Springer Series in Opti

MW00814532

Volume 224

Springer Series in Optical Sciences is led by Editor-in-Chief William T. Rhodes, Florida Atlantic University, USA, and provides an expanding selection of research monographs in all major areas of optics:

– lasers and quantum optics
– ultrafast phenomena
– optical spectroscopy techniques
– optoelectronics
– information optics
– applied laser technology
– industrial applications and
– other topics of contemporary interest.

With this broad coverage of topics the series is useful to research scientists and engineers who need up-to-date reference books.

More information about this series at http://www.springer.com/series/624

Kurt E. Oughstun

Electromagnetic and Optical Pulse Propagation

Volume 1: Spectral Representations in
Temporally Dispersive Media

Second Edition

 Springer

Kurt E. Oughstun
College of Engineering
and Mathematical Sciences
University of Vermont
Burlington, VT, USA

ISSN 0342-4111   ISSN 1556-1534 (electronic)
Springer Series in Optical Sciences
ISBN 978-3-030-20837-0   ISBN 978-3-030-20835-6 (eBook)
https://doi.org/10.1007/978-3-030-20835-6

This Springer imprint is published by the registered company Springer Nature Switzerland AG.
The registered company address is: Gewerbestrasse 11, 6330 Cham, Switzerland

The cautious guest
who comes to the table
speaks sparingly.
Listens with ears
learns with eyes.
Such is the seeker of
knowledge.
Hávamál
The Sayings of the Vikings
Translated by Björn Jónasson

This volume is dedicated to
Professor John B. Bulman
Professor George C. Sherman
Professor Emil Wolf
Professor Kenneth I. Golden
Professor Gagan Mirchandani
Dr. Walter J. Fader
Dr. Edward A. Sziklas
Dr. Richard Albanese
My teachers, mentors, and valued colleagues.

Preface to the Second Revised Edition

This revised second edition of the first volume of this two-volume text on time- and frequency-domain electromagnetics in dispersive attenuative media contains most of the same material presented in the first edition with minor editing and correction where necessary, as well as expanded development of several deserving topics. These include sections on the aberration of light and the relativistic Doppler effect, the conjugate electromagnetic fields and invariants, the near field of an extended time-harmonic linear dipole source, the molecular theory of reflection and refraction, the Ewald–Oseen extinction theorem, the Fresnel reflection and transmission coefficients for complex lossy media, the properties of metamaterials, and the Sherman expansion of source-free wave fields. In addition, the majority of the material presented in Chap. 9 of Volume 2 has been moved to Volume 1 as it is more appropriate for it to be included with the topics presented in this volume.

As now rewritten, Chaps. 1–5 provide a detailed development of the fundamental properties of and interrelationship between microscopic and macroscopic electromagnetic field theories. There is sufficient material here for a one-semester senior or first year graduate-level course in electromagnetic theory, including a variety of challenging problems at the end of each chapter. The remaining Chaps. 6–9, taken together with selected topics from Volume 2, would then form the basis for an advanced graduate-level course in electromagnetic wave theory.

My research on this topic began in the early 1970s when I was a graduate student at The Institute of Optics of The University of Rochester with George C. Sherman as my advisor and fellow, albeit more senior graduate students, Anthony J. Devaney and Jakob J. Stamnes as mentors. Constructive criticisms regarding several topics appearing in the first edition of this volume from Richard Albanese, Natalie Cartwright, and Christopher Palombini are gratefully acknowledged. In addition, I am indebted to the physics and engineering students who have successfully taken my undergraduate and graduate electromagnetic theory courses at the University of Vermont for pointing out some of the all too many typographical errors appearing in the first edition of this text. In spite of this, I blissfully remain a "two-finger" typist. Dyslexia doesn't help.

Burlington, VT, USA   Kurt E. Oughstun
January 2017

Preface to the First Edition

This two-volume text presents a systematic theoretical treatment of the radiation and propagation of pulsed electromagnetic and optical fields through temporally dispersive attenuative media. Although such fields are often referred to as transient, they may be short-lived only in the sense of an observation made at some fixed point in space. Because of their unique properties when the pulse spectrum is ultrawideband with respect to the material dispersion, specific features of the propagated pulse are found to persist in time long after the main body of the pulse has become exponentially small. Therein lies both their interest and usefulness.

The subject matter divides naturally into two volumes. Volume 1 presents a detailed development of the fundamental theory of pulsed electromagnetic radiation and wave propagation in causal linear media that are homogeneous and isotropic but which otherwise have rather general dispersive, absorptive properties. The analysis is specialized in Volume 2 to the propagation of electromagnetic and optical pulses in homogeneous, isotropic, locally linear media whose temporal frequency dispersion is described by a specific causal model. Dielectric, conducting, and semi-conducting material models are considered. These two volumes present sufficient material to cover a two-semester graduate sequence in electromagnetic and optical wave theory in physics and electrical engineering as well as in applied mathematics. Prerequisite material includes undergraduate courses in electromagnetic field theory and complex variable theory. Challenging problems are given throughout the text.

The development presented in Volume 1 provides a rigorous description of the fundamental time-domain electromagnetics and optics in linear temporally dispersive, absorptive media. The analysis begins with a detailed review of the classical Maxwell–Lorentz theory and the invariance of the field equations in the special theory of relativity. The macroscopic theory is then obtained from the microscopic theory through a well-defined spatial-averaging procedure. A general description of macroscopic electromagnetics and the role that causality plays in the constitutive (or material) relations are then developed. The description of electromagnetic energy flow in causally dispersive media is then developed based on the conservation laws for the electromagnetic field. The angular spectrum of plane wave representation of the pulsed radiation field in homogeneous, isotropic, locally linear, temporally dispersive media is derived and expressed in the form of the classical integral representations of Weyl, Sommerfeld, and Ott. The theory is then applied to the general description of pulsed electromagnetic and optical beam fields in dispersive materials, showing how the effects of temporal dispersion and spatial diffraction are coupled.

The theory presented in Volume 2 provides the necessary mathematical and physical basis to describe and explain the detailed dynamical evolution of a pulse as it travels through a linear temporally dispersive, absorptive medium. This is the subject of a classic theory with origins in the seminal research by Arnold Sommerfeld and Leon Brillouin in the early 1900s for a Lorentz model dielectric and described in modern textbooks on advanced electrodynamics. This classic theory has been carefully reexamined and extended by George Sherman and myself over the time period from 1974 to 1997 using modern asymptotic methods of approximation. We have developed a physical model that provides a straightforward quantitative algorithm that not only describes the entire dynamical field evolution in the mature dispersion regime but also explains each feature in the propagated field in simple physical terms. This model reduces to the group velocity description in the weak dispersion limit as the material loss approaches zero. The uniform asymptotic description was then completed by Natalie Cartwright and myself in 2007. Finally, the persistent controversy regarding the question of superluminal pulse propagation in dispersive media is examined in light of the recent results establishing the domain of applicability of the group velocity approximation.

My research in this area began in the early 1970s when I was a graduate student at The Institute of Optics of The University of Rochester in Rochester, New York. I am grateful for the financial support during that critical period by The Institute of Optics, the Corning Glass Works Foundation, the National Science Foundation, and the Center for Naval Research. This research continued while I was at the United Technologies Research Center, the University of Wisconsin at Madison, and, finally, the University of Vermont with an extended sabbatical at the Universitetet i Bergen in Norway that was generously supported by the Norwegian Research Council for Science and the Humanities. The critical, long-term support of this research by both Dr. Arje Nachmann of the United States Air Force Office of Scientific Research and Dr. Richard Albanese of the Air Force Research Laboratory is gratefully acknowledged.

A diverse variety of textbooks has contributed to my understanding of both electromagnetic and optical field and wave theory. Chief among these are J. D. Jackson's *Classical Electrodynamics*, M. Born and E. Wolf's *Principles of Optics: Electromagnetic Theory of Propagation, Interference and Diffraction of Light*, and J. M. Stone's *Radiation and Optics: An Introduction to the Classical Theory*. Each has had a fundamental influence on my own research.

Burlington, VT, USA   Kurt E. Oughstun
January 2010

Contents - Volume I

Contents - Volume II

Chapter 1
Introduction

"The beginning is the most important part of the work." Plato

1.1 Motivation

The dynamical evolution of an electromagnetic pulse as it propagates through
a linear, temporally dispersive medium (such as water) or system (such as a
dielectric waveguide) is a classical problem in both electromagnetics and optics.
With Maxwell's unifying theory [1, 2] of electromagnetism and optics, Lorentz's
classical model [3] of dielectric dispersion, and Einstein's special theory of relativity
[4], the stage was set for a long-standing problem of some controversy in classical
physics, engineering, and applied mathematics. If the system was nondispersive, an
arbitrary plane wave pulse would propagate unaltered in shape at the phase velocity
of the wave field in the medium. For example, for distortionless wave propagation
along the z-direction of a cartesian coordinate system, a one-dimensional wave is
described by a single-valued function of the form $f(z \pm vt)$, where the argument
$\varphi = z \pm vt$ is called the phase of the wave function. Any fixed value of this phase (for
example, the value at the temporal center of the pulse) then propagates undistorted
in shape along the $\pm z$-direction with the velocity $dz/dt = \pm v$. The first partial
derivatives of the wave function $f(z \pm vt)$ with respect to the independent variables
z and t are then given by $\partial f/\partial z = f'$ and $\partial f/\partial t = \pm v f'$, where $f' = df(\zeta)/d\zeta$.
These two expressions then show that

$$v = \mp \frac{\partial f/\partial t}{\partial f/\partial z},\tag{1.1}$$

assuming that these derivatives exist. A second partial differentiation of the wave
function then gives $\partial^2 f/\partial z^2 = f''$ and $\partial^2 f/\partial t^2 = v^2 f''$, so that

$$\frac{\partial^2 f}{\partial z^2} = \frac{1}{v^2}\frac{\partial^2 f}{\partial t^2},\tag{1.2}$$

© Springer Nature Switzerland AG 2019
K. E. Oughstun, *Electromagnetic and Optical Pulse Propagation*, Springer Series
in Optical Sciences 224, https://doi.org/10.1007/978-3-030-20835-6_1

1

provided that these second derivatives exist. This is the *one-dimensional wave equation for distortionless wave propagation* along the $\pm z$-direction. Notice that periodicity is not an inherent feature of wave motion [5]. An important feature of this wave equation is its *linearity*: if $f_1, f_2, \ldots, f_n$ are all solutions of this wave equation, then the linear combination $a_1 f_1 + a_2 f_2 + \cdots + a_n f_n$ with constant coefficients $a_1, a_2, \ldots, a_n$ is also a solution. Solutions of the wave equation (2) are then said to satisfy the *principle of superposition*.

For propagation in an attenuative medium, consider a time-harmonic solution to the wave equation (1.2) of the form

$$f(z - vt) = Ae^{i(kz-\omega t)}, \tag{1.3}$$

where A is a constant and where k and ω are independent of both z and t. Substitution of this expression into the wave equation (1.2) then shows that k and ω must be related by the elementary *dispersion relation*

$$k^2 v^2 = \omega^2 \tag{1.4}$$

which may be solved for k in terms of ω as $k(\omega) = \pm\omega/v$ or for ω in terms of k as $\omega(k) = \pm vk$. If v is a real-valued quantity, then it represents the velocity at which the phase $\varphi = kz - \omega t$ advances in the $+z$-direction and is accordingly called the *phase velocity*

$$v_p = \frac{\omega}{k}. \tag{1.5}$$

The velocity v in Eq. (1.1) is then seen to be this phase velocity.

In an attenuative medium, either the wavenumber k or the frequency ω is complex-valued. Consider first the case when k is complex-valued with

$$k = \beta + i\alpha, \tag{1.6}$$

where β is the real part and α the imaginary part of k, respectively. If ω is real-valued, then v must also be complex-valued, with

$$v = v_r + iv_i, \tag{1.7}$$

where v_r and v_i are the real and imaginary parts of v, respectively. The time-harmonic solution (1.3) then becomes

$$f(z - vt) = Ae^{-\alpha z}e^{i(\beta z-\omega t)}, \tag{1.8}$$

which represents a time-harmonic wave propagating in the $+z$-direction unaltered in form (i.e. undistorted) with amplitude decreasing exponentially with propagation distance, referred to as a *spatially attenuated wave* [5] with phase velocity

$$v_p = \frac{\omega}{\beta}. \tag{1.9}$$

Solutions of the wave equation that are in the form (1.8) physically represent *progressive harmonic waves*. Substitution of Eqs. (1.6) and (1.7) into the elementary dispersion relation (1.4) shows that

$$\beta = \omega \frac{v_r}{|v|^2} \quad \& \quad \alpha = \omega \frac{v_i}{|v|^2}, \tag{1.10}$$

where $|v|^2 = v_r^2 + v_i^2$. With this result, the phase velocity (1.9) becomes

$$v_p = \frac{|v|^2}{v_r} = v_r \left(1 + \frac{v_i^2}{v_r^2} \right), \tag{1.11}$$

which states that the phase velocity increases when material loss is included. Notice that the dependence of the phase velocity on v_i is a second-order effect, whereas the dependence of α on v_i is a first-order effect.

Alternatively, if k is real-valued and ω is complex-valued with

$$\omega = \omega_r + i\omega_i, \tag{1.12}$$

where ω_r is the real part and ω_i the imaginary part of ω, then v must be complex-valued, as given in Eq. (1.7). The time-harmonic solution (1.3) then becomes

$$f(z - vt) = Ae^{+\omega_i t} e^{i(kz - \omega_r t)}, \tag{1.13}$$

which represents a time-harmonic wave along the z-axis with amplitude decreasing exponentially in time for $\omega_i < 0$, referred to here as a *temporally attenuated wave* with phase velocity

$$v_p = \frac{\omega_r}{k}. \tag{1.14}$$

For $\omega_i > 0$, Eq. (1.13) represents a *temporally amplified wave*. Solutions of the wave equation that are in the form (1.13) physically represent *standing harmonic waves*. Substitution of Eqs. (1.7) and (1.12) into the dispersion relation (1.4) shows that

$$\omega_r = kv_r \quad \& \quad \omega_i = kv_i, \tag{1.15}$$

so that $v_i < 0$ for a temporally attenuated wave. The phase velocity (1.14) then becomes

$$v_p = v_r, \tag{1.16}$$

which, unlike the phase velocity (1.11) for a spatially attenuated progressive wave, is independent of v_i.

Notice that neither the spatially attenuated progressive wave solution (1.8) nor the temporally attenuated standing wave solution (1.13) satisfies the wave equation (1.2) for real v, as each requires v to be complex-valued. Wave propagation in a dispersive attenuative medium must then be based on a more general formulation that allows complex-valued quantities. This is accomplished in the frequency domain with Helmholtz' equation

$$\frac{\partial^2 \tilde{f}(z, \omega)}{\partial z^2} + \frac{\omega^2}{v^2(\omega)} \tilde{f}(z, \omega) = 0 \tag{1.17}$$

where

$$f(z, t) = \frac{1}{2\pi} \int_{-\infty}^{+\infty} \tilde{f}(z, \omega) e^{-i\omega t} d\omega \tag{1.18}$$

is the Fourier integral representation of the real-valued wave function $f(z, t)$ with inverse

$$\tilde{f}(z, \omega) = \int_{-\infty}^{+\infty} f(z, t) e^{i\omega t} dt. \tag{1.19}$$

If v is independent of ω, then $f(z, t)$, as given by Eq. (1.18) with $\tilde{f}(z, \omega)$ satisfying the Helmholtz equation (1.17), is found to satisfy the wave equation (1.2). The *complex wavenumber* is then defined as

$$\tilde{k}(\omega) = \beta(\omega) + i\alpha(\omega) \equiv \frac{\omega}{v(\omega)} \tag{1.20}$$

with real $\beta(\omega) = \Re\{\tilde{k}(\omega)\}$ and imaginary $\alpha(\omega) = \Im\{\tilde{k}(\omega)\}$ parts. With this identification, the Helmholtz equation (1.17) becomes

$$\frac{\partial^2 \tilde{f}(z, \omega)}{\partial z^2} + \tilde{k}^2(\omega) \tilde{f}(z, \omega) = 0, \tag{1.21}$$

with solution

$$\tilde{f}(z, \omega) = \tilde{f}_0(\omega) e^{i\tilde{k}(\omega)\Delta z} = \tilde{f}_0(\omega) e^{-\alpha(\omega)\Delta z} e^{i\beta(\omega)\Delta z}, \tag{1.22}$$

where $\Delta z = z - z_0$ and $\tilde{f}_0(\omega) = \tilde{f}(z_0, \omega)$ is the Fourier spectrum of the initial temporal wave behavior at $z = z_0$. The propagated wave is then given by the Fourier integral representation

$$f(z, t) = \frac{1}{2\pi} \int_{-\infty}^{+\infty} \tilde{f}_0(\omega) e^{-\alpha(\omega)\Delta z} e^{i(\beta(\omega)\Delta z - \omega t)} d\omega, \qquad (1.23)$$

provided that these integrals exist. This result is the generalization of the *spatially attenuated wave* solution given in Eq. (1.8) that properly includes the effects of dispersion.

Hence, in a dispersive medium, the pulse is modified as it propagates due to two fundamentally interconnected effects, as described by Eq. (1.23). First of all, each monochromatic spectral component of the initial pulse propagates through the dispersive system with its own phase velocity $v_p(\omega) = \omega/\beta(\omega)$, so that the phasal relationship between the various spectral components of the pulse changes with propagation distance. Secondly, but not of secondary importance, each monochromatic spectral component is absorbed with increasing propagation distance at its own rate, as specified by $\alpha(\omega)$, so that the relative amplitudes between the spectral components of the pulse also change with propagation distance. These two simple effects then result in a complicated change in the dynamical structure of the propagated pulsed wave field. The rigorous analysis of dispersive pulse propagation phenomena is further complicated by the simple fact that the phasal and absorptive parts of the system response are connected through the physical requirement of causality [6, 7]. For an initial pulse with a sufficiently rapid rise-time, fall-time, phase change, or amplitude change within the body of the pulse, these effects manifest themselves through the formation of well-defined precursor fields [8–11] whose evolution has been shown [12] to be completely determined by the interrelated dispersive and absorptive properties of the system. The precursor fields are readily distinguished in the dynamical evolution of the propagated wave field by the fact that the range of their oscillation frequency is typically quite different from that of the input pulse and their attenuation is typically much less than that at the input pulse carrier frequency.

The precursor fields (or forerunners, as they were originally called) were first described by Sommerfeld [9] and Brillouin [10] in their seminal analysis of optical signal propagation in a locally linear, isotropic, causally dispersive dielectric medium with frequency dispersion described by the single resonance Lorentz model [3]. Unfortunately, their analysis errantly concluded that the amplitudes of these precursor fields were, for the most part, negligible in comparison to the main signal evolution and that the main signal arrival occurred with a sudden rise in amplitude of the field. Because of this, it was further concluded that the signal arrival could not be given an unambiguous physical definition. Regrettably, these misconceptions have settled into some of the standard literature on electromagnetic wave theory [13]. The more recent analysis [14–22] of linear dispersive pulse propagation that is based upon modern asymptotic techniques [23–29] has provided a complete, rigorous description of the dynamical field evolution in both single and multiple resonance Lorentz model dielectrics. In particular, this analysis has shown that the precursor fields that result from an input Heaviside unit step-function modulated signal are a dominant feature of the field evolution in the *mature dispersion regime*. The mature dispersion regime has been shown [30–32] to typically include all propagation distances that are greater than one absorption depth $z_d(\omega_c) = \alpha^{-1}(\omega_c)$ in the medium at the signal frequency ω_c of the initial wave field. In addition, this modern asymptotic description [14, 15, 17] has also provided both a precise definition and physical interpretation [30] of the signal velocity in the dispersive medium. This proper description of the signal velocity is critically dependent upon the correct description and interpretation of the precursor fields. Most importantly, this signal velocity is shown to be bounded below by zero and above by the speed of light c in vacuum, in complete agreement with the special theory of relativity [4].

The central importance that the precursor fields hold in both the analysis and interpretation of linear dispersive pulse propagation phenomena is also realized in the study of ultrashort pulse dynamics. The asymptotic theory clearly shows that the resultant pulse distortion that an input rectangular envelope modulated signal undergoes as it propagates through a causally dispersive medium is primarily due to the precursor fields that are associated with the leading and trailing edges of the input pulse envelope regardless of the initial temporal pulse width $T > 0$ [18]. The interference between these two sets of precursor fields increases with the propagation distance in the dispersive medium and naturally leads to asymmetric pulse distortion. Similar results are obtained for a trapezoidal envelope pulse (of central interest in ultrawideband radar) provided that either the initial rise-time or fall-time is faster than the oscillation period $T_c = 1/f_c$ of the carrier wave. The situation is quite different for a pulse whose initial envelope function is infinitely smooth, such as that for a Gaussian envelope pulse (of central interest in ultrashort optics). In that case, the entire pulse evolves into a single set of precursor fields provided that the initial pulse width is shorter than the characteristic relaxation time of the medium [19, 33–35], reinforcing the fundamental role that the precursor fields play in dispersive pulse dynamics.

The subject of electromagnetic pulse propagation in dispersive media has been and continues to be of considerable practical importance in several areas of contemporary optics and engineering electromagnetics. For example, the effects of dispersion are prevalent in all fiber optic communication and integrated optics systems [36, 37]. As data rates continue to increase and enter the terabit rate domain, the temporal pulse widths will begin to exceed the characteristic optical relaxation times for typical fiber materials and the precursor fields will then dominate the field evolution. In addition, with the current experimental capabilities of producing both ultrashort (femtosecond) optical pulses [38] and digitally coded ultrawideband microwave signals, a new technology is rapidly developing in which the importance of the causal effects of dispersion is greatly magnified.

1.2   A Critical History of Previous Research

The history of published research in the area of dispersive wave propagation is extensive and varied in both content and depth of mathematical rigor. This section provides an overview of the most important published literature on the subject. This includes papers that are concerned with such closely related topics as electromagnetic wave propagation in waveguiding systems and fiber optics, as well as research on nonelectromagnetic pulse propagation phenomena that may only be peripherally related to the central topic of this work. Much of this research was and continues to be motivated by the fundamental question regarding the velocity of information transfer.

Early considerations of the wave theory of light represented the optical wave field as a coherent superposition of monochromatic scalar wave disturbances. Dispersive wave propagation was first considered in this manner by Sir William R. Hamilton [39] in 1839 where the concept of group velocity was first introduced. In that paper, Hamilton compared the phase and group velocities of light, stating that [39]

> the velocity with which such vibration spreads into those portions of the vibratory medium which were previously undisturbed, is in general different from the velocity of a passage of a given phase from one particle to another within that portion of the medium which is already fully agitated; since we have velocity of transmission of phase $= s/k$, but velocity of propagation of vibratory motion $= ds/dk$,

where s denotes the angular frequency and k the wavenumber of the disturbance in Hamilton's notation. Subsequent to this definition, Stokes [40] posed the concept of group velocity as a "Smith's Prize examination" question in 1876. Lord Rayleigh[1] then mistakenly attributed the original definition of the group velocity to Stokes, stating that [41]

> when a group of waves advances into still water, the velocity of the group is less than that of the individual waves of which it is composed; the waves appear to advance through the group, dying away as they approach its anterior limit. This phenomenon was, I believe, first explained by Stokes, who regarded the group as formed by the superposition of two infinite trains of waves, of equal amplitudes and of nearly equal wave-lengths, advancing in the same direction.

Rayleigh applied these results to explain the difference between the phase and group velocities of light with respect to their observability, arguing that [42]

> Unless we can deal with phases, a simple train of waves presents no mark by which its parts can be identified. The introduction of such a mark necessarily involves a departure from the original simplicity of a single train, and we have to consider how in accordance with Fourier's theorem the new state of things is to be represented. The only case in which we can expect a simple result is when the mark is of such a character that it leaves a considerable number of consecutive waves still sensibly of the given harmonic type, though the wave-length and amplitude may vary within moderate limits at points whose distance amounts to a very large multiple of λ ... From this we see that ... the deviations from the simple harmonic type travel with the velocity dn/dk and not with the velocity n/k,

---

[1] John William Strutt, 3rd Baron Rayleigh (1842–1919).

where n denotes the frequency and k the wavenumber in Rayleigh's notation.

These early considerations may best be illustrated by the coherent superposition of two time-harmonic waves with equal amplitudes and nearly equal wave numbers (k and $k + \delta k$) and angular frequencies (ω and $\omega + \delta\omega$, respectively) travelling in the positive z-direction. The linear superposition of these two wave functions then yields the waveform [5, 43]

$$U(z, t) = a \cos(kz - \omega t) + a \cos((k + \delta k)z - (\omega + \delta\omega)t)$$

$$= 2a \cos\left[\frac{1}{2}(z\delta k - t\delta\omega)\right] \cos(\bar{k}z - \bar{\omega}t). \tag{1.24}$$

This superposition results in an amplitude modulated wave with mean wavenumber $\bar{k} = k + \delta k/2$ and mean angular frequency $\bar{\omega} = \omega + \delta\omega/2$. The surfaces of constant phase propagate with the *phase velocity*

$$v_p \equiv \frac{\bar{\omega}}{\bar{k}}, \tag{1.25}$$

while the surfaces of constant amplitude propagate with the *group velocity*

$$v_g \equiv \frac{\delta\omega}{\delta k}. \tag{1.26}$$

Notice that these results are exact for the waveform given in Eq. (1.24). If the medium is nondispersive, then $\bar{k} = \bar{\omega}/c$, $\delta k = \delta\omega/c$, and the phase and group velocities are equal. However, if the medium exhibits temporal dispersion so that $k(\omega) = (\omega/c)n(\omega)$ where $n(\omega)$ is frequency-dependent, then the phase and group velocities will, in general, be different. In particular, if $n(\omega) > 0$ increases with increasing $\omega \geq 0$, then $v_p \geq v_g > 0$ and the phase fronts will advance through the wave group as described by Rayleigh [41]. This elementary phenomenon is illustrated in Fig. 1.1 for the simple wave group described in Eq. (1.24). Each wave pattern illustrated in this figure describes a "snapshot" of the wave group at a fixed instant of time. In the upper wave pattern the coincidence at $z = 0$ of a particular peak amplitude point in the envelope (marked with a G) with a peak amplitude point in the waveform (marked with a P) is indicated. As time increases from this initial instant of time ($t = 0$), these two points become increasingly separated in time, as illustrated in the middle ($t = \delta t$) and bottom ($t = 2\delta t$) wave patterns, showing that the phase velocity of the wave is greater than the group velocity of the envelope in this case. This phenomenon is readily observable in nature to the discerning observer.

The result given in Eq. (1.24) may be generalized [43] to obtain the Fourier–Laplace integral representation [44] [see the development leading to Eq. (1.23)]

$$U(z, t) = \frac{1}{2\pi} \int_C \tilde{f}(\omega) e^{i\left(\tilde{k}(\omega)z - \omega t\right)} d\omega \tag{1.27}$$

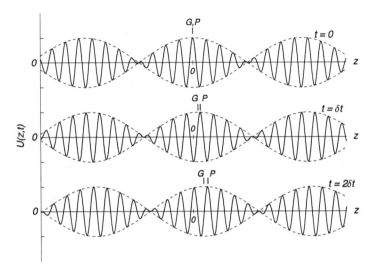

Fig. 1.1 Evolution of a simple wave group in a temporally dispersive medium with normal frequency dispersion. The point P marks a specific peak amplitude point in the wave form (solid curve) and the point G marks a specific peak amplitude point in the envelope of the wave group (dashed curves)

for plane wave pulse propagation in the $+z$-direction, where the temporal Fourier spectrum $\tilde{U}(z, \omega)$ of $U(z, t)$ satisfies the *Helmholtz equation*

$$\left(\nabla^2 + \tilde{k}^2(\omega)\right) \tilde{U}(z, \omega) = 0 \qquad (1.28)$$

with boundary value given by the initial pulse $U(0, t) = f(t)$ at the plane $z = 0$ with Fourier spectrum $\tilde{f}(\omega) = \int_{-\infty}^{\infty} f(t)e^{i\omega t} dt$, where

$$\tilde{k}(\omega) \equiv \frac{\omega}{c} n(\omega) \qquad (1.29)$$

describes the *complex wave number* with *propagation factor* $\beta(\omega) \equiv \Re\{\tilde{k}(\omega)\}$ and *attenuation factor* $\alpha(\omega) \equiv \Im\{\tilde{k}(\omega)\}$ in the medium with complex index of refraction $n(\omega) = n_r(\omega) + in_i(\omega)$. Here C denotes the straight-line contour extending from $ia - \infty$ to $ia + \infty$ where a is greater than the abscissa of absolute convergence [13] for the initial pulse function $f(t)$. For the special case when $f(t) = u(t) \sin(\omega_c t + \psi)$ with envelope function $u(t)$ and phase constant ψ [where $\psi = 0$ for a sine wave carrier and $\psi = \pi/2$ for a cosine wave carrier], Eq. (1.27) becomes

$$U(z, t) = \frac{1}{2\pi} \Re \left\{ i e^{-i\psi} \int_C \tilde{u}(\omega - \omega_c) e^{i\left(\tilde{k}(\omega)z - \omega t\right)} d\omega \right\}, \qquad (1.30)$$

where $\tilde{u}(\omega) = \int_{-\infty}^{\infty} u(t)e^{i\omega t} dt$ is the Fourier spectrum of the initial pulse envelope function. This exact integral representation forms the basis for much of the research on linear dispersive wave propagation phenomena.

With Maxwell's theory[2] [1, 2] of electromagnetic wave propagation firmly in place at the beginning of the twentieth century, the development of a theory of the dispersive properties of dielectric media was begun in terms of a classical atomistic model that culminated in Lorentz's classical work [3]. Drude [45] indicates in a footnote that Maxwell (1869) was the first to base the theory of anomalous dispersion upon such an atomistic model. In research independent of Maxwell's, Sellmeier, v. Helmholtz, and Ketteler also used this model as a basis for a theory of material dispersion.[3]

The distinction between the signal and group velocities originated in the early research by Voigt [48, 49] and Ehrenfest [50] on elementary dispersive waves, and by Laue [51] who first considered the problem of dispersive wave propagation in a region of anomalous dispersion where the absorption is both large and strongly dependent upon the frequency. Subsequently, the distinction between the front and signal velocities was considered by Sommerfeld [8, 9] who showed that no signal could travel faster than the vacuum speed of light c and that the signal front progressed with the velocity c in a dispersive medium, as well as by Brillouin [10, 11] who provided a detailed description of the signal evolution in a Lorentz model dielectric. In his 1907 paper, Sommerfeld [8] stated that (as translated by Brillouin [11]):

> It can be proven that the signal velocity is exactly equal to c, if we assume the observer to be equipped with a detector of infinite sensitivity, and this is true for normal or anomalous dispersion, for isotropic or anisotropic medium, that may or may not contain conduction electrons. The signal velocity has absolutely nothing to do with the phase velocity. There is nothing, in this problem, in the way of Relativity theory.

The "signal velocity" referred to here by Sommerfeld has since become properly identified as the front velocity, the signal velocity being described by Brillouin [10, 11] in terms of the moment of transition from the forerunner evolution to the signal evolution in the dynamical field evolution due to an initial Heaviside step-function modulated signal. Brillouin's asymptotic analysis, based upon the then newly developed method of steepest descent due to Debye [52], provided the first detailed description of the frequency dispersion of the signal velocity in a single resonance Lorentz model dielectric. Based upon this seminal analysis, Brillouin concluded that [10, 11]

---

[2]Maxwell also played a defining role in introducing the nabla symbol ∇, so named because of its similarity with the harp, the Greek word for the Hebrew or Egyptian harp being "nabla". This image arose in relation to a mathematical discourse between Maxwell and the Scottish mathematical physicist Peter Guthrie Tait regarding a problem on orthogonal surfaces. In his January 23, 1871 letter to Tait, Maxwell opened with "Still harping on that Nabla?"

[3]For a more complete discussion of this early work, see E. Mach [46] as well as the well-known undergraduate optics text by Jenkins and White [47].

The signal velocity does not differ from the group velocity, except in the region of anomalous dispersion. There the group velocity becomes greater than the velocity in vacuum if the reciprocal $c/U < 1$; it even becomes negative ...Naturally, the group velocity has a meaning only so long as it agrees with the signal velocity. The negative parts of the group velocity have no physical meaning ...The signal velocity is always less than or at most equal to the velocity of light in vacuum.

Sommerfeld's now classic analysis [8, 9] was the first to prove that the signal arrival in a dispersive medium (described by the Lorentz model) does not always propagate with the group velocity and that, even though the group velocity may exceed the vacuum speed of light c in a region of anomalous dispersion, the wave front arrives with a positive velocity that is always less than or equal to c. In a fundamental extension of Sommerfeld's results, Brillouin [10] also employed a Fourier integral representation of a Heaviside unit step-function modulated plane wave signal $U(0, t) = u_H(t) \sin(\omega_c t)$ with finite $\omega_c > 0$, where $u_H(t) = 0$ for $t < 0$ and $u_H(t) = 1$ for $t > 0$, as it propagates through a semi-infinite, single resonance Lorentz medium. This exact integral representation is given by [10]

$$U(z, t) = \frac{1}{2\pi} \Re \int_{ia-\infty}^{ia+\infty} \frac{e^{(z/c)\phi(\omega,\theta)}}{\omega - \omega_c} \, d\omega, \tag{1.31}$$

with constant $a > 0$ [cf. Eq. (1.30)]. The complex phase function $\phi(\omega, \theta)$ appearing in the exponential of the integrand in Eq. (1.31) is given by

$$\phi(\omega, \theta) \equiv i\frac{c}{z}(\tilde{k}(\omega)z - \omega t)$$

$$= i\omega(n(\omega) - \theta), \tag{1.32}$$

where $\theta \equiv ct/z$ is a dimensionless space–time parameter for all $z > 0$ and

$$n(\omega) = \left(1 - \frac{b^2}{\omega^2 - \omega_0^2 + 2i\delta\omega}\right)^{1/2} \tag{1.33}$$

is the complex index of refraction of the Lorentz medium with resonance frequency ω_0, plasma frequency b, and damping constant δ. Because $\phi(\omega, \theta)$ is complex-valued, Brillouin [10] applied the method of steepest descent [52] in order to obtain the asymptotic behavior of the propagated signal for large propagation distances $z \gg 0$. In using this asymptotic method, the original contour of integration extending from negative to positive infinity in the upper-half of the complex ω-plane is deformed through the appropriate saddle points $\omega_{SP_j}(\theta)$, $j = S, B$, of $\phi(\omega, \theta)$ along the path of steepest descent, where $\phi'(\omega_{SP_j}, \theta) = 0$ with $\phi' = d\phi/d\omega$. The dominant contribution to the asymptotic approximation of Eq. (1.31) is then due to the behavior in the immediate neighborhood of the dominant saddle point (the saddle point with the least exponential attenuation) together with the pole contribution at $\omega = \omega_c$ as the observation point moves off to infinity with the field. This asymptotic representation due to Brillouin may then be expressed as [14, 15]

$$U(z, t) \sim U_S(z, t) + U_B(z, t) + U_c(z, t) \qquad (1.34)$$

where

$$U_j(z, t) = a_j \left(\frac{c}{2\pi z} \right)^{1/2} \Re \left\{ \left(\frac{-1}{\omega_{SP_j} - \omega_c} \right) \frac{\exp\left[\frac{z}{c}\phi(\omega_{SP_j}, \theta)\right]}{[-\phi''(\omega_{SP_j}, \theta)]^{1/2}} \right\} \qquad (1.35)$$

for $j = S, B$, where $a_S = 2$ for all $\theta > 1$, $a_B = 1$ for $1 < \theta < \theta_1$, and $a_B = 2$ for $\theta > \theta_1$. Here $\theta_1 \approx \theta_0 + 2\delta^2 b^2 / 3\theta_0 \omega_0^4$ with $\theta_0 \equiv n(0)$, as originally described by Brillouin [10, 11]. With a first-order approximation of the behavior of the complex phase function $\phi(\omega, \theta)$ in the complex ω-plane, the approximate behavior of the saddle point locations as a function of θ together with their relative importance was deduced, giving rise to the following succession of events illustrated in Fig. 1.2. First of all, in complete agreement with the principle of relativistic causality [7], the propagated field identically vanishes for all $\theta = ct/z < 1$; this important result is due to the behavior of $\phi(\omega, \theta)$ at $|\omega| = \infty$ and the fact that the integrand appearing in the integral representation of the propagated field is analytic in the upper-half of the complex ω-plane. Between $\theta = ct/z \geq 1$ and larger values of this parameter, two sets of forerunners [$U_S(z, t)$ and $U_B(z, t)$] are present for a single resonance Lorentz medium. From a physical point of view, these forerunners arise from those Fourier components comprising the initial pulse shape whose velocities of propagation through the dispersive medium are greater than the velocity of propagation of the Fourier component at the applied signal frequency ω_c. The main signal at the finite applied signal frequency $\omega_c > 0$ will then arrive at

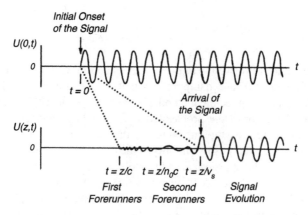

Fig. 1.2 Brillouin's description of the evolution of a Heaviside step-function signal $U(0, t) = u_H(t) \sin(\omega_c t)$ in a single resonance Lorentz medium, where $u_H(t) = 0$ for $t < 0$ and $u_H(t) = 1$ for $t > 0$, $n_0 = n(0)$, and where v_s denotes the signal velocity at the fixed signal frequency ω_c

some later space–time point $\theta = \theta_s \geq 1$ during the evolution of these forerunners. From a purely mathematical point of view, this main signal arrival is dependent upon the crossing of the deformed contour of integration with the simple pole singularity at $\omega = \omega_c$ that appears in the integrand of the integral representation (1.31) of the propagated field. At such a crossing, the integral may then be evaluated through use of the residue theorem with result

$$U_c(z, t) = -e^{-\alpha(\omega_c)z} \sin\left(\beta(\omega_c)z - \omega_c t\right). \tag{1.36}$$

As a consequence, the main signal is found to arrive with a mathematically well-defined (but physically incorrect) signal velocity $v_s = c/\theta_s$ which may, in certain special cases, coincide with the group velocity. Thus, in an involved analysis, Brillouin presented a basic asymptotic description of signal propagation in a Lorentz medium that is now a classic topic in electromagnetic theory.[4] This research then established the asymptotic description of pulse propagation in dispersive absorptive media. An essential feature of this description is its strict adherence to relativistic causality through careful treatment of the dispersive properties of both the real and imaginary parts of the complex index of refraction $n(\omega)$.

Based upon this asymptotic description, the transmission of an obliquely incident plane wave step-function signal across a planar interface from vacuum into a dispersive Lorentz model half-space was addressed by Colby [54] in 1915. The focus of this paper was on the evolution of the wavefront in both the reflected and transmitted signals, with the conclusion that

> an expression has been found for the refracted 'forerunners' reducible to the form given in Sommerfeld's paper for normal incidence and small values of time. The forerunners are found in general not to differ from the type described by Sommerfeld except in magnitude of amplitude and period, both quantities decreasing with increasing obliquity... the direction of the refracted ray is... a function of the time, varying from the incident direction at the first instant continuously toward the normal.

As for the reflected signal, Colby concluded that the "reflected ray has also a train of forerunners of extremely small amplitude and period."

At approximately the same time, Havelock [55, 56] completed his research on wave propagation in a dispersive medium based upon Kelvin's stationary phase method [57]. Havelock was the first to employ the Taylor series expansion of the wave number (κ in Havelock's notation) about a given wavenumber value κ_0 that the spectrum of the wave group is clustered about, referring to this approach as the *group method*. In addition, Havelock stated that [56]

> The range of integration is supposed to be small and the amplitude, phase and velocity of the members of the group are assumed to be continuous, slowly varying, functions of κ.

---

[4]The role that this research played in Brillouin's scientific career may be found in the biographical article by Mosseri [53].

This research then established the group velocity method for dispersive wave propagation. Because the method of stationary phase [58] requires that the wavenumber be real-valued, this method cannot properly treat causally dispersive attenuative media. Furthermore, notice that Havelock's group velocity method is a significant departure from Kelvin's stationary phase method with regard to the wavenumber value κ_0 about which the Taylor series expansion is taken. In Kelvin's method, κ_0 is the stationary phase point of the wavenumber κ, whereas in Havelock's method κ_0 describes the wavenumber about which the wave group spectrum is peaked. This apparently subtle change in the value of κ_0 results in significant consequences for the accuracy of the resulting group velocity description. Finally, notice that the fundamental hyperbolic character of the underlying wave equation is approximated as parabolic in this formulation, the characteristics then propagating instantaneously [59] instead of at the vacuum speed of light c.

There were then two different approaches to the problem of dispersive pulse propagation: the asymptotic approach (based upon Debye's method [52] of steepest descent) which provided a proper accounting of causality but was considered to be mathematically unwieldy without any simple physical interpretation, and Havelock's group velocity approximation (based upon Havelock's reformulation [55, 56] of Kelvin's asymptotic method [57] of stationary phase) which violates causality through its neglect of dispersive attenuation, but possesses a simple, physically appealing interpretation. Notice that both methods are based upon an asymptotic approximation but with two very different approaches, the method of stationary phase relying upon coherent interference and the method of steepest descent relying upon attenuation.

Brillouin's asymptotic solution [10] was revisited by Baerwald [60] in 1930 who reconsidered the signal velocity in a Lorentz medium. Because of the unnecessary constraint imposed on the deformed contour of integration by the method of steepest descent in Brillouin's analysis, the signal arrival was defined to occur when the path of steepest descent moved across the simple pole singularity at $\omega = \omega_c$ appearing in the integral representation (1.8). This misconception resulted in a frequency dependence of the signal velocity that erroneously peaks to the vacuum speed of light c near to the medium resonance frequency ω_0 and that is incomplete in its description when $\omega > \omega_0$. Baerwald [60] was the first to show that the signal velocity is at a minimum near the resonance frequency. A comparison of the frequency dependence of the signal velocity in a single resonance Lorentz medium as first described by Brillouin [10] and then by Baerwald [60] is depicted in Fig. 1.3. The asymptotic description was also revisited by Stratton [13] in 1941, who reformulated the problem in terms of the Laplace transform and derived an alternate contour integral representation of the propagated signal. Stratton [13] appears to have first referred to the forerunners described by Sommerfeld [9] and Brillouin [10] as precursors.

The first experimental measurement of the signal velocity was attempted by Shiren in 1962 using pulsed ultrasonic waves within a narrow absorption band. His experimental results [61] were "found to lie within theoretical limits established by calculations of Brillouin and Baerwald." A more detailed analysis of these results by

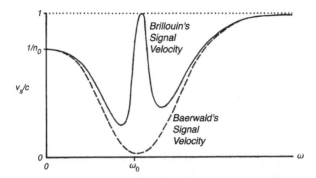

Fig. 1.3 Comparison of Brillouin's and Baerwald's descriptions of the angular frequency dependence of the relative signal velocity v_s/c in a single-resonance Lorentz medium with resonance frequency ω_0

Weber and Trizna [62] indicated that the velocity measured by Shiren was in reality that for the first precursor and not the signal. Subsequent research by Handelsman and Bleistein [27] in 1969 provided a uniform asymptotic description of the arrival and initial evolution of the signal front. The first experimental measurements of the precursor fields described by Sommerfeld [9] and Brillouin [10] were then published by Pleshko and Palócz [63] in 1969, who first referred to the first and second precursors as the *Sommerfeld* and *Brillouin precursors*, respectively. Although their experiments were conducted in the microwave domain on waveguiding structures with dispersion characteristics that are similar to that described by a single resonance Lorentz model dielectric, their results established the physical propriety of the asymptotic *Sommerfeld–Brillouin theory*.

The signal velocity of sound in superfluid ^3He-B was measured by Avenel, Rouff, Varoquaux and Williams [64, 65] in 1986 for moderate material damping, with results in agreement with Brillouin's description [10] in which the signal velocity peaks to a maximum value near the material resonance frequency (see Fig. 1.3). However, their experiment did not use a step-function modulated signal for which the signal velocity was defined. Rather, they used a continuous envelope pulse for which the signal velocity is undefined. The observation of a "precursory" motion that is similar to the Sommerfeld (or first) precursor was later observed by Varoquaux, Williams and Avenel [65] in superfluid ^3He-B.

The group velocity approximation was also refined and extended during this same period, most notably by Eckart [66] who considered the close relationship between the method of stationary phase and Hamilton–Jacobi ray theory in dispersive but nonabsorptive media. Of equal importance are the papers by Whitham [67] and Lighthill [68] on the general mathematical properties of three-dimensional wave propagation and the group velocity for ship-wave patterns and magnetohydrodynamic waves. The appropriate boundary value problem is solved in both papers through application of the method of stationary phase to a plane-wave expansion representation. Their approach, however, is useful only for nonabsorbing media, thereby limiting the types of dispersion relations that may be considered. The equivalence between the group velocity and the energy-transport velocity in loss-free media and systems was also established [67–71], thereby providing a physical basis for the group velocity in lossless systems with an inconclusive extension to dissipative media [72, 73]. In addition, the quasimonochromatic or slowly varying envelope approximation was precisely formulated by Born and Wolf [43] in the context of partial coherence theory. This then completed the mathematical and physical basis for the group velocity approximation.

This analysis naturally led to the asymptotic description of propagation in a dispersive medium by means of ray techniques. The first is the *direct-ray method* [74–79] which is applicable to solving partial differential equations with appropriately specified boundary or initial conditions. This is accomplished by assuming an asymptotic series for the solution which is then substituted into the partial differential equation. Families of rays are introduced along which the functional terms of the assumed series solution are described by ordinary differential equations which can then be directly solved. Alternatively, there is the *space–time ray theory* [80–83] which employs the plotting of rays and dispersion surfaces along with the initial or boundary values of the wave-field in order to demonstrate the propagation phenomena and develop the asymptotic representation.

Such ray techniques are useful in certain applications; however, they are heuristic in origin as each requires additional hypotheses about the solution form. Moreover, their applicability is limited because the theory does not provide error terms for the resultant asymptotic representations. Further evidence of these limitations is noted by Felsen and coworkers [84–86] who have applied space–time ray and dispersion surface techniques to the problems of propagation in dispersive media with applications to isotropic (cold) plasma media. Here Felsen noted the existence of certain transition regions wherein the ray-optic technique fails and one must resort to an exact integral representation and subsequent rigorous asymptotic analysis. This approach has recently been generalized by Heyman [87] and Felsen [88–90] and Melamed [91–94] using complex-source and complex-spectrum representations for pulsed-beam propagation in media that exhibit both spatial inhomogeneity and temporal dispersion. However, the formulation remains restricted to the case when the material attenuation is nondispersive.

Consider next the integral representation techniques that have been employed in connection with the one-dimensional boundary value problem for pulse propagation in metallic wave-guiding structures. The general method utilized in this type of problem, along with its limitations, is presented in Sect. 7.8 of Jackson [95]. Elliott [96], Forrer [97], and Wanselow [98] considered the problem of one-dimensional microwave pulse distortion in a dielectric filled metallic waveguide with propagation factor

$$\tilde{k}(\omega) = \frac{1}{v} \left(\omega^2 - \omega_{co}^2 \right)^{1/2}, \qquad (1.37)$$

where $v = (\epsilon\mu)^{-1/2}$ is the phase velocity in the dielectric filling the waveguide with real-valued dielectric permittivity ϵ and magnetic permeability μ, and where ω_{co} denotes the cutoff frequency for the particular waveguide mode under consideration. Direct application of Fourier transform techniques results in an integral representation for the pulse analogous to that given in Eq. (1.30) which is then solved by approximating the propagation factor $\tilde{k}(\omega) = \beta(\omega)$ for $|\omega| \geq \omega_{co}$ by the first three terms in its Taylor series expansion about the pulse frequency $\omega_c \gg \omega_{co}$ as

$$\beta(\omega) \simeq \beta_0 + \beta_1(\omega - \omega_c) + \beta_2(\omega - \omega_c)^2, \tag{1.38}$$

where $\beta_j = \beta^{(j)}(\omega_c)/j!$, $j = 0, 1, 2$ are the respective Taylor series coefficients with $\beta^{(j)} \equiv \partial^j\beta/\partial\omega^j$. Knop and Cohn [99] pointed out that this quadratic approximation applies only in the quasimonochromatic case and that if the initial pulse rise-time is short, this assumption no longer applies and the resulting system transfer function yields an output before an input is applied, violating the principle of primitive causality.[5] Nevertheless, approximate numerical solutions for this problem with the exact form of the propagation factor indicate that for times $t > L/c$ (L being the propagation length of the waveguide), the approximately degraded waveforms obtained by Elliott (and corrected by Knop and Cohn) are good approximations to the actual output pulse shapes of the waveguide. In order to overcome these difficulties, Knop [100] derived an exact solution to the problem of the propagation of a rectangular-pulse-modulated carrier input in a simple plasma medium. By employing transfer function techniques and the Laplace transform, a closed form analytic solution for the output was obtained in the form of a series summation over odd-order Bessel functions. Case and Haskell [101] simplified the result somewhat by indicating that this series summation could be written as the sum of two Lommel functions. Finally, the paper by Vogler [102] generalized these results to arbitrary waveforms in ideal (lossless) waveguides. However, these results apply only to such idealized situations.

A problem of related interest is the transient response treated in the papers by Wait [103] and Haskell and Case [104]. Wait deals with approximate methods for determining pulse distortion in a dispersive channel (such as a metallic waveguide) based upon the integral representation

$$E(z, t) = \frac{1}{2\pi}\int_{-\infty}^{\infty} R(z, \omega)e^{i\varphi(z,\omega)}g(\omega)e^{-i\omega t}\,d\omega, \tag{1.39}$$

where $R(z, \omega)e^{i\varphi(z,\omega)}$ is the *transfer function* for the linear dispersive system, $R(z, \omega)$ describing the slowly-varying (with respect to ω) amplitude and phase of

---

[5]The *causality condition* referred to as *primitive causality* states that the effect cannot precede the cause, where "cause" and "effect" must be appropriately defined for the particular situation considered. The *causality condition* known as *relativistic causality* states that any signal cannot propagate with a velocity greater than the speed of light c in vacuum [7].

the system, and where $g(\omega) = \int_{-\infty}^{\infty} E_0(t)e^{-i\omega t}\, dt$ is the spectrum of the source pulse $E_0(t) = E(0, t)$. For the approximate evaluation of this integral representation using asymptotic methods, it is first expressed as

$$J = \int_{-\infty}^{\infty} G(\omega)e^{-izF(\omega)}\, d\omega, \tag{1.40}$$

where $F(\omega)$ is real and contains all of the rapidly varying terms from the integrand in Eq. (1.39), whereas $G(\omega) = |G(\omega)|e^{i\hat{g}(\omega)}$ is a slowly-varying function of ω that may be complex-valued with phase $\hat{g}(\omega)$. The stationary phase method is then applied to evaluate this J integral for a general phase function $F(\omega)$ that describes the rapidly-varying part of the system dispersion. In this approach, the real-valued phase function $F(\omega)$ is first expanded in a Taylor series about each *stationary phase point* ω_s defined by

$$F'(\omega_s) = \left.\frac{\partial F(\omega)}{\partial \omega}\right|_{\omega=\omega_s} = 0, \tag{1.41}$$

so that

$$F(\omega) = F(\omega_s) + \frac{1}{2!}F''(\omega_s)(\omega - \omega_s)^2 + \frac{1}{3!}F'''(\omega_s)(\omega - \omega_s)^3 + \cdots . \tag{1.42}$$

For a symmetric pair of stationary phase points $\pm\omega_s$ with $G(-\omega_s) = G^*(\omega_s)$ and with $F(\omega) = -F(-\omega_s)$ approximated by the first two terms in Eq. (1.42) with $F''(\omega_s) \neq 0$, the asymptotic behavior of J is found to be given by[6]

$$J \sim 2|G(\omega_s)| \left[\frac{2\pi}{zF''(\omega_s)}\right]^{1/2} \cos\left(zF(\omega_s) - \hat{g}(\omega_s) \pm \pi/4\right) \tag{1.43}$$

as $z \to \infty$, where $+\pi/4$ is used when $F''(\omega_s) > 0$ and $-\pi/4$ when $F''(\omega_s) < 0$. When $F(\omega)$ has additional pairs of stationary phase points $\pm\omega_{sj}$ with $F''(\pm\omega_{sj}) \neq 0$, each contribution is of the form given in Eq. (1.43) and the asymptotic behavior of J is given by their superposition. The stationary phase method in mode theory is then developed in an application to a more realistic waveguide mode model for which $F''(\omega_s) = 0$. Instead of expanding $F(\omega)$ about the stationary phase point, Wait [103] expands about the point at which the group velocity has an extremum, as required by the condition

$$\left.\frac{\partial v_g(\omega)}{\partial \omega}\right|_{\omega=\omega_g} = 0. \tag{1.44}$$

---

[6]Notice that Wait does not explicitly display the asymptotic parameter z in his analysis [103] as is done here for clarity.

The resulting asymptotic expansion is then expressed in terms of the *Airy integral*

$$A(\zeta) = \frac{1}{\sqrt{\pi}} \int_0^\infty \cos{(\zeta s + s^3/3)} ds. \tag{1.45}$$

Finally, in a more generalized description of the leading-edge of the propagated signal (the transient solution in Wait's terminology) in an idealized waveguide, it was shown that this leading edge may be described in terms of the *Fresnel integral*

$$\mathcal{C}(\zeta) + i\mathcal{S}(\zeta) = \int_0^\zeta e^{i\frac{\pi}{2}t^2} dt, \tag{1.46}$$

where $\mathcal{C}(\zeta) = \int_0^\zeta \cos{(\frac{\pi}{2}t^2)} dt$ and $\mathcal{S}(\zeta) = \int_0^\zeta \sin{(\frac{\pi}{2}t^2)} dt$ are the well-known *cosine and sine Fresnel integrals* that appear in diffraction theory.

In a reexamination of Sommerfeld's analysis of the signal arrival [8, 9], Haskell and Case [104–106] presented the complete analysis of the arrival of the main signal in a lossless, isotropic plasma medium with dispersion relation

$$k(s) = \frac{1}{c}\left(s^2 + \omega_p^2\right)^{1/2}, \tag{1.47}$$

where ω_p denotes the plasma frequency. Here $s = -i\omega$ is the variable in the Laplace transform representation

$$E(z, t) = \frac{1}{2\pi i} \int_{s_0-\infty}^{s_0+\infty} \tilde{E}(0, s) e^{-[k(s)z-st]} ds, \tag{1.48}$$

where $\tilde{E}(0, s) = \int_0^\infty E(0, t) e^{-st} dt$ is the Laplace transform of the initial plane wave pulse [cf. Eq. (1.27) with $\tilde{k}(\omega) = -ik(-i\omega)$]. For a Heaviside unit step-function modulated signal $E(0, t) = u_H(t) \sin{(\omega_c t)}$ with Laplace transform $\tilde{E}(0, s) = \omega_c/(s^2 + \omega_c^2)$, this integral representation becomes

$$E(\eta, \zeta) = \frac{1}{2\pi i} \int_{x_0-\infty}^{x_0+\infty} \frac{1}{\Omega^2 + 1} \exp\left[\eta\Omega\left(\zeta - (1 + \Pi^2/\Omega^2)^{1/2}\right)\right] d\Omega, \tag{1.49}$$

where $\eta \equiv \omega_c z/c$, $\zeta \equiv \tau/\eta$, $\tau \equiv \omega_c t$, $\Pi \equiv \omega_p/\omega_c$, and $\Omega \equiv s/\omega_c$ are dimensionless. As in Sommerfeld's analysis, the integral in Eq. (1.49) identically vanishes for all $t < z/c$, as required by the principle of relativistic causality. Haskell and Case [104–106] then examined the uniform asymptotic behavior of this integral representation as $\eta \to \infty$ with $t > z/c$ using the saddle point method of integration. In their analysis, the asymptotic behavior of the signal arrival is described in three successive regions: a region before the saddle point crosses the pole (the *anterior transient*)

$$E(\eta, \zeta) \sim \left| \left(\frac{2\Pi}{\pi\eta} \right)^{1/2} \frac{(\zeta^2 - 1)^{1/4}}{\zeta^2(1 - \Pi^2) - 1} \right| \cos \left(\eta\Pi\sqrt{\zeta^2 - 1} + \pi/4 \right) \qquad (1.50)$$

as $\eta \to \infty$, a region when the saddle point is in the neighborhood of the pole (the *main signal buildup*)

$$E(\eta, \zeta) \sim \frac{1}{\sqrt{2}} \left[\left(\tfrac{1}{2} + \mathcal{C}(v) \right)^2 + \left(\tfrac{1}{2} + \mathcal{S}(v) \right)^2 \right]^{1/2} \sin \left[\eta \left(\zeta - \sqrt{1 - \Pi^2} \right) + \Theta_0 \right]$$

$$(1.51)$$

as $\eta \to \infty$ with $v \equiv (\eta/\pi\Pi)^{1/2} \left(1 - \Pi\zeta/\sqrt{\zeta^2 - 1} \right) (\zeta^2 - 1)^{3/4}$ and $\Theta_0 \equiv$ arctan $[(\mathcal{C}(v) - \mathcal{S}(v))/(1 + \mathcal{C}(v) + \mathcal{S}(v))]$, and a region after the saddle point has crossed the pole (the *posterior transient*)

$$E(\eta, \zeta) \sim \left| \left(\frac{2\Pi}{\pi\eta} \right)^{1/2} \frac{(\zeta^2 - 1)^{1/4}}{\zeta^2(1 - \Pi^2) - 1} \right| \sin \left(\eta\Pi\sqrt{\zeta^2 - 1} - \pi/4 \right)$$

$$+ \sin \left(\eta \left(\zeta - \sqrt{1 - \Pi^2} \right) \right) \qquad (1.52)$$

as $\eta \to \infty$. This set of results then provide a partial asymptotic solution of the transient response in an isotropic plasma medium. A review of the various applications of such uniform asymptotic expansion techniques for radiation and diffraction problems may be found in the paper by Ludwig [107].

A paper of significant importance and influence on optical pulse research was published by M. D. Crisp [108] in 1970. In this paper, Crisp showed that "small-area coherent pulses... exactly obey an area theorem which, in the case of an attenuating medium, requires that the pulse area drop to zero exponentially with increasing distance of propagation into the medium" while at the same time the pulse energy may not decrease exponentially. In particular [108], "for pulses short compared with the transverse relaxation time T_2 (including both homogeneous and inhomogeneous broadening), the electric-field envelope oscillates between positive and negative values in just such a way that the area theorem is satisfied but the pulses lose little energy after their initial shaping." The area of a pulse with envelope function $\mathcal{E}(z, t)$ is taken there as being given by the integral

$$\theta(z) \equiv \frac{p}{\hbar} \int_{-\infty}^{\infty} \mathcal{E}(z, t) dt, \qquad (1.53)$$

as defined by McCall and Hahn [109], where p denotes the electric dipole moment of the resonant transition and $\hbar \equiv h/2\pi$ where h is Planck's constant. Crisp [108] then shows that "the formation of these 'zero-degree pulses' gives rise to significant deviations from Beer's law."[7] However, because Crisp's analysis [108] relies upon the slowly-varying envelope approximation through the ansatz of separating the pulse into slowly-varying amplitude $\mathcal{E}(z, t)$ and phase $\varphi(z, t)$ functions as $E(z, t) = \mathcal{E}(z, t) \cos{[\omega(t - nz/c) - \varphi(z, t)]}$, where n denotes a background refractive index that is constant over the pulse spectrum, his results are not completely valid in the ultrashort pulse regime where these unique effects are observed. To this end, Crisp [108] states that

> on the assumption that the host medium is characterized by an index of refraction n and that $\mathcal{E}(z, t)$ and $\varphi(z, t)$ vary little in the time $2\pi/\omega$, the interesting phenomenon of precursors has been neglected.

In order to fully establish these reported results, a more rigorous description must be used, as provided, for example, by the modern asymptotic theory. Nevertheless, Crisp's description of the physical basis of dispersive pulse propagation is remarkably revealing. In this regard, Crisp [108] states that

> As a small-area pulse enters the resonant medium, its leading edge excites a macroscopic polarization in the thin slice of medium located at the surface. This polarization radiates 180° out of phase with respect to the input pulse for a time of the order of T_2 after the pulse has passed. If the trailing edge of the pulse drops off faster than the decay of the macroscopic polarization, then the envelope of the pulse leaving this slice will go through zero and become negative. The next slice of resonant material now sees a field envelope whose trailing edge drops off faster than before and then becomes negative. As a result, the polarization induced in the second slice by the positive lobe of the pulse envelope radiates 180° out of phase with respect to that field and adds to the negative lobe. In this way, a pulse can develop a field envelope which has negative-area regions that subtract from the total-pulse area. Thus, the total-pulse area can go to zero while the pulse energy remains finite.

A fundamental extension of the general topic of dispersive pulse propagation is provided by the problem of the reflection and transmission of a pulsed electromagnetic beam field that is incident upon the planar interface separating two half-spaces containing different dispersive media. Early treatments of this problem have either focused on time-harmonic beam fields in lossless media [110, 111] with emphasis on the Goos–Hänchen effect [112] or on pulsed plane wave fields when the incident medium is vacuum and the second medium is dispersive [54, 113–115]. The formation of a forerunner upon transmission into a plasma medium across a planar vacuum–plasma interface appears to have first been treated by Skrotskaya et al. [113] in 1969. The asymptotic description of the formation of precursors

---

[7]The Beer–Lambert–Bouger law (more commonly referred to as "Beer's law") was originally discovered by Pierre Bouger, as published in his *Essai d'Optique sur la Gradation de la Lumiere* (Claude Jombert, Paris, 1729) and subsequently cited by Johann Heinrich Lambert in *Photometri* (V. E. Klett, Augsburg, 1760). The result was then extended by August Beer in *Einleitung in die höhere Optik* (Friedrich Viewig, Braunschweig, 1853) [see also *Ann. Chem. Phys.* **86**, 78 (1852)] to include the concentration of solutions in the expression of the absorption coefficient for the intensity of light.

upon transmission into a Lorentz model dielectric was then given by Gitterman and Gitterman [114] in 1976. However, each of these descriptions considers only normal plane wave incidence upon the dispersive half-space. An extension to oblique plane wave incidence upon a dispersive half-space has been given by Blaschak and Franzen [115] using numerical methods and by Cartwright [116] using asymptotic methods. The formulation for the general situation in which both media are dispersive and absorptive and the incident field is a pulsed electromagnetic beam field has been given by Marozas and Oughstun [117] based on the form of the Fresnel reflection and transmission coefficients given by Stone [118]. A more compact matrix description has since been used [119] with significant computational advantage. The results of this analysis have direct application in integrated and fiber optic device technologies, the analysis and design of low-observable surfaces, and the analysis of bioelectromagnetic effects in stratified tissue.

A somewhat different application of the problem of dispersive pulse distortion is presented in the paper by Wait [120] which applies Laplace transform techniques to electromagnetic pulse propagation in a geological medium described by the frequency-dependent relative complex permittivity

$$\epsilon_c(\omega) \equiv \epsilon(\omega) + i\frac{\sigma(\omega)}{\omega} = \epsilon_\infty \left(1 + \frac{B}{1 - i\omega\tau}\right)^2, \tag{1.54}$$

which differs from the Debye model [121] of a polar dielectric with relaxation time τ through the square of the factor $\left(1 + \frac{B}{1-i\omega\tau}\right)$. Although no closed-form solutions are obtained and only numerical evaluations of the integral representation are presented, this paper does indicate the applicability of this theory to geological sensing through analysis of the resulting pulse distortion. With use of Laplace transform techniques, Wait [122] also considered the exact (causal) solution for the electromagnetic field radiated by a step-function modulated dipole source in a simple conducting medium whose complex dielectric permittivity has the more general form $\epsilon(\omega) = K(1 + ia/\omega)(1 + ib/\omega)$, where K, a, and b are constants.

A generalization of these integral representation techniques for the analytic description of pulse distortion in a linear dispersive system is given by Jones [123]. By employing the Fourier integral representation of the pulse, as given in Eq. (1.27), and assuming that the system dispersion may be expanded in a Taylor series about the carrier frequency ω_c as

$$\tilde{k}(\omega) = \tilde{k}(\omega_c) + \tilde{k}'(\omega_c)(\omega - \omega_c) + \frac{1}{2!}\tilde{k}''(\omega_c)(\omega - \omega_c)^2 + \frac{1}{3!}\tilde{k}'''(\omega_c)(\omega - \omega_c)^3 + \cdots, \tag{1.55}$$

where $\tilde{k}' = \partial\tilde{k}/\partial\omega$, $\tilde{k}'' = \partial^2\tilde{k}/\partial\omega^2$, $\tilde{k}''' = \partial^3\tilde{k}/\partial\omega^3$, and so on for higher-order terms. If the complex wave number is approximated by the *linear dispersion relation*

$$\tilde{k}(\omega) \approx \tilde{k}^{(1)}(\omega) \equiv \tilde{k}(\omega_c) + \tilde{k}'(\omega_c)(\omega - \omega_c), \tag{1.56}$$

the integral representation (1.30) of the propagated pulse yields

$$U(z,t) \approx \Re \left\{ u\left(t - \tilde{k}'(\omega_c)z\right) e^{i\left(\tilde{k}(\omega_c)z - \omega_c t + \pi/2 - \psi\right)} \right\}, \tag{1.57}$$

in which case the pulse propagates undistorted in shape at the complex group velocity. If the complex wave number is approximated by the *quadratic dispersion relation* (which is equivalent to the quasimonochromatic approximation, yielding noncausal results if the initial pulse rise-time is sufficiently short)

$$\tilde{k}(\omega) \approx \tilde{k}^{(2)}(\omega) \equiv \tilde{k}(\omega_c) + \tilde{k}'(\omega_c)(\omega - \omega_c) + \frac{1}{2}\tilde{k}''(\omega_c)(\omega - \omega_c)^2, \tag{1.58}$$

Jones [123] showed that the resultant integral can be expressed as

$$U(z,t) \approx \Re \left\{ \frac{1}{(2\pi \tilde{k}''(\omega_c))^{(1/2)}} e^{i[\tilde{k}(\omega_c)z - \omega_c t + 3\pi/4 - \psi]} \right.$$
$$\left. \times \int_{-\infty}^{\infty} u(t') \exp\left[-i \frac{(\tilde{k}'(\omega_c)z + t' - t)^2}{2\tilde{k}''(\omega_c)z} \right] dt' \right\}. \tag{1.59}$$

At this second level of approximation, the pulse envelope is again found to propagate at the complex group velocity at the input pulse carrier frequency, but the pulse shape is now found to be proportional to the Fresnel transform of the input pulse shape and, after the pulse has propagated sufficiently far through the dispersive system, the pulse shape becomes proportional to the Fourier transform of the initial pulse envelope shape.

More recently published treatments concerned with the propagation of wave packets in dispersive and absorptive media [124–127] have employed Havelock's technique of expanding the phase function appearing in the integral representation of the field in a Taylor series about some fixed characteristic frequency of the initial pulse. This approach may also be coupled with a recursive technique in order to obtain purported correction terms of arbitrary dispersive and absorptive orders for the resultant envelope function. This analysis again relies upon the quasimonochromatic approximation, and hence, can only be applied to study the evolution of pulses with slowly-varying envelope functions in weakly dispersive systems. This approximate approach has since been adopted as the standard in both fiber optics [36] and nonlinear optics in general [128, 129] with little regard for its accuracy.

A *phase-space asymptotic description* of wave propagation in homogeneous, dispersive, dissipative media has also been introduced by Hoc, Besieris and Sockell [130]. This phase-space approach uses a combined space-wavevector–time domain representation in which wave propagation is realized through the evolution of the Wigner distribution function [131]. However, the dispersive properties of the medium are approximated by an appropriate power series expansion in their analysis, thereby limiting the method to narrowband pulses.

A variety of purely numerical techniques have also been developed for the depiction of ultrashort pulse dynamics in temporally dispersive media and systems. In general, there are three computational approaches in time-domain electromagnetics and optics [132]: time-domain, frequency-domain, and hybrid time- and frequency-domain methods. Because of its natural representation in the frequency domain, as exhibited in Eqs. (1.27) and (1.30), dispersive pulse propagation phenomena are most amenable to computational methods that are based upon the Fourier–Laplace transform. Discrete Fourier transform methods have been successfully applied to a variety of problems in dispersive pulse propagation, most notably by Veghte and Balanis [133] for transient signals on microstrip transmission lines, Moten, Durney and Stockham [134], Albanese, Penn and Medina [135], and Blaschak and Franzen [115] for trapezoidal envelope pulses in Debye and Lorentz model dielectrics with application to bioelectromagnetics [136]. However, because the discrete Fourier transform can only be numerically computed (e.g., using the FFT algorithm) over a finite frequency domain $[-\omega_{max}, \omega_{max}]$, the complete evolution of the Sommerfeld precursor can never be completely described using this numerical approach alone. For this purpose, an efficient Laplace transform algorithm was developed by Hosono [137–139] and later updated by Wyns et al. [140]. Comparison of numerical results using Hosono's Laplace transform algorithm with the uniform asymptotic description of the Sommerfeld precursor due to Handelsman and Bleistein [27] established the accuracy of the modern asymptotic theory for the propagated signal front [140]. A hybrid asymptotic-FFT algorithm has also been developed [141] where the high-frequency structure $|\omega| > \omega_{max}$ of the propagated field is determined using the uniform asymptotic description. A similar approach has been described by Ziolkowski, Dudley and Dvorak [142] who have introduced an extraction technique that dramatically reduces the required number of sample points for an FFT simulation of pulse propagation through lossy plasma media. The extracted term contains the high-frequency information of the propagated pulse which may then be evaluated analytically. An alternate approach to the numerical inversion of the Laplace transform using the Dubner–Abate algorithm [143] with application to dispersive signal propagation in a Lorentz model dielectric has been given by Barakat [144]. However, the numerical results presented there do not resolve the high-frequency structure that is present at the onset of the Sommerfeld precursor.

Time-domain methods, on the other hand, involve the solution of a set of integro-differential equations that result from the differential form of Maxwell's equations taken together with the appropriate constitutive relation for the material dispersion. For example, for a linearly polarized plane wave pulse propagating in the positive z-direction through a locally linear, temporally dispersive dielectric medium, *Maxwell's equations* simplify to

$$\frac{\partial E_y(z,t)}{\partial z} = \frac{\partial B_x(z,t)}{\partial t}, \tag{1.60}$$

$$\frac{\partial H_x(z,t)}{\partial z} = \frac{\partial D_y(z,t)}{\partial t}, \tag{1.61}$$

where $H_x(z,t) = \mu^{-1}B_x(z,t)$ with constant magnetic permeability μ and where $D_y(z,t) = \epsilon_0 E_y(z,t) + P_y(z,t)$ with the *induced polarization* in the dielectric given by the causal relation $P_y(z,t) = \epsilon_0 \int_{-\infty}^{t} \hat{\chi}^{(1)}(t-t')E_y(z,t')dt'$, where $\hat{\chi}^{(1)}(t)$ is the first-order (or linear) electric susceptibility response function. The importance of this representation comes from the fact that the induced polarization can be directly obtained from a differential equation describing the causal dynamical response of the material's atomic and molecular structure to the local electromagnetic field. For example, for a single-resonance Lorentz-model dielectric [cf. Eq. (1.33)],

$$\frac{d^2 P_y}{dt^2} + 2\delta \frac{d P_y}{dt} + \omega_0^2 P_y = b^2 \epsilon_0 E_y, \tag{1.62}$$

where $P_y(z,t)$ is the resultant polarization from the locally induced molecular dipole moments from harmonically bound electrons with undamped resonance frequency ω_0, damping constant $\delta \geq 0$, and plasma frequency b. The set of relations given in Eqs. (1.60)–(1.62) provide a complete system of coupled differential relations for the description of plane-wave pulse propagation in a Lorentz-model dielectric. Analogous expressions can be obtained for multiple-resonance Lorentz-model dielectrics, Debye-model dielectrics, and Drude-model conductors.

The numerical solution of the coupled system of differential equations given above may be accomplished with the *finite-difference time-domain* (FDTD) method [145–147] which discretizes space and time into cells of size $(\Delta x, \Delta y, \Delta z, \Delta t)$. With application of central differencing in both time and space, Eqs. (1.60) and (1.61) can be expressed in *finite-difference form* as

$$H_x(j+\tfrac{1}{2}, k+\tfrac{1}{2}) = H_x(j+\tfrac{1}{2}, k-\tfrac{1}{2}) + \frac{\Delta t}{\mu \Delta z}\left[E_y(j+\tfrac{1}{2}, k) - E_y(j, k)\right],$$

$$\tag{1.63}$$

$$D_y(j, k+1) = D_y(j, k) + \frac{\Delta t}{\Delta z}\left[H_x(j+\tfrac{1}{2}, k+\tfrac{1}{2}) - H_x(j-\tfrac{1}{2}, k+\tfrac{1}{2})\right].$$

$$\tag{1.64}$$

Finally, with the substitution $P_y = D_y - \epsilon_0 E_y$ in Eq. (1.62), the finite-difference form of this second-order differential equation is found as

$$\begin{aligned}
E_y(j, k+1) = \Bigg\{ & \frac{1}{\epsilon_0}\Big[\left(\omega_0^2 (\Delta t)^2 + 2\delta \Delta t + 2\right) D_y(j, k+1) - D_y(j, k) \\
& + \left(\omega_0^2 (\Delta t)^2 - 2\delta \Delta t + 2\right) D_y(j, k-1)\Big] \\
& + 4 E_y(j, k) - \left[\left(\omega_0^2 + b^2\right)(\Delta t)^2 - 2\delta \Delta t + 2\right] E_y(j, k-1)\Bigg\} \\
& \times \left[\left(\omega_0^2 + b^2\right)(\Delta t)^2 + 2\delta \Delta t + 2\right]^{-1}.
\end{aligned} \tag{1.65}$$

This second-order dispersion relation completes the set of difference relations given in Eqs. (1.63) and (1.64) for the direct time integration of the linearly polarized plane-wave pulsed field equations in a single-resonance Lorentz-model dielectric.

The computational stability of the FDTD method is set by the *Courant condition* [148] which is based on the minimum time $\Delta t = \Delta z/c$ required to propagate one cell length Δz, for example, in the z-direction. The FDTD grid is then causally connected when the Courant condition is satisfied. The cell size is determined by the smallest sampled wavelength $\lambda_{min} = 2\pi c/\omega_{max}$ required by the physical problem, which is inversely related to the maximum angular frequency ω_{max} in the corresponding frequency-domain representation. Although this method can be used to model very complex propagation geometries, it suffers from numerical dispersion [148, 149] that is due to the nonzero spatial and time grid spacing which results in a frequency-dependent phase error. Joseph, Hagness and Taflove [146] used the FTDT method to numerically calculate the high-frequency leading edge of the Sommerfeld precursor in a Lorentz model dielectric, comparing it with uniform asymptotic results. When approximate saddle point locations were used in the uniform asymptotic description, they found a maximum error of approximately 10 percent. However, when precise numerical saddle point locations are used in the uniform asymptotic description, the quantitative agreement between the two results is exceptionally good with the rms error between them decreasing with increasing propagation distance.

Hybrid time- and frequency-domain methods are derived from the fact that the angular frequency ω and time t for a wave are conjugate physical variables that satisfy the *indeterminacy principle*

$$\Delta\omega\Delta t \geq \frac{1}{2},\tag{1.66}$$

where $\Delta\omega$ denotes the indeterminacy in the angular frequency measured in the time interval Δt. Any physically realizable function of time $f(t)$ can only be known over a finite time interval of duration τ, whereas its frequency structure can only be known up to some maximum angular frequency value ω_{max}. This function can then be described in terms of the conjugate physical variables (t, ω) in a rectangular region with sides τ and ω_{max}, respectively, known as an *information diagram*, illustrated in Fig. 1.4. The principle of indeterminacy then describes elementary cells in the (t, ω) plane with area $S_{el} = \Delta\omega\Delta t \sim 1/2$, where an elementary cell's particular shape is arbitrary. This information diagram can be divided into square elementary cells, called *Gabor cells* [150], with sides $\Delta t = 1$ s and $\Delta\omega = 1$ r/s. The function $f(t)$ can then be described by its values f_n that are specified in each cell labeled by the index n. Because the values f_n are complex-valued, having both amplitude and phase with some specified reference value, the total number of independent parameters which completely describe the physical process recorded by an instrument with spectral resolving power ω_{max} in a finite observation time τ is then given by $N_{max} = 2\tau\omega_{max}+1$.

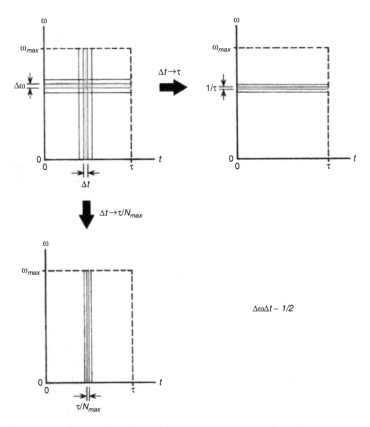

Fig. 1.4 Information diagram partitioned into Gabor cells (center) of dimension $\Delta t \times \Delta \omega$ for a physically realizable function of time measured over a finite time interval τ by an instrument with maximum frequency response $\omega_{\max}$, where $\Delta \omega$ denotes the indeterminacy in the frequency measured in the time interval Δt. As $\Delta t \to \tau$, the Gabor cells become rectangles of duration τ and spectral width $1/\tau$ (right), corresponding to a Fourier series expansion, while as $\Delta t \to \tau/N_{max}$ (bottom), the Gabor cells become rectangles of duration τ/N_{max} and spectral width ω_{max}, where $N_{max} = 2\tau\omega_{max} + 1$, corresponding to the sampling method

Unit area elementary cells can be formed by dividing the information rectangle into horizontal strips of length τ and width $1/\tau$, resulting in the Fourier series expansion

$$f(t) = \sum_{n=-N_{\max}}^{N_{\max}} c_n e^{in\omega_0 t} \tag{1.67}$$

with $\omega_0 \equiv 1/\tau$. This then corresponds to the temporal frequency domain representation of an optical pulse. Unit area elementary cells can also be formed by dividing the information rectangle into vertical strips of width $\tau/N_{\max}$ and length $\omega_{\max}$. The

Whittaker–Shannon–Kotelńikov sampling theorem [151–153] then states that the function $f(t)$ is completely defined over the time interval $t \in [0, \tau]$ by the set of sampled values $f_n \equiv f(n\tau/N_{max})$ if $N_{max} = 2\tau\omega_{max} + 1$. This then corresponds to the time-domain representation of an optical pulse. The signal that occupies the minimum area in the information diagram (i.e., a single Gabor cell) is the Gaussian pulse

$$f_{el}(t) \equiv e^{-(\kappa^2/2)(t-t_0)^2} e^{i(\omega_0 t + \phi_0)} \tag{1.68}$$

centered at the time t_0 with angular frequency $\omega_0 = 1/\tau$, where κ describes the signal duration and ϕ_0 is a phase constant. The Fourier transform of this elementary signal is

$$\tilde{f}_{el} = (\sqrt{2\pi}/\kappa) e^{-(\omega-\omega_0)^2/(2\kappa^2)} e^{i((\omega-\omega_0)t_0 - \phi_0)}. \tag{1.69}$$

The rms widths of these two functions are given by $\Delta t = 1/(\sqrt{2}\kappa)$ and $\Delta\omega = \kappa/\sqrt{2}$, respectively, so that $\Delta t \Delta\omega = 1/2$, which is the minimum allowed by the indeterminacy principle. A given temporal signal can then be decomposed in terms of elementary signals of the form given in Eq. (1.68) in a manner depending upon the value of the parameter κ. In the limit as $\kappa \to 0$, the elementary signal becomes a pure sinusoid and the decomposition is just the Fourier series expansion of the signal (i.e., a temporal frequency-domain representation is obtained). In the opposite limit as $\kappa \to \infty$, the elementary signal approaches a Dirac delta function and the decomposition yields the sampling method (i.e., a time-domain representation is obtained). The general series decomposition in terms of the elementary functions $f_{el}(t)$ results in a hybrid time-frequency-domain representation, which may be ideally suited for specific problems. One promising approach based on this decomposition is provided by the application of wavelets to electromagnetics [154].

Numerical methods have had immediate application to inverse problems in time-domain electromagnetics and optics. Beezley and Krueger [155] used a time-domain technique to derive a nonlinear integro-differential equation that relates the complex permittivity (through its susceptibility kernel) to the reflection operator for a one-dimensional homogeneous slab of an unknown material. The susceptibility kernel can then be determined from measured reflection data. An extension of this technique to stratified media was then given by Kristensson and Krueger [156, 157]. A Green's function technique developed by Krueger and Ochs [158] for nondispersive inhomogeneous media, which is related to the invariant embedding technique developed by Corones, Davison, Ammicht, Kristensson and Krueger [159–163], has also been applied to both direct- and inverse-scattering problems. The generaliztion of this method by Karlsson, Otterheim, and Stewart [164] has general applicability to one-dimensional dispersive media.

An equally important problem in dispersive wave theory, which has also received considerable attention in the open literature, concerns the propagation velocities of light in dispersive and absorptive media. The description of the velocity of energy transport through a causally dispersive medium, originally considered by Brillouin [10] in 1914 and again [165] in 1932, was reinvestigated by Schulz-DuBois [166] in 1969, Askne and Lind [167] and Loudon [168] in 1970, as well as by Anderson and Askne [169] in 1972. General conservation properties for waves in dispersive, dissipative media were also considered by Censor and Brandstatter [170]. By determining the total energy density associated with a time-harmonic, plane wave electromagnetic field in the dispersive medium as the sum of the energy density of the wave plus the energy stored in the medium, Loudon [168] arrived at a simple, closed form expression for the energy transport velocity for propagation of a monochromatic wave field in a classical single resonance Lorentz medium, given by

$$v_E(\omega) = \frac{c}{n_r(\omega) + n_i(\omega)/\delta}, \tag{1.70}$$

which is relativistically causal (i.e., is both positive and less than or equal to the vacuum speed of light c for all real angular frequencies $\omega \geq 0$). This description showed that the energy and group velocities are different in a region of anomalous dispersion in causally dispersive dielectrics whereas the energy velocity approaches the group velocity in regions of normal dispersion as the material loss becomes vanishingly small. A generalization of Loudon's energy velocity to a multiple resonance Lorentz model dielectric was then given by Oughstun and Shen [171] in 1988. The direct generalization of this result to an arbitrary dispersive medium has been shown by Barash and Ginzburg [172] to require a specific microscopic model of the material dispersion. Nevertheless, a general formulation for a material with arbitrary dispersive properties has been proposed [173–175] by Glasgow and co-workers.

The 1970 review paper by Smith [176] provides a general discussion of the definitions, physical significances, interrelationships, and observabilities of seven different velocities of light with an outlook to determine the proper velocity to be used in describing dispersive pulse propagation. Smith states here that "there is no observable physical quantity associated with the phase of a light wave," and that "a phase velocity cannot be attributed to a wave packet or to any wave except a monochromatic wave." As a consequence, the concept of the phase velocity is useful only in determining the phase of a monochromatic wave in space and time given the phase at some other position and time. Nevertheless, this elementary concept is indeed an essential ingredient in the mathematical description of dispersive pulse dynamics, as is evident in the Fourier–Laplace integral representation given in Eq. (1.27). Smith also showed that the standard definition of the group velocity fails to describe the motion of the peak in the envelope of an arbitrary pulse when the pulse frequency is within a region of anomalous dispersion. In order to correct this, a generalized group velocity defined as the velocity of motion of the temporal center of gravity of the amplitude of the wave packet has been proposed. The velocity of energy transport, defined as the ratio of the time-average Poynting vector to the time-average total electromagnetic energy density, is incorrectly criticized as not corresponding to any real observable physical quantity, as is the signal velocity introduced by Sommerfeld [8, 9] and Brillouin [10] and refined by Baerwald [60] as well as by Trizna and Weber [177, 178]. For completeness, Smith also briefly considered the relativistic velocity constant and the ratio of units velocity appearing in Maxwell's equations. Finally, Smith introduced a new definition for the velocity of light, called the *centrovelocity*, which is defined as the velocity of motion of the temporal center of gravity of the intensity of the pulse. It is important to note here that both the generalized definition of the group velocity and the centrovelocity are undefined for a step-function modulated signal. An "eighth" velocity of light has also been proposed by Bloch [179] that is based upon the cross-correlation of the original and received pulses.

Of further interest in this matter is the large number of experimental papers that report measurements of superluminal pulse velocities in apparent violation of the special theory of relativity. Early measurements by Basov et al. [180] and Faxvog et al. [181] report laboratory measurements of superluminal group velocity values in inverted media. Based upon the rate equations for resonant radiative energy transfer in a two-level active medium, Basov et al. [180] point out that as soon as the energy density in the leading edge of the pulse reaches a sufficient level, all of the active particles will produce stimulated emission into that leading edge, while the trailing edge of the pulse travels through the medium with either a much lower amplification or even attenuation. As a result, the peak of the pulse will undergo an additional shift forward, thereby resulting in an effective peak velocity that exceeds c. However, this does not contradict the relativistic principle of causality since such a motion of the peak amplitude point of the pulse occurs due to the deformation of the initially weak leading edge of nonzero intensity. That is, only the pulse shape has propagated with a velocity exceeding c and not the energy of the pulse. Later experimental results by Chu and Wong [182] in an absorbing medium are similarly flawed [183] because they do not adequately consider the effects of pulse distortion.

The proper (i.e., physically correct) description of these experimental results must then be obtained through a careful analysis of Gaussian pulse dynamics in a dispersive, absorptive medium. Early research by Garrett and McCumber [184] in 1970 on Gaussian pulse propagation in an anomalous dispersion medium employed the complex index of refraction

$$n(\omega) = n_\infty - \frac{\omega_0 \omega_p}{\omega(\omega - \omega_0 + i\gamma)}, \tag{1.71}$$

which is valid in the weak dispersion limit when $|\omega_p/\gamma| \ll n_\infty$. Here ω_0 denotes the resonance frequency, ω_p the plasma frequency, and γ describes the atomic linewidth of the medium whose refractive index has the large frequency limit $n_\infty \geq 1$. Their analysis showed that when a Gaussian pulse is incident upon a thin slab of such a weakly dispersive material,

> the power spectrum of the emerging pulse is still substantially Gaussian, and the peak of the pulse emerges at the instant given by the classical group velocity expression, even if that instant is earlier than the instant at which the peak of the input pulse entered the slab.

However, as the slab thickness increases, the pulse spectrum becomes sufficiently distorted such that the classical group velocity approximation is no longer applicable because [184] "the concept of a dominant central frequency $\bar{\omega}$ eventually fails." A more detailed analysis of Gaussian pulse dynamics in a Lorentz model dielectric that is not restricted to the weak dispersion limit was then given by Tanaka, Fujiwara, and Ikegami [185] in 1986. They found that

> the velocity of the wave packet, defined as the traveling distance of the peak amplitude divided by its flight time, decreases in the absorption range of frequency, although the group velocity become infinite in the same range. Fast pulse propagation, which was observed by Chu and Wong and is characterized by a packet velocity faster than the light velocity, turns out to be a characteristic in the early stage of the flight and is understood in terms of packet distortion due to damping of Fourier-component waves in an anomalous dispersion region. It also turns out that slow pulse propagation characterized by a packet velocity less than the light velocity appears for long travelling distance.

A uniform asymptotic description of Gaussian pulse dynamics in a Lorentz model dielectric has been given by Balictsis and Oughstun [19, 33, 35], beginning in 1993. This asymptotic description shows that a Gaussian pulse evolves into a set of precursor fields that are a characteristic of the dispersive properties of the medium. In a single-resonance Lorentz model dielectric, an ultrashort Gaussian pulse will then evolve into a pair of pulse components that travel at different velocities. The first component is a generalized Sommerfeld precursor whose peak amplitude travels near the vacuum speed of light c with near zero attenuation and dominates the total propagated field when the input pulse carrier angular frequency ω_c is well above the medium resonance frequency ω_0. The second component is a generalized Brillouin precursor whose peak amplitude travels near the zero frequency velocity $c/n(0)$ and only attenuates algebraically with propagation distance $z > 0$ as $1/\sqrt{z}$, thereby dominating the total propagated field when ω_c is either below or near resonance. Furthermore, the peak amplitude point in each pulse component is found [34] to evolve with the propagation distance $z \geq 0$ in such a manner that the pulse dynamics evolve from the classical group velocity description to the energy velocity description [186, 187] as the propagation distance increases into the mature dispersion regime.[8] The extension of these results to a joint time-frequency analysis of the precursor fields associated with gaussian pulse propagation in one-dimensional photonic crystals by Safian, Sarris, and Mojahedi [188] has shown that

> the precursor fields associated with superluminal pulse propagation travel at subluminal speeds, and thus, the arrival of these precursor fields must be associated with the arrival of "genuine information." ... abnormal group velocities do not contradict Einstein causality.

Asymptotic analysis of the electromagnetic Brillouin precursor in a one-dimensional photonic crystal has been given by Uitham and Hoenders [189].

The energy velocity description [186, 187] due to Sherman and Oughstun provided the first detailed physical explanation of ultrashort pulse dynamics in

---

[8]The mature dispersion regime typically includes propagation distances greater than an absorption depth at some characteristic frequency of the initial pulse.

dispersive, attenuative systems. The description is based upon the energy velocity and attenuation of time-harmonic waves in the dispersive system and is derived from the modern asymptotic theory of dispersive pulse propagation [14–19]. This new description reduces to the group velocity description in dispersive systems that are lossless. In particular, Sherman and Oughstun have shown that [186], in the mature dispersion regime,

> the field is dominated by a single real frequency at each space–time point, That frequency ω_E is the frequency of the time-harmonic wave with the least attenuation that has energy velocity equal to z/t."

The modern asymptotic description of dispersive pulse propagation has provided the first accurate, analytical description of ultrawideband pulse dynamics in causal systems. This description is based upon Olver's saddle point method [23] together with those uniform asymptotic techniques [22, 24–27] that are necessary to obtain a continuous evolution of the field at some fixed observation distance that is within the mature dispersion regime. Similar asymptotic results were also published at the same time by Vasilev, Kelbert, Sazonov, and Chaban [190] using saddle point techniques. This analysis has shown that the precursor fields that result from an input Heaviside unit step-function modulated signal are a dominant feature of the field evolution in the mature dispersion regime, as illustrated in Fig. 1.5 (compare with Fig. 1.2). The appearance of these precursor fields is critically dependent upon the rise-time of the initial pulse which must be faster than the characteristic relaxation time of the medium [20]. In addition, this modern asymptotic description has also provided both a precise definition [14, 15, 17] and physical interpretation [30] of the signal velocity in the dispersive medium. This proper description of the signal velocity is critically dependent upon the correct description and interpretation of the precursor fields. Furthermore, this signal velocity is shown [15] to be bounded below by zero and above by Loudon's energy transport velocity [168] for monochromatic waves, as illustrated in Fig. 1.6 (compare with Fig. 1.3). A comparison of the original Sommerfeld–Brillouin theory with the modern asymptotic theory is given in the pair of review papers [191, 192] by Cartwright and Oughstun. An extension of this

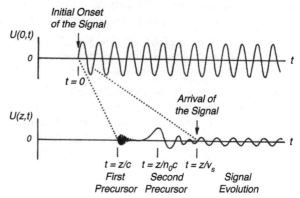

Fig. 1.5 Modern asymptotic description of the evolution of a Heaviside step-function signal $U(0, t) = u_H(t) \sin(\omega_c t)$ in a single-resonance Lorentz medium, where $n_0 = n(0)$ and v_s denotes the signal velocity at the signal frequency ω_c

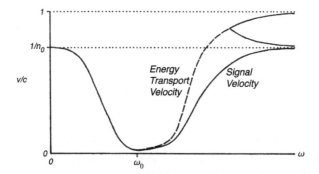

Fig. 1.6 Comparison of the frequency dependence of the relative energy transport velocity v_E/c for a monochromatic signal with the relative signal velocity v_s/c of a Heaviside step function signal in a single-resonance Lorentz model dielectric with resonance frequency ω_0

asymptotic theory to an active Lorentz medium has been given by Safian, Mojahedi, and Sarris [193].

The analysis of an ultrawideband, double-exponential electromagnetic pulse that is propagating through either a hollow metallic waveguide or a cold plasma medium has been described in detail by Dvorak and Dudley [194]. The dispersion relation is then given either by Eq. (1.37) or (1.47), respectively. Closed-form solutions are obtained in terms of incomplete Lipschitz–Hankel integrals which provide an accurate description of the leading edge of the Sommerfeld precursor that is characteristic of this dispersion relation type.

The central importance of the precursor fields in both the analysis and interpretation of linear dispersive pulse propagation phenomena is also realized in the study of ultrashort pulse dynamics. The asymptotic theory shows that the resultant pulse distortion due to an input rectangular envelope modulated pulse is primarily due to the precursor fields that are associated with the leading and trailing edges of the input pulse envelope regardless of the (positive-definite) initial temporal pulse width [18]. The interference between these two sets of precursor fields increases with the propagation distance in the dispersive medium and naturally leads to asymmetric pulse distortion. The situation is quite different for a pulse whose initial envelope function is infinitely smooth, such as that for a Gaussian envelope pulse. In that case, the entire pulse evolves into a single set of precursor fields provided that the initial pulse width is shorter than the characteristic relaxation time of the medium [19, 33–35], reinforcing the fundamental role that the precursor fields play in dispersive pulse dynamics. Optical precursors have also been shown by Gagnon [195] to be interpreted in terms of self-similar solutions of specific evolution equations. Predicted new precursors in materials exhibiting spatial dispersion have also been described by Birman and Frankel [196, 197], the effects of spatial dispersion increasing the signal velocity of the pulse. A Sommerfeld "precursor-like" effect has also been described [198] for pulse diffraction in vacuum. The diffraction of electromagnetic pulses in a Lorentz model dielectric has been described by Solhaug,

Stamnes, and Oughstun [199]. It is shown there that the edge diffraction process is itself dispersive and adds to the effects of material dispersion. The combined effects of edge diffraction and temporal dispersion then result in a singularity in the Brillouin precursor evolution that is due to the geometric focusing of the quasi-static portion of the transient field component.

Recently published research [31, 32] by Xiao and Oughstun has identified the space–time domain within which the group velocity approximation is valid. This group velocity description of dispersive pulse propagation is based on both the slowly varying envelope approximation and the Taylor series approximation of the complex wavenumber about some characteristic angular frequency ω_c of the initial pulse at which the temporal pulse spectrum is peaked, as originally described by Havelock [55, 56]. The *slowly varying envelope approximation* is a hybrid time and frequency domain representation [129] in which the temporal field behavior is separated into the product of a slowly varying temporal envelope function and an exponential phase term whose angular frequency is centered about ω_c. The envelope function is assumed to be slowly varying on the time scale $\Delta t_c \sim 1/\omega_c$, which is equivalent [200] to the quasimonochromatic assumption that its spectral bandwidth $\Delta \omega$ is sufficiently narrow that the inequality $\Delta \omega / \omega_c \ll 1$ is satisfied. Under these approximations, the frequency dependence of the wavenumber may then be approximated by the first few terms of its Taylor series expansion about the characteristic pulse frequency ω_c with the unfounded assumption [124, 125, 129] that improved accuracy can always be obtained through the inclusion of higher-order terms. This assumption has been proven incorrect [31, 32] in the ultrashort pulse, ultrawideband signal regime, optimal results being obtained using either the quadratic or the cubic dispersion approximation of the wavenumber.

Because of the slowly varying envelope approximation together with the neglect of the frequency dispersion of the material attenuation, the group velocity approximation is invalid in the ultrashort pulse regime in a causally dispersive material or system, its accuracy decreasing as the propagation distance $z \geq 0$ increases. This is in contrast with the modern asymptotic description whose accuracy increases in the sense of Poincaré [58] as the propagation distance increases. There is then a critical propagation distance $z_c > 0$ such that the group velocity description using either the quadratic or cubic dispersion approximation provides an accurate description of the pulse dynamics when $0 \leq z < z_c$, the accuracy increasing as $z \to 0$, while the modern asymptotic theory provides an accurate description when $z > z_c$, the accuracy increasing as $z \to \infty$. This critical distance z_c depends upon both the dispersive material and the input pulse characteristics including the pulse shape, temporal width, and characteristic angular frequency ω_c. For example, $z_c = \infty$ for the trivial case of vacuum for all pulse shapes, whereas $z_c \sim z_d$ for an ultrashort, ultrawideband pulse in a causally dispersive dielectric with e^{-1} amplitude penetration depth z_d at the characteristic oscillation frequency ω_c of the input pulse. An extension of the asymptotic theory to the weak dispersion, weak absorption limit when the absorption depth becomes very large has been given by Oughstun, Cartwtright, Gauthier, and Jeong [201].

In an attempt to overcome these critical difficulties, Brabec and Krausz [202] have proposed to replace the slowly varying envelope approximation with a slowly evolving wave approach that is supposed to be "applicable to the single-cycle regime of nonlinear optics." As with the slowly varying envelope approximation, the difficulty with the slowly evolving wave approach is twofold. First, the fundamental hyperbolic character of the underlying wave equation is approximated as parabolic. The characteristics then propagate instantaneously [59]. Second, the subsequent imposed Taylor series expansion of the complex wavenumber $\tilde{k}(\omega)$ about ω_c approximates the material dispersion by its local behavior about some characteristic angular frequency of the initial pulse. Because this approximation is incapable of correctly describing the precursor fields, it is then incapable of correctly describing the dynamical evolution of any ultrashort pulse and its accuracy monotonically decreases [31] as z exceeds the absorption depth z_d in the dispersive medium.

Recent research has been focused on the contentious topic of superluminal pulse propagation [203–210] in both linear and nonlinear optics. Again, the origin of this controversy may be found in the group velocity approximation which is typically favored by experimentalists. In response, Landauer [204] has argued for more careful analysis of experimental measurements reporting superluminal motions. Diener [205] then showed that "the group velocity cannot be interpreted as a velocity of information transfer" in those situations in which it exceeds the vacuum speed of light c. This analysis is in fact based upon an extension of Sommerfeld's now classic proof [8, 9] that the signal arrival cannot exceed c in a causally dispersive medium. Kuzmich, Dogariu, Wang, Milonni, and Chiao [208] defined a signal velocity that is operationally based upon the optical signal-to-noise ratio and showed that, in those cases when the group velocity is negative, "quantum fluctuations limit the signal velocity to values less than c." In addition, they argue that a more general definition of the "signal" velocity of a light pulse must satisfy two fundamental criteria: "First, it must be directly related to a known and practical way of detecting a signal. Second, it should refer to the fastest practical way of communicating information." In contrast, Nimtz and Haibel [209] argue regarding superluminal tunneling phenomena that

the principle of causality has not been violated by superluminal signals as a result of the finite signal duration and the corresponding narrow frequency-band width. But, amazingly enough, the time span between the cause and effect is reduced by a superluminal signal velocity compared with the time span in the case of light propagation from cause to effect.

In addition, Winful [210] argues that

distortionless tunneling of electromagnetic pulses through a barrier is a quasistatic process in which the slowly varying envelope of the incident pulse modulates the amplitude of a standing wave. For pulses longer than the barrier width, the barrier acts as a lumped element with respect to the pulse envelope. The envelopes of the transmitted and reflected fields can adiabatically follow the incident pulse with only a small delay that originates from energy storage.

Unfortunately, each of these arguments neglects the frequency-dependent attenuation of the material comprising the barrier. When material attenuation is properly included, the possibility of evanescent waves is replaced by inhomogeneous waves [211, 212], thereby rendering the accuracy of this superluminal tunneling analysis as questionable at best.

It is indeed clear that a more physically meaningful pulse velocity measure needs to be considered in order to accurately describe the complicated pulse evolution that occurs in ultrashort dispersive pulse dynamics when the initial pulse envelope is continuous (e.g., Gaussian). One possible velocity measure considers the packet velocity that is described by the temporal moments of the pulse [213]. A related measure would track the temporal centroid of the Poynting vector of the pulse. This pulse centroid velocity of the Poynting vector was first introduced by Lisak [214] in 1976. Recent descriptions of its properties by Peatross, Glasgow, and Ware [215, 216] and Cartwright and Oughstun [217, 218] have established its efficacy in describing the evolution of the pulse velocity with propagation distance in a single-resonance Lorentz model dielectric. In particular, it has been shown [218] that the instantaneous centroid velocity of the pulse Poynting vector approaches the classical group velocity in the limit as the propagation distance approaches zero from above when the input pulse carrier frequency lies within the normal dispersion region either above or below the medium absorption band, but in general not in the region of anomalous dispersion. However, the classical group velocity is always obtained at any fixed propagation distance in the limit as the initial pulse width is allowed to increase indefinitely. This limiting behavior reinforces the previously established [31, 32] region of applicability of the group velocity approximation.

The experimental observability of optical precursors has been proposed by Aaviksoo, Lippmaa, and Kuhl [219] in 1988 using the transient response of excitonic resonances to picosecond pulse excitation. The experimental observation of the Sommerfeld precursor was then reported [220] in 1991. The experimental observation of the Brillouin precursor in a linear medium exhibiting resonant dispersion using an ultrashort optical pulse has proved to be more difficult [221] with questionable results. Recent experimental observations [222] of optical precursors when the input ultrashort pulse is in the region of anomalous dispersion of a weakly dispersive medium has been reported by Jeong, Dawes, and Gauthier. However, their measurements have proved to be of the late-time behavior of the precursor evolution as described, for example, by Crisp [108] in 1970. Observation of optical precursors (comprised of the Sommerfeld and Brillouin precursors) at the biphoton level has also been reported by Du et al. [223]. Quite fortunately, these difficulties are not present in the microwave domain for Debye-type dielectrics. Measurements of the sub-exponential decay of the Brillouin precursor generated by an ultrawideband microwave pulse penetrating into deionized water have been reported by Pieraccini et al. [224] with similar results reported by Alejos, Dawood, and Medina [225] for foliage and by Mohammed, Dawood, and Alejos [226] for soil. Of significant related interest is the unpublished report [227] from the Directed Energy Technology Office of the Dahlgren Division of the Naval Surface Warfare Center which presents detailed measurements of the leading- and trailing-edge Brillouin precursors in water. These experimental results provide a complete, independent verification of the asymptotic description presented in [228]. Taken together with the experimental results of Pleshko and Palócz [63], as well as with the reported observation of magnetostatic wave precursors in thin ferrite films by Stancil [229], these experimental results provide a complete verification of the modern asymptotic theory in its description of ultrashort dispersive pulse dynamics.

1.3   Organization of the Book

The subject of pulsed electromagnetic radiation and propagation in a linear, temporally dispersive medium is considered here in two major movements. For convenience, the subject matter has been partitioned into two separate volumes in order to adequately represent these two movements. These are Volume 1 on the fundamental electromagnetic theory and Volume 2 on the ensuing analytical description of the propagated field through the use of mathematically well-defined asymptotic expansion techniques that are ideally suited for this type of problem.

The fundamental electromagnetic theory presented in volume 1 is developed for a general, linear dispersive medium with both a frequency-dependent dielectric permittivity $\epsilon(\omega)$, a frequency-dependent magnetic permeability $\mu(\omega)$, and a frequency-dependent conductivity $\sigma(\omega)$. This then provides the underlying theory for pulse propagation in purely dielectric or purely conductive media, as well as in semiconducting materials and metamaterials. The analysis presented in the final chapter of this volume is based upon the Advanced Physical Optics lecture notes on free-fields in homogeneous, isotropic, locally linear (but nondispersive) media given by George C. Sherman in 1973 at The Institute of Optics of the University of Rochester.

The asymptotic theory of the propagated pulse dynamics is developed in volume 2 for the multiple resonance Lorentz model dielectric, the Rocard–Powles extension of a Debye model dielectric, and the Drude model conductor as well as for composite models of semiconducting materials (i.e., a material that is a "good conductor" in one frequency domain and a "good dielectric" in another). This description then provides the framework for the description of pulse propagation phenomena in other types of linear dispersive media such as so-called metamaterials (i.e., materials with frequency-dependent dielectric permittivity and magnetic permeability such that the resultant complex index of refraction takes on negative values over a finite frequency domain).

For convenience, both mksa and Gaussian (cgs) units are used throughout this two volume work. This is accomplished through the use of a conversion factor $*$ that appears in double brackets $\| * \|$ in each equation (if required). If that factor is included in the affected equation, it is then in cgs units, while if that factor is replaced by unity, it is in mksa units. If no such factor is present, then that equation is correct in either system of units.

Problems

1.1 Derive the expression given in Eq. (1.24). *Hint:* Make use of the trigonometric identity $\cos\alpha + \cos\beta = 2\cos((\alpha+\beta)/2)\cos((\alpha-\beta)/2)$.

1.2 Extend the result given in Eq. (1.24) to the linear superposition of three time-harmonic waves with equal amplitudes and nearly equal wavenumbers k, $k-\delta k$, and $k+\delta k$ and associated angular frequencies ω, $\omega-\delta\omega$, and $\omega+\delta\omega$, respectively. Show that the surfaces of constant phase propagate with the phase velocity $v_p = \bar{\omega}/\bar{k}$ with mean wavenumber $\bar{k} = k$ and mean angular frequency $\bar{\omega} = \omega$, while the surfaces of constant amplitude propagate with the group velocity $v_g = \delta\omega/\delta k$. With $\omega_1 = 2\pi$ and $\delta\omega = \omega_1/10$, compute the superposition of these three waveforms both directly and using the derived expression when the system dispersion is given by $k(\omega) = (\omega/c)(1.5 + \omega/10)$.

1.3 Extend the result given in Eq. (1.24) to the linear superposition of two time-harmonic waves with unequal amplitudes a_1 and a_2 and nearly equal wavenumbers k and $k - \delta k$ and associated angular frequencies ω and $\omega - \delta\omega$, respectively. *Hint:* Express the resulting superposed waveform in the form $U(z,t) = a_1\cos(\psi - \varphi) + a_2\cos(\psi + \varphi)$, where $\varphi = \varphi(\delta k, \delta\omega)$, and use the angle sum and angle difference relations for the cosine function.

1.4 Show that the temporal Fourier transform of the Fourier–Laplace integral representation of $U(z,t)$ given in Eq. (1.27) satisfies the Helmholtz equation $\left(\nabla^2 + \tilde{k}^2(\omega)\right)\tilde{U}(z,\omega) = 0$ with boundary value $U(0,t) = u(t)\sin(\omega_c t + \psi)$.

1.5 Use the quadratic dispersion relation (1.38) to approximate the dispersion relation given in Eq. (1.37) about the angular frequency $\omega_c = 2\omega_{co}$. Determine the rms error of this quadratic approximation over the angular frequency domain $\omega \in [\omega_c - \delta\omega, \omega_c + \delta\omega]$ as $\delta\omega$ increases from 0 to ω_{co}. Describe the behavior of the quadratic approximation as ω_c decreases to the cutoff frequency ω_{co}.

1.6 With the expression $n(\omega) = \left(1 - \sum_{j=0}^{N}\omega_{pj}^2/(\omega^2 - \omega_j^2 + 2i\gamma_j\omega)\right)^{1/2}$ for the complex refractive index of a multiple-resonance Lorentz model dielectric as a starting point, derive the expression given in Eq. (1.71) for the behavior about the resonance frequency ω_0 in the weak dispersion limit when $|\omega_{p0}/\gamma_0| \ll n_\infty$, where n_∞ describes the limiting behavior of the index of refraction above ω_0. Explain any additional approximations that may need to be made.

1.7 Derive the results given in Eqs. (1.57) and (1.59).

1.8 Show that the temporal Fourier transform of the Gaussian pulse given in Eq. (1.68) is given by Eq. (1.69) and that the respective rms widths Δt and $\Delta\omega$ for these two functions satisfy $\Delta t\,\Delta\omega = 1/2$, the minimum allowed by the indeterminacy principle.

References

1. J. C. Maxwell, "A dynamical theory of the electromagnetic field," *Phil. Trans. Roy. Soc. (London)*, vol. 155, pp. 450–521, 1865.
2. J. C. Maxwell, *A Treatise on Electricity and Magnetism*. Oxford: Oxford University Press, 1873.
3. H. A. Lorentz, *The Theory of Electrons*. Leipzig: Teubner, 1906. Ch. IV.
4. A. Einstein, "Zur elektrodynamik bewegter körper," *Ann. Phys.*, vol. 17, pp. 891–921, 1905.
5. R. B. Lindsay, *Mechanical Radiation*. New York: McGraw-Hill, 1960. Ch. 1.
6. J. S. Toll, "Causality and the dispersion relation: Logical foundations," *Phys. Rev.*, vol. 104, no. 6, pp. 1760–1770, 1956.
7. H. M. Nussenzveig, *Causality and Dispersion Relations*. New York: Academic, 1972. Ch. 1.
8. A. Sommerfeld, "Ein einwand gegen die relativtheorie der elektrodynamok und seine beseitigung," *Phys. Z.*, vol. 8, p. 841, 1907.
9. A. Sommerfeld, "Über die fortpflanzung des lichtes in disperdierenden medien," *Ann. Phys.*, vol. 44, pp. 177–202, 1914.
10. L. Brillouin, "Über die fortpflanzung des licht in disperdierenden medien," *Ann. Phys.*, vol. 44, pp. 204–240, 1914.
11. L. Brillouin, *Wave Propagation and Group Velocity*. New York: Academic, 1960.
12. K. E. Oughstun, "Dynamical evolution of the precursor fields in linear dispersive pulse propagation in lossy dielectrics," in *Ultra-Wideband, Short-Pulse Electromagnetics 2* (L. Carin and L. B. Felsen, eds.), pp. 257–272, New York: Plenum, 1994.
13. J. A. Stratton, *Electromagnetic Theory*. New York: McGraw-Hill, 1941.
14. K. E. Oughstun, *Propagation of Optical Pulses in Dispersive Media*. PhD thesis, The Institute of Optics, University of Rochester, 1978.
15. K. E. Oughstun and G. C. Sherman, "Propagation of electromagnetic pulses in a linear dispersive medium with absorption (the Lorentz medium)," *J. Opt. Soc. Am. B*, vol. 5, no. 4, pp. 817–849, 1988.
16. S. Shen and K. E. Oughstun, "Dispersive pulse propagation in a double-resonance Lorentz medium," *J. Opt. Soc. Am. B*, vol. 6, pp. 948–963, 1989.
17. K. E. Oughstun and G. C. Sherman, "Uniform asymptotic description of electromagnetic pulse propagation in a linear dispersive medium with absorption (the Lorentz medium)," *J. Opt. Soc. Am. A*, vol. 6, no. 9, pp. 1394–1420, 1989.
18. K. E. Oughstun and G. C. Sherman, "Uniform asymptotic description of ultrashort rectangular optical pulse propagation in a linear, causally dispersive medium," *Phys. Rev. A*, vol. 41, no. 11, pp. 6090–6113, 1990.
19. C. M. Balictsis and K. E. Oughstun, "Uniform asymptotic description of ultrashort Gaussian pulse propagation in a causal, dispersive dielectric," *Phys. Rev. E*, vol. 47, no. 5, pp. 3645–3669, 1993.
20. K. E. Oughstun, "Noninstantaneous, finite rise-time effects on the precursor field formation in linear dispersive pulse propagation," *J. Opt. Soc. Am. A*, vol. 12, pp. 1715–1729, 1995.
21. J. A. Solhaug, K. E. Oughstun, J. J. Stamnes, and P. Smith, "Uniform asymptotic description of the Brillouin precursor in a single-resonance Lorentz model dielectric," *Pure Appl. Opt.*, vol. 7, no. 3, pp. 575–602, 1998.
22. N. A. Cartwright and K. E. Oughstun, "Uniform asymptotics applied to ultrawideband pulse propagation," *SIAM Rev.*, vol. 49, no. 4, pp. 628–648, 2007.
23. F. W. J. Olver, "Why steepest descents," *SIAM Rev.*, vol. 12, no. 2, pp. 228–247, 1970.
24. C. Chester, B. Friedman, and F. Ursell, "An extension of the method of steepest descents," *Proc. Cambridge Phil. Soc.*, vol. 53, pp. 599–611, 1957.
25. N. Bleistein, "Uniform asymptotic expansions of integrals with stationary point near algebraic singularity," *Com. Pure and Appl. Math.*, vol. XIX, no. 4, pp. 353–370, 1966.
26. N. Bleistein, "Uniform asymptotic expansions of integrals with many nearby stationary points and algebraic singularities," *J. Math. Mech*, vol. 17, no. 6, pp. 533–559, 1967.

27. R. A. Handelsman and N. Bleistein, "Uniform asymptotic expansions of integrals that arise in the analysis of precursors," *Arch. Rat. Mech. Anal.*, vol. 35, pp. 267–283, 1969.
28. L. Felsen and N. Marcuvitz, *Radiation and Scattering of Waves*. Englewood Cliffs, NJ: Prentice-Hall, 1973.
29. N. Bleistein and R. Handelsman, *Asymptotic Expansions of Integrals*. New York: Dover, 1975.
30. K. E. Oughstun, P. Wyns, and D. P. Foty, "Numerical determination of the signal velocity in dispersive pulse propagation," *J. Opt. Soc. Am. A*, vol. 6, no. 9, pp. 1430–1440, 1989.
31. K. E. Oughstun and H. Xiao, "Failure of the quasimonochromatic approximation for ultrashort pulse propagation in a dispersive, attenuative medium," *Phys. Rev. Lett.*, vol. 78, no. 4, pp. 642–645, 1997.
32. H. Xiao and K. E. Oughstun, "Failure of the group velocity description for ultrawideband pulse propagation in a double resonance Lorentz model dielectric," *J. Opt. Soc. Am. B*, vol. 16, no. 10, pp. 1773–1785, 1999.
33. C. M. Balictsis and K. E. Oughstun, "Uniform asymptotic description of Gaussian pulse propagation of arbitrary initial pulse width in a linear, causally dispersive medium," in *Ultra-Wideband, Short-Pulse Electromagnetics 2* (L. Carin and L. B. Felsen, eds.), pp. 273–283, New York: Plenum, 1994.
34. K. E. Oughstun and C. M. Balictsis, "Gaussian pulse propagation in a dispersive, absorbing dielectric," *Phys. Rev. Lett.*, vol. 77, no. 11, pp. 2210–2213, 1996.
35. C. M. Balictsis and K. E. Oughstun, "Generalized asymptotic description of the propagated field dynamics in Gaussian pulse propagation in a linear, causally dispersive medium," *Phys. Rev. E*, vol. 55, no. 2, pp. 1910–1921, 1997.
36. G. P. Agrawal, *Nonlinear Fiber Optics*. Academic, 1989.
37. M. N. Islam, *Ultrafast Fiber Switching Devices and Systems*. Cambridge: Cambridge University Press, 1992.
38. C. Rullière, ed., *Femtosecond Laser Pulses: Principles and Experiments*. Berlin: Springer-Verlag, 1998.
39. W. R. Hamilton, "Researches respecting vibration, connected with the theory of light," *Proc. Royal Irish Academy*, vol. 1, pp. 341–349, 1839.
40. G. G. Stokes, "Smith's prize examination question no. 11," in *Mathematical and Physical Papers*, vol. 5, Cambridge University Press, 1905. pg. 362.
41. L. Rayleigh, "On progressive waves," *Proc. London Math. Soc.*, vol. IX, pp. 21–26, 1877.
42. L. Rayleigh, "On the velocity of light," *Nature*, vol. XXIV, pp. 52–55, 1881.
43. M. Born and E. Wolf, *Principles of Optics*. Cambridge: Cambridge University Press, seventh (expanded) ed., 1999.
44. K. E. Oughstun and G. C. Sherman, *Pulse Propagation in Causal Dielectrics*. Berlin: Springer-Verlag, 1994.
45. P. Drude, *Lehrbuch der Optik*. Leipzig: Teubner, 1900. Ch. V.
46. E. Mach, *The Principles of Physical Optics: An Historical and Philosophical Treatment*. New York: First German edition 1913. English translation 1926, reprinted by Dover, 1953. Ch. VII.
47. F. A. Jenkins and H. E. White, *Fundamentals of Optics*. New York: McGraw-Hill, third ed., 1957. Ch. 23.
48. W. Voigt, "Über die änderung der schwingungsform des lichtes beim fortschreiten in einem dispergirenden oder absorbirenden mittel," *Ann. Phys. und Chem. (Leipzig)*, vol. 68, pp. 598–603, 1899.
49. W. Voigt, "Weiteres zur änderung der schwingungsform des lichtes beim fortschreiten in einem dispergirenden oder absorbirenden mittel," *Ann. Phys. (Leipzig)*, vol. 4, pp. 209–214, 1901.
50. P. Ehrenfest, "Mißt der aberrationswinkel in fall einer dispersion des äthers die wellengeschwindigkeit?," *Ann. Phys. (Leipzig)*, vol. 33, p. 1571, 1910.
51. A. Laue, "Die fortpflanzung der strahlung in dispergirenden und absorbirenden medien," *Ann. Phys.*, vol. 18, p. 523, 1905.

52. P. Debye, "Näherungsformeln für die zylinderfunktionen für grosse werte des arguments und unbeschränkt verander liche werte des index," *Math Ann.*, vol. 67, pp. 535–558, 1909.

53. R. Mosseri, "Léon Brillouin: une vie à la croisée des ondes," *Sciences et Vie*, vol. numero special "200 ans de sciences 1789-1989", pp. 256–261, 1989.

54. W. Colby, "Signal propagation in dispersive media," *Phys. Rev.*, vol. 5, no. 3, pp. 253–265, 1915.

55. T. H. Havelock, "The propagation of groups of waves in dispersive media," *Proc. Roy. Soc. A*, vol. LXXXI, p. 398, 1908.

56. T. H. Havelock, *The Propagation of Disturbances in Dispersive Media.* Cambridge: Cambridge University Press, 1914.

57. L. Kelvin, "On the waves produced by a single impulse in water of any depth, or in a dispersive medium," *Proc. Roy. Soc.*, vol. XLII, p. 80, 1887.

58. E. T. Copson, *Asymptotic Expansions.* London: Cambridge University Press, 1965.

59. P. M. Morse and H. Feshbach, *Methods of Theoretical Physics.* New York: McGraw-Hill, 1953. Vol. I.

60. H. Baerwald, "Über die fortpflanzung von signalen in disperdierenden medien," *Ann. Phys.*, vol. 7, pp. 731–760, 1930.

61. N. S. Shiren, "Measurement of signal velocity in a region of resonant absorption by ultrasonic paramagnetic resonance," *Phys. Rev.*, vol. 128, pp. 2103–2112, 1962.

62. T. A. Weber and D. B. Trizna, "Wave propagation in a dispersive and emissive medium," *Phys. Rev.*, vol. 144, pp. 277–282, 1966.

63. P. Pleshko and I. Palócz, "Experimental observation of Sommerfeld and Brillouin precursors in the microwave domain," *Phys. Rev. Lett.*, vol. 22, pp. 1201–1204, 1969.

64. O. Avenel, M. Rouff, E. Varoquaux, and G. A. Williams, "Resonant pulse propagation of sound in superfluid $^3He - B$," *Phys. Rev. Lett.*, vol. 50, no. 20, pp. 1591–1594, 1983.

65. E. Varoquaux, G. A. Williams, and O. Avenel, "Pulse propagation in a resonant medium: Application to sound waves in superfluid $^3He - B$," *Phys. Rev. B*, vol. 34, no. 11, pp. 7617–7640, 1986.

66. C. Eckart, "The approximate solution of one-dimensional wave equations," *Rev. Modern Physics*, vol. 20, pp. 399–417, 1948.

67. G. B. Whitham, "Group velocity and energy propagation for three-dimensional waves," *Comm. Pure Appl. Math.*, vol. XIV, pp. 675–691, 1961.

68. M. J. Lighthill, "Group velocity," *J. Inst. Maths. Applics.*, vol. 1, pp. 1–28, 1964.

69. L. J. F. Broer, "On the propagation of energy in linear conservative waves," *Appl. Sci. Res.*, vol. A2, pp. 329–344, 1950.

70. C. O. Hines, "Wave packets, the Poynting vector, and energy flow: Part I – Non-dissipative (anisotropic) homogeneous media," *J. Geophysical Research*, vol. 56, no. 1, pp. 63–72, 1951.

71. M. A. Biot, "General theorems on the equivalence of group velocity and energy velocity," *Phys. Rev.*, vol. 105, pp. 1129–1137, 1957.

72. C. O. Hines, "Wave packets, the Poynting vector, and energy flow: Part II – Group propagation through dissipative isotropic media," *J. Geophysical Research*, vol. 56, no. 2, pp. 197–220, 1951.

73. C. O. Hines, "Wave packets, the Poynting vector, and energy flow: Part III – Poynting and Macdonald velocities in dissipative anisotropic media (conclusions)," *J. Geophysical Research*, vol. 56, no. 4, pp. 535–544, 1951.

74. R. M. Lewis, "Asymptotic theory of wave-propagation," *Archive Rational Mech. Analysis*, vol. 20, pp. 191–250, 1965.

75. R. M. Lewis, "Asymptotic theory of transients," in *Electromagnetic Wave Theory* (J. Brown, ed.), pp. 845–869, New York: Pergamon, 1967.

76. Y. A. Kravtsov, L. A. Ostrovsky, and N. S. Stepanov, "Geometrical optics of inhomogeneous and nonstationary dispersive media," *Proc. IEEE*, vol. 62, pp. 1492–1510, 1974.

77. D. Censor, "The group Doppler effect," *Journal Franklin Inst.*, vol. 299, pp. 333–338, 1975.

78. D. Censor, "Fermat's principle and real space-time rays in absorbing media," *J. Phys. A*, vol. 10, pp. 1781–1790, 1977.

79. D. Censor, "Wave packets and localized pulses – A dual approach," *Phys. Rev. A*, vol. 24, pp. 1452–1459, 1981.
80. V. Krejči and L. Pekárek, "Determination of dispersion curve parameters from transient wave," *Czech. J. Phys.*, vol. 17, 1967.
81. L. B. Felsen, "Rays, dispersion surfaces and their uses for radiation and diffraction problems," *SIAM Rev.*, vol. 12, pp. 424–448, 1970.
82. K. A. Connor and L. B. Felsen, "Complex space-time rays and their application to pulse propagation in lossy dispersive media," *Proc. IEEE*, vol. 62, pp. 1586–1598, 1974.
83. J. Arnaud, "A theory of Gaussian pulse propagation," *Opt. Quantum Electron.*, vol. 16, pp. 125–130, 1984.
84. L. B. Felsen, "Transients in dispersive media-I. Theory," *IEEE Trans. Antennas Prop.*, vol. 17, pp. 191–200, 1969.
85. G. M. Whitham and L. B. Felsen, "Transients in dispersive media-Part II: Excitation of space waves in a bounded cold magnetoplasma," *IEEE Trans. Antennas Prop.*, vol. 17, pp. 200–208, 1969.
86. L. B. Felsen, "Asymptotic theory of pulse compression in dispersive media," *IEEE Trans. Antennas Prop.*, vol. 19, pp. 424–432, 1971.
87. E. Heyman, "Complex source pulsed beam expansion of transient radiation," *Wave Motion*, vol. 11, pp. 337–349, 1989.
88. E. Heyman and L. B. Felsen, "Weakly dispersive spectral theory of transients (SST). Part I: Formulation and interpretation," *IEEE Trans. Antennas Prop.*, vol. 35, pp. 80–86, 1987.
89. E. Heyman and L. B. Felsen, "Weakly dispersive spectral theory of transients (SST). Part II: Evaluation of the spectral integral," *IEEE Trans. Antennas Prop.*, vol. 35, pp. 574–580, 1987.
90. E. Heyman and L. B. Felsen, "Gaussian beam and pulsed-beam dynamics: complex-source and complex-spectrum formulations within and beyond paraxial asymptotics," *J. Opt. Soc. Am. A*, vol. 18, pp. 1588–1611, 2001.
91. T. Melamed, E. Heyman, and L. B. Felsen, "Local spectral analysis of short-pule-excited scattering from weakly inhomogeneous media. Part I: Forward scattering," *IEEE Trans. Antennas Prop.*, vol. 47, pp. 1208–1217, 1999.
92. T. Melamed, E. Heyman, and L. B. Felsen, "Local spectral analysis of short-pule-excited scattering from weakly inhomogeneous media. Part I: Inverse scattering," *IEEE Trans. Antennas Prop.*, vol. 47, pp. 1218–1227, 1999.
93. T. Melamed and L. B. Felsen, "Pulsed-beam propagation in lossless dispersive media. I. A numerical example," *J. Opt. Soc. Am. A*, vol. 15, pp. 1277–1284, 1998.
94. T. Melamed and L. B. Felsen, "Pulsed-beam propagation in lossless dispersive media. II. Theory," *J. Opt. Soc. Am. A*, vol. 15, pp. 1268–1276, 1998.
95. J. D. Jackson, *Classical Electrodynamics*. New York: John Wiley & Sons, third ed., 1999.
96. R. S. Elliott, "Pulse waveform degradation due to dispersion in waveguide," *IRE Trans. Microwave Theory Tech.*, vol. 5, pp. 254–257, 1957.
97. M. P. Forrer, "Analysis of millimicrosecond RF pulse transmission," *Proc. IRE*, vol. 46, pp. 1830–1835, 1958.
98. R. D. Wanselow, "Rectangular pulse distortion due to a nonlinear complex transmission propagation constant," *J. Franklin Inst.*, vol. 274, pp. 178–184, 1962.
99. C. M. Knop and G. I. Cohn, "Comments on pulse waveform degradation due to dispersion in waveguide," *IEEE Trans. Microwave Theory Tech.*, vol. 11, pp. 445–447, 1963.
100. C. M. Knop, "Pulsed electromagnetic wave propagation in dispersive media," *IEEE Trans. Antennas Prop.*, vol. 12, pp. 494–496, 1964.
101. C. T. Case and R. E. Haskell, "On pulsed electromagnetic wave propagation in dispersive media," *IEEE Trans. Antennas Prop.*, vol. 14, p. 401, 1966.
102. L. E. Vogler, "An exact solution for wave-form distortion of arbitrary signals in ideal wave guides," *Radio Sci.*, vol. 5, pp. 1469–1474, 1970.
103. J. R. Wait, "Propagation of pulses in dispersive media," *Radio Sci.*, vol. 69D, pp. 1387–1401, 1965.

104. R. E. Haskell and C. T. Case, "Transient signal propagation in lossless, isotropic plasmas," *IEEE Trans. Antennas Prop.*, vol. 15, no. 3, pp. 458–464, 1967.

105. R. E. Haskell and C. T. Case, "Transient signal propagation in lossless, isotropic plasmas: Volume I," Tech. Rep. No. 212, Physical Sciences Research Papers, Microwave Physics Laboratory, Air Force Cambridge Research Laboratories, 1966.

106. R. E. Haskell and C. T. Case, "Transient signal propagation in lossless, isotropic plasmas: Volume II," Tech. Rep. No. 241, Physical Sciences Research Papers, Microwave Physics Laboratory, Air Force Cambridge Research Laboratories, 1966.

107. D. Ludwig, "Uniform asymptotic expansions for wave propagation and diffraction problems," *SIAM Rev.*, vol. 12, pp. 325–331, 1970.

108. M. D. Crisp, "Propagation of small-area pulses of coherent light through a resonant medium," *Phys. Rev. A*, vol. 1, no. 6, pp. 1604–1611, 1970.

109. S. L. McCall and E. L. Hahn, "Self-induced transparency by pulsed coherent light," *Phys. Rev. Lett.*, vol. 18, no. 21, pp. 908–912, 1967.

110. B. R. Horowitz and T. Tamir, "Unified theory of total reflection phenomena at a dielectric interface," *Applied Phys.*, vol. 1, pp. 31–38, 1973.

111. C. C. Chan and T. Tamir, "Beam phenomena at and near critical incidence upon a dielectric interface," *J. Opt. Soc. Am. A*, vol. 4, pp. 655–663, 1987.

112. F. Goos and H. Hänchen, "Ein neuer und fundamentaler Versuch zur Totalreflexion," *Ann. Physik*, vol. 6, no. 1, pp. 333–345, 1947.

113. E. G. Skrotskaya, A. N. Makhlin, V. A. Kashin, and G. V. Skrotskiĭ, "Formation of a forerunner in the passage of the front of a light pulse through a vacuum-medium interface," *Zh. Eksp. Teor. Fiz.*, vol. 56, no. 1, pp. 220–226, 1969. [English translation: Sov. Phys. JETP vol. 29, 123–125 (1969)].

114. E. Gitterman and M. Gitterman, "Transient processes for incidence of a light signal on a vacuum-medium interface," *Phys. Rev. A*, vol. 13, pp. 763–776, 1976.

115. J. G. Blaschak and J. Franzen, "Precursor propagation in dispersive media from short-rise-time pulses at oblique incidence," *J. Opt. Soc. Am. A*, vol. 12, no. 7, pp. 1501–1512, 1995.

116. N. A. Cartwright, "Electromagnetic plane-wave pulse transmission into a Lorentz half-space," *J. Opt. Soc. Am. A*, vol. 28, no. 12, pp. 2647–2654, 2011.

117. J. A. Marozas and K. E. Oughstun, "Electromagnetic pulse propagation across a planar interface separating two lossy, dispersive dielectrics," in *Ultra-Wideband, Short-Pulse Electromagnetics 3* (C. Baum, L. Carin, and A. P. Stone, eds.), pp. 217–230, New York: Plenum, 1996.

118. J. M. Stone, *Radiation and Optics, An Introduction to the Classical Theory*. New York: McGraw-Hill, 1963.

119. C. L. Palombini and K. E. Oughstun, "Reflection and transmission of pulsed electromagnetic fields through multilayered biological media," in *Proc. 2011 International Conference on Electromagnetics in Advanced Applications*, 2011. paper #216.

120. J. R. Wait, "Electromagnetic-pulse propagation in a simple dispersive medium," *Elect. Lett.*, vol. 7, pp. 285–286, 1971.

121. P. Debye, *Polar Molecules*. New York: Dover, 1929.

122. J. R. Wait, "Electromagnetic fields of a pulsed dipole in dissipative and dispersive media," *Radio Sci.*, vol. 5, pp. 733–735, 1970.

123. J. Jones, "On the propagation of a pulse through a dispersive medium," *Am. J. Physics*, vol. 42, pp. 43–46, 1974.

124. D. G. Anderson and J. I. H. Askne, "Wave packets in strongly dispersive media," *Proc. IEEE*, vol. 62, pp. 1518–1523, 1974.

125. D. Anderson, J. Askne, and M. Lisak, "Wave packets in an absorptive and strongly dispersive medium," *Phys. Rev. A*, vol. 12, pp. 1546–1552, 1975.

126. L. A. Vainshtein, "Propagation of pulses," *Usp. Fiz. Nauk.*, vol. 118, pp. 339–367, 1976. [English translation: Sov. Phys.-Usp. vol.19, 189–205, (1976)].

127. S. A. Akhmanov, V. A. Yysloukh, and A. S. Chirkin, "Self-action of wave packets in a nonlinear medium and femtosecond laser pulse generation," *Usp. Fiz. Nauk.*, vol. 149, pp. 449–509, 1986. [English translation: Sov. Phys.-Usp. vol.29, 642–677 (1986)].

128. S. A. Akhamanov, V. A. Vysloukh, and A. S. Chirkin, *Optics of Femtosecond Laser Pulses*. New York: American Institute of Physics, 1992.

129. P. N. Butcher and D. Cotter, *The Elements of Nonlinear Optics*. Cambridge: Cambridge University Press, 1990.

130. N. D. Hoc, I. M. Besieris, and M. E. Sockell, "Phase-space asymptotic analysis of wave propagation in homogeneous dispersive and dissipative media," *IEEE Trans. Antennas Prop.*, vol. 33, no. 11, pp. 1237–1248, 1985.

131. E. Wigner, "On the quantum correction for thermodynamic equilibrium," *Phys. Rev.*, vol. 40, pp. 749–759, 1932.

132. K. E. Oughstun, "Computational methods in ultrafast time-domain optics," *Computing in Science & Engineering*, vol. 5, no. 6, pp. 22–32, 2003.

133. R. L. Veghte and C. A. Balanis, "Dispersion of transient signals in microstrip transmission lines," *IEEE Trans. Antennas Prop.*, vol. 34, no. 12, pp. 1427–1436, 1986.

134. K. Moten, C. H. Durney, and T. G. Stockham, "Electromagnetic pulse propagation in dispersive planar dielectrics," *Bioelectromagnetics*, vol. 10, pp. 35–49, 1989.

135. R. Albanese, J. Penn, and R. Medina, "Short-rise-time microwave pulse propagation through dispersive biological media," *J. Opt. Soc. Am. A*, vol. 6, pp. 1441–1446, 1989.

136. R. Albanese, J. Blaschak, R. Medina, and J. Penn, "Ultrashort electromagnetic signals: Biophysical questions, safety issues, and medical opportunities," *Aviation. Space and Environmental Medicine*, vol. 65, no. 5, pp. 116–120, 1994.

137. T. Hosono, "Numerical inversion of Laplace transform," *Trans. IEE of Japan*, vol. 99, no. 10, pp. 44–50, 1979.

138. T. Hosono, "Numerical inversion of Laplace transform and some applications to wave optics," in *Proceedings of the URSI Symposium on Electromagnetic Wave Theory*, (München), pp. C1–C4, 1980.

139. T. Hosono, "Numerical inversion of Laplace transform and some applications to wave optics," *Radio Science*, vol. 16, no. 6, pp. 1015–1019, 1981.

140. P. Wyns, D. P. Foty, and K. E. Oughstun, "Numerical analysis of the precursor fields in dispersive pulse propagation," *J. Opt. Soc. Am. A*, vol. 6, no. 9, pp. 1421–1429, 1989.

141. H. Xiao and K. E. Oughstun, "Hybrid numerical-asymptotic code for dispersive pulse propagation calculations," *J. Opt. Soc. Am. A*, vol. 15, no. 5, pp. 1256–1267, 1998.

142. S. L. Dvorak, D. G. Dudley, and R. W. Ziolkowski, "Propagation of UWB electromagnetic pulses through lossy plasmas," in *Ultra-Wideband, Short-Pulse Electromagnetics 3* (C. Baum, L. Carin, and A. P. Stone, eds.), pp. 247–254, New York: Plenum, 1997.

143. H. Dubner and J. Abate, "Numerical inversion of Laplace transforms by relating them to the finite Fourier cosine transform," *J. Assoc. Comput. Mach.*, vol. 15, pp. 115–123, 1968.

144. R. Barakat, "Ultrashort optical pulse propagation in a dispersive medium," *J. Opt. Soc. Am. B*, vol. 3, no. 11, pp. 1602–1604, 1986.

145. T. Kashiwa and I. Fukai, "A treatment of the dispersive characteristics associated with electronic polarization," *Microwave Opt. Technol. Lett.*, vol. 3, pp. 203–205, 1990.

146. R. Joseph, S. Hagness, and A. Taflove, "Direct time integration of Maxwell's equations in linear dispersive media with absorption for scattering and propagation of femtosecond electromagnetic pulses," *Opt. Lett.*, vol. 16, pp. 1412–1414, 1991.

147. R. J. Luebbers and F. Hunsberger, "FD-TD for n-th order dispersive media," *IEEE Trans. Antennas Propag.*, vol. 40, pp. 1297–1301, 1992.

148. A. F. Peterson, S. L. Ray, and R. Mittra, *Computational Methods for Electromagnetics*. Piscataway, NJ: IEEE Press, 1998.

149. L. A. Trefethen, "Group velocity in finite difference schemes," *SIAM Review*, vol. 24, no. 2, pp. 113–136, 1982.

150. D. Gabor, "Theory of communication," *J. Inst. Electrical Eng.*, vol. 93, no. 26, pp. 429–457, 1946.

151. E. T. Whittaker, "On the functions which are represented by the expansions of the interpolation theory," *Proc. Roy. Soc. Edinburgh, Sect. A*, vol. 35, p. 181, 1915.

152. C. E. Shannon, "Communication in the presence of noise," *Proc. IRE*, vol. 37, p. 10, 1949.

153. L. M. Soroko, *Holography and Coherent Optics*. New York: Plenum, 1980. Ch. 5.

154. G. Kaiser, *A Friendly Guide to Wavelets*. Boston: Birkhäuser, 1994.

155. R. S. Beezley and R. J. Krueger, "An electromagnetic inverse problem for dispersive media," *J. Math. Phys.*, vol. 26, no. 2, pp. 317–325, 1985.

156. G. Kristensson and R. J. Krueger, "Direct and inverse scattering in the time domain for a dissipative wave equation. I. Scattering operators," *J. Math. Phys.*, vol. 27, no. 6, pp. 1667–1682, 1986.

157. G. Kristensson and R. J. Krueger, "Direct and inverse scattering in the time domain for a dissipative wave equation. II. Simultaneous reconstruction of dissipation and phase velocity profiles," *J. Math. Phys.*, vol. 27, no. 6, pp. 1683–1693, 1986.

158. R. J. Krueger and R. L. Ochs, "A Green's function approach to the determination of internal fields," *Wave Motion*, vol. 11, pp. 525–543, 1989.

159. J. P. Corones, M. E. Davison, and R. J. Krueger, "Direct and inverse scattering in the time domain via invariant embedding equations," *J. Acoustic Soc. Am.*, vol. 74, pp. 1535–1541, 1983.

160. G. Kristensson and R. J. Krueger, "Direct and inverse scattering in the time domain for a dissipative wave equation. Part I. Scattering operators," *J. Math. Phys.*, vol. 27, pp. 1683–1693, 1986.

161. G. Kristensson and R. J. Krueger, "Direct and inverse scattering in the time domain for a dissipative wave equation. Part II. Simultaneous reconstruction of dissipation and phase velocity profiles," *J. Math. Phys.*, vol. 27, pp. 1667–1682, 1986.

162. G. Kristensson and R. J. Krueger, "Direct and inverse scattering in the time domain for a dissipative wave equation. Part III. Scattering operators in the presence of a phase velocity mismatch," *J. Math. Phys.*, vol. 28, pp. 360–370, 1987.

163. E. Ammicht, J. P. Corones, and R. J. Krueger, "Direct and inverse scattering for viscoelestic media," *J. Acoustic Soc. Am.*, vol. 81, pp. 827–834, 1987.

164. A. Karlsson, H. Otterheim, and R. Stewart, "Transient wave propagation in composite media: Green's function approach," *J. Opt. Soc. Am. A*, vol. 10, no. 5, pp. 886–895, 1993.

165. L. Brillouin, "Propagation of electromagnetic waves in material media," in *Congrès International d'Electricité*, vol. 2, pp. 739–788, Paris: Gauthier-Villars, 1933.

166. E. O. Schulz-DuBois, "Energy transport velocity of electromagnetic propagation in dispersive media," *Proc. IEEE*, vol. 57, pp. 1748–1757, 1969.

167. J. Askne and B. Lind, "Energy of electromagnetic waves in the presence of absorption and dispersion," *Phys. Rev. A*, vol. 2, no. 6, pp. 2335–2340, 1970.

168. R. Loudon, "The propagation of electromagnetic energy through an absorbing dielectric," *Phys. A*, vol. 3, pp. 233–245, 1970.

169. D. Anderson and J. Askne, "Energy relations for waves in strongly dispersive media," *Proc. IEEE Lett.*, vol. 60, pp. 901–902, 1972.

170. D. Censor and J. J. Brandstatter, "Conservation and balance equations for waves in dissipative media," *Appl. Sci. Res.*, vol. 30, pp. 291–303, 1975.

171. K. E. Oughstun and S. Shen, "Velocity of energy transport for a time-harmonic field in a multiple-resonance Lorentz medium," *J. Opt. Soc. Am. B*, vol. 5, no. 11, pp. 2395–2398, 1988.

172. Y. S. Barash and V. L. Ginzburg, "Expressions for the energy density and evolved heat in the electrodynamics of a dispersive and absorptive medium," *Usp. Fiz. Nauk.*, vol. 118, pp. 523–530, 1976. [English translation: Sov. Phys.-Usp. vol. 19, 163–270 (1976)].

173. C. Broadbent, G. Hovhannisyan, M. Clayton, J. Peatross, and S. A. Glasgow, "Reversible and irreversible processes in dispersive/dissipative optical media: Electro-magnetic free energy and heat production," in *Ultra-Wideband, Short-Pulse Electromagnetics 6* (E. L. Mokole, M. Kragalott, and K. R. Gerlach, eds.), pp. 131–142, New York: Kluwer Academic, 2003.

174. S. Glasgow and M. Ware, "Real-time dissipation of optical pulses in passive dielectrics," *Phys. Rev. A*, vol. 80, pp. 043817–043827, 2009.

175. S. Glasgow, J. Corson, and C. Verhaaren, "Dispersive dielectrics and time reversal: Free energies, orthogonal spectra, and parity in dissipative media," *Phys. Rev. E*, vol. 82, p. 011115, 2010.

176. R. L. Smith, "The velocities of light," *Am. J. Phys.*, vol. 38, no. 8, pp. 978–984, 1970.

177. D. B. Trizna and T. A. Weber, "Time delays of electromagnetic pulses due to molecular resonances in the atmosphere and the interstellar medium," *Astrophysical J.*, vol. 159, no. 1, pp. 309–317, 1970.

178. D. B. Trizna and T. A. Weber, "Brillouin revisited: Signal velocity definition for pulse propagation in a medium with resonant anomalous dispersion," *Radio Science*, vol. 17, no. 5, pp. 1169–1180, 1982.

179. S. C. Bloch, "Eighth velocity of light," *Am. J. Phys.*, vol. 45, no. 6, pp. 538–549, 1977.

180. N. G. Basov, R. V. Ambartsumyan, V. S. Zuev, P. G. Kryukov, and V. S. Letokhov, "Propagation velocity of an intense light pulse in a medium with inverted population," *Doklady Akademii Nauk SSSR*, vol. 165, pp. 58–60, 1965. [English translation: Sov. Phys.-Doklady vol. 10, 1039–1040 (1966)].

181. F. R. Faxvog, C. N. Y. Chow, T. Bieber, and J. A. Carruthers, "Measured pulse velocity greater than c in a neon absorption cell," *Appl. Phys. Lett.*, vol. 17, pp. 192–193, 1970.

182. S. Chu and S. Wong, "Linear pulse propagation in an absorbing medium," *Phys. Rev. Lett.*, vol. 48, pp. 738–741, 1982.

183. A. Katz and R. R. Alfano, "Pulse propagation in an absorbing medium," *Phys. Rev. Lett.*, vol. 49, p. 1292, 1982.

184. C. G. B. Garrett and D. E. McCumber, "Propagation of a Gaussian light pulse through an anomalous dispersion medium," *Phys. Rev. A*, vol. 1, pp. 305–313, 1970.

185. M. Tanaka, M. Fujiwara, and H. Ikegami, "Propagation of a Gaussian wave packet in an absorbing medium," *Phys. Rev. A*, vol. 34, pp. 4851–4858, 1986.

186. G. C. Sherman and K. E. Oughstun, "Description of pulse dynamics in Lorentz media in terms of the energy velocity and attenuation of time-harmonic waves," *Phys. Rev. Lett.*, vol. 47, pp. 1451–1454, 1981.

187. G. C. Sherman and K. E. Oughstun, "Energy velocity description of pulse propagation in absorbing, dispersive dielectrics," *J. Opt. Soc. Am. B*, vol. 12, pp. 229–247, 1995.

188. R. Safian, C. D. Sarris, and M. Mojahedi, "Joint time-frequency and finite-difference time-domain analysis of precursor fields in dispersive media," *Phys. Rev. E*, vol. 73, pp. 066602–1–9, 2006.

189. R. Uitham and B. J. Hoenders, "The electromagnetic Brillouin precursor in one-dimensional photonic crystals," *Opt. Comm.*, vol. 281, pp. 5910–5918, 2008.

190. V. A. Vasilev, M. Y. Kelbert, I. A. Sazonov, and I. A. Chaban, "Propagation of ultrashort light pulses in a resonant absorbing medium," *Opt. Spektrosk.*, vol. 64, no. 4, pp. 862–868, 1988.

191. N. A. Cartwright and K. E. Oughstun, "Precursors and dispersive pulse dynamics, a century after the Sommerfeld-Brillouin theory: the original theory," in *Progress in Optics* (E. Wolf, ed.), vol. 59, pp. 209–265, Amsterdam, The Netherlands: Elsevier, 2014.

192. N. A. Cartwright and K. E. Oughstun, "Precursors and dispersive pulse dynamics, a century after the Sommerfeld-Brillouin theory: the modern asymptotic theory," in *Progress in Optics* (E. Wolf, ed.), vol. 60, pp. 263–344, Amsterdam, The Netherlands: Elsevier, 2015.

193. R. Safian, M. Mojahedi, and C. D. Sarris, "Asymptotic description of wave propagation in an active Lorentzian medium," *Phys. Rev. E*, vol. 75, pp. 066611–1–8, 2007.

194. S. Dvorak and D. Dudley, "Propagation of ultra-wide-band electromagnetic pulses through dispersive media," *IEEE Trans. Elec. Comp.*, vol. 37, no. 2, pp. 192–200, 1995.

195. L. Gagnon, "Similarity properties and nonlinear effects on optical precursors," *Phys. Lett. A*, vol. 148, no. 8, pp. 452–456, 1990.

196. J. L. Birman and M. J. Frankel, "Predicted new electromagnetic precursors and altered signal velocity in dispersive media," *Opt. Comm.*, vol. 13, no. 3, pp. 303–306, 1975.

197. M. J. Frankel and J. L. Birman, "Transient optical response of a spatially dispersive medium," *Phys. Rev. A*, vol. 15, no. 5, pp. 2000–2008, 1977.
198. E. M. Belenov and A. V. Nazarkin, "Transient diffraction and precursorlike effects in vacuum," *J. Opt. Soc. Am. A*, vol. 11, no. 1, pp. 168–172, 1994.
199. J. A. Solhaug, J. J. Stamnes, and K. E. Oughstun, "Diffraction of electromagnetic pulses in a single-resonance Lorentz model dielectric," *Pure Appl. Opt.*, vol. 7, no. 5, pp. 1079–1101, 1998.
200. L. Mandel and E. Wolf, *Optical Coherence and Quantum Optics*. Cambridge: Cambridge University Press, 1995. Ch. 3.
201. K. E. Oughstun, N. A. Cartwright, D. J. Gauthier, and H. Jeong, "Optical precursors in the singular and weak dispersion limits," *J. Opt. Soc. Am. B*, vol. 27, no. 8, pp. 1664–1670, 2010.
202. T. Brabec and F. Krausz, "Nonlinear optical pulse propagation in the single-cycle regime," *Phys. Rev. Lett.*, vol. 78, no. 17, pp. 3282–3285, 1997.
203. A. Steinberg, P. G. Kwiat, and R. Y. Chiao, "Measurement of the single-photon tunneling time," *Phys. Rev. Lett.*, vol. 71, no. 5, pp. 708–711, 1993.
204. R. Landauer, "Light faster than light," *Nature*, vol. 365, pp. 692–693, 1993.
205. G. Diener, "Superluminal group velocities and information transfer," *Phys. Lett. A*, vol. 223, pp. 327–331, 1996.
206. P. W. Milonni, K. Furuya, and R. Y. Chiao, "Quantum theory of superluminal pulse propagation," *Optics Express*, vol. 8, no. 2, pp. 59–65, 2001.
207. A. Dogariu, A. Kuzmich, H. Cao, and L. J. Wang, "Superluminal light pulse propagation via rephasing in a transparent anomalously dispersive medium," *Optics Express*, vol. 8, no. 6, pp. 344–350, 2001.
208. A. Kuzmich, A. Dogariu, L. J. Wang, P. W. Milonni, and R. Y. Chiao, "Signal velocity, causality, and quantum noise in superluminal light pulse propagation," *Phys. Rev. Lett.*, vol. 86, no. 18, pp. 3925–3929, 2001.
209. G. Nimtz and A. Haibel, "Basics of superluminal signals," *Ann. Phys. (Leipzig)*, vol. 11, no. 2, pp. 163–171, 2002.
210. H. Winful, "Nature of "superluminal" barrier tunneling," *Phys. Rev. Lett.*, vol. 90, no. 2, pp. 239011–239014, 2003.
211. K. E. Oughstun, "Asymptotic description of pulse ultrawideband electromagnetic beam field propagation in dispersive, attenuative media," *J. Opt. Soc. Am. A*, vol. 18, no. 7, pp. 1704–1713, 2001.
212. K. E. Oughstun, "Asymptotic description of ultrawideband, ultrashort pulsed electromagnetic beam field propagation in a dispersive, attenuative medium," in *Ultra-Wideband, Short-Pulse Electromagnetics 5* (P. D. Smith and S. R. Cloude, eds.), pp. 687–696, New York: Kluwer Academic, 2002.
213. D. Anderson, J. Askne, and M. Lisak, "The velocity of wave packets in dispersive and slightly absorptive media," *Proc. IEEE Lett.*, vol. 63, no. 4, pp. 715–717, 1975.
214. M. Lisak, "Energy expressions and energy velocity for wave packets in an absorptive and dispersive medium," *J. Phys. A: Math. Gen.*, vol. 9, pp. 1145–1158, 1976.
215. J. Peatross, S. A. Glasgow, and M. Ware, "Average energy flow of optical pulses in dispersive media," *Phys. Rev. Lett.*, vol. 84, no. 11, pp. 2370–2373, 2000.
216. M. Ware, S. A. Glasgow, and J. Peatross, "Role of group velocity in tracking field energy in linear dielectrics," *Opt. Exp.*, vol. 9, no. 10, pp. 506–518, 2001.
217. K. E. Oughstun and N. A. Cartwright, "Dispersive pulse dynamics and associated pulse velocity measures," *Pure Appl. Opt.*, vol. 4, no. 5, pp. S125–S134, 2002.
218. N. A. Cartwright and K. E. Oughstun, "Pulse centroid velocity of the Poynting vector," *J. Opt. Soc. Am. A*, vol. 21, no. 3, pp. 439–450, 2004.
219. J. Aaviksoo, J. Lippmaa, and J. Kuhl, "Observability of optical precursors," *J. Opt. Soc. Am. B*, vol. 5, no. 8, pp. 1631–1635, 1988.
220. J. Aaviksoo, J. Kuhl, and K. Ploog, "Observation of optical precursors at pulse propagation in GaAs," *Phys. Rev. A*, vol. 44, no. 9, pp. 5353–5356, 1991.

221. S.-H. Choi and U. Österberg, "Observation of optical precursors in water," *Phys. Rev. Lett.*, vol. 92, no. 19, pp. 1939031–1939033, 2004.
222. H. Jeong, A. M. C. Dawes, and D. J. Gauthier, "Direct observation of optical precursors in a region of anomalous dispersion," *Phys. Rev. Lett.*, vol. 96, no. 14, p. 143901, 2006.
223. S. Du, C. Belthangady, P. Kolchin, G. Yin, and S. Harris, "Observation of optical precursors at the biphoton level," *Opt. Lett.*, vol. 33, no. 18, pp. 2149–2151, 2008.
224. M. Pieraccini, A. Bicci, D. Mecatti, G. Macaluso, and C. Atzeni, "Propagation of large bandwidth microwave signals in water," *IEEE Trans. Ant. Prop.*, vol. 57, no. 11, pp. 3612–3618, 2009.
225. A. V. Alejos, M. Dawood, and L. Medina, "Experimental dynamical evolution of the Brillouin precursor for broadband wireless communication through vegetation," in *Progress in Electromagnetics Research* (J. A. Kong, ed.), vol. 111, pp. 291–309, Cambridge, MA: EMW Publishing, 2011.
226. H. Mohammed, M. Dawood, and A. V. Alejos, "Experimental detection and characterization of Brillouin precursor through loamy soil at microwave frequencies," *IEEE Trans. Geoscience and Remote Sensing*, vol. 50, no. 2, pp. 436–445, 2012.
227. D. C. Stoudt, F. E. Peterkin, and B. J. Hankla, "Transient RF and microwave pulse propagation in a Debye medium (water)," Interaction Note 622, Dahlgren Division, Naval Surface Warfare Center, Directed Technology Office, Electromagnetic and Solid State Technologies Division, Code B20/Bldg 1470, Dahlgren, VA, 4 July 2011.
228. K. E. Oughstun, "Dynamical evolution of the Brillouin precursor in Rocard-Powles-Debye model dielectrics," *IEEE Trans. Ant. Prop.*, vol. 53, no. 5, pp. 1582–1590, 2005.
229. D. D. Stancil, "Magnetostatic wave precursors in thin ferrite films," *J. Appl. Phys.*, vol. 53, no. 3, pp. 2658–2660, 1982.

Chapter 2
Microscopic Electromagnetics

"War es ein Gott der diese Gleichungen schrieb?"

Ludwig Boltzmann, quoting from Goethe's *Faust,* on Maxwell's equations.

A mathematically rigorous, physically based development of the classical theory of electromagnetism is introduced here through a consideration of the microscopic Maxwell–Lorentz theory [1–4]. Although the Lorentz theory of electrons is a purely classical, heuristic model that is incapable of analyzing many fundamental problems associated with the atomic constituency of matter, it is nevertheless an expedient model in providing the proper source terms for the microscopic Maxwell equations. Indeed, the classical Lorentz theory does yield many results connected with the electromagnetic properties of matter that agree in functional form with that given by quantum theory. In particular, the Lorentz theory assumes additional forces of just the right nature such that qualitatively correct expressions are obtained and, by empirical adjustment of the parameters appearing in these ad hoc force relations, quantitatively correct predictions may also be obtained. Even though quantum theory justifies the assumption of these additional forces and shows them to be of electrical origin, the Lorentz theory is incapable of arriving at this fundamental level of understanding.

The starting point of the theory developed here is the microscopic formulation of Maxwell's equations. The microscopic field vectors are related to the microscopic properties of matter through the elementary source terms, these being the microscopic charge and convective current densities. The fundamental electromagnetic interaction between the source terms appearing in the microscopic theory and a test particle employed to measure the constituent field vectors then occurs in vacuum and is completely specified in the microscopic Lorentz–Maxwell theory through the Lorentz force relation. Based upon this microscopic field formalism, the macroscopic properties of material media in their interaction with an electromagnetic field may then be developed in a consistent fashion. An important consequence of this approach is that the relationship of the macroscopic electromagnetic field

© Springer Nature Switzerland AG 2019
K. E. Oughstun, *Electromagnetic and Optical Pulse Propagation*, Springer Series in Optical Sciences 224, https://doi.org/10.1007/978-3-030-20835-6_2

vectors to their microscopic counterparts may be clearly defined with respect to the relationship between the macroscopic and microscopic properties of the particular material medium. As stated by Einstein regarding Maxwell's theory of electromagnetism in his 1936 article on *Physics and Reality* [5],

> the electric field theory of Faraday and Maxwell represents probably the most profound transformation which has been experienced by the foundations of physics since Newton's time... The existence of the field manifests itself, indeed, only when electrically charged bodies are introduced into it. The differential equations of Maxwell connect the spatial and temporal differential coefficients of the electric and magnetic fields. The electric masses are nothing more than places of non-disappearing divergency of the electric field. Light waves appear as undulatory electromagnetic field processes in space.

The important role that this microscopic theory holds in the advancement of theoretical physics, as well as its limitations, were aptly described by Einstein who stated that Lorentz's theory was based on the following hypothesis [5]:

> Everywhere (including the interior of ponderable bodies) the seat of the field is the empty space. The participation of matter in electromagnetic phenomena has its origin only in the fact that the elementary particles of matter carry unalterable electric charges, and, on this account are subject on the one hand to the actions of ponderomotive[1] forces and on the other hand possess the property of generating a field. The elementary particles obey Newton's law of motion for the material point.
>
> This is the basis on which H. A. Lorentz obtained his synthesis of Newton's mechanics and Maxwell's field theory. The weakness of this theory lies in the fact that it tried to determine the phenomena by a combination of partial differential equations (Maxwell's field equations for empty space) and total differential equations (equations of motion of points), which procedure was obviously unnatural. The unsatisfactory part of the theory showed up externally by the necessity of assuming finite dimensions for the particles in order to prevent the electromagnetic field existing at their surfaces from becoming infinitely great. The theory failed moreover to give any explanation concerning the tremendous forces which hold the electric charges on the individual particles. H. A. Lorentz accepted these weaknesses of his theory, which were well known to him, in order to explain the phenomena correctly at least as regards their general lines.
>
> What appears certain to me... is that, in the foundations of any consistent field theory, there shall not be, in addition to the concept of field, any concept concerning particles. The whole theory must be based solely on partial differential equations and their singularity-free solutions.

Throughout the remainder of this book the analysis is presented in both cgs (Gaussian) units and mksa units. This is accomplished by writing each equation such that it can be interpreted in mksa units provided that the factor * in the double brackets $\| * \|$ is omitted, whereas the inclusion of this factor yields the appropriate expression in cgs units. If there isn't any double bracketed quantity present, the expression can be interpreted equally in Gaussian or mksa units unless specifically noted otherwise.

---

[1]The adjective "ponderomotive" describes the tendency to produce movement of a ponderable body. By a "ponderable body" is meant one having nonzero mass.

2.1   The Microscopic Maxwell–Lorentz Theory

In accordance with the classical Lorentz theory [3, 4, 6] it is assumed that matter is composed of three basic physical entities: mass, positive electric charge, and negative electric charge. The spatial distributions of the two types of charge at any given instant of time t are specified at each point of space by two non-negative scalar fields: the density of positive charge $\rho_+(\mathbf{r}, t)$ and the density of negative charge $\rho_-(\mathbf{r}, t)$. These two scalar fields are formally defined through use of the following limiting procedure. Let $Q_+(t)$ denote the total positive charge contained in the interior of a simply connected closed region of space $\mathcal{V}$ with volume V and let P be an interior point of the region with position vector $\mathbf{r}$. The ratio Q_+/V then gives the average volume density of positive charge in $\mathcal{V}$ and in the limit as the region $\mathcal{V}$ shrinks to the point P, the *density of positive charge* at P is obtained, where

$$\rho_+(\mathbf{r}, t) \equiv \lim_{V \to 0} \left(\frac{Q_+(t)}{V} \right). \tag{2.1}$$

In a similar fashion, the *density of negative charge* at P is specified by the limit

$$\rho_-(\mathbf{r}, t) \equiv \lim_{V \to 0} \left(\frac{Q_-(t)}{V} \right), \tag{2.2}$$

where $Q_-(t)$ is the total negative charge contained in the region $\mathcal{V}$. By definition, $\rho_+(\mathbf{r}, t)$ is the positive charge per unit volume at any point of space, and $\rho_-(\mathbf{r}, t)$ is the negative charge per unit volume at any point of space and is taken to be a non-negative quantity. The *charge density* $\rho(\mathbf{r}, t)$, given by

$$\rho(\mathbf{r}, t) = \rho_+(\mathbf{r}, t) - \rho_-(\mathbf{r}, t), \tag{2.3}$$

is the net charge per unit volume at any point of space and time defined by the limit

$$\rho(\mathbf{r}, t) \equiv \lim_{V \to 0} \left(\frac{Q(t)}{V} \right), \tag{2.4}$$

where $Q(t) = Q_+(t) - Q_-(t)$ is the net charge contained in the region $\mathcal{V}$. Each of these charge densities is assumed to be finite and to vary continuously from point to point in space. This mathematical technique of defining a field through a limiting ratio must be used (and interpreted) with caution in order to achieve results that correspond in some sense to physical reality. Quite fortunately, it does provide a convenient means for introducing discontinuities into the formalism. In particular, the concept of a point charge may be introduced in a generalized function sense as an appropriate limit of a continuous charge distribution. With this generalized interpretation, all results derived for a continuous distribution remain valid in the discrete case.

The kinematics of the two types of charge at a given instant are assumed to be specified at each point of space by two vector fields, the convective current densities of positive and negative charge. If, at a given point $\mathbf{r}$ and instant of time t, the positive charge is moving with velocity $\mathbf{v}_+(\mathbf{r}, t)$, then the *convective current density of positive charge* is defined by

$$\mathbf{j}_+(\mathbf{r}, t) \equiv \rho_+(\mathbf{r}, t)\mathbf{v}_+(\mathbf{r}, t). \tag{2.5}$$

In a similar fashion, the *convective current density of negative charge* is defined by

$$\mathbf{j}_-(\mathbf{r}, t) \equiv \rho_-(\mathbf{r}, t)\mathbf{v}_-(\mathbf{r}, t). \tag{2.6}$$

Let da be an infinitesimal regular surface element with unit normal vector $\hat{\mathbf{n}}$, as illustrated in Fig. 2.1. If $\mathbf{j}_+$ is evaluated at an interior point of this surface element, the quantity $\mathbf{j}_+ \cdot \hat{\mathbf{n}} da$ is seen to be the rate at which positive charge flows across da into the region of space that $\hat{\mathbf{n}}$ is directed. The scalar quantity $\mathbf{j}_+ \cdot \hat{\mathbf{n}}$ is then seen to be the rate per unit area of positive charge flow and $\mathbf{j}_+(\mathbf{r}, t)$ itself can be interpreted as the flow of positive charge per unit area per unit time through a surface element oriented perpendicular to and in the direction of the vector $\mathbf{j}_+$. In a similar fashion, $\mathbf{j}_-(\mathbf{r}, t)$ can be interpreted as the flow of negative charge per unit area per unit time through a surface element oriented perpendicular to and in the direction of the vector $\mathbf{j}_-$. The *net convective current density* $\mathbf{j} = \mathbf{j}(\mathbf{r}, t)$ is then given by

$$\mathbf{j}(\mathbf{r}, t) = \mathbf{j}_+(\mathbf{r}, t) - \mathbf{j}_-(\mathbf{r}, t), \tag{2.7}$$

which is simply called the *convective current density*. Consequently, through a surface that is oriented perpendicular to the vector $\mathbf{j}$, a net charge $j = |\mathbf{j}|$ flows per unit area per unit time, so that across an arbitrarily oriented surface element, the net flow of charge per unit area per unit time is given by

$$\mathcal{I}(\mathbf{r}, t) = \mathbf{j}(\mathbf{r}, t) \cdot \hat{\mathbf{n}} da. \tag{2.8}$$

The total convective current flowing across a regular surface $\mathcal{S}$ into the region of space that the unit normal vector $\hat{\mathbf{n}}$ to $\mathcal{S}$ is directed is then given by the surface

Fig. 2.1 Depiction of the microscopic convective current density vector $\mathbf{j}(\mathbf{r}, t)$ passing through an infinitesimal regular surface element da with unit normal vector $\hat{\mathbf{n}}$

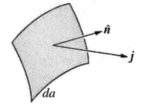

integral

$$\mathcal{I}_{tot}(\mathbf{r}, t) = \int_S \mathbf{j}(\mathbf{r}, t) \cdot \hat{\mathbf{n}} da. \tag{2.9}$$

Suppose now that the surface S appearing in Eq. (2.9) is closed and that $\hat{\mathbf{n}}$ denotes the positive (or outward) unit normal vector to the closed surface. The surface integral of the normal component of the net convective current density $\mathbf{j}(\mathbf{r}, t)$ over the closed surface S then measures the loss of charge from the region of space that is enclosed by S. From *conservation of charge*,[2]

$$\oint_S \mathbf{j}(\mathbf{r}, t) \cdot \hat{\mathbf{n}} da = -\frac{d}{dt} \int_V \rho(\mathbf{r}, t) d^3 r, \tag{2.10}$$

where V is the region enclosed by S. If the net current flow is out of the region, then the net charge in V decreases ($dQ/dt < 0$) and both sides of Eq. (2.10) are positive, whereas if the net current flow is into the region then the net charge in V increases ($dQ/dt > 0$) and both sides of Eq. (2.10) are negative, where

$$Q(t) = \int_V \rho(\mathbf{r}, t) d^3 r \tag{2.11}$$

is the total charge contained in V.

The flow of charge across S originates in two different ways: the surface S may be fixed in space and the net charge density $\rho(\mathbf{r}, t)$ may vary in time, or the net charge density may be invariable with time while the surface S moves in some prescribed manner. Of course, a combination of these two situations may also occur. In either of the latter two cases the integrals in Eq. (2.10) are functions of time by virtue of their variable limits of integration. If S is fixed in space and the volume integral of $\rho(\mathbf{r}, t)$ over the region V enclosed by S is convergent for all time t, then Leibniz' rule applies to the right-hand side of Eq. (2.10) and

$$\oint_S \mathbf{j}(\mathbf{r}, t) \cdot \hat{\mathbf{n}} da = -\int_V \frac{\partial \rho(\mathbf{r}, t)}{\partial t} d^3 r. \tag{2.12}$$

Application of the divergence theorem results in the expression

$$\int_V \left(\nabla \cdot \mathbf{j}(\mathbf{r}, t) + \frac{\partial \rho(\mathbf{r}, t)}{\partial t} \right) d^3 r = 0. \tag{2.13}$$

The integrand here is a continuous function of the coordinates so that there must exist small regions of space within which the integrand does not change sign. If the

---

[2]Conservation laws in nature originated with Democritus, a fifth century B.C. atomist who asserted that matter was indestructible.

integral is to vanish for arbitrarily small regions of space $\mathcal{V}$, it is then necessary that the integrand be identically zero, viz.

$$\nabla \cdot \mathbf{j}(\mathbf{r}, t) + \frac{\partial \rho(\mathbf{r}, t)}{\partial t} = 0, \tag{2.14}$$

which is the *equation of continuity*.

2.1.1   Differential Form of the Microscopic Maxwell Equations

The *microscopic Maxwell–Lorentz theory* [1–4] asserts that charges and currents give rise to electric and magnetic fields, and these fields in turn exert forces on the charges and influence their motion that at all times satisfy the equation of continuity (2.14). This interplay of source and field is governed by a set of equations which, (i) relate the fields to the charges and currents that produce them, (ii) relate the fields to the forces that they exert on the charges, and (iii) specify the law of motion of the charges under the influence of all forces acting on them, including non-electromagnetic as well as those forces that are electromagnetic in origin.

The source terms appearing in the microscopic Maxwell equations are the net charge $\rho(\mathbf{r}, t) = \rho_+(\mathbf{r}, t) - \rho_-(\mathbf{r}, t)$ and current $\mathbf{j}(\mathbf{r}, t) = \mathbf{j}_+(\mathbf{r}, t) - \mathbf{j}_-(\mathbf{r}, t)$ densities. In the classical theory the source fields $\rho(\mathbf{r}, t)$ and $\mathbf{j}(\mathbf{r}, t)$ represent the net charge and current distributions in minute detail. For example, if the net charge of a molecule is zero, the net microscopic charge density $\rho(\mathbf{r}, t)$ will not be zero within the region of space occupied by the molecule if there is a separation of positive and negative charges.

The net charge and current densities give rise to a microscopic electric field vector $\mathbf{e} = \mathbf{e}(\mathbf{r}, t)$ and a microscopic magnetic field vector $\mathbf{b} = \mathbf{b}(\mathbf{r}, t)$. At every point in space and time the microscopic field vectors $\mathbf{e}(\mathbf{r}, t)$ and $\mathbf{b}(\mathbf{r}, t)$ satisfy the microscopic Maxwell equations

$$\nabla \times \mathbf{e}(\mathbf{r}, t) = - \left\| \frac{1}{c} \right\| \frac{\partial \mathbf{b}(\mathbf{r}, t)}{\partial t}, \tag{2.15}$$

$$\nabla \times \mathbf{b}(\mathbf{r}, t) = \left\| \frac{4\pi}{c} \right\| \mu_0 \mathbf{j}(\mathbf{r}, t) + \left\| \frac{1}{c} \right\| \epsilon_0 \mu_0 \frac{\partial \mathbf{e}(\mathbf{r}, t)}{\partial t}, \tag{2.16}$$

where c is the speed of light in vacuum, and where ϵ_0 is the dielectric permittivity and μ_0 the magnetic permeability of free space. In Gaussian (or cgs) units, $\epsilon_0 = \mu_0 = 1$.

Two additional conditions that are satisfied by the electric and magnetic vector fields $\mathbf{e}$ and $\mathbf{b}$ may be directly deduced from Maxwell's equations by noting that the divergence of the curl of any vector vanishes identically. Upon taking the divergence of Eq. (2.15) there results $\nabla \cdot (\partial \mathbf{b}/\partial t) = \partial(\nabla \cdot \mathbf{b})/\partial t = 0$. The commutation of the differential operators $\nabla\cdot$ and $\partial/\partial t$ is admissible in the above relation because the field vector $\mathbf{b}(\mathbf{r}, t)$ and all of its space and time derivatives are assumed to be continuous. From this result it is seen that at every point in space, the divergence of $\mathbf{b}(\mathbf{r}, t)$ is a constant. Because this constant is zero in the static case, one then has that

$$\nabla \cdot \mathbf{b}(\mathbf{r}, t) = 0 \tag{2.17}$$

and the magnetic field is solenoidal. In a similar fashion, the divergence of Eq. (2.16) results in $\partial(\nabla \cdot \mathbf{e})/\partial t + (\|4\pi\|/\epsilon_0)\nabla \cdot \mathbf{j} = 0$, which, with the equation of continuity (2.14), may be rewritten as $\partial[\nabla \cdot \mathbf{e} - (\|4\pi\|/\epsilon_0)\rho]/\partial t = 0$. The quantity $[\nabla \cdot \mathbf{e} - (\|4\pi\|/\epsilon_0)\rho]$ is then seen to be a constant. Because this constant is zero in the static case, one then has that

$$\nabla \cdot \mathbf{e}(\mathbf{r}, t) = \frac{\|4\pi\|}{\epsilon_0}\rho(\mathbf{r}, t), \tag{2.18}$$

and the charge distributed with net density $\rho(\mathbf{r}, t)$ constitutes a source term for the microscopic electric field vector $\mathbf{e}(\mathbf{r}, t)$.

The divergence relations appearing in Eqs. (2.17) and (2.18) are frequently included as part of Maxwell's equations. However, they are not independent relations if the conservation of charge is assumed, as has been done here. On the other hand, if Eqs. (2.15)–(2.18) are postulated as Maxwell's equations, then the conservation of charge as expressed by the equation of continuity (2.14) is a direct consequence of this set of differential field relations.

A certain asymmetry is apparent in Eqs. (2.15)–(2.18) that has important physical significance. The densities $\rho(\mathbf{r}, t)$ and $\mathbf{j}(\mathbf{r}, t)$ that are electrical in origin appear in Eqs. (2.16) and (2.18), but in the corresponding places in Eqs. (2.15) and (2.17) where one would expect similar densities of magnetic charge and current to appear, there are none. The assertion that the Maxwell equations as they stand are entirely adequate to account for all classical electromagnetic phenomena has (so far) withstood the test of time. There has yet to be found any experimental evidence of magnetic charge, and the microscopic Maxwell equations are written in such a way so as to explicitly display this fact.

Electric and magnetic fields are fundamentally fields of force that ultimately originate from electric charges. Whether such a force field may be termed electric, magnetic, or electromagnetic hinges upon the motional state of the electric charges relative to the point at which the field observation is made. Electric charges at rest relative to an observation point give rise to an electrostatic field at that point. A relative motion of the charges yields a convective current and provides an additional magnetic force field at the observation point. This additional field is a magnetostatic field if the charges are all moving at constant velocities relative to the observation point. Finally, if the charges undergo accelerated motions, both time-varying electric and magnetic fields are produced that are coupled through the microscopic Maxwell equations (2.15) and (2.16), and are termed electromagnetic fields.

The microscopic Maxwell theory is completed by the equation for the force per unit volume (i.e., the microscopic force density) that is exerted on the charges and convective currents by an electromagnetic field. The microscopic force density $\mathbf{f}_+ = \mathbf{f}_+(\mathbf{r}, t)$ acting on the positive charge and convective current densities $\rho_+(\mathbf{r}, t)$ and $\mathbf{j}_+(\mathbf{r}, t)$ is given by the Lorentz force

$$\mathbf{f}_+(\mathbf{r}, t) = \rho_+(\mathbf{r}, t)\mathbf{e}(\mathbf{r}, t) + \left\|\frac{1}{c}\right\| \mathbf{j}_+(\mathbf{r}, t) \times \mathbf{b}(\mathbf{r}, t), \qquad (2.19)$$

and the microscopic force density $\mathbf{f}_- = \mathbf{f}_-(\mathbf{r}, t)$ acting on the negative charge and convective current densities $\rho_-(\mathbf{r}, t)$ and $\mathbf{j}_-(\mathbf{r}, t)$ is given by

$$\mathbf{f}_-(\mathbf{r}, t) = -\rho_-(\mathbf{r}, t)\mathbf{e}(\mathbf{r}, t) - \left\|\frac{1}{c}\right\| \mathbf{j}_-(\mathbf{r}, t) \times \mathbf{b}(\mathbf{r}, t). \qquad (2.20)$$

The addition of Eqs. (2.19) and (2.20) yields the net microscopic force density $\mathbf{f}(\mathbf{r}, t) = \mathbf{f}_+(\mathbf{r}, t) + \mathbf{f}_-(\mathbf{r}, t)$, which is given by the Lorentz force relation [3]

$$\mathbf{f}(\mathbf{r}, t) = \rho(\mathbf{r}, t)\mathbf{e}(\mathbf{r}, t) + \left\|\frac{1}{c}\right\| \mathbf{j}(\mathbf{r}, t) \times \mathbf{b}(\mathbf{r}, t), \qquad (2.21)$$

in terms of the net microscopic charge and current densities.

The physical significance of the net force density $\mathbf{f} = \mathbf{f}(\mathbf{r}, t)$ may be described in the following manner. Consider an infinitesimal region of space with volume ΔV in which the net charge density is $\rho = \rho(\mathbf{r}, t)$ and the net convective current density is $\mathbf{j} = \mathbf{j}(\mathbf{r}, t)$ at some instant of time. At that particular instant, let the microscopic field vectors in the region ΔV have the spatially averaged values $\langle \mathbf{e} \rangle$ and $\langle \mathbf{b} \rangle$. The Lorentz theory then asserts that the net force exerted on the charged mass in ΔV arising from the action of the electric and magnetic fields is given by

$$\Delta \mathbf{F} = (\rho \Delta V)\langle \mathbf{e} \rangle + \left\|\frac{1}{c}\right\| (\mathbf{j}\Delta V) \times \langle \mathbf{b} \rangle. \qquad (2.22)$$

Let P be an interior point of the region ΔV, as illustrated in Fig. 2.2. In the limit as ΔV shrinks to the point P, the average field values $\langle \mathbf{e} \rangle$ and $\langle \mathbf{b} \rangle$ go over to their respective values $\mathbf{e}(\mathbf{r}, t)$ and $\mathbf{b}(\mathbf{r}, t)$ at the point P, and the limiting ratio

$$\mathbf{f}(\mathbf{r}, t) \equiv \lim_{\Delta V \to 0} \frac{\Delta \mathbf{F}}{\Delta V} \qquad (2.23)$$

then defines $\mathbf{f} = \mathbf{f}(\mathbf{r}, t)$ as the net electromagnetic force per unit volume acting at the point $\mathbf{r}$ at the instant of time t.

Consider now an elementary charged particle of small but nonvanishing spatial extent that is part of some physical system. If it is sufficiently well removed from any other charges or currents in the system, then the external fields $\mathbf{e}_0(\mathbf{r}, t)$ and $\mathbf{b}_0(\mathbf{r}, t)$

Fig. 2.2 Illustration of the
Lorentz force vector $\Delta\mathbf{F}$
exerted on an infinitesimal
volume element ΔV with net
charge density ρ and
convective current density $\mathbf{j}$

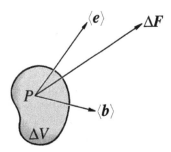

arising from these other charges and currents are essentially constant over the spatial
extent of the particle and the net force acting upon the particle is readily obtained by
integrating the Lorentz force relation given in Eq. (2.21) over the particle volume,
with the result

$$\mathbf{F}(\mathbf{r}, t) = q\left(\mathbf{e}_0(\mathbf{r}, t) + \left\|\frac{1}{c}\right\| \mathbf{v}(\mathbf{r}, t) \times \mathbf{b}_0(\mathbf{r}, t)\right). \qquad (2.24)$$

Here q is the total charge of the elementary particle (e.g., $q = -q_e$ for an electron
and $q = +q_e$ for a proton) and $\mathbf{v}(\mathbf{r}, t)$ is the instantaneous velocity of the particle.
The following six comments then apply to this equation [4].

1. The electromagnetic field vectors $\mathbf{e}_0$ and $\mathbf{b}_0$ are the microscopic fields arising
   from all charges and convective currents in the system with the exception of
   those of the charged particle itself.
2. Because of the finiteness of the charge q, its influence on neighboring charges
   and currents cannot be neglected. It is thus assumed that the external fields $\mathbf{e}_0$
   and $\mathbf{b}_0$ have been determined with this influence taken into account.
3. If the elementary charged particle is part of or is in collision with another
   elementary particle or system of particles (such as an atom or molecule),
   additional forces of a nonelectromagnetic nature must be included.[3]
4. Equation (3.24) applies to a rotating particle if $\mathbf{v}$ is taken as the velocity of the
   charge centroid of the elementary particle.[4]
5. If the motion of the particle involves any acceleration, there is then a small force
   of radiation reaction[5] that needs to be added to Eq. (2.24).

---

[3]This occurs, for example, in the collisional broadening of radiation from a Lorentz atom. See
Chap. 12 of Stone [4] for a detailed development of collisional broadening effects in a system of
Lorentz atoms.

[4]The electron spin is a purely quantum-mechanical effect that is not susceptible to a complete
analysis in classical physics. For a detailed historical discussion of this point, see Chap. VI of
Kramers' classic treatise *Quantum Mechanics* [7].

[5]For a complete rigorous development of radiation reaction in the Lorentz–Abraham model, see
Yaghjian's treatise *Relativistic Dynamics of a Charged Sphere* [8].

6. If the magnitude of the velocity of the particle is very small in comparison to the vacuum speed of light c (i.e., in the nonrelativistic limit $|\mathbf{v}| \ll c$), then the second term appearing on the right-hand side of Eq. (2.24) is typically small when compared with the first term (see Problem 2.4).

A fundamental property of the microscopic Maxwell–Lorentz equations is their linearity as expressed by the *principle of superposition*. Let $\rho_1(\mathbf{r}, t)$ and $\mathbf{j}_1(\mathbf{r}, t)$ be any given distribution of microscopic charge and current densities satisfying the equation of continuity (2.14) and let the microscopic field vectors due to them be denoted by $\mathbf{e}_1(\mathbf{r}, t)$ and $\mathbf{b}_1(\mathbf{r}, t)$. Let the field vectors $\mathbf{e}_2(\mathbf{r}, t)$ and $\mathbf{b}_2(\mathbf{r}, t)$ be due to an independent distribution of microscopic charge and current densities $\rho_2(\mathbf{r}, t)$ and $\mathbf{j}_2(\mathbf{r}, t)$, also satisfying the equation of continuity. Due to the linearity of the microscopic Maxwell equations and the Lorentz force relation, it then follows that the combined charge and current distributions $\rho(\mathbf{r}, t) = \rho_1(\mathbf{r}, t) + \rho_2(\mathbf{r}, t)$ and $\mathbf{j}(\mathbf{r}, t) = \mathbf{j}_1(\mathbf{r}, t) + \mathbf{j}_2(\mathbf{r}, t)$ satisfy the equation of continuity and produce the field vectors $\mathbf{e}(\mathbf{r}, t) = \mathbf{e}_1(\mathbf{r}, t) + \mathbf{e}_2(\mathbf{r}, t)$ and $\mathbf{b}(\mathbf{r}, t) = \mathbf{b}_1(\mathbf{r}, t) + \mathbf{b}_2(\mathbf{r}, t)$. The principle of superposition is of particular importance in the solution of many electromagnetic problems. In its most direct application, it allows one to calculate the field vectors due to a given fixed distribution of charged particles by first determining the separate fields produced by each particle alone and then adding the individual results to give the total field. However, if the electromagnetic field is due to an unspecified motion of some system of interacting charged particles, then the Maxwell–Lorentz equations (with proper account taken for radiation reaction, collisional effects, etc.) must be solved in a self-consistent fashion in which the fields, in part, determine the motion of each charged particle in the system, and each charged particle is, in turn, a source term for the electromagnetic field. In this more complicated (and more realistic) situation, the principle of superposition applies in each separate part of the dynamical computation. An example of this self-consistent procedure may be found in plasma electrodynamics [9–11].

All that remains now is to show how the microscopic electric and magnetic fields can be measured (at least in principle) and their units defined in terms of the elementary sources appearing in the theory. For that purpose, the classical concept of a charged *test particle* is employed. Such a particle is defined to consist of a distribution of charge, each element of which is in some way held rigidly fixed with respect to the other elements regardless of any electromagnetic forces that may be acting on it. The part of these electromagnetic forces due to the charges and currents in the test particle itself is assumed to be exactly balanced by internal stabilizing forces of the particle (which remain unspecified in the classical theory). The remainder of the electromagnetic force that is exerted on the test particle is due to the fields arising from external charges and other fields whose origin is not due to the test particle itself, and these are just the fields that are to be measured by introducing the test particle into the system. Furthermore, in order that the system under consideration be undisturbed by the presence of the test particle when it is introduced, it is assumed that the latter is of infinitesimal size and has infinitesimal charge dq. Moreover, it is assumed that the test particle can be controlled to such a degree that it can be held fixed in space at any point or given a definite velocity at any point, and in all cases the force exerted upon it due to the interacting fields under study can always be measured at any instant of time with infinitesimal accuracy. The plausible existence of such an infinitesimal charged test particle along with its complete controllability is possible only within the framework of classical physics. Such is not the case, however, in the physics of quantum field theory [12]. In the purely classical framework of the Maxwell–Lorentz theory, the concept of field measurements through the use of infinitesimally small test charges is pushed to the limit of physical conceptualization by asserting that the field vectors can be measured with microscopic precision even inside the basic entities of matter. It is in this sense that the microscopic field vectors $\mathbf{e}(\mathbf{r}, t)$ and $\mathbf{b}(\mathbf{r}, t)$ have meaning at all points of space.

If the test particle dq moves with the velocity $\mathbf{v}$ without acceleration or rotation, it then gives rise to a current density $\mathbf{j}(\mathbf{r}, t) = \rho(\mathbf{r}, t)\mathbf{v}$ defined at each point of space, and Eq. (2.24) gives the total force acting on the test particle as[6]

$$d\mathbf{F}(\mathbf{r}, t) = dq \left(\mathbf{e}(\mathbf{r}, t) + \left\| \frac{1}{c} \right\| \mathbf{v} \times \mathbf{b}(\mathbf{r}, t) \right). \tag{2.25}$$

Let this infinitesimal test charge be at initially rest at a fixed point in the field so that $\mathbf{v} = \mathbf{0}$. The force exerted on the particle is then given by $d\mathbf{F}(\mathbf{r}, t) = \mathbf{e}(\mathbf{r}, t)dq$ and

$$\mathbf{e}(\mathbf{r}, t) = \frac{d\mathbf{F}(\mathbf{r}, t)}{dq} = \lim_{\Delta q \to 0} \frac{\Delta \mathbf{F}(\mathbf{r}, t)}{\Delta q}, \tag{2.26}$$

so that knowing Δq and measuring $\Delta \mathbf{F}(\mathbf{r}, t)$ gives the value of the microscopic field vector $\mathbf{e}(\mathbf{r}, t)$ at that point in space and time. Next, refer space to a fixed Cartesian coordinate system and let the infinitesimal test particle move in the y-coordinate direction with velocity $\mathbf{v} = \hat{\mathbf{1}}_y v_y$ and no acceleration. From Eq. (2.25), the force exerted on this test particle is now given by

$$d\mathbf{F}(\mathbf{r}, t) = \mathbf{e}(\mathbf{r}, t)dq + \left\| \frac{1}{c} \right\| dq \left(\hat{\mathbf{1}}_x v_y b_z(\mathbf{r}, t) - \hat{\mathbf{1}}_z v_y b_x(\mathbf{r}, t) \right). \tag{2.27}$$

Subtract out the component of the force that is due to the electric field $\mathbf{e}(\mathbf{r}, t)$, known from the measurement specified by Eq. (2.26); the x-component of the remaining force is then given by $dF'_x = \|1/c\| v_y b_z dq$, so that

$$b_z(\mathbf{r}, t) = \|c\| \frac{1}{v_y} \frac{dF'_x(\mathbf{r}, t)}{dq} = \|c\| \frac{1}{v_y} \lim_{\Delta q \to 0} \frac{\Delta F'_x(\mathbf{r}, t)}{\Delta q}. \tag{2.28}$$

---

[6]Notice that no distinction need be made between the external field vectors $\mathbf{e}_0$ and $\mathbf{b}_0$ employed in Eq. (2.24) and the total field vectors $\mathbf{e}(\mathbf{r}, t)$ and $\mathbf{b}(\mathbf{r}, t)$ appearing in Eq. (2.25) at the fixed observation point P at which the test particle is introduced because dq is infinitesimally small.

Similarly, the components $b_x(\mathbf{r}, t)$ and $b_y(\mathbf{r}, t)$ can be obtained by first taking the test particle velocity in the z-direction and then in the x-direction and measuring the components $dF_y'(\mathbf{r}, t)$ and $dF_z'(\mathbf{r}, t)$, respectively, of the force remaining after that due to the electric field has been subtracted out. The results are given by

$$b_x(\mathbf{r}, t) = \|c\| \frac{1}{v_z} \frac{dF_y'(\mathbf{r}, t)}{dq} = \|c\| \frac{1}{v_z} \lim_{\Delta q \to 0} \frac{\Delta F_y'(\mathbf{r}, t)}{\Delta q}, \qquad (2.29)$$

$$b_y(\mathbf{r}, t) = \|c\| \frac{1}{v_x} \frac{dF_z'(\mathbf{r}, t)}{dq} = \|c\| \frac{1}{v_x} \lim_{\Delta q \to 0} \frac{\Delta F_z'(\mathbf{r}, t)}{\Delta q}. \qquad (2.30)$$

The limiting expressions appearing in Eqs. (2.26) and (2.28)–(2.30) are not only of a formal mathematical nature but are also connected with the physical definition of the electric and magnetic fields, respectively. Finally, notice that only two measurements of the magnetic force $\mathbf{F}_m(\mathbf{r}, t) = \left\| \frac{1}{c} \right\| q\mathbf{v} \times \mathbf{b}(\mathbf{r}, t)$ at a point are sufficient to uniquely determine the magnetic induction field vector $\mathbf{b}(\mathbf{r}, t)$ at that point (see Problem 2.3).

It is convenient to introduce the auxiliary field vectors defined by

$$\mathbf{d}(\mathbf{r}, t) \equiv \epsilon_0 \mathbf{e}(\mathbf{r}, t), \qquad (2.31)$$

$$\mathbf{h}(\mathbf{r}, t) \equiv \frac{1}{\mu_0} \mathbf{b}(\mathbf{r}, t), \qquad (2.32)$$

where $\mathbf{d} = \mathbf{d}(\mathbf{r}, t)$ is the *microscopic electric displacement vector* and $\mathbf{h} = \mathbf{h}(\mathbf{r}, t)$ is the *microscopic magnetic intensity vector*. With these identifications, the differential form of the microscopic Maxwell equations becomes

$$\nabla \times \mathbf{e}(\mathbf{r}, t) = - \left\| \frac{1}{c} \right\| \frac{\partial \mathbf{b}(\mathbf{r}, t)}{\partial t}, \qquad (2.33)$$

$$\nabla \times \mathbf{h}(\mathbf{r}, t) = \left\| \frac{4\pi}{c} \right\| \mathbf{j}(\mathbf{r}, t) + \left\| \frac{1}{c} \right\| \frac{\partial \mathbf{d}(\mathbf{r}, t)}{\partial t}, \qquad (2.34)$$

with

$$\nabla \cdot \mathbf{d}(\mathbf{r}, t) = \|4\pi\| \rho(\mathbf{r}, t), \qquad (2.35)$$

$$\nabla \cdot \mathbf{b}(\mathbf{r}, t) = 0. \qquad (2.36)$$

Because $\epsilon_0 = \mu_0 = 1$ in cgs or Gaussian units, then $\mathbf{d}(\mathbf{r}, t) = \mathbf{e}(\mathbf{r}, t)$ and $\mathbf{h}(\mathbf{r}, t) = \mathbf{b}(\mathbf{r}, t)$ and there is no real distinction between these microscopic field vectors in that system. The importance of these auxiliary field vectors is fully realized in the macroscopic theory of the electromagnetic field in either polarizable, conducting, or magnetizable materials, presented in Chap. 4.

2.1.1.1 Gaussian (cgs) Units

In *Gaussian units* (centimeters, grams, seconds), the basic unit of charge is the statcoulomb so that the volume charge density ρ is in units of statcoulombs per cubic centimeter. The fundamental unit of current is then that current which transports charge at the rate of 1 statcoulomb per second and is called the statampere. The unit of current density $\mathbf{j}$ is then the statampere per square centimeter.

The unit of force in Gaussian units is the dyne so that, from Eq. (2.26), the unit of measure of the microscopic electric field intensity is the dyne per statcoulomb, called the esu (electrostatic unit) of electric field intensity, viz.

$$1 \text{ esu of electric field intensity} \equiv 1 \text{ dyne/statcoulomb.} \tag{2.37}$$

In terms of energy, 1 esu of electric field intensity = 1 statvolt/cm, so that

$$1 \text{ statvolt} = 1 \text{ erg/statcoulomb.} \tag{2.38}$$

The unit of measure of the microscopic magnetic induction field is the gauss, so that from Eqs. (2.28)–(2.30)

$$1 \text{ gauss} = 1 \text{ dyne/statcoulomb.} \tag{2.39}$$

2.1.1.2 MKSA (SI) Units

The principal factor relating mksa and cgs units is the vacuum speed of light c, where $c = 2.9979 \times 10^8$ m/s. In *mksa* or *SI (Système Internationale) units* (meters, kilograms, seconds, amperes) the basic unit of charge is the coulomb, where 1 coulomb $= 2.9979 \times 10^9$ statcoulombs. The volume charge density ρ is accordingly in units of coulombs per cubic meter. Furthermore, the basic unit of current is that current which transports charge at the rate of 1 coulomb per second and is called the ampere, where 1 ampere $= 2.9979 \times 10^9$ statamperes, and the unit of current density $\mathbf{j}$ is the ampere per square meter. In the conventional mksa system of electromagnetic units, the electric current is arbitrarily chosen as the fourth fundamental dimension.

The unit of force in mksa units is the newton so that, from Eq. (2.26), the unit of measure of the microscopic electric field intensity is the newton per coulomb. In terms of energy, 1 volt/m = 1 newton/coulomb, so that

$$1 \text{ volt} = 1 \text{ joule/coulomb,} \tag{2.40}$$

where (since 1 newton $= 1 \times 10^5$ dynes), 1 volt = (1/299.79) statvolt. The unit of measure of the microscopic magnetic induction field in mksa units is the tesla, so that, from Eqs. (2.28)–(2.30),

$$1 \text{ tesla} = 1 \text{ newton/coulomb}, \tag{2.41}$$

where [noting the factor c in Eqs. (2.28)–(2.30) when cgs units are used] 1 tesla = 1×10^4 gauss. Finally, the mksa values of the dielectric permittivity and magnetic permeability of free space are given respectively by $\epsilon_0 = (1 \times 10^7$ farad $\cdot$ m/s$^2)/(4\pi c^2) = 8.854 \times 10^{-12}$ farad/m and $\mu_0 = 4\pi \times 10^{-7}$ weber/(ampere$\cdot$m), where 1 farad = 1 coulomb/volt is the unit of capacitance and 1 weber = 1 volt $\cdot$ s = 1 joule/ampere is the unit of magnetic flux. The displacement vector **d** is then in units of coulombs/m^2 and the magnetic intensity vector **h** is in units of amperes/m.

2.1.2   Integral Form of the Microscopic Maxwell Equations

The electromagnetic field properties described by the set of differential equations given in Eqs. (2.33)–(2.36) can also be expressed by an equivalent set of integral equations. Consider first the microscopic differential relation given in Eq. (2.33). Upon integrating this equation over any regular open surface S that is bounded by a closed, regular (i.e. nonintersecting) contour C, followed by application of Stokes' theorem to the surface integral of $\nabla \times \mathbf{e}(\mathbf{r}, t)$, one obtains

$$\oint_C \mathbf{e}(\mathbf{r}, t) \cdot d\mathbf{l} + \left\| \frac{1}{c} \right\| \int_S \frac{\partial \mathbf{b}(\mathbf{r}, t)}{\partial t} \cdot \hat{\mathbf{n}} da = 0, \tag{2.42}$$

where the integration around the contour C is traversed in the positive sense. Because the electromagnetic field vectors and their partial derivatives are assumed to be continuous, if the surface S is fixed in space, then by Leibniz' rule, the differential operator $\partial/\partial t$ may be brought out from under the integration, with the result

$$\oint_C \mathbf{e}(\mathbf{r}, t) \cdot d\mathbf{l} = -\left\| \frac{1}{c} \right\| \frac{d}{dt} \int_S \mathbf{b}(\mathbf{r}, t) \cdot \hat{\mathbf{n}} da, \tag{2.43}$$

which is known as *Faraday's law*. The scalar quantity defined by

$$\Phi_m(\mathbf{R}, t) \equiv \int_S \mathbf{b}(\mathbf{r}, t) \cdot \hat{\mathbf{n}} da \tag{2.44}$$

is the flux of the magnetic induction vector $\mathbf{b}(\mathbf{r}, t)$ through the surface S, and is appropriately called the *magnetic flux*. In cgs units the unit measure of Φ_m is in gauss $\cdot$ cm^2 which is called the *maxwell*, while in mksa units the unit of measure of Φ_m is in tesla $\cdot$ m^2, which is called the *weber*, where 1 weber = 1×10^8 maxwells. Notice that the vector **R** appearing in the argument of the magnetic flux in Eq. (2.44) depends upon both the position vector **r** appearing in the argument of the magnetic induction vector as well as upon the shape and orientation of the surface of integration S appearing in that equation.

According to Eq. (2.43), the circulation of the microscopic electric field vector $\mathbf{e}(\mathbf{r}, t)$ about any closed, regular contour C that is situated in that field is determined by the time rate of change of the magnetic flux through any open regular surface bounded by that contour. The experiments of Faraday [13] demonstrated that this relation holds whatever the cause of the magnetic flux variation. The total time derivative appearing outside the surface integral in Eq. (2.43) was obtained from Eq. (2.42) under the assumption of a variable magnetic flux density threading a fixed contour C, but the total magnetic flux Φ_m will also change through any deformation or motion of the contour C. This time derivative of the magnetic flux must then be interpreted in this more general sense. This more general interpretation of Faraday's law as given in Eq. (2.43) is indeed a consequence of the differential field equations, but the general proof must be based on the electrodynamics of moving media [14].

It is important to note that the microscopic electric field $\mathbf{e}(\mathbf{r}, t)$ appearing in Eq. (2.43) is the electric field intensity at the differential path element $d\mathbf{l}$ along the contour C in the reference frame in which $d\mathbf{l}$ is at rest. If the contour C is moving with a constant velocity **V**, as illustrated in Fig. 2.3, then the total time derivative appearing in Eq. (2.43) must take this translational motion into account. The magnetic flux Φ_m linking the circuit then changes because both the flux changes with time at a given point and the translation of the circuit changes the location of the boundary. If the ordered-triple (x_1, x_2, x_3) denotes the coordinates of the boundary C, then the rate of change with time of some property of the field that is associated with the point $(x_1, x_2, x_3) = (x_1(t), x_2(t), x_3(t))$ at the time t is given by

$$\frac{d}{dt} = \frac{\partial}{\partial t} + \sum_{k=1}^{3} \frac{\partial x_k}{\partial t} \frac{\partial}{\partial x_k} = \frac{\partial}{\partial t} + \sum_{k=1}^{3} V_k \frac{\partial}{\partial x_k}. \tag{2.45}$$

Fig. 2.3 Closed, oriented contour C moving with a constant translational velocity **V** relative to an inertial reference frame

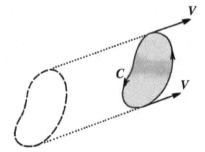

Because $\mathbf{V} = (V_1, V_2, V_3)$, the total time derivative is seen to be given by the *convective derivative*

$$\frac{d}{dt} = \frac{\partial}{\partial t} + \mathbf{V} \cdot \nabla, \tag{2.46}$$

where $\nabla = (\partial/\partial x_1, \partial/\partial x_2, \partial/\partial x_3)$.

The total time derivative of the magnetic flux linking the circuit $\mathcal{C}$ is then given by (where the subscript L denotes the *laboratory frame of reference* in which the field measurements are made)

$$\frac{d}{dt} \int_S \mathbf{b}_L \cdot \hat{\mathbf{n}} da = \int_S \frac{\partial \mathbf{b}_L}{\partial t} \cdot \hat{\mathbf{n}} da + \int_S [(\mathbf{V} \cdot \nabla) \mathbf{b}_L] \cdot \hat{\mathbf{n}} da$$

$$= \int_S \frac{\partial \mathbf{b}_L}{\partial t} \cdot \hat{\mathbf{n}} da + \int_S \nabla \times (\mathbf{b}_L \times \mathbf{V}) \cdot \hat{\mathbf{n}} da,$$

since $\nabla \times (\mathbf{b}_L \times \mathbf{V}) = (\mathbf{V} \cdot \nabla)\mathbf{b}_L - \mathbf{V}(\nabla \cdot \mathbf{b}_L) - (\mathbf{b}_L \cdot \nabla)\mathbf{V} + \mathbf{b}_L(\nabla \cdot \mathbf{V})$, where the last three terms appearing on the right-hand side of this identity vanish by virtue of both the divergenceless character of the magnetic induction vector $\mathbf{b}_L$ and the constant value of the velocity vector $\mathbf{V}$. Application of Stokes' theorem to the second surface integral on the right-hand side of the above equation then results in the expression

$$\frac{d}{dt} \int_S \mathbf{b}_L \cdot \hat{\mathbf{n}} da = \int_S \frac{\partial \mathbf{b}_L}{\partial t} \cdot \hat{\mathbf{n}} da + \oint_{\mathcal{C}} (\mathbf{b}_L \times \mathbf{V}) \cdot d\mathbf{l}. \tag{2.47}$$

With this result, Faraday's law (2.43) may then be written in the form

$$\oint_{\mathcal{C}} [\mathbf{e} - \|1/c\|(\mathbf{V} \times \mathbf{b}_L)] \cdot d\mathbf{l} = - \left\| \frac{1}{c} \right\| \int_S \frac{\partial \mathbf{b}_L}{\partial t} \cdot \hat{\mathbf{n}} da. \tag{2.48}$$

Equation (2.48) is an equivalent statement of Faraday's law as applied to a uniformly moving circuit $\mathcal{C}$ with constant velocity $\mathbf{V}$. The electric field intensity $\mathbf{e}$ appearing in Eq. (2.48) is measured with respect to the moving reference frame in which the contour $\mathcal{C}$ is at rest. An alternate interpretation of this situation, however, is to view the contour $\mathcal{C}$ and surface S as being instantaneously located at a certain position in space in the laboratory frame of reference. Application of Faraday's law (2.43) to that circuit results in the expression

$$\oint_{\mathcal{C}} \mathbf{e}_L \cdot d\mathbf{l} = - \left\| \frac{1}{c} \right\| \frac{d}{dt} \int_S \mathbf{b}_L \cdot \hat{\mathbf{n}} da, \tag{2.49}$$

where $\mathbf{e}_L$ is the microscopic electric field vector in the laboratory reference frame.

The assumption of *Galilean invariance* (i.e., that physical laws should be invariant under a Galilean transformation[7]) then implies that the left-hand sides of Eqs. (2.48) and (2.49) must be equal. This in turn implies that the electric field intensity **e** in the moving coordinate system of the circuit $\mathcal{C}$ is given by

$$\mathbf{e}(\mathbf{r}, t) = \mathbf{e}_L(\mathbf{r}, t) + \left\| \frac{1}{c} \right\| \mathbf{V} \times \mathbf{b}_L(\mathbf{r}, t). \tag{2.50}$$

This is then the *Galilean transformation of the electric field intensity*. A charged test particle q at rest in the moving coordinate frame of the circuit $\mathcal{C}$ then experiences the force $\mathbf{F} = q\mathbf{e}$. When viewed from the laboratory frame of reference, the charge represents a microscopic current density vector $\mathbf{j} = q\mathbf{V}\delta(\mathbf{r} - \mathbf{r}_0)$, where $\delta(\mathbf{r})$ is the three-dimensional Dirac delta function (see Appendix A) and where $\mathbf{r}_0 = \mathbf{r}_0(t)$ denotes the position of the test particle in the laboratory frame; it then experiences the Lorentz force density

$$\mathbf{f} = q\delta(\mathbf{r} - \mathbf{r}_0)\left[\mathbf{e}_L(\mathbf{r}, t) + \left\| \frac{1}{c} \right\| \mathbf{V} \times \mathbf{b}_L(\mathbf{r}, t) \right], \tag{2.51}$$

in agreement with the result implied by Eq. (2.50) taken together with the force relation $\mathbf{f} = q\mathbf{e}\delta(\mathbf{r} - \mathbf{r}_0)$.

Consider now the second curl relation (2.34) of the microscopic Maxwell's equations. Upon integration of this relation over any given regular open surface $\mathcal{S}$ that is bounded by a closed regular contour $\mathcal{C}$, followed by application of Stokes' theorem to the surface integral of $\nabla \times \mathbf{h}(\mathbf{r}, t)$, there results

$$\oint_{\mathcal{C}} \mathbf{h}(\mathbf{r}, t) \cdot d\mathbf{l} = \left\| \frac{4\pi}{c} \right\| \int_{\mathcal{S}} \mathbf{j}(\mathbf{r}, t) \cdot \hat{\mathbf{n}} da + \left\| \frac{1}{c} \right\| \int_{\mathcal{S}} \frac{\partial \mathbf{d}(\mathbf{r}, t)}{\partial t} \cdot \hat{\mathbf{n}} da, \tag{2.52}$$

where the integration about the contour $\mathcal{C}$ is traversed in the positive sense. The total current $\mathcal{J}(\mathbf{R}, t)$ linking the contour $\mathcal{C}$ is given by [see Eq. (2.9)]

$$\mathcal{J}(\mathbf{R}, t) = \int_{\mathcal{S}} \mathbf{j}(\mathbf{r}, t) \cdot \hat{\mathbf{n}} da, \tag{2.53}$$

where the vector **R** appearing in the argument of the total current depends upon both the position vector **r** appearing in the argument of the current density vector $\mathbf{j}(\mathbf{r}, t)$ as well as upon the shape and orientation of the surface of integration $\mathcal{S}$. As in the preceding analysis, the partial differential operator $\partial/\partial t$ may be brought out from under the surface integral in Eq. (2.52) and replaced by the total time derivative

---

[7] *Galilean invariance* states that physical phenomena are the same when viewed by two observers moving with a constant velocity **V** relative to each other, provided that the coordinates in space and time are related by the *Galilean transformation* $\mathbf{r}' = \mathbf{r} + \mathbf{V}t, \, t' = t$.

(Leibniz' rule), resulting in the final expression

$$\oint_{\mathcal{C}} \mathbf{h}(\mathbf{r}, t) \cdot d\mathbf{l} = \left\| \frac{4\pi}{c} \right\| \int_{\mathcal{S}} \mathbf{j}(\mathbf{r}, t) \cdot \hat{\mathbf{n}} da + \left\| \frac{1}{c} \right\| \frac{d}{dt} \int_{\mathcal{S}} \mathbf{d}(\mathbf{r}, t) \cdot \hat{\mathbf{n}} da, \qquad (2.54)$$

which is known as *Ampère's law* [15]. This expression is valid whatever the cause of the electric flux variation. Again, a rigorous proof of the complete generality of this result must be based upon the electrodynamics of moving media [14].

In the steady state, the time derivative of the surface integral appearing on the right-hand side of Ampère's circuital law (2.54) is zero and the total convection current $\mathcal{J}$ through any open regular surface $\mathcal{S}$ is equal to $\|c/4\pi\|$ times the circulation of the magnetic intensity vector $\mathbf{h}$ about the contour $\mathcal{C}$ that forms the boundary of the open surface $\mathcal{S}$. However, if the microscopic electric field is variable in time, the vector $\|1/4\pi\|\partial\mathbf{d}/\partial t$ has associated with it a magnetic intensity vector $\mathbf{h}(\mathbf{r}, t)$ exactly equal to that which would be produced by a microscopic current distribution with density $\mathbf{j}'(\mathbf{r}, t) = \|1/4\pi\|\partial\mathbf{d}(\mathbf{r}, t)/\partial t$; this term is accordingly called *Maxwell's displacement current.*

The microscopic magnetic intensity vector $\mathbf{h}$ appearing in Eq. (2.54) is the magnetic field at the path element $d\mathbf{l}$ along the contour $\mathcal{C}$ in the coordinate system in which $d\mathbf{l}$ is at rest. Let the contour $\mathcal{C}$ move with a constant velocity $\mathbf{V}$; the total time derivative appearing in Eq. (2.54) is then given by the convective derivative defined in Eq. (2.46). The total time derivative of the electric flux linking the contour $\mathcal{C}$ is then given by (the subscript L denoting the laboratory frame of reference in which the field measurements are made)

$$\frac{d}{dt} \int_{\mathcal{S}} \mathbf{d}_L \cdot \hat{\mathbf{n}} da = \int_{\mathcal{S}} \frac{\partial \mathbf{d}_L}{\partial t} \cdot \hat{\mathbf{n}} da + \int_{\mathcal{S}} [(\mathbf{V} \cdot \nabla) \mathbf{d}_L] \cdot \hat{\mathbf{n}} da$$

$$= \int_{\mathcal{S}} \frac{\partial \mathbf{d}_L}{\partial t} \cdot \hat{\mathbf{n}} da + \oint_{\mathcal{C}} (\mathbf{d}_L \times \mathbf{V}) \cdot d\mathbf{l} + \|4\pi\| \int_{\mathcal{S}} \rho \mathbf{V} \cdot \hat{\mathbf{n}} da,$$

where the divergence relation given in Eq. (2.35) has been employed. In addition, the total current $\mathcal{J}$ linking the contour $\mathcal{C}$ as measured in the moving reference frame is augmented by the convection current with density $\mathbf{j}_V = \rho \mathbf{V}$, so that

$$\mathcal{J} = \mathcal{J}_L - \int_{\mathcal{S}} \rho \mathbf{V} \cdot \hat{\mathbf{n}} da, \qquad (2.55)$$

where $\mathcal{J}_L$ is the current measured in the laboratory frame of reference. With these two results, Eq. (2.54) may then be written in the form

$$\oint_{\mathcal{C}} \left[\mathbf{h} - \left\| \frac{1}{c} \right\| (\mathbf{d}_L \times \mathbf{V}) \right] \cdot d\mathbf{l} = \left\| \frac{4\pi}{c} \right\| \mathcal{J}_L + \left\| \frac{1}{c} \right\| \int_{\mathcal{S}} \frac{\partial \mathbf{d}_L}{\partial t} \cdot \hat{\mathbf{n}} da. \qquad (2.56)$$

This equation is an equivalent statement of the integral representation (2.53) of the microscopic Maxwell equation (2.34) applied to a uniformly moving contour $\mathcal{C}$. The magnetic intensity vector $\mathbf{h} = \mathbf{h}(\mathbf{r}, t)$ is measured here in the moving frame in which the contour $\mathcal{C}$ is at rest. An alternate interpretation of this situation, however, is to view the contour $\mathcal{C}$ and the associated surface $\mathcal{S}$ as being instantaneously located at a certain position in space in the laboratory coordinate system. Application of Eq. (2.52) to that fixed contour then yields the expression

$$\oint_{\mathcal{C}} \mathbf{h}_L \cdot d\mathbf{l} = \left\| \frac{4\pi}{c} \right\| \| \mathfrak{J}_L + \left\| \frac{1}{c} \right\| \int_{\mathcal{S}} \frac{\partial \mathbf{d}_L}{\partial t} \cdot \hat{\mathbf{n}} da, \qquad (2.57)$$

where $\mathbf{h}_L$ is the microscopic magnetic field intensity in the laboratory reference frame.

The assumption of *Galilean invariance* implies that the left-hand sides of Eqs. (2.56) and (2.57) must be equal. This in turn implies that the microscopic magnetic intensity vector $\mathbf{h}$ in the moving coordinate system of the circuit is related to that in the laboratory frame through the relation

$$\mathbf{h}(\mathbf{r}, t) = \mathbf{h}_L(\mathbf{r}, t) - \left\| \frac{1}{c} \right\| \mathbf{V} \times \mathbf{d}_L(\mathbf{r}, t). \qquad (2.58)$$

This is then the *Galilean transformation of the magnetic field intensity.*

In order to illustrate the inadequacy of Galilean relativity, consider a convective current density $\mathbf{j} = \rho \mathbf{V}_c$ in the moving coordinate frame. In the laboratory frame of reference this corresponds to the convective current density $\mathbf{j}_L = \rho(\mathbf{V}_c + \mathbf{V})$ and the Lorentz force density acting upon it in the laboratory frame is given by

$$\mathbf{f}_L = \rho \mathbf{e}_L + \left\| \frac{1}{c} \right\| \mathbf{j}_L \times \mathbf{b}_L. \qquad (2.59)$$

In the moving coordinate frame the current density experiences the force density

$$\mathbf{f} = \rho \mathbf{e} + \left\| \frac{1}{c} \right\| \rho \mathbf{V}_c \times \mathbf{b}, \qquad (2.60)$$

which, with substitution from the Galilean transformation relations given in Eqs. (2.50) and (2.58), becomes

$$\begin{aligned}
\mathbf{f} &= \rho \left(\mathbf{e}_L + \left\| \frac{1}{c} \right\| \mathbf{V} \times \mathbf{b}_L \right) + \left\| \frac{1}{c} \right\| \rho \mathbf{V}_c \times \left(\mathbf{b}_L - \left\| \frac{1}{c} \right\| \mathbf{V} \times \mathbf{e}_L \right) \\
&= \rho \mathbf{e}_L + \left\| \frac{1}{c} \right\| \rho(\mathbf{V}_c + \mathbf{V}) \times \mathbf{b}_L - \left\| \frac{1}{c^2} \right\| \rho \mathbf{V}_c \times (\mathbf{V} \times \mathbf{e}_L) \\
&= \rho \mathbf{e}_L + \left\| \frac{1}{c} \right\| \mathbf{j}_L \times \mathbf{b}_L - \left\| \frac{1}{c^2} \right\| \rho \mathbf{V}_c \times (\mathbf{V} \times \mathbf{e}_L).
\end{aligned} \qquad (2.61)$$

These two expressions [(2.59) and (2.61)] for the force density differ by an amount that is on the order of $V_c V/c^2$, an inconsistency that is a direct consequence of the assumption of Galilean invariance.

The violation of invariance of the force relation in the Maxwell–Lorentz theory with the assumption of Galilean relativity is resolved by the special theory of relativity. Classical electrodynamics is indeed consistent with special relativity wherein the Maxwell–Lorentz equations are invariant under a *Poincaré-Lorentz transformation*. Indeed, Henri Poincaré [16–18], followed by H. A. Lorentz [19], originally arrived at this transformation by requiring the invariance of the Maxwell–Lorentz equations between any two inertial reference frames. Inconsistencies such as that raised above by calculating the microscopic force density due to a magnetic field acting upon a convective current density in two different inertial reference frames that are moving with a relative velocity **V** are resolved by the fact, shown so clearly in the special theory of relativity, that magnetic and electric fields have no separate meaning; instead, special relativity implies the single unified concept of an electromagnetic field. In general, a field that is purely electric, or purely magnetic, in one inertial frame will have both electric and magnetic field components in another inertial reference frame.

The two additional differential field relations given in Eqs. (2.35) and (2.36) can easily be expressed in an equivalent integral form with the aid of the divergence theorem. Accordingly, let $\mathcal{R}$ be any regular region of space that is bounded by a closed surface $\mathcal{S}$; integration of Eq. (2.36) over the region $\mathcal{R}$ and application of the divergence theorem to the resultant volume integral of $\nabla \cdot \mathbf{b}$ results in the expression

$$\oint_{\mathcal{S}} \mathbf{b}(\mathbf{r}, t) \cdot \hat{\mathbf{n}} da = 0, \qquad (2.62)$$

which is *Gauss' law* [20] for the magnetic field, where $\hat{\mathbf{n}}$ denotes the positive (outward directed) unit normal vector to the surface element da. Hence, the total flux of the microscopic magnetic induction vector $\mathbf{b}(\mathbf{r}, t)$ crossing any closed regular surface is zero. Similarly, upon integration of Eq. (2.35) over the region $\mathcal{R}$ and application of the divergence theorem to the resultant volume integral of $\nabla \cdot \mathbf{d}$, one obtains the expression

$$\oint_{\mathcal{S}} \mathbf{d}(\mathbf{r}, t) \cdot \hat{\mathbf{n}} da = \|4\pi\| \int_{\mathcal{R}} \rho(\mathbf{r}, t) d^3 r = \|4\pi\| q(t). \qquad (2.63)$$

which is *Gauss' law* [20] for the electric field. Hence, the total flux of the microscopic electric displacement vector $\mathbf{d}(\mathbf{r}, t)$ crossing any given closed regular surface $\mathcal{S}$ is equal to the total charge $q(t)$ contained within the region $\mathcal{R}$ that is bounded by that surface.

The *time-domain integral form of the microscopic Maxwell equations* are then given by

$$\oint_{\mathcal{C}} \mathbf{e}(\mathbf{r}, t) \cdot d\mathbf{l} = - \left\| \frac{1}{c} \right\| \frac{d}{dt} \int_{\mathcal{S}} \mathbf{b}(\mathbf{r}, t) \cdot \hat{\mathbf{n}} da, \tag{2.64}$$

$$\oint_{\mathcal{C}} \mathbf{h}(\mathbf{r}, t) \cdot d\mathbf{l} = \left\| \frac{4\pi}{c} \right\| \int_{\mathcal{S}} \mathbf{j}(\mathbf{r}, t) \cdot \hat{\mathbf{n}} da + \left\| \frac{1}{c} \right\| \frac{d}{dt} \int_{\mathcal{S}} \mathbf{d}(\mathbf{r}, t) \cdot \hat{\mathbf{n}} da, \tag{2.65}$$

and

$$\oint_{\mathcal{S}} \mathbf{d}(\mathbf{r}, t) \cdot \hat{\mathbf{n}} da = \| 4\pi \| \int_{\mathcal{R}} \rho(\mathbf{r}, t) d^3 r, \tag{2.66}$$

$$\oint_{\mathcal{S}} \mathbf{b}(\mathbf{r}, t) \cdot \hat{\mathbf{n}} da = 0. \tag{2.67}$$

In the first pair of these equations, $\mathcal{S}$ is any regular open surface that is bounded by a closed regular contour $\mathcal{C}$, while in the second pair of equations $\mathcal{R}$ is any regular region of space that is bounded by a closed surface $\mathcal{S}$. Unlike the differential (or point) form of Maxwell's equations [given in Eqs. (2.33)–(2.36)] which require that the field vectors are piecewise continuous with piecewise continuous first partial derivatives, the integral (or space) form of Maxwell's equations are valid over any region of space. They are then the more general form because they no longer require that the field properties be everywhere well-behaved.

2.2   Invariance of the Maxwell–Lorentz Equations in the Special Theory of Relativity

The failure of the expected invariance of the Maxwell–Lorentz theory under a Galilean transformation in Newtonian relativity eventually led to the development of the special theory of relativity. This special theory describes physical phenomena as measured by different observers in reference frames that are moving at a constant velocity relative to each other, appropriately referred to as inertial reference frames. This special theory then did away with the "Newtonian" idea of an absolute frame of reference with respect to which the absolute dynamical state of any physical system could be specified. This concept of an absolute reference frame resulted in the introduction of a so-called "luminiferous ether" through which electromagnetic waves propagated. Originally introduced by Thomas Young [21] in order to explain Bradley's discovery of stellar aberration [22] in terms of the wave theory of light, the idea of a "luminiferous ether" was widely accepted at that time and was adopted by Maxwell [1] as an essential part of his theory of electromagnetic waves.

The null result of the landmark Michelson–Morley experiment [23] in 1887 which set out to measure the relative motion of the Earth with respect to the "luminiferous ether" led Lorentz in 1895 to propose that a moving body experienced a length contraction along the direction of the relative motion. In a later response in 1901 to this "ad hoc" hypothesis by Lorentz, Poincáre wrote [16]

> I am not satisfied with the explanation (physical contraction) of the negative result of the Michelson experiment by the Lorentz theory, I would say that the laws of optics are only depending on the relative motion of the involved bodies.

This then led to an amicable controversy with Lorentz regarding the physics of absolute space and time. In his 1902 book *La Science et l'hypothèse* [17], Poincáre devoted an entire chapter to this *principle of relativity*, stating that

> There is no absolute uniform motion, no physical experience can therefore detect any inertial motion (no force felt), there is no absolute time, saying that two events have the same duration is conventional, as well as saying they are simultaneous is purely conventional as they occur in different places.

In addition, Poincáre stated that the concept of a "luminiferous ether" was a metaphysical rather than a physical concept. His derivation of the transformation equations between two inertial reference frames proceeded from symmetry considerations and the observation that the transformation between inertial reference frames Σ to Σ' whose coordinate axes are parallel with relative uniform motion along their x-axes is the same as the successive transformations between inertial reference frames Σ to Σ'' to Σ' with parallel coordinate axes and relative uniform motion along their respective x-axes. The resulting transformation relations between the primed and unprimed coordinate systems was found by Poincáre to be given by

$$x' = \frac{x - Vt}{\sqrt{1 - V^2/K^2}}, \tag{2.68}$$

$$t' = \frac{t - Vx/K^2}{\sqrt{1 - V^2/K^2}}, \tag{2.69}$$

with $y' = y$ and $z' = z$, where K is left (in Poincáre's derivation) as an unspecified parameter with the dimension of velocity that provides an upper bound to the magnitude of the relative velocity V in order that the transformation relations are real-valued. Notice that the classical Galilean transformation relations $x' \to x - Vt$, $t' \to t$ result in the limit as $K \to \infty$.

Based upon the negative experimental results of Trouton and Noble [24] in 1903 to detect the effects of the velocity of the Earth through this 'luminiferous ether' and following the mathematical arguments of Lorentz [19] and Poincaré [25], Einstein [26] was led to the conclusion that

> the same laws of electrodynamics and optics will be valid for all frames of reference for which the equations of mechanics hold good.

He then went on to

raise this conjecture (the purport of which will hereafter be called the 'Principle of Relativity') to the status of a postulate, and also introduce another postulate, which is only apparently irreconcilable with the former, namely, that light is always propagated in empty space with a definite velocity c which is independent of the state of motion of the emitting body. These two postulates suffice for the attainment of a simple and consistent theory.

Finally, he concluded that the

introduction of a 'luminiferous ether' will prove to be superfluous inasmuch as the view here to be developed will not require an 'absolutely stationary space' provided with special properties...

Einstein summarized these conjectures in the following two *postulates of special relativity theory*:

1. The Fundamental Postulate of Special Relativity: *The laws of physics are the same in all inertial reference frames.*

2. The Postulate of the Constancy of the Speed of Light: *The speed of light in free space has the same value c in all inertial systems.*

As an immediate consequence of the first postulate, no preferred inertial reference system exists. Because of this, it is then physically impossible to detect the uniform motion of one inertial frame of reference from observations that are made entirely within that reference frame.

With regard to the fundamental postulate of special relativity, the identical formulation of the laws of physics in all inertial reference frames leads to two fundamental concepts that depend upon the manner in which the equations of physics transform from one inertial reference frame to another. Equations that do not change in form under a transformation from one inertial reference frame to another are said to be *Lorentz invariant*, whereas equations that remain valid because their terms, which are not all invariant, transform according to identical laws are said to be *Lorentz covariant*. Lorentz covariance then states that, in two different inertial reference frames located at the same event in space–time, all nongravitational laws of physics must yield the same results for identical experiments.

These two postulates immediately lead to a re-examination of the concept of simultaneity. Because the first postulate requires that the concept of a universal time must be abandoned, the second postulate then means that the only way that simultaneity can be defined is in terms of the velocity of light c. This then elevates c to a level of significance that is much more fundamental than its property of being the propagation velocity of an electromagnetic wave in free space. It is in fact the maximum velocity of propagation of any interaction and its finite value then excludes instantaneous action at a distance. The fundamental concept of *simultaneity* may then be defined in the following operational manner [27]: two instants of time t_1 and t_2 that are observed at two points P_1 and P_2 in a particular inertial reference frame are said to be *simultaneous* if an impulse of light emitted at the geometric midpoint of the line connecting P_1 and P_2 arrives at the point P_1

at the time t_1 and at the point P_2 at the time t_2. This definition guarantees that an impulse of light that is emitted from a point source will reach all equidistant points from that source point simultaneously in a given inertial reference frame so that its phase surface is spherical in that reference frame. Notice that the simultaneity of two events at two different points in an inertial reference frame has no significance independent of the frame.

2.2.1 Transformation Laws in Special Relativity

The proper relationship between the space–time coordinates $(\mathbf{r}, t) = (x, y, z, t)$ of some event as observed in an inertial coordinate frame Σ and the space–time coordinates $(\mathbf{r}', t') = (x', y', z', t')$ of that same event as observed in another inertial coordinate frame Σ' provides the fundamental transformation laws of special relativity. This relationship is provided by the equations of transformation as formally expressed by the functional expressions $x' = x'(x, y, z, t)$, $y' = y'(x, y, z, t)$, $z' = z'(x, y, z, t)$, and $t' = t'(x, y, z, t)$ which relate the space–time coordinates of an event in inertial frame Σ with those of the same event as observed in another inertial frame Σ'.

With the assumption that both *space and time are homogeneous* (i.e., that all points in space–time are equivalent), these equations of transformation must then be linear. The connection between space–time points in the two inertial reference frames is obtained from the postulate of the constancy of the velocity of light c. Consider an impulse of light that is emitted as a spherical wave from the origin O in the inertial reference frame Σ at the instant $t = 0$ and assume that the origin O' of the inertial reference frame Σ' coincides with O at that precise instant of time. The emitted spherical light wave then propagates with the velocity c in all directions in each inertial frame Σ and Σ', so that

$$c^2 t^2 - \left(x^2 + y^2 + z^2 \right) = c^2 t'^2 - \left(x'^2 + y'^2 + z'^2 \right). \tag{2.70}$$

This relationship then provides the connection between the coordinates of any given event as observed in the two inertial reference frames Σ and Σ'.

When reference frame Σ' is moving at a constant velocity $\mathbf{V}$ with magnitude $V = |\mathbf{V}|$ in some fixed direction relative to the inertial reference frame Σ, the most general form of these transformation equations is found to be [28]

$$\mathbf{r}' = \mathbf{r} + \mathbf{V} \left[\frac{\mathbf{V} \cdot \mathbf{r}}{V^2} (\gamma - 1) - \gamma t \right], \tag{2.71}$$

$$ct' = \gamma \left(ct - \frac{\mathbf{V} \cdot \mathbf{r}}{c} \right), \tag{2.72}$$

where

$$\gamma \equiv \frac{1}{\sqrt{1-\beta^2}} \qquad (2.73)$$

is the so-called *dilation factor*, with $\beta \equiv V/c$ denoting the magnitude of the *normalized velocity*. The inverse coordinate transformation is obtained by interchanging primed and unprimed quantities and by replacing $\mathbf{V}$ with $-\mathbf{V}$ (the velocity of Σ relative to Σ'), with the result

$$\mathbf{r} = \mathbf{r}' + \mathbf{V}\left[\frac{\mathbf{V}\cdot\mathbf{r}'}{V^2}(\gamma-1) + \gamma t'\right], \qquad (2.74)$$

$$ct = \gamma\left(ct' + \frac{\mathbf{V}\cdot\mathbf{r}'}{c}\right). \qquad (2.75)$$

These equations are known as the *Poincaré–Lorentz transformation relations*. Notice that the transformation relation given in Eq. (2.71) may be separated into components parallel ($\|$) and perpendicular ($\perp$) to the velocity vector $\mathbf{V}$ as

$$\mathbf{r}'_\| = \gamma(\mathbf{r}_\| - \mathbf{V}t), \qquad (2.76)$$

$$\mathbf{r}'_\perp = \mathbf{r}_\perp. \qquad (2.77)$$

Similar expressions may also be written for the inverse transformation given in Eq. (2.74). Finally, notice that the transformation relations given in Eqs. (2.68) and (2.69) are a special case of these relations when the parameter K is set equal to c.

The specific form of the Poincaré–Lorentz transformation equations is a result of the invariance of the relation given in Eq. (2.70) describing a spherical light wave whose radius increases with time at a rate given by the invariant vacuum speed of light c. As a consequence, the quadratic form

$$ds^2 = c^2 dt^2 - (dx^2 + dy^2 + dz^2) \qquad (2.78)$$

for the world distance ds between two neighboring events with space–time coordinates $(\mathbf{r}, t) = (x, y, z, t)$ and $(\mathbf{r} + d\mathbf{r}, t + dt) = (x + dx, y + dy, z + dz, t + dt)$ is invariant under the Poincaré–Lorentz transformation. In fact, the Poincaré–Lorentz transformation relations are the only nonsingular transformation relations that possess a unique inverse and which maintain the invariance of the quadratic form for the world distance [29]. The world distance s between two events is accordingly said to be either *time-like* ($s^2 > 0$), *light-like* ($s = 0$), or *space-like* ($s^2 < 0$) depending on the value of s^2.

A simple test of the physical propriety of the Poincaré–Lorentz transformations is that they reduce to the approximate Galilean transformation equations in the *nonrelativistic limit* when $V/c \ll 1$. In that limit the transformation equation (2.76) becomes

$$\mathbf{r}'_\| = \left(\mathbf{r}_\| - \mathbf{V}t\right)\left(1 + \frac{V^2}{2c^2} + \cdots\right) \rightarrow \mathbf{r}_\| - \mathbf{V}t, \qquad (2.79)$$

and the transformation equation (2.72), when applied to the motion of the origin O' of Σ', for example, which is given by $\mathbf{r} = \mathbf{V}t$, becomes

$$t' = t\left(1 - \frac{V^2}{c^2}\right)\left(1 + \frac{V^2}{2c^2} + \cdots\right) \rightarrow t. \qquad (2.80)$$

These are just the classical Galilean transformation equations.

2.2.1.1 Time Dilation and The Convective Derivative Operator

The first important consequence of the Poincaré–Lorentz transformations is the phenomenon of time dilation. Consider a clock that is at rest ($\Delta x' = 0$) at the point x' in the Σ' reference frame which is moving with velocity $\mathbf{V} = \hat{\mathbf{1}}_x V$ along the common $x - x'$ axis of the Σ and Σ' inertial frames. Let two successive events at that point in the Σ' frame span the time interval from t' to $t' + \Delta t'$. An observer in the Σ frame observes these two successive events as occurring at the two instances of time given by [cf. Eq. (2.75)]

$$t_1 = \frac{t' + (V/c^2)x'}{\sqrt{1 - \beta^2}} \quad \& \quad t_2 = \frac{(t' + \Delta t') + (V/c^2)x'}{\sqrt{1 - \beta^2}}.$$

A single clock at the fixed position x' in the Σ' frame is used to measure the time interval from t' to $t' + \Delta t'$ while the corresponding time instances t_1 and $t_2 = t_1 + \Delta t$ in the Σ frame are measured by two different clocks that are synchronized, one stationary Σ-clock that is coincident with the moving Σ'-clock at the beginning of the time interval and another stationary Σ-clock that is coincident with the moving Σ'-clock at the end of the time interval. The two time intervals are then related by

$$\Delta t = \frac{\Delta t'}{\sqrt{1 - \beta^2}}. \qquad (2.81)$$

As a consequence, the duration of a measured time interval is a minimum in the inertial reference frame in which the clock is at rest relative to the observer. When a clock moves with a velocity $\mathbf{V}$ relative to the observer, its rate is measured by that observer to have slowed down by the factor $\sqrt{1 - \beta^2}$ because any measured time interval is increased by the factor $1/\sqrt{1 - \beta^2}$.

If $d\tau$ represents a *proper differential time interval*, then the *non-proper differential time interval* dt is given by the *time dilation relation*

$$dt = \frac{d\tau}{\sqrt{1-\beta^2}} = \gamma d\tau. \tag{2.82}$$

It is then seen that $(d\tau)^2 = (dt)^2 - \frac{1}{c^2}(dx^2 + dy^2 + dz^2)$ which may, in turn, be written in the more suggestive form $(icd\tau)^2 = dx^2 + dy^2 + dz^2 + (icdt)^2$. The transformation between two inertial reference frames is thus seen to be described by an imaginary rotation in space–time (x, y, z, ct) by an amount proportional to the relative velocity V between them (see Sect. 1.7 of [30]).

The general relativistic transformation between partial time differentiation in the Σ and Σ' inertial coordinate systems is obtained from Eqs. (2.72) and (2.75) as

$$\frac{\partial}{\partial t} = \frac{\sqrt{1-\beta^2}}{1 + \mathbf{u}' \cdot \mathbf{V}/c^2} \frac{\partial}{\partial t'}, \tag{2.83}$$

with inverse

$$\frac{\partial}{\partial t'} = \frac{\sqrt{1-\beta^2}}{1 - \mathbf{u} \cdot \mathbf{V}/c^2} \frac{\partial}{\partial t}. \tag{2.84}$$

This pair of differential transformation relations reduces to that obtained from the time dilation relation (2.82) when either $\mathbf{u}' = \mathbf{0}$ (in which case the proper time differential is dt') or when $\mathbf{u} = \mathbf{0}$ (in which case the proper time differential is dt).

The convective or substantial derivative operator given in Eq. (2.46) relates the rate of change of any given variable along a fixed contour C in an inertial reference frame Σ' that is moving with velocity $\mathbf{V}$ relative to another inertial reference frame Σ, as illustrated in Fig. 2.3. From the time dilation expression given in Eq. (2.81), one obtains the general expression

$$\frac{d}{dt} = \frac{1}{\sqrt{1-\beta^2}} \left(\frac{\partial}{\partial t} + \mathbf{V} \cdot \nabla \right), \tag{2.85}$$

which reduces to the classical expression in Eq. (2.46) as $\beta \to 0^+$.

2.2.1.2 Lorentz–Fitzgerald Contraction

The next important consequence of the Poincaré–Lorentz transformations is the phenomenon of Lorentz–Fitzgerald contraction. Let the Σ' reference frame move with velocity $\mathbf{V} = \hat{\mathbf{1}}_x V$ along the common $x - x'$ axis of the Σ and Σ' inertial frames. Consider a measuring rod that is stationary in the Σ' inertial reference frame and is lying along the x'-axis. Let its end-points be at the coordinate positions x'_1

and $x_2' \geq x_1'$ so that its length in this proper frame, referred to as its *proper length*, is given by the difference $\Delta x' = x_2' - x_1'$. From the Poincaré–Lorentz transformation in Eq. (2.76), $x_j' = \gamma(x_j - Vt_j)$ for $j = 1, 2$ so that their difference is given by $x_2' - x_1' = \gamma[(x_2 - x_1) - V(t_2 - t_1)]$. The length of the rod in the Σ frame is given by the distance between the coordinates x_2 and x_1 of the two end points measured at the same instant of time in that inertial frame. With $t_2 = t_1$, the preceding relation yields the *Lorentz–Fitzgerald contraction*

$$\Delta x = \sqrt{1 - \beta^2}\,(\Delta x')\,. \tag{2.86}$$

The measured length Δx of the moving rod is then contracted from its proper length $\Delta x'$ by the factor $\sqrt{1 - \beta^2}$. This contraction of length was put forth independently by both FitzGerald and Lorentz [31] (who acknowledged FitzGerald's unpublished speculation on this effect) in order to explain the negative result of the Michelson–Morley experiment. For the dimensions of the rod along the y'- and z'-coordinate directions, which are perpendicular to the relative motion of the two inertial reference frames, the transformation relation in Eq. (2.77) states that the measurement of these dimensions is the same in both frames. As a consequence, an object's length is a maximum when it is measured by an observer at rest with respect to it. When the object moves with a velocity $\mathbf{V}$ relative to the observer, its measured length is contracted in the direction of its motion by the factor $\sqrt{1 - \beta^2}$ while its dimensions perpendicular to the direction of the motion remain unaltered.

2.2.1.3 Relativistic Velocity and Acceleration Transformations

The relativistic velocity transformation relation of a rigid body moving with velocity $\mathbf{u}'$ in the Σ' inertial frame that is moving with fixed velocity $\mathbf{V}$ relative to the inertial frame Σ is obtained from the Poincaré–Lorentz transformation relations (2.71) and (2.72) as (see Problem 2.8)

$$\mathbf{u} = \frac{\mathbf{u}'\sqrt{1 - \beta^2} + \mathbf{V}\left[(\mathbf{u}' \cdot \mathbf{V}/V^2)\left(1 - \sqrt{1 - \beta^2}\right) + 1\right]}{1 + \mathbf{u}' \cdot \mathbf{V}/c^2}\,. \tag{2.87}$$

The inverse transformation is obtained by interchanging primed and unprimed quantities and replacing $\mathbf{V}$ with $-\mathbf{V}$ in the above relation to yield

$$\mathbf{u}' = \frac{\mathbf{u}\sqrt{1 - \beta^2} + \mathbf{V}\left[(\mathbf{u} \cdot \mathbf{V}/V^2)\left(1 - \sqrt{1 - \beta^2}\right) - 1\right]}{1 - \mathbf{u} \cdot \mathbf{V}/c^2}\,. \tag{2.88}$$

The transformation relation in Eq. (2.87) may be separated into components parallel
($\parallel$) and perpendicular ($\perp$) to the velocity vector $\mathbf{V}$ as

$$u_{\parallel} = \frac{u'_{\parallel} + V}{1 + \mathbf{u}' \cdot \mathbf{V}/c^2},$$

(2.89)

$$\mathbf{u}_{\perp} = \frac{\sqrt{1 - \beta^2}}{1 + \mathbf{u}' \cdot \mathbf{V}/c^2} \mathbf{u}'_{\perp},$$

(2.90)

with similar expressions for the inverse transformation. Each perpendicular (or
transverse) component of an object's velocity as measured in the Σ frame is then
seen to be related to the corresponding transverse velocity component as well as to
the parallel velocity component of the object in the Σ' frame. The complexity of the
transformation relations exhibited in Eqs. (2.87) and (2.88) is a consequence of the
fact that neither reference frame is proper.

A special case occurs when an inertial reference frame Σ' is chosen such that
$u'_{\parallel} = 0$. The velocity transformation equations then become $u_{\parallel} = V$ and $\mathbf{u}_{\perp} =$
$\sqrt{1 - \beta^2}\mathbf{u}'_{\perp}$. In this case there is no length contraction involved and the transverse
velocity transformation is due to the effects of time-dilation alone.

The relativistic acceleration transformation equations are obtained from the
velocity transformation relations by direct time differentiation as (see Problem 2.9)

$$a_{\parallel} = \frac{(1 - \beta^2)^{3/2}}{(1 + \mathbf{u}' \cdot \mathbf{V}/c^2)^3} a'_{\parallel},$$

(2.91)

$$\mathbf{a}_{\perp} = \frac{1 - \beta^2}{(1 + \mathbf{u}' \cdot \mathbf{V}/c^2)^2} \left(\mathbf{a}'_{\perp} - \frac{\mathbf{u}'_{\perp}\beta/c}{1 + \mathbf{u}' \cdot \mathbf{V}/c^2} a'_{\parallel} \right),$$

(2.92)

with inverse

$$a'_{\parallel} = \frac{(1 - \beta^2)^{3/2}}{(1 - \mathbf{u} \cdot \mathbf{V}/c^2)^3} a_{\parallel},$$

(2.93)

$$\mathbf{a}'_{\perp} = \frac{1 - \beta^2}{(1 - \mathbf{u} \cdot \mathbf{V}/c^2)^2} \left(\mathbf{a}_{\perp} + \frac{\mathbf{u}_{\perp}\beta/c}{1 - \mathbf{u} \cdot \mathbf{V}/c^2} a_{\parallel} \right).$$

(2.94)

Unlike the Galilean transformation result which states that the acceleration is
independent of the inertial reference frame, the relativistic transformation relations
given in Eqs. (2.91) and (2.92) states that the acceleration of a rigid body depends
upon the inertial reference frame in which it is measured. The classical result that
$\mathbf{a} = \mathbf{a}'$ is obtained in the non-relativistic limit as $\beta \to 0$.

2.2.2 *Transformation of Dynamical Quantities*

Classical Newtonian mechanics is found to be inconsistent with the special theory
of relativity because, although its laws are invariant under a Galilean transformation, they are not invariant under a Poincaré–Lorentz transformation. In addition,
Newton's third law of motion requiring that action and reaction forces must be
equal has no meaning in special relativity for action-at-a-distance forces (allowed in
the classical theory of Newtonian mechanics) because the simultaneity of spatially
separated events depends upon the inertial reference frame from which they are
observed. Finally, from Newton's second law of motion, the action of a force on a
particle in Newtonian mechanics can accelerate that particle to an unlimited velocity,
in violation of the second fundamental postulate of the special theory of relativity.

2.2.2.1 Energy and Momentum in Special Relativity

Consider a body of fixed mass at rest in an inertial reference frame Σ. Because the
mass of this body must then be the ordinary mass in non-relativistic Newtonian
mechanics, it is referred to as the *rest mass* or *proper mass* m_0 of the body. If
this body is now moving with velocity $\mathbf{u}$ in the Σ frame, conservation of linear
momentum requires that its mass, as observed by a stationary observer in Σ, be
given by its *relativistic mass*

$$m = \frac{m_0}{\sqrt{1 - u^2/c^2}}. \tag{2.95}$$

Here u is the absolute value of the velocity of the body as observed in the inertial
reference frame Σ. Notice that this expression for the relativistic mass preserves the
form $\mathbf{p} = m\mathbf{u}$ for linear momentum while implicitly assuming that the relativistic
mass of a body is independent of its acceleration. In addition, $m \to m_0$ in the nonrelativistic limit as $u \to 0$, whereas in the relativistic limit as $u \to c$ from below,
the relativistic mass becomes unbounded. The *relativistic form of Newton's second
law of motion* is then given by

$$\mathbf{F} = \frac{d\mathbf{p}}{dt} = \frac{d}{dt}\left(\frac{m_0\mathbf{u}}{\sqrt{1 - u^2/c^2}}\right). \tag{2.96}$$

The conservation law of relativistic momentum then states that in the absence of
external forces ($\mathbf{F} = \mathbf{0}$), $\mathbf{p} = m_0\mathbf{u}/\sqrt{1 - u^2/c^2}$ is conserved.

The kinetic energy U_k of a particle is given by the work expended by an external
force $\mathbf{F}$ to increase its velocity from $\mathbf{0}$ up to some final value $\mathbf{u}$, so that

$$U_k = \int_0^{\mathbf{u}} \mathbf{F} \cdot d\mathbf{r} = \int_0^{\mathbf{u}} \frac{d}{dt}(m\mathbf{u}) \cdot d\mathbf{r}. \tag{2.97}$$

The integrand appearing in the final form of this expression may be rewritten as $d/dt(m\mathbf{u}) \cdot d\mathbf{r} = d(m\mathbf{u}) \cdot \mathbf{u} = m\mathbf{u} \cdot d\mathbf{u} + u^2 dm$. From the expression (2.95) for the relativistic mass, one obtains the equation $m^2 c^2 - m^2 \mathbf{u} \cdot \mathbf{u} = m_0^2 c^2$ with total differential $2mc^2 dm - 2mu^2 dm - 2m^2 \mathbf{u} \cdot d\mathbf{u} = 0$, which may be rewritten as $m\mathbf{u} \cdot d\mathbf{u} + u^2 dm = c^2 dm$. With these substitutions, Eq. (2.97) becomes

$$U_k = c^2 \int_{m_0}^{m} dm = mc^2 - m_0 c^2. \tag{2.98}$$

With substitution from Eq. (2.95) for the relativistic mass, the final expression for the *relativistic kinetic energy* of a particle with rest mass m_0 is found as

$$U_k = m_0 c^2 \left(\frac{1}{\sqrt{1 - u^2/c^2}} - 1 \right). \tag{2.99}$$

In the nonrelativistic limit $\beta \to 0$, the relativistic kinetic energy becomes $U_k = m_0 c^2 \left[(1 - \beta^2)^{-1/2} - 1 \right] = m_0 c^2 \left(1 + \frac{1}{2}\beta^2 + \frac{3}{8}\beta^4 + \cdots - 1 \right) \to \frac{1}{2} m_0 u^2$, which is just the classical result for the kinetic energy of a body with mass m_0 and velocity $\mathbf{u}$. Notice that the omitted terms are $\mathcal{O}(\beta^2)$ to the nonrelativistic result.[8] In the opposite limit as $\beta \to 1^-$, $U_k \to \infty$ so that an infinite amount of work needs to be expended in order to accelerate a particle with nonzero rest mass up to the vacuum speed of light c.

With the *total energy* of the particle given by the celebrated *Einstein mass–energy relation*[26]

$$E = mc^2 = m_0 c^2 + U_k \tag{2.100}$$

so that $U_k = (m - m_0)c^2$, it is then seen that any change in the kinetic energy of a particle may be related to the change in its inertial mass. The quantity $m_0 c^2$ is accordingly referred to as the *rest energy* of the particle.

From the expression $\mathbf{p} = m_0 \mathbf{u}/\sqrt{1 - u^2/c^2}$ for the relativistic momentum of a particle with rest mass m_0, one finds that $p^2 = m_0^2 u^2/(1 - u^2/c^2)$ which may be solved for the square of the magnitude of the velocity with the result $u^2 = p^2/(m_0^2 + p^2/c^2)$. Substitution of this result into Eq. (2.99) then yields $(U_k + m_0 c^2)^2 = m_0^2 c^4 + p^2 c^2$ which may be written in terms of the total energy $E = U_k + m_0 c^2$ as

$$E^2 = \left(m_0 c^2 \right)^2 + (pc)^2 . \tag{2.101}$$

---

[8]The order symbol $\mathcal{O}$ is defined as follows. Let $f(z)$ and $g(z)$ be two functions of the complex variable z that possess limits as $z \to z_0$ in some domain $\mathcal{D}$. Then $f(z) = \mathcal{O}(g(z))$ as $z \to z_0$ iff there exist positive constants K and δ such that $|f(z)| \le K|g(z)|$ whenever $0 < |z - z_0| < \delta$.

Upon expanding the equation $(U_k + m_0c^2)^2 = m_0^2c^4 + p^2c^2$ one obtains the quadratic relation $U_k^2 + 2m_0c^2U_k = p^2c^2$ for the kinetic energy of the particle. In the nonrelativistic limit when $u/c \ll 1$, the kinetic energy U_k of the particle will be much less than its rest energy m_0c^2 so that the quadratic term U_k^2 becomes negligible in comparison to the term linear in U_k, thereby yielding the classical result that $p \rightarrow \sqrt{2m_oU_k}$ as $u/c \rightarrow 0$.

2.2.2.2   The Mass of a Photon

A *photon* is a special type of elementary particle called a *luxon* whose velocity relative to any inertial reference frame is always c. Luxons do not have a rest mass because Eq. (2.95) would then become infinite. Because of this, their momentum must be carefully defined as the relativistically correct expression $p = m_0u/\sqrt{1 - u^2/c^2} \rightarrow 0/0$ is now undefined. From Eq. (2.101), the energy E and momentum p of a photon are related by $E = pc$, and from the quantum mechanical description of the photoelectric effect, the energy of a photon is given by $E = h\nu$, where $h \simeq 6.62 \times 10^{-34}$ J · s is *Planck's constant* and ν is the oscillation frequency of the photon, where $\lambda\nu = c$ with λ denoting the wavelength. The magnitude of the momentum of a photon is then given by

$$p = \frac{E}{c} = \frac{h\nu}{c}, \qquad (2.102)$$

which is nonzero if $\nu > 0$. The *De Broglie wavelength*[32] of a photon is then given by

$$\lambda = \frac{h}{p}. \qquad (2.103)$$

A photon has a measurable wavelength λ which, in turn, provides a measure of its momentum p and energy E through these two fundamental relations. Hence, a photon has all the attributes of a particle with a non-zero rest mass, although its own rest mass m_γ is zero.[9]

2.2.2.3   Acceleration and Force in Special Relativity

Beginning with Eq. (2.101) rewritten as $E = c\sqrt{m_0^2c^2 + p^2}$, the derivative of the total energy with respect to the magnitude of the momentum p is found as

---

[9]While it is practically impossible to measure a zero rest mass, upper limits on the rest mass of a photon have been obtained experimentally by Fischbach et al. [33]. Their results provide a geomagnetic limit given by $m_\gamma \leq 1 \times 10^{-48}$ g, eighteen orders of magnitude smaller than the rest mass m_e of an electron.

$dE/dp = pc^2/E$. With $E = mc^2$ and $p^2 = m^2u^2$, this expression reduces to the rather elegant result

$$\frac{dE}{dp} = u \tag{2.104}$$

so that the magnitude of the velocity of a particle is given by the change in total energy with respect to the momentum of the particle.

Consider now obtaining a relativistically correct expression for the acceleration of a particle of relativistic mass m due to the application of a force, where

$$\mathbf{F} = \frac{d}{dt}(m\mathbf{u}) = m\frac{d\mathbf{u}}{dt} + \mathbf{u}\frac{dm}{dt}. \tag{2.105}$$

Because $m = E/c^2$ where $E = U_k + m_0c^2$, then $dm/dt = (1/c^2)dU_k/dt$, and because $dU_k = \mathbf{F} \cdot d\mathbf{r}$ [cf. Eq. (2.97)], then $dU_k/dt = \mathbf{F} \cdot d\mathbf{r}/dt = \mathbf{F} \cdot \mathbf{u}$ so that $dm/dt = (1/c^2)\mathbf{F} \cdot \mathbf{u}$. Substitution of this expression into Eq. (2.103) then yields

$$\mathbf{F} = m\frac{d\mathbf{u}}{dt} + \mathbf{u}\frac{\mathbf{F} \cdot \mathbf{u}}{c^2}. \tag{2.106}$$

The acceleration of the particle is then given by the general expression

$$\mathbf{a} \equiv \frac{d\mathbf{u}}{dt} = \frac{1}{m}\mathbf{F} - \left(\frac{\mathbf{F} \cdot \mathbf{u}}{c^2}\right)\mathbf{u}, \tag{2.107}$$

and the acceleration is, in general, not parallel to the applied force.

Two special cases of Eq. (2.105) immediately arise. The first is when the applied force $\mathbf{F}$ is parallel to the velocity $\mathbf{u}$ of the particle. Because the vectors $\mathbf{a}$, $\mathbf{F}$, and $\mathbf{u}$ are then all parallel, Eq. (2.105) may then be written in the scalar form

$$F = m\frac{du}{dt} + u\frac{dm}{dt} = \frac{m_0}{\sqrt{1 - u^2/c^2}}\frac{du}{dt} + u\frac{d}{dt}\left(\frac{m_0}{\sqrt{1 - u^2/c^2}}\right)$$

$$= \frac{m_0}{\sqrt{1 - u^2/c^2}}\left(1 + \frac{u^2/c^2}{1 - u^2/c^2}\right)\frac{du}{dt} = \frac{m_0}{(1 - u^2/c^2)^{3/2}}\frac{du}{dt}.$$

Because $a = du/dt$, the above result may then be written as

$$\mathbf{F}_\parallel = \frac{m_0}{(1 - u^2/c^2)^{3/2}}\mathbf{a}_\parallel, \tag{2.108}$$

where $m_0/(1 - u^2/c^2)^{3/2}$ is referred to as the *longitudinal mass*. The other special case is realized when the applied force $\mathbf{F}$ is perpendicular to the velocity $\mathbf{u}$, in which case $\mathbf{F} \cdot \mathbf{u} = 0$. Equation (2.107) then yields the expression

$$\mathbf{F}_\perp = \frac{m_0}{\sqrt{1 - u^2/c^2}} \mathbf{a}_\perp, \tag{2.109}$$

where $m_0/\sqrt{1 - u^2/c^2}$ is referred to as the *transverse mass* and is equal to the relativistic mass m of the particle.

Finally, from the relation following Eq. (2.105) it is seen that

$$\frac{dU_k}{dt} = c^2 \frac{dm}{dt}, \tag{2.110}$$

which states that a change in the kinetic energy of a body is equal to a proportionate change in its relativistic mass. *Einstein's mass–energy relation*

$$E = mc^2 \tag{2.111}$$

is then seen to be an expression of the equivalence of mass and energy. Based upon his derivation of this fundamental expression, Einstein [26] hypothesized that mass and energy form a single invariant quantity that is referred to as the *mass–energy*.

As a consequence of the mass–energy relation, because the rest mass of a body is internal energy, then a body with zero rest mass has no internal energy, its energy being all external. If such a body with zero rest mass moved with a velocity that was less than c in any inertial reference frame, then another inertial reference frame could always be found in which it is at rest. However, if it travels at the velocity c in any inertial reference frame, then it will travel at the velocity c in every inertial reference frame. Thus, a body with zero rest mass must travel at the speed of light and can never be at rest in any inertial reference frame. Any particle moving at the speed of light is called a *luxon*, an important example of which is the photon (see Sect. 2.2.2.2). Because of time dilation, as expressed in Eq. (2.81), a photon is emitted (created) and absorbed (destroyed) at the same instant. In contrast, any particle that has an initial velocity that is less than c is called either a *bradyon* or a *tardyon*; such a particle can be accelerated to a velocity that approaches c but can never exceed this value. Finally, any particle whose velocity is larger than c during its entire lifetime is referred to as a *tachyon* [34, 35]. Because of the unifying notion of space–time in the special theory of relativity, as time slows down as the relative velocity u of a bradyon increases relative to a stationary observer, approaching zero as $u \to c$, one may argue that motion through three-dimensional space takes place at the expense of "motion" through time.

2.2.2.4   Linear Momentum and Energy in Special Relativity

Consider now the transformation relations for the linear momentum and energy of a body with rest mass m_0. These quantities are defined in the inertial frame Σ by the pair of relations

$$\mathbf{p} = \frac{m_0 \mathbf{u}}{\sqrt{1 - u^2/c^2}}, \quad E = \frac{m_0 c^2}{\sqrt{1 - u^2/c^2}},$$

whereas in the inertial frame Σ' the corresponding quantities are defined by

$$\mathbf{p}' = \frac{m_0 \mathbf{u}'}{\sqrt{1 - u'^2/c^2}}, \quad E' = \frac{m_0 c^2}{\sqrt{1 - u'^2/c^2}},$$

where the Σ' frame is moving with the velocity $\mathbf{V}$ with respect to Σ. Application of the velocity transformation relation given in Eq. (2.87) together with the identity [see Problem 2.10]

$$\frac{1}{\sqrt{1 - u^2/c^2}} = \frac{1 + \mathbf{u}' \cdot \mathbf{V}/c^2}{\sqrt{1 - u'^2/c^2}\sqrt{1 - V^2/c^2}} \tag{2.112}$$

then yields the general transformation relations

$$\mathbf{p} = \mathbf{p}' + (\gamma - 1)\frac{\mathbf{p}' \cdot \mathbf{V}}{V^2}\mathbf{V} + \gamma \frac{E'}{c^2}\mathbf{V} \tag{2.113}$$

$$E = \gamma(E' + \mathbf{V} \cdot \mathbf{p}'), \tag{2.114}$$

where $\gamma = 1/\sqrt{1 - \beta^2}$ with $\beta = V/c$. The inverse transformation relations are obtained by replacing $\mathbf{V}$ with $-\mathbf{V}$ and interchanging primed and unprimed quantities, yielding

$$\mathbf{p}' = \mathbf{p} + (\gamma - 1)\frac{\mathbf{p} \cdot \mathbf{V}}{V^2}\mathbf{V} - \gamma \frac{E}{c^2}\mathbf{V} \tag{2.115}$$

$$E' = \gamma(E - \mathbf{V} \cdot \mathbf{p}). \tag{2.116}$$

The transformation relations for the parallel and perpendicular components of the momentum (relative to the direction of $\mathbf{V}$) are then given by the relations $p'_{\parallel} = (p_{\parallel} - EV/c^2)/\sqrt{1 - V^2/c^2}$ and $\mathbf{p}'_{\perp} = \mathbf{p}_{\perp}$, with similar expressions for the inverse transformation. Notice that the quantities $\mathbf{p}$ and E/c^2 transform in precisely the same manner as do the space–time coordinates $\mathbf{r}$ and t of a particle [cf. Eqs. (2.74) and (2.75)].

On a more fundamental level, notice the interdependence of momentum and energy in the special theory of relativity. If the energy and momentum are conserved in an interaction that is observed in one inertial reference frame, then these two quantities must be conserved in every inertial reference frame. In addition, if momentum is conserved then energy must also be conserved, and conversely, if energy is conserved then so also must momentum be conserved.

The transformation relations for mass follow directly from the energy and momentum transformation relations given in Eqs. (2.113)–(2.116). Because $E = mc^2$ with $m = m_0/\sqrt{1 - u^2/c^2}$ and because $E' = m'c^2$ with $m' = m_0/\sqrt{1 - u'^2/c^2}$, then

$$mc^2 = \frac{1}{\sqrt{1 - V^2/c^2}}\left(m'c^2 + \mathbf{V}\cdot m'\mathbf{u}'\right) = \frac{m'}{\sqrt{1 - V^2/c^2}}\left(c^2 + \mathbf{u}'\cdot\mathbf{V}\right).$$

The transformation relation for the mass is then seen to be given by

$$m = m'\frac{1 + \mathbf{u}'\cdot\mathbf{V}/c^2}{\sqrt{1 - V^2/c^2}}, \tag{2.117}$$

with inverse

$$m' = m\frac{1 - \mathbf{u}\cdot\mathbf{V}/c^2}{\sqrt{1 - V^2/c^2}}. \tag{2.118}$$

The transformation relation for the force $\mathbf{F} = d\mathbf{p}/dt$ is obtained directly from those for the linear momentum $\mathbf{p}$ as

$$\mathbf{F}' = \frac{\sqrt{1 - V^2/c^2}\,\mathbf{F} + \mathbf{V}\left[\left(1 - \sqrt{1 - V^2/c^2}\right)\mathbf{F}\cdot\mathbf{V}/V^2 - \mathbf{F}\cdot\mathbf{u}/c^2\right]}{1 - \mathbf{u}\cdot\mathbf{V}/c^2}, \tag{2.119}$$

where the inverse transformation relation is obtained by replacing $\mathbf{V}$ with $-\mathbf{V}$ and interchanging primed and unprimed quantities, so that

$$\mathbf{F} = \frac{\sqrt{1 - V^2/c^2}\,\mathbf{F}' + \mathbf{V}\left[\left(1 - \sqrt{1 - V^2/c^2}\right)\mathbf{F}'\cdot\mathbf{V}/V^2 + \mathbf{F}'\cdot\mathbf{u}'/c^2\right]}{1 + \mathbf{u}'\cdot\mathbf{V}/c^2}. \tag{2.120}$$

The transformation relations for the parallel and perpendicular components of the force (relative to the direction of $\mathbf{V}$) are then given by the pair of relations

$$F'_{\parallel} = \frac{F_{\parallel} - \mathbf{F}\cdot\mathbf{u}V/c^2}{1 - \mathbf{u}\cdot\mathbf{V}/c^2},$$

$$\mathbf{F}'_{\perp} = \frac{\sqrt{1 - V^2/c^2}}{1 - \mathbf{u}\cdot\mathbf{V}/c^2}\mathbf{F}_{\perp}.$$

Notice that the force $\mathbf{F}$ in one inertial frame is related to the power $\mathbf{F}'\cdot\mathbf{u}'$ developed by the force in another inertial frame. This then implies that power and force are interrelated in the special theory of relativity. In particular, the transformation relation (2.119) and its inverse (2.120) are completed by the pair of equations (see

Problem 2.14)

$$\mathbf{u}' \cdot \mathbf{F}' = \frac{(\mathbf{u} - \mathbf{V}) \cdot \mathbf{F}}{(1 - \mathbf{u} \cdot \mathbf{V}/c^2)^2}, \tag{2.121}$$

$$\mathbf{u} \cdot \mathbf{F} = \frac{(\mathbf{u}' + \mathbf{V}) \cdot \mathbf{F}'}{(1 + \mathbf{u}' \cdot \mathbf{V}/c^2)^2}. \tag{2.122}$$

Notice that these transformation relations for the force reduce to the Newtonian limits $\mathbf{F}' = \mathbf{F}$ and $\mathbf{u}' \cdot \mathbf{F}' = \mathbf{u} \cdot \mathbf{F}$ as $V/c \to 0$.

As a case of special interest, consider a body that is at rest at some instant of time $t' = t'_0$ in the inertial frame Σ' where it is then subjected to the force $\mathbf{F}'$. Because $\mathbf{u}' = \mathbf{0}$ at $t' = t'_0$ in this proper frame, then the force transformation relation given in Eq. (2.120) yields $F_\parallel = F'_\parallel$ and $\mathbf{F}_\perp = \sqrt{1 - V^2/c^2}\mathbf{F}'_\perp$. It is then seen that *the force measured in the proper reference frame for the body is greater than the corresponding force measured in any other inertial reference frame.*

2.2.2.5 The Aberration of Light and the Doppler Effect

Of considerable importance to the special theory of relativity is the complete explanation of *stellar aberration*, an astronomical phenomenon reported by Bradley [36] in 1727. With reference to a stationary light source S' in the Σ' inertial reference frame with coordinate axes parallel to those in the Σ inertial reference frame and moving with velocity $\mathbf{V} = \hat{\mathbf{1}}_z V = \hat{\mathbf{1}}'_z V$ relative to Σ, consider a photon emitted by S' and propagating in the $y'z'$-plane at an angle θ' to the z'-axis. The velocity component of this photon along the z'-direction is then $u'_z = c \cos \theta'$ and its component along the y'-direction is $u'_y = c \sin \theta'$. From the velocity transformation relations given in Eqs. (2.89) and (2.90), the observed velocity components of this photon along the parallel z and perpendicular y coordinate axes of the Σ reference frame are given by

$$u_z = c \frac{\cos \theta' + \beta}{1 + \beta \cos \theta'}, \quad \& \quad u_y = c \frac{\sqrt{1 - \beta^2} \sin \theta'}{1 + \beta \cos \theta'},$$

where $\beta = V/c$. Because $u_z = c \cos \theta$ and $u_y = c \sin \theta$ in the Σ reference frame, then

$$\tan \theta = \frac{u_y}{u_z} = \frac{\sqrt{1 - \beta^2} \sin \theta'}{\cos \theta' + \beta}, \tag{2.123}$$

which is the *relativistic equation for the aberration of light*, with inverse

$$\tan \theta' = \frac{u_y}{u_z} = \frac{\sqrt{1 - \beta^2} \sin \theta}{\cos \theta - \beta}, \tag{2.124}$$

relating the observed directions of light propagation from two different inertial reference frames. For a star directly overhead to a stationary observer on the surface of the Earth, $\theta = 3\pi/2$ and the observed angle is then given by $\tan \theta' = \sqrt{1/\beta^2 - 1}$ so that, in the nonrelativistic limit $\beta = V/c \ll 1$, the stellar aberration angle becomes $\tan \theta' = 1/\beta = c/V$, in agreement with the classical result.

The importance of this relativistic description in the explanation of stellar aberration was summarized by Max Born [37] as follows:

> This result is particularly remarkable because all the other theories have considerable difficulty in explaining aberration. From the Galilean transformation one obtains no deflection at all of the wave plane and the wave direction, and to explain aberration one has to introduce the concept 'ray', which in moving systems need not coincide with the direction of propagation. In Einstein's theory this difficulty disappears. In every inertial system S the direction of the ray (that is, the direction along which the energy is transported) coincides with the perpendicular to the wave planes, and the aberration results, in the same way as the Doppler effect and Fresnel's convection coefficient, from the concept of a wave with the help of the Lorentz transformation. This method of deriving the fundamental laws of the optics of moving bodies shows very strikingly that Einstein's theory of relativity is superior to all other theories.

For the relativistic treatment of the Doppler effect, consider a photon traveling in the $y'z'$-plane at an angle θ' to the positive z'-direction of the Σ' inertial frame where its frequency is ν'. Its energy, as measured in Σ', is then given by $h\nu'$. Let Σ' be moving with velocity $\mathbf{V} = \hat{\mathbf{1}}_z V$ with respect to the parallel reference frame Σ. From Eq. (2.114), the energy $E = h\nu$ of this photon measured in the Σ reference frame is related to its values in the Σ' frame by $h\nu = h\nu'(1 + \beta \cos\theta')/\sqrt{1 - \beta^2}$ so that the frequency ν of the photon, as observed in the Σ frame, is given by

$$\nu = \frac{1 + \beta \cos\theta'}{\sqrt{1 - \beta^2}} \nu', \tag{2.125}$$

with inverse

$$\nu' = \frac{1 - \beta \cos\theta}{\sqrt{1 - \beta^2}} \nu, \tag{2.126}$$

where θ and θ' are related through the aberration equations (2.123) and (2.124). This *relativistic Doppler effect* reduces to the classical Doppler frequency shift $\nu \simeq (1 + \beta \cos\alpha)\nu'$ when $\beta \ll 1$. The classical (Galilean) result is then modified by the relativistic time dilation factor γ appearing in Eqs. (2.125) and (1.26) due to the time dilation $T' = T\gamma = T/\sqrt{1 - \beta^2}$ of the oscillation period $T = 1/\nu$. Notice that all frequencies experience the same Doppler shift in vacuum.

2.2.2.6 Charge and Current in Special Relativity

Consider the electronic charge contained within a microscopic cubic element of volume whose edges have rest length l_0. If there are N electrons in the cube, then its total charge is Nq_e and the volume charge density is $\rho_0 = Nq_e/l_0^3$. Let the charges be at rest in the inertial frame Σ' so that their convective current density in that frame vanishes ($\mathbf{j}_0 = \mathbf{0}$). From the perspective of an inertial frame Σ that is moving with constant velocity $\mathbf{V}$ relative to Σ' along a coordinate direction situated along one of the edges of the cube, that edge will have the measured length $l_0\sqrt{1 - V^2/c^2}$ in Σ whereas the measured lengths of the edges along the directions transverse to the motion remain unchanged at l_0. The volume of this "cube" as observed in the reference frame Σ is then given by $l_0^3\sqrt{1 - V^2/c^2}$. However, the number of electrons contained in this "cube" remains unchanged and the charge on each electron remains fixed to an observer in Σ so that the observed charge density is given by $\rho = Nq_e/(l_0^3\sqrt{1 - V^2/c^2})$, which may be expressed in terms of the rest value $\rho_0 = Nq_e/l_0^3$ as

$$\rho = \frac{\rho_0}{\sqrt{1 - V^2/c^2}} = \frac{\rho_0}{m_0}m, \tag{2.127}$$

where m is the relativistic mass given in Eq. (2.95). Notice that this result is independent of the direction of motion between the two inertial reference frames. As observed from the reference frame Σ, the charges move with velocity $\mathbf{V}$ so that there is an observed convective current density $\mathbf{j} = \rho_0\mathbf{V}/\sqrt{1 - V^2/c^2}$, which may be expressed as

$$\mathbf{j} = \frac{\rho_0}{m_0}\mathbf{p}, \tag{2.128}$$

where $\mathbf{p}$ is the relativistic momentum of the moving charged particle. Equations (2.127) and (2.128) then show that the charge and current densities ρ and $\mathbf{j}$ transform in the same manner as do the mass m and momentum $\mathbf{p}$, respectively.

Notice that the relation between the current density and the charge density is similar to that between space and time coordinates as well as that between momentum and energy. As a consequence, just as both of the quantities

$$c^2 t^2 - |\mathbf{r}|^2 = c^2 \tau^2 \tag{2.129}$$

and

$$c^2 m^2 - |\mathbf{p}|^2 = c^2 m_0^2 \tag{2.130}$$

are invariant, the quantity

$$c^2 \rho^2 - |\mathbf{j}|^2 = c^2 \rho_0^2 \tag{2.131}$$

is also invariant.

Consider now a simplified model of the convection current flowing in a metal wire whose axis is situated along the x-axis of an inertial frame Σ in which the positive ions of the metal are at rest. It is assumed here that the free electrons travel with an average drift velocity $\mathbf{u} = \hat{\mathbf{1}}_x u$ and that the average number N of free electrons per unit volume is equal to the number of positive ions per unit volume so that the average net charge density is zero throughout the entire volume of the wire. Although the spatial distribution of free electrons in the metal is random on a microscopic scale, because the average charge density is zero, the spatial distribution of free electrons on a macroscopic average scale must be uniform and identical to that of the positive ions. The average negative charge density is then $\rho_- = Nq_e$, where q_e denotes the magnitude of the electronic charge, the positive charge density is $\rho_+ = Nq_e$, and the net charge density is then $\rho = \rho_+ - \rho_- = 0$. However, the average negative and positive current densities are $\mathbf{j}_- = Nq_e\mathbf{u}$ and $\mathbf{j}_+ = \mathbf{0}$, respectively, so that the net current density is then $\mathbf{j} = \mathbf{j}_+ - \mathbf{j}_- = -Nq_e\mathbf{u}$.

From the viewpoint of an observer in a different inertial frame Σ' that is moving with velocity $\mathbf{V} = \hat{\mathbf{1}}_x V$ relative to Σ, the negative and positive charge densities are, from Eqs. (2.127) and (2.128) with Eqs. (2.118) and (2.115), given by

$$\rho_-' = \frac{1 - Vu/c^2}{\sqrt{1 - V^2/c^2}}\rho_-,$$

$$\rho_+' = \frac{1}{\sqrt{1 - V^2/c^2}}\rho_+,$$

respectively, where $\rho_- = \rho_+ = Nq_e$. With this substitution, the net charge density observed in Σ' is found as

$$\rho' = \rho_+' - \rho_-' = \frac{Nq_e Vu/c^2}{\sqrt{1 - V^2/c^2}}, \tag{2.132}$$

and the observer in Σ' finds the wire to be positively charged. The origin of this observed net positive charge density in Σ' is simply due to the relativity of simultaneity [27]; observers in the two different inertial frames disagree on the simultaneity of measurements made on the end point positions of adjacent like charge pairs, thereby measuring different average distances of separation between adjacent pairs of positive ions and adjacent pairs of conduction electrons. This then leads to different measured values of the positive and negative charge densities, resulting in a measured net positive charge density in Σ' when the net charge density is zero in Σ.

In the inertial frame Σ where the positive ions are stationary, the net charge density is zero whereas the current density is nonzero. There is then no observed electric field whereas the magnetic field about the conductor is nonvanishing and the field observed in Σ is purely magnetic. In the inertial reference frame Σ', however, there is a positive net charge density together with a net current density so that both an electric and a magnetic field are observed.

As a consequence, *whether an electromagnetic field is purely magnetic or purely electric, or part electric and magnetic, is dependent upon the inertial reference frame from which the charge and current sources for that electromagnetic field are observed.*

2.2.3 Interdependence of Electric and Magnetic Fields

The fact that electric and magnetic fields have no separate meaning is eminently demonstrated in the special theory of relativity. In particular, a field that is purely electric or magnetic in one inertial frame will, in general, have both electric and magnetic field components in another inertial frame. The two "separate" concepts of an electric and a magnetic field are then subsumed by the single unifying concept of an electromagnetic field.

2.2.3.1 Transformation Relations for Electric and Magnetic Fields

The Lorentz force relation for a point charge q that is moving with velocity $\mathbf{u}$ at a particular space–time point at which the microscopic electric field is $\mathbf{e}$ and the microscopic magnetic field is $\mathbf{b}$ is given by [cf. Eq. (2.24)]

$$\mathbf{F} = q \left(\mathbf{e} + \left\| \frac{1}{c} \right\| \mathbf{u} \times \mathbf{b} \right). \tag{2.133}$$

From Eq. (2.120), the set of force transformation relations between an inertial frame Σ and another inertial frame Σ' in which the particle experiencing the force is instantaneously at rest ($\mathbf{u}' = \mathbf{0}$) is given by $F_{\parallel} = F_{\parallel}'$ and $\mathbf{F}_{\perp} = \mathbf{F}_{\perp}'/\gamma$ with $\gamma = 1/\sqrt{1 - V^2/c^2}$, where Σ' moves with velocity $\mathbf{V}$ with respect to Σ. Because the charged particle is assumed to be initially at rest in the inertial reference frame Σ', then the observed Lorentz force on the particle is given by $\mathbf{F}' = q\mathbf{e}'$, whereas in the inertial reference frame Σ the corresponding Lorentz force at that instant is given by $\mathbf{F} = q(\mathbf{e} + \|1/c\|\mathbf{V} \times \mathbf{b})$.

Application of the transformation relations for the parallel and perpendicular components of the force then yields the set of transformation relations for the parallel and perpendicular components of the electric field vector as $e_{\parallel}' = e_{\parallel}$ and

$$\mathbf{e}_{\perp}' = \gamma \left(\mathbf{e}_{\perp} + \left\| \frac{1}{c} \right\| \mathbf{V} \times \mathbf{b} \right), \tag{2.134}$$

with inverse

$$\mathbf{e}_\perp = \gamma \left(\mathbf{e}'_\perp - \left\| \frac{1}{c} \right\| \mathbf{V} \times \mathbf{b}' \right). \tag{2.135}$$

These *relativistically correct transformation relations* state that the component of the electric field vector that is parallel to the relative velocity between the two inertial reference frames is unchanged, whereas the components of the electric field vector that are perpendicular to that relative velocity vector are transformed into both electric and magnetic field components.

Consider next the transformation properties of the magnetic field vector. Assume again that the inertial frame Σ' moves with velocity $\mathbf{V}$ with respect to Σ. Let the point charge q be moving with velocity $\mathbf{u}'$ in Σ' so that the Lorentz force is given by $\mathbf{F}' = q(\mathbf{e}' + \|1/c\|\mathbf{u}' \times \mathbf{b}')$. In the Σ inertial frame the velocity $\mathbf{u}$ of the charged particle has components $u_\| = (u'_\| + V)/(1 + \mathbf{u}' \cdot \mathbf{V}/c^2)$ and $u_\perp = \sqrt{1 - V^2/c^2}\mathbf{u}'_\perp/(1 + \mathbf{u}' \cdot \mathbf{V}/c^2)$ parallel and perpendicular to the relative velocity $\mathbf{V}$ between the two inertial reference frames [cf. Eqs. (2.89) and (2.90)]. With the Lorentz force relation $\mathbf{F} = q(\mathbf{e} + \|1/c\|\mathbf{u} \times \mathbf{b})$ in Σ, substitution of the above results into the general force transformation relation and use of the transformation relation (2.134) for the electric field then results in the set of transformation relations for the parallel and perpendicular components of the magnetic field vector as $b'_\| = b_\|$ and

$$\mathbf{b}'_\perp = \gamma \left(\mathbf{b}_\perp - \|c\| \frac{1}{c^2} \mathbf{V} \times \mathbf{e} \right), \tag{2.136}$$

with inverse

$$\mathbf{b}_\perp = \gamma \left(\mathbf{b}'_\perp + \|c\| \frac{1}{c^2} \mathbf{V} \times \mathbf{e}' \right). \tag{2.137}$$

These *relativistically correct transformation relations* state that the component of the magnetic field vector that is parallel to the relative velocity between the two inertial frames is unchanged, whereas the components of the magnetic field vector that are perpendicular to that relative velocity vector are transformed into both electric and magnetic field components.

Notice the similarity between the transformation relations (2.134)–(2.135) and (2.136)–(2.137) for the electric and magnetic field vectors, respectively; the only differences are in the sign change and the additional $1/c^2$ factor. Because these transformation relations involve all components of both the electric and magnetic field vectors, it is once again seen that electric and magnetic fields are interdependent.

Finally, in vector form, the transformation relations given in Eqs. (2.134) and (2.135) become [compare with the Galilean transformation relations given in

Eqs. (2.50) and (2.58), respectively]

$$\mathbf{e}' = (1 - \gamma)\frac{\mathbf{e} \cdot \mathbf{V}}{V} + \gamma\left(\mathbf{e} + \left\|\frac{1}{c}\right\|\mathbf{V} \times \mathbf{b}\right), \tag{2.138}$$

$$\mathbf{b}' = (1 - \gamma)\frac{\mathbf{b} \cdot \mathbf{V}}{V} + \gamma\left(\mathbf{b} - \|c\|\frac{1}{c^2}\mathbf{V} \times \mathbf{e}\right). \tag{2.139}$$

The corresponding inverse transformations are given by interchange of primed and unprimed quantities and replacement of $\mathbf{V}$ with $-\mathbf{V}$.

2.2.3.2 Invariance of Maxwell's Equations

The invariance of the differential form of the microscopic Maxwell's equations

$$\nabla \cdot \mathbf{e}(\mathbf{r}, t) = \frac{\|4\pi\|}{\epsilon_0}\rho(\mathbf{r}, t), \tag{2.140}$$

$$\nabla \cdot \mathbf{b}(\mathbf{r}, t) = 0, \tag{2.141}$$

$$\nabla \times \mathbf{e}(\mathbf{r}, t) = -\left\|\frac{1}{c}\right\|\frac{\partial \mathbf{b}(\mathbf{r}, t)}{\partial t}, \tag{2.142}$$

$$\nabla \times \mathbf{b}(\mathbf{r}, t) = \left\|\frac{4\pi}{c}\right\|\mu_0\mathbf{j}(\mathbf{r}, t) + \left\|\frac{1}{c}\right\|\epsilon_0\mu_0\frac{\partial \mathbf{e}(\mathbf{r}, t)}{\partial t}, \tag{2.143}$$

is now finally considered. These coupled partial differential equations for the microscopic electric and magnetic field vectors hold at any given space–time point $(\mathbf{r}, t)$ in the inertial frame Σ. The same form of these relations must also hold in an inertial frame Σ' that is moving at a fixed velocity $\mathbf{V}$ relative to Σ so that

$$\nabla' \cdot \mathbf{e}'(\mathbf{r}', t') = \frac{\|4\pi\|}{\epsilon_0}\rho'(\mathbf{r}', t'), \tag{2.144}$$

$$\nabla' \cdot \mathbf{b}'(\mathbf{r}', t') = 0, \tag{2.145}$$

$$\nabla' \times \mathbf{e}'(\mathbf{r}', t') = -\left\|\frac{1}{c}\right\|\frac{\partial \mathbf{b}'(\mathbf{r}', t')}{\partial t'}, \tag{2.146}$$

$$\nabla' \times \mathbf{b}'(\mathbf{r}', t') = \left\|\frac{4\pi}{c}\right\|\mu_0\mathbf{j}'(\mathbf{r}', t') + \left\|\frac{1}{c}\right\|\epsilon_0\mu_0\frac{\partial \mathbf{e}'(\mathbf{r}', t')}{\partial t'}, \tag{2.147}$$

where the space–time coordinates $(\mathbf{r}', t') \in \Sigma'$ and $(\mathbf{r}, t) \in \Sigma$ are related through the Poincaré–Lorentz transformation.

For simplicity, let Σ' move with fixed velocity $\mathbf{V} = \hat{\mathbf{1}}_x V$ relative to Σ along their common $x - x'$ axes. In this case, the Poincaré–Lorentz transformation relations become (in component form) $x' = \gamma(x - Vt)$, $y' = y$, $z' = z$, and $t' = \gamma(t -$

xV/c^2). Application of the chain rule with these transformation relations then yields the set of differential relations

$$\frac{\partial}{\partial x} = \frac{\partial x'}{\partial x}\frac{\partial}{\partial x'} + \frac{\partial y'}{\partial x}\frac{\partial}{\partial y'} + \frac{\partial z'}{\partial x}\frac{\partial}{\partial z'} + \frac{\partial t'}{\partial x}\frac{\partial}{\partial t'} = \gamma\left(\frac{\partial}{\partial x'} - \frac{V}{c^2}\frac{\partial}{\partial t'}\right),$$

$$\frac{\partial}{\partial y} = \frac{\partial x'}{\partial y}\frac{\partial}{\partial x'} + \frac{\partial y'}{\partial y}\frac{\partial}{\partial y'} + \frac{\partial z'}{\partial y}\frac{\partial}{\partial z'} + \frac{\partial t'}{\partial y}\frac{\partial}{\partial t'} = \frac{\partial}{\partial y'},$$

$$\frac{\partial}{\partial z} = \frac{\partial x'}{\partial z}\frac{\partial}{\partial x'} + \frac{\partial y'}{\partial z}\frac{\partial}{\partial y'} + \frac{\partial z'}{\partial z}\frac{\partial}{\partial z'} + \frac{\partial t'}{\partial z}\frac{\partial}{\partial t'} = \frac{\partial}{\partial z'},$$

$$\frac{\partial}{\partial t} = \frac{\partial x'}{\partial t}\frac{\partial}{\partial x'} + \frac{\partial y'}{\partial t}\frac{\partial}{\partial y'} + \frac{\partial z'}{\partial t}\frac{\partial}{\partial z'} + \frac{\partial t'}{\partial t}\frac{\partial}{\partial t'} = \gamma\left(\frac{\partial}{\partial t'} - V\frac{\partial}{\partial x'}\right).$$

Substitution of these relations into the y-component of Eq. (2.142), given by $\partial e_x/\partial z - \partial e_z/\partial x = -\|1/c\|\partial b_y/\partial t$, results in the relation

$$\frac{\partial e_x}{\partial z'} - \gamma\left(\frac{\partial e_z}{\partial x'} - \frac{V}{c^2}\frac{\partial e_z}{\partial t'}\right) = -\left\|\frac{1}{c}\right\|\gamma\left(\frac{\partial b_y}{\partial t'} - V\frac{\partial b_y}{\partial x'}\right),$$

which may be rewritten as

$$\frac{\partial e_x}{\partial z'} - \frac{\partial}{\partial x'}\left(\gamma\left(e_z + \left\|\frac{1}{c}\right\|Vb_y\right)\right) = -\left\|\frac{1}{c}\right\|\frac{\partial}{\partial t'}\left(\gamma\left(b_y + \|c\|\frac{V}{c^2}e_z\right)\right).$$

Comparison of this relation with the y'-component of Eq. (2.146) shows that these components of the electric and magnetic field vectors transform as $e'_x = e_x$ and

$$e'_z = \gamma\left(e_z + \|1/c\|\,Vb_y\right),$$

$$b'_y = \gamma\left(b_y + \|c\|(V/c^2)e_z\right),$$

in agreement with the corresponding relations given in Eqs. (2.144) and (2.146). Similarly, substitution of the above transformation relations for the partial differential operators into the z-component of Eq. (2.142), given by $\partial e_y/\partial x - \partial e_x/\partial y = -\|1/c\|\partial b_z/\partial t$, results in the relation

$$\gamma\left(\frac{\partial e_y}{\partial x'} - \frac{V}{c^2}\frac{\partial e_y}{\partial t'}\right) - \frac{\partial e_x}{\partial y'} = -\left\|\frac{1}{c}\right\|\gamma\left(\frac{\partial b_z}{\partial t'} - V\frac{\partial b_z}{\partial x'}\right),$$

which may be rewritten as

$$\frac{\partial}{\partial x'}\left(\gamma\left(e_y - \left\|\frac{1}{c}\right\|Vb_z\right)\right) - \frac{\partial e_x}{\partial y'} = -\left\|\frac{1}{c}\right\|\frac{\partial}{\partial t'}\left(\gamma\left(b_z - \|c\|\frac{V}{c^2}e_y\right)\right).$$

Comparison of this relation with the z'-component of Eq. (2.146) then gives

$$e'_y = \gamma \left(e_y - \left\| \frac{1}{c} \right\| V b_z \right),$$

$$b'_z = \gamma \left(b_z - \|c\| \frac{V}{c^2} e_y \right),$$

in agreement with the corresponding relations in Eqs. (2.134) and (2.136). Finally, substitution of the transformation relations for the partial differential operators into Eq. (2.141), given by $\partial b_x / \partial x + \partial b_y / \partial y + \partial b_z / \partial z = 0$, gives

$$\gamma \left(\frac{\partial b_x}{\partial x'} - \frac{V}{c^2} \frac{\partial b_x}{\partial t'} \right) + \frac{\partial b_y}{\partial y'} + \frac{\partial b_z}{\partial z'} = 0.$$

Substitution of the inverses of the above transformation relations for b_y and b_z then gives

$$\frac{\partial b_x}{\partial x'} - \frac{V}{c^2} \frac{\partial b_x}{\partial t'} + \frac{\partial}{\partial y'} \left(b'_y - \|c\| \frac{V}{c^2} e'_z \right) + \frac{\partial}{\partial z'} \left(b'_z + \|c\| \frac{V}{c^2} e'_y \right) = 0,$$

which may be rewritten as

$$\frac{\partial b_x}{\partial x'} + \frac{\partial b'_y}{\partial y'} + \frac{\partial b'_z}{\partial z'} - \|c\| \frac{V}{c^2} \left(\frac{\partial e'_z}{\partial y'} - \frac{\partial e'_y}{\partial z'} + \left\| \frac{1}{c} \right\| \frac{\partial b_x}{\partial t'} \right) = 0.$$

Comparison of this expression with that given in Eq. (2.145) then gives the final transformation relation, $b'_x = b_x$.

The same set of transformation relations for the components of the electric and magnetic field vectors is obtained using Eqs. (2.140) and (2.143) when the appropriate transformation relations for the source terms are employed. Thus, *the microscopic Maxwell equations are invariant in form under a Poincaŕe–Lorentz transformation* when the field vectors are transformed by the relations specified in Eqs. (2.138) and (2.139) and the charge and current sources are transformed in the manner required by Eqs. (2.127) and (2.128).

2.3   Conservation Laws for the Electromagnetic Field

The mathematical formulation of the laws of conservation of energy and momentum for the microscopic electromagnetic field are of fundamental importance to the physical interpretation of electromagnetic phenomena. Of particular interest here is the classical interpretation of these microscopic conservation laws for the combined system of charged particles and fields in the Maxwell–Lorentz theory.

2.3.1 Conservation of Energy: The Poynting–Heaviside Theorem

As described in Sect. 2.1, matter is assumed to be composed of both positive and negative charged particles that are characterized by their rest mass m_0 and their microscopic charge density, ρ_+ for a positive charge and ρ_- for a negative charge. The kinematics of these two types of charged particles are respectively described by their convective current densities $\mathbf{j}_+ = \rho_+ \mathbf{v}_+$ and $\mathbf{j}_- = \rho_- \mathbf{v}_-$. The electromagnetic forces that act upon these charges are the Lorentz forces with densities at each point of space and time given by

$$\mathbf{f}_+(\mathbf{r}, t) = \rho_+(\mathbf{r}, t)\mathbf{e}(\mathbf{r}, t) + \left\|\frac{1}{c}\right\| \mathbf{j}_+(\mathbf{r}, t) \times \mathbf{b}(\mathbf{r}, t),$$

$$\mathbf{f}_-(\mathbf{r}, t) = -\rho_-(\mathbf{r}, t)\mathbf{e}(\mathbf{r}, t) - \left\|\frac{1}{c}\right\| \mathbf{j}_-(\mathbf{r}, t) \times \mathbf{b}(\mathbf{r}, t).$$

There are also additional unspecified forces of a non-electromagnetic nature.

Let $\mathcal{V}$ be any simply connected region of space containing charges that are specified by the microscopic densities $\rho_+(\mathbf{r}, t)$, $\rho_-(\mathbf{r}, t)$, $\mathbf{j}_+(\mathbf{r}, t)$, and $\mathbf{j}_-(\mathbf{r}, t)$. The rate at which the Lorentz forces acting inside the region $\mathcal{V}$ do work on the enclosed charged particles is then given by

$$\int_{\mathcal{V}} (\mathbf{f}_+ \cdot \mathbf{v}_+ + \mathbf{f}_- \cdot \mathbf{v}_-) \, d^3r = \int_{\mathcal{V}} (\rho_+ \mathbf{v}_+ - \rho_- \mathbf{v}_-) \cdot \mathbf{e} \, d^3r$$

$$= \int_{\mathcal{V}} \mathbf{j} \cdot \mathbf{e} \, d^3r,$$

where $\mathbf{j} = \mathbf{j}_+ - \mathbf{j}_-$ is the total convective current density. The microscopic magnetic field is absent in this expression because terms involving $\mathbf{b}$ contain scalar triple products with repeated factors and accordingly vanish. If the particles in the region $\mathcal{V}$ are enumerated by the index k, and if the velocity of the kth particle is $\mathbf{v}_k$ and the nonelectromagnetic force acting on it is $\mathbf{F}_k$, then the rate at which all forces acting in the region $\mathcal{V}$ do work (i.e., the power generated by them) is given by

$$P = \sum_k \mathbf{F}_k \cdot \mathbf{v}_k + \int_{\mathcal{V}} \mathbf{j} \cdot \mathbf{e} \, d^3r. \tag{2.148}$$

This equation provides a *local form of the energy theorem* in the sense that it involves the forces acting on each individual particle within a region $\mathcal{V}$.

Consider now obtaining an alternate form of the volume integral appearing in Eq. (2.148) that involves only the microscopic electromagnetic field vectors within the region $\mathcal{V}$. From the differential form of Ampére's law in Eq. (2.34)

$$\mathbf{j} \cdot \mathbf{e} = \left\| \frac{c}{4\pi} \right\| \mathbf{e} \cdot (\nabla \times \mathbf{h}) - \left\| \frac{1}{4\pi} \right\| \mathbf{e} \cdot \frac{\partial \mathbf{d}}{\partial t}, \tag{2.149}$$

and from the differential form of Faraday's law in Eq. (2.33)

$$0 = - \left\| \frac{c}{4\pi} \right\| \mathbf{h} \cdot (\nabla \times \mathbf{e}) - \left\| \frac{1}{4\pi} \right\| \mathbf{h} \cdot \frac{\partial \mathbf{b}}{\partial t}. \tag{2.150}$$

Addition of these two equations then yields the expression

$$\mathbf{j} \cdot \mathbf{e} = -\frac{\partial}{\partial t} \left(\left\| \frac{1}{4\pi} \right\| \frac{\epsilon_0 e^2 + \mu_0 h^2}{2} \right) - \nabla \cdot \left(\left\| \frac{c}{4\pi} \right\| \mathbf{e} \times \mathbf{h} \right), \tag{2.151}$$

which is a direct mathematical consequence of the microscopic field equations for the electromagnetic field. If one defines the *electromagnetic energy density* as

$$u(\mathbf{r}, t) \equiv \frac{1}{2} \left\| \frac{1}{4\pi} \right\| \left(\epsilon_0 e^2(\mathbf{r}, t) + \mu_0 h^2(\mathbf{r}, t) \right) \tag{2.152}$$

so that the *total field energy* in a region $\mathcal{V}$ is given by

$$U(t) \equiv \int_{\mathcal{V}} u(\mathbf{r}, t) d^3 r = \frac{1}{2} \left\| \frac{1}{4\pi} \right\| \int_{\mathcal{V}} \left(\epsilon_0 e^2(\mathbf{r}, t) + \mu_0 h^2(\mathbf{r}, t) \right) d^3 r, \tag{2.153}$$

and the *microscopic Poynting vector* as

$$\mathbf{s}(\mathbf{r}, t) \equiv \left\| \frac{c}{4\pi} \right\| \mathbf{e}(\mathbf{r}, t) \times \mathbf{h}(\mathbf{r}, t), \tag{2.154}$$

the relation appearing in Eq. (2.151) takes the form

$$\mathbf{j} \cdot \mathbf{e} = -\frac{\partial u}{\partial t} - \nabla \cdot \mathbf{s}. \tag{2.155}$$

Substitution of this expression into Eq. (2.148) and application of the divergence theorem to the volume integral of $\nabla \cdot \mathbf{s}$ over the regular region $\mathcal{V}$ finally gives

$$P = \sum_k \mathbf{F}_k \cdot \mathbf{v}_k - \frac{dU}{dt} - \oint_{\mathcal{S}} \mathbf{s} \cdot \hat{\mathbf{n}} d^2 r, \tag{2.156}$$

where $\hat{\mathbf{n}}$ is the unit outward normal vector to the surface $\mathcal{S}$ enclosing $\mathcal{V}$.

Equation (2.156) is a form of the local energy theorem that is mathematically equivalent to the expression given in Eq. (2.148). The scalar quantity $u(\mathbf{r}, t)$ that is defined in Eq. (2.152) may be interpreted as a measure of the density of energy

in the microscopic electromagnetic field and is, in general, a function of both
position and time (i.e., it is a local instantaneous quantity). The scalar quantity $U(t)$
that is defined in Eq. (2.153) may then be interpreted as the total electromagnetic
field energy in the region $\mathcal{V}$ and is a function only of the time for that region.
The surface integral appearing in Eq. (2.156) is then interpreted as representing
a flow of electromagnetic energy through the regular, closed surface $\mathcal{S}$; the vector
quantity $s(r, t)$ that is defined in Eq. (2.154) must then represent the microscopic
flux density of the electromagnetic field energy. If that surface integral is positive,
there must then be a net loss of electromagnetic energy from the region $\mathcal{V}$, whereas
if it is negative there must be a net gain of electromagnetic energy within the
region $\mathcal{V}$. The vector quantity $s(r, t)$ is called the *Poynting vector of the microscopic
electromagnetic field* and is, in general, a function of both position and time at each
space–time point at which the electromagnetic field is defined. The relation that
appears in Eq. (2.156) is commonly referred to as *Poynting's theorem*[38] as well as
the *Poynting–Heaviside theorem*[10] which, with the aid of the local energy theorem
as expressed in Eq. (2.148), may be written in the alternate form

$$\int_{\mathcal{V}} \mathbf{j} \cdot \mathbf{e}\, d^3r = -\frac{dU}{dt} - \oint_{\mathcal{S}} \mathbf{s} \cdot \hat{\mathbf{n}}\, d^2r, \qquad (2.157)$$

where $\mathbf{j} = \rho\mathbf{v}$ is the total convective current density in the region $\mathcal{V}$. The Poynting–
Heaviside theorem for the microscopic electromagnetic field is a mathematical
statement of the conservation of energy of the combined system of charged particles
and fields in the Maxwell–Lorentz theory.

The total energy $U(t)$ of a system is an *extensive variable* because it refers to
the system as a whole. Associated with it is the *intensive variable* $u(\mathbf{r}, t)$ defined
in Eq. (2.152) which describes the local (microscopic) behavior. The differential
form of Poynting's theorem given in Eq. (2.155) is an example of the balance
equation for an intensive variable [39]. In general, balance equations of this type
naturally lead to a classification of physical quantities into two general categories.
The first category consists of conserved quantities where the source term vanishes.
An important example of a conserved quantity is given by the charge density $\rho(\mathbf{r}, t)$
which satisfies the equation of continuity [cf. Eq. (2.14)]

$$\nabla \cdot \mathbf{j}(\mathbf{r}, t) + \frac{\partial \rho(\mathbf{r}, t)}{\partial t} = 0,$$

where the current density $\mathbf{j}(\mathbf{r}, t)$ describes the flow of the conserved quantity. When
$\mathbf{j}(\mathbf{r}, t) = 0$, the differential form of Poynting's theorem becomes

$$\nabla \cdot \mathbf{s}(\mathbf{r}, t) + \frac{\partial u(\mathbf{r}, t)}{\partial t} = 0,$$

[10]This result was also derived by Oliver Heaviside in the same year.

and the energy density $u(\mathbf{r}, t)$ is the conserved quantity whereas the Poynting vector $\mathbf{s}(\mathbf{r}, t)$ describes the flow of this quantity. The other category consists of nonconserved quantities where the source term does not vanish, as occurs in Eq. (2.155) when $\mathbf{j}(\mathbf{r}, t) \neq \mathbf{0}$. The scalar quantity $\mathbf{j}(\mathbf{r}, t) \cdot \mathbf{e}(\mathbf{r}, t)$ describes the work done by the electromagnetic field on the local charge and represents a loss term in Eqs. (2.155) and (2.157).

The interpretation of the Poynting vector $\mathbf{s}(\mathbf{r}, t)$ is that at any point of observation of an electromagnetic field at which $\mathbf{s}(\mathbf{r}, t)$ is different from zero one can assert that electromagnetic energy is flowing in the direction of $\mathbf{s}(\mathbf{r}, t)$ such that across an elemental plane surface perpendicular to the direction of $\mathbf{s}(\mathbf{r}, t)$ at that point, the rate of flow of energy in the field is given by the quantity $|\mathbf{s}(\mathbf{r}, t)|$. However, it must be emphasized that the local quantities $u(\mathbf{r}, t)$ and $\mathbf{s}(\mathbf{r}, t)$ merely provide an interpretation (albeit a useful one) of the density and transfer of energy in a given electromagnetic field. The physically proper statement of the transfer of electromagnetic energy is embodied in Poynting's theorem as given in either of Eq. (2.156) or (2.157). As was first pointed out by Thompson,[11] even though the total flow of electromagnetic energy through a closed surface S may be correctly represented by the surface integral of the normal component of the Poynting vector, it cannot be definitely concluded that the time rate of energy flow at any given point is uniquely specified by the Poynting vector $\mathbf{s}(\mathbf{r}, t)$ at that point, for one may add to the Poynting vector any solenoidal vector field (which then integrates to zero over any closed regular surface) without affecting the statement of conservation of energy that is expressed by Poynting's theorem. Hence, there is no strictly valid justification for the accepted interpretation of the Poynting vector except that it is useful and seldom leads to erroneous results provided that proper care is taken in its application and interpretation. When using this interpretation, however, one should keep in mind that the energy flow implied by the value of the Poynting vector $\mathbf{s}(\mathbf{r}, t)$ need not coincide with any intuitive notion of energy flow. In particular, because $\mathbf{s}(\mathbf{r}, t)$ need not vanish for a static electromagnetic field, this interpretation can lead to a nonzero value of the energy flow even in a static field.

The classical *Poynting–Heaviside interpretation* of Poynting's theorem as a statement of the conservation of energy is then seen to depend to a considerable degree on hypothesis. Various critiques [40] as well as alternative forms [41–43] have been offered but none has the advantage of greater plausibility so as to supersede the Poynting–Heaviside interpretation. Poynting's theorem as embodied in Eq. (2.157) is a direct mathematical consequence of the Maxwell–Lorentz theory and the associated hypothesis of an energy density and a flow of energy in the classical electromagnetic field has proved to be extremely useful. As stated by Stratton [44] in this context,

> A theory is not an absolute truth but a self-consistent analytical formulation of the relations governing a group of natural phenomena. By this standard there is every reason to retain the Poynting–Heaviside viewpoint until a clash with new experimental evidence shall call for its revision.

---

[11] See J. J. Thompson, *Recent Researches*, pp. 251–387.

With the acceptance of the validity of the Maxwell–Lorentz theory, Poynting's theorem is indeed a valid self-consistent relationship.

2.3.2 Conservation of Linear Momentum

The conservation of linear momentum in a combined system of charged particles and fields can be considered in a similar manner in the microscopic Maxwell–Lorentz theory. From the Lorentz force relation given in Eq. (2.21) and Newton's second law of motion $\mathbf{F} = d\mathbf{p}_{mech}/dt$, the time rate of change of the total mechanical momenta $\mathbf{p}_{mech}$ of all the charged particles in a region $\mathcal{V}$ is given by (with the assumption that no particles enter or leave $\mathcal{V}$)

$$\frac{d\mathbf{p}_{mech}}{dt} = \int_{\mathcal{V}} \left(\rho\mathbf{e} + \left\| \frac{1}{c} \right\| \mathbf{j} \times \mathbf{b} \right) d^3r. \tag{2.158}$$

In the same manner as taken in the derivation of Poynting's theorem, the microscopic Maxwell's equations given in Eqs. (2.34) and (2.35) are now employed for the purpose of eliminating the microscopic source terms $\rho(\mathbf{r}, t)$ and $\mathbf{j}(\mathbf{r}, t)$ from Eq. (2.158), where $\rho = \left\| \frac{1}{4\pi} \right\| \nabla \cdot \mathbf{d}$ and $\mathbf{j} = \left\| \frac{c}{4\pi} \right\| \nabla \times \mathbf{h} - \left\| \frac{1}{4\pi} \right\| \frac{\partial \mathbf{d}}{\partial t}$. With these two substitutions, the integrand appearing in Eq. (2.158) becomes

$$\rho\mathbf{e} + \left\| \frac{1}{c} \right\| \mathbf{j} \times \mathbf{b} = \left\| \frac{1}{4\pi} \right\| \left((\nabla \cdot \mathbf{d})\mathbf{e} + \left\| \frac{1}{c} \right\| \mathbf{b} \times \frac{\partial \mathbf{d}}{\partial t} - \mathbf{b} \times (\nabla \times \mathbf{h}) \right).$$

Because $\mathbf{b} \times \partial \mathbf{d}/\partial t = -\partial(\mathbf{d} \times \mathbf{b})/\partial t + \mathbf{d} \times \partial \mathbf{b}/\partial t$ and because $(\nabla \cdot \mathbf{b})\mathbf{h} = 0$, this expression becomes

$$\| 4\pi \| \, \rho\mathbf{e} + \left\| \frac{4\pi}{c} \right\| \mathbf{j} \times \mathbf{b}$$

$$= (\nabla \cdot \mathbf{d})\mathbf{e} + (\nabla \cdot \mathbf{b})\mathbf{h} - \left\| \frac{1}{c} \right\| \frac{\partial}{\partial t}(\mathbf{d} \times \mathbf{b}) + \left\| \frac{1}{c} \right\| \mathbf{d} \times \frac{\partial \mathbf{b}}{\partial t} - \mathbf{b} \times (\nabla \times \mathbf{h}).$$

Furthermore, because $\|1/c\|\partial \mathbf{b}/\partial t = -\nabla \times \mathbf{e}$ from Eq. (2.33), one finally obtains

$$\| 4\pi \| \, \rho\mathbf{e} + \left\| \frac{4\pi}{c} \right\| \mathbf{j} \times \mathbf{b}$$

$$= (\nabla \cdot \mathbf{d})\mathbf{e} + (\nabla \cdot \mathbf{b})\mathbf{h} - \mathbf{d} \times (\nabla \times \mathbf{e}) - \mathbf{b} \times (\nabla \times \mathbf{h}) - \left\| \frac{1}{c} \right\| \frac{\partial}{\partial t}(\mathbf{d} \times \mathbf{b}).$$

With this result, the expression in Eq. (2.154) can now be written as

$$\frac{d\mathbf{p}_{mech}}{dt} + \frac{d}{dt}\int_{\mathcal{V}}\left\|\frac{1}{4\pi c}\right\|(\mathbf{d} \times \mathbf{b})d^3r$$
$$= \left\|\frac{1}{4\pi}\right\|\int_{\mathcal{V}}[(\nabla \cdot \mathbf{d})\mathbf{e} - \mathbf{d} \times (\nabla \times \mathbf{e}) + (\nabla \cdot \mathbf{b})\mathbf{h} - \mathbf{b} \times (\nabla \times \mathbf{h})]d^3r.$$

$$(2.159)$$

The volume integral appearing on the left-hand side of this equation may be formally defined as the *total electromagnetic momentum* $\mathbf{p}_{em}$ of the electromagnetic field in the region $\mathcal{V}$, so that [45]

$$\mathbf{p}_{em} \equiv \epsilon_0\mu_0\left\|\frac{1}{4\pi c}\right\|\int_{\mathcal{V}}(\mathbf{e} \times \mathbf{h})d^3r. \qquad (2.160)$$

The integrand of this expression, viz.

$$\mathfrak{p}_{em}(\mathbf{r}, t) \equiv \epsilon_0\mu_0\left\|\frac{1}{4\pi c}\right\|(\mathbf{e}(\mathbf{r}, t) \times \mathbf{h}(\mathbf{r}, t)), \qquad (2.161)$$

may then be interpreted as a *density of electromagnetic momentum* in the microscopic field. Notice that this momentum density is proportional to the energy flux density $\mathbf{s}(\mathbf{r}, t)$, viz.

$$\mathfrak{p}_{em}(\mathbf{r}, t) = \frac{1}{c^2}\mathbf{s}(\mathbf{r}, t), \qquad (2.162)$$

where $\mathbf{s}(\mathbf{r}, t)$ is the Poynting vector of the microscopic electromagnetic field.

In order to complete the identification of the volume integral of $\mathfrak{p}_{em}(\mathbf{r}, t)$ as the total electromagnetic momentum, and further, to establish Eq. (2.159) as a statement of the conservation of linear momentum, the volume integral appearing on the right-hand side of that expression needs to be converted into a surface integral of the normal component of a linear momentum flow. With use of the identity $\nabla(\mathbf{U} \cdot \mathbf{V}) = \mathbf{U} \cdot \nabla\mathbf{V} + \mathbf{V} \cdot \nabla\mathbf{U} + \mathbf{U} \times (\nabla \times \mathbf{V}) + \mathbf{V} \times (\nabla \times \mathbf{U})$ which yields, with $\mathbf{V} = \mathbf{U}$,

$$\mathbf{U} \times (\nabla \times \mathbf{U}) = -\mathbf{U} \cdot \nabla\mathbf{U} + \frac{1}{2}\nabla\left(U^2\right),$$

the terms involving the magnetic induction vector $\mathbf{b}$ in Eq. (2.159) can be written as

$$\mathbf{b}(\nabla \cdot \mathbf{b}) - \mathbf{b} \times (\nabla \times \mathbf{b}) = \mathbf{b}(\nabla \cdot \mathbf{b}) + \mathbf{b} \cdot \nabla\mathbf{b} - \frac{1}{2}\nabla\left(b^2\right)$$
$$= \nabla \cdot \left(\mathbf{b}\mathbf{b} - \frac{1}{2}\mathfrak{I}b^2\right),$$

where $\underline{\mathfrak{I}} = \hat{\mathbf{1}}_x\hat{\mathbf{1}}_x + \hat{\mathbf{1}}_y\hat{\mathbf{1}}_y + \hat{\mathbf{1}}_z\hat{\mathbf{1}}_z$ is the unit dyadic or *idemfactor*. Similarly, the terms involving the electric field intensity vector $\mathbf{e}$ in Eq. (2.159) can be written as

$$\mathbf{e}(\nabla \cdot \mathbf{e}) - \mathbf{e} \times (\nabla \times \mathbf{e}) = \nabla \cdot \left(\mathbf{ee} - \frac{1}{2}\underline{\mathfrak{I}}e^2\right).$$

With these two substitutions the expression appearing in Eq. (2.159) for the *conservation of linear momentum* for the combined system of charged particles and fields in a region $\mathcal{V}$ becomes

$$\frac{d}{dt}(\mathbf{p}_{mech} + \mathbf{p}_{em}) = \int_{\mathcal{V}} \nabla \cdot \underline{\mathfrak{I}} d^3r = \oint_{\mathcal{S}} \hat{\mathbf{n}} \cdot \underline{\mathfrak{I}} d^2r, \qquad (2.163)$$

where $\mathcal{S}$ is the closed boundary surface for the regular region $\mathcal{V}$ and where $\hat{\mathbf{n}}$ is the unit outward normal vector to $\mathcal{S}$. The dyadic (second-rank tensor) quantity $\underline{\mathfrak{I}}$ defined by

$$\underline{\mathfrak{I}} \equiv \left\|\frac{1}{4\pi}\right\| \left[\epsilon_0\mathbf{ee} + \mu_0\mathbf{hh} - \frac{1}{2}\underline{\mathfrak{I}}\left(\epsilon_0 e^2 + \mu_0 h^2\right)\right] \qquad (2.164)$$

is the *microscopic Maxwell stress tensor* of the electromagnetic field. In terms of the electromagnetic energy density defined in Eq. (2.152), the Maxwell stress tensor becomes

$$\underline{\mathfrak{I}}(\mathbf{r}, t) = \left\|\frac{1}{4\pi}\right\| \left(\epsilon_0\mathbf{e}(\mathbf{r}, t)\mathbf{e}(\mathbf{r}, t) + \mu_0\mathbf{h}(\mathbf{r}, t)\mathbf{h}(\mathbf{r}, t)\right) - \underline{\mathfrak{I}}u(\mathbf{r}, t). \qquad (2.165)$$

It is seen that $\underline{\mathfrak{I}}$ is a symmetric tensor with elements

$$\mathfrak{I}_{ij} = \left\|\frac{1}{4\pi}\right\| \left[\epsilon_0 e_i e_j + \mu_0 h_i h_j - \frac{1}{2}\left(\epsilon_0 e^2 + \mu_0 h^2\right)\delta_{ij}\right], \qquad (2.166)$$

where δ_{ij} is the Kronecker-delta function[12] ($\delta_{ii} = 1$, $\delta_{ij} = 0$ when $i \neq j$).

If the region $\mathcal{V}$ is taken to be all of space and if all of the components of the stress tensor $\underline{\mathfrak{I}}$ go to zero with sufficient rapidity such that the surface integral of the normal component of $\underline{\mathfrak{I}}$ vanishes as $\mathcal{S}$ recedes to infinity, Eq. (2.163) becomes

$$\frac{d}{dt}(\mathbf{p}_{mech} + \mathbf{p}_{em}) = 0 \qquad (2.167)$$

---

[12]Leopold Kronecker (1823–1891) introduced this delta function in his 1866 paper "Über bilineare Formen" presented to the Royal Prussian Academy of Sciences in Berlin that was later published in the *Monthly Bulletin of the Academy* in 1868.

for all space. Hence, it is the total linear momentum in all of space that does not change with time (and hence is conserved) rather than just the mechanical linear momentum of the system of charged particles in all of space (which then need not be conserved). The quantity $\hat{\mathbf{n}} \cdot \mathcal{T}$ appearing in the integrand of the surface integral in Eq. (2.158) is then seen to represent the normal flow of linear momentum per unit area across the boundary surface S into the enclosed volume $\mathcal{V}$. That is, the quantity $\hat{\mathbf{n}} \cdot \mathcal{T}$ is the force per unit area transmitted across the surface S by the microscopic electromagnetic field.[13]

Just as was found for the Heaviside–Poynting interpretation of Poynting's theorem, the Poynting vector, and the electromagnetic energy density, the concept of electromagnetic momentum just presented need not coincide with any intuitive notion of momentum flow in the field. Even though the total flow of electromagnetic momentum across a closed surface may be correctly represented by the surface integral in Eq. (2.163), it cannot be definitely concluded that the time rate of electromagnetic momentum flux in the direction specified by the unit vector $\hat{\mathbf{n}}$ at any given point in space is uniquely specified by the component $\hat{\mathbf{n}} \cdot \mathcal{T}$ of Maxwell's stress tensor at that point. Indeed, one may always add to the vector $\hat{\mathbf{n}} \cdot \mathcal{T}$ any other vector field that integrates to zero over a closed surface without affecting the statement of conservation of linear momentum that is expressed in Eq. (2.163). The classical interpretation of Eq. (2.163) as a statement of the conservation of linear momentum in the combined system of charged particles and fields is then seen to depend to a considerable degree on hypothesis. Nevertheless, that relation is a direct mathematical consequence of the Maxwell–Lorentz theory and, just as for Poynting's theorem, is a valid self-consistent relationship.

2.3.3   Conservation of Angular Momentum

The conservation of angular momentum in a combined system of charged particles and fields follows directly from the preceding derivation of the conservation of linear momentum. The mechanical angular momentum of a system of point particles (labeled by the index i) of mass m_i taken about a fixed point O as illustrated in Fig. 2.4 is given by

$$\mathbf{l}_{mech} \equiv \mathbf{R} \times M\mathbf{v} + \sum_i \mathbf{r}_i \times \mathbf{p}_{mech}^{(i)},$$

where $\mathbf{R}$ is the position vector (relative to the fixed point O) of the center of mass of the system of particles of total mass $M = \sum_i m_i$, $\mathbf{v}$ is the velocity of the center

---

[13]For an electrostatic field this force is given by $\mathcal{T}_{ij} n_j da$ at the surface S of a conductor, the sign change accounting for the fact that the unit normal vector $\hat{\mathbf{n}}$ is directed outwards from the conductor body. Because the electric field is normal on the surface of a perfect conductor, then $d\mathbf{f}(\mathbf{r})/da = \|\frac{1}{4\pi}\| \frac{1}{2}\epsilon_0 e^2 \hat{\mathbf{n}} = \frac{\|4\pi\|}{2\epsilon_0} \rho_s^2(\mathbf{r}) \hat{\mathbf{n}} = \frac{1}{2}\rho_s(\mathbf{r})\mathbf{e}(\mathbf{r})$ when $\mathbf{r} \in S$.

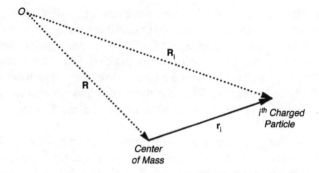

Fig. 2.4 Center of mass of a closed system of charged particles q_i and the associated position vectors for the determination of the system angular momentum

of mass of the system relative to O, and where $\mathbf{r}_i$ is the position vector and $\mathbf{p}_{mech}^{(i)}$ is the linear momentum of the ith particle relative to the center of mass of the system. It is assumed here that the center of mass of the system of charged particles is at rest with respect to O so that $\mathbf{v} = \mathbf{0}$ and the total mechanical angular momentum is then given by

$$\mathbf{l}_{mech} = \sum_i \mathbf{r}_i \times \mathbf{p}_{mech}^{(i)}. \qquad (2.168)$$

The total time derivative of this expression then yields

$$\frac{d\mathbf{l}_{mech}}{dt} = \sum_i \mathbf{r}_i \times \frac{d\mathbf{p}_{mech}^{(i)}}{dt}. \qquad (2.169)$$

because $(d\mathbf{r}_i/dt) \times \mathbf{p}_{mech}^{(i)} = \mathbf{v}_i \times (m_i \mathbf{v}_i) = \mathbf{0}$.

From Eq. (2.158), the time rate of change of the sum of the mechanical angular momenta of the total system of charged particles in a given region $\mathcal{V}$ is given by (assuming that no particles enter or leave the region $\mathcal{V}$)

$$\frac{d\mathbf{l}_{mech}}{dt} = \int_{\mathcal{V}} \mathbf{r} \times \left(\rho \mathbf{e} + \left\| \frac{1}{c} \right\| \mathbf{j} \times \mathbf{b} \right) d^3 r. \qquad (2.170)$$

From the relation preceding Eq. (2.159) the integrand appearing on the right-hand side of this expression may be written as

$$\mathbf{r} \times \left(\rho \mathbf{e} + \left\| \frac{1}{c} \right\| \mathbf{j} \times \mathbf{b} \right) = \mathbf{r} \times (\nabla \cdot \underline{\mathfrak{T}}) - \mathbf{r} \times \frac{\partial \mathfrak{p}_{em}}{\partial t}$$

$$= -\nabla \cdot \left(\underline{\mathfrak{T}} \times \mathbf{r} \right) - \frac{\partial}{\partial t} \left(\mathbf{r} \times \mathfrak{p}_{em} \right), \qquad (2.171)$$

where the fact that $\partial \mathbf{r}/\partial t = \mathbf{0}$ (because $\mathbf{r}$ now denotes the position vector from the fixed origin O to the differential volume element d^3r) and the dyadic identity [46] $\mathbf{r} \times (\nabla \cdot \mathfrak{T}) = -\nabla \cdot (\mathfrak{T} \times \mathbf{r})$ for the microscopic stress tensor have been employed. Substitution of Eq. (2.171) into Eq. (2.170) then yields

$$\frac{d\mathbf{l}_{mech}}{dt} + \frac{d}{dt}\int_V (\mathbf{r} \times \mathbf{p}_{em})\, d^3r = -\int_V \nabla \cdot (\mathfrak{T} \times \mathbf{r})\, d^3r$$

$$= -\oint_S \hat{\mathbf{n}} \cdot (\mathfrak{T} \times \mathbf{r})\, d^2r, \qquad (2.172)$$

where S is the boundary surface enclosing V and $\hat{\mathbf{n}}$ is the outward unit normal vector to S. The volume integral appearing on the left-hand side of this expression is identified as the *total electromagnetic angular momentum* $\mathbf{l}_{em}$ in the region V, so that

$$\mathbf{l}_{em} \equiv \int_V (\mathbf{r} \times \mathbf{p}_{em})\, d^3r. \qquad (2.173)$$

The integrand of this expression, viz.

$$\mathbf{l}_{em} \equiv \mathbf{r} \times \mathbf{p}_{em} = \frac{1}{c^2}\mathbf{r} \times \mathbf{s}, \qquad (2.174)$$

is then interpreted as the *microscopic density of electromagnetic angular momentum*. The *flux of angular momentum* of the microscopic electromagnetic field is then described by the tensor

$$\mathfrak{M} \equiv \mathfrak{T} \times \mathbf{r}. \qquad (2.175)$$

Notice that this quantity may be written as a third-rank tensor as

$$\mathfrak{M}_{ijk} = \mathfrak{T}_{ij}r_k - \mathfrak{T}_{ik}r_j \qquad (2.176)$$

where the indices i, j, k take on the integer values $1, 2, 3$. Because this tensor is antisymmetric in the indices j and k (i.e., $\mathfrak{M}_{ijk} = -\mathfrak{M}_{ikj}$), it only has three independent elements. With the inclusion of the index i, the tensor $\mathfrak{M}_{ijk}$ is seen to have nine components and can then be written as a pseudotensor of the second rank (i.e., as a pseudodyadic), as has been done in Eq. (2.175).

With these identifications, the expression given in Eq. (2.172) for the conservation of angular momentum for the combined system of charged particles and fields in a region V of space becomes

$$\frac{d}{dt}(\mathbf{l}_{mech} + \mathbf{l}_{em}) = -\oint_S \hat{\mathbf{n}} \cdot \mathfrak{M}\, d^2r. \qquad (2.177)$$

If the region $\mathcal{V}$ is taken to be all of space and if the angular momentum flux tensor $\underline{\mathbb{M}}$ goes to zero with sufficient rapidity such that the surface integral of the normal component of $\underline{\mathbb{M}}$ vanishes as the boundary surface $\mathcal{S}$ recedes to infinity, then Eq. (2.177) for the conservation of angular momentum for the combined system of charged particles and the electromagnetic field becomes

$$\frac{d}{dt}\left(\mathbf{l}_{mech}+\mathbf{l}_{em}\right)=\mathbf{0} \tag{2.178}$$

for all space. Hence, it is the total angular momentum in all of space that is conserved rather than just the total mechanical angular momentum of the entire system of charged particles (which need not be conserved). The quantity $\hat{\mathbf{n}}\cdot\underline{\mathbb{M}}$ appearing in the surface integral of Eq. (2.177) is then seen to represent the normal flow of electromagnetic angular momentum per unit area out of the region $\mathcal{V}$ across the boundary surface $\mathcal{S}$.

2.4   Conjugate Electromagnetic Fields and Invariants

The microscopic Maxwell's equations in source-free regions of space take the rather suggestive symmetric form

$$\nabla\cdot\mathbf{e}(\mathbf{r},t)=\nabla\cdot\mathbf{h}(\mathbf{r},t)=0, \tag{2.179}$$

$$\nabla\times\mathbf{e}(\mathbf{r},t)=-\frac{1}{c}\frac{\partial\mathbf{h}(\mathbf{r},t)}{\partial t}, \tag{2.180}$$

$$\nabla\times\mathbf{h}(\mathbf{r},t)=\frac{1}{c}\frac{\partial\mathbf{e}(\mathbf{r},t)}{\partial t}, \tag{2.181}$$

in cgs units. If one then defines the *complex vector field* introduced by Silberstein [47, 48] and Bateman [49] as

$$\bar{\mathbf{k}}(\mathbf{r},t)\equiv\mathbf{h}(\mathbf{r},t)\pm i\mathbf{e}(\mathbf{r},t), \tag{2.182}$$

then Maxwell's equations (2.179)–(2.181) are replaced by the pair of equations

$$\nabla\cdot\bar{\mathbf{k}}(\mathbf{r},t)=0,\quad\nabla\times\bar{\mathbf{k}}(\mathbf{r},t)=\mp\frac{i}{c}\frac{\partial\bar{\mathbf{k}}(\mathbf{r},t)}{\partial t}. \tag{2.183}$$

An electromagnetic field is then defined by Bateman [49] as follows:

Definition 2.1 (Electromagnetic Field) A solution of the set of vector differential equations (2.183) which provides single-valued vector functions $\mathbf{e}(\mathbf{r},t)$ and $\mathbf{h}(\mathbf{r},t)$ for each space–time point $(\mathbf{r},t)$ belonging to a certain domain $\mathcal{D}$ defines an electromagnetic field in $\mathcal{D}$.

Because Maxwell's equations are linear, the superposition of any number of solutions of Eq. (2.183) is also a solution.

Definition 2.2 (Conjugate Electromagnetic Fields) Two electromagnetic fields $(\mathbf{e}_j(\mathbf{r}, t), \mathbf{h}_j(\mathbf{r}, t))$, $j = 1, 2$ in a given space–time domain $\mathcal{D}$ are said to be *conjugate* in $\mathcal{D}$ if and only if their associated complex vectors $\bar{\mathbf{k}}_j(\mathbf{r}, t) \equiv \mathbf{h}_j(\mathbf{r}, t) \pm i\mathbf{e}_j(\mathbf{r}, t)$ are orthogonal throughout $\mathcal{D}$, viz.

$$\bar{\mathbf{k}}_1(\mathbf{r}, t) \cdot \bar{\mathbf{k}}_2(\mathbf{r}, t) = 0 \qquad (2.184)$$

for all $(\mathbf{r}, t) \in \mathcal{D}$.

The magnitude squared of the complex vector for an electromagnetic field in some domain $\mathcal{D}$, given by

$$\bar{k}^2 \equiv \bar{\mathbf{k}} \cdot \bar{\mathbf{k}} = h^2 - e^2 \pm 2i\mathbf{e} \cdot \mathbf{h} \qquad (2.185)$$

may be written as

$$\bar{k}^2(\mathbf{r}, t) = I_1(\mathbf{r}, t) \pm 2i I_2(\mathbf{r}, t) \qquad (2.186)$$

where

$$I_1(\mathbf{r}, t) \equiv h^2(\mathbf{r}, t) - e^2(\mathbf{r}, t), \qquad (2.187)$$

$$I_2(\mathbf{r}, t) \equiv \mathbf{e}(\mathbf{r}, t) \cdot \mathbf{h}(\mathbf{r}, t), \qquad (2.188)$$

are *invariants* for the group of linear transformations under which the source-free electromagnetic field equations remain unaltered [49].

Let $(\mathbf{e}_1, \mathbf{h}_1)$ and $(\mathbf{e}_2, \mathbf{h}_2)$ be conjugate electromagnetic fields in a region $\mathcal{D}$, so that $\bar{\mathbf{k}}_1 \cdot \bar{\mathbf{k}}_2 = (\mathbf{h}_1 \pm i\mathbf{e}_1) \cdot (\mathbf{h}_2 \pm i\mathbf{e}_2) = \mathbf{h}_1 \cdot \mathbf{h}_2 - \mathbf{e}_1 \cdot \mathbf{e}_2 \pm i(\mathbf{h}_1 \cdot \mathbf{e}_2 + \mathbf{e}_1 \cdot \mathbf{h}_2) = 0$ by definition, and hence $\mathbf{h}_1 \cdot \mathbf{h}_2 - \mathbf{e}_1 \cdot \mathbf{e}_2 = 0$ and $\mathbf{h}_1 \cdot \mathbf{e}_2 + \mathbf{e}_1 \cdot \mathbf{h}_2 = 0$. The invariants for the superposed electromagnetic field $\mathbf{e} = \mathbf{e}_1 + \mathbf{e}_2$, $\mathbf{h} = \mathbf{h}_1 + \mathbf{h}_2$ in $\mathcal{D}$ are then given by

$$\begin{aligned} I_1 &= h^2 - e^2 = (\mathbf{h}_1 + \mathbf{h}_2) \cdot (\mathbf{h}_1 + \mathbf{h}_2) - (\mathbf{e}_1 + \mathbf{e}_2) \cdot (\mathbf{e}_1 + \mathbf{e}_2) \\ &= (h_1^2 - e_1^2) + (h_2^2 - e_2^2) + 2(\mathbf{h}_1 \cdot \mathbf{h}_2 - \mathbf{e}_1 \cdot \mathbf{e}_2) \\ &= I_1^{(1)} + I_1^{(2)} \end{aligned} \qquad (2.189)$$

and

$$\begin{aligned} I_2 &= \mathbf{e} \cdot \mathbf{h} = (\mathbf{e}_1 + \mathbf{e}_2) \cdot (\mathbf{h}_1 + \mathbf{h}_2) \\ &= \mathbf{e}_1 \cdot \mathbf{h}_1 + \mathbf{e}_2 \cdot \mathbf{h}_2 + \mathbf{e}_1 \cdot \mathbf{h}_2 + \mathbf{e}_2 \cdot \mathbf{h}_1 \\ &= I_2^{(1)} + I_2^{(2)}. \end{aligned} \qquad (2.190)$$

Thus, when two conjugate electromagnetic fields in a region $\mathcal{D}$ are superposed, the invariants I_j for the total field in $\mathcal{D}$ is given by the sum of the respective invariants $I_j^{(k)}$ for the two component fields $k = 1, 2$.

Definition 2.3 (Self-Conjugate Field) If the invariants I_1 and I_2 of an electromagnetic field vanish throughout a region $\mathcal{D}$, the field is said to be self-conjugate over this region.

Consider finally the flow of energy in the electromagnetic field in a region $\mathcal{D}$ where the source terms ρ and $\mathbf{j} = \rho \mathbf{v}_c$ vanish. In that case, Poynting's theorem (2.155) becomes

$$\frac{\partial u}{\partial t} + \nabla \cdot \mathbf{s} = 0, \tag{2.191}$$

where $u = \frac{1}{8\pi}(e^2 + h^2)$ and $\mathbf{s} = \frac{c}{4\pi} \mathbf{e} \times \mathbf{h}$ are the electromagnetic energy density and Poynting vector, respectively, in cgs units. The general form of the *equation of continuity* for some physical entity with volume density ρ that is conserved moving with velocity $\mathbf{v}$ is given by

$$\frac{\partial \rho}{\partial t} + \nabla \cdot (\rho \mathbf{v}) = 0. \tag{2.192}$$

Application of this general expression to the conservation of charge is obvious. For Poynting's theorem, $\rho \to u$ is the energy density that is the conserved entity and $\rho \mathbf{v} \to \mathbf{s}$ is the rate at which this entity (electromagnetic energy) flows in the field. Following the analysis due to Bateman [49], consider the quantity

$$(4\pi)^2 u^2 (c^2 - v^2) = \frac{1}{4} c^2 (e^2 + h^2)^2 - c^2 s^2.$$

Because $(4\pi/c)^2 s^2 = (4\pi/c)^2 \mathbf{s} \cdot \mathbf{s} = (\mathbf{e} \times \mathbf{h}) \cdot (\mathbf{e} \times \mathbf{h}) = e^2 h^2 - (\mathbf{e} \cdot \mathbf{h})^2$, the above expression then becomes

$$(4\pi)^2 u^2 (c^2 - v^2) = \frac{1}{4} c^2 (e^2 - h^2)^2 + c^2 (\mathbf{e} \cdot \mathbf{h})^2. \tag{2.193}$$

Because the right-hand side of this equation is, in general, nonzero, it is then seen that electromagnetic energy travels with a velocity that is less than the speed of light c in vacuum. Indeed, as stated by Bateman [49], *"the velocity c is attained only in the case of a self-conjugate field."*

2.5 Time-Reversal Invariance of the Microscopic Field Equations

Time-reversal symmetry (or *T-symmetry*) is defined as the analytical symmetry of a particular physical law under the *time-reversal transformation* $t \overset{T}{\mapsto} -t$. The key word here is "analytical" because the second law of thermodynamics dictates that, with the exception of isolated systems in thermodynamic equilibrium, physically observable phenomena do not exhibit T-symmetry, at least in the classical sense.

Because the elementary source for the microscopic electric field $\mathbf{e}(\mathbf{r}, t)$ is the microscopic charge density $\rho(\mathbf{r}, t)$ which is T-symmetric $[\rho(\mathbf{r}, -t) = \rho(\mathbf{r}, t)]$, then $\mathbf{e}(\mathbf{r}, t)$ is also T-symmetric; that is

$$\rho(\mathbf{r}, -t) = \rho(\mathbf{r}, t) \Longrightarrow \mathbf{e}(\mathbf{r}, -t) = \mathbf{e}(\mathbf{r}, t). \tag{2.194}$$

However, because the microscopic current density $\mathbf{j}(\mathbf{r}, t) = \rho(\mathbf{r}, t)\mathbf{v}(\mathbf{r}, t)$ changes sign under a time-reversal transformation due to the fact that the direction of the charge velocity $\mathbf{v}(\mathbf{r}, t)$ is reversed under this transformation $[\mathbf{v}(\mathbf{r}, -t) = -\mathbf{v}(\mathbf{r}, t)]$, and because this is the elementary source term for the microscopic magnetic induction field $\mathbf{b}(\mathbf{r}, t)$, one then has that

$$\mathbf{j}(\mathbf{r}, -t) = -\mathbf{j}(\mathbf{r}, t) \Longrightarrow \mathbf{b}(\mathbf{r}, -t) = -\mathbf{b}(\mathbf{r}, t) \tag{2.195}$$

and the magnetic field is not T-symmetric, changing sign under a time-reversal transformation.

Because $\partial/\partial t \overset{T}{\mapsto} -\partial/\partial t$ under a time-reversal transformation, Faraday's law (2.33) transforms as

$$\nabla \times \mathbf{e}(\mathbf{r}, t) = -\frac{\partial \mathbf{b}(\mathbf{r}, t)}{\partial t} \overset{T}{\mapsto} \nabla \times \mathbf{e}(\mathbf{r}, -t) = -\frac{\partial \mathbf{b}(\mathbf{r}, -t)}{\partial t},$$

which returns to the correct form of Faraday's law with the substitutions $\mathbf{e}(\mathbf{r}, -t) = \mathbf{e}(\mathbf{r}, t)$ and $\mathbf{b}(\mathbf{r}, -t) = -\mathbf{b}(\mathbf{r}, t)$. The time-reversal transformation of Ampére's law (2.34) gives

$$\nabla \times \mathbf{h}(\mathbf{r}, t) = \mathbf{j}(\mathbf{r}, t) + \frac{\partial \mathbf{d}(\mathbf{r}, t)}{\partial t} \overset{T}{\mapsto} \nabla \times \mathbf{h}(\mathbf{r}, -t) = \mathbf{j}(\mathbf{r}, -t) - \frac{\partial \mathbf{d}(\mathbf{r}, -t)}{\partial t},$$

which returns to the correct form of Ampére's law with the substitutions $\mathbf{h}(\mathbf{r}, -t) = -\mathbf{h}(\mathbf{r}, t)$, $\mathbf{j}(\mathbf{r}, -t) = -\mathbf{j}(\mathbf{r}, t)$, and $\mathbf{d}(\mathbf{r}, -t) = \mathbf{d}(\mathbf{r}, t)$. As the two Gauss' laws $\nabla \cdot \mathbf{d}(\mathbf{r}, t) = \rho$ and $\nabla \cdot \mathbf{b}(\mathbf{r}, t) = 0$ are time-translation invariant, it is then seen that the *microscopic Maxwell's equations are time-translation invariant* [50], as are the equation of continuity $\nabla \cdot \mathbf{j}(\mathbf{r}, t) + \partial\rho(\mathbf{r}, t)/\partial t = 0$ and Lorentz force relation $\mathbf{f}(\mathbf{r}, t) = \rho(\mathbf{r}, t)\mathbf{e}(\mathbf{r}, t) + \mathbf{j}(\mathbf{r}, t) \times \mathbf{b}(\mathbf{r}, t)$.

2.6 Uniqueness of Solution

Consider now determining the conditions that must be satisfied by the microscopic electromagnetic field vectors obtained as solutions of the microscopic Maxwell equations in order that these solutions are unique. Helmholtz' theorem (Appendix B) shows that a given vector field may be uniquely expressed in a region V in terms of a scalar and vector potential that are determined, respectively, from the divergence and curl of that vector field in V together with its normal and tangential components on the closed surface S bounding V. Such is the case for both the electrostatic and magnetostatic fields when either Dirichlet, Neumann, or mixed boundary conditions are specified. However, the microscopic Maxwell equations

$$\nabla \cdot \mathbf{e}(\mathbf{r}, t) = \frac{\|4\pi\|}{\epsilon_0} \rho(\mathbf{r}, t), \quad \nabla \cdot \mathbf{b}(\mathbf{r}, t) = 0,$$

$$\nabla \times \mathbf{e}(\mathbf{r}, t) = - \left\| \frac{1}{c} \right\| \frac{\partial \mathbf{b}(\mathbf{r}, t)}{\partial t}, \quad \nabla \times \mathbf{b}(\mathbf{r}, t) = \left\| \frac{4\pi}{c} \right\| \mu_0 \mathbf{j}(\mathbf{r}, t) + \left\| \frac{1}{c} \right\| \epsilon_0 \mu_0 \frac{\partial \mathbf{e}(\mathbf{r}, t)}{\partial t},$$

are a set of differential relations that specify the divergence and curl of both the microscopic electric $\mathbf{e}(\mathbf{r}, t)$ and magnetic $\mathbf{b}(\mathbf{r}, t)$ field vectors in any region of space V in terms of the fundamental charge $\rho(\mathbf{r}, t)$ and current $\mathbf{j}(\mathbf{r}, t)$ source densities as well as upon the time rate of change of the other field quantity. Hence, the uniqueness of each field vector is intimately connected to the uniqueness of the other. As a consequence, the sufficient conditions for the uniqueness of solution for the time-dependent electromagnetic field vectors $\mathbf{e}(\mathbf{r}, t)$ and $\mathbf{b}(\mathbf{r}, t)$ must be obtained simultaneously, and this is most conveniently accomplished through Poynting's theorem.

Consider then a finite region V of space that is bounded by a closed regular surface S. Assume that there are two sets of solutions $\{\mathbf{e}_1(\mathbf{r}, t), \mathbf{b}_1(\mathbf{r}, t)\}$ and $\{\mathbf{e}_2(\mathbf{r}, t), \mathbf{b}_2(\mathbf{r}, t)\}$ to the microscopic Maxwell equations that describe the electromagnetic field behavior within V and which are equal at some time $t = t_0$. It is then desired to determine the conditions under which these two sets of solutions remain equal for all time $t > t_0$ in V. Because the microscopic field equations are linear, then the difference field defined by $\mathbf{e}(\mathbf{r}, t) \equiv \mathbf{e}_1(\mathbf{r}, t) - \mathbf{e}_2(\mathbf{r}, t)$ and $\mathbf{b}(\mathbf{r}, t) \equiv \mathbf{b}_1(\mathbf{r}, t) - \mathbf{b}_2(\mathbf{r}, t)$ is also a solution of the field equations within the region V. The field vectors $\mathbf{e}(\mathbf{r}, t)$ and $\mathbf{b}(\mathbf{r}, t)$ therefore satisfy Poynting's theorem (2.157), which may be written as

$$\frac{1}{2} \left\| \frac{1}{4\pi} \right\| \frac{d}{dt} \int_V \left(\epsilon_0 e^2 + \mu_0 h^2 \right) d^3 r + \oint_S \mathbf{s} \cdot \hat{\mathbf{n}} d^2 r = - \int_V \mathbf{j} \cdot \mathbf{e} \, d^3 r, \qquad (2.196)$$

where $\mathbf{s} = \|c/4\pi\| \mathbf{e} \times \mathbf{h}$ is the Poynting vector for the difference field and where $\hat{\mathbf{n}}$ is the unit outward normal vector to the surface S. In order that the surface integral of the Poynting vector appearing in Eq. (2.196) vanish, it is only necessary that either the tangential components of $\mathbf{e}_1(\mathbf{r}, t)$ and $\mathbf{e}_2(\mathbf{r}, t)$, or else the tangential components of $\mathbf{b}_1(\mathbf{r}, t)$ and $\mathbf{b}_2(\mathbf{r}, t)$ be identical on $\mathcal{S}$ for all time $t \geq t_0$; for then either $\hat{\mathbf{n}} \times \mathbf{e} = \mathbf{0}$ or $\hat{\mathbf{n}} \times \mathbf{b} = \mathbf{0}$ and $\mathbf{s}$ has no normal component over the entire surface $\mathcal{S}$. Consequently, there is no net flow of electromagnetic energy associated with the difference field across the closed surface $\mathcal{S}$ and the form (2.196) of Poynting's theorem becomes

$$\frac{1}{2} \left\| \frac{1}{4\pi} \right\| \frac{d}{dt} \int_V \left(\epsilon_0 e^2 + \mu_0 h^2 \right) d^3r = - \int_V \mathbf{j} \cdot \mathbf{e} \, d^3r. \qquad (2.197)$$

The volume integral of $\mathbf{j} \cdot \mathbf{e}$ over $\mathcal{V}$ is the rate at which the Lorentz forces acting inside the region $\mathcal{V}$ do work and hence is a non-negative quantity for a dissipative system. The right-hand side of Eq. (2.197) is consequently always less than or equal to zero. On the other hand, the electromagnetic field energy that is given by the volume integral on the left-hand side of Eq. (2.197) is always greater than or equal to zero and, by construction, vanishes at $t = t_0$. Because the right-hand side of Eq. (2.197) is always negative or zero, the time derivative of the total electromagnetic field energy in the region $\mathcal{V}$ must also always be negative or zero and hence does not ever increase with time. Because it vanishes at $t = t_0$, it must then vanish for all $t \geq t_0$. Because each of the two terms appearing in the integrand of the electromagnetic field energy integral are non-negative, Eq. (2.197) can only be satisfied if both $\mathbf{e}(\mathbf{r}, t) = \mathbf{e}_1(\mathbf{r}, t) - \mathbf{e}_2(\mathbf{r}, t) = \mathbf{0}$ and $\mathbf{b}(\mathbf{r}, t) = \mathbf{b}_1(\mathbf{r}, t) - \mathbf{b}_2(\mathbf{r}, t) = \mathbf{0}$ for all $t \geq t_0$ in $\mathcal{V}$. The following uniqueness theorem has therefore been established:

Theorem 2.1 (Uniqueness) *A microscopic electromagnetic field is uniquely determined within a bounded, regular region $\mathcal{V}$ for all time $t \geq t_0$ by both the initial values of the microscopic electric and magnetic field vectors throughout the region $\mathcal{V}$ and the values of the tangential component of either the electric or magnetic field vector over the closed boundary surface $\mathcal{S}$ for all $t \geq t_0$.*

Notice that this does not prove the existence of any field vectors $\mathbf{e}(\mathbf{r}, t)$ and $\mathbf{b}(\mathbf{r}, t)$ that satisfy the imposed conditions; the theorem only states that if such a field did exist, it is then the only such field.

It has thus been established that the values of the microscopic electric and magnetic field vectors are uniquely determined throughout any closed (bounded) regular region $\mathcal{V}$ at any given time t by a tangential boundary condition on either of the field vectors on the boundary surface $\mathcal{S}$ and by the initial values of both field vectors everywhere in the region $\mathcal{V}$. If the boundary surface $\mathcal{S}$ recedes to infinity, the region $\mathcal{V}$ is externally unbounded and one must ensure the vanishing of the surface integral of the Poynting vector over an infinitely remote surface. If the field was established in the finite past, this difficulty may be circumvented by the assumption that the boundary surface $\mathcal{S}$ lies beyond the spatial zone reached at the time t by a field that is propagated with a finite velocity c. The above uniqueness theorem does not take into full account this finiteness of propagation of the electromagnetic field. Because the field is propagated with a finite velocity, only those elements of the region $\mathcal{V}$ whose distance from the point of observation is less than or equal to the quantity $c(t - t_0)$ need be accounted for. The classical uniqueness theorem given above has been extended in this sense by Rubinowicz [51].

2.7   Synopsis

With the rejection of a "luminiferous ether" as the seat of electromagnetism, the Maxwell–Lorentz theory does not provide any meaningful replacement for it beyond the hypothesis that [5] "the seat of the field is the empty space. The participation of matter in electromagnetic phenomena has its origin only in the fact that the elementary particles of matter carry unalterable electric charges, and, on this account, are subject on the one hand to the actions of ponderomotive forces and on the other hand possess the property of generating a field." The classical theory can go no further, can go no deeper, as that requires both quantum theory and general relativity. String theory [52], for example, suggests that both electromagnetic and gravitational waves are distortions in the fabric of space–time. As far as this development is concerned, this is sufficient.

Problems

2.1 Show that charge is conserved using only the time-domain integral form of Maxwell's equations.

2.2 Consider a charged point particle of mass m and charge q that is moving through an externally applied electromagnetic field $\{\mathbf{e}, \mathbf{b}\}$ with velocity $\mathbf{v} = d\mathbf{r}/dt$, where $\mathbf{r} = \mathbf{r}(t)$ denotes the position vector of the charged particle with respect to a fixed origin of coordinates. From the Lorentz force relation (2.24) and Newton's second law of motion, the equation of motion for the charged particle is found to be given by

$$m\frac{d^2\mathbf{r}}{dt^2} = q\mathbf{e} + \left\|\frac{1}{c}\right\| q\mathbf{v} \times \mathbf{b}. \tag{2.198}$$

(It should be noted that neither side of this equation is strictly correct: the left-hand side is relativistically incorrect but is a good approximation for small particle velocities such that $v/c \ll 1$, whereas the right-hand side of this equation does not include the radiation damping that is due to the electromagnetic field radiated by the charged particle when it undergoes an acceleration, the energy of which must be drawn from the particle's kinetic energy).

(a) Prove that the magnetic field vector b has no influence on the magnitude $v = |\mathbf{v}|$ of the velocity of the particle, and hence, on it's kinetic energy.

(b) Consider the motion in a uniform steady magnetic field $\mathbf{b} = \hat{\mathbf{1}}_z b$ alone that is directed along the positive z-axis of the chosen system of coordinates. The equations of motion then become $m(d^2x/dt^2) = \|1/c\|qb(dy/dt)$, $m(d^2y/dt^2) = -\|1/c\|qb(dx/dt)$, and $m(d^2z/dt^2) = 0$ in component form, where $\mathbf{r} = \hat{\mathbf{1}}_x x + \hat{\mathbf{1}}_y y + \hat{\mathbf{1}}_z z$. The third equation shows that the component of motion in the direction of $\mathbf{b}$ is unaffected by the field so that one need only consider the projection of the motion onto the xy-plane (the component of $\mathbf{r}$ perpendicular to $\mathbf{b}$). Show that this projection describes an orbit with radius $R = \|c\|mv/|qb|$ and that the projected motion has the constant angular velocity $\omega = \|1/c\|qb/m$.

(c) Write a computer program that computes and plots the three-dimensional trajectory followed by an electron in a uniform, steady magnetic field that is directed along the z-axis.

2.3 Beginning with the expression $\mathbf{F}_m = \|1/c\|q\mathbf{v} \times \mathbf{B}$ for the magnetic force, show that two measurements $\mathbf{F}_1(\mathbf{r})$ and $\mathbf{F}_2(\mathbf{r})$ of $\mathbf{F}_m$ that are made with mutually perpendicular velocities $\mathbf{v}_1$ and $\mathbf{v}_2$, respectively, at the same point $\mathbf{r}$ result in the unique determination of the magnetic induction vector at that point as

$$\mathbf{B}(\mathbf{r}) = \frac{\|c\|}{q}\left[\frac{\mathbf{F}_1 \times \mathbf{v}_1}{v_1^2} + \frac{(\mathbf{F}_2 \times \mathbf{v}_2)\cdot\mathbf{v}_1}{v_1^2 v_2^2}\mathbf{v}_1\right]. \tag{2.199}$$

2.4 Compare the electric and magnetic forces between two moving point charges: charge q moving with velocity $\mathbf{v}$ and charge q_1 moving with velocity $\mathbf{v}_1$. In particular, show that the magnetic force $\mathbf{F}_m$ acting on q due to q_1 may be expressed as

$$\mathbf{F}_m = \frac{\mathbf{v}}{c} \times \left(\frac{\mathbf{v}_1}{c} \times \mathbf{F}_e\right), \tag{2.200}$$

where $\mathbf{F}_e$ is the electric force exerted on q by q_1.

2.5 Two events occur at the same spatial point in an inertial reference frame Σ, but not simultaneously. Show that the temporal sequence of these two events remains unchanged in any (all) other inertial reference frame(s) Σ'.

2.6 Show that the Poincaré–Lorentz transformation relations given in Eqs. (3.71) and (3.72) may be expressed in matrix form as

$$\mathbf{r}^{\Sigma'} = \underline{\underline{Q}}_{\Sigma}^{\Sigma'}\mathbf{r}^{\Sigma}, \tag{2.201}$$

where $\mathbf{r}^{\Sigma} = (x^1, x^2, x^3, x^4) \equiv (x, y, z, ct)$ and $\mathbf{r}^{\Sigma'} = (x'^1, x'^2, x'^3, x'^4) \equiv (x', y', z', ct')$ are the row vector forms of the column vectors appearing in the above equation with coordinate transformation matrix

$$
\underline{\underline{Q}}_{\Sigma}^{\Sigma'} = \begin{pmatrix}
1 + \frac{V_x^2}{V^2}(\gamma - 1) & \frac{V_x V_y}{V^2}(\gamma - 1) & \frac{V_x V_z}{V^2}(\gamma - 1) & -\frac{V_x}{c}\gamma \\
\frac{V_y V_x}{V^2}(\gamma - 1) & 1 + \frac{V_y^2}{V^2}(\gamma - 1) & \frac{V_y V_z}{V^2}(\gamma - 1) & -\frac{V_y}{c}\gamma \\
\frac{V_z V_x}{V^2}(\gamma - 1) & \frac{V_z V_y}{V^2}(\gamma - 1) & 1 + \frac{V_z^2}{V^2}(\gamma - 1) & -\frac{V_z}{c}\gamma \\
-\gamma \frac{V_x}{c} & -\gamma \frac{V_y}{c} & -\gamma \frac{V_z}{c} & \gamma
\end{pmatrix}. \tag{2.202}
$$

The inverse transformation relation is then given by

$$
\mathbf{r}^{\Sigma} = \underline{\underline{Q}}_{\Sigma'}^{\Sigma} \mathbf{r}^{\Sigma'}, \tag{2.203}
$$

where $\underline{\underline{Q}}_{\Sigma'}^{\Sigma} = (\underline{\underline{Q}}_{\Sigma}^{\Sigma'})^{-1}$ is the inverse of the transformation matrix given above. Show that

$$
\underline{\underline{Q}}_{\Sigma}^{\Sigma'} \underline{\underline{Q}}_{\Sigma'}^{\Sigma} = \underline{\underline{Q}}_{\Sigma'}^{\Sigma} \underline{\underline{Q}}_{\Sigma}^{\Sigma'} = \underline{\underline{J}}, \tag{2.204}
$$

where $\underline{\underline{J}} \equiv (\delta_{ij})$ is the *idemfactor*. Show that $\underline{\underline{Q}}_{\Sigma}^{\Sigma'} \underline{\underline{Q}}_{\Sigma'}^{\Sigma} = \underline{\underline{Q}}_{\Sigma'}^{\Sigma} \underline{\underline{Q}}_{\Sigma}^{\Sigma'} = \underline{\underline{J}}$.

2.7 Determine the relativistic motion of a charged particle of charge q and rest mass m_0 as it passes through a uniform, time-independent magnetic field $\mathbf{b} = \hat{\mathbf{1}}_z b$. Assume that the initial velocity of the particle is along the x-direction.

2.8 Derive the relativistic velocity transformation relation given in Eq. (2.87).

2.9 Derive the relativistic acceleration transformation relation given in Eqs. (2.91) and (2.92).

2.10 Derive the identity given in Eq. (2.112). (Hint: Beginning with the relativistic velocity transformation relations in Eqs. (2.89) and (2.90), determine the transformation relation for the quantity $c^2 - u^2$, where $u^2 = u_\parallel^2 + u_\perp^2$.

2.11 Derive the general transformation relation for linear momentum given in Eq. (2.114).

2.12 Derive the general transformation relation for energy given in Eq. (2.115).

2.13 Derive the general transformation relation for the force given in Eq. (2.120).

2.14 Verify the expression for the force given in Eq. (2.121).

2.15 Derive the appropriate transformation relations for the microscopic charge density ρ and current density $\mathbf{j}$.

2.16 Derive the transformation relations given in Eq. (2.130) for the rectangular components of the microscopic electric field vector $\mathbf{e}$.

2.17 Derive the transformation relations given in Eq. (2.132) for the rectangular components of the microscopic magnetic field vector $\mathbf{b}$.

2.18 Show that the microscopic Maxwell equations (2.136) and (2.139) are invariant in form under a Poincaŕe–Lorentz transformation when the field vectors are transformed by the relations specified in Eqs. (2.134)–(2.135) and the charge and current sources are transformed by the relations obtained in Problem 2.15.

2.19 Show that the relativistically correct transformation relations given in Eqs. (2.134) and (2.135) for the electric and magnetic field vectors, respectively, reduce to the corresponding Galilean transformation relations given in Eqs. (2.50) and (2.58) in the small velocity limit as $V/c \to 0$. Determine the first higher-order correction term to these Galilean transformation relations.

2.20 A static distribution of charge and current can set up time-independent electric and magnetic fields in a common region such that the Poynting vector is nonvanishing, but there is no net power flow. Show that under these conditions

$$\oint_S \mathbf{s} \cdot \hat{\mathbf{n}} d^2 r = 0,$$

for any closed surface S in the region.

2.21 Is uniqueness obtained when the volume integral of the quantity $\mathbf{j} \cdot \mathbf{e}$ over the region V appearing on the right-hand side of Eq. (2.195) is either zero or negative?

References

1. J. C. Maxwell, "A dynamical theory of the electromagnetic field," *Phil. Trans. Roy. Soc. (London)*, vol. 155, pp. 450–521, 1865.
2. J. C. Maxwell, *A Treatise on Electricity and Magnetism*. Oxford: Oxford University Press, 1873.
3. H. A. Lorentz, *The Theory of Electrons*. Leipzig: Teubner, 1906. Ch. IV.
4. J. M. Stone, *Radiation and Optics, An Introduction to the Classical Theory*. New York: McGraw-Hill, 1963.
5. A. Einstein, *Out of My Later Years*. New York: Philosophical Library, 1950. pp. 76–77.
6. L. Rosenfeld, *Theory of Electrons*. Amsterdam: North-Holland, 1951.
7. H. A. Kramers, *Quantum Mechanics*. Amsterdam: North-Holland, 1957.
8. A. D. Yaghjian, *Relativistic Dynamics of a Charged Sphere*. Berlin-Heidelberg: Springer-Verlag, 1992.
9. K. I. Golden and G. Kalman, "Phenomenological electrodynamics of two-dimensional Coulomb systems," *Phys. Rev. B*, vol. 45, no. 11, pp. 5834–5837, 1992.
10. K. I. Golden and G. Kalman, "Phenomenological electrodynamics of electronic superlattices," *Phys. Rev. B*, vol. 52, no. 20, pp. 14719–14727, 1995.
11. K. I. Golden and G. J. Kalman, "Quasilocalized charge approximation in strongly coupled plasma physics," *Physics of Plasmas*, vol. 7, no. 1, pp. 14–32, 2000.
12. N. Bohr and L. Rosenfeld, "Field and charge measurements in quantum electrodynamics," *Phys. Rev.*, vol. 78, no. 6, pp. 794–798, 1950.
13. M. Faraday, *Experimental Researches in Electricity*. London: Bernard Quaritch, 1855.
14. P. Penfield and H. A. Haus, *Electrodynamics of Moving Media*. Cambridge, MA: M.I.T. Press, 1967.

15. A. M. Ampère, "Memoir on the mutual action of two electric currents," *Annales de Chimie et Physique*, vol. 15, pp. 59–76, 1820.
16. H. Poincaré, *Electricité et Optique*. Paris: Carré Nadaud, 1901.
17. H. Poincaré, *La Science et l'hypothése*. Paris: Flammarion, 1902.
18. H. Poincaré, "Sur la dynamique de l'électron," *Comptes rendus Acad. Sci. Paris*, vol. 140, pp. 1504–1508, 1905.
19. H. A. Lorentz, "Electromagnetic phenomena in a system moving with any velocity less than that of light," *Proc. Acad. Sci. Amsterdam*, vol. 6, pp. 809–832, 1904.
20. K. F. Gauss, "Theoria Attractionis Corporum Sphaeroidicorum Ellipticorum Homogeneorum," in *Werke*, vol. 5, pp. 1–22, Göttingen: Royal Society of Science, 1870.
21. T. Young, "Experiments and calculations relative to physical optics," in *Miscellaneous Works* (G. Peacock, ed.), vol. 1, pp. 179–191, London: John Murray Publishers, 1855. p.188.
22. J. Bradley, "An account of a new discovered motion of the fix'd stars," *Phil. Trans. Roy. Soc. (London)*, vol. 35, pp. 637–660, 1728.
23. A. A. Michelson and E. W. Morley, "On the relative motion of the Earth and the luminiferous ether," *Am. J. Sci.*, no. 203, pp. 333–345, 1887.
24. F. T. Trouton and H. R. Noble, "Forces acting on a charged condenser moving through space," *Proc. Roy. Soc. (London)*, vol. 72, pp. 132–133, 1903.
25. J. H. Poincaré, "L'etat actuel et l'avenir de la physique mathématique," *Bull. Sci. Math.*, vol. 28, pp. 302–324, 1904. English translation in *Monist*, vol. 15, 1 (1905).
26. A. Einstein, "Zur elektrodynamik bewegter körper," *Ann. Phys.*, vol. 17, pp. 891–921, 1905.
27. R. Resnick, *Introduction to Special Relativity*. New York: John Wiley & Sons, 1968.
28. J. V. Bladel, *Relativity and Engineering*. Berlin-Heidelberg: Springer-Verlag, 1984.
29. A. S. Eddington, *The Mathematical Theory of Relativity*. New York: Chelsea, third ed., 1975. Section 5.
30. P. M. Morse and H. Feshbach, *Methods of Theoretical Physics*. New York: McGraw-Hill, 1953. Vol. I.
31. H. A. Lorentz, *Versuch einer Theorie der elektrischen und optischen Erscheinungen in bewegten Körpern*. Leiden: E. J. Brill, 1895. Sections 89–92. English translation: "Michelson's Interference Experiment," in *The Principle of Relativity. A Collection of Original Memoirs on the Special and General Theory of Relativity* by A. Einstein, H. A. Lorentz, H. Minkowski, and H. Weyl, New York: Dover, 1958.
32. L. D. Broglie, *Matière et Lumière*. Paris: Albin Michel, 1937.
33. E. Fischbach, H. Kloor, R. A. Langel, A. T. Y. Lui, and M. Peredo, "New geomagnetic limits on the photon mass and on long-range forces coexisting with electromagnetism," *Phys. Rev. Lett.*, vol. 73, no. 4, pp. 514–517, 1994.
34. G. Feinberg, "Possibility of faster-than-light particles," *Phys. Rev.*, vol. 159, no. 5, pp. 1089–1105, 1967.
35. O.-M. Bilaniuk and E. C. G. Sudarshan, "Particles beyond the light barrier," *Physics Today*, vol. 22, no. 5, pp. 43–51, 1969.
36. A. Stewart, "The discovery of stellar aberration," *Scientific American*, vol. 210, no. 3, p. 100, 1964.
37. M. Born, *Einstein's Theory of Relativity*. New York: Dover, 1962.
38. J. H. Poynting, "Transfer of energy in the electromagnetic field," *Phil. Trans.*, vol. 175, pp. 343–361, 1884.
39. G. Nicolis, *Introduction to Nonlinear Science*. Cambridge: Cambridge University Press, 1995. Section 2.2.
40. W. S. Franklin, "Poynting's theorem and the distribution of electric field inside and outside of a conductor carrying electric current," *Phys. Rev.*, vol. 13, no. 3, pp. 165–181, 1901.
41. MacDonald, *Electric Waves*. Cambridge: Cambridge University Press, 1902.
42. Livens, *The Theory of Electricity*. Cambridge: Cambridge University Press, 1926. pp. 238 ff.
43. Mason and Weaver, *The Electromagnetic Field*. Chicago: University of Chicago Press, 1929. pp. 264 ff.
44. J. A. Stratton, *Electromagnetic Theory*. New York: McGraw-Hill, 1941.

45. M. Abraham, "Prinzipien der Dynamik der Elektrons," *Ann. Physik*, vol. 10, pp. 105–179, 1903.
46. H. B. Phillips, *Vector Analysis*. New York: John Wiley & Sons, 1933.
47. L. Silberstein, "Electromagnetische Grundgleichungen in bivectorielle Behandlung," *Ann. Phys.*, vol. 22, pp. 579–586, 1907.
48. L. Silberstein, "Electromagnetische Grundgleichungen in bivectorielle Behandlung," *Ann. Phys.*, vol. 24, pp. 783–784, 1907.
49. H. Bateman, *The Mathematical Analysis of Electrical & Optical Wave-Motion*. Cambridge: Cambridge University Press, 1905.
50. D. B. Malament, "On the time reversal invariance of classical electromagnetic theory," *Studies in History and Philosophy of Modern Physics*, vol. 35, pp. 295–315, 2004.
51. A. Rubinowicz, "Uniqueness of solution of Maxwell's equations," *Phys. Zeits.*, vol. 27, pp. 707–710, 1926.
52. B. Greene, *The Elegant Universe: Superstrings, Hidden Dimensions, and the Quest for the Ultimate Theory*. New York: W. W. Norton & Company, 1999.

Chapter 3
Microscopic Potentials and Radiation

> "A real field is a mathematical function we use for avoiding the idea of action at a distance." Richard Feynman, *The Feynman Lectures on Physics.*

The microscopic Maxwell equations consist of a set of coupled first-order partial differential equations relating the electric and magnetic field vectors that comprise the electromagnetic field to each other as well as to their charge and current sources. Their solution is often facilitated by the introduction of auxiliary fields known as potentials. These potentials have their origin in the two homogeneous equations $\nabla \cdot \mathbf{b} = 0$ and $\nabla \times \mathbf{e} = -\|1/c\| \partial \mathbf{b}/\partial t$ which indicate that not all of the components of the field vectors $\mathbf{e}$ and $\mathbf{b}$ are entirely independent.

3.1 The Microscopic Electromagnetic Potentials

The vector and scalar potentials[1] for an electromagnetic field are introduced through the differential properties of the electric and magnetic field vectors as described by the microscopic Maxwell equations given in Eqs. (2.33)–(2.36). Consider a microscopic electromagnetic field that is due to a given microscopic distribution of charge density $\rho(\mathbf{r}, t)$ and convective current density $\mathbf{j}(\mathbf{r}, t)$. The spatiotemporal properties of this field are then described by

$$\nabla \times \mathbf{e}(\mathbf{r}, t) + \left\| \frac{1}{c} \right\| \frac{\partial \mathbf{b}(\mathbf{r}, t)}{\partial t} = \mathbf{0}, \tag{3.1}$$

$$\nabla \times \mathbf{h}(\mathbf{r}, t) - \left\| \frac{1}{c} \right\| \frac{\partial \mathbf{d}(\mathbf{r}, t)}{\partial t} = \left\| \frac{4\pi}{c} \right\| \mathbf{j}(\mathbf{r}, t), \tag{3.2}$$

---

[1]George Green introduced the concept of the potential function into the theory of electricity and magnetism in 1828. Franz Neumann (the father of Karl Neumann) introduced the vector potential in 1845.

© Springer Nature Switzerland AG 2019
K. E. Oughstun, *Electromagnetic and Optical Pulse Propagation*, Springer Series in Optical Sciences 224, https://doi.org/10.1007/978-3-030-20835-6_3

with $\nabla \cdot \mathbf{d}(\mathbf{r}, t) = \|4\pi\| \rho(\mathbf{r}, t)$ and $\nabla \cdot \mathbf{b}(\mathbf{r}, t) = 0$, where the charge and current densities are related by the equation of continuity $\nabla \cdot \mathbf{j}(\mathbf{r}, t) = -\partial \rho(\mathbf{r}, t)/\partial t$. Due to the divergenceless character of the magnetic induction vector $\mathbf{b}$ throughout all space–time, the magnetic $\mathbf{b}$-field vector is solenoidal so that it can always be expressed as the curl of another vector field $\mathbf{a_0}$ as

$$\mathbf{b}(\mathbf{r}, t) = \nabla \times \mathbf{a_0}(\mathbf{r}, t). \qquad (3.3)$$

The subsidiary vector field $\mathbf{a_0}$ is not uniquely specified by this equation, for $\mathbf{b}$ may also be given by the curl of some other vector field $\mathbf{a}$,

$$\mathbf{b}(\mathbf{r}, t) = \nabla \times \mathbf{a}(\mathbf{r}, t), \qquad (3.4)$$

with

$$\mathbf{a}(\mathbf{r}, t) = \mathbf{a_0}(\mathbf{r}, t) - \nabla \psi(\mathbf{r}, t) \qquad (3.5)$$

where ψ is any arbitrary scalar function of both position and time. If $\mathbf{b}$ is replaced in Eq. (3.1) by either of the expressions given in Eqs. (3.3) or (3.4), there results, respectively

$$\nabla \times \left(\mathbf{e}(\mathbf{r}, t) + \left\| \frac{1}{c} \right\| \frac{\partial \mathbf{a_0}(\mathbf{r}, t)}{\partial t} \right) = \mathbf{0},$$

$$\nabla \times \left(\mathbf{e}(\mathbf{r}, t) + \left\| \frac{1}{c} \right\| \frac{\partial \mathbf{a}(\mathbf{r}, t)}{\partial t} \right) = \mathbf{0}.$$

The vector fields $(\mathbf{e} + \|1/c\| \partial \mathbf{a_0}/\partial t)$ and $(\mathbf{e} + \|1/c\| \partial \mathbf{a}/\partial t)$ are therefore everywhere irrotational and each can then be expressed as the gradient of some scalar function, so that

$$\mathbf{e}(\mathbf{r}, t) = -\left\| \frac{1}{c} \right\| \frac{\partial \mathbf{a_0}(\mathbf{r}, t)}{\partial t} - \nabla \phi_0(\mathbf{r}, t), \qquad (3.6)$$

$$\mathbf{e}(\mathbf{r}, t) = -\left\| \frac{1}{c} \right\| \frac{\partial \mathbf{a}(\mathbf{r}, t)}{\partial t} - \nabla \phi(\mathbf{r}, t). \qquad (3.7)$$

From Eq. (3.5), the scalar functions ϕ and ϕ_0 are seen to be related as

$$\phi(\mathbf{r}, t) = \phi_0(\mathbf{r}, t) + \left\| \frac{1}{c} \right\| \frac{\partial \psi(\mathbf{r}, t)}{\partial t}. \qquad (3.8)$$

Notice that arbitrary scalar constants may always be added to both of the scalar functions ψ and ϕ without altering the field vectors $\mathbf{e}$ and $\mathbf{b}$.

The vector functions $\mathbf{a}(\mathbf{r}, t)$ are called the *vector potentials* and the scalar functions $\phi(\mathbf{r}, t)$ are called the *scalar potentials* of the microscopic electromagnetic field. The functions $\mathbf{a_0}$ and ϕ_0 designate one specific pair of potentials from which the electromagnetic field vectors may be determined through application of Eqs. (3.3) and (3.6). An infinite number of potentials that describe the same

Fig. 3.1 Network structure of the gauge transformation for the microscopic vector and scalar potential fields

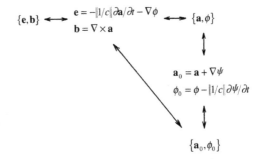

electromagnetic field can then be constructed from Eqs. (3.5) and (3.8) through choice of the function $\psi(\mathbf{r}, t)$ while the field vectors themselves are uniquely determined (provided that the conditions stated in Theorem 2.1 of Sect. 2.5 are satisfied).

The homogeneous field equations are naturally satisfied by the scalar and vector potentials. The remaining equations then provide the pair of inhomogeneous partial differential equations for the potential fields as

$$\nabla^2 \phi + \left\| \frac{1}{c} \right\| \frac{\partial}{\partial t} (\nabla \cdot \mathbf{a}) = - \frac{\|4\pi\|}{\epsilon_0} \rho, \tag{3.9}$$

$$\nabla^2 \mathbf{a} - \frac{1}{c^2} \frac{\partial^2 \mathbf{a}}{\partial t^2} - \nabla \left(\nabla \cdot \mathbf{a} + \left\| \frac{1}{c} \right\| \epsilon_0 \mu_0 \frac{\partial \phi}{\partial t} \right) = - \left\| \frac{4\pi}{c} \right\| \mu_0 \mathbf{j}. \tag{3.10}$$

Hence, with the introduction of these two auxiliary functions, the set of four Maxwell's equations have been reduced to a system of two equations that are, however, still coupled. They may, nevertheless, be uncoupled by exploiting the arbitrariness involved in the definition of the potential functions through the scalar function $\psi(\mathbf{r}, t)$ appearing in Eqs. (3.5) and (3.8). The transformation relation pair

$$\mathbf{a}(\mathbf{r}, t) \quad \longrightarrow \quad \mathbf{a}'(\mathbf{r}, t) = \mathbf{a}(\mathbf{r}, t) + \nabla \psi(\mathbf{r}, t), \tag{3.11}$$

$$\phi(\mathbf{r}, t) \quad \longrightarrow \quad \phi'(\mathbf{r}, t) = \phi(\mathbf{r}, t) - \left\| \frac{1}{c} \right\| \frac{\partial \psi(\mathbf{r}, t)}{\partial t}, \tag{3.12}$$

is called a *gauge transformation* and the invariance of the electromagnetic field vectors under such a transformation is called *gauge invariance*.[2] Finally, the scalar function $\psi(\mathbf{r}, t)$ is called the *gauge function*. The interrelationship between the electric and magnetic field vectors and the scalar and vector potential fields under a gauge transformation is illustrated in Fig. 3.1.

---

[2]Gauge invariance has its origin in quantum theory where gauge transformations possess the freedom to choose unobservable reference frames required for the description of certain types of elementary particles. The scalar and vector potentials in electromagnetics are necessary for the description of the dynamics of these particles.

3.1.1 The Lorenz Condition and the Lorenz Gauge

The freedom of choice implied by the pair of relations in Eqs. (3.11) and (3.12) means that one can always choose a pair of potentials $\{\mathbf{a}(\mathbf{r}, t), \phi(\mathbf{r}, t)\}$ such that they satisfy the differential relation

$$\nabla \cdot \mathbf{a} + \left\| \frac{1}{c} \right\| \epsilon_0 \mu_0 \frac{\partial \phi}{\partial t} = 0 \qquad (3.13)$$

which is known[3] as the *Lorenz condition* [2]. In order to show that a pair of potentials can always be found such that they satisfy the Lorenz condition, suppose that the potentials $\{\mathbf{a}(\mathbf{r}, t), \phi(\mathbf{r}, t)\}$ which satisfy the pair of relations given in Eqs. (3.9) and (3.10) do not satisfy the Lorenz condition. One can then undertake a gauge transformation to a new pair of potential functions $\{\mathbf{a}'(\mathbf{r}, t), \phi'(\mathbf{r}, t)\}$ and demand that this set does satisfy the Lorenz condition; hence

$$\nabla \cdot \mathbf{a}' + \frac{\epsilon_0 \mu_0}{\|c\|} \frac{\partial \phi'}{\partial t} = \nabla \cdot (\mathbf{a} + \nabla \psi) + \frac{\epsilon_0 \mu_0}{\|c\|} \frac{\partial}{\partial t} \left(\phi - \frac{1}{\|c\|} \frac{\partial \psi}{\partial t} \right)$$

$$= \left(\nabla \cdot \mathbf{a} + \frac{\epsilon_0 \mu_0}{\|c\|} \frac{\partial \phi}{\partial t} \right) + \left(\nabla^2 \psi - \frac{1}{c^2} \frac{\partial^2 \psi}{\partial t^2} \right) \equiv 0.$$

Thus, provided that a gauge function $\psi(\mathbf{r}, t)$ can be found to satisfy the inhomogeneous wave equation

$$\nabla^2 \psi - \frac{1}{c^2} \frac{\partial^2 \psi}{\partial t^2} = -\nabla \cdot \mathbf{a} - \frac{\epsilon_0 \mu_0}{\|c\|} \frac{\partial \phi}{\partial t}, \qquad (3.14)$$

the new pair of potentials $\{\mathbf{a}'(\mathbf{r}, t), \phi'(\mathbf{r}, t)\}$ will satisfy the Lorenz condition.

For a pair of potentials $\{\mathbf{a}(\mathbf{r}, t), \phi(\mathbf{r}, t)\}$ that satisfy the Lorenz condition the pair of differential equations given in Eqs. (3.9) and (3.10) for the potentials become uncoupled as

$$\nabla^2 \phi - \frac{1}{c^2} \frac{\partial^2 \phi}{\partial t^2} = -\frac{\|4\pi\|}{\epsilon_0} \rho, \qquad (3.15)$$

$$\nabla^2 \mathbf{a} - \frac{1}{c^2} \frac{\partial^2 \mathbf{a}}{\partial t^2} = -\left\| \frac{4\pi}{c} \right\| \mu_0 \mathbf{j}, \qquad (3.16)$$

---

[3] Although the erroneous attribution of this gauge condition to H. A. Lorentz was corrected by E. T. Whittaker [1] in 1951, the majority of texts have regrettably continued to attribute it to Lorentz and not to Lorenz who introduced the retarded potentials in a series of three articles beginning in 1867 [2]; see J. Van Bladel [3].

and one is left with two separate inhomogeneous wave equations. However, they are not completely independent as the charge and current densities are related through the equation of continuity $\nabla \cdot \mathbf{j} + \partial \rho / \partial t = 0$.

Even for a pair of potentials that satisfy the Lorenz condition (3.13) there still remains a certain degree of arbitrariness in their determination. In particular, the *restricted gauge transformation*

$$\mathbf{a}(\mathbf{r}, t) \quad \longrightarrow \quad \mathbf{a}'(\mathbf{r}, t) = \mathbf{a}(\mathbf{r}, t) + \nabla \psi(\mathbf{r}, t), \tag{3.17}$$

$$\phi(\mathbf{r}, t) \quad \longrightarrow \quad \phi'(\mathbf{r}, t) = \phi(\mathbf{r}, t) - \left\| \frac{1}{c} \right\| \frac{\partial \psi(\mathbf{r}, t)}{\partial t}, \tag{3.18}$$

where the gauge function $\psi(\mathbf{r}, t)$ satisfies the homogeneous wave equation

$$\nabla^2 \psi - \frac{1}{c^2} \frac{\partial^2 \psi}{\partial t^2} = 0, \tag{3.19}$$

preserves the Lorenz condition provided that the pair of potential functions $\{\mathbf{a}(\mathbf{r}, t), \phi(\mathbf{r}, t)\}$ satisfy it initially. All pairs of potentials in this restricted class are said to belong to the *Lorenz gauge*. The Lorenz gauge is commonly employed in both the physics and engineering communities; first, because it yields the pair of inhomogeneous wave equations (3.15) and (3.16) which treat the scalar and vector potentials on an equal level, and second, because it is independent of the particular coordinate system chosen, it fits naturally into the framework of the special theory of relativity.

3.1.2 The Coulomb Gauge

In the *Coulomb gauge* the vector potential is chosen to be solenoidal, so that

$$\nabla \cdot \mathbf{a}(\mathbf{r}, t) = 0. \tag{3.20}$$

This is also known as either the *transverse gauge* or as the *radiation gauge*. In this gauge the scalar potential satisfies Poisson's equation

$$\nabla^2 \phi = - \frac{\| 4\pi \|}{\epsilon_0} \rho \tag{3.21}$$

with solution

$$\phi(\mathbf{r}, t) = \frac{\| 4\pi \|}{4\pi \epsilon_0} \int \frac{\rho(\mathbf{r}', t)}{|\mathbf{r} - \mathbf{r}'|} d^3 r', \tag{3.22}$$

where the integration is taken over all space. Hence, the scalar potential in the Coulomb gauge is just the instantaneous Coulomb potential due to the microscopic charge distribution with density $\rho(\mathbf{r}, t)$.

The vector potential in the Coulomb gauge satisfies the inhomogeneous wave equation

$$\nabla^2 \mathbf{a} - \frac{1}{c^2}\frac{\partial^2 \mathbf{a}}{\partial t^2} = - \left\| \frac{4\pi}{c} \right\| \mu_0 \mathbf{j} + \frac{\epsilon_0 \mu_0}{\|c\|}\nabla\frac{\partial \phi}{\partial t}. \tag{3.23}$$

The apparent "current" term that appears in Eq. (3.23) can, in principle, be determined from the instantaneous Coulomb potential given in Eq. (3.22). With use of the equation of continuity [cf. Eq. (2.14)] in Eq. (3.22), there results

$$\nabla\frac{\partial \phi}{\partial t} = \frac{\|4\pi\|}{4\pi\epsilon_0}\nabla\int\frac{\partial \rho(\mathbf{r}', t)/\partial t}{|\mathbf{r}-\mathbf{r}'|}d^3 r'$$

$$= -\frac{\|4\pi\|}{4\pi\epsilon_0}\nabla\int\frac{\nabla'\cdot\mathbf{j}(\mathbf{r}', t)}{|\mathbf{r}-\mathbf{r}'|}d^3 r'. \tag{3.24}$$

Because this current term is irrotational, it may then cancel a corresponding term that is contained in the microscopic current density $\mathbf{j}(\mathbf{r}, t)$. From Helmholtz' theorem [4] (see Appendix B), the current density may be expressed as the sum of an irrotational and a solenoidal component as

$$\mathbf{j}(\mathbf{r}, t) = \mathbf{j}_\ell(\mathbf{r}, t) + \mathbf{j}_t(\mathbf{r}, t), \tag{3.25}$$

where $\mathbf{j}_\ell(\mathbf{r}, t)$ is referred to as the *longitudinal* or *irrotational current density* with $\nabla\times\mathbf{j}_\ell = \mathbf{0}$ and $\mathbf{j}_t(\mathbf{r}, t)$ as the *transverse* or *solenoidal current density* with $\nabla\cdot\mathbf{j}_t = 0$. From Helmholtz' theorem, these two current densities are uniquely determined from the microscopic current density $\mathbf{j}(\mathbf{r}, t)$ as

$$\mathbf{j}_\ell(\mathbf{r}, t) = -\frac{1}{4\pi}\nabla\int\frac{\nabla'\cdot\mathbf{j}(\mathbf{r}', t)}{|\mathbf{r}-\mathbf{r}'|}d^3 r', \tag{3.26}$$

$$\mathbf{j}_t(\mathbf{r}, t) = \frac{1}{4\pi}\nabla\times\nabla\times\int\frac{\mathbf{j}(\mathbf{r}', t)}{|\mathbf{r}-\mathbf{r}'|}d^3 r'. \tag{3.27}$$

Comparison of Eqs. (3.24) and (3.26) then yields the identification

$$\nabla\frac{\partial \phi(\mathbf{r}, t)}{\partial t} = \frac{\|4\pi\|}{\epsilon_0}\mathbf{j}_\ell(\mathbf{r}, t). \tag{3.28}$$

As a consequence, the source term for the inhomogeneous wave equation (3.23) of the vector potential in the Coulomb gauge may be expressed entirely in terms of the transverse current density given in Eq. (3.27), so that

$$\nabla^2 \mathbf{a} - \frac{1}{c^2}\frac{\partial^2 \mathbf{a}}{\partial t^2} = -\left\|\frac{4\pi}{c}\right\|\mu_0\mathbf{j}_t. \tag{3.29}$$

This result is the origin of the name "transverse gauge." The term "radiation gauge" has its origin in the fact that transverse radiation fields are given by the vector potential alone, the instantaneous Coulomb potential given in Eq. (3.22) contributing only to the near-field behavior. This gauge is of particular importance in quantum electrodynamics because its description of photons would then need to quantize only the vector potential. Finally, notice that Eq. (3.28) states that the scalar potential is determined by the longitudinal current density. In analogy with wave motion in elastic solids, *scalar potential waves are longitudinal waves* and *vector potential waves are transverse waves*.

The Coulomb or transverse gauge is particularly useful when no charge or current sources are present. In that case the instantaneous Coulomb potential vanishes everywhere [$\phi(\mathbf{r}, t) = 0$] and the vector potential $\mathbf{a}(\mathbf{r}, t)$ satisfies the homogeneous wave equation

$$\nabla^2 \mathbf{a} - \frac{1}{c^2} \frac{\partial^2 \mathbf{a}}{\partial t^2} = \mathbf{0}. \tag{3.30}$$

The microscopic electric and magnetic field vectors are then given by the pair of expressions

$$\mathbf{e}(\mathbf{r}, t) = - \left\| \frac{1}{c} \right\| \frac{\partial \mathbf{a}(\mathbf{r}, t)}{\partial t}, \tag{3.31}$$

$$\mathbf{b}(\mathbf{r}, t) = \nabla \times \mathbf{a}(\mathbf{r}, t). \tag{3.32}$$

It is important at this point to note an interesting peculiarity of the Coulomb gauge. The special theory of relativity requires that the electromagnetic field in vacuum have a propagation velocity that is at most the speed of light c in free space. However, Eq. (3.22) indicates that in the Coulomb gauge the scalar potential for the electromagnetic disturbance is "propagated" instantaneously everywhere in space. The vector potential in the Coulomb gauge, on the other hand, satisfies the wave equation given in Eq. (3.23) with its associated finite velocity of propagation c in free space. Notice further that in the Lorenz gauge, both potential functions satisfy a wave equation with finite propagation velocity c. The incongruity appearing in the scalar potential of the Coulomb gauge is then seen to be a mere consequence of the gauge choice and hence, has no bearing upon the resultant electromagnetic field which is independent of the particular choice of gauge. Indeed, it is not the potentials but the field vectors themselves (and the forces they exert on any given test particle) that are the physical quantities of interest in the theory, and these propagate with a finite velocity that does not exceed the speed of light c in free space.

As a final point of interest, return to the Lorenz gauge and consider any region of space where the charge density $\rho(\mathbf{r}, t)$ is zero. The scalar potential $\phi(\mathbf{r}, t)$ then satisfies the homogeneous wave equation

$$\nabla^2 \phi - \frac{1}{c^2} \frac{\partial^2 \phi}{\partial t^2} = 0$$

and a gauge function $\psi(\mathbf{r}, t)$ may then be chosen such that the scalar potential vanishes throughout the region. In particular, from Eqs. (3.12) and (3.19), it is only necessary to take the gauge function to be given by

$$\psi(\mathbf{r}, t) = \|c\| \int \phi(\mathbf{r}, t)\, dt, \tag{3.33}$$

because then

$$\phi(\mathbf{r}, t) \quad\longrightarrow\quad \phi'(\mathbf{r}, t) = \phi(\mathbf{r}, t) - \left\|\frac{1}{c}\right\| \frac{\partial \psi(\mathbf{r}, t)}{\partial t}$$

$$= \phi(\mathbf{r}, t) - \phi(\mathbf{r}, t) = 0. \tag{3.34}$$

The electromagnetic field vectors can then be derived from the vector potential alone, as given by the pair of relations in Eqs. (3.31) and (3.32), and the Lorenz condition (3.13) reduces to $\nabla \cdot \mathbf{a} = 0$, which is simply the condition given in Eq. (3.20) for the Coulomb gauge. Hence, in charge-free regions of space the Lorenz and Coulomb gauges become identical when the gauge function $\psi(\mathbf{r}, t)$ is chosen such that the scalar potential (in the Lorenz gauge) vanishes throughout the region.

3.1.3   The Retarded Potentials

The general solutions of the inhomogeneous wave equations given in Eqs. (3.15) and (3.16) for the scalar and vector potentials in the Lorenz gauge are given by the *retarded potentials*

$$\phi(\mathbf{r}, t) = \frac{\|4\pi\|}{4\pi\epsilon_0} \int_V \frac{\rho(\mathbf{r}', t - R/c)}{R} d^3 r', \tag{3.35}$$

$$\mathbf{a}(\mathbf{r}, t) = \frac{\|4\pi/c\|}{4\pi}\mu_0 \int_V \frac{\mathbf{j}(\mathbf{r}', t - R/c)}{R} d^3 r'. \tag{3.36}$$

Here $R \equiv |\mathbf{r}-\mathbf{r}'| = \sqrt{(x - x')^2 + (y - y')^2 + (z - z')^2}$ is the distance between the observation point $\mathbf{r}$ and the source point $\mathbf{r}'$ where the volume element of integration $d^3 r'$ is located, the integration being carried out throughout the volume V that includes all charge and current sources that produce the electromagnetic field under consideration. In the above expressions the argument $t - R/c$ denotes that the pair of Lorenz gauge potentials $\{\mathbf{a}(\mathbf{r}, t), \phi(\mathbf{r}, t)\}$ are due to the charge and current distributions at the *retarded time* $t - R/c$, where c is the vacuum speed of light. Because of this retardation effect, the pair of potentials $\{\mathbf{a}(\mathbf{r}, t), \phi(\mathbf{r}, t)\}$ are called *retarded potentials*.

Consider now proving[4] that these retarded potentials are indeed the special solutions to the inhomogeneous wave equations in the Lorenz gauge. Because of the similarity of the wave equations for the scalar and vector potentials, it is necessary to only consider the proof for the scalar potential that satisfies

$$\nabla^2 \phi - \frac{1}{c^2} \frac{\partial^2 \phi}{\partial t^2} = -\frac{\|4\pi\|}{\epsilon_0} \rho. \tag{3.37}$$

Let the integration region $\mathcal{V}$ appearing in Eq. (3.35) be divided into two distinct regions $\mathcal{V}_1$ and $\mathcal{V}_2$ such that $\mathcal{V} = \mathcal{V}_1 \cup \mathcal{V}_2$ and $\mathcal{V}_1 \cap \mathcal{V}_2 = \emptyset$, where $\emptyset$ denotes the empty set, and where $\mathcal{V}_1$ is an arbitrarily small volume surrounding the point $\mathbf{r}$ at which the potential is to be observed (the observation or field point). The scalar potential $\phi(\mathbf{r}, t)$ is then composed of two parts

$$\phi(\mathbf{r}, t) = \phi_1(\mathbf{r}, t) + \phi_2(\mathbf{r}, t), \tag{3.38}$$

where

$$\phi_i(\mathbf{r}, t) = \frac{\|4\pi\|}{4\pi\epsilon_0} \int_{\mathcal{V}_i} \frac{\rho(\mathbf{r}', t - R/c)}{R} d^3 r' \tag{3.39}$$

for $i = 1, 2$. Let the region $\mathcal{V}_1$ containing the field point $\mathbf{r}$ be sufficiently small such that retardation effects are negligible for all source points within that region. In the limit as $\mathcal{V}_1 \to 0$ one then has that

$$\rho(\mathbf{r}'.t - R/c) \to \rho(\mathbf{r}', t); \quad \mathbf{r}' \in \mathcal{V}_1 \tag{3.40}$$

and the expression (3.39) for $\phi_1(\mathbf{r}, t)$ becomes

$$\phi_1(\mathbf{r}, t) = \frac{\|4\pi\|}{4\pi\epsilon_0} \int_{\mathcal{V}_1} \frac{\rho(\mathbf{r}', t)}{R} d^3 r'. \tag{3.41}$$

Because this equation is identical with the instantaneous Coulomb potential [cf. Eq. (3.22)], it follows that $\phi_1(\mathbf{r}, t)$ is the solution of Poisson's equation

$$\nabla^2 \phi_1 = -\frac{\|4\pi\|}{\epsilon_0} \rho. \tag{3.42}$$

In the region $\mathcal{V}_2$, $R > 0$ and it is permissible to differentiate $\phi_2(\mathbf{r}, t)$ under the integral sign. Because the Laplacian operator ∇^2 acts only upon the unprimed coordinates and because the integrand $\rho(\mathbf{r}', t - R/c)/R$ depends upon these coordinates only through the radial distance R from the observation point to the

---

[4]This proof was first given by Georg Friedrich Bernhard Riemann in 1858.

source point, one may then use spherical coordinates to evaluate the Laplacian of $\rho(\mathbf{r}', t - R/c)/R$ as

$$\nabla^2 \left(\frac{\rho}{R} \right) = \frac{1}{R^2} \frac{\partial}{\partial R} \left[R^2 \frac{\partial}{\partial R} \left(\frac{\rho}{R} \right) \right] = \frac{1}{R} \frac{\partial^2 \rho}{\partial R^2}.$$

Because

$$\frac{\partial^2 \rho(t - R/c)}{\partial R^2} = \frac{1}{c^2} \frac{\partial^2 \rho(t - R/c)}{\partial t^2},$$

one finally has that

$$\nabla^2 \left(\frac{\rho(t - R/c)}{R} \right) = \frac{1}{Rc^2} \frac{\partial^2 \rho(t - R/c)}{\partial t^2}.$$

With these expressions there then results

$$\nabla^2 \phi_2 - \frac{1}{c^2} \frac{\partial^2 \phi_2}{\partial t^2}$$

$$= \frac{\|4\pi\|}{4\pi \epsilon_0} \int_{V_2} \left[\nabla^2 \left(\frac{\rho(\mathbf{r}', t - R/c)}{R} \right) - \frac{1}{Rc^2} \frac{\partial^2 \rho(\mathbf{r}', t - R/c)}{\partial t^2} \right] d^3 r'$$

$$= 0. \tag{3.43}$$

Hence, in the limit as $V_1 \to 0$, Eqs. (3.38), (3.41), and (3.43) yield

$$\nabla^2 \phi(\mathbf{r}, t) = \nabla^2 \left(\phi_1(\mathbf{r}, t) + \phi_2(\mathbf{r}, t) \right) = -\frac{\|4\pi\|}{\epsilon_0} \rho(\mathbf{r}, t) + \frac{1}{c^2} \frac{\partial^2 \phi(\mathbf{r}, t)}{\partial t^2},$$

so that

$$\nabla^2 \phi(\mathbf{r}, t) - \frac{1}{c^2} \frac{\partial^2 \phi(\mathbf{r}, t)}{\partial t^2} = -\frac{\|4\pi\|}{\epsilon_0} \rho(\mathbf{r}, t), \tag{3.44}$$

and the retarded potential given in Eq. (3.35) satisfies the inhomogeneous wave equation (3.15) for the scalar potential in the Lorenz gauge. Notice that the term $-(\|4\pi\|/\epsilon_0)\rho(\mathbf{r}, t)$ arises from the observation point whereas the term $(1/c^2)\partial^2 \phi(\mathbf{r}, t)/\partial t^2$ arises from everywhere else but the observation point. In a strictly analogous manner, one can show that each Cartesian component of the vector potential appearing in Eq. (3.36) satisfies the appropriate component of the inhomogeneous wave equation for the vector potential in the Lorenz gauge given in Eq. (3.16). Hence, Eq. (3.36) satisfies the inhomogeneous wave equation for the vector potential in the Lorenz gauge. Moreover, on account of the equation of continuity [Eq. (2.14)], these solutions also satisfy the Lorenz condition given in Eq. (3.13).

The expressions appearing in Eqs. (3.35) and (3.36) show that one may regard the vector and scalar potentials $\{\mathbf{a}(\mathbf{r}, t), \phi(\mathbf{r}, t)\}$ in the Lorenz gauge as arising from contributions from each volume element of space, a typical element d^3r' contributing the amounts $\|4\pi/c\|(\mu_0/4\pi)\mathbf{j}(\mathbf{r}', t - R/c)$ and $(\|4\pi\|/4\pi\epsilon_0)\rho(\mathbf{r}', t - R/c)$ to $\mathbf{a}(\mathbf{r}, t)$ and $\phi(\mathbf{r}, t)$, respectively. The quantity R/c is precisely the time needed for an electromagnetic disturbance to propagate from the source point at $\mathbf{r}'$ to the observation (or field) point at $\mathbf{r}$ to contribute to the field at the required time t. For this reason, these expressions [Eqs. (3.35) and (3.36)] are called *retarded potentials*. It is also possible to construct solutions in the form of *advanced potentials* with the time argument $t + R/c$ appearing in the integrand. Both types of potential satisfy the inhomogeneous wave equations (3.15) and (3.16) in the Lorenz gauge and are mathematically admissible; however, only the retarded potentials that refer to charge and current sources at earlier times correspond to the physical law of causality so that the advanced potentials are not considered.

The retarded potentials given in Eqs. (3.35) and (3.36) represent a special solution of the inhomogeneous wave equations (3.15) and (3.16) in the Lorenz gauge, namely that which arises from the given distribution of charge and current sources with densities $\rho(\mathbf{r}, t)$ and $\mathbf{j}(\mathbf{r}, t) = \rho(\mathbf{r}, t)\mathbf{v}(\mathbf{r}, t)$. The general solution is obtained by adding to these the general solutions of the homogeneous wave equations

$$\nabla^2\phi - \frac{1}{c^2}\frac{\partial^2\phi}{\partial t^2} = 0, \tag{3.45}$$

$$\nabla^2\mathbf{a} - \frac{1}{c^2}\frac{\partial^2\mathbf{a}}{\partial t^2} = \mathbf{0}, \tag{3.46}$$

where $\phi(\mathbf{r}, t)$ and $\mathbf{a}(\mathbf{r}, t)$ are again subject to the Lorenz condition.

For slowly-varying systems the retarded potentials in the Lorenz gauge reduce to the static potentials. The fields due to a localized charge or current distribution about the origin then fall off as R^{-2} or faster in the limit as $R \to \infty$ because the potentials behave asymptotically as

$$\phi(\mathbf{r}, t) \sim \frac{\|4\pi\|}{4\pi\epsilon_0 R}\int_\mathcal{V}\rho(\mathbf{r}', t)d^3r', \tag{3.47}$$

$$\mathbf{a}(\mathbf{r}, t) \sim \frac{\|4\pi/c\|}{4\pi R}\mu_0\int_\mathcal{V}\mathbf{j}(\mathbf{r}', t)d^3r', \tag{3.48}$$

where the radial distance R from the origin to the field point is much larger than the maximum radial extent of the source region $\mathcal{V}$ from the origin. On the other hand, if the charge distribution in $\mathcal{V}$ varies rapidly with time, then although the vector potential $\mathbf{a}(\mathbf{r}, t)$ still falls off as R^{-1} for large R, the electric field vector $\mathbf{e}(\mathbf{r}, t)$ contains a term proportional to $-\partial\mathbf{a}/\partial t$ that also decays as R^{-1}. There is a similar term in the magnetic field vector $\mathbf{h}(\mathbf{r}, t)$ as required by the second of Maxwell's equations [Eq. (3.2)]. Thus, in a system that is oscillating with a period

T the asymptotic behavior of the microscopic $\mathbf{e}$ and $\mathbf{h}$ fields at sufficiently large distances R varies as $(RT)^{-1}$; this part of the field is responsible for radiation.

3.2 The Hertz Potential and Elemental Dipole Radiation

The analysis of the preceding section has shown that the microscopic electromagnetic field vectors may be specified by the scalar and vector potentials, thereby reducing the component dimensionality of the theory from six $\{\mathbf{e}, \mathbf{b}\}$ to four $\{\mathbf{a}, \phi\}$ components. However, because these potential functions are, for example, connected by the Lorenz condition (3.13) when they are expressed in the Lorenz gauge, it then follows that it is possible to express the electromagnetic field vectors in terms of a single vector function, known as the Hertz potential,[5] thereby further reducing the component dimensionality of the theory to three.

3.2.1 The Hertz Potential

Consider the vector and scalar potential fields in the Lorenz gauge, viz.

$$\nabla^2 \mathbf{a} - \frac{1}{c^2}\frac{\partial^2 \mathbf{a}}{\partial t^2} = -\left\|\frac{4\pi}{c}\right\| \mu_0 \mathbf{j}, \tag{3.49}$$

$$\nabla^2 \phi - \frac{1}{c^2}\frac{\partial^2 \phi}{\partial t^2} = -\frac{\|4\pi\|}{\epsilon_0}\rho, \tag{3.50}$$

where the microscopic current and charge densities satisfy the continuity equation

$$\nabla \cdot \mathbf{j} + \frac{\partial \rho}{\partial t} = 0. \tag{3.51}$$

As an immediate consequence of this relation, there exists a vector function $\mathbf{p}(\mathbf{r}, t)$ such that

$$\mathbf{j}(\mathbf{r}, t) = \frac{\partial \mathbf{p}(\mathbf{r}, t)}{\partial t}, \tag{3.52}$$

$$\rho(\mathbf{r}, t) = -\nabla \cdot \mathbf{p}(\mathbf{r}, t), \tag{3.53}$$

---

[5]The Hertz potential is a generalization of a potential function that was introduced for the electromagnetic field of an oscillating dipole by H. Hertz [5] in 1889. The vector character of the Hertz potential was first noted by A. Righi [6] in 1901.

in which case Eq. (3.51) is identically satisfied. This vector field $\mathbf{p}(\mathbf{r}, t)$ is called the *electric moment*, in terms of which the above pair of wave equations become

$$\nabla^2 \mathbf{a} - \frac{1}{c^2} \frac{\partial^2 \mathbf{a}}{\partial t^2} = - \left\| \frac{4\pi}{c} \right\| \mu_0 \frac{\partial \mathbf{p}}{\partial t}, \tag{3.54}$$

$$\nabla^2 \phi - \frac{1}{c^2} \frac{\partial^2 \phi}{\partial t^2} = \frac{\|4\pi\|}{\epsilon_0} \nabla \cdot \mathbf{p}. \tag{3.55}$$

In a similar manner, the vector and scalar potentials may be expressed as

$$\mathbf{a}(\mathbf{r}, t) = \frac{\mu_0}{\|c\|} \frac{\partial \mathbf{\Pi}(\mathbf{r}, t)}{\partial t}, \tag{3.56}$$

$$\phi(\mathbf{r}, t) = -\frac{1}{\epsilon_0} \nabla \cdot \mathbf{\Pi}(\mathbf{r}, t), \tag{3.57}$$

which automatically satisfy the Lorenz condition (3.13). With this substitution, the inhomogeneous wave equations (3.54) and (3.55) become

$$\frac{\partial}{\partial t} \left[\nabla^2 \mathbf{\Pi} - \frac{1}{c^2} \frac{\partial^2 \mathbf{\Pi}}{\partial t^2} + \|4\pi\| \mathbf{p} \right] = \mathbf{0},$$

$$\nabla \cdot \left[\nabla^2 \mathbf{\Pi} - \frac{1}{c^2} \frac{\partial^2 \mathbf{\Pi}}{\partial t^2} + \|4\pi\| \mathbf{p} \right] = 0.$$

This pair of equations will be satisfied if

$$\nabla^2 \mathbf{\Pi} - \frac{1}{c^2} \frac{\partial^2 \mathbf{\Pi}}{\partial t^2} = -\|4\pi\| \mathbf{p}. \tag{3.58}$$

The vector field $\mathbf{\Pi}(\mathbf{r}, t)$ is called the *Hertz vector* or *Hertz potential*. [5]

The above analysis shows that the Hertz vector $\mathbf{\Pi}(\mathbf{r}, t)$ determines an electromagnetic field through its vector and scalar potential functions, but it does not show that a given electromagnetic field can be represented in terms of the vector field $\mathbf{\Pi}(\mathbf{r}, t)$. In order to accomplish this, it must be proven that given the vector and scalar potentials for an electromagnetic field, the vector field $\mathbf{\Pi}(\mathbf{r}, t)$ is "uniquely defined." To that end, let $\phi_0(\mathbf{r}) \equiv \phi(\mathbf{r}, 0)$ be the value of the scalar potential $\phi(\mathbf{r}, t)$ when $t = 0$ and let $\mathbf{\Pi}_0(\mathbf{r})$ be any vector function of position that is independent of the time t such that $\nabla \cdot \mathbf{\Pi}_0 = -\epsilon_0 \phi_0$. Define the time-dependent vector field $\mathbf{\Pi}'(\mathbf{r}, t)$ by the relation

$$\mathbf{\Pi}'(\mathbf{r}, t) \equiv \mathbf{\Pi}_0(\mathbf{r}) + \frac{\|c\|}{\mu_0} \int_0^t \mathbf{a}(\mathbf{r}, t') \, dt', \tag{3.59}$$

130

so that

$$\mathbf{a}(\mathbf{r}, t) = \frac{\mu_0}{\|c\|} \frac{\partial \mathbf{\Pi}'(\mathbf{r}, t)}{\partial t} \tag{3.60}$$

by straightforward differentiation of Eq. (3.59), and

$$\nabla \cdot \mathbf{\Pi}'(\mathbf{r}, t) = \nabla \cdot \mathbf{\Pi}_0(\mathbf{r}) + \frac{\|c\|}{\mu_0} \int_0^t \nabla \cdot \mathbf{a}(\mathbf{r}, t')\, dt'$$

$$= -\epsilon_0 \phi_0(\mathbf{r}) - \epsilon_0 \int_0^t \frac{\partial \phi(\mathbf{r}, t')}{\partial t'} dt' = -\epsilon_0 \phi(\mathbf{r}, t), \tag{3.61}$$

where the Lorenz condition (3.13) has been invoked. From the pair of relations in Eqs. (3.54) and (3.55) one then obtains, by direct substitution,

$$\frac{\partial}{\partial t}\left[\nabla^2 \mathbf{\Pi}' - \frac{1}{c^2}\frac{\partial^2 \mathbf{\Pi}'}{\partial t^2}\right] = -\|4\pi\|\frac{\partial \mathbf{p}}{\partial t},$$

$$\nabla \cdot \left[\nabla^2 \mathbf{\Pi}' - \frac{1}{c^2}\frac{\partial^2 \mathbf{\Pi}'}{\partial t^2}\right] = -\|4\pi\|\nabla \cdot \mathbf{p},$$

and consequently

$$\nabla^2 \mathbf{\Pi}' - \frac{1}{c^2}\frac{\partial^2 \mathbf{\Pi}'}{\partial t^2} + \|4\pi\|\mathbf{p} = \nabla \times \mathbf{\Xi}, \tag{3.62}$$

where $\mathbf{\Xi}(\mathbf{r})$ is a vector function of position that is independent of the time t. Let

$$\mathbf{\Pi}'(\mathbf{r}, t) = \mathbf{\Pi}(\mathbf{r}, t) + \nabla \times \mathbf{\Psi}(\mathbf{r}), \tag{3.63}$$

where $\mathbf{\Psi}(\mathbf{r})$ is a time-independent vector function of position such that

$$\nabla^2 \mathbf{\Psi}(\mathbf{r}) = \mathbf{\Xi}(\mathbf{r}) \tag{3.64}$$

as required by Eq. (3.62). The vector field $\mathbf{\Pi}'(\mathbf{r}, t)$ then satisfies Eq. (3.58), and because

$$\nabla \cdot \mathbf{\Pi}'(\mathbf{r}, t) = \nabla \cdot \mathbf{\Pi}(\mathbf{r}, t), \tag{3.65}$$

$$\frac{\partial \mathbf{\Pi}'(\mathbf{r}, t)}{\partial t} = \frac{\partial \mathbf{\Pi}(\mathbf{r}, t)}{\partial t}, \tag{3.66}$$

then Eqs. (3.60) and (3.61) show that $\mathbf{\Pi}'(\mathbf{r}, t)$ satisfies the two relations given in Eqs. (3.56) and (3.57). Hence, for a given electromagnetic field one can always determine a vector field $\mathbf{\Pi}(\mathbf{r}, t)$ such that

$$\mathbf{e}(\mathbf{r}, t) = \frac{1}{\epsilon_0}\nabla(\nabla \cdot \mathbf{\Pi}(\mathbf{r}, t)) - \frac{\mu_0}{\|c^2\|}\frac{\partial^2 \mathbf{\Pi}(\mathbf{r}, t)}{\partial t^2}, \tag{3.67}$$

$$\mathbf{b}(\mathbf{r}, t) = \frac{\mu_0}{\|c\|} \nabla \times \frac{\partial \mathbf{\Pi}(\mathbf{r}, t)}{\partial t}, \tag{3.68}$$

where $\mathbf{\Pi}(\mathbf{r}, t)$ satisfies the inhomogeneous wave equation given in Eq. (3.58) and is the Hertz potential for the microscopic electromagnetic field. This Hertz potential is a superpotential in the sense that the electromagnetic field, when described by $\mathbf{\Pi}(\mathbf{r}, t)$, is given by a three-component equation, whereas the vector and scalar potential pair $\{\mathbf{a}(\mathbf{r}, t), \phi(\mathbf{r}, t)\}$ amount to a four-component potential.

The Hertz potential $\mathbf{\Pi}(\mathbf{r}, t)$ is not uniquely determined because the vector and scalar potentials are invariant under the *gauge transformation*

$$\mathbf{\Pi}(\mathbf{r}, t) \rightarrow \mathbf{\Pi}'(\mathbf{r}, t) = \mathbf{\Pi}(\mathbf{r}, t) + \nabla \times \mathbf{\Gamma}(\mathbf{r}), \tag{3.69}$$

where $\mathbf{\Gamma}(\mathbf{r})$ is an arbitrary vector function of position that is independent of the time t. An even greater degree of arbitrariness is present here because $\phi(\mathbf{r}, t)$ and $\mathbf{a}(\mathbf{r}, t)$ are themselves not uniquely determined in the chosen Lorenz gauge because of the restricted gauge transformation given in Eqs. (3.17) and (3.18). As a consequence, the microscopic electromagnetic field vectors are invariant under the more general *gauge transformation* [7]

$$\mathbf{\Pi}(\mathbf{r}, t) \rightarrow \mathbf{\Pi}'(\mathbf{r}, t) = \mathbf{\Pi}(\mathbf{r}, t) + \nabla \times \mathbf{\Gamma}(\mathbf{r}) - \frac{1}{c} \frac{\partial \mathbf{\Lambda}(\mathbf{r}, t)}{\partial t} - \nabla \lambda(\mathbf{r}, t), \tag{3.70}$$

where $\mathbf{\Gamma}(\mathbf{r})$, $\mathbf{\Lambda}(\mathbf{r}, t)$, and $\lambda(\mathbf{r}, t)$ are arbitrary functions. Under this transformation the vector and scalar potentials given in Eqs. (3.56) and (3.57) undergo the gauge transformation

$$\mathbf{a}(\mathbf{r}, t) \rightarrow \mathbf{a}'(\mathbf{r}, t) = \mathbf{a}(\mathbf{r}, t) - \frac{\mu_0}{\|c\|c} \frac{\partial^2 \mathbf{\Lambda}(\mathbf{r}, t)}{\partial t^2} - \frac{\mu_0}{\|c\|} \nabla \frac{\partial \lambda(\mathbf{r}, t)}{\partial t}, \tag{3.71}$$

$$\phi(\mathbf{r}, t) \rightarrow \phi'(\mathbf{r}, t) = \phi(\mathbf{r}, t) + \frac{1}{\epsilon_0 c} \nabla \cdot \frac{\partial \mathbf{\Lambda}(\mathbf{r}, t)}{\partial t} + \frac{1}{\epsilon_0} \nabla^2 \lambda(\mathbf{r}, t). \tag{3.72}$$

The Lorenz condition (3.13) is satisfied by the pair of transformed potentials $\{\mathbf{a}'(\mathbf{r}, t), \phi'(\mathbf{r}, t)\}$ if it is satisfied by the pair $\{\mathbf{a}(\mathbf{r}, t), \phi(\mathbf{r}, t)\}$, because

$$\nabla \cdot \mathbf{a}' + \frac{\epsilon_0 \mu_0}{\|c\|} \frac{\partial \phi'}{\partial t} = \nabla \cdot \mathbf{a} + \frac{\epsilon_0 \mu_0}{\|c\|} \frac{\partial \phi}{\partial t} - \frac{\mu_0}{\|c\|c} \left(\nabla \cdot \frac{\partial^2 \mathbf{\Lambda}}{\partial t^2} - \frac{\partial^2 (\nabla \cdot \mathbf{\Lambda})}{\partial t^2} \right)$$

$$- \frac{\mu_0}{\|c\|} \left(\nabla^2 \frac{\partial \lambda}{\partial t} - \frac{\partial (\nabla^2 \lambda)}{\partial t} \right) = 0. \tag{3.73}$$

Upon comparison of this equation with the expressions given in Eqs. (3.71) and (3.72), the gauge function $\psi(\mathbf{r}, t)$ appearing in Eqs. (3.11) and (3.12) is seen to

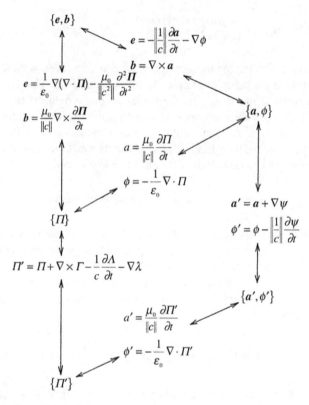

Fig. 3.2 Network structure of the restricted gauge transformation for the vector and scalar potentials in the Lorenz gauge and the Hertz vector

satisfy the pair of relations

$$\nabla \psi = -\frac{\mu_0}{\|c\|} \left(\frac{1}{c} \frac{\partial^2 \mathbf{\Lambda}}{\partial t^2} + \nabla \frac{\partial \lambda}{\partial t} \right), \tag{3.74}$$

$$\frac{\partial \psi}{\partial t} = -\frac{\|c\|}{\epsilon_0} \left(\frac{1}{c} \nabla \cdot \frac{\partial \mathbf{\Lambda}}{\partial t} + \nabla^2 \lambda \right), \tag{3.75}$$

from which it is seen that $\psi(\mathbf{r}, t)$ satisfies the homogeneous wave equation (3.19). The interrelationship among the electric and magnetic field vectors, the scalar and vector potentials, and the Hertz vector under these various gauge transformations within the Lorenz gauge is illustrated in Fig. 3.2. Because the gauge transformation for the Hertz potential leaves the scalar and vector potentials in the Lorenz gauge, it then corresponds to a restricted gauge transformation for the scalar and vector potential functions.

3.2.2   Radiation from an Elemental Hertzian Dipole

As a direct application of the Hertz potential, consider the microscopic electromagnetic field produced by an elemental linear electric dipole situated at the fixed point $\mathbf{r} = \mathbf{r}_0$ and oscillating along a fixed direction specified by the unit vector $\hat{\mathbf{d}}$. Such an *ideal point dipole* is characterized by the electric moment

$$\mathbf{p}(\mathbf{r}, t) = p(t)\delta(\mathbf{r} - \mathbf{r}_0)\hat{\mathbf{d}}, \tag{3.76}$$

where $\delta(\mathbf{r})$ denotes the Dirac delta function (see Appendix A). The particular solution of the inhomogeneous wave equation (3.58) for the Hertz potential may be expressed in the form given in Eqs. (3.35) and (3.36) for the retarded potentials as

$$\mathbf{\Pi}(\mathbf{r}, t) = \frac{\|4\pi\|}{4\pi} \int \frac{\mathbf{p}(\mathbf{r}', t - R/c)}{|\mathbf{r} - \mathbf{r}'|} d^3 r', \tag{3.77}$$

so that, with substitution from Eq. (3.76), the Hertz vector for a point dipole is found as

$$\mathbf{\Pi}(\mathbf{r}, t) = \frac{\|4\pi\|}{4\pi R} p(t - R/c)\hat{\mathbf{d}}, \tag{3.78}$$

where $R = |\mathbf{r} - \mathbf{r}_0|$. The microscopic electric and magnetic field vectors are then given by Eqs. (3.67) and (3.68) where [making note of the mathematical equivalence $\nabla p(t - R/c) = (\partial p(t - R/c)/\partial t)(-\nabla R/c)$ with $\nabla R = \mathbf{R}/R$]

$$\nabla \cdot \left(\frac{1}{R} p(t - R/c)\hat{\mathbf{d}} \right) = p(t - R/c) \left(\nabla \frac{1}{R} \right) \cdot \hat{\mathbf{d}} + \frac{1}{R} \left(\nabla p(t - R/c) \right) \cdot \hat{\mathbf{d}}$$

$$= - \left(\frac{p(t - R/c)}{R^3} + \frac{1}{cR^2} \frac{\partial p(t - R/c)}{\partial t} \right) (\mathbf{R} \cdot \hat{\mathbf{d}})$$

so that

$$\frac{4\pi}{\|4\pi\|} \nabla (\nabla \cdot \mathbf{\Pi}) = \left\{ \frac{3}{R^5} p + \frac{3}{cR^4} \frac{\partial p}{\partial t} + \frac{1}{c^2 R^3} \frac{\partial^2 p}{\partial t^2} \right\} (\hat{\mathbf{d}} \cdot \mathbf{R}) \mathbf{R}$$

$$- \left\{ \frac{1}{R^3} p + \frac{1}{cR^2} \frac{\partial p}{\partial t} \right\} \hat{\mathbf{d}},$$

and

$$\frac{4\pi}{\|4\pi\|} \nabla \times \mathbf{\Pi} = \left\{ \frac{1}{R^3} p + \frac{1}{cR^2} \frac{\partial p}{\partial t} \right\} (\hat{\mathbf{d}} \times \mathbf{R}),$$

where all quantities on the right-hand side of the above equations are evaluated at the retarded time $t - R/c$. With these substitutions, Eqs. (3.67) and (3.68) become

$$\mathbf{e}(\mathbf{r}, t) = \frac{\|4\pi\|}{4\pi \epsilon_0} \left\{ \frac{3}{R^5} p + \frac{3}{cR^4} \frac{\partial p}{\partial t} + \frac{1}{c^2 R^3} \frac{\partial^2 p}{\partial t^2} \right\} (\hat{\mathbf{d}} \cdot \mathbf{R}) \mathbf{R}$$

$$- \frac{\|4\pi\| \mu_0}{4\pi c^2} \left\{ \frac{1}{R^3} p + \frac{1}{cR^2} \frac{\partial p}{\partial t} + \frac{1}{c^2 R} \frac{\partial^2 p}{\partial t^2} \right\} \hat{\mathbf{d}}, \qquad (3.79)$$

$$\mathbf{b}(\mathbf{r}, t) = \left\| \frac{4\pi}{c} \right\| \frac{\mu_0}{4\pi} \left\{ \frac{1}{R^3} \frac{\partial p}{\partial t} + \frac{1}{cR^2} \frac{\partial^2 p}{\partial t^2} \right\} (\hat{\mathbf{d}} \times \mathbf{R}). \qquad (3.80)$$

These are then the electromagnetic field vectors for an elemental Hertzian dipole, where $p = p(t - R/c)$.

For a time-harmonic point dipole with electric moment $p(t) = \Re\left\{ \tilde{p}_0 e^{-i\omega t} \right\}$ one finds that $p(t - R/c) = \Re\left\{ \tilde{p}_0 e^{i(kR - \omega t)} \right\}$. The field vectors are then also time-harmonic with angular frequency $\omega = 2\pi f$ and wavenumber $k = \omega/c$ with wavelength $\lambda = 2\pi/k = c/f$. The evolution of the electric field streamlines in the near-field region $kr \overset{<}{\sim} 1$ is illustrated in Fig. 3.3 at the successive instants

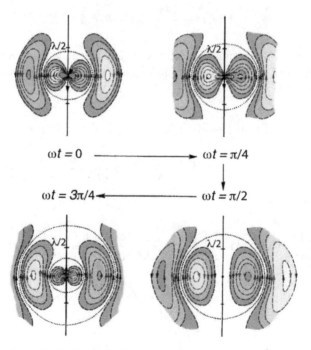

Fig. 3.3 Streamlines of the electric field vector in the near-field ($kr \overset{<}{\sim} 1$) of an ideal time-harmonic linear electric dipole oscillator at successive instants of time. The dotted circles indicate the contours of zero field strength

$\omega t = 0, \pi/4, \pi/2, 3\pi/4$ when the dipole axis is oriented along the polar axis. This same figure sequence with the streamline directions reversed also applies when $\omega t = \pi, 5\pi/4, 3\pi/2, 7\pi/4$, so that a complete cycle in time is obtained. The detailed properties of this radiation field are described in Sect. 3.4 following a more general derivation.

3.3   The Liénard–Wiechert Potentials and the Field of a Moving Charged Particle

Specific attention is now turned to the fundamental source of all electromagnetic radiation, that being the accelerated motion of charged particles. It is ultimately desired to be able to calculate the electric and magnetic field vectors produced by charged particles whose relative motion is described in a given inertial reference frame. In order to accomplish this one may employ the retarded potentials given in Eqs. (3.35) and (3.36); however, in order to perform the required integrations it is necessary to know the detailed motion of the charged particles because the calculations depend explicitly upon their positions and velocities at the retarded time $t - R/c$. It is therefore desirable to obtain expressions for the vector and scalar potentials $\{\mathbf{a}(\mathbf{r}, t), \phi(\mathbf{r}, t)\}$ that explicitly display their dependence on the velocity and acceleration of a given charged particle, and this is provided by the Liénard–Wiechert potentials [8, 9].

3.3.1   The Liénard–Wiechert Potentials

Consider a single point particle of charge q whose motion traces out a trajectory described by the radius vector $\mathbf{r}_q(t')$ relative to some fixed origin O in an inertial reference frame, as depicted in Fig. 3.4. The calculation of the resultant scalar potential function at any given fixed observation point P with position vector $\mathbf{r}$ according to the expressions given in Eq. (3.35) requires a retarded time integration over the entire region of space containing charge that contributes to $\phi(\mathbf{r}, t)$. For the single point charge q one may express the retarded time calculation in terms of the Dirac delta function as

$$\phi(\mathbf{r}, t) = \frac{\|4\pi\|}{4\pi\epsilon_0} q \int_{-\infty}^{\infty} \frac{\delta(t' - t + |\mathbf{r} - \mathbf{r}_q(t')|/c)}{|\mathbf{r} - \mathbf{r}_q(t')|} dt'. \tag{3.81}$$

In order to evaluate this integral one needs to perform a change of variable so that the variable of integration is the same as the argument of the delta function. This variable change is necessary because the position vector $\mathbf{r}_q(t')$ that describes the

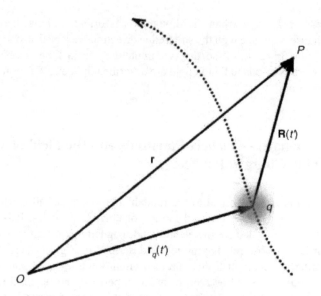

Fig. 3.4 Trajectory of a single point charge q that is described by the position vector $\mathbf{r}_q(t')$ with fixed origin O. The field produced by this charge is observed at the fixed observation (or field) point P with position vector $\mathbf{r}$

path of the point charge q is an explicit function of the time variable. Define then the new variable of integration

$$t'' \equiv t' - t + \frac{1}{c}|\mathbf{r} - \mathbf{r}_q(t')|. \tag{3.82}$$

Upon differentiation of this expression and taking note of the fact that $dt = 0$ as t is the fixed time of observation, one obtains

$$dt'' = \left(1 + \frac{1}{c}\frac{d}{dt'}|\mathbf{r} - \mathbf{r}_q(t')|\right) dt'. \tag{3.83}$$

The scalar quantity $|\mathbf{r} - \mathbf{r}_q(t')|$ is given by the distance expression

$$|\mathbf{r} - \mathbf{r}_q(t')| = \left[\sum_{i=1}^{3} (x_i - x_{qi}(t'))^2\right]^{1/2}, \tag{3.84}$$

where $x_{qi} = x_{qi}(t')$, $i = 1, 2, 3$ are the cartesian coordinates of the source point at time t' and where the coordinates x_i are constant as they represent the position of the fixed observation (or field) point P. Notice that the quantity $|\mathbf{r} - \mathbf{r}_q(t')|$ is only an implicit function of t' through each of the coordinate values $x_{qi} = x_{qi}(t')$ which are themselves explicit functions of t'. The derivative of $|\mathbf{r} - \mathbf{r}_q(t')|/c$ with respect to t' is then given by

$$\frac{1}{c}\frac{d}{dt'}|\mathbf{r} - \mathbf{r}_q(t')| = \frac{1}{c}\sum_{i=1}^{3}\left(\frac{\partial}{\partial x_{qi}}|\mathbf{r} - \mathbf{r}_q(t')|\right)\frac{dx_{qi}}{dt'}$$

$$= \frac{1}{c}\left(\nabla_q|\mathbf{r} - \mathbf{r}_q(t')|\right)\cdot\frac{d\mathbf{r}_q}{dt'}, \qquad (3.85)$$

where the subscript q appearing on the gradient operator in this expression indicates that the partial differentiation operations are to be taken with respect to the coordinates of the charged particle, so that

$$\nabla_q|\mathbf{r} - \mathbf{r}_q(t')| = -\frac{\mathbf{r} - \mathbf{r}_q(t')}{|\mathbf{r} - \mathbf{r}_q(t')|} = -\frac{\mathbf{R}(t')}{R(t')}. \qquad (3.86)$$

The derivative of the position vector $\mathbf{r}_q(t')$ with respect to the time variable t' is just the velocity $\mathbf{v}(t')$ of the point charge. With the definition of the *normalized velocity* as

$$\mathbf{\Upsilon}(t') \equiv \frac{\mathbf{v}(t')}{c} \qquad (3.87)$$

with magnitude $\beta(t') \equiv |\mathbf{\Upsilon}(t')| = v(t')/c$, Eq. (3.85) becomes

$$\frac{1}{c}\frac{d}{dt'}|\mathbf{r} - \mathbf{r}_q(t')| = -\frac{\mathbf{\Upsilon}(t')\cdot\mathbf{R}(t')}{R(t')}. \qquad (3.88)$$

Substitution of this expression in Eq. (3.83) then yields

$$dt'' = \left(1 - \frac{\mathbf{\Upsilon}(t')\cdot\mathbf{R}(t')}{R(t')}\right)dt' \qquad (3.89)$$

so that

$$dt' = \frac{R(t')}{R(t') - \mathbf{\Upsilon}(t')\cdot\mathbf{R}(t')}dt''. \qquad (3.90)$$

With the change of variable defined in Eq. (3.82), the expression given in Eq. (3.81) for the scalar potential becomes

$$\phi(\mathbf{r}, t) = \frac{\|4\pi\|}{4\pi\epsilon_0}q\int_{-\infty}^{\infty}\delta(t'')\frac{1}{R(t')}\left(\frac{R(t')}{R(t') - \mathbf{\Upsilon}(t')\cdot\mathbf{R}(t')}\right)dt''$$

$$= \frac{\|4\pi\|}{4\pi\epsilon_0}\left[\frac{q}{R(t') - \mathbf{\Upsilon}(t')\cdot\mathbf{R}(t')}\right]_{t''=0}.$$

Because $t'' = 0$ implies that $t' = t - R(t')/c$, then

$$\phi(\mathbf{r}, t) = \frac{\|4\pi\|}{4\pi\epsilon_0} \frac{q}{R(t') - \mathbf{\Upsilon}(t') \cdot \mathbf{R}(t')}, \tag{3.91}$$

where t' denotes the retarded time here and throughout the remainder of this section. Because the current density $\mathbf{j}$ is just equal to the charge density multiplied by the velocity of the particle, an analogous calculation can be carried out for the vector potential with the result

$$\mathbf{a}(\mathbf{r}, t) = \left\|\frac{4\pi}{c}\right\| \frac{\mu_0 c}{4\pi} \frac{q\mathbf{\Upsilon}(t')}{R(t') - \mathbf{\Upsilon}(t') \cdot \mathbf{R}(t')}. \tag{3.92}$$

Equations (3.91) and (3.92) for the scalar and vector potentials, which explicitly exhibit the dependence of the potentials on the velocity of the charged particle, are called the *Liénard–Wiechert potentials*.

Some physical insight concerning the form of the Liénard–Wiechert potentials for a moving charge may be gained through a consideration of the following construction due to Panofsky and Phillips [10]. Consider the calculation of the scalar potential $\phi(\mathbf{r}, t)$ at a given fixed observation point P at some specific instant of time t. Surround the point P with a spherical shell of radius R which is sufficiently large that it contains all of the charge that contributes to the potential at the time t. At the initial instant of time the shell is allowed to collapse to the center P with the velocity c. As the collapsing shell sweeps through the charge distribution it gathers information regarding the charge density. The spherical shell will then have collapsed to the observation point P at the prescribed instant of time t and will have gathered all of the information necessary to compute the value of the scalar potential $\phi(\mathbf{r}, t)$ at that particular space–time point, where $\mathbf{r}$ is the position vector for the point P. For a distribution of stationary charges this construct just yields the retarded potentials given in Eqs. (3.35) and (3.36). However, if the charge distribution has a net outward (inward) velocity, the volume integral of the charge density that is measured by the collapsing sphere will yield a result that is smaller (larger) than the total charge of the system.

Consider then a differential element of charge dq that is distributed uniformly with a charge density ρ throughout the volume element dV illustrated in Fig. 3.5. Let $\mathbf{R}$ be the vector from dV to the observation point P. If the charge dq is stationary, then the amount of charge that the spherical shell crosses as it contracts by an amount dR in the time interval dt is given by $\rho(t')da\,dR = \rho(t')dV$. On the other hand, if the charge moves with velocity $\mathbf{v}$, then the amount of charge crossed by the spherical shell in the time interval dt will be altered from its stationary value to the value

$$dq = \rho(t')dV - \rho(t')\frac{\mathbf{v(t')} \cdot \mathbf{R(t')}}{R(t')}da\,dt.$$

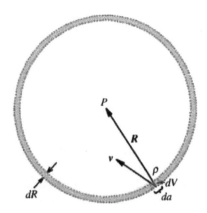

Fig. 3.5 Spherical "information shell" collapsing to the fixed observation point P with velocity c so that it arrives at precisely the time t at which the field at P is determined

If the velocity $\mathbf{v}$ of this differential element of charge is directed inward then $\mathbf{v} \cdot \mathbf{R} > 0$ and dq is reduced from its stationary value; if $\mathbf{v}$ is directed outward then $\mathbf{v} \cdot \mathbf{R} < 0$ and dq is increased from its stationary value. Because $dV = (da)(dR)$ and $dR = cdt$, then $(da)(dt) = (dV/dR)(dR/c) = (1/c)dV$ and

$$dq = \left(1 - \frac{\boldsymbol{\Upsilon}(t') \cdot \mathbf{R}(t')}{R(t')}\right) \rho(t')dV = \left[R(t') - \boldsymbol{\Upsilon}(t') \cdot \mathbf{R}(t')\right] \frac{\rho(t')}{R(t')} dV,$$

so that

$$\frac{\rho(t')}{R(t')} dV = \frac{dq}{R(t') - \boldsymbol{\Upsilon}(t') \cdot \mathbf{R}(t')}.$$

Substitution of this expression in Eq. (3.35) for the retarded scalar potential then gives

$$\phi(\mathbf{r}, t) = \frac{\|4\pi\|}{4\pi \epsilon_0} \int_V \frac{dq}{R(t') - \boldsymbol{\Upsilon}(t') \cdot \mathbf{R}(t')}. \tag{3.93}$$

If the complete charge distribution contains a total charge q that is confined to an infinitesimally small region of space, one can then perform the required integration by neglecting the variation of the denominator over the charge region and Eq. (3.93) reduces to the expression given in Eq. (3.91). A similar argument for the vector potential gives

$$\mathbf{a}(\mathbf{r}, t) = \left\| \frac{4\pi}{c} \right\| \frac{\mu_0 c}{4\pi} \int_V \frac{\boldsymbol{\Upsilon}(t')dq}{R(t') - \boldsymbol{\Upsilon}(t') \cdot \mathbf{R}(t')}, \tag{3.94}$$

which reduces to the expression given in Eq. (3.92) for the case of a single point charge. Once again, notice that $\mathbf{R}$, $\boldsymbol{\Upsilon} = \mathbf{v}/c$, and $R = |\mathbf{R}|$ are all functions of the retarded time $t' = t - R(t')/c$.

The preceding derivation provides some further insight into the physical interpretation of the concept of a point charge as well as to the physical applicability
of the results obtained for such a mathematical idealization. All that is required to
obtain the Liénard–Wiechert potentials (3.91) and (3.92) for the scalar and vector
potentials is that the quantities $(R(t') - \mathbf{R}(t') \cdot \mathbf{\Upsilon}(t'))$ and $\mathbf{\Upsilon}(t')$ have a negligible
variation over the spatial extent of the charge q.

3.3.2   The Field Produced by a Moving Charged Particle

With the expressions given in Eqs. (3.93) and (3.94) for the scalar and vector
potentials in the Lorenz gauge for the microscopic electromagnetic field that is
produced by a moving point charge q in free space, the microscopic field vectors
themselves are given by

$$\mathbf{e}(\mathbf{r}, t) = -\nabla \phi(\mathbf{r}, t) - \left\| \frac{1}{c} \right\| \frac{\partial \mathbf{a}(\mathbf{r}, t)}{\partial t}, \tag{3.95}$$

$$\mathbf{b}(\mathbf{r}, t) = \nabla \times \mathbf{a}(\mathbf{r}, t). \tag{3.96}$$

Because the time derivative of the vector potential $\mathbf{a}(\mathbf{r}, t)$, and hence of the
normalized velocity $\mathbf{\Upsilon} = \mathbf{v}/c$, is involved, it is then seen that the electric and
magnetic field vectors will be functions not only of the velocity $\mathbf{v}$, but also of the
acceleration $\breve{\mathbf{a}} = d\mathbf{v}/dt$ of the charged particle.[6] The electric and magnetic field
vectors may then each be separated into two components as

$$\mathbf{e}(\mathbf{r}, t) = \mathbf{e}_v(\mathbf{r}, t) + \mathbf{e}_a(\mathbf{r}, t), \tag{3.97}$$

$$\mathbf{b}(\mathbf{r}, t) = \mathbf{b}_v(\mathbf{r}, t) + \mathbf{b}_a(\mathbf{r}, t). \tag{3.98}$$

Both $\mathbf{e}_a(\mathbf{r}, t)$ and $\mathbf{b}_a(\mathbf{r}, t)$ involve the particle acceleration and are accordingly called
the *acceleration fields*, both of which go to zero when $\breve{\mathbf{a}} = \mathbf{0}$, whereas both $\mathbf{e}_v(\mathbf{r}, t)$
and $\mathbf{b}_v(\mathbf{r}, t)$ involve the particle velocity and are accordingly called the *velocity
fields*, both of which reduce to the static field for a point charge with velocity $\mathbf{v} = \mathbf{0}$.

Let the coordinates of the field point $P = P(x_1, x_2, x_3)$ be denoted by x_α, with
$\alpha = 1, 2, 3$, and let the retarded source point coordinates be denoted by $x'_\alpha = x'_\alpha(t')$, where t' denotes the retarded time. The field and source point variables are
connected by the *retardation condition*

$$R(x_\alpha, x'_\alpha) \equiv \left[\sum_{\alpha=1}^{3} (x_\alpha - x'_\alpha)^2 \right]^{1/2} = c(t - t'). \tag{3.99}$$

---

[6]This holds for the magnetic field vector because the spatial derivatives involved in the curl
operation are to be applied to the retarded quantities appearing in the Liénard–Wiechert vector
potential.

The components of the vector differential operator ∇ appearing in Eqs. (3.95) and (3.96) are partial derivatives at a fixed time t, and therefore not at constant retarded time t'. Partial differentiation with respect to the field coordinates x_α compares the potentials at neighboring points at the same time t, but the signals producing these potentials originated from the source charge at different retarded times t'. Similarly, the partial derivative with respect to the time t implies constant field point coordinates x_α, and hence refers to the comparison of the potentials at a given field point over an interval of time during which the coordinates x'_α of the source charge will have changed. Because only the time variation with respect to t' is given, it is then necessary to transform the differential operators $\partial/\partial t|_{x_\alpha}$ and $\nabla|_t$ to expressions involving the partial differential operator $\partial/\partial t'|_{x_\alpha}$ in order to compute the fields.

The retarded values of the velocity and acceleration of the point charge q, given by

$$v_\alpha = \frac{dx'_\alpha}{dt'},\tag{3.100}$$

$$\breve{a}_\alpha = \frac{dv_\alpha}{dt'} = \frac{d^2 x'_\alpha}{dt'^2},\tag{3.101}$$

are assumed known. In terms of the "position" vector $\mathbf{R}$ these expressions may be written in vector form as

$$\mathbf{v} = -\frac{d\mathbf{R}}{dt'},\tag{3.102}$$

$$\breve{\mathbf{a}} = \frac{d\mathbf{v}}{dt'} = -\frac{d^2\mathbf{R}}{dt'^2},\tag{3.103}$$

where the minus sign appears because $\mathbf{R}$ is the radius vector from the charge q to the observation point P, and not the reverse. Furthermore, differentiation of the identity $R^2 = \mathbf{R}\cdot\mathbf{R}$ with respect to t' with fixed field coordinates x_α yields

$$2R\left(\frac{\partial R}{\partial t'}\right)_{x_\alpha} = 2\mathbf{R}\cdot\left(\frac{\partial \mathbf{R}}{\partial t'}\right)_{x_\alpha}\tag{3.104}$$

so that

$$\left(\frac{\partial R}{\partial t'}\right)_{x_\alpha} = -\frac{\mathbf{R}\cdot\mathbf{v}}{R}.\tag{3.105}$$

Upon differentiation of the retardation condition (3.99) with respect to t, one obtains

$$\frac{\partial R}{\partial t} = c\left(1 - \frac{\partial t'}{\partial t}\right) = \frac{\partial R}{\partial t'}\frac{\partial t'}{\partial t} = -\frac{\mathbf{R}\cdot\mathbf{v}}{R}\frac{\partial t'}{\partial t},$$

so that

$$\frac{\partial t'}{\partial t} = \frac{1}{1 - (\mathbf{R} \cdot \mathbf{v}/(Rc))} = \frac{1}{1 - (\mathbf{R} \cdot \mathbf{\Upsilon})/R},$$

or

$$\frac{\partial}{\partial t} = \frac{R}{R - \mathbf{R} \cdot \mathbf{\Upsilon}} \frac{\partial}{\partial t'} = \frac{R}{s} \frac{\partial}{\partial t'}, \tag{3.106}$$

which is the desired transformation for the time derivatives, where

$$s \equiv R - \mathbf{R} \cdot \mathbf{\Upsilon}. \tag{3.107}$$

Similarly, differentiation of the same expression with respect to the spatial coordinates x_α gives

$$\nabla R = -c \nabla t' = \nabla_1 R + \frac{\partial R}{\partial t'} \nabla t' = \frac{\mathbf{R}}{R} - \frac{\mathbf{R} \cdot \mathbf{v}}{R} \nabla t',$$

so that

$$\nabla t' = -\frac{\mathbf{R}}{c(R - \mathbf{R} \cdot \mathbf{\Upsilon})} = -\frac{\mathbf{R}}{sc}, \tag{3.108}$$

and hence, in general

$$\nabla = \nabla_1 + (\nabla t') \frac{\partial}{\partial t'} = \nabla_1 - \frac{\mathbf{R}}{sc} \frac{\partial}{\partial t'}. \tag{3.109}$$

The operator ∇_1 implies differentiation with respect to the first argument of the function $f(x_\alpha, t') = R(x_\alpha, x'_\alpha(t'))$ in Eq. (3.99), that is, differentiation at fixed retarded time t'. The expressions in Eqs. (3.106) and (3.109) constitute the required transformation relations for the differential operators from the coordinates of the field point to those of the moving source point.

The computation of the electric field vector from the Liénard–Wiechert potentials given in Eqs. (3.91) and (3.92) then proceeds as

$$\mathbf{e} = -\nabla \phi - \left\|\frac{1}{c}\right\| \frac{\partial \mathbf{a}}{\partial t} = -\frac{\|4\pi\|}{4\pi\epsilon_0} q \left\{ \nabla \left(\frac{1}{s}\right) + \frac{1}{c^2} \frac{\partial}{\partial t} \frac{\mathbf{v}}{s} \right\}$$

$$= \frac{\|4\pi\|}{4\pi\epsilon_0} q \left\{ \frac{1}{s^2} \nabla s - \frac{1}{c^2} \frac{\partial}{\partial t} \frac{\mathbf{v}}{s} \right\}. \tag{3.110}$$

With Eqs. (3.106) and (3.109) one then obtains

$$\frac{4\pi\epsilon_0}{\|4\pi\|q}\mathbf{e} = \frac{1}{s^2}\nabla_1 s - \frac{\mathbf{R}}{cs^3}\frac{\partial s}{\partial t'} - \frac{\mathbf{R}}{c^2 s}\frac{\partial}{\partial t'}\left(\frac{\mathbf{v}}{s}\right)$$

$$= \frac{1}{s^2}\nabla_1 s - \frac{\mathbf{R}}{cs^3}\frac{\partial s}{\partial t'} - \frac{\mathbf{R}}{c^2 s}\left(\frac{\mathbf{\check{a}}}{s} - \frac{\mathbf{v}}{s^2}\frac{\partial s}{\partial t'}\right)$$

$$= \frac{1}{s^2}\nabla_1 s - \frac{\mathbf{R}}{cs^3}\frac{\partial s}{\partial t'} - \frac{\mathbf{R}}{c^2 s^2}\mathbf{\check{a}} + \frac{\mathbf{R}\mathbf{v}}{c^2 s^3}\frac{\partial s}{\partial t'}. \tag{3.111}$$

Differentiation of the retardation relation (3.107) yields

$$\nabla_1 s = \nabla_1\left(R - \frac{1}{c}\mathbf{R}\cdot\mathbf{v}\right) = \frac{\mathbf{R}}{R} - \frac{\mathbf{v}}{c}, \tag{3.112}$$

$$\frac{\partial s}{\partial t'} = \frac{\partial R}{\partial t'} - \frac{1}{c}\frac{\partial}{\partial t'}(\mathbf{R}\cdot\mathbf{v}) = -\frac{\mathbf{R}\cdot\mathbf{v}}{R} - \frac{1}{c}\frac{\partial\mathbf{R}}{\partial t'}\cdot\mathbf{v} - \frac{1}{c}\mathbf{R}\cdot\frac{\partial\mathbf{v}}{\partial t'}$$

$$= -\frac{\mathbf{R}\cdot\mathbf{v}}{R} + \frac{v^2}{c} - \frac{\mathbf{R}\cdot\mathbf{\check{a}}}{c}, \tag{3.113}$$

which, when substituted in Eq. (3.111), results in the expression

$$\frac{4\pi\epsilon_0}{\|4\pi\|q}\mathbf{e} = \frac{1}{s^2}\left(\frac{\mathbf{R}}{R} - \frac{\mathbf{v}}{c}\right) - \frac{\mathbf{R}}{cs^3}\left(\frac{v^2}{c} - \frac{\mathbf{R}\cdot\mathbf{v}}{R} - \frac{\mathbf{R}\cdot\mathbf{\check{a}}}{c}\right)$$

$$- \frac{\mathbf{R}}{c^2 s^2}\mathbf{\check{a}} + \frac{\mathbf{R}\mathbf{v}}{c^2 s^3}\left(\frac{v^2}{c} - \frac{\mathbf{R}\cdot\mathbf{v}}{R} - \frac{\mathbf{R}\cdot\mathbf{\check{a}}}{c}\right). \tag{3.114}$$

The *velocity and acceleration parts of the electric field vector* are then given by

$$\mathbf{e}_v(\mathbf{r}, t) = \frac{\|4\pi\|}{4\pi\epsilon_0}\frac{q}{s^3}(\mathbf{R} - R\mathbf{\Upsilon})(1 - \beta^2), \tag{3.115}$$

$$\mathbf{e}_a(\mathbf{r}, t) = \frac{\|4\pi\|}{4\pi\epsilon_0}\frac{q}{c^2 s^3}\mathbf{R}\times\left((\mathbf{R} - R\mathbf{\Upsilon})\times\mathbf{\check{a}}\right), \tag{3.116}$$

respectively, where $\beta = |\mathbf{\Upsilon}| = v/c$.

The computation of the magnetic induction vector from the Liénard–Wiechert potentials follows along similar lines as

$$\frac{4\pi\epsilon_0 c^2}{\|4\pi c\|q}\mathbf{b} = \nabla\times\left(\frac{\mathbf{v}}{s}\right) = \nabla\left(\frac{1}{s}\right)\times\mathbf{v} + \frac{1}{s}\nabla\times\mathbf{v} = -\frac{1}{s^2}(\nabla s)\times\mathbf{v} + \frac{1}{s}\nabla\times\mathbf{v}$$

$$= -\frac{1}{s^2}\left(\nabla_1 s - \frac{\mathbf{R}}{sc}\frac{\partial s}{\partial t'}\right)\times\mathbf{v} + \frac{1}{s}\left(\nabla_1\times\mathbf{v} - \frac{\mathbf{R}}{sc}\times\frac{\partial\mathbf{v}}{\partial t'}\right)$$

$$= -\frac{1}{s^2}\left[\frac{\mathbf{R}}{R} - \frac{\mathbf{v}}{c} - \frac{\mathbf{R}}{sc}\left(-\frac{\mathbf{R}\cdot\mathbf{v}}{R} + \frac{v^2}{c} - \frac{\mathbf{R}\cdot\mathbf{\check{a}}}{c}\right)\right]\times\mathbf{v} - \frac{\mathbf{R}\times\mathbf{\check{a}}}{s^2 c},$$

because $\nabla_1 \times \mathbf{v} = \mathbf{0}$, so that

$$\frac{4\pi\epsilon_0 c^2}{\|4\pi c\| q}\mathbf{b} = -\frac{\mathbf{R}\times\breve{\mathbf{a}}}{cs^2} + \frac{\mathbf{v}\times\mathbf{R}}{s^2}\left[\frac{1}{R} + \frac{1}{s}\left(\frac{\mathbf{R}\cdot\mathbf{v}}{cR} + \frac{\mathbf{R}\cdot\breve{\mathbf{a}}}{c^2} - \frac{v^2}{c^2}\right)\right]. \qquad (3.117)$$

The *velocity and acceleration parts of the magnetic induction field vector* are then given by

$$\mathbf{b}_v(\mathbf{r}, t) = \frac{\|4\pi c\|}{4\pi\epsilon_0 c^2}\frac{q}{s^3}(\mathbf{v}\times\mathbf{R})(1 - \beta^2), \qquad (3.118)$$

$$\mathbf{b}_a(\mathbf{r}, t) = \frac{\|4\pi c\|}{4\pi\epsilon_0 c^3}\frac{q}{s^3}\frac{\mathbf{R}}{R}\times\left\{\mathbf{R}\times\left[(\mathbf{R} - R\Upsilon)\times\breve{\mathbf{a}}\right]\right\}. \qquad (3.119)$$

Taken together, Eqs. (3.115)–(3.116) and (3.118)–(3.119) constitute the set of velocity and acceleration fields for the electromagnetic field produced by a moving charged particle q. Once again, the quantities R, $\mathbf{R}$, Υ, and $s = R - \mathbf{R}\cdot\Upsilon$ appearing on the right-hand side of each of these expressions are all evaluated at the retarded time $t' = t - R(t')/c$. Upon comparison of Eqs. (3.118) and (3.119) with Eqs. (3.115) and (3.116), it is seen that

$$\mathbf{b}(\mathbf{r}, t) = \frac{\|c\|}{c}\frac{\mathbf{R}}{R}\times\mathbf{e}(\mathbf{r}, t). \qquad (3.120)$$

Hence, *the magnetic field vector is always perpendicular to both the electric field vector and the retarded radius vector* $\mathbf{R}$. Furthermore, $\mathbf{b}_v$ is perpendicular to $\mathbf{e}_v$, and $\mathbf{b}_a$ is perpendicular to $\mathbf{e}_a$. The retarded radius vector $\mathbf{R}$ is perpendicular to $\mathbf{e}_a$, $\mathbf{b}_v$, and $\mathbf{b}_a$, but is not in general perpendicular to $\mathbf{e}_v$.

It is important to point out that the above set of equations for the electromagnetic field vectors is relativistically correct. Indeed, the velocity $\mathbf{v} = c\Upsilon$ may simply be considered as the relative velocity between the point charge and the observer. However, the Liénard–Wiechert potentials were derived prior to the advent of the special theory of relativity; it was only after the same results were obtained relativistically that it became apparent that $\mathbf{v}$ could indeed be interpreted as the relative velocity.

The velocity part of the electric field, given in Eq. (3.115), is seen to vary as R^{-2} for large distances R between the source and field point, and is formally identical with the convective field produced by a uniformly moving charged particle [10]. With regard to this field one may define the quantity

$$\mathbf{R}_v \equiv \mathbf{R} - R\Upsilon = \mathbf{R} - R\frac{\mathbf{v}}{c} \qquad (3.121)$$

as the *"virtual present radius vector"*, that is, the position vector that would describe the position the point charge would occupy "at present" if it had continued with a

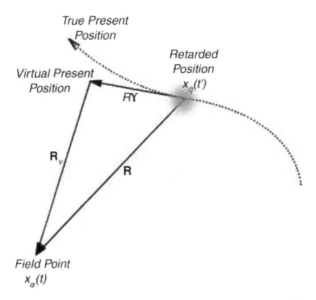

Fig. 3.6 Geometric relationship between the retarded position, the true present position, and the virtual present position of a point charge q moving along a (dotted) curved path in space

uniform velocity from the point $x_q(t')$, as illustrated in Fig. 3.6. In terms of this quantity the velocity part of the field vector becomes

$$\mathbf{e}_v(\mathbf{r}, t) = \frac{\|4\pi\|}{4\pi\epsilon_0}\frac{q}{s^3}\mathbf{R}_v(1 - \beta^2), \tag{3.122}$$

and this is just the electric field produced by a charge in uniform motion.

3.3.3   Radiated Energy from a Moving Charged Particle

The radiated energy from a moving charged particle q is obtained from the Poynting vector of the electromagnetic field. From Eqs. (3.97) and (3.98), the microscopic Poynting vector may be expressed as

$$\mathbf{s}(\mathbf{r}, t) = \mathbf{s}_v(\mathbf{r}, t) + \mathbf{s}_a(\mathbf{r}, t) + \mathbf{s}_\times(\mathbf{r}, t), \tag{3.123}$$

where $\mathbf{s}_v(\mathbf{r}, t) \equiv \|c/4\pi\|\mathbf{e}_v(\mathbf{r}, t)\times\mathbf{b}_v(\mathbf{r}, t)$ is the Poynting vector associated with the velocity part of the field, $\mathbf{s}_a(\mathbf{r}, t) \equiv \|c/4\pi\|\mathbf{e}_a(\mathbf{r}, t) \times \mathbf{b}_a(\mathbf{r}, t)$ is the Poynting vector associated with the acceleration part of the field, and $\mathbf{s}_\times(\mathbf{r}, t)$ is due to the cross terms from both the velocity and acceleration parts of the field. From Eqs. (3.115) and (3.118), the *microscopic Poynting vector associated with the velocity part of the field* is found to be given by

$$\mathbf{s}_v = \frac{\|4\pi\|c}{(4\pi)^2\epsilon_0}\frac{q^2}{s^6}(1-\beta^2)(\mathbf{R}-R\mathbf{\Upsilon})\times(\mathbf{\Upsilon}\times\mathbf{R})$$

$$= \frac{\|4\pi\|c}{(4\pi)^2\epsilon_0}\frac{q^2}{s^6}(1-\beta^2)\Big[R(R-\mathbf{\Upsilon}\cdot\mathbf{R})+(R\beta^2-\mathbf{\Upsilon}\cdot\mathbf{R})\mathbf{R}\Big]. \quad (3.124)$$

From Eqs. (3.116) and (3.119), the *microscopic Poynting vector associated with the acceleration part of the field* is found to be given by

$$\mathbf{s}_a = \frac{\|4\pi\|c}{(4\pi)^2\epsilon_0}\frac{q^2}{Rs^6}\left\{\big[\mathbf{R}\times((\mathbf{R}-R\mathbf{\Upsilon})\times\check{\mathbf{a}})\big]\times\big[\mathbf{R}\times(\mathbf{R}\times((\mathbf{R}-R\mathbf{\Upsilon})\times\check{\mathbf{a}}))\big]\right\}.$$

In order to evaluate the complicated cross product appearing in this expression, let $\mathbf{\Gamma}\equiv(\mathbf{R}-R\mathbf{\Upsilon})\times\check{\mathbf{a}}=\mathbf{R}_v\times\check{\mathbf{a}}$, so that

$$(\mathbf{R}\times\mathbf{\Gamma})\times(\mathbf{R}\times(\mathbf{R}\times\mathbf{\Gamma})) = (\mathbf{R}\times\mathbf{\Gamma})\times[(\mathbf{R}\cdot\mathbf{\Gamma})\mathbf{R}-R^2\mathbf{\Gamma}]$$

$$= R^2\mathbf{\Gamma}\times(\mathbf{R}\times\mathbf{\Gamma})-(\mathbf{R}\cdot\mathbf{\Gamma})\mathbf{R}\times(\mathbf{R}\times\mathbf{\Gamma})$$

$$= R^2[\Gamma^2\mathbf{R}-(\mathbf{\Gamma}\cdot\mathbf{R})\mathbf{\Gamma}]-(\mathbf{R}\cdot\mathbf{\Gamma})[(\mathbf{R}\cdot\mathbf{\Gamma})\mathbf{R}-R^2\mathbf{\Gamma}]$$

$$= \Big[R^2\Gamma^2-(\mathbf{R}\cdot\mathbf{\Gamma})^2\Big]\mathbf{R}.$$

Hence,

$$\mathbf{s}_a = \frac{\|4\pi\|c}{(4\pi)^2\epsilon_0}\frac{q^2}{s^6}\left[\frac{R^2(\mathbf{R}_v\times\check{\mathbf{a}})^2-(\mathbf{R}\cdot(\mathbf{R}_v\times\check{\mathbf{a}}))^2}{R}\right]\mathbf{R}, \quad (3.125)$$

which is along the direction of the retarded radius vector from the point charge q to the observation point P.

The radial dependencies of the Poynting vectors for the velocity and acceleration fields are therefore given by

$$s_v\propto\frac{1}{R^4}, \quad s_a\propto\frac{1}{R^2},$$

whereas the cross-term $s_\times\propto R^{-3}$. In order to determine the energy radiated by a moving charged particle, the normal component of the Poynting vector $\mathbf{s}(\mathbf{r},t)$ is integrated over the surface of a sphere which varies as R^2. As R becomes large, the surface integral involving $\mathbf{s}_v$ vanishes as $1/R^2$, whereas the integral involving $\mathbf{s}_a$ remains finite and nonvanishing in general. Hence, a charged particle that moves with a constant velocity, and so has $\mathbf{s}_a = \mathbf{0}$, cannot radiate energy; only a charged particle that is undergoing acceleration can produce radiation. The fact that a charged particle moving with a uniform velocity cannot radiate energy is consistent with the relativistic nature of the field quantities, for if $\mathbf{v}$ is the relative velocity between the charged particle and the observation point, an inertial reference frame exists in which the particle is at rest and the observation point is in uniform motion. That is, if it is possible to find an inertial reference frame with respect to which the charged particle is at rest, then radiation cannot occur.

3.4   The Radiation Field Produced by a General Dipole Oscillator

The radiation field that is most fundamental to the microscopic Maxwell–Lorentz theory is that of an atomic dipole oscillator. Although classical physics fails to provide a correct description of electronic motion in an atomic system, the quantum theory shows that in many cases the atomic radiation field is primarily due to the dipole moment of the atomic system so that the radiation field can be thought of as being produced by a classical dipole oscillator. Consider then an atomic system in which an electron of charge $q_1 = -q_e$ moves about a fixed nucleus of charge $q_2 = +q_e$ (any other nuclear charges are assumed to be effectively screened by the other bound electrons of the electrically neutral system) such that its velocity v is small in comparison to the velocity of light c in free space and its position is always in close proximity to the nucleus. Let $\mathbf{d} = \mathbf{d}(t)$ denote[7] the temporally varying position vector of the electron relative to the nucleus, let $\mathbf{r}$ denote the position vector from the nucleus to the fixed observation point P, and let $\mathbf{r}_e$ denote the vector from the electron position to the point P, as illustrated in Fig. 3.7. The radiation field at any observation point P that is not too near the source atomic site ($r \gg d$) is then given to a high degree of accuracy by just the first few terms in an expansion of the exact field in a power series in terms of the dimensionless quantities d/r and v/c. However, instead of performing this expansion on the exact field, it is more

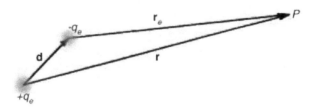

Fig. 3.7 Atomic dipole oscillator comprised of a fixed nucleus and a bound electron of equal and opposite charge

---

[7]The position vector $\mathbf{d}(t)$ should not be confused here with the microscopic electric displacement vector $\mathbf{d}(\mathbf{r}, t) \equiv \epsilon_0 \mathbf{e}(\mathbf{r}, t)$.

appropriate to expand the Liénard–Wiechert potentials in a power series in terms of the same quantities and then determine the radiation field that results from the lowest order terms that appear in this expansion. This latter approach is taken in the analysis presented here which follows that given by Stone [11].

3.4.1   The Field Vectors Produced by a General Dipole Oscillator

The scalar and vector potentials at the stationary field observation point P at the time t consist of the superposition of the Liénard–Wiechert potentials due to the fixed nucleus (fixed relative to P) and the bound electron. This physical situation then involves the following three interrelated time variables:

- $t \equiv$ the time at which both the fields and potentials at the observation point P are to be determined.
- $t' \equiv t - r/c =$ the retarded time at the fixed position of the nucleus corresponding to the time t at the observation point P.
- $t'_e \equiv t - r_e(t'_e)/c =$ the retarded time at the variable position of the bound electron corresponding to the time t at the observation point P.

In addition to these specific time variables, it is convenient to introduce an arbitrary time variable τ that may be assigned any particular value. It is assumed that the position of the bound electron is a known vector function $\mathbf{d}(\tau)$ relative to the fixed position of the nucleus, with velocity $\dot{\mathbf{d}}(\tau)$ and acceleration $\ddot{\mathbf{d}}(\tau)$. The bound electron's distance from the nucleus is assumed to always be less than some finite maximum distance d_{max}, so that

$$|\mathbf{d}(\tau)| \leq d_{max} \tag{3.126}$$

for all τ, and the distance $r = |\mathbf{r}|$ of the field observation point P from the nucleus is assumed to be much greater than that same maximum distance d_{max}, so that

$$r \gg d_{max}. \tag{3.127}$$

In addition, the velocity of the bound electron is assumed to always be less than some maximum value v_{max}, where

$$\left|\dot{\mathbf{d}}(\tau)\right| \leq v_{max} \ll c \tag{3.128}$$

for all τ. These three inequalities constitute the basic approximations under which the analysis of the radiation field due to an atomic dipole oscillator is undertaken in this development.

From Eqs. (3.91) and (3.92), the Liénard–Wiechert potentials due to the charge $+q_e$ of the (stationary) nucleus are given by

$$\phi_n(\mathbf{r}, t) = \frac{\|4\pi\|}{4\pi\epsilon_0} \frac{q_e}{r}, \tag{3.129}$$

$$\mathbf{a}_n(\mathbf{r}, t) = \mathbf{0}, \tag{3.130}$$

and those due to the bound electron of charge $-q_e$ are given by

$$\phi_e(\mathbf{r}, t) = -\frac{\|4\pi\|}{4\pi\epsilon_0} \frac{q_e}{r_e} \left(1 - \frac{\dot{\mathbf{d}} \cdot \mathbf{r}_e}{c r_e}\right)^{-1}, \tag{3.131}$$

$$\mathbf{a}_e(\mathbf{r}, t) = -\frac{\|4\pi\|\mu_0}{4\pi\|c\|} q_e \frac{\dot{\mathbf{d}}}{r_e} \left(1 - \frac{\dot{\mathbf{d}} \cdot \mathbf{r}_e}{c r_e}\right)^{-1}. \tag{3.132}$$

Because of the inequality in Eq. (3.128), these expressions for the scalar and vector potentials of the bound electron may be expanded as

$$\phi_e(\mathbf{r}, t) = -\frac{\|4\pi\|}{4\pi\epsilon_0} \frac{q_e}{r_e} \left(1 + \frac{\dot{\mathbf{d}} \cdot \mathbf{r}_e}{c r_e} + \cdots\right), \tag{3.133}$$

$$\mathbf{a}_e(\mathbf{r}, t) = -\frac{\|4\pi\|\mu_0}{4\pi\|c\|} q_e \frac{\dot{\mathbf{d}}}{r_e} \left(1 + \frac{\dot{\mathbf{d}} \cdot \mathbf{r}_e}{c r_e} + \cdots\right), \tag{3.134}$$

respectively, where the sum of the remaining (unwritten) terms in each series is $\mathcal{O}((v_{\max}/c)^2)$.

The position vector $\mathbf{r}_e$ from the bound electron position to the observation point P is given by the vector difference $\mathbf{r}_e = \mathbf{r} - \mathbf{d}$ so that its magnitude is

$$r_e \equiv |\mathbf{r}_e| = \left(r^2 - 2\mathbf{r} \cdot \mathbf{d} + d^2\right)^{1/2} = r\left(1 - 2\frac{\mathbf{r} \cdot \mathbf{d}}{r^2} + \frac{d^2}{r^2}\right)^{1/2},$$

and consequently

$$\frac{1}{r_e} = \frac{1}{r}\left(1 - 2\frac{\mathbf{r} \cdot \mathbf{d}}{r^2} + \frac{d^2}{r^2}\right)^{-1/2}$$

$$= \frac{1}{r} + \frac{\mathbf{r} \cdot \mathbf{d}}{r^3} - \frac{d^2}{2r^3} + \frac{3(\mathbf{r} \cdot \mathbf{d})^2}{2r^5} + \cdots, \tag{3.135}$$

where the sum of the remaining (unwritten) terms in the series is $\mathcal{O}(d^3/r^4)$.

The vector $\mathbf{r}$ and its magnitude r are to be evaluated at the retarded time $t' = t - r/c$ in the Liénard–Wiechert potentials appearing in Eqs. (3.133) and (3.134) whereas the vector $\mathbf{d}$ and its magnitude d are to be evaluated at the retarded time $t_e' = t - r_e(t_e')/c$. It is possible (and quite advantageous) to express $\mathbf{d}$, d, and their various time derivatives in terms of the retarded time t' which is simply proportional to the observation time t and is only slightly different from t_e'. The relation between these two retarded times is

$$t_e' = t' + \frac{\ell(t_e')}{c}, \tag{3.136}$$

where $\ell(t_e') \equiv r - r_e(t_e')$. By *Lagrange's theorem* [12], if $\mathbf{x}(\tau)$ denotes either $\mathbf{d}(\tau)$ or one of its derivatives with respect to τ, then

$$\mathbf{x}(t_e') = \mathbf{x}(t') + \sum_{n=1}^{\infty} \frac{1}{n! c^n} \frac{d^{n-1}}{dt'^{n-1}} \left[\dot{\mathbf{x}}(t') \left(\ell(t') \right)^n \right] \tag{3.137}$$

$$= \mathbf{x}(t') + \frac{1}{c} \dot{\mathbf{x}}(t') \ell(t') + \cdots . \tag{3.138}$$

Because

$$\ell(t_e') = r - r_e(t_e') = r - r \left(1 - 2\frac{\mathbf{r} \cdot \mathbf{d}(t_e')}{r^2} + \frac{d^2(t_e')}{r^2} \right)^{1/2}$$

$$= r - r \left(1 - \frac{\mathbf{d}(t_e') \cdot \mathbf{r}}{r^2} + \frac{d^2(t_e')}{2r^2} - \frac{\left(\mathbf{d}(t_e') \cdot \mathbf{r} \right)^2}{2r^4} + \cdots \right)$$

$$= \frac{\mathbf{d}(t_e') \cdot \mathbf{r}}{r} - \frac{d^2(t_e')}{2r} + \frac{\left(\mathbf{d}(t_e') \cdot \mathbf{r} \right)^2}{2r^3} + \cdots$$

then, with this result evaluated at t', Eq. (3.138) becomes

$$\mathbf{x}(t_e') = \mathbf{x}(t') + \frac{1}{c} \dot{\mathbf{x}}(t') \left(\frac{\mathbf{d}(t') \cdot \mathbf{r}}{r} - \frac{d^2(t')}{2r} + \frac{\left(\mathbf{d}(t') \cdot \mathbf{r} \right)^2}{2r^3} + \cdots \right) + \cdots . \tag{3.139}$$

The relative position vector and velocity of the bound electron at the retarded time t_e' may then be expressed as a function of the retarded time t' at the fixed position of the nucleus as

$$\mathbf{d}(t_e') = \mathbf{d}(t') + \frac{1}{c} \dot{\mathbf{d}}(t') \left(\frac{\mathbf{d}(t') \cdot \mathbf{r}}{r} - \frac{d^2(t')}{2r} + \frac{\left(\mathbf{d}(t') \cdot \mathbf{r} \right)^2}{2r^3} + \cdots \right) + \cdots , \tag{3.140}$$

$$\dot{\mathbf{d}}(t_e') = \dot{\mathbf{d}}(t') + \frac{1}{c} \ddot{\mathbf{d}}(t') \left(\frac{\mathbf{d}(t') \cdot \mathbf{r}}{r} - \frac{d^2(t')}{2r} + \frac{\left(\mathbf{d}(t') \cdot \mathbf{r} \right)^2}{2r^3} + \cdots \right) + \cdots . \tag{3.141}$$

With these expansions the Liénard–Wiechert potentials due to the bound electron of charge $q = -q_e$ may be expressed in terms of quantities that are evaluated at the retarded time t' in the following manner. Substitution of Eq. (3.135) into the expansion given in Eq. (3.133) for the scalar potential gives

$$\phi_e(\mathbf{r}, t) = -\frac{\|4\pi\|}{4\pi\epsilon_0} \frac{q_e}{r_e(t'_e)} \left(1 + \frac{\dot{\mathbf{d}}(t'_e) \cdot \mathbf{r}_e(t'_e)}{c r_e(t'_e)} + \cdots \right)$$

$$= -\frac{\|4\pi\|}{4\pi\epsilon_0} q_e \left(\frac{1}{r} + \frac{\mathbf{d}(t'_e) \cdot \mathbf{r}}{r^3} - \frac{d^2(t'_e)}{2r^3} + \cdots \right)$$

$$\times \left(1 + \frac{\dot{\mathbf{d}}(t'_e) \cdot \mathbf{r}_e(t'_e)}{c} \left(\frac{1}{r} + \frac{\mathbf{d}(t'_e) \cdot \mathbf{r}}{r^3} - \frac{d^2(t'_e)}{2r^3} + \cdots \right) + \cdots \right),$$

so that, with the further substitution of the relation $\mathbf{r}_e(t'_e) = \mathbf{r} - \mathbf{d}(t'_e)$, there results

$$\phi_e(\mathbf{r}, t) = -\frac{\|4\pi\|}{4\pi\epsilon_0} q_e \left(\frac{1}{r} + \frac{\mathbf{d}(t'_e) \cdot \mathbf{r}}{r^3} - \frac{d^2(t'_e)}{2r^3} + \cdots \right)$$

$$\times \left[1 + \frac{\dot{\mathbf{d}}(t'_e) \cdot \mathbf{r}}{c} \left(\frac{1}{r} + \frac{\mathbf{d}(t'_e) \cdot \mathbf{r}}{r^3} - \frac{d^2(t'_e)}{2r^3} + \cdots \right) \right.$$

$$\left. - \frac{\dot{\mathbf{d}}(t'_e) \cdot \mathbf{d}(t'_e)}{c} \left(\frac{1}{r} + \frac{\mathbf{d}(t'_e) \cdot \mathbf{r}}{r^3} - \frac{d^2(t'_e)}{2r^3} + \cdots \right) + \cdots \right].$$

Substitution of the pair of relations appearing in Eqs. (3.140) and (3.141) into the above expansion of the scalar potential then yields (where all quantities on the right-hand side of the equal sign are now evaluated at the retarded time t' at the stationary position of the nucleus)

$$\phi_e(\mathbf{r}, t) = -\frac{\|4\pi\|}{4\pi\epsilon_0} q_e \left[\frac{1}{r} + \frac{\mathbf{r}}{r^3} \cdot \left(\mathbf{d} + \frac{\dot{\mathbf{d}}}{c} \left(\frac{\mathbf{d} \cdot \mathbf{r}}{r} - \frac{d^2}{2r} + \frac{(\mathbf{d} \cdot \mathbf{r})^2}{2r^3} + \cdots \right) + \cdots \right) \right.$$

$$- \frac{1}{2r^3} \left(s^2 + \frac{2}{c} \mathbf{d} \cdot \dot{\mathbf{d}} \left(\frac{\mathbf{d} \cdot \mathbf{r}}{r} - \frac{d^2}{2r} + \frac{(\mathbf{d} \cdot \mathbf{r})^2}{2r^3} + \cdots \right) \right.$$

$$\left. + \frac{\dot{d}^2}{c^2} \left(\frac{\mathbf{d} \cdot \mathbf{r}}{r} - \frac{d^2}{2r} + \frac{(\mathbf{d} \cdot \mathbf{r})^2}{2r^3} + \cdots \right)^2 + \cdots \right) + \cdots \right]$$

$$\times \left\{ 1 + \frac{\mathbf{r}}{c} \cdot \left[\dot{\mathbf{d}} + \frac{\ddot{\mathbf{d}}}{c} \left(\frac{\mathbf{d} \cdot \mathbf{r}}{r} - \frac{d^2}{2r} + \frac{(\mathbf{d} \cdot \mathbf{r})^2}{2r^3} + \cdots \right) + \cdots \right] \right.$$

$$\times \left[\frac{1}{r} + \frac{\mathbf{r}}{r^3} \cdot \left(\mathbf{d} + \frac{\dot{\mathbf{d}}}{c} \left(\frac{\mathbf{d} \cdot \mathbf{r}}{r} - \frac{d^2}{2r} + \frac{(\mathbf{d} \cdot \mathbf{r})^2}{2r^3} + \cdots \right) + \cdots \right) \right.$$

$$-\frac{1}{2r^3}\left(d^2 + \frac{2}{c}\mathbf{d}\cdot\dot{\mathbf{d}}\left(\frac{\mathbf{d}\cdot\mathbf{r}}{r} - \frac{d^2}{2r} + \frac{(\mathbf{d}\cdot\mathbf{r})^2}{2r^3} + \cdots\right)\right.$$

$$\left. + \frac{\dot{s}^2}{c^2}\left(\frac{\mathbf{d}\cdot\mathbf{r}}{r} - \frac{d^2}{2r} + \frac{(\mathbf{d}\cdot\mathbf{r})^2}{2r^3} + \cdots\right)^2 + \cdots\right) + \cdots\right]$$

$$-\frac{1}{c}\left[\dot{\mathbf{d}} + \frac{\ddot{\mathbf{d}}}{c}\left(\frac{\mathbf{d}\cdot\mathbf{r}}{r} - \frac{d^2}{2r} + \frac{(\mathbf{d}\cdot\mathbf{r})^2}{2r^3} + \cdots\right) + \cdots\right]$$

$$\cdot\left[\mathbf{d} + \frac{\dot{\mathbf{d}}}{c}\left(\frac{\mathbf{d}\cdot\mathbf{r}}{r} - \frac{d^2}{2r} + \frac{(\mathbf{d}\cdot\mathbf{r})^2}{2r^3} + \cdots\right) + \cdots\right]$$

$$\times\left[\frac{1}{r} + \frac{\mathbf{r}}{r^3}\cdot\left(\mathbf{d} + \frac{\dot{\mathbf{d}}}{c}\left(\frac{\mathbf{d}\cdot\mathbf{r}}{r} - \frac{d^2}{2r} + \frac{(\mathbf{d}\cdot\mathbf{r})^2}{2r^3} + \cdots\right) + \cdots\right)\right.$$

$$-\frac{1}{2r^3}\left(d^2 + \frac{2}{c}\mathbf{d}\cdot\dot{\mathbf{d}}\left(\frac{\mathbf{d}\cdot\mathbf{r}}{r} - \frac{d^2}{2r} + \frac{(\mathbf{d}\cdot\mathbf{r})^2}{2r^3} + \cdots\right)\right.$$

$$\left.\left. + \frac{\dot{d}^2}{c^2}\left(\frac{\mathbf{d}\cdot\mathbf{r}}{r} - \frac{d^2}{2r} + \frac{(\mathbf{d}\cdot\mathbf{r})^2}{2r^3} + \cdots\right)^2 + \cdots\right) + \cdots\right] + \cdots\right\}.$$

$$(3.142)$$

Upon collecting terms that are ordered according to the number of times the electronic position vector $\mathbf{d}$ or any of its time derivatives or combinations thereof appears, one finally obtains

$$\phi_e(\mathbf{r}, t) = -\frac{\|4\pi\|}{4\pi\epsilon_0}q_e\left[\frac{1}{r} + \frac{\mathbf{r}\cdot\mathbf{d}}{r^3} + \frac{(\mathbf{r}\cdot\dot{\mathbf{d}})(\mathbf{r}\cdot\mathbf{d})}{cr^4} - \frac{d^2}{2r^3} + O(d^3)\right]$$

$$\times\left[1 + \frac{\mathbf{r}\cdot\dot{\mathbf{d}}}{cr} + \frac{(\mathbf{r}\cdot\dot{\mathbf{d}})(\mathbf{r}\cdot\mathbf{d})}{cr^3} + \frac{(\mathbf{r}\cdot\ddot{\mathbf{d}})(\mathbf{r}\cdot\mathbf{d})}{c^2r^2} - \frac{\mathbf{d}\cdot\dot{\mathbf{d}}}{cr} + O(d^3)\right]$$

$$= -\frac{\|4\pi\|}{4\pi\epsilon_0}q_e\left(\frac{1}{r} + \frac{\mathbf{d}\cdot\hat{\mathbf{r}}}{r^2} + \frac{\dot{\mathbf{d}}\cdot\hat{\mathbf{r}}}{cr} - \frac{d^2}{2r^3}\right.$$

$$\left. + \frac{3(\mathbf{d}\cdot\hat{\mathbf{r}})(\dot{\mathbf{d}}\cdot\hat{\mathbf{r}})}{c^2r} + \frac{(\mathbf{d}\cdot\hat{\mathbf{r}})(\ddot{\mathbf{d}}\cdot\hat{\mathbf{r}})}{c^2r} + O(d^3)\right), \quad (3.143)$$

where $\hat{\mathbf{r}} \equiv \mathbf{r}/r$ is the unit vector in the direction of the observation point P.

From the pair of relations given in Eqs. (3.133) and (3.134), the vector potential $\mathbf{a}_e(\mathbf{r}, t)$ that is associated with the scalar potential $\phi_e(\mathbf{r}, t)$ is seen to be given by

$$\mathbf{a}_e(\mathbf{r}, t) = \frac{\epsilon_0\mu_0}{\|c\|}\dot{\mathbf{d}}(t'_e)\phi_e(\mathbf{r}, t) \quad (3.144)$$

so that, with substitution from Eqs. (3.141) and (3.143), there results

$$\mathbf{a}_e(\mathbf{r}, t) = -\left|\left|\frac{4\pi}{c}\right|\right|\frac{\mu_0}{4\pi}q_e\left(\dot{\mathbf{d}} + \frac{\mathbf{d}\cdot\hat{\mathbf{r}}}{c}\ddot{\mathbf{d}} + O(d^3)\right)\left(\frac{1}{r} + \frac{\mathbf{d}\cdot\hat{\mathbf{r}}}{r^2} + \frac{\dot{\mathbf{d}}\cdot\hat{\mathbf{r}}}{cr} + O(d^2)\right)$$

$$= -\left|\left|\frac{4\pi}{c}\right|\right|\frac{\mu_0}{4\pi}q_e\left(\frac{1}{r}\dot{\mathbf{d}} + \frac{\mathbf{d}\cdot\hat{\mathbf{r}}}{r^2}\dot{\mathbf{d}} + \frac{\dot{\mathbf{d}}\cdot\hat{\mathbf{r}}}{cr}\dot{\mathbf{d}} + \frac{\mathbf{d}\cdot\hat{\mathbf{r}}}{cr}\ddot{\mathbf{d}} + O(d^3)\right). \quad (3.145)$$

The above pair of expressions (3.143) and (3.145) for the scalar and vector Liénard–
Wiechert potentials for the bound electron of charge $q = -q_e$ are both in the form
of a series of terms involving the position vector $\mathbf{r} = r\hat{\mathbf{r}}$ of the field observation
point relative to the fixed position of the nucleus, the position vector $\mathbf{d}(t')$ of
the bound electron relative to the position of the nucleus, and the derivatives of
$\mathbf{d}(t')$ with respect to the retarded time $t' = t - r/c$ at the fixed position of the
nucleus corresponding to the time t at the stationary observation point P. The terms
in each series expansion have been ordered into groups that represent successive
approximations to each of the potentials in the following manner: in the zeroth
or lowest-order group the electronic position vector $\mathbf{d}$ does not appear (the scalar
potential alone has this term); in the first-order group the electronic position vector
$\mathbf{d}$ appears once in each term (either as $\mathbf{d}$ or as $\dot{\mathbf{d}}$); in the second-order group the
electronic position vector $\mathbf{d}$ appears twice in each term (either as $\mathbf{d}\cdot\mathbf{d} = d^2$, $\mathbf{d}\cdot\dot{\mathbf{d}}$,
$(\mathbf{d}\cdot\hat{\mathbf{r}})(\dot{\mathbf{d}}\cdot\hat{\mathbf{r}})$, $(\mathbf{d}\cdot\hat{\mathbf{r}})\dot{\mathbf{d}}$, etc.); and so on for higher-order terms. The zeroth-order group in
the scalar potential represents the *monopole contribution* due to the electron whose
position is taken at the fixed position of the nucleus (this contribution to the total
scalar potential of the atomic dipole oscillator will then be identically canceled by
the monopole contribution due to the fixed nucleus of charge $q = +q_e$). The first-
order groups in both the scalar and vector potentials represent the *electric dipole
contribution* to the total field because the electric dipole moment $\mathbf{p} = q_e\mathbf{d}$ and its
time derivative $\dot{\mathbf{p}} = q_e\dot{\mathbf{d}}$ only appear. The second-order groups in both the scalar
and vector potentials represent both the *electric quadrupole and magnetic dipole
contributions* to the total field. In general, the nth-order groups represent both the
electric 2^n-pole and the *magnetic 2^{n-1}-pole* contributions to the total field.

3.4.2   The Electric Dipole Approximation

Equations (3.143) and (3.145) show that any $(n + 1)$th-order group provides a
negligible contribution to the total radiated field when compared to the contribution
from the nth-order group provided that the pair of inequalities in Eqs. (3.127) and
(3.128) are both satisfied and that the electric 2^n-pole and the magnetic 2^{n-1}-pole
contributions are non-vanishing. As a consequence, an accurate approximation to
the field produced by an atomic dipole oscillator may be obtained by considering
only the terms up through the electric dipole contribution (the first group) provided

that these inequalities are satisfied and that the electric dipole moment does not vanish nor is vanishingly small. If the electronic motion is such that the resultant electric dipole moment $\mathbf{p} = q_e\mathbf{d}$ and its time derivative both vanish for all time, the first non-vanishing higher-order multipole contribution must then be considered.

The first-order or electric dipole approximation to the Liénard–Wiechert potentials due to the motion of a bound electron about the nucleus in an atom with a stationary nuclear site is then given by

$$\phi(\mathbf{r}, t) = \phi_n(\mathbf{r}, t) + \phi_e(\mathbf{r}, t)$$

$$\cong \frac{\|4\pi\|}{4\pi\,\epsilon_0}\frac{q_e}{r} - \frac{\|4\pi\|}{4\pi\,\epsilon_0}q_e\left(\frac{1}{r} + \frac{\mathbf{d}(t')\cdot\hat{\mathbf{r}}}{r^2} + \frac{\dot{\mathbf{d}}(t')\cdot\hat{\mathbf{r}}}{cr}\right)$$

$$= -\frac{\|4\pi\|}{4\pi\,\epsilon_0}q_e\left(\frac{\mathbf{d}(t')\cdot\hat{\mathbf{r}}}{r^2} + \frac{\dot{\mathbf{d}}(t')\cdot\hat{\mathbf{r}}}{cr}\right), \qquad (3.146)$$

$$\mathbf{a}(\mathbf{r}, t) = \mathbf{a}_n(\mathbf{r}, t) + \mathbf{a}_e(\mathbf{r}, t) \cong -\left\|\frac{4\pi}{c}\right\|\frac{\mu_0}{4\pi}q_e\frac{\dot{\mathbf{d}}(t')}{r}, \qquad (3.147)$$

where $\hat{\mathbf{r}} = \mathbf{r}/r$ is independent of the time and where $\mathbf{d}(t')$ and $\dot{\mathbf{d}}(t')$ are both evaluated at the retarded time $t' = t - r/c$ at the stationary position of the nucleus.

The electric and magnetic field vectors of the radiated field are obtained from the scalar and vector Liénard–Wiechert potentials through application of the differential relations

$$\mathbf{e}(\mathbf{r}, t) = -\nabla\phi(\mathbf{r}, t) - \left\|\frac{1}{c}\right\|\frac{\partial\mathbf{a}(\mathbf{r}, t)}{\partial t}, \qquad (3.148)$$

$$\mathbf{b}(\mathbf{r}, t) = \nabla \times \mathbf{a}(\mathbf{r}, t). \qquad (3.149)$$

Because r is independent of the time, then $\partial/\partial t' = \partial/\partial t$. However, care must be taken in evaluating the required spatial derivatives because of the presence of r in the retarded time. If one lets the ordered triple (x, y, z) denote the coordinates of the field observation point P and (x', y', z') be the coordinates of the atomic nucleus, then

$$\frac{\partial\mathbf{d}(t')}{\partial x} = \frac{d\mathbf{d}(t')}{dt'}\frac{\partial t'}{\partial x} = \dot{\mathbf{d}}(t')\frac{\partial}{\partial x}\left(t - \frac{r}{c}\right)$$

$$= -\frac{1}{c}\dot{\mathbf{d}}(t')\frac{\partial}{\partial x}\left((x-x')^2 + (y-y')^2 + (z-z')^2\right)^{1/2}$$

$$= -\frac{1}{c}\dot{\mathbf{d}}(t')\frac{x-x'}{r}, \qquad (3.150)$$

with analogous expressions for the partial derivatives with respect to y and z. With Eqs. (3.146) and (3.147), Eq. (3.148) for the electric field vector yields

$$\mathbf{e}(\mathbf{r}, t) \cong \frac{\|4\pi\|}{4\pi\epsilon_0} q_e \left[\nabla \left(\frac{\mathbf{d}(t') \cdot \hat{\mathbf{r}}}{r^2} + \frac{\dot{\mathbf{d}}(t') \cdot \hat{\mathbf{r}}}{cr} \right) + \frac{1}{c^2 r} \ddot{\mathbf{d}}(t') \right].$$

The vector identity $\nabla(\mathbf{U} \cdot \mathbf{V}) = \mathbf{U} \cdot \nabla\mathbf{V} + \mathbf{V} \cdot \nabla\mathbf{U} + \mathbf{U} \times (\nabla \times \mathbf{V}) + \mathbf{V} \times (\nabla \times \mathbf{U})$ applied to the two gradient quantities in the above equation then gives

$$\mathbf{e}(\mathbf{r}, t) \cong \frac{\|4\pi\|}{4\pi\epsilon_0} q_e \left[\mathbf{d}(t') \cdot \nabla \left(\frac{\hat{\mathbf{r}}}{r^2} \right) + \frac{\hat{\mathbf{r}}}{r^2} \cdot \mathbf{d}(t') + \frac{\hat{\mathbf{r}}}{r^2} \times (\nabla \times \mathbf{d}(t')) \right.$$

$$\left. + \frac{1}{c}\dot{\mathbf{d}}(t') \cdot \nabla \left(\frac{\hat{\mathbf{r}}}{r} \right) + \frac{\hat{\mathbf{r}}}{cr} \cdot \nabla\dot{\mathbf{d}}(t') + \frac{\hat{\mathbf{r}}}{cr} \times (\nabla \times \dot{\mathbf{d}}(t')) + \frac{1}{c^2 r}\ddot{\mathbf{d}}(t') \right],$$

$$(3.151)$$

because $\nabla \times (\hat{\mathbf{r}}/r^n) = \mathbf{0}$ for $n = 0, 1, 2, \ldots$.

The various spatial derivatives appearing in Eq. (3.151) are now evaluated. First of all, for any vector field $\mathbf{U}$ and integer value n,

$$\mathbf{U} \cdot \nabla \left(\frac{\hat{\mathbf{r}}}{r^n} \right) = \frac{\mathbf{U}}{r^{n+1}} - (n+1)\frac{\mathbf{U} \cdot \hat{\mathbf{r}}}{r^{n+1}}\hat{\mathbf{r}}, \qquad (3.152)$$

so that

$$\mathbf{d}(t') \cdot \nabla \left(\frac{\hat{\mathbf{r}}}{r^2} \right) = \frac{\mathbf{d}(t')}{r^3} - 3\frac{\mathbf{d}(t') \cdot \hat{\mathbf{r}}}{r^3}\hat{\mathbf{r}}, \qquad (3.153)$$

$$\dot{\mathbf{d}}(t') \cdot \nabla \left(\frac{\hat{\mathbf{r}}}{r^2} \right) = \frac{\dot{\mathbf{d}}(t')}{r^3} - 3\frac{\dot{\mathbf{d}}(t') \cdot \hat{\mathbf{r}}}{r^3}\hat{\mathbf{r}}. \qquad (3.154)$$

With the expression given in Eq. (3.150) for $\partial\mathbf{d}(t')/\partial x$ with analogous expressions for $\partial\mathbf{d}(t')/\partial y$ and $\partial\mathbf{d}(t')/\partial z$, there then results

$$\hat{\mathbf{r}} \cdot \nabla\mathbf{d}(t') = -\frac{1}{c}\dot{\mathbf{d}}(t'), \qquad (3.155)$$

$$\hat{\mathbf{r}} \cdot \nabla\dot{\mathbf{d}}(t') = -\frac{1}{c}\ddot{\mathbf{d}}(t'), \qquad (3.156)$$

and

$$\nabla \times \mathbf{d}(t') = -\frac{1}{c}\hat{\mathbf{r}} \times \dot{\mathbf{d}}(t'), \qquad (3.157)$$

$$\nabla \times \dot{\mathbf{d}}(t') = -\frac{1}{c}\hat{\mathbf{r}} \times \ddot{\mathbf{d}}(t'). \qquad (3.158)$$

Substitution of these expressions into Eq. (3.151) for the electric field vector then gives

$$\mathbf{e}(\mathbf{r}, t) \cong \frac{\|4\pi\|}{4\pi\epsilon_0} q_e \left\{ \frac{\mathbf{d}(t')}{r^3} - \left[3\frac{\mathbf{d}(t')\cdot\hat{\mathbf{r}}}{r^3} + 2\frac{\dot{\mathbf{d}}(t')\cdot\hat{\mathbf{r}}}{cr^2} \right]\hat{\mathbf{r}} \right.$$

$$\left. -\hat{\mathbf{r}}\times\left[\hat{\mathbf{r}}\times\left(\frac{\dot{\mathbf{d}}(t')}{cr^2} + \frac{\ddot{\mathbf{d}}(t')}{c^2r} \right) \right] \right\}. \quad (3.159)$$

Because any vector $\mathbf{U}$ may be resolved into components parallel and perpendicular to the unit vector $\hat{\mathbf{r}}$ as $\mathbf{U} = (\mathbf{U}\cdot\hat{\mathbf{r}})\hat{\mathbf{r}} - \hat{\mathbf{r}}\times(\hat{\mathbf{r}}\times\mathbf{U})$, then the above expression for the electric field vector may accordingly be resolved as

$$\mathbf{e}(\mathbf{r}, t) \cong \frac{\|4\pi\|}{4\pi\epsilon_0} q_e \left\{ \left[2\left(\frac{\mathbf{d}(t')}{r^3} + \frac{\dot{\mathbf{d}}(t')}{cr^2} \right)\cdot\hat{\mathbf{r}} \right]\hat{\mathbf{r}} \right.$$

$$\left. +\hat{\mathbf{r}}\times\left[\hat{\mathbf{r}}\times\left(\frac{\mathbf{d}(t')}{r^3} + \frac{\dot{\mathbf{d}}(t')}{cr^2} + \frac{\ddot{\mathbf{d}}(t')}{c^2r} \right) \right] \right\}. \quad (3.160)$$

For the magnetic field vector, Eqs. (3.147) and (3.149) give

$$\mathbf{b}(\mathbf{r}, t) \cong \left\|\frac{4\pi}{c}\right\| \frac{\mu_0}{4\pi} q_e \hat{\mathbf{r}}\times\left(\frac{\dot{\mathbf{d}}(t')}{r^2} + \frac{\ddot{\mathbf{d}}(t')}{cr} \right), \quad (3.161)$$

where Eq. (3.158) has been used. These two expressions constitute the field vectors of the radiation field produced by an atomic dipole oscillator in the electric dipole approximation, where $t' = t - r/c$ is the retarded time at the stationary position of the atomic nucleus.

3.4.3 The Field Produced by a Monochromatic Dipole Oscillator in the Electric Dipole Approximation

Consider a single atom whose (screened) nucleus with charge $+q_e$ is at rest in the laboratory frame of reference, and where a single bound electron of charge $-q_e$ oscillates at a constant angular frequency ω in an elliptical orbit about the nucleus, as illustrated in Fig. 3.8. The real-valued position vector $\mathbf{d}(t')$ extending from the center of the nucleus to the center of the orbiting electron may then be represented as

$$\mathbf{d}(t') = \Re\left\{ \mathbf{d}_0 e^{-i\omega t'} \right\}$$

$$= \Re\left\{ (\mathbf{d}_0' + i\mathbf{d}_0'') e^{-i\omega t'} \right\}, \quad (3.162)$$

Fig. 3.8 Elliptical orbit of a
bound electron about a fixed
nucleus

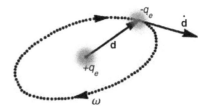

where $\mathbf{d}_0 = \mathbf{d}'_0 + i\mathbf{d}''_0$ is a fixed complex-valued vector with real and imaginary parts
$\mathbf{d}'_0 = \Re\{\mathbf{d}_0\}$ and $\mathbf{d}''_0 = \Im\{\mathbf{d}_0\}$, respectively, that specify the elliptical orbit. Here
$t' = t - r/c$ is the retarded time at the stationary nuclear position, so that

$$e^{-i\omega t'} = e^{-i\omega(t - r/c)}$$

$$= e^{i(kr - \omega t)}, \tag{3.163}$$

where

$$k \equiv \frac{\omega}{c} \equiv \frac{2\pi}{\lambda} \tag{3.164}$$

is the *wavenumber* of the time-harmonic wave motion with *wavelength* λ. This is
an extremely satisfying result, stating that the phenomenon of wave propagation is
a direct consequence of the retardation condition $t' = t - r/c$. The relevant time
derivatives of the electronic position vector $\mathbf{d}(t')$ are then given by

$$\dot{\mathbf{d}}(t') = \Re\left\{-i\omega\mathbf{d}_0 e^{-i\omega t'}\right\},$$

$$\ddot{\mathbf{d}}(t') = \Re\left\{-\omega^2\mathbf{d}_0 e^{-i\omega t'}\right\}.$$

Substitution of these expressions into the pair of relations given in Eqs. (3.160) and
(3.161) then yields the pair of expressions

$$\mathbf{e}(\mathbf{r}, t) \cong -\frac{\|4\pi\|}{4\pi\,\epsilon_0} q_e k^3 \Re\left\{\left[\left(\frac{2}{(kr)^3} - \frac{2i}{(kr)^2}\right)(\mathbf{d}_0 \cdot \hat{\mathbf{r}})\,\hat{\mathbf{r}}\right.\right.$$

$$\left.\left. + \left(\frac{1}{(kr)^3} - \frac{i}{(kr)^2} - \frac{1}{kr}\right)(\hat{\mathbf{r}} \times (\hat{\mathbf{r}} \times \mathbf{d}_0))\right] e^{i(kr - \omega t)}\right\}, \tag{3.165}$$

$$\mathbf{b}(\mathbf{r}, t) \cong -\left\|\frac{4\pi}{c}\right\|\frac{\mu_0 c}{4\pi} q_e k^3 \Re\left\{\left(\frac{i}{(kr)^2} + \frac{1}{kr}\right)(\hat{\mathbf{r}} \times \mathbf{d}_0)e^{i(kr - \omega t)}\right\}, \tag{3.166}$$

for the electric dipole radiation field vectors. From the inequalities given in
Eqs. (3.127) and (3.128), this pair of expressions for the radiation field produced

by a time-harmonic electric dipole are good approximations of the actual radiated field provided that both of the inequalities

$$d_{max} \ll r \quad \Longrightarrow \quad d_0 \ll r \tag{3.167}$$

$$\left. |\dot{\mathbf{d}}| \right|_{max} \ll c \quad \Longrightarrow \quad d_0 \ll \frac{c}{\omega} = \frac{1}{k} \tag{3.168}$$

are satisfied, where $d_0 \equiv |\mathbf{d}_0|$ is the maximum displacement of the bound electron from the nucleus. The pair of relations given in Eqs. (3.165) and (3.166) constitutes the *electric dipole approximation of the radiation field produced by a monochromatic dipole oscillator*.

The space–time evolution of the radiation field vectors given in Eqs. (3.165) and (3.166) is best illustrated for the special case of a linear oscillation of the bound electron. For that special case, choose a spherical coordinate system with polar axis along the $\hat{\mathbf{1}}_z$-direction and let $\mathbf{d}_0 = d_0\hat{\mathbf{1}}_z$. The *electric dipole moment* is then given by

$$\mathbf{p} = p\hat{\mathbf{1}}_z = q_e d_0\hat{\mathbf{1}}_z. \tag{3.169}$$

With this chosen geometry one obtains

$$\mathbf{d}_0 \cdot \hat{\mathbf{r}} = d_0 \cos\theta,$$

$$\hat{\mathbf{r}} \times \mathbf{d}_0 = -\hat{\mathbf{1}}_\phi d_0 \sin\theta,$$

$$\hat{\mathbf{r}} \times (\hat{\mathbf{r}} \times \mathbf{d}_0) = \hat{\mathbf{1}}_\theta d_0 \sin\theta, \tag{3.170}$$

where θ is the angle of declination from the positive z-axis. The polar coordinate representation of the linear dipole radiation field vectors is then seen to be given by

$$\mathbf{e}(\mathbf{r}, t) = \hat{\mathbf{1}}_r e_r(r, \theta, t) + \hat{\mathbf{1}}_\theta e_\theta(r, \theta, t), \tag{3.171}$$

$$\mathbf{b}(\mathbf{r}, t) = \hat{\mathbf{1}}_\phi b_\phi(r, \theta, t), \tag{3.172}$$

with

$$e_r(r, \theta, t) \cong -\frac{\|4\pi\|}{4\pi\epsilon_0} \left(pk^3 \cos\theta \right) \Re\left\{ \left(\frac{2}{(kr)^3} - \frac{2i}{(kr)^2} \right) e^{i(kr-\omega t)} \right\}, \tag{3.173}$$

$$e_\theta(r, \theta, t) \cong -\frac{\|4\pi\|}{4\pi\epsilon_0} \left(pk^3 \sin\theta \right) \Re\left\{ \left(\frac{1}{(kr)^3} - \frac{i}{(kr)^2} - \frac{1}{kr} \right) e^{i(kr-\omega t)} \right\}, \tag{3.174}$$

$$b_\phi(r, \theta, t) \cong \left\|\frac{4\pi}{c}\right\| \frac{\mu_0 c}{4\pi} \left(pk^3 \sin\theta \right) \Re\left\{ \left(\frac{i}{(kr)^2} + \frac{1}{kr} \right) e^{i(kr-\omega t)} \right\}. \tag{3.175}$$

The streamlines of the magnetic field vector are then comprised of concentric circles about the polar axis and lie in planes perpendicular to the polar axis. The streamlines[8] of the electric field vector are independent of the azimuthal angle ϕ and are contained in planes that pass through (or contain) the polar axis. Maps of these streamlines at various instants of time, depicted in Fig. 3.3, were first given by H. Hertz [13] in 1893.

The above set of equations also applies to Hertzian dipole antennas with an appropriately specified dipole moment $\mathbf{p}$ [compare with Eqs. (3.79) and (3.80) for an elemental Hertzian dipole]. Because of the complicated radial dependence indicated in Eqs. (3.173)–(3.175), the field structure naturally separates into three radial zones (the static zone, the intermediate zone, and the wave zone), each of which is now described in some detail.

3.4.3.1 The Static Zone

The *static zone* of the dipole radiation field is defined by the inequality

$$kr \ll 1, \tag{3.176}$$

so that $r \ll \lambda/2\pi$, provided that the additional pair of restrictions $r \gg d_0$ and $d_0 \ll \lambda/2\pi$ are also satisfied [as required by the pair of inequalities given in Eq. (3.167) and (3.168)]. When this inequality is satisfied for an ideal point dipole (obtained in the limit as $d_0 \to 0$ with $p = q_e d_0$ held fixed), all terms appearing in the expressions given in Eqs. (3.173)–(3.175) for the field vectors that contain the factors $(kr)^{-2}$ and $(kr)^{-1}$ are negligible in comparison to the term containing the factor $(kr)^{-3}$. Furthermore, the propagation phase delay is negligible when $kr \ll 1$ so that the approximation $e^{ikr} \approx 1$ is valid throughout the static zone. With these simplifications, the set of relations in Eqs. (3.173)–(3.175) reduce to the pair of expressions

$$\mathbf{e}(\mathbf{r}, t) \cong -\frac{\|4\pi\|}{4\pi\epsilon_0} \frac{p}{r^3} \left(\hat{\mathbf{1}}_r 2\cos\theta + \hat{\mathbf{1}}_\theta \sin\theta \right) \cos(\omega t), \tag{3.177}$$

$$\mathbf{b}(\mathbf{r}, t) \cong \mathbf{0}. \tag{3.178}$$

Hence, in the static zone of an ideal point dipole, the radiation field is essentially the field due to an electrostatic dipole of variable moment $p\cos(\omega t)$.

The relative error that results in approximating Eqs. (3.173)–(3.175) by Eqs. (3.177) and (3.178), given by kr for the r-component and by $kr\sqrt{1 + (kr)^2}$ for the θ-component of the electric field vector, is illustrated in Fig. 3.9. Notice that

---

[8]The streamlines of the electric field vector are obtained from the vector differential relation $d\boldsymbol{\ell} = K\mathbf{e}$, where K is a constant and where $d\boldsymbol{\ell}$ is a differential element of directed length in the chosen coordinate system.

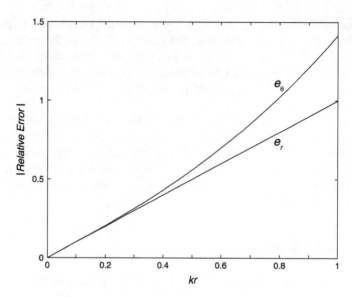

Fig. 3.9 Magnitude of the relative error in the static zone approximation of the radial e_r and angular e_θ components of the radiation field produced by an ideal point dipole at the origin

the error exceeds 10% when $kr \approx 0.1$, in agreement with the inequality $kr \ll 1$ defining the static zone. Although this error goes to zero as $kr \to 0$, this is valid only for the special case of an ideal point dipole. Even in that idealized case, this static zone approximation given in Eqs. (3.177) and (3.178) still breaks down whenever the oscillatory factor $\cos(\omega t)$ approaches zero. Instead of the electric streamlines contracting into and disappearing at the origin, they instead pinch off into closed loops which then propagate outwards from the dipole location, as illustrated by the field streamline sequence presented in Fig. 3.3.

For a time-harmonic linear dipole with finite but sufficiently small size d_0 such that $t'_e \to t' \to t$ throughout the static zone where $kr \ll 1$, the scalar potential may be obtained from a direct application of Coulomb's law in spherical coordinates as

$$\phi(r, \theta, t) \cong \frac{\|4\pi\|}{4\pi \epsilon_0} q_e \left(\frac{1}{r} - \frac{1}{r_e(t)} \right)$$

$$\cong \frac{\|4\pi\|}{4\pi \epsilon_0} \frac{q_e}{r} \left[1 - \left(1 - 2\frac{d(t)}{r} \cos \theta + \frac{d^2(t)}{r^2} \right)^{-1/2} \right] \quad (3.179)$$

when the dipole is chosen to lie along the positive z-axis with the fixed positive charge $+q_e$ located at the origin and the bound negative charge $-q_e$ located at the variable distance $d(t) = d_0 \cos(\omega t)$ from the origin along the z-axis. The field observation point $P(r, \theta)$ is assumed to lie within close proximity to the origin such that $r < 2d_0$, where θ denotes the angle of declination from the positive z-axis. The scalar potential contribution to the static zone electric field is then given by

$$\mathbf{e}(r,\theta,t) = -\nabla\phi(r,\theta,t)$$

$$\cong -\frac{\|4\pi\|}{4\pi\epsilon_0}q_e\left(\hat{\mathbf{1}}_r\frac{\partial}{\partial r}+\hat{\mathbf{1}}_\theta\frac{1}{r}\frac{\partial}{\partial\theta}\right)\left(\frac{1}{r}-\frac{1}{\left(r^2-2rd(t)\cos\theta+d^2(t)\right)^{1/2}}\right)$$

$$\cong -\frac{\|4\pi\|}{4\pi\epsilon_0}\frac{q_e}{\left(r^2-2rd(t)\cos\theta+d^2(t)\right)^{3/2}}$$

$$\times\left\{\hat{\mathbf{1}}_r\left[r-d\cos\theta-r\left(1-2\frac{d}{r}\cos\theta+\frac{d^2}{r^2}\right)^{3/2}\right]+\hat{\mathbf{1}}_\theta d\sin\theta\right\}.$$

$$(3.180)$$

In the limiting case of an ideal point dipole, obtained when $d_0 \to 0$ with the magnitude of the electric dipole moment $p_0 = q_e d_0$ held fixed, this expression reduces to that given in Eq. (3.177).

The corresponding vector potential $\mathbf{a}(r,t) = \hat{\mathbf{1}}_z a(r,\theta,t)$ may be directly obtained from Eq. (3.179) for the scalar potential through the Lorenz condition (3.13) with $\nabla\cdot\mathbf{a}=\partial a/\partial z$ as

$$\frac{\partial a}{\partial z}\cong -\frac{\|4\pi/c\|}{4\pi}\mu_0\frac{q_e}{r}\frac{\partial}{\partial t}\left[1-\left(1-2\frac{d(t)}{r}\cos\theta+\frac{d^2(t)}{r^2}\right)^{-1/2}\right]$$

$$\cong\frac{\|4\pi/c\|}{4\pi}\mu_0\omega p_0\sin(\omega t)\frac{(d(t)-z)r}{\left(r^2-2zd(t)+d^2(t)\right)^{3/2}}.\qquad(3.181)$$

Integration of this expression results in a vector potential yielding both an electric field contribution and a magnetic field that vary as $1/r$ as $r \to 0$, and hence, may be neglected for sufficiently small r.

3.4.3.2 The Intermediate Zone

The *intermediate zone* of the dipole radiation field is roughly defined by the expression

$$kr\approx 1,\qquad(3.182)$$

so that $r \approx \lambda/2\pi$. The intermediate zone occupies the region of space about the dipole oscillator that is just beyond the static zone ($kr \ll 1$) but is not so large that the opposite inequality ($kr \gg 1$) is satisfied, where this latter inequality defines the wave zone of the radiation field. The evolution of the electric field streamlines in the intermediate zone is illustrated in Fig. 3.3. In this intermediate zone, the radial component $e_r(r, \theta, t)$ is appreciable in comparison to the transverse component $e_\theta(r, \theta, t)$ of the electric field vector of the radiation field and the electromagnetic field has not yet settled into a well-defined fixed wavelength. Finally, notice that the intermediate zone is distinguished from the static zone by the formation of closed loops in the electric field streamlines which begin to occur about $r \approx \lambda/2$.

3.4.3.3 The Wave Zone

By following a few sets of closed electric field streamlines as they propagate outward from the dipole oscillator, they are found to rapidly settle into the simple pattern illustrated in Fig. 3.10. Inspection of this figure shows that the radial component of the electric field vector is rapidly becoming negligible in comparison with the θ-component, especially in any small angular region about the $\theta = \pi/2$ direction. Throughout any such angular region the electric field vector is found to rapidly approach (with increasing propagation distance) the condition of being completely transverse to the radial direction from the dipole source location. In addition, the distance between successive zeroes of $e_\theta(r, \theta, t)$ approaches the constant wavelength

$$\lambda = \frac{2\pi}{k} = 2\pi \frac{c}{\omega} \tag{3.183}$$

associated with the constant angular frequency of oscillation ω of the source dipole.

Fig. 3.10 Streamlines of the electric field vector near the onset of the wave zone $(kr \gg 1)$ for a time-harmonic linear electric dipole oscillator

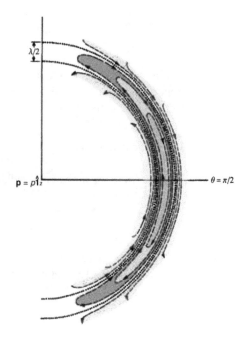

The *wave zone* of the dipole radiation field is defined by the inequality

$$kr \gg 1, \tag{3.184}$$

so that $r \gg \lambda/2\pi$. In that case, the field vectors given in Eqs. (3.171)–(3.175) simplify to the pair of expressions

$$\mathbf{e}(\mathbf{r}, t) \cong \hat{\mathbf{1}}_\theta e_\theta(r, \theta, t)$$

$$\cong \hat{\mathbf{1}}_\theta \frac{\|4\pi\|}{4\pi\epsilon_0} \frac{pk^2}{r} \sin(\theta) \cos(kr - \omega t), \tag{3.185}$$

$$\mathbf{b}(\mathbf{r}, t) \cong \hat{\mathbf{1}}_\phi b_\phi(r, \theta, t)$$

$$\cong \hat{\mathbf{1}}_\phi \left\| \frac{4\pi}{c} \right\| \frac{\mu_0 c}{4\pi} \frac{pk^2}{r} \sin(\theta) \cos(kr - \omega t), \tag{3.186}$$

which become increasingly accurate as $kr \to \infty$. Throughout the wave zone, the dipole radiation field is essentially transverse to the radial direction from the dipole source with wavelength $\lambda = 2\pi c/\omega$. These general features of the wave zone field are readily evident in the field graphs presented in Figs. 3.11 and 3.12. The graph in Fig. 3.11 describes the radial dependence of the transverse field components e_θ and b_ϕ (in Gaussian units) in the equatorial plane ($\theta = \pi/2$) at several successive values of the temporal phase quantity ωt. The radial dependence of these field components

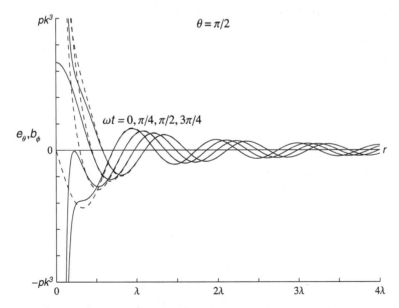

Fig. 3.11 Radial dependence of the transverse field components e_θ and b_ϕ (in gaussian units) in the equatorial plane ($\theta = \pi/2$) at several successive values of the temporal phase quantity ωt

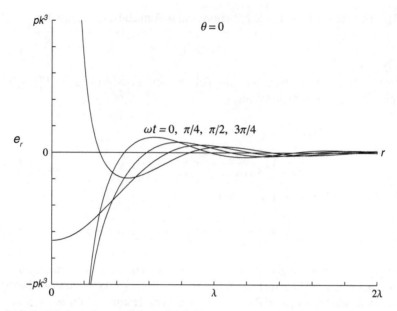

Fig. 3.12 Radial dependence of the radial field components e_r (in gaussian units) along the positive dipole axis ($\theta = 0$) at several successive values of the temporal phase quantity ωt

along a line at any other angle of declination θ may then be obtained from these graphs by multiplying the ordinate values by $\sin(\theta)$. The outward propagation of the radiation field as the time variable t increases is evident in this figure. The graph in Fig. 3.12 describes the radial dependence of the radial electric field component e_r along the polar axis ($\theta = 0$) at several successive values of the temporal phase quantity ωt. The radial dependence of e_r along a line at any other angle θ may then be obtained from this graph by multiplying the ordinate values by $\cos(\theta)$. In addition, this pair of graphs shows that the radial component e_r of the electric field vector becomes negligibly small in comparison to e_θ in the region between $r = 2\lambda$ and $r = 3\lambda$, and that a definite fixed wavelength has appeared in the radiated field structure at about $r = 2\lambda$. As a consequence, the approximate boundary between the intermediate and wave zones is at $r \cong 2\lambda$.

The Poynting vector for the dipole radiation field in the wave zone is directly obtained from Eqs. (3.185) and (3.186) as

$$\mathbf{s}(\mathbf{r}, t) = \left\| \frac{c}{4\pi} \right\| \mathbf{e}(\mathbf{r}, t) \times \mathbf{h}(\mathbf{r}, t) = \left\| \frac{c}{4\pi} \right\| \hat{\mathbf{1}}_r e_\theta(r, \theta, t) h_\phi(r, \theta, t)$$

$$= \hat{\mathbf{1}}_r \frac{\|4\pi\| c}{4\pi \epsilon_0} \frac{p^2 k^4}{4\pi r^2} \sin^2(\theta) \cos^2(kr - \omega t). \tag{3.187}$$

Because $4\pi r^2$ is the surface area of a sphere of radius r centered on the point dipole, the magnitude of the dipole Poynting vector at the radial distance r from the dipole times the surface area of the sphere of radius r centered on the point dipole is independent of r.

3.5   The Complex Potential and the Scalar Optical Field

In source-free regions of space (i.e., in vacuum) the Lorenz and Coulomb gauges become identical when the gauge function is chosen such that the scalar potential in the Lorenz gauge vanishes throughout the region [cf. Eqs. (3.33) and (3.34)]. In that situation the electromagnetic field is completely specified by the vector potential $\mathbf{a}(\mathbf{r}, t)$ that satisfies the subsidiary condition

$$\nabla \cdot \mathbf{a}(\mathbf{r}, t) = 0. \tag{3.188}$$

Based upon the analysis of Green and Wolf [14] in 1953 and Wolf [15] in 1959, let this vector potential possess the spatial Fourier integral representation

$$\mathbf{a}(\mathbf{r}, t) = \frac{1}{(2\pi)^3} \int_K \mathbf{C}(\mathbf{k}, t) e^{i\mathbf{k} \cdot \mathbf{r}} d^3 k, \tag{3.189}$$

the integration being taken over all of real $\mathbf{k}$-space, indicated by the symbol K. Application of the subsidiary condition given in Eq. (3.188) to this representation then yields the orthogonality condition

$$\mathbf{k} \cdot \mathbf{C}(\mathbf{k}, t) = 0 \tag{3.190}$$

that is satisfied for all $\mathbf{k} \in K$. Furthermore, because $\mathbf{a}(\mathbf{r}, t)$ is real-valued, the complex conjugate of Eq. (3.189) gives

$$\begin{aligned} \mathbf{a}(\mathbf{r}, t) &= \frac{1}{(2\pi)^3} \int_K \mathbf{C}^*(\mathbf{k}, t) e^{-i\mathbf{k} \cdot \mathbf{r}} d^3 k \\ &= \frac{1}{(2\pi)^3} \int_K \mathbf{C}(-\mathbf{k}, t) e^{-i\mathbf{k} \cdot \mathbf{r}} d^3 k \end{aligned}$$

so that the symmetry relation

$$\mathbf{C}(-\mathbf{k}, t) = \mathbf{C}^*(\mathbf{k}, t) \tag{3.191}$$

is satisfied for all $\mathbf{k} \in K$.

In order to fully utilize the orthogonality relation given in Eq. (3.190), Wolf [15] introduced a pair of real, mutually orthogonal unit vectors $\hat{\mathbf{l}}_1$ and $\hat{\mathbf{l}}_2$ defined as

$$\hat{\mathbf{l}}_1(\mathbf{k}) \equiv \frac{\mathbf{n} \times \mathbf{k}}{|\mathbf{n} \times \mathbf{k}|}, \quad \hat{\mathbf{l}}_2(\mathbf{k}) \equiv \frac{\mathbf{k} \times \hat{\mathbf{l}}_1}{|\mathbf{k} \times \hat{\mathbf{l}}_1|}, \tag{3.192}$$

where $\mathbf{n}$ is an arbitrary but fixed real vector. In terms of these unit vectors, Wolf [15] then defined a set of *complex basis vectors* as

$$\mathbf{L}(\mathbf{k}) \equiv \hat{\mathbf{l}}_1(\mathbf{k}) + i\hat{\mathbf{l}}_2(\mathbf{k}) \tag{3.193}$$

for $k_z \geq 0$, with

$$\mathbf{L}(-\mathbf{k}) \equiv \mathbf{L}(\mathbf{k}). \tag{3.194}$$

The *complex potential* of the vector potential field $\mathbf{a}(\mathbf{r}, t)$ is then defined as

$$V(\mathbf{r}, t) \equiv \frac{1}{(2\pi)^3} \int_K \gamma(\mathbf{k}, t) e^{i\mathbf{k}\cdot\mathbf{r}} d^3k, \tag{3.195}$$

where

$$\gamma(\mathbf{k}, t) \equiv \mathbf{L}(\mathbf{k}) \cdot \mathbf{C}(\mathbf{k}, t). \tag{3.196}$$

The Fourier inverse of Eq. (3.189), viz.

$$\mathbf{C}(\mathbf{k}, t) = \int_{-\infty}^{\infty} \mathbf{a}(\mathbf{r}, t) e^{-i\mathbf{k}\cdot\mathbf{r}} d^3r, \tag{3.197}$$

then gives, with Eq. (3.196),

$$\gamma(\mathbf{k}, t) = \mathbf{L}(\mathbf{k}) \cdot \int_{-\infty}^{\infty} \mathbf{a}(\mathbf{r}, t) e^{-i\mathbf{k}\cdot\mathbf{r}} d^3r. \tag{3.198}$$

Substitution of this expression into Eq. (3.195) for the complex potential then yields the result

$$V(\mathbf{r}, t) = \int_{-\infty}^{\infty} \mathbf{a}(\mathbf{r}', t) \cdot \mathbf{M}(\mathbf{r}' - \mathbf{r}) d^3r', \tag{3.199}$$

where

$$\mathbf{M}(\mathbf{r}) \equiv \frac{1}{(2\pi)^3} \int_K \mathbf{L}(\mathbf{k}) e^{-i\mathbf{k}\cdot\mathbf{r}} d^3k. \tag{3.200}$$

This result shows that the complex potential $V(\mathbf{r}, t)$ is obtained from a linear transformation of the real-valued vector potential $\mathbf{a}(\mathbf{r}, t)$ with transform kernel $\mathbf{M}$ given by the Fourier transform of the complex basis vectors $\mathbf{L}(\mathbf{k})$. A similar representation in source-free regions has been given by Whittaker [16] in terms of two real scalar wave functions.

Because $\mathbf{k} \cdot \mathbf{C}(\mathbf{k}, t) = 0$, the complex vector $\mathbf{C}(\mathbf{k}, t)$ lies in the plane formed by the two orthogonal unit vectors $\hat{\mathbf{l}}_1(\mathbf{k})$ and $\hat{\mathbf{l}}_2(\mathbf{k})$. The symmetry relations given in Eqs. (3.191) and (3.194), together with the definition in Eq. (3.196), then yield the result

$$\mathbf{C}(\mathbf{k}, t) = \frac{1}{2}\left[\gamma(\mathbf{k}, t)\mathbf{L}^*(\mathbf{k}) + \gamma^*(-\mathbf{k}, t)\mathbf{L}(\mathbf{k})\right]. \tag{3.201}$$

With the Fourier inverse of Eq. (3.195), this relation becomes

$$\mathbf{C}(\mathbf{k}, t) = \frac{1}{2}\left\{\mathbf{L}^*(\mathbf{k}) \int_{-\infty}^{\infty} V(\mathbf{r}, t)e^{-i\mathbf{k}\cdot\mathbf{r}}d^3r + \mathbf{L}(\mathbf{k}) \int_{-\infty}^{\infty} V^*(\mathbf{r}, t)e^{-i\mathbf{k}\cdot\mathbf{r}}d^3r\right\}.$$
$$\tag{3.202}$$

Substitution of this expression into the Fourier integral representation given in Eq. (3.189) then gives

$$\mathbf{a}(\mathbf{r}, t) = \Re \int_{-\infty}^{\infty} V(\mathbf{r}', t)\mathbf{M}^*(\mathbf{r}' - \mathbf{r})d^3r', \tag{3.203}$$

where the symbol $\Re$ denotes the real part. This then is the inverse transformation to that given in Eq. (3.199).

3.5.1 The Wave Equation for the Complex Potential

Because the vector potential $\mathbf{a}(\mathbf{r}, t)$ satisfies the homogeneous vector wave equation

$$\nabla^2 \mathbf{a}(\mathbf{r}, t) - \frac{1}{c^2}\frac{\partial^2 \mathbf{a}(\mathbf{r}, t)}{\partial t^2} = \mathbf{0}, \tag{3.204}$$

in source-free regions [cf. Eq. (3.30)], then each spatial Fourier component $\mathbf{C}(\mathbf{k}, t)$ of $\mathbf{a}(\mathbf{r}, t)$ must satisfy the differential equation

$$\frac{\partial^2 \mathbf{C}(\mathbf{k}, t)}{\partial t^2} + k^2 c^2 \mathbf{C}(\mathbf{k}, t) = \mathbf{0}.$$

Correspondingly, each spatial Fourier component $\gamma(\mathbf{k}, t)$ of the complex potential $V(\mathbf{r}, t)$ must satisfy the differential equation

$$\frac{\partial^2 \gamma(\mathbf{k}, t)}{\partial t^2} + k^2 c^2 \gamma(\mathbf{k}, t) = 0.$$

As a consequence, the complex potential must then satisfy the homogeneous scalar wave equation

$$\nabla^2 V(\mathbf{r}, t) - \frac{1}{c^2}\frac{\partial^2 V(\mathbf{r}, t)}{\partial t^2} = 0, \tag{3.205}$$

The electromagnetic field vectors are then uniquely determined by this single scalar complex potential $V(\mathbf{r}, t)$ in source-free regions of space.

3.5.2 Electromagnetic Energy and Momentum Densities

The electric and magnetic field vectors are given in terms of the vector potential as

$$\mathbf{e}(\mathbf{r}, t) = -\left\|\frac{1}{c}\right\|\frac{\partial \mathbf{a}(\mathbf{r}, t)}{\partial t},$$

$$\mathbf{b}(\mathbf{r}, t) = \nabla \times \mathbf{a}(\mathbf{r}, t),$$

where $\mathbf{a}(\mathbf{r}, t)$ is given by Eq. (3.203) in terms of the complex potential. The total electromagnetic momentum of the field is then given by [cf. Eq. (2.160)]

$$\mathbf{p}_{em}(t) = \frac{\epsilon_0 \mu_0}{\|4\pi c\|}\int_{-\infty}^{\infty}(\mathbf{e}(\mathbf{r}, t) \times \mathbf{h}(\mathbf{r}, t))\, d^3 r$$

$$= -\left\|\frac{1}{4\pi}\right\|\frac{1}{c^2}\int_{-\infty}^{\infty}\frac{\partial \mathbf{a}(\mathbf{r}, t)}{\partial t} \times (\nabla \times \mathbf{a}(\mathbf{r}, t))\, d^3 r. \tag{3.206}$$

Substitution from Eq. (3.203) then yields [14, 15]

$$\mathbf{p}_{em}(t) = -\left\|\frac{1}{4\pi}\right\|\frac{1}{2c^2}\int_{-\infty}^{\infty}\left(\frac{\partial V^*(\mathbf{r}, t)}{\partial t}\nabla V(\mathbf{r}, t) + \frac{\partial V^{(}\mathbf{r}, t)}{\partial t}\nabla V^*(\mathbf{r}, t)\right)d^3 r,$$

$$\tag{3.207}$$

and the electromagnetic momentum density is then given by

$$\mathbf{p}_{em}(\mathbf{r}, t) = -\frac{1}{\|4\pi\|2c^2}\left(\frac{\partial V^*(\mathbf{r}, t)}{\partial t}\nabla V(\mathbf{r}, t) + \frac{\partial V(\mathbf{r}, t)}{\partial t}\nabla V^*(\mathbf{r}, t)\right) \tag{3.208}$$

in terms of the complex potential. The Poynting vector $\mathbf{s} = c^2\bar{\mathbf{p}}_{em}$ is then

$$\mathbf{s}(\mathbf{r}, t) = -\frac{1}{2\|4\pi\|}\left(\frac{\partial V^*(\mathbf{r}, t)}{\partial t}\nabla V(\mathbf{r}, t) + \frac{\partial V^{(}\mathbf{r}, t)}{\partial t}\nabla V^*(\mathbf{r}, t)\right). \tag{3.209}$$

The other relevant quantity appearing in energy conservation and flow in the microscopic electromagnetic field is the electromagnetic energy density [cf. Eq. (2.152)]

$$u(\mathbf{r}, t) = \frac{1}{2\|4\pi\|} \left(\epsilon_0 e^2(\mathbf{r}, t) + \mu_0 h^2(\mathbf{r}, t) \right)$$

$$= \frac{1}{2\|4\pi\|\mu_0} \left[\frac{1}{c^2} \left(\frac{\partial \mathbf{a}(\mathbf{r}, t)}{\partial t} \right)^2 + (\nabla \times \mathbf{a}(\mathbf{r}, t))^2 \right].$$ (3.210)

Substitution from Eq. (3.203) then yields the expression [14, 15]

$$u(\mathbf{r}, t) = \frac{1}{2\|4\pi\|\mu_0} \left(\frac{1}{c^2} \frac{\partial V(\mathbf{r}, t)}{\partial t} \frac{\partial V^*(\mathbf{r}, t)}{\partial t} + \nabla V(\mathbf{r}, t) \cdot \nabla V^*(\mathbf{r}, t) \right).$$ (3.211)

In agreement with Poynting's theorem in source-free regions [cf. Eq. (2.155)], these quantities are found to satisfy the conservation law

$$\nabla \cdot \mathbf{s}(\mathbf{r}, t) + \frac{\partial u(\mathbf{r}, t)}{\partial t} = 0.$$ (3.212)

These expressions for the electromagnetic energy density $u(\mathbf{r}, t)$ and Poynting vector $\mathbf{s}(\mathbf{r}, t)$ in terms of the complex potential $V(\mathbf{r}, t)$ for the microscopic electromagnetic field in source-free regions may be formally identified with the respective quantum mechanical expressions for the probability density and current [15, 17].

3.5.3 A Scalar Representation of the Optical Field

Because $V(\mathbf{r}, t)$ is a complex function, it may be expressed as

$$V(\mathbf{r}, t) \equiv \mathcal{V}(\mathbf{r}, t)e^{i\varphi(\mathbf{r}, t)},$$ (3.213)

where $\mathcal{V}(\mathbf{r}, t)$ is the real amplitude and $\varphi(\mathbf{r}, t)$ the phase of the complex potential. Upon substitution of this representation into the wave equation (3.205) for the complex potential and separation of the result into real and imaginary parts then yields the pair of wave equations

$$\nabla^2 \mathcal{V} - \mathcal{V}(\nabla\varphi)^2 - \frac{1}{c^2} \left(\frac{\partial^2 \mathcal{V}}{\partial t^2} - \mathcal{V} \left(\frac{\partial\varphi}{\partial t} \right)^2 \right) = 0,$$ (3.214)

$$2\nabla\mathcal{V} \cdot \nabla\varphi + \mathcal{V}\nabla^2\varphi - \frac{1}{c^2} \left(2\frac{\partial\mathcal{V}}{\partial t}\frac{\partial\varphi}{\partial t} + \mathcal{V}\frac{\partial^2\varphi}{\partial t^2} \right) = 0.$$ (3.215)

The electromagnetic momentum density given in Eq. (3.208) and the electromagnetic energy density given in Eq. (3.211) may be expressed in terms of the real-valued functions $\mathcal{V}(\mathbf{r}, t)$ and $\varphi(\mathbf{r}, t)$ as

$$\mathfrak{p}_{em}(\mathbf{r}, t) = -\frac{1}{\|4\pi\|c^2}\left[\dot{\mathcal{V}}(\mathbf{r}, t)\nabla\mathcal{V}(\mathbf{r}, t) + \mathcal{V}^2(\mathbf{r}, t)\dot{\varphi}(\mathbf{r}, t)\nabla\varphi(\mathbf{r}, t)\right], \quad (3.216)$$

$$u(\mathbf{r}, t) = \frac{1}{2\|4\pi\|\mu_0}\left[\frac{1}{c^2}\left(\left(\dot{\mathcal{V}}(\mathbf{r}, t)\right)^2 + (\mathcal{V}(\mathbf{r}, t)\dot{\varphi}(\mathbf{r}, t))^2\right)\right.$$

$$\left. + (\nabla\mathcal{V}(\mathbf{r}, t))^2 + (\mathcal{V}(\mathbf{r}, t)\nabla\varphi(\mathbf{r}, t))^2\right]. \quad (3.217)$$

Of particular interest here is the case of a monochromatic (or time-harmonic) wave field where the time enters only through the exponential factor[9] $e^{-i\omega t}$. The amplitude of the complex potential function is then independent of time ($\dot{\mathcal{V}} = 0$) and the phase function is of the form

$$\varphi(\mathbf{r}, t) = k\mathfrak{S}(\mathbf{r}) - \omega t, \quad (3.218)$$

where $k = \omega/c$ is the wavenumber. With this substitution, the pair of wave equations appearing in Eqs. (3.214) and (3.215) become

$$(\nabla\mathfrak{S})^2 - \frac{1}{k^2\mathcal{V}}\nabla^2\mathcal{V} = 1, \quad (3.219)$$

$$\nabla\mathcal{V}\cdot\nabla\mathfrak{S} + \frac{1}{2}\mathcal{V}\nabla^2\mathfrak{S} = 0, \quad (3.220)$$

and Eqs. (3.216) and (3.217) for the electromagnetic momentum and energy densities become

$$\mathfrak{p}_{em}(\mathbf{r}, t) = \frac{k^2}{\|4\pi\|c}\mathcal{V}^2(\mathbf{r})\nabla\mathfrak{S}(\mathbf{r}), \quad (3.221)$$

$$u(\mathbf{r}, t) = \frac{1}{2\|4\pi\|\mu_0}\mathcal{V}^2(\mathbf{r})\left[1 + (\nabla\mathfrak{S}(\mathbf{r}))^2 + \frac{1}{k^2}\left[\nabla\ln(\mathcal{V}(\mathbf{r}))\right]^2\right]. \quad (3.222)$$

Equation (3.221) shows that both the electromagnetic momentum and energy flow are orthogonal to the *cophasal surfaces* defined by

$$\mathfrak{S}(\mathbf{r}) = \text{constant}. \quad (3.223)$$

---

[9]A general monochromatic wave must be represented by a complex potential function of the form $V(\mathbf{r}, t) = V_1(\mathbf{r})e^{-i\omega t} + V_2(\mathbf{r})e^{i\omega t}$, where V_1 and V_2 are complex functions of the position. Wolf [15] has shown that the case considered here in which $V_2 = 0$ corresponds to a monochromatic wave of arbitrary shape but with a spatial average circular polarization.

In addition, Eq. (3.220) may be written in the form

$$\nabla \cdot \left(\mathcal{V}^2(\mathbf{r}) \nabla \mathfrak{S}(\mathbf{r}) \right) = 0, \tag{3.224}$$

so that, from Eq. (3.221), it is found that

$$\nabla \cdot \mathfrak{p}_{em}(\mathbf{r}) = 0. \tag{3.225}$$

Hence, the momentum density of a monochromatic electromagnetic wave field is solenoidal [this result also follows from the conservation law given in Eq. (3.212) because $\partial u / \partial t = 0$]. Finally, Eq. (3.219) may be written in the form of a *generalized free-space eikonal equation*

$$(\nabla \mathfrak{S}(\mathbf{r}))^2 = \mathcal{N}^2(\mathbf{r}) \tag{3.226}$$

where

$$\mathcal{N}(\mathbf{r}) \equiv \left(1 + \frac{1}{k^2 \mathcal{V}(\mathbf{r})} \nabla^2 \mathcal{V}(\mathbf{r}) \right)^{1/2} \tag{3.227}$$

is a *modified refractive index function* that depends upon the amplitude of the complex potential [14].

Consider finally the implication of Eq. (3.220). With the definition of the differential operator [18]

$$\frac{\partial}{\partial \tau} \equiv \nabla \mathfrak{S} \cdot \nabla, \tag{3.228}$$

so that τ specifies the position along the orthogonal trajectories to the cophasal surfaces $\mathfrak{S}(\mathbf{r}) = \text{constant}$, and hence is along the direction of energy flow, Eq. (3.220) becomes

$$\frac{\partial \mathcal{V}}{\partial \tau} + \frac{1}{2} \mathcal{V} \left(\nabla^2 \mathfrak{S} \right) = 0. \tag{3.229}$$

The formal solution to this *transport equation* for the amplitude of the complex potential along these orthogonal trajectories is then given by

$$\mathcal{V}(\tau) = \mathcal{V}(\tau_0) \exp \left\{ -\frac{1}{2} \int_{\tau_0}^{\tau} \nabla^2 \mathfrak{S}(\tau) d\tau \right\}. \tag{3.230}$$

In the infinite frequency limit as $k \to \infty$, Eq. (3.226) reduces to the free-space eikonal equation of geometrical optics [19]

$$(\nabla \mathfrak{S})^2 = 1. \tag{3.231}$$

172   3 Microscopic Potentials and Radiation

In this limiting case the phase function $\mathfrak{S}(\mathbf{r})$ is completely specified by its boundary conditions alone. The cophasal surfaces $\mathfrak{S}(\mathbf{r}) = $ constant are then the wavefronts of geometrical optics and their orthogonal trajectories are the familiar rays of geometrical optics.

3.6   The Four-Potential and Lorentz Invariance

Of final importance to this development of the vector and scalar potentials for the microscopic electromagnetic field is their compatibility with special relativity; in particular, their behavior under a Lorentz transformation between two inertial reference frames. The most direct way to investigate this is through the invariance properties of four-vectors. Consider first the *four-dimensional gradient operator*

$$\Box \equiv \left(\frac{\partial}{\partial x}, \frac{\partial}{\partial y}, \frac{\partial}{\partial z}, \frac{1}{ic} \frac{\partial}{\partial t} \right), \tag{3.232}$$

in gaussian (cgs) units, which leads to the *d'Alembertian operator*

$$\Box^2 \equiv \Box \cdot \Box = \nabla^2 - \frac{1}{c^2} \frac{\partial^2}{\partial t^2}, \tag{3.233}$$

which is *Lorentz invariant* (see Sect. 2.2). With this in mind, define the four-dimensional *"charge current"* vector as

$$\tilde{\mathbf{I}} \equiv (\mathbf{j}, i\rho c) = (j_x, j_y, j_z, i\rho c). \tag{3.234}$$

In order to show that this four-dimensional vector is Lorentz invariant, that is, to show that it is a *four-vector*, all one need do is to show that its "scalar product" with a known four-vector is a Lorentz invariant quantity. Application of the four-dimensional gradient operator (3.232) to this charge–current vector yields the equation of continuity

$$\Box \cdot \tilde{\mathbf{I}} = \nabla \cdot \mathbf{j} + \frac{\partial \rho}{\partial t} = 0, \tag{3.235}$$

and hence, $\tilde{\mathbf{I}}$ is Lorentz invariant.

In a similar manner, the four-dimensional microscopic potential defined by (in cgs units)

$$\tilde{\mathbf{V}} \equiv (\mathbf{a}, i\phi) = (a_x, a_y, a_z, i\phi) \tag{3.236}$$

is a four-vector in the Lorenz gauge because of the Lorenz condition [see Eq. (3.13)], as

$$\square \cdot \tilde{\mathbf{V}} = \nabla \cdot \mathbf{a} + \frac{1}{c} \frac{\partial \phi}{\partial t} = 0, \tag{3.237}$$

but only in the Lorenz gauge. As a consequence, equations for the vector $\mathbf{a}(\mathbf{r}, t)$ and scalar $\phi(\mathbf{r}, t)$ potentials in the Lorenz gauge have the same form in any inertial reference frame. However, under the gauge transformation $\mathbf{a}' = \mathbf{a} - \nabla \psi$, $\phi' = \phi + (1/c)\partial \psi / \partial t$, where ψ is a solution of the wave equation, the transformed four-dimensional potential $\tilde{\mathbf{V}}' = (\mathbf{a}', i\phi')$ does not necessarily form a four-vector (the Coulomb gauge is an example), even though the electromagnetic field vectors they describe are the same as before the gauge transformation is made. Finally, the wave equation for the *four-potential* is

$$\square^2 \tilde{\mathbf{V}} = -\frac{4\pi}{c} \tilde{\mathbf{I}}, \tag{3.238}$$

which is Lorentz invariant.

3.7   The Lagrangian for a System of Charged Particles in an Electromagnetic Field

The microscopic Lorentz force relation $\mathbf{f}(\mathbf{r}, t) = q[\mathbf{e}(\mathbf{r}, t) + \|1/c\|\mathbf{v} \times \mathbf{b}(\mathbf{r}, t)]$ for a charged particle q moving with velocity $\mathbf{v}$ through an electromagnetic field may be expressed in terms of the scalar and vector potentials as

$$\mathbf{f}(\mathbf{r}, t) = q \left[-\nabla \phi(\mathbf{r}, t) - \frac{1}{\|c\|} \left(\frac{\partial \mathbf{a}(\mathbf{r}, t)}{\partial t} - \mathbf{v} \times \nabla \times \mathbf{a}(\mathbf{r}, t) \right) \right]. \tag{3.239}$$

For the x-component of this force, $(\nabla \phi)_x = \partial \phi / \partial x$ and

$$(\mathbf{v} \times \nabla \times \mathbf{a})_x = \frac{\partial}{\partial x}(\mathbf{v} \cdot \mathbf{a}) - \frac{da_x}{dt} + \frac{\partial a_x}{\partial t},$$

where $da_x/dt = \partial a_x/\partial t + \partial(\mathbf{v} \cdot \mathbf{a})/\partial x$ is the x-component of the convective derivative of a_x. The x-component of the Lorentz force (3.239) is then given by

$$f_x(\mathbf{r}, t) = q \left[-\frac{\partial}{\partial x} \left(\phi(\mathbf{r}, t) - \frac{1}{\|c\|} \mathbf{v} \cdot \mathbf{a}(\mathbf{r}, t) \right) - \frac{1}{\|c\|} \frac{d}{dt} \left(\frac{\partial}{\partial v_x} (\mathbf{v} \cdot \mathbf{a}(\mathbf{r}, t)) \right) \right]. \tag{3.240}$$

Because the scalar potential $\phi(\mathbf{r}, t)$ is independent of the velocity $\mathbf{v}$ of the charged particle, this expression may be rewritten as [20]

$$f_x(\mathbf{r}, t) = -\frac{U(\mathbf{r}, t)}{\partial x} + \frac{d}{dt}\frac{\partial U(\mathbf{r}, t)}{\partial v_x}, \tag{3.241}$$

where

$$U(\mathbf{r}, t) \equiv q\phi(\mathbf{r}, t) - \frac{1}{\|c\|}q\mathbf{v} \cdot \mathbf{a}(\mathbf{r}, t), \tag{3.242}$$

defines a *generalized potential* for the microscopic electromagnetic field. The *Lagrangian* for a charged particle moving with velocity $\mathbf{v}(\mathbf{r}, t)$ through an electromagnetic field is then given by [20]

$$L = T - U = T - q\phi + \frac{1}{\|c\|}q\mathbf{v} \cdot \mathbf{a}, \tag{3.243}$$

where $T = \frac{1}{2}mv^2$ is the kinetic energy of the charged particle.

If not all of the forces acting on the charged particle are derivable from a potential, then *Lagrange's equations* may be expressed as [20]

$$\frac{d}{dt}\left(\frac{\partial L}{\partial \dot{\xi}_j}\right) - \frac{\partial L}{\partial \xi_j} = Q_j, \tag{3.244}$$

where the Lagrangian L includes the potential of the conservative forces, as given here by Eq. (3.243), Q_j represents those forces not arising from a potential, and where the ξ_j denote the *generalized coordinates* of the charged particle q and the $\dot{\xi}_j$ are the *Lagrange conjugate momenta*. Once the Lagrangian of a given system is known, Lagrange's equations (3.244) provide the equations of motion for that system.

For a continuous system the Lagrangian L is replaced by the *Lagrangian density* $\mathcal{L}$, where

$$L = \int\int\int \mathcal{L} \, d^3r, \tag{3.245}$$

and *Lagrange's equations* become

$$\frac{d}{dt}\left(\frac{\partial \mathcal{L}}{\partial \dot{\xi}_j}\right) - \frac{\partial \mathcal{L}}{\partial \xi_j} = \mathcal{Q}, \tag{3.246}$$

where $\mathcal{Q}$ is a force density not arising from a potential.

For example, a Lagrangian density leading to the pair of inhomogeneous relations in Maxwell's equations (the two homogeneous equations directly following from the scalar and vector potentials) is given by [20]

$$\mathcal{L} = \frac{1/2}{\|4\pi\|}(\epsilon_0 e^2 - \mu_0 b^2) - \rho\phi + \frac{1}{\|c\|}\mathbf{j} \cdot \mathbf{a}, \tag{3.247}$$

where $\mathbf{e} = -\nabla\phi - (1/\|c\|)\partial\mathbf{a}/\partial t$ and $\mathbf{b} = \nabla \times \mathbf{a}$, and where $\mathbf{j} = \rho\mathbf{v}$ is the microscopic current density. By using ϕ and $\mathbf{a} = (a_1, a_2, a_3)$ as the generalized coordinates of the field, one first finds that

$$\frac{\partial\mathcal{L}}{\partial\phi} = -\rho \quad \& \quad \frac{\partial\mathcal{L}}{\partial(\partial\phi/\partial\xi_k)} = \frac{1}{\|4\pi\|}\epsilon_0 e_k \frac{\partial e_k}{\partial(\partial\phi/\partial\xi_k)} = -\frac{1}{\|4\pi\|}\epsilon_0 e_k$$

and Lagrange's equation (3.246) for ϕ becomes

$$\frac{\epsilon_0}{\|4\pi\|}\sum_k \frac{\partial e_k}{\partial\xi_k} - \rho = 0 \quad \Longrightarrow \quad \nabla \cdot \mathbf{e} = \frac{\|4\pi\|}{\epsilon_0}\rho,$$

which is Gauss' law for the microscopic electric field. For the equations of motion associated with the components of the vector potential $\mathbf{a} = (a_1, a_2, a_3)$, any one component will suffice, say a_1, in which case one finds that

$$\frac{\partial\mathcal{L}}{\partial a_1} = \frac{1}{\|c\|}a_1 \quad \& \quad \frac{\partial\mathcal{L}}{\partial\dot{a}_1} = \frac{1}{\|4\pi\|}\epsilon_0 e_1 \frac{\partial e_1}{\partial\dot{a}_1} = -\frac{1}{\|4\pi c\|}\epsilon_0 e_1,$$

and

$$\frac{\partial\mathcal{L}}{\partial(\partial a_1/\partial\xi_2)} = -\frac{1}{\|4\pi\|\mu_0}b_3 \frac{\partial b_3}{\partial(\partial a_1/\partial\xi_2)} = \frac{1}{\|4\pi\|\mu_0}b_3,$$

$$\frac{\partial\mathcal{L}}{\partial(\partial a_1/\partial\xi_3)} = -\frac{1}{\|4\pi\|\mu_0}b_2.$$

Lagrange's equation (3.246) for a_1 then becomes

$$\frac{1}{\|4\pi\|\mu_0}\left(\frac{\partial b_3}{\partial\xi_2} - \frac{\partial b_2}{\partial\xi_3}\right) - \frac{\epsilon_0}{\|4\pi c\|}\frac{\partial e_1}{\partial t} - \frac{1}{\|c\|}j_1 = 0,$$

which is the 1-component of Ampère's law [cf. Eq. (2.16)]

$$\nabla \times \mathbf{b} = \left\|\frac{4\pi}{c}\right\|\mu_0\mathbf{j} + \frac{\epsilon_0\mu_0}{\|c\|}\frac{\partial\mathbf{e}}{\partial t}.$$

Hence, the four Lagrange's equations with the Lagrangian density defined by Eq. (3.247) using the scalar potential ϕ and the three components of the vector potential $\mathbf{a} = (a_1, a_2, a_3)$ as generalized coordinates are identical with the microscopic Maxwell's equations given in Eqs. (2.15)–(2.18).

The Lagrangian for the electromagnetic field is then given by

$$L = \int\int\int \left[\frac{1/2}{\|4\pi\|} \left(\epsilon_0 e^2 - \mu_0 b^2 \right) - \rho \left(\phi - \frac{1}{\|c\|} \mathbf{v} \cdot \mathbf{a} \right) \right] d^3r. \qquad (3.248)$$

Notice that the quantity $\rho(\phi - \mathbf{v}\cdot\mathbf{a}/\|c\|)$ appearing in the integrand of this Lagrangian also appears as $q(\phi - \mathbf{v} \cdot \mathbf{a}/\|c\|)$ in the Lagrangian (3.243) for a charged particle q moving through an electromagnetic field. With this in mind, consider the volume integral $\int\int\int \rho(\mathbf{r}) \left[\phi(\mathbf{r}) - \frac{1}{\|c\|} \mathbf{v} \cdot \mathbf{a}(\mathbf{r}) \right] d^3r$ with microscopic charge density

$$\rho(\mathbf{r}) = \sum_{i=1}^{n} q_i \delta(\mathbf{r} - \mathbf{r}_i) \qquad (3.249)$$

for a system of n charged particles with charge q_i and position vector $\mathbf{r}_i$, so that

$$\int\int\int \rho(\mathbf{r}) \left[\phi(\mathbf{r}) - \frac{1}{\|c\|} \mathbf{v} \cdot \mathbf{a}(\mathbf{r}) \right] d^3r = \sum_{i=1}^{n} q_i \left[\phi(\mathbf{r}_i) - \frac{1}{\|c\|} \mathbf{v}_i \cdot \mathbf{a}(\mathbf{r}_i) \right].$$

With this result, Eq. (3.248) for the electromagnetic field Lagrangian becomes

$$L = \frac{1/2}{\|4\pi\|} \int\int\int \left(\epsilon_0 e^2 - \mu_0 b^2 \right) d^3r - \sum_{i=1}^{n} q_i \left[\phi(\mathbf{r}_i) - \frac{1}{\|c\|} \mathbf{v}_i \cdot \mathbf{a}(\mathbf{r}_i) \right].$$

$$(3.250)$$

The summation appearing in this expression is recognized as a direct extension of the generalized potential given in Eq. (3.242) to that for a system of n charged particles q_i that produce the electromagnetic forces acting on these same charged particles. With the same generalization of Eq. (3.243) to this system of n charged particles, the total Lagrangian is found to be given by [20]

$$L = \frac{1/2}{\|4\pi\|} \int\int\int \left(\epsilon_0 e^2 - \mu_0 b^2 \right) d^3r$$

$$- \sum_{i=1}^{n} q_i \left[\phi(\mathbf{r}_i) - \frac{1}{\|c\|} \mathbf{v}_i \cdot \mathbf{a}(\mathbf{r}_i) \right] + \frac{1}{2} \sum_{i=1}^{n} m_i v_i^2, \qquad (3.251)$$

the additional summation accounting for the total kinetic energy of the particles. As stated by Goldstein [20], in Eq. (3.251)

we have a total Lagrangian which describes both the electromagnetic field on the one hand, and the mechanical motion of the n particles on the other. It is a function of the generalized coordinates $\phi, \mathbf{a}$ with continuous space indices x_1, x_2, x_3, and the particle generalized coordinates $\mathbf{r}_i$ distinguished by a discrete index i. A single Hamilton's principle thus suffices for both systems! Variation with respect to the potentials produces the Maxwell's equations of the field, while variation with respect to the particle coordinates results in the particle equations of motion.

Finally, because the first term $\frac{1}{2\|4\pi\|}\int\int\int\left(\epsilon_0 e^2 - \mu_0 b^2\right) d^3 r$ in Eq. (3.251) represents the Lagrangian for the electromagnetic field in the absence of the system of charged particles, whereas the last term $\frac{1}{2}\sum_{i=1}^{n} m_i v_i^2$ represents the Lagrangian for the system of charged particles in the absence of the electromagnetic field, the middle term

$$\sum_{i=1}^{n} q_i \left[\phi(\mathbf{r}_i) - \frac{1}{\|c\|}\mathbf{v}_i \cdot \mathbf{a}(\mathbf{r}_i)\right] = \int\int\int \sum_{i=1}^{n} q_i \delta(\mathbf{r} - \mathbf{r}_i)\left[\phi(\mathbf{r}) - \frac{1}{\|c\|}\mathbf{v}_i \cdot \mathbf{a}(\mathbf{r})\right]d^3 r$$

(3.252)

is then seen to provide the mutual interaction between the system of particles and the electromagnetic field that they produce. This formulation therefore completes the classical Maxwell–Lorentz theory of electromagnetism.

3.8   Concluding Remarks

The Lorentz invariance of the scalar and vector potentials in the Lorenz gauge is of some importance to electromagnetic field theory as it shows that the Liénard–Wiechert potentials are relativistically correct. However, because the Liénard–Wiechert potentials require detailed knowledge of the position and velocity of each charged particle source, the results are not consistent with the Heisenberg uncertainty relations for position and momentum ($\Delta x \Delta p \geq \hbar$) as well as for energy and time ($\Delta E \Delta t \geq \hbar$), where $\hbar = h/2\pi$, h being Planck's constant. The first inequality results in an uncertainty in the position and orientation of the dipole source and hence of the radiation pattern, and the second inequality results in an uncertainty in the period of oscillation of the dipole and hence in the frequency and wavelength of the radiated electromagnetic wave. In order to properly address these criticisms, one must turn to quantum theory based upon the classical Lagrangian formulation of Sect. 3.7. As this is outside the scope of this text, the interested reader must accordingly consult other sources in their pursuit of this topic.

Problems

3.1 Prove that the instantaneous Coulomb potential given in Eq. (3.22) is the solution to Poisson's equation $\nabla^2 \phi = -(\|4\pi\|/\epsilon_0)\rho$. Notice that, in part, this entails proving that

$$\nabla^2 \left(\frac{1}{|\mathbf{r} - \mathbf{r}'|}\right) = -4\pi \delta(\mathbf{r} - \mathbf{r}'),$$

where $\delta(\mathbf{r})$ is the three-dimensional Dirac delta function. In order to accomplish this, first prove that $\nabla^2(|\mathbf{r} - \mathbf{r}'|^{-1}) = 0$ when $\mathbf{r} \neq \mathbf{r}'$. When $\mathbf{r} = \mathbf{r}'$, translate the origin of coordinates to $\mathbf{r}'$ and consider the quantity $\nabla^2(1/r)$ where $r = \mathbf{r}$. Upon integration of the quantity $\nabla^2(1/r)$ over an arbitrary volume V that contains the origin, show that

$$\int_V \nabla^2 \left(\frac{1}{r}\right) d^3r = -4\pi,$$

thereby proving the above identity for the three-dimensional delta function.

3.2 Show that the retarded potentials given in Eqs. (3.35) and (3.36) satisfy the Lorenz condition (3.13). Notice that the vector $\mathbf{R}$ is the radius vector from the position of the elementary source element $\rho(\mathbf{r}', t - R/c)d^3r'$ to the fixed observation or field point, so that $\mathbf{v} = -\partial\mathbf{R}/\partial t$, where $\mathbf{j} = \rho\mathbf{v}$.

3.3 From Eqs. (3.62) and (3.63), prove that $\nabla^2\Psi = \Xi$, Notice that, in part, this entails proving that $\nabla^2(\nabla \times \Psi) = \nabla \times (\nabla^2\Psi)$ for any differentiable vector field $\Psi(\mathbf{r})$.

3.4 Derive Eqs. (3.67) and (3.68).

3.5 Obtain an expression for that part $[\mathbf{s}_\times(\mathbf{r}, t)]$ of the Poynting vector that is associated with the cross-terms between the velocity and acceleration parts of the electromagnetic field produced by a moving charged point particle. Show that the radial dependence of this term varies as R^{-3} as $R \to \infty$.

3.6 Through use of the general retarded expressions given in Eqs. (3.115)–(3.116) and (3.118)–(3.119) for the electromagnetic field produced by a moving charged particle, determine the field vectors and the Poynting vector produced by a point charge q that is moving with a fixed velocity $\mathbf{v}$ in terms of the present position vector $\mathbf{R}(t)$ of the particle. From these results, describe the field behavior in each of the two limiting cases obtained as $v \to 0$ and as $v \to c$.

3.7 Prove the differential vector identity $\nabla \times (\hat{\mathbf{r}}/r^n) = \mathbf{0}$ for $n = 0, 1, 2, \ldots$.

3.8 Derive Eq. (3.152).

3.9 Derive Eqs. (3.155)–(3.158).

3.10 Derive the equation for the electric field streamlines for the monochromatic dipole radiation field given in Eqs. (3.171) and (3.173)–(3.174). Hint: Let $e_r \equiv (1/r\sin(\theta))\partial/\partial\theta(f(r,\theta)\sin(\theta))$ and $e_\theta \equiv -(1/r)\partial/\partial r(rf(r,\theta))$.

3.11 Under what conditions, if any, is the dipole radiation field given in Eqs. (3.173)–(3.175) a self-conjugate field.

References

1. E. T. Whittaker, *A History of the Theories of the Aether and Electricity*. London: T. Nelson & Sons, 1951.
2. L. Lorenz, "On the identity of the vibrations of light with electrical currents," *Philos. Mag.*, vol. 34, pp. 287–301, 1867.
3. J. V. Bladel, "Lorenz or Lorentz?," *IEEE Antennas Prop. Mag.*, vol. 33, p. 69, 1991.
4. P. M. Morse and H. Feshbach, *Methods of Theoretical Physics*. New York: McGraw-Hill, 1953. Vol. I.
5. H. Hertz, "Die Kräfte electrischer Schwingungen, behandelt nach der Maxwell'schen Theorie," *Ann. Phys.*, vol. 36, pp. 1–22, 1889.
6. A. Righi, "Electromagnetic fields," *Nuovo Cimento*, vol. 2, pp. 104–121, 1901.
7. A. Nisbet, "Hertzian electromagnetic potentials and associated gauge transformations," *Proc. Roy. Soc. A*, vol. 231, pp. 250–263, 1955.
8. A. Liénard, "Electric and magnetic field produced by a moving charged particle," *L'Éclairage Électrique*, vol. 16, pp. 5–14, 53–59, 106–112, 1898.
9. E. Wiechert, "Electrodynamical laws," *Arch. Néerland.*, vol. 5, pp. 549–573, 1900.
10. W. K. H. Panofsky and M. Phillips, *Classical Electricity and Magnetism*. Reading, MA: Addison-Wesley, 1955. Ch. 19–20.
11. J. M. Stone, *Radiation and Optics, An Introduction to the Classical Theory*. New York: McGraw-Hill, 1963.
12. E. T. Whittaker and G. N. Watson, *Modern Analysis*. London: Cambridge University Press, fourth ed., 1963. p. 133.
13. H. Hertz, *Electric Waves*. London: Macmillan, 1893. English translation.
14. H. S. Green and E. Wolf, "A scalar representation of electromagnetic fields," *Proc. Phys. Soc. A*, vol. 66, no. 12, pp. 1129–1137, 1953.
15. E. Wolf, "A scalar representation of electromagnetic fields: III," *Proc. Phys. Soc.*, vol. 74, pp. 281–289, 1959.
16. E. T. Whittaker, "On an expression of the electromagnetic field due to electrons by means of two scalar potential functions," *Proc. Lond. Math. Soc.*, vol. 1, pp. 367–372, 1904.
17. P. Roman, "A scalar representation of electromagnetic fields: II," *Proc. Phys. Soc.*, vol. 74, pp. 269–280, 1959.
18. A. Nisbet and E. Wolf, "On linearly polarized electromagnetic waves of arbitrary form," *Proc. Camb. Phil. Soc.*, vol. 50, pp. 614–622, 1954.
19. M. Born and E. Wolf, *Principles of Optics*. Cambridge: Cambridge University Press, seventh (expanded) ed., 1999.
20. H. Goldstein, *Classical Mechanics*. Reading: Addison-Wesley, 1950. Sections 1–5 and 11–5.

Chapter 4
Macroscopic Electromagnetics

> *"A curious case of coincidence."* H. A. Lorentz on the
> Lorentz–Lorenz relation.

In the classical Maxwell–Lorentz theory [1–3], matter is regarded as being composed of point charges (e.g., point electrons and point protons and nuclei) that produce microscopic electric and magnetic fields. The microscopic equations of electromagnetics given in Eqs. (2.33)–(2.36) together with the Lorentz force relation given in Eq. (2.21) describe the detailed classical behavior of the charged particles and fields, as presented in Chaps. 2 and 3. The macroscopic equations of electromagnetics, in turn, describe the average behavior of the charged particles and fields. It is then expected that, through a suitable averaging procedure, the macroscopic electromagnetic field equations may be derived from the microscopic equations, a viewpoint that was initially developed by H. A. Lorentz [3] in 1906 and has since been extended by J. H. van Vleck [4], R. Russakoff [5], and F. N. H. Robinson [6].

4.1 Correlation of Microscopic and Macroscopic Electromagnetics

A macroscopically small volume of matter at rest in the laboratory frame typically contains on the order of between 10^{18} and 10^{28} electrons and nuclei that are all in dynamical motion due to, for example, thermal agitation, zero point vibration, or orbital motion. The microscopic electromagnetic fields that are produced by these fundamental charge sources are then seen to vary with extreme rapidity in both space and time. The spatial variations occur over distances with an upper limit set by intermolecular spacing which is of the order of 10^{-8} cm or less, and the temporal fluctuations occur with typical average oscillation periods ranging from 10^{-13} s for nuclear vibrations to 10^{-17} s for electronic orbital motion. Macroscopic measurement instruments typically average over both space and time intervals that

© Springer Nature Switzerland AG 2019
K. E. Oughstun, *Electromagnetic and Optical Pulse Propagation*, Springer Series
in Optical Sciences 224, https://doi.org/10.1007/978-3-030-20835-6_4

are much larger than these characteristic dimensions so that these microscopic fluctuations are usually averaged out. It is then necessary to obtain a macroscopic electromagnetic theory that is complementary to this type of measurement process.

The type of averaging that is appropriate to carry the theory from the microscopic domain to the macroscopic domain must first be carefully considered. First of all, it is well known that the propagation of light in dielectric materials is adequately described by the Maxwell equations with a continuous dielectric permittivity, whereas X-ray diffraction exhibits the atomistic nature of matter. As a consequence, it is reasonable to take the length $\ell_0 \approx 10^{-6}$ cm as the absolute lower limit to the macroscopic domain. The period of oscillation associated with electromagnetic radiation of this wavelength is $\ell_0/c \approx 3 \times 10^{-17}$ s. In a volume of material of size $\ell_0^3 \approx 10^{-18}$ cm^3 there are typically on the order of 10^6 electrons and nuclei. Hence, in any macroscopically small region of ponderable media with linear dimensions $\ell \gg \ell_0$, there are so many microscopic charged particles that their random fluctuations will be completely washed out by a spatial averaging procedure because, in the absence of any special preparation and the establishment of ordering over macroscopic distances, the temporal variations of the microscopic fields are uncorrelated over distances of order ℓ. All that survives from a spatial averaging are the frequency components that correspond to oscillations that are driven at the external, applied frequencies.

4.1.1   Spatial Average of the Microscopic Field Equations

The spatial average of a microscopic function $f(\mathbf{r}, t)$ of position and time is defined as

$$\langle\langle f(\mathbf{r}, t)\rangle\rangle \equiv \int w(\mathbf{r}')f(\mathbf{r} - \mathbf{r}', t)d^3r', \tag{4.1}$$

where $w(\mathbf{r})$ is a real-valued, positive, sufficiently well-behaved function that is nonzero only in some nonvanishing region of space surrounding the point $\mathbf{r} = \mathbf{0}$. This "weighting" function is normalized to unity over all of space as

$$\int_{-\infty}^{\infty} w(\mathbf{r})d^3r = 1, \tag{4.2}$$

and varies sufficiently slowly such that the first three terms in the local series expansion [5]

$$w(\mathbf{r} + \mathbf{d}) = w(\mathbf{r}) + (\mathbf{d} \cdot \nabla)w(\mathbf{r}) + \frac{1}{2}(\mathbf{d} \cdot \nabla)^2 w(\mathbf{r}) + \cdots \tag{4.3}$$

provides a sufficiently accurate approximation to this weighting function when $d = |\mathbf{d}|$ is of the order of the molecular size.

The operations of space and time differentiation commute with this spatial-averaging operation, as

$$\frac{\partial}{\partial x_j} \langle\langle f(\mathbf{r}, t)\rangle\rangle = \int w(\mathbf{r}') \frac{\partial f(\mathbf{r} - \mathbf{r}', t)}{\partial x_j} d^3 r'$$

$$= \left\langle\left\langle \frac{\partial f(\mathbf{r}, t)}{\partial x_j} \right\rangle\right\rangle, \tag{4.4}$$

and

$$\frac{\partial}{\partial t} \langle\langle f(\mathbf{r}, t)\rangle\rangle = \int w(\mathbf{r}') \frac{\partial f(\mathbf{r} - \mathbf{r}', t)}{\partial t} d^3 r'$$

$$= \left\langle\left\langle \frac{\partial f(\mathbf{r}, t)}{\partial t} \right\rangle\right\rangle. \tag{4.5}$$

It then follows that the spatial differential operator ∇ also commutes with the spatial-averaging operation in the usual gradient, divergence, curl and Laplacian operations.

The macroscopic electric $\mathbf{E}(\mathbf{r}, t)$ and magnetic $\mathbf{B}(\mathbf{r}, t)$ field vectors are defined as the spatial averages of their respective microscopic field vectors $\mathbf{e}(\mathbf{r}, t)$ and $\mathbf{b}(\mathbf{r}, t)$ as

$$\mathbf{E}(\mathbf{r}, t) \equiv \langle\langle \mathbf{e}(\mathbf{r}, t)\rangle\rangle, \tag{4.6}$$

$$\mathbf{B}(\mathbf{r}, t) \equiv \langle\langle \mathbf{b}(\mathbf{r}, t)\rangle\rangle. \tag{4.7}$$

The spatial average of the microscopic field equations given in Eqs. (2.33)–(2.36) then gives

$$\nabla \cdot \mathbf{E}(\mathbf{r}, t) = \frac{\|4\pi\|}{\epsilon_0} \langle\langle \rho(\mathbf{r}, t)\rangle\rangle, \tag{4.8}$$

$$\nabla \cdot \mathbf{B}(\mathbf{r}, t) = 0, \tag{4.9}$$

$$\nabla \times \mathbf{E}(\mathbf{r}, t) = -\left\|\frac{1}{c}\right\| \frac{\partial \mathbf{B}(\mathbf{r}, t)}{\partial t}, \tag{4.10}$$

$$\nabla \times \mathbf{B}(\mathbf{r}, t) = \left\|\frac{4\pi}{c}\right\| \mu_0 \langle\langle \mathbf{j}(\mathbf{r}, t)\rangle\rangle + \left\|\frac{1}{c}\right\| \epsilon_0 \mu_0 \frac{\partial \mathbf{E}(\mathbf{r}, t)}{\partial t}. \tag{4.11}$$

The two spatially-averaged homogeneous equations (4.9) and (4.10) remain the same as their microscopic counterparts. For the spatially-averaged inhomogeneous equations the spatial averages $\langle\langle \rho(\mathbf{r}, t)\rangle\rangle$ and $\langle\langle \mathbf{j}(\mathbf{r}, t)\rangle\rangle$ remain to be determined, the derived macroscopic field vectors $\mathbf{D}(\mathbf{r}, t)$ and $\mathbf{H}(\mathbf{r}, t)$ being introduced by the extraction from $\langle\langle \rho(\mathbf{r}, t)\rangle\rangle$ and $\langle\langle \mathbf{j}(\mathbf{r}, t)\rangle\rangle$ of certain contributions that can be identified with the bulk properties of the material medium [7–9].

4.1.2   Spatial Average of the Charge Density

Consider a material medium that is comprised of molecules composed of nuclei and electrons that are bound together and, in addition, that contains "free" charges that are not localized around any particular molecule of the material. The microscopic charge density may then be separated into two groups: *bound charge density* $\rho_b(\mathbf{r}, t)$ belonging to the molecular structure of the material and *free charge density* $\rho_f(\mathbf{r}, t)$ accounting for the conduction current in the material. The microscopic charge density is then given by

$$\rho(\mathbf{r}, t) = \rho_b(\mathbf{r}, t) + \rho_f(\mathbf{r}, t), \tag{4.12}$$

where

$$\rho_f(\mathbf{r}, t) = \sum_{\substack{j \\ (\text{free})}} q_j \delta\left(\mathbf{r} - \mathbf{r}_j(t)\right) \tag{4.13}$$

is the *microscopic free charge density*, and

$$\rho_b(\mathbf{r}, t) = \sum_{\substack{n \\ (\text{mol})}} \rho_n(\mathbf{r}, t) \tag{4.14}$$

is the *microscopic bound charge density*, where

$$\rho_n(\mathbf{r}, t) = \sum_{j(n)} q_j \delta\left(\mathbf{r} - \mathbf{r}_j(t)\right), \tag{4.15}$$

the summation extending over the bound charge in the nth molecule. Consider first taking the spatial average of the microscopic charge density in the nth molecule and then summing the contributions from all of the molecules in the material. In order to accomplish this, it is appropriate to first express the coordinates of the charges in the nth molecule with respect to an origin O' that is at rest with respect to that molecule (for example, at the center of mass of the molecule). Let the position vector of that fixed point be denoted by $\mathbf{r}_n(t)$ relative to the fixed origin O and let the position vector of the jth charge in the molecule be denoted by $\mathbf{r}_j(t)$ relative to O as well as by $\mathbf{r}_{jn}(t)$ relative to O', as illustrated in Fig. 4.1, where

$$\mathbf{r}_j(t) = \mathbf{r}_n(t) + \mathbf{r}_{jn}(t). \tag{4.16}$$

The spatial average of the microscopic charge density in the nth molecule is then given by

$$\langle\langle \rho_n(\mathbf{r}, t) \rangle\rangle = \int w(\mathbf{r}') \rho_n(\mathbf{r} - \mathbf{r}', t) d^3 r'$$

Fig. 4.1 Position vectors $\mathbf{r}_n$ for the nth molecule and $\mathbf{r}_j$ for the jth electron in that molecule with respect to the fixed origin O, and $\mathbf{r}_{jn} = \mathbf{r}_j - \mathbf{r}_n$ for the relative position of the jth electron with respect to the position O' of the molecule

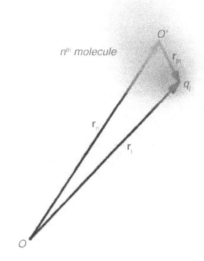

n^{th} molecule

$$= \sum_{j(n)} q_j \int w(\mathbf{r}')\delta(\mathbf{r} - \mathbf{r}' - \mathbf{r}_{jn}(t) - \mathbf{r}_n(t), t)d^3r'$$

$$= \sum_{j(n)} q_j w(\mathbf{r} - \mathbf{r}_n(t) - \mathbf{r}_{jn}(t)). \tag{4.17}$$

Because $|\mathbf{r}_{jn}|$ is on the order of atomic dimensions, the terms in the above summation have arguments that differ only slightly from the vector $\mathbf{r} - \mathbf{r}_n$ on the spatial scale over which the weighting function $w(\mathbf{r})$ changes appreciably. One may then expand the function $w(\mathbf{r} - \mathbf{r}_n - \mathbf{r}_{jn})$ about the point $\mathbf{r} - \mathbf{r}_n$ for each term in the summation appearing in Eq. (4.17) so that, with application of Eq. (4.3), there results

$$\langle\langle \rho_n(\mathbf{r}, t) \rangle\rangle = \sum_{j(n)} q_j \big[w(\mathbf{r} - \mathbf{r}_n) - (\mathbf{r}_{jn} \cdot \nabla)w(\mathbf{r} - \mathbf{r}_n)$$

$$+ \frac{1}{2}(\mathbf{r}_{jn} \cdot \nabla)^2 w(\mathbf{r} - \mathbf{r}_n) + \cdots \big]$$

$$= q_n w(\mathbf{r} - \mathbf{r}_n) - \mathbf{p}_n \cdot \nabla w(\mathbf{r} - \mathbf{r}_n) + \frac{1}{6}\underline{Q}'_n \cdot \nabla \cdot \nabla w(\mathbf{r} - \mathbf{r}_n) + \cdots . \tag{4.18}$$

The quantities introduced here, which entail summations over the charges in the molecule, are the *molecular multipole moments*

$$\text{Molecular Charge (Monopole Moment):} \quad q_n \equiv \sum_{j(n)} q_j, \tag{4.19}$$

Molecular Dipole Moment Vector: $\quad \mathbf{p}_n \equiv \sum_{j(n)} q_j \mathbf{r}_{jn},$ $\qquad$ (4.20)

Molecular Quadrupole Moment Tensor: $\quad \underline{Q}'_n \equiv \sum_{j(n)} q_j \mathbf{r}_{jn} \mathbf{r}_{jn},$ $\qquad$ (4.21)

where the elements of the molecular quadrupole moment tensor are given by $(\underline{Q}'_n)_{\alpha\beta} = 3\sum_{j(n)} q_j (\mathbf{r}_{jn})_\alpha (\mathbf{r}_{jn})_\beta$, the indices α and β taking on the values $1, 2, 3$, and so-on for higher-order moments.

When Eq. (4.18) is viewed in light of the definition of the spatial average given in Eq. (4.1), the first term is seen to be the spatial average of a point charge density located at $\mathbf{r} = \mathbf{r}_n$, the second term as the divergence of the average of a point dipole density at $\mathbf{r} = \mathbf{r}_n$, and the third term as the second spatial derivative of the average of a point quadrupole density at $\mathbf{r} = \mathbf{r}_n$. The expression appearing in Eq. (4.18) for the spatial average of the microscopic charge density may then be written as [5, 10]

$$\langle\langle \rho_n(\mathbf{r},t) \rangle\rangle = \langle\langle q_n \delta(\mathbf{r} - \mathbf{r}_n) \rangle\rangle - \nabla \cdot \langle\langle \mathbf{p}_n \delta(\mathbf{r} - \mathbf{r}_n) \rangle\rangle$$

$$+ \frac{1}{6} \nabla \cdot \nabla \cdot \langle\langle \underline{Q}'_n \delta(\mathbf{r} - \mathbf{r}_n) \rangle\rangle + \cdots . \qquad (4.22)$$

Through the process of spatial averaging, each molecule is viewed as a collection of point multipoles located at a single fixed point O' in the molecule (see Fig. 4.1). Although the detailed structure of the molecular charge distribution is important at the microscopic level, at the macroscopic level it is replaced by a sum of point multipoles at each molecular site.

The total microscopic charge density given in Eq. (4.12) consists of both free and bound charges. Upon taking the summation of Eq. (4.22) over all of the molecules in the material (which may be of different species) and combining this result with the spatial average of the free charge density, the spatial average of the (total) microscopic charge density is obtained as

$$\langle\langle \rho(\mathbf{r},t) \rangle\rangle = \langle\langle \rho_f(\mathbf{r},t) \rangle\rangle + \langle\langle \rho_b(\mathbf{r},t) \rangle\rangle$$

$$= \left\langle\left\langle \sum_{j \atop (\text{free})} q_j \delta(\mathbf{r} - \mathbf{r}_j(t)) \right\rangle\right\rangle + \left\langle\left\langle \sum_{n \atop (\text{mol})} \rho_n(\mathbf{r},t) \right\rangle\right\rangle$$

$$= \varrho(\mathbf{r},t) - \nabla \cdot \mathbf{P}(\mathbf{r},t) + \nabla \cdot \nabla \cdot \underline{Q}'(\mathbf{r},t) + \cdots , \qquad (4.23)$$

where $\nabla \cdot \nabla \cdot \underline{Q}'(\mathbf{r},t) = \sum_{\alpha=1}^{3} \sum_{\beta=1}^{3} \frac{\partial^2}{\partial x_\alpha \partial x_\beta} Q'_{\alpha\beta}(\mathbf{r},t)$. Here $\varrho(\mathbf{r},t)$ is the *macroscopic charge density*

$$\varrho(\mathbf{r},t) = \varrho_f(\mathbf{r},t) + \varrho_b(\mathbf{r},t) \qquad (4.24)$$

that is comprised of the *macroscopic free charge density*

$$\varrho_f(\mathbf{r}, t) \equiv \left\langle\!\!\left\langle \sum_{\substack{j \\ (\text{free})}} q_j \delta(\mathbf{r} - \mathbf{r}_j(t)) \right\rangle\!\!\right\rangle \tag{4.25}$$

plus the *macroscopic bound charge density*

$$\varrho_b(\mathbf{r}, t) \equiv \left\langle\!\!\left\langle \sum_{\substack{n \\ (\text{mol})}} \sum_{j(n)} q_j \delta(\mathbf{r} - \mathbf{r}_n(t)) \right\rangle\!\!\right\rangle . \tag{4.26}$$

In addition, $\mathbf{P}(\mathbf{r}, t)$ is the *macroscopic polarization density*

$$\mathbf{P}(\mathbf{r}, t) \equiv \left\langle\!\!\left\langle \sum_{\substack{n \\ (\text{mol})}} \sum_{j(n)} q_j \mathbf{r}_{jn} \delta(\mathbf{r} - \mathbf{r}_n(t)) \right\rangle\!\!\right\rangle , \tag{4.27}$$

and $\underline{Q}'(\mathbf{r}, t)$ is the *macroscopic quadrupole moment density tensor*

$$\underline{Q}'(\mathbf{r}, t) \equiv \frac{1}{2} \left\langle\!\!\left\langle \sum_{\substack{n \\ (\text{mol})}} \sum_{j(n)} q_j \mathbf{r}_{jn} \mathbf{r}_{jn} \delta(\mathbf{r} - \mathbf{r}_n(t)) \right\rangle\!\!\right\rangle \tag{4.28}$$

with elements

$$Q'_{\alpha\beta}(\mathbf{r}, t) = \frac{1}{2} \left\langle\!\!\left\langle \sum_{\substack{n \\ (\text{mol})}} \sum_{j(n)} q_j (\mathbf{r}_{jn})_\alpha (\mathbf{r}_{jn})_\beta \delta(\mathbf{r} - \mathbf{r}_n(t)) \right\rangle\!\!\right\rangle , \tag{4.29}$$

the indices α and β taking on the values 1, 2, 3.

Because not all components of the quadrupole moment tensor are independent, a traceless microscopic molecular quadrupole moment density is defined as [8, 10]

$$\begin{aligned} \left(\underline{Q}_n\right)_{\alpha\beta} &\equiv \left(\underline{Q}'_n\right)_{\alpha\beta} - \sum_{j(n)} q_j \left(\mathbf{r}_{jn}\right)^2 \delta_{\alpha\beta} \\ &= \sum_{j(n)} q_j \left(3(\mathbf{r}_{jn})_\alpha \mathbf{r}_{jn})_\alpha - \left(\mathbf{r}_{jn}\right)^2 \delta_{\alpha\beta} \right) . \end{aligned} \tag{4.30}$$

Define the *mean square charge radius* $\overline{r_n^2}$ of the molecular charge distribution for the nth molecule as

$$\hat{q}\, \overline{r_n^2} \equiv \sum_{j(n)} q_j \left(\mathbf{r}_{jn}\right)^2 , \tag{4.31}$$

where $\hat{q}$ is some convenient unit of (positive) charge such as that of a proton. With this substitution, Eq. (4.30) may be written as

$$(\underline{Q}'_n)_{\alpha\beta} = (\underline{Q}_n)_{\alpha\beta} + \hat{q}\,\overline{r_n^2}\,\delta_{\alpha\beta}. \tag{4.32}$$

The elements of the macroscopic quadrupole moment density given in Eq. (4.29) then become

$$Q'_{\alpha\beta}(\mathbf{r}, t) = Q_{\alpha\beta}(\mathbf{r}, t) + \frac{1}{6}\left\langle\!\left\langle \sum_{\substack{n \\ \text{(mol)}}} \hat{q}\,\overline{r_n^2}\,\delta_{\alpha\beta}\delta(\mathbf{r} - \mathbf{r}_n(t)) \right\rangle\!\right\rangle. \tag{4.33}$$

The net result is that, in Eq. (4.23) for the spatial average of the microscopic charge density, the traceless macroscopic quadrupole moment density tensor $\underline{Q}(\mathbf{r}, t)$ replaces $\underline{Q}'(\mathbf{r}, t)$ and the macroscopic bound charge density $\varrho_b(\mathbf{r}, t)$ is augmented by an additional term so that Eq. (4.26) is replaced by[1]

$$\varrho_b(\mathbf{r}, t) = \left\langle\!\left\langle \sum_{\substack{n \\ \text{(mol)}}} q_n\delta(\mathbf{r} - \mathbf{r}_n(t)) \right\rangle\!\right\rangle + \frac{1}{6}\nabla^2\left\langle\!\left\langle \sum_{\substack{n \\ \text{(mol)}}} \hat{q}r_n^2\delta(\mathbf{r} - \mathbf{r}_n(t)) \right\rangle\!\right\rangle. \tag{4.34}$$

The trace of the macroscopic quadrupole tensor $\underline{Q}'$ is grouped with the macroscopic charge density ϱ because it is an $\ell = 0$ contribution from the multipole expansion of the molecular charge distribution of the medium [10]. Taken together, the molecular charge and mean square radius terms represent the first two terms in an expansion of the $\ell = 0$ molecular multipole as one goes beyond the infinite wavelength-zero wavenumber limit of electrostatics [6].

4.1.3 Spatial Average of the Current Density

The microscopic current density can also be separated into two groups: a *microscopic "bound" current density* $\mathbf{j}_b(\mathbf{r}, t)$ due to the motion of bound charge belonging to the molecular structure of the material, and the *microscopic free current density* $\mathbf{j}_f(\mathbf{r}, t)$ due to the motion of free charge in the material giving rise to the conduction current. The microscopic current density is then given by

$$\mathbf{j}(\mathbf{r}, t) = \mathbf{j}_b(\mathbf{r}, t) + \mathbf{j}_f(\mathbf{r}, t), \tag{4.35}$$

---

[1]Notice that the quantity $\nabla \cdot \mathbf{P}(\mathbf{r}, t)$ is sometimes referred to as the bound charge density. In order to avoid confusion with the bound charge density described in Eq. (4.34), the quantity $\nabla \cdot \mathbf{P}(\mathbf{r}, t)$ is referred to here by its proper name, the polarization charge density.

where

$$\mathbf{j}_f(\mathbf{r}, t) = \sum_{\substack{j \\ \text{(free)}}} q_j \mathbf{v}_j \delta(\mathbf{r} - \mathbf{r}_j(t)) \tag{4.36}$$

is the microscopic free current density, and

$$\mathbf{j}_b(\mathbf{r}, t) = \sum_{\substack{n \\ \text{(mol)}}} \mathbf{j}_n(\mathbf{r}, t) \tag{4.37}$$

is the microscopic bound current density with

$$\mathbf{j}_n(\mathbf{r}, t) = \sum_{j(n)} q_j \mathbf{v}_j \delta(\mathbf{r} - \mathbf{r}_j(t)) \tag{4.38}$$

describing the microscopic current density of the nth molecule in the material. Here $\mathbf{v}_j = d\mathbf{r}_j/dt$ denotes the velocity of the jth charge (bound or free).

The spatial average of the current density in the nth molecule is obtained in the same manner as that in the derivation of Eq. (4.17) for the charge density. For the current density

$$\langle\langle \mathbf{j}_n(\mathbf{r}, t) \rangle\rangle = \int w(\mathbf{r}')\mathbf{j}_n(\mathbf{r} - \mathbf{r}', t) d^3 r'$$

$$= \sum_{j(n)} q_j \int w(\mathbf{r}')\mathbf{v}_j \delta(\mathbf{r} - \mathbf{r}' - \mathbf{r}_j(t)) d^3 r'$$

$$= \sum_{j(n)} q_j \left(\mathbf{v}_{jn} + \mathbf{v}_n\right) \int w(\mathbf{r}')\delta(\mathbf{r} - \mathbf{r}' - \mathbf{r}_n - \mathbf{r}_{jn}) d^3 r'$$

$$= \sum_{j(n)} q_j \left(\mathbf{v}_{jn} + \mathbf{v}_n\right) w(\mathbf{r} - \mathbf{r}_n - \mathbf{r}_{jn}), \tag{4.39}$$

where $\mathbf{r}_j(t) = \mathbf{r}_n(t) + \mathbf{r}_{jn}(t)$ so that (assuming nonrelativistic motion of the charged particles) $\mathbf{v}_j(t) = \mathbf{v}_n(t) + \mathbf{v}_{jn}(t)$, where $\mathbf{v}_n(t)$ describes the velocity of the origin O' in the nth molecule (see Fig. 4.1) and where $\mathbf{v}_{jn}$ denotes the internal relative velocity of the jth charged particle in that molecule. Upon expanding the weighting function $w(\mathbf{r} - \mathbf{r}_n - \mathbf{r}_{jn})$ about the point $\mathbf{r} - \mathbf{r}_n$ in each term of the above series, there results

$$\langle\langle \mathbf{j}_n(\mathbf{r}, t) \rangle\rangle = \sum_{j(n)} q_j \left(\mathbf{v}_{jn} + \mathbf{v}_n\right) \times \left[w(\mathbf{r} - \mathbf{r}_n) - (\mathbf{r}_{jn} \cdot \nabla)w(\mathbf{r} - \mathbf{r}_n) \right.$$

$$\left. + \frac{1}{2}(\mathbf{r}_{jn} \cdot \nabla)^2 w(\mathbf{r} - \mathbf{r}_n) + \cdots \right]$$

$$= q_n \mathbf{v}_n w(\mathbf{r} - \mathbf{r}_n) + \frac{d\mathbf{p}_n}{dt} w(\mathbf{r} - \mathbf{r}_n) - \mathbf{v}_n (\mathbf{p}_n \cdot \nabla) w(\mathbf{r} - \mathbf{r}_n)$$

$$+ \frac{1}{6} \mathbf{v}_n \left(\underline{Q}'_n \cdot \nabla \cdot \nabla \right) w(\mathbf{r} - \mathbf{r}_n) - \sum_{j(n)} q_j \mathbf{v}_{jn} (\mathbf{r}_{jn} \cdot \nabla) w(\mathbf{r} - \mathbf{r}_n)$$

$$+ \cdots , \tag{4.40}$$

where q_n is the molecular charge, $\mathbf{p}_n$ the molecular dipole moment, and $\underline{Q}'_n$ the molecular quadrupole moment, as defined in Eqs. (4.19)–(4.21).

The *molecular magnetic moment* is defined as

$$\mathbf{m}_n \equiv \frac{1}{2\|c\|} \sum_{j(n)} q_j (\mathbf{r}_{jn} \times \mathbf{v}_{jn}). \tag{4.41}$$

In addition, it is seen that

$$\frac{d}{dt} (\mathbf{p}_n w(\mathbf{r} - \mathbf{r}_n)) + \nabla \times (\mathbf{p}_n \times \mathbf{v}_n w(\mathbf{r} - \mathbf{r}_n))$$

$$= \frac{d\mathbf{p}_n}{dt} w(\mathbf{r} - \mathbf{r}_n) - \mathbf{v}_n (\mathbf{v}_n \cdot \nabla) w(\mathbf{r} - \mathbf{r}_n),$$

and

$$\nabla \times (\|c\| \mathbf{m}_n w(\mathbf{r} - \mathbf{r}_n)) - \frac{1}{6} \frac{d}{dt} \left(\underline{Q}'_n \cdot \nabla w(\mathbf{r} - \mathbf{r}_n) \right)$$

$$- \frac{1}{6} \nabla \times \nabla \cdot \left(\underline{Q}'_n \times \nabla w(\mathbf{r} - \mathbf{r}_n) \right)$$

$$= \frac{1}{6} \mathbf{v}_n \left(\underline{Q}'_n \cdot \nabla \cdot \nabla \right) w(\mathbf{r} - \mathbf{r}_n) - \sum_{j(n)} q_j \mathbf{v}_{jn} (\mathbf{r}_{jn} \cdot \nabla) w(\mathbf{r} - \mathbf{r}_n).$$

With these substitutions, Eq. (4.40) becomes

$$\langle \langle \mathbf{j}_n(\mathbf{r}, t) \rangle \rangle = q_n \mathbf{v}_n w(\mathbf{r} - \mathbf{r}_n) + \frac{d}{dt} (\mathbf{p}_n w(\mathbf{r} - \mathbf{r}_n)) + \nabla \times (\mathbf{p}_n \times \mathbf{v}_n w(\mathbf{r} - \mathbf{r}_n))$$

$$+ \frac{1}{6} \frac{d}{dt} \left(\underline{Q}'_n \cdot \nabla w(\mathbf{r} - \mathbf{r}_n) \right) + \|c\| \nabla \times \mathbf{m}_n w(\mathbf{r} - \mathbf{r}_n))$$

$$- \frac{1}{6} \nabla \times \nabla \cdot \left[\underline{Q}'_n \times \mathbf{v}_n w(\mathbf{r} - \mathbf{r}_n)) \right] + \cdots$$

$$= \langle q_n \mathbf{v}_n \delta(\mathbf{r} - \mathbf{r}_n) \rangle + \frac{d}{dt} \left[\langle \mathbf{p}_n \delta(\mathbf{r} - \mathbf{r}_n) \rangle - \frac{1}{6} \nabla \cdot \langle \underline{Q}'_n \delta(\mathbf{r} - \mathbf{r}_n) \rangle \right]$$

$$+ \|c\| \nabla \times \langle \mathbf{m}_n \delta(\mathbf{r} - \mathbf{r}_n) \rangle$$

$$+ \nabla \times \left[\langle \mathbf{p}_n \times \mathbf{v}_n \delta(\mathbf{r} - \mathbf{r}_n) \rangle - \frac{1}{6} \nabla \cdot \left(\underline{Q}'_n \times \mathbf{v}_n \delta(\mathbf{r} - \mathbf{r}_n) \right) \right]$$

$$+ \cdots . \tag{4.42}$$

Upon taking the summation of the above expression over all of the molecules in the material and then combining this result with the spatial average of the free current density, one finally obtains the spatial average of the microscopic current density as

$$\langle\langle \mathbf{j}(\mathbf{r}, t) \rangle\rangle = \langle\langle \mathbf{j}_f(\mathbf{r}, t) \rangle\rangle + \langle\langle \mathbf{j}_b(\mathbf{r}, t) \rangle\rangle$$

$$= \left\langle\!\!\left\langle \sum_{\substack{j \\ (\text{free})}} q_j \mathbf{v}_j \delta(\mathbf{r} - \mathbf{r}_j(t)) \right\rangle\!\!\right\rangle + \left\langle\!\!\left\langle \sum_{\substack{n \\ (\text{mol})}} \mathbf{j}_n(\mathbf{r}, t) \right\rangle\!\!\right\rangle$$

$$= \mathbf{J}(\mathbf{r}, t) + \frac{d}{dt} \left(\mathbf{P}(\mathbf{r}, t) - \nabla \cdot \underline{Q}'(\mathbf{r}, t) \right) + \|c\| \nabla \times \mathbf{M}(\mathbf{r}, t)$$

$$+ \nabla \times \left[\left\langle\!\!\left\langle \sum_{\substack{n \\ (\text{mol})}} \mathbf{p}_n \times \mathbf{v}_n \delta(\mathbf{r} - \mathbf{r}_n) \right\rangle\!\!\right\rangle \right.$$

$$\left. - \frac{1}{6} \nabla \cdot \left\langle\!\!\left\langle \sum_{\substack{n \\ (\text{mol})}} \underline{Q}'_n \times \mathbf{v}_n \delta(\mathbf{r} - \mathbf{r}_n) \right\rangle\!\!\right\rangle \right]$$

$$+ \cdots . \tag{4.43}$$

Here $\mathbf{J}(\mathbf{r}, t)$ is the *macroscopic current density*

$$\mathbf{J}(\mathbf{r}, t) = \mathbf{J}_f(\mathbf{r}, t) + \mathbf{J}_b(\mathbf{r}, t) \tag{4.44}$$

that is comprised of the *macroscopic free current density*

$$\mathbf{J}_f(\mathbf{r}, t) \equiv \left\langle\!\!\left\langle \sum_{\substack{j \\ (\text{free})}} q_j \mathbf{v}_j \delta(\mathbf{r} - \mathbf{r}_j(t)) \right\rangle\!\!\right\rangle \tag{4.45}$$

plus the *macroscopic bound current density*

$$\mathbf{J}_b(\mathbf{r}, t) \equiv \left\langle\!\!\left\langle \sum_{\substack{n \\ (\text{mol})}} \sum_{j(n)} q_n \mathbf{v}_j \delta(\mathbf{r} - \mathbf{r}_j(t)) \right\rangle\!\!\right\rangle , \tag{4.46}$$

and $\mathbf{M}(\mathbf{r}, t)$ is the *macroscopic magnetization*

$$\mathbf{M}(\mathbf{r}, t) \equiv \left\langle\!\!\left\langle \sum_{\substack{n \\ (\text{mol})}} \mathbf{m}_n \delta(\mathbf{r} - \mathbf{r}_n(t)) \right\rangle\!\!\right\rangle . \tag{4.47}$$

If the free charges also possess intrinsic magnetic moments, then the spatial average of these microscopic moments must also be included in the defining relation (4.47) for the macroscopic magnetization vector.

4.1.4   The Macroscopic Maxwell Equations

Substitution of Eq. (4.23) for the spatial average of the microscopic charge density into Eq. (4.8) yields

$$\nabla \cdot \mathbf{E}(\mathbf{r}, t) = \frac{\|4\pi\|}{\epsilon_0} \left(\varrho(\mathbf{r}, t) - \nabla \cdot \mathbf{P}(\mathbf{r}, t) + \nabla \cdot \nabla \cdot \underline{Q}'(\mathbf{r}, t) + \cdots \right),$$

which may be rewritten as

$$\nabla \cdot \left(\epsilon_0 \mathbf{E}(\mathbf{r}, t) + \|4\pi\| \mathbf{P}(\mathbf{r}, t) - \|4\pi\| \nabla \cdot \underline{Q}'(\mathbf{r}, t) + \cdots \right) = \|4\pi\| \varrho(\mathbf{r}, t). \quad (4.48)$$

In addition, substitution of Eq. (4.43) for the spatial average of the microscopic current density into Eq. (4.11) yields

$$\nabla \times \mathbf{B}(\mathbf{r}, t) = \frac{\epsilon_0 \mu_0}{\|c\|} \frac{\partial \mathbf{E}(\mathbf{r}, t)}{\partial t}$$

$$+ \left\| \frac{4\pi}{c} \right\| \mu_0 \left[\mathbf{J}(\mathbf{r}, t) + \frac{\partial}{\partial t} \left(\mathbf{P}(\mathbf{r}, t) - \nabla \cdot \underline{Q}'(\mathbf{r}, t) \right) \right.$$

$$+ \|c\| \nabla \times \mathbf{M}(\mathbf{r}, t)$$

$$+ \nabla \times \left(\left\langle\!\!\left\langle \sum_{\substack{n \\ (\text{mol})}} \mathbf{p}_n \times \mathbf{v}_n \delta(\mathbf{r} - \mathbf{r}_n) \right\rangle\!\!\right\rangle \right)$$

$$\left. - \frac{1}{6} \nabla \cdot \left(\left\langle\!\!\left\langle \sum_{\substack{n \\ (\text{mol})}} \underline{Q}'_n \times \mathbf{v}_n \delta(\mathbf{r} - \mathbf{r}_n) \right\rangle\!\!\right\rangle \right) + \cdots \right],$$

which may be rewritten as

$$\nabla \times \left(\frac{1}{\mu_0} \mathbf{B}(\mathbf{r}, t) - \|4\pi\| \mathbf{M}(\mathbf{r}, t) \right.$$

$$- \left\| \frac{4\pi}{c} \right\| \left\langle\!\!\left\langle \sum_{\substack{n \\ (\text{mol})}} \mathbf{p}_n \times \mathbf{v}_n \delta(\mathbf{r} - \mathbf{r}_n) \right\rangle\!\!\right\rangle$$

$$+ \left\| \frac{4\pi}{c} \right\| \frac{1}{6} \nabla \cdot \left\langle \left\langle \sum_{\substack{n \\ (\text{mol})}} \underline{Q}'_n \times \mathbf{v}_n \delta(\mathbf{r} - \mathbf{r}_n) \right\rangle \right\rangle + \cdots \right)$$

$$= \left\| \frac{4\pi}{c} \right\| \mathbf{J}(\mathbf{r}, t)$$

$$+ \left\| \frac{1}{c} \right\| \frac{\partial}{\partial t} \Big(\epsilon_0 \mathbf{E}(\mathbf{r}, t) + \|4\pi\| \mathbf{P}(\mathbf{r}, t) - \|4\pi\| \nabla \cdot \underline{Q}'(\mathbf{r}, t) + \cdots \Big). \quad (4.49)$$

Define the *macroscopic electric displacement vector* $\mathbf{D}(\mathbf{r}, t)$ as the vector field given by

$$\mathbf{D}(\mathbf{r}, t) = \epsilon_0 \mathbf{E}(\mathbf{r}, t) + \|4\pi\| \mathbf{P}(\mathbf{r}, t) - \|4\pi\| \nabla \cdot \underline{Q}'(\mathbf{r}, t) + \cdots, \quad (4.50)$$

and define the *macroscopic magnetic intensity vector* $\mathbf{H}(\mathbf{r}, t)$ as the vector field given by

$$\mathbf{H}(\mathbf{r}, t) = \frac{1}{\mu_0} \mathbf{B}(\mathbf{r}, t) - \|4\pi\| \mathbf{M}(\mathbf{r}, t)$$

$$- \left\| \frac{4\pi}{c} \right\| \left\langle \left\langle \sum_{\substack{n \\ (\text{mol})}} \mathbf{p}_n \times \mathbf{v}_n \delta(\mathbf{r} - \mathbf{r}_n) \right\rangle \right\rangle$$

$$+ \frac{1}{6} \nabla \cdot \left\langle \left\langle \sum_{\substack{n \\ (\text{mol})}} \underline{Q}'_n \times \mathbf{v}_n \delta(\mathbf{r} - \mathbf{r}_n) \right\rangle \right\rangle + \cdots. \quad (4.51)$$

Notice that any approximation due to neglected higher-order terms has been placed in these two derived field vectors. With these definitions, the *macroscopic Maxwell equations* assume the familiar form

$$\nabla \cdot \mathbf{D}(\mathbf{r}, t) = \|4\pi\| \varrho(\mathbf{r}, t), \quad (4.52)$$

$$\nabla \cdot \mathbf{B}(\mathbf{r}, t) = 0, \quad (4.53)$$

$$\nabla \times \mathbf{E}(\mathbf{r}, t) = - \left\| \frac{1}{c} \right\| \frac{\partial \mathbf{B}(\mathbf{r}, t)}{\partial t}, \quad (4.54)$$

$$\nabla \times \mathbf{H}(\mathbf{r}, t) = \left\| \frac{4\pi}{c} \right\| \mathbf{J}(\mathbf{r}, t) + \left\| \frac{1}{c} \right\| \frac{\partial \mathbf{D}(\mathbf{r}, t)}{\partial t}. \quad (4.55)$$

It is seen from this derivation that, because the macroscopic field vectors $\mathbf{E}$ and $\mathbf{B}$ are the respective spatial averages of the microscopic field vectors $\mathbf{e}$ and $\mathbf{b}$, then $\mathbf{E}$ and $\mathbf{B}$ are of more fundamental significance than the derived field vectors $\mathbf{D}$ and $\mathbf{H}$. The derived field vectors $\mathbf{D}$ and $\mathbf{H}$ should be regarded as a shorthand notation for the expressions appearing on the right-hand sides of Eqs. (4.50) and (4.51), respectively.

For the conservation of charge, the spatial average of the microscopic equation of continuity given in Eq. (2.14) directly yields the macroscopic equation of continuity

$$\nabla \cdot \mathbf{J}(\mathbf{r}, t) + \frac{\partial \varrho(\mathbf{r}, t)}{\partial t} = 0. \tag{4.56}$$

This result also follows from the divergence of Eq. (4.55) with substitution from Eq. (4.52), as it must for internal consistency.

Finally, consider the spatial average of the microscopic Lorentz force relation (2.21)

$$\langle\langle \mathbf{f}(\mathbf{r}, t) \rangle\rangle = \langle\langle \rho(\mathbf{r}, t) \mathbf{e}(\mathbf{r}, t) \rangle\rangle + \left\| \frac{1}{c} \right\| \langle\langle \mathbf{j}(\mathbf{r}, t) \times \mathbf{b}(\mathbf{r}, t) \rangle\rangle. \tag{4.57}$$

With the *macroscopic Lorentz force density* defined by the spatial average $\mathbf{F}(\mathbf{r}, t) \equiv \langle \mathbf{f}(\mathbf{r}, t) \rangle$, application of the spatial averaging process (4.1) to the above expression yields

$$\mathbf{F}(\mathbf{r}, t) = \int w(\mathbf{r}') \rho(\mathbf{r} - \mathbf{r}', t) \mathbf{e}(\mathbf{r} - \mathbf{r}', t) d^3 r'$$

$$+ \left\| \frac{1}{c} \right\| \int w(\mathbf{r}') \mathbf{j}(\mathbf{r} - \mathbf{r}', t) \times \mathbf{b}(\mathbf{r} - \mathbf{r}', t) d^3 r'. \tag{4.58}$$

Let $\mathbf{e}'(\mathbf{r}, t) \equiv \mathbf{E}(\mathbf{r}, t) - \mathbf{e}(\mathbf{r}, t)$ and $\mathbf{b}'(\mathbf{r}, t) \equiv \mathbf{B}(\mathbf{r}, t) - \mathbf{b}(\mathbf{r}, t)$ denote the difference fields between the macroscopic and microscopic field vectors and assume that $\mathbf{E}(\mathbf{r} - \mathbf{r}', t) \approx \mathbf{E}(\mathbf{r}, t)$ and $\mathbf{B}(\mathbf{r} - \mathbf{r}', t) \approx \mathbf{B}(\mathbf{r}, t)$ when $\mathbf{r}'$ varies over the macroscopically small spatial size of the weighting function $w(\mathbf{r}')$. The above expression for the macroscopic force density then becomes

$$\mathbf{F}(\mathbf{r}, t) \approx \langle\langle \rho(\mathbf{r}, t) \rangle\rangle \mathbf{E}(\mathbf{r}, t) + \left\| \frac{1}{c} \right\| \langle\langle \mathbf{j}(\mathbf{r}, t) \rangle\rangle \times \mathbf{B}(\mathbf{r}, t)$$

$$- \langle\langle \rho(\mathbf{r}, t) \mathbf{e}'(\mathbf{r}, t) \rangle\rangle - \left\| \frac{1}{c} \right\| \langle\langle \mathbf{j}(\mathbf{r}, t) \times \mathbf{b}'(\mathbf{r}, t) \rangle\rangle$$

$$\approx \varrho(\mathbf{r}, t) \mathbf{E}(\mathbf{r}, t) + \left\| \frac{1}{c} \right\| \mathbf{J}(\mathbf{r}, t) \times \mathbf{B}(\mathbf{r}, t)$$

$$- (\nabla \cdot \mathbf{P}(\mathbf{r}, t)) \mathbf{E}(\mathbf{r}, t) + (\nabla \cdot \nabla \cdot \underline{Q}'(\mathbf{r}, t)) \mathbf{E}(\mathbf{r}, t)$$

$$+ \left\| \frac{1}{c} \right\| \left[\frac{d}{dt} \left(\mathbf{P}(\mathbf{r}, t) - \nabla \cdot \underline{Q}'(\mathbf{r}, t) \right) \right] \times \mathbf{B}(\mathbf{r}, t)$$

$$+ (\nabla \times \mathbf{M}(\mathbf{r}, t)) \times \mathbf{B}(\mathbf{r}, t)$$

$$- \langle\langle \rho(\mathbf{r}, t) \mathbf{e}'(\mathbf{r}, t) \rangle\rangle - \left\| \frac{1}{c} \right\| \langle\langle \mathbf{j}(\mathbf{r}, t) \times \mathbf{b}'(\mathbf{r}, t) \rangle\rangle,$$

and it is seen that there is no simple expression of the spatially averaged Lorentz force relation in a ponderable medium. Nevertheless, the microscopic Lorentz force relation (2.21) holds everywhere and so can always be applied to determine the electromagnetic force in microscopic detail.

4.1.5 Electromagnetics in Moving Media

Suppose now that the medium as a whole is moving with the translational velocity **v**. If one neglects any other motion of the molecules comprising the medium, then $\mathbf{v}_n \approx \mathbf{v}$ for all n. Equation (4.51) then becomes (neglecting quadrupole and higher-order terms)

$$\frac{1}{\mu_0}\mathbf{B}(\mathbf{r},t) - \mathbf{H}(\mathbf{r},t) \approx \|4\pi\|\mathbf{M}(\mathbf{r},t) + \left\|\frac{4\pi}{c}\right\| \mathbf{P}(\mathbf{r},t) \times \mathbf{v}, \qquad (4.59)$$

where, from Eq. (4.50)

$$\mathbf{P}(\mathbf{r},t) \approx \frac{1}{\|4\pi\|}\left(\mathbf{D}(\mathbf{r},t) - \epsilon_0 \mathbf{E}(\mathbf{r},t)\right), \qquad (4.60)$$

so that

$$\frac{1}{\mu_0}\mathbf{B}(\mathbf{r},t) - \mathbf{H}(\mathbf{r},t) \approx \|4\pi\|\mathbf{M}(\mathbf{r},t) + \left\|\frac{1}{c}\right\|(\mathbf{D}(\mathbf{r},t) - \epsilon_0 \mathbf{E}(\mathbf{r},t)) \times \mathbf{v}. \quad (4.61)$$

Hence, the electric polarization **P** (and any higher-order moments, if included) enters the expression for the effective magnetization for a moving medium. Equation (4.61) is the nonrelativistic limit of one of the equations in the Minkowski formulation of the electrodynamics of moving media [11].

The phase velocity of an electromagnetic wave is also effected by the motion of the medium it is traveling through, an experimentally observed phenomenon known as "Fresnel drag" [12]. Let the phase velocity of the time-harmonic electromagnetic wave in the moving medium be given by $u' = c/n$, where $n = n(\omega)$ is the real index of refraction of the medium at the frequency ω of the wave, and let the medium be moving along a parallel direction at the constant velocity V with respect to the laboratory reference frame. From Eq. (2.89), the velocity of the electromagnetic wave relative to the laboratory reference frame is then given by

$$u = \frac{c/n + V}{1 + V/nc} \approx \frac{c}{n} + V\left(1 - \frac{1}{n^2}\right), \qquad (4.62)$$

where the final approximation is valid for sufficiently small $V/c \ll 1$. The factor $(1 - 1/n^2)$ is known as the *Fresnel drag coefficient*. A detailed description of this relativistic effect is given by J. Van Bladel [13].

4.2   Constitutive Relations in Linear Electromagnetics and Optics

The macroscopic Maxwell's equations given in Eqs. (4.52)–(4.55), together with the macroscopic equation of continuity (4.56), are completed by the appropriate constitutive (or material) relations relating the induced electric displacement vector $\mathbf{D}(\mathbf{r}, t)$, magnetic intensity vector $\mathbf{H}(\mathbf{r}, t)$, and conduction current density $\mathbf{J}_c(\mathbf{r}, t)$, these three field vectors being referred to as the *induction fields*, to both the electric field intensity $\mathbf{E}(\mathbf{r}, t)$ and the magnetic induction field $\mathbf{B}(\mathbf{r}, t)$, these latter two field vectors being referred to as the *primitive fields*. For many physical situations, magnetic field effects are entirely negligible in comparison with the effects produced by the electric field for both the electric displacement field and the conduction current density, and electric field effects are likewise negligible in comparison with magnetic field effects for the magnetic induction vector.[2] With both this and the additional assumption that the field strengths considered are sufficiently small that nonlinear effects are negligible, each material relation may be expressed in the form of the *general constitutive relation*

$$\mathbf{G}(\mathbf{r}, t) = \int_{-\infty}^{\infty} d^3 r' \int_{-\infty}^{t} dt' \, \hat{\underline{\zeta}}(\mathbf{r}', t', \mathbf{r}, t) \cdot \mathbf{F}(\mathbf{r}', t'), \qquad (4.63)$$

where the upper limit of integration in the time integral is imposed by primitive causality [16]. Here $\hat{\underline{\zeta}}(\mathbf{r}', t', \mathbf{r}, t)$ denotes either the dielectric permittivity response tensor $\hat{\underline{\epsilon}}(\mathbf{r}', t', \mathbf{r}, t)$ when $\mathbf{F}(\mathbf{r}, t) = \mathbf{E}(\mathbf{r}, t)$ and $\mathbf{G}(\mathbf{r}, t) = \mathbf{D}(\mathbf{r}, t)$, the inverse of the magnetic permeability response tensor $\hat{\underline{\mu}}(\mathbf{r}', t', \mathbf{r}, t)$ when $\mathbf{F}(\mathbf{r}, t) = \mathbf{B}(\mathbf{r}, t)$ and $\mathbf{G}(\mathbf{r}, t) = \mathbf{H}(\mathbf{r}, t)$, or the electric conductivity response tensor $\hat{\underline{\sigma}}(\mathbf{r}', t', \mathbf{r}, t)$ when $\mathbf{F}(\mathbf{r}, t) = \mathbf{E}(\mathbf{r}, t)$ and $\mathbf{G}(\mathbf{r}, t) = \mathbf{J}_c(\mathbf{r}, t)$, where $\mathbf{J}_c(\mathbf{r}, t)$ denotes the macroscopic conduction current density whose origin is derived from the macroscopic free current density defined in Eq. (4.45). That the inverse of the magnetic permeability tensor is used in the magnetic constitutive relation is a consequence of the fact that the magnetic intensity vector $\mathbf{H}(\mathbf{r}, t)$ is the induced field whereas the magnetic induction vector $\mathbf{B}(\mathbf{r}, t)$ is the primitive field [17]. The linear properties of the material are then determined by the analytical properties of these material tensors.

The linearity property expressed in the general constitutive relation (4.63) is due to the independence of the dielectric permittivity, magnetic permeability, and

---

[2]Notable exceptions are bi-anisotropic and bi-isotropic materials that exhibit chirality [14, 15]. A brief description of their general formulation is included at the end of this section.

conductivity tensors on the applied field strength. For example, this relation states that both $\mathbf{D}(\mathbf{r}, t)$ and $\mathbf{J}_c(\mathbf{r}, t)$ are doubled if $\mathbf{E}(\mathbf{r}, t)$ is doubled for all $\mathbf{r}$ and t, while $\mathbf{H}(\mathbf{r}, t)$ is doubled if $\mathbf{B}(\mathbf{r}, t)$ is doubled for all $\mathbf{r}$ and t.

The material medium is said to be *spatially locally linear* if and only if $\hat{\underline{\zeta}}(\mathbf{r}', t', \mathbf{r}, t) = \hat{\underline{\zeta}}(\mathbf{r}, t', t)\delta(\mathbf{r} - \mathbf{r}')$. With this substitution, the general constitutive relation given in Eq. (4.63) simplifies to

$$\mathbf{G}(\mathbf{r}, t) = \int_{-\infty}^{t} \hat{\underline{\zeta}}(\mathbf{r}, t', t) \cdot \mathbf{F}(\mathbf{r}, t')dt', \tag{4.64}$$

so that, for example, $\mathbf{D}(\mathbf{r}, t)$ is doubled at a particular point $\mathbf{r} = \mathbf{r}_0$ if $\mathbf{E}(\mathbf{r}, t)$ is doubled at that same point $\mathbf{r} = \mathbf{r}_0$ for all time t. The mathematical statement of spatial local linearity then corresponds in an intuitive way to the physical property that each molecule in the material is uncoupled from every other molecule in the material.

A material is said to be *spatially homogeneous* if and only if its properties vary with position within the material as $\hat{\underline{\zeta}}(\mathbf{r}', t', \mathbf{r}, t) = \hat{\underline{\zeta}}(\mathbf{r} - \mathbf{r}', t', t)$. If this mathematical property is not satisfied, then the material is said to be *spatially inhomogeneous*. Similarly, a material is said to be *temporally homogeneous* if and only if its properties vary with time as $\hat{\underline{\zeta}}(\mathbf{r}', t', \mathbf{r}, t) = \hat{\underline{\zeta}}(\mathbf{r}', \mathbf{r}, t - t')$. If this mathematical property is not satisfied, then the material is said to be *temporally inhomogeneous*.

For a *spatially and temporally homogeneous material*, the general constitutive relation given in Eq. (4.63) becomes

$$\mathbf{G}(\mathbf{r}, t) = \int_{-\infty}^{\infty} d^3r' \int_{-\infty}^{t} dt' \, \hat{\underline{\zeta}}(\mathbf{r} - \mathbf{r}', t - t') \cdot \mathbf{F}(\mathbf{r}', t'). \tag{4.65}$$

The spatio-temporal Fourier transform of this relation then yields, with application of the convolution theorem

$$\tilde{\mathbf{G}}(\mathbf{k}, \omega) = \tilde{\underline{\zeta}}(\mathbf{k}, \omega) \cdot \tilde{\mathbf{F}}(\mathbf{k}, \omega). \tag{4.66}$$

Here

$$\tilde{\mathbf{F}}(\mathbf{k}, \omega) = \int_{-\infty}^{\infty} d^3r \int_{-\infty}^{\infty} dt \, \mathbf{F}(\mathbf{r}, t)e^{-i(\mathbf{k}\cdot\mathbf{r}-\omega t)} \tag{4.67}$$

is the spatio-temporal Fourier transform of $\mathbf{F}(\mathbf{r}, t)$, provided that the spatial and temporal integrals exist, with inverse transform

$$\mathbf{F}(\mathbf{r}, t) = \frac{1}{(2\pi)^4} \int_{-\infty}^{\infty} d^3k \int_{-\infty}^{\infty} d\omega \, \tilde{\mathbf{F}}(\mathbf{k}, \omega)e^{i(\mathbf{k}\cdot\mathbf{r}-\omega t)}. \tag{4.68}$$

Hence, a spatially and temporally homogeneous material is linear in Fourier space in the sense that if $\tilde{\mathbf{F}}(\mathbf{k}, \omega)$ is doubled at $\mathbf{k} = \mathbf{k}_0$, $\omega = \omega_0$, then $\tilde{\mathbf{G}}(\mathbf{k}_0, \omega_0)$ is doubled. If the material tensor $\tilde{\underline{\zeta}}(\mathbf{k}, \omega)$ depends on the wave vector $\mathbf{k}$, then the medium is said to be *spatially dispersive*, whereas if it depends upon the temporal angular frequency ω, it is *temporally dispersive*. Spatial dispersion is typically considered in connection with crystal optics [18], whereas temporal dispersion appears in all ponderable media.

If the material tensor is independent of both $\mathbf{k}$ and ω, the medium is said to be *nondispersive*. In that case, Eq. (4.66) simplifies to the expression $\tilde{\mathbf{G}}(\mathbf{k}, \omega) = \underline{\tilde{\zeta}} \cdot \tilde{\mathbf{F}}(\mathbf{k}, \omega)$, whose inverse transform then yields the result $\mathbf{G}(\mathbf{r}, t) = \underline{\tilde{\zeta}} \cdot \mathbf{F}(\mathbf{r}, t)$, which is seen to be a special case of Eq. (4.64). Consequently, a linear nondispersive medium is both spatially and temporally locally linear. Notice that the material response of a temporally locally linear medium is instantaneous, which is clearly nonphysical (with the single exception of vacuum) and is not considered any further here.

For a material that is locally linear, spatially inhomogeneous, and temporally dispersive, the response tensor assumes the form $\underline{\hat{\zeta}}(\mathbf{r}', t', \mathbf{r}, t) = \underline{\hat{\zeta}}(\mathbf{r}, t-t')\delta(\mathbf{r}-\mathbf{r}')$. With this substitution, the general constitutive relation given in Eq. (4.63) becomes

$$\mathbf{G}(\mathbf{r}, t) = \int_{-\infty}^{t} \underline{\hat{\zeta}}(\mathbf{r}, t - t') \cdot \mathbf{F}(\mathbf{r}, t')dt', \qquad (4.69)$$

with temporal Fourier transform $\tilde{\mathbf{G}}(\mathbf{r}, \omega) = \underline{\zeta}(\mathbf{r}, \omega) \cdot \tilde{\mathbf{F}}(\mathbf{r}, \omega)$. The appropriate frequency domain constitutive relation for a locally linear, spatially homogeneous, spatially nondispersive, temporally dispersive material is then seen to be given by $\tilde{\mathbf{G}}(\mathbf{r}, \omega) = \underline{\zeta}(\omega) \cdot \tilde{\mathbf{F}}(\mathbf{r}, \omega)$, where the material properties are now completely independent of position within the material.

Finally, consider the property of *isotropy*. The material is said to be *isotropic* if and only if the components of the response tensor satisfy the relation $\hat{\zeta}_{ij}(\mathbf{r}', t', \mathbf{r}, t) = \hat{\zeta}(\mathbf{r}', t', \mathbf{r}, t)\delta_{ij}$, where δ_{ij} is the Kronecker delta function, and where $\hat{\zeta}(\mathbf{r}', t', \mathbf{r}, t)$ is independent of the indices i, j. With this substitution, the general constitutive relation (4.63) becomes

$$\mathbf{G}(\mathbf{r}, t) = \int_{-\infty}^{\infty} d^3r' \int_{-\infty}^{t} dt' \, \hat{\zeta}(\mathbf{r}', t', \mathbf{r}, t)\mathbf{F}(\mathbf{r}', t'). \qquad (4.70)$$

If the material is not isotropic, it is then said to be *anisotropic*.

For a spatially and temporally homogeneous, anisotropic medium, Eq. (4.66) applies and the properties of the material are described by the tensor $\tilde{\underline{\zeta}}(\mathbf{k}, \omega)$ which has the dyadic and matrix representations

$$\tilde{\underline{\zeta}}(\mathbf{k}, \omega) = \zeta_{xx}\hat{\mathbf{1}}_x\hat{\mathbf{1}}_x + \zeta_{xy}\hat{\mathbf{1}}_x\hat{\mathbf{1}}_y + \zeta_{xz}\hat{\mathbf{1}}_x\hat{\mathbf{1}}_z$$

$$+\zeta_{yx}\hat{\mathbf{1}}_y\hat{\mathbf{1}}_x + \zeta_{yy}\hat{\mathbf{1}}_y\hat{\mathbf{1}}_y + \zeta_{yz}\hat{\mathbf{1}}_y\hat{\mathbf{1}}_z$$

$$+\zeta_{zx}\hat{1}_z\hat{1}_x + \zeta_{zy}\hat{1}_z\hat{1}_y + \zeta_{zz}\hat{1}_z\hat{1}_z \qquad (4.71)$$

$$= \begin{pmatrix} \zeta_{xx}(\mathbf{k}, \omega) & \zeta_{xy}(\mathbf{k}, \omega) & \zeta_{xz}(\mathbf{k}, \omega) \\ \zeta_{yx}(\mathbf{k}, \omega) & \zeta_{yy}(\mathbf{k}, \omega) & \zeta_{yz}(\mathbf{k}, \omega) \\ \zeta_{zx}(\mathbf{k}, \omega) & \zeta_{zy}(\mathbf{k}, \omega) & \zeta_{zz}(\mathbf{k}, \omega) \end{pmatrix}, \qquad (4.72)$$

respectively. If this anisotropic medium is lossless, then the law of conservation of energy in the electromagnetic field requires [14] that this material tensor be complex symmetric, so that $\zeta_{ij} = \zeta_{ji}^*$. If these elements are real, the matrix can then be reduced to diagonal form, in which case Eqs. (4.71) and (4.72) may be simplified as

$$\tilde{\underline{\zeta}}(\mathbf{k}, \omega) = \zeta_{x'}\hat{1}_{x'}\hat{1}_{x'} + \zeta_{y'}\hat{1}_{y'}\hat{1}_{y'} + \zeta_{z'}\hat{1}_{z'}\hat{1}_{z'} \qquad (4.73)$$

$$= \begin{pmatrix} \zeta_{x'}(\mathbf{k}, \omega) & 0 & 0 \\ 0 & \zeta_{y'}(\mathbf{k}, \omega) & 0 \\ 0 & 0 & \zeta_{z'}(\mathbf{k}, \omega) \end{pmatrix} \qquad (4.74)$$

in the orthogonal (x', y', z') *principal axis coordinate system* formed by the eigenvectors of $\tilde{\underline{\zeta}}(\mathbf{k}, \omega)$ with eigenvalues $\zeta_{x'}, \zeta_{y'}, \zeta_{z'}$. If this anisotropic material is *uniaxial*, then two of these diagonal elements are equal and different from the third. One typically sets $\zeta_{x'} = \zeta_{y'} = \zeta^o$ with $\zeta_{z'} = \zeta^e \neq \zeta^o$, where the superscript o stands for *ordinary* and the superscript e for *extraordinary*. If all three of the diagonal elements in Eq. (4.74) are unequal (i.e. if $\zeta_{x'} \neq \zeta_{y'} \neq \zeta_{z'}$), then the anisotropic medium is said to be *biaxial*.

The material response tensor for a spatially homogeneous, isotropic, locally linear, temporally dispersive material is obtained from Eqs. (4.69) and (4.70) with the additional constraint of spatial homogeneity. In that case $\hat{\underline{\zeta}}(\mathbf{r}', t', \mathbf{r}, t) = \hat{\underline{\zeta}}(t - t')\delta(\mathbf{r} - \mathbf{r}')$ and the constitutive relation given in Eq. (4.70) becomes

$$\mathbf{G}(\mathbf{r}, t) = \int_{-\infty}^{t} \hat{\underline{\zeta}}(t - t')\mathbf{F}(\mathbf{r}, t')dt', \qquad (4.75)$$

with temporal Fourier transform $\tilde{\mathbf{G}}(\mathbf{r}, \omega) = \zeta(\omega)\tilde{\mathbf{F}}(\mathbf{r}, \omega)$. Hence, $\tilde{\mathbf{G}}(\mathbf{r}, \omega)$ is parallel to $\tilde{\mathbf{F}}(\mathbf{r}, \omega)$; however, this does not necessarily imply that $\mathbf{G}(\mathbf{r}, t)$ is parallel to $\mathbf{F}(\mathbf{r}, t)$ for all $\mathbf{r}$ and t. For this to occur, the homogeneous, linear, second-order integral equation

$$\gamma\mathbf{F}(\mathbf{r}, t) = \int_{-\infty}^{t} \hat{\underline{\zeta}}(t - t')\mathbf{F}(\mathbf{r}, t')dt' \qquad (4.76)$$

must be satisfied.

It is of special interest here to examine in more detail the constitutive relation appearing in Eq. (4.75) for a spatially homogeneous, isotropic, locally linear, temporally dispersive material. Because the field $\mathbf{G}(\mathbf{r}, t)$ depends upon the past history of both the field $\mathbf{F}(\mathbf{r}, t')$ and the material response $\hat{\zeta}(t')$, it is appropriate to express $\mathbf{F}(\mathbf{r}, t')$ in a Taylor series expansion about the instant $t' = t$ as

$$\mathbf{F}(\mathbf{r}, t') = \sum_{j=0}^{\infty} \frac{1}{j!} \frac{\partial^j \mathbf{F}(\mathbf{r}, t)}{\partial t^j} (t' - t)^j, \tag{4.77}$$

which is valid provided that $\mathbf{F}(\mathbf{r}, t')$ and all of its time derivatives exist at each instant $t' \leq t$. Substitution of this expression into the constitutive relation given in Eq. (4.75) then gives

$$\mathbf{G}(\mathbf{r}, t') = \sum_{j=0}^{\infty} \hat{\zeta}^{(j)} \frac{\partial^j \mathbf{F}(\mathbf{r}, t)}{\partial t^j}, \tag{4.78}$$

where

$$\hat{\zeta}^{(j)} \equiv \frac{(-1)^j}{j!} \int_0^{\infty} \hat{\zeta}(\tau) \tau^j d\tau \tag{4.79}$$

is proportional to the jth moment of the material response function. The temporal Fourier transform of Eq. (4.78) then shows that the temporal frequency spectrum of the material response function is given by

$$\zeta(\omega) = \sum_{j=0}^{\infty} \hat{\zeta}^{(j)} (-i\omega)^j. \tag{4.80}$$

The real part of $\zeta(\omega)$ is then seen to be an even function of ω whereas the imaginary part is an odd function of ω. Furthermore, Eq. (4.80) shows that the material expansion coefficients are given by

$$\hat{\zeta}^{(j)} = \frac{i^j}{j!} \frac{\partial^j \zeta(\omega)}{\partial \omega^j} \bigg|_{\omega=0}. \tag{4.81}$$

Because of uniqueness, Eq. (4.80) is seen to be the Maclaurin's series expansion of the function $\zeta(\omega)$.

As a final point of interest here, consider the more general situation first considered by Tellegen [19] in which both of the induction fields $\mathbf{D}(\mathbf{r}, t)$ and $\mathbf{H}(\mathbf{r}, t)$ are functions of both primitive fields $\mathbf{E}(\mathbf{r}, t)$ and $\mathbf{B}(\mathbf{r}, t)$. In that case, the material is said to be *bianisotropic* and the corresponding constitutive relation must be written as [14, 15, 19]

$$\begin{pmatrix} \tilde{\mathbf{D}}(\mathbf{k}, \omega) \\ \tilde{\mathbf{H}}(\mathbf{k}, \omega) \end{pmatrix} = \begin{pmatrix} \tilde{\underline{\mathcal{P}}}(\mathbf{k}, \omega) & \tilde{\underline{\mathcal{L}}}(\mathbf{k}, \omega) \\ \tilde{\underline{\mathcal{M}}}(\mathbf{k}, \omega) & \tilde{\underline{\mathcal{Q}}}(\mathbf{k}, \omega) \end{pmatrix} \begin{pmatrix} \tilde{\mathbf{E}}(\mathbf{k}, \omega) \\ \tilde{\mathbf{B}}(\mathbf{k}, \omega) \end{pmatrix} \tag{4.82}$$

so that

$$\tilde{\mathbf{D}}(\mathbf{k}, \omega) = \tilde{\underline{M}}(\mathbf{k}, \omega) \cdot \tilde{\mathbf{E}}(\mathbf{k}, \omega) + \tilde{\underline{L}}(\mathbf{k}, \omega) \cdot \tilde{\mathbf{B}}(\mathbf{k}, \omega), \qquad (4.83)$$

$$\tilde{\mathbf{H}}(\mathbf{k}, \omega) = \tilde{\underline{P}}(\mathbf{k}, \omega) \cdot \tilde{\mathbf{E}}(\mathbf{k}, \omega) + \tilde{\underline{Q}}(\mathbf{k}, \omega) \cdot \tilde{\mathbf{B}}(\mathbf{k}, \omega). \qquad (4.84)$$

Each element $\tilde{\underline{P}}$, $\tilde{\underline{L}}$, $\tilde{\underline{M}}$, and $\tilde{\underline{Q}}$ may be expressed either in the dyadic form (4.71) or in the equivalent matrix form (4.72), each of whose elements may be spatially inhomogeneous, spatially homogeneous, spatially dispersive, spatially nondispersive, and temporally dispersive or nondispersive. An isotropic material body at rest in one inertial reference frame will then appear to be bianisotropic to a stationary observer in a different inertial reference frame when one is moving with a constant velocity $\mathbf{v}$ relative to the other as, for example, exhibited in Eq. (4.61).

If the elements $\tilde{\underline{P}}$, $\tilde{\underline{L}}$, $\tilde{\underline{M}}$, and $\tilde{\underline{Q}}$ are all scalars, then the material is said to be *bi-isotropic*., abbreviated as *BI*. A Chiral material (i.e. a material exhibiting chirality in which its composite molecules are distinguishable from their mirror image) is an important example of a *BI*-medium [15].

4.3   Causality and Dispersion Relations

Causality is an essential feature of electromagnetic wave theory. At the most fundamental level is the *principle of primitive causality* [16, 20] which simply states that the effect cannot precede the cause. As trivially obvious as this is, the events of "cause" and "effect" must be carefully defined in order to properly apply this principle [21]. This is modified in the special theory of relativity to what is referred to as the *principle of relativistic causality* which states that no signal can propagate with a velocity greater than the speed of light c in vacuum. These two fundamental physical principles have direct bearing on the problem of ultrashort dispersive pulse propagation with regard to the models used in the description of the material dispersion.

The most direct derivation [22] of the dispersion relations that must be satisfied by any linear system begins by expressing the angular frequency spectrum of the output signal $f_{\text{out}}(t)$—the effect—in terms of the spectrum of the input signal $f_{\text{in}}(t)$—the cause—through the linear relationship

$$\tilde{f}_{\text{out}}(\omega) = \chi(\omega) \tilde{f}_{\text{in}}(\omega), \qquad (4.85)$$

where

$$\chi(\omega) = \int_{-\infty}^{\infty} \hat{\chi}(t) e^{i\omega t} \, dt \qquad (4.86)$$

describes the linear system response at the angular frequency ω. Because of primitive causality, the system response may be expressed as

$$\hat{\chi}(t) = U(t)\hat{\Psi}(t), \qquad (4.87)$$

where $U(t) = 0$ for $t < 0$ and $U(t) = 1$ for $t > 0$ denotes the Heaviside unit step function (see Appendix A), and where $\hat{\Psi}(t) \equiv \hat{\chi}(t)$ for all $t > 0$. The Fourier transform of the Heaviside unit step function is given by

$$\tilde{U}(\omega) = \int_0^\infty e^{i\omega t}\,dt = \lim_{\Delta \to 0^+}\left(\frac{i}{\omega + i\Delta}\right)$$

$$= \mathcal{P}\left\{\frac{i}{\omega}\right\} + \pi\delta(\omega), \qquad (4.88)$$

where $\mathcal{P}$ indicates that the Cauchy principal value is to be taken. Equation (4.86) with (4.87) then yields, with application of the convolution theorem and the above result

$$\chi(\omega) = \frac{1}{2\pi}\int_{-\infty}^\infty \Psi(\omega')\tilde{U}(\omega - \omega')d\omega'$$

$$= \frac{i}{2\pi}\mathcal{P}\int_{-\infty}^\infty \frac{\Psi(\omega')}{\omega - \omega'}d\omega' + \frac{1}{2}\Psi(\omega). \qquad (4.89)$$

Because $\hat{\chi}(t) = U(t)\hat{\Psi}(t)$, the behavior of the function $\hat{\Psi}(t)$ for $t < 0$ may be freely chosen.

As the first choice, let $\hat{\Psi}(-|t|) = \hat{\Psi}(|t|)$ so that $\hat{\Psi}(t)$ is an even function of t. Its Fourier spectrum $\Psi(\omega)$ is then purely real and Eq. (4.89) yields $\Psi(\omega) = 2\Re\{\chi(\omega)\}$, so that

$$\Im\{\chi(\omega)\} = -\frac{1}{\pi}\mathcal{P}\int_{-\infty}^\infty \frac{\Re\{\chi(\omega')\}}{\omega' - \omega}d\omega'. \qquad (4.90)$$

For the second choice, let $\hat{\Psi}(-|t|) = -\hat{\Psi}(|t|)$ so that $\hat{\Psi}(t)$ is an odd function of t. Its Fourier spectrum $\Psi(\omega)$ is then purely imaginary and Eq. (4.89) yields $\Psi(\omega) = 2i\Im\{\chi(\omega)\}$, so that

$$\Re\{\chi(\omega)\} = \frac{1}{\pi}\mathcal{P}\int_{-\infty}^\infty \frac{\Im\{\chi(\omega')\}}{\omega' - \omega}d\omega'. \qquad (4.91)$$

The real and imaginary parts of a causal system response function then form a Hilbert transform pair, as expressed in Eqs. (4.90) and (4.91). The relationships contained in this pair of integral relations, referred to as either the *Plemelj formulae* [23] or the *dispersion relations*, are precisely stated by Titchmarsh's theorem [16, 24]:

Theorem 4.1 (Titchmarsh's Theorem) *Any square-integrable function $\chi(\omega)$ with inverse Fourier transform $\hat{\chi}(t)$ that satisfies one of the following four conditions satisfies all four of them:*

1. *$\hat{\chi}(t) = 0$ for all $t < 0$.*
2. *$\chi(\omega') = \lim_{\omega''\to 0^+}\{\chi(\omega' + i\omega'')\}$ for almost all ω', where $\chi(\omega)$ is holomorphic in the upper-half of the complex $\omega = \omega' + i\omega''$ plane and is square-integrable over any line parallel to the ω'-axis in the upper-half plane.*
3. *$\Re\{\chi(\omega)\}$ and $\Im\{\chi(\omega)\}$ satisfy the first Plemelj formula (4.90).*
4. *$\Re\{\chi(\omega)\}$ and $\Im\{\chi(\omega)\}$ satisfy the second Plemelj formula (4.91).*

The restriction to almost all ω' appearing in condition 2 of this theorem refers to the fact that the Fourier transform of a function remains unchanged when that function is changed over a set of measure zero.

4.3.1   The Dielectric Permittivity

For a spatially homogeneous, isotropic, locally linear, temporally dispersive dielectric, the constitutive relation (4.69) for the electric displacement vector becomes

$$\mathbf{D}(\mathbf{r}, t) = \int_{-\infty}^{t} \hat{\epsilon}(t - t')\mathbf{E}(\mathbf{r}, t')dt' \tag{4.92}$$

with temporal Fourier transform

$$\tilde{\mathbf{D}}(\mathbf{r}, \omega) = \epsilon(\omega)\tilde{\mathbf{E}}(\mathbf{r}, \omega), \tag{4.93}$$

where

$$\epsilon(\omega) = \int_{-\infty}^{\infty} \hat{\epsilon}(t)e^{i\omega t}\,dt \tag{4.94}$$

is the *dielectric permittivity*. In addition, the constitutive relation for the macroscopic polarization density is given by

$$\mathbf{P}(\mathbf{r}, t) = \epsilon_0 \int_{-\infty}^{t} \hat{\chi}_e(t - t')\mathbf{E}(\mathbf{r}, t')dt' \tag{4.95}$$

with temporal Fourier transform

$$\tilde{\mathbf{P}}(\mathbf{r}, \omega) = \epsilon_0 \chi_e(\omega)\tilde{\mathbf{E}}(\mathbf{r}, \omega), \tag{4.96}$$

where

$$\chi_e(\omega) = \int_{-\infty}^{\infty} \hat{\chi}_e(t)e^{i\omega t}\,dt \tag{4.97}$$

is the *electric susceptibility*. From the Fourier transform of Eq. (4.50) for a *simple polarizable dielectric*[3]

$$\tilde{\mathbf{D}}(\mathbf{r}, \omega) = \epsilon_0 \left(1 + \|4\pi\| \chi_e(\omega)\right) \tilde{\mathbf{E}}(\mathbf{r}, \omega), \tag{4.98}$$

so that the dielectric permittivity and electric susceptibility are related by

$$\epsilon(\omega) = \epsilon_0 \left(1 + \|4\pi\| \chi_e(\omega)\right). \tag{4.99}$$

Because $\mathbf{E}(\mathbf{r}, t)$, $\mathbf{P}(\mathbf{r}, t)$, and $\mathbf{D}(\mathbf{r}, t)$ are all real-valued vector fields, then each of their temporal frequency spectra satisfies the respective symmetry property

$$\tilde{\mathbf{E}}^*(\mathbf{r}, \omega) = \tilde{\mathbf{E}}(\mathbf{r}, -\omega), \tag{4.100}$$

$$\tilde{\mathbf{P}}^*(\mathbf{r}, \omega) = \tilde{\mathbf{P}}(\mathbf{r}, -\omega), \tag{4.101}$$

$$\tilde{\mathbf{D}}^*(\mathbf{r}, \omega) = \tilde{\mathbf{D}}(\mathbf{r}, -\omega), \tag{4.102}$$

for real ω. As a consequence, the real part of each spectral vector field is an even function of ω whereas the imaginary part is an odd function of ω. Furthermore, Eq. (4.101) with (4.96) and Eq. (4.102) with (4.93) yields the symmetry relations

$$\chi_e^*(\omega) = \chi_e(-\omega), \tag{4.103}$$

$$\epsilon^*(\omega) = \epsilon(-\omega) \tag{4.104}$$

for real ω. These results also follow respectively from Eqs. (4.94) and (4.97) because the dielectric response functions $\hat{\epsilon}(t)$ and $\hat{\chi}(t)$ are both real-valued. As a consequence, the real part of $\epsilon(\omega)$ is an even function of ω whereas its imaginary part is an odd function of ω when ω is real-valued. For complex $\omega = \omega' + i\omega''$ with $\omega' \equiv \Re\{\omega\}$ and $\omega'' \equiv \Im\{\omega\}$ these symmetry relations become

$$\chi_e^*(\omega) = \chi_e(-\omega^*), \tag{4.105}$$

$$\epsilon^*(\omega) = \epsilon(-\omega^*), \tag{4.106}$$

as seen from Eqs. (4.97) and (4.94), respectively. Hence, along the imaginary axis, $\chi_e^*(i\omega'') = \chi_e(i\omega'')$ and $\epsilon^*(i\omega'') = \epsilon(i\omega'')$ so that both $\chi_e(\omega)$ and $\epsilon(\omega)$ are real-valued on the imaginary axis.

Because Eq. (4.95) represents the most fundamental cause and effect relation for the dielectric response, the electric susceptibility then satisfies Titchmarsh's theorem. In particular, the Plemelj formulae for the electric susceptibility are given

---

[3] A simple polarizable dielectric is defined here as one for which the quadrupole and all higher-order moments of the molecular charge distribution identically vanish.

by the pair of Hilbert transform relations

$$\Re\{\chi_e(\omega)\} = \frac{1}{\pi}\mathcal{P}\int_{-\infty}^{\infty}\frac{\Im\{\chi_e(\omega')\}}{\omega'-\omega}d\omega', \tag{4.107}$$

$$\Im\{\chi_e(\omega)\} = -\frac{1}{\pi}\mathcal{P}\int_{-\infty}^{\infty}\frac{\Re\{\chi_e(\omega')\}}{\omega'-\omega}d\omega'. \tag{4.108}$$

Because of the symmetry relation expressed in Eq. (4.103), this transform pair may be expressed over the positive real frequency axis alone as

$$\Re\{\chi_e(\omega)\} = \frac{2}{\pi}\mathcal{P}\int_{0}^{\infty}\frac{\omega'\Im\{\chi_e(\omega')\}}{\omega'^2-\omega^2}d\omega', \tag{4.109}$$

$$\Im\{\chi_e(\omega)\} = -\frac{2\omega}{\pi}\mathcal{P}\int_{0}^{\infty}\frac{\Re\{\chi_e(\omega')\}}{\omega'^2-\omega^2}d\omega'. \tag{4.110}$$

With Eq. (4.99), the appropriate dispersion relations for the dielectric permittivity are then seen to be given by the *Kramers–Kronig relations*

$$\Re\{\epsilon(\omega)-\epsilon_0\} = \frac{1}{\pi}\mathcal{P}\int_{-\infty}^{\infty}\frac{\Im\{\epsilon(\omega')\}}{\omega'-\omega}d\omega', \tag{4.111}$$

$$\Im\{\epsilon(\omega)\} = -\frac{1}{\pi}\mathcal{P}\int_{-\infty}^{\infty}\frac{\Re\{\epsilon(\omega')-\epsilon_0\}}{\omega'-\omega}d\omega', \tag{4.112}$$

first derived by Kramers [25] and Kronig [26] in 1927 and 1926, respectively. As a consequence of Titchmarsh's theorem, $\epsilon(\omega)$ is holomorphic in the upper-half of the complex ω-plane. In addition, $\epsilon(\omega)/\epsilon_0 \to 1$ in the limit as $|\omega| \to \infty$ in the upper-half plane ($\omega'' > 0$). By continuity, $\epsilon(\omega')/\epsilon_0 \to 1$ as $\omega' \to \pm\infty$ along the real ω'−axis, so that $\epsilon''(\omega') \to 0$ as $\omega' \to \pm\infty$. In addition, because $\epsilon''(\omega')$ is an odd function along the real ω'-axis, then $\epsilon''(0) = 0$.

Upon returning to the fundamental constitutive relation given in Eq. (4.92) it is seen that at any fixed point in space the displacement vector depends upon the past history of the electric field intensity at that point through the dielectric permittivity response function $\hat{\epsilon}(t - t')$. From a physical point of view, it seems reasonable to expect that the sensitivity of the medium response decreases as the past time t' extends further into the past from the present time t at which the field vector $\mathbf{D}(\mathbf{r}, t)$ is evaluated. It is then seen appropriate to expand the electric field intensity $\mathbf{E}(\mathbf{r}, t')$ in a Taylor series about the instant $t' = t$ as

$$\mathbf{E}(\mathbf{r}, t') = \sum_{n=0}^{\infty}\frac{1}{n!}\frac{\partial^n\mathbf{E}(\mathbf{r}, t)}{\partial t^n}(t' - t)^n, \tag{4.113}$$

provided that $\mathbf{E}(\mathbf{r}, t')$ and all of its time derivatives exist at each instant $t' \le t$. This Taylor series expansion represents $\mathbf{E}(\mathbf{r}, t')$ for all $t' \le t$ if and only if the remainder

$R_N(t')$ after N terms approaches zero as $N \to \infty$. Substitution of this expansion into the constitutive relation (4.92) then gives

$$\mathbf{D}(\mathbf{r}, t) = \sum_{n=0}^{\infty} \epsilon^{(n)} \frac{\partial^n \mathbf{E}(\mathbf{r}, t)}{\partial t^n} \tag{4.114}$$

with

$$\epsilon^{(n)} \equiv \frac{(-1)^n}{n!} \int_0^{\infty} \hat{\epsilon}(\tau) \tau^n d\tau. \tag{4.115}$$

Notice that each coefficient $\epsilon^{(n)}$ is proportional to the *nth-order moment of the dielectric permittivity response function*. This form of the constitutive relation explicitly exhibits the local linear dependence of the derived field vector $\mathbf{D}(\mathbf{r}, t)$ on the temporal derivatives of the electric field intensity vector $\mathbf{E}(\mathbf{r}, t)$. For a quasi-static field this form of the constitutive relation may approximated as $\mathbf{D}(\mathbf{r}, t) \simeq \epsilon^{(0)} \mathbf{E}(\mathbf{r}, t) + \epsilon^{(1)} \partial \mathbf{E}(\mathbf{r}, t)/\partial t$ from which it is seen that the zeroth-order moment $\epsilon^{(0)}$ is just the static dielectric permittivity of the medium; that is, $\epsilon^{(0)} = \epsilon(0)$.

The temporal Fourier transform of the constitutive relation given in Eq. (4.114) yields, with Eq. (4.93),

$$\tilde{\mathbf{D}}(\mathbf{r}, \omega) = \epsilon(\omega) \tilde{\mathbf{E}}(\mathbf{r}, \omega) = \left(\sum_{n=0}^{\infty} (-i\omega)^n \epsilon^{(n)} \right) \tilde{\mathbf{E}}(\mathbf{r}, \omega),$$

so that the temporal frequency spectrum of the dielectric permittivity is given by

$$\epsilon(\omega) = \sum_{n=0}^{\infty} \epsilon^{(n)} (-i\omega)^n. \tag{4.116}$$

The real and imaginary parts of this expression then yield the pair of relations

$$\epsilon_r(\omega) \equiv \Re\{\epsilon(\omega)\} = \epsilon^{(0)} - \epsilon^{(2)} \omega^2 + \epsilon^{(4)} \omega^4 - \epsilon^{(6)} \omega^6 + \cdots, \tag{4.117}$$

$$\epsilon_i(\omega) \equiv \Im\{\epsilon(\omega)\} = -\epsilon^{(1)} \omega + \epsilon^{(3)} \omega^3 - \epsilon^{(5)} \omega^5 + \cdots. \tag{4.118}$$

Once again it is seen that the real part of $\epsilon(\omega)$ is an even function of real ω and the imaginary part is an odd function of ω. Furthermore, it is seen that

$$\epsilon^{(n)} = \frac{i^n}{n!} \frac{\partial^n \epsilon(\omega)}{\partial \omega^n} \bigg|_{\omega=0}, \tag{4.119}$$

so that the nth-order moment of the dielectric response function is proportional to the nth-order derivative of the dielectric permittivity at the origin.

4.3.2   The Electric Conductivity

For a spatially homogeneous, isotropic, locally linear, temporally dispersive conducting or semiconducting material, the constitutive relation (4.69) for the conduction current vector becomes

$$\mathbf{J}_c(\mathbf{r}, t) = \int_{-\infty}^{t} \hat{\sigma}(t - t') \mathbf{E}(\mathbf{r}, t') dt' \qquad (4.120)$$

with temporal Fourier transform

$$\tilde{\mathbf{J}}_c(\mathbf{r}, \omega) = \sigma(\omega) \tilde{\mathbf{E}}(\mathbf{r}, \omega), \qquad (4.121)$$

where

$$\sigma(\omega) = \int_{-\infty}^{\infty} \hat{\sigma}(t) e^{i\omega t} dt \qquad (4.122)$$

is the *electric conductivity*. Because both $\mathbf{J}_c(\mathbf{r}, t)$ and $\mathbf{E}(\mathbf{r}, t)$ are real-valued vector fields, the conductivity then satisfies the symmetry relation

$$\sigma^*(\omega) = \sigma(-\omega^*) \qquad (4.123)$$

for complex ω.

The manner in which the electric conductivity enters the dispersive material properties is obtained from the temporal Fourier transform of the macroscopic form (4.55) of Ampére's law in a semiconducting material, given by $\nabla \times \tilde{\mathbf{H}}(\mathbf{r}, \omega) = \|4\pi/c\| \tilde{\mathbf{J}}(\mathbf{r}, \omega) - \|1/c\| i\omega \tilde{\mathbf{D}}(\mathbf{r}, \omega)$ with total current density $\tilde{\mathbf{J}}(\mathbf{r}, \omega) = \tilde{\mathbf{J}}_{ext}(\mathbf{r}, \omega) + \tilde{\mathbf{J}}_c(\mathbf{r}, \omega)$, where $\tilde{\mathbf{J}}_{ext}(\mathbf{r}, \omega)$ describes any externally applied current source. Substitution of the constitutive relations given in Eqs. (4.93) and (4.121) then yields

$$\nabla \times \tilde{\mathbf{H}}(\mathbf{r}, \omega) = \left\|\frac{4\pi}{c}\right\| \tilde{\mathbf{J}}_{ext}(\mathbf{r}, \omega) - \left\|\frac{1}{c}\right\| i\omega \epsilon_c(\omega) \tilde{\mathbf{E}}(\mathbf{r}, \omega), \qquad (4.124)$$

where

$$\epsilon_c(\omega) \equiv \epsilon(\omega) + i\|4\pi\| \frac{\sigma(\omega)}{\omega} \qquad (4.125)$$

is defined as the *complex permittivity* of the material. For a nonconducting material, $\sigma(\omega) = 0$ and $\epsilon_c(\omega) = \epsilon(\omega)$, whereas for a purely conducting material, $\epsilon(\omega) = \epsilon_0$ and $\epsilon_c(\omega) = \epsilon_0 + i\|4\pi\|\sigma(\omega)/\omega$.

Because of the simple pole singularity at $\omega = 0$ appearing in Eq. (4.125), the first part of condition 2 in Titchmarsh's theorem is not satisfied. The Hilbert transform of the complex permittivity then includes a boundary contribution from this pole.

The dispersion relations given in Eqs. (4.111) and (4.112) must then be modified [27, 28] for the complex permittivity as

$$\Re\{\epsilon_c(\omega) - \epsilon_0\} = \frac{1}{\pi}\mathcal{P}\int_{-\infty}^{\infty}\frac{\Im\{\epsilon_c(\omega')\}}{\omega' - \omega}d\omega', \tag{4.126}$$

$$\Im\{\epsilon_c(\omega)\} - \|4\pi\|\frac{\sigma(0)}{\omega} = -\frac{1}{\pi}\mathcal{P}\int_{-\infty}^{\infty}\frac{\Re\{\epsilon_c(\omega') - \epsilon_0\}}{\omega' - \omega}d\omega', \tag{4.127}$$

where the pole contribution at $\omega = 0$ has been subtracted. This pair of relations reduces to the pair of Kramers–Kronig relations given in Eqs. (4.11) and (4.112) in the limit of zero static conductivity. For a purely conducting material they yield the pair of dispersion relations

$$\Re\left\{\frac{\sigma(\omega)}{\omega}\right\} - \frac{\sigma(0)}{\omega} = \frac{1}{\pi}\mathcal{P}\int_{-\infty}^{\infty}\frac{\Im\{\sigma(\omega')/\omega'\}}{\omega' - \omega}d\omega', \tag{4.128}$$

$$\Im\left\{\frac{\sigma(\omega)}{\omega}\right\} = -\frac{1}{\pi}\mathcal{P}\int_{-\infty}^{\infty}\frac{\Re\{\sigma(\omega')/\omega'\}}{\omega' - \omega}d\omega', \tag{4.129}$$

for the electric conductivity.

Consider again the fundamental constitutive relation given in Eq. (4.120) for the conduction current. This relation becomes, with substitution of the Taylor series expansion of the electric field intensity about the instant $t' = t$,

$$\mathbf{J}_c(\mathbf{r}, t) = \sum_{n=0}^{\infty}\sigma^{(n)}\frac{\partial^n \mathbf{E}(\mathbf{r}, t)}{\partial t^n} \tag{4.130}$$

with

$$\sigma^{(n)} \equiv \frac{(-1)^n}{n!}\int_0^{\infty}\hat{\sigma}(\tau)\tau^n d\tau. \tag{4.131}$$

This form of the constitutive relation explicitly displays the local linear dependence of the derived conduction current vector $\mathbf{J}_c(\mathbf{r}, t)$ on the temporal derivatives of the electric field intensity vector $\mathbf{E}(\mathbf{r}, t)$. Each coefficient $\sigma^{(n)}$ is then seen to be proportional to the *nth-order moment of the electric conductivity response function*. In particular, the zeroth-order moment is seen to be identical with the static conductivity of the medium; that is, $\sigma^{(0)} = \sigma(0)$.

The temporal Fourier transform of the constitutive relation given in Eq. (4.130) yields, with Eq. (4.121)

$$\tilde{\mathbf{J}}_c(\mathbf{r}, \omega) = \sigma(\omega)\tilde{\mathbf{E}}(\mathbf{r}, \omega) = \left(\sum_{n=0}^{\infty}(-i\omega)^n\sigma^{(n)}\right)\tilde{\mathbf{E}}(\mathbf{r}, \omega),$$

so that the temporal frequency spectrum of the electric conductivity is given by

$$\sigma(\omega) = \sum_{n=0}^{\infty} \sigma^{(n)}(-i\omega)^n. \tag{4.132}$$

The real and imaginary parts of this expression then yield the pair of relations

$$\sigma_r(\omega) \equiv \Re\{\sigma(\omega)\} = \sigma^{(0)} - \sigma^{(2)}\omega^2 + \sigma^{(4)}\omega^4 - \sigma^{(6)}\omega^6 + \cdots, \tag{4.133}$$

$$\sigma_i(\omega) \equiv \Im\{\sigma(\omega)\} = -\sigma^{(1)}\omega + \sigma^{(3)}\omega^3 - \sigma^{(5)}\omega^5 + \cdots. \tag{4.134}$$

Once again it is seen that the real part of $\sigma(\omega)$ is an even function of real ω and the imaginary part is an odd function of ω. Furthermore, it is seen that

$$\sigma^{(n)} = \frac{i^n}{n!} \frac{\partial^n \sigma(\omega)}{\partial \omega^n}\bigg|_{\omega=0}, \tag{4.135}$$

so that the nth-order moment of the electric conductivity response function is proportional to the nth-order derivative of the conductivity at the origin.

4.3.3 The Magnetic Permeability

For a spatially homogeneous, isotropic, locally linear, temporally dispersive, non-ferromagnetic material, the constitutive relation (4.69) for the magnetic intensity vector becomes

$$\mathbf{H}(\mathbf{r}, t) = \int_{-\infty}^{t} \hat{\mu}^{-1}(t - t')\mathbf{B}(\mathbf{r}, t')dt', \tag{4.136}$$

with temporal Fourier transform

$$\tilde{\mathbf{H}}(\mathbf{r}, \omega) = \mu^{-1}(\omega)\tilde{\mathbf{B}}(\mathbf{r}, \omega), \tag{4.137}$$

where

$$\mu^{-1}(\omega) = \int_{-\infty}^{\infty} \hat{\mu}^{-1}(t)e^{i\omega t} dt \tag{4.138}$$

is the *inverse of the magnetic permeability* $\mu(\omega)$. The constitutive relation given in Eq. (4.137) may then be expressed in the more familiar form

$$\tilde{\mathbf{B}}(\mathbf{r}, \omega) = \mu(\omega)\tilde{\mathbf{H}}(\mathbf{r}, \omega). \tag{4.139}$$

In addition, the constitutive relation for the macroscopic magnetization vector is given by

$$\mathbf{M}(\mathbf{r}, t) = \frac{1}{\mu_0} \int_{-\infty}^{t} \hat{\chi}_b(t - t') \mathbf{B}(\mathbf{r}, t') dt' \tag{4.140}$$

with temporal Fourier transform

$$\tilde{\mathbf{M}}(\mathbf{r}, \omega) = \frac{1}{\mu_0} \chi_b(\omega) \tilde{\mathbf{B}}(\mathbf{r}, \omega), \tag{4.141}$$

where

$$\chi_b(\omega) = \int_{-\infty}^{\infty} \hat{\chi}_b(t) e^{i\omega t} dt. \tag{4.142}$$

For a *simple magnetizable medium*, the magnetization vector is given by [cf. Eq.(4.51)][4]

$$\mathbf{M}(\mathbf{r}, t) = \frac{1}{\|4\pi\|} \left(\frac{1}{\mu_0} \mathbf{B}(\mathbf{r}, t) - \mathbf{H}(\mathbf{r}, t) \right). \tag{4.143}$$

The temporal Fourier transform of this relation then yields, with substitution from Eqs. (4.139) and (4.141)

$$\tilde{\mathbf{B}}(\mathbf{r}, \omega) = \mu_0 \left(1 + \|4\pi\| \chi_m(\omega) \right) \tilde{\mathbf{H}}(\mathbf{r}, \omega), \tag{4.144}$$

where $1 + \|4\pi\| \chi_m(\omega) = 1/(1 - \|4\pi\| \chi_b(\omega))$, so that

$$\chi_m(\omega) = \frac{\chi_b(\omega)}{1 - \|4\pi\| \chi_b(\omega)} \tag{4.145}$$

is the *magnetic susceptibility* of the material. Comparison of Eqs. (4.139) and (4.144) then shows that the magnetic permeability and magnetic susceptibility are related by

$$\mu(\omega) = \mu_0 \left(1 + \|4\pi\| \chi_m(\omega) \right). \tag{4.146}$$

This is the familiar textbook relation that is commonly employed.

The magnetic susceptibility of a simple magnetizable material is typically much smaller than the electric susceptibility for that material. This is due to the fact that the magnetization of a *nonferromagnetic material* is a relativistic effect and so is on the order of v^2/c^2, where v is the velocity of the atomic electrons [27]. For

---

[4]This equation is taken here as the definition of a simple magnetizable medium.

diamagnetic materials, $\chi_m(\omega)$ is slightly less than zero ($\chi_m \sim -1 \times 10^{-5} \rightarrow -1 \times 10^{-8}$) so that $\mu(\omega)/\mu_0$ is slightly smaller than unity, whereas for *paramagnetic materials*, $\chi_m(\omega)$ is slightly larger than zero ($\chi_m \sim 1 \times 10^{-4} \rightarrow 1 \times 10^{-7}$) so that $\mu(\omega)/\mu_0$ is slightly greater than unity. In either case, $|\chi_m(\omega)| \ll 1$ and Eq. (4.145), which may be rewritten as $1 - \|4\pi\|\chi_b(\omega) = 1/(1 + \|4\pi\|\chi_m(\omega)) \approx 1 - \|4\pi\|\chi_m(\omega)$, shows that

$$\chi_m(\omega) \approx \chi_b(\omega) \tag{4.147}$$

over the frequency domain of interest. Within the accuracy of this approximate equivalence, the constitutive relation given in Eq. (4.136) may be inverted as

$$\mathbf{B}(\mathbf{r}, t) = \int_{-\infty}^{t} \hat{\mu}(t - t')\mathbf{H}(\mathbf{r}, t')dt', \tag{4.148}$$

which, in effect, reverses the proper roles of cause and effect in simple magnetizable materials.

Because $\mathbf{B}(\mathbf{r}, t)$, $\mathbf{M}(\mathbf{r}, t)$, and $\mathbf{H}(\mathbf{r}, t)$ are all real-valued vector fields, then each of their temporal frequency spectra satisfies the respective symmetry property

$$\tilde{\mathbf{B}}^*(\mathbf{r}, \omega) = \tilde{\mathbf{B}}(\mathbf{r}, -\omega^*), \tag{4.149}$$

$$\tilde{\mathbf{M}}^*(\mathbf{r}, \omega) = \tilde{\mathbf{M}}(\mathbf{r}, -\omega^*), \tag{4.150}$$

$$\tilde{\mathbf{H}}^*(\mathbf{r}, \omega) = \tilde{\mathbf{H}}(\mathbf{r}, -\omega^*), \tag{4.151}$$

for complex $\omega = \omega' + i\omega''$. As a consequence, the real part of each spectral vector field is an even function of real $\omega = \omega'$ and the imaginary part is an odd function of ω'. Furthermore, Eqs. (4.138), (4.142), and (4.145) yield the symmetry relations

$$\chi_b^*(\omega) = \chi_b(-\omega^*), \tag{4.152}$$

$$\chi_m^*(\omega) = \chi_m(-\omega^*), \tag{4.153}$$

$$\mu^*(\omega) = \mu(-\omega^*). \tag{4.154}$$

Hence, along the imaginary axis, $\chi_j^*(i\omega'') = \chi_j(i\omega'')$ for $j = m, b$ and $\mu^*(i\omega'') = \mu(i\omega'')$ so that both $\chi_j(\omega)$ and $\mu(\omega)$ are real-valued on the imaginary axis. Furthermore, the real part of $\mu(\omega)$ is an even function and the imaginary part of $\mu(\omega)$ is an odd function along the real ω'-axis.

As Eq. (4.140) represents the most fundamental cause and effect relation for the magnetic response, the susceptibility $\chi_b(\omega)$ then satisfies Titchmarsh's theorem. In particular, the Plemelj formulae for this susceptibility function are given by the pair of Hilbert transform relations

$$\Re\{\chi_j(\omega)\} = \frac{1}{\pi}\mathcal{P}\int_{-\infty}^{\infty} \frac{\Im\{\chi_j(\omega')\}}{\omega' - \omega}d\omega', \tag{4.155}$$

$$\Im\{\chi_j(\omega)\} = -\frac{1}{\pi}\mathcal{P}\int_{-\infty}^{\infty}\frac{\Re\{\chi_j(\omega')\}}{\omega'-\omega}d\omega', \tag{4.156}$$

with $j = b$. Because of the approximate equality stated in Eq. (4.147), the above pair of dispersion relations also holds for the magnetic susceptibility $\chi_m(\omega)$ with $j = m$. Because of the symmetry relations expressed in Eqs. (4.152) and (4.153), this transform pair may be expressed over the positive real frequency axis alone as

$$\Re\{\chi_j(\omega)\} = \frac{2}{\pi}\mathcal{P}\int_0^{\infty}\frac{\omega'\Im\{\chi_j(\omega')\}}{\omega'^2-\omega^2}d\omega', \tag{4.157}$$

$$\Im\{\chi_j(\omega)\} = -\frac{2\omega}{\pi}\mathcal{P}\int_0^{\infty}\frac{\Re\{\chi_j(\omega')\}}{\omega'^2-\omega^2}d\omega', \tag{4.158}$$

for $j = m, b$. With Eq. (4.146), the appropriate dispersion relations for the magnetic permeability are then seen to be given by

$$\Re\{\mu(\omega)\} - \mu_0 = \frac{1}{\pi}\mathcal{P}\int_{-\infty}^{\infty}\frac{\Im\{\mu(\omega')\}}{\omega'-\omega}d\omega', \tag{4.159}$$

$$\Im\{\mu(\omega)\} = -\frac{1}{\pi}\mathcal{P}\int_{-\infty}^{\infty}\frac{\Re\{\mu(\omega')\} - \mu_0}{\omega'-\omega}d\omega'. \tag{4.160}$$

Notice that these dispersion relations should be used with appropriate caution as the magnetic permeability function $\mu(\omega)$ loses physical meaning as ω exceeds some angular frequency value ω_μ [27].

Consider again the derived constitutive relation given in Eq. (4.148). Expansion of the magnetic intensity vector $\mathbf{H}(\mathbf{r}, t')$ in a Taylor series about the instant $t' = t$ results in the expression

$$\mathbf{H}(\mathbf{r}, t') = \sum_{n=0}^{\infty}\frac{1}{n!}\frac{\partial^n\mathbf{H}(\mathbf{r}, t)}{\partial t^n}(t'-t)^n, \tag{4.161}$$

provided that $\mathbf{H}(\mathbf{r}, t')$ and all of its time derivatives exist at each instant $t' \le t$. Substitution of this expansion in Eq. (4.148) then gives

$$\mathbf{B}(\mathbf{r}, t) = \sum_{n=0}^{\infty}\mu^{(n)}\frac{\partial^n\mathbf{H}(\mathbf{r}, t)}{\partial t^n} \tag{4.162}$$

with

$$\mu^{(n)} \equiv \frac{(-1)^n}{n!}\int_0^{\infty}\hat{\mu}(\tau)\tau^n d\tau. \tag{4.163}$$

The zeroth-order moment $\mu^{(0)}$ is then seen to be identical with the static permeability of the medium; that is, $\mu^{(0)} = \mu(0)$. The temporal Fourier transform of the constitutive relation given in Eq. (4.162) yields, with the Fourier transform of Eq. (4.148),

$$\tilde{\mathbf{B}}(\mathbf{r}, \omega) = \mu(\omega)\tilde{\mathbf{H}}(\mathbf{r}, \omega) = \left(\sum_{n=0}^{\infty}(-i\omega)^n\mu^{(n)}\right)\tilde{\mathbf{H}}(\mathbf{r}, \omega),$$

so that the temporal frequency spectrum of the magnetic permeability is given by

$$\mu(\omega) = \sum_{n=0}^{\infty}\mu^{(n)}(-i\omega)^n. \tag{4.164}$$

The real and imaginary parts of this expression then yield the pair of relations

$$\mu_r(\omega) \equiv \Re\{\mu(\omega)\} = \mu^{(0)} - \mu^{(2)}\omega^2 + \mu^{(4)}\omega^4 - \mu^{(6)}\omega^6 + \cdots, \tag{4.165}$$

$$\mu_i(\omega) \equiv \Im\{\mu(\omega)\} = -\mu^{(1)}\omega + \mu^{(3)}\omega^3 - \mu^{(5)}\omega^5 + \cdots. \tag{4.166}$$

Once again it is seen that the real part of $\mu(\omega)$ is an even function of real ω and the imaginary part is an odd function of ω. Furthermore, it is seen that

$$\mu^{(n)} = \frac{i^n}{n!}\frac{\partial^n\mu(\omega)}{\partial\omega^n}\bigg|_{\omega=0}, \tag{4.167}$$

so that the *nth-order moment of the effective magnetic permeability response function* is proportional to the nth-order derivative of the permeability at the origin.

4.4 Causal Models of the Material Dispersion

The theory of temporal frequency dispersion in homogeneous, isotropic, locally linear materials in either the solid, liquid, or gaseous states has its origins in the classical theories due to Drude [29] for free-electron conductors, Lorentz [3] for high-frequency resonant dispersion phenomena in dielectrics, and Debye [30] for low-frequency orientational polarization phenomena in dielectrics. Although these models are phenomenological in their origin, they do provide reasonably accurate expressions for describing the dispersive properties of such media over the appropriate frequency domain of interest. Composite models may then be constructed from appropriate generalizations [9, 31–33] of these individual models in order to more accurately describe the classical frequency dispersion over the entire frequency domain. It is essential that each of these physical models be causal for obvious reasons.

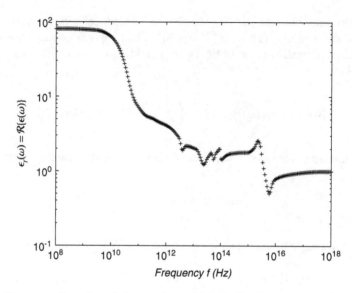

Fig. 4.2 Frequency dispersion of the real part of the relative dielectric permittivity data for triply distilled water at 25 °C. (Numerical data supplied by the School of Aerospace Medicine, Armstrong Laboratory, Brooks Air Force Base.)

A dielectric material of central importance to both electromagnetics and optics is water [34–38] as it is so prevalent in our natural environment. Representative data points for the computed frequency dispersion of the real and imaginary parts of the dielectric permittivity of triply-distilled (or deionized) water along the positive real frequency axis are presented in Figs. 4.2 and 4.3, respectively, where $f = \omega'/2\pi$ is the oscillation frequency in Hz. These computed data points have been obtained from representative experimental data points for the real index of refraction $n_r(\omega') \equiv \Re\{n(\omega')\}$ and attenuation coefficient $\alpha(\omega') \equiv (\omega'/c)n_i(\omega')$ of triply distilled water at 25 °C with $n_i(\omega') \equiv \Im\{n(\omega')\}$ denoting the imaginary part of the complex index of refraction $n(\omega) \equiv \sqrt{(\mu/\mu_0)(\epsilon(\omega)/\epsilon_0)}$, where $\mu/\mu_0 = 1$ for triply-distilled water. At zero frequency, $\epsilon_i(0) = 0$ and [because $\epsilon(0) = \epsilon_r(0)$]

$$\epsilon(0) = \epsilon_0 + \frac{2}{\pi}\mathcal{P}\int_0^\infty \frac{\epsilon_i(\omega')}{\omega'}d\omega', \qquad (4.168)$$

so that the static permittivity is given by the weighted sum of the contributions from all of the absorption mechanisms in the medium. Because of the Kramers–Kronig relations [cf. Eqs. 4.107)–(4.112)], each instance of rapid variation in $\epsilon_r(\omega')$ over some frequency interval is accompanied by a peak in $\epsilon_i(\omega')$ in that same frequency interval. The low frequency dependence of $\epsilon_r(\omega')$ depicted in Fig. 4.2 for $0 \leq f \lesssim 10^{13}$ Hz is characteristic of rotational polarization phenomena in polar dielectrics, whereas the high frequency dependence for $f \gtrsim 10^{13}$ Hz is more characteristic of resonance polarization phenomena, first molecular (due to

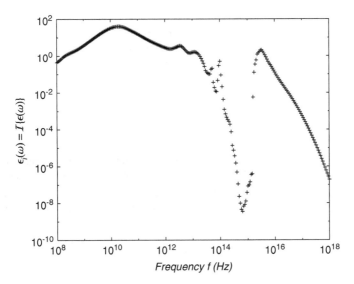

Fig. 4.3 Frequency dispersion of the imaginary part of the relative dielectric permittivity data for triply distilled water at 25 °C. (Numerical data supplied by the School of Aerospace Medicine, Armstrong Laboratory, Brooks Air Force Base.)

vibrational modes through the infrared region of the spectrum) and then atomic (due to electronic modes through the ultraviolet region of the spectrum). Additional polarization modes due to atomic core electrons may also appear in the x-ray region of the spectrum as relatively weak variations in the dielectric permittivity [39]. Similar remarks apply to the frequency dependence of the imaginary part of the dielectric permittivity depicted in Fig. 4.3.

Causal model-based analysis of the frequency dependence of the dielectric permittivity that is presented in the following subsections shows that as the frequency increases above zero, a critical frequency value is reached at which the permanent molecular dipoles of H_2O can no longer follow the oscillations of the driving wave field and the real permittivity rapidly decreases from its near static value $\epsilon^{(0)} = \epsilon(0)$ to a value $\epsilon_V^{(0)}$ that is given by the low-frequency limit of the combined vibrational and electronic polarization modes of H_2O. As the frequency f increases through each vibrational polarization mode of H_2O, the dielectric permittivity exhibits characteristic resonance responses as $\epsilon_r(\omega')$ approaches a value $\epsilon_e^{(0)}$ that is given by the low-frequency limit of its electronic polarization modes. As f increases further through each electronic polarization mode, the dielectric permittivity exhibits characteristic resonance responses as the dielectric permittivity approaches the high-frequency limits $\epsilon_r \to 1$, $\epsilon_i \to 0$ as $f \to \infty$.

Application of these analytical models to the dielectric frequency dispersion illustrated in Figs. 4.2 and 4.3 reveals the error present in this data for frequencies in and above the EHF radio frequency band (i.e. for frequencies $f > 1 \times 10^{11}$ Hz). Because of this, a numerical application of the Kramers–Kronig relations was performed on this data [40], revealing that it is inconsistent over certain frequency domains above $f \sim 1 \times 10^{11}$ Hz, particularly so in the infrared and optical regions of the electromagnetic spectrum. Both this inconsistency and its correction are more fully addressed in Sect. 4.4.5 of this chapter.

4.4.1   The Lorentz–Lorenz Relation

Because appropriate physical models for the material response originate at the microscopic level, the relationship between the electromagnetic response at the molecular level and its macroscopic equivalent needs to be established first. The macroscopic polarization $\mathbf{P}(\mathbf{r}, t)$ due to a material comprised of molecular species labeled by the index j with number densities N_j is given by

$$\mathbf{P}(\mathbf{r}, t) = \sum_j N_j \langle\langle \mathbf{p}_j(\mathbf{r}, t) \rangle\rangle, \qquad (4.169)$$

where $\langle\langle \mathbf{p}_j(\mathbf{r}, t) \rangle\rangle$ is the *macroscopic spatial average of the microscopic molecular dipole moment*, given by the causal linear relation

$$\langle\langle \mathbf{p}_j(\mathbf{r}, t) \rangle\rangle = \int_{-\infty}^{t} \hat{\alpha}_j(t - t') \langle\langle \mathbf{E}_{\text{eff}}(\mathbf{r}, t') \rangle\rangle dt'. \qquad (4.170)$$

Here $\mathbf{E}_{\text{eff}}(\mathbf{r}, t)$ denotes the *effective electric field* at the local molecular site. The temporal Fourier transform of Eq. (4.163) gives

$$\langle\langle \tilde{\mathbf{p}}_j(\mathbf{r}, \omega) \rangle\rangle = \alpha_j(\omega) \langle\langle \tilde{\mathbf{E}}_{\text{eff}}(\mathbf{r}, \omega) \rangle\rangle, \qquad (4.171)$$

where $\alpha_j(\omega)$ is the *mean polarizability* for the j-type molecules at the angular frequency ω. The term "mean" is used here to indicate a spatial average over molecular sites.

For a simple polarizable medium, the spatial average of the effective electric field is given by (see Appendix C)

$$\langle\langle \mathbf{E}_{\text{eff}}(\mathbf{r}, t) \rangle\rangle = \mathbf{E}(\mathbf{r}, t) + \frac{\|4\pi\|}{3\epsilon_0} \mathbf{P}(\mathbf{r}, t), \qquad (4.172)$$

where $\mathbf{E}(\mathbf{r}, t)$ is the external, applied electric field. Substitution of Eq. (4.171) with the temporal Fourier transform of Eq. (4.172) into the temporal Fourier transform of Eq. (4.169) then yields the expression

$$\tilde{\mathbf{P}}(\mathbf{r}, \omega) = \frac{\sum_j N_j \alpha_j(\omega)}{1 - (\|4\pi\|/3\epsilon_0) \sum_j N_j \alpha_j(\omega)} \tilde{\mathbf{E}}(\mathbf{r}, \omega) \qquad (4.173)$$

relating the macroscopic induced polarization to the external, applied electric field. Comparison of this expression with that given in Eq. (4.96) shows that the *electric susceptibility* of the material is given by

$$\chi_e(\omega) = \frac{\sum_j N_j \alpha_j(\omega)}{\epsilon_0 - (\|4\pi\|/3) \sum_j N_j \alpha_j(\omega)}, \tag{4.174}$$

so that the *dielectric permittivity* $\epsilon(\omega) = \epsilon_0 (1 + \|4\pi\| \chi_e(\omega))$ is given by

$$\epsilon(\omega) = \epsilon_0 \left[1 + \|4\pi\| \frac{\sum_j N_j \alpha_j(\omega)}{\epsilon_0 - (\|4\pi\|/3) \sum_j N_j \alpha_j(\omega)} \right]. \tag{4.175}$$

Upon solving this relation for the summation over the molecular polarizabilities of the medium, one finally obtains the *Lorentz–Lorenz formula* [41, 42]

$$\sum_j N_j \alpha_j(\omega) = \frac{3\epsilon_0}{\|4\pi\|} \frac{\epsilon(\omega)/\epsilon_0 - 1}{\epsilon(\omega)/\epsilon_0 + 2}, \tag{4.176}$$

which is also referred to as the *Clausius–Mossotti relation*.[5]. This simple relation provides the required connection between the phenomenological macroscopic Maxwell theory and the microscopic atomic theory of matter [44].

4.4.2 The Debye Model of Orientational Polarization

The *microscopic orientational (or dipolar) polarization* due to permanent molecular dipole moments may be described by the relaxation equation due to Debye [30]

$$\frac{d\mathbf{p}(\mathbf{r}, t)}{dt} + \frac{1}{\tau_m} \mathbf{p}(\mathbf{r}, t) = a \mathbf{E}_{\text{eff}}(\mathbf{r}, t), \tag{4.177}$$

where a is a constant in time and where τ_m denotes the characteristic exponential relaxation time of the molecular dipole moment in the absence of an externally applied electromagnetic field. Rotational Brownian motion theory [31] shows that the *dipolar relaxation time* is given by

$$\tau_m = \frac{\zeta}{2 K_B T}, \tag{4.178}$$

---

[5]Lorentz [3] attributed Eq. (4.176) with $j = 1$ to the earlier work by R. Clausius (1879) and O. F. Mossotti (1850); see p. 50 of [43].

where K_B is Boltzmann's constant, T is the absolute temperature, and where ζ is a frictional constant describing the resistance to dipolar rotation in the medium. The temporal frequency Fourier transform of the spatial average of Eq. (4.177) then yields

$$\langle\langle \tilde{\mathbf{p}}(\mathbf{r}, \omega)\rangle\rangle = \frac{a\tau_m}{1 - i\omega\tau}\langle\langle \tilde{\mathbf{E}}_{\text{eff}}(\mathbf{r}, \omega)\rangle\rangle. \tag{4.179}$$

Comparison of this expression with that given in Eq. (4.171) then shows that the *molecular polarizability for the Debye model* is given by

$$\alpha(\omega) = \frac{a\tau_m}{1 - i\omega\tau_m}. \tag{4.180}$$

With this expression substituted in Eq. (4.174), the *electric susceptibility* is found to be given by

$$\chi_e(\omega) = \frac{1}{\epsilon_0}\frac{Na\tau}{1 - i\omega\tau},$$

where $\tau \equiv \tau_m/(1 - (\|4\pi\|/3\epsilon_0)Na\tau_m)$, so that the *dielectric permittivity* is

$$\epsilon(\omega) = \epsilon_0 + \|4\pi\|\frac{Na\tau}{1 - i\omega\tau}.$$

Evaluation of this expression at $\omega = 0$ then shows that

$$a = \frac{1}{\|4\pi\|}\frac{\epsilon_s - \epsilon_0}{N\tau}, \tag{4.181}$$

where $\epsilon_s \equiv \epsilon(0)$ denotes the static permittivity of the medium. With this substitution, the single relaxation time Debye model susceptibility and permittivity expressions become

$$\chi_e(\omega) = \frac{1}{\|4\pi\|}\frac{\epsilon_s/\epsilon_0 - 1}{1 - i\omega\tau}, \tag{4.182}$$

$$\epsilon(\omega) = \epsilon_0\left(1 + \frac{\epsilon_s/\epsilon_0 - 1}{1 - i\omega\tau}\right), \tag{4.183}$$

where

$$\tau \equiv \frac{\epsilon_s/\epsilon_0 + 2}{3}\tau_m \tag{4.184}$$

is the *effective relaxation time*.

When considered as a function of complex $\omega = \omega' + i\omega''$, the Debye model dielectric permittivity given in Eq. (4.183) is found to have a single simple pole singularity at

$$\omega_p \equiv -\frac{i}{\tau} \qquad (4.185)$$

along the imaginary axis in the lower half of the complex ω-plane. The Debye model dielectric permittivity is then analytic in the upper-half of the complex ω-plane. In addition,

$$\epsilon(\omega') = \lim_{\omega'' \to 0} \epsilon(\omega' + i\omega'')$$

for all ω'. Finally, because

$$|\epsilon(\omega)/\epsilon_0 - 1|^2 = \frac{(\epsilon_s/\epsilon_0 - 1)^2}{1 + 2\tau\omega'' + \tau^2|\omega|^2}$$

is integrable over any line parallel to the ω'-axis for all $\omega'' > 0$, condition 2 of Titchmarsh's theorem is then satisfied, thereby proving that the Debye model permittivity is causal.

A simple generalization of the Debye model expression given in Eq. (4.183) in order to accommodate the influence of higher frequency ($\omega \gg 1/\tau$) polarization mechanisms on the lower frequency behavior is given by

$$\epsilon(\omega)/\epsilon_0 = \epsilon_\infty + \frac{\epsilon_{sr} - \epsilon_\infty}{1 - i\omega\tau}, \qquad (4.186)$$

where $\epsilon_\infty \geq 1$ denotes the large frequency limit of the relative dielectric permittivity due to the Debye model alone. The real and imaginary parts of this expression are then given by

$$\epsilon_r(\omega)/\epsilon_0 = \epsilon_\infty + \frac{\epsilon_{sr} - \epsilon_\infty}{1 + \tau^2\omega^2},$$

$$\epsilon_i(\omega)/\epsilon_0 = (\epsilon_{sr} - \epsilon_\infty)\frac{\tau\omega}{1 + \tau^2\omega^2},$$

so that

$$\left(\epsilon_r(\omega)/\epsilon_0 - (\epsilon_{sr} + \epsilon_\infty)/2\right)^2 + \left(\epsilon_i(\omega)/\epsilon_0\right)^2 = \left((\epsilon_{sr} - \epsilon_\infty)/2\right)^2. \qquad (4.187)$$

The locus of points $(\epsilon_r(\omega), \epsilon_i(\omega))$ then lie on a semicircle with center along the ϵ'-axis at $\epsilon' = (\epsilon_{sr} + \epsilon_\infty)/2$ and radius $(\epsilon_{sr} - \epsilon_\infty)/2$, as illustrated by the solid curve in Fig. 4.4 for triply distilled water. The numerical data for triply distilled water presented in Figs. 4.2 and 4.3 is also presented by the $+$ signs in this figure, where

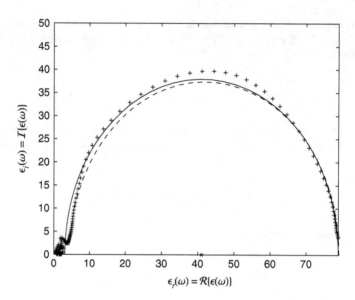

Fig. 4.4 Cole–Cole plot of the single (solid curve) and double (dashed curve) relaxation time Debye models compared with the numerical data (+ signs) for the real and imaginary parts of the relative dielectric permittivity of triply distilled water at 25 °C. The single relaxation time Debye model parameters used here are $\epsilon_\infty = 3.1$, $\epsilon_{sr} = 79.0$, $a = 75.9$, and $\tau = 7.88 \times 10^{-12}$ s, and the double relaxation time parameters are $\epsilon_\infty = 1.7$, $a_1 = 74.7$, $\tau_1 = 8.44 \times 10^{-12}$ s, $a_2 = 2.8$, and $\tau_2 = 4.77 \times 10^{-14}$ s. The point marked with an × along the abscissa locates the center of the semicircular Cole–Cole plot for the single relaxation time Debye model

increasing angular frequency values progress from right to left (notice the high-frequency structure near the origin of this plot that is due to resonance polarization contributions). Such a graphical representation is referred to as a *Cole–Cole plot* [32]. From Eq. (4.187), the maximum value of $\epsilon_i(\omega)/\epsilon_0$ occurs when $\epsilon_i(\omega)/\epsilon_0 = (\epsilon_{sr} + \epsilon_\infty)/2$ which in turn occurs when $\omega\tau = 1$. The relaxation time τ for a single relaxation time Debye model may then be estimated from the experimentally determined angular frequency value at which $\epsilon_i(\omega)$ reaches a maximum. The Cole–Cole plot readily shows whether or not the experimental data points for a given material can be described either by a single relaxation time, multiple relaxation times, or a distribution of relaxation times.

In order to account for additional orientational polarization modes as well as for any polarization mechanisms at higher frequencies (due, for example, to resonance polarization effects) the Debye model expression given in Eq. (4.186) may be generalized as

$$\epsilon(\omega)/\epsilon_0 = \epsilon_\infty + \sum_j \frac{a_j}{1 - i\omega\tau_j}, \qquad (4.188)$$

where ϵ_∞ is the large frequency limit of the relative permittivity due to orientational polarization effects alone, and where

$$\sum_j a_j = \epsilon_{sr} - \epsilon_\infty. \tag{4.189}$$

with $\epsilon_{sr} \equiv \epsilon_s/\epsilon_0$ denoting the relative static permittivity of the material. The summation appearing in the above two equations is taken over each orientational mode present in the material, each mode characterized by its individual relaxation time τ_j and strength a_j. Because of linearity, the resultant model is causal. Its Cole–Cole plot for triply distilled water, represented by the dashed curve in Fig. 4.4, possesses a better fit to the numerical data at the higher frequency values while maintaining the accuracy at very low frequencies.

The frequency dependence of a double relaxation time Debye model that has been numerically fitted to the numerical data for the relative dielectric permittivity of triply distilled water (see Figs. 4.2 and 4.3) is presented in Fig. 4.5 over the frequency domain from $f = 100\,\text{MHz}$ to $f = 1\,\text{THz}$. The frequency behavior illustrated here is typical of orientational polarization phenomena in polar dielectrics. Discrepancies between the model curves and data points may be reduced by employing appropriate extensions of the Debye model.

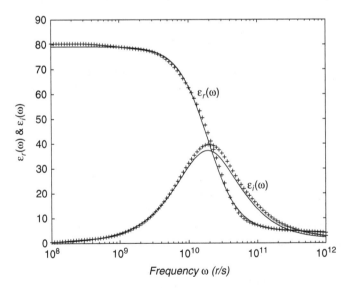

Fig. 4.5 Comparison of the double relaxation time Debye model results with $\epsilon_\infty = 1.7$, $a_1 = 74.7$, $\tau_1 = 8.44 \times 10^{-12}$ s, $a_2 = 2.8$, and $\tau_2 = 4.77 \times 10^{-14}$ s (solid curves) and numerical data (+ signs) for the real and imaginary parts of the relative dielectric permittivity of triply distilled water at 25 °C

4.4.3 Generalizations of the Debye Model

A major shortcoming of the Debye model is that it yields a nonvanishing absorption coefficient[6] $\alpha(\omega) = \Im\{\tilde{k}(\omega)\} = (\omega/c)n(\omega)$ at sufficiently high frequencies such that $\omega \gg 1/\tau_1$, assuming that the relaxation times τ_j are ordered in decreasing value. This deficiency is evident in Fig. 4.6 which presents a comparison of the double relaxation time Debye model curve for the imaginary part of the dielectric permittivity with the numerical data for water over the frequency domain of interest. Notice that the entire visible window for water has been effectively cut off by the Debye model curve. This implies that the early time ($t \ll \tau$) spontaneous relaxation of orientational molecular polarization cannot be exponential in time. Because such an exponential time decay can result only from a macroscopic averaging process taken over the collisions between a polar molecule and the random thermal motion of the surrounding molecules, it then becomes necessary to include the effects of rotational Brownian motion in the dynamical theory [31].

4.4.3.1 The Rocard–Powles–Debye Model

The complex molecular polarizability including, to first-order, the effects of rotational Brownian motion on the dipolar molecular relaxation process, is given by [31]

$$\alpha(\omega) = \frac{a\tau_m}{(1 - i\omega\tau_m)(1 - i\omega\tau_{mf})}. \qquad (4.190)$$

This first-order correction to the Debye model expression (4.180) was first given by Rocard [45] and Powles [46] through the inclusion of inertial effects in the Debye model. Here

$$\tau_{mf} = \frac{I}{\zeta} \qquad (4.191)$$

is the *associated friction time*, where I is the moment of inertia of the polar molecule and ζ is the same frictional constant appearing in Eq. (4.178) for the dipolar relaxation time. With this substitution in Eqs. (4.174) and (4.175), the single relaxation time *Rocard–Powles–Debye model susceptibility and permittivity* expressions are found to be given by

$$\chi_e(\omega) = \frac{1}{\|4\pi\|} \frac{\epsilon_s/\epsilon_0 - 1}{(1 - i\omega\tau)(1 - i\omega\tau_f)}, \qquad (4.192)$$

---

[6]See Eq. (1.20) with complex phase velocity given by $v(\omega) = c/n(\omega)$, where $n(\omega) \equiv (\epsilon(\omega)/\epsilon_0)^{1/2}$ is the complex index of refraction with $\mu/\mu_0 = 1$. This notation for the absorption coefficient should not be confused with that for the molecular polarizability, as each is determined by the context it is being used in.

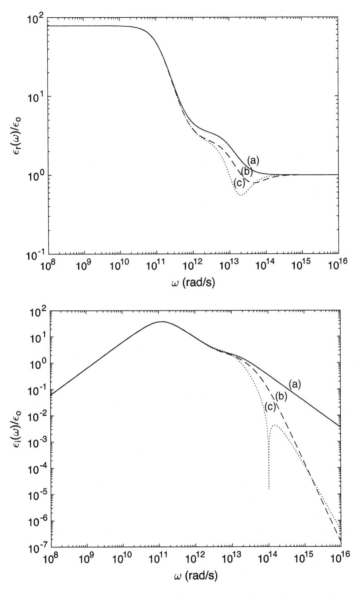

Fig. 4.6 Comparison of (a) the double relaxation time Debye, (b) the Rocard–Powles–Debye, and (c) the Cole–Cole extension of the Rocard–Powles–Debye models of the real (upper graph) and imaginary (lower graph) parts of the relative dielectric permittivity of triply distilled water at 25 °C

$$\epsilon(\omega) = \epsilon_0 \left(1 + \frac{\epsilon_s/\epsilon_0 - 1}{(1 - i\omega\tau)(1 - i\omega\tau_f)} \right), \tag{4.193}$$

where the effective relaxation time τ is given in Eq. (4.184), and where τ_f denotes an *effective friction time* for the model.

When considered as a function of complex $\omega = \omega' + i\omega''$, the Rocard–Powles–Debye model dielectric permittivity given in Eq. (4.193) is found to have isolated simple pole singularities at

$$\omega_{p1} \equiv -\frac{i}{\tau}, \qquad \omega_{p2} \equiv -\frac{i}{\tau_f}, \qquad (4.194)$$

which are situated along the imaginary axis in the lower-half of the complex ω–plane. The Rocard–Powles–Debye model dielectric permittivity is then analytic in the upper-half of the complex ω-plane. In addition,

$$\epsilon(\omega') = \lim_{\omega'' \to 0} \epsilon(\omega' + i\omega'')$$

for all ω'. Finally, because

$$|\epsilon(\omega)/\epsilon_0 - 1|^2 = \frac{(\epsilon_s/\epsilon_0 - 1)^2}{(1 + 2\tau\omega'' + \tau^2|\omega|^2)(1 + 2\tau_f\omega'' + \tau_f^2|\omega|^2)}$$

is integrable over any line parallel to the ω'-axis for all $\omega'' > 0$, condition 2 of Titchmarsh's theorem is then satisfied. The Rocard–Powles–Debye model permittivity is then causal.

As in Eq. (4.186), the Rocard–Powles model expression (4.193) may be generalized as

$$\epsilon(\omega)/\epsilon_0 = \epsilon_\infty + \frac{\epsilon_{sr} - \epsilon_\infty}{(1 - i\omega\tau)(1 - i\omega\tau_f)} \qquad (4.195)$$

in order to account for the influence of higher frequency ($\omega \gg 1/\tau$) polarization mechanisms on the lower frequency behavior, where $\epsilon_\infty \geq 1$ denotes the large frequency limit of the relative dielectric permittivity due to the Rocard–Powles–Debye model alone. The real and imaginary parts of this expression are then given by

$$\epsilon_r(\omega)/\epsilon_0 = \epsilon_\infty + (\epsilon_{sr} - \epsilon_\infty)\frac{(1 - \tau\tau_f\omega^2)}{(1 + \tau^2\omega^2)(1 + \tau_f^2\omega^2)},$$

$$\epsilon_i(\omega)/\epsilon_0 = (\epsilon_{sr} - \epsilon_\infty)\frac{(\tau + \tau_f)\omega}{(1 + \tau^2\omega^2)(1 + \tau_f^2\omega^2)},$$

so that

$$\left(\epsilon_r(\omega)/\epsilon_0 - (\epsilon_{sr} + \epsilon_\infty)/2\right)^2 + \left(\epsilon_i(\omega)/\epsilon_0\right)^2 \leq \left((\epsilon_{sr} - \epsilon_\infty)/2\right)^2, \qquad (4.196)$$

where the equality results when $\tau_f = 0$. The locus of points $(\epsilon_r(\omega), \epsilon_i(\omega))$ for the Rocard–Powles extension then lies within the semicircle of the Cole–Cole plot for the Debye model limit (as $\tau_f \to 0$).

In order to account for additional orientational polarization modes as well as for any polarization mechanisms at higher frequencies, the Rocard–Powles–Debye model expression given in Eq. (4.195) may be generalized as

$$\epsilon(\omega)/\epsilon_0 = \epsilon_\infty + \sum_j \frac{a_j}{(1 - i\omega\tau_j)(1 - i\omega\tau_{fj})} \tag{4.197}$$

with the sum rule given in Eq. (4.189), where ϵ_∞ is the large frequency limit of the relative permittivity due to orientational polarization effects alone. The summation is taken over each orientational mode present in the material with relaxation time τ_j, friction time τ_{fj}, and strength a_j.

Comparison of the frequency dependence of a double relaxation time Rocard–Powles–Debye model with the corresponding Debye model for the relative dielectric permittivity of triply distilled water, presented in Fig. 4.6, shows that they are practically indistinguishable over the frequency domain from the static ($\omega = 0\,\mathrm{r/s}$) up to approximately $2\,\mathrm{GHz}$ ($\omega = 2 \times 10^{11}\,\mathrm{r/s}$). Above that frequency, both the real and imaginary parts diverge for each model. While the real parts re-converge at a higher frequency ($\omega \sim 5 \times 10^{14}\,\mathrm{r/s}$), the imaginary parts do not. Specifically, the Rocard–Powles extension of the Debye model results in an expression for the imaginary part of the dielectric permittivity that decreases as $1/\omega^3$ as compared to $1/\omega$ for the Debye model for sufficiently higher frequencies ($\omega \gg 1/\tau_1$).

4.4.3.2   Fröhlich's Generalization to a Continuous Distribution of Relaxation Times

A further generalization of the Debye model, as well as of the Rocard–Powles extension of the Debye model, is obtained through the introduction of a continuous distribution of relaxation times. As introduced by Fröhlich [33] for the Debye model permittivity and generalized here to the Rocard–Powles–Debye model,

$$\epsilon(\omega)/\epsilon_0 - \epsilon_\infty = \int_0^\infty \frac{f(\tau)}{(1 - i\omega\tau)(1 - i\omega\tau_f)}d\tau, \tag{4.198}$$

where $\tau_f = \gamma(\tau)$ [31]. The function $f(\tau)$ is the distribution function for the relaxation time. Evaluation of this expression at $\omega = 0$ gives

$$\epsilon_{sr} - \epsilon_\infty = \int_0^\infty f(\tau)d\tau \tag{4.199}$$

so that the quantity $f(\tau)d\tau$ measures the relative contribution to the permittivity difference $\epsilon_{sr} - \epsilon_\infty$ from those components of the dielectric response with relaxation times in the interval $[\tau, \tau + d\tau]$, where $\epsilon_{sr} \equiv \epsilon(0)/\epsilon_0$ is the relative static permittivity and ϵ_∞ is the high frequency limit of the relative permittivity due to orientational polarization effects alone. Because the distribution function $f(\tau)$ is nonnegative for all $\tau \in [0, \infty]$, $\epsilon_i(\omega)$ is then comprised of a superposition of absorption curves with varying positions of their corresponding maxima. The half-width of the resulting absorption curve is then broadened from that for a single Rocard–Powles–Debye model. This distribution can also result in a multiplicity of absorption peaks. The inversion of Eq. (4.198) with $\tau_f = 0$ when $\epsilon(\omega)/\epsilon_0$ is described by an analytic function of ω has been studied by Stieltjes, as ascribed by Titchmarsh [47].

Fröhlich [48] introduced the distribution function

$$f_F(\tau) \equiv \frac{\epsilon_{sr} - \epsilon_\infty}{\ln(\tau_2/\tau_1)\tau}, \qquad \tau_1 < \tau < \tau_2 \tag{4.200}$$

that incorporates a finite range of relaxation times between τ_1 and τ_2 and is zero elsewhere. Substitution of this distribution function in Eq. (4.198) with τ_f independent of τ then results in the expression [49]

$$\epsilon(\omega)/\epsilon_0 = \epsilon_\infty + \frac{\epsilon_{sr} - \epsilon_\infty}{\ln(\tau_2/\tau_1)(1 - i\omega\tau_f)} \left\{ \ln\left[\frac{\tau_2(1 + (\omega\tau_1)^2)^{1/2}}{\tau_1(1 + (\omega\tau_2)^2)^{1/2}}\right] \right.$$

$$\left. + i\,(\arctan(\omega\tau_2) - \arctan(\omega\tau_1)) \right\}. \tag{4.201}$$

Because $\arctan(\omega\tau) \to \pi/2$ as $\omega\tau \to \infty$, the attenuation at high frequencies ($\omega \gg 1/\tau$) for the Fröhlich distribution function with $\tau_f = 0$ is due solely to the difference between two similar curves.

4.4.3.3 Cole–Cole Extension of the Rocard–Powles–Debye Model

Perhaps a more appropriate distribution function is that due to Cole and Cole [32], given by

$$f_{CC}(\tau) \equiv (\epsilon_{sr} - \epsilon_\infty)\frac{\sin(\pi\nu)/(\pi\tau)}{(\tau/\tau_1)^{1-\nu} - 2\cos(\pi\nu) + (\tau_1/\tau)^{1-\nu}}, \tag{4.202}$$

which is continuous for all $\tau > 0$. Substitution of this distribution function in Eq. (4.198) with τ_f independent of τ results in the expression, [49]

$$\epsilon(\omega)/\epsilon_0 = \epsilon_\infty + \frac{\epsilon_{sr} - \epsilon_\infty}{(1 - i\omega\tau_1)^{1-\nu}(1 - i\omega\tau_f)} \tag{4.203}$$

with $\nu < 1$. When considered as a function of complex $\omega = \omega' + i\omega''$, this *Cole–Cole extension of the Rocard–Powles–Debye model dielectric permittivity* with a noninteger value of ν is found to have a branch point singularity ω_b and a simple pole singularity ω_p at

$$\omega_b \equiv -\frac{i}{\tau_1}, \qquad \omega_p \equiv -\frac{i}{\tau_f}, \qquad (4.204)$$

situated along the negative imaginary axis in the lower-half of the complex ω-plane. For a negative integer value of ν, the branch point ω_b becomes a pole singularity and the causality analysis following Eq. (4.194) applies. With a branch cut chosen along the negative imaginary axis extending from ω_b to $-i\infty$, the Cole–Cole model dielectric permittivity is then analytic in the upper-half of the complex ω-plane. In addition,

$$\epsilon(\omega') = \lim_{\omega'' \to 0} \epsilon(\omega' + i\omega'')$$

for all ω'. Finally, because

$$|\epsilon(\omega)/\epsilon_0 - \epsilon_\infty|^2 = \frac{(\epsilon_{sr} - \epsilon_\infty)^2}{(1 + 2\tau_1\omega'' + \tau_1^2|\omega|^2)^{1-\nu}(1 + 2\tau_f\omega'' + \tau_f^2|\omega|^2)}$$

is integrable over any line parallel to the ω'-axis for all $\omega'' > 0$ when $\nu < 1$, condition 2 of Titchmarsh's theorem is then satisfied. The Cole–Cole extension of the Rocard–Powles–Debye model permittivity given in Eq. (4.203) is then causal. Notice that the Cole–Cole model is typically stated [49] as being restricted to values of ν satisfying the inequality $0 \le \nu < 1$, the lower limit $\nu = 0$ resulting in the Rocard–Powles model extension whereas the upper limit $\nu \to 1^-$ results in the original Debye model with τ_f replaced by τ_1; however, the lower limit does not appear to be necessary and may then be omitted so that negative values of ν may be allowed, as is done here.

The generalization of the Cole–Cole expression given in Eq. (4.203) in order to account for additional distributions of orientational polarization modes as well as for any higher frequency polarization mechanisms is given by [cf. Eq. (4.197)]

$$\epsilon(\omega)/\epsilon_0 = \epsilon_\infty + \sum_j \frac{a_j}{(1 - i\omega\tau_j)^{1-\nu_j}(1 - i\omega\tau_{fj})} \qquad (4.205)$$

together with the sum rule given in Eq. (4.189). The summation appearing here is taken over each orientational mode distribution present in the material, each mode characterized by a representative relaxation time τ_j, friction time τ_{fj}, strength a_j, and distribution parameter ν_j with $\nu_j < 1$.

The real and imaginary parts of the frequency dependence of a double relaxation time Cole–Cole extended Rocard–Powles–Debye model for the relative dielectric permittivity of triply distilled water with $\nu_1 = 0$ and $\nu_2 = -0.5$, viz.

$$\epsilon(\omega)/\epsilon_0 = \epsilon_\infty + \frac{a_1}{(1 - i\omega\tau_1)(1 - i\omega\tau_{f1})} + \frac{a_2}{(1 - i\omega\tau_2)^{3/2}(1 - i\omega\tau_{f2})}, \quad (4.206)$$

is represented by the dotted curves in Fig. 4.6. For sufficiently high frequencies ($\omega \gg 1/\tau_1$), this Cole–Cole model extension results in an expression for the imaginary part of the dielectric permittivity that initially decreases as $1/\omega^4$ as compared to $1/\omega^3$ for the Rocard–Powles–Debye model alone, as evident in Fig. 4.6, becoming negative over a small frequency interval before increasing back above that for the Rocard–Powles–Debye model.

4.4.4 The Classical Lorentz Model of Resonance Polarization

The classical Lorentz model [3] of resonance polarization phenomena in dielectrics describes the material as a collection of neutral atoms with elastically bound electrons to the nucleus (i.e., as a collection of *Lorentz oscillators*), where each electron is bound to the nucleus by a Hooke's law restoring force. Under the action of an applied electromagnetic field, the equation of motion of a typical bound electron is given by

$$m\left(\frac{d^2\mathbf{r}_j}{dt^2} + 2\delta_j\frac{d\mathbf{r}_j}{dt} + \omega_j^2\mathbf{r}_j\right) = -q_e\mathbf{E}_{\text{eff}}, \quad (4.207)$$

where m is the mass of the electron and q_e is the magnitude of the electronic charge. The quantity ω_j is the *undamped resonance frequency* of the jth oscillator type and δ_j is the associated *phenomenological damping constant* of the oscillator. The same dynamical equation of motion also applies to molecular vibration modes when m is the ionic mass and ω_0 is the undamped resonance frequency of the transverse vibrational mode of the ionic lattice structure [39]. The field $\mathbf{E}_{\text{eff}} = \mathbf{E}_{\text{eff}}(\mathbf{r}, t)$ is the *effective local electric field intensity* that acts on the electron (or ion) as a driving force. The additional force $-\|1/c\|q_e\mathbf{v}_j \times \mathbf{B}_{\text{eff}}$ arising from the interaction of the bound electron (or ion) with the effective local magnetic field is assumed to be negligible in comparison to the electric field interaction (due to the smallness of the velocity of the electron in comparison with the vacuum speed of light c) and is consequently neglected in the classical theory; its relative influence on the atomic polarizability is presented in Appendix D. Finally, the term $2m\delta_j d\mathbf{r}_j/dt$ represents a *phenomenological damping* term for the electronic motion of the jth Lorentz oscillator type. The actual loss mechanism for a free atom is radiation damping [50], but it arises from a variety of scattering mechanisms in both solid and liquids [39].

The temporal frequency transform of Eq. (4.207) directly yields the frequency-domain solution

$$\tilde{\mathbf{r}}_j(\mathbf{r}, \omega) = \frac{q_e/m}{\omega^2 - \omega_j^2 + 2i\delta_j\omega}\tilde{\mathbf{E}}_{\text{eff}}(\mathbf{r}, \omega), \quad (4.208)$$

and the local induced dipole moment $\tilde{\mathbf{p}}_j = -q_e \tilde{\mathbf{r}}_j$ is then given by

$$\tilde{\mathbf{p}}_j(\mathbf{r}, \omega) = \frac{-q_e^2/m}{\omega^2 - \omega_j^2 + 2i\delta_j\omega} \tilde{\mathbf{E}}_{\text{eff}}(\mathbf{r}, \omega). \tag{4.209}$$

If there are N_j Lorentz oscillators per unit volume that are characterized by the undamped resonance frequencies ω_j and damping constants δ_j, then the macroscopic polarization induced in the medium is given by the summation over all oscillator types of the spatially averaged locally induced dipole moments as

$$\tilde{\mathbf{P}}(\mathbf{r}, \omega) = \sum_j N_j \langle\!\langle \tilde{\mathbf{p}}_j(\mathbf{r}, \omega) \rangle\!\rangle$$

$$= \langle\!\langle \tilde{\mathbf{E}}_{\text{eff}}(\mathbf{r}, \omega) \rangle\!\rangle \sum_j N_j \alpha_j(\omega). \tag{4.210}$$

Here

$$\alpha_j(\omega) = \frac{-q_e^2/m}{\omega^2 - \omega_j^2 + 2i\delta_j\omega} \tag{4.211}$$

is the *atomic polarizability* of the Lorentz oscillator type characterized by ω_j and δ_j with number density N_j. In addition, $N = \sum_j N_j$ is the total number of electrons per unit volume that interact with the effective local applied electromagnetic field. Substitution of this result into the Lorentz–Lorenz formula (4.176) then yields the expression [see Eq. (4.175)]

$$\epsilon(\omega)/\epsilon_0 = \frac{1 - (2/3)\sum_j b_j^2/(\omega^2 - \omega_j^2 + 2i\delta_j\omega)}{1 + (1/3)\sum_j b_j^2/(\omega^2 - \omega_j^2 + 2i\delta_j\omega)} \tag{4.212}$$

for the relative dielectric permittivity, where

$$b_j \equiv \sqrt{\frac{\|4\pi\|}{\epsilon_0} \frac{N_j q_e^2}{m}} \tag{4.213}$$

is the plasma frequency for the type j Lorentz oscillators with number density N_j.

When the inequality $b_j^2/(6\delta_j\omega_j) \ll 1$ is satisfied, the denominator in Eq. (4.212) may be approximated by the first two terms in its power series expansion, so that

$$\epsilon(\omega)/\epsilon_0 \approx \left(1 - \frac{2}{3}\sum_j \frac{b_j^2}{\omega^2 - \omega_j^2 + 2i\delta_j\omega}\right)\left(1 - \frac{1}{3}\sum_j \frac{b_j^2}{\omega^2 - \omega_j^2 + 2i\delta_j\omega}\right)$$

$$\approx 1 - \sum_j \frac{b_j^2}{\omega^2 - \omega_j^2 + 2i\delta_j\omega}, \tag{4.214}$$

which is the usual expression [3, 44, 51] for the frequency dispersion of a classical Lorentz model dielectric.

As an example, consider the single resonance Lorentz model material parameters $\omega_0 = 4 \times 10^{16}$ r/s, $\delta_0 = 0.28 \times 10^{16}$ r/s, $b_0 = \sqrt{20} \times 10^{16}$ r/s that were chosen by Brillouin [52, 53] in his now classic analysis of dispersive absorptive signal propagation. These values correspond to a highly absorptive dielectric. Furthermore, because $b_0^2/(6\delta_0\omega_0) = 2.976$, the approximation made in Eq. (4.214) does not strictly apply and the Lorentz–Lorenz modified expression (4.212) must be used for these particular model parameters. The frequency dispersion of the relative dielectric permittivity for the Lorentz model alone is illustrated by the dashed curves in Fig. 4.7 for the real part $\epsilon_r(\omega)/\epsilon_0 = \Re\{\epsilon(\omega)/\epsilon_0\}$ and in Fig. 4.8 for the imaginary part $\epsilon_i(\omega)/\epsilon_0 = \Im\{\epsilon(\omega)/\epsilon_0\}$. The corresponding solid curves in this pair of figures describe the resultant frequency dispersion for this Lorentz model dielectric when the Lorentz–Lorenz relation (4.212) is used. The Lorentz–Lorenz modified material dispersion is seen to primarily shift the resonance frequency to a lower value while increasing both the absorption and the below resonance real dielectric permittivity. However, if the plasma frequency is decreased to the value $b_0 = \sqrt{2} \times 10^{16}$ r/s so that $b_0^2/(6\delta_0\omega_0) = 0.2976$, then the modification of the Lorentz model by the Lorentz–Lorenz relation is relatively small (and the dispersion is also weak), as exhibited by the second set of curves in Figs. 4.7 and 4.8. The approximate expression given in Eq. (4.214) may then be used in this latter case, but not in the former.

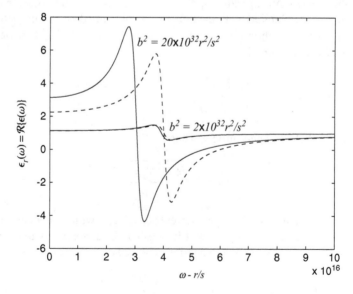

Fig. 4.7 Angular frequency dependence of the real part of the relative dielectric permittivity $\epsilon(\omega)/\epsilon_0$ for a single-resonance Lorentz model dielectric with (solid curves) and without (dashed curves) the Lorentz–Lorenz formula correction for two different values of the material plasma frequency

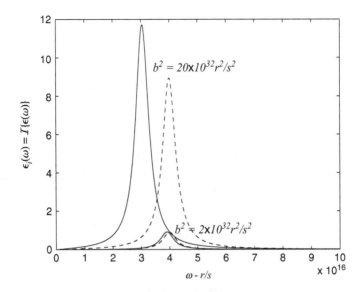

Fig. 4.8 Angular frequency dependence of the imaginary part of the relative dielectric permittivity $\epsilon(\omega)/\epsilon_0$ for a single-resonance Lorentz model dielectric with (solid curves) and without (dashed curves) the Lorentz–Lorenz formula correction for two different values of the material plasma frequency

Because the primary effect of the Lorentz–Lorenz formula on the Lorentz model is to downshift the effective resonance frequency and increase the static permittivity value, consider then determining the resonance frequency ω_* appearing in the Lorentz–Lorenz formula for a single-resonance Lorentz model dielectric that will yield the same value for $\epsilon_s/\epsilon_0 \equiv \epsilon(0)/\epsilon_0$ as that given by the Lorentz model alone with resonance frequency ω_0. Equations (4.212) and (4.214) then result in the equivalence relation [54, 55]

$$1 + b_0^2/\omega_0^2 = \frac{1 + (2/3)(b_0^2/\omega_*^2)}{1 - (1/3)(b_0^2/\omega_*^2)}$$

with solution

$$\omega_* = \sqrt{\omega_0^2 + b_0^2/3}. \tag{4.215}$$

This equivalence relation then allows the construction of a dielectric model satisfying the Lorentz–Lorenz relation with a resonance frequency ω_* that is upshifted to match a given Lorentz model resonance frequency ω_0.

The branch points of the complex index of refraction $n(\omega) \equiv \sqrt{\epsilon(\omega)/\epsilon_0}$ for a single-resonance Lorentz model dielectric with relative magnetic permeability $\mu/\mu_0 = 1$ and dielectric permittivity described by Eq. (4.214) are given by

$$\omega_p^\pm = -i\delta_0 \pm \sqrt{\omega_0^2 - \delta_0^2}, \qquad \omega_z^\pm = -i\delta_0 \pm \sqrt{\omega_0^2 + b_0^2 - \delta_0^2},$$

whereas the branch points of the complex index of refraction for the Lorentz–Lorenz modified Lorentz model dielectric with dielectric permittivity described by Eq. (4.212) are given by

$$\omega_{p'}^{\pm} = -i\delta_0 \pm \sqrt{\omega_*^2 - b_0^2/3 - \delta_0^2}, \qquad \omega_{z'}^{\pm} = -i\delta_0 \pm \sqrt{\omega_*^2 + 2b_0^2/3 - \delta_0^2}.$$

If $\omega_* = \omega_0$, then the branch points $\omega_{p'}^{\pm}$ and $\omega_{z'}^{\pm}$ of $n(\omega)$ for the Lorentz–Lorenz modified Lorentz model are shifted inward toward the imaginary axis from the respective branch point locations $\omega_p^{\pm}$ and $\omega_z^{\pm}$ for the Lorentz model alone provided that the inequality $\omega_*^2 - b_0^2/3 - \delta_0^2 \geq 0$ is satisfied. If the opposite inequality $\omega_*^2 - b_0^2/3 - \delta_0^2 < 0$ is satisfied, then the branch points $\omega_{p'}^{\pm}$ are located along the imaginary axis. However, if ω_* is given by the equivalence relation (4.215), then the locations of the branch points of the complex index of refraction $n(\omega)$ for the Lorentz–Lorenz modified Lorentz model and the Lorentz model alone are exactly the same. It then follows that the analyticity properties for these two models of the dielectric permittivity are also the same.

A comparison of the angular frequency dependence of the real and imaginary parts of the complex index of refraction $n(\omega) = \sqrt{\epsilon(\omega)/\epsilon_0}$ with dielectric permittivity described by Eqs. (4.212) and (4.214) for a single-resonance Lorentz model dielectric with ω_* given by the equivalence relation (4.215) is presented in Figs. 4.9 and 4.10 for Brillouin's choice of the material parameters ($\omega_0 = 4 \times 10^{16}$ r/s, $\delta_0 = 0.28 \times 10^{16}$ r/s, $b_0 = \sqrt{20} \times 10^{16}$ r/s). The rms error between the

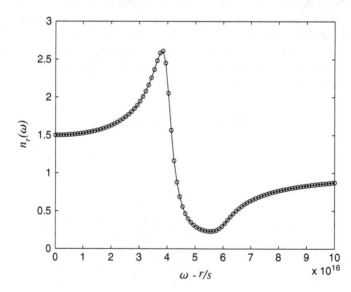

Fig. 4.9 Comparison of the angular frequency dependence of the real part of the complex index of refraction $n(\omega) = \sqrt{\epsilon(\omega)/\epsilon_0}$ for a single-resonance Lorentz model dielectric alone (solid curve) and for the equivalent Lorentz–Lorenz formula modified Lorentz model (circles)

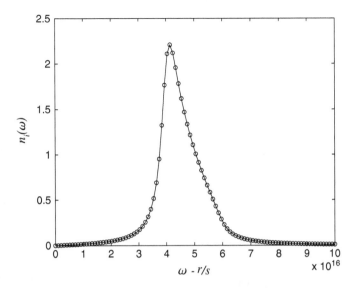

Fig. 4.10 Comparison of the angular frequency dependence of the imaginary part of the complex index of refraction $n(\omega) = \sqrt{\epsilon(\omega)/\epsilon_0}$ for a single-resonance Lorentz model dielectric alone (solid curve) and for the equivalent Lorentz–Lorenz formula modified Lorentz model (circles)

two sets of data points presented here is approximately 2.3×10^{-16} for the real part and 2.0×10^{-16} for the imaginary part of the complex index of refraction, with a maximum single point rms error of $\approx 2.5 \times 10^{-16}$. The corresponding rms error for the relative dielectric permittivity is $\approx 1.1 \times 10^{-15}$ for both the real and imaginary parts with a maximum single point rms error of $\approx 1 \times 10^{-14}$. Variation of any of the remaining material parameters in the equivalent Lorentz–Lorenz modified Lorentz model dielectric, including the value of the plasma frequency from that specified in Eq. (4.215), results in an increase in the rms error. This approximate equivalence relation between the Lorentz–Lorenz formula modified Lorentz model dielectric and the Lorentz model dielectric alone is then seen to provide a "best fit" in the rms sense between the frequency dependence of the two models. Finally, for a multiple-resonance Lorentz model dielectric, the equivalence relation (4.215) becomes

$$\omega_{j*} = \sqrt{\omega_j^2 + b_j^2/3} \qquad (4.216)$$

for each resonance line j present in the dielectric.

The behavior of the real and imaginary parts of the complex index of refraction $n(\omega)$ along the positive real frequency axis illustrated in Figs. 4.9 and 4.10 is typical of a single resonance Lorentz model dielectric. The frequency regions wherein the real index of refraction $n_r(\omega) \equiv \Re\{n(\omega)\}$ increases with increasing ω [i.e., where $n_r(\omega)$ has a positive slope] are said to be *normally dispersive*, whereas each frequency region where $n_r(\omega)$ decreases with increasing ω [i.e., where $n_r(\omega)$ has a negative slope] is said to exhibit *anomalous dispersion*. Notice that the real index of refraction $n_r(\omega)$ varies rapidly with ω within the region of anomalous dispersion and that this region essentially coincides with the region of strong absorption for the medium, the angular frequency width of the absorption line increasing with increasing values of the phenomenological damping constant δ_j for that line.

When considered as a function of complex $\omega = \omega' + i\omega''$, the *Lorentz model dielectric permittivity*

$$\epsilon(\omega)/\epsilon_0 = 1 - \frac{b_0^2}{\omega^2 - \omega_0^2 + 2i\delta_0\omega} \tag{4.217}$$

is found to have isolated simple pole singularities at

$$\omega_p^\pm \equiv -i\delta_0 \pm \sqrt{\omega_0^2 - \delta_0^2}, \tag{4.218}$$

which are symmetrically located about the imaginary axis in the lower half-plane. The Lorentz model dielectric permittivity is then analytic in the upper half-plane. In addition,

$$\epsilon(\omega') = \lim_{\omega'' \to 0} \epsilon(\omega' + i\omega'')$$

for all ω'. Finally, because

$$|\epsilon(\omega)/\epsilon_0 - 1|$$

$$= \frac{b_0^4}{(\omega'^2 + \omega''^2)^2 - 2\omega_0^2(\omega'^2 - \omega''^2) + \omega_0^4 + 4\delta_0\omega''(\omega'^2 + \omega''^2) + 4\delta_0\omega_0^2\omega''}$$

is integrable over any line parallel to the ω'-axis for all $\omega'' > 0$, condition 2 of Titchmarsh's theorem is then satisfied. The Lorentz model of resonance polarization in dielectrics is then a causal model.

From a quantum mechanical analysis [56] it is found that the dielectric response function of a multiple resonance dielectric material is given by

$$\epsilon(\omega)/\epsilon_0 = 1 - \|4\pi\| \frac{Nq_e^2}{m} \sum_j \frac{f_j}{\omega^2 - \omega_j^2 + 2i\gamma_j\omega}. \tag{4.219}$$

Although there is a formal similarity between this equation and the classical Lorentz model expression given in Eq. (4.214), the physical interpretation of some of the terms is quite different. Classically, ω_j is the undamped resonance frequency of a bound electron, whereas quantum mechanically it is the transition frequency of a bound electron between two atomic states that are separated in energy by the amount $\hbar\omega_j$, where $\hbar \equiv h/2\pi$, h denoting Planck's constant. The parameter f_j, called the *oscillator strength* of the dipole line, is a measure of the relative probability of a quantum mechanical dipole transition taking place with energy change $\hbar\omega_j$, which satisfies the sum rule [compare with the sum rule $N = \sum_j N_j$ following Eq. (4.211)]

$$\sum_j f_j = 1. \tag{4.220}$$

Finally, the quantity δ_j in the classical theory is a phenomenological damping constant that appears in the damping term $2\delta_j \dot{\mathbf{r}}_j$ of the equation of motion (4.207). This term represents the loss of energy due to electromagnetic radiation and is the classical analogue of spontaneous emission in the quantum theory [56]. Consequently, although the physical theory upon which the Lorentz model is based is rather simplistic, the proper functional form of the dielectric response is obtained for dielectric media. The model parameters b_j, ω_j, and δ_j cannot be obtained from the classical theory, but they can be obtained by fitting them to experimental measurements of the dispersive dielectric under consideration.

4.4.5 Composite Model of the Dielectric Permittivity

A general composite *Cole–Cole–Rocard–Powles–Debye–Lorentz* model for the frequency dispersion of the dielectric permittivity of a (nonconducting) dielectric material is given by

$$\epsilon(\omega)/\epsilon_0 = 1 + \sum_{j=1}^{J_o} \frac{a_j}{(1 - i\omega\tau_j)^{1-\nu_j}(1 - i\omega\tau_{fj})} - \sum_{\substack{(j\,\text{even})\\j=0}}^{J_r} \frac{b_j^2}{\omega^2 - \omega_j^2 + 2i\delta_j\omega}, \tag{4.221}$$

where the material exhibits J_o orientational polarization modes and J_r resonance lines. The static relative permittivity for this composite is then given by

$$\epsilon(0)/\epsilon_0 = 1 + \sum_{j=1}^{J_o} a_j + \sum_{\substack{(j\,\text{even})\\j=0}}^{J_r} \frac{b_j^2}{\omega_j^2}, \tag{4.222}$$

and its high-frequency limit is

$$\lim_{\omega\to\infty} \epsilon(\omega)/\epsilon_0 = 1. \tag{4.223}$$

The model parameters a_j, τ_j, τ_{fj}, ν_j, b_j, ω_j, and δ_j may then be determined through a numerical fit of the real and imaginary parts of Eq. (4.221) to experimental data. However, because this is a causal model, before any 'best-fit' numerical procedure is performed for a given dielectric material in order to determine these model parameters, the experimental data for $\epsilon_r(\omega)/\epsilon_0$ and $\epsilon_i(\omega)/\epsilon_0$ must first be checked in order to see if it is compliant with the Kramers–Kronig relations given in Eqs. (4.111) and (4.112).

4.4.5.1 Dispersion Relation Analysis of the Dielectric Permittivity Data for H_2O

For numerical evaluation of the principal value integrals appearing in the Kramers–Kronig relations (4.111) and (4.112), these dispersion relations relating the real and imaginary parts of the dielectric permittivity are rewritten as [40]

$$\epsilon_r(\omega)/\epsilon_0 - 1 = -\frac{1}{\pi} \lim_{\delta \to 0+} \int_\delta^\infty \frac{\epsilon_i(\omega+\omega')/\epsilon_0 - \epsilon_i(\omega-\omega')/\epsilon_0}{\omega'} d\omega', \quad (4.224)$$

$$\epsilon_i(\omega)/\epsilon_0 = \frac{1}{\pi} \lim_{\delta \to 0+} \int_\delta^\infty \frac{\epsilon_r(\omega+\omega')/\epsilon_0 - \epsilon_r(\omega-\omega')/\epsilon_0}{\omega'} d\omega'. \quad (4.225)$$

Each principal value integral may then be numerically evaluated from the "experimental data" illustrated in Figs. 4.2 and 4.3 for triply-distilled H_2O over the frequency domain $[\omega_{min}, \omega_{max}] \sim [1 \times 10^8 \text{ r/s}, 1 \times 10^{18} \text{ r/s}]$ using the even and odd symmetry properties of $\epsilon_r(\omega)$ and $\epsilon_i(\omega)$, respectively, provided that this "experimental data" can be analytically continued into both the low frequency domain $\omega' = 0 \to \omega_{min}$ and the high frequency domain $\omega' = \omega_{max} \to +\infty$.

For continuation into the low frequency domain below $\omega_{min} \sim 1 \times 10^8$ r/s, the Debye model relative permittivity [see Eq. (4.186)]

$$\epsilon(\omega)/\epsilon_0 = \epsilon_\infty + \frac{\epsilon_s - \epsilon_\infty}{1 - i\tau\omega} \qquad (4.226)$$

is employed, where $\epsilon_s \equiv \epsilon(0)/\epsilon_0$ is the relative static permittivity and ϵ_∞ is the limiting behavior of $\epsilon_r(\omega)/\epsilon_0$ as ω increases above $\omega \sim \tau^{-1}$. For sufficiently small $\omega < \omega_{min} < 1/\tau$, $\epsilon_r(\omega)/\epsilon_0 \simeq \epsilon_s - (\epsilon_s - \epsilon_\infty)\tau^2\omega^2$ and $\epsilon_i(\omega)/\epsilon_0 \simeq (\epsilon_s - \epsilon_\infty)\tau\omega(1 - \tau^2\omega^2)$. With the measured static value $\epsilon_s \simeq 79.0$, the Debye-model material parameters ϵ_∞ and τ may be estimated from the "experimental data" for $\epsilon_r(\omega)/\epsilon_0$ and $\epsilon_i(\omega)/\epsilon_0$ by requiring a continuously smooth transition between the analytical and "experimental" values at ω_{min}.

In a similar manner, continuation of this "experimental data" into the high frequency domain above ω_{max} is based on the Lorentz model permittivity

$$\epsilon(\omega)/\epsilon_0 = 1 - \frac{b_\gamma^2}{\omega^2 - \omega_\gamma^2 + 2i\delta_\gamma\omega}, \qquad (4.227)$$

with plasma frequency b_γ, resonance frequency ω_γ, and damping $\delta_\gamma \geq 0$ determined, in part, by the uppermost resonance line in water. For large $\omega > \omega_{max} \gg \omega_\gamma$,
$\epsilon_r(\omega) \simeq 1 - (b_\gamma^2/\omega^2)(1 - 4\delta_\gamma^2/\omega^2)$ and $\epsilon_i(\omega) \simeq 2\delta_\gamma b_\gamma^2/\omega^3$.

A numerical evaluation [40] of the dispersion relation pair in Eqs. (4.224) and
(4.225) for the extended "experimental data" for triply-distilled H_2O presented in
Figs. 4.2 and 4.3 over the extended frequency domain[7] from $\omega_{MIN} = 0$ to $\omega_{MAX} = 2\pi \times 10^{40}$ r/s is presented in Fig. 4.11(a) for the real part and in Fig. 4.11(b) for the
imaginary part of the relative dielectric permittivity. The green curve in each graph
describes the "experimental data", denoted here by $\epsilon_{ro}(\omega)/\epsilon_0$ and $\epsilon_{io}(\omega)/\epsilon_0$. The
violet curve in each graph describes the numerically computed Hilbert transform
of the "experimental data" in the other graph, denoted here by $\epsilon_{rH}(\omega)$ and $\epsilon_{iH}(\omega)$,
where

$$\epsilon_{rH}(\omega)/\epsilon_0 - 1 = \frac{1}{\pi}\mathcal{P}\int_{-\infty}^{+\infty}\frac{\epsilon_{io}(\omega')/\epsilon_0}{\omega' - \omega}d\omega', \qquad (4.228)$$

$$\epsilon_{iH}(\omega)/\epsilon_0 = -\frac{1}{\pi}\mathcal{P}\int_{-\infty}^{+\infty}\frac{\epsilon_{ro}(\omega')/\epsilon_0 - 1}{\omega' - \omega}d\omega'. \qquad (4.229)$$

The green curve in each graph then forms a Hilbert transform pair with the violet
curve in the other graph. Either pair may then be used in numerical studies. The
inverse transforms for each are also satisfied so that

$$\epsilon_{io}(\omega)/\epsilon_0 = -\frac{1}{\pi}\mathcal{P}\int_{-\infty}^{+\infty}\frac{\epsilon_{rH}(\omega')/\epsilon_0 - 1}{\omega' - \omega}d\omega', \qquad (4.230)$$

$$\epsilon_{ro}(\omega)/\epsilon_0 - 1 = \frac{1}{\pi}\mathcal{P}\int_{-\infty}^{+\infty}\frac{\epsilon_{iH}(\omega')/\epsilon_0}{\omega' - \omega}d\omega', \qquad (4.231)$$

These numerical results then show that the "experimental data" presented in
Figs. 4.2 and 4.3 only approximately satisfies the Kramers–Kronig relations above
$\sim 100\,\text{GHz}$.

Because the Hilbert transform is linear in ϵ, an average of the "experimental
data" $\epsilon_{ro}(\omega)/\epsilon_0$ and $\epsilon_{io}(\omega)/\epsilon_0$ with the respective computed values $\epsilon_{rH}(\omega)/\epsilon_0$ and
$\epsilon_{iH}(\omega)/\epsilon_0$ over the angular frequency domain $[\omega_{min}, \omega_{max}]$ will result in a causal

---

[7]This maximum frequency is below the frequency $f_P \equiv 1/t_P \simeq 1.855 \times 10^{43}\,\text{s}^{-1}$ associated
with the *Planck time* $t_P \equiv \sqrt{\hbar G/c^5} \simeq 5.39 \times 10^{-44}$ s, a unique combination of the gravitational
constant G, the speed of light constant c in special relativity, and the rationalized Planck constant
$\hbar \equiv h/2\pi$ from quantum theory. The Planck time supposedly provides an estimate of the quantized
time scale at which quantum gravitational effects may appear. Smaller time intervals would then be
meaningless, implying that frequencies above f_P could not possibly exist. A more realistic bound
is obtained from the realization that macroscopic quantities such as the index of refraction become
meaningless when the wavelength decreases below intermolecular distances, so that $f_{max} \approx 1 \times 10^{20}\,\text{s}^{-1}$ for water. .

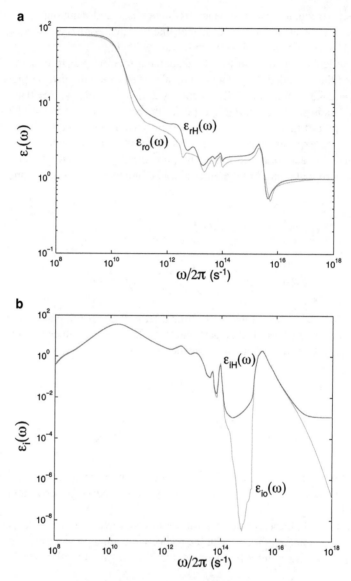

Fig. 4.11 Frequency dependence of the "experimental data" (green curves) and its computed Hilbert transform (violet curves) for the real (upper graph) and imaginary (lower graph) parts of the relative dielectric permittivity for triply-distilled H_2O at 25 °C

set of data that compares well with experimental data over this entire frequency domain. This causal set may then be used as the basis for finalizing an accurate analytical model, constructed from multiple relaxation time Rocard–Powles–Debye and multiple resonance Lorentz models, over the entire frequency domain $[0, \infty]$.

4.4.5.2 Composite Model of the Dielectric Permittivity of H_2O

Careful examination of Figs. 4.2, 4.3 and 4.11 depicting the frequency dependence of the real and imaginary parts of the relative dielectric permittivity of triply distilled water shows that there are five significant resonance features above 1 THz. With Eq. (4.206) for the observed rotational polarization phenomena, a *composite Rocard–Powles–Debye–Lorentz model* for the frequency dispersion of the dielectric permittivity of triply distilled water is given by

$$\epsilon(\omega)/\epsilon_0 = 1 + \sum_{j=1}^{2} \frac{a_j}{(1 - i\omega\tau_j)(1 - i\omega\tau_{fj})} - \sum_{\substack{(j \text{ even}) \\ j=0}}^{8} \frac{b_j^2}{\omega^2 - \omega_j^2 + 2i\delta_j\omega}.$$

(4.232)

An approximate fit of the parameters appearing in this model results in the Rocard–Powles–Debye model parameter set given in Table 4.1 and the Lorentz model parameter set given in Table 4.2.

A comparison of the frequency dependence of the relative dielectric permittivity described by the composite model given in Eq. (4.232) with the parameters listed in Tables 4.1 and 4.2 and the Hilbert transform averaged numerical data for triply distilled water is presented in Fig. 4.12 for the real part and in Fig. 4.13 for the imaginary part. Notice that the $j = 0$ resonance feature appears in close proximity to the $j = 1$ relaxation feature. The composite model (4.232) with material parameters given in Tables 4.1 and 4.2 is thus seen to provide a reasonably accurate description of both the real and imaginary parts of the dielectric permittivity of triply distilled water at 25 °C over the entire angular frequency domain $\omega \in [0, \infty]$.

Table 4.1 Estimated rms fit Rocard–Powles–Debye model parameters for triply distilled water at 25 °C

j	a_j	τ_j	τ_{fj}
1	76.59	8.14×10^{-12} s	6.7×10^{-14} s
2	2.4	1.1×10^{-13} s	4.0×10^{-15} s

Table 4.2 Estimated Lorentz model parameters for triply distilled water at 25 °C

j	ω_j	δ_j	b_j
0	2.31×10^{13} r/s	8.2×10^{12} r/s	2.4×10^{13} r/s
2	9.5×10^{13} r/s	4.5×10^{13} r/s	9.2×10^{13} r/s
4	3.0×10^{14} r/s	1.8×10^{13} r/s	3.4×10^{13} r/s
6	6.19×10^{14} r/s	2.9×10^{13} r/s	1.2×10^{14} r/s
8	2.2×10^{16} r/s	5.4×10^{15} r/s	2.1×10^{16} r/s

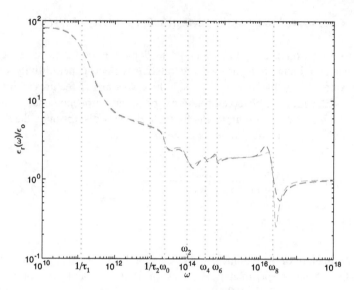

Fig. 4.12 Comparison of the angular frequency dependence of the real part of the dielectric permittivity given by the composite Rocard–Powles–Debye model (green dashed curve) with the Hilbert transform averaged numerical data (blue dashed curve) for triply-distilled water at 25 °C

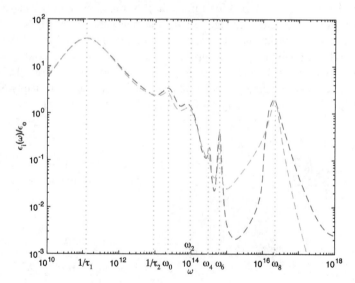

Fig. 4.13 Comparison of the angular frequency dependence of the imaginary part of the dielectric permittivity given by the composite Rocard–Powles–Debye model (green dashed curve) with the Hilbert transform averaged numerical data (blue dashed curve) for triply distilled water at 25 °C

4.4.6 Composite Model of the Magnetic Permeability

The frequency dependence of the magnetic permeability $\mu(\omega)$ may also, in some circumstances [57], be described by a composite Rocard–Powles–Debye–Lorentz model, where

$$\mu(\omega)/\mu_0 = 1 + \sum_{j=1}^{N} \frac{m_j}{(1 - i\omega\Upsilon_j)(1 - i\omega\Upsilon_{fj})} - \sum_{\substack{(j\ \text{even}) \\ j=0}} \frac{\Omega_{mj}^2}{\omega^2 - \Omega_j^2 + 2i\,\Gamma_j\omega}.$$

$$(4.233)$$

The first summation on the right-hand side of this expression describes the low-frequency orientational response of the material magnetization, where Υ_j is the macroscopic relaxation time and Υ_{fj} is the associated friction time. The second summation describes the high-frequency resonance response of the material magnetization [15], where Ω_j denotes the undamped resonance frequency, Ω_{mj} is the magnetic equivalent to the plasma frequency in the Lorentz model for frequency dispersion in dielectrics, and Υ_j is the phenomenological damping constant associated with the jth resonance frequency.

4.4.7 The Drude Model of Free Electron Metals

The classical Drude model [29] treats conduction electrons that are near the Fermi level as essentially free electrons [39] with equation of motion

$$m\left(\frac{d^2\mathbf{r}}{dt^2} + \gamma\frac{d\mathbf{r}}{dt}\right) = -q_e\mathbf{E}_{\text{eff}}.$$

$$(4.234)$$

The relative dielectric permittivity is then given by

$$\epsilon(\omega)/\epsilon_0 = 1 - \frac{\omega_p^2}{\omega(\omega + i\gamma)},$$

$$(4.235)$$

where $\gamma = 1/\tau_c$ is a damping constant given by the inverse of the relaxation time τ_c associated with the mean-free path for free electrons in the material, and $\omega_p \equiv \sqrt{(\|4\pi\|/\epsilon_0)(Nq_e^2/m)}$ is the plasma frequency. Here N is the number density of free electrons with effective mass m. Because the Lorentz model is causal, the Drude model is then also a causal model.

The physical significance of the plasma frequency ω_p is derived from the theory of *plasma oscillations* [58], the simplest type of collective behavior in a plasma medium. In order to describe this motion, consider a fully ionized gas uniformly filling all of space with a sufficiently low absolute temperature such that the random thermal motions of the ions and electrons comprising the plasma can be neglected. In addition, assume that the ratio of the electron mass m_e to that of the ion mass m_i satisfies the inequality $m_e/m_i \ll 1$ so that the collective motion of the ions is negligible in comparison to that for the electrons. For simplicity, consider the one-dimensional motion along the x-direction of a plane lamina of electrons in the yz-plane with equilibrium position x that is displaced by the (small) distance ξ to the position $x + \xi$ such that the electron laminae near $x = \pm\infty$ are left undisturbed and the ordering of these electron laminae along the x-axis remains unchanged. This then results in the production of a dipole layer due to an excess of positive charge $Nq_e\xi$ per unit area on one side of the electron lamina and a deficit of charge $-Nq_e\xi$ on the other side, where N denotes the number density of electrons. The electric field experienced by this displaced electron lamina is then given by the linearly increasing field in the interior of the plane dipole layer as

$$\mathbf{E}(\xi) = \frac{\|4\pi\|}{\epsilon_0} Nq_e\xi \hat{\mathbf{1}}_x.$$

The equation of motion of each electron in this displaced lamina is then given by $m_e d^2\xi/dt^2 = -q_e E$, which may be rewritten as

$$\frac{d^2\xi}{dt^2} + \omega_p^2\xi = 0. \tag{4.236}$$

The laminar displacement then oscillates at the plasma frequency ω_p.

Problems

4.1 Show that the spatially averaged charge is conserved using the expressions given in Eqs. (4.23) and (4.43) for the spatial averages of the microscopic charge and current densities.

4.2 Show that $\mathcal{F}\{sgn(t)\} = \mathcal{P}\{2i/\omega\}$ where $\mathcal{F}\{f(t)\} \equiv \int_{-\infty}^{\infty} f(t)e^{i\omega t}dt$ denotes the Fourier transform of the function $f(t)$ and where $sgn(t) \equiv -1$ for $t < 0$ and $sgn(t) \equiv 1$ for $t > 0$ is the sign function. From this result, show that $\mathcal{F}\{U(t)\} = \mathcal{P}\{i/\omega\} + \pi\delta(\omega)$, where $U(t)$ denotes the Heaviside unit step function.

4.3 Show that $\hat{\epsilon}(t) = \epsilon_0\left(\delta(t) + \|4\pi\|\hat{\chi}(t)\right)$ for a simple polarizable dielectric.

4.4 Show that $\int_{-\infty}^{t}\hat{\mu}^{-1}(t-t')\hat{\mu}_m(t'-t'')dt' = \delta(t''-t)$, where $\hat{\mu}^{-1}(t)$ is as defined in Eq. (4.136) and where $\hat{\mu}_m(t)$ is defined through the linear relation $\mathbf{B}(\mathbf{r}, t) = \int_{-\infty}^{\infty}\hat{\mu}_m(t-t')\mathbf{H}(\mathbf{r}, t')dt'$.

4.5 Derive the approximate expression given in Eq. (4.146) for the magnetic susceptibility in either diamagnetic or paramagnetic materials.

4.6 Explain in physical terms why the effective local field given in Eq. (4.172) [see also Appendix C] is larger in magnitude than the macroscopic applied electric field $\tilde{\mathbf{E}}(\mathbf{r}, \omega)$.

4.7 With the molecular polarizability given by Eq. (4.180) for the single relaxation time Debye model, derive Eqs. (4.182) and (4.183) for the electric susceptibility and dielectric permittivity.

4.8 With the molecular polarizability given by Eq. (4.190) for the single relaxation time Rocard–Powles–Debye model, derive Eqs. (4.192) and (4.193) for the electric susceptibility and dielectric permittivity.

4.9 Establish the inequality given in Eq. (4.196) for the Rocard–Powles extension of the Debye model.

References

1. J. C. Maxwell, "A dynamical theory of the electromagnetic field," *Phil. Trans. Roy. Soc. (London)*, vol. 155, pp. 450–521, 1865.
2. J. C. Maxwell, *A Treatise on Electricity and Magnetism*. Oxford: Oxford University Press, 1873.
3. H. A. Lorentz, *The Theory of Electrons*. Leipzig: Teubner, 1906. Ch. IV.
4. J. H. van Vleck, *Electric and Magnetic Susceptibilities*. London: Oxford University Press, 1932.
5. R. Russakoff, "A derivation of the macroscopic Maxwell equations," *Am. J. Phys.*, vol. 38, no. 10, pp. 1188–1195, 1970.
6. F. N. H. Robinson, *Macroscopic Electromagnetism*. Oxford: Pergamon, 1973.
7. E. M. Purcell, *Electricity and Magnetism*. New York: McGraw-Hill, 1965. pp. 344–347.
8. C. J. F. Böttcher, *Theory of Electric Polarization: Volume I. Dielectrics in Static Fields*. Amsterdam: Elsevier, second ed., 1973.
9. C. J. F. Böttcher and P. Bordewijk, *Theory of Electric Polarization: Volume II. Dielectrics in Time-Dependent Fields*. Amsterdam: Elsevier, second ed., 1978.
10. J. D. Jackson, *Classical Electrodynamics*. New York: John Wiley & Sons, third ed., 1999.
11. P. Penfield and H. A. Haus, *Electrodynamics of Moving Media*. Cambridge, MA: M.I.T. Press, 1967.
12. R. Resnick, *Introduction to Special Relativity*. New York: John Wiley & Sons, 1968.
13. J. V. Bladel, *Relativity and Engineering*. Berlin-Heidelberg: Springer-Verlag, 1984.
14. J. A. Kong, *Theory of Electromagnetic Waves*. Wiley, 1975. Chapter 1.
15. I. V. Lindell, A. H. Sihvola, S. A. Tretyakov, and A. J. Viitanen, *Electromagnetic Waves in Chiral and Bi-Isotropic Media*. Boston: Artech House, 1994. Ch. 6.
16. H. M. Nussenzveig, *Causality and Dispersion Relations*. New York: Academic, 1972. Ch. 1.
17. R. W. Whitworth and H. V. Stopes-Roe, "Experimental demonstration that the couple on a bar magnet depends on H, not B," *Nature*, vol. 234, pp. 31–33, 1971.
18. V. M. Agranovich and V. L. Ginzburg, *Crystal Optics with Spatial Dispersion, and Excitons*. Berlin-Heidelberg: Springer-Verlag, second ed., 1984.
19. B. D. H. Tellegen, "The gyrator, a new electric network element," *Philips Res. Rep.*, vol. 3, pp. 81–101, 1948.
20. J. S. Toll, "Causality and the dispersion relation: Logical foundations," *Phys. Rev.*, vol. 104, no. 6, pp. 1760–1770, 1950.

21. A. Grünbaum, "Is preacceleration of particles in Dirac's electrodynamics a case of backward causation? The myth of retrocausation in classical electrodynamics," *Philosophy of Science*, vol. 43, pp. 165–201, 1976.

22. B. Y. Hu, "Kramers-Kronig in two lines," *Am. J. Phys.*, vol. 57, no. 9, p. 821, 1989.

23. J. Plemelj, "Ein Ergänzungssatz zur Cauchyschen Integraldarstellung analytischer Funktionen, Randwerte betreffend," *Monatshefte für Mathematik und Physik*, vol. 19, pp. 205–210, 1908.

24. E. C. Titchmarsh, *Introduction to the Theory of Fourier Integrals*. London: Oxford University Press, 1939. Ch. I.

25. H. A. Kramers, "La diffusion de la lumière par les atomes," in *Estratto dagli Atti del Congresso Internazional de Fisici Como*, pp. 545–557, Bologna: Nicolo Zonichelli, 1927.

26. R. de L. Kronig, "On the theory of dispersion of X-Rays," *J. Opt. Soc. Am. & Rev. Sci. Instrum.*, vol. 12, no. 6, pp. 547–557, 1926.

27. L. D. Landau and E. M. Lifshitz, *Electrodynamics of Continuous Media*. Oxford: Pergamon, 1960. Ch. IX.

28. V. Lucarini, F. Bassani, K.-E. Peiponen, and J. J. Saarinen, "Dispersion theory and sum rules in linear and nonlinear optics," *Riv. Nuovo Cimento*, vol. 26, pp. 1–127, 2003.

29. P. Drude, *Lehrbuch der Optik*. Leipzig: Teubner, 1900. Ch. V.

30. P. Debye, *Polar Molecules*. New York: Dover, 1929.

31. J. McConnel, *Rotational Brownian Motion and Dielectric Theory*. London: Academic, 1980.

32. K. S. Cole and R. H. Cole, "Dispersion and absorption in dielectrics. I. Alternating current characteristics," *J. Chem. Phys.*, vol. 9, pp. 341–351, 1941.

33. H. Fröhlich, *Theory of Dielectrics: Dielectric Constant and Dielectric Loss*. Oxford: Oxford University Press, 1949.

34. K. Moten, C. H. Durney, and T. G. Stockham, "Electromagnetic pulse propagation in dispersive planar dielectrics," *Bioelectromagnetics*, vol. 10, pp. 35–49, 1989.

35. R. Albanese, J. Penn, and R. Medina, "Short-rise-time microwave pulse propagation through dispersive biological media," *J. Opt. Soc. Am. A*, vol. 6, pp. 1441–1446, 1989.

36. K. Moten, C. H. Durney, and T. G. Stockham, "Electromagnetic pulsed-wave radiation in spherical models of dispersive biological substances," *Bioelectromagnetics*, vol. 12, p. 319, 1991.

37. J. E. K. Laurens and K. E. Oughstun, "Electromagnetic impulse response of triply-distilled water," in *Ultra-Wideband, Short-Pulse Electromagnetics 4* (E. Heyman, B. Mandelbaum, and J. Shiloh, eds.), pp. 243–264, New York: Plenum, 1999.

38. P. D. Smith and K. E. Oughstun, "Ultrawideband electromagnetic pulse propagation in triply-distilled water," in *Ultra-Wideband, Short-Pulse Electromagnetics 4* (E. Heyman, B. Mandelbaum, and J. Shiloh, eds.), pp. 265–276, New York: Plenum, 1999.

39. C. F. Bohren and D. R. Huffman, *Absorption and Scattering of Light by Small Particles*. New York: Wiley-Interscience, 1984. Ch. 9.

40. C. D. MacDonald and K. E. Oughstun, "Dispersion relation analysis of $\epsilon(\omega)$ data for H_2O," in *2016 IEEE Antennas & Propagation Soc. International Symposium*, 2016. Paper #TUP-A2.2P.5.

41. H. A. Lorentz, "Über die Beziehungzwischen der Fortpflanzungsgeschwindigkeit des Lichtes der Körperdichte," *Ann. Phys.*, vol. 9, pp. 641–665, 1880.

42. L. Lorenz, "Über die Refractionsconstante," *Ann. Phys.*, vol. 11, pp. 70–103, 1880.

43. J. W. F. Brown, "Dielectrics," in *Handbuch der Physik: Dielektrika*, vol. XVII, Berlin: Springer-Verlag, 1956.

44. M. Born and E. Wolf, *Principles of Optics*. Cambridge: Cambridge University Press, seventh (expanded) ed., 1999.

45. M. Y. Rocard, "x," *J. Phys. Radium*, vol. 4, pp. 247–250, 1933.

46. J. G. Powles, "x," *Trans. Faraday Soc.*, vol. 44, pp. 802–806, 1948.

47. E. C. Titchmarsh, *Introduction to the Theory of Fourier Integrals*. London: Oxford University Press, second ed., 1948. Sect. 11.8.

48. H. Fröhlich, "Remark on the calculation of the static dielectric constant," *Physica*, vol. 22, pp. 898–904, 1956.

49. B. K. P. Scaife, *Principles of Dielectrics*. Oxford: Oxford University Press, 1989.

50. A. D. Yaghjian, *Relativistic Dynamics of a Charged Sphere*. Berlin-Heidelberg: Springer-Verlag, 1992.

51. J. M. Stone, *Radiation and Optics, An Introduction to the Classical Theory*. New York: McGraw-Hill, 1963.

52. L. Brillouin, "Über die fortpflanzung des licht in disperdierenden medien," *Ann. Phys.*, vol. 44, pp. 204–240, 1914.

53. L. Brillouin, *Wave Propagation and Group Velocity*. New York: Academic, 1960.

54. K. E. Oughstun and N. A. Cartwright, "On the Lorentz-Lorenz formula and the Lorentz model of dielectric dispersion," *Opt. Exp.*, vol. 11, no. 13, pp. 1541–1546, 2003.

55. K. E. Oughstun and N. A. Cartwright, "On the Lorentz-Lorenz formula and the Lorentz model of dielectric dispersion: addendum," *Opt. Exp.*, vol. 11, no. 21, pp. 2791–2792, 2003.

56. R. Loudon, *The Quantum Theory of Light*. London: Oxford University Press, 1973.

57. J. B. Pendry, A. J. Holden, D. J. Robbins, and W. J. Stewart, "Magnetism from conductors and enhanced nonlinear phenomena," *IEEE Trans. Microwave Theory & Tech.*, vol. 47, pp. 2075–2084, 1999.

58. P. A. Sturrock, *Plasma Physics*. Cambridge: Cambridge University Press, 1994.

Chapter 5
Fundamental Field Equations in Temporally Dispersive Media

"We can scarcely avoid the inference that light consists in the transverse undulations of the same medium which is the cause of electric and magnetic phenomena."

James Clerk Maxwell (1862).

The fundamental macroscopic electromagnetic field equations and elementary plane wave solutions in linear, temporally dispersive absorptive media are developed in this chapter with particular emphasis on homogeneous, isotropic, locally linear (HILL), temporally absorptive dispersive media. The general frequency dependence of the dielectric permittivity, magnetic permeability, and electric conductivity is included in the analysis so that the resultant field equations rigorously apply to both perfect and imperfect dielectrics, conductors and semiconducting materials, as well as to metamaterials, over the entire frequency domain.

5.1 Macroscopic Electromagnetic Field Equations in Temporally Dispersive Media

The macroscopic electromagnetic field behavior in an isotropic, locally linear, temporally dispersive medium with no externally supplied charge or current sources is described by the macroscopic Maxwell's equations [cf. Eqs. (4.52)–(4.55)]

$$\nabla \cdot \mathbf{D}(\mathbf{r}, t) = \|4\pi\| \varrho_c(\mathbf{r}, t), \tag{5.1}$$

$$\nabla \times \mathbf{E}(\mathbf{r}, t) = -\left\|\frac{1}{c}\right\| \frac{\partial \mathbf{B}(\mathbf{r}, t)}{\partial t}, \tag{5.2}$$

$$\nabla \cdot \mathbf{B}(\mathbf{r}, t) = 0, \tag{5.3}$$

$$\nabla \times \mathbf{H}(\mathbf{r}, t) = \left\|\frac{1}{c}\right\| \frac{\partial \mathbf{D}(\mathbf{r}, t)}{\partial t} + \left\|\frac{4\pi}{c}\right\| \mathbf{J}_c(\mathbf{r}, t), \tag{5.4}$$

© Springer Nature Switzerland AG 2019
K. E. Oughstun, *Electromagnetic and Optical Pulse Propagation*, Springer Series in Optical Sciences 224, https://doi.org/10.1007/978-3-030-20835-6_5

taken together with the constitutive relations

$$\mathbf{D}(\mathbf{r}, t) = \int_{-\infty}^{t} \hat{\epsilon}(\mathbf{r}, t - t')\mathbf{E}(\mathbf{r}, t')dt', \tag{5.5}$$

$$\mathbf{H}(\mathbf{r}, t) = \int_{-\infty}^{t} \hat{\mu}^{-1}(\mathbf{r}, t - t')\mathbf{B}(\mathbf{r}, t')dt', \tag{5.6}$$

$$\mathbf{J}_c(\mathbf{r}, t) = \int_{-\infty}^{t} \hat{\sigma}(\mathbf{r}, t - t')\mathbf{E}(\mathbf{r}, t')dt', \tag{5.7}$$

as described in Eq. (4.69). Here $\hat{\epsilon}(\mathbf{r}, t)$ denotes the (real-valued) dielectric permittivity, $\hat{\mu}(\mathbf{r}, t)$ denotes the (real-valued) magnetic permeability, and $\hat{\sigma}(\mathbf{r}, t)$ denotes the (real-valued) electric conductivity response of the spatially inhomogeneous, temporally dispersive medium. By causality, $\hat{\epsilon}(\mathbf{r}, t - t') = 0$, $\hat{\mu}^{-1}(\mathbf{r}, t - t') = 0$, and $\hat{\sigma}(\mathbf{r}, t - t') = 0$ for $t' > t$, as exhibited in the upper limit of integration in the above three relations. The dependence of the conduction current density $\mathbf{J}_c(\mathbf{r}, t)$ on the magnetic field through the $(\mathbf{J} \times \mathbf{B})$ Lorentz force term, known as the *Hall effect*, is negligible in all but very special materials and is not considered here.

The conduction current density $\mathbf{J}_c(\mathbf{r}, t)$ and charge density $\varrho_c(\mathbf{r}, t)$ are related by the equation of continuity [cf. Eq. (4.56)],

$$\nabla \cdot \mathbf{J}_c(\mathbf{r}, t) + \frac{\partial \varrho_c(\mathbf{r}, t)}{\partial t} = 0, \tag{5.8}$$

which also follows from the divergence of Eq. (5.4) with substitution from Eq. (5.1). The temporal Fourier transform (see Appendix E) of this relation yields

$$\nabla \cdot \tilde{\mathbf{J}}_c(\mathbf{r}, \omega) = i\omega\tilde{\varrho}_c(\mathbf{r}, \omega), \tag{5.9}$$

and the temporal Fourier transform of the constitutive relation (5.7) yields $\tilde{\mathbf{J}}_c(\mathbf{r}, \omega) = \sigma(\mathbf{r}, \omega)\tilde{\mathbf{E}}(\mathbf{r}, \omega)$, with application of the convolution theorem, so that

$$\tilde{\varrho}_c(\mathbf{r}, \omega) = \frac{1}{i\omega} \left(\sigma(\mathbf{r}, \omega)\nabla \cdot \tilde{\mathbf{E}}(\mathbf{r}, \omega) + (\nabla\sigma(\mathbf{r}, \omega)) \cdot \tilde{\mathbf{E}}(\mathbf{r}, \omega) \right).$$

Furthermore, the temporal Fourier transform of the constitutive relation (5.5) yields $\tilde{\mathbf{D}}(\mathbf{r}, \omega) = \epsilon(\omega)\tilde{\mathbf{E}}(\mathbf{r}, \omega)$, with application of the convolution theorem, so that the temporal Fourier transform of Gauss' law (5.1) gives

$$\|4\pi\|\tilde{\varrho}_c(\mathbf{r}, \omega) = \nabla \cdot \tilde{\mathbf{D}}(\mathbf{r}, \omega)$$

$$= \epsilon(\mathbf{r}, \omega)\nabla \cdot \tilde{\mathbf{E}}(\mathbf{r}, \omega) + (\nabla\epsilon(\mathbf{r}, \omega)) \cdot \tilde{\mathbf{E}}(\mathbf{r}, \omega).$$

The solution of this pair of equations for the spectrum of the conduction charge density then yields

$$\tilde{\varrho}_c(\mathbf{r}, \omega) = \frac{\epsilon(\mathbf{r}, \omega)\nabla\sigma(\mathbf{r}, \omega) - \sigma(\mathbf{r}, \omega)\nabla\epsilon(\mathbf{r}, \omega)}{i\omega\epsilon(\mathbf{r}, \omega) - \|4\pi\|\sigma(\mathbf{r}, \omega)} \cdot \tilde{\mathbf{E}}(\mathbf{r}, \omega), \tag{5.10}$$

provided that $i\omega\epsilon(\mathbf{r}, \omega) \neq \|4\pi\|\sigma(\mathbf{r}, \omega)$. The conduction charge density $\varrho(\mathbf{r}, t)$ then vanishes for $\mathbf{E}(\mathbf{r}, t) \neq \mathbf{0}$ when any of the following conditions is satisfied:

1. $\nabla\sigma(\mathbf{r}, \omega) = \nabla\epsilon(\mathbf{r}, \omega) = 0$ so that the electrical properties of the material are spatially homogeneous (the magnetic permeability $\mu(\mathbf{r}, \omega)$ may still be spatially inhomogeneous),
2. $(\nabla\epsilon(\mathbf{r}, \omega)) \cdot \tilde{\mathbf{E}}(\mathbf{r}, \omega) = (\nabla\sigma(\mathbf{r}, \omega)) \cdot \tilde{\mathbf{E}}(\mathbf{r}, \omega) = 0$ so that the spectrum of the electric field vector $\tilde{\mathbf{E}}(\mathbf{r}, \omega)$ is orthogonal to both $\nabla\epsilon(\mathbf{r}, \omega)$ and $\nabla\sigma(\mathbf{r}, \omega)$,
3. $\nabla\sigma(\mathbf{r}, \omega) = \sigma(\mathbf{r}, \omega)\boldsymbol{\zeta}$ and $\nabla\epsilon(\mathbf{r}, \omega) = \epsilon(\mathbf{r}, \omega)\boldsymbol{\zeta}$ so that the spatial variation of $\epsilon(\mathbf{r}, \omega)$ and $\sigma(\mathbf{r}, \omega)$ is in the same direction $\boldsymbol{\zeta}$ with each equally proportional to its value at that point,
4. the medium is nonconducting, viz. $\sigma(\mathbf{r}, \omega) = 0$.

The electromagnetic field equations when the first (spatial homogeneity) and the last (nonconducting) conditions are satisfied are now separately considered.

5.1.1   Temporally Dispersive HILL Media

Maxwell's equations in a homogeneous, isotropic, locally linear (HILL), temporally dispersive medium with no externally supplied charge or current sources are then given by

$$\nabla \cdot \mathbf{D}(\mathbf{r}, t) = 0, \tag{5.11}$$

$$\nabla \times \mathbf{E}(\mathbf{r}, t) = -\left\|\frac{1}{c}\right\|\frac{\partial \mathbf{B}(\mathbf{r}, t)}{\partial t}, \tag{5.12}$$

$$\nabla \cdot \mathbf{B}(\mathbf{r}, t) = 0, \tag{5.13}$$

$$\nabla \times \mathbf{H}(\mathbf{r}, t) = \left\|\frac{1}{c}\right\|\frac{\partial \mathbf{D}(\mathbf{r}, t)}{\partial t} + \left\|\frac{4\pi}{c}\right\|\mathbf{J}_c(\mathbf{r}, t), \tag{5.14}$$

with the constitutive relations given in Eqs. (5.5)–(5.7) and the equation of continuity

$$\nabla \cdot \mathbf{J}_c(\mathbf{r}, t) = 0. \tag{5.15}$$

Hence, the electric displacement vector $\mathbf{D}(\mathbf{r}, t)$, the electric field intensity vector $\mathbf{E}(\mathbf{r}, t)$, the magnetic induction vector $\mathbf{B}(\mathbf{r}, t)$, the magnetic field intensity vector $\mathbf{H}(\mathbf{r}, t)$, and the conduction current density vector $\mathbf{J}_c(\mathbf{r}, t)$ are all solenoidal vector fields in the source-free medium. Because the material properties are spatially continuous, it then follows, for example, that

$$\oint_S \mathbf{J}_c(\mathbf{r}, t) \cdot \hat{\mathbf{n}} d^2 r = 0 \tag{5.16}$$

for any closed surface S, with similar expressions for the field vectors.

5.1.1.1 Temporal Frequency Domain Representation

The temporal frequency spectra of the electric field intensity and magnetic induction field vectors are defined by the Fourier transform integrals

$$\tilde{\mathbf{E}}(\mathbf{r}, \omega) = \int_{-\infty}^{\infty} \mathbf{E}(\mathbf{r}, t) e^{i\omega t} dt, \tag{5.17}$$

$$\tilde{\mathbf{B}}(\mathbf{r}, \omega) = \int_{-\infty}^{\infty} \mathbf{B}(\mathbf{r}, t) e^{i\omega t} dt, \tag{5.18}$$

with corresponding inverse transformations

$$\mathbf{E}(\mathbf{r}, t) = \frac{1}{2\pi} \int_{-\infty}^{\infty} \tilde{\mathbf{E}}(\mathbf{r}, \omega) e^{-i\omega t} d\omega, \tag{5.19}$$

$$\mathbf{B}(\mathbf{r}, t) = \frac{1}{2\pi} \int_{-\infty}^{\infty} \tilde{\mathbf{B}}(\mathbf{r}, \omega) e^{-i\omega t} d\omega. \tag{5.20}$$

Because both of the electromagnetic field vectors $\mathbf{E}(\mathbf{r}, t)$ and $\mathbf{B}(\mathbf{r}, t)$ are real-valued vector functions of both position and time, then their corresponding temporal frequency spectra satisfy the *symmetry property*

$$\tilde{\mathbf{E}}^*(\mathbf{r}, \omega) = \tilde{\mathbf{E}}(\mathbf{r}, -\omega^*), \tag{5.21}$$

$$\tilde{\mathbf{B}}^*(\mathbf{r}, \omega) = \tilde{\mathbf{B}}(\mathbf{r}, -\omega^*), \tag{5.22}$$

for complex ω. The temporal frequency spectra of the electric displacement vector $\mathbf{D}(\mathbf{r}, t)$, magnetic intensity vector $\mathbf{H}(\mathbf{r}, t)$, and conduction current vector $\mathbf{J}_c(\mathbf{r}, t)$ also satisfy this symmetry property, so that [cf. Eqs. (4.106), (4.123), and (4.154)]

$$\epsilon^*(\omega) = \epsilon(-\omega^*), \quad \mu^*(\omega) = \mu(-\omega^*), \quad \sigma^*(\omega) = \sigma(-\omega^*), \tag{5.23}$$

for complex ω. Finally, the *temporal frequency transform of the source-free Maxwell's equations* (5.11)–(5.14) yields

$$\nabla \cdot \tilde{\mathbf{E}}(\mathbf{r}, \omega) = 0, \tag{5.24}$$

$$\nabla \times \tilde{\mathbf{E}}(\mathbf{r}, \omega) = \left\| \frac{1}{c} \right\| i\omega \tilde{\mathbf{B}}(\mathbf{r}, \omega), \tag{5.25}$$

$$\nabla \cdot \tilde{\mathbf{B}}(\mathbf{r}, \omega) = 0, \tag{5.26}$$

$$\nabla \times \tilde{\mathbf{B}}(\mathbf{r}, \omega) = - \left\| \frac{1}{c} \right\| i \omega \mu(\omega) \epsilon_c(\omega) \tilde{\mathbf{E}}(\mathbf{r}, \omega), \tag{5.27}$$

where

$$\epsilon_c(\omega) \equiv \epsilon(\omega) + i \|4\pi\| \frac{\sigma(\omega)}{\omega} \tag{5.28}$$

defines the *complex permittivity* $\epsilon_c(\omega)$ of the dispersive HILL medium.

Upon taking the curl of Eq. (5.25) and substituting Eq. (5.27) there results

$$\nabla^2 \tilde{\mathbf{E}}(\mathbf{r}, \omega) + \left\| \frac{1}{c^2} \right\| \omega^2 \mu(\omega) \epsilon_c(\omega) \tilde{\mathbf{E}}(\mathbf{r}, \omega) = \mathbf{0}, \tag{5.29}$$

after application of the vector identity $\nabla \times \nabla \times = \nabla(\nabla \cdot) - \nabla^2$ for the curl-curl operator and substitution from Eq. (5.24). In a similar fashion, upon taking the curl of Eq. (5.27) and substituting Eq. (5.25) one obtains

$$\nabla^2 \tilde{\mathbf{B}}(\mathbf{r}, \omega) + \left\| \frac{1}{c^2} \right\| \omega^2 \mu(\omega) \epsilon_c(\omega) \tilde{\mathbf{B}}(\mathbf{r}, \omega) = \mathbf{0}, \tag{5.30}$$

where the solenoidal character of the magnetic field has been used. Define the *vacuum wavenumber* of the field as

$$k_0 \equiv \frac{\omega}{c}, \tag{5.31}$$

and the *complex index of refraction* of the medium as

$$n(\omega) \equiv \left(\frac{\mu(\omega)\epsilon_c(\omega)}{\mu_0 \epsilon_0} \right)^{1/2}. \tag{5.32}$$

Notice that the complex index of refraction combines the frequency dispersion of each individual dispersive material parameter $\epsilon(\omega)$, $\mu(\omega)$, and $\sigma(\omega)$ into a single, physically meaningful parameter. With these substitutions, the wave equations given in Eqs. (5.29) and (5.30) become

$$\nabla^2 \tilde{\mathbf{E}}(\mathbf{r}, \omega) + k_0^2 n^2(\omega) \tilde{\mathbf{E}}(\mathbf{r}, \omega) = \mathbf{0}, \tag{5.33}$$

$$\nabla^2 \tilde{\mathbf{B}}(\mathbf{r}, \omega) + k_0^2 n^2(\omega) \tilde{\mathbf{B}}(\mathbf{r}, \omega) = \mathbf{0}. \tag{5.34}$$

These are the *Helmholtz equations* for the electromagnetic field vectors in homogeneous, isotropic, locally linear (HILL), temporally dispersive media.

5.1.1.2 Complex Time-Harmonic Form of the Field Quantities

The complex (phasor) form of the electric and magnetic field vectors for a strictly time-harmonic field with fixed angular frequency ω is obtained from the representation

$$\mathbf{E}_\omega(\mathbf{r}, t) = \tilde{\mathbf{E}}(\mathbf{r})e^{-i\omega t}, \tag{5.35}$$

$$\mathbf{B}_\omega(\mathbf{r}, t) = \tilde{\mathbf{B}}(\mathbf{r})e^{-i\omega t}, \tag{5.36}$$

where each complex field vector is related to its real-valued counterpart by

$$\mathbf{E}(\mathbf{r}, t) = \Re\{\mathbf{E}_\omega(\mathbf{r}, t)\} = \Re\left\{\tilde{\mathbf{E}}(\mathbf{r})e^{-i\omega t}\right\}, \tag{5.37}$$

$$\mathbf{B}(\mathbf{r}, t) = \Re\{\mathbf{B}_\omega(\mathbf{r}, t)\} = \Re\left\{\tilde{\mathbf{B}}(\mathbf{r})e^{-i\omega t}\right\}. \tag{5.38}$$

With these substitutions, the constitutive relations given in Eqs. (5.5)–(5.7) become

$$\mathbf{D}(\mathbf{r}, t) = \Re\left\{\epsilon(\omega)\tilde{\mathbf{E}}(\mathbf{r})e^{-i\omega t}\right\}, \tag{5.39}$$

$$\mathbf{H}(\mathbf{r}, t) = \Re\left\{\mu^{-1}(\omega)\tilde{\mathbf{B}}(\mathbf{r})e^{-i\omega t}\right\}, \tag{5.40}$$

$$\mathbf{J}_c(\mathbf{r}, t) = \Re\left\{\sigma(\omega)\tilde{\mathbf{E}}(\mathbf{r})e^{-i\omega t}\right\}. \tag{5.41}$$

The appropriate complex form of these induced field vectors is then given by

$$\mathbf{D}_\omega(\mathbf{r}, t) = \tilde{\mathbf{D}}(\mathbf{r})e^{-i\omega t} = \epsilon(\omega)\tilde{\mathbf{E}}(\mathbf{r})e^{-i\omega t}, \tag{5.42}$$

$$\mathbf{H}_\omega(\mathbf{r}, t) = \tilde{\mathbf{H}}(\mathbf{r})e^{-i\omega t} = \mu^{-1}(\omega)\tilde{\mathbf{B}}(\mathbf{r})e^{-i\omega t}, \tag{5.43}$$

$$\mathbf{J}_{c\omega}(\mathbf{r}, t) = \tilde{\mathbf{J}}_c(\mathbf{r})e^{-i\omega t} = \sigma(\omega)\tilde{\mathbf{E}}(\mathbf{r})e^{-i\omega t}, \tag{5.44}$$

so that $\tilde{\mathbf{D}}(\mathbf{r}) = \epsilon(\omega)\tilde{\mathbf{E}}(\mathbf{r})$, $\tilde{\mathbf{B}}(\mathbf{r}) = \mu(\omega)\tilde{\mathbf{H}}(\mathbf{r})$, and $\tilde{\mathbf{J}}_c(\mathbf{r}) = \sigma(\omega)\tilde{\mathbf{E}}(\mathbf{r})$.

Consider determining the relationship between the complex phasor form of the field vectors and the corresponding temporal frequency spectra of the real-valued field vectors. Substitution of Eq. (5.37) into Eq. (5.17) for the temporal frequency spectrum of the electric field intensity vector results in

$$\tilde{\mathbf{E}}(\mathbf{r}, \omega') = \int_{-\infty}^{\infty} \Re\left\{\tilde{\mathbf{E}}(\mathbf{r})e^{-i\omega t}\right\}e^{i\omega' t}\,dt$$

$$= \frac{1}{2}\left\{\tilde{\mathbf{E}}(\mathbf{r})\int_{-\infty}^{\infty} e^{i(\omega'-\omega)t}\,dt + \tilde{\mathbf{E}}^*(\mathbf{r})\int_{-\infty}^{\infty} e^{i(\omega'+\omega)t}\,dt\right\},$$

so that

$$\tilde{\mathbf{E}}(\mathbf{r}, \omega') = \pi \left\{ \tilde{\mathbf{E}}(\mathbf{r})\delta(\omega' - \omega) + \tilde{\mathbf{E}}^*(\mathbf{r})\delta(\omega' + \omega) \right\} \tag{5.45}$$

with similar expressions for the other field vectors. Hence, the frequency spectrum of the complex monochromatic field given in Eqs. (5.35) and (5.36) with real-valued field vectors given in Eqs. (5.37) and (5.38) is comprised of a symmetric pair of spectral lines at $\omega' = \pm\omega$.

Substitution of Eqs. (5.37)–(5.41) into the source-free form of Maxwell's equations given in Eqs. (5.11)–(5.14) results in the set of equations

$$\Re \left\{ \epsilon(\omega)\nabla \cdot \tilde{\mathbf{E}}(\mathbf{r})e^{-i\omega t} \right\} = 0, \tag{5.46}$$

$$\Re \left\{ \nabla \times \tilde{\mathbf{E}}(\mathbf{r})e^{-i\omega t} \right\} = \left\| \frac{1}{c} \right\| \Re \left\{ i\omega \tilde{\mathbf{B}}(\mathbf{r})e^{-i\omega t} \right\}, \tag{5.47}$$

$$\Re \left\{ \nabla \cdot \tilde{\mathbf{B}}(\mathbf{r})e^{-i\omega t} \right\} = 0, \tag{5.48}$$

$$\Re \left\{ \mu^{-1}(\omega)\nabla \times \tilde{\mathbf{B}}(\mathbf{r})e^{-i\omega t} \right\} = - \left\| \frac{1}{c} \right\| \Re \left\{ i\omega \epsilon_c(\omega)\tilde{\mathbf{E}}(\mathbf{r})e^{-i\omega t} \right\}, \tag{5.49}$$

where $\epsilon_c(\omega)$ is the complex permittivity defined in Eq. (5.28). Because it is only the real parts of the above relations that enter into the determination of the real-valued field vectors through Eqs. (5.37)–(5.41), one is free to impose the additional condition that the imaginary parts of these quantities also satisfy the same set of relations. One then obtains the *complex time-harmonic differential (phasor) form of Maxwell's equations*

$$\nabla \cdot \tilde{\mathbf{E}}(\mathbf{r}) = 0, \tag{5.50}$$

$$\nabla \times \tilde{\mathbf{E}}(\mathbf{r}) = \left\| \frac{1}{c} \right\| i\omega \tilde{\mathbf{B}}(\mathbf{r}), \tag{5.51}$$

$$\nabla \cdot \tilde{\mathbf{B}}(\mathbf{r}) = 0, \tag{5.52}$$

$$\nabla \times \tilde{\mathbf{B}}(\mathbf{r}) = - \left\| \frac{1}{c} \right\| i\omega\mu(\omega)\epsilon_c(\omega)\tilde{\mathbf{E}}(\mathbf{r}), \tag{5.53}$$

which are identical with the temporal frequency domain form given in Eqs. (5.24)–(5.27); the difference between these two sets of equations then lies entirely in the interpretation of the field vectors. Furthermore, the complex time-harmonic field vectors satisfy the *Helmholtz equations*

$$\nabla^2 \tilde{\mathbf{E}}(\mathbf{r}) + k_0^2 n^2(\omega)\tilde{\mathbf{E}}(\mathbf{r}) = \mathbf{0}, \tag{5.54}$$

$$\nabla^2 \tilde{\mathbf{B}}(\mathbf{r}) + k_0^2 n^2(\omega)\tilde{\mathbf{B}}(\mathbf{r}) = \mathbf{0}, \tag{5.55}$$

in temporally dispersive HILL media, where $n(\omega)$ is the complex index of refraction defined in Eq. (5.32).

5.1.1.3 The Harmonic Electromagnetic Plane Wave Field

The simplest electromagnetic wave field is that of a time-harmonic electromagnetic plane wave for which the complex field vectors appearing in Eqs. (5.37) and (5.38) are given by

$$\tilde{\mathbf{E}}(\mathbf{r}) = \mathbf{E}_0 e^{i\tilde{\mathbf{k}}\cdot\mathbf{r}}, \tag{5.56}$$

$$\tilde{\mathbf{B}}(\mathbf{r}) = \mathbf{B}_0 e^{i\tilde{\mathbf{k}}\cdot\mathbf{r}}, \tag{5.57}$$

where $\mathbf{E}_0$ and $\mathbf{B}_0$ are fixed complex-valued vectors, independent of both time and the coordinate position $\mathbf{r}$. The surfaces of constant phase for this wave-field are then given by $\Re\{\tilde{\mathbf{k}} \cdot \mathbf{r}\}$ = constant which yields a family of parallel plane surfaces with normal vector parallel to the vector $\Re\{\tilde{\mathbf{k}}\}$. Substitution of either of these expressions for the complex field vectors into the appropriate vector Helmholtz equation given in Eqs. (5.54) and (5.55) then results in the identification

$$\tilde{\mathbf{k}} \cdot \tilde{\mathbf{k}} = k_0^2 n^2(\omega),$$

where $k_0 = \omega/c$ is the *vacuum wavenumber* of the field. Let

$$\tilde{\mathbf{k}} \equiv \tilde{k}(\omega)\hat{\mathbf{s}}, \tag{5.58}$$

where $\hat{\mathbf{s}}$ is a real-valued unit vector in the direction of the wave vector $\tilde{\mathbf{k}}$, and where

$$\tilde{k}(\omega) \equiv k_0 n(\omega) = \frac{\omega}{c} n(\omega) \tag{5.59}$$

is the *complex wavenumber* in the temporally dispersive HILL medium.

Substitution of each complex field vector given in Eqs. (5.56) and (5.57) into its appropriate divergence relation given either in Eq. (5.50) or (5.52) of the complex time-harmonic differential form of Maxwell's equations then yields the *transversality relation*

$$\hat{\mathbf{s}} \cdot \mathbf{E}_0 = \hat{\mathbf{s}} \cdot \mathbf{B}_0 = 0. \tag{5.60}$$

Furthermore, the curl relation in Eq. (5.51) gives $\tilde{\mathbf{k}} \times \mathbf{E}_0 = \|1/c\|\omega\mathbf{B}_0$, so that

$$\mathbf{B}_0 = \|c\|\frac{n(\omega)}{c}\hat{\mathbf{s}} \times \mathbf{E}_0, \tag{5.61}$$

and the curl relation in Eq. (5.53) yields $\tilde{\mathbf{k}} \times \mathbf{B}_0 = \|1/c\| \omega \mu(\omega) \epsilon_c(\omega) \mathbf{E}_0$, so that

$$\mathbf{E}_0 = -\frac{1}{\|c\|} \frac{c}{n(\omega)} \hat{\mathbf{s}} \times \mathbf{B}_0. \tag{5.62}$$

These three relations [Eqs. (5.60)–(5.62)] then show that the ordered triple of vectors $\{\mathbf{E}_0, \mathbf{B}_0, \hat{\mathbf{s}}\}$ forms a right-handed orthogonal triad at any point of space when $\Re\{n(\omega)\} > 0$. The field vectors $\mathbf{E}_0$ and $\mathbf{B}_0$ then lie in the surfaces of constant phase and their complex amplitudes are related by

$$B_0 = \|c\| \frac{n(\omega)}{c} E_0 \tag{5.63}$$

at each point of space. Notice that the harmonic plane wave field in vacuum is a self-conjugate field [see Eqs. (2.189)–(2.190) and Definition 2.3]. The *complex intrinsic impedance* of the dispersive medium is defined as the ratio of the complex electric field intensity to the complex magnetic field intensity $H_0 = B_0/\mu(\omega)$, so that, with Eqs. (5.63) and (5.32),

$$\eta(\omega) \equiv \frac{E_0}{H_0} = \left[\frac{\mu(\omega)}{\epsilon_c(\omega)}\right]^{1/2}. \tag{5.64}$$

For free space the intrinsic impedance is given by $\eta_0 = \sqrt{\mu_0/\epsilon_0} = 376.73\Omega$ in MKSA units whereas it is unity in cgs units.

From Eqs. (5.28), (5.32), and (5.59), the *complex wavenumber* of the harmonic electromagnetic plane wave field is found to be given by

$$\tilde{k}(\omega) = \frac{\omega}{\|c\|} \left\{\mu(\omega)\left[\epsilon(\omega) + i\|4\pi\|\frac{\sigma(\omega)}{\omega}\right]\right\}^{1/2}. \tag{5.65}$$

Let

$$\tilde{k}(\omega) \equiv \beta(\omega) + i\alpha(\omega), \tag{5.66}$$

where $\beta(\omega) \equiv \Re\{\tilde{k}(\omega)\}$ is the *propagation factor* and $\alpha(\omega) \equiv \Im\{\tilde{k}(\omega)\}$ is the *attenuation factor* for the harmonic plane wave of angular frequency ω in the dispersive medium. Because $\epsilon(\omega) = \epsilon_r(\omega) + i\epsilon_i(\omega)$ and $\sigma(\omega) = \sigma_r(\omega) + i\sigma_i(\omega)$, then the *real and imaginary parts of the complex permittivity* are given by

$$\epsilon'_c(\omega) \equiv \Re\{\epsilon_c(\omega)\} = \epsilon_r(\omega) - \|4\pi\|\frac{\sigma_i(\omega)}{\omega}, \tag{5.67}$$

$$\epsilon''_c(\omega) \equiv \Im\{\epsilon_c(\omega)\} = \epsilon_i(\omega) + \|4\pi\|\frac{\sigma_r(\omega)}{\omega}, \tag{5.68}$$

respectively. The real and imaginary parts of the complex wavenumber given in Eq. (5.65) are then determined from the relation

$$\beta^2 - \alpha^2 + 2i\alpha\beta = \left\|\frac{1}{c^2}\right\| \omega^2 \left(\mu' + i\mu''\right)\left(\epsilon'_c + i\epsilon''_c\right),$$

which, upon equating real and imaginary parts, yields the pair of relations

$$\beta^2 - \alpha^2 = \left\|\frac{1}{c^2}\right\| \omega^2 \left(\mu'\epsilon'_c - \mu''\epsilon''_c\right),$$

$$2\alpha\beta = \left\|\frac{1}{c^2}\right\| \omega^2 \left(\mu'\epsilon''_c + \mu''\epsilon'_c\right).$$

Upon solving the second equation for β and substituting in the first equation, one obtains the fourth-order polynomial equation

$$\alpha^4 + \left\|\frac{1}{c^2}\right\| \omega^2 \left(\mu'\epsilon'_c - \mu''\epsilon''_c\right)\alpha^2 - \left\|\frac{1}{c^4}\right\| \frac{1}{4}\omega^4 \left(\mu'\epsilon''_c + \mu''\epsilon'_c\right)^2 = 0$$

with solution (assuming that $\mu'\epsilon'_c - \mu''\epsilon''_c > 0$)

$$\alpha^2 = \left\|\frac{1}{c^2}\right\| \frac{1}{2}\omega^2 \left(\mu'\epsilon'_c - \mu''\epsilon''_c\right)\left\{\left[1 + \left(\frac{\mu'\epsilon''_c + \mu''\epsilon'_c}{\mu'\epsilon'_c - \mu''\epsilon''_c}\right)^2\right]^{1/2} - 1\right\},$$

and consequently

$$\beta^2 = \left\|\frac{1}{c^2}\right\| \omega^2 \left(\mu'\epsilon'_c - \mu''\epsilon''_c\right) + \alpha^2$$

$$= \left\|\frac{1}{c^2}\right\| \frac{1}{2}\omega^2 \left(\mu'\epsilon'_c - \mu''\epsilon''_c\right)\left\{\left[1 + \left(\frac{\mu'\epsilon''_c + \mu''\epsilon'_c}{\mu'\epsilon'_c - \mu''\epsilon''_c}\right)^2\right]^{1/2} + 1\right\}.$$

The plane wave *propagation and attenuation factors* in a temporally dispersive HILL medium are then given by

$$\beta = \left\|\frac{1}{c}\right\| \omega \left[\frac{1}{2}\left(\mu'\epsilon'_c - \mu''\epsilon''_c\right)\right]^{1/2}\left\{\left[1 + \left(\frac{\mu'\epsilon''_c + \mu''\epsilon'_c}{\mu'\epsilon'_c - \mu''\epsilon''_c}\right)^2\right]^{1/2} + 1\right\}^{1/2}, \quad (5.69)$$

$$\alpha = \left\|\frac{1}{c}\right\| \omega \left[\frac{1}{2}\left(\mu'\epsilon'_c - \mu''\epsilon''_c\right)\right]^{1/2}\left\{\left[1 + \left(\frac{\mu'\epsilon''_c + \mu''\epsilon'_c}{\mu'\epsilon'_c - \mu''\epsilon''_c}\right)^2\right]^{1/2} - 1\right\}^{1/2}, \quad (5.70)$$

when $\mu' \epsilon'_c - \mu'' \epsilon''_c > 0$. If $\mu' \epsilon'_c - \mu'' \epsilon''_c < 0$, then the quantity $(\mu' \epsilon'_c - \mu'' \epsilon''_c)$ is replaced by its absolute value and the above pair of expressions for β and α are interchanged. Finally, if $\mu' \epsilon'_c - \mu'' \epsilon''_c = 0$, then

$$\beta = \alpha = \left\| \frac{1}{c} \right\| \omega \left(\frac{1}{2} \mu'' \epsilon'_c \right)^{1/2} \left(1 + \frac{\epsilon''^2_c}{\epsilon'^2_c} \right)^{1/2}, \qquad (5.71)$$

where $\mu' / \mu'' = \epsilon''_c / \epsilon'_c$ in this special case. These relations may then be used to classify dispersive materials according to the origin of the loss over the frequency domain of interest. Several specific cases of interest are now separately considered in more detail.

5.1.1.4 Near-Ideal Dielectrics

For nonmagnetic materials with $\mu(\omega) = \mu$ where μ is real-valued, Eqs. (5.69) and (5.70) simplify to the pair of relations

$$\beta(\omega) = \left\| \frac{1}{c} \right\| \omega \left[\frac{1}{2} \left(\mu \epsilon'_c(\omega) \right) \right]^{1/2} \left\{ \left[1 + \left(\frac{\epsilon''_c(\omega)}{\epsilon'_c(\omega)} \right)^2 \right]^{1/2} + 1 \right\}^{1/2}, \qquad (5.72)$$

$$\alpha(\omega) = \left\| \frac{1}{c} \right\| \omega \left[\frac{1}{2} \left(\mu \epsilon'_c(\omega) \right) \right]^{1/2} \left\{ \left[1 + \left(\frac{\epsilon''_c(\omega)}{\epsilon'_c(\omega)} \right)^2 \right]^{1/2} - 1 \right\}^{1/2}, \qquad (5.73)$$

when $\epsilon'_c > 0$. For low-loss materials in a given spectral domain, which may be referred to as *near-ideal dielectrics*,[1] the inequality

$$\frac{\epsilon''_c(\omega)}{\epsilon'_c(\omega)} \ll 1 \qquad (5.74)$$

is satisfied over the angular frequency domain of interest. The plane wave propagation and attenuation factors may then be approximated as

$$\beta(\omega) \approx k_n(\omega) \left[1 + \frac{1}{8} \left(\frac{\epsilon''_c(\omega)}{\epsilon'_c(\omega)} \right)^2 \right], \qquad (5.75)$$

$$\alpha(\omega) \approx \frac{1}{2} k_n(\omega) \frac{\epsilon''_c(\omega)}{\epsilon'_c(\omega)}, \qquad (5.76)$$

---

[1] Some texts refer to such low-loss materials as being imperfect dielectrics. This terminology is not used here as all dispersive dielectric media have loss over some frequency domain and consequently are imperfect.

where

$$k_n(\omega) \equiv \left\| \frac{1}{c} \right\| c\sqrt{\mu\epsilon_c'(\omega)}k_0 = n_r(\omega)k_0 \tag{5.77}$$

is the wavenumber in a fictitious lossless dispersive nonmagnetic medium (i.e., an "ideal dielectric") with real-valued refractive index $n_r(\omega)$. The complex intrinsic impedance of the harmonic plane wave field is, from Eq. (5.64), then found as

$$\eta(\omega) = \eta_n(\omega) \left[1 + i\frac{\epsilon_c''(\omega)}{\epsilon_c'(\omega)} \right]^{-1/2}$$

$$\approx \eta_n(\omega) \left\{ \left[1 - \frac{3}{8}\left(\frac{\epsilon_c''(\omega)}{\epsilon_c'(\omega)} \right)^2 \right] - i\frac{\epsilon_c''(\omega)}{2\epsilon_c'(\omega)} \right\}, \tag{5.78}$$

where

$$\eta_n(\omega) \equiv \sqrt{\frac{\mu}{\epsilon_c'(\omega)}} \tag{5.79}$$

is the intrinsic impedance of the fictitious lossless medium with refractive index $n_r(\omega)$. The phase difference δ_f between the complex electric and magnetic field amplitudes is then given by

$$\tan\delta_f \approx \frac{\epsilon_c''(\omega)}{2\epsilon_c'(\omega)}, \tag{5.80}$$

where $\tilde{E} = |\eta_n|\tilde{H}e^{-i\delta_f}$. Finally, the *phase velocity* of a time-harmonic plane wave field in a near-ideal dielectric is given by

$$v_p(\omega) = \frac{\omega}{\beta(\omega)} \approx \frac{c/n_r(\omega)}{1 + \frac{1}{8}\left(\frac{\epsilon_c''(\omega)}{\epsilon_c'(\omega)} \right)^2}, \tag{5.81}$$

where $c/n_r(\omega)$ is the phase velocity in the fictitious lossless medium. Once again, notice that the above expressions given in Eqs. (5.75)–(5.81) are valid approximations only within those frequency domains where the inequality in Eq. (5.74) is satisfied and the medium behaves as a near-ideal dielectric.

5.1.1.5 Semiconducting Media

For a material such as a semiconductor where the losses are primarily conductive, the complex wavenumber given in Eq. (5.65) for a time-harmonic plane wave field may be approximated as

$$\tilde{k}(\omega) \approx \left\| \frac{1}{c} \right\| \omega \left\{ \mu \left[\epsilon_c'(\omega) + i\|4\pi\|\frac{\sigma_r(\omega)}{\omega} \right] \right\}^{1/2}, \tag{5.82}$$

where $\epsilon'_c(\omega) = \epsilon_r(\omega) - \|4\pi\|\sigma_i(\omega)/\omega$. For $\omega \gg \sigma_r(\omega)/\epsilon'_c(\omega)$ the loss term in the above expression is sufficiently small that the previous approximations apply with $\epsilon''_c = \|4\pi\|\sigma_r(\omega)/\omega$ and the semiconductor behaves like a near-ideal dielectric. On the other hand, for $\omega \ll \sigma_r(\omega)/\epsilon'_c(\omega)$ the semiconductor material behavior is dominated by the real part $\sigma_r(\omega)$ of the conductivity and may then be considered as a good conductor. In that low-frequency domain the complex wavenumber given in Eq. (5.82) may be further approximated as

$$\tilde{k}(\omega) \approx \frac{1+i}{d_p} \quad \text{when } \omega \ll \sigma_r(\omega)/\epsilon'_c(\omega), \tag{5.83}$$

where

$$d_p \equiv \left\|\frac{c}{2\sqrt{\pi}}\right\| \left[\frac{2}{\omega\mu\sigma_r(\omega)}\right]^{1/2} \tag{5.84}$$

is the *penetration depth* of the conductor. It is then seen from Eq. (5.83) with Eqs. (5.61) and (5.62) that the electric and magnetic fields are $\pi/4$ out of phase in a good conductor. The complex intrinsic impedance of a time-harmonic plane wave field with oscillation frequency in the low-frequency domain of a semiconductor material may be approximated as

$$\eta(\omega) \approx \left\|\frac{1}{2\sqrt{\pi}}\right\| e^{-i\pi/4} \left(\frac{\mu\omega}{\sigma_r(\omega)}\right)^{1/2} \quad \text{when } \omega \ll \sigma_r(\omega)/\epsilon'_c(\omega). \tag{5.85}$$

Because $\sigma_r(\omega) \approx \sigma^{(0)} - \sigma^{(2)}\omega^2$ as $\omega \to 0$, then $\eta(\omega) \to 0$ as $\omega \to 0$. Finally, the phase velocity in the low frequency domain of a semiconductor is approximately given by

$$v_p(\omega) \approx \omega d_p = \left\|\frac{c}{2\sqrt{\pi}}\right\| \left(\frac{2\omega}{\mu\sigma_r(\omega)}\right)^{1/2} \quad \text{when } \omega \ll \sigma_r(\omega)/\epsilon'_c(\omega), \tag{5.86}$$

which is very small in a good conductor.

An important example of a semiconducting material is provided by sea-water.[2] From Eq. (4.235) describing the dielectric permittivity for the Drude model of free electron conductors and Eq. (5.28) for the complex permittivity, the *frequency dispersion of the electric conductivity* is found to be given by

$$\sigma(\omega) = i\frac{\gamma\sigma_0}{\omega + i\gamma}, \tag{5.87}$$

where $\sigma_0 \equiv (\epsilon_0/\|4\pi\|)\omega_p^2/\gamma$ denotes the static conductivity of the material and where $\tau_c = 1/\gamma$ is the relaxation time associated with the mean free path for

---

[2]In sea-water, the ionic bonds in *NaCl* are weakened by the large relative static dielectric permittivity of H_2O ($\epsilon_s \approx 79.2$), thereby effectively "dissolving" the molecule into positive and negative ions, thereby giving rise to its electrical conductivity.

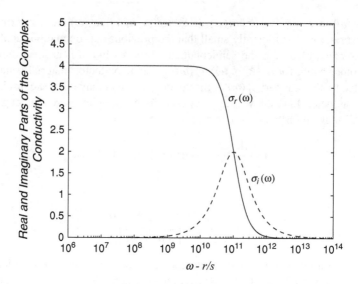

Fig. 5.1 Angular frequency dispersion of the real (solid curve) and imaginary (dashed curve) parts of the complex conductivity described by the Drude model for sea-water with static conductivity $\sigma_0 \approx 4\,\text{mhos/m}$ and mean relaxation time $\tau_c \approx 1 \times 10^{-11}\text{s}$

free electrons in the material. Estimates of these parameters for sea-water are $\sigma_0 \approx 4\,\text{mho/m}$ and $\gamma \approx 1 \times 10^{11}/\text{s}$. The resultant angular frequency dispersion of the real and imaginary parts of the Drude model conductivity for sea-water is presented in Fig. 5.1. Notice that any effects due to the conductivity rapidly diminish as the frequency increases above γ and that the imaginary part of the conductivity is negligible in comparison to the real part for $|\omega| \ll 1/\gamma$.

The relative complex dielectric permittivity with the Drude model is given by

$$\epsilon_c(\omega)/\epsilon_0 = \epsilon(\omega)/\epsilon_0 - \omega_p^2 \frac{\omega - i\gamma}{\omega(\omega^2 + \gamma^2)}, \tag{5.88}$$

where $\epsilon(\omega)/\epsilon_0$ is given in Eq. (4.232) for triply-distilled water with parameters given in Tables 4.1 and 4.2. The influence of the Drude model conductivity on the imaginary part of the complex permittivity of water is depicted in Fig. 5.2 when the numerical value of the coefficient a_1 appearing in the Rocard–Powles contribution is changed from $a_1 = 74.65$ to $a_1 = 79.17$ in order to reflect the addition of the contribution $-\omega_p^2/\gamma^2$ to the relative static value from the Drude model. Notice that the imaginary part of the complex permittivity now increases monotonically as the frequency decreases to zero below $\omega \sim 1/\gamma$ whereas the real part is essentially unaltered from its zero conductivity behavior. The angular frequency dispersion of the real and imaginary parts of the corresponding complex index of refraction are illustrated in Fig. 5.3, where $\mu = \mu_0$. Notice that the introduction of conductivity has an equal effect on both the real and imaginary parts of $n(\omega)$.

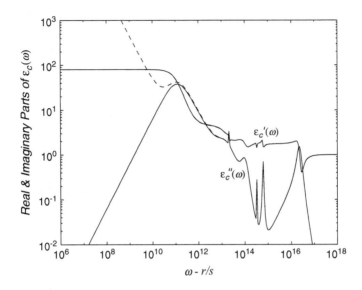

Fig. 5.2 Angular frequency dispersion of the real and imaginary parts of the complex dielectric permittivity for water both without (solid curves) and with (dashed curves) conductivity appropriate for sea-water

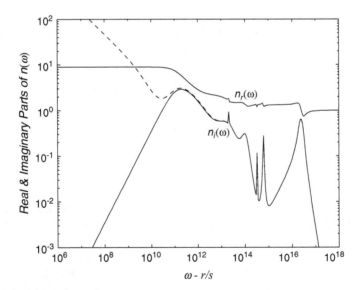

Fig. 5.3 Angular frequency dispersion of the real and imaginary parts of the complex index of refraction for water both without (solid curves) and with (dashed curves) conductivity appropriate for sea-water

For small $\omega \to 0$ the dielectric permittivity $\epsilon(\omega)$ of sea-water is dominated by the Debye model orientational polarization contribution and this, in turn, is dominated by the effects of Drude model conductivity $\sigma(\omega)$ in the complex permittivity $\epsilon_c(\omega) = \epsilon(\omega) + i\|4\pi\|\sigma(\omega)/\omega$, so that

$$\epsilon_c(\omega)/\epsilon_0 \approx \epsilon_\infty + \frac{\epsilon_s - \epsilon_\infty}{1 - i\omega\tau} - \frac{\omega_p^2}{\omega(\omega + i\gamma)}$$

$$\approx \epsilon_s + i(\epsilon_s - \epsilon_\infty)\tau\omega + i\frac{\omega_p^2}{\gamma\omega} \tag{5.89}$$

as $\omega \to 0$. With $\mu/\mu_0 \simeq 1$ for sea-water, the complex index of refraction $n(\omega) = (\mu/\mu_0)^{1/2}(\epsilon_c(\omega)/\epsilon_0)^{1/2}$ is found to be given by

$$n(\omega) \approx \frac{\omega_p}{\gamma^{1/2}\omega^{1/2}}e^{i\pi/4}\left(1 - i\frac{\epsilon_s\gamma}{2\omega_p^2}\omega\right), \tag{5.90}$$

and the complex wavenumber $\tilde{k}(\omega) = (\omega/c)n(\omega)$ becomes

$$\tilde{k}(\omega) \approx \frac{\omega_p}{c\gamma^{1/2}}e^{i\pi/4}\omega^{1/2} + \frac{\epsilon_s\gamma^{1/2}}{2c\omega_p}e^{-i\pi/4}\omega^{3/2} \tag{5.91}$$

as $\omega \to 0$, showing that the attenuation coefficient $\alpha(\omega) = \Im\{\tilde{k}(\omega)\}$ for sea-water varies as

$$\alpha(\omega) \approx \frac{\omega_p}{2^{1/2}c\gamma^{1/2}}\omega^{1/2} - \frac{\epsilon_s\gamma^{1/2}}{2^{3/2}c\omega_p}\omega^{3/2} \tag{5.92}$$

as $\omega \to 0$, where $\sigma_0 = (\epsilon_0/\|4\pi\|)\omega_p^2/\gamma$ is the static conductivity. Hence, to a first approximation in MKSA units [cf. Eq. (5.83)],

$$\alpha(\omega) \approx \sqrt{\pi\mu_0\sigma_0 f}, \tag{5.93}$$

where $f = \omega/2\pi$. For example, for $f = 1.0\,\text{kHz} = 1.0 \times 10^3\,\text{Hz}$ in sea-water, $\alpha \approx 0.126\,\text{Np/m}$ and the penetration depth is found to be $d_p = 1/\alpha \approx 7.94\,\text{m}$ so that the distance z at which the initial electric field amplitude is reduced to 1% of its initial value, given by $e^{-z/d_p} = 0.01$ is $z \approx 36.5\,\text{m}$. This increases to $d_p \approx 25.1\,\text{m}$ when the frequency is decreased by a factor of 10 to $f = 100\,\text{Hz}$, the 1% field decay propagation distance increasing to $z \approx 115.6\,\text{m}$. Communication through sea-water, either between two submersibles or between a surface (on land or sea) and submersible vessel, must then be conducted with the smallest frequency possible,[3] resulting in a painfully slow data rate.

5.1.1.6 Double Negative (DNG) Media

Over certain frequency domains the real part of the dielectric permittivity can be negative [as described, for example, by the Lorentz model in Eq. (4.217)], as can the real part of the magnetic permeability. If both $\epsilon_r(\omega)$ and $\mu_r(\omega)$ are negative over some frequency domain (given by the nonempty intersection Ω of their individual frequency domains of negative values), then the real part of the index of refraction may also be negative, as was originally proposed by V. G. Veselago [3] in 1968. A so-called *metamaterial* having this rather unusual property is said to be a *double negative* (DNG) medium.

Based upon the analysis of Ziolkowski and Heyman [4], the relative dielectric permittivity $\tilde{\epsilon}(\omega) \equiv \epsilon(\omega)/\epsilon_0$ and magnetic permeability $\tilde{\mu}(\omega) \equiv \mu(\omega)/\mu_0$ in a DNG medium must be expressed as

$$\tilde{\epsilon}(\omega) = |\tilde{\epsilon}(\omega)| \, e^{i\varphi_\epsilon(\omega)}, \quad \varphi_\epsilon \in (\pi/2, \pi], \tag{5.94}$$

$$\tilde{\mu}(\omega) = |\tilde{\mu}(\omega)| \, e^{i\varphi_\mu(\omega)}, \quad \varphi_\mu \in (\pi/2, \pi], \tag{5.95}$$

over the frequency domain Ω. With the complex index of refraction and intrinsic impedance expressed as

$$n(\omega) = |n(\omega)| e^{i\varphi_n(\omega)} \quad \& \quad \eta(\omega) = |\eta(\omega)| e^{i\varphi_\eta(\omega)}, \tag{5.96}$$

there are then two possible principal choices for each phase quantity $\varphi_{n,\eta}$ corresponding to the phase branches given in Eqs. (5.94) and (5.95). The first choice is given by

$$\varphi_n(\omega) = \frac{1}{2}(\varphi_\mu(\omega) + \varphi_\epsilon(\omega)) \in (\pi/2, \pi], \tag{5.97}$$

$$\varphi_\eta(\omega) = \frac{1}{2}(\varphi_\mu(\omega) - \varphi_\epsilon(\omega)) \in (-\pi/4, \pi/4], \tag{5.98}$$

---

[3]This was the driving force behind the U. S. Navy's *Project Sanguine* [1, 2] which developed an extremely low frequency (ELF) 2.6 MW radio wave transmitter facility operating at 76 Hz. Because of public opposition incited by concerns of adverse health effects caused by ELF radiation, Project Sanguine was abandoned and replaced by a series of scaled-down versions, beginning with *Project Seafarer* in 1975, *Austere ELF* in 1978, and finally *Project ELF* in 1981 that was constructed and deployed at Clam Lake, Wisconsin and Republic, Michigan. A similar ELF system is reportedly operated by the Russian Navy at Murmansk.

and the second is

$$\varphi_n(\omega) = \frac{1}{2}(\varphi_\mu(\omega) + \varphi_\epsilon(\omega)) - \pi \in (-\pi/2, 0], \tag{5.99}$$

$$\varphi_\eta(\omega) = \frac{1}{2}(\varphi_\mu(\omega) - \varphi_\epsilon(\omega)) + \pi \in (3\pi/4, 5\pi/4]. \tag{5.100}$$

The complex index of refraction

$$n(\omega) = \left[(\tilde{\mu}_r + i\tilde{\mu}_i)(\tilde{\epsilon}_r + i\tilde{\epsilon}_i) \right]^{1/2}$$

$$= (\tilde{\mu}_r\tilde{\epsilon}_r)^{1/2} \left[1 - \frac{\tilde{\mu}_i\tilde{\epsilon}_i}{\tilde{\mu}_r\tilde{\epsilon}_r} + i\left(\frac{\tilde{\mu}_i}{\tilde{\mu}_r} + \frac{\tilde{\epsilon}_i}{\tilde{\epsilon}_r} \right) \right]^{1/2},$$

so that in the small loss limit $0 \leq \tilde{\mu}_i \ll \tilde{\mu}_r$ and $0 \leq \tilde{\epsilon}_i \ll \tilde{\epsilon}_r$ with the principal phase choices given either in Eqs. (5.97)–(5.98) or Eqs. (5.99)–(5.100),

$$n(\omega) \simeq \mp\sqrt{|\tilde{\mu}_r\tilde{\epsilon}_r|} \left[1 - \frac{i}{2}\left(\frac{\tilde{\mu}_i}{|\tilde{\mu}_r|} + \frac{\tilde{\epsilon}_i}{|\tilde{\epsilon}_r|} \right) \right], \tag{5.101}$$

where the upper sign choice corresponds to the phase choice in Eqs. (5.97)–(5.98) and the lower sign choice to that in Eqs. (5.99) and (5.100). Similarly, the complex intrinsic impedance

$$\eta(\omega) = \left(\frac{\tilde{\mu}_r + i\tilde{\mu}_i}{\tilde{\epsilon}_r + i\tilde{\epsilon}_i} \right)^{1/2} = \left(\frac{\tilde{\mu}_r}{\tilde{\epsilon}_r} \right)^{1/2} \left[1 + i\left(\frac{\tilde{\mu}_i}{\tilde{\mu}_r} - \frac{\tilde{\epsilon}_i}{\tilde{\epsilon}_r} \right) \right]^{1/2},$$

so that in the small loss limit

$$\eta(\omega) \simeq \pm\sqrt{\left| \frac{\tilde{\mu}_r}{\tilde{\epsilon}_r} \right|} \left[1 - \frac{i}{2}\left(\frac{\tilde{\mu}_i}{|\tilde{\mu}_r|} - \frac{\tilde{\epsilon}_i}{|\tilde{\epsilon}_r|} \right) \right], \tag{5.102}$$

where the upper sign choice corresponds to the phase choice in Eqs. (5.97)–(5.98) and the lower sign choice to that in Eqs. (5.99) and (5.100).

Consequently, with the phase choice given in Eqs. (5.97) and (5.98) one finds that $\Re\{\eta(\omega)\} > 0$, $\Re\{n(\omega)\} < 0$, and $\Im\{n(\omega)\} \geq 0$, whereas with the phase choice given in Eqs. (5.99) and (5.100) one finds that $\Re\{\eta(\omega)\} < 0$, $\Re\{n(\omega)\} > 0$, and $\Im\{n(\omega)\} \leq 0$. The set of relations given in Eqs. (5.58)–(5.62) then show that for a DNG medium an electromagnetic plane wave is left-handed (LH) with respect to the direction $\hat{s}$ of the wave vector $\tilde{k}$ and right-handed (RH) with respect to the direction of advance of the plane wavefront for the phase choice given in Eqs. (5.97) and (5.98), whereas it is LH with respect to both the direction of the wave vector and the direction of propagation for the phase choice given in Eqs. (5.99) and (5.100). More will be said about these two choices in connection with Poynting's theorem in Sect. 5.2.

5.1.2 Nonconducting Spatially Inhomogeneous, Temporally Dispersive Media

The source-free time-harmonic form of Maxwell's equations in a nonconducting spatially inhomogeneous, locally linear, temporally dispersive medium are given by

$$\nabla \cdot \left(\epsilon(\mathbf{r}, \omega) \tilde{\mathbf{E}}(\mathbf{r}, \omega) \right) = 0, \tag{5.103}$$

$$\nabla \cdot \left(\mu(\mathbf{r}, \omega) \tilde{\mathbf{H}}(\mathbf{r}, \omega) \right) = 0, \tag{5.104}$$

and

$$\nabla \times \tilde{\mathbf{E}}(\mathbf{r}, \omega) = i\omega\mu(\mathbf{r}, \omega)\tilde{\mathbf{H}}(\mathbf{r}, \omega), \tag{5.105}$$

$$\nabla \times \tilde{\mathbf{H}}(\mathbf{r}, \omega) = -i\omega\epsilon(\mathbf{r}, \omega)\tilde{\mathbf{E}}(\mathbf{r}, \omega). \tag{5.106}$$

Upon taking the curl of Faraday's law (5.105) after dividing through by $\mu(\mathbf{r}, \omega)$, one obtains

$$\nabla \times \left(\frac{1}{\mu(\mathbf{r}, \omega)} \nabla \times \tilde{\mathbf{E}}(\mathbf{r}, \omega) \right) = i\omega \nabla \times \tilde{\mathbf{H}}(\mathbf{r}, \omega)$$

$$= \omega^2 \epsilon(\mathbf{r}, \omega) \tilde{\mathbf{E}}(\mathbf{r}, \omega)$$

after substitution from Ampére's law (5.106). Upon expanding the left-hand side of this equation as

$$\nabla \times \left(\frac{1}{\mu} \nabla \times \tilde{\mathbf{E}} \right) = \frac{1}{\mu} \nabla \times \nabla \times \tilde{\mathbf{E}} + \nabla \left(\frac{1}{\mu} \right) \times \left(\nabla \times \tilde{\mathbf{E}} \right)$$

and using the curl identity $\nabla \times \nabla \times \tilde{\mathbf{E}} = \nabla(\nabla \cdot \tilde{\mathbf{E}}) - \nabla^2 \tilde{\mathbf{E}}$, one obtains

$$\nabla^2 \tilde{\mathbf{E}} + \tilde{k}^2(\mathbf{r}, \omega)\tilde{\mathbf{E}} + \nabla(\ln(\mu(\mathbf{r}, \omega))) \times (\nabla \times \tilde{\mathbf{E}}) - \nabla(\nabla \cdot \tilde{\mathbf{E}}) = \mathbf{0}, \tag{5.107}$$

where $\nabla(\ln(\mu)) = -\mu\nabla(1/\mu)$, and where

$$\tilde{k}(\mathbf{r}, \omega) = \frac{\omega}{c} n(\mathbf{r}, \omega) \tag{5.108}$$

is the complex wavenumber of the spatially inhomogeneous, temporally dispersive medium with complex index of refraction $n(\mathbf{r}, \omega) = [\tilde{\mu}(\mathbf{r}, \omega)\tilde{\epsilon}(\mathbf{r}, \omega)]^{1/2}$, $\tilde{\mu} = \mu/\mu_0$ and $\tilde{\epsilon} = \epsilon/\epsilon_0$ being the relative permeability and permittivity, respectively. Finally, from Gauss' law (5.103), $\epsilon\nabla \cdot \tilde{\mathbf{E}} + (\nabla\epsilon) \cdot \tilde{\mathbf{E}} = 0$ so that

$$\nabla \cdot \tilde{\mathbf{E}} = -\frac{\nabla\epsilon}{\epsilon} \cdot \tilde{\mathbf{E}} = -(\nabla \ln(\epsilon)) \cdot \tilde{\mathbf{E}},$$

which is then substituted into Eq. (5.107), yielding

$$\nabla^2 \tilde{\mathbf{E}} + \tilde{k}^2(\mathbf{r}, \omega) + \nabla \left[\ln(\mu(\mathbf{r}, \omega)) \right] \times (\nabla \times \tilde{\mathbf{E}}) + \left[\nabla \ln(\epsilon(\mathbf{r}, \omega)) \cdot \tilde{\mathbf{E}} \right] = \mathbf{0}. \quad (5.109)$$

In a similar manner, the curl of Ampére's law (5.106) after dividing through by $\epsilon(\mathbf{r}, \omega)$ gives

$$\nabla \times \left(\frac{1}{\epsilon(\mathbf{r}, \omega)} \nabla \times \tilde{\mathbf{H}}(\mathbf{r}, \omega) \right) = -i\omega \nabla \times \tilde{\mathbf{E}}(\mathbf{r}, \omega)$$

$$= \omega^2 \mu(\mathbf{r}, \omega) \tilde{\mathbf{H}}(\mathbf{r}, \omega)$$

after substitution from Faraday's law (5.105). Upon expanding the left-hand side of this equation as

$$\nabla \times \left(\frac{1}{\epsilon} \nabla \times \tilde{\mathbf{H}} \right) = \frac{1}{\epsilon} \nabla \times \nabla \times \tilde{\mathbf{H}} + \nabla \left(\frac{1}{\epsilon} \right) \times \left(\nabla \times \tilde{\mathbf{H}} \right)$$

$$= \frac{1}{\epsilon} \nabla(\nabla \cdot \tilde{\mathbf{H}}) - \frac{1}{\epsilon} \nabla^2 \tilde{\mathbf{H}} + \nabla \left(\frac{1}{\epsilon} \right) \times \left(\nabla \times \tilde{\mathbf{H}} \right),$$

so that

$$\nabla^2 \tilde{\mathbf{H}} + \tilde{k}^2(\mathbf{r}, \omega) \tilde{\mathbf{H}} + \nabla(\ln(\epsilon(\mathbf{r}, \omega))) \times (\nabla \times \tilde{\mathbf{H}}) - \nabla(\nabla \cdot \tilde{\mathbf{H}}) = \mathbf{0}. \quad (5.110)$$

Finally, from Gauss' law (5.104), $\mu\nabla \cdot \tilde{\mathbf{H}} + \tilde{\mathbf{H}} \cdot \nabla\mu = 0$ so that

$$\nabla \cdot \tilde{\mathbf{H}} = -\frac{\nabla\mu}{\mu} \cdot \tilde{\mathbf{H}} = -(\nabla \ln(\mu)) \cdot \tilde{\mathbf{H}},$$

which is then substituted into Eq. (5.110), yielding

$$\nabla^2 \tilde{\mathbf{H}} + \tilde{k}^2(\mathbf{r}, \omega) \tilde{\mathbf{H}} + \nabla \left[\ln(\epsilon(\mathbf{r}, \omega)) \times (\nabla \times \tilde{\mathbf{H}}) \right] + \nabla \left[\tilde{\mathbf{H}} \cdot \nabla(\ln(\mu(\mathbf{r}, \omega))) \right] = \mathbf{0}.$$
$$(5.111)$$

Equations (5.109) and (5.111) are the inhomogeneous Helmholtz equations for the electric and magnetic field intensity vectors in source-free regions of a nonconducting, isotropic, spatially inhomogeneous, temporally dispersive, locally linear medium. These two equations reduce directly to Eqs. (5.54) and (5.55), respectively, in the special case when the medium is spatially homogeneous.

5.1.2.1 The High-Frequency Limit of Maxwell's Equations

With the exception of special forms of the spatial inhomogeneity, such as that in dielectric optical waveguides [5], the solution of the inhomogeneous Helmholtz equations (5.109) and (5.111) is a rather formidable problem. Nevertheless, much can be learned about electromagnetic wave propagation behavior through a spatially inhomogeneous medium in the high-frequency limit as $\omega \to \infty$. Because the wavelength goes to zero ($\lambda \to 0$) in this high-frequency limit, it is also referred to as the *geometrical optics limit*.

With the source-free time-harmonic form of Maxwell's equations given in Eqs. (5.103)–(5.106) as a starting point, let

$$\tilde{\mathbf{E}}(\mathbf{r}, \omega) = \tilde{\mathbf{e}}(\mathbf{r}, \omega)e^{ik_0 \mathcal{S}(\mathbf{r},\omega)}, \tag{5.112}$$

$$\tilde{\mathbf{H}}(\mathbf{r}, \omega) = \tilde{\mathbf{h}}(\mathbf{r}, \omega)e^{ik_0 \mathcal{S}(\mathbf{r},\omega)}, \tag{5.113}$$

where $k_0 = 2\pi/\lambda_0 = \omega/c$ is the free-space wavenumber, λ_0 is the free-space wavelength, and where $\mathcal{S}(\mathbf{r}, \omega) = \mathcal{S}_r(\mathbf{r}, \omega) + i\mathcal{S}_i(\mathbf{r}, \omega)$ is a complex-valued scalar function of position $\mathbf{r}$ and frequency ω with real $\mathcal{S}_r(\mathbf{r}, \omega) \equiv \Re\{\mathcal{S}(\mathbf{r}, \omega)\}$ and imaginary $\mathcal{S}_i(\mathbf{r}, \omega) \equiv \Im\{\mathcal{S}(\mathbf{r}, \omega)\}$ parts. In addition, $\tilde{\mathbf{e}}(\mathbf{r}, \omega)$ and $\tilde{\mathbf{h}}(\mathbf{r}, \omega)$ are complex-valued vector functions of position and frequency.

Notice that the time-harmonic phase term $(k_0 \mathcal{S}_r(\mathbf{r}, \omega) - \omega t)$ will be the same at the two neighboring space–time points $(\mathbf{r}, t)$ and $(\mathbf{r} + d\mathbf{r}, t + dt)$ provided that $k_0 \nabla \mathcal{S}_r(\mathbf{r}, \omega) \cdot d\mathbf{r} - \omega dt = 0$. Let $\hat{\mathbf{s}}$ denote the unit vector in the direction of $d\mathbf{r}$ so that $d\mathbf{r} = \hat{\mathbf{s}} ds$. With this substitution, the above equation yields

$$\frac{ds}{dt} = \frac{\omega}{k_0 \hat{\mathbf{s}} \cdot \nabla \mathcal{S}_r(\mathbf{r}, \omega)} = \frac{c}{\hat{\mathbf{s}} \cdot \nabla \mathcal{S}_r(\mathbf{r}, \omega)}.$$

This expression will be a minimum when the unit vector $\hat{\mathbf{s}}$ is parallel to $\nabla \mathcal{S}_r(\mathbf{r}, \omega)$ and hence normal to the *co-phasal surface* $\mathcal{S}_r(\mathbf{r}, \omega) = $ constant of the wavefront, so that

$$\hat{\mathbf{s}} = \frac{\nabla \mathcal{S}_r(\mathbf{r}, \omega)}{|\nabla \mathcal{S}_r(\mathbf{r}, \omega)|}. \tag{5.114}$$

With this identification, the expression given above for the speed ds/dt of the co-phasal surfaces becomes

$$v_p(\mathbf{r}, \omega) = \frac{c}{|\nabla \mathcal{S}_r(\mathbf{r}, \omega)|}, \tag{5.115}$$

which is otherwise known as the *phase velocity*.

Returning to the pair of expressions in Eqs. (5.112) and (5.113), because

$$\nabla \cdot (\epsilon \tilde{\mathbf{E}}) = \left(\epsilon \nabla \cdot \tilde{\mathbf{e}} + \tilde{\mathbf{e}} \cdot \nabla \epsilon + i k_0 \epsilon \tilde{\mathbf{e}} \cdot \nabla \mathcal{S}\right) e^{i k_0 \mathcal{S}},$$

$$\nabla \cdot (\mu \tilde{\mathbf{H}}) = \left(\mu \nabla \cdot \tilde{\mathbf{h}} + \tilde{\mathbf{h}} \cdot \nabla \mu + i k_0 \mu \tilde{\mathbf{h}} \cdot \nabla \mathcal{S}\right) e^{i k_0 \mathcal{S}},$$

and

$$\nabla \times \tilde{\mathbf{E}} = \left(\nabla \times \tilde{\mathbf{e}} + i k_0 \nabla \mathcal{S} \times \tilde{\mathbf{e}}\right) e^{i k_0 \mathcal{S}},$$

$$\nabla \times \tilde{\mathbf{H}} = \left(\nabla \times \tilde{\mathbf{h}} + i k_0 \nabla \mathcal{S} \times \tilde{\mathbf{h}}\right) e^{i k_0 \mathcal{S}},$$

Maxwell's equations (5.103)–(5.106) become

$$\tilde{\mathbf{e}} \cdot \nabla \mathcal{S} = -\frac{1}{i k_0} \left(\tilde{\mathbf{e}} \cdot \nabla(\ln \epsilon) + \nabla \cdot \tilde{\mathbf{e}}\right), \tag{5.116}$$

$$\tilde{\mathbf{h}} \cdot \nabla \mathcal{S} = -\frac{1}{i k_0} \left(\tilde{\mathbf{h}} \cdot \nabla(\ln \mu) + \nabla \cdot \tilde{\mathbf{h}}\right), \tag{5.117}$$

and

$$\nabla \mathcal{S} \times \tilde{\mathbf{e}} - \frac{c}{\|c\|} \mu \tilde{\mathbf{h}} = -\frac{1}{i k_0} \nabla \times \tilde{\mathbf{e}}, \tag{5.118}$$

$$\nabla \mathcal{S} \times \tilde{\mathbf{h}} + \frac{c}{\|c\|} \epsilon \tilde{\mathbf{e}} = -\frac{1}{i k_0} \nabla \times \tilde{\mathbf{h}}. \tag{5.119}$$

In the high-frequency limit[4] as $\omega \to \infty$, Eqs. (5.118) and (5.119) become

$$\nabla \mathcal{S} \times \tilde{\mathbf{e}} - \frac{c}{\|c\|} \mu \tilde{\mathbf{h}} = \mathbf{0}, \tag{5.120}$$

$$\nabla \mathcal{S} \times \tilde{\mathbf{h}} + \frac{c}{\|c\|} \epsilon \tilde{\mathbf{e}} = \mathbf{0}. \tag{5.121}$$

The other pair of equations (5.116) and (5.117) become $\tilde{\mathbf{e}} \cdot \nabla \mathcal{S} = 0$ and $\tilde{\mathbf{h}} \cdot \nabla \mathcal{S} = 0$, which are redundant as they directly follow from Eqs. (5.120) and (5.121) upon scalar multiplication with $\nabla \mathcal{S}$. The pair of relations in Eqs. (5.120) and (5.121) are thus the *high-frequency limiting form of Maxwell's equations* in a nonconducting, isotropic, spatially inhomogeneous, temporally dispersive, locally linear medium. Substitution of $\tilde{\mathbf{h}} = (\|c\|/\mu c) \nabla \mathcal{S} \times \tilde{\mathbf{e}}$ from Eq. (5.120) into Eq. (1.121) yields

$$\frac{\|c\|}{\mu c} \nabla \mathcal{S} \times \left(\nabla \mathcal{S} \times \tilde{\mathbf{e}}\right) + \frac{c}{\|c\|} \epsilon \tilde{\mathbf{e}} = \mathbf{0},$$

---

[4]For optical wave fields, $\omega \sim 10^{15} \to 10^{16}$ r/s and $k_0 \sim 10^7 \to 10^8$ m^{-1}.

and because $\nabla S \times (\nabla S \times \tilde{\mathbf{e}}) = (\nabla S \cdot \tilde{\mathbf{e}})\nabla S - (\nabla S)^2 \tilde{\mathbf{e}} = -(\nabla S)^2 \tilde{\mathbf{e}}$, this relation becomes

$$\left[(\nabla S)^2 - \epsilon \mu c^2 / \|c^2\|\right]\tilde{\mathbf{e}} = \mathbf{0}.$$

Because $\tilde{\mathbf{e}}(\mathbf{r}, \omega)$ does not vanish everywhere for a nontrivial field, one must then have that

$$(\nabla S(\mathbf{r}, \omega))^2 = n^2(\mathbf{r}, \omega), \qquad (5.122)$$

where $n(\mathbf{r}, \omega) = (\tilde{\epsilon}(\mathbf{r}, \omega)\tilde{\mu}(\mathbf{r}, \omega))^{1/2}$ is the complex index of refraction of the spatially inhomogeneous medium. The same expression is also obtained from Eq. (5.120) with substitution from Eq. (5.121). The real and imaginary parts of $S(\mathbf{r}, \omega)$ then satisfy the pair of coupled equations

$$(\nabla S_r(\mathbf{r}, \omega))^2 - (\nabla S_i(\mathbf{r}, \omega))^2 = n_r^2(\mathbf{r}, \omega) - n_i^2(\mathbf{r}, \omega), \qquad (5.123)$$

$$\nabla S_r(\mathbf{r}, \omega) \cdot \nabla S_i(\mathbf{r}, \omega) = n_r(\mathbf{r}, \omega)n_i(\mathbf{r}, \omega). \qquad (5.124)$$

If $n(\mathbf{r}, \omega)$ is real-valued, in which case the medium is lossless and $n_i(\mathbf{r}, \omega) = 0$, then $S(\mathbf{r}, \omega)$ is, in turn, also real-valued. In that idealized, lossless case, the function $S(\mathbf{r}, \omega)$ is called the *eikonal*, from the Greek word for image, and Eq. (5.122), known as the *eikonal equation*, forms the foundation of geometrical optics. In the more general case of a dispersive attenuative medium, the function $S(\mathbf{r}, \omega)$ may be referred to as the *complex eikonal*. The surfaces

$$S_r(\mathbf{r}, \omega) = \text{constant} \qquad (5.125)$$

then define the *geometrical wave fronts*. The eikonal equation may be regarded as the equation for the characteristics of the wave equations (5.109) and (5.111) and rigorously describes the propagation of discontinuities in the electromagnetic wave field [6].

This high-frequency approximation may be extended by expanding the electric and magnetic field vectors in asymptotic series of the form

$$\tilde{\mathbf{E}}(\mathbf{r}, \omega) = e^{ik_0 S(\mathbf{r}, \omega)} \sum_{m=0}^{\infty} \frac{\tilde{\mathbf{e}}^{(m)}(\mathbf{r}, \omega)}{(ik_o)^m},$$

$$\tilde{\mathbf{H}}(\mathbf{r}, \omega) = e^{ik_0 S(\mathbf{r}, \omega)} \sum_{m=0}^{\infty} \frac{\tilde{\mathbf{h}}^{(m)}(\mathbf{r}, \omega)}{(ik_o)^m}.$$

Geometrical optics then corresponds to the leading term in each of these two expressions. This generalized formulation forms the basis for Keller's [7] *geometrical theory of diffraction* (GTD).

5.1.2.2 Differential Equations for the Light Rays

On account of the complex eikonal equation (5.122), $\nabla S(\mathbf{r}, \omega)/n(\mathbf{r}, \omega)$ defines a unit vector $\hat{\mathbf{s}}$ such that $\nabla S(\mathbf{r}, \omega) = n(\mathbf{r}, \omega)\hat{\mathbf{s}}$. The real and imaginary parts of this expression satisfy $\nabla S_r(\mathbf{r}, \omega) = n_r(\mathbf{r}, \omega)\hat{\mathbf{s}}$ and $\nabla S_i(\mathbf{r}, \omega) = n_i(\mathbf{r}, \omega)\hat{\mathbf{s}}$, respectively, which also satisfy the pair of coupled equations (5.123) and (5.124), as they must, so that $\hat{\mathbf{s}}$ is orthogonal to the wave fronts given by Eq. (5.125). This unit vector is then given by [cf. Eq. (5.114)]

$$\hat{\mathbf{s}} = \frac{\nabla S_r(\mathbf{r}, \omega)}{n_r(\mathbf{r}, \omega)} = \frac{\nabla S_i(\mathbf{r}, \omega)}{n_i(\mathbf{r}, \omega)}, \tag{5.126}$$

as well as by $\hat{\mathbf{s}} = \nabla S_r(\mathbf{r}, \omega)/|\nabla S_r(\mathbf{r}, \omega)| = \nabla S_i(\mathbf{r}, \omega)/|\nabla S_i(\mathbf{r}, \omega)|$. Geometrical light rays are then defined as the orthogonal trajectories to the geometrical wave fronts $S_r(\mathbf{r}, \omega) = $ constant. Let $\mathbf{r}(s)$ denote the position vector of a point P on a ray, considered as a function of the arc length s along the ray measured from some convenient starting point P_0. Then $d\mathbf{r}/ds = \hat{\mathbf{s}}$ and the *equation of the ray* may be written as

$$n_r(\mathbf{r}, \omega)\frac{d\mathbf{r}}{ds} = \nabla S_r(\mathbf{r}, \omega). \tag{5.127}$$

Because $\tilde{\mathbf{e}} \cdot \nabla S = 0$ and $\tilde{\mathbf{h}} \cdot \nabla S = 0$, the real and imaginary parts of the electric and magnetic field vectors are found to be orthogonal to the ray direction $\hat{\mathbf{s}}(\mathbf{r}, \omega)$ at every point along the ray trajectory (see Problem 5.3). Upon differentiating this expression with respect to the arc length s along the ray, one obtains

$$\frac{d}{ds}\left(n_r\frac{d\mathbf{r}}{ds}\right) = \frac{d}{ds}\nabla S_r = \frac{d\mathbf{r}}{ds} \cdot \nabla(\nabla S_r) = \frac{1}{n_r}\nabla S_r \cdot \nabla(\nabla S_r)$$

$$= \frac{1}{2n_r}\nabla\left[(\nabla S_r)^2\right] = \frac{1}{2n_r}\nabla\left(n_r^2\right),$$

so that

$$\frac{d}{ds}\left(n_r\frac{d\mathbf{r}}{ds}\right) = \nabla n_r(\mathbf{r}, \omega). \tag{5.128}$$

This is the *vector form of the differential equations of the light rays*. An analogous expression also holds when $n_r(\mathbf{r}, \omega)$ is replaced by $n_i(\mathbf{r}, \omega)$. Notice that this does no imply that $n_r(\mathbf{r}, \omega)$ and $n_i(\mathbf{r}, \omega)$ are equal, but only that their spatial dependence is the same.

For a spatially homogeneous medium, $n(\mathbf{r}, \omega) = n(\omega)$ is independent of position and Eq. (5.128) then simplifies to $d^2\mathbf{r}/ds^2 = 0$ with solution

$$\mathbf{r}(s) = \mathbf{a}s + \mathbf{b}.$$

Each light ray then travels along a straight line through such a spatially homogenous medium.

For a more interesting example, consider the cylindrically symmetric refractive index profile

$$n^2(r) = n_0^2 \pm b^2 r^2,$$

where $r^2 = x^2 + y^2$ with the direction of propagation into the half-space $z > 0$. Because the parabolic index of refraction $n(r)$ is independent of z, the z-component of Eq. (5.128) is given by

$$\frac{d}{ds}\left(n(r)\frac{dz}{ds}\right) = 0 \implies n(r)\frac{dz}{ds} = n_0 \cos\theta_0,$$

where $n_0 \equiv n(0)$ and θ_0 is the initial angle of the ray with respect to the z-axis. This relation then implies that $nd/ds = \ell_0 d/dz$ where $\ell_0 \equiv n_0 \cos\theta_0$. The x-component of the ray equation (5.128), given by

$$n\frac{d}{ds}\left(n\frac{dx}{ds}\right) = n\frac{\partial n}{\partial x} = \frac{1}{2}\frac{\partial n^2}{\partial x} = \pm b^2 x$$

then becomes

$$\frac{d^2 x}{dz^2} = \pm\left(\frac{b}{\ell_0}\right)^2 x$$

with the same equation for the y-component of the ray. With the negative sign choice in the refractive index $[n^2(r) = n_0^2 - b^2 r^2]$ this equation becomes

$$\frac{d^2 x}{dz^2} + \left(\frac{b}{\ell_0}\right)^2 x = 0$$

which has the stable solution $x(z) = A\cos(bx/\ell_0) + B\sin(bx/\ell_0)$ with constants A and B determined by the initial conditions, where stability is defined here in the sense that the ray oscillates about the z-axis. This solution then provides the motivation for the use of parabolic index profiles in fiber optics [5]. Finally, notice that the other sign choice in the refractive index profile $[n^2(r) = n_0^2 + b^2 r^2]$ results in an unstable solution.

5.1.3 Anisotropic, Locally Linear Media with Spatial and Temporal Dispersion

For a spatially and temporally homogeneous, dispersive, anisotropic dielectric medium, of central interest, for example, in crystal optics, the dielectric properties of the material are different in different directions, as described by the *complex dielectric tensor* $\tilde{\underline{\epsilon}}(\mathbf{k}, \omega)$ which has the dyadic and matrix representations [see Eqs. (4.71) and (4.72)]

$$
\begin{aligned}
\tilde{\underline{\epsilon}}(\mathbf{k}, \omega) = {}& \epsilon_{xx}\hat{1}_x\hat{1}_x + \epsilon_{xy}\hat{1}_x\hat{1}_y + \epsilon_{xz}\hat{1}_x\hat{1}_z \\
& + \epsilon_{yx}\hat{1}_y\hat{1}_x + \epsilon_{yy}\hat{1}_y\hat{1}_y + \epsilon_{yz}\hat{1}_y\hat{1}_z \\
& + \epsilon_{zx}\hat{1}_z\hat{1}_x + \epsilon_{zy}\hat{1}_z\hat{1}_y + \epsilon_{zz}\hat{1}_z\hat{1}_z
\end{aligned}
\tag{5.129}
$$

$$
= \begin{pmatrix}
\epsilon_{xx}(\mathbf{k}, \omega) & \epsilon_{xy}(\mathbf{k}, \omega) & \epsilon_{xz}(\mathbf{k}, \omega) \\
\epsilon_{yx}(\mathbf{k}, \omega) & \epsilon_{yy}(\mathbf{k}, \omega) & \epsilon_{yz}(\mathbf{k}, \omega) \\
\epsilon_{zx}(\mathbf{k}, \omega) & \epsilon_{zy}(\mathbf{k}, \omega) & \epsilon_{zz}(\mathbf{k}, \omega)
\end{pmatrix},
\tag{5.130}
$$

respectively. The electric displacement vector is then given by

$$
\tilde{\mathbf{D}}(\mathbf{k}, \omega) = \tilde{\underline{\epsilon}}(\mathbf{k}, \omega) \cdot \tilde{\mathbf{E}}(\mathbf{k}, \omega)
\tag{5.131}
$$

with inverse transform [see Eq. (4.65)]

$$
\mathbf{D}(\mathbf{r}, t) = \int_{-\infty}^{t} d^3t' \int_{-\infty}^{\infty} dt' \, \hat{\underline{\epsilon}}(\mathbf{r} - \mathbf{r}', t - t') \cdot \mathbf{E}(\mathbf{r}', t'),
\tag{5.132}
$$

where (see Appendix E)

$$
\tilde{\underline{\epsilon}}(\mathbf{k}, \omega) = \int_{-\infty}^{\infty} d^3r \int_{-\infty}^{\infty} dt \, \hat{\underline{\epsilon}}(\mathbf{r}, t) e^{-i(\mathbf{k}\cdot\mathbf{r} - \omega t)},
\tag{5.133}
$$

$$
\hat{\underline{\epsilon}}(\mathbf{r}, t) = \frac{1}{(2\pi)^4} \int_{-\infty}^{\infty} d^3k \int_{-\infty}^{\infty} d\omega \, \tilde{\underline{\epsilon}}(\mathbf{k}, \omega) e^{i(\mathbf{k}\cdot\mathbf{r} - \omega t)},
\tag{5.134}
$$

with similar relations for the electric field and displacement vectors. In addition, the macroscopic polarization density is given by [cf. Eqs. (4.97)–(4.99)]

$$
\mathbf{P}(\mathbf{r}, t) = \int_{-\infty}^{t} d^3t' \int_{-\infty}^{\infty} dt' \, \hat{\underline{\chi}}_e(\mathbf{r} - \mathbf{r}', t - t') \cdot \mathbf{E}(\mathbf{r}', t'),
\tag{5.135}
$$

so that

$$
\tilde{\mathbf{P}}(\mathbf{k}, \omega) = \tilde{\underline{\chi}}_e(\mathbf{k}, \omega) \cdot \tilde{\mathbf{E}}(\mathbf{k}, \omega),
\tag{5.136}
$$

where $\tilde{\underline{\chi}}_e(\mathbf{k}, \omega)$ is the *electric susceptibility tensor* which, in turn, determines the dielectric permittivity tensor as [see Eq. (4.99)]

$$
\tilde{\underline{\epsilon}}(\mathbf{k}, \omega) = \epsilon_0 \left(\underline{\mathbb{1}} + \|4\pi\|\tilde{\underline{\chi}}_e(\mathbf{k}.\omega) \right),
\tag{5.137}
$$

where $\underline{\mathfrak{I}} \equiv \hat{\mathbf{1}}_x\hat{\mathbf{1}}_x + \hat{\mathbf{1}}_y\hat{\mathbf{1}}_y + \hat{\mathbf{1}}_z\hat{\mathbf{1}}_z$ is the unit dyad or *idemfactor*. The inverse of this relationship then yields (see Problem 4.3)

$$\hat{\underline{\epsilon}}(\mathbf{r}, t) = \epsilon_0 \left(\delta(\mathbf{r})\delta(t)\underline{\mathfrak{I}} + \|4\pi\|\hat{\underline{\chi}}(\mathbf{r}, t) \right). \tag{5.138}$$

Unlike the polarization response tensor $\hat{\underline{\chi}}(\mathbf{r}, t)$ which has no singularities, the dielectric response tensor $\hat{\underline{\epsilon}}(\mathbf{r}, t)$ possesses a delta function singularity at the space–time point $(\mathbf{r}, t)$ corresponding to the instantaneous response of the medium to the electric field vector at that point.

The temporal frequency dependence of the dielectric permittivity tensor $\tilde{\underline{\epsilon}}(\mathbf{k}, \omega)$ corresponds to *temporal dispersion*, whereas its dependence on the wave vector $\mathbf{k}$ corresponds to the *spatial dispersion* of the medium. The variables $\mathbf{k} = (k_x, k_y, k_z)$ and ω are, in general, independent variables. Whereas the effect of temporal dispersion on wave propagation through such an anisotropic dielectric material can be pronounced under rather ordinary circumstances, the effect of spatial dispersion may typically be regarded as being weak when the inequality $a/\lambda \ll 1$ is satisfied, where a is some molecular dimension characteristic of the medium such as the lattice constant for a crystal. One notable exception is provided by a material medium exhibiting *gyrotropy* (the medium is then said to be *gyrotropic*), in which case the dielectric permittivity tensor depends upon an externally applied quasi-static magnetic field intensity $\tilde{\mathbf{H}}_{ext}(\mathbf{k}, \omega)$ through its *magneto-optical susceptibility* $\chi_m^{(0)}$ in such a manner that

$$\tilde{\mathbf{D}}(\mathbf{k}, \omega) = \tilde{\underline{\epsilon}}^{(0)}(\mathbf{k}, \omega) \cdot \tilde{\mathbf{E}}(\mathbf{k}, \omega) + i\tilde{\mathbf{E}}(\mathbf{k}, \omega) \times \tilde{\mathbf{g}}(\mathbf{k}, \omega), \tag{5.139}$$

where (to a first-order approximation) $\tilde{\mathbf{g}}(\mathbf{k}, \omega) = \epsilon_0\chi_m^{(0)}\tilde{\mathbf{H}}_{ext}(\mathbf{k}, \omega)$ is a pseudo-vector called the *gyration vector*. In such a gyrotropic medium, left- and right-handed elliptically polarized light can propagate with different phase velocities, an optical phenomenon known as the *Faraday effect* in which the plane of polarization is rotated. As this text is primarily interested in the effects of dispersion, unless specifically stated otherwise, the remaining analysis is restricted to nongyrotropic media where Eq. (5.131) is satisfied.

The inverse of the dielectric permittivity tensor expressed in matrix form is given by

$$\tilde{\underline{\epsilon}}^{-1}(\mathbf{k}, \omega) = \frac{\text{adj}\left\{\tilde{\underline{\epsilon}}(\mathbf{k}, \omega)\right\}}{\left|\tilde{\underline{\epsilon}}(\mathbf{k}, \omega)\right|}, \tag{5.140}$$

provided that the matrix is nonsingular, where adj$\{\underline{A}\}$ denotes the adjoint of the matrix $\underline{A}$. If this inverse does exist, then the inverse of Eq. (5.131) is

$$\tilde{\mathbf{E}}(\mathbf{k}, \omega) = \tilde{\underline{\epsilon}}^{-1}(\mathbf{k}, \omega) \cdot \tilde{\mathbf{D}}(\mathbf{k}, \omega). \tag{5.141}$$

Interestingly enough, the dielectric permittivity tensor completely describes both the electric and magnetic properties of the medium in the following manner [8].

The spatio-temporal frequency transform of Faraday's law (5.2) yields $\tilde{\mathbf{B}}(\mathbf{k}, \omega) = (\|c\|/\omega)\mathbf{k} \times \tilde{\mathbf{E}}(\mathbf{k}, \omega) = (\|c\|/\omega)\mathbf{k} \times \left[\tilde{\underline{\epsilon}}^{-1}(\mathbf{k}, \omega) \cdot \tilde{\mathbf{D}}(\mathbf{k}, \omega)\right]$ so that the relation between $\tilde{\mathbf{B}}(\mathbf{k}, \omega)$ and $\tilde{\mathbf{D}}(\mathbf{k}, \omega)$ is uniquely determined provided that $\omega \neq 0$.[5] However, if spatial dispersion is entirely negligible, one may not, in general, set $\tilde{\underline{\epsilon}}(\mathbf{0}, \omega) = \underline{\epsilon}(\omega)$, in which case the *magnetic permeability tensor* $\tilde{\mu}(\omega)$ needs to be introduced. Nevertheless, the approximation of classical crystal optics is formally obtained in the limit [8]

$$\lim_{\mathbf{k} \to 0} \tilde{\underline{\epsilon}}(\mathbf{k}, \omega) = \tilde{\underline{\epsilon}}(\mathbf{0}, \omega) = \underline{\epsilon}(\omega), \tag{5.142}$$

where $\underline{\epsilon}(\omega)$ is the permittivity tensor in the absence of spatial dispersion.

Because the field vectors in Eq. (5.132) are real-valued, the dielectric permittivity tensor satisfies the symmetry relation

$$\tilde{\underline{\epsilon}}(\mathbf{k}, \omega) = \tilde{\underline{\epsilon}}^*(-\mathbf{k}^*, -\omega^*) \tag{5.143}$$

for complex-valued $\mathbf{k} = \mathbf{k}' + i\mathbf{k}''$ and $\omega = \omega' + i\omega''$. In addition, because of time-reversal symmetry (see Sect. 2.5), which corresponds to the transformation $\omega \overset{T}{\mapsto} -\omega$ in the temporal frequency domain, the dielectric permittivity tensor satisfies the *Onsager relation* [9]

$$\tilde{\underline{\epsilon}}(\mathbf{k}, \omega)\big|_{\mathbf{B}} = \tilde{\underline{\epsilon}}^T(-\mathbf{k}, \omega)\big|_{-\mathbf{B}}, \tag{5.144}$$

where the dependence on the sign of the magnetic induction field under the time-reversal transformation $t \overset{T}{\mapsto} -t$ is explicitly indicated.

In general, the dielectric permittivity tensor $\tilde{\underline{\epsilon}}(\mathbf{k}, \omega)$ is neither symmetric[6] nor Hermitian.[7] It may then be written as the sum of two Hermitian matrices as

$$\tilde{\underline{\epsilon}}(\mathbf{k}, \omega) = \tilde{\underline{\epsilon}}_s(\mathbf{k}, \omega) + i\tilde{\underline{\epsilon}}_a(\mathbf{k}, \omega), \tag{5.145}$$

---

[5]Notice that the zero frequency limit of the spatio-temporal Fourier transform of Faraday's law $\tilde{\mathbf{B}}(\mathbf{k}, \omega) = (\|c\|/\omega)\mathbf{k} \times \tilde{\mathbf{E}}(\mathbf{k}, \omega)$ as $\omega \to 0$ exists provided that one lets $\mathbf{k} \to \mathbf{0}$ simultaneously. In this idealized static limit, Maxwell's equations are reduced into separate equations governing the electrostatic and steady-state magnetic fields.

[6]A square matrix $\underline{A} = (a_{ij})$ is said to be *symmetric* if and only if $a_{ij} = a_{ji}$ for all i, j, in which case the matrix is equal to its transpose, viz. $\underline{A} = \underline{A}^T$. For a *skew-symmetric* square matrix $\underline{B} = (b_{ij})$, $b_{ij} = -b_{ji}$ for all i, j. The diagonal elements of a skew-symmetric matrix must then all be zeroes.

[7]A square matrix $\underline{A} = (a_{ij})$ is said to be *Hermitian* if and only if $a_{ij} = a_{ji}^*$ for all i, j, so that $\underline{A} = \underline{A}^\dagger$, where the superscript † notation indicates the complex conjugate of the transpose of the matrix. The diagonal elements a_{jj} of an Hermitian matrix must then be real-valued. For a *skew-Hermitian* (or *anti-Hermitian*) matrix $\underline{B} = (b_{ij})$, $b_{ij} = -b_{ji}^*$ for all i, j, so that $\underline{B} = -\underline{B}^\dagger$. The diagonal elements of a skew-Hermitian matrix are then either zeroes or are pure imaginary. Finally, notice that a skew-Hermitian matrix $\underline{B} = (b_{ij})$ may be expressed in terms of a Hermitian matrix $\underline{C} = (c_{ij})$ as $\underline{B} = i\underline{C}$ so that $c_{ij} = -ib_{ij}$ for all i, j.

with

$$\tilde{\underline{\underline{\epsilon}}}_s(\mathbf{k}, \omega) \equiv \frac{1}{2} \left(\tilde{\underline{\underline{\epsilon}}}(\mathbf{k}, \omega) + \tilde{\underline{\underline{\epsilon}}}^\dagger(\mathbf{k}, \omega) \right), \tag{5.146}$$

$$\tilde{\underline{\underline{\epsilon}}}_a(\mathbf{k}, \omega) \equiv \frac{1}{2i} \left(\tilde{\underline{\underline{\epsilon}}}(\mathbf{k}, \omega) - \tilde{\underline{\underline{\epsilon}}}^\dagger(\mathbf{k}, \omega) \right), \tag{5.147}$$

where $i\tilde{\underline{\underline{\epsilon}}}_a(\mathbf{k}, \omega)$ forms a skew-Hermitian tensor, $\tilde{\underline{\underline{\epsilon}}}_a(\mathbf{k}, \omega)$ itself being Hermitian. The dielectric permittivity tensor may also be separated into real and imaginary parts as

$$\tilde{\underline{\underline{\epsilon}}}(\mathbf{k}, \omega) = \tilde{\underline{\underline{\epsilon}}}'(\mathbf{k}, \omega) + i\tilde{\underline{\underline{\epsilon}}}''(\mathbf{k}, \omega) \tag{5.148}$$

where $\tilde{\underline{\underline{\epsilon}}}'(\mathbf{k}, \omega) \equiv \Re\left\{\tilde{\underline{\underline{\epsilon}}}(\mathbf{k}, \omega)\right\}$ and $\tilde{\underline{\underline{\epsilon}}}''(\mathbf{k}, \omega) \equiv \Im\left\{\tilde{\underline{\underline{\epsilon}}}(\mathbf{k}, \omega)\right\}$, as may their respective Hermitian parts,

$$\tilde{\underline{\underline{\epsilon}}}_s(\mathbf{k}, \omega) = \tilde{\underline{\underline{\epsilon}}}'_s(\mathbf{k}, \omega) + i\tilde{\underline{\underline{\epsilon}}}''_s(\mathbf{k}, \omega), \tag{5.149}$$

$$\tilde{\underline{\underline{\epsilon}}}_a(\mathbf{k}, \omega) = \tilde{\underline{\underline{\epsilon}}}'_a(\mathbf{k}, \omega) + i\tilde{\underline{\underline{\epsilon}}}''_a(\mathbf{k}, \omega), \tag{5.150}$$

where $\tilde{\underline{\underline{\epsilon}}}'(\mathbf{k}, \omega) = \tilde{\underline{\underline{\epsilon}}}'_s(\mathbf{k}, \omega) - \tilde{\underline{\underline{\epsilon}}}''_a(\mathbf{k}, \omega)$ and $\tilde{\underline{\underline{\epsilon}}}''(\mathbf{k}, \omega) = \tilde{\underline{\underline{\epsilon}}}''_s(\mathbf{k}, \omega) + \tilde{\underline{\underline{\epsilon}}}'_a(\mathbf{k}, \omega)$. If spatial dispersion is entirely negligible, the dielectric permittivity tensor is then described by a symmetric matrix, in which case $\tilde{\underline{\underline{\epsilon}}}_s(\mathbf{k}, \omega) = \tilde{\underline{\underline{\epsilon}}}'(\mathbf{k}, \omega)$ and $\tilde{\underline{\underline{\epsilon}}}_a(\mathbf{k}, \omega) = \tilde{\underline{\underline{\epsilon}}}''(\mathbf{k}, \omega)$ and there is no real distinction between Eqs. (5.145) and (5.148). Finally, the elements of the real and imaginary parts of the dielectric permittivity tensor are interrelated through the dispersion relations [8]

$$\epsilon'_{ij}(\mathbf{k}, \omega)/\epsilon_0 - \delta_{ij} = \frac{1}{\pi} \mathcal{P} \int_{-\infty}^{\infty} \frac{\epsilon''_{ij}(\mathbf{k}, \omega)/\epsilon_0}{\omega' - \omega} d\omega', \tag{5.151}$$

$$\epsilon''_{ij}(\mathbf{k}, \omega)/\epsilon_0 = -\frac{1}{\pi} \mathcal{P} \int_{-\infty}^{\infty} \frac{\epsilon'_{ij}(\mathbf{k}, \omega)/\epsilon_0 - \delta_{ij}}{\omega' - \omega} d\omega', \tag{5.152}$$

the integration being taken along the real ω'-axis [cf. Eqs. (4.111) and (4.112)]. Notice that all of the above symmetry properties for the permittivity tensor $\tilde{\underline{\underline{\epsilon}}}(\mathbf{k}, \omega)$ also apply to its inverse $\tilde{\underline{\underline{\epsilon}}}^{-1}(\mathbf{k}, \omega)$, as well as to the permeability tensor $\tilde{\underline{\underline{\mu}}}(\mathbf{k}, \omega)$ and its inverse $\tilde{\underline{\underline{\mu}}}^{-1}(\mathbf{k}, \omega)$, as well as to the conductivity tensor $\tilde{\underline{\underline{\sigma}}}(\mathbf{k}, \omega)$. In analogy with Eq. (5.28), a *complex permittivity tensor* may then be defined as

$$\tilde{\underline{\underline{\epsilon}}}_c(\mathbf{k}, \omega) \equiv \tilde{\underline{\underline{\epsilon}}}(\mathbf{k}, \omega) + i\frac{\|4\pi\|}{\omega} \tilde{\underline{\underline{\sigma}}}(\mathbf{k}, \omega), \tag{5.153}$$

which also satisfies the above symmetry relations.

The response of an isotropic medium, by definition, does not possess any preferential direction [see Eq. (4.70)]. For an isotropic medium that is temporally dispersive but not spatially dispersive (i.e., is spatially nondispersive), the dielectric permittivity tensor is independent of the wave vector $\mathbf{k}$ in both its magnitude $k = |\mathbf{k}|$ and direction $\hat{k} \equiv \mathbf{k}/k$ (recalling that $\mathbf{k}$ and ω are, in general, independent variables), in which case

$$\epsilon_{ij}(\omega) = \epsilon(\omega)\delta_{ij}, \tag{5.154}$$

where δ_{ij} is the Kronecker delta function. For an isotropic, temporally and spatially dispersive medium, the dielectric permittivity tensor can be separated into transverse and longitudinal components as

$$\epsilon_{ij}(\mathbf{k}, \omega) = \epsilon_T(\mathbf{k}, \omega)(\delta_{ij} - \kappa_i\kappa_j) + \epsilon_L(\mathbf{k}, \omega)\kappa_i\kappa_j, \tag{5.155}$$

so that

$$\tilde{\tilde{\epsilon}}(\mathbf{k}, \omega) = \begin{pmatrix} \epsilon_T(\mathbf{k}, \omega) + \Delta\epsilon\kappa_1^2 & \Delta\epsilon(\mathbf{k}, \omega)\kappa_1\kappa_2 & \Delta\epsilon(\mathbf{k}, \omega)\kappa_1\kappa_3 \\ \Delta\epsilon(\mathbf{k}, \omega)\kappa_2\kappa_1 & \epsilon_T(\mathbf{k}, \omega) + \Delta\epsilon\kappa_2^2 & \Delta\epsilon(\mathbf{k}, \omega)\kappa_2\kappa_3 \\ \Delta\epsilon(\mathbf{k}, \omega)\kappa_3\kappa_1 & \Delta\epsilon(\mathbf{k}, \omega)\kappa_3\kappa_2 & \epsilon_T(\mathbf{k}, \omega) + \Delta\epsilon\kappa_3^2 \end{pmatrix}, \tag{5.156}$$

where $\Delta\epsilon(\mathbf{k}, \omega) \equiv \epsilon_L(\mathbf{k}, \omega) - \epsilon_T(\mathbf{k}, \omega)$ is the difference between the longitudinal and transverse components of the permittivity tensor. Notice that Eq. (5.156) is such that the longitudinal permittivity $\epsilon_L(\mathbf{k}, \omega)$ is always the value along the direction $\hat{k}$ of the wave vector and that the transverse permittivity $\epsilon_T(\mathbf{k}, \omega)$ is always the value in the direction orthogonal to $\hat{k}$, as it should be for an isotropic medium. With the electric field vector $\tilde{\mathbf{E}}(\mathbf{k}, \omega)$ decomposed into transverse $\tilde{\mathbf{E}}_\perp(\mathbf{k}, \omega)$ and longitudinal $\tilde{\mathbf{E}}_\parallel(\mathbf{k}, \omega)$ components with respect to the direction $\hat{k}$ of the wave vector $\mathbf{k}$ as

$$\tilde{\mathbf{E}}(\mathbf{k}, \omega) = \tilde{\mathbf{E}}_\perp(\mathbf{k}, \omega) + \tilde{\mathbf{E}}_\parallel(\mathbf{k}, \omega) \tag{5.157}$$

where

$$\hat{k} \cdot \tilde{\mathbf{E}}_\perp(\mathbf{k}, \omega) = 0 \quad \& \quad \tilde{\mathbf{E}}_\parallel(\mathbf{k}, \omega) = \tilde{E}_\parallel(\mathbf{k}, \omega)\hat{k}, \tag{5.158}$$

then Eq. (5.155) states that

$$\epsilon_{ij}(\mathbf{k}, \omega)\tilde{E}_{\perp,j}(\mathbf{k}, \omega) = \epsilon_T(\mathbf{k}, \omega)\tilde{E}_{\perp,i}(\mathbf{k}, \omega), \tag{5.159}$$

$$\epsilon_{ij}(\mathbf{k}, \omega)\tilde{E}_{\parallel,j}(\mathbf{k}, \omega) = \epsilon_L(\mathbf{k}, \omega)\tilde{E}_{\parallel,i}(\mathbf{k}, \omega), \tag{5.160}$$

where the summation convention applies. Analogous expressions hold for the complex permittivity and magnetic permeability tensors.

In the approximation of classical crystal optics, formally obtained in the limit as $\mathbf{k} \rightarrow \mathbf{0}$ given in Eq. (5.141), the symmetry relation (5.148) simplifies to $\tilde{\underline{\epsilon}}^*(\omega^*) = \tilde{\underline{\epsilon}}(-\omega)$, and the Onsager relation (5.149) becomes $\tilde{\underline{\epsilon}}(\omega) = \tilde{\underline{\epsilon}}^T(\omega)$ so that the dielectric response tensor is symmetric. Because any symmetric matrix can be diagonalized and because such a diagonalization procedure corresponds to a rotation of the coordinate axes along preferential directions in the medium, these preferential directions are called the *principal axes* of the anisotropic medium. If the medium has no preferential direction, then it is isotropic. If there is one preferential direction the medium is said to be *uniaxial* and the diagonalized form of the dielectric permittivity tensor is given by

$$\tilde{\underline{\epsilon}}(\omega) = \begin{pmatrix} \epsilon_\perp(\omega) & 0 & 0 \\ 0 & \epsilon_\perp(\omega) & 0 \\ 0 & 0 & \epsilon_\parallel(\omega) \end{pmatrix}, \qquad (5.161)$$

where $\epsilon_\perp(\omega)$ and $\epsilon_\parallel(\omega)$ are, in general, complex-valued. If there are two preferential directions the medium is said to be *biaxial* and the diagonalized form of the dielectric permittivity tensor is given by

$$\tilde{\underline{\epsilon}}(\omega) = \begin{pmatrix} \epsilon_1(\omega) & 0 & 0 \\ 0 & \epsilon_2(\omega) & 0 \\ 0 & 0 & \epsilon_3(\omega) \end{pmatrix}, \qquad (5.162)$$

where $\epsilon_1(\omega)$, $\epsilon_2(\omega)$, and $\epsilon_3(\omega)$ are all complex-valued, in general, and are called the *principal dielectric permittivities*.[8] If the magnetic permeability tensor $\tilde{\underline{\mu}}(\omega)$ is diagonalized along the same preferential direction, then the *principal indices of refraction* of the biaxial crystal are given by

$$n_j(\omega) = \left[(\epsilon_j(\omega)/\epsilon_0)(\mu_j(\omega)/\mu_0) \right]^{1/2}, \quad j = 1, 2, 3. \qquad (5.163)$$

Similar remarks hold for the conductivity tensor $\tilde{\underline{\sigma}}(\omega)$ as well as for the complex permittivity tensor $\tilde{\underline{\epsilon}}_c(\omega) = \tilde{\underline{\epsilon}}(\omega) + i \|4\pi\| \tilde{\underline{\sigma}}(\omega)/\omega$. For a uniaxial crystal

$$n_1(\omega) = n_2(\omega) \equiv n_o(\omega) \quad \& \quad n_3(\omega) \equiv n_e(\omega) \neq n_o(\omega),$$

where $n_o(\omega) = \left[(\epsilon_\perp(\omega)/\epsilon_0)(\mu_\perp(\omega)/\mu_0) \right]^{1/2}$ is the *ordinary refractive index* and $n_e(\omega) = \left[(\epsilon_\parallel(\omega)/\epsilon_0)(\mu_\parallel(\omega)/\mu_0) \right]^{1/2}$ is the *extraordinary refractive index*.

In a *chiral* or *optically active medium*, the plane of polarization of a linearly polarized electromagnetic wave is rotated with propagation distance. At the micro-

---

[8]The standard terminology of 'principal dielectric constants' is not adopted here because ϵ_1, ϵ_2, and ϵ_3 are, in general, frequency dependent and hence, not constant.

scopic level, a molecule (or any other geometric shape or object) is said to be *chiral*, or to exhibit *chirality*, if it cannot be mapped into its mirror image through application of translations and rotations alone. A nonchiral object, such as a sphere, is said to be *achiral*. Taken together in combination, a chiral object and its mirror image are said to be *enantiomorphic*.[9] A chiral molecule then possesses a certain handedness such that a chiral material can alter the plane of polarization of a linearly polarized electromagnetic wave passing through it. The material relation between the electric displacement vector $\tilde{\mathbf{D}}(\mathbf{k}, \omega)$ and the electric field vector $\tilde{\mathbf{E}}(\mathbf{k}, \omega)$ is accordingly given by [10]

$$\tilde{\mathbf{D}}(\mathbf{k}, \omega) = \tilde{\underline{\epsilon}}^{(0)}(\mathbf{k}, \omega) \cdot \tilde{\mathbf{E}}(\mathbf{k}, \omega) + i \left(\tilde{\boldsymbol{\delta}}(\mathbf{k}, \omega) \times \tilde{\mathbf{E}}(\mathbf{k}, \omega) \right) \cdot \underline{\mathfrak{I}}, \qquad (5.164)$$

where $\underline{\mathfrak{I}} = (\delta_{ij})$ is the idemfactor. Notice the similarity between this constitutive relation and that given in Eq. (5.139) for a gyrotropic medium. This expression may be expressed in the usual form $\tilde{\mathbf{D}}(\mathbf{k}, \omega) = \tilde{\underline{\epsilon}}(\mathbf{k}, \omega) \cdot \tilde{\mathbf{E}}(\mathbf{k}, \omega)$ [see Eq. (5.131)] with the dielectric permittivity tensor

$$\tilde{\underline{\epsilon}}(\mathbf{k}, \omega) = \begin{pmatrix} \epsilon_x(\mathbf{k}, \omega) & -i\delta_z(\mathbf{k}, \omega) & i\delta_y(\mathbf{k}, \omega) \\ i\delta_z(\mathbf{k}, \omega) & \epsilon_y(\mathbf{k}, \omega) & -i\delta_x(\mathbf{k}, \omega) \\ -i\delta_y(\mathbf{k}, \omega) & i\delta_x(\mathbf{k}, \omega) & \epsilon_z(\mathbf{k}, \omega) \end{pmatrix}. \qquad (5.165)$$

Because optical activity is a small effect, the (equally small) material relations between the electric and magnetic field vectors should also be included in its description for completeness. A more general form of the constitutive relations for a chiral medium is then given by [11]

$$\begin{pmatrix} \tilde{\mathbf{D}}(\mathbf{k}, \omega) \\ \tilde{\mathbf{H}}(\mathbf{k}, \omega) \end{pmatrix} = \begin{pmatrix} \tilde{\underline{\epsilon}}(\mathbf{k}, \omega) & i\tilde{\underline{\gamma}}(\mathbf{k}, \omega) \\ i\tilde{\underline{\gamma}}(\mathbf{k}, \omega) & \tilde{\underline{\mu}}^{-1}(\mathbf{k}, \omega) \end{pmatrix} \cdot \begin{pmatrix} \tilde{\mathbf{E}}(\mathbf{k}, \omega) \\ \tilde{\mathbf{B}}(\mathbf{k}, \omega) \end{pmatrix}. \qquad (5.166)$$

Because $\tilde{\mathbf{D}}$ and $\tilde{\mathbf{E}}$ are "ordinary" (or *polar*) vectors whereas $\tilde{\mathbf{H}}$ and $\tilde{\mathbf{B}}$ are *pseudovectors*,[10] then for an isotropic, lossless chiral medium that is reciprocal, the tensors $\tilde{\underline{\epsilon}}(\mathbf{k}, \omega)$ and $\tilde{\underline{\mu}}(\mathbf{k}, \omega)$ reduce to real scalars $\epsilon(\mathbf{k}, \omega)$ and $\mu(\mathbf{k}, \omega)$, respectively, whereas $\tilde{\underline{\gamma}}(\mathbf{k}, \omega)$ reduces to a pseudoscalar.

---

[9]From the Greek for opposite (enantio) form (morphe).

[10]A *polar vector* (an ordinary vector such as the position vector $\mathbf{r}$) reverses sign when the coordinate axes are reversed, whereas a *pseudovector* (an *axial vector*) does not reverse sign. The cross product of two polar vectors produces a pseudovector, whereas the dot product of a polar vector with a pseudovector produces a *pseudoscalar*, a scalar which changes sign under an inversion.

5.2 Electromagnetic Energy and Energy Flow in Temporally Dispersive HILL Media

Of considerable importance to the physical interpretation of electromagnetic wave propagation in dispersive media are the related concepts of electromagnetic energy and energy flow in the dispersive host medium. The statement of the conservation of energy in the coupled field–medium system, as embodied in Poynting's theorem, is readily obtained as a direct mathematical consequence of the real space–time form of Maxwell's field equations, just as was derived for the microscopic field equations in Sect. 2.3.1, but now with the appropriate constitutive relations included. Although subject to interpretation, this relation provides a mathematically consistent formulation of energy flow in the electromagnetic field that is extremely useful in the analysis and physical interpretation of electromagnetic information transmission in temporally dispersive media and systems [12–14].

5.2.1 Poynting's Theorem and the Conservation of Energy

The analysis begins with the pair of curl relations (5.12) and (5.14) of the real space–time form of Maxwell's equations with no externally supplied charge or current sources. Upon taking the scalar product of the first equation (5.12) with $\mathbf{H}(\mathbf{r}, t)$ and the second equation (5.14) with $\mathbf{E}(\mathbf{r}, t)$ and then taking their difference, one obtains

$$\mathbf{E} \cdot \nabla \times \mathbf{H} - \mathbf{H} \cdot \nabla \times \mathbf{E} = \left\| \frac{1}{c} \right\| \left(\mathbf{E} \cdot \frac{\partial \mathbf{D}}{\partial t} + \mathbf{H} \cdot \frac{\partial \mathbf{B}}{\partial t} \right) + \left\| \frac{4\pi}{c} \right\| \mathbf{J}_c \cdot \mathbf{E},$$

which may be rewritten as

$$\mathbf{J}_c \cdot \mathbf{E} = - \left\| \frac{c}{4\pi} \right\| \nabla \cdot (\mathbf{E} \times \mathbf{H}) - \left\| \frac{1}{4\pi} \right\| \left(\mathbf{E} \cdot \frac{\partial \mathbf{D}}{\partial t} + \mathbf{H} \cdot \frac{\partial \mathbf{B}}{\partial t} \right). \tag{5.167}$$

The Poynting vector for the macroscopic electromagnetic field is defined at each space–time point as [cf. Eq. (2.154)]

$$\mathbf{S}(\mathbf{r}, t) \equiv \left\| \frac{c}{4\pi} \right\| \mathbf{E}(\mathbf{r}, t) \times \mathbf{H}(\mathbf{r}, t), \tag{5.168}$$

which has the dimensional units of power per unit area. In addition, one may define the scalar quantities $\mathcal{U}_e'(\mathbf{r}, t)$ and $\mathcal{U}_m'(\mathbf{r}, t)$ that are associated with the electric and magnetic fields, respectively, through the general relations

$$\frac{\partial \mathcal{U}_e'(\mathbf{r}, t)}{\partial t} \equiv \left\| \frac{1}{4\pi} \right\| \mathbf{E}(\mathbf{r}, t) \cdot \frac{\partial \mathbf{D}(\mathbf{r}, t)}{\partial t}, \tag{5.169}$$

$$\frac{\partial \mathcal{U}'_m(\mathbf{r}, t)}{\partial t} \equiv \left\| \frac{1}{4\pi} \right\| \mathbf{H}(\mathbf{r}, t) \cdot \frac{\partial \mathbf{B}(\mathbf{r}, t)}{\partial t}, \tag{5.170}$$

which have the dimensions of energy per unit volume, with sum

$$\mathcal{U}'(\mathbf{r}, t) = \mathcal{U}'_e(\mathbf{r}, t) + \mathcal{U}'_m(\mathbf{r}, t). \tag{5.171}$$

With these definitions, Eq. (5.167) becomes

$$\mathbf{J}_c(\mathbf{r}, t) \cdot \mathbf{E}(\mathbf{r}, t) = -\nabla \cdot \mathbf{S}(\mathbf{r}, t) - \frac{\partial \mathcal{U}'(\mathbf{r}, t)}{\partial t}, \tag{5.172}$$

which is the differential form of the Heaviside–Poynting theorem for the macroscopic electromagnetic field [compare with its microscopic counterpart in Eq. (2.155)]. Integration of this expression over an arbitrary volume V bounded by a simply connected closed surface S followed by application of the divergence theorem then yields

$$\int_V \mathbf{J}_c(\mathbf{r}, t) \cdot \mathbf{E}(\mathbf{r}, t) d^3r = -\oint_S \mathbf{S}(\mathbf{r}, t) \cdot \hat{\mathbf{n}} d^2r - \frac{d}{dt} \int_V \mathcal{U}'(\mathbf{r}, t) d^3r, \tag{5.173}$$

where $\hat{\mathbf{n}}$ denotes the unit outward normal vector to S. This integral form of the Heaviside–Poynting theorem may be taken as a mathematical expression of the conservation of energy in the coupled field–medium system. Because of this physical interpretation, the volume integral on the left-hand side of this relation is interpreted as the time rate of work being done on the conduction current in the region V by the electromagnetic field, the volume integral on the right-hand side of this relation is interpreted as the time rate of energy loss by the electromagnetic field in V, and the surface integral is interpreted as the time rate of electromagnetic energy flow out of V across the boundary surface S.

The generally accepted physical interpretation of the scalar quantity $\mathcal{U}'(\mathbf{r}, t)$ is that it represents the total electromagnetic energy density of the coupled field–medium system at the space–time point $(\mathbf{r}, t)$, measured in ergs/cm^3 in Gaussian (cgs) units and joules/m^3 in SI (mksa) units. The separate scalar quantities $\mathcal{U}'_e(\mathbf{r}, t)$ and $\mathcal{U}'_m(\mathbf{r}, t)$ then represent the electric and magnetic energy densities of the coupled field–medium system.

If one assumes that, in the limit as $t \to -\infty$, $\mathbf{E}(-\infty) = \mathbf{B}(-\infty) = \mathbf{0}$ and that $\mathcal{U}'(-\infty) = 0$ everywhere in space, then Eqs. (5.169) and (5.170) yield

$$\mathcal{U}'_e(\mathbf{r}, t) = \left\| \frac{1}{4\pi} \right\| \int \mathbf{E}(\mathbf{r}, t) \cdot \frac{\partial \mathbf{D}(\mathbf{r}, t)}{\partial t} dt, \tag{5.174}$$

$$\mathcal{U}'_m(\mathbf{r}, t) = \left\| \frac{1}{4\pi} \right\| \int \mathbf{H}(\mathbf{r}, t) \cdot \frac{\partial \mathbf{B}(\mathbf{r}, t)}{\partial t} dt. \tag{5.175}$$

Because the conditions leading to these expressions are not satisfied by a strictly monochromatic field, these expressions do not have unrestricted applicability. However, they do apply to physically realizable pulsed electromagnetic fields.

The general interpretation of the Poynting vector $\mathbf{S}(\mathbf{r}, t)$ is that at any point $\mathbf{r}$ at which $\mathcal{S} \equiv |\mathbf{S}|$ is different from zero one can assert that electromagnetic energy is flowing in the direction $\hat{\mathbf{s}} \equiv \mathbf{S}/\mathcal{S}$ such that across an elemental plane surface oriented perpendicular to $\hat{\mathbf{s}}$ at that point the rate of flow of energy in the field is $\mathcal{S}$ ergs/s/cm^2 (in Gaussian units) or $\mathcal{S}$ joules/s/m^2 (in SI units). In the absence of any conductive current density (i.e., in a nonconducting medium), the differential form (5.172) of Poynting's theorem becomes

$$\nabla \cdot \mathbf{S}(\mathbf{r}, t) + \frac{\partial \mathcal{U}'(\mathbf{r}, t)}{\partial t} = 0, \qquad (5.176)$$

which has exactly the same form as the equation of continuity (4.56) for the conservation of charge. By analogy with this equation, the total energy density $\mathcal{U}'(\mathbf{r}, t)$ of the coupled electromagnetic field–medium system is the conserved quantity and the Poynting vector $\mathbf{S}(\mathbf{r}, t)$ then represents the density flow of this conserved quantity.

The conservation of electromagnetic energy in the coupled field–medium system is now illustrated through a careful consideration of the explicit form that the energy density takes in a temporally dispersive HILL medium. For any general homogeneous, isotropic, locally linear, temporally dispersive medium satisfying the constitutive relations given in Eqs. (5.5) and (5.6), one has the multipole expansions [cf. Eqs. (4.50) and (4.51)]

$$\mathbf{D}(\mathbf{r}, t) = \epsilon_0 \mathbf{E}(\mathbf{r}, t) + \|4\pi\| \mathbf{P}(\mathbf{r}, t) - \|4\pi\| \nabla \cdot \underline{Q}'(\mathbf{r}, t) + \cdots, \qquad (5.177)$$

$$\mathbf{B}(\mathbf{r}, t) = \mu_0 \mathbf{H}(\mathbf{r}, t) + \|4\pi\| \mu_0 \mathbf{M}(\mathbf{r}, t) + \cdots, \qquad (5.178)$$

where $\mathbf{P}(\mathbf{r}, t)$ is the macroscopic polarization density vector [see Eq. (4.27)], $\underline{Q}'(\mathbf{r}, t)$ is the macroscopic quadrupole moment density tensor [see Eqs. (4.28)–(4.33)], and $\mathbf{M}(\mathbf{r}, t)$ is the macroscopic magnetization vector [see Eq. (4.47)] of the temporally dispersive medium. If the dielectric response is dominated by the material polarizabilty, then Eqs. (5.5) and (5.177) yield

$$\epsilon_0 \mathbf{E}(\mathbf{r}, t) + \|4\pi\| \mathbf{P}(\mathbf{r}, t) = \int_{-\infty}^{t} \hat{\epsilon}(t - t') \mathbf{E}(\mathbf{r}, t') dt', \qquad (5.179)$$

to a high degree of approximation, so that

$$\mathbf{P}(\mathbf{r}, t) = \frac{1}{\|4\pi\|} \int_{-\infty}^{\infty} \left[\hat{\epsilon}(t - t') - \epsilon_0 \delta(t - t') \right] \mathbf{E}(\mathbf{r}, t') dt'$$

$$= \int_{-\infty}^{\infty} \hat{\chi}_e(t - t') \mathbf{E}(\mathbf{r}, t') dt', \qquad (5.180)$$

where the upper limit of integration has been extended from t to $+\infty$ because $\hat{\epsilon}(t-t')$ vanishes when $t' > t$. Here

$$\hat{\chi}_e(t) = \frac{1}{\|4\pi\|}\left[\hat{\epsilon}(t) - \epsilon_0\delta(t)\right] \tag{5.181}$$

is the electric susceptibility of the polarizable medium. Similarly, if the magnetic response is dominated by the material magnetization, then Eqs. (5.6) and (5.178) yield

$$\frac{1}{\mu_0}\mathbf{B}(\mathbf{r}, t) - \|4\pi\|\mathbf{M}(\mathbf{r}, t) = \int_{-\infty}^{t} \hat{\mu}^{-1}(t-t')\mathbf{B}(\mathbf{r}, t')dt', \tag{5.182}$$

to a high degree of approximation, so that

$$\mathbf{M}(\mathbf{r}, t) = \frac{1}{\|4\pi\|}\int_{-\infty}^{\infty}\left[\mu_0^{-1}\delta(t-t') - \|4\pi\|\hat{\mu}^{-1}(t-t')\right]\mathbf{B}(\mathbf{r}, t')dt'$$

$$= \frac{1}{\mu_0}\int_{-\infty}^{\infty}\hat{\chi}_b(t-t')\mathbf{B}(\mathbf{r}, t')dt', \tag{5.183}$$

where the upper limit of integration has been extended from t to $+\infty$ because $\hat{\mu}^{-1}(t-t')$ and $\hat{\chi}_b(t-t')$ both vanish when $t' > t$. Here

$$\hat{\chi}_b(t) = \frac{1}{\|4\pi\|}\left[\delta(t) - \|4\pi\|\mu_0\hat{\mu}^{-1}(t)\right] \tag{5.184}$$

is related to the magnetic susceptibility $\hat{\chi}_m(t)$ through the frequency domain relation [cf. Eq. (4.145)] $(1 - \|4\pi\|\chi_b(\omega)) = 1/(1 + \|4\pi\|\chi_m(\omega))$, so that

$$\hat{\chi}_m(t) = \frac{1}{\|4\pi\|}\left[\frac{\hat{\mu}(t)}{\mu_0} - \delta(t)\right]. \tag{5.185}$$

The electric and magnetic energy densities of the coupled field–medium system are then given by

$$\mathcal{U}'_e(\mathbf{r}, t) = \left\|\frac{1}{4\pi}\right\|\frac{\epsilon_0}{2}\mathbf{E}(\mathbf{r}, t)\cdot\mathbf{E}(\mathbf{r}, t) + \int \mathbf{E}(\mathbf{r}, t)\cdot\frac{\partial\mathbf{P}(\mathbf{r}, t)}{\partial t}dt, \tag{5.186}$$

$$\mathcal{U}'_m(\mathbf{r}, t) = \left\|\frac{1}{4\pi}\right\|\frac{\mu_0}{2}\mathbf{H}(\mathbf{r}, t)\cdot\mathbf{H}(\mathbf{r}, t) + \mu_0\int \mathbf{H}(\mathbf{r}, t)\cdot\frac{\partial\mathbf{M}(\mathbf{r}, t)}{\partial t}dt, \tag{5.187}$$

respectively. This pair of expressions explicitly states that a portion of the electromagnetic field energy resides in the dispersive medium. As for Eqs. (5.174) and (5.175), these two equations do not have unrestricted applicability; however, they are valid if the electromagnetic field vanishes as $t \to -\infty$.

Substitution of Eqs. (5.186) and (5.187) combined with Eq. (5.171) into the differential form (5.172) of the Heaviside–Poynting theorem yields

$$\nabla \cdot \mathbf{S}(\mathbf{r}, t) = -\frac{1}{\|4\pi\|} \left[\mu_0 \mathbf{H}(\mathbf{r}, t) \cdot \frac{\partial \mathbf{H}(\mathbf{r}, t)}{\partial t} + \epsilon_0 \mathbf{E}(\mathbf{r}, t) \cdot \frac{\partial \mathbf{E}(\mathbf{r}, t)}{\partial t} \right]$$

$$+ \mathbf{E}(\mathbf{r}, t) \cdot \left[\mathbf{J}_c(\mathbf{r}, t) + \frac{\partial \mathbf{P}(\mathbf{r}, t)}{\partial t} \right] + \mu_0 \mathbf{H}(\mathbf{r}, t) \cdot \frac{\partial \mathbf{M}(\mathbf{r}, t)}{\partial t}.$$

$$(5.188)$$

This is the appropriate differential form of Poynting's theorem for a homogeneous, isotropic, locally linear, temporally dispersive semiconducting medium whose dielectric response is described by the polarization density vector $\mathbf{P}(\mathbf{r}, t)$ and whose magnetic response is described by the macroscopic magnetization vector $\mathbf{M}(\mathbf{r}, t)$. Because its derivation does not rely on any assumption regarding the field behavior as $t \to -\infty$, this expression then has general applicability. Furthermore, it is straightforward to show that the same expression holds with an external current source if the conduction current density $\mathbf{J}_c(\mathbf{r}, t)$ is replaced by the total current density $\mathbf{J}(\mathbf{r}, t) = \mathbf{J}_c(\mathbf{r}, t) + \mathbf{J}_{ext}(\mathbf{r}, t)$.

It is evident that one may associate the first two terms appearing on the right-hand side of Eq. (5.188) with the rate of change of the energy density in the magnetic field alone, the next two terms with the electric energy density, and the final term with the magnetic energy density that is interacting with the dispersive medium. A certain portion of this electric and magnetic interaction energy is reactively stored in the medium and the remainder is absorbed, acting as a source for the evolved heat in the dispersive absorptive medium. The total energy density of the field is then given by the sum of the first two terms plus the reactively stored portion of the remaining three terms.

The open literature is littered with criticisms of the Heaviside–Poynting theorem [15] as well as with revised versions of energy balance and flow in the electromagnetic field [16, 17]. An alternate form of the Heaviside–Poynting theorem has been more recently proposed [18] based upon a redefinition of the energy flux vector in terms of the time derivatives of the vector and scalar potentials for the field. This alternate definition of power flow in the electromagnetic field would then yield an identically zero value in the static field case. However, as shown in Problem 2.20, although the Poynting vector may not vanish in specially constructed static field cases, it does indeed yield the physically correct result of zero net energy flow in those exceptional cases [19]. A detailed critique of this alternate form of the energy flux density has been given by F. N. H. Robinson [20], but not without reply [21]. With the support of a wealth of experimental verification together with the lack of any definitive experiment proving otherwise, the physical validity of the Heaviside–Poynting interpretation of electromagnetic energy flow is maintained throughout this work.

5.2.2 The Energy Density and Evolved Heat in a Dispersive and Absorptive Medium

Attention is now given to the precise physical definition of the expressions for the electric and magnetic energy densities $\mathcal{U}_e(\mathbf{r}, t)$ and $\mathcal{U}_m(\mathbf{r}, t)$, respectively, as well as to the evolved heat (or dissipation) $\mathcal{Q}(\mathbf{r}, t)$ in the electrodynamics of a dispersive, absorptive HILL medium. Without the formal definitions given in Eqs. (5.169) and (5.170), the differential form (5.172) of the Heaviside–Poynting theorem may be written as

$$-\nabla \cdot \mathbf{S}(\mathbf{r}, t) = \frac{1}{\|4\pi\|} \left[\mathbf{E}(\mathbf{r}, t) \cdot \frac{\partial \mathbf{D}(\mathbf{r}, t)}{\partial t} + \mathbf{H}(\mathbf{r}, t) \cdot \frac{\partial \mathbf{B}(\mathbf{r}, t)}{\partial t} \right] + \mathbf{J}(\mathbf{r}, t) \cdot \mathbf{E}(\mathbf{r}, t),$$

$$(5.189)$$

where $\mathbf{J}(\mathbf{r}, t) = \mathbf{J}_c(\mathbf{r}, t) + \mathbf{J}_{ext}(\mathbf{r}, t)$ denotes the total current density.

In the simplest case of a nonmagnetic and nondispersive HILL medium, the constitutive relations given in Eqs. (5.5) and (5.6) yield $\mathbf{D}(\mathbf{r}, t) = \epsilon \mathbf{E}(\mathbf{r}, t)$ and $\mathbf{H}(\mathbf{r}, t) = \mu^{-1} \mathbf{B}(\mathbf{r}, t)$, respectively, where μ and ϵ are real scalar constants. In that case the quantity

$$
\begin{aligned}
\mathcal{U}_m(\mathbf{r}, t) &= \frac{1}{\|4\pi\|} \int \mathbf{H}(\mathbf{r}, t) \cdot \frac{\partial \mathbf{B}(\mathbf{r}, t)}{\partial t} dt \\
&= \frac{1}{\|4\pi\|} \frac{1}{2\mu} \int \frac{\partial}{\partial t} \left(\mathbf{B}(\mathbf{r}, t) \cdot \mathbf{B}(\mathbf{r}, t) \right) dt \\
&= \frac{1}{\|4\pi\|} \frac{1}{2} \mathbf{B}(\mathbf{r}, t) \cdot \mathbf{H}(\mathbf{r}, t)
\end{aligned}
$$
$$(5.190)$$

is immediately identified with the energy density of the magnetic field and is the same as that given by Eq. (5.175), and the quantity

$$
\begin{aligned}
\mathcal{U}_e(\mathbf{r}, t) &= \frac{1}{\|4\pi\|} \int \mathbf{E}(\mathbf{r}, t) \cdot \frac{\partial \mathbf{D}(\mathbf{r}, t)}{\partial t} dt \\
&= \frac{1}{\|4\pi\|} \frac{\epsilon}{2} \int \frac{\partial}{\partial t} \left(\mathbf{E}(\mathbf{r}, t) \cdot \mathbf{E}(\mathbf{r}, t) \right) dt \\
&= \frac{1}{\|4\pi\|} \frac{1}{2} \mathbf{E}(\mathbf{r}, t) \cdot \mathbf{D}(\mathbf{r}, t)
\end{aligned}
$$
$$(5.191)$$

is immediately identified with the energy density of the electric field and is the same as that given by Eq. (5.174). The total electromagnetic energy density is then given by the sum of these two quantities as

$$
\begin{aligned}
\mathcal{U}(\mathbf{r}, t) &= \mathcal{U}_e(\mathbf{r}, t) + \mathcal{U}_m(\mathbf{r}, t) \\
&= \frac{1}{\|4\pi\|} \frac{1}{2} \left(\epsilon |\mathbf{E}(\mathbf{r}, t)|^2 + \mu |\mathbf{H}(\mathbf{r}, t)|^2 \right),
\end{aligned}
$$
$$(5.192)$$

which, according to Landau and Lifshitz [22], has the "exact thermodynamic significance" of representing "the difference between the internal energy per unit volume with and without the field, the density and entropy remaining unchanged." The Poynting vector $\mathbf{S}(\mathbf{r}, t) = \|c/4\pi\| \mathbf{E}(\mathbf{r}, t) \times \mathbf{H}(\mathbf{r}, t)$ has the physical meaning of representing the total electromagnetic energy flux across a unit area normal to the direction of $\mathbf{S}(\mathbf{r}, t)$.

For a homogeneous, isotropic, locally linear, temporally dispersive medium with frequency-dependent dielectric permittivity $\epsilon(\omega) = \epsilon_r(\omega) + i\epsilon_i(\omega)$, magnetic permeability $\mu(\omega) = \mu_r(\omega) + i\mu_i(\omega)$, and electric conductivity $\sigma(\omega) = \sigma_r(\omega) + i\sigma_i(\omega)$, the physical interpretation of the Poynting vector $\mathbf{S}(\mathbf{r}, t) = \|c/4\pi\| \mathbf{E}(\mathbf{r}, t) \times \mathbf{H}(\mathbf{r}, t)$ as the total electromagnetic energy flux density remains intact [22]. However, the expressions given in Eqs. (5.190) and (5.1191) for the quantities $\mathcal{U}_m(\mathbf{r}, t)$ and $\mathcal{U}_e(\mathbf{r}, t)$ no longer represent the magnetic and electric energy densities, respectively. Furthermore, the quantities $\mathcal{U}'_e(\mathbf{r}, t)$ and $\mathcal{U}'_m(\mathbf{r}, t)$ given in Eqs. (5.174) and (5.175) may now be interpreted as generalized electric and magnetic energy densities, respectively, of the coupled field–medium system, as is evident in Eqs. (5.186) and (5.187) for a dispersive medium whose dielectric response is dominated by the medium polarizability and whose magnetic response is dominated by the medium magnetization. A formal separation of these coupled field–medium quantities can be made by rewriting Eqs. (5.169) and (5.170) as [23]

$$\frac{\partial \mathcal{U}_e(\mathbf{r}, t)}{\partial t} + \mathcal{Q}_e(\mathbf{r}, t) = \left\| \frac{1}{4\pi} \right\| \mathbf{E}(\mathbf{r}, t) \cdot \frac{\partial \mathbf{D}(\mathbf{r}, t)}{\partial t} + \mathbf{J}_c(\mathbf{r}, t) \cdot \mathbf{E}(\mathbf{r}, t), \quad (5.193)$$

$$\frac{\partial \mathcal{U}_m(\mathbf{r}, t)}{\partial t} + \mathcal{Q}_m(\mathbf{r}, t) = \left\| \frac{1}{4\pi} \right\| \mathbf{H}(\mathbf{r}, t) \cdot \frac{\partial \mathbf{B}(\mathbf{r}, t)}{\partial t}, \quad (5.194)$$

where $\mathcal{U}_e(\mathbf{r}, t)$ now represents just that part of the electric energy density residing both in the field and reactively stored in the dispersive medium, $\mathcal{U}_m(\mathbf{r}, t)$ now represents just that part of the magnetic energy density residing both in the field and reactively stored in the dispersive medium, and where

$$\mathcal{Q}(\mathbf{r}, t) \equiv \mathcal{Q}_e(\mathbf{r}, t) + \mathcal{Q}_m(\mathbf{r}, t) \quad (5.195)$$

represents the *evolved heat* or *dissipation* in the medium. Comparison of Eq. (5.193) with Eq. (5.169) and comparison of Eq. (5.117) with Eq. (5.170) shows that $\mathcal{U}_e = \mathcal{U}'_e$ and $\mathcal{U}_m = \mathcal{U}'_m$ only in the absence of all loss mechanisms in the medium. With Eqs. (5.193) and (5.194), the Heaviside–Poynting theorem (5.189) becomes

$$\frac{\partial \mathcal{U}(\mathbf{r}, t)}{\partial t} + \mathcal{Q}(\mathbf{r}, t) = -\nabla \cdot \mathbf{S}(\mathbf{r}, t) - \mathbf{J}_{ext}(\mathbf{r}, t) \cdot \mathbf{E}(\mathbf{r}, t), \quad (5.196)$$

where $\mathcal{U}(\mathbf{r}, t) \equiv \mathcal{U}_e(\mathbf{r}, t) + \mathcal{U}_m(\mathbf{r}, t)$. All that remains is to obtain separate expressions for $\mathcal{U}(\mathbf{r}, t)$ and $\mathcal{Q}(\mathbf{r}, t)$.

The sum of the expressions given in Eqs. (5.193) and (5.194) yields, with substitution of the series expansions of the constitutive relations given in Eqs. (4.114),

(4.130), and (4.162) for homogeneous, isotropic, locally linear, temporally dispersive media,

$$\frac{\partial \mathcal{U}}{\partial t} + \mathcal{Q} = \left\| \frac{1}{4\pi} \right\| \left(\mathbf{E} \cdot \frac{\partial \mathbf{D}}{\partial t} + \mathbf{H} \cdot \frac{\partial \mathbf{B}}{\partial t} \right) + \mathbf{J}_c \cdot \mathbf{E}$$

$$= \left\| \frac{1}{4\pi} \right\| \left(\mathbf{E} \cdot \sum_{n=0}^{\infty} \epsilon^{(n)} \frac{\partial^{n+1} \mathbf{E}}{\partial t^{n+1}} + \mathbf{H} \cdot \sum_{n=0}^{\infty} \mu^{(n)} \frac{\partial^{n+1} \mathbf{H}}{\partial t^{n+1}} \right)$$

$$+ \mathbf{E} \cdot \sum_{n=0}^{\infty} \sigma^{(n)} \frac{\partial^{n} \mathbf{E}}{\partial t^{n}} .$$

Upon collecting terms together with the same order time derivatives and field vectors, there results

$$\frac{\partial \mathcal{U}}{\partial t} + \mathcal{Q} = \sigma^{(0)} E^2 + \frac{1}{2} \left(\epsilon^{(0)} / \| 4\pi \| + \sigma^{(1)} \right) \frac{\partial E^2}{\partial t} + \frac{1}{2} \left((\mu^{(0)} / \| 4\pi \|) \right) \frac{\partial H^2}{\partial t}$$

$$+ \frac{1}{2} \left(\epsilon^{(1)} / \| 4\pi \| + \sigma^{(2)} \right) \frac{\partial^2 E^2}{\partial t^2} + \frac{1}{2} \left((\mu^{(1)} / \| 4\pi \|) \right) \frac{\partial^2 H^2}{\partial t^2}$$

$$- \left(\epsilon^{(1)} / \| 4\pi \| + \sigma^{(2)} \right) \left(\frac{\partial^2 E}{\partial t} \right)^2 - \left((\mu^{(1)} / \| 4\pi \|) \right) \left(\frac{\partial H^2}{\partial t} \right)^2$$

$$+ \cdots . \tag{5.197}$$

In general, one then has that

$$\mathcal{U}(\mathbf{r}, t) = \frac{1}{2} \left\| \frac{1}{4\pi} \right\| \left(\alpha_1 E^2(\mathbf{r}, t) + \beta_1 H^2(\mathbf{r}, t) \right) + \cdots , \tag{5.198}$$

$$\mathcal{Q}(\mathbf{r}, t) = \sigma^{(0)} E^2(\mathbf{r}, t) + \frac{1}{2} \left\| \frac{1}{4\pi} \right\| \left(\xi_1 \frac{\partial E^2}{\partial t} + \zeta_1 \frac{\partial H^2}{\partial t} \right) + \cdots , \tag{5.199}$$

where

$$\alpha_1 + \beta_1 + \xi_1 + \zeta_1 = \epsilon^{(0)} + \mu^{(0)} + \| 4\pi \| \sigma^{(1)}, \tag{5.200}$$

and so on for higher-order coefficients. Unfortunately, the first-order coefficients α_1, β_1, ξ_1, and ζ_1 cannot be separately expressed in terms of the zeroth-order dielectric permittivity and magnetic permeability moments $\epsilon^{(0)}$ and $\mu^{(0)}$ [see Eqs. (4.119) and (4.167)] and the first-order electric conductivity moment $\sigma^{(1)}$ [see Eq. (4.135)].

In the simplest situation of a nondispersive medium, in which case $\epsilon = \epsilon^{(0)}$, $\mu = \mu^{(0)}$, and $\sigma = \sigma^{(0)}$, the above expression becomes

$$\alpha_1 + \beta_1 = \epsilon + \mu,$$

because $\xi_1 = \zeta_1 = 0$ when a causal medium is nondispersive. It is then seen that the natural choice is to take $\alpha_1 = \epsilon$ and $\beta_1 = \mu$, in which case the expression given in Eq. (5.198) simplifies to that in Eq. (5.192), as required.

Such a natural separation is not possible in the general dispersive case. For example, if the electric and magnetic field vectors are assumed to vary sufficiently slowly with time such that $|\xi_1|/T \ll \sigma^{(0)}$ and $|\zeta_1|/T \ll \sigma^{(0)}$, where T is the characteristic temporal period describing variations in the electromagnetic field, then Eqs. (5.198) and (5.199) may be approximated as

$$\mathcal{U}(\mathbf{r}, t) \cong \frac{1}{2} \left\| \frac{1}{4\pi} \right\| \left(\epsilon^{(0)} E^2(\mathbf{r}, t) + \mu^{(0)} H^2(\mathbf{r}, t) \right),$$

$$\mathcal{Q}(\mathbf{r}, t) \cong \sigma^{(0} E^2(\mathbf{r}, t), \qquad (5.201)$$

when $|\alpha_1| \gg |\xi_1|$ and $|\beta_1| \gg |\zeta_1|$. However, it is by no means certain that the conditions $|\alpha_1| \gg |\xi_1|$ and $|\beta_1| \gg |\zeta_1|$ will be satisfied and it is entirely possible that either $|\alpha_1| \overset{<}{\sim} |\xi_1|$ or $|\beta_1| \overset{<}{\sim} |\zeta_1|$. Based upon these results, Barash and Ginzburg [23] concluded that[11]

> one cannot, in general, express the electromagnetic energy density $\mathcal{U}(\mathbf{r}, t)$ and dissipation $\mathcal{Q}(\mathbf{r}, t)$ separately in terms of the dielectric permittivity, magnetic permeability and electric conductivity of a general causal, temporally dispersive medium.

Consequently, in order to unambiguously determine these quantities, it may be necessary to employ a specific physical model of the dispersive medium through, for example, the equation of motion at the microscopic level. Further discussion of this conclusion from an opposing viewpoint is presented in Appendix F.

5.2.3 Complex Time-Harmonic Form of Poynting's Theorem

For a completely time-harmonic electromagnetic field with fixed, real-valued angular frequency ω, each real-valued field quantity may be expressed in the form [cf. Eqs. (5.37)–(5.41)]

$$\mathcal{A}(\mathbf{r}, t) = \Re\left\{ \tilde{\mathbf{A}}(\mathbf{r}) e^{-i\omega t} \right\} = \frac{1}{2} \left[\tilde{\mathbf{A}}(\mathbf{r}) e^{-i\omega t} + \tilde{\mathbf{A}}^*(\mathbf{r}) e^{i\omega t} \right], \qquad (5.202)$$

---

[11]This result has been criticized in an analysis [24] based upon a revised formulation of electromagnetic energy conservation [18]. However, this analysis relies, in part, on the revised constitutive relations [compare with the relations given in Eqs. (5.5) and (5.6)] $\mathbf{D} = \epsilon * \partial \mathbf{E}/\partial t$ and $\mathbf{B} = \mu * \partial \mathbf{H}/\partial t$, where $*$ denotes the convolution operation. This then results in the real parts of both $\epsilon(\omega)$ and $\mu(\omega)$ being odd functions of real ω and their imaginary parts are now even, in disagreement with experimental results (see Figs. 4.2 and 4.3).

where $\tilde{\mathbf{A}}(\mathbf{r})$ is in general complex-valued and is related to the positive frequency component of the temporal frequency spectrum of $\mathcal{A}(\mathbf{r}, t)$, as described in Eq. (5.45). The scalar product of two such real-valued vector fields with the same oscillation frequency ω is then given by

$$
\begin{aligned}
\mathcal{A}(\mathbf{r}, t) \cdot \mathcal{B}(\mathbf{r}, t) &= \frac{1}{2} \Re \left\{ \tilde{\mathbf{A}}(\mathbf{r}) \cdot \tilde{\mathbf{B}}^*(\mathbf{r}) + \tilde{\mathbf{A}}(\mathbf{r}) \cdot \tilde{\mathbf{B}}(\mathbf{r}) e^{-i2\omega t} \right\} \\
&= \frac{1}{2} \Re \left\{ \tilde{\mathbf{A}}^*(\mathbf{r}) \cdot \tilde{\mathbf{B}}(\mathbf{r}) + \tilde{\mathbf{A}}(\mathbf{r}) \cdot \tilde{\mathbf{B}}(\mathbf{r}) e^{-i2\omega t} \right\},
\end{aligned} \tag{5.203}
$$

and their vector product is given by

$$
\begin{aligned}
\mathcal{A}(\mathbf{r}, t) \times \mathcal{B}(\mathbf{r}, t) &= \frac{1}{2} \Re \left\{ \tilde{\mathbf{A}}(\mathbf{r}) \times \tilde{\mathbf{B}}^*(\mathbf{r}) + \tilde{\mathbf{A}}(\mathbf{r}) \times \tilde{\mathbf{B}}(\mathbf{r}) e^{-i2\omega t} \right\} \\
&= \frac{1}{2} \Re \left\{ \tilde{\mathbf{A}}^*(\mathbf{r}) \times \tilde{\mathbf{B}}(\mathbf{r}) + \tilde{\mathbf{A}}(\mathbf{r}) \times \tilde{\mathbf{B}}(\mathbf{r}) e^{-i2\omega t} \right\}. \tag{5.204}
\end{aligned}
$$

The time-average of a periodic function $\mathcal{A}(\mathbf{r}, t)$ is simply defined by the integral

$$
\langle \mathcal{A}(\mathbf{r}, t) \rangle \equiv \frac{1}{T} \int_{-T/2}^{T/2} \mathcal{A}(\mathbf{r}, t) dt, \tag{5.205}
$$

where $T > 0$ is the fundamental period of oscillation of the function $\mathcal{A}(\mathbf{r}, t)$; for an aperiodic function one takes the limit as $T \to \infty$ in the above expression. If $\mathcal{A}(\mathbf{r}, t)$ is a simple harmonic function of time with fixed angular frequency ω, then $T = 2\pi/\omega$ and $\langle \mathcal{A}(\mathbf{r}, t) \rangle = 0$. The time-average of the scalar and vector products of two such time-harmonic vector fields with the same oscillation frequency do not vanish, however, as seen from Eqs. (5.203) and (5.204), and are given by

$$
\langle \mathcal{A}(\mathbf{r}, t) \cdot \mathcal{B}(\mathbf{r}, t) \rangle = \frac{1}{2} \Re \left\{ \tilde{\mathbf{A}}(\mathbf{r}) \cdot \tilde{\mathbf{B}}^*(\mathbf{r}) \right\} = \frac{1}{2} \Re \left\{ \tilde{\mathbf{A}}^*(\mathbf{r}) \cdot \tilde{\mathbf{B}}(\mathbf{r}) \right\}, \tag{5.206}
$$

$$
\langle \mathcal{A}(\mathbf{r}, t) \times \mathcal{B}(\mathbf{r}, t) \rangle = \frac{1}{2} \Re \left\{ \tilde{\mathbf{A}}(\mathbf{r}) \times \tilde{\mathbf{B}}^*(\mathbf{r}) \right\} = \frac{1}{2} \Re \left\{ \tilde{\mathbf{A}}^*(\mathbf{r}) \times \tilde{\mathbf{B}}(\mathbf{r}) \right\}. \tag{5.207}
$$

The *time-average Poynting vector* for a time-harmonic field is then given by

$$
\begin{aligned}
\langle \mathbf{S}(\mathbf{r}, t) \rangle &= \left\| \frac{c}{4\pi} \right\| \langle \tilde{\mathbf{E}}(\mathbf{r}, t) \times \tilde{\mathbf{H}}(\mathbf{r}, t) \rangle \\
&= \frac{1}{2} \left\| \frac{c}{4\pi} \right\| \Re \left\{ \tilde{\mathbf{E}}(\mathbf{r}) \times \tilde{\mathbf{H}}^*(\mathbf{r}) \right\} = \Re \left\{ \tilde{\mathbf{S}}(\mathbf{r}) \right\}, \tag{5.208}
\end{aligned}
$$

where

$$
\tilde{\mathbf{S}}(\mathbf{r}) \equiv \frac{1}{2} \left\| \frac{c}{4\pi} \right\| \tilde{\mathbf{E}}(\mathbf{r}) \times \tilde{\mathbf{H}}^*(\mathbf{r}) \tag{5.209}
$$

is the *complex Poynting vector* of the time-harmonic field.[12]

Consider now the complex, time-harmonic differential form of Maxwell's equations, as given in Eqs. (5.50)–(5.53), whose curl relations are

$$\nabla \times \tilde{\mathbf{E}}(\mathbf{r}) = \left\|\frac{1}{c}\right\| i\omega \tilde{\mathbf{B}}(\mathbf{r}),$$

$$\nabla \times \tilde{\mathbf{H}}(\mathbf{r}) = -\left\|\frac{1}{c}\right\| i\omega\epsilon(\omega)\tilde{\mathbf{E}}(\mathbf{r}) + \left\|\frac{4\pi}{c}\right\| \tilde{\mathbf{J}}_c(\mathbf{r}),$$

in the absence of external current sources. The complex conjugate of the second curl relation gives

$$\tilde{\mathbf{J}}_c^*(\mathbf{r}) = \left\|\frac{c}{4\pi}\right\| \nabla \times \tilde{\mathbf{H}}^*(\mathbf{r}) - \left\|\frac{1}{4\pi}\right\| i\omega\epsilon^*(\omega)\tilde{\mathbf{E}}^*(\mathbf{r}),$$

where ω is assumed here to be real-valued. The scalar product of this expression with the complex (phasor) electric field vector $\tilde{\mathbf{E}}(\mathbf{r})$ yields, with the first of the above curl relations,

$$\tilde{\mathbf{J}}_c^* \cdot \tilde{\mathbf{E}} = \left\|\frac{c}{4\pi}\right\| \tilde{\mathbf{E}} \cdot \left(\nabla \times \tilde{\mathbf{H}}^*\right) - \left\|\frac{1}{4\pi}\right\| i\omega \tilde{\mathbf{E}} \cdot \tilde{\mathbf{D}}^*$$

$$= -\left\|\frac{c}{4\pi}\right\| \nabla \cdot \left(\tilde{\mathbf{E}} \times \tilde{\mathbf{H}}^*\right) + \left\|\frac{c}{4\pi}\right\| \tilde{\mathbf{H}}^* \cdot \left(\nabla \times \tilde{\mathbf{E}}\right) - \left\|\frac{1}{4\pi}\right\| i\omega \tilde{\mathbf{E}} \cdot \tilde{\mathbf{D}}^*$$

$$= -\left\|\frac{c}{4\pi}\right\| \nabla \cdot \left(\tilde{\mathbf{E}} \times \tilde{\mathbf{H}}^*\right) + \left\|\frac{1}{4\pi}\right\| i\omega \left(\tilde{\mathbf{B}} \cdot \tilde{\mathbf{H}}^* - \tilde{\mathbf{E}} \cdot \tilde{\mathbf{D}}^*\right).$$

With the definition (5.209) of the complex Poynting vector, one finally obtains

$$\frac{1}{2}\tilde{\mathbf{J}}_c^*(\mathbf{r}) \cdot \tilde{\mathbf{E}}(\mathbf{r}) = -\nabla \cdot \tilde{\mathbf{S}}(\mathbf{r}) - 2i\omega\left(\tilde{u}_e(\mathbf{r}) - \tilde{u}_m(\mathbf{r})\right), \tag{5.210}$$

which is the *time-harmonic* analogue of the *differential form* (5.172) of the *Heaviside–Poynting theorem*. The scalar quantity

$$\tilde{u}_e(\mathbf{r}) \equiv \frac{1}{4}\left\|\frac{1}{4\pi}\right\| \tilde{\mathbf{E}}(\mathbf{r}) \cdot \tilde{\mathbf{D}}^*(\mathbf{r}) \tag{5.211}$$

is referred to as the *harmonic electric energy density*, and the quantity

$$\tilde{u}_m(\mathbf{r}) \equiv \frac{1}{4}\left\|\frac{1}{4\pi}\right\| \tilde{\mathbf{B}}(\mathbf{r}) \cdot \tilde{\mathbf{H}}^*(\mathbf{r}) \tag{5.212}$$

---

[12]Notice that some authors choose to leave the factor $\frac{1}{2}$ in Eq. (5.208) when mksa units are employed, in which case $\tilde{\mathbf{S}} = \tilde{\mathbf{E}} \times \tilde{\mathbf{H}}^*$ and $\langle \mathbf{S} \rangle = \frac{1}{2}\mathfrak{R}\{\tilde{\mathbf{S}}\}$.

as the *harmonic magnetic energy density*. However, neither of these two defined quantities $\tilde{u}_e(\mathbf{r})$ and $\tilde{u}_m(\mathbf{r})$ can be rigorously related to the respective electric and magnetic energy densities $\mathcal{U}_e(\mathbf{r}, t)$ and $\mathcal{U}_m(\mathbf{r}, t)$ of the coupled field–medium system because Eqs. (5.174) and (5.75) are not strictly applicable for a time-harmonic field. Nevertheless, Eqs. (5.169) and (5.170) are applicable in the time-harmonic case and do provide a partial connection of the complex densities defined in Eqs. (5.211) and (5.212) with physically meaningful quantities. The time-average of Eq. (5.169) yields

$$
\begin{aligned}
\left\langle \frac{\partial \mathcal{U}'_e(\mathbf{r}, t)}{\partial t} \right\rangle &= \left\| \frac{1}{4\pi} \right\| \left\langle \mathbf{E}(\mathbf{r}, t) \cdot \frac{\partial \mathbf{D}(\mathbf{r}, t)}{\partial t} \right\rangle \\
&= \left\| \frac{1}{4\pi} \right\| \frac{1}{4} \left\langle \left(\tilde{\mathbf{E}} e^{-i\omega t} + \tilde{\mathbf{E}}^* e^{i\omega t} \right) \cdot \left(-i\omega \tilde{\mathbf{D}} e^{-i\omega t} + i\omega \tilde{\mathbf{D}}^* e^{i\omega t} \right) \right\rangle \\
&= \left\| \frac{1}{4\pi} \right\| \frac{1}{4} \left(i\omega \tilde{\mathbf{E}} \cdot \tilde{\mathbf{D}}^* - i\omega \tilde{\mathbf{E}}^* \cdot \tilde{\mathbf{D}} \right) \\
&= \left\| \frac{1}{4\pi} \right\| \frac{1}{2} \Re \left\{ i\omega \tilde{\mathbf{E}}(\mathbf{r}) \cdot \tilde{\mathbf{D}}^*(\mathbf{r}) \right\} = 2\Re \left\{ i\omega \tilde{u}_e(\mathbf{r}) \right\},
\end{aligned}
$$

so that

$$
\Im \{ \tilde{u}_e(\mathbf{r}) \} = -\frac{1}{2\omega} \left\langle \frac{\partial \mathcal{U}'_e(\mathbf{r}, t)}{\partial t} \right\rangle, \tag{5.213}
$$

and the imaginary part of the harmonic electric energy density is related to the time-average of the time-rate of change of the electric energy density of the coupled field–medium system. In a similar manner, the time-average of Eq. (5.170) yields

$$
\begin{aligned}
\left\langle \frac{\partial \mathcal{U}'_m(\mathbf{r}, t)}{\partial t} \right\rangle &= \left\| \frac{1}{4\pi} \right\| \left\langle \mathbf{H}(\mathbf{r}, t) \cdot \frac{\partial \mathbf{B}(\mathbf{r}, t)}{\partial t} \right\rangle \\
&= \left\| \frac{1}{4\pi} \right\| \frac{1}{4} \left(i\omega \tilde{\mathbf{H}} \cdot \tilde{\mathbf{B}}^* - i\omega \tilde{\mathbf{H}}^* \cdot \tilde{\mathbf{B}} \right) \\
&= \left\| \frac{1}{4\pi} \right\| \frac{1}{2} \Re \left\{ i\omega \tilde{\mathbf{H}}(\mathbf{r}) \cdot \tilde{\mathbf{B}}^*(\mathbf{r}) \right\} = 2\Re \left\{ -i\omega \tilde{u}_m(\mathbf{r}) \right\},
\end{aligned}
$$

so that

$$
\Im \{ \tilde{u}_m(\mathbf{r}) \} = \frac{1}{2\omega} \left\langle \frac{\partial \mathcal{U}'_m(\mathbf{r}, t)}{\partial t} \right\rangle, \tag{5.214}
$$

and the imaginary part of the harmonic magnetic energy density is related to the time-average of the time-rate of change of the magnetic energy density of the coupled field–medium system. Unfortunately, such a simple, physically meaningful relationship does not result for the real parts of $\tilde{u}_e(\mathbf{r})$ and $\tilde{u}_m(\mathbf{r})$.

Integration of Eq. (5.210) over an arbitrary fixed volume V bounded by a simple closed surface S and application of the divergence theorem yields

$$\frac{1}{2} \int_V \tilde{\mathbf{J}}^*(\mathbf{r}) \cdot \tilde{\mathbf{E}}(\mathbf{r}) \, d^3r = - \oint_S \tilde{\mathbf{S}}(\mathbf{r}) \cdot \hat{\mathbf{n}} \, d^2r - 2i\omega \int_V (\tilde{u}_e(\mathbf{r}) - \tilde{u}_m(\mathbf{r})) \, d^3r,$$

(5.215)

which is the analogue of the *integral form* (5.173) of the *Heaviside–Poynting theorem for time-harmonic fields*. The time-harmonic forms (5.210) and (5.215) of the Heaviside–Poynting theorem are, just as are their general counterparts in Eqs. (5.172) and (5.173), a mathematically rigorous consequence of Maxwell's equations and are therefore self-consistent relationships within the framework of classical electrodynamics. As in the general case, it is the physical interpretation of the quantities appearing in these relations that depend to a certain degree on hypothesis.

The real part of the time-harmonic Heaviside–Poynting theorem given in Eq. (5.215) is, with substitution from Eqs. (5.208), (5.213), and (5.214),

$$\frac{1}{2} \int_V \Re \left\{ \tilde{\mathbf{J}}^*(\mathbf{r}) \cdot \tilde{\mathbf{E}}(\mathbf{r}) \right\} d^3r$$
$$= - \oint_S \langle \mathcal{S}(\mathbf{r}, t) \rangle \cdot \hat{\mathbf{n}} \, d^2r + 2\omega \int_V \Im \{ \tilde{u}_e(\mathbf{r}) - \tilde{u}_m(\mathbf{r}) \} \, d^3r$$
$$= - \oint_S \langle \mathcal{S}(\mathbf{r}, t) \rangle \cdot \hat{\mathbf{n}} \, d^2r - \int_V \left\{ \left\langle \frac{\partial \mathcal{U}'_e(\mathbf{r}, t)}{\partial t} \right\rangle + \left\langle \frac{\partial \mathcal{U}'_m(\mathbf{r}, t)}{\partial t} \right\rangle \right\} d^3r,$$

(5.216)

which is just a statement of the conservation of energy in the coupled field–medium system. The imaginary part of Eq. (5.215) is

$$\frac{1}{2} \int_V \Im \left\{ \tilde{\mathbf{J}}^*(\mathbf{r}) \cdot \tilde{\mathbf{E}}(\mathbf{r}) \right\} d^3r$$
$$= - \oint_S \Im \left\{ \tilde{\mathbf{S}}(\mathbf{r}) \right\} \cdot \hat{\mathbf{n}} \, d^2r - 2\omega \int_V \Re \{ \tilde{u}_e(\mathbf{r}) - \tilde{u}_m(\mathbf{r}) \} \, d^3r. \quad (5.217)$$

Because $\tilde{\mathbf{J}}_c(\mathbf{r}) = \sigma(\omega)\tilde{\mathbf{E}}(\mathbf{r})$ so that $\tilde{\mathbf{J}}_c^*(\mathbf{r}) \cdot \tilde{\mathbf{E}}(\mathbf{r}) = \sigma(\omega)\tilde{\mathbf{E}}^*(\mathbf{r}) \cdot \tilde{\mathbf{E}}(\mathbf{r})$, then the real and imaginary parts are expressible as

$$\Re \left\{ \tilde{\mathbf{J}}_c^*(\mathbf{r}) \cdot \tilde{\mathbf{E}}(\mathbf{r}) \right\} = \sigma_r(\omega)|\tilde{\mathbf{E}}(\mathbf{r}|^2,$$

$$\Im \left\{ \tilde{\mathbf{J}}_c^*(\mathbf{r}) \cdot \tilde{\mathbf{E}}(\mathbf{r}) \right\} = -\sigma_i(\omega)|\tilde{\mathbf{E}}(\mathbf{r}|^2.$$

In addition, because $(4\|4\pi\|)\tilde{u}_e = \epsilon^*(\omega)|\tilde{\mathbf{E}}|^2 = (\epsilon_r(\omega) - i\epsilon_i(\omega))|\tilde{\mathbf{E}}|^2$, then [from Eq. (5.213)]

$$\left\langle \frac{\partial \mathcal{U}_e'(\mathbf{r}, t)}{\partial t} \right\rangle = -2\omega \Im\{\tilde{u}_e(\mathbf{r})\} = \frac{1}{2} \left\| \frac{1}{4\pi} \right\| \omega\epsilon_i(\omega)|\tilde{\mathbf{E}}(\mathbf{r})|^2, \qquad (5.218)$$

and because $(4\|4\pi\|)\tilde{u}_m = \mu(\omega)|\tilde{\mathbf{H}}|^2 = (\mu_r(\omega) + i\mu_i(\omega))|\tilde{\mathbf{H}}|^2$, then [from Eq. (5.214)]

$$\left\langle \frac{\partial \mathcal{U}_m'(\mathbf{r}, t)}{\partial t} \right\rangle = 2\omega \Im\{\tilde{u}_m(\mathbf{r})\} = \frac{1}{2} \left\| \frac{1}{4\pi} \right\| \omega\mu_i(\omega)|\tilde{\mathbf{H}}(\mathbf{r})|^2. \qquad (5.219)$$

Substitution of these expressions into Eq. (5.216) then results in

$$-\oint_S \langle \mathcal{S}(\mathbf{r}, t) \rangle \cdot \hat{\mathbf{n}} d^2 r$$

$$= \frac{\omega}{2} \left\| \frac{1}{4\pi} \right\| \int_V \left\{ \left[\epsilon_i(\omega) + \|4\pi\| \frac{\sigma_r(\omega)}{\omega} \right] |\tilde{\mathbf{E}}(\mathbf{r})|^2 + \mu_i(\omega)|\tilde{\mathbf{H}}(\mathbf{r})|^2 \right\} d^3 r$$

$$= \frac{\omega}{2} \left\| \frac{1}{4\pi} \right\| \int_V \Im\left\{ \epsilon_c(\omega)|\tilde{\mathbf{E}}(\mathbf{r})|^2 + \mu(\omega)|\tilde{\mathbf{H}}(\mathbf{r})|^2 \right\} d^3 r. \qquad (5.220)$$

The left-hand side of this equation is interpreted as the time-average electromagnetic power flow into the region V across the simple closed boundary surface S, so that the right-hand side may then be interpreted as the time-average rate of electromagnetic energy that is dissipated in the temporally dispersive HILL medium contained within the region V. In addition, substitution of the above expressions into Eq. (5.217) results in

$$\oint_S \Im\left\{ \tilde{\mathbf{S}}(\mathbf{r}) \right\} \cdot \hat{\mathbf{n}} d^2 r$$

$$= -\frac{\omega}{2} \left\| \frac{1}{4\pi} \right\| \int_V \left\{ \left[\epsilon_r(\omega) - \|4\pi\| \frac{\sigma_i(\omega)}{\omega} \right] |\tilde{\mathbf{E}}(\mathbf{r})|^2 - \mu_r(\omega)|\tilde{\mathbf{H}}(\mathbf{r})|^2 \right\} d^3 r$$

$$= -\frac{\omega}{2} \left\| \frac{1}{4\pi} \right\| \int_V \Re\left\{ \epsilon_c(\omega)|\tilde{\mathbf{E}}(\mathbf{r})|^2 - \mu(\omega)|\tilde{\mathbf{H}}(\mathbf{r})|^2 \right\} d^3 r. \qquad (5.221)$$

The right-hand side of this expression may be interpreted as the time-average of the reactive or stored energy within the coupled field–medium system enclosed by the surface S. It would seem that these physical interpretations of the various quantities appearing in Eqs. (5.220) and (5.221) provide separate identifications of both the time-average electromagnetic energy density $\mathcal{U}(\mathbf{r}, t)$ and dissipation $\mathcal{Q}(\mathbf{r}, t)$ for a time-harmonic field in terms of the dielectric permittivity, magnetic permeability, and electric conductivity of the dispersive medium. However, there is no unambiguous connection between the integrands appearing on the right-hand sides of Eqs. (5.220) and (5.221) and the time-average quantities $\langle \mathcal{U}_e(\mathbf{r}, t) \rangle$, $\langle \mathcal{U}_m(\mathbf{r}, t) \rangle$, and $\langle \mathcal{Q}(\mathbf{r}, t) \rangle$.

For an attenuative medium, the time-average electromagnetic power flow into any given region V in the medium across the simple closed boundary surface S of V must be positive for any nonzero field, so that

$$-\oint_S \langle \mathcal{S}(\mathbf{r}, t) \rangle \cdot \hat{\mathbf{n}} \, d^2 r > 0,$$

and Eq. (5.220) then yields the inequality

$$\omega \int_V \Im\left\{ \epsilon_c(\omega) |\tilde{\mathbf{E}}(\mathbf{r})|^2 + \mu(\omega) |\tilde{\mathbf{H}}(\mathbf{r})|^2 \right\} d^3 r > 0,$$

which is valid for all nonstatic fields[13] ($\omega \neq 0$). Because this inequality holds for any region V in the medium, then $\Im\{\epsilon_c(\omega)|\tilde{\mathbf{E}}(\mathbf{r})|^2 + \mu(\omega)|\tilde{\mathbf{H}}(\mathbf{r})|^2\} > 0$ and consequently

$$\left(\epsilon_i(\omega) + \|4\pi\| \frac{\sigma_r(\omega)}{\omega} \right) |\tilde{\mathbf{E}}(\mathbf{r})|^2 + \mu_i(\omega) |\tilde{\mathbf{H}}(\mathbf{r})|^2 > 0, \tag{5.222}$$

where $\omega > 0$ is the real-valued angular frequency of the (nonstatic) time-harmonic field. The first term in this inequality is the electric loss and the second term is the magnetic loss. From the time-average of Eq. (5.196) with $\mathbf{J}_{ext}(\mathbf{r}, t) = \mathbf{0}$ one obtains

$$\left\langle \frac{\partial \mathcal{U}(\mathbf{r}, t)}{\partial t} \right\rangle + \langle \mathcal{Q}(\mathbf{r}, t) \rangle$$
$$= \frac{\omega}{2} \left\| \frac{1}{4\pi} \right\| \left\{ \left(\epsilon_i(\omega) + \|4\pi\| \frac{\sigma_r(\omega)}{\omega} \right) |\tilde{\mathbf{E}}(\mathbf{r})|^2 + \mu_i(\omega)|\tilde{\mathbf{H}}(\mathbf{r})|^2 \right\},$$

$$\tag{5.223}$$

so that the sum of the time-average time rate of change of the electromagnetic energy density reactively stored in the medium and the time-average dissipation of electromagnetic energy in the medium is determined by the imaginary parts of $\epsilon(\omega)$ and $\mu(\omega)$ together with the real part of $\sigma(\omega)$. Because of the second law of thermodynamics (the law of increase of entropy), the dissipation of electromagnetic energy must be accompanied by the evolution of heat [22] so that $\langle \mathcal{Q}(\mathbf{r}, t) \rangle > 0$. However, the sign of the time-average quantity $\langle \partial \mathcal{U}(\mathbf{r}, t)/\partial t \rangle$ is left undetermined. The general inequality appearing in Eq. (5.222) is then seen to allow for amplification in one or two of the medium parameters provided that it is exceeded by the loss associated with the remaining medium parameter(s). If the dielectric permittivity, magnetic

---

[13] There is no loss for a strictly static field, as is evident from Eq. (5.221).

permeability, and electric conductivity of the medium are separately attenuative, then

$$\epsilon_i(\omega) > 0,$$
$$\mu_i(\omega) > 0, \quad (5.224)$$
$$\sigma_r(\omega) > 0,$$

for all finite positive $\omega > 0$. However, the signs for the real parts $\epsilon_r(\omega)$ of the dielectric permittivity and $\mu_r(\omega)$ of the magnetic permeability and the imaginary part $\sigma_i(\omega)$ of the medium conductivity are not subject to any physical restriction beyond that imposed by causality.

5.2.4 Electromagnetic Energy in the Harmonic Plane Wave Field in Positive Index and DNG Media

As an example, consider the time-harmonic energy densities and Poynting vector for a plane wave electromagnetic field in a temporally dispersive HILL medium with dielectric permittivity $\epsilon(\omega)$, magnetic permeability $\mu(\omega)$, and electric conductivity $\sigma(\omega)$. From Eqs. (5.211) and (5.212) with substitution from Eqs. (5.56) and (5.57), respectively, the time-harmonic electric and magnetic energy densities are found as

$$\tilde{u}_e(\mathbf{r}) = \frac{1}{4\|4\pi\|}\epsilon^*(\omega)|E_0|^2 e^{-2\alpha(\omega)\hat{s}\cdot\mathbf{r}}, \quad (5.225)$$

$$\tilde{u}_m(\mathbf{r}) = \frac{1}{4\|4\pi\|}\mu(\omega)|H_0|^2 e^{-2\alpha(\omega)\hat{s}\cdot\mathbf{r}}. \quad (5.226)$$

Because $H_0 = (\epsilon_c(\omega)/\mu(\omega))^{1/2}E_0$, from Eq. (5.64), where E_0 and H_0 are complex-valued, then $|H_0|^2 = |\epsilon_c(\omega)/\mu(\omega)||E_0|^2$, so that

$$\tilde{u}_m(\mathbf{r}) = \frac{1}{4\|4\pi\|}|\epsilon_c(\omega)||E_0|^2 e^{i\varphi_\mu(\omega)} e^{-2\alpha(\omega)\hat{s}\cdot\mathbf{r}}, \quad (5.227)$$

where $\varphi_\mu(\omega) \equiv arg\{\mu(\omega)\}$ is the phase angle of $\mu(\omega) = |\mu(\omega)|e^{i\varphi_\mu(\omega)}$.

From Eq. (5.209) with substitution from Eqs. (5.56), (5.57), and (5.61) with (5.64), the complex Poynting vector for a time-harmonic plane electromagnetic wave field in a temporally dispersive HILL medium is given by

$$\tilde{S}(\mathbf{r}) = \left\|\frac{c}{4\pi}\right\|\frac{1}{2\eta^*(\omega)}|E_0|^2\hat{s}e^{-2\alpha(\omega)\hat{s}\cdot\mathbf{r}}, \quad (5.228)$$

where $\eta(\omega) = [\mu(\omega)/\epsilon_c(\omega)]^{1/2}$ denotes the complex intrinsic impedance of the material. The time-average Poynting vector in the dispersive HILL medium is then

$$\langle \mathbf{S}(\mathbf{r}, t) \rangle = \left\| \frac{c}{4\pi} \right\| \frac{\eta_r(\omega)}{2|\eta(\omega)|^2} |E_0|^2 \hat{\mathbf{s}} e^{-2\alpha(\omega)\hat{\mathbf{s}} \cdot \mathbf{r}}$$

$$= \left\| \frac{c}{4\pi} \right\| \frac{\cos(\varphi_\eta(\omega))}{2|\eta(\omega)|} |E_0|^2 \hat{\mathbf{s}} e^{-2\alpha(\omega)\hat{\mathbf{s}} \cdot \mathbf{r}}, \tag{5.229}$$

where $\varphi_\eta(\omega) \equiv arg\{\eta(\omega)\}$ is the phase angle of the complex intrinsic impedance $\eta(\omega) = |\eta(\omega)|e^{i\varphi_\eta(\omega)}$.

In a positive index, nonconducting medium, either one or the other of $\tilde{\epsilon}_r(\omega) \equiv \Re\{\tilde{\epsilon}(\omega)\}$ and $\tilde{\mu}_r(\omega) \equiv \Re\{\tilde{\mu}(\omega)\}$ may be negative over some frequency domain, but not both such that $\varphi_n(\omega) \in [0, \pi/2)$, where $\varphi_n(\omega) \equiv arg\{n(\omega)\}$ is the phase angle of the complex index of refraction $n(\omega) = [\tilde{\epsilon}(\omega)\tilde{\mu}(\omega)]^{1/2}$ with relative permittivity $\tilde{\epsilon}(\omega) = \epsilon(\omega)/\epsilon_0$ and permeability $\tilde{\mu}(\omega) = \mu(\omega)/\mu_0$. In that case, $\beta(\omega) > 0$ and $\alpha(\omega) \geq 0$ so that the plane wave phase fronts advance in the positive wave vector direction $+\hat{\mathbf{s}}$. In addition, because $\cos(\varphi_\eta(\omega)) > 0$, then the time-average Poynting vector, and hence, electromagnetic energy flow, is also in the positive wave vector direction $+\hat{\mathbf{s}}$.

For a DNG medium with the branch choice given in Eqs. (5.97) and (5.98), $\beta(\omega) < 0$ and $\alpha(\omega) > 0$ so that the plane wave phase fronts advance in the negative wave vector direction $-\hat{\mathbf{s}}$. However, because $\cos(\varphi_\eta(\omega)) > 0$, the time-average Poynting vector is in the positive wave vector direction $+\hat{\mathbf{s}}$.

Just the opposite occurs for a DNG medium with the branch choice given in Eqs. (5.99) and (5.100). In that case, $\beta(\omega) > 0$ and $\alpha(\omega) < 0$ so that the plane wave phase fronts advance in the positive wave vector direction $\hat{\mathbf{s}}$. However, because $\cos(\varphi_\eta(\omega)) < 0$, the time-average Poynting vector is in the negative wave vector direction $-\hat{\mathbf{s}}$.

In order to better understand this rather marked difference between these two branch choices for a double negative metamaterial, consider a temporally finite pulsed electromagnetic plane wave propagating in the positive z-direction with $\hat{\mathbf{s}} = \hat{\mathbf{1}}_z$ normally incident upon a DNG medium filling the half-space $z > 0$ with surface S occupying the entire xy-plane at $z = 0$ and outward unit normal $\hat{\mathbf{n}} = -\hat{\mathbf{1}}_z$. The branch choice given in Eqs. (5.97) and (5.98) results in a left-handed (LH) plane wave with respect to the direction $\hat{\mathbf{s}} = \hat{\mathbf{1}}_z$ of the wave vector $\tilde{\mathbf{k}}(\omega)$ and a positive power flow into the half-space $z > 0$ with a negative Poynting flux $\int \int_S \mathbf{S} \cdot \hat{\mathbf{n}} d^2 r = -\int \int_S \mathbf{S} \cdot \hat{\mathbf{1}}_z dx dy$, whereas the branch choice given in Eqs. (5.99) and (5.100) results in a left-handed (LH) plane wave with respect to the wave vector direction $\hat{\mathbf{s}} = \hat{\mathbf{1}}_z$ and a negative power flow into the half-space $z > 0$ with a positive Poynting flux. One may then conclude that the first branch choice given in Eqs. (5.97) and (5.98), resulting in a negative refractive index and a positive impedance, is the physically proper choice, as borne out by detailed numerical analysis [4].

5.2.5   Light Rays and the Intensity Law of Geometrical Optics in Spatially Inhomogeneous Media

Attention is now returned to the high-frequency (or geometrical optics) limit of electromagnetic wave propagation in a nonconducting spatially inhomogeneous, temporally dispersive medium that was begun in Sect. 5.1.2.1. From Eq. (5.211) and the representation given in Eq. (5.112) for the time-harmonic electric field vector $\tilde{\mathbf{E}}(\mathbf{r}, \omega) = \tilde{\mathbf{e}}(\mathbf{r}, \omega)e^{ik_0 \mathcal{S}(\mathbf{r}, \omega)}$ in terms of the complex eikonal $\mathcal{S}(\mathbf{r}, \omega) = \mathcal{S}_r(\mathbf{r}, \omega) + i\mathcal{S}_i(\mathbf{r}, \omega)$, the harmonic electric energy density becomes

$$\tilde{u}_e(\mathbf{r}, \omega) = \frac{1}{4} \left\| \frac{1}{4\pi} \right\| \epsilon^*(\mathbf{r}, \omega)\tilde{\mathbf{e}}(\mathbf{r}, \omega) \cdot \tilde{\mathbf{e}}^*(\mathbf{r}, \omega)e^{-2k_0 \mathcal{S}_i(\mathbf{r}, \omega)},$$

where $\mathcal{S}_i \equiv \Im\{\mathcal{S}\}$. From Eq. (5.121) one finds that $\epsilon^* \tilde{\mathbf{e}}^* = -(\|c\|/c)\nabla \mathcal{S}^* \times \tilde{\mathbf{h}}^*$ in the limit as $k_0 \to \infty$, so that

$$\tilde{u}_e(\mathbf{r}, \omega) = \frac{1}{4c} \left\| \frac{c}{4\pi} \right\| \left[\tilde{\mathbf{e}}(\mathbf{r}, \omega), \tilde{\mathbf{h}}^*(\mathbf{r}, \omega), \nabla \mathcal{S}^*(\mathbf{r}, \omega) \right] e^{-2k_0 \mathcal{S}_i(\mathbf{r}, \omega)}, \qquad (5.230)$$

where $[\mathbf{a}, \mathbf{b}, \mathbf{c}] \equiv \mathbf{a} \cdot \mathbf{b} \times \mathbf{c}$ denotes the scalar triple product. In a similar manner, from Eq. (5.212) and the representation given in Eq. (5.113) for the time-harmonic magnetic field vector $\tilde{\mathbf{H}}(\mathbf{r}, \omega) = \tilde{\mathbf{h}}(\mathbf{r}, \omega)e^{ik_0 \mathcal{S}(\mathbf{r}, \omega)}$ in terms of the complex eikonal $\mathcal{S}(\mathbf{r}, \omega)$, the harmonic magnetic energy density becomes

$$\tilde{u}_m(\mathbf{r}, \omega) = \frac{1}{4} \left\| \frac{1}{4\pi} \right\| \mu(\mathbf{r}, \omega)\tilde{\mathbf{h}}(\mathbf{r}, \omega) \cdot \tilde{\mathbf{h}}^*(\mathbf{r}, \omega)e^{-2k_0 \mathcal{S}_i(\mathbf{r}, \omega)}.$$

From Eq. (5.120) one finds that $\mu\tilde{\mathbf{h}} = (\|c\|/c)\nabla \mathcal{S} \times \tilde{\mathbf{e}}$ in the limit as $k_0 \to \infty$, so that

$$\tilde{u}_m(\mathbf{r}, \omega) = \frac{1}{4c} \left\| \frac{c}{4\pi} \right\| \left[\tilde{\mathbf{e}}(\mathbf{r}, \omega), \tilde{\mathbf{h}}^*(\mathbf{r}, \omega), \nabla \mathcal{S}(\mathbf{r}, \omega) \right] e^{-2k_0 \mathcal{S}_i(\mathbf{r}, \omega)}. \qquad (5.231)$$

Hence, within the high-frequency limit of geometric optics, the harmonic electric and magnetic energy densities differ only through the complex conjugate of the complex eikonal $\mathcal{S}(\mathbf{r}, \omega)$ and hence, are equal only when the medium is lossless. The *total harmonic energy density of the electromagnetic wave field* in the geometric optics limit as $k_0 \to \infty$, given by the sum of Eqs. (5.230) and (5.231), is then found to be given by

$$\tilde{u}(\mathbf{r}, \omega) = \frac{1}{2c} \left\| \frac{c}{4\pi} \right\| \left[\tilde{\mathbf{e}}(\mathbf{r}, \omega), \tilde{\mathbf{h}}^*(\mathbf{r}, \omega), \Re\{\nabla \mathcal{S}(\mathbf{r}, \omega)\} \right] e^{-2k_0 \mathcal{S}_i(\mathbf{r}, \omega)}. \qquad (5.232)$$

From Eq. (5.208) with substitution from Eqs. (5.112) and (5.113), the time-average Poynting vector is found to be given by

$$\langle \mathbf{S}(\mathbf{r}, t) \rangle = \frac{1}{2} \left\| \frac{c}{4\pi} \right\| \Re \left\{ \tilde{\mathbf{e}}(\mathbf{r}, \omega) \times \tilde{\mathbf{h}}^*(\mathbf{r}, \omega) \right\} e^{-2k_0 \mathcal{S}_i(\mathbf{r}, \omega)}.$$

With direct substitution from the complex conjugate of Eq. (5.120), given by $\tilde{\mathbf{h}}^* = (\|c\|/\mu^* c) \nabla \mathcal{S}^* \times \tilde{\mathbf{e}}^*$, the above expression for the time-average Poynting vector becomes

$$\langle \mathbf{S}(\mathbf{r}, t) \rangle = \frac{1}{2c} \left\| \frac{c^2}{4\pi} \right\| \Re \left\{ \frac{1}{\mu^*(\mathbf{r}, \omega)} \tilde{\mathbf{e}}(\mathbf{r}, \omega) \times \left(\nabla \mathcal{S}^*(\mathbf{r}, \omega) \times \tilde{\mathbf{e}}^*(\mathbf{r}, \omega) \right) \right\} e^{-2k_0 \mathcal{S}_i(\mathbf{r}, \omega)}$$

$$= \frac{1}{2c} \left\| \frac{c^2}{4\pi} \right\| \Re \left\{ \frac{1}{\mu^*(\mathbf{r}, \omega)} \left[\left(\tilde{\mathbf{e}}(\mathbf{r}, \omega) \cdot \tilde{\mathbf{e}}^*(\mathbf{r}, \omega) \right) \nabla \mathcal{S}^*(\mathbf{r}, \omega) \right. \right.$$

$$\left. \left. - \left(\tilde{\mathbf{e}}(\mathbf{r}, \omega) \cdot \nabla \mathcal{S}^*(\mathbf{r}, \omega) \right) \tilde{\mathbf{e}}^*(\mathbf{r}, \omega) \right] \right\} e^{-2k_0 \mathcal{S}_i(\mathbf{r}, \omega)}$$

(5.233)

which may be expressed as $\langle \mathbf{S}(\mathbf{r}, t) \rangle = \langle \mathbf{S}(\mathbf{r}, t) \rangle_1 - \langle \mathbf{S}(\mathbf{r}, t) \rangle_2$ where $\langle \mathbf{S}(\mathbf{r}, t) \rangle_1$ is the first term on the right in Eq. (5.233) and $\langle \mathbf{S}(\mathbf{r}, t) \rangle_2$ the second.

At this point, it is instructive to consider the zero loss limit of Eq. (5.233). In that limit, $\mathcal{S}_i = 0$, $\mathcal{S}^* = \mathcal{S} \equiv \mathcal{S}_0$, and $\mu^* = \mu$ so that $\langle \mathbf{S}(\mathbf{r}, t) \rangle_2 = 0$ and hence

$$\langle \mathbf{S}(\mathbf{r}, t) \rangle \to \langle \mathbf{S}(\mathbf{r}, t) \rangle_0 = \frac{1}{2c} \left\| \frac{c^2}{4\pi} \right\| \frac{1}{\mu(\mathbf{r}, \omega)} \left(\tilde{\mathbf{e}}(\mathbf{r}, \omega) \cdot \tilde{\mathbf{e}}^*(\mathbf{r}, \omega) \right) \nabla \mathcal{S}_0(\mathbf{r}, \omega).$$

In addition, the complex electric and magnetic energy densities are equal and the time-average electromagnetic energy density is then given by twice the real part of the complex electric energy density, so that

$$\langle \mathcal{U}(\mathbf{r}, t) \rangle_0 = \frac{1}{2} \left\| \frac{1}{4\pi} \right\| \epsilon(\mathbf{r}, \omega) \Re \left\{ \tilde{\mathbf{e}}(\mathbf{r}, \omega) \cdot \tilde{\mathbf{e}}^*(\mathbf{r}, \omega) \right\}.$$

The time-average Poynting vector in a lossless, spatially inhomogeneous, isotropic, dispersive medium is then given by [25]

$$\langle \mathbf{S}(\mathbf{r}, t) \rangle_0 = v_p(\mathbf{r}, \omega) \langle \mathcal{U}(\mathbf{r}, t) \rangle_0 \hat{\mathbf{s}}_0, \qquad (5.234)$$

where $v_p(\mathbf{r}, \omega) = c/n(\mathbf{r}, \omega)$ is the (real-valued) phase velocity. This rather satisfying result states that, in the geometrical optics limit, the time-average electromagnetic energy density propagates at the phase velocity $v_p(\mathbf{r}, \omega)$ in the direction of the normal $\hat{\mathbf{s}}_0 = n(\mathbf{r}, \omega) \nabla \mathcal{S}_0(\mathbf{r}, \omega)$ to the geometrical wave front when the medium is lossless. In addition, because of the high-frequency limiting form

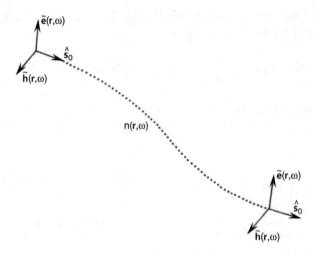

Fig. 5.4 Optical ray path through an isotropic, spatially inhomogeneous, dispersive medium with index of refraction $n(\mathbf{r}, \omega)$

of Eqs. (5.116) and (5.117), $\tilde{\mathbf{e}}(\mathbf{r}, \omega) \cdot \mathcal{S}_0(\mathbf{r}, \omega) = \tilde{\mathbf{h}}(\mathbf{r}, \omega) \cdot \mathcal{S}_0(\mathbf{r}, \omega) = 0$ and the electric and magnetic field vectors are everywhere orthogonal to $\hat{\mathbf{s}}_0$ in this idealized case, as well as being everywhere orthogonal to each other, as depicted in Fig. 5.4. Furthermore, in this zero loss limit, the vector differential equation of the light rays given in Eq. (5.128) becomes

$$n(\mathbf{r}, \omega)\frac{d\mathbf{r}}{ds} = \nabla \mathcal{S}_0(\mathbf{r}, \omega), \tag{5.235}$$

where the scalar s is the arc length along the ray.

With material loss included Eq. (5.234) is no longer strictly valid and one must return to Eq. (5.233). The first term in that equation is given by

$$\langle \mathbf{S}(\mathbf{r}, t) \rangle_1 \equiv \frac{1}{2c} \left\| \frac{c^2}{4\pi} \right\| \Re\left\{ \frac{1}{\mu^*(\mathbf{r}, \omega)} \left[\left(\tilde{\mathbf{e}}(\mathbf{r}, \omega) \cdot \tilde{\mathbf{e}}^*(\mathbf{r}, \omega)\right) \nabla \mathcal{S}^*(\mathbf{r}, \omega)\right] \right\} e^{-2k_0 \mathcal{S}_i(\mathbf{r}, \omega)}. \tag{5.236}$$

From the unnumbered equation preceding Eq. (5.230) for the harmonic electric energy density $\tilde{u}_e(\mathbf{r}, \omega)$, one finds that

$$\tilde{\mathbf{e}}(\mathbf{r}, \omega) \cdot \tilde{\mathbf{e}}^*(\mathbf{r}, \omega) = 4\|4\pi\| \frac{\tilde{u}_e(\mathbf{r}, \omega)}{\epsilon^*(\mathbf{r}, \omega)} e^{+2k_0 \mathcal{S}_i(\mathbf{r}, \omega)}.$$

With this substitution, Eq. (5.236) simplifies to

$$\langle \mathbf{S}(\mathbf{r}, t) \rangle_1 = \frac{2}{c} \left\| c^2 \right\| \Re \left\{ \frac{\tilde{u}_e(\mathbf{r}, \omega)}{\mu^*(\mathbf{r}, \omega) \epsilon^*(\mathbf{r}, \omega)} \nabla S^*(\mathbf{r}, \omega) \right\}. \tag{5.237}$$

Because

$$\frac{\nabla S^*(\mathbf{r}, \omega)}{\mu^*(\mathbf{r}, \omega) \epsilon^*(\mathbf{r}, \omega)} = \frac{c^2}{\|c^2\|} \frac{1}{n^*(\mathbf{r}, \omega)} \hat{\mathbf{s}}$$

Eq. (5.237) then becomes

$$\langle \mathbf{S}(\mathbf{r}, t) \rangle_1 = 2c \Re \left\{ \frac{\tilde{u}_e(\mathbf{r}, \omega)}{n^*(\mathbf{r}, \omega)} \right\} \hat{\mathbf{s}}. \tag{5.238}$$

With the association $\tilde{\mathbf{e}}(\mathbf{r}, \omega) \cdot \tilde{\mathbf{e}}^*(\mathbf{r}, \omega) = 2\|4\pi\| \langle \mathcal{U}(\mathbf{r}, t) \rangle_0 / \epsilon_r(\mathbf{r}, \omega)$ connecting the lossy to the lossless case, the time-harmonic electric energy density in a lossy medium becomes

$$\tilde{u}_e(\mathbf{r}, \omega) = \frac{\epsilon^*(\mathbf{r}, \omega)}{2\epsilon_r(\mathbf{r}, \omega)} \langle \mathcal{U}(\mathbf{r}, t) \rangle_0 e^{-2k_0 S_i(\mathbf{r},\omega)}, \tag{5.239}$$

which decays with the value of the imaginary part of the complex eikonal. With this substitution, Eq. (5.237) takes the more physically appealing form

$$\langle \mathbf{S}(\mathbf{r}, t) \rangle_1 = c \Re \left\{ \frac{\epsilon^*(\mathbf{r}, \omega)}{\epsilon_r(\mathbf{r}, \omega) n^*(\mathbf{r}, \omega)} \right\} \langle \mathcal{U}(\mathbf{r}, t) \rangle_0 \hat{\mathbf{s}} e^{-2k_0 S_i(\mathbf{r},\omega)}, \tag{5.240}$$

which then also decays with the value of the imaginary part of the complex eikonal. Let

$$\epsilon(\mathbf{r}, \omega) = |\epsilon(\mathbf{r}, \omega)| e^{i\varphi_\epsilon(\mathbf{r},\omega)} \quad \& \quad n(\mathbf{r}, \omega) = |n(\mathbf{r}, \omega)| e^{i\varphi_n(\mathbf{r},\omega)} \tag{5.241}$$

so that

$$\Re \left\{ \frac{\epsilon^*(\mathbf{r}, \omega)}{\epsilon_r(\mathbf{r}, \omega) n^*(\mathbf{r}, \omega)} \right\} = \frac{\cos(\varphi_n)}{n_r(\mathbf{r}, \omega) \cos(\varphi_\epsilon)} \cos(\varphi_n - \varphi_\epsilon).$$

With this final substitution and comparison of the resulting expression with Eq. (5.234) for the time-average Poynting vector in the idealized lossless medium case, one finally obtains

$$\langle \mathbf{S}(\mathbf{r}, t) \rangle_1 = \langle \mathbf{S}(\mathbf{r}, t) \rangle_0 \frac{\cos(\varphi_n)}{\cos(\varphi_\epsilon)} \cos(\varphi_n - \varphi_\epsilon) e^{-2k_0 S_i(\mathbf{r},\omega)} \tag{5.242}$$

for the first term in Eq. (5.233). Notice that $\varphi_n = (\varphi_\epsilon + \varphi_\mu)/2$ so that $\varphi_n - \varphi_\epsilon = (\varphi_\mu - \varphi_\epsilon)/2$, where $\mu(\mathbf{r}, \omega) = |\mu(\mathbf{r}, \omega)|e^{i\varphi_\mu(\mathbf{r},\omega)}$. If the medium is either nonmagnetic, paramagnetic, or diamagnetic, then $\varphi_\mu \simeq 0$ so that $(\cos(\varphi_n)/\cos(\varphi_\epsilon))\cos(\varphi_n - \varphi_\epsilon) \simeq \cos^2(\varphi_\epsilon/2/)\cos(\varphi_\epsilon) \geq 1$. This magnifying factor is compensated for by both the exponential decay factor $e^{-2k_0 S_i(\mathbf{r},\omega)}$ and the second term in Eq. (5.233), given by

$$\langle \mathbf{S}(\mathbf{r}, t)\rangle_2 = \frac{1}{2c}\left\|\frac{c^2}{4\pi}\right\|\Re\left\{\left[\frac{\tilde{\mathbf{e}}(\mathbf{r}, \omega)\cdot\nabla S^*(\mathbf{r}, \omega)}{\mu^*(\mathbf{r}, \omega)}\right]\tilde{\mathbf{e}}^*(\mathbf{r}, \omega)\right\}e^{-2k_0 S_i(\mathbf{r},\omega)}.$$

Because $\tilde{\mathbf{e}} \cdot \nabla S = 0$, then $\tilde{\mathbf{e}} \cdot \nabla S^* = 2\tilde{\mathbf{e}} \cdot \Re\{\nabla S\}$ and the above equation becomes

$$\langle \mathbf{S}(\mathbf{r}, t)\rangle_2 = \frac{1}{c}\left\|\frac{c^2}{4\pi}\right\|\Re\left\{\left[\frac{\tilde{\mathbf{e}}(\mathbf{r}, \omega)\cdot\Re\{\nabla S(\mathbf{r}, \omega)\}}{\mu^*(\mathbf{r}, \omega)}\right]\tilde{\mathbf{e}}^*(\mathbf{r}, \omega)\right\}e^{-2k_0 S_i(\mathbf{r},\omega)}.$$

$$(5.243)$$

This term, although typically small in comparison to that given in Eq. (5.242), nonetheless reduces the magnitude of the time-average Poynting vector in an isotropic, spatially inhomogeneous, lossy dispersive medium given in Eq. (5.233). This topic is returned to in Chap. 11 of Volume 2.

5.2.6 Energy Velocity of a Time-Harmonic Field in a Multiple-Resonance Lorentz Model Dielectric

The analysis of the previous subsection shows that a useful phenomenological point of view is to consider the electromagnetic field as having an associated electromagnetic energy that flows through space with an energy velocity $\mathbf{v}_E$ which, in the idealized lossless case of geometrical optics (i.e., in the limit as $\omega \to \infty$), should be equal to the phase velocity $\mathbf{v}_p(\omega) = \hat{\mathbf{s}}c/n(\omega)$. The direction and rate of flow of electromagnetic energy per unit area through a surface normal to the direction of flow is taken to be given by the time-average of the Poynting vector $\mathbf{S}(\mathbf{r}, t) = \|c/4\pi\|\mathbf{E}(\mathbf{r}, t) \times \mathbf{H}(\mathbf{r}, t)$, where $\mathbf{E}(\mathbf{r}, t)$ and $\mathbf{H}(\mathbf{r}, t)$ denote the real-valued electric and magnetic field vectors, respectively, of the electromagnetic field. By analogy with Eq. (5.234), the time-average velocity of propagation of electromagnetic energy is then defined by

$$\mathbf{v}_E \equiv \frac{\langle \mathbf{S}(\mathbf{r}, t)\rangle}{\langle \mathcal{U}_{tot}(\mathbf{r}, t)\rangle},\qquad(5.244)$$

where $\mathcal{U}_{tot}(\mathbf{r}, t)$ is the volume density of the total energy associated with the electromagnetic field. When the electromagnetic field propagates in a nondispersive HILL medium with real-valued relative dielectric permittivity $\tilde{\epsilon}$ and magnetic permeability

$\tilde{\mu}$, the velocity of energy propagation is found to have magnitude $v_E = |\mathbf{v}_E|$ that is given by $c/\sqrt{\tilde{\epsilon}\tilde{\mu}}$, where $\tilde{\epsilon} = 1$ and $\tilde{\mu} = 1$ in the special case when the medium is free-space (i.e., vacuum). The calculation is much more difficult, however, when the field propagates through a temporally dispersive dielectric medium because it is then impossible to explicitly express the electric energy density and the dissipation separately in terms of the general dielectric permittivity of a general dispersive dielectric, as described in Sects. 5.2.1–5.2.3. Nevertheless, the calculation is possible if a specific dynamical model for the medium response is given, such as that provided by the Lorentz model (see Sect. 4.4.4). Loudon's [26, 27] original derivation of the energy velocity for a single-resonance Lorentz model dielectric as well as its generalization [28] to a multiple-resonance dielectric is now presented for a monochromatic electromagnetic field. The connection between the resulting energy velocity for a time-harmonic wave and the signal velocity of a pulse in a Lorentz model dielectric is not apparent from the theory presented here. The fact that there is a close connection has been established by the modern asymptotic theory [29, 30] and is carefully developed in Volume 2. One of the most spectacular results [31] of this modern asymptotic theory is that the frequency dependence of the energy velocity of a monochromatic wave with angular frequency ω_c is shown to be an upper envelope to the frequency dependence of the signal velocity of a pulse with signal frequency ω_c.

The analysis here begins with the differential form of Poynting's theorem which may be written as [cf. Eq. (5.188)]

$$\nabla \cdot \mathbf{S} = -\left\| \frac{1}{4\pi} \right\| \left(\mu_0 \mathbf{H} \cdot \frac{\partial \mathbf{H}}{\partial t} + \epsilon_0 \mathbf{E} \cdot \frac{\partial \mathbf{E}}{\partial t} + \|4\pi\| \mathbf{E} \cdot \frac{\partial \mathbf{P}}{\partial t} \right) \tag{5.245}$$

for a nonconducting, nonmagnetic, dispersive dielectric with $\mu = \mu_0$. First, consider obtaining an expression for the interaction term $\mathbf{E} \cdot \partial \mathbf{P}/\partial t$ for a general, multiple-resonance Lorentz model dielectric. From Eq. (4.207),

$$\mathbf{E} = -\frac{m}{q_e} \left(\frac{d^2 \mathbf{r}_j}{dt^2} + 2\delta_j \frac{d \mathbf{r}_j}{dt} + \omega_j^2 \mathbf{r}_j \right), \tag{5.246}$$

and from Eqs. (4.208)–(4.211) the macroscopic polarization vector is found to be given by

$$\mathbf{P} = -\sum_j N_j q_e \mathbf{r}_j, \tag{5.247}$$

so that

$$\mathbf{E} \cdot \frac{\partial \mathbf{P}}{\partial t} = \sum_j N_j m \left[\frac{1}{2} \frac{d}{dt} \left(\frac{d\mathbf{r}_j}{dt} \right)^2 + 2\delta_j \left(\frac{d\mathbf{r}_j}{dt} \right)^2 + \frac{1}{2} \omega_j^2 \frac{d}{dt} \left(\mathbf{r}_j \right)^2 \right]. \tag{5.248}$$

The quantity appearing on the right-hand side of Poynting's theorem (5.245) is then seen to be the sum of a perfect differential in time and a term in δ_j which corresponds to the dissipation mechanism. If one then defines the *total electromagnetic energy density* as [compare with the discussion involving Eqs. (5.193) and (5.194)]

$$\mathcal{U}_{\text{tot}} \equiv \mathcal{U}_{\text{em}} + \mathcal{U}_{\text{rev}}, \tag{5.249}$$

where

$$\mathcal{U}_{\text{em}} \equiv \left\| \frac{1}{4\pi} \right\| \frac{1}{2} \left(\epsilon_0 E^2 + \mu_0 H^2 \right) \tag{5.250}$$

is identified as the *energy density stored in the electromagnetic field* alone, and where

$$\mathcal{U}_{\text{rev}} \equiv \frac{1}{2} \sum_j N_j m \left[\left(\frac{d\mathbf{r}_j}{dt} \right)^2 + \omega_j^2 (\mathbf{r}_j)^2 \right] \tag{5.251}$$

is identified as the thermodynamically *reversible energy density* stored in the multiple resonance Lorentz medium, then the differential form (5.245) of Poynting's theorem becomes

$$\nabla \cdot \mathbf{S} + 2m \sum_j N_j \delta_j \left(\frac{d\mathbf{r}_j}{dt} \right)^2 = -\frac{d\mathcal{U}_{\text{tot}}}{dt}. \tag{5.252}$$

Integration of this expression over an arbitrary region V bounded by a simple closed surface S followed by application of the divergence theorem then yields

$$\oint_S \mathbf{S} \cdot \hat{\mathbf{n}} d^2 r + 2m \sum_j N_j \delta_j \int_V \left(\frac{d\mathbf{r}_j}{dt} \right)^2 d^3 r = -\frac{d}{dt} \int_V \mathcal{U}_{\text{tot}} d^3 r, \tag{5.253}$$

where $\hat{\mathbf{n}}$ is the outward unit normal vector to the surface S. This relation expresses the conservation of energy in the dispersive Lorentz model dielectric, where the two terms on the left-hand side represent the rate of energy loss in the region V through flow across its surface S and by dissipation in the medium contained in V, respectively, whereas the integral on the right-hand side represents the rate of change of the total electromagnetic energy stored within the region V.

For a time-harmonic field with angular frequency ω, the time-average energy density stored in the multiple-resonance Lorentz model dielectric is found from Eq. (5.251) with (4.208) as

$$\langle \mathcal{U}_{\text{rev}} \rangle = \left\| \frac{1}{4\pi} \right\| \frac{\epsilon_0}{4} E_0^2 e^{-2\alpha(\omega)\hat{s}\cdot\mathbf{r}} \sum_j \frac{b_j^2(\omega^2 + \omega_j^2)}{(\omega^2 - \omega_j^2)^2 + 4\delta_j^2 \omega^2}. \tag{5.254}$$

The time-average value of the energy density stored in the monochromatic electromagnetic field is obtained from Eq. (5.250) with substitution of the relation $H_0/E_0 = \sqrt{\epsilon(\omega)/\mu_0}$ [cf. Eq. (5.64)]

$$\langle \mathcal{U}_{em} \rangle = \left\| \frac{1}{4\pi} \right\| \frac{\epsilon_0}{4} \left(n_r^2(\omega) + n_i^2(\omega) + 1 \right) E_0^2 e^{-2\alpha(\omega)\hat{s}\cdot r}. \tag{5.255}$$

With use of the first of the pair of relations [cf. Eqs. (4.214) and (5.32)]

$$n_r^2(\omega) - n_i^2(\omega) = 1 - \sum_j \frac{b_j^2(\omega^2 - \omega_j^2)}{(\omega^2 - \omega_j^2)^2 + 4\delta_j^2\omega^2}, \tag{5.256}$$

$$n_r(\omega)n_i(\omega) = \sum_j \frac{b_j^2 \delta_j \omega}{(\omega^2 - \omega_j^2)^2 + 4\delta_j^2\omega^2}, \tag{5.257}$$

the total time-average electromagnetic energy density $\langle \mathcal{U}_{tot} \rangle = \langle \mathcal{U}_{em} \rangle + \langle \mathcal{U}_{rev} \rangle$ stored in both the field and the medium is found to be given by

$$\langle \mathcal{U}_{tot} \rangle = \left\| \frac{1}{4\pi} \right\| \frac{\epsilon_0}{2} E_0^2 e^{-2\alpha(\omega)\hat{s}\cdot r} \left[n_r^2(\omega) + \sum_j \frac{b_j^2 \omega^2}{(\omega^2 - \omega_j^2)^2 + 4\delta_j^2\omega^2} \right], \tag{5.258}$$

where the summation extends over all of the medium resonances. Finally, from Eq. (5.229), the time-average value of the magnitude of the Poynting vector for a time-harmonic plane wave field is found to be given by

$$\langle |\mathbf{S}(\mathbf{r}, t)| \rangle = \left\| \frac{c^2}{4\pi} \right\| \frac{n_r(\omega)}{2\mu_0 c} E_0^2 e^{-2\alpha(\omega)\hat{s}\cdot r}, \tag{5.259}$$

where $n_r(\omega)$ denotes the real part of the complex index of refraction.

For a single-resonance Lorentz model dielectric, Eq. (5.254) simplifies to

$$\langle \mathcal{U}_{rev} \rangle = \left\| \frac{1}{4\pi} \right\| \frac{\epsilon_0}{4} E_0^2 e^{-2\alpha(\omega)\hat{s}\cdot r} \frac{b^2 \left(\omega^2 + \omega_0^2 \right)}{\left(\omega^2 - \omega_0^2 \right)^2 + 4\delta^2\omega^2}. \tag{5.260}$$

Furthermore, the magnitude of the macroscopic polarization vector [given in Eq. (5.247) with substitution from Eq. (4.208)] becomes

$$P = -\frac{b^2}{\sqrt{(\omega^2 - \omega_0^2)^2 + 4\delta^2\omega^2}} E_0 e^{i\tilde{k}\cdot r} \tag{5.261}$$

with time derivative

$$\dot{P} = -\frac{b^2\omega}{\sqrt{(\omega^2 - \omega_0^2)^2 + 4\delta^2\omega^2}} E_0 e^{i\tilde{\mathbf{k}}\cdot\mathbf{r}}. \tag{5.262}$$

With these results, Eq. (5.260) for the time-average energy density stored in a single-resonance Lorentz model dielectric becomes

$$\langle \mathcal{U}_{rev} \rangle = \left\|\frac{1}{4\pi}\right\| \frac{\epsilon_0}{4} \left[\frac{\omega_0^2}{b^2} P^2 + \frac{1}{b^2}\dot{P}^2\right] e^{-2\alpha(\omega)\hat{\mathbf{s}}\cdot\mathbf{r}}. \tag{5.263}$$

In addition, Eq. (5.258) becomes, with substitution from Eq. (5.257)

$$\langle \mathcal{U}_{tot} \rangle = \left\|\frac{1}{4\pi}\right\| \frac{\epsilon_0}{2} E_0^2 e^{-2\alpha(\omega)\hat{\mathbf{s}}\cdot\mathbf{r}} \left[n_r^2(\omega) + \frac{n_r(\omega)n_i(\omega)}{\delta}\right], \tag{5.264}$$

where $n_i(\omega)$ denotes the imaginary part of the complex index of refraction. The time-average velocity of energy transport in a single-resonance Lorentz model dielectric is then given by the ratio of Eqs. (5.259) and (5.264) as

$$v_E(\omega) \equiv \frac{\langle|\mathbf{S}|\rangle}{\langle\mathcal{U}_{tot}\rangle} = \frac{c}{n_r(\omega) + \omega n_i(\omega)/\delta}, \tag{5.265}$$

which is Loudon's now classic result [26]. Brillouin's derivation [32] of the energy velocity is in error due to the neglect of the electromagnetic energy reactively stored in the Lorentz oscillators. Because $n_r(\omega) + \omega n_i(\omega)/\delta \geq 1$ for all $\omega \geq 0$, then the energy transport velocity satisfies the inequality

$$0 \leq v_E(\omega) \leq c \tag{5.266}$$

for all $\omega \geq 0$, in agreement with the relativistic principle of causality.

The angular frequency dependence of the relative time-average energy velocity $v_E(\omega)/c$ in a single-resonance Lorentz model dielectric is illustrated in Fig. 5.5 for a highly lossy medium. For comparison, the relative phase and group velocities are also depicted. Although the phase velocity $v_p(\omega) = \omega/\beta(\omega)$ yields superluminal values when $\omega > \omega_0$, and the group velocity $v_g(\omega) = (d\beta(\omega)/d\omega)^{-1}$ yields both superluminal and negative values in the region of anomalous dispersion that extends from the angular frequency value at which the phase velocity is a minimum to the angular frequency value at which it is a maximum, the energy transport velocity $v_E(\omega)$ is subluminal for all $\omega \in [0, +\infty)$, approaching c from below together with the group velocity in the limit as $\omega \rightarrow +\infty$ from below, whereas the phase velocity approaches c from above in this high-frequency limit. Because of this, the phase velocity appearing in Eq. (5.234) may be replaced by either the group velocity or, more precisely, the energy velocity, as has been expressed in Eq. (5.244).

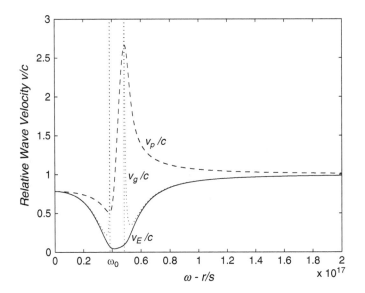

Fig. 5.5 Angular frequency dependence of the relative phase velocity v_p/c (dashed curve), group velocity v_g/c (dotted curve), and energy transport velocity v_E/c (solid curve) for a plane wave field in a single-resonance Lorentz model dielectric with medium parameters $\omega_0 = 4 \times 10^{16}$ r/s, $b = \sqrt{20} \times 10^{16}$ r/s, and $\delta = 0.28 \times 10^{16}$ r/s

Notice that the energy velocity attains a minimum value just above the resonance angular frequency ω_0 near the point where $n_i(\omega)$ attains its maximum value and that v_E remains relatively small throughout the region of anomalous dispersion that approximately extends from ω_0 to $\omega_1 \equiv \sqrt{\omega_0^2 + b^2}$.

For a multiple-resonance Lorentz model dielectric, the ratio of the expression (5.259) for the time-average magnitude of the Poynting vector to the expression given in Eq. (5.258) for the total time-average electromagnetic energy density results in the general expression [28]

$$v_E(\omega) = \frac{c}{n_r(\omega) + \frac{1}{n_r(\omega)} \sum_j \frac{b_j^2 \omega^2}{(\omega^2 - \omega_j^2)^2 + 4\delta_j^2 \omega^2}} \tag{5.267}$$

for the time-average velocity of energy transport. This result is an important generalization of the single-resonance expression (5.265) due to Loudon [26] and reduces to that result in that special case. Because the denominator appearing in Eq. (5.267) is greater than or equal to unity for all $\omega \in [0, +\infty)$, then for all physically realizable values of the angular frequency of oscillation of the monochromatic wave field, this expression for the energy transport velocity yields results that are in agreement with relativistic causality. The angular frequency dependence of the relative time-average energy velocity $v_E(\omega)/c$ in a multiple-resonance Lorentz

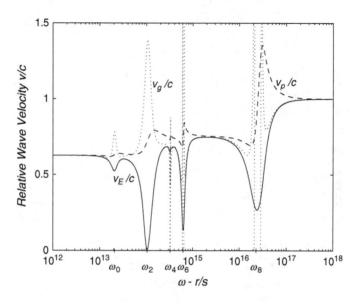

Fig. 5.6 Angular frequency dependence of the relative phase velocity v_p/c (dashed curve), group velocity v_g/c (dotted curve), and energy transport velocity v_E/c (solid curve) for a plane wave field in a multiple-resonance Lorentz model of triply distilled water with medium parameters given in Table 4.2

model of triply distilled water is illustrated in Fig. 5.6. For comparison, the relative phase and group velocities are also depicted. Notice that the relative phase velocity $v_p(\omega)/c$ remains subluminal until the angular frequency of oscillation exceeds the uppermost resonance frequency ω_8. Notice also that the relative group velocity $v_g(\omega)/c$ remains positive throughout the first two absorption bands and that it approaches the relative energy velocity in each region of normal dispersion, but differs from it near each resonance frequency ω_j where the energy velocity attains a local minimum.

5.3   Boundary Conditions

Consider a continuous, smooth surface S that forms the boundary (or interface) between two separate homogeneous, isotropic, locally linear, temporally dispersive materials, across which there may occur rapid changes in the values of the constitutive parameters $\epsilon(\omega)$, $\mu(\omega)$, and $\sigma(\omega)$ at a fixed frequency ω.[14] On a macroscopic

[14]Because the electric and magnetic material properties are temporally dispersive, there may be specific angular frequency values at which one or more of the material parameters do not change across S.

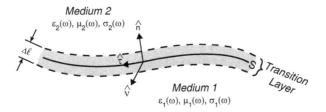

Fig. 5.7 Transition layer about the interface S separating two temporally dispersive HILL media. At each point on S, the unit vector $\hat{\mathbf{n}}$ is normal to S and directed from medium 1 into medium 2, and the mutually orthogonal unit vectors $\hat{\boldsymbol{\tau}}$ and $\hat{\boldsymbol{\upsilon}}$ are tangent to S such that $\hat{\boldsymbol{\tau}} = \hat{\boldsymbol{\upsilon}} \times \hat{\mathbf{n}}$

scale, these changes in the material properties may typically be considered to be discontinuous so that the electromagnetic field vectors themselves are expected to exhibit corresponding discontinuities across S. The discontinuous change in the constitutive parameters across S may be considered as the limiting case of the more physically realistic situation in which S is replaced by an infinitesimally thin transition layer within which the material parameters $\epsilon(\omega)$, $\mu(\omega)$, and $\sigma(\omega)$ at fixed ω vary rapidly, but continuously, from their (complex) values $\epsilon_1(\omega)$, $\mu_1(\omega)$, $\sigma_1(\omega)$ in medium 1 to their (complex) values $\epsilon_2(\omega)$, $\mu_2(\omega)$, $\sigma_2(\omega)$ in medium 2, as depicted in Fig. 5.7. The thickness of the transition layer is denoted by $\Delta\ell$ and the appropriate limit to the discontinuous change across S is obtained as $\Delta\ell \to 0$. The variation in material properties across the transition layer is assumed to be sufficiently smooth such that the electromagnetic field vectors and their first derivatives are continuous bounded functions of both position and time.

Consider first Gauss' law for the electric and magnetic fields, given respectively by the pair of integral relations

$$\oint_{\Sigma} \tilde{\mathbf{D}}(\mathbf{r}, \omega) \cdot d\mathbf{s} = \|4\pi\| \int_{R} \tilde{\varrho}(\mathbf{r}, \omega) d^3 r, \qquad (5.268)$$

$$\oint_{\Sigma} \tilde{\mathbf{B}}(\mathbf{r}, \omega) \cdot d\mathbf{s} = 0 \qquad (5.269)$$

in the temporal frequency domain, where the simply connected closed surface Σ encloses the region R, and where $d\mathbf{s}$ denotes the outward-oriented differential element of surface area on Σ. Choose this surface Σ to be a vanishingly small right circular cylinder whose generators are normal to the interface S and whose end-caps (each with surface area Δa) respectively lie in the upper and lower surfaces of the transition layer so that they are separated by the layer thickness $\Delta\ell$. For the magnetic field behavior across the interface S, let $\tilde{\mathbf{B}}_1(\mathbf{r}, \omega)$ denote the magnetic field vector at the center of the cylinder base in medium 1 with outward unit normal vector $\hat{\mathbf{n}}_1$, and let $\tilde{\mathbf{B}}_2(\mathbf{r}, \omega)$ denote the magnetic field vector at the center of the cylinder base in medium 2 with outward unit normal vector $\hat{\mathbf{n}}_2$. In the limit as both $\Delta\ell \to 0$ and

$\Delta a \to 0$, the magnetic field vectors $\tilde{\mathbf{B}}_1(\mathbf{r}, \omega)$ and $\tilde{\mathbf{B}}_2(\mathbf{r}, \omega)$ come to be evaluated at the same point but on opposite sides of S, and $\hat{\mathbf{n}}_1 \to -\hat{\mathbf{n}}$ and $\hat{\mathbf{n}}_2 \to \hat{\mathbf{n}}$, where the unit vector $\hat{\mathbf{n}}$ is normal to S and directed from medium 1 into medium 2, as depicted in Fig. 5.7. In that limit, Eq. (5.269) yields the boundary condition

$$\hat{\mathbf{n}} \cdot \left(\tilde{\mathbf{B}}_2(\mathbf{r}, \omega) - \tilde{\mathbf{B}}_1(\mathbf{r}, \omega) \right) = 0, \quad \mathbf{r} \in S, \tag{5.270}$$

stating that *the normal component of the magnetic induction field vector is continuous across any surface of discontinuity in the material parameters* $\epsilon(\omega)$, $\mu(\omega)$, *and* $\sigma(\omega)$.

For the behavior of the normal component of the electric displacement vector across the interface S, a similar analysis applied to Eq. (5.268) results in the limiting expression

$$\hat{\mathbf{n}} \cdot \left(\tilde{\mathbf{D}}_2(\mathbf{r}, \omega) - \tilde{\mathbf{D}}_1(\mathbf{r}, \omega) \right) = \|4\pi\| \lim_{\Delta\ell \to 0} \lim_{\Delta a \to 0} \int_R \tilde{\varrho}(\mathbf{r}, \omega) d^3 r.$$

In the limit as $\Delta\ell \to 0$, the charge enclosed in the region R approaches the value $\tilde{\varrho}_S(\mathbf{r}, \omega)\Delta a$, where $\tilde{\varrho}_S(\mathbf{r}, \omega)$ denotes the *free surface charge density* on the interface S, given by

$$\tilde{\varrho}_S(\mathbf{r}, \omega) \equiv \lim_{\Delta\ell \to 0} \{\tilde{\varrho}(\mathbf{r}, \omega)\Delta\ell\}. \tag{5.271}$$

With this identification, the boundary condition for the electric displacement vector becomes

$$\hat{\mathbf{n}} \cdot \left(\tilde{\mathbf{D}}_2(\mathbf{r}, \omega) - \tilde{\mathbf{D}}_1(\mathbf{r}, \omega) \right) = \|4\pi\|\tilde{\varrho}_S(\mathbf{r}, \omega), \quad \mathbf{r} \in S, \tag{5.272}$$

provided that the order of the limiting process as $\Delta\ell \to 0$ and the integration over the region R can be interchanged. Thus, *the normal component of the electric displacement vector is discontinuous across any surface of discontinuity in the material parameters* $\epsilon(\omega)$, $\mu(\omega)$, *and* $\sigma(\omega)$, *the amount of the discontinuity being proportional to the free surface charge density* $\tilde{\varrho}_S(\mathbf{r}, \omega)$ *at that point.*

Consider next Faraday's and Ampére's laws in the temporal frequency domain, given respectively by the pair of integral relations

$$\oint_C \tilde{\mathbf{E}}(\mathbf{r}, \omega) \cdot d\mathbf{l} = \frac{i\omega}{\|c\|} \int_\Sigma \tilde{\mathbf{B}}(\mathbf{r}, \omega) \cdot \hat{v} d^2 r, \tag{5.273}$$

$$\oint_C \tilde{\mathbf{H}}(\mathbf{r}, \omega) \cdot d\mathbf{l} = -\frac{i\omega}{\|c\|} \int_\Sigma \tilde{\mathbf{D}}(\mathbf{r}, \omega) \cdot \hat{v} d^2 r + \left\|\frac{4\pi}{c}\right\| \int_\Sigma \tilde{\mathbf{J}}(\mathbf{r}, \omega) \cdot \hat{v} d^2 r,$$

$$\tag{5.274}$$

where Σ denotes the open surface region enclosed by the contour C, and where $\hat{\boldsymbol{v}}$ denotes the positive unit normal vector to Σ, the direction of which is determined by the direction of integration about C. Choose the contour C to be an infinitesimally small plane rectangular loop whose sides (each with length $\Delta\ell$) are perpendicular to the interface S and whose top and bottom (each with length Δs) respectively lie in the upper and lower surfaces of the transition layer about S. Let $\hat{\boldsymbol{\tau}}_1$ and $\hat{\boldsymbol{\tau}}_2$ denote unit vectors in the direction of circulation about the contour C along the lower (medium 1) and upper (medium 2) sides, respectively, of the rectangular contour at the edge of each transition layer. In addition, let $\tilde{\mathbf{E}}_1(\mathbf{r}, \omega)$ denote the electric field value at the midpoint of the lower side of C in medium 1 with circulation vector $\hat{\boldsymbol{\tau}}_1$, and let $\tilde{\mathbf{E}}_2(\mathbf{r}, \omega)$ denote the electric field value at the midpoint of the upper side of C in medium 2 with circulation vector $\hat{\boldsymbol{\tau}}_2$. In the limit as both $\Delta\ell \to 0$ and $\Delta s \to 0$, the electric field vectors $\tilde{\mathbf{E}}_1(\mathbf{r}, \omega)$ and $\tilde{\mathbf{E}}_2(\mathbf{r}, \omega)$ come to be evaluated at the same point but on opposite sides of S, whereas $\hat{\boldsymbol{\tau}}_1 \to -\hat{\boldsymbol{\tau}}$ and $\hat{\boldsymbol{\tau}}_2 \to \hat{\boldsymbol{\tau}}$, where (see Fig. 5.7) $\hat{\boldsymbol{\tau}} \equiv \hat{\boldsymbol{v}} \times \hat{\mathbf{n}}$ defines the unit tangent vector to the surface S at that point. In this limit, Faraday's law [Eq. (5.273)] becomes

$$\hat{\boldsymbol{v}} \cdot \left[\hat{\mathbf{n}} \times \left(\tilde{\mathbf{E}}_2(\mathbf{r}, \omega) - \tilde{\mathbf{E}}_1(\mathbf{r}, \omega) \right) - \frac{i\omega}{\|c\|} \lim_{\Delta\ell \to 0} \left(\tilde{\mathbf{B}}(\mathbf{r}, \omega)\Delta\ell \right) \right] = 0, \quad \mathbf{r} \in S,$$

because $(\hat{\boldsymbol{v}} \times \hat{\mathbf{n}}) \cdot \tilde{\mathbf{E}} = \hat{\boldsymbol{v}} \cdot (\hat{\mathbf{n}} \times \tilde{\mathbf{E}})$. Because the orientation of the contour C, and hence the direction of the unit vector $\hat{\boldsymbol{v}}$, is entirely arbitrary, then

$$\hat{\mathbf{n}} \times \left(\tilde{\mathbf{E}}_2(\mathbf{r}, \omega) - \tilde{\mathbf{E}}_1(\mathbf{r}, \omega) \right) = \frac{i\omega}{\|c\|} \lim_{\Delta\ell \to 0} \left(\tilde{\mathbf{B}}(\mathbf{r}, \omega)\Delta\ell \right), \quad \mathbf{r} \in S.$$

Finally, because the field vectors and their first derivatives are assumed to be bounded, the right-hand side of this relation vanishes with $\Delta\ell$, resulting in the boundary condition

$$\hat{\mathbf{n}} \times \left(\tilde{\mathbf{E}}_2(\mathbf{r}, \omega) - \tilde{\mathbf{E}}_1(\mathbf{r}, \omega) \right) = \mathbf{0}, \quad \mathbf{r} \in S. \tag{5.275}$$

Hence, *the tangential component of the electric field intensity vector is continuous across any surface of discontinuity in the material parameters $\epsilon(\omega)$, $\mu(\omega)$, and $\sigma(\omega)$.*

For the behavior of the tangential component of the magnetic intensity vector across the interface S, a similar analysis applied to Ampére's law [Eq. (5.333)] results in the limiting expression

$$\hat{\mathbf{n}} \times \left(\tilde{\mathbf{H}}_2(\mathbf{r}, \omega) - \tilde{\mathbf{H}}_1(\mathbf{r}, \omega) \right) = \left\| \frac{4\pi}{c} \right\| \lim_{\Delta\ell \to 0} \left(\tilde{\mathbf{J}}(\mathbf{r}, \omega)\Delta\ell \right)$$

$$- \frac{i\omega}{\|c\|} \lim_{\Delta\ell \to 0} \left(\tilde{\mathbf{D}}(\mathbf{r}, \omega)\Delta\ell \right), \quad \mathbf{r} \in S.$$

Because $\tilde{\mathbf{D}}(\mathbf{r}, \omega)$ and its first derivatives are bounded, the second term on the right-hand side of this expression vanishes in the limit as $\Delta\ell \to 0$. However, the first term on the right-hand side of this expression does not necessarily vanish as it is proportional to the current $\mathcal{J} = \mathbf{J}(\mathbf{r}, t) \cdot \hat{\mathbf{v}} \Delta s \Delta \ell$ flowing through the rectangular loop C. In the limit as $\Delta\ell \to 0$, this current $\mathcal{J}$ approaches the value $\mathbf{J}_S(\mathbf{r}, t) \cdot \hat{\mathbf{n}} \Delta s$, where

$$\tilde{\mathbf{J}}_S(\mathbf{r}, \omega) \equiv \lim_{\Delta\ell \to 0} \left\{ \tilde{\mathbf{J}}(\mathbf{r}, \omega) \Delta\ell \right\} \tag{5.276}$$

is the temporal frequency spectrum of the *surface current density* on the interface S. With this identification, the above boundary condition becomes

$$\hat{\mathbf{n}} \times \left(\tilde{\mathbf{H}}_2(\mathbf{r}, \omega) - \tilde{\mathbf{H}}_1(\mathbf{r}, \omega) \right) = \left\| \frac{4\pi}{c} \right\| \tilde{\mathbf{J}}_S(\mathbf{r}, \omega) \quad \mathbf{r} \in S. \tag{5.277}$$

Hence, *the presence of a surface current on the interface S across which the material parameters $\epsilon(\omega)$, $\mu(\omega)$, and $\sigma(\omega)$ change discontinuously, results in a discontinuous change in the tangential component of the magnetic field intensity vector, the amount of the discontinuity being proportional to the surface current density $\tilde{\mathbf{J}}_S(\mathbf{r}, \omega)$ at that point.*

The surface charge and current densities are not independent, but rather are related by the equation of continuity given in Eq. (4.56), taken in the limit as $\Delta\ell \to 0$, whose temporal Fourier transform yields

$$\nabla \cdot \tilde{\mathbf{J}}_S(\mathbf{r}, \omega) - i\omega \tilde{\varrho}_S(\mathbf{r}, \omega) = 0. \tag{5.278}$$

Because of this relationship, it is necessary only to apply the boundary conditions given in Eqs. (5.275) and (5.277) on the tangential components of the electric and magnetic field vectors, the boundary conditions given in Eqs. (5.270) and (5.272) for the normal components then being automatically satisfied.

5.3.1 Boundary Conditions for Nonconducting Dielectric Media

For a pure dielectric material, the conductivity is identically zero at all frequencies; viz.,

$$\sigma(\omega) = 0 \quad \forall \omega. \tag{5.279}$$

Such a dielectric cannot then furnish free charge, so that if no excess charge is externally supplied to the interface between two such dielectric materials, then the surface charge density must vanish; viz.,

$$\tilde{\varrho}_S(\mathbf{r}, \omega) = 0 \quad \forall \omega, \quad \mathbf{r} \in S. \tag{5.280}$$

The equation of continuity (5.278) for the surface charge and current densities then requires that the surface current density be solenoidal. However, because the conductivity for each material is identically zero, the constitutive relation given in Eq. (4.113) then requires that the conduction current density vanish in each medium, so that

$$\tilde{\mathbf{J}}_S(\mathbf{r}, \omega) = 0 \quad \forall \omega, \quad \mathbf{r} \in S \qquad (5.281)$$

on the interface between these two media. The boundary conditions for the electric and magnetic field vectors across an interface surface S separating two nonconducting dielectric media with material properties $\epsilon_1(\omega)$, $\mu_1(\omega)$ in medium 1 and $\epsilon_2(\omega)$, $\mu_2(\omega)$ in medium 2 are then given by the set of relations

$$\hat{\mathbf{n}} \cdot \left(\tilde{\mathbf{D}}_2(\mathbf{r}, \omega) - \tilde{\mathbf{D}}_1(\mathbf{r}, \omega) \right) = 0, \quad \mathbf{r} \in S, \qquad (5.282)$$

$$\hat{\mathbf{n}} \cdot \left(\tilde{\mathbf{B}}_2(\mathbf{r}, \omega) - \tilde{\mathbf{B}}_1(\mathbf{r}, \omega) \right) = 0, \quad \mathbf{r} \in S, \qquad (5.283)$$

$$\hat{\mathbf{n}} \times \left(\tilde{\mathbf{E}}_2(\mathbf{r}, \omega) - \tilde{\mathbf{E}}_1(\mathbf{r}, \omega) \right) = \mathbf{0}, \quad \mathbf{r} \in S, \qquad (5.284)$$

$$\hat{\mathbf{n}} \times \left(\tilde{\mathbf{H}}_2(\mathbf{r}, \omega) - \tilde{\mathbf{H}}_1(\mathbf{r}, \omega) \right) = \mathbf{0}, \quad \mathbf{r} \in S. \qquad (5.285)$$

This set of relations can then be used to determine the bending of the electric and magnetic lines of force across the interface S.

5.3.2 Boundary Conditions for a Dielectric–Conductor Interface

Consider first the case in which medium 1 is a perfect conductor and medium 2 is nonconducting, so that $\sigma_1(\omega) = \infty$ and $\sigma_2(\omega) = 0$ for all ω. Let the unit normal vector $\hat{\mathbf{n}}$ to the interface S be directed from the perfectly conducting medium 1 into the nonconducting medium 2, as depicted in Fig. 5.7. In order that the conduction current density in a perfect conductor have, at most, finite values, the electric field inside such an idealized conductor must identically vanish and the boundary condition given in Eq. (5.272) becomes

$$\hat{\mathbf{n}} \cdot \tilde{\mathbf{D}}_2(\mathbf{r}, \omega) = \|4\pi\| \tilde{\varrho}_S(\mathbf{r}, \omega), \quad \mathbf{r} \in S. \qquad (5.286)$$

The free charge in a perfect conductor then moves "instantaneously" in response to any applied time-varying electromagnetic field, producing the correct surface charge density to identically cancel the externally applied electric field within the body of the ideal conductor. In turn, Faraday's law then requires that the magnetic field also identically vanish inside a perfect conductor so that the boundary condition given in Eq. (5.277) becomes

$$\hat{n} \times \tilde{\mathbf{H}}_2(\mathbf{r}, \omega) = \left\| \frac{4\pi}{c} \right\| \tilde{\mathbf{J}}_S(\mathbf{r}, \omega), \quad \mathbf{r} \in S. \tag{5.287}$$

The surface charge on a perfect conductor then moves "instantaneously" in response to the tangential magnetic field to identically cancel the externally applied magnetic field within the body of the ideal conductor. The remaining two boundary conditions appearing in Eqs. (5.270) and (5.275) for the normal component of the magnetic induction field vector and the tangential component of the electric field vector, respectively, then become

$$\hat{n} \cdot \tilde{\mathbf{B}}_2(\mathbf{r}, \omega) = 0, \quad \mathbf{r} \in S, \tag{5.288}$$

$$\hat{n} \times \tilde{\mathbf{E}}_2(\mathbf{r}, \omega) = \mathbf{0}, \quad \mathbf{r} \in S. \tag{5.289}$$

The boundary conditions in Eqs. (5.286)–(5.289) then state that *only normal* **E** *and tangential* **H** *fields can exist immediately outside the surface of a perfect conductor, the field magnitudes dropping discontinuously to zero inside the perfect conductor body.*

If the conducting medium has finite conductivity, then Ohm's law [see Eqs. (4.121) and (4.122)] for a homogeneous, isotropic, locally linear conductor applies and

$$\tilde{\mathbf{J}}_1(\mathbf{r}, \omega) = \sigma_1(\omega)\tilde{\mathbf{E}}_1(\mathbf{r}, \omega) \tag{5.290}$$

in the absence of any externally supplied current source. The divergence of the temporal frequency domain form of Ampére's law [see Eq. (5.27)] yields

$$\nabla \cdot \left(\nabla \times \tilde{\mathbf{H}}_2(\mathbf{r}, \omega) \right) = 0 = -\frac{1}{\|c\|} i \omega \epsilon_{c2}(\omega) \nabla \cdot \tilde{\mathbf{E}}(\mathbf{r}, \omega),$$

where $\epsilon_{c2}(\omega) = \epsilon_2(\omega) + i\|4\pi\|\sigma_2(\omega)/\omega$ is the complex permittivity of the conducting material, so that with Gauss' law

$$\epsilon_2(\omega)\nabla \cdot \tilde{\mathbf{E}}_2(\mathbf{r}, \omega) = \tilde{\varrho}_2(\mathbf{r}, \omega), \tag{5.291}$$

one finds that $\tilde{\varrho}_2(\mathbf{r}, \omega) = 0$ in the conducting material. This result, taken together with the fact that the nonconducting dielectric medium 1 cannot furnish free charge, then leads to the conclusion that the surface charge density vanishes on S, viz.

$$\tilde{\varrho}_S(\mathbf{r}, \omega) = 0, \quad \mathbf{r} \in S. \tag{5.292}$$

The equation of continuity given in Eq. (5.278) for the surface charge and current densities then requires that

$$\nabla \cdot \tilde{\mathbf{J}}_S(\mathbf{r}, \omega) = 0, \quad \mathbf{r} \in S, \tag{5.293}$$

and the idealized surface current density is solenoidal.

5.4 Penetration Depth in a Conducting Medium

Inside the conducting medium the electric and magnetic fields are attenuated exponentially with the normal distance ξ from the interface as $e^{-\xi/d_p}$, where [see Eq. (5.84)]

$$d_p \approx \left\| \frac{c}{2\sqrt{\pi}} \right\| \left[\frac{2}{\omega \mu_1'(\omega) \sigma_1'(\omega)} \right]^{1/2} \tag{5.294}$$

is the *penetration depth* or *skin depth* of the conductor. Here $\mu_1'(\omega) \equiv \Re\{\mu_1(\omega)\}$ and $\sigma_1'(\omega) \equiv \Re\{\sigma_1(\omega)\}$ denote the real parts of the magnetic permeability and electric conductivity, respectively, in the conducting medium. Notice that this expression for the penetration depth is valid when the overall material loss is dominated by the conductive loss over the frequency domain of interest. This distance then defines [33] a physical transition layer between the electromagnetic field quantities in medium 2 just outside the conductor body above the surface S and the vanishing field quantities sufficiently deep inside the conductor body (medium 1). As a consequence, the boundary condition appearing in Eq. (5.287) for the tangential component of the magnetic field vector is replaced here by [cf. Eq. (5.285)]

$$\hat{\mathbf{n}} \times \left(\tilde{\mathbf{H}}_2(\mathbf{r}, \omega) - \tilde{\mathbf{H}}_1(\mathbf{r}, \omega) \right) = 0, \quad \mathbf{r} \in S. \tag{5.295}$$

The other tangential boundary condition given in Eq. (5.275) remains valid.

If there exists a tangential component $\tilde{\mathbf{H}}_{2\|}(\mathbf{r}, \omega)$ of the magnetic field just outside the conducting body above the surface S, then the boundary condition given in Eq. (5.295) states that the same tangential component exists just inside the conducting body below the surface S. With this boundary value, the electromagnetic field behavior in the conducting material can then be constructed from the appropriate solution to the pair of field equations

$$\nabla \times \tilde{\mathbf{E}}_1(\mathbf{r}, \omega) = \left\| \frac{1}{c} \right\| i\omega \mu_1(\omega) \tilde{\mathbf{H}}_1(\mathbf{r}, \omega), \tag{5.296}$$

$$\nabla \times \tilde{\mathbf{H}}_1(\mathbf{r}, \omega) = -\left\| \frac{1}{c} \right\| i\omega \epsilon_{c1}(\omega) \tilde{\mathbf{E}}_1(\mathbf{r}, \omega), \tag{5.297}$$

where $\epsilon_{c1}(\omega) = \epsilon_1(\omega) + i\,\|4\pi\|\,\sigma_1(\omega)/\omega$ is the complex permittivity and $\mu_1(\omega)$ the magnetic permeability of the conducting material. For sufficiently small real-valued angular frequency values ω such that [where $\epsilon_1''(\omega) \equiv \Im\{\epsilon_1(\omega)\}$]

$$\epsilon_1''(\omega) \ll \|4\pi\|\frac{\sigma_1'(\omega)}{\omega}, \tag{5.298}$$

the conductive loss dominates the dielectric loss, and because the dielectric loss is typically assumed to dominate the magnetic loss [i.e., $\mu_1''(\omega) \ll \epsilon_1''(\omega)$], the above pair of curl relations may then be approximated as

$$\tilde{\mathbf{E}}_1(\mathbf{r}, \omega) \cong \left\|\frac{c}{4\pi}\right\| \frac{1}{\sigma_1'(\omega)} \nabla \times \tilde{\mathbf{H}}_1(\mathbf{r}, \omega), \tag{5.299}$$

$$\tilde{\mathbf{H}}_1(\mathbf{r}, \omega) \cong -\|c\|\frac{i}{\omega\mu_1'(\omega)} \nabla \times \tilde{\mathbf{E}}_1(\mathbf{r}, \omega). \tag{5.300}$$

Let the variable $\xi \geq 0$ denote the normal coordinate distance to the interface surface S into the conducting medium, where $\xi = 0$ on S. Then

$$\nabla \approx -\hat{\mathbf{n}}\frac{\partial}{\partial\xi}, \tag{5.301}$$

the remaining spatial derivatives of the field quantities being assumed negligible within the conducting medium [33]. With this substitution, the pair of curl relations in Eqs. (5.299) and (5.300) becomes

$$\tilde{\mathbf{E}}_1(\xi, \omega) \approx -\left\|\frac{c}{4\pi}\right\| \frac{1}{\sigma_1'(\omega)} \hat{\mathbf{n}} \times \frac{\partial\tilde{\mathbf{H}}_1(\xi, \omega)}{\partial\xi}, \tag{5.302}$$

$$\tilde{\mathbf{H}}_1(\xi, \omega) \approx \|c\|\frac{i}{\omega\mu_1'(\omega)} \hat{\mathbf{n}} \times \frac{\partial\tilde{\mathbf{E}}_1(\xi, \omega)}{\partial\xi}, \tag{5.303}$$

leading immediately to the (approximate) transversality relation

$$\hat{\mathbf{n}} \cdot \tilde{\mathbf{E}}_1(\xi, \omega) \approx \hat{\mathbf{n}} \cdot \tilde{\mathbf{H}}_1(\xi, \omega) \approx 0, \tag{5.304}$$

so that both $\tilde{\mathbf{E}}_1(\mathbf{r}, \omega)$ and $\tilde{\mathbf{H}}_1(\mathbf{r}, \omega)$ are (approximately) parallel to the interface surface S. Substitution of Eq. (5.302) into Eq. (5.303) then yields

$$\tilde{\mathbf{H}}_1(\xi, \omega) \approx -\left\|\frac{c^2}{4\pi}\right\| \frac{i}{\omega\mu_1'(\omega)\sigma_1'(\omega)} \frac{\partial^2}{\partial\xi^2}\left[\left(\hat{\mathbf{n}} \cdot \tilde{\mathbf{H}}_1(\xi, \omega)\right)\hat{\mathbf{n}} - \tilde{\mathbf{H}}_1(\xi, \omega)\right],$$

so that, after application of the approximate transversality relation (5.363),

$$\frac{\partial^2}{\partial \xi^2}\tilde{\mathbf{H}}_1(\xi, \omega) + \frac{2i}{d_p^2}\tilde{\mathbf{H}}_1(\xi, \omega) \approx \mathbf{0}, \tag{5.305}$$

where d_p is the penetration depth defined in Eq. (5.294). The solution of this partial differential equation that is consistent with the boundary condition given in Eq. (5.295) is then

$$\tilde{\mathbf{H}}_1(\xi, \omega) \approx \tilde{\mathbf{H}}_{2\|}(\mathbf{r}, \omega)e^{-\xi/d_p}e^{i\xi/d_p}, \quad \mathbf{r} \in S. \tag{5.306}$$

From Eq. (5.299), the electric field in the conducting material corresponding to this magnetic field solution is then given by

$$\tilde{\mathbf{E}}_1(\xi, \omega) \approx \left\| \frac{1}{2\sqrt{\pi}} \right\| \left[\frac{\omega\mu_1'(\omega)}{\sigma_1'(\omega)} \right]^{1/2} \left(\hat{\mathbf{n}} \times \tilde{\mathbf{H}}_{2\|}(\mathbf{r}, \omega) \right) e^{-\xi/d_p}e^{i\xi/d_p}e^{-i\pi/4}, \quad \mathbf{r} \in S, \tag{5.307}$$

which is shifted in phase by $-\pi/4$ from the magnetic field vector. For sufficiently small angular frequencies such that the inequality $\omega \ll \sigma_1'(\omega)/\mu_1'(\omega)$ is satisfied, the electric field strength will be negligible in comparison to the magnetic field strength in the conductor, whereas in the opposite limit when $\omega \gg \sigma_1'(\omega)/\mu_1'(\omega)$, keeping in mind the inequality in Eq. (5.298), the magnetic field strength will be negligible in comparison to the electric field strength.

5.4.1 First-Order Correction to the Normal Boundary Conditions for a Dielectric–Conductor Interface

From the boundary condition (5.275) for the tangential component of the electric field vector, there exists a tangential component of the electric field just outside of the conductor body and above the interface surface S that is given by the expression in Eq. (5.307) evaluated at $\xi = 0$, so that

$$\tilde{\mathbf{E}}_{2\|}(\mathbf{r}, \omega) \approx \left\| \frac{1}{2\sqrt{\pi}} \right\| \left[\frac{\omega\mu_1'(\omega)}{\sigma_1'(\omega)} \right]^{1/2} \left(\hat{\mathbf{n}} \times \tilde{\mathbf{H}}_{2\|}(\mathbf{r}, \omega) \right) e^{-i\pi/4}, \quad \mathbf{r} \in S. \tag{5.308}$$

Faraday's law [$\tilde{\mathbf{H}}(\mathbf{r}, \omega) = (\|c\|/i\omega\mu(\omega)) \nabla \times \tilde{\mathbf{E}}(\mathbf{r}, \omega)$] then shows that there exists a small normal component $\tilde{\mathbf{H}}_{2\perp}(\mathbf{r}, \omega)$ of the magnetic field vector just above the surface of the conducting medium. With the decomposition $\tilde{\mathbf{E}}(\mathbf{r}, \omega) = \tilde{\mathbf{E}}_\perp(\mathbf{r}, \omega) +$

$\tilde{\mathbf{E}}_{\|}(\mathbf{r}, \omega)$ into normal and longitudinal components at the interface and the identity $\nabla = \nabla_{\perp} - \hat{\mathbf{n}}\partial/\partial\xi$, Faraday's law becomes

$$\tilde{\mathbf{H}}_2(\mathbf{r}, \omega) = \frac{\|c\|}{i\omega\mu_2(\omega)}\left[\nabla_{\perp} \times \left(\tilde{\mathbf{E}}_{2\perp}(\mathbf{r}, \omega) + \tilde{\mathbf{E}}_{2\|}(\mathbf{r}, \omega)\right) - \hat{\mathbf{n}} \times \frac{\partial\tilde{\mathbf{E}}_{2\|}(\mathbf{r}, \omega)}{\partial\xi}\right].$$

The $\nabla_{\perp} \times \tilde{\mathbf{E}}_{2\perp}(\mathbf{r}, \omega)$ and $\hat{\mathbf{n}} \times \partial\tilde{\mathbf{E}}_{2\|}(\mathbf{r}, \omega)/\partial\xi$ terms appearing on the right-hand side of this expression yield the tangential magnetic field component $\tilde{\mathbf{H}}_{2\|}(\mathbf{r}, \omega)$ on the interface surface S, the second of these two terms giving the zeroth-order boundary value for the magnetic intensity vector appearing in Eqs. (5.307) and (5.308) so that the first of these two terms provides the first-order correction to this prescribed boundary value. The $\nabla_{\perp} \times \tilde{\mathbf{E}}_{2\|}(\mathbf{r}, \omega)$ term yields the desired expression for the normal component of the magnetic field vector on the interface S as $\tilde{\mathbf{H}}_{2\perp}(\mathbf{r}, \omega) = (\|c\|/i\omega\mu_2(\omega)) \nabla_T \times \tilde{\mathbf{E}}_{2\|}(\mathbf{r}, \omega)$, so that, with the result in Eq. (5.308), one obtains the *first-order correction*

$$\tilde{\mathbf{H}}_{2\perp}(\mathbf{r}, \omega) \approx \left\|\frac{c}{2\sqrt{\pi}}\right\| \left[\frac{\mu_1'(\omega)}{\omega\mu_2'(\omega)\sigma_1'(\omega)}\right]^{1/2} \hat{\mathbf{n}} \times \left(\nabla_T \times \tilde{\mathbf{H}}_{2\|}(\mathbf{r}, \omega)\right) e^{i\pi/4}, \quad \mathbf{r} \in S$$

$$(5.309)$$

to the boundary value on S. If required, an iterative procedure may be used to numerically determine the precise boundary values for a given problem.

5.4.2 Dielectric–Conductor Surface Current Density

The conduction current density $\tilde{\mathbf{J}}_c(\mathbf{r}, \omega) \approx \sigma_1'(\omega)\tilde{\mathbf{E}}_1(\mathbf{r}, \omega)$ in the conducting material is, with substitution from Eq. (5.308), given by

$$\tilde{\mathbf{J}}_c(\xi, \omega) \approx \left\|\frac{c}{4\pi}\right\| \frac{\sqrt{2}}{d_p(\omega)}\left(\hat{\mathbf{n}} \times \tilde{\mathbf{H}}_{2\|}(\mathbf{r}, \omega)\right) e^{-i\pi/4} e^{-(1-i)\xi/d_p}, \quad \mathbf{r} \in S, \quad (5.310)$$

for $\xi \geq 0$. Because this current density is effectively confined to a thin layer just inside the conductor body below the interface surface S, it is equivalent to an *effective surface current density*

$$\tilde{\mathbf{J}}_{S_{eff}}(\mathbf{r}, \omega) \equiv \int_0^{\infty} \tilde{\mathbf{J}}_c(\xi, \omega)d\xi \approx \left\|\frac{c}{4\pi}\right\| \left(\hat{\mathbf{n}} \times \tilde{\mathbf{H}}_{2\|}(\mathbf{r}, \omega)\right), \quad \mathbf{r} \in S. \quad (5.311)$$

Thus, *a good conductor behaves effectively as a perfect conductor with the idealized surface current replaced by an effective surface current that is exponentially distributed beneath the surface of the conducting body* [33].

5.4.3 Ohmic Power Loss Across a Conductor Surface

The existence of non-vanishing tangential components of the electric and magnetic field vectors at the interface surface S implies that electromagnetic power is coupled into the conducting medium across S. From the time-harmonic form of Poynting's theorem (see Sect. 5.2.3), the time-average electromagnetic power absorbed per unit area on the surface S is given by the real part of the complex Poynting vector as

$$\left\langle \frac{d P_{loss}}{da} \right\rangle \approx -\frac{1}{2} \left\| \frac{c}{4\pi} \right\| \Re \left\{ \hat{\mathbf{n}} \cdot \left(\tilde{\mathbf{E}}_{2\|}(\mathbf{r}, \omega) \times \tilde{\mathbf{H}}_{2\|}^*(\mathbf{r}, \omega) \right) \right\},$$

for $\mathbf{r} \in S$. With substitution from Eq. (5.308), this expression becomes

$$\left\langle \frac{d P_{loss}}{da} \right\rangle \approx \left\| \frac{1}{4\pi} \right\| \frac{1}{4} \omega \mu_1'(\omega) d_p(\omega) \left| \tilde{\mathbf{H}}_{2\|}(\mathbf{r}, \omega) \right|^2, \quad \mathbf{r} \in S, \tag{5.312}$$

which should be equal to the time-average ohmic power loss in the conductor [see Eq. (5.215)] per unit surface area on S. This latter quantity may be obtained from the expression

$$\left\langle \frac{d P_\Omega}{dV} \right\rangle = \frac{1}{2} \tilde{\mathbf{J}}_c(\xi, \omega) \cdot \tilde{\mathbf{E}}_1^*(\xi, \omega) \approx \frac{1}{2\sigma_1'(\omega)} \left| \tilde{\mathbf{J}}_c(\xi, \omega) \right|^2$$

$$\approx \left\| \left(\frac{c}{4\pi} \right)^2 \right\| \frac{1}{\sigma_1'(\omega) d_p^2(\omega)} \left| \tilde{\mathbf{H}}_{2\|}(\mathbf{r}, \omega) \right|^2 e^{-2\xi/d_p}, \tag{5.313}$$

for the time-average ohmic power loss per unit volume in the conductor, where $\mathbf{r} \in S$. The ohmic power loss in the conducting medium per unit area on the interface surface S is then given by

$$\left\langle \frac{d P_\Omega}{da} \right\rangle = \int_0^\infty \left\langle \frac{d P_\Omega}{dV} \right\rangle d\xi \approx \left\| \left(\frac{c}{4\pi} \right)^2 \right\| \frac{1}{\sigma_1'(\omega) d_p(\omega)} \left| \tilde{\mathbf{H}}_{2\|}(\mathbf{r}, \omega) \right|^2$$

$$= \left\| \frac{1}{4\pi} \right\| \frac{1}{4} \omega \mu_1'(\omega) d_p(\omega) \left| \tilde{\mathbf{H}}_{2\|}(\mathbf{r}, \omega) \right|^2, \quad \mathbf{r} \in S, \tag{5.314}$$

which is just that given in Eq. (5.312).

5.4.4 Current Density in a Homogeneous, Isotropic Cylindrical Conductor

A problem of considerable practical importance in both electrical and power system engineering involving low-frequency electromagnetic field effects concerns the distributed resistance and internal inductance of a transmission line conductor and the associated radial dependence of the current density in the conductor. Because this is primarily of interest in engineering, MKSA units alone are used.

For the elementary case of a solid homogeneous, isotropic wire with circular cross-section of radius a and static conductivity σ_0, the resistance per unit length at zero frequency is given by $R_0 = (\sigma_0 \pi a^2)^{-1}$. The inductance per unit length, however, is another matter. As described by Chipman [34], the distributed inductance of a conductor (either isolated or not) is comprised of two parts: that due to internal flux-current linkages, designated as $L_{int}(\omega)$, and that due to external flux-current linkages, designated as $L_{ext}(\omega)$, where $\omega = 2\pi f$ is the angular oscillation frequency of the current source.

Consider first a uniform conductor with magnetic permeability $\mu \approx \mu_0$ and circular cross-section of radius a carrying a dc ($f = 0$) current I. The differential amount of current flowing in an infinitesimal annular tube of radius $r' \le a$ and radial width dr' is then given by $(I/\pi a^2)(2\pi r' dr')$. With the definition that inductance is the magnetic flux linking a circuit per unit current flowing in the circuit, this cylindrical differential element constitutes a fraction $(2\pi r' dr')/(\pi a^2)$ of the conductor as a circuit element. Because the magnetic flux lines due to each cylindrical current element in the wire are circles concentric with the conductor, this fractional circuit element is linked by the magnetic flux (in Wb/m^2 or Tesla T)

$$B(r) = \frac{\mu I}{2\pi r} \left(\frac{\pi r^2}{\pi a^2} \right) \tag{5.315}$$

inside the conductor between r' and a (i.e., $r' \le r \le a$). The contribution to the distributed dc internal inductance $L_{int}(0)$ per unit length due to the fractional circuit consisting of the cylindrical tube with thickness dr' at radius r' is given by (in H/m)

$$dL_{int}(0) = \frac{2\pi r' dr'}{\pi a^2 I} \int_{r'}^{a} \frac{\mu I}{2\pi r} \frac{r^2}{a^2} dr = \mu \left(a^2 - r'^2 \right) r' dr' \tag{5.316}$$

The total internal dc inductance is then obtained by integrating this expression over the radial extent of the wire as

$$L_{int}(0) = \int_{0}^{a} dL_i(0) = \frac{\mu}{8\pi}, \tag{5.317}$$

and is thus independent of the conductor radius.

From Eq. (5.316), the incremental internal dc inductance per unit annular cross-sectional area is given by $dL_i(0)/(2\pi r' dr') = (\mu/2\pi)(a^2 - r'^2)$ so that the distributed internal inductance decreases monotonically as r' increases from the center $r' = 0$ to the periphery $r' = a$ of the cylindrical conductor. This result implies that, at a fixed frequency, the distributed internal reactance $X_L = i\omega L$ also decreases as r' increases from 0 to a. If an ac voltage is applied across the ends of a section of the cylindrical conductor, less current per unit area will flow in the high reactance region near the center of the conductor than in the low reactance region nearer to the periphery of the conductor. Unlike the uniform current density obtained in the zero frequency limit, the current density is now nonuniform and this effect becomes increasingly pronounced as the frequency increases. At a sufficiently high frequency the current flow will be essentially confined to a very thin layer at the conductor surface, a phenomenon known as the *skin effect*. Based upon the analysis given by Chipman [34], a more detailed description of this effect is now given for the important case of cylindrical symmetry about the z-axis wherein both the electric field and current density are functions of the radius r alone.

For a time-harmonic field with angular frequency $\omega = 2\pi f$, the phasor form of the conduction current density $J_z(r, t) = \Re \left\{ \tilde{J}_z(r) e^{-i\omega t} \right\}$ is given by Ohm's law as

$$\tilde{J}_z(r) = \sigma(\omega)\tilde{E}_z(r), \tag{5.318}$$

where $\tilde{E}_z(r)$ is the complex phasor of the electric field intensity along the cylinder axis of the conductor. The phasor form of the azimuthally-directed magnetic induction field at a radial distance r from the conductor axis produced by the total conduction current $\tilde{I}(r) = \int_0^r \tilde{J}_z(r')2\pi r'dr'$ flowing along the z-direction within that cylinder of radius r is obtained from Eq. (2.141) as

$$\tilde{B}_\varphi(r) = \frac{\mu \tilde{I}(r)}{2\pi r} = \frac{\mu}{r} \int_0^r \tilde{J}_z(r')r'dr'. \tag{5.319}$$

The phasor electric $\tilde{E}_z(r)$ and magnetic $\tilde{B}_\varphi(r)$ fields are related through the phasor form of Faraday's law as

$$\oint_C \tilde{\mathbf{E}} \cdot d\boldsymbol{\ell} = i\omega \iint_S \tilde{\mathbf{B}} \cdot \hat{\mathbf{n}} da. \tag{5.320}$$

Application of this law to a rectangular contour situated in the rz-plane of the conductor at a distance r from the conductor axis with length Δz and radial width Δr then gives

$$\left\{ \left[\tilde{E}_z(r) + \frac{\partial \tilde{E}_z(r)}{\partial r} \right] \Delta z - \tilde{E}_z(r)\Delta z \right\} \Delta r = i\omega \frac{\mu}{r} \Delta z \Delta r \int_0^r \tilde{J}_z(r')r'dr'.$$

With Ohm's law (5.318), one then obtains the integro-differential equation

$$r\frac{d\tilde{J}_z(r)}{dr} = i\omega\mu\sigma(\omega) \int_0^r \tilde{J}_z(r')r'dr' \tag{5.321}$$

for the phasor conduction current density in the conductor. Differentiation of this expression with respect to r then gives the second-order differential equation

$$\frac{d^2 \tilde{J}_z(r)}{dr^2} + \frac{1}{r}\frac{d\tilde{J}_z(r)}{dr} - i\omega\mu\sigma(\omega)\tilde{J}_z(r) = 0, \qquad (5.322)$$

which is a modified form of Bessel's equation of order zero, with solution

$$\tilde{J}_z(r) = a_1 J_0\left(\sqrt{-2i}\,(r/d_p(\omega))\right) + a_2 Y_0\left(\sqrt{-2i}\,(r/d_p(\omega))\right), \qquad (5.323)$$

where a_1 and a_2 are constants of integration, and where [cf. Eq. (5.84)]

$$d_p(\omega) \equiv \sqrt{\frac{2}{\omega\mu\sigma(\omega)}} \qquad (5.324)$$

is the *skin depth* of the conductor at angular frequency ω. Because the argument of the Bessel functions appearing in Eq. (5.323) is complex-valued, one introduces the real and imaginary parts through the *Kelvin functions*

$$ber(\xi) \equiv \Re\left\{J_0\left(\sqrt{-i}\,\xi\right)\right\}, \quad bei(\xi) \equiv \Im\left\{J_0\left(\sqrt{-i}\,\xi\right)\right\}, \qquad (5.325)$$

$$ker(\xi) \equiv \Re\left\{Y_0\left(\sqrt{-i}\,\xi\right)\right\}, \quad kei(\xi) \equiv \Im\left\{Y_0\left(\sqrt{-i}\,\xi\right)\right\}, \qquad (5.326)$$

where ξ is real-valued. With these substitutions, Eq. (5.323) becomes

$$\tilde{J}_z(r) = a_1\left[ber\left(\sqrt{2}\,r/d_p\right) + i\,bei\left(\sqrt{2}\,r/d_p\right)\right]$$
$$+ a_2\left[ker\left(\sqrt{2}\,r/d_p\right) + i\,kei\left(\sqrt{2}\,r/d_p\right)\right]. \qquad (5.327)$$

This represents the general solution for the current density for a hollow tubular conductor with r bounded away from zero. However, because $\lim_{\xi\to 0} ker(\xi) = \infty$, the coefficient a_2 must be zero for a solid tubular conductor, in which case the solution becomes

$$\tilde{J}_z(r) = a_1\left[ber\left(\sqrt{2}\,r/d_p\right) + i\,bei\left(\sqrt{2}\,r/d_p\right)\right], \quad 0 \le r \le a. \qquad (5.328)$$

Because the current penetrates into the conductor due to the external electric field at the conductors surface ($r = a$), the quantity of interest here is the current density $\tilde{J}_z(r)$ relative to the current density $\tilde{J}_z(a)$ at the surface of the conductor, given by

$$\frac{\tilde{J}_z(r)}{\tilde{J}_z(a)} = \frac{ber\left(\sqrt{2}\,r/d_p\right) + i\,bei\left(\sqrt{2}\,r/d_p\right)}{ber\left(\sqrt{2}\,a/d_p\right) + i\,bei\left(\sqrt{2}\,a/d_p\right)}. \qquad (5.329)$$

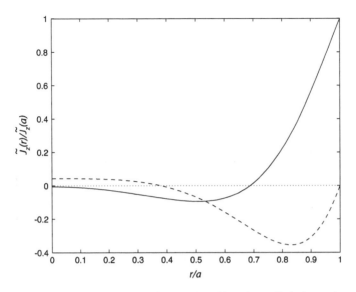

Fig. 5.8 Radial dependence of the real (solid curve) and imaginary (dashed curve) parts of the relative current density $\tilde{J}_z(r)/\tilde{J}_z(a)$ in a solid cylindrical conductor of radius a at a frequency for which $a/d_p = 5$

The radial dependence of both the real and imaginary parts of this expression for the relative current density is depicted in Figs. 5.8 and 5.9 for angular frequency values at which $a/d_p = 5$ and $a/d_p = 10$, respectively. Notice that as the skin depth d_p decreases (i.e., as a/d_p increases), the current density becomes increasingly concentrated near the surface of the conductor. In particular, since $\sigma(0) = \sigma_0$, where σ_0 denotes the static conductivity, then $d_p \to \infty$ and $a/d_p \to 0$ as $\omega \to 0$ so that the current density $\tilde{J}_z(r)$ becomes uniformly distributed (i.e. independent of r) throughout the cross-section of the wire under static conditions.

The *distributed internal impedance* Z_{int} of a solid cylindrical conductor is related to the distributed resistance R and internal inductance L_{int} of the conductor by $Z_{int} = R + i\omega L_{int}$. This complex impedance is defined as the ratio of the longitudinal phasor potential difference per unit length of the conductor at the conductor's surface, obtained from Ohm's law (5.378) as $\tilde{E}_z(a) = \tilde{J}_z(a)/\sigma(\omega)$, to the total phasor current $\tilde{I}_z$ in the conductor, so that

$$Z_{int}(\omega) = R(\omega) + i\omega L_{int}(\omega) = \frac{\tilde{J}_z(a)}{\sigma(\omega)\tilde{I}_z}. \qquad (5.330)$$

It now remains to determine an expression for the total phasor current $\tilde{I}_z$ in the cylindrical conductor. Because of the cylindrical symmetry of the conductor geometry, $\tilde{\mathbf{E}}(\mathbf{r}) = \hat{\mathbf{1}}_z \tilde{E}_z(r)$ and $\tilde{\mathbf{H}}(\mathbf{r}) = \hat{\mathbf{1}}_\varphi \tilde{H}_\varphi(r)$. From the integral form of

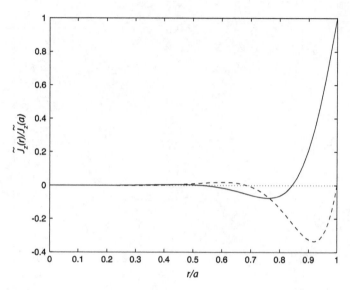

Fig. 5.9 Radial dependence of the real (solid curve) and imaginary (dashed curve) parts of the relative current density $\tilde{J}_z(r)/\tilde{J}_z(a)$ in a solid cylindrical conductor of radius a at a frequency for which $a/d_p = 10$

Ampère's law, $\tilde{I}_z$ is given by the contour integral of $\tilde{H}_\varphi(r)$ around the periphery $(r = a)$ of the cylindrical conductor, with the result

$$\tilde{I}_z = 2\pi a \tilde{H}_\varphi(a). \tag{5.331}$$

From the phasor form of Faraday's law, given by $\nabla \times \tilde{\mathbf{E}} = i\omega \tilde{\mathbf{B}}$, with $\tilde{\mathbf{B}} = \mu \tilde{\mathbf{H}}$, one obtains

$$\tilde{H}_\varphi(r) = \frac{1}{i\mu\omega} \frac{\partial \tilde{E}_z(r)}{\partial r} = \frac{1}{i\mu\sigma\omega} \frac{\partial \tilde{J}_z(r)}{\partial r}. \tag{5.332}$$

With substitution from Eq. (5.328), this expression becomes

$$\tilde{H}_\varphi(r) = \frac{\sqrt{2}}{i d_p \mu\sigma\omega} a_1 \left[ber'(\sqrt{2}r/d_p) + i bei'(\sqrt{2}r/d_p) \right],$$

for $0 \le r \le a$, where the prime denotes differentiation with respect to the argument. With this result, Eq. (5.331) for the total phasor current becomes $\tilde{I}_z = \frac{2^{3/2}\pi a}{i d_p \mu\sigma\omega} a_1 \left[ber'(\sqrt{2}a/d_p) + i bei'(\sqrt{2}a/d_p) \right]$. With a_1 given by Eq. (5.328) evaluated at $r = a$, one finally obtains

$$\tilde{I}_z = \frac{2^{3/2}\pi a}{i d_p \mu\sigma\omega} \tilde{J}_z(a) \left[\frac{ber'(\sqrt{2}a/d_p) + i bei'(\sqrt{2}a/d_p)}{ber(\sqrt{2}a/d_p) + i bei(\sqrt{2}a/d_p)} \right], \tag{5.333}$$

so that the expression (5.330) for the *distributed internal impedance* becomes

$$Z_{int}(\omega) = i \frac{R_s(\omega)}{\sqrt{2}\pi a} \left[\frac{ber\left(\sqrt{2}\,a/d_p\right) + i\,bei\left(\sqrt{2}\,a/d_p\right)}{ber'\left(\sqrt{2}\,a/d_p\right) + i\,bei'\left(\sqrt{2}\,a/d_p\right)} \right], \qquad (5.334)$$

where

$$R_s(\omega) \equiv \frac{1}{d_p(\omega)\sigma(\omega)} = \sqrt{\frac{\mu\omega}{2\sigma(\omega)}} \qquad (5.335)$$

is the *surface resistivity* (also known as the *skin resistivity*), in Ω/m^2.

5.5 Concluding Remarks

The results presented in this chapter provide a detailed description of the fundamental electromagnetic properties that appear in causal, temporally dispersive HILL media. In particular, the general properties of time-harmonic plane wave propagation in a causally dispersive HILL medium have been developed, including energy flow and dissipation. The description is valid for homogeneous, isotropic, locally linear dielectric, conducting, or semiconducting materials. These results are fundamental to both the mathematical formulation and physical interpretation of pulsed electromagnetic beam field propagation in such media. In addition, the approximate boundary conditions for a dielectric–conductor interface with finite conductivity provide the basis for analyzing the electromagnetic mode properties of dielectric-filled metallic waveguides. The analysis of waveguide modes has been extensively studied for ideal metallic waveguides [35] where mode guidance is achieved by near ideal reflection from the highly conducting waveguide walls, as well as for lossless dielectric waveguides [36] where mode guidance is achieved by total internal reflection from the lossless dielectric interface. The extension of this waveguide mode theory to fully include dispersive attenuative materials remains to be completed.

Of fundamental importance to this macroscopic theory are the conservation laws of energy and momentum. The Heaviside–Poynting theorem for the microscopic electromagnetic field as expressed by the conservation of energy relation [see Eq. (2.155)]

$$\nabla \cdot \mathbf{s}(\mathbf{r}, t) + \frac{\partial u(\mathbf{r}, t)}{\partial t} = -\mathbf{j}(\mathbf{r}, t) \cdot \mathbf{e}(\mathbf{r}, t) \qquad (5.336)$$

is a an exact mathematical consequence of the microscopic Maxwell's equations, where

$$u(\mathbf{r}, t) \equiv \frac{1}{2} \left\| \frac{1}{4\pi} \right\| \left(\epsilon_0 e^2(\mathbf{r}, t) + \mu_0 h^2(\mathbf{r}, t) \right) \qquad (5.337)$$

is the electromagnetic energy density at the space–time point $(\mathbf{r}, t)$ and where $\mathbf{s}(\mathbf{r}, t)$ is the microscopic Poynting vector defined by [see Eq. (2.154)]

$$\mathbf{s}(\mathbf{r}, t) \equiv \left\| \frac{c}{4\pi} \right\| \mathbf{e}(\mathbf{r}, t) \times \mathbf{h}(\mathbf{r}, t), \qquad (5.338)$$

which describes the flow of this electromagnetic energy density. The nonnegative quantity $\mathbf{j}(\mathbf{r}, t) \cdot \mathbf{e}(\mathbf{r}, t)$ then describes the microscopic work done per unit volume by the microscopic electromagnetic field on the local charge density $\rho(\mathbf{r}, t)$ at that space–time point through the convective current density $\mathbf{j}(\mathbf{r}, t) = \rho(\mathbf{r}, t)\mathbf{v}(\mathbf{r}, t)$.

By comparison, the conservation of energy for the macroscopic electromagnetic field in a temporally dispersive HILL medium is described by [see Eq. (5.95)]

$$\nabla \cdot \mathbf{S}(\mathbf{r}, t) + \frac{\partial \mathcal{U}'(\mathbf{r}, t)}{\partial t} = -\mathbf{J}_c(\mathbf{r}, t) \cdot \mathbf{E}(\mathbf{r}, t), \qquad (5.339)$$

where

$$\mathbf{S}(\mathbf{r}, t) \equiv \left\| \frac{c}{4\pi} \right\| \mathbf{E}(\mathbf{r}, t) \times \mathbf{H}(\mathbf{r}, t) \qquad (5.340)$$

is the macroscopic Poynting vector, and where

$$\frac{\partial \mathcal{U}'(\mathbf{r}, t)}{\partial t} = \left\| \frac{1}{4\pi} \right\| \left(\mathbf{E}(\mathbf{r}, t) \cdot \frac{\partial \mathbf{D}(\mathbf{r}, t)}{\partial t} + \mathbf{H}(\mathbf{r}, t) \cdot \frac{\partial \mathbf{B}(\mathbf{r}, t)}{\partial t} \right) \qquad (5.341)$$

is the time rate of change of the electromagnetic energy density of the coupled field–medium system. The primitive macroscopic electric and magnetic field vectors are related to their primitive microscopic counterparts through the spatial averages $\mathbf{E}(\mathbf{r}, t) \equiv \langle\langle \mathbf{e}(\mathbf{r}, t) \rangle\rangle$ and $\mathbf{B}(\mathbf{r}, t) \equiv \langle\langle \mathbf{b}(\mathbf{r}, t) \rangle\rangle$, as defined in Eqs. (4.6) and (4.7), whereas the induced macroscopic displacement and magnetic intensity vectors are related to their primitive microscopic counterparts in a temporally dispersive HILL medium through the respective convolution relations [see Eqs. (4.85) and (4.129)]

$$\mathbf{D}(\mathbf{r}, t) = \frac{1}{\epsilon_0} \int_{-\infty}^{t} \hat{\epsilon}(t - t') \langle\langle \mathbf{d}(\mathbf{r}, t') \rangle\rangle dt', \qquad (5.342)$$

$$\mathbf{H}(\mathbf{r}, t) = \mu_0 \int_{-\infty}^{t} \hat{\mu}^{-1}(t - t') \langle\langle \mathbf{h}(\mathbf{r}, t') \rangle\rangle dt'. \qquad (5.343)$$

In addition, the conduction current density is related to the microscopic electric field vector through the convolution relation [see Eq. (4.113)]

$$\mathbf{J}_c(\mathbf{r}, t) = \int_{-\infty}^{t} \hat{\sigma}(t - t') \langle\langle \mathbf{e}(\mathbf{r}, t') \rangle\rangle dt'. \qquad (5.344)$$

Because $\langle\langle f^2(\mathbf{r})\rangle\rangle \geq \langle\langle f(\mathbf{r})\rangle\rangle^2$ (see Problem 5.11), it is then seen that

$$\mathbf{S}(\mathbf{r}, t) \leq \langle\langle \mathbf{s}(\mathbf{r}, t)\rangle\rangle, \tag{5.345}$$

with similar remarks for the other terms in Poynting's theorem. Consequently, although it is tempting to say it does, the spatial average of the microscopic Heaviside–Poynting theorem (5.336) does not, in general, lead directly to the macroscopic form of the energy theorem with a direct mapping of microscopic quantities into their macroscopic equivalents. One obvious exception occurs when the medium is vacuum, in which case there is no distinction between microscopic and macroscopic quantities. Nevertheless, conservation of energy in the microscopic domain does indeed imply energy conservation in the macroscopic domain, the form of the conservation equation remaining the same in dispersive attenuative HILL media.

A similar analysis for the conservation of linear momentum in macroscopic electromagnetics for any medium except the vacuum is more complicated. This is, in part, due to the fact that the spatial average of the Lorentz force density relation (2.21) for the microscopic electromagnetic field does not yield its macroscopic counterpart in ponderable media, as [cf. Eqs. (4.57) and (4.58)]

$$\mathbf{F}_{em}(\mathbf{r}, t) \equiv \langle\langle \mathbf{f}_L(\mathbf{r}, t)\rangle\rangle = \langle\langle \rho(\mathbf{r}, t)\mathbf{e}(\mathbf{r}, t)\rangle\rangle + \left\|\frac{1}{c}\right\| \langle\langle \mathbf{j}(\mathbf{r}, t) \times \mathbf{b}(\mathbf{r}, t)\rangle\rangle$$

$$\neq \varrho(\mathbf{r}, t)\mathbf{E}(\mathbf{r}, t) + \left\|\frac{1}{c}\right\| \mathbf{J}(\mathbf{r}, t) \times \mathbf{B}(\mathbf{r}, t). \tag{5.346}$$

Nevertheless, with the substitutions $\varrho = \nabla \cdot \mathbf{D}/\|4\pi\|$ and $\mathbf{J} = \|c/4\pi\| \nabla \times \mathbf{H} - \|1/4\pi\| \partial \mathbf{D}/\partial t$, the macroscopic quantity corresponding to the microscopic Lorentz force density becomes

$$\varrho\mathbf{E} + \left\|\frac{1}{4\pi}\right\| \mathbf{J} \times \mathbf{B} = \left\|\frac{1}{4\pi}\right\| \left\{ (\nabla \cdot \mathbf{D})\mathbf{E} + \left\|\frac{1}{c}\right\| \mathbf{B} \times \frac{\partial \mathbf{D}}{\partial t} - \mathbf{B} \times (\nabla \times \mathbf{H}) \right\}, \tag{5.347}$$

where the left-hand side, referred to in the literature as the Minkowski force density [37–39] $\mathbf{F}_M = d\mathbf{p}_M/dt$ may be identified as some part of the mechanical force density acting on the macroscopic charge and current densities at the space–time point $(\mathbf{r}, t)$. Because $(\nabla \cdot \mathbf{B})\mathbf{H} = 0$ and

$$\mathbf{B} \times \frac{\partial \mathbf{D}}{\partial t} = -\frac{\partial}{\partial t}(\mathbf{D} \times \mathbf{B}) + \mathbf{D} \times \frac{\partial \mathbf{B}}{\partial t},$$

the right-hand side of the preceding equation may be rewritten as

$$\|4\pi\|\frac{d\mathbf{p}_M}{dt} = (\nabla \cdot \mathbf{D})\mathbf{E} + (\nabla \cdot \mathbf{B})\mathbf{H}$$

$$-\left\|\frac{1}{c}\right\|\frac{\partial}{\partial t}(\mathbf{D} \times \mathbf{B}) + \left\|\frac{1}{c}\right\|\mathbf{D} \times \frac{\partial \mathbf{B}}{\partial t} - \mathbf{B} \times (\nabla \times \mathbf{H}).$$

With substitution from Faraday's law that $\|1/c\|\partial\mathbf{B}/\partial t = -\nabla \times \mathbf{E}$, the above expression finally becomes

$$\|4\pi\|\frac{d\mathbf{p}_M}{dt} = (\nabla \cdot \mathbf{D})\mathbf{E} + (\nabla \cdot \mathbf{B})\mathbf{H}$$

$$-\mathbf{D} \times (\nabla \times \mathbf{E}) - \mathbf{B} \times (\nabla \times \mathbf{H}) - \left\|\frac{1}{c}\right\|\frac{\partial}{\partial t}(\mathbf{D} \times \mathbf{B}).$$

$$(5.348)$$

By analogy with its microscopic counterpart, define the *macroscopic electromagnetic momentum density* as

$$\mathbf{p}_{em}(\mathbf{r}, t) \equiv \left\|\frac{1}{4\pi c}\right\| \mathbf{D}(\mathbf{r}, t) \times \mathbf{B}(\mathbf{r}, t)$$

$$= \left\|\frac{1}{4\pi c}\right\| \int_{-\infty}^{t} \hat{\epsilon}(t - t') \int_{-\infty}^{t} \hat{\mu}(t - t'')\mathbf{E}(\mathbf{r}, t') \times \mathbf{H}(\mathbf{r}, t'')dt'dt'',$$

$$(5.349)$$

which may be rewritten as

$$\mathbf{p}_{em}(\mathbf{r}, t) = \frac{1}{c^2}\int_{-\infty}^{t} \hat{\epsilon}_{rel}(t - t') \int_{-\infty}^{t} \hat{\mu}_{rel}(t - t'')\mathbf{S}(\mathbf{r}, t', t'')dt'dt'', \qquad (5.350)$$

where $\hat{\epsilon}_{rel}(t) = \hat{\epsilon}(t)/\epsilon_0$ and $\hat{\mu}_{rel}(t) = \hat{\mu}(t)/\mu_0$ are the relative dielectric permittivity and magnetic permeability response functions, respectively, and where

$$\mathbf{S}(\mathbf{r}, t', t'') \equiv \left\|\frac{c}{4\pi}\right\| \mathbf{E}(\mathbf{r}, t') \times \mathbf{H}(\mathbf{r}, t'') \qquad (5.351)$$

is a generalized, two-time form of the macroscopic Poynting vector with temporal frequency transform $\tilde{\mathbf{S}}(\mathbf{r}, \omega) = \|c/4\pi\|\tilde{\mathbf{E}}(\mathbf{r}, \omega) \times \tilde{\mathbf{H}}(\mathbf{r}, \omega)$, which differs from the complex Poynting vector of a time-harmonic field [cf. Eq. (5.209)].

Upon returning to Eq. (5.348), one finds with the vector differential identity $\nabla(\mathbf{B} \cdot \mathbf{H}) = \mathbf{H} \cdot \nabla \mathbf{B} + \mathbf{B} \cdot \nabla \mathbf{H} + \mathbf{H} \times (\nabla \times \mathbf{B}) + \mathbf{B} \times (\nabla \times \mathbf{H})$, with an analogous relation for $\nabla(\mathbf{E} \cdot \mathbf{D})$, that

$$(\nabla \cdot \mathbf{B})\mathbf{H} - \mathbf{B} \times (\nabla \times \mathbf{H})$$
$$= \nabla \cdot \int_{-\infty}^{t} \hat{\mu}(t - t') \left[\mathbf{H}(r, t)\mathbf{H}(\mathbf{r}, t') - \frac{1}{2} \left(\mathbf{H}(r, t) \cdot \mathbf{H}(\mathbf{r}, t') \right) \underline{\mathfrak{I}} \right] dt'$$

and

$$(\nabla \cdot \mathbf{D})\mathbf{E} - \mathbf{D} \times (\nabla \times \mathbf{E})$$
$$= \nabla \cdot \int_{-\infty}^{t} \hat{\epsilon}(t - t') \left[\mathbf{E}(r, t)\mathbf{E}(\mathbf{r}, t') - \frac{1}{2} \left(\mathbf{E}(r, t) \cdot \mathbf{E}(\mathbf{r}, t') \right) \underline{\mathfrak{I}} \right] dt',$$

where $\underline{\mathfrak{I}} = \hat{\mathbf{1}}_x \hat{\mathbf{1}}_x + \hat{\mathbf{1}}_y \hat{\mathbf{1}}_y + \hat{\mathbf{1}}_z \hat{\mathbf{1}}_z$ is the unit dyadic or *idemfactor*. With these results, Eq. (5.348) becomes

$$\frac{d\mathbf{p}_M}{dt} + \frac{d\mathbf{p}_{em}}{dt}$$
$$= \frac{1}{\|4\pi\|} \nabla \cdot \int_{-\infty}^{t} \left\{ \hat{\epsilon}(t - t')\mathbf{E}(\mathbf{r}, t')\mathbf{E}(\mathbf{r}, t) + \hat{\mu}(t - t')\mathbf{H}(\mathbf{r}, t')\mathbf{H}(\mathbf{r}, t) \right.$$
$$\left. - \frac{1}{2} \left[\hat{\epsilon}(t - t')\mathbf{E}(\mathbf{r}, t') \cdot \mathbf{E}(\mathbf{r}, t) + \hat{\mu}(t - t')\mathbf{H}(\mathbf{r}, t') \cdot \mathbf{H}(\mathbf{r}, t) \right] \underline{\mathfrak{I}} \right\} dt'.$$

$$(5.352)$$

The *macroscopic Minkowski stress tensor* is defined here as

$$\underline{\mathfrak{I}}_M(\mathbf{r}, t) \equiv \frac{1}{\|4\pi\|} \int_{-\infty}^{t} \left\{ \hat{\epsilon}(t - t')\mathbf{E}(\mathbf{r}, t')\mathbf{E}(\mathbf{r}, t) + \hat{\mu}(t - t')\mathbf{H}(\mathbf{r}, t')\mathbf{H}(\mathbf{r}, t) \right.$$
$$\left. - \frac{1}{2} \left[\hat{\epsilon}(t - t')\mathbf{E}(\mathbf{r}, t') \cdot \mathbf{E}(\mathbf{r}, t) + \hat{\mu}(t - t')\mathbf{H}(\mathbf{r}, t') \cdot \mathbf{H}(\mathbf{r}, t) \right] \underline{\mathfrak{I}} \right\} dt'$$
$$= \frac{1}{\|4\pi\|} \left\{ \mathbf{D}(\mathbf{r}, t)\mathbf{E}(\mathbf{r}, t) + \mathbf{B}(\mathbf{r}, t)\mathbf{H}(\mathbf{r}, t) \right.$$
$$\left. - \frac{1}{2} \underline{\mathfrak{I}} \left[\mathbf{D}(\mathbf{r}, t) \cdot \mathbf{E}(\mathbf{r}, t) + \mathbf{B}(\mathbf{r}, t) \cdot \mathbf{H}(\mathbf{r}, t) \right] \right\}, \qquad (5.353)$$

which is seen to be the obvious generalization of the microscopic Maxwell stress tensor $\underline{\mathfrak{I}}$ given in Eq. (2.164). However, it cannot be written in the form given in Eq. (2.165) because the quantity $\frac{1}{2}[\mathbf{D}(\mathbf{r}, t) \cdot \mathbf{E}(\mathbf{r}, t) + \mathbf{B}(\mathbf{r}, t) \cdot \mathbf{H}(\mathbf{r}, t)]$ is not equal to the macroscopic electromagnetic energy density in a dispersive attenuative medium. With the above definition of the macroscopic Maxwell stress tensor $\underline{\mathfrak{T}}(\mathbf{r}, t)$, Eq. (5.352) then assumes the suggestive form

$$\frac{d\mathbf{p}_M}{dt} + \frac{d\mathbf{p}_{em}}{dt} = \nabla \cdot \underline{\mathfrak{T}}_M \tag{5.354}$$

which, when integrated over a volume V enclosed by the surface S, yields

$$\frac{d}{dt}\left(\mathbf{P}_M + \mathbf{P}_{em}\right) = \int_V \nabla \cdot \underline{\mathfrak{T}}_M d^3 r = \oint_S \hat{\mathbf{n}} \cdot \underline{\mathfrak{T}}_M d^2 r. \tag{5.355}$$

This is the macroscopic counterpart of the conservation of momentum relation given in Eq. (2.163) for the microscopic electromagnetic field. However, it is not equal to the spatial average of that microscopic relation. Nevertheless, with the definition of the Minkowski force density and stress tensor, Eqs. (5.354) and (5.355) are precise mathematical statements that directly follow from the macroscopic Maxwell's equations in a dispersive attenuative medium.

Problems

5.1 Derive the differential equation satisfied by the complex charge density $\varrho_{c\omega}(\mathbf{r}, t)$ in a temporally dispersive semiconducting material. From the solution to this equation, determine the relaxation time of the charge density in the medium and show that $\varrho_c(\mathbf{r}, t)$ must vanish in the absence of external sources.

5.2 From the symmetry relations given in Eqs. (4.104), (4.123), and (4.154) for real ω, obtain the possible forms of the symmetry relation for the complex index of refraction $n(\omega) = [\mu(\omega)\epsilon_c(\omega)/(\mu_0\epsilon_0)]^{1/2}$, where $\epsilon_c(\omega) = \epsilon(\omega) + i\|4\pi\|\sigma(\omega)/\omega$ is the complex permittivity.

5.3 From the relations $\tilde{\mathbf{e}} \cdot \nabla S = 0$ and $\tilde{\mathbf{h}} \cdot \nabla S = 0$, valid in the geometric optics limit as $k_0 \to \infty$, show that the real and imaginary parts of the electric $\tilde{\mathbf{e}}(\mathbf{r}, \omega)$ and magnetic $\tilde{\mathbf{h}}(\mathbf{r}, \omega)$ field vectors are orthogonal to the ray direction $\hat{\mathbf{s}}(\mathbf{r}, \omega)$ at every point along the ray trajectory.

5.4 Consider a homogeneous, bianisotropic, locally linear, nondispersive, nonconducting medium that is characterized by the constitutive relations

$$\mathbf{D}(\mathbf{r}) = \underline{\epsilon} \cdot \mathbf{E}(\mathbf{r}) + \underline{\xi} \cdot \mathbf{H}(\mathbf{r}),$$

$$\mathbf{B}(\mathbf{r}) = \underline{\zeta} \cdot \mathbf{E}(\mathbf{r}) + \underline{\mu} \cdot \mathbf{H}(\mathbf{r}),$$

where $\underline{\epsilon}$, $\underline{\xi}$, $\underline{\zeta}$, and $\underline{\mu}$ are complex-constant dyadics. Determine the conditions that these material tensors must satisfy if the medium is lossless.

5.5 Obtain appropriate expressions for the time-harmonic plane wave propagation and attenuation factors $\beta(\omega)$ and $\alpha(\omega)$, respectively, that are valid when $\epsilon_c'(\omega) < 0$ and when $\epsilon_c'(\omega) = 0$.

5.6 Derive the expression given in Eq. (5.152).

5.7 Derive the expression given in Eq. (5.259) for the time-average value of the magnitude of the Poynting vector for a time-harmonic plane wave field in a temporally dispersive HILL dielectric.

5.8 Show that the general expression given in Eq. (5.189) for the irreversible electromagnetic energy yields the proper loss term for a single resonance Lorentz model dielectric.

5.9 Show that the energy and group velocities are approximately equal in the normal dispersion regions of a single resonance Lorentz model dielectric.

5.10 Determine how the electric and magnetic streamlines change across an interface S separating two dissimilar lossy dielectric media.

5.11 Show that the spatial average of the square of a function of position $f(\mathbf{r})$ is greater than or equal to the square of the spatial average of that function; that is, show that

$$\langle\langle f^2(\mathbf{r})\rangle\rangle \geq \langle\langle f(\mathbf{r})\rangle\rangle^2.$$

References

1. J. R. Wait, "Project Sanguine," *Science*, vol. 178, pp. 272–275, 1972.
2. J. R. Wait, "Propagation of ELF electromagnetic waves and Project Sanguine/Seafarer," *IEEE Journal Oceanic Engineering*, vol. 2, no. 2, pp. 161–172, 1977.
3. V. G. Veselago, "The electrodynamics of substances with simultaneously negative values of ϵ and μ," *Sov. Phys. Uspekhi*, vol. 10, no. 4, pp. 509–514, 1968.
4. R. W. Ziolkowski and E. Heyman, "Wave propagation in media having negative permittivity and permeability," *Phys. Rev. E*, vol. 64, no. 5, pp. 056625–1–056625–15, 2001.
5. M. S. Sodha and A. K. Ghatak, *Inhomogeneous Optical Waveguides*. New York: Plenum Press, 1977.
6. M. Kline and I. W. Kay, *Electromagnetic Theory and Geometrical Optics*. New York: Wiley, 1965.
7. J. B. Keller, "Geometrical theory of diffraction," *J. Opt. Soc. Am.*, vol. 52, no. 2, pp. 116–130, 1962.
8. V. M. Agranovich and V. L. Ginzburg, *Crystal Optics with Spatial Dispersion, and Excitons*. Berlin-Heidelberg: Springer-Verlag, second ed., 1984.
9. D. B. Melrose and R. C. McPhedran, *Electromagnetic Processes in Dispersive Media: A Treatment Based on the Dielectric Tensor*. Cambridge: Cambridge University Press, 1991.
10. J. M. Stone, *Radiation and Optics, An Introduction to the Classical Theory*. New York: McGraw-Hill, 1963.
11. S. Bassiri, C. H. Papas, and N. Engheta, "Electromagnetic wave propagation through a dielectric-chiral interface and through a chiral slab," *J. OPt. Soc. Am. A*, vol. 5, no. 9, pp. 1450–1459, 1988.

12. G. C. Sherman and K. E. Oughstun, "Description of pulse dynamics in Lorentz media in terms of the energy velocity and attenuation of time-harmonic waves," *Phys. Rev. Lett.*, vol. 47, pp. 1451–1454, 1981.

13. G. C. Sherman and K. E. Oughstun, "Energy velocity description of pulse propagation in absorbing, dispersive dielectrics," *J. Opt. Soc. Am. B*, vol. 12, pp. 229–247, 1995.

14. N. A. Cartwright and K. E. Oughstun, "Pulse centroid velocity of the Poynting vector," *J. Opt. Soc. Am. A*, vol. 21, no. 3, pp. 439–450, 2004.

15. W. S. Franklin, "Poynting's theorem and the distribution of electric field inside and outside of a conductor carrying electric current," *Phys. Rev.*, vol. 13, no. 3, pp. 165–181, 1901.

16. J. Neufeld, "Revised formulation of the macroscopic Maxwell theory. I. Fundamentals of the proposed formulation," *Il Nuovo Cimento*, vol. LXV B, no. 1, pp. 33–68, 1970.

17. J. Neufeld, "Revised formulation of the macroscopic Maxwell theory. II. Propagation of an electromagnetic disturbance in dispersive media," *Il Nuovo Cimento*, vol. LXVI B, no. 1, pp. 51–76, 1970.

18. C. Jeffries, "A new conservation law for classical electrodynamics," *SIAM Review*, vol. 34, no. 4, pp. 386–405, 1992.

19. H. G. Schantz, "On the localization of electromagnetic energy," in *Ultra-Wideband, Short-Pulse Electromagnetics 5* (P. D. Smith and S. R. Cloude, eds.), pp. 89–96, New York: Kluwer Academic, 2002.

20. F. N. H. Robinson, "Poynting's vector: Comments on a recent paper by Clark Jeffries," *SIAM Review*, vol. 36, no. 4, pp. 633–637, 1994.

21. C. Jeffries, "Response to a commentary by F. N. H. Robinson," *SIAM Review*, vol. 36, no. 4, pp. 638–641, 1994.

22. L. D. Landau and E. M. Lifshitz, *Electrodynamics of Continuous Media*. Oxford: Pergamon, 1960. Ch. IX.

23. Y. S. Barash and V. L. Ginzburg, "Expressions for the energy density and evolved heat in the electrodynamics of a dispersive and absorptive medium," *Usp. Fiz. Nauk.*, vol. 118, pp. 523–530, 1976. [English translation: Sov. Phys.-Usp. vol. 19, 163–270 (1976)].

24. J. M. Carcione, "On energy definition in electromagnetism: An analogy with viscoelasticity," *J. Acoust. Soc. Am.*, vol. 105, no. 2, pp. 626–632, 1999.

25. M. Born and E. Wolf, *Principles of Optics*. Cambridge: Cambridge University Press, seventh (expanded) ed., 1999.

26. R. Loudon, "The propagation of electromagnetic energy through an absorbing dielectric," *Phys. A*, vol. 3, pp. 233–245, 1970.

27. R. Loudon, *The Quantum Theory of Light*. London: Oxford University Press, 1973.

28. K. E. Oughstun and S. Shen, "Velocity of energy transport for a time-harmonic field in a multiple-resonance Lorentz medium," *J. Opt. Soc. Am. B*, vol. 5, no. 11, pp. 2395–2398, 1988.

29. K. E. Oughstun and G. C. Sherman, "Propagation of electromagnetic pulses in a linear dispersive medium with absorption (the Lorentz medium)," *J. Opt. Soc. Am. B*, vol. 5, no. 4, pp. 817–849, 1988.

30. S. Shen and K. E. Oughstun, "Dispersive pulse propagation in a double-resonance Lorentz medium," *J. Opt. Soc. Am. B*, vol. 6, pp. 948–963, 1989.

31. K. E. Oughstun and G. C. Sherman, "Optical pulse propagation in temporally dispersive Lorentz media," *J. Opt. Soc. Am.*, vol. 65, no. 10, p. 1224A, 1975.

32. L. Brillouin, *Wave Propagation and Group Velocity*. New York: Academic, 1960.

33. J. D. Jackson, *Classical Electrodynamics*. New York: John Wiley & Sons, third ed., 1999.

34. R. A. Chipman, *Theory and Problems of Transmission Lines*. New York: McGraw-Hill, 1968.

35. R. E. Collin, *Field Theory of Guided Waves*. Piscataway, NJ: IEEE, second ed., 1991.

36. D. Marcuse, *Theory of Dielectric Optical Waveguides*. New York: Academic, 1974.

37. H. Minkowski, "Die Grundgleichungen für die elektromagnetischen Vorgänge in bewegten Körpen," *Nachr. Königl. Ges. Wiss. Göttingen*, pp. 53–111, 1908. Reprinted in Math. Ann, **68**, pages 472–525 (1910).

38. S. M. Barnett and R. Loudon, "On the electromagnetic force on a dielectric medium," *Journal of Physics B: Atomic, Molecular and Optical Physics*, vol. 39, no. 15, p. S671, 2006.
39. A. Shevchenko and M. Kaivola, "Electromagnetic force density and energy-momentum tensor in an arbitrary continuous medium," *Journal of Physics B: Atomic, Molecular and Optical Physics*, vol. 44, no. 17, p. 175401, 2011.

Chapter 6
Plane Wave Reflection and Refraction

"Do not Bodies and Light act mutually upon one another; that is to say, Bodies upon Light in emitting, reflecting, refracting and inflecting it, and Light upon Bodies for heating them, and putting their parts into a vibrating motion wherein heat consists?"

Sir Isaac Newton (1704).

A practical problem of fundamental importance in electromagnetic wave theory concerns the reflection and transmission of an electromagnetic wave at an interface separating two different material media. As this problem can become rather complicated for a general pulsed wave field incident upon the surface S of a dispersive body immersed in a dispersive medium, it is best to consider first the much simpler case of a time-harmonic (monochromatic) plane wave field incident upon an infinitely extended plane surface. The general solution to this problem will then form the basis for the analysis of the more general problem of a pulsed electromagnetic beam field incident upon a planar interface separating two different dispersive media. This more general problem is treated in some detail in Volume 2.

6.1 General Formulation

Let the homogeneous, isotropic, locally linear medium in which the incident and reflected wave fields reside be described by the frequency-dependent complex valued dielectric permittivity $\epsilon_1(\omega)$, electric conductivity $\sigma_1(\omega)$, and magnetic permeability $\mu_1(\omega)$, and let the homogeneous, isotropic, locally linear medium in which the transmitted (or refracted) wave field resides be described by the frequency-dependent complex valued dielectric permittivity $\epsilon_2(\omega)$, electric conductivity $\sigma_2(\omega)$, and magnetic permeability $\mu_2(\omega)$. Take the incident electromagnetic wave to be propagating along the direction specified by the unit vector $\hat{\mathbf{1}}_w$ which is

© Springer Nature Switzerland AG 2019
K. E. Oughstun, *Electromagnetic and Optical Pulse Propagation*, Springer Series in Optical Sciences 224, https://doi.org/10.1007/978-3-030-20835-6_6

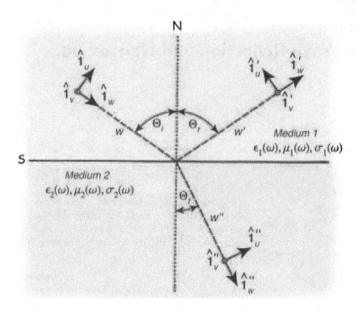

Fig. 6.1 Orientation of the incident (u, v, w), reflected (u', v', w'), and refracted (u'', v'', w'') coordinate systems in the plane of incidence at a planar interface S with normal N separating two homogeneous, isotropic, locally linear, dispersive half-spaces $j = 1, 2$ with dielectric permittivity $\epsilon_j(\omega)$, magnetic permeability $\mu_j(\omega)$, and electric conductivity $\sigma_j(\omega)$

at the angle Θ_i with respect to the normal N to the interface S, take the reflected wave to be propagating along the direction specified by the unit vector $\hat{\mathbf{1}}_w'$ which is at the angle Θ_r with respect to the surface normal N, and take the transmitted wave to be propagating along the direction specified by the unit vector $\hat{\mathbf{1}}_w''$ which is at the angle Θ_t with respect to N, as depicted in Fig. 6.1. The right-handed rectangular coordinate systems $\left(\hat{\mathbf{1}}_u, \hat{\mathbf{1}}_v, \hat{\mathbf{1}}_w\right)$, $\left(\hat{\mathbf{1}}_u', \hat{\mathbf{1}}_v', \hat{\mathbf{1}}_w'\right)$, and $\left(\hat{\mathbf{1}}_u'', \hat{\mathbf{1}}_v'', \hat{\mathbf{1}}_w''\right)$ are then defined along each of these wave directions such that the unit vectors $\hat{\mathbf{1}}_v$, $\hat{\mathbf{1}}_v'$, and $\hat{\mathbf{1}}_v''$ are each directed out of the *plane of incidence* that is defined by the unit vector $\hat{\mathbf{1}}_w$ of the incident wave and the normal N to the interface, as indicated in Fig. 6.1. If the unit vector $\hat{\mathbf{1}}_w$ is along the normal N to the interface, then the plane of incidence is not uniquely defined; in that case, any plane containing the normal N may be chosen as the plane of incidence.

Because plane wave reflection and transmission at a planar interface S is confined to the plane of incidence defined by the incident wave propagation direction and the normal N to S, the corresponding electromagnetic field analysis may be undertaken in the two-dimensional uw-space defined by the unit vectors $\hat{\mathbf{1}}_u$ and $\hat{\mathbf{1}}_w$, the electric and magnetic field vectors being independent of the independent coordinate variable v. The phasor form of the source-free Maxwell's equations given in Eqs. (5.50)–(5.53) then yield the set of relations

$$\frac{\partial \tilde{H}_u}{\partial u} + \frac{\partial \tilde{H}_w}{\partial w} = 0 \qquad\qquad \frac{\partial \tilde{E}_u}{\partial u} + \frac{\partial \tilde{E}_w}{\partial w} = 0$$

$$\frac{\partial \tilde{E}_v}{\partial w} = -\left\| \frac{1}{c} \right\| i\omega\mu(\omega)\tilde{H}_u \qquad\qquad \frac{\partial \tilde{E}_u}{\partial w} - \frac{\partial \tilde{E}_w}{\partial u} = \left\| \frac{1}{c} \right\| i\omega\mu(\omega)\tilde{H}_v$$

$$\frac{\partial \tilde{E}_v}{\partial u} = \left\| \frac{1}{c} \right\| i\omega\mu(\omega)\tilde{H}_w \qquad\qquad \frac{\partial \tilde{H}_v}{\partial w} = \left\| \frac{1}{c} \right\| i\omega\epsilon_c(\omega)\tilde{E}_u$$

$$\frac{\partial \tilde{H}_u}{\partial w} - \frac{\partial \tilde{H}_w}{\partial u} = -\left\| \frac{1}{c} \right\| i\omega\epsilon_c(\omega)\tilde{E}_v \qquad\qquad \frac{\partial \tilde{H}_v}{\partial u} = -\left\| \frac{1}{c} \right\| i\omega\epsilon_c(\omega)\tilde{E}_w.$$

Careful inspection of this set of relations reveals that they have been separated into two mutually orthogonal sets, each satisfying the orthogonality relation $\tilde{\mathbf{E}} \cdot \tilde{\mathbf{H}} = 0$. The first set is given by

$$\tilde{\mathbf{E}} = \hat{\mathbf{1}}_v \tilde{E}_v \quad \& \quad \tilde{\mathbf{H}} = \hat{\mathbf{1}}_u \tilde{H}_u + \hat{\mathbf{1}}_w \tilde{H}_w \tag{6.1}$$

satisfying the set of equations listed in the left column of equations above, and the second set is given by

$$\tilde{\mathbf{E}} = \hat{\mathbf{1}}_u \tilde{E}_u + \hat{\mathbf{1}}_w \tilde{E}_w \quad \& \quad \tilde{\mathbf{H}} = \hat{\mathbf{1}}_v \tilde{H}_v \tag{6.2}$$

satisfying the set of equations listed in the right column of equations above. Because the electric field vector in Eq. (6.1) does not have a component along the propagation direction, it is referred to as a *transverse electric* or *TE* wave field, whereas it is the magnetic field vector in Eq. (6.2) that does not have a component along the propagation direction and is consequently referred to as a *transverse magnetic* or *TM* wave field. A *transverse electromagnetic* or *TEM* wave field is one whose electric and magnetic field vectors are both orthogonal to the propagation direction, such as a plane wave [see Eqs. (5.60)–(5.62)].

For the TE wave field (6.1), the v-component of the electric field vector is found to satisfy the two-dimensional Helmholtz equation

$$\frac{\partial^2 \tilde{E}_v}{\partial u^2} + \frac{\partial^2 \tilde{E}_v}{\partial w^2} + \tilde{k}^2(\omega)\tilde{E}_v = 0, \tag{6.3}$$

with

$$\tilde{H}_u = \|c\| \frac{i}{\omega\mu(\omega)} \frac{\partial \tilde{E}_v}{\partial w}, \tag{6.4}$$

$$\tilde{H}_w = -\|c\| \frac{i}{\omega\mu(\omega)} \frac{\partial \tilde{E}_v}{\partial u}, \tag{6.5}$$

where $\tilde{k}(\omega) = (\omega/c)n(\omega)$ is the complex wave number of the host medium. For the TM wave field (6.2), the v-component of the magnetic field vector is found to satisfy the two-dimensional Helmholtz equation

$$\frac{\partial^2 \tilde{H}_v}{\partial u^2} + \frac{\partial^2 \tilde{H}_v}{\partial w^2} + \tilde{k}^2(\omega)\tilde{E}_v = 0, \tag{6.6}$$

with

$$\tilde{E}_u = -\|c\| \frac{i}{\omega \epsilon_c(\omega)} \frac{\partial \tilde{H}_v}{\partial w}, \tag{6.7}$$

$$\tilde{E}_w = \|c\| \frac{i}{\omega \epsilon_c(\omega)} \frac{\partial \tilde{H}_v}{\partial u}. \tag{6.8}$$

With regard to plane wave reflection and refraction, the boundary conditions presented in Sect. 5.3 show that two separate cases need be considered based upon the orientation of the electric and magnetic field vectors with respect to the plane of incidence depicted in Fig. 6.1. For TE-polarization, the electric field vector is linearly polarized along the v-coordinate direction, as described by Eq. (6.1) with $\tilde{H}_w = 0$ for a plane wave. This is also referred to as s (for *senkrecht*) or *perpendicular-polarization* with the electric field vector perpendicular to the plane of incidence and the magnetic field vector in the plane of incidence. For TM-polarization, the magnetic field vector is linearly polarized along the v-coordinate direction, as described by Eq. (6.2) with $\tilde{E}_w = 0$ for a plane wave. This is also referred to as p or *parallel-polarization* with the magnetic field vector perpendicular to the plane of incidence and the electric field vector in the plane of incidence. Any other state of polarization may then be described by a superposition of these two orthogonal states.

6.2 Perfect Dielectric–Conductor Interface

Consider a time-harmonic plane wave propagating in a perfect (i.e., lossless) dielectric medium and incident upon a perfect conductor ($\sigma_2 = \infty$). Let the $+z$-axis be taken along the normal N to the planar interface S at $z = 0$ separating the perfect dielectric half-space (medium 1) from the perfect conductor half-space (medium 2), directed from medium 1 into medium 2. The boundary conditions given in Eqs. (5.288) and (5.289) then apply with the subscript 2 interchanged with 1. Because the field vectors are identically zero in a perfect conductor, there is only an incident and reflected plane wave field in medium 1. For definiteness, let the plane of incidence be the xz-plane.

6.2.1   *T M-Polarization*

For a TM-polarized plane wave field in medium 1, the superposition of the incident and reflected phasor electric field vectors at any point (x, z) in the half-space $z \leq 0$ is given by

$$\tilde{\mathbf{E}}_1(x, z) = \tilde{\mathbf{E}}_u e^{ik_1 w} + \tilde{\mathbf{E}}_{u'} e^{-ik_1 w'}, \tag{6.9}$$

where $k_1 = (\omega/c)n_1(\omega)$ is the real-valued wave number in medium 1 with (real-valued) index of refraction $n_1(\omega) = \sqrt{(\epsilon_1(\omega)/\epsilon_0)(\mu_1(\omega)/\mu_0)}$. From the geometry depicted in Fig. 6.1,

$$w = x \sin \Theta_i + z \cos \Theta_i \quad \& \quad w' = -x \sin \Theta_r + z \cos \Theta_r$$

for $z \leq 0$, so that

$$\tilde{E}_x(x, z) = \tilde{E}_u \cos \Theta_i e^{ik_1(x \sin \Theta_i + z \cos \Theta_i)} - \tilde{E}_{u'} \cos \Theta_r e^{-ik_1(-x \sin \Theta_r + z \cos \Theta_r)},$$
$$\tag{6.10}$$

$$\tilde{E}_z(x, z) = -\tilde{E}_u \sin \Theta_i e^{ik_1(x \sin \Theta_i + z \cos \Theta_i)} - \tilde{E}_{u'} \sin \Theta_r e^{-ik_1(-x \sin \Theta_r + z \cos \Theta_r)}.$$
$$\tag{6.11}$$

The total magnetic field in the half-space $z \leq 0$ is then given by

$$\tilde{H}_y(x, z) = \tilde{H}_v e^{ik_1(x \sin \Theta_i + z \cos \Theta_i)} + \tilde{H}_{v'} e^{-ik_1(-x \sin \Theta_r + z \cos \Theta_r)}, \tag{6.12}$$

where $\tilde{E}_u/\tilde{H}_v = \tilde{E}_{u'}/\tilde{H}_{v'} = \eta_1(\omega)$, where $\eta_1(\omega) = \sqrt{\mu(\omega)/\epsilon(\omega)}$ is the (real-valued) intrinsic impedance of medium 1.

Application of the boundary condition in Eq. (5.289) at the interface S (the surface of the perfect conductor), which states here that $\tilde{E}_x(x, 0) = 0$ for all x, Eq. (6.10) yields the expression

$$\tilde{E}_u \cos \Theta_i e^{ik_1 x \sin \Theta_i} = \tilde{E}_{u'} \cos \Theta_r e^{ik_1 x \sin \Theta_r}.$$

This equation can be satisfied for all x only if the phase factors in the two terms are equal, and this, in turn, requires that the *law of reflection*

$$\Theta_r = \Theta_i \tag{6.13}$$

is satisfied. The above boundary condition then yields

$$\tilde{E}_{u'} = \tilde{E}_u \tag{6.14}$$

and the reflection is said to be "perfect", the reflected phasor field perfectly imaging the incident phasor field.

With these results, Eqs. (6.10)–(6.12) simplify as

$$\tilde{E}_x(x, z) = 2i\,\tilde{E}_u \cos\Theta_i \sin(k_1 z \cos\Theta_i) e^{ik_1 x \sin\Theta_i},$$

$$\tilde{E}_z(x, z) = -2\tilde{E}_u \sin\Theta_i \cos(k_1 z \cos\Theta_i) e^{ik_1 x \sin\Theta_i},\qquad (6.15)$$

$$\tilde{H}_y(x, z) = 2\frac{\tilde{E}_u}{\eta_1} \cos(k_1 z \cos\Theta_i) e^{ik_1 x \sin\Theta_i},$$

for $z \leq 0$. Notice that this total field in medium 1 has the character of a traveling wave in the x-direction, but that of a standing wave in the z-direction. The x-component of the electric field standing wave vanishes when $k_1 z \cos\Theta_i = -m\pi$, $m = 0, 1, 2, \ldots$, located at

$$z_{E_{min}} = -m\frac{\lambda_1}{2\cos\Theta_i}, \qquad m = 0, 1, 2, \ldots, \qquad (6.16)$$

where λ_1 is the wavelength in medium 1. Notice that these zero points in the electric field standing wave pattern correspond to the maxima in the magnetic field standing wave whose zeros are given by $k_1 z \cos\Theta_i = -(2m + 1)\pi/2$, $m = 0, 1, 2, \ldots$, so that

$$z_{H_{min}} = -(m + \tfrac{1}{2})\frac{\lambda_1}{2\cos\Theta_i}, \qquad m = 0, 1, 2, \ldots. \qquad (6.17)$$

Notice, in turn, that these zero points in the magnetic field standing wave pattern correspond the maxima in the electric field standing wave.

6.2.2   T E-Polarization

For a TE-polarized plane wave field in medium 1, the superposition of the incident and reflected phasor electric field vectors at any point (x, z) in the half-space $z \leq 0$ is given by

$$\tilde{\mathbf{E}}_1(x, z) = \tilde{\mathbf{E}}_v e^{ik_1 w} + \tilde{\mathbf{E}}_{v'} e^{-ik_1 w'}, \qquad (6.18)$$

so that

$$\tilde{E}_y(x, z) = \tilde{E}_v e^{ik_1(x \sin\Theta_i + z \cos\Theta_i)} + \tilde{E}_{v'} e^{ik_1(x \sin\Theta_i - z \cos\Theta_i)}, \qquad (6.19)$$

$$\tilde{H}_x(x, z) = -\frac{\tilde{E}_v}{\eta_1} \cos\Theta_i e^{ik_1(x \sin\Theta_i + z \cos\Theta_i)} + \frac{\tilde{E}_v}{\eta_1} \cos\Theta_r e^{ik_1(x \sin\Theta_i - z \cos\Theta_i)},$$

$$(6.20)$$

$$\tilde{H}_z(x, z) = \frac{\tilde{E}_v}{\eta_1} \sin\Theta_i e^{ik_1(x \sin\Theta_i + z \cos\Theta_i)} + \frac{\tilde{E}_v}{\eta_1} \sin\Theta_r e^{ik_1(x \sin\Theta_i - z \cos\Theta_i)}.$$

$$(6.21)$$

Application of the boundary condition (5.289) at the interface S, which states that $\tilde{E}_v(x, 0) = 0$ for all x then yields the law of reflection given in Eq. (6.13) as well as the relation

$$\tilde{E}_{v'} = -\tilde{E}_v \tag{6.22}$$

between the reflected and incident phasor electric fields. The reflection is again said to be "perfect", but now with a π phase-shift between the incident and reflected electric field phasors.

6.2.3 Phase Velocity with Respect to the Interface

The incident phasor electric field propagating along the w-axis is given by

$$\tilde{\mathbf{E}}_{inc}(x, z) = \tilde{\mathbf{E}}_j e^{ik_1 w} = \tilde{\mathbf{E}}_j e^{ik_1(x \sin \Theta_i + z \cos \Theta_i)}, \tag{6.23}$$

with instantaneous form

$$\mathbf{E}_{inc}(x, z, t) = \Re \left\{ \tilde{\mathbf{E}}_j e^{i(k_1 w - \omega t)} \right\}, \tag{6.24}$$

where $j = u$ for TM-polarization and $j = v$ for TE-polarization. The surfaces of constant phase $k_1 w - \omega t = C$ for this incident plane wave field in medium 1 then travel with the *phase velocity* $v_p = dw/dt$ along the positive w-direction, where $k_1 dw - \omega dt = 0$, so that

$$v_p(\omega) = \frac{\omega}{k_1(\omega)} = \frac{c}{n_1(\omega)}, \tag{6.25}$$

where $n_1(\omega)$ is the real-valued index of refraction of medium 1 [see the discussion following Eq. (6.9)].

From Eq. (6.23), the propagation factors in the x and z coordinate directions are given by

$$\beta_x(\omega) = k_1(\omega) \sin \Theta_i \quad \& \quad \beta_z(\omega) = k_1(\omega) \cos \Theta_i.$$

The instantaneous form of the incident electric field vector (6.24) is then given by

$$\mathbf{E}_{inc}(x, z, t) = \Re \left\{ \tilde{\mathbf{E}}_j e^{i(\beta_x x + \beta_z z - \omega t)} \right\}. \tag{6.26}$$

The *phase velocity* in the x-coordinate direction *parallel to the interface surface S* is then given by the velocity of propagation of the constant phase surface $\beta_x x - \omega t = C$ as

$$v_{p_x}(\omega) = \frac{\omega}{\beta_x(\omega)} = \frac{v_p}{\sin \Theta_i}, \tag{6.27}$$

and the *phase velocity* in the z-coordinate direction *perpendicular to the interface surface S* is given by the velocity of propagation of the constant phase surface $\beta_z z - \omega t = C$ as

$$v_{p_z}(\omega) = \frac{\omega}{\beta_z(\omega)} = \frac{v_p}{\cos \Theta_i}. \tag{6.28}$$

In vacuum, $v_p = c$ and the phase velocities parallel and perpendicular to the interface surface S both exceed the speed of light c in vacuum, v_{p_x} becoming infinite at normal incidence ($\Theta_i = 0$) and v_{p_z} becoming infinite at grazing incidence ($\Theta_i = \pi/2$). This is one in a series of examples purporting to demonstrate the appearance of a superluminal velocity in classical electromagnetic field theory as described by the Maxwell–Lorentz theory. Each such example does not violate Special Relativity because the wave property considered is not a physically measurable quantity, the quantity in the present example being the phase of the traveling wave.

6.3   Perfect Dielectric-Dielectric Interface

Consider a time-harmonic plane wave propagating in a perfect dielectric medium 1 with real-valued dielectric permittivity $\epsilon_1(\omega)$ and magnetic permeability $\mu_1(\omega)$ and incident upon another perfect dielectric medium 2 with real-valued dielectric permittivity $\epsilon_2(\omega)$ and magnetic permeability $\mu_2(\omega)$, as illustrated in Fig. 6.1. Define a position vector $\mathbf{r}$ with origin at a fixed point O on the interface S and let $\hat{\mathbf{n}}$ be the unit vector normal to the interface S and directed from medium 1 into medium 2.[1]

The total phasor field vectors in medium 1 are given by the superposition of the incident and reflected wave fields as

$$\tilde{\mathbf{E}}_1(\mathbf{r}) = \tilde{\mathbf{E}}_i e^{i\mathbf{k}_i \cdot \mathbf{r}} + \tilde{\mathbf{E}}_r e^{i\mathbf{k}_r \cdot \mathbf{r}}, \tag{6.29}$$

$$\tilde{\mathbf{H}}_1(\mathbf{r}) = \tilde{\mathbf{H}}_i e^{i\mathbf{k}_i \cdot \mathbf{r}} + \tilde{\mathbf{H}}_r e^{i\mathbf{k}_r \cdot \mathbf{r}}, \tag{6.30}$$

---

[1]Unlike the analysis presented in Sect. 6.2 for reflection from a perfect conductor, the present analysis is conducted in a more general framework.

and the total phasor field vectors in medium 2 are given by the transmitted wave field as

$$\tilde{\mathbf{E}}_2(\mathbf{r}) = \tilde{\mathbf{E}}_t e^{i\tilde{\mathbf{k}}_t \cdot \mathbf{r}}, \tag{6.31}$$

$$\tilde{\mathbf{H}}_2(\mathbf{r}) = \tilde{\mathbf{H}}_t e^{i\tilde{\mathbf{k}}_t \cdot \mathbf{r}}, \tag{6.32}$$

where $\tilde{\mathbf{k}}_t = \mathbf{k}_t + i\mathbf{a}_t$ for complete generality. The incident wave vector $\mathbf{k}_i = \mathbf{k}_i(\omega)$ has magnitude $k_i(\omega) = n_1(\omega)k_0$ and is directed along the positive w-axis with $k_0 \equiv \omega/c$, where $n_1(\omega)$ is the real-valued index of refraction of medium 1. The incident electric field vector phasor $\tilde{\mathbf{E}}_i$ is perpendicular to the incident wave vector $\mathbf{k}_i$ but is otherwise left arbitrary and is in general complex-valued so that the incident plane wave field can have any polarization (either TE or TM polarization or any linear combination of these two orthogonal polarizations). The remaining reflected and transmitted wave vectors and phasor field vectors are then to be determined from the boundary conditions given in Eqs. (5.282)–(5.285).

Application of these boundary conditions for any position vector $\mathbf{r}_s$ on the interface S results in the set of equations

$$\hat{\mathbf{n}} \cdot \left(\epsilon_1 \tilde{\mathbf{E}}_i e^{i\mathbf{k}_i \cdot \mathbf{r}_s} + \epsilon_1 \tilde{\mathbf{E}}_r e^{i\mathbf{k}_r \cdot \mathbf{r}_s} - \epsilon_2 \tilde{\mathbf{E}}_t e^{i\tilde{\mathbf{k}}_t \cdot \mathbf{r}_s} \right) = 0, \tag{6.33}$$

$$\hat{\mathbf{n}} \cdot \left(\mu_1 \tilde{\mathbf{H}}_i e^{i\mathbf{k}_i \cdot \mathbf{r}_s} + \mu_1 \tilde{\mathbf{H}}_r e^{i\mathbf{k}_r \cdot \mathbf{r}_s} - \mu_2 \tilde{\mathbf{H}}_t e^{i\tilde{\mathbf{k}}_t \cdot \mathbf{r}_s} \right) = 0, \tag{6.34}$$

$$\hat{\mathbf{n}} \times \left(\tilde{\mathbf{E}}_i e^{i\mathbf{k}_i \cdot \mathbf{r}_s} + \tilde{\mathbf{E}}_r e^{i\mathbf{k}_r \cdot \mathbf{r}_s} - \tilde{\mathbf{E}}_t e^{i\tilde{\mathbf{k}}_t \cdot \mathbf{r}_s} \right) = \mathbf{0}, \tag{6.35}$$

$$\hat{\mathbf{n}} \times \left(\tilde{\mathbf{H}}_i e^{i\mathbf{k}_i \cdot \mathbf{r}_s} + \tilde{\mathbf{H}}_r e^{i\mathbf{k}_r \cdot \mathbf{r}_s} - \tilde{\mathbf{H}}_t e^{i\tilde{\mathbf{k}}_t \cdot \mathbf{r}_s} \right) = \mathbf{0}. \tag{6.36}$$

Because these equations must be satisfied for all points $\mathbf{r}_s \in S$, one must then have that

$$\mathbf{k}_i \cdot \mathbf{r}_s = \mathbf{k}_r \cdot \mathbf{r}_s = \mathbf{k}_t \cdot \mathbf{r}_s \quad \& \quad \mathbf{a}_t \cdot \mathbf{r}_s = 0, \tag{6.37}$$

upon equating real and imaginary parts. Furthermore, because $\mathbf{r}_s = \hat{\mathbf{n}} \times \mathbf{r}$ for each point $\mathbf{r}_s \in S$, where the position vector $\mathbf{r}$ describes any point in space, then, for example, the first pair of equations above yields

$$(\mathbf{k}_i - \mathbf{k}_r) \cdot (\hat{\mathbf{n}} \times \mathbf{r}) = (\mathbf{k}_i \times \hat{\mathbf{n}} - \mathbf{k}_r \times \hat{\mathbf{n}}) \cdot \mathbf{r} = 0,$$

stating that $\mathbf{k}_i \times \hat{\mathbf{n}} - \mathbf{k}_r \times \hat{\mathbf{n}} = \mathbf{0}$. By this and the analogous argument applied to the remaining pairs in Eq. (6.37), one obtains

$$\mathbf{k}_i \times \hat{\mathbf{n}} = \mathbf{k}_r \times \hat{\mathbf{n}} = \mathbf{k}_t \times \hat{\mathbf{n}} \quad \& \quad \mathbf{a}_t \times \hat{\mathbf{n}} = \mathbf{0}. \tag{6.38}$$

The first part of this expression then states that each wave vector $\mathbf{k}_i$, $\mathbf{k}_r$, and $\mathbf{k}_t$ and the unit normal vector $\hat{\mathbf{n}}$ lie in one and the same plane, called the *plane of incidence*, and that this plane is unambiguously defined if the incident wave vector $\mathbf{k}_i$ is not normally incident upon S, and hence, is not parallel to $\hat{\mathbf{n}}$. This result then justifies the tacit assumption made in connection with Fig. 6.1 that the reflected and transmitted wave vectors are in the plane of incidence defined by the incident wave vector and normal N to the interface S. The second part of Eq. (6.38) states that the imaginary part $\mathbf{a}_t$ of the complex transmitted wave vector $\tilde{\mathbf{k}}_t = \mathbf{k}_t + i\mathbf{a}_t$ is parallel to $\hat{\mathbf{n}}$.

Because the incident, reflected, and transmitted wave vectors satisfy the relations [see Eqs. (5.58) and (5.59)]

$$\mathbf{k}_i \cdot \mathbf{k}_i = \mathbf{k}_r \cdot \mathbf{k}_r = k_0^2 n_1^2(\omega) \quad \& \quad \tilde{\mathbf{k}}_t \cdot \tilde{\mathbf{k}}_t = k_0^2 n_2^2(\omega) \tag{6.39}$$

where $k_0 \equiv \omega/c$, then the first pair of relations in Eq. (6.38) state that $k_0 n_1 \sin \Theta_i = k_0 n_1 \sin \Theta_r$, resulting in the *law of reflection*

$$\Theta_r = \Theta_i. \tag{6.40}$$

In addition, the equality between the first and third parts of Eq. (6.38) states that $k_0 n_1 \sin \Theta_i = k_t \sin \Theta_t$, which is only part of the law of refraction. In order to relate the real and imaginary parts of the transmitted wave vector $\tilde{\mathbf{k}}_t = \mathbf{k}_t + ia_t\hat{\mathbf{n}}$ to the incident wave vector $\mathbf{k}_i$, one first resolves each wave vector into components parallel and perpendicular to $\hat{\mathbf{n}}$ as

$$\mathbf{k}_i = (\mathbf{k}_i \cdot \hat{\mathbf{n}})\hat{\mathbf{n}} - \hat{\mathbf{n}} \times (\hat{\mathbf{n}} \times \mathbf{k}_i) = k_0 n_1 \cos \Theta_i \hat{\mathbf{n}} - \hat{\mathbf{n}} \times (\hat{\mathbf{n}} \times \mathbf{k}_i),$$

and

$$\mathbf{k}_t + i\mathbf{a}_t = (\mathbf{k}_t \cdot \hat{\mathbf{n}} + i\mathbf{a}_t \cdot \hat{\mathbf{n}})\hat{\mathbf{n}} - \hat{\mathbf{n}} \times (\hat{\mathbf{n}} \times \mathbf{k}_t + i\hat{\mathbf{n}} \times \mathbf{a}_t)$$

$$= (k_t \cos \Theta_t + ia_t)\hat{\mathbf{n}} - \hat{\mathbf{n}} \times (\hat{\mathbf{n}} \times \mathbf{k}_t),$$

because $\hat{\mathbf{n}} \times \mathbf{a}_t = 0$. Subtraction of the first from the second of these two expressions, noting from Eq. (6.38) that $\hat{\mathbf{n}} \times \mathbf{k}_i = \hat{\mathbf{n}} \times \mathbf{k}_t$, results in

$$(k_t \cos \Theta_t + ia_t)\hat{\mathbf{n}} = \mathbf{k}_t + i\mathbf{a}_t - \mathbf{k}_i + \hat{\mathbf{n}}k_0 n_1 \cos \Theta_i,$$

so that, upon taking the scalar product of this expression with itself,

$$(k_t \cos \Theta_t + ia_t)^2 = (\mathbf{k}_t + i\mathbf{a}_t) \cdot (\mathbf{k}_t + i\mathbf{a}_t) - 2\mathbf{k}_i \cdot (\mathbf{k}_t + i\mathbf{a}_t)$$

$$+2k_0 n_1 \cos \Theta_i (\mathbf{k}_t + i\mathbf{a}_t) \cdot \hat{\mathbf{n}} - 2k_0 n_1 \cos \Theta_i \mathbf{k}_i \cdot \hat{\mathbf{n}}$$

$$+\mathbf{k}_i \cdot \mathbf{k}_i + k_0^2 n_1^2 \cos^2 \Theta_i$$

$$= k_0^2 n_2^2 - 2(k_0 n_1 \cos \Theta_i \hat{\mathbf{n}} - \hat{\mathbf{n}} \times (\hat{\mathbf{n}} \times \mathbf{k}_i)) \cdot (\mathbf{k}_t + i\mathbf{a}_t)$$

$$+2k_0n_1 \cos \Theta_i k_t \cos \Theta_t + 2ik_0n_1 \cos \Theta_i a_t$$
$$-2k_0^2n_1^2 \cos^2 \Theta_i + k_0^2n_1^2 + k_0^2n_1^2 \cos^2 \Theta_i$$
$$= k_0^2n_2^2 + k_0^2n_1^2(1 - \cos^2 \Theta_i)$$
$$-2(\hat{\mathbf{n}} \times (\hat{\mathbf{n}} \times \mathbf{k}_t)) \cdot (\hat{\mathbf{n}} \times (\hat{\mathbf{n}} \times \mathbf{k}_t)).$$

Because [see Eq. (6.38)] $(\hat{\mathbf{n}} \times (\hat{\mathbf{n}} \times \mathbf{k}_t)) \cdot (\hat{\mathbf{n}} \times (\hat{\mathbf{n}} \times \mathbf{k}_t)) = (\hat{\mathbf{n}} \times (\hat{\mathbf{n}} \times \mathbf{k}_i)) \cdot (\hat{\mathbf{n}} \times (\hat{\mathbf{n}} \times \mathbf{k}_i)) = k_0^2n_1^2 \sin^2 \Theta_i$, the above expression becomes

$$(k_t \cos \Theta_t + ia_t)^2 = k_0^2n_2^2 + k_0^2n_1^2(1 - \cos^2 \Theta_i) - 2k_0^2n_1^2 \sin^2 \Theta_i$$
$$= k_0^2n_2^2 - k_0^2n_1^2 \sin^2 \Theta_i.$$

The *law of refraction* is then described by the pair of equations

$$k_t \sin \Theta_t = k_0n_1 \sin \Theta_i, \tag{6.41}$$

$$k_t \cos \Theta_t + ia_t = k_0 \left[n_2^2 - n_1^2 \sin^2 \Theta_i \right]^{1/2}. \tag{6.42}$$

The solution of this pair of equations separates into two distinct cases that depend upon the sign of the quantity appearing in the square root in Eq. (6.42):

1. If $n_2^2 - n_1^2 \sin^2 \Theta_i \geq 0$, then $a_t = 0$ and $k_t = k_0n_2$ and the law of refraction simplifies to *Snell's law*

$$n_2 \sin \Theta_t = n_1 \sin \Theta_i. \tag{6.43}$$

2. If $n_2^2 - n_1^2 \sin^2 \Theta_i < 0$, then $\Theta_t = \frac{\pi}{2}$ and

$$k_t = k_0n_1 \sin \Theta_i \quad \& \quad a_t = k_0 \left[n_1^2 \sin^2 \Theta_i - n_2^2 \right]^{1/2}. \tag{6.44}$$

These two cases suggest that a *critical angle* of incidence $\Theta_i = \Theta_c$ be defined as

$$\Theta_c \equiv \arcsin (n_2/n_1), \tag{6.45}$$

which occurs when incidence is upon the optically rarer medium ($n_1 > n_2$).

With these identifications, the tangential boundary conditions given in Eqs. (6.35) and (6.36) become

$$\left(\tilde{\mathbf{E}}_i + \tilde{\mathbf{E}}_r \right) \times \hat{\mathbf{n}} = \tilde{\mathbf{E}}_t \times \hat{\mathbf{n}}, \tag{6.46}$$

$$\left(\tilde{\mathbf{H}}_i + \tilde{\mathbf{H}}_r \right) \times \hat{\mathbf{n}} = \tilde{\mathbf{H}}_t \times \hat{\mathbf{n}}. \tag{6.47}$$

Each of these phasor field vectors may be resolved along the three orthogonal unit vectors $\hat{\mathbf{n}}$, $\hat{\boldsymbol{\tau}}$, and $\hat{\boldsymbol{\upsilon}}$ which form the basis for a right-handed coordinate system in $\mathbb{R}^3$, where $\hat{\mathbf{n}} = \hat{\boldsymbol{\tau}} \times \hat{\boldsymbol{\upsilon}}$ is normal to the interface and directed from medium 1 into medium 2, $\hat{\boldsymbol{\tau}}$ is on the interface S and parallel ($\parallel$) to the plane of incidence, and $\hat{\boldsymbol{\upsilon}} = \hat{\mathbf{1}}_\upsilon$ is on S and perpendicular ($\perp$) to the plane of incidence (see Fig. 6.1). In terms of the incident wave vector $\mathbf{k}_i$ and the unit normal vector $\hat{\mathbf{n}}$ to S, the other two unit vectors are unambiguously given by

$$\hat{\boldsymbol{\upsilon}} = \frac{\hat{\mathbf{n}} \times \mathbf{k}_i}{|\hat{\mathbf{n}} \times \mathbf{k}_i|} \quad \& \quad \hat{\boldsymbol{\tau}} = \hat{\boldsymbol{\upsilon}} \times \hat{\mathbf{n}}. \tag{6.48}$$

For notational convenience, the interface S may be situated in the xy-plane at $z = 0$ with $\hat{\mathbf{n}} = \hat{\mathbf{1}}_z$ and the plane of incidence chosen to be parallel to the xz-plane so that $\hat{\boldsymbol{\tau}} = \hat{\mathbf{1}}_x$ and $\hat{\boldsymbol{\upsilon}} = \hat{\mathbf{1}}_y$; one may always change back to the $\hat{\boldsymbol{\tau}}$, $\hat{\boldsymbol{\upsilon}}$, $\hat{\mathbf{n}}$ unit vector notation through a simple substitution. The desired field vector resolutions are then given by

$$
\begin{aligned}
\tilde{\mathbf{E}}_i &= \tilde{E}_{in}\hat{\mathbf{n}} + \tilde{E}_{ix}\hat{\mathbf{1}}_x + \tilde{E}_{iy}\hat{\mathbf{1}}_y, & \tilde{\mathbf{H}}_i &= \tilde{H}_{in}\hat{\mathbf{n}} + \tilde{H}_{ix}\hat{\mathbf{1}}_x + \tilde{H}_{iy}\hat{\mathbf{1}}_y, \\
\tilde{\mathbf{E}}_r &= \tilde{E}_{rn}\hat{\mathbf{n}} + \tilde{E}_{rx}\hat{\mathbf{1}}_x + \tilde{E}_{ry}\hat{\mathbf{1}}_y, & \tilde{\mathbf{H}}_r &= \tilde{H}_{rn}\hat{\mathbf{n}} + \tilde{H}_{rx}\hat{\mathbf{1}}_x + \tilde{H}_{ry}\hat{\mathbf{1}}_y, \\
\tilde{\mathbf{E}}_t &= \tilde{E}_{tn}\hat{\mathbf{n}} + \tilde{E}_{tx}\hat{\mathbf{1}}_x + \tilde{E}_{ty}\hat{\mathbf{1}}_y, & \tilde{\mathbf{H}}_t &= \tilde{H}_{tn}\hat{\mathbf{n}} + \tilde{H}_{tx}\hat{\mathbf{1}}_x + \tilde{H}_{ty}\hat{\mathbf{1}}_y,
\end{aligned}
\tag{6.49}
$$

with the corresponding wave vector decompositions

$$
\begin{aligned}
\mathbf{k}_i &= k_{in}\hat{\mathbf{n}} + k_{ix}\hat{\mathbf{1}}_x, \\
\mathbf{k}_r &= k_{rn}\hat{\mathbf{n}} + k_{rx}\hat{\mathbf{1}}_x, \\
\mathbf{k}_t &= k_{tn}\hat{\mathbf{n}} + k_{tx}\hat{\mathbf{1}}_x \quad \& \quad \mathbf{a}_t = a_t\hat{\mathbf{n}},
\end{aligned}
\tag{6.50}
$$

where a_t is determined by the appropriate solution of Eq. (6.42) given in Eqs. (6.43) and (6.44). The relationship between the incident, reflected, and transmitted field vectors given in Eq. (6.49) then naturally separates into two mutually orthogonal fields: a *transverse electric* (TE) field whose electric field vector is perpendicular to the plane of incidence and whose magnetic field vector is then parallel to the plane of incidence, and a *transverse magnetic* (TM) field whose magnetic field vector is perpendicular to the plane of incidence and whose electric field is then parallel to the plane of incidence, as described earlier.

6.3.1 TE-Polarization

For *TE-polarization*, the incident phasor electric field vector in medium 1 is directed along the $\hat{\boldsymbol{\upsilon}} = \hat{\mathbf{1}}_y$ direction as

$$\tilde{\mathbf{E}}_i = \hat{\mathbf{1}}_y E_s e^{i\mathbf{k}_i \cdot \mathbf{r}}, \tag{6.51}$$

so that, from the time-harmonic form (5.51) of Faraday's law, the corresponding magnetic field vector is given by [see Eqs. (5.61)–5.64)]

$$\tilde{\mathbf{H}}_i = \frac{\|c\|}{\mu_1 \omega} \mathbf{k}_i \times \hat{\mathbf{1}}_y E_s e^{i\mathbf{k}_i \cdot \mathbf{r}}$$

$$= \frac{\|c\|}{\mu_1 \omega} \left(-k_{in} \hat{\mathbf{1}}_x + k_{ix} \hat{\mathbf{1}}_z \right) E_s e^{i\mathbf{k}_i \cdot \mathbf{r}}, \tag{6.52}$$

where E_s is, in general, complex-valued. Here

$$\mathbf{k}_i = n_1 k_0 (\hat{\mathbf{1}}_x \sin \Theta_i + \hat{\mathbf{1}}_z \cos \Theta_i) \tag{6.53}$$

is the incident wave vector with $k_0 = \omega/c$ the free-space wave number, and where $k_{in} = n_1 k_0 \cos \Theta_i$ and $k_{ix} = n_1 k_0 \sin \Theta_i$. The reflected plane wave phasor field vectors in medium 1 are then given by

$$\tilde{\mathbf{E}}_r = \hat{\mathbf{1}}_y \Gamma_{TE} E_s e^{i\mathbf{k}_r \cdot \mathbf{r}}, \tag{6.54}$$

$$\tilde{\mathbf{H}}_r = \frac{\|c\|}{\mu_1 \omega} \mathbf{k}_r \times \hat{\mathbf{1}}_y \Gamma_{TE} E_s e^{i\mathbf{k}_r \cdot \mathbf{r}}$$

$$= \frac{\|c\|}{\mu_1 \omega} \left(k_{rn} \hat{\mathbf{1}}_x + k_{rx} \hat{\mathbf{1}}_z \right) \Gamma_{TE} E_s e^{i\mathbf{k}_r \cdot \mathbf{r}}, \tag{6.55}$$

where $\Gamma_{TE} \equiv E_r/E_s$ is the *reflection coefficient* for TE-polarization, and

$$\mathbf{k}_r = n_1 k_0 (\hat{\mathbf{1}}_x \sin \Theta_r - \hat{\mathbf{1}}_z \cos \Theta_r) \tag{6.56}$$

is the reflected wave vector with $k_{rn} = n_1 k_0 \cos \Theta_r$ and $k_{rx} = n_1 k_0 \sin \Theta_r$, where $\Theta_r = \Theta_i$ by the law of reflection given in Eq. (6.40). Finally, the transmitted plane wave phasor field vectors in medium 2 are given by

$$\tilde{\mathbf{E}}_t = \hat{\mathbf{1}}_y \tau_{TE} E_s e^{i\tilde{\mathbf{k}}_t \cdot \mathbf{r}}, \tag{6.57}$$

$$\tilde{\mathbf{H}}_t = \frac{\|c\|}{\mu_2 \omega} \tilde{\mathbf{k}}_t \times \hat{\mathbf{1}}_y \tau_{TE} E_s e^{i\tilde{\mathbf{k}}_t \cdot \mathbf{r}}$$

$$= \frac{\|c\|}{\mu_2 \omega} \left(-(k_{tn} + i a_t) \hat{\mathbf{1}}_x + k_{tx} \hat{\mathbf{1}}_z \right) \tau_{TE} E_s e^{i\tilde{\mathbf{k}}_t \cdot \mathbf{r}}, \tag{6.58}$$

where $\tau_{TE} \equiv E_t/E_s$ is the *transmission coefficient* for TE-polarization, and

$$\tilde{\mathbf{k}}_t = \mathbf{k}_t + i a_t \hat{\mathbf{n}} = n_2 k_0 (\hat{\mathbf{1}}_x \sin \Theta_t + \hat{\mathbf{1}}_z \cos \Theta_t) + i a_t \hat{\mathbf{1}}_z \tag{6.59}$$

is the transmitted complex wave vector with components $k_{tn} = n_2 k_0 \cos \Theta_t$, $k_{tx} = n_2 k_0 \sin \Theta_t$, and $a_t = 0$ when $n_2^2 - n_1^2 \sin^2 \Theta_i \geq 0$, while $k_{tn} = 0$, $k_{tx} = k_0 n_1 \sin \Theta_i$, and $a_t = k_0 \left[n_1^2 \sin^2 \Theta_i - n_2^2 \right]^{1/2}$ when $n_2^2 - n_1^2 \sin^2 \Theta_i < 0$, as described by the law of refraction in Eqs. (6.41)–(6.44).

The total field in medium 1 at the interface S is given by the superposition of the incident and reflected TE-polarized wave fields as

$$\tilde{\mathbf{E}}_1 = \tilde{\mathbf{E}}_i + \tilde{\mathbf{E}}_r = \hat{\mathbf{1}}_y E_s (1 + \Gamma_{TE}) e^{i \mathbf{k}_i \cdot \mathbf{r}_s}, \tag{6.60}$$

$$\tilde{\mathbf{H}}_1 = \tilde{\mathbf{H}}_i + \tilde{\mathbf{H}}_r$$

$$= \frac{E_s}{\eta_1} \left(\hat{\mathbf{1}}_x (\Gamma_{TE} - 1) \cos \Theta_i + \hat{\mathbf{1}}_z (\Gamma_{TE} + 1) \sin \Theta_i \right) e^{i \mathbf{k}_i \cdot \mathbf{r}_s}, \tag{6.61}$$

where $\eta_1 = (\mu_1/\epsilon_1)^{1/2}$ is the intrinsic impedance of medium 1, and the total field in medium 2 at the interface S is just the transmitted field given in Eqs. (6.57) and (6.58) evaluated at $\mathbf{r} = \mathbf{r}_s$. For both *critical* and *subcritical angles of incidence* $\Theta_i \le \Theta_c$, Eq. (6.58) becomes

$$\tilde{\mathbf{H}}_t = \tau_{TE} \frac{E_s}{\eta_2} \left(-\hat{\mathbf{1}}_x \cos \Theta_t + \hat{\mathbf{1}}_z \sin \Theta_t \right) e^{i \tilde{\mathbf{k}}_t \cdot \mathbf{r}_s}, \tag{6.62}$$

where $\eta_2 = (\mu_2/\epsilon_2)^{1/2}$ is the intrinsic impedance of medium 2. The tangential boundary conditions in Eqs. (6.46) and (6.47) then yield

$$1 + \Gamma_{TE} = \tau_{TE},$$

$$\frac{1}{\eta_1} (\Gamma_{TE} - 1) \cos \Theta_i = -\frac{1}{\eta_2} \tau_{TE} \cos \Theta_t,$$

with solution

$$\Gamma_{TE} = \frac{\frac{1}{\eta_1} \cos \Theta_i - \frac{1}{\eta_2} \cos \Theta_t}{\frac{1}{\eta_1} \cos \Theta_i + \frac{1}{\eta_2} \cos \Theta_t}, \tag{6.63}$$

$$\tau_{TE} = 1 + \Gamma_{TE} = \frac{\frac{2}{\eta_1} \cos \Theta_i}{\frac{1}{\eta_1} \cos \Theta_i + \frac{1}{\eta_2} \cos \Theta_t}, \tag{6.64}$$

provided that $\Theta_i \le \Theta_c$. This pair of equations constitutes the *TE-polarization reflection and transmission coefficients*.

At normal incidence ($\Theta_i = \Theta_t = 0$), $\Gamma_{TE} = (\eta_2 - \eta_1)/(\eta_2 + \eta_1)$ and $\tau_{TE} = 2\eta_2/(\eta_2 + \eta_1)$. Notice then that $|\Gamma_{TE}| \le 1$ always and that $|\tau_{TE}| \le 1$ when $\eta_1 \ge \eta_2$ whereas $|\tau_{TE}| > 1$ when $\eta_2 > \eta_1$, all at normal incidence. The result that the magnitude of the transmission coefficient will be greater than unity at normal incidence when $\eta_2 > \eta_1$ does not violate any conservation rule as the field amplitude is not a conserved quantity. The incident, reflected, and transmitted electromagnetic power must, however, be conserved; this aspect of the problem is addressed later on in this section through consideration of the reflectivity and transmissivity for the Poynting vector per unit area.

For *supercritical angles of incidence* $\Theta_i > \Theta_c$, Eq. (5.393) becomes

$$
\begin{aligned}
\tilde{\mathbf{H}}_t &= \frac{\|c\|}{\mu_2 \omega} \tau_{TE} E_s \left(-\hat{\mathbf{1}}_x i a_t + \hat{\mathbf{1}}_z k_0 n_1 \sin \Theta_i \right) e^{i \mathbf{k}_t \cdot \mathbf{r}_s} \\
&= \tau_{TE} \frac{E_s}{n_2} \left[-\hat{\mathbf{1}}_x i \left(\frac{n_1^2}{n_2^2} \sin^2 \Theta_i - 1 \right)^{1/2} + \hat{\mathbf{1}}_z \frac{n_1}{n_2} \sin \Theta_i \right],
\end{aligned}
\tag{6.65}
$$

and the tangential boundary condition (5.382) becomes

$$
\frac{1}{\eta_1} (\Gamma_{TE} - 1) \cos \Theta_i = -i \frac{1}{\eta_2} \tau_{TE} \left(\frac{n_1^2}{n_2^2} \sin^2 \Theta_i - 1 \right)^{1/2},
\tag{6.66}
$$

the tangential boundary condition for the electric field vector remaining the same as before so that $1 + \Gamma_{TE} = \tau_{TE}$. Substitution of this expression in the above equation then leads to the solution

$$
\Gamma_{TE} = \frac{\frac{1}{\eta_1} \cos \Theta_i - \frac{i}{\eta_2} \left(\frac{n_1^2}{n_2^2} \sin^2 \Theta_i - 1 \right)^{1/2}}{\frac{1}{\eta_1} \cos \Theta_i + \frac{i}{\eta_2} \left(\frac{n_1^2}{n_2^2} \sin^2 \Theta_i - 1 \right)^{1/2}},
$$

so that

$$
\Gamma_{TE} = e^{-i\phi_s},
\tag{6.67}
$$

$$
\tau_{TE} = 1 + \Gamma_{TE} = 1 + e^{-i\phi_s},
\tag{6.68}
$$

when $\Theta_i > \Theta_c$, where

$$
\phi_s \equiv 2 \arctan \left(\frac{\mu_1 \left(\sin^2 \Theta_i - \frac{n_2^2}{n_1^2} \right)^{1/2}}{\mu_2 \cos \Theta_i} \right).
\tag{6.69}
$$

Hence, at supercritical angles of incidence, $|\Gamma_{TE}| = 1$ and the phenomenon of *total internal reflection* occurs, an essential ingredient in low-loss fiber optics communication systems. At the *critical angle of incidence* $\Theta_i = \Theta_c$, $\Theta_t = \pi/2$, $\Gamma_{TE} = 1$ and $\tau_{TE} = 2$. Furthermore, in the limit of *grazing incidence* as $\Theta_i \to \pi/2$, $\Gamma_{TE} \to -1$ and $\tau_{TE} \to 0$.

6.3.2 TM-Polarization

For *TM-polarization*, the phasor magnetic field vector in medium 1 is directed along the $\hat{v} = \hat{1}_y$ direction as

$$\tilde{\mathbf{H}}_i = \hat{1}_y H_p e^{i\mathbf{k}_i \cdot \mathbf{r}}, \tag{6.70}$$

so that, from the time-harmonic form (5.53) of Ampére's law, the corresponding electric field vector is given by

$$\tilde{\mathbf{E}}_i = -\frac{\eta_1}{n_1 k_0} \mathbf{k}_i \times \hat{1}_y H_p e^{i\mathbf{k}_i \cdot \mathbf{r}} \ ,$$

$$= \frac{\eta_1}{n_1 k_0} \left(k_{in}\hat{1}_x - k_{ix}\hat{1}_z \right) H_p e^{i\mathbf{k}_i \cdot \mathbf{r}}, \tag{6.71}$$

where H_p is, in general, complex-valued, and where the incident wave vector is given in Eq. (5.388). The reflected phasor magnetic field vector is directed along the $-\hat{v} = -\hat{1}_y$ direction so that the reflected plane wave phasor field vectors in medium 1 are given by

$$\tilde{\mathbf{H}}_r = -\hat{1}_y \Gamma_{TM} H_p e^{i\mathbf{k}_r \cdot \mathbf{r}}, \tag{6.72}$$

$$\tilde{\mathbf{E}}_r = \frac{\eta_1}{n_1 k_0} \mathbf{k}_r \times \hat{1}_y \Gamma_{TM} H_p e^{i\mathbf{k}_r \cdot \mathbf{r}}$$

$$= \frac{\eta_1}{n_1 k_0} \left(k_{rn}\hat{1}_x + k_{rx}\hat{1}_z \right) \Gamma_{TM} H_p e^{i\mathbf{k}_i \cdot \mathbf{r}}, \tag{6.73}$$

where $\Gamma_{TM} \equiv H_r / H_p$ is the *reflection coefficient* for *TM*-polarization, and where the reflected wave vector $\mathbf{k}_r$ is given in Eq. (6.56). Finally, the transmitted plane wave phasor field vectors in medium 2 are given by

$$\tilde{\mathbf{H}}_t = \hat{1}_y \tau_{TM} H_p e^{i\mathbf{k}_t \cdot \mathbf{r}}, \tag{6.74}$$

$$\tilde{\mathbf{E}}_t = -\frac{\eta_2}{n_2 k_0} \mathbf{k}_t \times \hat{1}_y \tau_{TM} H_p e^{i\mathbf{k}_t \cdot \mathbf{r}}$$

$$= \frac{\eta_2}{n_2 k_0} \left((k_{tn} + i a_t)\hat{1}_x - k_{tx}\hat{1}_z \right) \tau_{TM} H_p e^{i\mathbf{k}_t \cdot \mathbf{r}}, \tag{6.75}$$

where $\tau_{TM} \equiv H_t / H_p$ is the *transmission coefficient* for *TM*-polarization, and where the transmitted wave vector $\mathbf{k}_t$ is given in Eq. (6.59).

The total field in medium 1 at the interface S is given by the superposition of the incident and reflected TM-polarized wave fields as

$$\tilde{\mathbf{H}}_1 = \tilde{\mathbf{H}}_i + \tilde{\mathbf{H}}_r = \hat{\mathbf{1}}_y (1 - \Gamma_{TM}) H_p e^{i\mathbf{k}_i \cdot \mathbf{r}_s}, \tag{6.76}$$

$$\tilde{\mathbf{E}}_1 = \tilde{\mathbf{E}}_i + \tilde{\mathbf{E}}_r$$

$$= \eta_1 \left(\hat{\mathbf{1}}_x (1 + \Gamma_{TM}) \cos \Theta_i - \hat{\mathbf{1}}_z (1 - \Gamma_{TM}) \sin \Theta_i \right) H_p e^{i\mathbf{k}_i \cdot \mathbf{r}_s}, \tag{6.77}$$

and the total field in medium 2 at the interface S is just the transmitted field given in Eqs. (6.74) and (6.75) evaluated at $\mathbf{r} = \mathbf{r}_s$. For both *critical* and *subcritical angles of incidence* $\Theta_i \leq \Theta_c$, Eq. (6.75) becomes

$$\tilde{\mathbf{E}}_t = \eta_2 \left(\hat{\mathbf{1}}_x \cos \Theta_t - \hat{\mathbf{1}}_z \sin \Theta_t \right) \tau_{TM} H_p e^{i\mathbf{k}_i \cdot \mathbf{r}_s}. \tag{6.78}$$

The tangential boundary conditions in Eqs. (6.46) and (6.47) become

$$1 - \Gamma_{TM} = \tau_{TM},$$

$$\eta_1 (1 + \Gamma_{TM}) \cos \Theta_i = \eta_2 \tau_{TM} \cos \Theta_t,$$

with solution

$$\Gamma_{TM} = \frac{\eta_2 \cos \Theta_t - \eta_1 \cos \Theta_i}{\eta_1 \cos \Theta_i + \eta_2 \cos \Theta_t}, \tag{6.79}$$

$$\tau_{TM} = 1 - \Gamma_{TM} = \frac{2\eta_1 \cos \Theta_i}{\eta_1 \cos \Theta_i + \eta_2 \cos \Theta_t}, \tag{6.80}$$

provided that $\Theta_i \leq \Theta_c$. This pair of equations constitutes the *TM-polarization reflection and transmission coefficients*. By comparison, the subcritical E-field referenced reflection and transmission coefficients (used in optics) for TM-polarization are given by Γ_{TM} and $(\eta_2/\eta_1)\tau_{TM}$, respectively.

At normal incidence ($\Theta_i = \Theta_t = 0$), $\Gamma_{TM} = (\eta_2 - \eta_1)/(\eta_2 + \eta_1)$ and $\tau_{TM} = 2\eta_1/(\eta_2 + \eta_1)$. Notice then that $|\Gamma_{TM}| \leq 1$ always and that $|\tau_{TM}| \leq 1$ when $\eta_1 < \eta_2$ whereas $|\tau_{TM}| > 1$ when $\eta_1 > \eta_2$, all at normal incidence.

For *supercritical angles of incidence* $\Theta_i > \Theta_c$, Eq. (6.75) becomes

$$\tilde{\mathbf{E}}_t = \frac{\eta_2}{n_2 k_0} \tau_{TM} H_p \left(\hat{\mathbf{1}}_x i \alpha_t - \hat{\mathbf{1}}_z k_0 n_1 \sin \Theta_i \right) e^{i\mathbf{k}_t \cdot \mathbf{r}}$$

$$= \frac{n_1}{n_2} \eta_2 \tau_{TM} H_p \left[\hat{\mathbf{1}}_x i \left(\sin^2 \Theta_i - \frac{n_2^2}{n_1^2} \right)^{1/2} - \hat{\mathbf{1}}_z \sin \Theta_i \right] e^{i\mathbf{k}_t \cdot \mathbf{r}}, \tag{6.81}$$

and the tangential boundary condition (6.46) becomes

$$\eta_1(1+\Gamma_{TM})\cos\Theta_i = i\frac{n_1}{n_2}\eta_2\tau_{TM}\left(\sin^2\Theta_i - \frac{n_2^2}{n_1^2}\right)^{1/2}, \tag{6.82}$$

the tangential boundary condition for the magnetic field vector remaining the same as before so that $\tau_{TM} = 1 - \Gamma_{TM}$. Substitution of this expression into the above equation then leads to the solution

$$\Gamma_{TM} = -\frac{\epsilon_2\cos\Theta_i - i\epsilon_1\left(\sin^2\Theta_i - (n_2^2/n_1^2)\right)^{1/2}}{\epsilon_2\cos\Theta_i + i\epsilon_1\left(\sin^2\Theta_i - (n_2^2/n_1^2)\right)^{1/2}},$$

so that

$$\Gamma_{TM} = -e^{-i\phi_p} = e^{i(\pi-\phi_p)}, \tag{6.83}$$

$$\tau_{TM} = 1 - \Gamma_{TM} = 1 + e^{-i\phi_p}, \tag{6.84}$$

when $\Theta_i > \Theta_c$, where

$$\phi_p \equiv 2\arctan\left(\frac{\epsilon_1\left(\sin^2\Theta_i - (n_2^2/n_1^2)\right)^{1/2}}{\epsilon_2\cos\Theta_i}\right). \tag{6.85}$$

By comparison, the supercritical E-field referenced reflection and transmission coefficients for TM-polarization are again given by Γ_{TM} and $(\eta_2/\eta_1)\tau_{TM}$.

Just as for TE-polarization, at supercritical angles of incidence, $|\Gamma_{TM}| = 1$ and the phenomenon of *total internal reflection* occurs. At the critical angle of incidence $\Theta_i = \Theta_c$, $\Theta_t = \pi/2$, $\Gamma_{TM} = -1$ and $\tau_{TM} = 2$. At grazing incidence $\Theta_i \to \pi/2$, $\Gamma_{TM} \to +1$ amd $\tau_{TM} \to 0$.

6.3.3 Reflectivity and Transmissivity

For the relation between the electromagnetic power that is incident, reflected, and transmitted across the planar interface S separating two ideal (i.e., lossless) dielectrics, one must consider the time-average or, equivalently, the complex Poynting vector for each plane wave field. From Eq. (5.228), the complex Poynting vector for a time-harmonic plane wave is given by

$$\tilde{\mathbf{S}}_j(\mathbf{r}) = \left\|\frac{c}{4\pi}\right\|\frac{1}{2n_jn_jk_0}\left|\tilde{\mathbf{E}}_j(\mathbf{r})\right|^2\mathbf{k}_j, \tag{6.86}$$

where $\eta_j = (\mu_j/\epsilon_j)^{1/2}$ is the impedance and $\mathbf{k}_j = n_j k_0 \hat{\mathbf{s}}$ the wave vector in medium j with refractive index $n_j = [(\epsilon_j/\epsilon_0)(\mu_j)(\mu_0)]^{1/2}$. From an elementary geometrical construction, the electromagnetic power in the plane wave incident per unit area on the interface S is given by [1]

$$J_i = S_i \cos \Theta_i = \left\| \frac{c}{4\pi} \right\| \frac{1}{2\eta_1} |E_i|^2 \cos \Theta_i, \tag{6.87}$$

while for the reflected and transmitted plane waves, the electromagnetic power per unit area leaving the interface S is given by

$$J_r = S_r \cos \Theta_r = \left\| \frac{c}{4\pi} \right\| \frac{1}{2\eta_1} |E_r|^2 \cos \Theta_i, \tag{6.88}$$

$$J_t = S_t \cos \Theta_t = \left\| \frac{c}{4\pi} \right\| \frac{1}{2\eta_2} |E_t|^2 \cos \Theta_t, \tag{6.89}$$

respectively. The plane wave *reflectivity* and *transmissivity* are then defined as

$$\mathcal{R} \equiv \frac{J_r}{J_i} = |r|^2 \tag{6.90}$$

and

$$\mathcal{T} \equiv \frac{J_t}{J_i} = \frac{\eta_1 \cos \Theta_t}{\eta_2 \cos \Theta_i} |t|^2, \tag{6.91}$$

respectively, where $r = E_r/E$ is the polarization appropriate reflection coefficient and $t = E_t/E_i$ is the corresponding transmission coefficient. In particular, for TE-polarization, $r = \Gamma_{TE}$ and $t = \tau_{TE}$, whereas for TM-polarization, $r = \Gamma_{TM}$ and $t = (\eta_2/\eta_1)\tau_{TM}$. Because $J_r + J_t = J_i$ for conservation of energy, then

$$\mathcal{R} + \mathcal{T} = 1. \tag{6.92}$$

This result may be directly verified for both TE and TM polarizations with substitution from the appropriate set of reflection and transmission coefficients. Notice that, for both critical and supercritical angles of incidence $\Theta_i \geq \Theta_c$, $\Theta_t = \pi/2$ and $\mathcal{R} = 1$ and $\mathcal{T} = 0$, independent of the polarization state. As a consequence, there is no net power transmission into medium 2 at supercritical angles of incidence, another key factor in low-loss fiber optic communication systems. For example, for TE-polarization with $\Theta_i > \Theta_c$, the transmitted phasor electric field vector (5.392) becomes

$$\tilde{\mathbf{E}}_t = \hat{\mathbf{1}}_y 2 E_s \cos (\phi_s/2) e^{-k_0 z (n_1^2 \sin^2 \Theta_i - n_2^2)^{1/2}} e^{i(n_1 k_0 x \sin \Theta_i - \phi_s/2)}$$

for $z > 0$, so that both the field vectors and the complex Poynting vector in medium 2 decay exponentially with increasing distance away from the interface. Such a wave field is said to be *evanescent*.

6.3.4 Nonmagnetic Media

For the case of two nonmagnetic, perfect (i.e., lossless) dielectric media with $\mu_1 = \mu_2 = \mu_0$ (the permeability of free-space), in which case $\eta_1/\eta_2 = n_2/n_1$, the E-field reflection and transmission coefficients given in Eqs. (6.63) and (6.64) for TE-polarization (s or $\perp$ polarization) become

$$r_\perp \equiv r_s = \frac{n_1 \cos\Theta_i - n_2 \cos\Theta_t}{n_1 \cos\Theta_i + n_2 \cos\Theta_t}, \tag{6.93}$$

$$t_\perp \equiv t_s = 1 + r_s = \frac{2n_1 \cos\Theta_i}{n_1 \cos\Theta_i + n_2 \cos\Theta_t}, \tag{6.94}$$

and the E-field reflection and transmission coefficients obtained from Eqs. (6.79) and (6.80) for TM-polarization (p or $\parallel$ polarization) are

$$r_\parallel \equiv r_p = \frac{n_1 \cos\Theta_t - n_2 \cos\Theta_i}{n_2 \cos\Theta_i + n_1 \cos\Theta_t}, \tag{6.95}$$

$$t_\parallel \equiv t_p = \frac{n_1}{n_2}(1 - r_p) = \frac{2n_1 \cos\Theta_i}{n_2 \cos\Theta_i + n_1 \cos\Theta_t}, \tag{6.96}$$

for critical and subcritical angles of incidence $\Theta_i \le \Theta_c$, where $n_j = \sqrt{\epsilon_j/\epsilon_0}$. These are known as the *Fresnel reflection and transmission coefficients* in both acoustics [2] and optics [3]. With substitution from the law of refraction that $n_1/n_2 = \sin\Theta_t/\sin\Theta_i$, these Fresnel reflection and transmission coefficients become

$$r_s = \frac{\sin(\Theta_t - \Theta_i)}{\sin(\Theta_t + \Theta_i)} \quad \& \quad t_s = \frac{2\sin\Theta_t \cos\Theta_i}{\sin(\Theta_t + \Theta_i)}, \tag{6.97}$$

$$r_p = \frac{\tan(\Theta_i - \Theta_t)}{\tan(\Theta_i + \Theta_t)} \quad \& \quad t_p = \frac{2\sin\Theta_t \cos\Theta_i}{\sin(\Theta_i + \Theta_t)\cos(\Theta_i - \Theta_t)}, \tag{6.98}$$

for $\Theta_i \le \Theta_c$. The reflectivity and transmissivity for s-polarization are then

$$\mathcal{R}_\perp \equiv \mathcal{R}_s = \frac{(n_1 \cos\Theta_i - n_2 \cos\Theta_t)^2}{(n_1 \cos\Theta_i + n_2 \cos\Theta_t)^2}, \tag{6.99}$$

$$\mathcal{T}_\perp \equiv \mathcal{T}_s = \frac{4n_1 n_2 \cos\Theta_i \cos\Theta_t}{(n_1 \cos\Theta_i + n_2 \cos\Theta_t)^2}, \tag{6.100}$$

and the reflectivity and transmissivity for p-polarization are

$$\mathcal{R}_\| \equiv \mathcal{R}_p = \frac{(n_2 \cos \Theta_i - n_1 \cos \Theta_t)^2}{(n_2 \cos \Theta_i + n_1 \cos \Theta_t)^2}, \tag{6.101}$$

$$\mathcal{T}_\| \equiv \mathcal{T}_p = \frac{4 n_1 n_2 \cos \Theta_i \cos \Theta_t}{(n_2 \cos \Theta_i + n_1 \cos \Theta_t)^2}. \tag{6.102}$$

6.3.4.1 Incidence on the Optically Denser Medium ($n_2 > n_1$)

When incidence is on the optically denser medium, then Eq. (6.43) always applies, $a_t = 0$ for all angles of incidence $\Theta_i \in [0, \pi/2)$ and the transmitted time-harmonic plane wave is of the homogeneous type with surfaces of constant phase and surfaces of constant amplitude both orthogonal to the wave vector direction. The angle of refraction Θ_t is then given by *Snell's law*[2]

$$n_2 \sin \Theta_t = n_1 \sin \Theta_i. \tag{6.103}$$

The angular dependence of the Fresnel reflection and transmission coefficients with $n_2 = (3/2)n_1$ is illustrated in Figs. 6.2 and 6.3, respectively. For comparison, the *H-field reflection and transmission coefficients* for TM-polarization are described by the dotted curves in these two figures, where

$$\Gamma_{TM} = r_p = \frac{n_1 \cos \Theta_t - n_2 \cos \Theta_i}{n_1 \cos \Theta_t + n_2 \cos \Theta_i}, \tag{6.104}$$

$$\tau_{TM} = 1 - \Gamma_{TM} = \frac{n_2}{n_1} t_p = \frac{2 n_2 \cos \Theta_i}{n_1 \cos \Theta_t + n_2 \cos \Theta_i}, \tag{6.105}$$

respectively. At *normal incidence* on the interface ($\Theta_i = \Theta_t = 0$), the Fresnel reflection and transmission coefficients for parallel and perpendicular polarizations are equal, where

$$r_{s,p} = \frac{n_1 - n_2}{n_1 + n_2} \quad \& \quad t_{s,p} = \frac{2 n_1}{n_1 + n_2}, \quad (\Theta_i = 0).$$

---

[2] Willebrord Snellius (1580–1626), a Dutch mathematician and astronomer, rediscovered the law of refraction in 1621, commonly referred to as *Snell's law* but also known as the *Snell–Descarte law* due to Descarte's independent heuristic derivation in his 1637 essay *Dioptrics*. This law of refraction was first described by the Greco-Egyptian mathematician Ptolemy (c. 100–170), was later described by the Persian mathematician Ibn Sahl (c. 940–1000) in his 984 treatise *On Burning Mirrors and Lenses*, and subsequently described by the Silesian friar and natural philosopher Erazmus Witelo in 1284.

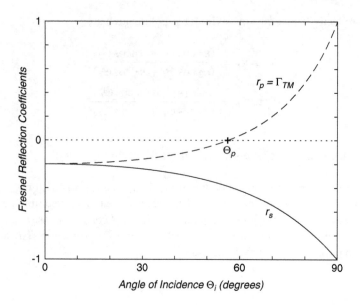

Fig. 6.2 Parallel (p) and perpendicular (s) Fresnel reflection coefficients $r_p = r_\parallel$ and $r_s = r_\perp$, respectively, when incidence is on the optically denser medium with $n_2 = (3/2)n_1$

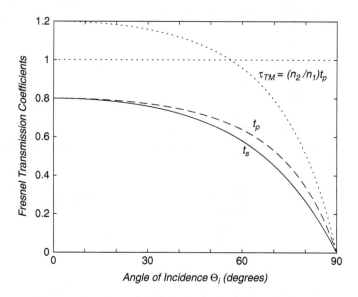

Fig. 6.3 Parallel (p) and perpendicular (s) Fresnel transmission coefficients $t_p = t_\parallel$ and $t_s = t_\perp$, respectively, when incidence is on the optically denser medium with $n_2 = (3/2)n_1$. The dotted curve describes the magnetic field transmission coefficient $\tau_{TM} = \frac{n_2}{n_1} t_p$

The Fresnel reflection coefficient r_s for perpendicular (s)-polarization then decreases with increasing angle of incidence Θ_i from its normal incidence value while the Fresnel reflection coefficient r_p for parallel (p)-polarization increases with increasing angle of incidence Θ_i from its normal incidence value, the perpendicular coefficient r_s approaching the value -1 as Θ_i approaches *grazing incidence* ($\Theta_i \to \pi/2$) while the parallel component r_p approaches the value $+1$ at grazing incidence, vanishing when $\Theta_i = \Theta_p$. Both transmission coefficients t_s and t_p [as well as $\tau_{TM} = (n_2/n_1)t_p$] decrease monotonically to 0 as Θ_i increases to $\pi/2$. In addition, the reflected s-polarized plane wave experiences a π-phase shift relative to the incident plane wave for all angles of incidence while the reflected p-polarized plane wave experiences a π-phase shift relative to the incident plane wave only for angles of incidence $\Theta_i < \Theta_p$. The transmission coefficients t_p and t_s, on the other hand, are positive for all angles of incidence Θ_i, vanishing only at grazing incidence ($\Theta_i = \pi/2$). The corresponding reflectivity and transmissivity for $TE(s)$ and $TM(p)$ polarizations are depicted in Figs. 6.4 and 6.5, respectively.

When the electric field is polarized parallel to the plane of incidence (p-polarization), the reflection coefficient r_p vanishes when $\Theta_i = \Theta_p$, where [from Eq. (6.95)]

$$n_1 \cos \Theta_t = n_2 \cos \Theta_p.$$

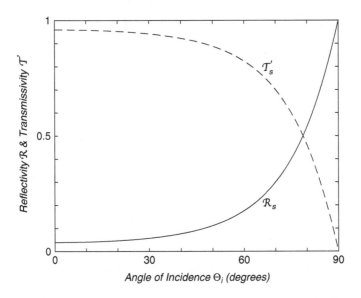

Fig. 6.4 Reflectivity $\mathcal{R}_s = \mathcal{R}_\perp$ and transmissivity $\mathcal{T}_s = \mathcal{T}_\perp$ for $TE(s)$-polarization when incidence is on the optically denser medium with $n_2 = (3/2)n_1$

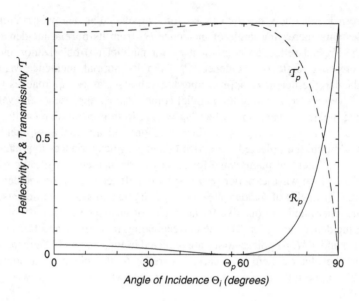

Fig. 6.5 Reflectivity $\mathcal{R}_p = \mathcal{R}_\parallel$ and transmissivity $\mathcal{T}_p = \mathcal{T}_\parallel$ for $TM(p)$-polarization when incidence is on the optically denser medium with $n_2 = (3/2)n_1$

Upon squaring this equation and using Snell's law, which states here that $n_1 \sin \Theta_p = n_2 \sin \Theta_t$, one obtains

$$\frac{n_2^2}{n_1^2} + \frac{n_1^2}{n_2^2} \tan^2 \Theta_p = \frac{1}{\cos^2 \Theta_p} = 1 + \tan^2 \Theta_p,$$

which may then be solved for the tangent of Θ_p, with the result

$$\tan \Theta_p = \frac{n_2}{n_1}. \tag{6.106}$$

This angle of incidence is known as *Brewster's angle* or the *polarizing angle* because the reflected wave due to an incident wave with any state of polarization is entirely *s*-polarized. Furthermore, upon multiplying the defining equation $n_1 \cos \Theta_t = n_2 \cos \Theta_p$ for the polarizing angle with the corresponding Snell's law of refraction $n_1 \sin \Theta_p = n_2 \sin \Theta_t$, one obtains

$$n_1 n_2 \sin \Theta_p \cos \Theta_p = n_1 n_2 \sin \Theta_t \cos \Theta_t \implies \sin(2\Theta_p) = \sin(2\Theta_t),$$

the pertinent solution of which yields

$$\Theta_p + \Theta_t = \frac{\pi}{2}. \tag{6.107}$$

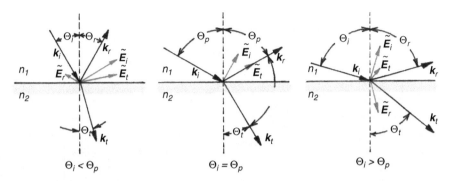

Fig. 6.6 Incident, reflected, and transmitted phasor electric field vectors $\tilde{\mathbf{E}}_j$ and wave vectors $\mathbf{k}_j$ when the incidence angle Θ_i is below, equal to, and greater than the polarizing angle Θ_p

Thus, at the polarizing angle, the reflected wave vector $\mathbf{k}_r$ is perpendicular to the transmitted wave vector $\mathbf{k}_t$, as depicted in Fig. 6.6.

6.3.4.2   Incidence on the Rarer Medium ($n_1 > n_2$)

When the time-harmonic electromagnetic plane wave field is incident upon the optically rarer medium ($n_1 > n_2$), then the right-hand side of Eq. (6.42) is real-valued only when $\Theta_i < \Theta_c$, where Θ_c is the *critical angle*

$$\Theta_c \equiv \arcsin\left(\frac{n_2}{n_1}\right). \tag{6.108}$$

For *subcritical angles of incidence* $\Theta_i < \Theta_c$, the transmitted wave vector is real-valued ($a_t = 0$) and the transmitted wave is of the homogeneous type, Snell's law (6.43) applies, and the reflection and transmission coefficients are given by the Fresnel equations (6.93) and (6.94) for *(s)*-polarization and by the Fresnel equations (6.95) and (6.96) for *(p)*-polarization. There is also a *polarizing angle* at which $r_p = 0$ given by *Brewster's law*

$$\Theta_p \equiv \arctan\left(\frac{n_2}{n_1}\right). \tag{6.109}$$

The angular dependence of the Fresnel reflection and transmission coefficients when $n_1 = (3/2)n_2$ is illustrated in Figs. 6.7 and 6.8, respectively. In this case of incidence on the optically rarer medium, both reflection coefficients r_s and r_p are positive at normal incidence ($\Theta_i = 0$) and are given by $r_{s,p} = (n_1 - n_2)/(n_1 + n_2)$. The reflection coefficient $r_s = r_\perp$ for perpendicular (*TE*) polarization remains positive and increases to $+1$ as the angle of incidence increases to the critical angle Θ_c, while the reflection coefficient $r_p = r_\parallel$ for parallel (*TM*) polarization decreases

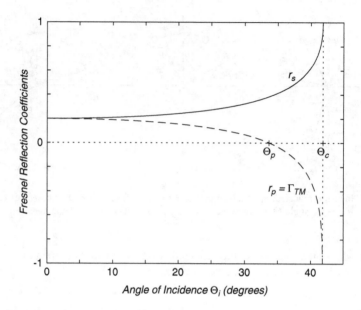

Fig. 6.7 Parallel (p) and perpendicular (s) Fresnel reflection coefficients $r_p = r_\parallel$ and $r_s = r_\perp$, respectively, when incidence is on the optically rarer medium with $n_1 = (3/2)n_2$

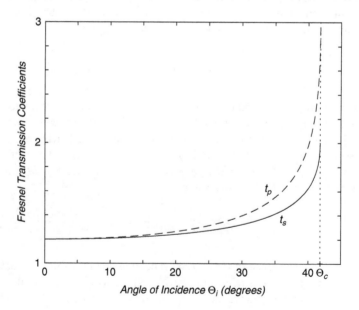

Fig. 6.8 Parallel (p) and perpendicular (s) Fresnel transmission coefficients $t_p = t_\parallel$ and $t_s = t_\perp$, respectively, when incidence is on the optically rarer medium with $n_1 = (3/2)n_2$

to zero as the angle of incidence increases to the polarizing angle Θ_p, vanishes at $\Theta_i = \Theta_p$, and then decreases to -1 as Θ_i increases above Θ_p to the critical angle Θ_c, as illustrated in Fig. 6.7. At the critical angle, $r_s(\Theta_c) = 1$ and $r_p(\Theta_c) = -1$. The reflected wave for p-polarization thus experiences a π-phase shift relative to the incident plane wave when $\Theta_p < \Theta_i \leq \Theta_c$. The transmission coefficients $t_p = t_\parallel$ and $t_s = t_\perp$, on the other hand, are positive for all subcritical angles of incidence with $t_{s,p}(0) = 2n_1/(n_1 + n_2)$, and $t_s(\Theta_c) = 2$ and $t_p(\Theta_c) = 2n_1/n_2$. The corresponding reflectivity $\mathcal{R}$ and transmissivity $\mathcal{T}$ for both (s) and (p) polarizations are given in Figs. 6.9 and 6.10, respectively.

For *supercritical angles of incidence* $\Theta_i > \Theta_c$, the transmitted wave vector is complex-valued and the transmitted wave is of the inhomogeneous type with different surfaces of constant phase and amplitude. In this case, Eq. (6.44) applies so that $\Theta_t = \pi/2$ and

$$k_t = k_0 n_1 \sin \Theta_i \quad \& \quad a_t = k_0 \left[n_1^2 \sin^2 \Theta_i - n_2^2 \right]^{1/2} . \tag{6.110}$$

Because $\Theta_t = \pi/2$, then $k_{tn} = \mathbf{k}_t \cdot \hat{\mathbf{n}} = 0$ and hence, $k_{t\tau} = k_t$, so that the transmitted complex wave vector is given by

$$\tilde{\mathbf{k}}_t = \mathbf{k}_t + i\mathbf{a}_t = k_t \hat{\mathbf{1}}_x + i a_t \hat{\mathbf{1}}_z . \tag{6.111}$$

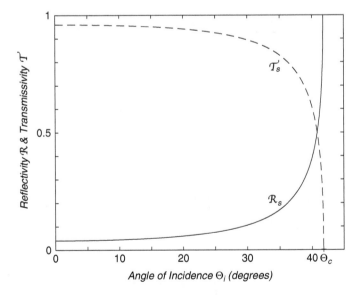

Fig. 6.9 Perpendicular (s) polarization reflectivity $\mathcal{R}_s$ and transmissivity $\mathcal{T}_s$ when incidence is on the optically rarer medium with $n_1 = (3/2)n_2$

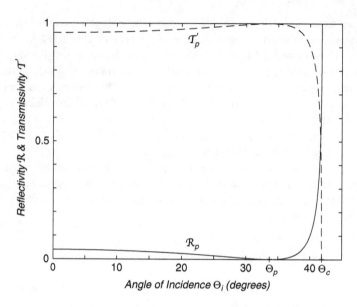

Fig. 6.10 Parallel (p) polarization reflectivity $\mathcal{R}_p$ and transmissivity $\mathcal{T}_p$ when incidence is on the optically rarer medium with $n_1 = (3/2)n_2$

The *reflection coefficient for perpendicular* (s)*-polarization* is then given by

$$r_\perp \equiv r_s = \frac{\cos\Theta_i - i\left[\sin^2\Theta_i - n_2^2/n_1^2\right]^{1/2}}{\cos\Theta_i + i\left[\sin^2\Theta_i - n_2^2/n_1^2\right]^{1/2}} = e^{-i\phi_s}, \qquad (6.112)$$

where

$$\phi_s = 2\arctan\left(\frac{\left(\sin^2\Theta_i - n_2^2/n_1^2\right)^{1/2}}{\cos\Theta_i}\right), \qquad (6.113)$$

and the *reflection coefficient for parallel* (p)*-polarization* is given by

$$r_\| \equiv r_p = -\frac{(n_2^2/n_1^2)\cos\Theta_i - i\left[\sin^2\Theta_i - n_2^2/n_1^2\right]^{1/2}}{(n_2^2/n_1^2)\cos\Theta_i + i\left[\sin^2\Theta_i - n_2^2/n_1^2\right]^{1/2}} = e^{i(\pi-\phi_p)}, \qquad (6.114)$$

where

$$\phi_p = 2\arctan\left(\frac{n_1^2\left(\sin^2\Theta_i - n_2^2/n_1^2\right)^{1/2}}{n_2^2\cos\Theta_i}\right). \qquad (6.115)$$

In the limit as the critical angle Θ_c is approached from above,

$$\lim_{\Theta_i \to \Theta_c^+} \phi_s = -2 \arctan(0) = 0,$$

$$\lim_{\Theta_i \to \Theta_c^+} (\pi - \phi_p) = \pi - 2 \arctan(0) = \pi,$$

and in the limit as grazing incidence is approached from below

$$\lim_{\Theta_i \to \pi/2} \phi_s = 2 \arctan(\infty) = \pi,$$

$$\lim_{\Theta_i \to \pi/2} (\pi - \phi_p) = \pi - 2 \arctan(\infty) = 0,$$

so that ϕ_s increases monotonically from 0 to π and $\pi - \phi_p$ decreases monotonically from π to 0 as the supercritical angle of incidence Θ_i increases from the critical angle Θ_c to grazing incidence, as illustrated in Fig. 6.11.

The imaginary part a_t of the transmitted wave vector $\tilde{\mathbf{k}}_t = k_t \hat{\boldsymbol{\tau}} + i a_t \hat{\mathbf{n}}$, given in Eq. (6.44) and illustrated in Fig. 6.12 relative to the free-space wave number k_0,

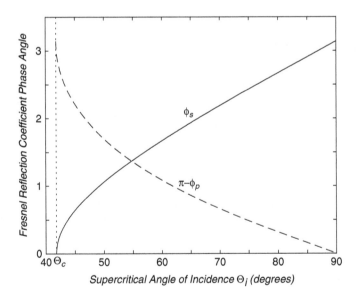

Fig. 6.11 Phase angles ϕ_p (dashed curve) and ϕ_s (solid curve) for the parallel (p) and perpendicular (s) Fresnel reflection coefficients $r_p = r_\parallel = e^{i(\phi_p - \pi)}$ and $r_s = r_\perp = e^{i\phi_s}$, respectively, for supercritical angles of incidence with $n_1 = (3/2)n_2$

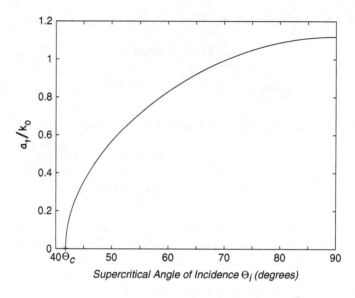

Fig. 6.12 Magnitude a_t of the imaginary part of the complex wave vector $\tilde{\mathbf{k}}_t = \mathbf{k}_t + i\mathbf{a}_t$ relative to the free-space wave number k_0 for supercritical angles of incidence with $n_1 = (3/2)n_2$

possesses the limiting behavior

$$\lim_{\Theta_i \to \Theta_c} \frac{a_t}{k_0} = \lim_{\Theta_i \to \Theta_c} \left[n_1^2 \sin^2 \Theta_i - n_2^2 \right]^{1/2} = 0,$$

$$\lim_{\Theta_i \to \pi/2} \frac{a_t}{k_0} = \lim_{\Theta_i \to \pi/2} \left[n_1^2 \sin^2 \Theta_i - n_2^2 \right]^{1/2} = \sqrt{n_1^2 - n_2^2}.$$

The attenuation factor of the transmitted inhomogeneous plane wave then increases monotonically from 0 to $\sqrt{n_1^2 - n_2^2}$ as the supercritical angle of incidence increases from Θ_c to $\pi/2$, as illustrated, the rate of increase being most rapid as Θ_i increases away from Θ_c and monotonically decreasing as Θ_i approaches grazing incidence.

6.4 Lossy Dispersive-Lossy Dispersive Interface

The generalization of the Fresnel coefficients to the more physically realistic case where the planar interface separates two different lossy dispersive media is not a trivial task. Not surprisingly, there are two different approaches to the solution of this problem (see Chap. IX of Stratton [4], Chap. 5 of Balanis [5], and the paper by Roy [6]). In one approach, complex-valued angles are introduced into Snell's law in order to accommodate for the material loss while real-valued angles are maintained in the other at the expense of complex-valued Fresnel coefficients. In his 2011 paper,

Besieris [7] showed that the criticism raised by Canning [8] with regard to these two different approaches to the solution of this problem was unnecessary as both were indeed correct in describing the reflection and transmission coefficients when one medium was lossless and the other lossy. The approach taken by Canning [8], as reformulated by Besieris [7], is generalized here to the case when both media are lossy.

Each temporally dispersive material is assumed to be homogeneous, isotropic, and locally linear, as described by the frequency-domain constitutive relations

$$\tilde{\mathbf{D}}_j(\mathbf{r}, \omega) = \epsilon_j(\omega)\tilde{\mathbf{E}}_j(\mathbf{r}, \omega),$$

$$\tilde{\mathbf{B}}_j(\mathbf{r}, \omega) = \mu_j(\omega)\tilde{\mathbf{H}}_j(\mathbf{r}, \omega), \qquad (6.116)$$

$$\tilde{\mathbf{J}}_{cj}(\mathbf{r}, \omega) = \sigma_j(\omega)\tilde{\mathbf{E}}_j(\mathbf{r}, \omega),$$

$j = 1, 2$, with complex-valued dielectric permittivity $\epsilon_j(\omega) = \epsilon'_j(\omega) + i\epsilon''_j(\omega)$, magnetic permeability $\mu_j(\omega) = \mu'_j(\omega) + i\mu''_j(\omega)$, and electric conductivity $\sigma_j(\omega) = \sigma'_j(\omega) + i\sigma''_j(\omega)$. The *complex permittivity* of each material is defined as

$$\epsilon_{cj}(\omega) \equiv \epsilon_j(\omega) + \frac{i}{\|4\pi\|} \frac{\sigma_j(\omega)}{\omega}, \qquad (6.117)$$

$j = 1, 2$, with real part $\epsilon'_{cj}(\omega) = \epsilon'_j(\omega) - (1/\|4\pi\|)\sigma''_j(\omega)/\omega$ and imaginary part $\epsilon''_{cj}(\omega) = \epsilon''_j(\omega) + (1/\|4\pi\|)\sigma'_j(\omega)/\omega$, for real ω. Finally, let $n'_j(\omega) \equiv \Re\{n_j(\omega)\}$ and $n''_j(\omega) \equiv \Im\{n_j(\omega)\}$ denote the real and imaginary parts of the *complex index of refraction*

$$n_j(\omega) = \left(\frac{\mu_j(\omega)\epsilon_{cj}(\omega)}{\mu_0\epsilon_0}\right)^{1/2} \qquad (6.118)$$

for each complex medium $j = 1, 2$.

The analysis is based on the frequency domain form of the tangential boundary conditions

$$\hat{\mathbf{n}} \times \left(\tilde{\mathbf{E}}_2(\mathbf{r}, \omega) - \tilde{\mathbf{E}}_1(\mathbf{r}, \omega)\right) = \mathbf{0}, \qquad (6.119)$$

$$\hat{\mathbf{n}} \times \left(\tilde{\mathbf{H}}_2(\mathbf{r}, \omega) - \tilde{\mathbf{H}}_1(\mathbf{r}, \omega)\right) = \mathbf{0}, \qquad (6.120)$$

for all $\mathbf{r} \in S$, where $\hat{\mathbf{n}}$ is the unit normal to the interface S directed from medium 1 into medium 2. A surface current density $\mathbf{J}_s(\mathbf{r}, \omega)$ is not included in Eq. (6.120) because it is nonzero only when one of the media is a perfect electric conductor (PEC) [5]. As in the previous subsections, let the interface S be situated in the xy-plane at $z = 0$ with unit normal $\hat{\mathbf{n}} = \hat{\mathbf{1}}_z$ directed from medium 1 $(\epsilon_1, \mu_1, \sigma_1)$ where the incident and reflected fields reside, into medium 2 $(\epsilon_2, \mu_2, \sigma_2)$ where the transmitted field resides. With the origin O fixed at a point on S so that $\hat{\mathbf{1}}_z \cdot \mathbf{r} = 0$ for all $\mathbf{r} \in S$, the interface position vector $\mathbf{r} \in S$ may then be expressed as

$$\mathbf{r} = -\hat{\mathbf{1}}_z \times (\hat{\mathbf{1}}_z \times \mathbf{r}), \tag{6.121}$$

where $\mathbf{r} = \hat{\mathbf{1}}_x x + \hat{\mathbf{1}}_y y$, the xz-plane defining the *plane of incidence*.

6.4.1 Complex $TE(s)$-Polarization Fresnel Coefficients

For a time-harmonic TE (s)-polarized plane electromagnetic wave incident in medium 1 upon the planar interface S with the electric field vector linearly polarized along the $\hat{\mathbf{1}}_y$-direction,

$$\tilde{\mathbf{E}}_i(\mathbf{r}, \omega) = \hat{\mathbf{1}}_y E_s e^{i\tilde{\mathbf{k}}_i(\omega)\cdot\mathbf{r}}, \tag{6.122}$$

$$\tilde{\mathbf{H}}_i(\mathbf{r}, \omega) = \frac{\|c\|}{i\mu_1\omega} \nabla \times \tilde{\mathbf{E}}_i(\mathbf{r}, \omega)$$

$$= \frac{\|c\|}{\mu_1\omega} \tilde{\mathbf{k}}_i(\omega) \times \hat{\mathbf{1}}_y E_s e^{i\tilde{\mathbf{k}}_i(\omega)\cdot\mathbf{r}}, \tag{6.123}$$

where

$$\tilde{\mathbf{k}}_i(\omega) = \boldsymbol{\beta}_i(\omega) + i\boldsymbol{\alpha}_i(\omega) \tag{6.124}$$

is the *incident wave vector*, with (for real-valued ω)

$$\beta_i(\omega) \equiv |\boldsymbol{\beta}_i(\omega)| = \beta_1(\omega) = \frac{\omega}{c} n_1'(\omega), \tag{6.125}$$

$$\alpha_i(\omega) \equiv |\boldsymbol{\alpha}_i(\omega)| = \alpha_1(\omega) = \frac{\omega}{c} n_1''(\omega). \tag{6.126}$$

In addition, if Θ_i is the *angle of incidence* of the incident plane wave phase front, then

$$\boldsymbol{\beta}_i(\omega) = \beta_1(\omega)\left(\hat{\mathbf{1}}_x \sin\Theta_i + \hat{\mathbf{1}}_z \cos\Theta_i\right). \tag{6.127}$$

The reflected plane wave field vectors in medium 1 are then given by

$$\tilde{\mathbf{E}}_r(\mathbf{r}, \omega) = \hat{\mathbf{1}}_y \Gamma_{TE}(\omega) E_s e^{i\tilde{\mathbf{k}}_r(\omega)\cdot\mathbf{r}}, \tag{6.128}$$

$$\tilde{\mathbf{H}}_r(\mathbf{r}, \omega) = \frac{\|c\|}{i\mu_1\omega} \nabla \times \tilde{\mathbf{E}}_r(\mathbf{r}, \omega)$$

$$= \frac{\|c\|}{\mu_1\omega} \tilde{\mathbf{k}}_r(\omega) \times \hat{\mathbf{1}}_y \Gamma_{TE}(\omega) E_s e^{i\tilde{\mathbf{k}}_r(\omega)\cdot\mathbf{r}}, \tag{6.129}$$

where $\Gamma_{TE}(\omega) \equiv E_r/E_s$ is the Fresnel reflection coefficient for TE (s)-polarization, and

$$\tilde{\mathbf{k}}_r(\omega) = \boldsymbol{\beta}_r(\omega) + i\boldsymbol{\alpha}_r(\omega) \tag{6.130}$$

is the *reflected wave vector*, with (for real-valued ω)

$$\beta_r(\omega) \equiv |\boldsymbol{\beta}_r(\omega)| = \beta_1(\omega) = \frac{\omega}{c} n_1'(\omega), \tag{6.131}$$

$$\alpha_r(\omega) \equiv |\boldsymbol{\alpha}_r(\omega)| = \alpha_1(\omega) = \frac{\omega}{c} n_1''(\omega). \tag{6.132}$$

If Θ_r is the *angle of reflection* of the reflected plane wave phase front, then

$$\boldsymbol{\beta}_r(\omega) = \beta_1(\omega) \left(\hat{\mathbf{1}}_x \sin \Theta_r - \hat{\mathbf{1}}_z \cos \Theta_r \right). \tag{6.133}$$

Finally, the transmitted plane wave field vectors in medium 2 are given by

$$\tilde{\mathbf{E}}_t(\mathbf{r}, \omega) = \hat{\mathbf{1}}_y \tau_{TE}(\omega) E_s e^{i\tilde{\mathbf{k}}_t(\omega)\cdot\mathbf{r}}, \tag{6.134}$$

$$\tilde{\mathbf{H}}_t(\mathbf{r}, \omega) = \frac{\|c\|}{i\mu_2\omega} \nabla \times \tilde{\mathbf{E}}_t(\mathbf{r}, \omega)$$

$$= \frac{\|c\|}{\mu_2\omega} \tilde{\mathbf{k}}_t(\omega) \times \hat{\mathbf{1}}_y \tau_{TE}(\omega) E_s e^{i\tilde{\mathbf{k}}_t(\omega)\cdot\mathbf{r}}, \tag{6.135}$$

where $\tau_{TE}(\omega) \equiv E_t/E_s$ is the Fresnel transmission coefficient for TE (s)-polarization, and

$$\tilde{\mathbf{k}}_t(\omega) = \boldsymbol{\beta}_t(\omega) + i\boldsymbol{\alpha}_t(\omega) \tag{6.136}$$

is the *transmitted wave vector*, with (for real-valued ω)

$$\beta_t(\omega) \equiv |\boldsymbol{\beta}_t(\omega)| = \beta_2(\omega) = \frac{\omega}{c} n_2'(\omega), \tag{6.137}$$

$$\alpha_t(\omega) \equiv |\boldsymbol{\alpha}_t(\omega)| = \alpha_2(\omega) = \frac{\omega}{c} n_2''(\omega). \tag{6.138}$$

If Θ_t is the *angle of refraction* of the transmitted plane wave phase front, then

$$\boldsymbol{\beta}_t(\omega) = \beta_2(\omega) \left(\hat{\mathbf{1}}_x \sin \Theta_t + \hat{\mathbf{1}}_z \cos \Theta_t \right). \tag{6.139}$$

The total field in medium 1 at the interface S is given by the superposition of the incident and reflected fields as $\tilde{\mathbf{E}}_1(\mathbf{r}, \omega) = \tilde{\mathbf{E}}_i(\mathbf{r}, \omega) + \tilde{\mathbf{E}}_r(\mathbf{r}, \omega)$ and $\tilde{\mathbf{H}}_1(\mathbf{r}, \omega) = \tilde{\mathbf{H}}_i(\mathbf{r}, \omega) + \tilde{\mathbf{H}}_r(\mathbf{r}, \omega)$, so that

$$\tilde{\mathbf{E}}_1(\mathbf{r}, \omega) = \hat{\mathbf{1}}_y E_s \left\{ e^{i\tilde{\mathbf{k}}_i \cdot \mathbf{r}} + \Gamma_{TE} e^{i\tilde{\mathbf{k}}_r \cdot \mathbf{r}} \right\}, \tag{6.140}$$

$$\tilde{\mathbf{H}}_1(\mathbf{r}, \omega) = E_s \frac{\|c\|}{\mu_1 \omega} \left\{ \tilde{\mathbf{k}}_i \times \hat{\mathbf{1}}_y e^{i\tilde{\mathbf{k}}_i \cdot \mathbf{r}} + \Gamma_{TE} \tilde{\mathbf{k}}_r \times \hat{\mathbf{1}}_y e^{i\tilde{\mathbf{k}}_r \cdot \mathbf{r}} \right\}, \tag{6.141}$$

where $\tilde{\mathbf{k}}_i \times \hat{\mathbf{1}}_y = \beta_1(\omega) \left(\hat{\mathbf{1}}_z \sin \Theta_i - \hat{\mathbf{1}}_x \cos \Theta_i \right) + i\alpha_i(\omega) \times \hat{\mathbf{1}}_y$ and where $\tilde{\mathbf{k}}_r \times \hat{\mathbf{1}}_y = \beta_1(\omega) \left(\hat{\mathbf{1}}_z \sin \Theta_r + \hat{\mathbf{1}}_x \cos \Theta_r \right) + i\alpha_r(\omega) \times \hat{\mathbf{1}}_y$. The total field in medium 2 is just the transmitted field. With these substitutions, the tangential boundary condition (6.119) yields

$$\tau_{TE} e^{i\tilde{\mathbf{k}}_t \cdot \mathbf{r}} = e^{i\tilde{\mathbf{k}}_i \cdot \mathbf{r}} + \Gamma_{TE} e^{i\tilde{\mathbf{k}}_r \cdot \mathbf{r}}, \tag{6.142}$$

while the tangential boundary condition (6.120) becomes

$$\frac{\tau_{TE}}{\mu_2} \left[-\hat{\mathbf{1}}_y \beta_2 \cos \Theta_t e^{i\tilde{\mathbf{k}}_t \cdot \mathbf{r}} + i\hat{\mathbf{1}}_z \times (\alpha_t \times \hat{\mathbf{1}}_y) e^{i\tilde{\mathbf{k}}_t \cdot \mathbf{r}} \right]$$

$$= -\hat{\mathbf{1}}_y \frac{\beta_1}{\mu_1} \left[\cos \Theta_i e^{i\tilde{\mathbf{k}}_i \cdot \mathbf{r}} - \Gamma_{TE} \cos \Theta_r e^{i\tilde{\mathbf{k}}_r \cdot \mathbf{r}} \right]$$

$$+ \frac{i}{\mu_1} \left[\hat{\mathbf{1}}_z \times (\alpha_i \times \hat{\mathbf{1}}_y) e^{i\tilde{\mathbf{k}}_i \cdot \mathbf{r}} + \Gamma_{TE} \hat{\mathbf{1}}_z \times (\alpha_r \times \hat{\mathbf{1}}_y) e^{i\tilde{\mathbf{k}}_r \cdot \mathbf{r}} \right]. \tag{6.143}$$

Because these two equations must be satisfied for all points $\mathbf{r} \in S$, the exponential factors must all be equal so that

$$\tilde{\mathbf{k}}_i(\omega) \cdot \mathbf{r} = \tilde{\mathbf{k}}_r(\omega) \cdot \mathbf{r} = \tilde{\mathbf{k}}_t(\omega) \cdot \mathbf{r}, \quad \mathbf{r} \in S. \tag{6.144}$$

With substitution from Eqs. (6.124), (6.130), and (6.136), the real parts of Eq. (6.144) when Θ_t is real-valued give

$$\beta_1(\omega) \left(\hat{\mathbf{1}}_x \sin \Theta_i + \hat{\mathbf{1}}_z \cos \Theta_i \right) \cdot \left(\hat{\mathbf{1}}_z \times \left(\hat{\mathbf{1}}_z \times \mathbf{r} \right) \right)$$

$$= \beta_1(\omega) \left(\hat{\mathbf{1}}_x \sin \Theta_r - \hat{\mathbf{1}}_z \cos \Theta_r \right) \cdot \left(\hat{\mathbf{1}}_z \times \left(\hat{\mathbf{1}}_z \times \mathbf{r} \right) \right)$$

$$= \beta_2(\omega) \left(\hat{\mathbf{1}}_x \sin \Theta_t + \hat{\mathbf{1}}_z \cos \Theta_t \right) \cdot \left(\hat{\mathbf{1}}_z \times \left(\hat{\mathbf{1}}_z \times \mathbf{r} \right) \right),$$

so that $\beta_1(\omega) \sin \Theta_i = \beta_1(\omega) \sin \Theta_r = \beta_2(\omega) \sin \Theta_t$, resulting in the relation

$$\Theta_r = \Theta_i, \tag{6.145}$$

and, with Eqs. (6.125) and (6.137),

$$n_1'(\omega) \sin \Theta_i = n_2'(\omega) \sin \Theta_t. \tag{6.146}$$

In addition, the imaginary parts of Eq. (6.144) when Θ_t is real-valued give

$$\boldsymbol{\alpha}_i(\omega) \cdot \left(\hat{\mathbf{1}}_z \times \left(\hat{\mathbf{1}}_z \times \mathbf{r}\right)\right) = \boldsymbol{\alpha}_r(\omega) \cdot \left(\hat{\mathbf{1}}_z \times \left(\hat{\mathbf{1}}_z \times \mathbf{r}\right)\right) = \boldsymbol{\alpha}_t(\omega) \cdot \left(\hat{\mathbf{1}}_z \times \left(\hat{\mathbf{1}}_z \times \mathbf{r}\right)\right),$$

so that $\alpha_i(\omega) \cos \psi_i = \alpha_r(\omega) \cos \psi_r = \alpha_t(\omega) \cos \psi_t$, where ψ_j $(j = i, r, t)$ is the angle between the attenuation vector $\boldsymbol{\alpha}_j$ and the interface plane S (taken here as the xy-plane). Hence, $n_1'' \cos \psi_i = n_1'' \cos \psi_r = n_2'' \cos \psi_t$, so that

$$\psi_r = \psi_i \tag{6.147}$$

and, with Eqs. (6.126) and (6.138),

$$n_1'' \cos \psi_i = n_2'' \cos \psi_t. \tag{6.148}$$

Taken together, Eqs. (6.145) and (6.147) constitute the *generalized law of reflection*, and Eqs. (6.146) and (6.148) constitute the *generalized law of refraction* for complex media with an incident inhomogeneous plane wave.

Notwithstanding its change in direction of propagation with respect to the normal direction from medium 1 to medium 2, the reflected plane wave is then seen to be unchanged in form from that of the incident plane wave. For the form of the transmitted plane wave in a lossy medium 2 $(n_2'' \neq 0)$, there are three distinct possibilities depending upon whether or not medium 1 is lossy and whether or not the incident plane wave field is homogeneous $(\Theta_i + \psi_i = \pi/2)$ or inhomogeneous $(\Theta_i + \psi_i \neq \pi/2)$:

1. If medium 1 is lossless $(n_1'' = 0)$, then $\psi_t = \pi/2$ and, except for the special case of normal incidence, the transmitted plane wave is inhomogeneous with $\boldsymbol{\alpha}_t(\omega) = \hat{\mathbf{1}}_z \alpha_2(\omega)$.
2. If medium 1 is lossy $(n_1'' > 0)$ and the incident plane wave is homogeneous $(\psi_i = \pi/2 - \Theta_i)$, then

$$\cos \psi_t = \frac{n_1''(\omega)}{n_2''(\omega)} \sin \Theta_i \tag{6.149}$$

and the transmitted plane wave is inhomogeneous unless either $\Theta_i = 0$ (normal incidence) or $n_1''(\omega) = n_2''(\omega)$, in which case it is also homogeneous.
3. If the incident plane wave is inhomogeneous, then the transmitted plane wave is also inhomogeneous unless $\psi_t = \pi/2 - \Theta_t$.

Hence, for the general case when medium 1 and medium 2 are both lossy, Eqs. (6.124) and (6.127), (6.130) and (6.133), and (6.136) and (6.139) yield

$$\tilde{\mathbf{k}}_i(\omega) = \beta_1(\omega)(\hat{\mathbf{1}}_x \sin \Theta_i + \hat{\mathbf{1}}_z \cos \Theta_i) + i\alpha_1(\omega)(\hat{\mathbf{1}}_x \cos \psi_i + \hat{\mathbf{1}}_z \sin \psi_i),$$

$$\tilde{\mathbf{k}}_r(\omega) = \beta_1(\omega)(\hat{\mathbf{1}}_x \sin \Theta_r - \hat{\mathbf{1}}_z \cos \Theta_r) + i\alpha_1(\omega)(\hat{\mathbf{1}}_x \cos \psi_r - \hat{\mathbf{1}}_z \sin \psi_r), \quad (6.150)$$

$$\tilde{\mathbf{k}}_t(\omega) = \beta_2(\omega)(\hat{\mathbf{1}}_x \sin \Theta_t + \hat{\mathbf{1}}_z \cos \Theta_t) + i\alpha_2(\omega)(\hat{\mathbf{1}}_x \cos \psi_t + \hat{\mathbf{1}}_z \sin \psi_t).$$

For the special case of an electromagnetic plane wave traveling through a layered stack of complex media, the incident plane wave is typically homogeneous and in near vacuum with zero attenuation. The transmitted wave fields in each complex layer then each have an attenuative part (the imaginary part) that is z-directed.

With these results, Eq. (6.142) becomes

$$\tau_{TE} = 1 + \Gamma_{TE}, \quad (6.151)$$

and Eq. (6.143) yields

$$\frac{\tau_{TE}}{\mu_2}(\beta_2 \cos \Theta_t + i\alpha_2 \sin \psi_t) = \frac{1}{\mu_1}(1 - \Gamma_{TE})(\beta_1 \cos \Theta_i + i\alpha_1 \sin \psi_i). \quad (6.152)$$

The solution of this pair of equations then yields the *generalized Fresnel reflection and transmission coefficients* for TE (s)-polarization, given by

$$\Gamma_{TE}(\omega) \equiv \Gamma_s(\omega) = \frac{\mu_2(n_1' \cos \Theta_i + in_1'' \sin \psi_i) - \mu_1(n_2' \cos \Theta_t + in_2'' \sin \psi_t)}{\mu_2(n_1' \cos \Theta_i + in_1'' \sin \psi_i) + \mu_1(n_2' \cos \Theta_t + in_2'' \sin \psi_t)}, \quad (6.153)$$

$$\tau_{TE}(\omega) \equiv \tau_s(\omega) = \frac{2\mu_2(n_1' \cos \Theta_i + in_1'' \sin \psi_i)}{\mu_2(n_1' \cos \Theta_i + in_1'' \sin \psi_i) + \mu_1(n_2' \cos \Theta_t + in_2'' \sin \psi_t)}, \quad (6.154)$$

respectively. Notice that both μ_1 and μ_2 may be complex-valued. Furthermore, because $|\mu|/\mu_0$ is slightly smaller than unity for diamagnetic materials whereas it is slightly greater than unity for paramagnetic materials, then the effect of the magnetic permeabilities $\mu_1(\omega)$ and $\mu_2(\omega)$ on both Γ_{TE} and τ_{TE} is negligible for such materials; however, this is not the case for metamaterials where the relative magnetic permeability may become large and negative. Notice that in the idealized lossless limit, these generalized Fresnel reflection and transmission coefficients reduce to the well-known results for perfect dielectrics given in Eqs. (6.63) and (6.64). Furthermore, for a homogeneous $TE(s)$-polarized plane wave at *normal incidence* ($\Theta_i = \Theta_t = 0$ and $\psi_i = \psi_t = \pi/2$), these generalized Fresnel equations

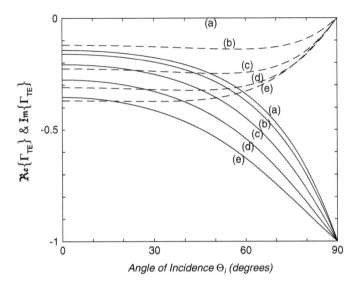

Fig. 6.13 Real (solid curves) and imaginary (dashed curves) parts of the TE (s)-polarization Fresnel reflection coefficient $\Gamma_{TE} = \Gamma_s$ as a function of the incidence angle Θ_i when $\psi_i = \pi/2 - \Theta_i$, $\mu_1 = \mu_2$, $n_1 = 1.5$ and (a) $n_2 = 2.0$, (b) $n_2 = 2.0 + 0.5i$, (c) $n_2 = 2.0 + i$, (d) $n_2 = 2.0 + 1.5i$, and (e) $n_2 = 2.0 + 2.0i$

simplify to the well-known expressions

$$\Gamma_{TE}(\omega)\big|_{\Theta_i=0} = \frac{\eta_2(\omega) - \eta_1(\omega)}{\eta_2(\omega) + \eta_1(\omega)} \quad \& \quad \tau_{TE}(\omega)\big|_{\Theta_i=0} = \frac{2\eta_2(\omega)}{\eta_2(\omega) + \eta_1(\omega)},$$

where $\eta_j(\omega) = [\mu_j(\omega)/\epsilon_{cj}(\omega)]^{1/2}$ is the *complex impedance*[3] of medium j.

The angular dependence of the Fresnel reflection coefficient Γ_{TE} when $\mu_1 = \mu_2$ and incidence is on the optically denser medium, as defined by the inequality $n_2' > n_1'$, is illustrated in Fig. 6.13 when the incident plane wave is homogeneous ($\psi_i = \pi/2 - \Theta_i$) and medium 1 is lossless with refractive index $n_1 = 1.5$ and medium 2 has complex refractive index $n_2 = 2.0 + in_2''$ with imaginary part varying from $n_2'' = 0 \rightarrow 2$ in 0.5 increments. In that case $\psi_t = \pi/2$ and, except for normal incidence, the transmitted plane wave is inhomogeneous. Notice that $\Re\{\Gamma_{TE}\} \rightarrow -1$ and $\Im\{\Gamma_{TE}\} \rightarrow 0$ as $\theta_i \rightarrow 90°$ independent of the loss in medium 2 when medium 1 is lossless, as seen from Eq. (6.153) when $n_1'' = 0$. The magnitude of Γ_{TE} then increases monotonically from its normal incidence value

---

[3]Notice that $n_j(\omega)/\mu_j(\omega) = (\epsilon_0\mu_0)^{-1/2}[\epsilon_{cj}(\omega)/\mu_j(\omega)]^{1/2} = (c/\|c\|)/\eta_j(\omega)$ and that $n_j(\omega)/\epsilon_{cj}(\omega) = (\epsilon_0\mu_0)^{-1/2}[\mu_j(\omega)/\epsilon_{cj}(\omega)]^{1/2} = (c/\|c\|)\eta_j(\omega)$ where $\eta_j(\omega)$ is the complex impedance and $n_j(\omega) = [(\epsilon_{cj}(\omega)/\epsilon_0)(\mu_j(\omega)/\mu_0)]^{1/2}$ is the complex index of refraction of medium j with complex permittivity $\epsilon_{cj}(\omega) = \epsilon(\omega) + (i/\|4\pi\|)\sigma_j(\omega)/\omega$.

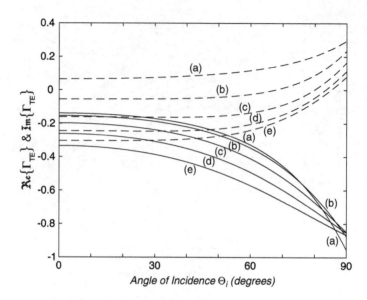

Fig. 6.14 Real (solid curves) and imaginary (dashed curves) parts of the TE (s)-polarization Fresnel reflection coefficient $\Gamma_{TE} = \Gamma_s$ as a function of the incidence angle Θ_i when $\psi_i = \psi_t = \pi/2$, $\mu_1 = \mu_2$, $n_1 = 1.5 + 0.2i$ and (a) $n_2 = 2.0$, (b) $n_2 = 2.0 + 0.5i$, (c) $n_2 = 2.0 + i$, (d) $n_2 = 2.0 + 1.5i$, and (e) $n_2 = 2.0 + 2.0i$

$\Gamma_{TE}|_{\Theta_i=0} = (\eta_2 - \eta_1)/(\eta_2 + \eta_1)$ to unity at grazing incidence while its radian phase angle decreases monotonically from its negative normal incidence value to $-\pi$ at grazing incidence. However, such is not the case when $\psi_i = \psi_t = \pi/2$ and medium 1 is also lossy, as illustrated in Figs. 6.14 and 6.15 when $n_1 = 1.5+0.2i$ and $n_2 = 2.0+in_2''$ with imaginary part varying from $n_2'' = 0 \to 2$ in 0.5 increments. At normal incidence with $\mu_1 = \mu_2$, $\Gamma_{TE}|_{\Theta_i=0} = (n_1 - n_2)/(n_1 + n_2)$ has magnitude that initially decreases as n_2'' increases above 0, reaches a minimum when $n_2'' \simeq n_1''$, and then increases towards unity as n_2'' increases above n_1''.

At *grazing incidence* $\Theta_i = \pi/2$, $\sin \Theta_t = n_1'/n_2'$, $\cos \Theta_t = \sqrt{1 - (n_1'/n_2')^2}$, and the Fresnel reflection coefficient (6.153) becomes

$$\Gamma_{TE}|_{\Theta_i=\pi/2} = -\frac{\frac{\mu_1}{\mu_2}\sqrt{n_2'^2 - n_1'^2} + i\left(\frac{\mu_1}{\mu_2}n_2'' \sin \psi_t - n_1'' \sin \psi_i\right)}{\frac{\mu_1}{\mu_2}\sqrt{n_2'^2 - n_1'^2} + i\left(\frac{\mu_1}{\mu_2}n_2'' \sin \psi_t + n_1'' \sin \psi_i\right)},$$

when incidence is on the optically denser medium. Notice that if the incident plane wave is homogeneous (in which case $\psi_i = \pi/2 - \Theta_i = 0$), then $\cos \psi_t = n_1''/n_2''$, $\sin \psi_t = \sqrt{1 - (n_1''/n_2'')^2}$ and $\Gamma_{TE}(\omega) = -1$, $\tau_{TE}(\omega) = 0$ independent of the loss in either medium. Hence: *Total reflection at grazing incidence on the optically denser medium ($n_2' > n_1'$) always occurs, independent of the loss in either medium,*

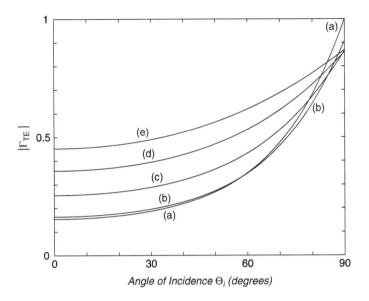

Fig. 6.15 Magnitude of the TE (s)-polarization Fresnel reflection coefficient $\Gamma_{TE} = \Gamma_s$ as a function of the incidence angle Θ_i when $\psi_i = \pi/2$, $\mu_1 = \mu_2$, $n_1 = 1.5 + 0.2i$ and (a) $n_2 = 2.0$, (b) $n_2 = 2.0 + 0.5i$, (c) $n_2 = 2.0 + i$, (d) $n_2 = 2.0 + 1.5i$, and (e) $n_2 = 2.0 + 2i$

when the incident plane wave is homogeneous. Notice further that if $n_1'' = 0$, then $\Gamma_{TE}(\omega) = -1$, $\tau_{TE}(\omega) = 0$ independent of the loss in medium 2 for both homogeneous and inhomogeneous plane wave incidence. Hence:
Total reflection at grazing incidence on the optically denser medium ($n_2' > n_1'$) always occurs if medium 1 is lossless ($n_1'' = 0$), independent of the loss in medium 2, for both homogeneous and inhomogeneous plane wave incidence.
In addition, if $n_2'' = 0$ and μ_2/μ_1 is real-valued, then $\Gamma_{TE}(\omega) = -e^{-i\phi_s(\omega)}$ with

$$\phi_s(\omega) = 2\arctan\left(\frac{\mu_2 n_1'' \sin\psi_i}{\mu_1\sqrt{n_2'^2 - n_1'^2}}\right),$$

and $\tau_{TE}(\omega) = 2i\,e^{-i\phi_s(\omega)/2}\sin(\phi_s/2)$. The magnitude $|\Gamma_{TE}(\omega)|$ at grazing incidence when medium 1 is lossy ($n_1'' > 0$) and the incident plane wave is inhomogeneous ($\psi_i \neq \pi/2 - \Theta_i$) is then seen to decrease from unity as the loss in medium 2 increases (i.e., as n_2'' increases above 0), reaching a minimum when $n_2''^2 \sin^2 \psi_t = n_2'^2 - n_1'^2 + (\mu_2/\mu_1)^2 n_1''^2 \sin^2 \psi_i$, and then increases as n_2'' increases above this value, determined as $n_2'' \simeq 1.338$ from this equation for the complex index of refraction values used in Fig. 6.15 with $\psi_t = \psi_i = \pi/2$, occurring between cases (d) and (e) in that figure.

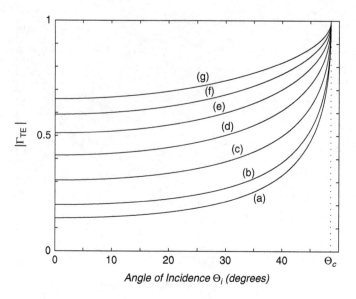

Fig. 6.16 Magnitude of the TE (s)-polarization Fresnel reflection coefficient Γ_{TE} as a function of the incidence angle Θ_i when $\psi_i = \pi/2 - \Theta_i$, $\mu_1 = \mu_2$, $n_1 = 2.0$ and (a) $n_2 = 1.5$, (b) $n_2 = 1.5 + 0.5i$, (c) $n_2 = 1.5 + i$, (d) $n_2 = 1.5 + 1.5i$, (e) $n_2 = 1.5 + 2i$, (f) $n_2 = 1.5 + 2.5i$, and (g) $n_2 = 1.5 + 3i$

Because the angle of refraction given by Eq. (6.146) does not involve the imaginary part of the complex index of refraction for either medium, the *critical angle* of incidence

$$\Theta_c \equiv \arcsin\left(\frac{n_2'(\omega)}{n_1'(\omega)}\right) \tag{6.155}$$

when incidence is on the optically rarer medium ($n_1' > n_2'$) is unaffected by the presence of material loss. However, the reflected and transmitted wave-fields when incidence is at or above the critical angle is effected by the presence of material loss. The angular dependence of the magnitude of the Fresnel reflection coefficient Γ_{TE} when $\psi_i = \pi/2 - \Theta_i$, $\mu_1 = \mu_2$ and incidence is on the optically rarer medium ($n_1' > n_2'$) is illustrated in Fig. 6.16 when $n_1 = 2.0$ and $n_2 = 1.5 + in_2''$ with imaginary part varying from $n_2'' = 0 \rightarrow 3.0$ in 0.5 increments. In this case $\sqrt{n_2'^2 - n_1'^2} = i\sqrt{n_1'^2 - n_2'^2}$ at the critical angle $\Theta_i = \Theta_c$ and

$$\Gamma_{TE}\big|_{\Theta_i=\Theta_c} = \frac{\sqrt{n_1'^2 - n_2'^2} + i\left(n_1'' \sin\psi_i - \frac{\mu_1}{\mu_2}n_2'' \sin\psi_t\right)}{\sqrt{n_1'^2 - n_2'^2} + i\left(n_1'' \sin\psi_i + \frac{\mu_1}{\mu_2}n_2'' \sin\psi_t\right)}, \tag{6.156}$$

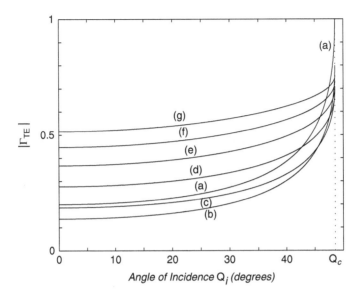

Fig. 6.17 Magnitude of the TE (s)-polarization Fresnel reflection coefficient Γ_{TE} as a function of the incidence angle Θ_i when $\psi_i = \pi/2$, $\mu_1 = \mu_2$, $n_1 = 2.0 + 0.5i$ and (a) $n_2 = 1.5$, (b) $n_2 = 1.5 + 0.5i$, (c) $n_2 = 1.5 + i$, (d) $n_2 = 1.5 + 1.5i$, (e) $n_2 = 1.5 + 2i$, (f) $n_2 = 1.5 + 2.5i$, and (g) $n_2 = 1.5 + 3i$

from which it is seen that $|\Gamma_{TE}| = 1$ at the critical angle when either $n_1'' = 0$ or $n_2'' = 0$. In particular, if $n_2'' = 0$, then $\Gamma_{TE} = 1$ and $\tau_{TE} = 2$, whereas if $n_1'' = 0$ and μ_1/μ_2 is real-valued, then $\Gamma_{TE} = e^{-i\phi_s(\omega)}$ where

$$\phi_s(\omega) = 2 \arctan \left(\frac{\mu_1 n_2'' \sin \psi_t}{\mu_2 \sqrt{n_1'^2 - n_2'^2}} \right).$$

However, such is not the case when both medium 1 and medium 2 are lossy, as illustrated in Fig. 6.17 when $\psi_t = \psi_i = \pi/2$, $n_1 = 2.0 + 0.5i$ and $n_2 = 1.5 + in_2''$ with imaginary part varying from $n_2'' = 0 \rightarrow 3$ in 0.5 increments. In that case the magnitude of the reflection coefficient reaches a minimum value at the critical angle of incidence when $n_2'' \sin \psi_i = (\mu_2/\mu_1)\sqrt{n_1'^2 - n_2'^2 + n_1''^2 \sin^2 \psi_i}$ (again assuming that μ_1/μ_2 is real-valued) and then increases back towards unity as n_2'' increases above that value. This then establishes the following result:

Total reflection does not occur at critical incidence on the optically rarer medium when both media exhibit finite nonzero loss.

At both critical and supercritical angles of incidence $\Theta_i \geq \Theta_c$, $\Theta_t = \pi/2$ and the transmitted wave vector in medium 2 becomes [cf. Eqs. (6.110) and (6.111)]

$$\tilde{\mathbf{k}}_t(\omega) = \mathbf{k}_t(\omega) + i\mathbf{a}_t(\omega)$$

$$= \frac{\omega}{c}\left\{\hat{\mathbf{1}}_x\left[n_1'(\omega)\sin\Theta_i + in_2''(\omega)\cos\psi_t\right]\right.$$

$$\left. + \hat{\mathbf{1}}_z i\left[n_2''(\omega)\sin\psi_t + \sqrt{n_1'^2\sin^2\Theta_i - n_2'^2}\right]\right\}. \quad (6.157)$$

Continuity of the tangential E-field components across the interface S then yields [cf. Eq. (6.119)]

$$\tau_{TE} = 1 + \Gamma_{TE},$$

while continuity of the tangential H-field components across S yields [cf. Eq. (6.120)]

$$i\frac{\tau_{TE}}{\mu_2}\left[n_2''\sin\psi_t + \left(n_1'^2\sin^2\Theta_i - n_2'^2\right)^{1/2}\right] = \frac{1}{\mu_1}(1 - \Gamma_{TE})\left(n_1'\cos\Theta_i + in_1''\sin\psi_i\right).$$

Solution of this pair of equations then yields the *critical and supercritical Fresnel reflection and transmission coefficients for TE (s)-polarization*

$$\Gamma_{TE} = \frac{\mu_2 n_1'\cos\Theta_i - i\left[\mu_1 n_2''\sin\psi_t - \mu_2 n_1''\sin\psi_i + \mu_1\left(n_1'^2\sin^2\Theta_i - n_2'^2\right)^{1/2}\right]}{\mu_2 n_1'\cos\Theta_i + i\left[\mu_1 n_2''\sin\psi_t + \mu_2 n_1''\sin\psi_i + \mu_1\left(n_1'^2\sin^2\Theta_i - n_2'^2\right)^{1/2}\right]},$$

$$(6.158)$$

$$\tau_{TE} = \frac{2\mu_2(n_1'\cos\Theta_i + in_1''\sin\psi_i)}{\mu_2 n_1'\cos\Theta_i + i\left[\mu_1 n_2''\sin\psi_t + \mu_2 n_1''\sin\psi_i + \mu_1\left(n_1'^2\sin^2\Theta_i - n_2'^2\right)^{1/2}\right]},$$

$$(6.159)$$

for $\Theta_i \geq \Theta_c$. For incidence at the critical angle $\Theta_i = \Theta_c \equiv \arcsin(n_2'/n_1')$, both Eqs. (6.153)–(6.154) and Eqs. (6.158)–(6.159) yield [cf. Eq. (6.156)]

$$\Gamma_{TE}(\Theta_c) = \frac{\mu_2\sqrt{n_1'^2 - n_2'^2} - i(\mu_1 n_2''\sin\psi_t - \mu_2 n_1''\sin\psi_i)}{\mu_2\sqrt{n_1'^2 - n_2'^2} + i(\mu_1 n_2''\sin\psi_t + \mu_2 n_1''\sin\psi_i)}, \quad (6.160)$$

$$\tau_{TE}(\Theta_c) = \frac{2\mu_2\left(\sqrt{n_1'^2 - n_2'^2} + in_1''\sin\psi_i\right)}{\mu_2\sqrt{n_1'^2 - n_2'^2} + i(\mu_1 n_2''\sin\psi_t + \mu_2 n_1''\sin\psi_i)}. \quad (6.161)$$

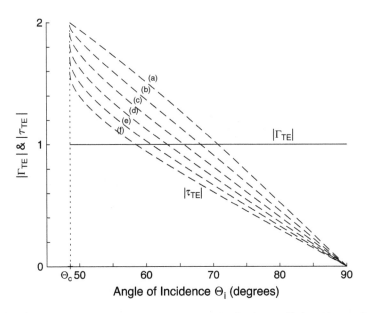

Fig. 6.18 Magnitude of the TE (s)-polarization Fresnel reflection coefficient $\Gamma_{TE} = \Gamma_s$ (solid curves) and transmission coefficient $\tau_{TE} = \tau_s$ (dashed curves) as a function of the critical to supercritical incidence angle $\Theta_i \geq \Theta_c$ when $\psi_i = \pi/2 - \Theta_i$, $\mu_1 = \mu_2$, $n_1 = 2.0$ and (a) $n_2 = 1.5$, (b) $n_2 = 1.5 + 0.2i$, (c) $n_2 = 1.5 + 0.4i$, (d) $n_2 = 1.5 + 0.6i$, (e) $n_2 = 1.5 + 0.8i$, and (f) $n_2 = 1.5 + 1.0i$

The magnitude and phase of the Fresnel reflection (solid curves) and transmission (dashed curves) coefficients for $TE(s)$-polarization when the incident plane wave is homogeneous ($\psi_i = \pi/2 - \Theta_i$) as a function of the critical to supercritical angle of incidence $\Theta_i \geq \Theta_c$ are illustrated in Figs. 6.18 and 6.19, respectively, when $\mu_1 = \mu_2$ and medium 1 is lossless with $n_1 = 2.0$ and the loss in medium 2 varies from $n_2'' = 0 \to 1.0$ in $\Delta n_2'' = 0.2$ increments. Notice that $|\Gamma_{TE}| = 1$ from critical to grazing angles of incidence $\Theta_i = \Theta_c \to 90°$ when medium 1 is lossless independent of the loss in medium 2, with phase angle $\phi_s \equiv \arg\{\Gamma_{TE}\}$ monotonically decreasing from $0 \to -\pi$. The magnitude $|\tau_{TE}|$ of the transmission coefficient, on the other hand, monotonically decreases from $2 \to 0$ as Θ_i increases from critical to grazing angles of incidence $\Theta_i = \Theta_c \to 90°$ with phase angle $\arg\{\tau_{TE}\}$ decreasing from $0 \to -\pi/2$. In each case illustrated here, the amount of decrease in the values of $\arg\{\Gamma_{TE}\}$, $|\tau_{TE}|$, and $\arg\{\tau_{TE}\}$ from its value at the critical angle of incidence to grazing incidence ($\Theta_i = \Theta_c \to 90°$) increases with increasing imaginary part $n_2'' \equiv \Im\{n_2\}$.

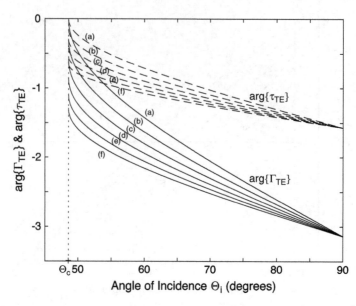

Fig. 6.19 Phase angle (in radians) of the TE (s)-polarization Fresnel reflection coefficient $\Gamma_{TE} = \Gamma_s$ (solid curves) and transmission coefficient $\tau_{TE} = \tau_s$ (dashed curves) as a function of the critical to supercritical incidence angle $\Theta_i \geq \Theta_c$ when $\psi_i = \pi/2 - \Theta_i$, $\mu_1 = \mu_2$, $n_1 = 2.0$ and (a) $n_2 = 1.5$, (b) $n_2 = 1.5 + 0.2i$, (c) $n_2 = 1.5 + 0.4i$, (d) $n_2 = 1.5 + 0.6i$, (e) $n_2 = 1.5 + 0.8i$, and (f) $n_2 = 1.5 + 1.0i$

For comparison, the magnitude and phase of the Fresnel reflection (solid curves) and transmission (dashed curves) coefficients for TE(s)-polarization with $\psi_i = \pi/2$ as a function of the critical to supercritical angle of incidence $\Theta_i \geq \Theta_c$ is illustrated in Figs. 6.20 and 6.21, respectively, when $\mu_1 = \mu_2$ and medium 1 is lossy with $n_1 = 2.0 + 0.4i$ and the loss in medium 2 varies from $n_2'' = 0 \rightarrow 1.0$ in $\Delta n_2'' = 0.2$ increments. The magnitude $|\Gamma_{TE}|$ of the reflection coefficient is no longer constant, decreasing with increasing critical to supercritical angles of incidence. Notice that $|\Gamma_{TE}|$ initially decreases with increasing imaginary part n_2'', but that this behavior is reversed for larger supercritical angles of incidence, as illustrated in Fig. 6.20. The magnitude $|\tau_{TE}|$ of the transmission coefficient, on the other hand, monotonically decreases with increasing supercritical angles of incidence as well as decreasing monotonically with increasing n_2'' at any fixed supercritical angle of incidence. The phase angle $\phi_s \equiv \arg\{\Gamma_{TE}\}$ for the reflection coefficient monotonically decreases from its value at the critical angle of incidence to $-\pi$ at grazing incidence, as well as decreasing with increasing imaginary part n_2'' at any fixed supercritical angle of incidence. By comparison, the phase angle $\arg\{\tau_{TE}\}$ first decreases with increasing $\Theta_i \geq \Theta_c$ before reaching a minimum value

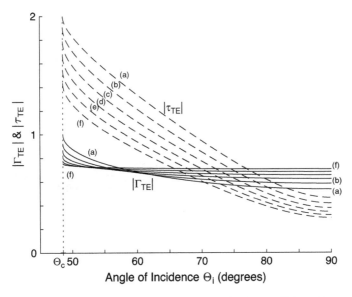

Fig. 6.20 Magnitude of the TE (s)-polarization Fresnel reflection coefficient $\Gamma_{TE} = \Gamma_s$ (solid curves) and transmission coefficient $\tau_{TE} = \tau_s$ (dashed curves) as a function of the critical to supercritical incidence angle $\Theta_i \geq \Theta_c$ when $\psi_i = \pi/2$, $\mu_1 = \mu_2$, $n_1 = 2.0 + 0.4i$ and (a) $n_2 = 1.5$, (b) $n_2 = 1.5 + 0.2i$, (c) $n_2 = 1.5 + 0.4i$, (d) $n_2 = 1.5 + 0.6i$, (e) $n_2 = 1.5 + 0.8i$, and (f) $n_2 = 1.5 + 1.0i$

and then increasing to 0 at grazing incidence. This phase angle for the transmitted inhomogeneous plane wave field also decreases with increasing imaginary part n_2'' at any fixed supercritical angle of incidence.

6.4.2 Complex $TM(p)$-Polarization Fresnel Coefficients

For a time-harmonic $TM(p)$-polarized plane electromagnetic wave propagating in medium 1 and incident upon the planar interface S separating medium 1 from medium 2 with the magnetic field vector linearly polarized along the $\hat{v} = \hat{1}_y$-direction (see Fig. 6.1)

$$\tilde{\mathbf{H}}_i(\mathbf{r}, \omega) = \hat{1}_y H_0 e^{i\tilde{\mathbf{k}}_i(\omega)\cdot\mathbf{r}}, \tag{6.162}$$

$$\tilde{\mathbf{E}}_i(\mathbf{r}, \omega) = \|c\| \frac{i}{\omega\epsilon_{c1}(\omega)} \nabla \times \tilde{\mathbf{H}}_i(\mathbf{r}, \omega)$$

$$= -\frac{\|c\|}{\omega\epsilon_{c1}(\omega)} \tilde{\mathbf{k}}_i(\omega) \times \hat{1}_y H_0 e^{i\tilde{\mathbf{k}}_i(\omega)\cdot\mathbf{r}}, \tag{6.163}$$

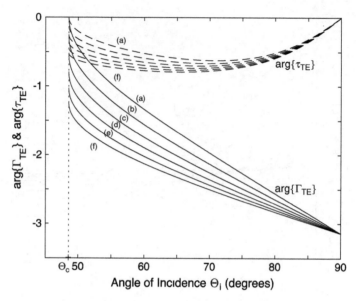

Fig. 6.21 Phase angle (in radians) of the TE (s)-polarization Fresnel reflection coefficient $\Gamma_{TE} = \Gamma_s$ (solid curves) and transmission coefficient $\tau_{TE} = \tau_s$ (dashed curves) as a function of the critical to supercritical incidence angle $\Theta_i \geq \Theta_c$ when $\psi_i = \pi/2$, $\mu_1 = \mu_2$, $n_1 = 2.0 + 0.4i$ and (a) $n_2 = 1.5$, (b) $n_2 = 1.5 + 0.2i$, (c) $n_2 = 1.5 + 0.4i$, (d) $n_2 = 1.5 + 0.6i$, (e) $n_2 = 1.5 + 0.8i$, and (f) $n_2 = 1.5 + 1.0i$

from the time-harmonic form (5.53) of Ampére's law, where the incident wave vector $\mathbf{k}_i(\omega) = \boldsymbol{\beta}_i(\omega) + i\boldsymbol{\alpha}_i(\omega)$ is described by Eqs. (6.125)–(6.127). The reflected wave field in medium 1 is given by

$$\tilde{\mathbf{H}}_r(\mathbf{r}, \omega) = -\hat{\mathbf{1}}_y \Gamma_{TM}(\omega) H_0 e^{i\tilde{\mathbf{k}}_r(\omega)\cdot\mathbf{r}}, \tag{6.164}$$

$$\tilde{\mathbf{E}}_r(\mathbf{r}, \omega) = \|c\| \frac{i}{\omega\epsilon_{c1}(\omega)} \nabla \times \tilde{\mathbf{H}}_r(\mathbf{r}, \omega)$$

$$= \frac{\|c\|}{\omega\epsilon_{c1}(\omega)} \tilde{\mathbf{k}}_r(\omega) \times \hat{\mathbf{1}}_y \Gamma_{TM}(\omega) H_0 e^{i\tilde{\mathbf{k}}_r(\omega)\cdot\mathbf{r}}, \tag{6.165}$$

where $\Gamma_{TM} \equiv H_r/H_0$ is defined as the Fresnel reflection coefficient for $TM(p)$-polarization, and where the reflected wave vector $\mathbf{k}_r(\omega) = \boldsymbol{\beta}_r(\omega) + i\boldsymbol{\alpha}_r(\omega)$ is described by Eqs. (6.130)–(6.133). Finally, the transmitted plane wave field in medium 2 is given by

$$\tilde{\mathbf{H}}_t(\mathbf{r}, \omega) = \hat{\mathbf{1}}_y \tau_{TM}(\omega) H_0 e^{i\tilde{\mathbf{k}}_t(\omega)\cdot\mathbf{r}}, \tag{6.166}$$

$$\tilde{\mathbf{E}}_t(\mathbf{r}, \omega) = \|c\| \frac{i}{\omega\epsilon_{c2}(\omega)} \nabla \times \tilde{\mathbf{H}}_t(\mathbf{r}, \omega)$$

$$= -\frac{\|c\|}{\omega\epsilon_{c2}(\omega)} \tilde{\mathbf{k}}_t(\omega) \times \hat{\mathbf{1}}_y \tau_{TM}(\omega) H_0 e^{i\tilde{\mathbf{k}}_t(\omega)\cdot\mathbf{r}}, \tag{6.167}$$

where $\tau_{TM} \equiv H_t/H_0$ is the Fresnel transmission coefficient for $TM(p)$-polarization, and where the transmitted wave vector $\mathbf{k}_t(\omega) = \boldsymbol{\beta}_t(\omega) + i\boldsymbol{\alpha}_t(\omega)$ is specified by Eqs. (6.136)–(6.139).

The total field in medium 1 at the interface S is given by the superposition of the incident and reflected fields as $\tilde{\mathbf{H}}_1(\mathbf{r}, \omega) = \tilde{\mathbf{H}}_i(\mathbf{r}, \omega) + \tilde{\mathbf{H}}_r(\mathbf{r}, \omega)$ and $\tilde{\mathbf{E}}_1(\mathbf{r}, \omega) = \tilde{\mathbf{E}}_i(\mathbf{r}, \omega) + \tilde{\mathbf{E}}_r(\mathbf{r}, \omega)$, so that

$$\tilde{\mathbf{H}}_1(\mathbf{r}, \omega) = \hat{\mathbf{1}}_y H_0 \left\{ e^{i\tilde{\mathbf{k}}_i \cdot \mathbf{r}} - \Gamma_{TM} e^{i\tilde{\mathbf{k}}_r \cdot \mathbf{r}} \right\}, \tag{6.168}$$

$$\tilde{\mathbf{E}}_1(\mathbf{r}, \omega) = -H_0 \frac{\|c\|}{\omega \epsilon_{c1}} \left\{ \tilde{\mathbf{k}}_i \times \hat{\mathbf{1}}_y e^{i\tilde{\mathbf{k}}_i \cdot \mathbf{r}} - \Gamma_{TM} \tilde{\mathbf{k}}_r \times \hat{\mathbf{1}}_y e^{i\tilde{\mathbf{k}}_r \cdot \mathbf{r}} \right\}, \tag{6.169}$$

where

$$\tilde{\mathbf{k}}_i \times \hat{\mathbf{1}}_y = \beta_1(\omega) \left(\hat{\mathbf{1}}_z \sin \Theta_i - \hat{\mathbf{1}}_x \cos \Theta_i \right) + i\boldsymbol{\alpha}_i(\omega) \times \hat{\mathbf{1}}_y,$$

$$\tilde{\mathbf{k}}_r \times \hat{\mathbf{1}}_y = \beta_1(\omega) \left(\hat{\mathbf{1}}_z \sin \Theta_r + \hat{\mathbf{1}}_x \cos \Theta_r \right) + i\boldsymbol{\alpha}_r(\omega) \times \hat{\mathbf{1}}_y,$$

and the total field in medium 2 is just the transmitted field given in Eqs. (6.166) and (6.167) with

$$\tilde{\mathbf{k}}_t \times \hat{\mathbf{1}}_y = \beta_2(\omega) \left(\hat{\mathbf{1}}_z \sin \Theta_t - \hat{\mathbf{1}}_x \cos \Theta_t \right) + i\boldsymbol{\alpha}_t(\omega) \times \hat{\mathbf{1}}_y.$$

With these substitutions, the tangential boundary condition (6.120) yields

$$\tau_{TM} e^{i\tilde{\mathbf{k}}_t \cdot \mathbf{r}} = e^{i\tilde{\mathbf{k}}_i \cdot \mathbf{r}} - \Gamma_{TM} e^{i\tilde{\mathbf{k}}_r \cdot \mathbf{r}}, \tag{6.170}$$

while the tangential boundary condition (6.119) gives

$$\frac{\tau_{TM}}{\epsilon_{c2}} \left[-\hat{\mathbf{1}}_y \beta_2 \cos \Theta_t e^{i\tilde{\mathbf{k}}_t \cdot \mathbf{r}} + i\hat{\mathbf{1}}_z \times (\boldsymbol{\alpha}_t \times \hat{\mathbf{1}}_y) e^{i\tilde{\mathbf{k}}_t \cdot \mathbf{r}} \right]$$

$$= -\hat{\mathbf{1}}_y \frac{\beta_1}{\epsilon_{c1}} \left[\cos \Theta_i e^{i\tilde{\mathbf{k}}_i \cdot \mathbf{r}} + \Gamma_{TM} \cos \Theta_r e^{i\tilde{\mathbf{k}}_r \cdot \mathbf{r}} \right]$$

$$+ \frac{i}{\epsilon_{c1}} \left[\hat{\mathbf{1}}_z \times (\boldsymbol{\alpha}_i \times \hat{\mathbf{1}}_y) e^{i\tilde{\mathbf{k}}_i \cdot \mathbf{r}} - \Gamma_{TM} \hat{\mathbf{1}}_z \times (\boldsymbol{\alpha}_r \times \hat{\mathbf{1}}_y) e^{i\tilde{\mathbf{k}}_r \cdot \mathbf{r}} \right]. \tag{6.171}$$

As for the $TE(s)$-polarized case presented in Sect. 6.4.1, these two expressions must be satisfied for all $\mathbf{r} \in S$, so that Eq. (6.144) is again obtained, leading to the law of reflection given in Eq. (6.145) and the law of refraction given in Eqs. (6.146) and (6.148), as well as to the set of relations in Eqs. (6.149)–(6.151) for the incident, reflected, and transmitted wave vectors.

With these results, Eq. (6.170) becomes

$$\tau_{TM} = 1 - \Gamma_{TM},$$

and Eq. (6.171) simplifies to

$$\frac{\tau_{TM}}{\epsilon_{c2}}(\beta_2 \cos \Theta_t + i\alpha_2 \sin \psi_t) = \frac{1}{\epsilon_{c1}}(1 + \Gamma_{TM})(\beta_1 \cos \Theta_i + i\alpha_1 \sin \psi_i).$$

The solution of this pair of equations then yields the *generalized Fresnel reflection and transmission coefficients for $TM(p)$-polarization*, given by

$$\Gamma_{TM}(\omega) \equiv \Gamma_p(\omega) = \frac{\epsilon_{c1}(n_2' \cos \Theta_t + in_2'' \sin \psi_t) - \epsilon_{c2}(n_1' \cos \Theta_i + in_1'' \sin \psi_i)}{\epsilon_{c1}(n_2' \cos \Theta_t + in_2'' \sin \psi_t) + \epsilon_{c2}(n_1' \cos \Theta_i + in_1'' \sin \psi_i)},$$

$$(6.172)$$

$$\tau_{TM}(\omega) \equiv \tau_p(\omega) = \frac{2\epsilon_{c2}(n_1' \cos \Theta_i + in_1'' \sin \psi_i)}{\epsilon_{c1}(n_2' \cos \Theta_t + in_2'' \sin \psi_t) + \epsilon_{c2}(n_1' \cos \Theta_i + in_1'' \sin \psi_i)},$$

$$(6.173)$$

respectively. For a homogeneous plane wave ($\psi_i = \psi_t = \pi/2$) at *normal incidence* ($\Theta_i = \Theta_t = 0$), these generalized Fresnel equations simplify to the familiar expressions

$$\Gamma_{TM}(\omega)\big|_{\Theta_i=0} = \frac{\eta_2(\omega) - \eta_1(\omega)}{\eta_2(\omega) + \eta_1(\omega)} \quad \& \quad \tau_{TM}(\omega)\big|_{\Theta_i=0} = \frac{2\eta_1(\omega)}{\eta_2(\omega) + \eta_1(\omega)},$$

so that $\Gamma_{TM}(\omega) = \Gamma_{TE}(\omega)$ and $\tau_{TM}(\omega) = (\eta_2/\eta_1)\tau_{TE}(\omega)$ at normal incidence. In the idealized lossless limit, the generalized Fresnel reflection and transmission coefficients given in Eqs. (6.172) and (6.173) reduce to the well-known results for perfect dielectrics given in Eqs. (6.78)–(6.80).

The angular dependence of the Fresnel reflection coefficient Γ_{TM} when a homogeneous plane wave ($\psi_i = \pi/2 - \Theta_i$) is incident on an optically denser medium ($n_2' > n_1'$) is illustrated in Fig. 6.22 when medium 1 is lossless with refractive index $n_1 = 1.5$ and medium 2 is lossy with complex refractive index $n_2 = 2.0 + in_2''$ with imaginary part varying from $n_2'' = 0 \to 2$ in 0.5 increments. Notice that $|\Gamma_{TM}| \to 1$ and $|\tau_{TM}| \to 0$ as $\Theta_i \to 90°$ independent of the loss in medium 2 when medium 1 is lossless, as seen from Eq. (6.172) when $n_1'' = 0$. However, such is not always the case when medium 1 is also lossy, as illustrated in Fig. 6.23 when $n_1 = 1.5 + 0.5i$ and $n_2 = 2.0 + in_2''$ with imaginary part varying from $n_2'' = 0 \to 0.5$ in 0.1 increments and in Fig. 6.24 with imaginary part varying from $n_2'' = 0.5 \to 3.0$ in 0.5 increments. For the special case when $n_1'' = 0$ and the ratio $\epsilon_{c1}/\epsilon_{c2}$ is real-valued, the $TM(p)$-reflection coefficient simplifies to

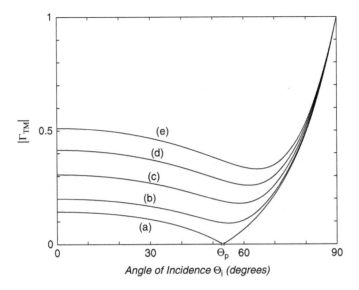

Fig. 6.22 Magnitude of the $TM(p)$-polarization Fresnel reflection coefficient $\Gamma_{TM} = \Gamma_p$ as a function of the incidence angle Θ_i when $\psi_i = \pi/2 - \Theta_i$, $\mu_1 = \mu_2$, $n_1 = 1.5$ and (a) $n_2 = 2.0$, (b) $n_2 = 2.0 + 0.5i$, (c) $n_2 = 2.0 + i$, (d) $n_2 = 2.0 + 1.5i$, and (e) $n_2 = 2.0 + 2i$

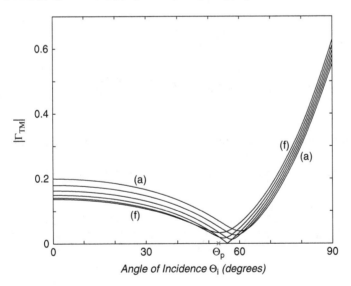

Fig. 6.23 Magnitude of the $TM(p)$-polarization Fresnel reflection coefficient $\Gamma_{TM} = \Gamma_p$ as a function of the incidence angle Θ_i when $\psi_i = \pi/2$, $\mu_1 = \mu_2$, $n_1 = 1.5 + 0.5i$ and (a) $n_2 = 2.0$, (b) $n_2 = 2.0 + 0.1i$, (c) $n_2 = 2.0 + 0.2i$, (d) $n_2 = 2.0 + 0.3i$, (e) $n_2 = 2.0 + 0.4i$, and (f) $n_2 = 2.0 + 0.5i$. Notice that the curves appear in alphabetic order from (a) to (f)

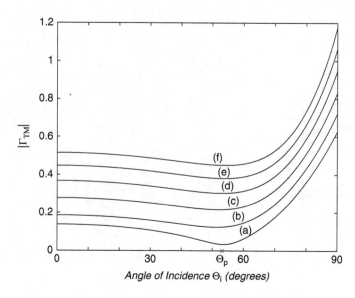

Fig. 6.24 Magnitude of the TM-polarization Fresnel reflection coefficient $\Gamma_{TM} = \Gamma_p$ as a function of the incidence angle Θ_i when $\psi_i = \pi/2$, $\mu_1 = \mu_2$, $n_1 = 1.5 + 0.5i$ and (a) $n_2 = 2.0 + 0.5i$, (b) $n_2 = 2.0 + i$, (c) $n_2 = 2.0 + 1.5i$, (d) $n_2 = 2.0 + 2i$, (e) $n_2 = 2.0 + 2.5i$, and (f) $n_2 = 2.0 + 3i$

$$\Gamma_{TM}(\omega)\big|_{\Theta_i = \pi/2} = -e^{-i\phi_p(\omega)} \quad \text{with}$$

$$\phi_p(\omega) = 2\arctan\left(\frac{\epsilon_{c1} n_2'' \sin\psi_t}{\epsilon_{c2}\sqrt{n_1'^2 - n_2'^2}}\right). \tag{6.174}$$

Finally, notice that $|\Gamma_{TM}|$ at grazing incidence on the optically denser medium ($n_2' > n_1'$) increases above unity when the imaginary part of $n_2 = n_2' + in_2''$ increases above the real part, as seen in Fig. 6.24. From Eq. (6.172) with $\Theta_i = \pi/2$ so that $\cos\Theta_t = (1 - n_1'^2/n_2'^2)^{1/2}$, and with $\psi_t = \psi_i = \pi/2$, one obtains

$$\Gamma_{TM}\big|_{\Theta_i = \pi/2} = \frac{\left(\epsilon_{c1}' \Delta n' + \epsilon_{c2}'' n_1'' - \epsilon_{c1}'' n_2''\right) + i\left(\epsilon_{c1}'' \Delta n' - \epsilon_{c2}' n_1'' + \epsilon_{c1}' n_2''\right)}{\left(\epsilon_{c1}' \Delta n' - \epsilon_{c2}'' n_1'' - \epsilon_{c1}'' n_2''\right) + i\left(\epsilon_{c1}'' \Delta n' + \epsilon_{c2}' n_1'' + \epsilon_1' n_2''\right)} \tag{6.175}$$

with $\Delta n' \equiv (n_2'^2 - n_1'^2)^{1/2}$. A maximum value of $|\Gamma_{TM}| \leq 2$ at grazing incidence is reached at some value of $n_2'' > n_2'$ after which $|\Gamma_{TM}|$ decreases to unity from above as $n_2'' \to \infty$, as illustrated in Fig. 6.25. If $\psi_i \neq \pi/2$, then n_1'' and n_2'' in Eq. (6.175) are replaced by $n_1'' \sin\psi_i$ and $n_2'' \sin\psi_t$, respectively.

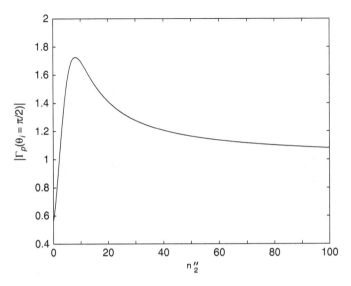

Fig. 6.25 Magnitude of the TM-polarization Fresnel reflection coefficient $\Gamma_{TM} = \Gamma_p$ with $\psi_i = \pi/2$ at grazing incidence ($\theta_i = \pi/2$) on the optically denser medium as a function of the imaginary part n_2'' of the complex index of refraction of medium 2 when $n_1 = 1.5 + 0.5i$ and $n_2 = 2.0 + in_2''$ with $\mu_1 = \mu_2 = 1$

The *polarizing* or *Brewster's* angle Θ_P at which Γ_{TM} vanishes, given by [1]

$$\tan \Theta_P = \frac{n_2'}{n_1'}, \quad n_1'' = n_2'' = 0, \tag{6.176}$$

when both media are lossless, indicated in Fig. 6.22 for curve (a), only occurs in the general case at a specific value of the material loss and at a slightly larger angle $\Theta_P' > \Theta_P$, as indicated in Fig. 6.23 for curve (d) when $n_2 = 2 + 2i$. The condition for the appearance of a polarizing angle is thus obtained from Eq. (6.172) by setting the real and imaginary parts of the numerator equal to zero, resulting in the pair of equations

$$\epsilon_{c2}' n_1' \cos \Theta_i - \epsilon_{c2}'' n_1'' \sin \psi_i = \epsilon_{c1}' n_2' \cos \Theta_t - \epsilon_{c1}'' n_2'' \sin \psi_t,$$

$$\epsilon_{c2}'' n_1' \cos \Theta_i + \epsilon_{c2}' n_1'' \sin \psi_i = \epsilon_{c1}'' n_2' \cos \Theta_t + \epsilon_{c1}' n_2'' \sin \psi_t,$$

where $\cos \Theta_t = \sqrt{1 - (n_1'^2/n_2'^2) \sin^2 \Theta_i}$ from the law of refraction. With this substitution, this pair of simultaneous equations can then be solved for $\cos \Theta_i$ when

$\Theta_i = \Theta'_p$, resulting in the pair of relations

$$\cos\Theta'_p = \frac{1}{n'_1 \Delta\epsilon'_c}\left[-\epsilon'_{c2}\Delta''_{\epsilon n} + \sqrt{\Delta''^2_{\epsilon n} + \Delta\epsilon'_c \Delta n'}\right], \qquad (6.177)$$

$$\cos\Theta'_p = \frac{1}{n'_1 \Delta\epsilon''_c}\left[-\epsilon''_{c2}\Delta'_{\epsilon n} + \epsilon''_{c1}\sqrt{\Delta'^2_{\epsilon n} + \Delta\epsilon''_c \Delta n'}\right], \qquad (6.178)$$

where the five inter-medium difference quantities $\Delta n' \equiv n'^2_2 - n'^2_1$, $\Delta\epsilon'_c \equiv \epsilon'^2_{c2} - \epsilon'^2_{c1}$, $\Delta\epsilon''_c \equiv \epsilon''^2_{c2} - \epsilon''^2_{c1}$, $\Delta'_{\epsilon n} \equiv \epsilon'_{c2}n''_1 \sin\psi_i - \epsilon'_{c1}n''_2 \sin\psi_t$, and lastly $\Delta''_{\epsilon n} \equiv \epsilon''_{c1}n''_2 \sin\psi_t - \epsilon''_{c2}n''_1 \sin\psi_i$, have been introduced for notational convenience. The polarizing angle Θ'_p is then given by the value satisfying both Eqs. (6.177) and (6.178). For example, for the situation depicted in Figs. 6.23 and 6.24 with $n'_1 = 1.5$ and $n'_2 = 2.0$, the loss-free polarizing angle is found from Eq. (6.176) to be $\Theta_P \simeq 53.1°$, whereas when both media are lossy with $n_1 = 1.5 + 0.5i$ and $n_2 = 2.0 + n''_2 i$, a polarizing angle is obtained only when $n''_2 \simeq 0.2898$ at which value $\Theta'_p \simeq 55.9°$, as occurs (approximately) for curve (c) in Fig. 6.23. This then establishes the result:

When both media are lossy, a polarizing angle occurs only for a specific pair of values for n''_1 and n''_2, and that polarizing angle Θ'_p is typically larger than the loss-free value Θ_p. Otherwise a pseudo-Brewster angle [9, 10] occurs at an incidence angle Θ_{pb} at which $|\Gamma_{TM}|$ is a minimum.

As a special case, when medium 1 is loss-free ($\epsilon''_{c1} = 0$) and medium 2 is lossy ($\epsilon''_{c2} > 0$), the second relation yields $\cos\Theta_i = \epsilon''_{c1}n''_2/\epsilon''_{c2}n'_1$ and the first relation yields $\cos\Theta_i = \sqrt{\epsilon'^2_{c1}\Delta n'/n'^2_1 \Delta\epsilon'_c}$, so that a polarizing angle exists and is given by

$$\cos\Theta'_p = \frac{\epsilon'_{c1}}{n'_1}\sqrt{\frac{\Delta n'}{\Delta\epsilon'_c}} \qquad (6.179)$$

when the condition $n''_2/\epsilon''_{c2} = \sqrt{\Delta n'/\Delta\epsilon'_c}$ is satisfied.

In the lossless media case ($n''_1 = n''_2 = 0$), the phase angle $\arg\{\Gamma_{TM}\}$ of the $TM(p)$-polarization reflection coefficient is equal to $-\pi$ for angles of incidence Θ_i below the polarizing angle Θ_p and then discontinuously jumps to 0 for incidence angles above Θ_p, as illustrated by curve (a) in Fig. 6.26. This discontinuous behavior only occurs in the idealized lossless media case, the reflection phase angle change being continuous for all $\Theta_i \in [0, \pi]$ when medium 2 possesses loss, as seen in the figure. Notice that the set of curves presented here for $\arg\{\Gamma_{TM}\}$ correspond to the set of curves illustrated in Fig. 6.22 for $|\Gamma_{TM}|$. In each case, $\arg\{\Gamma_{TM}\} \to 0$ as $\Theta_i \to 90°$. For comparison, the angular dependence of the reflection coefficient phase angle when both media are lossy is illustrated in Fig. 6.27 for the same set of cases presented in Fig. 6.24 for $|\Gamma_{TM}|$. Notice that the phase angle at grazing incidence ($\Theta_i \to 90°$) decreases as the imaginary part of n_2 increases.

Just as for the $TE(s)$-polarization case, the critical angle of incidence Θ_c defined in Eq. (6.155) when incidence is on the optically rarer medium ($n'_1 > n'_2$)

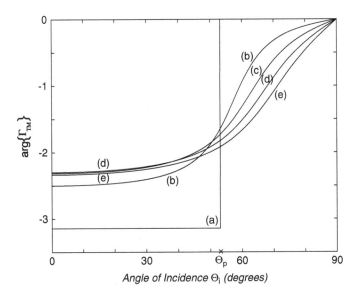

Fig. 6.26 Phase angle (in radians) of the $TM(p)$-polarization Fresnel reflection coefficient $\Gamma_{TM} = \Gamma_p$ (see Fig. 6.22 for $|\Gamma_{TM}|$) as a function of the incidence angle Θ_i when $\psi_i = \pi/2 - \Theta_i$, $\mu_1 = \mu_2 = 1$, $n_1 = 1.5$ and (a) $n_2 = 2.0$, (b) $n_2 = 2.0 + 0.5i$, (c) $n_2 = 2.0 + i$, (d) $n_2 = 2.0 + 1.5i$, and (e) $n_2 = 2.0 + 2i$

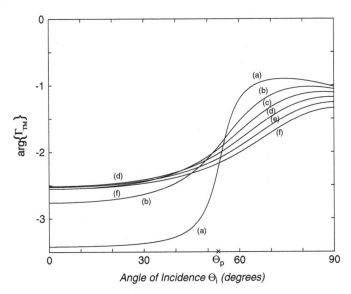

Fig. 6.27 Phase angle (in radians) of the $TM(p)$-polarization Fresnel reflection coefficient $\Gamma_{TM} = \Gamma_p$ (see Fig. 6.33 for $|\Gamma_{TM}|$) as a function of the incidence angle Θ_i when $\psi_i = \pi/2$, $\mu_1 = \mu_2 = 1$, $n_1 = 1.5 + 0.5i$ and (a) $n_2 = 2.0 + 0.5i$, (b) $n_2 = 2.0 + i$, (c) $n_2 = 2.0 + 1.5i$, (d) $n_2 = 2.0 + 2i$, (e) $n_2 = 2.0 + 2.5i$, and (f) $n_2 = 2.0 + 3i$

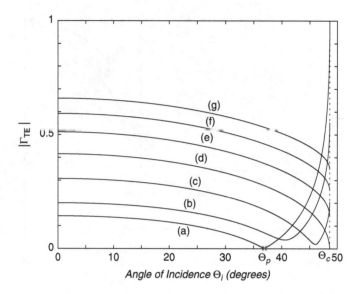

Fig. 6.28 Magnitude of the $TM(p)$-polarization Fresnel reflection coefficient $\Gamma_{TM} = \Gamma_p$ as a function of the incidence angle Θ_i when $\psi_i = \pi/2 - \Theta_i$, $\mu_1 = \mu_2 = 1$, $n_1 = 2.0$ and (a) $n_2 = 1.5$, (b) $n_2 = 1.5 + 0.5i$, (c) $n_2 = 1.5 + i$, (d) $n_2 = 1.5 + 1.5i$, (e) $n_2 = 1.5 + 2i$, (f) $n_2 = 1.5 + 2.5i$, and (g) $n_2 = 1.5 + 3i$

is unaffected by the presence of material loss. The angular dependence of the magnitude of the Fresnel reflection coefficient Γ_{TM} in that case with homogeneous plane wave incidence ($\psi_i = \pi/2 - \Theta_i$) is illustrated in Fig. 6.28 when $n_1 = 2.0$ and $n_2 = 1.5 + in_2''$ with imaginary part varying from $n_2'' = 0 \rightarrow 3.0$ in 0.5 increments. For comparison, the corresponding $TE(s)$-polarization case is given in Fig. 6.16. Notice that the polarizing angle Θ_p only occurs for the zero loss case described by curve (a) and that as the loss in medium 2 increases, the incidence angle at which the minimum in the magnitude of the reflection coefficient occurs is shifted upwards towards the critical angle Θ_c. For incidence at the critical angle $\Theta_i = \Theta_c$ where $\sin \Theta_c = n_2'/n_1'$, $\Theta_t = \pi/2$ and $\cos \Theta_c = \sqrt{1 - (n_2'/n_1')^2}$ so that [cf. Eq. (6.156)]

$$\Gamma_{TM}\big|_{\Theta_i=\Theta_c} = -\frac{\sqrt{n_1'^2 - n_2'^2} + i\left(n_1'' \sin \psi_i - \frac{\epsilon_{c1}}{\epsilon_{c2}} n_2'' \sin \psi_t\right)}{\sqrt{n_1'^2 - n_2'^2} + i\left(n_1'' \sin \psi_i + \frac{\epsilon_{c1}}{\epsilon_{c2}} n_2'' \sin \psi_t\right)}, \qquad (6.180)$$

and

$$\tau_{TM}\big|_{\Theta_i=\Theta_c} = \frac{2\left[\sqrt{n_1'^2 - n_2'^2} + in_1'' \sin \psi_i\right]}{\sqrt{n_1'^2 - n_2'^2} + i\left(n_1'' \sin \psi_i + \frac{\epsilon_{c1}}{\epsilon_{c2}} n_2'' \sin \psi_t\right)}. \qquad (6.181)$$

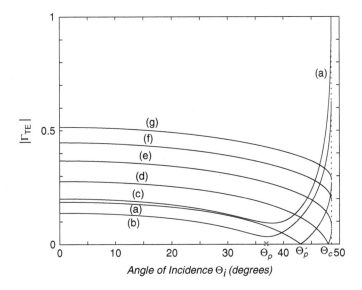

Fig. 6.29 Magnitude of the $TM(p)$-polarization Fresnel reflection coefficient $\Gamma_{TM} = \Gamma_p$ as a function of the incidence angle Θ_i when $\psi_i = \pi/2$, $\mu_1 = \mu_2 = 1$, $n_1 = 2.0 + 0.5i$ and (a) $n_2 = 1.5$, (b) $n_2 = 1.5 + 0.5i$, (c) $n_2 = 1.5 + i$, (d) $n_2 = 1.5 + 1.5i$, (e) $n_2 = 1.5 + 2i$, (f) $n_2 = 1.5 + 2.5i$, and (g) $n_2 = 1.5 + 3i$

Hence, if medium 2 is lossless ($n_2'' = 0$), then $\Gamma_{TM}(\omega) = -1$ and $\tau_{TM}(\omega) = 2$ at critical incidence, as seen for curve (a) in Fig. 6.29, whereas if medium 1 is lossless ($n_1'' = 0$), then $\Gamma_{TM}(\omega) = -e^{-i\phi_p(\omega)}$ with

$$\phi_p(\omega) = 2 \arctan \left(\frac{\epsilon_{c1}(\omega) n_2''(\omega) \sin \psi_t}{\epsilon_{c2}(\omega) \sqrt{n_1'^2(\omega) - n_2'^2(\omega)}} \right), \tag{6.182}$$

provided that $\epsilon_{c1}/\epsilon_{c2}$ is real-valued.

At both critical and supercritical angles of incidence $\Theta_i \geq \Theta_c$, $\Theta_t = \pi/2$, and the transmitted wave vector in medium 2 is given by Eq. (6.157). With the total field in medium 1 (incident plus reflected wave fields) given by Eqs. (6.168)–(6.169) and the total field in medium2 (the transmitted wave field) given by Eqs. (6.166) and (6.167), the continuity of the tangential component of the $\tilde{\mathbf{H}}$-field, as specified by the boundary condition in Eq. (6.47), yields

$$\tau_{TM} = 1 - \Gamma_{TM},$$

and the continuity of the tangential component of the $\tilde{\mathbf{E}}$-field, as specified by the boundary condition given in Eq. (6.46), yields

$$i \frac{\tau_{TM}}{\epsilon_{c2}} \left(n_2'' \sin \psi_t + \sqrt{n_1'^2 \sin^2 \Theta_i - n_2'^2} \right) = \frac{(1 + \Gamma_{TM})}{\epsilon_{c1}} \left(n_1' \cos \Theta_i + i n_1'' \sin \psi_i \right).$$

The solution of this simultaneous pair of algebraic equations then yields the *critical and supercritical Fresnel reflection and transmission coefficients for TM (p)-polarization*

$$\Gamma_{TM} = -\frac{\epsilon_{c2}(n_1' \cos \Theta_i + i n_1'' \sin \psi_i) - i\epsilon_{c1} \left(n_2'' \sin \psi_t + \sqrt{n_1'^2 \sin^2 \Theta_i - n_2'^2} \right)}{\epsilon_{c2}(n_1' \cos \Theta_i + i n_1'' \sin \psi_i) + i\epsilon_{c1} \left(n_2'' \sin \psi_t + \sqrt{n_1'^2 \sin^2 \Theta_i - n_2'^2} \right)},$$

$$\tag{6.183}$$

$$\tau_{TM} = \frac{2\epsilon_{c2}(n_1' \cos \Theta_i + i n_1'' \sin \psi_i)}{\epsilon_{c2}(n_1' \cos \Theta_i + i n_1'' \sin \psi_i) + i\epsilon_{c1} \left(n_2'' \sin \psi_t + \sqrt{n_1'^2 \sin^2 \Theta_i - n_2'^2} \right)},$$

$$\tag{6.184}$$

for $\Theta_i \geq \Theta_c$. At the critical angle of incidence $\Theta_i = \Theta_c = \arcsin(n_2'/n_1')$, $\cos \Theta_c = \sqrt{n_1'^2 - n_2'^2}/n_1'$ and Eqs. (6.183) and (6.184) reduce to those given in Eqs. (6.180) and (6.181), respectively. At grazing incidence, $\Theta_i = \pi/2$ and these two coefficients simplify to

$$\Gamma_{TM}\big|_{\Theta_i=\pi/2} = -\frac{\epsilon_{c2}n_1'' \sin \psi_i - \epsilon_{c1} \left(n_2'' \sin \psi_t + \sqrt{n_1'^2 - n_2'^2} \right)}{\epsilon_{c2}n_1'' \sin \psi_i + \epsilon_{c1} \left(n_2'' \sin \psi_t + \sqrt{n_1'^2 - n_2'^2} \right)}, \tag{6.185}$$

$$\tau_{TM}\big|_{\Theta_i=\pi/2} = \frac{2\epsilon_{c2}n_1'' \sin \psi_i}{\epsilon_{c2}n_1'' \sin \psi_i + \epsilon_{c1} \left(n_2'' \sin \psi_t + \sqrt{n_1'^2 - n_2'^2} \right)}, \tag{6.186}$$

so that $\Gamma_{TM} = 1$ and $\tau_{TM} = 0$ at grazing incidence provided that $n_1'' = 0$; otherwise $|\Gamma_{TM}| < 1$ at grazing incidence.

The magnitude and phase of the $TM(p)$-polarization Fresnel reflection and transmission coefficients as a function of critical to supercritical angles of incidence $\Theta_i \geq \Theta_c$ are illustrated in Figs. 6.30 and 6.31, respectively, for homogeneous plane wave incidence ($\psi_i = \pi/2 - \Theta_i$) when $\mu_1 = \mu_2$, medium 1 is lossless ($n_1 = 2.0$), and the loss in medium 2 ($n_2 = 1.5 + i n_2''$) varies from $n_2'' = 0 \to 1$ in $\Delta n_2'' = 0.2$ increments. Notice that $|\Gamma_{TM}|$ decreases from critical to grazing angles

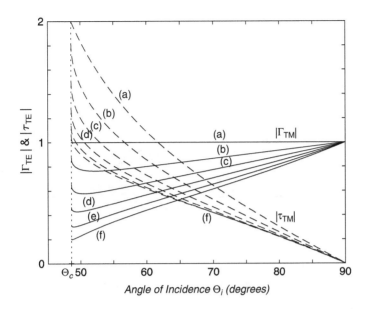

Fig. 6.30 Magnitude of the TM (p)-polarization Fresnel reflection coefficient $\Gamma_{TM} = \Gamma_p$ (solid curves) and transmission coefficient $\tau_{TM} = \tau_p$ (dashed curves) as a function of the critical to supercritical incidence angle $\Theta_i \geq \Theta_c$ when $\psi_i = \pi/2 - \Theta_i$, $\mu_1 = \mu_2 = 1$, $n_1 = 2.0$ and (a) $n_2 = 1.5$, (b) $n_2 = 1.5 + 0.2i$, (c) $n_2 = 1.5 + 0.4i$, (d) $n_2 = 1.5 + 0.6i$, (e) $n_2 = 1.5 + 0.8i$, and (f) $n_2 = 1.5 + i$

of incidence with increasing loss in medium 2 with phase angle $\phi_p \equiv \arg\{\Gamma_{TM}\}$ monotonically decreasing from $\pi \to 0$. In addition, $|\tau_{TM}|$ monotonically decreases from $2 \to 0$ from critical to grazing angles of incidence with phase angle $\arg\{\tau_{TM}\}$ monotonically decreasing from 0 with grazing incidence value increasing from $-\pi/2$ as the loss in medium 2 increases. The corresponding behavior for the TE-polarization case given in Figs. 6.28 and 6.29 shows that their behavior is quite dissimilar, this being due to the appearance of $\epsilon_{cj}(\omega)$ in place of μ_j.

For comparison, the magnitude and phase of the $TM(p)$-polarization Fresnel reflection and transmission coefficients as a function of critical to supercritical angles of incidence $\Theta_i \geq \Theta_c$ is illustrated in Figs. 6.32 and 6.33, respectively, when $\psi_t = \psi_i = \pi/2$, $\mu_1 = \mu_2$, and medium 1 is lossy with $n_1 = 2.0 + 0.4i$ with the loss in medium 2 again varying from $n_2'' = 0 \to 1.0$ in $\Delta n_2'' = 0.2$ increments. Notice that $|\Gamma_{TM}|$ decreases with increasing loss n_2'' for supercritical incidence angles increasing away from Θ_c, but that this ordering reverses beyond some angle before grazing incidence, as illustrated in Fig. 6.32. This behavior is partially reflected in the behavior of $|\tau_{TM}| = |1 + \Gamma_{TM}|$. The phase angle $\phi_p \equiv \arg\{\Gamma_{TM}\}$ for the reflection coefficient monotonically decreases from its value at the critical angle of incidence towards 0 and below at grazing incidence, as well as decreasing with increasing imaginary part n_2'' at any fixed supercritical angle of incidence. By comparison, the phase angle $\arg\{\tau_{TM}\}$ first decreases with increasing $\Theta_i \geq \Theta_c$ before reaching a

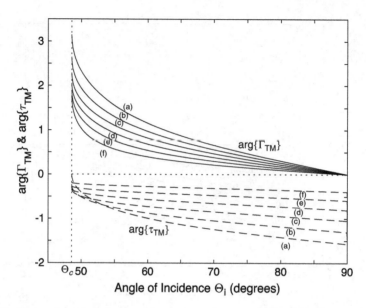

Fig. 6.31 Phase angle (in radians) of the TM (p)-polarization Fresnel reflection coefficient $\Gamma_{TM} = \Gamma_p$ (solid curves) and transmission coefficient $\tau_{TM} = \tau_p$ (dashed curves) as a function of the critical to supercritical incidence angle $\Theta_i \geq \Theta_c$ when $\psi_i = \pi/2 - \Theta_i$, $\mu_1 = \mu_2 = 1$, $n_1 = 2.0$ and (a) $n_2 = 1.5$, (b) $n_2 = 1.5 + 0.2i$, (c) $n_2 = 1.5 + 0.4i$, (d) $n_2 = 1.5 + 0.6i$, (e) $n_2 = 1.5 + 0.8i$, and (f) $n_2 = 1.5 + i$

minimum value and then increasing towards 0 and above at grazing incidence. The corresponding behavior for the TE-polarization case given in Figs. 6.20 and 6.21 shows that, as before, their behavior is quite dissimilar, this also being due to the appearance of $\epsilon_{cj}(\omega)$ in place of μ_j.

Finally, notice that the reflection and transmission behavior for the two orthogonal $TE(s)$ and $TM(p)$ polarization cases become increasingly similar when loss is also included in the magnetic permeability and not just in the dielectric permittivity. This will occur, for example, in DNG meta-materials.

6.5 Lossy Dispersive-Lossy DNG Meta-Material Interface

Consider first a time-harmonic plane electromagnetic wave traveling in an ordinary, naturally occurring dispersive lossy medium 1 with complex index of refraction $n_1(\omega) = n_1'(\omega) + in_1''(\omega)$, $n_1'(\omega) > 0$, $n_1''(\omega) \geq 0$, that is incident upon a planar interface S separating this medium from a dispersive lossy double negative index (DNG) meta-material (medium 2) with complex index of refraction $n_2(\omega) = n_2'(\omega) + in_2''(\omega)$, $n_2'(\omega) < 0$, $n_2''(\omega) \geq 0$ over the frequency domain of interest. The tangential boundary conditions given in Eqs. (6.119) and (6.120) then apply as is,

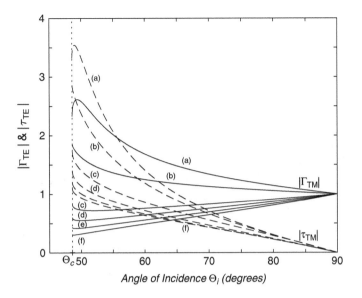

Fig. 6.32 Magnitude of the TM (p)-polarization Fresnel reflection coefficient $\Gamma_{TM} = \Gamma_p$ (solid curves) and transmission coefficient $\tau_{TM} = \tau_p$ (dashed curves) as a function of the critical to supercritical incidence angle $\Theta_i \geq \Theta_c$ when $\mu_1 = \mu_2 = 1$, $n_1 = 2.0 + 0.4i$ and (a) $n_2 = 1.5$, (b) $n_2 = 1.5 + 0.2i$, (c) $n_2 = 1.5 + 0.4i$, (d) $n_2 = 1.5 + 0.6i$, (e) $n_2 = 1.5 + 0.8i$, and (f) $n_2 = 1.5 + i$

stating that the tangential components of the **E** and **H** field vectors are continuous across the interface.

For TE(s)-polarization, the field vectors incident upon and reflected from the interface S are given by Eqs. (6.122)–(6.127) and (6.128)–(6.133), respectively. For the transmitted field, Eqs. (6.37) and (6.38) apply with Eqs. (6.39) and (6.40) replaced here by

$$\tilde{\mathbf{k}}_t(\omega) = -\boldsymbol{\beta}_t(\omega) + i\boldsymbol{\alpha}_t(\omega), \tag{6.187}$$

with

$$\beta_t(\omega) \equiv |-\boldsymbol{\beta}_t(\omega)| = -\beta_2(\omega) = \frac{\omega}{c} |n_2'(\omega)|, \tag{6.188}$$

$$\alpha_t(\omega) \equiv |\boldsymbol{\alpha}_t(\omega)| = \alpha_2(\omega) = \frac{\omega}{c} n_2''(\omega), \tag{6.189}$$

for real-valued ω. Notice that $\beta_2(\omega) = (\omega/c)n_2'(\omega)$ with $n_2'(\omega) < 0$ over the (real-valued) frequency domain of interest. If Θ_i is the *angle of incidence*, Θ_r the *angle of reflection*, and Θ_t the *angle of refraction*, each measured from the normal to the

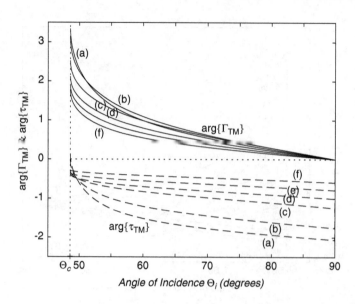

Fig. 6.33 Phase angle (in radians) of the TM (p)-polarization Fresnel reflection coefficient $\Gamma_{TM} = \Gamma_p$ (solid curves) and transmission coefficient $\tau_{TM} = \tau_p$ (dashed curves) as a function of the critical to supercritical incidence angle $\Theta_i \geq \Theta_c$ when $\mu_1 = \mu_2 = 1$, $n_1 = 2.0 + 0.4i$ and (a) $n_2 = 1.5$, (b) $n_2 = 1.5 + 0.2i$, (c) $n_2 = 1.5 + 0.4i$, (d) $n_2 = 1.5 + 0.6i$, (e) $n_2 = 1.5 + 0.8i$, and (f) $n_2 = 1.5 + i$

interface surface S, then [cf. Eqs. (6.53), (6.56), and (6.59)]

$$\boldsymbol{\beta}_i(\omega) = \beta_1(\omega)\left(\hat{\mathbf{1}}_x \sin \Theta_i + \hat{\mathbf{1}}_z \cos \Theta_i\right), \qquad (6.190)$$

$$\boldsymbol{\beta}_r(\omega) = \beta_1(\omega)\left(\hat{\mathbf{1}}_x \sin \Theta_r - \hat{\mathbf{1}}_z \cos \Theta_r\right), \qquad (6.191)$$

$$\boldsymbol{\beta}_t(\omega) = \beta_2(\omega)\left(\hat{\mathbf{1}}_x \sin \Theta_t + \hat{\mathbf{1}}_z \cos \Theta_t\right). \qquad (6.192)$$

Because Eq. (6.38) must be satisfied for all $\mathbf{r} \in S$, one again finds that

$$\beta_1(\omega)\left(\hat{\mathbf{1}}_x \sin \Theta_i + \hat{\mathbf{1}}_z \cos \Theta_i\right) \cdot \left(\hat{\mathbf{1}}_z \times \left(\hat{\mathbf{1}}_z \times \mathbf{r}\right)\right)$$
$$= \beta_1(\omega)\left(\hat{\mathbf{1}}_x \sin \Theta_r - \hat{\mathbf{1}}_z \cos \Theta_r\right) \cdot \left(\hat{\mathbf{1}}_z \times \left(\hat{\mathbf{1}}_z \times \mathbf{r}\right)\right)$$
$$= \beta_2(\omega)\left(\hat{\mathbf{1}}_x \sin \Theta_t + \hat{\mathbf{1}}_z \cos \Theta_t\right) \cdot \left(\hat{\mathbf{1}}_z \times \left(\hat{\mathbf{1}}_z \times \mathbf{r}\right)\right),$$

for the real part, so that $\beta_1(\omega)\sin\Theta_i = \beta_1(\omega)\sin\Theta_r = \beta_2(\omega)\sin\Theta_t$, and from the imaginary part that $\alpha_i(\omega)\cos\psi_i = \alpha_r(\omega)\cos\psi_r = \alpha_t(\omega)\cos\psi_t$, where ψ_j ($j = i, r, t$) is the angle between the attenuation vector $\boldsymbol{\alpha}_j$ and the interface plane S (taken here as the xy-plane). Hence, one again finds that $n_1'' \cos \psi_i = n_1'' \cos \psi_r = n_2'' \cos \psi_t$. Upon combining these results, one obtains the *generalized law of reflection*

$$\Theta_r = \Theta_i \quad \& \quad \psi_r = \psi_i \tag{6.193}$$

and the *generalized law of refraction*

$$n_1'(\omega) \sin \Theta_i = n_2'(\omega) \sin \Theta_t, \tag{6.194}$$

$$n_1''(\omega) \cos \psi_i = n_2''(\omega) \cos \psi_t. \tag{6.195}$$

Because $n_2'(\omega) < 0$ while $n_1'(\omega) > 0$, then $\Theta_t < 0$ and *negative refraction* occurs. Negative refraction also occurs when the situation is reversed and the incident wave field is in the DNG meta-material and the transmitted wave is in the ordinary, naturally occurring dispersive medium. If both media are DNG meta-materials, then both $n_1'(\omega) < 0$ and $n_2'(\omega) < 0$, in which case normal refraction occurs (i.e., Θ_i and Θ_t have the same sign). The same results apply for $TM(p)$-polarization.

As for the form of the transmitted wave, the reflected plane wave remaining unchanged in form from that of the incident plane wave, there are again three distinct possibilities depending upon whether or not medium 1 is lossy and whether or not the incident plane wave field is homogeneous ($\Theta_i + \psi_i = \pi/2$) or inhomogeneous ($\Theta_i + \psi_i \neq \pi/2$). These are [see the discussion following Eq. (6.148)]:

1. If medium 1 is lossless ($n_1''(\omega) = 0$), then $\psi_t = \pi/2$ and, except for the special case of normal incidence, the transmitted plane wave is inhomogeneous with $\boldsymbol{\alpha}_t(\omega) = \mathbf{1}_z \alpha_2(\omega)$.
2. If medium 1 is lossy ($n_1'' > 0$) and the incident plane wave is homogeneous ($\psi_i = \pi/2 - \Theta_i$), then

$$\cos \psi_t = \frac{n_1''(\omega)}{n_2''(\omega)} \sin \Theta_i, \tag{6.196}$$

   and the transmitted plane wave is inhomogeneous unless either $\Theta_i = 0$ (normal incidence) or $n_1''(\omega) = n_2''(\omega)$, in which case it is also homogeneous.
3. If the incident plane wave is inhomogeneous, then the transmitted plane wave is also inhomogeneous unless $\psi_t = \pi/2 + \Theta_t = \pi/2 - |\Theta_t|$.

A *critical angle of incidence*

$$\Theta_c \equiv \arcsin \left| \frac{n_2'(\omega)}{n_1'(\omega)} \right| \tag{6.197}$$

also occurs for both $TE(s)$ and $TM(p)$-polarizations when incidence is on an optically rarer DNG meta-material ($|n_1'(\omega)| > |n_2'(\omega)|$). Because both $\epsilon_2(\omega) < 0$ and $\mu_2(\omega) < 0$ for a DNG meta-material, it is found best to express the incident, reflected, and transmitted field vectors in terms of the *complex intrinsic impedance* of the host medium, given by [cf. Eqs. (5.63)–(5.64) and footnote 3]

$$\eta_j(\omega) \equiv \frac{|\tilde{\mathbf{E}}_j|}{|\tilde{\mathbf{H}}_j|} = \left(\frac{\mu_j(\omega)}{\epsilon_{cj}(\omega)}\right)^{1/2} = \frac{c}{\|c\|}\frac{\mu_j(\omega)}{n_j(\omega)}, \tag{6.198}$$

$j = 1, 2$, where $\eta_j(\omega) > 0$ for both ordinary and DNG meta-materials (see Sect. 5.2.4). In addition,

$$\frac{n_j(\omega)}{\epsilon_{cj}(\omega)} = \frac{\|c\|}{c}\left(\frac{\mu_j(\omega)}{\epsilon_{cj}(\omega)}\right)^{1/2} = \frac{c}{\|c\|}\eta_j(\omega), \tag{6.199}$$

$j = 1, 2$. Let $\mu_j(\omega) = |\mu_j|e^{i\varphi_{\mu j}}$, $\epsilon_{cj}(\omega) = |\epsilon_{cj}|e^{i\varphi_{\epsilon cj}}$, $n_j(\omega) = |n_j|e^{i\varphi_{n j}}$, and $\eta_j(\omega) = |\eta_j|e^{i\varphi_{\eta j}}$. Notice that, for notational convenience, the frequency dependence of each magnitude and phase term has been omitted; this is not to imply that either or both are frequency independent (quite the contrary, both may be rapidly-varying functions of ω). One then finds that, for example,

$$\frac{n_j'(\omega)}{\mu_j(\omega)} = \frac{|n_j|\cos\varphi_{n j}}{|\mu_j|e^{i\varphi_{\mu j}}} = \frac{c}{\|c\|}\frac{\cos\varphi_{n j}}{|\eta_j|}e^{-i\varphi_{\mu j}},$$

$$\frac{n_j''(\omega)}{\mu_j(\omega)} = \frac{|n_j|\sin\varphi_{n j}}{|\mu_j|e^{i\varphi_{\mu j}}} = \frac{c}{\|c\|}\frac{\sin\varphi_{n j}}{|\eta_j|}e^{-i\varphi_{\mu j}},$$

so that

$$n_j'(\omega) = \frac{c}{\|c\|}\frac{|\mu_j(\omega)|}{|\eta_j(\omega)|}\cos\varphi_{n j}, \tag{6.200}$$

$$n_j''(\omega) = \frac{c}{\|c\|}\frac{|\mu_j(\omega)|}{|\eta_j(\omega)|}\sin\varphi_{n j}, \tag{6.201}$$

and so on. The law of refraction $[n_1'(\omega)\sin\Theta_i = n_2'(\omega)\sin\Theta_t]$ given in Eq. (6.194) then takes the somewhat more complicated form

$$\frac{|\mu_1(\omega)|}{|\eta_1(\omega)|}\cos\varphi_{n_1}\sin\Theta_i = \frac{|\mu_2(\omega)|}{|\eta_2(\omega)|}\cos\varphi_{n_2}\sin\Theta_t \tag{6.202}$$

when expressed in terms of the material impedances, where $\varphi_{n_j} = \varphi_{\mu j} - \varphi_{\eta j}$.

Expressed in terms of the material impedances, the incident, reflected, and transmitted complex wave vectors given in Eq. (6.150) for subcritical angles of incidence become

$$\tilde{\mathbf{k}}_i(\omega) = \frac{\omega}{\|c\|} \frac{|\mu_1(\omega)|}{|\eta_1(\omega)|} \Big[\cos(\varphi_{n_1}) \Big(\hat{\mathbf{1}}_x \sin \Theta_i + \hat{\mathbf{1}}_z \cos \Theta_i \Big)$$

$$+ i \sin(\varphi_{n_1}) \Big(\hat{\mathbf{1}}_x \cos \psi_i + \hat{\mathbf{1}}_z \sin \psi_i \Big) \Big], \quad (6.203)$$

$$\tilde{\mathbf{k}}_r(\omega) = \frac{\omega}{\|c\|} \frac{|\mu_1(\omega)|}{|\eta_1(\omega)|} \Big[\cos(\varphi_{n_1}) \Big(\hat{\mathbf{1}}_x \sin \Theta_r - \hat{\mathbf{1}}_z \cos \Theta_r \Big)$$

$$+ i \sin(\varphi_{n_1}) \Big(\hat{\mathbf{1}}_x \cos \psi_r - \hat{\mathbf{1}}_z \sin \psi_r \Big) \Big], \quad (6.204)$$

$$\tilde{\mathbf{k}}_t(\omega) = \frac{\omega}{\|c\|} \frac{|\mu_2(\omega)|}{|\eta_2(\omega)|} \Big[- \cos(\varphi_{n_1}) \Big(\hat{\mathbf{1}}_x \sin \Theta_t + \hat{\mathbf{1}}_z \cos \Theta_t \Big)$$

$$+ i \sin(\varphi_{n_2}) \Big(\hat{\mathbf{1}}_x \cos \psi_t - \hat{\mathbf{1}}_z \sin \psi_t \Big) \Big], \quad (6.205)$$

respectively, where the (real-valued) angles Θ_j and ψ_j are related by the generalized laws of reflection and refraction given in Eqs. (6.193) and (6.194).

For a $TE(s)$-polarized plane wave incident upon a DNG meta-material, the incident electromagnetic field vectors are given by

$$\tilde{\mathbf{E}}_i(\mathbf{r}, \omega) = \hat{\mathbf{1}}_y E_s e^{i \tilde{\mathbf{k}}_i \cdot \mathbf{r}}, \quad (6.206)$$

$$\tilde{\mathbf{H}}_i(\mathbf{r}, \omega) = \frac{\|c\|}{\omega \mu_1(\omega)} \tilde{\mathbf{k}}_i(\omega) \times \hat{\mathbf{1}}_y E_s e^{i \tilde{\mathbf{k}}_i \cdot \mathbf{r}}$$

$$= \frac{E_s}{|\eta_1(\omega)|} e^{-i \varphi_{\mu_1}} \Big[\cos(\varphi_{n_1}) \Big(- \hat{\mathbf{1}}_x \cos \Theta_i + \hat{\mathbf{1}}_z \sin \Theta_i \Big)$$

$$+ i \sin(\varphi_{n_1}) \Big(- \hat{\mathbf{1}}_x \sin \psi_i + \hat{\mathbf{1}}_z \cos \psi_i \Big) \Big] e^{i \tilde{\mathbf{k}}_i \cdot \mathbf{r}}$$

$$(6.207)$$

when $\mathbf{r} \in S$, the reflected $TE(s)$-polarized plane wave is given by

$$\tilde{\mathbf{E}}_r(\mathbf{r}, \omega) = \hat{\mathbf{1}}_y \Gamma_{TE}(\omega) E_s e^{i \tilde{\mathbf{k}}_r \cdot \mathbf{r}}, \quad (6.208)$$

$$\tilde{\mathbf{H}}_i(\mathbf{r}, \omega) = \frac{\Gamma_{TE}(\omega) E_s}{|\eta_1(\omega)|} e^{-i \varphi_{\mu_1}} \Big[\cos(\varphi_{n_1}) \Big(\hat{\mathbf{1}}_x \cos \Theta_r + \hat{\mathbf{1}}_z \sin \Theta_r \Big)$$

$$+ i \sin(\varphi_{n_1}) \Big(- \hat{\mathbf{1}}_x \sin \psi_r + \hat{\mathbf{1}}_z \cos \psi_r \Big) \Big] e^{i \tilde{\mathbf{k}}_r \cdot \mathbf{r}},$$

$$(6.209)$$

when $\mathbf{r} \in S$ where $\Gamma_{TE} \equiv E_r/E_s$ is the reflection coefficient, and the transmitted $TE(s)$-polarized plane wave in the DNG meta-material is given by

$$\tilde{\mathbf{E}}_t(\mathbf{r}, \omega) = \hat{\mathbf{1}}_y \tau_{TE}(\omega) E_s e^{i\tilde{\mathbf{k}}_t \cdot \mathbf{r}}, \tag{6.210}$$

$$\tilde{\mathbf{H}}_t(\mathbf{r}, \omega) = \frac{\tau_{TE}(\omega) E_s}{|\eta_2(\omega)|} e^{-i\varphi_{\mu_2}} \Big[\cos(\varphi_{n_2}) \left(-\hat{\mathbf{1}}_x \cos \Theta_t + \hat{\mathbf{1}}_z \sin \Theta_t \right)$$
$$+ i \sin(\varphi_{n_2}) \left(-\hat{\mathbf{1}}_x \sin \psi_t + \hat{\mathbf{1}}_z \cos \psi_t \right) \Big] e^{i\tilde{\mathbf{k}}_t \cdot \mathbf{r}}, \tag{6.211}$$

when $\mathbf{r} \in S$ where $\tau_{TE} \equiv E_t/E_s$ is the transmission coefficient. The total field in medium 1 is then given by the vector sums $\tilde{\mathbf{E}}_1(\mathbf{r}, \omega) = \tilde{\mathbf{E}}_i(\mathbf{r}, \omega) + \tilde{\mathbf{E}}_r(\mathbf{r}, \omega)$ and $\tilde{\mathbf{H}}_1(\mathbf{r}, \omega) = \tilde{\mathbf{H}}_i(\mathbf{r}, \omega) + \tilde{\mathbf{H}}_r(\mathbf{r}, \omega)$, whereas the total field in medium 2 is just that given by the transmitted field vectors in Eqs. (6.210) and (6.211). Continuity of tangential E [see Eq. (6.119)] across the interface S then results in the expression

$$\tau_{TE} = 1 + \Gamma_{TE}, \tag{6.212}$$

whereas continuity of tangential H [see Eq. (6.120)] across S results in the expression

$$-\tau_{TE} \frac{e^{-i\varphi_{\mu_2}}}{|\eta_2(\omega)|} \Big[\cos \varphi_{n_2} \cos \Theta_t + i \sin \varphi_{n_2} \sin \psi_t \Big]$$
$$= -\frac{e^{-i\varphi_{\mu_1}}}{|\eta_1(\omega)|} (1 - \Gamma_{TE}) \Big[\cos \varphi_{n_1} \cos \Theta_i + i \sin \varphi_{n_1} \sin \psi_i \Big]. \tag{6.213}$$

Solution of this pair of equations results in the *generalized $TE(s)$-polarized Fresnel reflection and transmission coefficients for DNG metamaterials*

$$\Gamma_{TE}(\omega) = \frac{\frac{e^{-i\varphi_{\mu_1}}}{|\eta_1(\omega)|}(\cos\varphi_{n_1}\cos\Theta_i + i\sin\varphi_{n_1}\sin\psi_i) - \frac{e^{-i\varphi_{\mu_2}}}{|\eta_2(\omega)|}(\cos\varphi_{n_2}\cos\Theta_t + i\sin\varphi_{n_2}\sin\psi_t)}{\frac{e^{-i\varphi_{\mu_1}}}{|\eta_1(\omega)|}(\cos\varphi_{n_1}\cos\Theta_i + i\sin\varphi_{n_1}\sin\psi_i) + \frac{e^{-i\varphi_{\mu_2}}}{|\eta_2(\omega)|}(\cos\varphi_{n_2}\cos\Theta_t + i\sin\varphi_{n_2}\sin\psi_t)}, \tag{6.214}$$

$$\tau_{TE}(\omega) = \frac{2\frac{e^{-i\varphi_{\mu_1}}}{|\eta_1(\omega)|}(\cos\varphi_{n_1}\cos\Theta_i + i\sin\varphi_{n_1}\sin\psi_i)}{\frac{e^{-i\varphi_{\mu_1}}}{|\eta_1(\omega)|}(\cos\varphi_{n_1}\cos\Theta_i + i\sin\varphi_{n_1}\sin\psi_i) + \frac{e^{-i\varphi_{\mu_2}}}{|\eta_2(\omega)|}(\cos\varphi_{n_2}\cos\Theta_t + i\sin\varphi_{n_2}\sin\psi_t)}. \tag{6.215}$$

As they are written, this pair of equations is applicable when either medium 1 or medium 2 is a DNG meta-material (the situation of most interest here), as well as when both are DNG meta-materials or when neither are. These two equations reduce to those given in Eqs. (6.153) and (6.154) provided that the real part of the complex index of refraction for the DNG meta-material is replaced by its absolute value.

When incidence is on the optically rarer medium ($|n_1'| > |n_2'|$) at critical and supercritical angles of incidence $\Theta_i \geq \Theta_c$, where [cf. Eq. (6.197)]

$$\Theta_c = \arcsin\left(\frac{|\mu_2|\,|\eta_1|\,\cos\varphi_{n_2}}{|\mu_1|\,|\eta_2|\,\cos\varphi_{n_1}}\right), \tag{6.216}$$

then $\Theta_t = \pi/2$, $\cos\Theta_c = \sqrt{1 - (|\mu_2|/|\mu_1|)^2(|\eta_1|/|\eta_2|)^2(\cos\varphi_{n_2}/\cos\varphi_{n_1})^2}$, and Eq. (6.205) for the transmitted complex wave vector must be replaced by

$$\tilde{\mathbf{k}}_t(\omega) = \frac{\omega}{\|c\|}\frac{|\mu_2(\omega)|}{|\eta_2(\omega)|}\left\{\hat{\mathbf{1}}_x\left[\frac{|\mu_1|\,|\eta_2|}{|\mu_2|\,|\eta_1|}\cos\varphi_{n_2}\sin\Theta_i + i\sin\varphi_{n_2}\cos\psi_t\right]\right.$$

$$\left.+\hat{\mathbf{1}}_z i\left[\sin\varphi_{n_2}\sin\psi_t + \sqrt{\frac{|\mu_1|^2\,|\eta_2|^2}{|\mu_2|^2\,|\eta_1|^2}\cos^2\varphi_{n_1}\sin^2\Theta_i - \cos^2\varphi_{n_2}}\right]\right\}. \tag{6.217}$$

The magnetic field vector for the transmitted $TE(s)$-polarized plane wave is then given by

$$\tilde{\mathbf{H}}_t(\mathbf{r},\omega) = \tau_{TE}(\omega)E_s\left\{\hat{\mathbf{1}}_z\left[\frac{\cos\varphi_{n_1}}{|\eta_1|}e^{-i\varphi_{\mu_1}}\sin\Theta_i + i\frac{\sin\varphi_{n_2}}{|\eta_2|}e^{-i\varphi_{\mu_2}}\cos\psi_t\right]\right.$$

$$-\hat{\mathbf{1}}_x i\left[\frac{\sin\varphi_{n_2}}{|\eta_2|}e^{-i\varphi_{\mu_2}}\sin\psi_t\right.$$

$$\left.\left.+\sqrt{\frac{\cos^2\varphi_{n_1}}{|\eta_1|^2}e^{-i2\varphi_{\mu_2}}\sin^2\Theta_i - \frac{\cos^2\varphi_{n_2}}{|\eta_2|^2}e^{-i2\varphi_{\mu_2}}}\right]\right\}e^{i\mathbf{k}_t\cdot\mathbf{r}}, \tag{6.218}$$

the other field vectors remaining unchanged from those given in Eqs. (6.206)–(6.210). Continuity of tangential E and H across the interface S [Eqs. (6.119) and (6.120)] then results in the *generalized Fresnel reflection and transmission coefficients for critical and supercritical angles of incidence on a DNG metamaterial* given by

$$\Gamma_{TE}(\omega)\big|_{\Theta_i \geq \Theta_c} = \frac{Z_-}{Z_+} \tag{6.219}$$

$$\tau_{TE}(\omega)\big|_{\Theta_i \geq \Theta_c} = 1 + \Gamma_{TE}(\omega)\big|_{\Theta_i \geq \Theta_c} = \frac{Z_- + Z_+}{Z_+}, \tag{6.220}$$

respectively, where

$$Z_{\pm} = |\eta_2|e^{i\varphi_{\mu 2}} \cos \varphi_{n_1} \cos \Theta_i$$

$$+i\Big[|\eta_2|e^{i\varphi_{\mu 2}} \sin \varphi_{n_1} \sin \psi_i \pm |\eta_1|e^{i\varphi_{\mu 1}} \sin \varphi_{n_2} \sin \psi_t$$

$$\pm \sqrt{ |\eta_2|^2 e^{i2\varphi_{\mu 1}} \cos^2 \varphi_{n_1} \sin^2 \Theta_1 - |\eta_1|^2 e^{i2\varphi_{\mu 1}} \cos^2 \varphi_{n_2} } \Big],$$

$$(6.221)$$

so that

$$Z_- + Z_+ = 2|\eta_2|e^{i\varphi_{\mu 2}} \left(\cos \varphi_{n_1} \cos \Theta_i + i \sin \varphi_{n_1} \sin \psi_i \right). \qquad (6.222)$$

These two equations reduce to those given in Eqs. (6.158) and (6.159) provided that the real part of the complex index of refraction for the DNG meta-material is replaced by its absolute value.

For a $TM(p)$-polarized plane wave incident upon a DNG meta-material, the incident electromagnetic field vectors are given by

$$\tilde{\mathbf{H}}_i(\mathbf{r}, \omega) = \hat{\mathbf{1}}_y H_0 e^{i\tilde{\mathbf{k}}_i \cdot \mathbf{r}}, \qquad (6.223)$$

$$\tilde{\mathbf{E}}_i(\mathbf{r}, \omega) = -\frac{\|c\|}{\omega \epsilon_{c1}(\omega)} \tilde{\mathbf{k}}_i(\omega) \times \hat{\mathbf{1}}_y H_0 e^{i\tilde{\mathbf{k}}_i \cdot \mathbf{r}}$$

$$= \eta_1(\omega) e^{-i\varphi_{n_1}} \big[\hat{\mathbf{1}}_x (\cos \varphi_{n_1} \cos \Theta_i + i \sin \varphi_{n_1} \sin \psi_i)$$

$$- \hat{\mathbf{1}}_z (\cos \varphi_{n_1} \sin \Theta_i + i \sin \varphi_{n_1} \cos \psi_i) \big] H_0 e^{i\tilde{\mathbf{k}}_i \cdot \mathbf{r}}, \qquad (6.224)$$

when $\mathbf{r} \in S$, the reflected $TM(p)$-polarized plane wave is given by

$$\tilde{\mathbf{H}}_r(\mathbf{r}, \omega) = -\hat{\mathbf{1}}_y \Gamma_{TM}(\omega) H_0 e^{i\tilde{\mathbf{k}}_r \cdot \mathbf{r}}, \qquad (6.225)$$

$$\tilde{\mathbf{E}}_r(\mathbf{r}, \omega) = -\frac{\|c\|}{\omega \epsilon_{c1}(\omega)} \tilde{\mathbf{k}}_r(\omega) \times \hat{\mathbf{1}}_y \Gamma_{TM}(\omega) H_0 e^{i\tilde{\mathbf{k}}_r \cdot \mathbf{r}}$$

$$= -\eta_1(\omega) e^{-i\varphi_{n_1}} \Gamma_{TM}(\omega) \big[\hat{\mathbf{1}}_x (\cos \varphi_{n_1} \cos \Theta_r + i \sin \varphi_{n_1} \sin \psi_r)$$

$$+ \hat{\mathbf{1}}_z (\cos \varphi_{n_1} \sin \Theta_r + i \sin \varphi_{n_1} \cos \psi_r) \big] H_0 e^{i\tilde{\mathbf{k}}_r \cdot \mathbf{r}}, \qquad (6.226)$$

when $\mathbf{r} \in S$, and the transmitted $TM(p)$-polarized plane wave is given by

$$\tilde{\mathbf{H}}_t(\mathbf{r}, \omega) = \hat{\mathbf{1}}_y \tau_{TM}(\omega) H_0 e^{i\tilde{\mathbf{k}}_t \cdot \mathbf{r}}, \qquad (6.227)$$

$$
\begin{aligned}
\tilde{\mathbf{E}}_t(\mathbf{r}, \omega) &= -\frac{\|c\|}{\omega \epsilon_{c2}(\omega)} \tilde{\mathbf{k}}_t(\omega) \times \hat{\mathbf{1}}_y \tau_{TM}(\omega) H_0 e^{i\tilde{\mathbf{k}}_t \cdot \mathbf{r}} \\
&= \eta_2(\omega) e^{-i\varphi_{n_2}} \tau_{TM}(\omega) \big[\hat{\mathbf{1}}_x (\cos\varphi_{n_2} \cos\Theta_t + i \sin\varphi_{n_2} \sin\psi_t) \\
&\quad + \hat{\mathbf{1}}_z (\cos\varphi_{n_2} \sin\Theta_t + i \sin\varphi_{n_2} \cos\psi_t) \big] H_0 e^{i\tilde{\mathbf{k}}_t \cdot \mathbf{r}},
\end{aligned}
$$
$$(6.228)$$

when $\mathbf{r} \in S$. Continuity of the total tangential H-field across the interface S then gives

$$\tau_{TM} = 1 - \Gamma_{TM}, \qquad (6.229)$$

whereas continuity of the total tangential E-field across S gives

$$
\begin{aligned}
\eta_2(\omega) e^{-i\varphi_{n_2}} \tau_{TM}(\omega) & \big(\cos\varphi_{n_2} \cos\Theta_t + i \sin\varphi_{n_2} \sin\psi_t \big) \\
&= -\eta_1(\omega) e^{-i\varphi_{n_1}} \big(1 + \Gamma_{TM}(\omega) \big) \big(\cos\varphi_{n_1} \cos\Theta_i + i \sin\varphi_{n_1} \sin\psi_i \big). \quad (6.230)
\end{aligned}
$$

Solution to this pair of equations gives the *generalized $TM(p)$-polarized Fresnel reflection and transmission coefficients for DNG meta-materials*

$$\Gamma_{TM} = \frac{\eta_2 e^{-i\varphi_{n_2}} \left(\cos\varphi_{n_2} \cos\Theta_t + i \sin\varphi_{n_2} \sin\psi_t \right) - \eta_1 e^{-i\varphi_{n_1}} \left(\cos\varphi_{n_1} \cos\Theta_i + i \sin\varphi_{n_1} \sin\psi_i \right)}{\eta_2 e^{-i\varphi_{n_2}} \left(\cos\varphi_{n_2} \cos\Theta_t + i \sin\varphi_{n_2} \sin\psi_t \right) + \eta_1 e^{-i\varphi_{n_1}} \left(\cos\varphi_{n_1} \cos\Theta_i + i \sin\varphi_{n_1} \sin\psi_i \right)},$$
$$(6.231)$$

$$\tau_{TM} = \frac{-2\eta_1 e^{-i\varphi_{n_1}} \left(\cos\varphi_{n_1} \cos\Theta_i + i \sin\varphi_{n_1} \sin\psi_i \right)}{\eta_2 e^{-i\varphi_{n_2}} \left(\cos\varphi_{n_2} \cos\Theta_t + i \sin\varphi_{n_2} \sin\psi_t \right) + \eta_1 e^{-i\varphi_{n_1}} \left(\cos\varphi_{n_1} \cos\Theta_i + i \sin\varphi_{n_1} \sin\psi_i \right)}.$$
$$(6.232)$$

Because $\eta_j(\omega) = (c/\|c\|) n_j(\omega)/\epsilon_{cj}(\omega)$, $j = 1, 2$, these two expressions are identical to those given in Eqs. (6.172) and (6.173).

For *critical and supercritical angles of incidence* the above pair of expressions is replaced by

$$\Gamma_{TM}(\omega)\big|_{\Theta_i \geq \Theta_c} = \frac{\Xi_-}{\Xi_+}, \qquad (6.233)$$

$$\tau_{TM}(\omega)\big|_{\Theta_i \geq \Theta_c} = 1 - \Gamma_{TM}(\omega)\big|_{\Theta_i \geq \Theta_c} = \frac{\Xi_+ - \Xi_-}{\Xi_+}, \qquad (6.234)$$

where

$$\Xi_{\pm} = \eta_1 e^{-i\varphi_{n_1}} \left(\cos \varphi_{n_1} \cos \Theta_i + i \sin \varphi_{n_1} \sin \psi_r\right)$$

$$\pm i |\eta_2| e^{-i\varphi_{\epsilon c2}} \left[\sin \varphi_{n_2} \sin \psi_t + \sqrt{\frac{|\mu_1|^2}{|\mu_2|^2} \frac{|\eta_2|^2}{|\eta_1|^2} \cos^2 \varphi_{n_1} \sin^2 \Theta_i - \cos^2 \varphi_{n_2}}\right]$$

$$(6.235)$$

These two equations reduce to those given in Eqs. (6.183) and (6.184) provided that the real part of the complex index of refraction for the DNG meta-material is replaced by its absolute value.

The generalized Fresnel reflection and transmission coefficients given in Eqs. (6.153) and (6.154) for $TE(s)$-polarization and by Eqs. (6.172) and (6.173) for $TM(p)$-polarization for subcritical incidence, and by Eqs. (6.158) and (6.159) for $TE(s)$-polarization and by Eqs. (6.183) and (6.184) for $TM(p)$-polarization for critical and supercritical incidence remain valid when medium 2 is a DNG metamaterial provided that $n_2'(\omega)$ replaced by its absolute value $|n_2'(\omega)|$. A *polarizing* or *Brewster's angle* Θ_P at which angle of incidence Γ_{TM} vanishes, given by

$$\tan \Theta_P = \left|\frac{n_2'(\omega)}{n_1'(\omega)}\right|, \quad n_1'' = n_2'' = 0, \qquad (6.236)$$

then occurs when both media are lossless.

6.6 Molecular Optics Formulation

The classical theory of molecular optics,[4] originally developed by M. Born [11], L. Rosenfeld [12], and É. Lalor & E. Wolf [13], leads to the fundamentally important Ewald–Oseen extinction theorem which shows that a time-harmonic electromagnetic plane wave incident upon a dielectric interface initially penetrates into the medium undisturbed, exciting the molecules that comprise the medium which, in turn, produce a scattered wave field that extinguishes (through destructive interference) the incident plane wave field and constructs the refracted plane wave field. This analysis thus reveals the fundamental physical processes involved in the refraction of light at a dielectric interface at the microscopic level, bringing this entire development full-circle back to its beginning in Chaps. 2–4.

---

[4]A simpler asymptotic description of molecular optics may be found in the essential optics text *Principles of Optics: Electromagnetic Theory of Propagation, Interference and Diffraction of Light* by M. Born and E. Wolf. [1].

6.6.1 Integral Equation Representation

Consider the propagation of an electromagnetic plane wave in a homogeneous, isotropic, locally linear, nonmagnetic, nonconducting HILL dielectric medium comprised of bound molecules that react to an incident field in the manner of ideal point dipoles. The electric and magnetic field vectors $\mathbf{E}'_j(\mathbf{r}, t)$ and $\mathbf{H}'_j(\mathbf{r}, t)$ which act on the jth molecular dipole in the interior of the medium can then be formally separated into the superposition of the incident electromagnetic field vectors $\mathbf{E}^i(\mathbf{r}, t)$ and $\mathbf{H}^i(\mathbf{r}, t)$ that are propagating as if they were in vacuum with phase velocity equal to the constant vacuum speed of light c and the contribution arising from all of the other molecular dipoles in the medium, so that

$$\mathbf{E}'_j(\mathbf{r}, t) = \mathbf{E}^i(\mathbf{r}, t) + \sum_\ell \mathbf{E}_{j\ell}(\mathbf{r}, t), \tag{6.237}$$

$$\mathbf{H}'_j(\mathbf{r}, t) = \mathbf{H}^i(\mathbf{r}, t) + \sum_\ell \mathbf{H}_{j\ell}(\mathbf{r}, t), \tag{6.238}$$

with $\ell \neq j$. At the point $\mathbf{r}_j$ where the jth dipole is situated, the field of the ℓth dipole is given [1] by the pair of expressions [see Eqs. (3.67)–(3.78)]

$$\mathbf{E}_{j\ell}(\mathbf{r}, t) = \frac{1}{\epsilon_0} \nabla \times \nabla \times \frac{\mathbf{p}_\ell(t - R_{j\ell}/c)}{R_{j\ell}}, \tag{6.239}$$

$$\mathbf{H}_{j\ell}(\mathbf{r}, t) = \frac{1}{\|c\|} \nabla \times \frac{\dot{\mathbf{p}}_\ell(t - R_{j\ell}/c)}{R_{j\ell}}, \tag{6.240}$$

where $\mathbf{p}_\ell(t)$ is the moment of the ℓth dipole, $R_{j\ell} \equiv |\mathbf{r}_j - \mathbf{r}_\ell|$, and where the spatial differentiation is taken with respect to the coordinates (x_j, y_j, z_j) of the jth dipole.

The following simplifying, but non-compromising, approximations are now made:

1. first, the spatial distribution of the molecular dipoles comprising the dielectric medium is treated as continuous so that $\mathbf{p}_\ell(t) \to \mathbf{p}(\mathbf{r}, t)$;
2. and second, the number density $\mathcal{N}(\mathbf{r})$ of dipolar molecules is assumed to be a constant $\mathcal{N}$ in the material body (spatial homogeneity and isotropy) and zero outside.

For a homogeneous, isotropic, locally linear dielectric exhibiting temporal dispersion, the total electric dipole moment per unit volume (the *macroscopic polarization density*) is obtained from the continuous extension of Eq. (4.169) with substitution from Eq. (4.170) as

$$\mathbf{P}(\mathbf{r}, t) = \mathcal{N} \int_{-\infty}^{t} \hat{\alpha}(t - t') \mathbf{E}'(\mathbf{r}, t') dt', \tag{6.241}$$

where $\mathbf{E}'(\mathbf{r}, t)$ is a spatial average of the effective local electric field intensity $\mathbf{E}'_j(\mathbf{r}, t')$. For a causal medium response, Titchmarsh's theorem [14] (see Sect. 4.3) requires that $\hat{\alpha}(t - t') = 0$ for all $t' > t$, in which case the upper limit of integration in Eq. (6.241) may be extended to $+\infty$. The temporal Fourier transform of the resulting convolution relation then yields

$$\tilde{\mathbf{P}}(\mathbf{r}, \omega) = \mathcal{N}\alpha(\omega)\tilde{\mathbf{E}}'(\mathbf{r}, \omega), \tag{6.242}$$

where $\alpha(\omega) = \int_{-\infty}^{\infty} \hat{\alpha}(t)e^{i\omega t}\,dt$ is the *mean molecular polarizability* of the dielectric medium. This mean molecular polarizability characterizes the frequency-dependent linear response of the molecules comprising the dielectric body to the applied electric field (see Appendix C). Notice that the 'simple' linear relation expressed in Eqs. (6.241) and (6.242) may be generalized to include spatial inhomogeneity by letting $\mathcal{N} \rightarrow \mathcal{N}(\mathbf{r})$ as well as anisotropy by generalizing the scalar quantity α to its dyadic form. Spatial dispersion may also be included by removing the condition imposed by spatial locality in a generalized version of Eq. (6.241) by letting $\hat{\alpha}(t) \rightarrow \hat{\alpha}(\mathbf{r} - \mathbf{r}', t - t')$. For the analysis presented here, only a 'simple' linear, causally dispersive dielectric is considered.

Upon going over to a continuous distribution of the molecular dipoles in the temporally dispersive HILL dielectric medium, Eqs. (6.237) and (6.238) become, with substitution from Eqs. (6.241) and (6.242),

$$\mathbf{E}'(\mathbf{r}, t) = \mathbf{E}^i(\mathbf{r}, t) + \frac{1}{\epsilon_0} \int \nabla \times \nabla \times \left[\frac{\mathcal{N}}{R} \int_{-\infty}^{t} \hat{\alpha}(t - t')\mathbf{E}'(\mathbf{r}', t - R/c)dt'\right] d^3 r', \tag{6.243}$$

$$\mathbf{H}'(\mathbf{r}, t) = \mathbf{H}^i(\mathbf{r}, t) + \frac{1}{\|c\|} \int \nabla \times \left[\frac{\mathcal{N}}{R} \int_{-\infty}^{t} \hat{\alpha}(t - t')\dot{\mathbf{E}}'(\mathbf{r}', t - R/c)dt'\right] d^3 r', \tag{6.244}$$

where $R \equiv |\mathbf{r} - \mathbf{r}'|$, and where the indicated spatial differentiations are taken with respect to the unprimed coordinates (x, y, z) of the field point $\mathbf{r}$. If the point of observation (or field point) $\mathbf{r}$ is outside the dielectric medium, then the integration is taken throughout the entire space occupied by the medium. However, if it is inside the medium, then a small domain $\mathcal{D}_\Delta$ occupied by the molecule at that observation point must be excluded from the integration domain. As a matter of course, one then takes the limit as $\mathcal{D}_\Delta \rightarrow 0$ after the analysis has been completed.

Equation (6.243) is an integro-differential equation for the electric field vector $\mathbf{E}'(\mathbf{r}, t)$. When it is solved, the magnetic intensity vector $\mathbf{H}'(\mathbf{r}, t)$ associated with that electromagnetic wave field is then obtained from Eq. (6.244). This pair of relations is essentially equivalent to Maxwell's equations for homogeneous, isotropic, locally linear, nonmagnetic materials that exhibit temporal dispersion. Their solution has been obtained [11–13] for the special case when the incident wave field is strictly monochromatic with angular frequency ω, in which case the mean polarizability may be treated as a constant with $\hat{\alpha} = \alpha \delta(t - t')$.

When the incident electromagnetic wave is pulsed, and especially when it is ultrawideband with respect to the material dispersion, special care must be taken in solving Eqs. (6.243) and (6.244). In that case, this pair of integro-differential equations may be somewhat simplified in the temporal frequency domain by taking the temporal Fourier–Laplace transform of each, with the result

$$\tilde{\mathbf{E}}'(\mathbf{r}, \omega) = \tilde{\mathbf{E}}^i(\mathbf{r}, \omega) + \frac{N\alpha(\omega)}{\epsilon_0} \int \nabla \times \nabla \times \left[\tilde{\mathbf{E}}'(\mathbf{r}', \omega) \frac{e^{i\omega R/c}}{R} \right] d^3 r',$$

(6.245)

$$\tilde{\mathbf{H}}'(\mathbf{r}, \omega) = \tilde{\mathbf{H}}^i(\mathbf{r}, \omega) - \frac{i\omega}{\|c\|} N\alpha(\omega) \int \nabla \times \left[\tilde{\mathbf{E}}'(\mathbf{r}', \omega) \frac{e^{i\omega R/c}}{R} \right] d^3 r'.$$

(6.246)

Let Σ denote the closed boundary surface of the dielectric body occupying the region $\mathcal{D} \in \mathcal{R}^3$. For any observation point $\mathbf{r} \in \mathcal{D}$ inside the dielectric, Eqs. (6.242) and (6.245) may be combined to yield the relation

$$\tilde{\mathbf{P}}(\mathbf{r}, \omega) = N\alpha(\omega) \left[\tilde{\mathbf{E}}^i(\mathbf{r}, \omega) + \tilde{\mathbf{E}}^d(\mathbf{r}, \omega) \right],$$

(6.247)

where

$$\tilde{\mathbf{E}}^d(\mathbf{r}, \omega) \equiv \int_{\mathcal{D} - \mathcal{D}_\Delta} \nabla \times \nabla \times \left[\tilde{\mathbf{P}}(\mathbf{r}', \omega) \frac{e^{i\omega R/c}}{R} \right] d^3 r'$$

(6.248)

denotes the contribution from the molecular dipoles in the dielectric body $\mathcal{D}$ excluding the small region $\mathcal{D}_\Delta$ about the observation (or field) point $\mathbf{r}$. The electric field vector

$$\mathbf{E}'(\mathbf{r}, t) = \mathbf{E}^i(\mathbf{r}, t) + \mathbf{E}^d(\mathbf{r}, t)$$

(6.249)

acting on the molecular dipoles in the interior of the dielectric body may then be obtained from the inverse Fourier transform of the expression [from Eq. (6.242)]

$$\tilde{\mathbf{E}}'(\mathbf{r}, \omega) = \frac{1}{N\alpha(\omega)} \tilde{\mathbf{P}}(\mathbf{r}, \omega).$$

(6.250)

The solution then depends upon the particular form of the macroscopic polarization density $\mathbf{P}(\mathbf{r}, t)$. For example, for the nondispersive case the dipole moment density is taken to satisfy the homogeneous wave equation $\nabla^2 \mathbf{P}(\mathbf{r}, t) - (\acute{n}^2/c^2) \partial^2 \mathbf{P}(\mathbf{r}, t)/\partial t^2 = \mathbf{0}$ where the constant n is an unknown quantity that is to be determined [1, 13].

6.6.2 Molecular Optics with the Lorentz Model

For a Lorentz model dielectric, the macroscopic polarization density $\mathbf{P}(\mathbf{r}, t)$ is derived from the microscopic equation of motion [see Eq. (4.207)]

$$m_e \left(\ddot{\mathbf{s}}_j + 2\delta \dot{\mathbf{s}}_j + \omega_0^2 \mathbf{s}_j \right) = -q_e \mathbf{E}'_j(\mathbf{r}, t), \qquad (6.251)$$

where m_e is the mass of the electron and q_e the magnitude of the electronic charge, $\mathbf{E}'_j(\mathbf{r}, t)$ being identified as the *effective local electric field intensity* acting on the jth molecular dipole as driving force. The additional force $-\|1/c\|q_e(\dot{\mathbf{s}}_j/c) \times \mathbf{B}'_j(\mathbf{r}, t)$, arising from the interaction of the moving charge with the effective local magnetic field, is assumed here to be negligible in comparison to the electric field interaction (due to the smallness of the magnitude of this charge velocity in comparison with the vacuum speed of light c), as described in Appendix D. Here ω_0 is the undamped resonance frequency, and δ is the associated phenomenological damping constant of the microscopic oscillator. The same dynamical equation of motion also applies to molecular vibration modes when m_e is replaced by the ionic mass and ω_0 is the undamped resonance frequency of the transverse vibrational mode of the ionic lattice structure [15].

The temporal frequency transform of Eq. (6.251) directly yields the frequency domain solution

$$\tilde{\mathbf{s}}_j(\mathbf{r}, \omega) = \frac{q_e/m_e}{\omega^2 - \omega_0^2 + 2i\delta\omega} \tilde{\mathbf{E}}'_j(\mathbf{r}, \omega), \qquad (6.252)$$

so that the locally induced dipole moment $\tilde{\mathbf{p}}_j(\mathbf{r}, \omega) = -q_e \tilde{\mathbf{s}}_j(\mathbf{r}, \omega)$ of the jth molecular dipole is given by

$$\tilde{\mathbf{p}}_j(\mathbf{r}, \omega) = \alpha(\omega)\tilde{\mathbf{E}}'_j(\mathbf{r}, \omega), \qquad (6.253)$$

where

$$\alpha(\omega) = \frac{-q_e^2/m_e}{\omega^2 - \omega_0^2 + 2i\delta\omega} \qquad (6.254)$$

is the *mean molecular polarizability*. The *macroscopic polarization density* is then given by the spatial average of the locally induced (microscopic) dipole moments as

$$\tilde{\mathbf{P}}(\mathbf{r}, \omega) = \mathcal{N}\langle\langle\tilde{\mathbf{p}}_j(\mathbf{r}, \omega)\rangle\rangle = \mathcal{N}\alpha(\omega)\langle\langle\tilde{\mathbf{E}}'_j(\mathbf{r}, \omega)\rangle\rangle. \qquad (6.255)$$

With the identification that

$$\tilde{\mathbf{E}}'(\mathbf{r}, \omega) = \langle\langle \tilde{\mathbf{E}}'_j(\mathbf{r}, \omega)\rangle\rangle, \tag{6.256}$$

the expression given in Eq. (6.255) directly reduces to that in Eq. (6.242).

Upon following the analytical tack taken by Lalor and Wolf [13], it is now assumed that this polarization density satisfies the Helmholtz equation

$$\left(\nabla^2 + n^2(\omega)k_0^2\right)\tilde{\mathbf{P}}(\mathbf{r}, \omega) = \mathbf{0}, \tag{6.257}$$

where $k_0 \equiv \omega/c$ denotes the wave number in vacuum, and where the complex-valued quantity $n(\omega)$ remains to be determined. As a trial solution, let

$$\tilde{\mathbf{P}}(\mathbf{r}, \omega) \equiv \left(n^2(\omega) - 1\right)k_0^2\tilde{\mathbf{Q}}(\mathbf{r}, \omega). \tag{6.258}$$

The spectral quantity $\tilde{\mathbf{Q}}(\mathbf{r}, \omega)$ then satisfies the same vector Helmholtz equation

$$\left(\nabla^2 + n^2(\omega)k_0^2\right)\tilde{\mathbf{Q}}(\mathbf{r}, \omega) = \mathbf{0} \tag{6.259}$$

as does the polarization density. The complex wave field $\tilde{\mathbf{Q}}(\mathbf{r}, \omega)$ then "travels" in the dielectric body with the complex phase velocity $c/n(\omega)$. In addition, it is assumed that $\tilde{\mathbf{Q}}(\mathbf{r}, \omega)$ has no sources[5] inside the medium, so that

$$\nabla \cdot \tilde{\mathbf{Q}}(\mathbf{r}, \omega) = 0 \tag{6.260}$$

for all $\mathbf{r} \in \mathcal{D}$. Combination of Eqs. (6.248) and (6.258) then gives

$$\tilde{\mathbf{Q}}(\mathbf{r}, \omega) = \mathcal{N}\alpha(\omega)\left\{\frac{1}{\left(n^2(\omega) - 1\right)k_0^2}\tilde{\mathbf{E}}^i(\mathbf{r}, \omega)\right.$$
$$\left. + \int_{\mathcal{D}-\mathcal{D}_\Delta} \nabla \times \nabla \times \left[\tilde{\mathbf{Q}}(\mathbf{r}', \omega)G(R, \omega)\right]d^3r'\right\}, \tag{6.261}$$

where

$$G(R, \omega) \equiv \frac{e^{i\omega R/c}}{R} \tag{6.262}$$

---

[5]Notice that this macroscopic polarization density is derived from the spatial average of the microscopic dipoles induced by the local effective electric field.

is the *free-space Green's function* with $R = |\mathbf{r} - \mathbf{r}'|$. With the definition [13]

$$\tilde{\mathbf{A}}^d(\mathbf{r}, \omega) \equiv \int_{\mathcal{D}-\mathcal{D}_\Delta} \nabla \times \nabla \times \left[\tilde{\mathbf{Q}}(\mathbf{r}', \omega) G(R, \omega) \right] d^3 r', \qquad (6.263)$$

the relation in Eq. (6.261) may be expressed in a more compact form as

$$\tilde{\mathbf{Q}}(\mathbf{r}, \omega) = N\alpha(\omega) \left\{ \frac{1}{\left(n^2(\omega) - 1 \right) k_0^2} \tilde{\mathbf{E}}^i(\mathbf{r}, \omega) + \tilde{\mathbf{A}}^d(\mathbf{r}, \omega) \right\}. \qquad (6.264)$$

Notice that the mathematical form of the spectral field quantity $\tilde{\mathbf{A}}^d(\mathbf{r}, \omega)$ is precisely that obtained in the classical description of molecular optics [13], which is restricted to a monochromatic wave field in a nondispersive dielectric.

6.6.3   The Ewald–Oseen Extinction Theorem

As the radius of the small spherical region $\mathcal{D}_\Delta$ surrounding the observation point $\mathbf{r}$ shrinks to zero, the integral appearing in Eq. (6.261) becomes (see Appendix V of Born and Wolf [1])

$$\int_{\mathcal{D}-\mathcal{D}_\Delta} \nabla \times \nabla \times \left[\tilde{\mathbf{Q}}(\mathbf{r}', \omega) G(R, \omega) \right] d^3 r'$$

$$= \nabla \times \nabla \times \int_{\mathcal{D}-\mathcal{D}_\Delta} \tilde{\mathbf{Q}}(\mathbf{r}', \omega) G(R, \omega) d^3 r' - \frac{8\pi}{3} \tilde{\mathbf{Q}}(\mathbf{r}, \omega).$$
$$(6.265)$$

Because the free-space Green's function $G(R, \omega)$ describes a monochromatic spherical wave of angular frequency ω and wave number $k_0 = \omega/c$ in free-space, it then satisfies Helmholtz' equation

$$\left(\nabla^2 + k_0^2 \right) G(R, \omega) = 0 \qquad (6.266)$$

in the region $\mathcal{D} - \mathcal{D}_\Delta$. From Eqs. (6.259) and (6.266) one finds that

$$\tilde{\mathbf{Q}}G = \frac{1}{(n^2 - 1)k_0^2} \left(\tilde{\mathbf{Q}}\nabla^2 G - G\nabla^2\tilde{\mathbf{Q}} \right), \qquad (6.267)$$

so that, with application of Green's theorem, the volume integral appearing on the right-hand side of Eq. (6.265) becomes

$$\int_{\mathcal{D}-\mathcal{D}_\Delta} \tilde{\mathbf{Q}}(\mathbf{r}', \omega) G(R, \omega) d^3 r' = \frac{1}{(n^2 - 1)k_0^2} \int_{\mathcal{D}-\mathcal{D}_\Delta} \left(\tilde{\mathbf{Q}} \nabla^2 G - G \nabla^2 \tilde{\mathbf{Q}} \right) d^3 r'$$

$$= \frac{1}{(n^2 - 1)k_0^2} \left\{ \oint_S \left(\tilde{\mathbf{Q}} \frac{\partial G}{\partial n'} - G \frac{\partial \tilde{\mathbf{Q}}}{\partial n'} \right) d^2 r' \right.$$

$$\left. - \oint_\Delta \left(\tilde{\mathbf{Q}} \frac{\partial G}{\partial R} - G \frac{\partial \tilde{\mathbf{Q}}}{\partial R} \right) d^2 r' \right\},$$

$$(6.268)$$

where $\partial/\partial n'$ denotes differentiation along the outward normal n' [not to be confused with the complex index of refraction $n = n(\omega)$] to the boundary surface of the region $\mathcal{D} - \mathcal{D}_\Delta$. Notice that a minus sign appears in the second surface integral of Eq. (6.268) because $\partial/\partial R$ is the inner-directed normal derivative to the spherical surface $\mathcal{D}_\Delta$. The evaluation of the limit of this inner surface integral over the spherical region $\mathcal{D}_\Delta$ as its radius R approaches zero closely follows that given in the derivation of the *integral theorem of Helmholtz and Kirchhoff* (see, for example, Sect. 8.3.1 of Born and Wolf [1]) with the result

$$\oint_\Delta \left(\tilde{\mathbf{Q}} \frac{\partial G}{\partial R} - G \frac{\partial \tilde{\mathbf{Q}}}{\partial R} \right) d^2 r' = -4\pi \tilde{\mathbf{Q}}(\mathbf{r}). \qquad (6.269)$$

With this substitution, Eq. (6.268) becomes

$$\int_{\mathcal{D}} \tilde{\mathbf{Q}}(\mathbf{r}', \omega) G(R, \omega) d^3 r' = \frac{1}{(n^2 - 1)k_0^2} \left\{ \oint_S \left(\tilde{\mathbf{Q}} \frac{\partial G}{\partial n'} - G \frac{\partial \tilde{\mathbf{Q}}}{\partial n'} \right) d^2 r' + 4\pi \tilde{\mathbf{Q}}(\mathbf{r}) \right\}. \qquad (6.270)$$

Substitution of Eqs. (6.265) and (6.270) into Eq. (6.263) then gives

$$\tilde{\mathbf{A}}^d(\mathbf{r}, \omega) = \frac{1}{(n^2 - 1) k_0^2} \left\{ 4\pi \nabla \times \nabla \times \tilde{\mathbf{Q}}(\mathbf{r}, \omega) \right.$$

$$\left. + \nabla \times \nabla \times \oint_S \left(\tilde{\mathbf{Q}} \frac{\partial G}{\partial n'} - G \frac{\partial \tilde{\mathbf{Q}}}{\partial n'} \right) d^2 r' \right\} - \frac{8\pi}{3} \tilde{\mathbf{Q}}(\mathbf{r}, \omega).$$

$$(6.271)$$

Because $\tilde{\mathbf{Q}}(\mathbf{r}, \omega)$ is solenoidal [see Eq. (6.260)], then, together with Eq. (6.259) it is found that

$$\nabla \times \nabla \times \tilde{\mathbf{Q}}(\mathbf{r}, \omega) = \nabla \left(\nabla \cdot \tilde{\mathbf{Q}}(\mathbf{r}, \omega) \right) - \nabla^2 \tilde{\mathbf{Q}}(\mathbf{r}, \omega) = n^2(\omega) k_0^2 \tilde{\mathbf{Q}}(\mathbf{r}, \omega), \quad (6.272)$$

and Eq. (6.271) becomes

$$\tilde{\mathbf{A}}^d(\mathbf{r}, \omega) = \frac{4\pi}{3} \left(\frac{n^2 + 2}{n^2 - 1} \right) \tilde{\mathbf{Q}}(\mathbf{r}, \omega)$$

$$+ \frac{1}{(n^2 - 1) k_0^2} \nabla \times \nabla \times \oint_S \left(\tilde{\mathbf{Q}} \frac{\partial G}{\partial n'} - G \frac{\partial \tilde{\mathbf{Q}}}{\partial n'} \right) d^2 r'.$$

$$(6.273)$$

Substitution of this expression into Eq. (6.264) then gives

$$\left[1 - \frac{4\pi}{3} \left(\frac{n^2(\omega) + 2}{n^2(\omega) - 1} \right) \mathcal{N}\alpha(\omega) \right] \tilde{\mathbf{Q}}(\mathbf{r}, \omega)$$

$$= \frac{\mathcal{N}\alpha(\omega)}{(n^2(\omega) - 1) k_0^2} \left\{ \tilde{\mathbf{E}}^i(\mathbf{r}, \omega) + \nabla \times \nabla \times \oint_S \left(\tilde{\mathbf{Q}} \frac{\partial G}{\partial n'} - G \frac{\partial \tilde{\mathbf{Q}}}{\partial n'} \right) d^2 r' \right\}.$$

$$(6.274)$$

Because the left-hand side of this equation describes a monochromatic electro-magnetic wave with complex phase velocity $c/n(\omega)$, whereas the right-hand side describes a wave that propagates with the velocity c, each side must then separately vanish. Thus

$$\frac{4\pi}{3} \mathcal{N}\alpha(\omega) = \frac{n^2(\omega) - 1}{n^2(\omega) + 2}, \quad (6.275)$$

which is known either as the *Lorentz–Lorenz relation* [16, 17] or the *Clausius–Mossotti relation* [18, 19], and

$$\tilde{\mathbf{E}}^i(\mathbf{r}, \omega) + \nabla \times \nabla \times \oint_S \left(\tilde{\mathbf{Q}} \frac{\partial G}{\partial n'} - G \frac{\partial \tilde{\mathbf{Q}}}{\partial n'} \right) d^2 r' = \mathbf{0}, \quad (6.276)$$

which is known as the *Ewald–Oseen extinction theorem* [20, 21]. Because the Lorentz–Lorenz relation (6.275) has been derived along a separate line of analysis in Sect. 4.4.1, the Ewald–Oseen extinction theorem may then also be viewed as a consequence of it as applied to Eq. (6.274).

The relation given in Eq. (6.276) expresses the *extinction of the incident spectral wave field component* $\tilde{\mathbf{E}}^i(\mathbf{r}, \omega)$ at any point within the dielectric body through destructive interference with part of the induced dipole wave field. Notice that Eq. (6.276) states that the extinction of this incident wave field is brought about entirely by the dipoles on the boundary surface of the dielectric body. The extinguished incident spectral wave field at the frequency ω is then replaced by the spectral wave field at the same frequency ω that is given by [see Eqs. (6.247) and (6.249)]

$$\tilde{\mathbf{E}}'(\mathbf{r}, \omega) = \tilde{\mathbf{E}}^i(\mathbf{r}, \omega) + \tilde{\mathbf{E}}^d(\mathbf{r}, \omega) = \frac{1}{\mathcal{N}\alpha(\omega)}\tilde{\mathbf{P}}(\mathbf{r}, \omega). \tag{6.277}$$

With substitution from Eq. (6.247) and the Lorentz–Lorenz relation (6.275), this frequency-domain expression for the electric field vector acting on the molecular dipoles in the dielectric body becomes

$$\tilde{\mathbf{E}}'(\mathbf{r}, \omega) = \frac{1}{\mathcal{N}\alpha(\omega)}\left(n^2(\omega) - 1\right)k_0^2\tilde{\mathbf{Q}}(\mathbf{r}, \omega)$$

$$= \frac{4\pi}{3}\left(n^2(\omega) + 2\right)\frac{\omega^2}{c^2}\tilde{\mathbf{Q}}(\mathbf{r}, \omega). \tag{6.278}$$

which "propagates" inside the dielectric body with the *complex phase velocity*

$$v_p \equiv \frac{c}{n(\omega)}. \tag{6.279}$$

In terms of the macroscopic polarization vector $\tilde{\mathbf{P}}(\mathbf{r}, \omega)$, this expression becomes

$$\tilde{\mathbf{E}}'(\mathbf{r}, \omega) = \frac{4\pi}{3}\frac{n^2(\omega) + 2}{n^2(\omega) - 1}\tilde{\mathbf{P}}(\mathbf{r}, \omega), \tag{6.280}$$

where $\tilde{\mathbf{P}}(\mathbf{r}, \omega)$ satisfies the Helmholtz equation (6.257).

Finally, because $\tilde{\mathbf{D}}(\mathbf{r}, \omega) = n^2(\omega)\tilde{\mathbf{E}}(\mathbf{r}, \omega)$ in a homogeneous, isotropic, locally linear, nonmagnetic medium, and $\tilde{\mathbf{D}}(\mathbf{r}, \omega) = \tilde{\mathbf{E}}(\mathbf{r}, \omega) + 4\pi\tilde{\mathbf{P}}(\mathbf{r}, \omega)$ in a simple polarizable dielectric, then the elimination of the macroscopic electric displacement vector spectrum from these two expressions yields

$$\tilde{\mathbf{E}}(\mathbf{r}, \omega) = \frac{4\pi}{n^2(\omega) - 1}\tilde{\mathbf{P}}(\mathbf{r}, \omega). \tag{6.281}$$

With Eq. (6.280) rewritten as

$$\tilde{\mathbf{E}}'(\mathbf{r}, \omega) = \frac{4\pi}{n^2(\omega) - 1}\tilde{\mathbf{P}}(\mathbf{r}, \omega) + \frac{4\pi}{3}\tilde{\mathbf{P}}(\mathbf{r}, \omega), \tag{6.282}$$

comparison with Eq. (6.281) then shows that the *effective field acting on the molecular dipoles in the dielectric body* is given by the well-known expression

$$\mathbf{E}'(\mathbf{r}, t) = \mathbf{E}(\mathbf{r}, t) + \frac{4\pi}{3}\mathbf{P}(\mathbf{r}, t) \qquad (6.283)$$

after a straightforward Fourier inversion into the space–time domain.

6.7 Résumé

A detailed analysis of the Fresnel reflection and transmission coefficients for a planar interface separating two complex materials has been presented for both $TE(s)$ and $TM(p)$ polarizations. Complex materials include dispersive attenuative dielectrics with complex-valued dielectric permittivity $\epsilon(\omega) = \epsilon'(\omega) + i\epsilon''(\omega)$ and semiconducting materials with complex-valued conductivity $\sigma(\omega) = \sigma'(\omega) + i\sigma''(\omega)$, brought together through the complex permittivity $\epsilon_c(\omega) \equiv \epsilon(\omega) + i(1/\|4\pi\|)\sigma(\omega)/\omega$, as well as lossy magnetic media through its complex-valued magnetic permeability $\mu(\omega) = \mu'(\omega) + i\mu''(\omega)$. Of final interest here is the frequency dependence of the reflection and transmission coefficients as a function of the incidence angle Θ_i of an inhomogeneous plane wave with attenuation along the normal direction to the interface, a typical situation occurring in layered media. As an illustration of the complex behavior that results, let the frequency dispersion of both dielectric media be described by a single relaxation time Rocard–Powles–Debye model, where [see Eq. (4.197)]

$$\epsilon_j(\omega)/\epsilon_0 = \epsilon_{\infty j} + \frac{a_j}{(1 - i\omega\tau_j)(1 - i\omega\tau_{fj})}, \qquad j = 1, 2. \qquad (6.284)$$

Let the model parameters for medium 1 (the so-called "incident medium") be representative of triply-distilled water at 20 °C with $\epsilon_{\infty 1} = 2.1$, $a_1 = 74.1$, $\tau_1 = 8.44 \times 10^{-12}$ s, $\tau_{f1} = 4.62 \times 10^{-14}$ s, and let those for medium 2 (the so-called "transmitted medium") have somewhat similar model parameters with $\epsilon_{\infty 2} = 14.1$, $a_2 = 71.8$, $\tau_2 = 1.71 \times 10^{-11}$ s, and $\tau_{f2} = 1.75 \times 10^{-14}$ s. The resultant frequency dependence of the complex index of refraction for both media, illustrated in Fig. 6.34, shows that incidence is on the optically rarer medium ($n_1' > n_2'$) for low frequencies $\omega < \Omega_c$ but then switches to incidence on the optically denser medium ($n_1' < n_2'$) for higher frequencies $\omega > \Omega_c$, where $\Omega_c \approx 1 \times 10^{12}$ r/s).

This is reflected in the three-dimensional plot of the angle of refraction Θ_t as a function of both the angle of incidence Θ_i and the angular frequency ω, illustrated in Fig. 6.35. As expected, supercritical incidence occurs when $\omega < \Omega_c$, but not for higher frequencies.

Three-dimensional plots of the angular frequency dependence of the magnitude and phase of the $TE(s)$-polarization Fresnel reflection coefficient $\Gamma_{TE}(\omega) = \Gamma_s(\omega)$ as a function of the incidence angle Θ_i for this interface example between two similar Debye-model dielectrics is presented in Figs. 6.36 and 6.37, respectively.

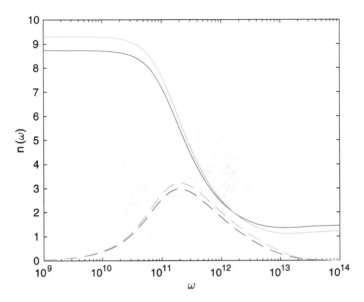

Fig. 6.34 Frequency dependence of the real (solid curves) and imaginary (dashed curves) parts of the complex index of refraction $n(\omega) = n'(\omega) + in''(\omega)$ described by the single relaxation time Rocard–Powles–Debye model of the dielectric permittivity for (a) triply-distilled H_2O with $\epsilon_{\infty 1} = 2.1$, $a_1 = 74.1$, $\tau_1 = 8.44 \times 10^{-12}$ s, $\tau_{f1} = 4.62 \times 10^{-14}$ s and (b) a slightly different dielectric with $\epsilon_{\infty 2} = 14.1$, $a_2 = 71.8$, $\tau_2 = 1.71 \times 10^{-11}$ s, and $\tau_{f2} = 1.75 \times 10^{-14}$ s

Notice that $|\Gamma_{TE}(\omega)|$ increases monotonically with increasing angle of incidence Θ_i for all frequencies, but only reaches total reflection for frequencies below Ω_c (see Fig. 6.35), the angular extent from the critical angle Θ_c to grazing incidence over which total reflection occurs decreasing as ω increases to Ω_c. The magnitude of the reflection coefficient reaches a minimum near Ω_c because $n'_1(\Omega_c) = n'_2(\Omega_c)$, but does not vanish because $n''_1(\Omega_c) \neq n''_2(\Omega_c)$, as seen in Fig. 6.36. The phase angle $\arg\{\Gamma_{TE}(\omega)\}$ behavior presented in Fig. 6.37 is seen to increase monotonically with increasing frequency for near normal angles of incidence but becomes increasingly complicated about Ω_c as the incidence angle increases towards grazing incidence.

The corresponding three-dimensional plots of the angular frequency dependence of the magnitude and phase of the $TE(s)$-polarization Fresnel transmission coefficient $\tau_{TE}(\omega) = \tau_s(\omega)$ as a function of the incidence angle Θ_i for this interface example between two similar Debye-model dielectrics is presented in Figs. 6.38 and 6.39, respectively. Both the magnitude and phase angle of $\tau_{TE}(\omega)$ vary smoothly with the frequency ω for near normal angles of incidence Θ_i, but as the incidence angle increases away from $0°$ and approaches $90°$ (grazing incidence), they both become increasingly variable, particularly in the transition region about Ω_c where incidence on the optically rarer medium for $\omega < \Omega_c$ switches to incidence on the optically denser medium for $\omega > \Omega_c$.

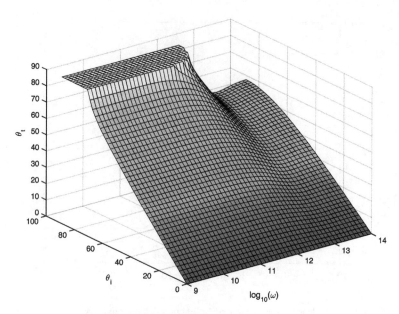

Fig. 6.35 Frequency dependence of the angle of refraction Θ_t vs. the angle of incidence Θ_i (both in degrees) for the lossy dispersive dielectric media used in the example presented in Fig. 6.43

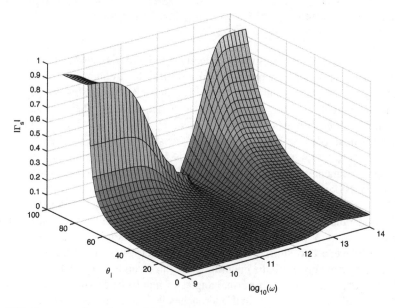

Fig. 6.36 Frequency dependence of the magnitude $|\Gamma(\omega)|$ of the $TE(s)$-polarization reflection coefficient $\Gamma_{TE}(\omega) = \Gamma_s(\omega)$ vs. the angle of incidence Θ_i (in degrees) for the example presented in Fig. 6.43

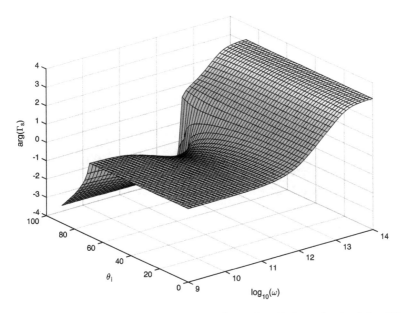

Fig. 6.37 Frequency dependence of the phase angle $\arg\{\Gamma(\omega)\}$ (in radians) of the $TE(s)$-polarization reflection coefficient $\Gamma_{TE}(\omega) = \Gamma_s(\omega)$ vs. the angle of incidence Θ_i (in degrees) for the example presented in Fig. 6.43

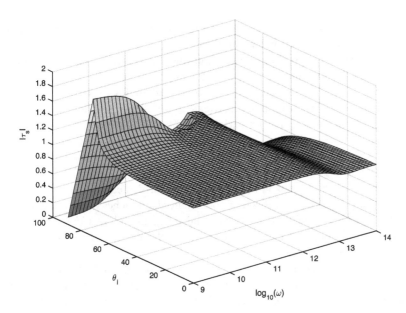

Fig. 6.38 Frequency dependence of the magnitude $|\tau(\omega)|$ of the $TE(s)$-polarization transmission coefficient $\tau_{TE}(\omega) = \tau_s(\omega)$ vs. the angle of incidence Θ_i (in degrees) for the example presented in Fig. 6.43.

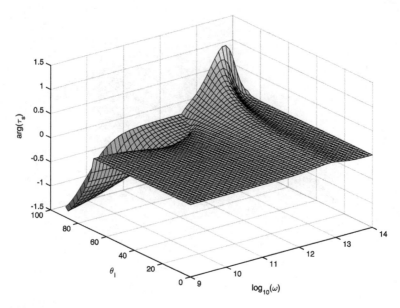

Fig. 6.39 Frequency dependence of the phase angle $\arg\{\tau(\omega)\}$ (in radians) of the $TE(s)$-polarization transmission coefficient $\tau_{TE}(\omega) = \tau_s(\omega)$ vs. the angle of incidence Θ_i (in degrees) for the example presented in Fig. 6.43

Three-dimensional plots of the angular frequency dependence of the magnitude and phase of the $TM(p)$-polarization Fresnel reflection coefficient $\Gamma_{TM}(\omega) = \Gamma_p(\omega)$ and the corresponding Fresnel transmission coefficient $\tau_{TM}(\omega) = \tau_p(\omega)$ as a function of the incidence angle Θ_i for this same interface example between two similar Debye-model dielectrics is presented in Figs. 6.40–6.41 and 6.42–6.43, respectively. The similarity between these plots and their corresponding counterparts for $TE(s)$ polarization is remarkable. As for the $TE(s)$-polarization case illustrated in Fig. 6.36, $|\Gamma_{TM}(\omega)|$ increases monotonically with increasing angle of incidence Θ_i for all frequencies, but only reaches total reflection for frequencies below Ω_c (see Fig. 6.35), the angular extent from the critical angle Θ_c to grazing incidence over which total reflection occurs again decreasing as ω increases to Ω_c. The magnitude of the reflection coefficient also reaches a minimum near Ω_c because $n'_1(\Omega_c) = n'_2(\Omega_c)$, but does not vanish because $n''_1(\Omega_c) \neq n''_2(\Omega_c)$. Notice further the appearance of near zero reflection at the Brewster angle Θ_p for lower frequencies where the loss is small.

The results presented in Figs. 6.36–6.39 and 6.40–6.43 indicate that the effects of frequency dispersion upon the reflected and transmitted fields from either an incident $TE(s)$-polarized or $TM(p)$-polarized ultrawideband (UWB) plane wave pulse whose frequency spectrum is centered about $\omega_c = 2\pi f_c$ in the EHF with $f_c = \Omega_c/2\pi \sim 100\,\mathrm{GHz}$ and extends from the UHF (1 GHz) to the infrared (10 THz) can be significant if the angle of incidence is either close to or exceeds the critical angle at the lower end of the incident pulse spectrum. In that case, the lower

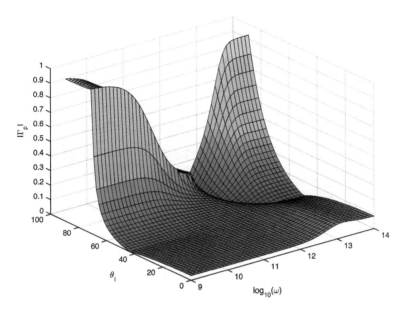

Fig. 6.40 Frequency dependence of the magnitude $|\Gamma(\omega)|$ of the $TM(p)$-polarization reflection coefficient $\Gamma_{TM}(\omega) = \Gamma_p(\omega)$ vs. the angle of incidence Θ_i (in degrees) for the example presented in Fig. 6.43

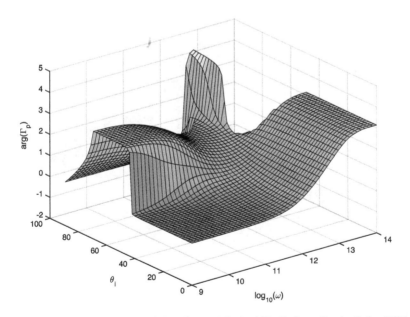

Fig. 6.41 Frequency dependence of the phase angle $\arg\{\Gamma(\omega)\}$ (in radians) of the $TM(p)$-polarization reflection coefficient $\Gamma_{TM}(\omega) = \Gamma_p(\omega)$ vs. the angle of incidence Θ_i (in degrees) for the example presented in Fig. 6.43

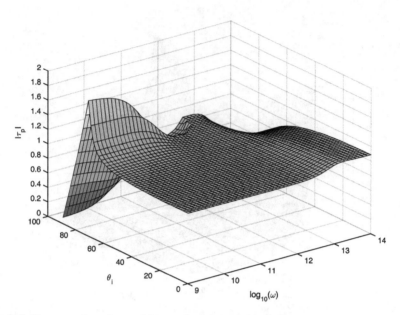

Fig. 6.42 Frequency dependence of the magnitude $|\tau(\omega)|$ of the $TM(p)$-polarization transmission coefficient $\tau_{TM}(\omega) = \tau_p(\omega)$ vs. the angle of incidence Θ_i (in degrees) for the example presented in Fig. 6.43

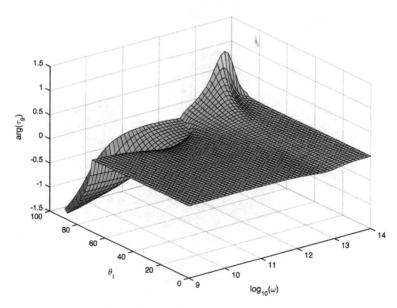

Fig. 6.43 Frequency dependence of the phase angle $\arg\{\tau(\omega)\}$ (in radians) of the $TM(p)$-polarization transmission coefficient $\tau_{TM}(\omega) = \tau_p(\omega)$ vs. the angle of incidence Θ_i (in degrees) for the example presented in this figure

frequency components of the incident pulse spectrum will undergo total internal reflection while the remaining higher frequency components will experience partial reflection. As a result, the reflected and transmitted pulses may be quite different from each other as well as from the incident pulse. This is considered in greater detail in Volume 2.

Finally, consider the behavior when medium 2 is a DNG metamaterial with resonant Lorentz model dielectric permittivity

$$\epsilon_2(\omega)/\epsilon_0 = 1 - \frac{\omega_p^2}{\omega^2 - \omega_0^2 + 2i\delta_p\omega} \qquad (6.285)$$

and magnetic permeability

$$\mu_2(\omega)/\mu_0 = 1 - \frac{\omega_m^2}{\omega^2 - \omega_1^2 + 2i\delta_m\omega}, \qquad (6.286)$$

where [from Eqs. (5.94) and (5.95)] $\epsilon_2(\omega) = |\epsilon_2(\omega)|e^{i\varphi_\epsilon(\omega)}$ with branch choice $\varphi_\epsilon(\omega) \in (\pi/2, \pi]$ when $\epsilon_2'(\omega) < 0$ and $\mu_2(\omega) = |\mu_2(\omega)|e^{i\varphi_\mu(\omega)}$ with branch choice $\varphi_\mu(\omega) \in (\pi/2, \pi]$ when $\mu_2'(\omega) < 0$. Such a material is referred to here as a *double Lorentz-model DNG metamaterial*. Define the *DNG medium domain* $\Omega'_{DNG}(\omega') = (\Omega_{min}, \Omega_{max})$ as the real angular frequency domain over which both $\epsilon'(\omega) < 0$ and $\mu'(\omega) < 0$. With $\omega_0 = 4.0 \times 10^{16}$ r/s, $\omega_1 = 4.2 \times 10^{16}$ r/s, $\omega_p = \omega_m = \sqrt{20} \times 10^{16}$ r/s, and $\delta_p = \delta_m = 0.28 \times 10^{16}$ r/s, one finds that $\Omega_{min} \simeq 4.1429 \times 10^{16}$ r/s and $\Omega_{max} \simeq 5.8415 \times 10^{16}$ r/s. It is over this finite real frequency domain that the real part of the complex index of refraction $n(\omega) = [(\epsilon(\omega)/\epsilon_0)(\mu(\omega)/\mu_0)]^{1/2}$ is negative; that is $n'(\omega) < 0$ when $\omega \in \Omega'_{DNG}$. Notice that this DNG frequency domain is a section through the two-dimensional domain $\Omega_{DNG}(\omega)$ in the complex ω-plane.

The real frequency dependence of the magnitudes of these quantities is illustrated in Fig. 6.44 and their associated phase angles are illustrated in Fig. 6.45. The resultant angular frequency dependence of the real and imaginary parts of the complex index of refraction $n(\omega) = [(\epsilon(\omega)/\epsilon_0)(\mu(\omega)/\mu_0)]^{1/2}$ is given in Fig. 6.46 with the resultant DNG domain $\Omega'_{DNG} = (\Omega_{min}, \Omega_{max})$ along the real frequency axis indicated. Notice that this branch choice for the phase arguments of $\epsilon(\omega)$ and $\mu(\omega)$ results in a continuous variation of the real and imaginary parts of the complex index of refraction with angular frequency over both the positive and negative index regions as well as at the transition points Ω_{min} and Ω_{max} between them.

Three-dimensional plots of the magnitude and phase of the relative dielectric permittivity $\epsilon(\omega)/\epsilon_0$ and magnetic permeability $\mu(\omega)/\mu_0$ in the complex frequency plane are given in Figs. 6.47, 6.48, 6.49 and 6.50. Notice in Figs. 6.48 and 6.50 the pair of symmetric branch cuts in the lower left and right-half planes for each phase angle $\arg\{\epsilon(\omega)\}$ and $\arg\{\mu(\omega)\}$. It is about each of these two regions that the two-dimensional DNG frequency domain $\Omega_{DNG}(\omega)$ for the complex index of refraction $n(\omega)$ appears in the complex ω-plane. The intersection of this two-

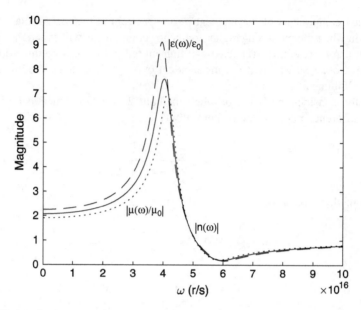

Fig. 6.44 Angular frequency dependence of the magnitude of the relative dielectric permittivity $|\epsilon(\omega)|/\epsilon_0$ (dashed curve), relative magnetic permeability $|\mu(\omega)|/\mu_0$ (dotted curve), and the complex index of refraction $n(\omega) = [(\epsilon(\omega)/\epsilon_0)(\mu(\omega)/\mu_0)]^{1/2}$ (solid curve) for a double Lorentz-model DNG metamaterial

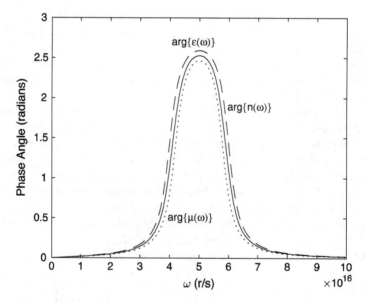

Fig. 6.45 Angular frequency dependence of the phase angle (in radians) of the relative dielectric permittivity $\arg\{\epsilon(\omega)\}$ (dashed curve), relative magnetic permeability $\arg\{\mu(\omega)\}$ (dotted curve), and the complex index of refraction $\arg\{n(\omega)\} = \arg\{[(\epsilon(\omega)/\epsilon_0)(\mu(\omega)/\mu_0)]^{1/2}\}$ (solid curve) for a double Lorentz-model DNG metamaterial

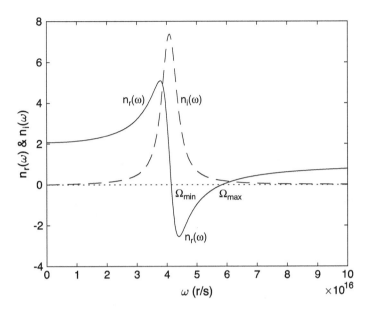

Fig. 6.46 Angular frequency dependence of the real (solid curve) and imaginary (dashed curve) parts of the complex index of refraction for a double Lorentz-model DNG metamaterial with $\omega_0 = 4.0 \times 10^{16}$ r/s, $\omega_1 = 4.2 \times 10^{16}$ r/s, $\omega_p = \omega_m = \sqrt{20} \times 10^{16}$ r/s, and $\delta_p = \delta_m = 0.28 \times 10^{16}$ r/s

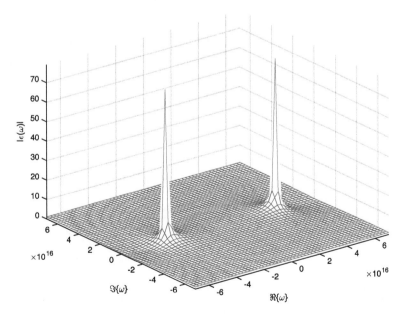

Fig. 6.47 Angular frequency dependence of the magnitude $|\epsilon(\omega)|/\epsilon_0$ of the relative dielectric permittivity in the complex ω-plane for a double Lorentz-model DNG metamaterial

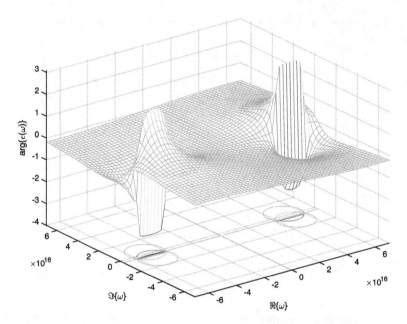

Fig. 6.48 Angular frequency dependence of the phase angle $\arg\{\epsilon(\omega)\}$ of the dielectric permittivity in the complex ω-plane for a double Lorentz-model DNG metamaterial

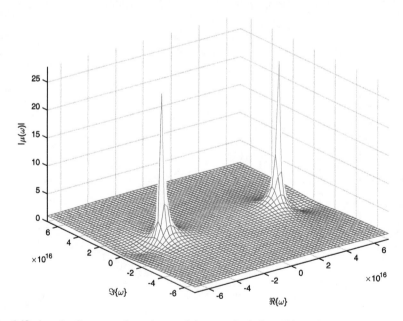

Fig. 6.49 Angular frequency dependence of the magnitude $|\mu(\omega)|/\mu_0$ of the relative magnetic permeability in the complex ω-plane for a double Lorentz-model DNG metamaterial

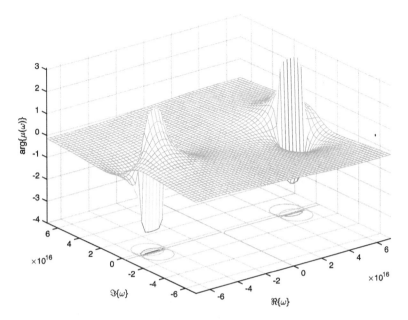

Fig. 6.50 Angular frequency dependence of the phase angle $\arg\{\mu(\omega)\}$ of the magnetic permeability in the complex ω-plane for a double Lorentz-model DNG metamaterial

dimensional region with the positive real frequency axis (the positive ω'-axis) is precisely the DNG frequency domain $\Omega'_{DNG}(\omega')$ defined above. Notice further that the magnitudes of both $\epsilon(\omega)$ and $\mu(\omega)$ possess relatively sharp peaks about the branch points

$$\omega_{\epsilon_\pm} \equiv \pm\sqrt{\omega_0^2 - \delta_p^2} - i\delta_p, \tag{6.287}$$

$$\omega_{\mu_\pm} \equiv \pm\sqrt{\omega_1^2 - \delta_m^2} - i\delta_m, \tag{6.288}$$

near their respective resonance frequencies, as seen in Figs. 6.47 and 6.49, respectively. The width of each resonance peak is primarily determined by the value of the phenomenological damping constant δ_j appearing in the Lorentz model for that particular medium property. The width of the resonance peak for the complex index of refraction

$$n(\omega) \equiv [(\epsilon(\omega)/\epsilon_0)(\mu(\omega)/\mu_0)]^{1/2}$$

$$= \left[\left(1 - \frac{\omega_p^2}{\omega^2 - \omega_0^2 + 2i\delta_p\omega}\right)\left(1 - \frac{\omega_m^2}{\omega^2 - \omega_1^2 + 2i\delta_m\omega}\right)\right]^{1/2} \tag{6.289}$$

then depends upon both values when $\omega_0 \approx \omega_1$; however, if the two resonance frequencies are widely separated, there will then be two individual peaks. In addition, the model exhibits the low- and high-frequency limiting values given by

$$\lim_{\omega \to 0} n(\omega) = n(0) = \sqrt{(1 + \omega_p^2/\omega_0^2)(1 + \omega_m^2/\omega_1^2)}, \qquad (6.290)$$

$$\lim_{\omega \to \infty} n(\omega) = 1, \qquad (6.291)$$

where $n(0) \simeq 2.1645$.

Consider finally the transmission of a plane electromagnetic wave from a normal, naturally occurring dielectric medium 1 into a DNG metamaterial medium 2. Let the metamaterial dielectric permittivity $\epsilon_2(\omega)/\epsilon_0$ and magnetic permeability $\mu_2(\omega)/\mu_0$ each be described by single resonance Lorentz models with complex index of refraction $n_2(\omega)$ as given in Eq. (6.289) with $\omega_0 = 4.0 \times 10^{16}$ r/s, $\omega_1 = 4.2 \times 10^{16}$ r/s, $\omega_p = \omega_m = \sqrt{20} \times 10^{16}$ r/s, and $\delta_p = \delta_m = 0.28 \times 10^{16}$ r/s as in the previous example. In addition, let medium 1 be described by the same dielectric permittivity as in medium 2 $[\epsilon_1(\omega) = \epsilon_2(\omega)]$ but with a constant magnetic permeability $\mu_1/\mu_0 = 1$. In a sense, the DNG metamaterial in medium 2 was engineered from medium 1 by modifying its magnetic permeability from the constant value $\mu_1 = \mu_0$ to a temporally dispersive one with a relative permeability $\mu_2(\omega)/\mu_0$ that is described by the Lorentz model in Eq. (6.286). The resultant angular frequency dependence of the angle of refraction $\Theta_i(\omega)$ is presented in the three-dimensional graph of Fig. 6.51. The discontinuity contour exhibited in this surface map describes the set of values (ω, Θ_i) at which $n_2' = \Re\{n_2\}$ vanishes. Points interior to this contour are in the negative refraction region that defines the DNG metamaterial, whereas points exterior to it lie in the region of normal dielectric refraction. Points on the zero-index contour where the discontinuity occurs are in neither region and so they assume the limiting value obtained from either side, negative if from the interior DNG region and positive if from the exterior normal dielectric region.

Problems

6.1 Using Gauss' law, show that an external electrostatic field $\mathbf{E}_{ext}(\mathbf{r})$ at the surface S of an ideal conductor ($\sigma = \infty$) produces an outward directed force $\mathbf{F}(\mathbf{r}) = \hat{\mathbf{n}} \Delta U_e(\mathbf{r})/\Delta \zeta$ per unit area given by

$$\frac{d\mathbf{F}(\mathbf{r})}{da} = \|4\pi\| \frac{\varrho_s(\mathbf{r})}{\epsilon(0)} \hat{\mathbf{n}} = u_e(\mathbf{r})\hat{\mathbf{n}} \qquad (6.292)$$

at the surface S of the conductor that is equal in magnitude to the electrostatic energy density at that point, thereby producing a negative pressure on the conductor that tends to pull it into the field. Here $\hat{\mathbf{n}}$ is the outward unit normal vector to S

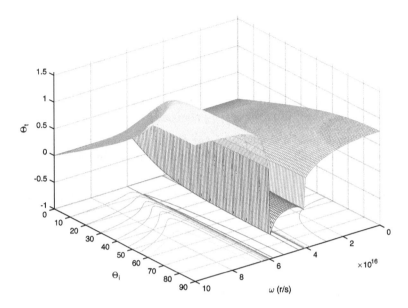

Fig. 6.51 Angular frequency dependence of the angle of refraction Θ_t vs. the angle of incidence Θ_i when medium 1 (a normal, naturally occurring dielectric) is described by the same complex dielectric permittivity $\epsilon_1(\omega)/\epsilon_0$ as medium 2 but with a non-dispersive magnetic permeability $\mu_1/\mu_0 = 1$, a metamaterial with frequency dependent dielectric permittivity $\epsilon_2(\omega)$ and magnetic permeability $\mu(\omega)$ whose Lorentz model frequency dispersion is presented in Figs. 6.44, 6.45, 6.46, 6.47, 6.48, 6.49, and 6.50

at the point $\mathbf{r} \in S$ directed from the conducting body and into the external medium characterized by the static dielectric permittivity $\epsilon(0)$, and where $\varrho_s(\mathbf{r})$ is the surface charge density at that point.

6.2 For a time-harmonic wave field [22], the electric field force acting on the surface of a perfect conductor body situated in a dielectric medium with permittivity $\epsilon(\omega)$ may be obtained through use of the Minkowski–Maxwell stress tensor defined in Eq. (5.353) with momentum flux density given by

$$- T_{ij} = \frac{1}{\|4\pi\|} \left(\frac{1}{2} D_i E_i \delta_{ij} - D_i E_j \right). \tag{6.293}$$

The force acting on a differential element of surface area $\hat{\mathbf{n}} da$ on the surface S of the conductor is given by the flux of momentum passing through it from the outside, given by $T_{ij} da_j = T_{ij} n_j da$, where the unit normal vector $\hat{\mathbf{n}}$ is directed outwards from the surface of the conductor body. Show that the force per unit area acting on the surface at a given point is given by

$$\frac{d\mathbf{F}(\mathbf{r}, t)}{da} = \left\| \frac{1}{4\pi} \right\| \frac{1}{2} \left(\mathbf{D}(\mathbf{r}, t) \cdot \mathbf{E}(\mathbf{r}, t) \right) \hat{\mathbf{n}}, \tag{6.294}$$

which reduces to that for an electrostatic field.

6.3 Provide the details for the derivation of the generalized law of refraction given in Eqs. (6.41)–(6.44).

6.4 Provide the details for the derivation of the generalized Fresnel reflection and transmission coefficients given in Eqs. (6.153) and (6.154) for $TE(s)$-polarization and Eqs. (6.158) and (6.159) for $TM(p)$-polarization.

6.5 Provide the details for the derivation of $\Gamma_{TE}(\omega)$ at normal incidence ($\Theta_i = 0$), at critical incidence ($\Theta_i = \Theta_c$), and approaching grazing incidence from below ($\Theta_i \to \pi/2^-$).

6.6 Derive the limiting expressions for $\Gamma_{TE}(\omega)$ and $\tau_{TE}(\omega)$ when (a) medium 1 is lossless ($n_1'' = 0$) and medium 2 may be lossy ($n_2'' \geq 0$), and (b) when both media are lossless ($n_1'' = n_2'' = 0$). Compare the latter results with the expressions given in Eqs. (6.63) and (6.64).

6.7 Repeat Problem 6.5 for $TM(p)$-polarization.

6.8 Repeat Problem 6.6 for $TM(p)$-polarization.

6.9 Derive Eq. (6.179) for the polarizing angle Θ_p at which $\Gamma_{TM}(\omega)$ vanishes.

6.10 Provide the details for the derivation of the generalized Fresnel reflection and transmission coefficients for incidence on a DNG metamaterial with loss, as given in Eqs. (6.214)–(6.215 and Eqs. (6.231)–(6.232).

6.11 Derive expressions for the reflection and transmission coefficients for the normal components of an incident $TE(s)$-polarized plane wave in terms of the Fresnel coefficients.

6.12 Consider a series of plane parallel interfaces S_j, $j = 1, 2, 3, \ldots, K$ between different dielectrics with corresponding indices of refraction n_j, where n_1 is the medium that the incident plane wave is situated in and n_K the medium where the final transmitted plane wave is situated. Show that the relation between the final exit angle Θ_K and the angle of incidence Θ_1 is governed by Snell's law and, as far as the indices of refraction of the media are concerned, depends only upon the indices of refraction of the entrance and exit dielectrics, provided that there is no intermediate surface at which total reflection occurs.

6.13 Show that the local change in angle of a geometrical ray of light in a spatially inhomogeneous medium with (real-valued) index of refraction $n(\mathbf{r}) = n(x, y, z)$ is given by

$$\Delta\Theta(\mathbf{r}) = -\frac{\nabla n(\mathbf{r}) \cdot \Delta\mathbf{r}}{n(\mathbf{r})} \tan\Theta(\mathbf{r}), \tag{6.295}$$

where $\Delta\mathbf{r} = \hat{\mathbf{1}}_x \Delta x + \hat{\mathbf{1}}_y \Delta y + \hat{\mathbf{1}}_z \Delta z$.

6.14 Consider a spatially inhomogeneous optical fiber with cylindrical symmetry (r, ϕ) about the z-axis; that is, $n = n(r)$ with $r = \sqrt{x^2 + y^2}$. With the position vector $\mathbf{r}(z)$ of a geometrical light ray expressed in cylindrical polar coordinates as

$$\mathbf{r}(z) = \hat{\mathbf{1}}_r(\phi(z))\mathbf{r}(z),$$

where the radial unit vector $\hat{\mathbf{1}}_r = \hat{\mathbf{1}}_r(\phi(z))$ depends upon the distance z along the fiber axis through the azimuthal angular dependence $\phi = \phi(z)$. From vector differential equation (5.128) of the light ray, determine the individual differential equations satisfied separately by the radial r- and azimuthal ϕ-components of the ray. Solve the azimuthal component equation for a general skew ray and sketch its behavior along the z-axis.

6.15 Consider a spatially inhomogeneous optical fiber with cylindrical symmetry about the z-axis (the fiber axis) and radially dependent parabolic index of refraction profile given by (illustrated below)

$$n^2(r) = n_0^2 - b^2 r^2, \quad 0 \le r \le r_0$$
$$n^2(r) = n_1^2, \qquad\quad r \ge r_0,$$

where $r = \sqrt{x^2 + y^2}$, and where n_0 is the on-axis refractive index value and n_1 is the constant refractive index of the surrounding fiber cladding. Consider a meridional ray in the xz-plane with initial ray height $x = 0$ and initial slope $dx/dz = \tan\gamma_0$ at $z = 0$. Determine the half-angle $\gamma_0|_{max}$ of the cone of rays that are guided (trapped) within the fiber and form stable solutions, expressing the answer solely in terms of the two refractive index values n_0 and n_1 that characterize the fiber. This value is related to the *numerical aperture* of the fiber (Fig. 6.52).

Fig. 6.52 Cross-sectional radial dependence of the real-valued (loss-free) index of refraction $n(r)$ for a parabolic profile index optical fiber

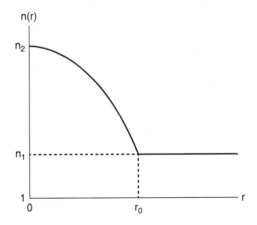

References

1. M. Born and E. Wolf, *Principles of Optics*. Cambridge: Cambridge University Press, seventh (expanded) ed., 1999.
2. G. Green, "On the reflexion and refraction of sound," *Trans. Cambridge Phil. Soc.*, vol. 6, p. 403, 1838.
3. A. Fresnel, "La réflexion de la lumière," in *Oeuvres*, vol. 1, p. 767, Paris: Imprimerie Imperiale, 1866.
4. J. A. Stratton, *Electromagnetic Theory*. New York: McGraw-Hill, 1941.
5. C. A. Balanis, *Advanced Engineering Electromagnetics*. Wiley, second ed., 2012.
6. J. E. Roy, "New results for the effective propagation constants of nonuniform plane waves at the planar interface of two lossy media," *IEEE Transactions on Antennas and Propagation*, vol. 51, pp. 1206–1215, 2003.
7. I. M. Besieris, "Comment on the 'Corrected Fresnel coefficients for lossy materials'," *IEEE Antennas and Propagation Magazine*, vol. 53, no. 4, pp. 161–164, 2011.
8. F. X. Canning, "Corrected Fresnel coefficients for lossy materials," in *Proceedings of the 2011 IEEE International Symposium on Antennas and Propagation*, pp. 2133–2136, 2011.
9. G. P. Ohman, "The pseudo-Brewster angle," *IEEE Trans. Antennas and Propagation*, vol. 25, no. 6, pp. 903–904, 1977.
10. S. Y. Kim and K. Vedam, "Analytic solution of the pseudo-Brewster angle," *J. Opt. Soc. Am. A*, vol. 3, no. 11, pp. 1772–1773, 1986.
11. M. Born, *Optik*. Berlin: Springer, 1933.
12. L. Rosenfeld, *Theory of Electrons*. Amsterdam: North-Holland, 1951.
13. É. Lalor and E. Wolf, "Exact solution of the equations of molecular optics for reflection and reflection of an electromagnetic wave on a semi-infinite dielectric," *J. Opt. Soc. Am.*, vol. 62, no. 10, pp. 1165–1174, 1972.
14. H. M. Nussenzveig, *Causality and Dispersion Relations*. New York: Academic, 1972. Ch. 1.
15. C. F. Bohren and D. R. Huffman, *Absorption and Scattering of Light by Small Particles*. New York: Wiley-Interscience, 1984. Ch. 9.
16. H. A. Lorentz, "Über die Beziehungzwischen der Fortpflanzungsgeschwindigkeit des Lichtes der Körperdichte," *Ann. Phys.*, vol. 9, pp. 641–665, 1880.
17. L. Lorenz, "Über die Refractionsconstante," *Ann. Phys.*, vol. 11, pp. 70–103, 1880.
18. P. F. Mossotti, "x," *Bibl. Univ. Modena*, vol. 6, p. 193, 1847.
19. R. Clausius, "Die Mechanische Wärmetheorie," vol. 2, pp. 62–97, Braunschweich: Vieweg, 1879.
20. P. P. Ewald, "Zur begründung der kristalloptik," *Ann. d. Physik*, vol. 48, p. 1, 1916.
21. C. W. Oseen, "About the interaction between two electric dipoles and the rotation of the polarization plane in crystals and liquids," *Ann. d. Physik*, vol. 49, p. 1, 1915.
22. L. D. Landau and E. M. Lifshitz, *Electrodynamics of Continuous Media*. Oxford: Pergamon, 1960. Ch. IX.

Chapter 7
The Angular Spectrum Representation of the Pulsed Radiation Field in Spatially and Temporally Dispersive Media

"Science is spectral analysis. Art is light synthesis."

Karl Kraus

Attention is now directed to the rigorous solution of the electromagnetic field that is radiated by a general current source in a homogeneous, anisotropic, locally linear, spatially and temporally dispersive medium occupying all of space that is characterized by the space and time-dependent dielectric permittivity response tensor $\hat{\underline{\epsilon}}(\mathbf{r} - \mathbf{r}', t - t')$, magnetic permeability response tensor $\hat{\underline{\mu}}(\mathbf{r} - \mathbf{r}', t - t')$, and electric conductivity response tensor $\hat{\underline{\sigma}}(\mathbf{r} - \mathbf{r}', t - t')$, where $(\overline{\mathbf{r}}', t')$ denotes the space–time point of the inducing field stimulus and $(\mathbf{r}, t)$ the space–time point of the induced field response. The applied current source $\mathbf{J}_0(\mathbf{r}, t)$ is assumed here to be a well-behaved known function of both position $\mathbf{r}$ and time t that identically vanishes for $|z| \geq Z$, where Z is a positive constant, as illustrated in Fig. 7.1. Furthermore, it is assumed that the current source is turned on at time $t = 0$, so that

$$\mathbf{J}_0(\mathbf{r}, t) = \mathbf{0}, \quad t \leq 0. \tag{7.1}$$

Because the present analysis is concerned only with the electromagnetic field that is radiated by this current source, it is then required that both field vectors $\mathbf{E}(\mathbf{r}, t)$ and $\mathbf{B}(\mathbf{r}, t)$ also vanish for $t \leq 0$; viz.,

$$\left.\begin{array}{l} \mathbf{E}(\mathbf{r}, t) = \mathbf{0} \\ \mathbf{B}(\mathbf{r}, t) = \mathbf{0} \end{array}\right\} \quad t \leq 0. \tag{7.2}$$

These requirements on the current source are not restrictive in any physical sense because all real radiation problems may be cast so as to satisfy them.

© Springer Nature Switzerland AG 2019   427
K. E. Oughstun, *Electromagnetic and Optical Pulse Propagation*, Springer Series in Optical Sciences 224, https://doi.org/10.1007/978-3-030-20835-6_7

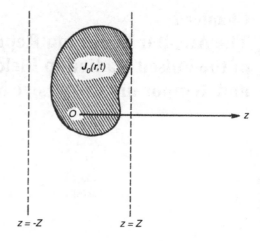

Fig. 7.1 General current source $\mathbf{J}_0(\mathbf{r}, t)$ embedded in a dispersive medium occupying all of space. The current source has finite spatial extent in the z-direction and identically vanishes for all time $t < 0$

7.1 The Fourier–Laplace Integral Representation of the Radiation Field

From the constitutive relations given in Eqs. (4.63)–(4.65) for a spatially and temporally dispersive HALL medium and the conditions given in Eqs. (7.1) and (7.2), the total current density in the medium after the current source has been turned on (i.e., for $t > 0$) is given by

$$\mathbf{J}(\mathbf{r}, t) = \mathbf{J}_0(\mathbf{r}, t) + \int_{-\infty}^{\infty}\int_0^t \hat{\underline{\sigma}}(\mathbf{r} - \mathbf{r}', t - t') \cdot \mathbf{E}(\mathbf{r}', t') d^3 r' dt', \qquad (7.3)$$

and the electric displacement and magnetic intensity vectors are

$$\mathbf{D}(\mathbf{r}, t) = \int_{-\infty}^{\infty}\int_0^t \hat{\underline{\epsilon}}(\mathbf{r} - \mathbf{r}', t - t') \mathbf{E}(\mathbf{r}', t') d^3 r' dt', \qquad (7.4)$$

$$\mathbf{H}(\mathbf{r}, t) = \int_{-\infty}^{\infty}\int_0^t \hat{\underline{\mu}}^{-1}(\mathbf{r} - \mathbf{r}', t - t') \mathbf{B}(\mathbf{r}', t') d^3 r' dt', \qquad (7.5)$$

respectively. The spatial Fourier and temporal Fourier–Laplace transform of these relations then gives, with application of the convolution theorem given in Eq. (E.15) for the Laplace transform,

$$\tilde{\tilde{\mathbf{J}}}(\mathbf{k}, \omega) = \tilde{\tilde{\mathbf{J}}}_0(\mathbf{k}, \omega) + \underline{\sigma}(\mathbf{k}, \omega) \cdot \tilde{\tilde{\mathbf{E}}}(\mathbf{k}, \omega), \qquad (7.6)$$

$$\tilde{\tilde{\mathbf{D}}}(\mathbf{k}, \omega) = \underline{\epsilon}(\mathbf{k}, \omega) \cdot \tilde{\tilde{\mathbf{E}}}(\mathbf{k}, \omega), \qquad (7.7)$$

$$\tilde{\tilde{\mathbf{H}}}(\mathbf{k}, \omega) = \underline{\mu}^{-1}(\mathbf{k}, \omega) \cdot \tilde{\tilde{\mathbf{B}}}(\mathbf{k}, \omega). \qquad (7.8)$$

The spatiotemporal evolution of the radiated electromagnetic field in a spatially and temporally dispersive HALL medium occupying all of three-dimensional space with charge and current sources is described by the macroscopic Maxwell's equations

$$\nabla \cdot \mathbf{D}(\mathbf{r}, t) = \|4\pi\| \varrho(\mathbf{r}, t), \tag{7.9}$$

$$\nabla \times \mathbf{E}(\mathbf{r}, t) = -\left\| \frac{1}{c} \right\| \frac{\partial \mathbf{B}(\mathbf{r}, t)}{\partial t}, \tag{7.10}$$

$$\nabla \cdot \mathbf{B}(\mathbf{r}, t) = 0, \tag{7.11}$$

$$\nabla \times \mathbf{H}(\mathbf{r}, t) = \left\| \frac{1}{c} \right\| \frac{\partial \mathbf{D}(\mathbf{r}, t)}{\partial t} + \left\| \frac{4\pi}{c} \right\| \mathbf{J}(\mathbf{r}, t). \tag{7.12}$$

The spatial Fourier and temporal Fourier–Laplace transform of these relations, with application of Eq. (E.13) and the initial conditions given in Eq. (7.2), then gives the set of vector algebraic equations, using Eqs. (E.19) and (E.20),

$$i\mathbf{k} \cdot \left(\underline{\epsilon}(\mathbf{k}, \omega) \cdot \tilde{\tilde{\mathbf{E}}}(\mathbf{k}, \omega) \right) = \|4\pi\| \tilde{\tilde{\varrho}}(\mathbf{k}, \omega), \tag{7.13}$$

$$i\mathbf{k} \times \tilde{\tilde{\mathbf{E}}}(\mathbf{k}, \omega) = \left\| \frac{1}{c} \right\| i\omega \tilde{\tilde{\mathbf{B}}}(\mathbf{k}, \omega), \tag{7.14}$$

$$i\mathbf{k} \cdot \tilde{\tilde{\mathbf{B}}}(\mathbf{k}, \omega) = 0, \tag{7.15}$$

$$i\mathbf{k} \times \left(\underline{\mu}^{-1}(\mathbf{k}, \omega) \cdot \tilde{\tilde{\mathbf{B}}}(\mathbf{k}, \omega) \right) = -\left\| \frac{1}{c} \right\| i\omega \left(\underline{\epsilon}(\mathbf{k}, \omega) \cdot \tilde{\tilde{\mathbf{E}}}(\mathbf{k}, \omega) \right) + \left\| \frac{4\pi}{c} \right\| \tilde{\tilde{\mathbf{J}}}(\mathbf{k}, \omega). \tag{7.16}$$

The spatial Fourier and temporal Fourier–Laplace transform of the equation of continuity [see Eq. (4.56)] yields

$$\tilde{\tilde{\varrho}}(\mathbf{k}, \omega) = \frac{1}{\omega} \left[\mathbf{k} \cdot \tilde{\tilde{\mathbf{J}}}_0(\mathbf{k}, \omega) + \mathbf{k} \cdot \left(\underline{\sigma}(\mathbf{k}, \omega) \cdot \tilde{\tilde{\mathbf{E}}}(\mathbf{k}, \omega) \right) \right],$$

so that with this substitution, Eq. (7.13) becomes

$$i\mathbf{k} \cdot \left(\underline{\epsilon}(\mathbf{k}, \omega) \cdot \tilde{\tilde{\mathbf{E}}}(\mathbf{k}, \omega) \right) = \frac{\|4\pi\|}{\omega} \mathbf{k} \cdot \left[\tilde{\tilde{\mathbf{J}}}_0(\mathbf{k}, \omega) + \left(\underline{\sigma}(\mathbf{k}, \omega) \cdot \tilde{\tilde{\mathbf{E}}}(\mathbf{k}, \omega) \right) \right].$$

This result then shows that Eq. (7.13) may be obtained directly from Eq. (7.16) simply by taking the scalar product of the latter with $\mathbf{k}$ and so is superfluous.

With Eq. (7.6), the spectral form of Ampère's law given in Eq. (7.16) may be written as

$$\mathbf{k} \times \left(\underline{\mu}^{-1}(\mathbf{k}, \omega) \cdot \tilde{\tilde{\mathbf{B}}}(\mathbf{k}, \omega) \right) = -\frac{\omega}{\|c\|} \underline{\epsilon}_c(\mathbf{k}, \omega) \cdot \tilde{\tilde{\mathbf{E}}}(\mathbf{k}, \omega) - i \left\| \frac{4\pi}{c} \right\| \tilde{\tilde{\mathbf{J}}}_0(\mathbf{k}, \omega),$$

where

$$\underline{\epsilon}_c(\mathbf{k}, \omega) \equiv \underline{\epsilon}(\mathbf{k}, \omega) + i \frac{\|4\pi\|}{\omega} \underline{\sigma}(\mathbf{k}, \omega) \tag{7.17}$$

is the *complex permittivity tensor*. With these results, the set of equations given in Eqs. (7.13)–(7.16) then becomes

$$\mathbf{k} \cdot \left(\underline{\epsilon}_c(\mathbf{k}, \omega) \cdot \tilde{\mathbf{E}}(\mathbf{k}, \omega) \right) = -\frac{\|4\pi\|}{\omega} \mathbf{k} \cdot \tilde{\mathbf{J}}_0(\mathbf{k}, \omega), \tag{7.18}$$

$$\mathbf{k} \times \tilde{\mathbf{E}}(\mathbf{k}, \omega) = \left\| \frac{1}{c} \right\| \omega \tilde{\mathbf{B}}(\mathbf{k}, \omega), \tag{7.19}$$

$$\mathbf{k} \cdot \tilde{\mathbf{B}}(\mathbf{k}, \omega) = 0, \tag{7.20}$$

$$\mathbf{k} \times \left(\underline{\mu}^{-1}(\mathbf{k}, \omega) \cdot \tilde{\mathbf{B}}(\mathbf{k}, \omega) \right) = -\frac{\omega}{\|c\|} \underline{\epsilon}_c(\mathbf{k}, \omega) \cdot \tilde{\mathbf{E}}(\mathbf{k}, \omega) - i \left\| \frac{4\pi}{c} \right\| \tilde{\mathbf{J}}_0(\mathbf{k}, \omega), \tag{7.21}$$

where Eq. (7.18) follows from Eq. (7.21). This is the general form of the *spatio-temporal spectral representation of Maxwell's equations in a spatially and temporally dispersive HALL medium*.

In order to proceed further, it is assumed that the magnetic permeability is isotropic, so that

$$\underline{\mu}^{-1}(\mathbf{k}, \omega) = \mu^{-1}(\mathbf{k}, \omega) \underline{\mathbb{I}}, \tag{7.22}$$

where $\underline{\mathbb{I}} = \hat{\mathbf{1}}_x \hat{\mathbf{1}}_x + \hat{\mathbf{1}}_y \hat{\mathbf{1}}_y + \hat{\mathbf{1}}_z \hat{\mathbf{1}}_z$ is the unit dyadic or idemfactor. With this assumption, Eq. (7.21) simplifies somewhat to

$$\mathbf{k} \times \tilde{\mathbf{B}}(\mathbf{k}, \omega) = -\frac{\omega}{\|c\|} \mu(\mathbf{k}, \omega) \underline{\epsilon}_c(\mathbf{k}, \omega) \cdot \tilde{\mathbf{E}}(\mathbf{k}, \omega) - i \left\| \frac{4\pi}{c} \right\| \mu(\mathbf{k}, \omega) \tilde{\mathbf{J}}_0(\mathbf{k}, \omega).$$

Upon taking the cross-product of Eq. (7.19) from the left with the wavenumber $\mathbf{k}$ and substituting from the equation above, there results

$$\mathbf{k} \times \left(\mathbf{k} \times \tilde{\mathbf{E}}(\mathbf{k}, \omega) \right) = -\frac{\omega}{\|c^2\|} \left[\mu(\mathbf{k}, \omega) \underline{\epsilon}_c(\mathbf{k}, \omega) \cdot \tilde{\mathbf{E}}(\mathbf{k}, \omega) + \mu(\mathbf{k}, \omega) \tilde{\mathbf{J}}_0(\mathbf{k}, \omega) \right],$$

which, after rearrangement, becomes

$$\left[k^2 \underline{\mathbb{I}} - \mathbf{k}\mathbf{k} - \frac{\omega^2}{\|c^2\|} \mu(\mathbf{k}, \omega) \underline{\epsilon}_c(\mathbf{k}, \omega) \right] \cdot \tilde{\mathbf{E}}(\mathbf{k}, \omega) = \left\| \frac{4\pi}{c^2} \right\| i\omega\mu(\mathbf{k}, \omega) \tilde{\mathbf{J}}_0(\mathbf{k}, \omega).$$

Following the analysis due to Stamnes[1], define the *propagation tensor*

$$\underline{\underline{A}}(\mathbf{k}, \omega) \equiv k^2 \underline{\underline{\mathfrak{I}}} - \mathbf{k}\mathbf{k} - \frac{\omega^2}{\|c^2\|} \mu(\mathbf{k}, \omega)\underline{\underline{\epsilon}}_c(\mathbf{k}, \omega) \tag{7.23}$$

with elements

$$\Lambda_{ij}(\mathbf{k}, \omega) \equiv k^2 \delta_{ij} - k_i k_j - \frac{\omega^2}{\|c^2\|} \mu(\mathbf{k}, \omega)\epsilon_{c_{ij}}(\mathbf{k}, \omega), \quad i, j = 1, 2, 3.$$

The above equation then becomes

$$\underline{\underline{A}}(\mathbf{k}, \omega) \cdot \tilde{\mathbf{E}}(\mathbf{k}, \omega) = \left\|\frac{4\pi}{c^2}\right\| i\omega\mu(\mathbf{k}, \omega)\tilde{\mathbf{J}}_0(\mathbf{k}, \omega),$$

with solution

$$\tilde{\mathbf{E}}(\mathbf{k}, \omega) = \left\|\frac{4\pi}{c^2}\right\| i\omega\mu(\mathbf{k}, \omega)\underline{\underline{A}}^{-1}(\mathbf{k}, \omega) \cdot \tilde{\mathbf{J}}_0(\mathbf{k}, \omega), \tag{7.24}$$

provided that the inverse dyadic $\underline{\underline{A}}^{-1}(\mathbf{k}, \omega)$ exists. The corresponding magnetic induction field vector is then immediately obtained from Eq. (7.19) as

$$\tilde{\mathbf{B}}(\mathbf{k}, \omega) = \left\|\frac{4\pi}{c}\right\| i\mu(\mathbf{k}, \omega)\mathbf{k} \times \left(\underline{\underline{A}}^{-1}(\mathbf{k}, \omega) \cdot \tilde{\mathbf{J}}_0(\mathbf{k}, \omega)\right). \tag{7.25}$$

The original set of partial differential relations for the real space–time form of the electric and magnetic field vectors radiated in spatially and temporally dispersive HALL medium have thus been replaced by a set of algebraic equations in which the spatio-temporal spectra of the electric and magnetic field vectors are the only remaining unknown quantities. The inverse spatio-temporal transform of this pair of equations will then yield space–time domain form of the radiated electromagnetic field.

At this juncture, one would typically specialize the analysis to a particular anisotropic structure such as that, for example, for a uniaxial crystal.[1] As the underlying purpose of this theoretical development is the space–time evolution of pulsed electromagnetic fields in temporally dispersive attenuative media, the analysis is now specialized to HILL medium with both spatial and temporal dispersion. The complex permittivity tensor given in Eq. (7.17) then simplifies to a complex scalar function of $\mathbf{k}$ and ω, as given in Eq. (5.28).

For a homogeneous, isotropic, locally linear (HILL) medium with both spatial and temporal dispersion, the spatio-temporal frequency form of Maxwell's equations given in Eqs. (7.18)–(7.21) take the somewhat simpler form

$$\mathbf{k} \cdot \tilde{\mathbf{E}}(\mathbf{k}, \omega) = -\|4\pi\| i \frac{\mathbf{k} \cdot \tilde{\mathbf{J}}_0(\mathbf{k}, \omega)}{\omega \epsilon_c(\mathbf{k}, \omega)}, \tag{7.26}$$

$$\mathbf{k} \times \tilde{\mathbf{E}}(\mathbf{k}, \omega) = \left\|\frac{1}{c}\right\| \omega \tilde{\mathbf{B}}(\mathbf{k}, \omega), \tag{7.27}$$

$$\mathbf{k} \cdot \tilde{\mathbf{B}}(\mathbf{k}, \omega) = 0, \tag{7.28}$$

$$\mathbf{k} \times \tilde{\mathbf{B}}(\mathbf{k}, \omega) = -\left\|\frac{1}{c}\right\| \omega \mu(\mathbf{k}, \omega)\epsilon_c(\mathbf{k}, \omega)\tilde{\mathbf{E}}(\mathbf{k}, \omega) - \left\|\frac{4\pi}{c}\right\| i\mu(\mathbf{k}, \omega)\tilde{\mathbf{J}}_0(\mathbf{k}, \omega). \tag{7.29}$$

Consider first the solution for the spatio-temporal frequency spectrum of the electric field vector. The vector product of $\mathbf{k}$ with the expression in Eq. (7.27) gives

$$\mathbf{k} \times \left(\mathbf{k} \times \tilde{\mathbf{E}}\right) = \left\|\frac{1}{c}\right\| \omega \mathbf{k} \times \tilde{\mathbf{B}}.$$

With the vector identity for the triple vector product, the left-hand side of this equation becomes

$$\mathbf{k} \times \left(\mathbf{k} \times \tilde{\mathbf{E}}\right) = \left(\mathbf{k} \cdot \tilde{\mathbf{E}}\right)\mathbf{k} - k^2\tilde{\mathbf{E}}$$

$$= -\|4\pi\| i \frac{\mathbf{k} \cdot \tilde{\mathbf{J}}_0}{\omega \epsilon_c(\mathbf{k}, \omega)}\mathbf{k} - k^2\tilde{\mathbf{E}},$$

after substitution from Eq. (7.26). Furthermore, the right-hand side of the equation resulting from the cross product of $\mathbf{k}$ with Eq. (7.27) can be expressed in terms of $\tilde{\mathbf{E}}(\mathbf{k}, \omega)$ with substitution from Eq. (7.29). With these substitutions, there then results

$$\left(k^2 - \frac{\mu(\mathbf{k}, \omega)\epsilon_c(\mathbf{k}, \omega)}{\|c^2\|}\omega^2\right)\tilde{\mathbf{E}} = \left\|\frac{4\pi}{c^2}\right\| i\omega\mu(\mathbf{k}, \omega)\tilde{\mathbf{J}}_0 - \|4\pi\| i \frac{\mathbf{k} \cdot \tilde{\mathbf{J}}_0}{\omega \epsilon_c(\mathbf{k}, \omega)}\mathbf{k}.$$

With the definition of the *complex velocity* in a temporally dispersive HILL medium with complex permittivity $\epsilon_c(\omega)$ and magnetic permeability $\mu(\omega)$ as the square root of the quantity

$$v^2(\mathbf{k}, \omega) \equiv \frac{\|c^2\|}{\mu(\mathbf{k}, \omega)\epsilon_c(\mathbf{k}, \omega)}, \tag{7.30}$$

one finally obtains the expression

$$\tilde{\mathbf{E}}(\mathbf{k}, \omega) = \|4\pi\| i \frac{\left\|\frac{1}{c^2}\right\| \omega\mu(\mathbf{k}, \omega)\tilde{\mathbf{J}}_0(\mathbf{k}, \omega) - \frac{\mathbf{k}\cdot\tilde{\mathbf{J}}_0(\mathbf{k},\omega)}{\omega\epsilon_c(\mathbf{k},\omega)}\mathbf{k}}{k^2 - \frac{\omega^2}{v^2(\mathbf{k},\omega)}}. \tag{7.31}$$

This equation can be somewhat simplified with the application of the vector identity $(\mathbf{k} \cdot \tilde{\mathbf{J}}_0)\mathbf{k} = \mathbf{k} \times (\mathbf{k} \times \tilde{\mathbf{J}}_0) + k^2\tilde{\mathbf{J}}_0$, with which the numerator in Eq. (7.31) becomes

$$\frac{\omega\mu(\mathbf{k}, \omega)}{\|c^2\|}\tilde{\mathbf{J}}_0 - \frac{\mathbf{k}\cdot\tilde{\mathbf{J}}_0}{\omega\epsilon_c(\mathbf{k}, \omega)}\mathbf{k} = \frac{1}{\omega\epsilon_c(\mathbf{k}, \omega)}\left[\left(\frac{\omega^2}{v^2(\mathbf{k}, \omega)} - k^2\right)\tilde{\mathbf{J}}_0 - \mathbf{k} \times \left(\mathbf{k} \times \tilde{\mathbf{J}}_0\right)\right].$$

With this substitution, Eq. (7.31) may then be rewritten as

$$\tilde{\mathbf{E}}(\mathbf{k}, \omega) = \|4\pi\|\frac{i}{\omega\epsilon_c(\mathbf{k}, \omega)}\left[\frac{\mathbf{k} \times \left(\mathbf{k} \times \tilde{\mathbf{J}}_0(\mathbf{k}, \omega)\right)}{\frac{\omega^2}{v^2(\mathbf{k},\omega)} - k^2} - \tilde{\mathbf{J}}_0(\mathbf{k}, \omega)\right]. \tag{7.32}$$

The solution for the spatio-temporal frequency spectrum of the magnetic induction field vector **B** is obtained by first taking the vector product of **k** with the expression in Eq. (7.29) and applying the relations that are in Eqs. (7.27) and (7.28) in the following way. Beginning with the vector identity $\mathbf{k}\times\left(\mathbf{k} \times \tilde{\mathbf{B}}\right) = \left(\mathbf{k} \cdot \tilde{\mathbf{B}}\right)\mathbf{k} - k^2\tilde{\mathbf{B}} = -k^2\tilde{\mathbf{B}}$, substitution from Eq. (7.29) then gives

$$-k^2\tilde{\mathbf{B}} = -\frac{\omega\mu(\mathbf{k}, \omega)\epsilon_c(\mathbf{k}, \omega)}{\|4\pi\|}\mathbf{k} \times \tilde{\mathbf{E}} - \left\|\frac{4\pi}{c}\right\| i\mu(\mathbf{k}, \omega)\mathbf{k} \times \tilde{\mathbf{J}}_0$$

$$= -\frac{\omega^2}{v^2(\mathbf{k}, \omega)}\tilde{\mathbf{B}} - \left\|\frac{4\pi}{c}\right\| i\mu(\mathbf{k}, \omega)\mathbf{k} \times \tilde{\mathbf{J}}_0,$$

and hence

$$\tilde{\mathbf{B}}(\mathbf{k}, \omega) = \left\|\frac{4\pi}{c}\right\| i\mu(\mathbf{k}, \omega)\frac{\mathbf{k} \times \tilde{\mathbf{J}}_0(\mathbf{k}, \omega)}{k^2 - \frac{\omega^2}{v^2(\mathbf{k},\omega)}}. \tag{7.33}$$

The inverse Fourier–Laplace transform of the vector solutions appearing in Eqs. (7.31)–(7.33) then yields the following Fourier–Laplace integral representation of the radiation field in a homogeneous, isotropic, locally linear (HILL) medium with spatial and temporal frequency-dependent dielectric permittivity $\epsilon(\mathbf{k}, \omega)$, magnetic permeability $\mu(\mathbf{k}, \omega)$, and electric conductivity $\sigma(\mathbf{k}, \omega)$:

$$\mathbf{E}(\mathbf{r}, t) = i \frac{\|4\pi\|}{(2\pi)^4} \int_C d\omega \int_{-\infty}^{\infty} d^3k \frac{\left\|\frac{1}{c^2}\right\| \omega\mu(\mathbf{k}, \omega)\tilde{\tilde{\mathbf{J}}}_0(\mathbf{k}, \omega) - \frac{\mathbf{k}\cdot\tilde{\mathbf{J}}_0(\mathbf{k}, \omega)}{\omega\epsilon_c(\mathbf{k}, \omega)}\mathbf{k}}{k^2 - \frac{\omega^2}{v^2(\mathbf{k}, \omega)}} e^{i(\mathbf{k}\cdot\mathbf{r}-\omega t)}$$

(7.34)

$$= -i \frac{\|4\pi\|}{(2\pi)^4} \int_C \frac{d\omega}{\omega} \int_{-\infty}^{\infty} \frac{d^3k}{\epsilon_c(\mathbf{k}, \omega)}$$

$$\times \left[\tilde{\tilde{\mathbf{J}}}_0(\mathbf{k}, \omega) + \frac{\mathbf{k} \times \left(\mathbf{k} \times \tilde{\tilde{\mathbf{J}}}_0(\mathbf{k}, \omega) \right)}{k^2 - \frac{\omega^2}{v^2(\mathbf{k}, \omega)}} \right] e^{i(\mathbf{k}\cdot\mathbf{r}-\omega t)},$$

(7.35)

$$\mathbf{B}(\mathbf{r}, t) = i \left\| \frac{4\pi}{c} \right\| \frac{1}{(2\pi)^4} \int_C d\omega \, \mu(\mathbf{k}, \omega) \int_{-\infty}^{\infty} d^3k \frac{\mathbf{k} \times \tilde{\tilde{\mathbf{J}}}_0(\mathbf{k}, \omega)}{k^2 - \frac{\omega^2}{v^2(\mathbf{k}, \omega)}} e^{i(\mathbf{k}\cdot\mathbf{r}-\omega t)},$$

(7.36)

where C is the Bromwich contour [2] given by $\omega = \omega' + ia$ where ω' varies from $-\infty$ to $+\infty$ and where a is the abscissa of absolute convergence for the integrand in each inverse Laplace transform (see Appendix E).

In order to apply contour integration techniques for the purpose of simplifying the set of expressions given in Eqs. (7.34)–(7.36), it is necessary to express the ω-integral in a form that involves only nonnegative real values of ω. Consider then the sum $\mathfrak{I} = \mathfrak{I}_+ + \mathfrak{I}_-$, with

$$\mathfrak{I}_\pm \equiv \int_{C_\pm} d\omega \int_{-\infty}^{\infty} d^3k \, \tilde{\tilde{\mathbf{U}}}(\mathbf{k}, \omega) e^{i(\mathbf{k}\cdot\mathbf{r}-\omega t)}, \tag{7.37}$$

where C_+ is the contour $\omega = \omega' + ia$ with ω' varying from 0 to $+\infty$, and where C_- is the contour $\omega = \omega' + ia$ with ω' varying from $-\infty$ to 0. If the spectral function $\tilde{\tilde{\mathbf{U}}}(\mathbf{k}, \omega)$ has the property that

$$\tilde{\tilde{\mathbf{U}}}(-\mathbf{k}, -\omega' + ia) = \tilde{\tilde{\mathbf{U}}}^*(\mathbf{k}, \omega' + ia), \tag{7.38}$$

then

$$\mathfrak{I}_+^* = \int_0^{\infty} d(\omega' + ia)^* \int_{-\infty}^{\infty} d^3k \, \tilde{\tilde{\mathbf{U}}}^*(\mathbf{k}, \omega' + ia) e^{-i(\mathbf{k}\cdot\mathbf{r}-(\omega'+ia)^*t)}$$

$$= \int_0^{\infty} d(\omega' - ia) \int_{-\infty}^{\infty} d^3k \, \tilde{\tilde{\mathbf{U}}}^*(\mathbf{k}, \omega' + ia) e^{-i(\mathbf{k}\cdot\mathbf{r}-(\omega'-ia)t)}.$$

Under the change of variable $\mathbf{k} \to -\mathbf{k}$, $\omega' \to -\omega'$, this integral becomes

$$\mathfrak{I}^*_+ = \int_0^{-\infty} d(-\omega' - ia) \int_{-\infty}^{\infty} d^3k \, \tilde{\mathbf{U}}^*(-\mathbf{k}, -\omega' + ia) e^{-i(-\mathbf{k}\cdot\mathbf{r}+(\omega'+ia)t)}$$

$$= \int_{-\infty}^{0} d(\omega' + ia) \int_{-\infty}^{\infty} d^3k \, \tilde{\mathbf{U}}(\mathbf{k}, \omega' + ia) e^{i(\mathbf{k}\cdot\mathbf{r}-(\omega'+ia)t)}$$

$$= \int_{ia-\infty}^{ia} d\omega \int_{-\infty}^{\infty} d^3k \, \tilde{\mathbf{U}}(\mathbf{k}, \omega) e^{i(\mathbf{k}\cdot\mathbf{r}-\omega t)} = \mathfrak{I}_-.$$

Hence, if the symmetry property specified in Eq. (7.38) is satisfied, then

$$\mathfrak{I} = \mathfrak{I}_+ + \mathfrak{I}_- = \mathfrak{I}_+ + \mathfrak{I}^*_+$$

$$= 2\Re\{\mathfrak{I}_+\}, \tag{7.39}$$

where the ω-integral in $\mathfrak{I}_+$ involves only nonnegative real values of ω.

In order to determine if the spectral functions appearing in Eqs. (7.34)–(7.36) satisfy the symmetry property given in Eq. (7.38), consider first the Fourier–Laplace transform $\tilde{\mathbf{V}}(\mathbf{k}, \omega)$ of a real-valued function $\mathbf{V}(\mathbf{r}, t)$, where

$$\tilde{\mathbf{V}}(\mathbf{k}, \omega) = \int_0^{\infty} dt \int_{-\infty}^{\infty} d^3r \, \mathbf{V}(\mathbf{r}, t) e^{-i(\mathbf{k}\cdot\mathbf{r}-\omega t)}, \tag{7.40}$$

in which case

$$\tilde{\mathbf{V}}(-\mathbf{k}, -\omega' + ia) = \int_0^{\infty} dt \int_{-\infty}^{\infty} d^3r \, \mathbf{V}(\mathbf{r}, t) e^{+i(\mathbf{k}\cdot\mathbf{r}-(\omega'-ia)t)}$$

$$= \tilde{\mathbf{V}}^*(\mathbf{k}, \omega' + ia) = \tilde{\mathbf{V}}^*(\mathbf{k}, \omega), \tag{7.41}$$

with $\omega = \omega' + ia$. Because $\mathbf{J}_0(\mathbf{r}, t)$, $\hat{\epsilon}(t)$, $\hat{\mu}(t)$, and $\hat{\sigma}(t)$ are all real-valued functions, their Fourier–Laplace spectra then satisfy the symmetry relations

$$\tilde{\mathbf{J}}_0(-\mathbf{k}, -\omega' + ia) = \tilde{\mathbf{J}}_0^*(\mathbf{k}, \omega), \tag{7.42}$$

and

$$\epsilon(-\mathbf{k}, -\omega' + ia) = \epsilon^*(\mathbf{k}, \omega), \tag{7.43}$$

$$\mu(-\mathbf{k}, -\omega' + ia) = \mu^*(\mathbf{k}, \omega), \tag{7.44}$$

$$\sigma(-\mathbf{k}, -\omega' + ia) = \sigma^*(\mathbf{k}, \omega), \tag{7.45}$$

so that the complex permittivity satisfies the same symmetry relation, as

$$\epsilon_c(-\mathbf{k}, -\omega' + ia) = \epsilon(-\mathbf{k}, -\omega' + ia) + i\|4\pi\|\frac{\sigma(-\mathbf{k}, -\omega' + ia)}{-\omega' + ia}$$

$$= \epsilon^*(\mathbf{k}, \omega) - i\|4\pi\|\frac{\sigma^*(\mathbf{k}, \omega)}{\omega' - ia}$$

$$= \left(\epsilon(\mathbf{k}, \omega) + i\|4\pi\|\frac{\sigma(\mathbf{k}, \omega)}{\omega}\right)^* = \epsilon_c^*(\mathbf{k}, \omega). \qquad (7.46)$$

As a consequence of this result it is then found that

$$i\left(-\omega' + ia\right)\epsilon_c(-\mathbf{k}, -\omega' + ia) = \left(i\left(\omega' + ia\right)\epsilon_c(\mathbf{k}, \omega)\right)^*$$

$$= \left(i\omega\epsilon_c(\mathbf{k}, \omega)\right)^*, \qquad (7.47)$$

and hence

$$\|c^2\|\frac{(-\omega' + ia)^2}{v^2(-\mathbf{k}, -\omega' + ia)} = \mu(-\mathbf{k}, -\omega' + ia)\epsilon_c(-\mathbf{k}, -\omega' + ia)\left(-\omega' + ia\right)^2$$

$$= \mu^*(\mathbf{k}, \omega)\epsilon_c^*(\mathbf{k}, \omega)\left(\omega' - ia\right)^2$$

$$= \left(\mu(\mathbf{k}, \omega)\epsilon_c(\mathbf{k}, \omega)\left(\omega' + ia\right)^2\right)^*$$

$$= \left(\omega^2\mu(\mathbf{k}, \omega)\epsilon_c(\mathbf{k}, \omega)\right)^*$$

$$= \|c^2\|\left(\frac{\omega^2}{v^2(\mathbf{k}, \omega)}\right)^*. \qquad (7.48)$$

It then follows from the relations given in Eqs. (7.42)–(7.48) that the spectral functions appearing in Eqs. (7.34)–(7.36) satisfy the condition specified in Eq. (7.38). The Fourier–Laplace integral representation of the radiation field in a temporally dispersive HILL medium with wave vector and frequency-dependent dielectric permittivity $\epsilon(\mathbf{k}, \omega)$, magnetic permeability $\mu(\mathbf{k}, \omega)$, and electric conductivity $\sigma(\mathbf{k}, \omega)$ may then be expressed as

$$\mathbf{E}(\mathbf{r}, t) = \frac{\|4\pi\|}{8\pi^4}\Re\left\{i\int_{C_+}d\omega\int_{-\infty}^{\infty}d^3k\frac{\frac{\omega\mu(\omega)}{\|c^2\|}\tilde{\mathbf{J}}_0(\mathbf{k}, \omega) - \frac{\mathbf{k}\cdot\tilde{\mathbf{J}}_0(\mathbf{k},\omega)}{\omega\epsilon_c(\omega)}\mathbf{k}}{k^2 - \frac{\omega^2}{v^2(\omega)}}e^{i(\mathbf{k}\cdot\mathbf{r}-\omega t)}\right\}$$

$$= -\frac{\|4\pi\|}{8\pi^4}\Re\left\{i\int_{C_+}d\omega\frac{1}{\omega\epsilon_c(\omega)}\right.$$

$$\times \int_{-\infty}^{\infty} d^3k \left[\tilde{\mathbf{J}}_0(\mathbf{k}, \omega) + \frac{\mathbf{k} \times \left(\mathbf{k} \times \tilde{\mathbf{J}}_0(\mathbf{k}, \omega) \right)}{k^2 - \frac{\omega^2}{v^2(\omega)}} \right] e^{i(\mathbf{k}\cdot\mathbf{r} - \omega t)} \Biggr\},$$

$$(7.49)$$

$$\mathbf{B}(\mathbf{r}, t) = \left\| \frac{4\pi}{c} \right\| \frac{1}{8\pi^4} \Re \left\{ i \int_{C_+} d\omega \, \mu(\omega) \int_{-\infty}^{\infty} d^3k \frac{\mathbf{k} \times \tilde{\mathbf{J}}_0(\mathbf{k}, \omega)}{k^2 - \frac{\omega^2}{v^2(\omega)}} e^{i(\mathbf{k}\cdot\mathbf{r} - \omega t)} \right\},$$

$$(7.50)$$

for $t > 0$, where C_+ is the contour $\omega = \omega' + ia$ with $\omega' \equiv \Re\{\omega\}$ varying from 0 to $+\infty$ and where a is fixed at a value greater than the abscissa of absolute convergence for the particular radiation problem under consideration. This final form of the Fourier–Laplace integral representation of the radiated electric and magnetic field vectors explicitly exhibits their real-valued character.

7.2   Scalar and Vector Potentials for the Radiation Field

Because the magnetic induction field $\mathbf{B}(\mathbf{r}, t)$ is a transverse (or solenoidal) vector field, it can always be expressed as the curl of another vector field $\mathbf{A}(\mathbf{r}, t)$ as

$$\mathbf{B}(\mathbf{r}, t) = \nabla \times \mathbf{A}(\mathbf{r}, t), \tag{7.51}$$

where $\mathbf{A}(\mathbf{r}, t)$ is the *macroscopic vector potential* (see Sect. 3.1). It is then seen from Eqs. (7.36) and (7.50) that such a vector potential for the radiation field is given by

$$\mathbf{A}(\mathbf{r}, t) = \left\| \frac{4\pi}{c} \right\| \frac{1}{(2\pi)^4} \int_C d\omega \int_{-\infty}^{\infty} d^3k, \mu(\mathbf{k}, \omega) \frac{\tilde{\mathbf{J}}_0(\mathbf{k}, \omega)}{k^2 - \frac{\omega^2}{v^2(\mathbf{k},\omega)}} e^{i(\mathbf{k}\cdot\mathbf{r} - \omega t)}$$

$$(7.52)$$

$$= \left\| \frac{4\pi}{c} \right\| \frac{1}{8\pi^4} \Re \left\{ \int_{C_+} d\omega \int_{-\infty}^{\infty} d^3k \, \mu(\omega) \frac{\tilde{\mathbf{J}}_0(\mathbf{k}, \omega)}{k^2 - \frac{\omega^2}{v^2(\mathbf{k},\omega)}} e^{i(\mathbf{k}\cdot\mathbf{r} - \omega t)} \right\}.$$

$$(7.53)$$

Comparison of Eq. (7.52) with Eq. (7.34) shows that the electric field vector $\mathbf{E}(\mathbf{r}, t)$ can be expressed as

$$\mathbf{E}(\mathbf{r}, t) = -i \frac{\|4\pi\|}{(2\pi)^4} \int_C \frac{d\omega}{\omega} \int_{-\infty}^{\infty} d^3k \frac{\mathbf{k} \cdot \tilde{\mathbf{J}}_0(\mathbf{k}, \omega)}{\epsilon_c(\mathbf{k}, \omega) \, k^2 - \frac{\omega^2}{v^2(\omega)}} \mathbf{k} e^{i(\mathbf{k}\cdot\mathbf{r} - \omega t)} - \left\| \frac{1}{c} \right\| \frac{\partial \mathbf{A}(\mathbf{r}, t)}{\partial t}$$

$$= -\frac{\|4\pi\|}{8\pi^4} \Re \left\{ i \int_{C_+} \frac{d\omega}{\omega} \int_{-\infty}^{\infty} \frac{d^3k}{\epsilon_c(\mathbf{k}, \omega)} \frac{\mathbf{k} \cdot \tilde{\mathbf{J}}_0(\mathbf{k}, \omega)}{k^2 - \frac{\omega^2}{v^2(\omega)}} \mathbf{k} e^{i(\mathbf{k}\cdot\mathbf{r} - \omega t)} \right\}$$

$$- \left\|\frac{1}{c}\right\| \frac{\partial \mathbf{A}(\mathbf{r}, t)}{\partial t},$$

which may be written in the form

$$\mathbf{E}(\mathbf{r}, t) = -\nabla \varphi(\mathbf{r}, t) - \left\|\frac{1}{c}\right\| \frac{\partial \mathbf{A}(\mathbf{r}, t)}{\partial t}, \tag{7.54}$$

where

$$\varphi(\mathbf{r}, t) = \frac{\|4\pi\|}{(2\pi)^4} \int_C \frac{d\omega}{\omega} \int_{-\infty}^{\infty} \frac{d^3k}{\epsilon_c(\mathbf{k}, \omega)} \frac{\mathbf{k} \cdot \tilde{\mathbf{J}}_0(\mathbf{k}, \omega)}{k^2 - \frac{\omega^2}{v^2(\mathbf{k}, \omega)}} e^{i(\mathbf{k}\cdot\mathbf{r} - \omega t)} \tag{7.55}$$

$$= \frac{\|4\pi\|}{8\pi^4} \Re \left\{ \int_{C_+} \frac{d\omega}{\omega} \int_{-\infty}^{\infty} \frac{d^3k}{\epsilon_c(\mathbf{k}, \omega)} \frac{\mathbf{k} \cdot \tilde{\mathbf{J}}_0(\mathbf{k}, \omega)}{k^2 - \frac{\omega^2}{v^2(\mathbf{k}, \omega)}} e^{i(\mathbf{k}\cdot\mathbf{r} - \omega t)} \right\}$$

$$\tag{7.56}$$

is the *macroscopic scalar potential* for the radiation field. Hence, by determining the vector and scalar potentials $\mathbf{A}(\mathbf{r}, t)$ and $\varphi(\mathbf{r}, t)$, respectively, for a given current source $\mathbf{J}_0(\mathbf{r}, t)$, the electric and magnetic field vectors of the radiation field produced by that current source are directly obtained from Eqs. (7.51) and (7.54). The problem of determining $\mathbf{E}(\mathbf{r}, t)$ and $\mathbf{B}(\mathbf{r}, t)$ for a given radiation problem specified by the current source $\mathbf{J}_0(\mathbf{r}, t)$ in a spatially and temporally dispersive HILL medium has thus been reduced to determining $\mathbf{A}(\mathbf{r}, t)$ and $\varphi(\mathbf{r}, t)$, which are specified by somewhat simpler expressions.

The gauge transformations described in Sect. 3.1 for the microscopic electromagnetic field vectors also apply to the macroscopic case. In particular, because the magnetic induction field vector $\mathbf{B}(\mathbf{r}, t)$ is defined through Eq. (7.51) in terms of the curl of the vector potential field $\mathbf{A}(\mathbf{r}, t)$, then this vector potential field is arbitrary to the extent that the gradient of some scalar function $\Lambda(\mathbf{r}, t)$ can be added to $\mathbf{A}(\mathbf{r}, t)$ without affecting $\mathbf{B}(\mathbf{r}, t)$. That is, the magnetic field vector $\mathbf{B}(\mathbf{r}, t)$ is left unchanged by the transformation [cf. Eq. (3.11)]

$$\mathbf{A}(\mathbf{r}, t) \longrightarrow \mathbf{A}'(\mathbf{r}, t) = \mathbf{A}(\mathbf{r}, t) + \nabla \Lambda(\mathbf{r}, t). \tag{7.57}$$

Under this transformation, Eq. (7.54) for the electric field vector becomes

$$\mathbf{E}(\mathbf{r}, t) = -\nabla \varphi(\mathbf{r}, t) - \left\|\frac{1}{c}\right\| \frac{\partial \mathbf{A}'(\mathbf{r}, t)}{\partial t}$$

$$= -\nabla\varphi(\mathbf{r}, t) - \left\|\frac{1}{c}\right\| \frac{\partial \mathbf{A}(\mathbf{r}, t)}{\partial t} - \left\|\frac{1}{c}\right\| \nabla\frac{\partial \Lambda(\mathbf{r}, t)}{\partial t}.$$

Hence, in order that the electric field vector remain unchanged under the vector potential transformation given in Eq. (7.57), the scalar potential $\varphi(\mathbf{r}, t)$ must be simultaneously transformed as

$$\varphi(\mathbf{r}, t) \longrightarrow \varphi'(\mathbf{r}, t) = \varphi(\mathbf{r}, t) - \left\|\frac{1}{c}\right\| \frac{\partial \Lambda(\mathbf{r}, t)}{\partial t}. \tag{7.58}$$

The complete transformation specified by the pair of relations in Eqs. (7.57) and (7.58) is called a *gauge transformation* and the invariance of the electric and magnetic field vectors under such a transformation is referred to as *gauge invariance*. The scalar function $\Lambda(\mathbf{r}, t)$ is called the *gauge function* of the transformation.

The freedom of choice implied by Eqs. (7.57) and (7.58) means that one can always choose a set of potentials $\{\mathbf{A}', \varphi'\}$ such that [cf. Eq. (3.13)]

$$\nabla \cdot \mathbf{A}'(\mathbf{r}, t) + \frac{\|c\|}{c^2} \frac{\partial \varphi'(\mathbf{r}, t)}{\partial t} = 0, \tag{7.59}$$

which is the *Lorenz condition* [3] for the macroscopic potentials. In order to prove that a pair of potentials $\{\mathbf{A}', \varphi'\}$ can always be found to satisfy the Lorenz condition, suppose that the original pair of potentials $\{\mathbf{A}, \varphi\}$ does not satisfy Eq. (7.59). One may then undertake a gauge transformation to the new pair of potentials $\{\mathbf{A}', \varphi'\}$ and demand that $\mathbf{A}'(\mathbf{r}, t)$ and $\varphi'(\mathbf{r}, t)$ satisfy the Lorenz condition, so that

$$\nabla^2\Lambda(\mathbf{r}, t) - \frac{1}{c^2}\frac{\partial^2\Lambda(\mathbf{r}, t)}{\partial t^2} = -\left(\nabla \cdot \mathbf{A}(\mathbf{r}, t) + \frac{\|c\|}{c^2}\frac{\partial\varphi(\mathbf{r}, t)}{\partial t}\right). \tag{7.60}$$

Thus, provided that a gauge function $\Lambda(\mathbf{r}, t)$ can be found that satisfies Eq. (7.60), the new pair of potentials $\{\mathbf{A}', \varphi'\}$ will satisfy the Lorenz condition.

Even for a pair of potentials that satisfy the Lorenz condition (7.59) there still remains a certain degree of arbitrariness. This is because the *restricted gauge transformation*

$$\mathbf{A}(\mathbf{r}, t) \longrightarrow \mathbf{A}'(\mathbf{r}, t) = \mathbf{A}(\mathbf{r}, t) + \nabla\psi(\mathbf{r}, t), \tag{7.61}$$

$$\varphi(\mathbf{r}, t) \longrightarrow \varphi'(\mathbf{r}, t) = \varphi(\mathbf{r}, t) - \left\|\frac{1}{c}\right\| \frac{\partial\psi(\mathbf{r}, t)}{\partial t}, \tag{7.62}$$

with

$$\nabla^2\psi(\mathbf{r}, t) - \frac{1}{c^2}\frac{\partial^2\psi(\mathbf{r}, t)}{\partial t^2} = 0 \tag{7.63}$$

preserves the Lorenz condition provided that the potential pair $\{\mathbf{A}, \varphi\}$ satisfies it initially, as is evident from Eq. (7.60). All pairs of potentials $\{\mathbf{A}, \varphi\}$ in this restricted class belong to the so-called Lorenz gauge.

The divergence of the vector potential for the radiation field given in Eq. (7.52) results in

$$\nabla \cdot \mathbf{A}(\mathbf{r}, t) = \left\|\frac{4\pi}{c}\right\| \frac{i}{(2\pi)^4} \int_C d\omega \int_{-\infty}^{\infty} d^3k\, \mu(\mathbf{k}, \omega) \frac{\mathbf{k} \cdot \tilde{\tilde{\mathbf{J}}}_0(\mathbf{k}, \omega)}{k^2 - \frac{\omega^2}{v^2(\mathbf{k},\omega)}} e^{i(\mathbf{k}\cdot\mathbf{r}-\omega t)},$$

and the partial time derivative of the scalar potential for the radiation field given in Eq. (7.56) is found as

$$\frac{\partial\varphi(\mathbf{r}, t)}{\partial t} = -\frac{\|4\pi\|}{(2\pi)^4} i \int_C d\omega \int_{-\infty}^{\infty} \frac{d^3k}{\epsilon_c(\mathbf{k}, \omega)} \frac{\mathbf{k} \cdot \tilde{\tilde{\mathbf{J}}}_0(\mathbf{k}, \omega)}{k^2 - \frac{\omega^2}{v^2(\mathbf{k},\omega)}} e^{i(\mathbf{k}\cdot\mathbf{r}-\omega t)}$$

$$= -i\frac{\|4\pi/c^2\|}{(2\pi)^4} \int_C d\omega \int_{-\infty}^{\infty} d^3k\, \mu(\mathbf{k}, \omega) v^2(\mathbf{k}, \omega) \frac{\mathbf{k} \cdot \tilde{\tilde{\mathbf{J}}}_0(\mathbf{k}, \omega)}{k^2 - \frac{\omega^2}{v^2(\mathbf{k},\omega)}} e^{i(\mathbf{k}\cdot\mathbf{r}-\omega t)},$$

where $v^2(\mathbf{k}, \omega)$ is the square of the complex velocity in the spatially and temporally dispersive HILL medium, defined in Eq. (7.30). With substitution of these two relations, the left-hand side of the Lorenz condition given in Eq. (7.60) is found to be

$$\nabla \cdot \mathbf{A}(\mathbf{r}, t) + \frac{\|c\|}{c^2}\frac{\partial\varphi(\mathbf{r}, t)}{\partial t}$$

$$= i\frac{\|4\pi/c\|}{(2\pi)^4} \int_C d\omega \int_{-\infty}^{\infty} d^3k\, \mu(\mathbf{k}, \omega) \left(1 - \frac{v^2(\mathbf{k}, \omega)}{c^2}\right) \frac{\mathbf{k} \cdot \tilde{\tilde{\mathbf{J}}}_0(\mathbf{k}, \omega)}{k^2 - \frac{\omega^2}{v^2(\omega)}} e^{i(\mathbf{k}\cdot\mathbf{r}-\omega t)},$$

$$(7.64)$$

which does not, in general, vanish. Hence, the vector and scalar potentials given in Eqs. (7.52) and (7.56) for the radiation field produced by the current source $\mathbf{J}_0(\mathbf{r}, t)$ in a spatially and temporally dispersive HILL medium do not, in general, satisfy the Lorenz condition. Nevertheless, a gauge function $\Lambda(\mathbf{r}, t)$ can always be determined from the solution of Eq. (7.60) so as to define a gauge transformation to a new pair of potentials $\{\mathbf{A}', \varphi'\}$ that do satisfy the Lorenz condition. However, the vector and scalar potentials given in Eqs. (7.52)–(7.53) and (7.55)–(7.56) are the most natural, and hence the simplest forms of the potentials for the given radiation problem.

7.2.1 The Nonconducting, Nondispersive Medium Case

It is because of the frequency dependence of the complex velocity appearing in Eq. (7.64) that the quantities $\nabla \cdot \mathbf{A}$ and $\partial \varphi / \partial t$ for the radiation problem cannot be directly related (without an appropriate gauge transformation) except in certain special cases. To that end, consider an idealized spatially and temporally nondispersive, nonconducting ($\sigma = 0$) dielectric medium with scalar constant, real-valued dielectric permittivity ϵ and magnetic permeability μ, so that $\epsilon_c(\omega) = \epsilon$ and $v(\omega) = v$, where

$$v^2 = \frac{\|c^2\|}{\mu \epsilon}. \tag{7.65}$$

The vector and scalar potentials given in Eqs. (7.52)–(7.53) and (7.55)–(7.56) for the radiation field are then seen to satisfy the *generalized Lorenz condition*

$$\nabla \cdot \mathbf{A}(\mathbf{r}, t) + \frac{\|c\|}{v^2} \frac{\partial \varphi(\mathbf{r}, t)}{\partial t} = 0, \tag{7.66}$$

so that $\varphi(\mathbf{r}, t)$ can be directly determined from $\mathbf{A}(\mathbf{r}, t)$.

Consider now determining the differential equations that are satisfied by the vector and scalar potentials in this generalized Lorenz gauge. From Eq. (7.52) one has that

$$\nabla^2 \mathbf{A}(\mathbf{r}, t) - \frac{1}{v^2} \frac{\partial^2 \mathbf{A}(\mathbf{r}, t)}{\partial t^2}$$

$$= \left\| \frac{4\pi}{c} \right\| \frac{\mu}{(2\pi)^4} \int_C d\omega \int_{-\infty}^{\infty} d^3k \, \frac{\tilde{\mathbf{J}}_0(\mathbf{k}, \omega)}{k^2 - \frac{\omega^2}{v^2}} \left(-k^2 + \frac{\omega^2}{v^2} \right) e^{i(\mathbf{k}\cdot\mathbf{r}-\omega t)}$$

$$= -\left\| \frac{4\pi}{c} \right\| \frac{\mu}{(2\pi)^4} \int_C d\omega \int_{-\infty}^{\infty} d^3k \, \tilde{\mathbf{J}}_0(\mathbf{k}, \omega) e^{i(\mathbf{k}\cdot\mathbf{r}-\omega t)}$$

$$= -\left\| \frac{4\pi}{c} \right\| \mu \mathscr{F}^{-1} \mathscr{L}^{-1} \left\{ \tilde{\mathbf{J}}_0(\mathbf{k}, \omega) \right\},$$

and the vector potential is seen to satisfy the inhomogeneous wave equation

$$\nabla^2 \mathbf{A}(\mathbf{r}, t) - \frac{1}{v^2} \frac{\partial^2 \mathbf{A}(\mathbf{r}, t)}{\partial t^2} = -\left\| \frac{4\pi}{c} \right\| \mu \mathbf{J}_0(\mathbf{r}, t) \tag{7.67}$$

when the medium is nonconducting and nondispersive. Similarly, from Eq. (7.56), one has that

$$\nabla^2 \varphi(\mathbf{r}, t) - \frac{1}{v^2} \frac{\partial^2 \varphi(\mathbf{r}, t)}{\partial t^2}$$

$$= \frac{\|4\pi\|}{(2\pi)^4 \epsilon} \int_C \frac{d\omega}{\omega} \int_{-\infty}^{\infty} d^3k \, \frac{\mathbf{k} \cdot \tilde{\mathbf{J}}_0(\mathbf{k}, \omega)}{k^2 - \frac{\omega^2}{v^2}} \left(-k^2 + \frac{\omega^2}{v^2} \right) e^{i(\mathbf{k} \cdot \mathbf{r} - \omega t)}$$

$$= -\frac{\|4\pi\|}{(2\pi)^4 \epsilon} \int_C d\omega \int_{-\infty}^{\infty} d^3k \, \frac{\mathbf{k} \cdot \tilde{\mathbf{J}}_0(\mathbf{k}, \omega)}{\omega} e^{i(\mathbf{k} \cdot \mathbf{r} - \omega t)}.$$

From the scalar simplification of the tensor equation following Eq. (7.16) with $\sigma = 0$ one finds that $\tilde{\varrho}_0(\mathbf{k}, \omega) = \mathbf{k} \cdot \tilde{\mathbf{J}}_0(\mathbf{k}, \omega)/\omega$, and the scalar potential is then seen to satisfy the inhomogeneous wave equation

$$\nabla^2 \varphi(\mathbf{r}, t) - \frac{1}{v^2} \frac{\partial^2 \varphi(\mathbf{r}, t)}{\partial t^2} = \frac{\|4\pi\|}{\epsilon} \varrho_0(\mathbf{r}, t), \qquad (7.68)$$

where

$$\varrho_0(\mathbf{r}, t) = \mathcal{F}^{-1} \mathcal{L}^{-1} \left\{ \frac{\mathbf{k} \cdot \tilde{\mathbf{J}}_0(\mathbf{k}, \omega)}{\omega} \right\} \qquad (7.69)$$

is the charge density associated with the externally supplied current source $\mathbf{J}_0(\mathbf{r}, t)$, as required by the equation of continuity in the nondispersive, nonconducting medium. The relations given in Eqs. (7.66)–(7.68) show that $\mathbf{A}(\mathbf{r}, t)$ and $\varphi(\mathbf{r}, t)$ are the standard macroscopic vector and scalar potentials, respectively, in the Lorenz gauge.

7.2.2   The Spectral Lorenz Condition for Temporally Dispersive HILL Media

The reason that the general vector and scalar potentials given in Eqs. (7.52) and (7.56) do not satisfy a Lorenz condition is due to the fact that the complex velocity is frequency dependent in any temporally dispersive medium. The temporal frequency transforms of these vector and scalar potentials for the radiation field in a temporally dispersive HILL medium without any spatial dispersion are found to be

$$\tilde{\mathbf{A}}(\mathbf{r}, \omega) = \left\| \frac{4\pi}{c} \right\| \frac{\mu(\omega)}{(2\pi)^3} \int_{-\infty}^{\infty} d^3k \, \frac{\tilde{\mathbf{J}}_0(\mathbf{k}, \omega)}{k^2 - \frac{\omega^2}{v^2(\omega)}} e^{i\mathbf{k} \cdot \mathbf{r}}, \qquad (7.70)$$

$$\tilde{\varphi}(\mathbf{r}, \omega) = \left\| \frac{4\pi}{c^2} \right\| \frac{\mu(\omega)v^2(\omega)}{(2\pi)^3\omega} \int_{-\infty}^{\infty} d^3k \, \frac{\mathbf{k} \cdot \tilde{\mathbf{J}}_0(\mathbf{k}, \omega)}{k^2 - \frac{\omega^2}{v^2(\omega)}} e^{i\mathbf{k}\cdot\mathbf{r}}. \tag{7.71}$$

Consider then the temporal spectral domain form of Eq. (7.67) with a temporal frequency-dependent complex velocity as given by Eq. (7.30) without the wave vector dependence. With Eqs. (7.70) and (7.71) one obtains the relation

$$\nabla \cdot \tilde{\mathbf{A}}(\mathbf{r}, \omega) - \left\| \frac{1}{c} \right\| i\omega\mu(\omega)\epsilon_c(\omega)\tilde{\varphi}(\mathbf{r}, \omega) = 0. \tag{7.72}$$

It is then seen that the temporal frequency spectra of the vector and scalar potentials for the radiation field in a temporally dispersive HILL medium satisfy a *generalized spectral Lorenz condition*. This condition reduces to the temporal frequency domain form given in Eq. (7.68) when the medium is nonconducting and nondispersive.

Consider now obtaining the individual wave equations satisfied by the temporal frequency spectra of the radiation potentials. From Eq. (7.71) one obtains

$$\nabla^2\tilde{\mathbf{A}}(\mathbf{r}, \omega) + \frac{\omega^2}{v^2(\omega)}\tilde{\mathbf{A}}(\mathbf{r}, \omega)$$

$$= \left\| \frac{4\pi}{c} \right\| \frac{\mu(\omega)}{(2\pi)^3} \int_{-\infty}^{\infty} d^3k \, \frac{\tilde{\mathbf{J}}_0(\mathbf{k}, \omega)}{k^2 - \frac{\omega^2}{v^2(\omega)}} \left(-k^2 + \frac{\omega^2}{v^2(\omega)} \right) e^{i\mathbf{k}\cdot\mathbf{r}}$$

$$= -\left\| \frac{4\pi}{c} \right\| \frac{\mu(\omega)}{(2\pi)^3} \int_{-\infty}^{\infty} d^3k \, \tilde{\mathbf{J}}_0(\mathbf{k}, \omega) e^{i\mathbf{k}\cdot\mathbf{r}}$$

$$= -\left\| \frac{4\pi}{c} \right\| \mu(\omega)\mathcal{F}^{-1}\left\{ \tilde{\mathbf{J}}_0(\mathbf{k}, \omega) \right\},$$

so that the temporal frequency spectrum of the vector potential satisfies the inhomogeneous, dispersive wave equation

$$\nabla^2\tilde{\mathbf{A}}(\mathbf{r}, \omega) + \frac{\omega^2}{v^2(\omega)}\tilde{\mathbf{A}}(\mathbf{r}, \omega) = -\left\| \frac{4\pi}{c} \right\| \mu(\omega)\tilde{\mathbf{J}}_0(\mathbf{r}, \omega). \tag{7.73}$$

Similarly, from Eq. (7.72) for the scalar potential one finds that

$$\nabla^2\tilde{\varphi}(\mathbf{r}, \omega) + \frac{\omega^2}{v^2(\omega)}\tilde{\varphi}(\mathbf{r}, \omega) = -\left\| \frac{4\pi}{c^2} \right\| \frac{\mu(\omega)v^2(\omega)}{(2\pi)^3\omega} \int_{-\infty}^{\infty} d^3k \left(\mathbf{k} \cdot \tilde{\mathbf{J}}_0(\mathbf{k}, \omega) \right) e^{i\mathbf{k}\cdot\mathbf{r}}.$$

The spatiotemporal spectrum of the charge density may be expressed in terms of the spatiotemporal spectrum of the current density through the use of the equation of continuity $\nabla \cdot \mathbf{J}(\mathbf{r}, t) = -\partial\varrho(\mathbf{r}, t)/\partial t$. The Fourier–Laplace transform of this conservation relation then results in the equation $i\mathbf{k} \cdot \tilde{\mathbf{J}}(\mathbf{k}, \omega) = i\omega\tilde{\varrho}(\mathbf{k}, \omega)$, where the fact that the initial value of the charge density $\varrho(\mathbf{r}, t)$ vanishes has been used. Substitution of Eq. (7.6) into this equation and solving for the spatiotemporal spectrum of the charge density be written as

$$\tilde{\tilde{\varrho}}(\mathbf{k}, \omega) = \tilde{\tilde{\varrho}}_0(\mathbf{k}, \omega) + \tilde{\tilde{\varrho}}_c(\mathbf{k}, \omega), \tag{7.74}$$

where

$$\tilde{\tilde{\varrho}}_0(\mathbf{k}, \omega) \equiv \frac{\mathbf{k} \cdot \tilde{\tilde{\mathbf{J}}}_0(\mathbf{k}, \omega)}{\omega} \tag{7.75}$$

is the *nonconductive charge density*, and where

$$\tilde{\tilde{\varrho}}_c(\mathbf{k}, \omega) \equiv \frac{\sigma(\omega)}{\omega} \mathbf{k} \cdot \tilde{\tilde{\mathbf{E}}}(\mathbf{k}, \omega) \tag{7.76}$$

is the *conductive charge density*. With this substitution the wave equation for the scalar potential becomes

$$\nabla^2 \tilde{\varphi}(\mathbf{r}, \omega) + \frac{\omega^2}{v^2(\omega)} \tilde{\varphi}(\mathbf{r}, \omega) = - \left\| \frac{4\pi}{c^2} \right\| \frac{\mu(\omega)v^2(\omega)}{(2\pi)^3\omega} \int_{-\infty}^{\infty} d^3k\, \tilde{\tilde{\varrho}}_0(\mathbf{k}, \omega) e^{i\mathbf{k}\cdot\mathbf{r}}$$

$$= - \left\| \frac{4\pi}{c^2} \right\| \mu(\omega)v^2(\omega) \tilde{\varrho}_0(\mathbf{r}, \omega),$$

so that the temporal frequency spectrum of the scalar potential satisfies the inhomogeneous, dispersive wave equation

$$\nabla^2 \tilde{\varphi}(\mathbf{r}, \omega) + \frac{\omega^2}{v^2(\omega)} \tilde{\varphi}(\mathbf{r}, \omega) = - \frac{\|4\pi\|}{\epsilon_c(\omega)} \tilde{\varrho}_0(\mathbf{r}, \omega). \tag{7.77}$$

Here

$$\tilde{\varrho}_0(\mathbf{r}, \omega) = \mathcal{F}^{-1}\left\{ \frac{\mathbf{k} \cdot \tilde{\tilde{\mathbf{J}}}_0(\mathbf{k}, \omega)}{\omega} \right\} \tag{7.78}$$

is the temporal frequency spectrum of the nonconductive charge density.

7.3   Angular Spectrum of Plane Waves Representation of the Radiation Field

Unless otherwise specifically stated, attention is henceforth focused on pulsed electromagnetic radiation and wave fields in temporally dispersive HILL media in the absence of spatial dispersion. Although both types of material dispersion are ubiquitous, the effects of temporal dispersion on both ultrawideband (UWB) and ultrashort electromagnetic pulse evolution is of greatest importance and hence of interest here. Any future continued research including the effects of spatial dispersion in HILL media may then pick up at the point following Eq. (7.64). For the inclusion of anisotropy too (HALL media), one needs to begin where the discussion ends following Eq. (7.25). Nevertheless, the analysis presented in the remainder of this text may serve as an outline on how best to proceed.

The Fourier–Laplace representations given in Eqs. (7.34)–(7.36) and (7.49)–(7.50) for the electromagnetic field vectors and in Eqs. (7.52)–(7.53) and (7.56)–(7.57) for the vector and scalar potentials for the radiation field in a temporally (but not spatially) dispersive HILL medium occupying all of space are each of the general form

$$\mathbf{G}(\mathbf{r}, t) = \frac{i}{(2\pi)^4} \int_C d\omega \int_{-\infty}^{\infty} d^3k \, \frac{\tilde{\tilde{\mathbf{F}}}(\mathbf{k}, \omega)}{k^2 - \frac{\omega^2}{v^2(\omega)}} e^{i(\mathbf{k}\cdot\mathbf{r}-\omega t)} \tag{7.79}$$

$$= \frac{1}{8\pi^4} \Re \left\{ i \int_{C_+} d\omega \int_{-\infty}^{\infty} d^3k \, \frac{\tilde{\tilde{\mathbf{F}}}(\mathbf{k}, \omega)}{k^2 - \frac{\omega^2}{v^2(\omega)}} e^{i(\mathbf{k}\cdot\mathbf{r}-\omega t)} \right\}, \tag{7.80}$$

where C is the contour $\omega = \omega' + ia$ with $\omega' = \Re\{\omega\}$ varying from $-\infty$ to $+\infty$, and C_+ is the contour $\omega = \omega' + ia$ with ω' varying from 0 to $+\infty$, and where the spectral function $\tilde{\tilde{\mathbf{F}}}(\mathbf{k}, \omega)$ satisfies the symmetry property [cf. Eq. (7.38)]

$$\tilde{\tilde{\mathbf{F}}}(-\mathbf{k}, -\omega' + ia) = \tilde{\tilde{\mathbf{F}}}^*(\mathbf{k}, \omega' + ia). \tag{7.81}$$

For each equation represented by the generic form given in Eq. (7.80) with Eq. (7.81), the spectral function $\tilde{\tilde{\mathbf{F}}}(\mathbf{k}, \omega)$ is given by either the scalar or vector product of $\mathbf{k}$ with $\tilde{\tilde{\mathbf{J}}}_0(\mathbf{k}, \omega) = \mathcal{F}\mathcal{L}\{\mathbf{J}_0(\mathbf{r}, t)\}$. Because $\mathbf{J}_0(\mathbf{r}, t)$ and its first-order spatial derivatives vanish for $|z| \geq Z$, where Z is a positive constant, it then follows that $\mathbf{F}(\mathbf{r}, t) = \mathcal{F}^{-1}\mathcal{L}^{-1}\left\{\tilde{\tilde{\mathbf{F}}}(\mathbf{k}, \omega)\right\}$ also vanishes for $|z| \geq Z$, where

$$\tilde{\tilde{\mathbf{F}}}(\mathbf{k}, \omega) = \int_0^{\infty} dt \int_{-\infty}^{\infty} d^3r \, \mathbf{F}(\mathbf{r}, t) e^{-i(\mathbf{k}\cdot\mathbf{r}-\omega t)}. \tag{7.82}$$

Let $k_z = k_z' + ik_1''$ and $\omega = \omega' + ia$ where k_z', k_z'', ω' and a are all real-valued variables. With these substitutions Eq. (7.82) becomes

$$\tilde{\tilde{\mathbf{F}}}(\mathbf{k}, \omega) = \int_{-Z}^{Z} dz \, e^{-ik_z'z} e^{k_z''z} \int_0^{\infty} dt \int_{-\infty}^{\infty} dx dy \, \mathbf{F}(\mathbf{r}, t) e^{-i(k_x x + k_y y - \omega' t)} e^{-at}$$

$$= \int_{-Z}^{Z} \mathbf{G}(k_x, k_y, \omega, z) e^{-ik_z'z} e^{k_z''z} dz. \tag{7.83}$$

Because the only integral involving $k_z = k_z' + ik_z''$ is taken over a bounded region, and because the integrand in Eq. (7.83) is an analytic function of k_z for all k_z and is also a continuous function of the variable of integration (provided that $\mathbf{F}(\mathbf{r}, t)$ is continuous), then $\tilde{\mathbf{F}}(\mathbf{k}, \omega)$ is an entire function of the complex variable k_z (i.e., it is analytic for all k_z) [4]. As a consequence, each component of $\tilde{\mathbf{F}}(\mathbf{k}, \omega)$ is bounded in magnitude as

$$\left| \tilde{F}_j(\mathbf{k}, \omega) \right| \leq \int_{-Z}^{Z} dz\, e^{k_z'' z} \int_0^\infty dt \int_{-\infty}^\infty dx\, dy\, \left| F_j(\mathbf{r}, t) \right|$$

$$\leq e^{|k_z''|Z} \int_0^\infty dt \int_{-\infty}^\infty dx\, dy\, \left| F_j(\mathbf{r}, t) \right|$$

$$\leq M_j e^{|k_z''|Z}, \tag{7.84}$$

where M_j is independent of k_z.

Because $\tilde{\mathbf{F}}(\mathbf{k}, \omega)$ is an entire function of complex k_z, the only singularities appearing in the integrand of Eq. (7.80) [as well as in Eq. (7.81)] are located where $k^2 - \omega^2/v^2(\omega) = 0$, and hence, this integrand has two simple pole singularities that are located at

$$k_z = \pm \left(\frac{\omega^2}{v^2(\omega)} - k_T^2 \right)^{1/2}, \tag{7.85}$$

where $k_T^2 \equiv k_x^2 + k_y^2$. Because $|\tilde{\mathbf{F}}(\mathbf{k}, \omega)|$ is bounded, the k_z-integrations in Eqs. (7.80) and (7.81) may then be evaluated through the use of Cauchy's residue theorem once the locations of the simple pole singularities given in Eq. (7.85) are determined.

In order to determine the location of the simple pole singularities satisfying Eq. (7.85) in the complex k_z-plane, it is necessary to determine the sign of the imaginary part of the quantity $\omega^2/v^2(\omega)$. From Eqs. (5.28) and the scalar, wave vector independent version of Eq. (7.30) one has that

$$\|c^2\| \frac{\omega^2}{v^2(\omega)} = \omega^2 \mu(\omega)\epsilon_c(\omega) = \omega^2 \mu(\omega) \left[\epsilon(\omega) + i\|4\pi\| \frac{\sigma(\omega)}{\omega} \right]$$

$$= \left[\mu_r \left(\omega^2 \epsilon_r - \|4\pi\| \omega\sigma_i \right) - \mu_i \left(\omega^2 \epsilon_i + \|4\pi\| \omega\sigma_r \right) \right]$$

$$+ i \left[\mu_i \left(\omega^2 \epsilon_r - \|4\pi\| \omega\sigma_i \right) + \mu_r \left(\omega^2 \epsilon_i + \|4\pi\| \omega\sigma_r \right) \right],$$

so that with $\omega = \omega' + ia$, where ω' and a are both real-valued,

$$
\begin{aligned}
\|c^2\| \frac{\omega^2}{v^2(\omega)} = & \left\{ \mu_r \left[\left(\omega'^2 - a^2 \right) \epsilon_r - 2a\omega'\epsilon_i - \|4\pi\| \left(\omega'\sigma_i + a\sigma_r \right) \right] \right. \\
& - \mu_i \left[\left(\omega'^2 - a^2 \right) \epsilon_i + 2a\omega'\epsilon_r + \|4\pi\| \left(\omega'\sigma_r - a\sigma_i \right) \right] \right\} \\
& + i \left\{ \mu_r \left[\left(\omega'^2 - a^2 \right) \epsilon_i + 2a\omega'\epsilon_r + \|4\pi\| \left(\omega'\sigma_r - a\sigma_i \right) \right] \right. \\
& + \mu_i \left[\left(\omega'^2 - a^2 \right) \epsilon_r - 2a\omega'\epsilon_i - \|4\pi\| \left(\omega'\sigma_i + a\sigma_r \right) \right] \right\},
\end{aligned}
$$

(7.86)

with $\epsilon_r(\omega) \equiv \Re\{\epsilon(\omega)\}$, $\epsilon_i(\omega) \equiv \Im\{\epsilon(\omega)\}$, $\mu_r(\omega) \equiv \Re\{\mu(\omega)\}$, $\mu_i(\omega) \equiv \Im\{\mu(\omega)\}$, $\sigma_r(\omega) \equiv \Re\{\sigma(\omega)\}$, and $\sigma_i(\omega) \equiv \Im\{\sigma(\omega)\}$. The only situation to be considered here is the one for which each of the inequalities

$$
\epsilon_i(\omega' + ia) \geq 0, \quad \mu_i(\omega' + ia) \geq 0, \quad \sigma_r(\omega' + ia) \geq 0, \tag{7.87}
$$

is satisfied for $\omega' \in [0, +\infty)$ for all $a \geq 0$, so that the medium is absorptive in each of its dielectric, magnetic, and conductive responses.[1] Along the contour C_+ in the representation $\mathfrak{J}^+$ given in Eq. (7.80), $\omega' = 0 \to +\infty$ with $a \geq 0$ a fixed constant that is greater than the abscissa of absolute convergence (see Appendix E) for the particular radiation problem under consideration. The inequality

$$
\Im \left\{ \frac{\omega^2}{v^2(\omega)} \right\} > 0, \tag{7.88}
$$

and hence (because k_x and k_y are both real-valued)

$$
\Im \left\{ \frac{\omega^2}{v^2(\omega)} - k_T^2 \right\} > 0, \tag{7.89}
$$

will then be satisfied along the contour C_+ in any of the following cases.

1. If $a = 0$, then

$$
\omega'\epsilon_i(\omega) + \|4\pi\|\sigma_r(\omega) + \frac{\mu_i(\omega)}{\mu_r(\omega)}\omega'\epsilon_r(\omega) > \|4\pi\|\frac{\mu_i(\omega)}{\mu_r(\omega)}\sigma_i(\omega). \tag{7.90}
$$

---

[1] A finite amount of gain in any one of these parameters may be allowed over a finite frequency domain provided that the inequality in Eq. (7.88) remains unchanged.

If $\mu_i(\omega) = 0$, then one must have that $\epsilon_i(\omega) > 0$ at every frequency value where $\sigma_r(\omega) = 0$ and $\sigma_r(\omega) > 0$ at every frequency value where $\epsilon_i(\omega) = 0$. If $\mu_i(\omega) \neq 0$, then there are two possibilities:

(a) If $\omega' \epsilon_r(\omega) \geq \|4\pi\| \sigma_i(\omega)$, then it is sufficient to require that $\epsilon_i(\omega) > 0$ at every frequency value where $\sigma_r(\omega) = 0$ and $\sigma_r(\omega) > 0$ at every frequency value where $\epsilon_i(\omega) = 0$.

(b) If $\omega' \epsilon_r(\omega) \leq \|4\pi\| \sigma_i(\omega)$ for sufficiently small nonnegative values of ω', then the general inequality given in Eq. (7.90) must be satisfied. This is easily satisfied for a simple (nonferromagnetic) magnetizable material for which $|\mu_i(\omega)/\mu_r(\omega)| \ll 1$.

2. If $\sigma_r(\omega) = \sigma_i(\omega) = 0$ so that the medium is nonconducting, then

$$\epsilon_r(\omega) > -\frac{(\omega'^2 - a^2) - 2a\omega' \frac{\mu_i(\omega)}{\mu_r(\omega)}}{2a\omega' + \frac{\mu_i(\omega)}{\mu_r(\omega)}(\omega'^2 - a^2)} \epsilon_i(\omega). \tag{7.91}$$

3. If $\epsilon_i(\omega) = 0$ so that the dielectric permittivity is real-valued, then

$$\left[2a\omega' + \frac{\mu_i(\omega)}{\mu_r(\omega)}(\omega'^2 - a^2) \right] \epsilon_r(\omega)$$

$$+ \|4\pi\| \left[(\omega' \sigma_r(\omega) - a\sigma_i(\omega)) - \frac{\mu_i(\omega)}{\mu_r(\omega)}(\omega' \sigma_i(\omega) + a\sigma_r(\omega)) \right] > 0. \tag{7.92}$$

If $a = 0$ this inequality requires that

$$\sigma_r(\omega) > -\frac{\mu_i(\omega)}{\mu_r(\omega)} \left(\frac{1}{\|4\pi\|} \omega' \epsilon_r(\omega) - \sigma_i(\omega) \right), \tag{7.93}$$

and the general inequality given in Eq. (7.92) must be satisfied when $a > 0$.

For a nondispersive medium, both ϵ_r and μ_r are positive constants and $\epsilon_i = \mu_i = 0$. If, in addition, $\sigma_r = \sigma_i = 0$ so that the medium is also nonconducting, then case 2 applies with $a > 0$, and the inequality in Eq. (7.91) becomes $\epsilon_r > 0$. On the other hand, if the medium is conducting so that $\sigma_r > 0$, then one can take $a = 0$ and the inequality in Eq. (7.93) of case 3 applies. Hence, for a nondispersive medium, the equivalent inequalities given in Eqs. (7.88) and (7.89) can always be satisfied.

For a causal, temporally dispersive medium both $\epsilon_r(\omega)$ and $\mu_r(\omega)$ can be negative over certain frequency intervals but both $\epsilon_i(\omega)$ and $\mu_i(\omega)$ are always nonnegative on C_+. In that case, $\Im\{\omega^2/v^2(\omega)\}$ will be negative for sufficiently small $\omega' < a$, as can be seen from the imaginary part of Eq. (7.86). One can then ensure that the equivalent inequalities given in Eqs. (7.88) and (7.89) are satisfied by taking $a = 0$, as in case 1. If this is not permissible because of a positive abscissa of absolute convergence for the particular radiation problem under consideration, one can always take the lower limit of the ω-integration in Eq. (7.80) to be $\omega' = a$ and proceed to the limit as $a \to 0$ after completion of the integration. In any event, ω' will always be considered to be positive when the k_z-integral in Eq. (7.80) is evaluated and then the ω-integral possesses a positive lower limit that one then lets pass to zero in a limiting process.

In order to proceed with the k_z-integration in Eq. (7.80) it is assumed that the equivalent inequalities in Eqs. (7.88) and (7.89) are satisfied. The complex quantity

$$\gamma(\omega) \equiv \left(\frac{\omega^2}{v^2(\omega)} - k_T^2 \right)^{1/2}, \tag{7.94}$$

where $k_T^2 \equiv k_x^2 + k_y^2$, is taken as the principal branch of the square root that is defined in the following manner: let

$$\frac{\omega^2}{v^2(\omega)} - k_T^2 = \Gamma e^{i\phi_\Gamma}; \quad 0 \leq \phi_\Gamma < 2\pi, \tag{7.95}$$

where Γ is the magnitude and ϕ_Γ the phase of the complex quantity $(\omega^2/v^2(\omega) - k_T^2)$, and define

$$\gamma(\omega) \equiv \sqrt{\Gamma} e^{i\phi_\Gamma/2}; \quad 0 \leq \phi_\Gamma < 2\pi. \tag{7.96}$$

Then $\gamma(\omega)$ always has a nonnegative imaginary part because $0 \leq \phi_\Gamma < \pi$. With reference to Eq. (7.85), the simple pole singularity at $k_z = \gamma(\omega)$ is then always in the upper-half of the complex k_z-plane and the pole at $k_z = -\gamma(\omega)$ is always in the lower-half of the complex k_z-plane.

Because of the inequality in Eq. (7.84), the jth component of the integrand in Eq. (7.80) is bounded as

$$\left| \frac{\tilde{F}_j(\mathbf{k}, \omega)}{k^2 - \frac{\omega^2}{v^2(\omega)}} e^{i(\mathbf{k} \cdot \mathbf{r} - \omega t)} \right| = \frac{\left| \tilde{F}_j(\mathbf{k}, \omega) \right|}{\left| k^2 - \frac{\omega^2}{v^2(\omega)} \right|} e^{-k_z'' z} e^{at}$$

$$\leq \frac{M_j}{\left| k^2 - \frac{\omega^2}{v^2(\omega)} \right|} e^{|k_z''| Z - k_z'' z}, \tag{7.97}$$

with $a \to 0$. Hence, in the upper-half of the complex k_z-plane the integrand of the integral in Eq. (7.80) decays exponentially and goes to zero as $|k_z''| \to \infty$ if $z > Z > 0$. The same exponential decay is present in the lower-half of the complex k_z-plane if $z < -Z$. It then follows from Jordan's lemma [5] that the k_z-integral in Eq. (7.80) may be evaluated by application of Cauchy's residue theorem either by completing the contour along the real axis into a closed circuit using the semicircular contour C_1 in the upper-half plane if $z > Z$, as illustrated in Fig. 7.2, or by using the semicircular contour C_2 in the lower-half plane if $z < -Z$, also illustrated in

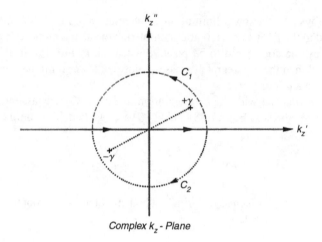

Complex k_z - Plane

Fig. 7.2 The semicircular contours C_1 and C_2 centered at the origin in the complex k_z-plane, where C_1 is contained entirely in the upper-half plane and C_2 is contained entirely in the lower-half plane

Fig. 7.2, each contour centered at the origin, where the radius of each is allowed to go to infinity. If $|z| < Z$, then the integral in Eq. (7.80) cannot be evaluated by completing the contour in this manner.

For the radiation field in the positive half-space $z > Z$ one completes the contour in the upper-half of the complex k_z-plane and the only contribution to the k_z-integral in Eq. (7.80) is due to the simple pole at $k_z = +\gamma(\omega)$. For the radiation field in the negative half-space $z < -Z$ one completes the contour in the lower-half of the complex k_z-plane and the only contribution to the k_z-integral in Eq. (7.80) is due to the simple pole at $k_z = -\gamma(\omega)$. Because

$$k^2 - \frac{\omega^2}{v^2(\omega)} = (k_z - \gamma(\omega))(k_z + \gamma(\omega)), \qquad (7.98)$$

then the residue of the simple pole at $k_z = +\gamma(\omega)$ is given by

Residue
$(k_z = +\gamma(\omega))$

$$= \frac{1}{8\pi^4} \lim_{k_z \to \gamma} (k_z - \gamma) \int_{C_+} d\omega \int_{-\infty}^{\infty} dk_x dk_y \frac{\tilde{\mathbf{F}}(\mathbf{k}, \omega)}{(k_z - \gamma)(k_z + \gamma)} e^{i(\mathbf{k} \cdot \mathbf{r} - \omega t)}$$

$$= \frac{1}{8\pi^4} \int_{C_+} d\omega \int_{-\infty}^{\infty} dk_x dk_y \frac{\tilde{\mathbf{F}}(\tilde{\mathbf{k}}^+, \omega)}{2\gamma(\omega)} e^{i(\tilde{\mathbf{k}}^+ \cdot \mathbf{r} - \omega t)}, \qquad (7.99)$$

and the residue of the simple pole at $k_z = -\gamma(\omega)$ is given by

$$
\begin{aligned}
\text{Residue} & \\
(k_z = -\gamma(\omega)) & \\
= \frac{1}{8\pi^4} & \lim_{k_z \to -\gamma} (k_z + \gamma) \int_{C_+} d\omega \int_{-\infty}^{\infty} dk_x dk_y \frac{\tilde{\mathbf{F}}(\mathbf{k}, \omega)}{(k_z - \gamma)(k_z + \gamma)} e^{i(\mathbf{k}\cdot\mathbf{r} - \omega t)} \\
= \frac{1}{8\pi^4} & \int_{C_+} d\omega \int_{-\infty}^{\infty} dk_x dk_y \frac{\tilde{\mathbf{F}}(\tilde{\mathbf{k}}^-, \omega)}{-2\gamma(\omega)} e^{i(\tilde{\mathbf{k}}^- \cdot \mathbf{r} - \omega t)},
\end{aligned}
\tag{7.100}
$$

Here

$$
\tilde{\mathbf{k}}^{\pm}(\omega) \equiv \hat{\mathbf{1}}_x k_x + \hat{\mathbf{1}}_y k_y \pm \hat{\mathbf{1}}_z \gamma(\omega)
\tag{7.101}
$$

is the *complex wave vector*, where

$$
\tilde{\mathbf{k}}^{\pm}(\omega) \cdot \tilde{\mathbf{k}}^{\pm}(\omega) = k_x^2 + k_y^2 + \gamma^2(\omega)
$$

$$
= \frac{\omega^2}{v^2(\omega)} = \tilde{k}^2(\omega),
\tag{7.102}
$$

so that

$$
\tilde{k}(\omega) = \left(\tilde{\mathbf{k}}^{\pm}(\omega) \cdot \tilde{\mathbf{k}}^{\pm}(\omega) \right)^{1/2}
$$

$$
= \frac{\omega}{v(\omega)} = \left\| \frac{1}{c} \right\| \omega \left(\mu(\omega)\epsilon_c(\omega) \right)^{1/2} = k_0 n(\omega)
\tag{7.103}
$$

is the complex wavenumber in the temporally dispersive HILL medium [cf. Eq. (5.58) for a time-harmonic plane wave field]. Here $k_0 = \omega/c$ is the vacuum wavenumber and $n(\omega)$ is the complex index of refraction of the dispersive medium, defined in Eq. (5.32). Consequently, by the residue theorem [6], Eq. (7.99) is multiplied by $2\pi i$ and Eq. (7.100) by $-2\pi i$ to obtain the final expressions for the k_z-integration, with the final result

$$
\begin{aligned}
\mathbf{G}(\mathbf{r}, t) &= \frac{1}{8\pi^4} \Re \left\{ i(\pm 2\pi i) \int_{C_+} d\omega \int_{-\infty}^{\infty} dk_x dk_y \frac{\tilde{\mathbf{F}}(\tilde{\mathbf{k}}^{\pm}, \omega)}{\pm 2\gamma(\omega)} e^{i(\tilde{\mathbf{k}}^{\pm} \cdot \mathbf{r} - \omega t)} \right\} \\
&= -\frac{1}{(2\pi)^3} \Re \left\{ \int_{C_+} d\omega \int_{-\infty}^{\infty} dk_x dk_y \frac{\tilde{\mathbf{F}}(\tilde{\mathbf{k}}^{\pm}, \omega)}{\gamma(\omega)} e^{i(\tilde{\mathbf{k}}^{\pm} \cdot \mathbf{r} - \omega t)} \right\}
\end{aligned}
\tag{7.104}
$$

for the generic integral in Eq. (7.80). This final expression is referred to as the *angular spectrum of plane waves representation* of the field $\mathbf{G}(\mathbf{r}, t)$; the origin of

452   7   The Angular Spectrum Representation of the Pulsed Radiation Field

this terminology is made clear in Sect. 7.4 when the spatial frequency integral in
Eq. (7.104) is transformed to polar coordinates in $k_x k_y$-space.

For the electric and magnetic field vectors for the radiation field, given in
Eqs. (7.49) and (7.50), the k_z-integration of each Fourier–Laplace representation
yields the pair of expressions

$$
\mathbf{E}(\mathbf{r}, t) = \frac{\|4\pi\|}{(2\pi)^3} \Re \left\{ \int_{C_+} \frac{d\omega}{\omega \epsilon_c(\omega)} \right.
$$

$$
\left. \int_{-\infty}^{\infty} dk_x dk_y \, \frac{\tilde{\mathbf{k}}^\pm \times \left(\tilde{\mathbf{k}}^\pm \times \tilde{\mathbf{J}}_0(\tilde{\mathbf{k}}^\pm, \omega) \right)}{\gamma(\omega)} e^{i\left(\tilde{\mathbf{k}}^\pm \cdot \mathbf{r} - \omega t\right)} \right\},
$$

(7.105)

$$
\mathbf{B}(\mathbf{r}, t) = -\frac{\|4\pi/c\|}{(2\pi)^3} \Re \left\{ \int_{C_+} d\omega \, \mu(\omega) \right.
$$

$$
\left. \int_{-\infty}^{\infty} dk_x dk_y \, \frac{\tilde{\mathbf{k}}^\pm \times \tilde{\mathbf{J}}_0(\tilde{\mathbf{k}}^\pm, \omega)}{\gamma(\omega)} e^{i\left(\tilde{\mathbf{k}}^\pm \cdot \mathbf{r} - \omega t\right)} \right\},
$$

(7.106)

for $t > 0$, where $\tilde{\mathbf{k}}^+(\omega)$ is used when $z > Z > 0$ and $\tilde{\mathbf{k}}^-(\omega)$ is used when $z < -Z$.
This pair of expressions is the *angular spectrum of plane waves representation of
the electromagnetic field vectors for the radiation field in a temporally dispersive
HILL medium* occupying all of space. For the vector and scalar potentials given in
Eqs. (7.53) and (7.56), respectively, for this radiated electromagnetic wave field, the
k_z-integration of each Fourier–Laplace representation yields the angular spectrum
of plane waves representations

$$
\mathbf{A}(\mathbf{r}, t) = \frac{\|4\pi/c\|}{(2\pi)^3} \Re \left\{ i \int_{C_+} d\omega \, \mu(\omega) \int_{-\infty}^{\infty} dk_x dk_y \, \frac{\tilde{\mathbf{J}}_0(\tilde{\mathbf{k}}^\pm, \omega)}{\gamma(\omega)} e^{i\left(\tilde{\mathbf{k}}^\pm \cdot \mathbf{r} - \omega t\right)} \right\},
$$

(7.107)

$$
\varphi(\mathbf{r}, t) = \frac{\|4\pi\|}{(2\pi)^3} \Re \left\{ i \int_{C_+} \frac{d\omega}{\omega \epsilon_c(\omega)} \int_{-\infty}^{\infty} dk_x dk_y \, \frac{\tilde{\mathbf{k}}^\pm \cdot \tilde{\mathbf{J}}_0(\tilde{\mathbf{k}}^\pm, \omega)}{\gamma(\omega)} e^{i\left(\tilde{\mathbf{k}}^\pm \cdot \mathbf{r} - \omega t\right)} \right\},
$$

(7.108)

for $t > 0$. The parameter a for the contour C_+ ($\omega = \omega' + ia$) can be taken to be zero
in Eqs. (7.106) and (7.107) because the points at which $\gamma(\omega) = 0$ are integrable
singularities [7], and can also be taken to be zero in Eqs. (7.105) and (7.108)
provided that the points at which $\omega \epsilon_c(\omega) = 0$ are also integrable singularities.

The exponential factor appearing in the angular spectrum of plane waves
representations given above in Eqs. (7.105)–(7.108) is, with substitution from
Eq. (7.101), given by

$$e^{i(\tilde{\mathbf{k}}^{\pm}\cdot\mathbf{r}-\omega t)} = e^{\pm i\gamma(\omega)z}e^{i(k_x x+k_y y-\omega t)}, \tag{7.109}$$

which represent either homogeneous or inhomogeneous plane waves depending upon whether $\gamma(\omega) \equiv \left(\omega^2/v^2(\omega) - k_T^2\right)^{1/2}$ is real, imaginary, or complex-valued. From Eqs. (5.28) and (7.30) one has that

$$\frac{\omega^2}{v^2(\omega)} = \left\|\frac{1}{c^2}\right\|\omega^2\mu(\omega)\epsilon(\omega) + \left\|\frac{4\pi}{c^2}\right\| i\omega\mu(\omega)\sigma(\omega). \tag{7.110}$$

Consider first the simplest case of a nondispersive nonconducting material where $\sigma = 0$ and ϵ and μ are both real-valued, so that $\omega^2/v^2 = \omega^2\mu\epsilon/\|c^2\|$. With ω taken to be real-valued (so that $a = 0$), then $\gamma = \left(\omega^2/v^2 - k_T^2\right)^{1/2}$ is either pure real or pure imaginary. If $k_T^2 \equiv k_x^2 + k_y^2 \leq \omega^2/v^2$, then γ is real and Eq. (7.109) represents pure oscillatory plane waves. On the other hand, if $k_T^2 > \omega^2/v^2$, then γ is pure imaginary with $\gamma = i|\gamma|$ so that $i\gamma$ is real and negative and the expression given in Eq. (7.109) possesses exponential decay in the z-direction for $|z| > Z$, because then $e^{\pm i\gamma z} = e^{-|\gamma||z|}$. Consequently, for $k_T^2 \leq \omega^2/v^2$, Eq. (6.109) represents homogeneous plane waves for which the surfaces of constant phase coincide with the surfaces of constant amplitude, whereas inhomogeneous plane waves are obtained when $k_T^2 > \omega^2/v^2$, where the phase is constant in planes parallel to the z-axis ($k_x x + k_y y = \kappa_1$) and the amplitude is constant in planes perpendicular to the z-axis ($\gamma z = \kappa_2$), where κ_j, $j = 1, 2$ are real constants. That the plane wave factors appearing in the angular spectrum of plane waves representations given in Eqs. (7.105)–(7.108) only decay along the z-direction has no real physical meaning and is simply a consequence of an imposed dependency due to the evaluation of the k_z-integral in the Fourier–Laplace representations.

It is then clear that, in general, the angular spectrum of plane waves representations for both the electromagnetic field vectors, given in Eqs. (7.105) and (7.106), and the vector and scalar potentials, given in Eqs. (7.107) and (7.108), combine both homogeneous and inhomogeneous plane waves in the 2π-solid angle about the positive z-axis for $z > Z$, as well as in the 2π-solid angle about the negative z-axis for $z < -Z$. These specific forms of the plane wave representation are known as *Weyl-type expansions* [8]; each is valid only in its particular half-space with plane waves propagating only within the 2π-solid angle into the respective half-space. When the source of the field ceases to radiate the problem reduces to an initial value problem and one can use a representation that possesses only homogeneous plane waves in the entire 4π-solid angle about the chosen origin. This specific form of the plane wave representation is known as a *Whittaker-type expansion* [9] and is deferred to Chap. 9. A detailed comparison of these two expansion types for the optical wave field in a nondispersive, nonconducting HILL medium may be found in the published papers by Sherman et al. [10, 11] and Devaney and Sherman [12]. Further discussion of the angular spectrum representation may be found in the research monographs by Clemmow [13], Stamnes [14], Nieto-Vesperinas [15], Hansen and Yaghjian [16], and Devaney [17] for a variety of other applications, including antenna analysis and synthesis, the field behavior in focal regions, diffraction theory, sensing, imaging and tomography.

7.4 Polar Coordinate Form of the Angular Spectrum Representation

Attention is now given to the spatial angular spectrum integral

$$u(\mathbf{r}) = \int_{-\infty}^{\infty} dk_x dk_y \frac{U(k_x, k_y)}{\gamma} e^{i\mathbf{k}^{\pm} \cdot \mathbf{r}} \tag{7.111}$$

that appears in the angular spectrum representation given in Eq. (7.105) for the electric field intensity vector with $U(k_x, k_y) = \tilde{\mathbf{k}}^{\pm} \times \left(\tilde{\mathbf{k}}^{\pm} \times \tilde{\mathbf{J}}_0(\tilde{\mathbf{k}}^{\pm}, \omega) \right)$, in the angular spectrum representation given in Eq. (7.106) for the magnetic induction field vector with $U(k_x, k_y) = \tilde{\mathbf{k}}^{\pm} \times \tilde{\mathbf{J}}_0(\tilde{\mathbf{k}}^{\pm}, \omega)$, in the angular spectrum representation given in Eq. (7.107) for the vector potential field with $U(k_x, k_y) = \tilde{\mathbf{J}}_0(\tilde{\mathbf{k}}^{\pm}, \omega)$, and in the angular spectrum representation given in Eq. (7.108) for the scalar potential field with $U(k_x, k_y) = \tilde{\mathbf{k}}^{\pm} \cdot \tilde{\mathbf{J}}_0(\tilde{\mathbf{k}}^{\pm}, \omega)$. Here $\mathbf{r} = \hat{\mathbf{1}}_x x + \hat{\mathbf{1}}_y y + \hat{\mathbf{1}}_z z$ is the position vector for the field, and

$$\tilde{\mathbf{k}}^{\pm}(\omega) \equiv \hat{\mathbf{1}}_x k_x + \hat{\mathbf{1}}_y k_y \pm \hat{\mathbf{1}}_z \gamma(\omega) \tag{7.112}$$

is the complex wave vector with

$$\gamma(\omega) \equiv \left(\tilde{k}^2(\omega) - k_x^2 - k_y^2 \right)^{1/2}, \tag{7.113}$$

where

$$\tilde{k}(\omega) = \frac{\omega}{v(\omega)} = \left\| \frac{1}{c} \right\| \omega \left(\mu(\omega) \epsilon_c(\omega) \right)^{1/2} \tag{7.114}$$

is the complex wave number. One employs $\tilde{\mathbf{k}}^{+}$ for $z > Z$ and $\tilde{\mathbf{k}}^{-}$ for $z < -Z$, where $Z > 0$ describes the maximum spatial extent of the applied current source in the z-coordinate direction (see Fig. 7.1). Finally,

$$U(k_x, k_y) = f(k_x, k_y, \gamma) \tag{7.115}$$

where $f(k_x, k_y, k_z)$ is an entire function of complex k_x, k_y, and k_z [see the discussion in connection with Eq. (7.83)]; however, $f(k_x, k_y, \gamma)$ is not necessarily an entire function of complex k_x and k_y.

Consider now the change of variable to polar coordinates that is defined by

$$k_x = \tilde{k}(\omega) \sin \alpha \cos \beta, \qquad (7.116)$$

$$k_y = \tilde{k}(\omega) \sin \alpha \sin \beta, \qquad (7.117)$$

where k_x and k_y are both real-valued and range over the domain from $-\infty$ to $+\infty$, $\tilde{k}(\omega)$ is the fixed complex number given by Eq. (7.114), and where α and β are, in general, complex-valued angles whose domains must be determined so as to yield the proper ranges for both k_x and k_y. The Jacobian of this coordinate transformation is

$$\frac{\partial(k_x, k_y)}{\partial(\alpha, \beta)} \equiv \begin{vmatrix} \partial k_x/\partial\alpha & \partial k_y/\partial\alpha \\ \partial k_x/\partial\beta & \partial k_y/\partial\beta \end{vmatrix} = \tilde{k}^2(\omega) \sin \alpha \cos \alpha. \qquad (7.118)$$

Furthermore, under this coordinate transformation

$$\gamma = \pm \tilde{k}(\omega) \cos \alpha, \qquad (7.119)$$

where the proper sign choice is determined by the domain of integration.

The derivation now proceeds with the determination of the integration contours along which the complex angles α and β vary in such a manner that both k_x and k_y are kept real-valued and varying from $-\infty$ to $+\infty$. The simplest case where $\tilde{k}$ is real-valued is considered first, followed by the general case where $\tilde{k}(\omega)$ is complex-valued. Clearly, the first case is a limiting situation of the latter and so lends important guidance to the latter, more general derivation.

Consider then the case where $\tilde{k} = k$ is real-valued. It is easily seen from Eqs. (7.116) and (7.117) that β must then be real-valued in order that both k_x and k_y are real-valued. The angle α must then be complex-valued such that $\sin \alpha$ is real-valued and ranges from 0 to $+\infty$. The real angle β ranging over 0 to 2π will then give k_x and k_y a range over $-\infty$ to $+\infty$. As a consequence, let $\alpha = \alpha' + i\alpha''$, where α' and α'' are both real-valued, so that $\sin \alpha = \sin (\alpha' + i\alpha'') = \sin \alpha' \cosh \alpha'' + i \cos \alpha' \sinh \alpha''$. Because $\sin \alpha$ must be real-valued, one then obtains the condition

$$\cos \alpha' \sinh \alpha'' = 0,$$

which is satisfied along either of the following two contours in the complex α-plane, illustrated in Fig. 7.3: the contour C where α varies from 0 to $\pi/2$ along the α'-axis, and then α varies from $\alpha = \pi/2$ to $\pi/2 - i\infty$ along the line $\alpha' = \pi/2$; the contour C' where α varies from 0 to $\pi/2$ along the α'-axis, and then α varies from $\alpha = \pi/2$ to $\pi/2 + i\infty$ along the line $\alpha' = \pi/2$. Along the $\alpha' = \Re\{\alpha\}$ axis, $\sin \alpha = \sin \alpha'$ so that

$$\gamma = k \left(1 - \sin^2 \alpha'\right)^{1/2} = \pm k \cos \alpha'$$

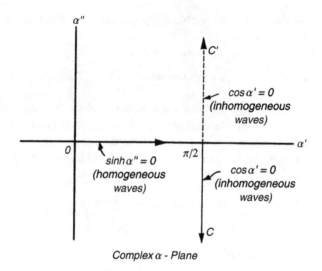

Complex α - Plane

Fig. 7.3 The contours C and C' in the complex α-plane for the special case when $\tilde{k} = k$ is real-valued

and the factor $e^{i\tilde{\mathbf{k}}^{\pm} \cdot \mathbf{r}}$ appearing in the integrand of Eq. (7.111) describes homogeneous plane waves along this portion of the contour for both C and C'. Along the line $\alpha' = \pi/2$, $\sin\alpha = \cosh\alpha''$ so that

$$\gamma = k\left(1 - \cosh^2\alpha''\right)^{1/2} = k\left(-\sinh^2\alpha''\right)^{1/2}$$

$$= \pm ik\sinh\alpha''$$

and the factor $e^{i\tilde{\mathbf{k}}^{\pm} \cdot \mathbf{r}}$ describes inhomogeneous plane waves along this portion of the contour for either C or C'. As α' varies from 0 to $\pi/2$ with $\alpha'' = 0$ along both C and C', $\sin\alpha = \sin\alpha'$ varies from 0 to 1, and as α varies from $\pi/2$ to $\pi/2 - i\infty$ with $\alpha' = \pi/2$ along C, $\sin\alpha = \cosh\alpha''$ varies from 1 to $+\infty$, and as α varies from $\pi/2$ to $\pi/2 + i\infty$ with $\alpha' = \pi/2$ along C', $\sin\alpha = \cosh\alpha''$ also varies from 1 to $+\infty$. Along the contour C, $\cos\alpha$ varies as follows:

$$\alpha'' = 0, \qquad \alpha' = 0 \to \frac{\pi}{2}: \qquad \cos\alpha = \cos\alpha' = 1 \to 0,$$

$$\alpha' = \frac{\pi}{2}, \qquad \alpha'' = 0 \to -\infty: \qquad \cos\alpha = -i\sinh\alpha'' = 0 \to +i\infty,$$

so that one must then choose the positive branch for γ when α varies along the contour C, viz.

$$\gamma = +k\cos\alpha; \quad \alpha \in C. \tag{7.120}$$

On the other hand, when α'' varies from 0 to $+\infty$ with $\alpha' = \pi/2$ along the contour C', $\cos\alpha = -i\sinh\alpha''$ varies from 0 to $-i\infty$ and one must then choose the negative branch for γ; that is, $\gamma = -k\cos\alpha$ when $\alpha \in C'$. For simplicity, the positive branch as specified in Eq. (7.120) is chosen here.

Consider now the case when $\tilde{k}(\omega)$ is complex-valued. In that case, let

$$\tilde{k}(\omega) \equiv k(\omega)e^{i\kappa(\omega)}, \tag{7.121}$$

where $k(\omega) \equiv |\tilde{k}(\omega)|$ is the magnitude and $\kappa(\omega) \equiv arg\{\tilde{k}(\omega)\}$ is the argument (or phase) of the complex wavenumber with $0 \le \kappa < \pi/2$, which will always be satisfied for an attenuative medium. Let $\alpha = \alpha' + i\alpha''$ and $\beta = \beta' + i\beta''$ with $\alpha' \equiv \Re\{\alpha\}$, $\alpha'' \equiv \Im\{\alpha\}$, $\beta' \equiv \Re\{\beta\}$, and $\beta'' \equiv \Im\{\beta\}$. With these substitutions, Eqs. (7.116) and (7.117) become

$$\frac{k_x}{k(\omega)} = e^{i\kappa(\omega)}\sin(\alpha' + i\alpha'')\cos(\beta' + i\beta'')$$

$$= (\cos\kappa + i\sin\kappa)(\sin\alpha'\cosh\alpha'' + i\cos\alpha'\sinh\alpha'')$$

$$\times (\cos\beta'\cosh\beta'' - i\sin\beta'\sinh\beta'')$$

$$= \left[(\sin\alpha'\cosh\alpha''\cos\beta'\cosh\beta'' + \cos\alpha'\sinh\alpha''\sin\beta'\sinh\beta'')\cos\kappa\right.$$

$$\left. - (\cos\alpha'\sinh\alpha''\cos\beta'\cosh\beta'' - \sin\alpha'\cosh\alpha''\sin\beta'\sinh\beta'')\sin\kappa\right]$$

$$+i\left[(\sin\alpha'\cosh\alpha''\cos\beta'\cosh\beta'' + \cos\alpha'\sinh\alpha''\sin\beta'\sinh\beta'')\sin\kappa\right.$$

$$\left. - (\cos\alpha'\sinh\alpha''\cos\beta'\cosh\beta'' - \sin\alpha'\cosh\alpha''\sin\beta'\sinh\beta'')\cos\kappa\right]$$

$$\tag{7.122}$$

and

$$\frac{k_y}{k(\omega)} = e^{i\kappa(\omega)}\sin(\alpha' + i\alpha'')\sin(\beta' + i\beta'')$$

$$= (\cos\kappa + i\sin\kappa)(\sin\alpha'\cosh\alpha'' + i\cos\alpha'\sinh\alpha'')$$

$$\times (\sin\beta'\cosh\beta'' + i\cos\beta'\sinh\beta'')$$

$$= \left[(\sin\alpha'\cosh\alpha''\sin\beta'\cosh\beta'' - \cos\alpha'\sinh\alpha''\cos\beta'\sinh\beta'')\cos\kappa\right.$$

$$\left. - (\sin\alpha'\cosh\alpha''\cos\beta'\sinh\beta'' + \cos\alpha'\sinh\alpha''\sin\beta'\cosh\beta'')\sin\kappa\right]$$

$$+i\left[(\sin\alpha'\cosh\alpha''\sin\beta'\cosh\beta'' - \cos\alpha'\sinh\alpha''\cos\beta'\sinh\beta'')\sin\kappa\right.$$

$$\left. + (\sin\alpha'\cosh\alpha''\cos\beta'\sinh\beta'' + \cos\alpha'\sinh\alpha''\sin\beta'\cosh\beta'')\cos\kappa\right], \tag{7.123}$$

respectively. Because the left-hand sides of both of these expressions are real-valued, then the imaginary parts appearing on the right-hand sides must both vanish and one then obtains the pair of expressions

$$\left(\sin\alpha'\cosh\alpha''\cos\beta'\cosh\beta'' + \cos\alpha'\sinh\alpha''\sin\beta'\sinh\beta''\right)\sin\kappa$$
$$-\left(\cos\alpha'\sinh\alpha''\cos\beta'\cosh\beta'' - \sin\alpha'\cosh\alpha''\sin\beta'\sinh\beta''\right)\cos\kappa = 0,$$
$$\left(\sin\alpha'\cosh\alpha''\sin\beta'\cosh\beta'' - \cos\alpha'\sinh\alpha''\cos\beta'\sinh\beta''\right)\sin\kappa$$
$$+\left(\sin\alpha'\cosh\alpha''\cos\beta'\sinh\beta'' + \cos\alpha'\sinh\alpha''\sin\beta'\cosh\beta''\right)\cos\kappa = 0.$$

Because $\sin\kappa$ and $\cos\kappa$ are known to exist, this pair of simultaneous linear equations in $\sin\kappa$ and $\cos\kappa$ must have a solution, which in turn requires that the determinant of its coefficients must vanish, which then results in the relation

$$\sinh\beta''\cosh\beta''\left(\sin^2\alpha'\cosh^2\alpha'' + \cos^2\alpha'\sinh^2\alpha''\right) = 0.$$

Because $\cosh\beta'' \neq 0$ for real β'' and $(\sin^2\alpha'\cosh^2\alpha'' + \cos^2\alpha'\sinh^2\alpha'') \neq 0$ for all real values of α' and α'', it then follows that $\sinh\beta'' = 0$ so that

$$\beta'' = 0 \qquad\qquad (7.124)$$

and $\beta = \beta'$ is real-valued. With this identification, the above pair of simultaneous linear equations becomes

$$\cos\beta\left(\sin\alpha'\cosh\alpha''\sin\kappa + \cos\alpha'\sinh\alpha''\cos\kappa\right) = 0,$$
$$\sin\beta\left(\sin\alpha'\cosh\alpha''\sin\kappa + \cos\alpha'\sinh\alpha''\cos\kappa\right) = 0,$$

and because $\cos\beta = \cos\beta'$ and $\sin\beta = \sin\beta'$ cannot both be equal to zero, one must then have that

$$\sin\alpha'\tan\kappa + \cos\alpha'\tanh\alpha'' = 0; \quad 0 \leq \kappa < \frac{\pi}{2}. \qquad (7.125)$$

Furthermore, Eqs. (7.122) and (7.123) become (with $\beta = \beta'$)

$$\frac{k_x}{k(\omega)} = \cos\beta\left(\sin\alpha'\cosh\alpha''\cos\kappa - \cos\alpha'\sinh\alpha''\sin\kappa\right), \qquad (7.126)$$

$$\frac{k_y}{k(\omega)} = \sin\beta\left(\sin\alpha'\cosh\alpha''\cos\kappa - \cos\alpha'\sinh\alpha''\sin\kappa\right), \qquad (7.127)$$

where α', α'', and $\beta = \beta'$ must vary in such a manner that k_x and k_y both vary continuously over the domain from $-\infty$ to $+\infty$.

Consider now the family of contours in the complex α-plane that are described by Eq. (7.125). In the limit as $\alpha'' \to +\infty$, $\tanh\alpha'' \to 1$ and this equation assumes the limiting form

$$\sin\alpha'\tan\kappa + \cos\alpha' = 0,$$

so that

$$\tan \alpha' = -\cot \kappa = \tan \left(\kappa - \frac{\pi}{2} \right),$$

and consequently

$$\alpha' \to \kappa - \frac{\pi}{2} (\text{mod } 2\pi) \quad \text{as} \quad \alpha'' \to +\infty. \tag{7.128}$$

On the other hand, in the limit as $\alpha'' \to -\infty$, $\tanh \alpha'' \to -1$ and Eq. (7.125) assumes the limiting form

$$\sin \alpha' \tan \kappa - \cos \alpha' = 0,$$

so that

$$\tan \alpha' = \cot \kappa = \tan \left(\frac{\pi}{2} - \kappa \right),$$

and consequently

$$\alpha' \to \frac{\pi}{2} - \kappa (\text{mod } 2\pi) \quad \text{as} \quad \alpha'' \to -\infty. \tag{7.129}$$

Furthermore, when $\alpha'' = 0$, $\tanh \alpha'' = 0$ and Eq. (7.125) becomes

$$\sin \alpha' \tan \kappa = 0,$$

and because $0 \leq \kappa < \pi/2$ for complex $\tilde{k}(\omega)$, then $\sin \alpha' = 0$ and

$$\alpha' = \pm m\pi; \quad m = 0, 1, 2, 3, \ldots \quad \text{when} \quad \alpha'' = 0. \tag{7.130}$$

Hence, each member C_m of the family of contours described by Eq. (7.125) begins at $\alpha' = m\pi - \pi/2 + \kappa$, $\alpha'' = +\infty$, passes across the α'-axis at $\alpha' = m\pi$, and ends at $\alpha' = m\pi + \pi/2 - \kappa$, $\alpha'' = -\infty$ for each positive and negative integer value of the index m, as illustrated in Fig. 7.4. The angle of slope as the curve crosses the α'-axis (i.e., at $\alpha'' = 0$) for each member C_m of this family of contours is obtained by differentiating Eq. (7.125) to obtain

$$\frac{d\alpha''}{d\alpha'} = \frac{\sin \alpha' \tanh \alpha'' - \tan \kappa \cos \alpha'}{\cos \alpha' \text{sech}^2 \alpha''},$$

so that

$$\tan^{-1} \left[\left(\frac{d\alpha''}{d\alpha'} \right)_{\alpha''=0} \right] = \tan^{-1} (-\tan \kappa) = -\kappa, \tag{7.131}$$

and the angle of slope is given by minus the phase of the complex wavenumber.

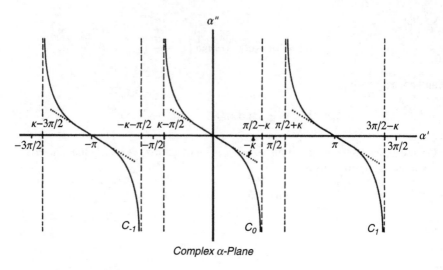

Fig. 7.4 Family of contours C_m in the complex α-plane

Fig. 7.5 The contour C in the complex α-plane when the wavenumber is, in general, complex-valued with $\tilde{k}(\omega) = k(\omega)e^{i\kappa(\omega)}$

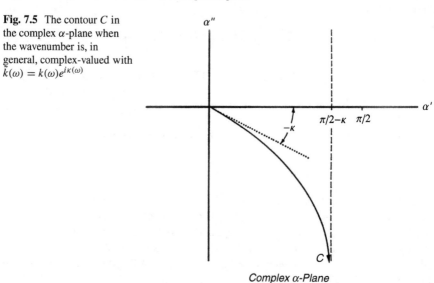

In the special case when $\kappa = 0$ and $\tilde{k}(\omega)$ is real-valued, Eq. (7.125) reduces to the equation $\cos\alpha' \sinh\alpha'' = 0$ that was obtained when the complex wave number was assumed to be real-valued [i.e., when $\tilde{k}(\omega) = k$ with real-valued k], and the lower-half of the contour C_0 reduces to the contour C that was obtained in that special case (see Fig. 7.3). Therefore, the general contour C to be employed here is given by the lower-half of C_0 and is directed from $\alpha = 0$ to $\alpha = \pi/2 - \kappa - i\infty$, as illustrated in Fig. 7.5. Along this contour the real and imaginary parts of $\alpha = \alpha' + i\alpha''$ vary as

$$\alpha' = 0 \to \frac{\pi}{2} - \kappa \Rightarrow \begin{cases} \sin \alpha' = 0 \to \sin(\pi/2 - \kappa) = \cos \kappa \geq 0, \\ \cos \alpha' = 1 \to \cos(\pi/2 - \kappa) = \sin \kappa \geq 0, \end{cases}$$

$$\alpha'' = 0 \to -\infty \quad \Rightarrow \begin{cases} \sinh \alpha'' = 0 \to -\infty, \\ \cosh \alpha' = 1 \to +\infty, \end{cases}$$

so that k_x and k_y vary along the contour C as

$$\frac{k_x}{k(\omega)} = A \cos \beta, \quad \frac{k_y}{k(\omega)} = A \sin \beta,$$

where A varies from 0 to $+\infty$. Thus, with $\alpha = \alpha' + i\alpha''$ varying along the contour C and with $\beta = \beta'$ varying between 0 and 2π, both k_x and k_y vary continuously from $-\infty$ to $+\infty$ and are real-valued, as desired. Furthermore, along the contour C, $\cos \alpha' \geq 0$, $\sin \alpha' \geq 0$, $\cosh \alpha'' \geq 0$, and $\sinh \alpha' \leq 0$, so that

$$\cos \alpha = \cos \alpha' \cosh \alpha'' + i \left| \sin \alpha' \sinh \alpha'' \right| \tag{7.132}$$

and the phase of $\cos \alpha$ is always between 0 and π. Once again, the positive branch of γ must be taken when α varies along C [see Eq. (7.120)], so that

$$\gamma(\omega) = +\tilde{k}(\omega) \cos \alpha; \quad \alpha \in C. \tag{7.133}$$

Hence, under the change of variable to polar coordinates given in Eqs. (7.116) and (7.117), the angular spectrum integral given in Eq. (7.111) becomes

$$u(\mathbf{r}) = \tilde{k}(\omega) \int_0^{2\pi} d\beta \int_C \sin \alpha \, d\alpha \, U(k_x, k_y) e^{i\tilde{\mathbf{k}}^\pm \cdot \mathbf{r}}, \tag{7.134}$$

where

$$\tilde{\mathbf{k}}^\pm(\omega) = \tilde{k}(\omega) \left[\hat{\mathbf{1}}_x \sin \alpha \cos \beta + \hat{\mathbf{1}}_y \sin \alpha \sin \beta \pm \hat{\mathbf{1}}_y \cos \alpha \right] \tag{7.135}$$

when the polar axis has been chosen to be along the k_z-axis.

7.4.1 Transformation to an Arbitrary Polar Axis

It is evident from a physical point of view that the choice of the polar axis to be along the k_z-axis is simply a matter of convenience that is unnecessary and the expression in Eq. (7.134) for the angular spectrum integral should be independent of this choice. The rigorous mathematical proof of the simple observation is, however, quite another matter. To that end, consider choosing some other polar axis that is

Fig. 7.6 Relation of the polar axis to the k_x, k_y, k_z coordinate axes when the wave vector **k** is real-valued

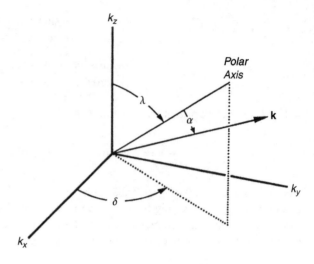

specified relative to the k_x, k_y, k_z coordinate axes by the (arbitrarily chosen) angles λ and δ, where λ is the angle of declination from the k_z-axis and δ is the azimuthal angle of the normal projection of the polar axis onto the $k_x k_y$-plane, measured from the k_x-axis, as illustrated in Fig. 7.6 (the wave vector **k** depicted here is real-valued strictly for the purpose of illustration). Under this transformation to a new polar axis, the expressions in Eqs. (7.116) and (7.117) become [retaining the complex variables α and β as the new angular coordinates for the complex wave vector $\tilde{\mathbf{k}}(\omega)$]

$$k_x = \tilde{k}(\omega)\left(\cos\alpha \sin\lambda \cos\delta \pm \sin\alpha \cos\beta \cos\lambda \cos\delta - \sin\alpha \sin\beta \sin\delta \right),$$

$$(7.136)$$

$$k_y = \tilde{k}(\omega)\left(\cos\alpha \sin\lambda \sin\delta \pm \sin\alpha \cos\beta \cos\lambda \sin\delta + \sin\alpha \sin\beta \cos\delta \right),$$

$$(7.137)$$

where the $\pm$ sign appearing in these two expressions corresponds to the $\pm$ sign appearing in $\tilde{\mathbf{k}}^{\pm}$ and takes into account the fact that for $z < -Z$, the polar axis is chosen in the opposite direction from that for $z > Z$, in which case $\lambda \rightarrow \pi - \lambda$, as illustrated in Fig. 7.7 (remembering that the polar axis is no longer fixed as it was in the preceding analysis).

For the Jacobian $\partial(k_x, k_y)/\partial(\alpha, \beta)$ of this transformation, the required partial differentials are given by

$$\frac{1}{\tilde{k}(\omega)} \frac{\partial k_x}{\partial \alpha} = -\sin\alpha \sin\lambda \cos\delta \pm \cos\alpha \cos\beta \cos\lambda \cos\delta - \cos\alpha \sin\beta \sin\delta,$$

$$\frac{1}{\tilde{k}(\omega)} \frac{\partial k_y}{\partial \beta} = \mp \sin\alpha \sin\beta \cos\lambda \sin\delta + \sin\alpha \cos\beta \cos\delta,$$

Fig. 7.7 Polar axis choices
when either $z > Z$ or $z < -Z$

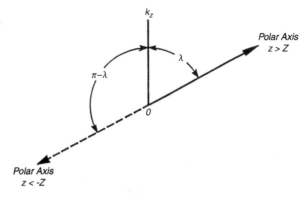

$$\frac{1}{\tilde{k}(\omega)}\frac{\partial k_y}{\partial \alpha} = -\sin\alpha \sin\lambda \sin\delta \pm \cos\alpha \cos\beta \cos\lambda \sin\delta + \cos\alpha \sin\beta \cos\delta,$$

$$\frac{1}{\tilde{k}(\omega)}\frac{\partial k_x}{\partial \beta} = \mp \sin\alpha \sin\beta \cos\lambda \cos\delta - \sin\alpha \cos\beta \sin\delta,$$

so that

$$\frac{1}{\tilde{k}^2(\omega)}\frac{\partial(k_x,k_y)}{\partial(\alpha,\beta)} = \begin{vmatrix} \tilde{k}^{-1}\partial k_x/\partial\alpha & \tilde{k}^{-1}\partial k_y/\partial\alpha \\ \tilde{k}^{-1}\partial k_x/\partial\beta & \tilde{k}^{-1}\partial k_y/\partial\beta \end{vmatrix}$$

$$= -\sin^2\alpha \cos\beta \sin\lambda \pm \sin\alpha \cos\alpha \cos\lambda. \qquad (7.138)$$

In addition, under the coordinate transformation given in Eqs. (7.136) and (7.137) one has that

$$1 - \frac{\gamma^2(\omega)}{\tilde{k}^2(\omega)} = \frac{k_x^2}{\tilde{k}^2(\omega)} + \frac{k_y^2}{\tilde{k}^2(\omega)}$$

$$= \left(\cos\alpha \sin\lambda \cos\delta \pm \sin\alpha \cos\beta \cos\lambda \cos\delta - \sin\alpha \sin\beta \sin\delta\right)^2$$

$$+ \left(\cos\alpha \sin\lambda \sin\delta \pm \sin\alpha \cos\beta \cos\lambda \sin\delta + \sin\alpha \sin\beta \cos\delta\right)^2$$

$$= \cos^2\alpha \sin^2\lambda + \sin^2\alpha \cos^2\beta \cos^2\lambda + \sin^2\alpha \sin^2\beta$$

$$\pm 2\sin\alpha \cos\alpha \cos\beta \sin\lambda \cos\lambda$$

$$= \cos^2\alpha - \cos^2\alpha \cos^2\lambda + \sin^2\alpha \cos^2\beta - \sin^2\alpha \cos^2\beta \sin^2\lambda$$

$$+ \sin^2\alpha \sin^2\beta \pm \pm 2\sin\alpha \cos\alpha \cos\beta \sin\lambda \cos\lambda$$

so that

$$1 - \frac{\gamma^2(\omega)}{\tilde{k}^2(\omega)} = 1 - \left(\cos\alpha \cos\lambda \mp \sin\alpha \cos\beta \sin\lambda\right)^2,$$

and hence

$$\gamma(\omega) = \pm \tilde{k}(\omega)\left(\cos\alpha\cos\lambda \mp \sin\alpha\cos\beta\sin\lambda\right).$$

When $\lambda = 0$ one must have $\gamma(\omega) = +\tilde{k}(\omega)\cos\alpha$ in accordance with Eq. (6.133), so that one finally obtains

$$\gamma(\omega) = \tilde{k}(\omega)\left(\cos\alpha\cos\lambda \mp \sin\alpha\cos\beta\sin\lambda\right) \tag{7.139}$$

and Eq. (7.138) then yields

$$\frac{\partial(k_x, k_y)}{\partial(\alpha, \beta)} = \pm \tilde{k}(\omega)\gamma(\omega)\sin\alpha \tag{7.140}$$

for the Jacobian of the transformation given in Eqs. (7.136) and (7.137), where the sign choice corresponds to the sign choice appearing in $\mathbf{k}^{\pm}$.

Hence, under the change of variable (7.136) and (7.137) to polar coordinates with an arbitrarily specified polar axis direction, the angular spectrum integral given in Eq. (7.111) becomes

$$u(\mathbf{r}) = \pm \tilde{k}(\omega) \int_{D(\alpha,\beta)} U(k_x, k_y) e^{i\tilde{\mathbf{k}}^{\pm}\cdot\mathbf{r}} \sin\alpha \, d\alpha d\beta, \tag{7.141}$$

where the polar coordinate variables $\alpha = \alpha' + i\alpha''$ and $\beta = \beta' + i\beta''$ must now both be complex-valued, in general, and where the domain of integration $D(\alpha, \beta)$ is some two-dimensional surface in the four-dimensional $\alpha', \alpha'', \beta', \beta''$-space. The determination of this integration domain in the general case is a very difficult problem. Nevertheless, it is shown in Sect. 7.4.2 that, given certain restrictions on the analyticity of the spectral function $U(k_x, k_y)$, the angular domain of integration $D(\alpha, \beta)$ can be taken to be the same domain of integration as that appearing in Eq. (7.134). Subject to these conditions on the analyticity of $U(k_x, k_y)$, the angular spectrum integral given in Eq. (7.141) becomes

$$u(\mathbf{r}) = \pm \tilde{k}(\omega) \int_0^{2\pi} d\beta \int_C \sin\alpha \, d\alpha \, U(k_x, k_y) e^{i\tilde{\mathbf{k}}^{\pm}\cdot\mathbf{r}}, \tag{7.142}$$

where

$$\begin{aligned}
\tilde{\mathbf{k}}^{\pm}(\omega) = \tilde{k}(\omega)\Big[&\hat{\mathbf{1}}_x\left(\cos\alpha\sin\lambda\cos\delta \pm \sin\alpha\cos\beta\cos\lambda\cos\delta - \sin\alpha\sin\beta\sin\delta\right) \\
&+ \hat{\mathbf{1}}_y\left(\cos\alpha\sin\lambda\sin\delta \pm \sin\alpha\cos\beta\cos\lambda\sin\delta + \sin\alpha\sin\beta\cos\delta\right) \\
&\pm \hat{\mathbf{1}}_z\left(\cos\alpha\cos\lambda \mp \sin\alpha\cos\beta\sin\lambda\right)\Big],
\end{aligned} \tag{7.143}$$

and where the contour C is that branch of the family of contours in the complex $\alpha = \alpha' + i\alpha''$ plane described by Eq. (7.125) that begins at $\alpha = 0$ and extends to $\alpha = \pi/2 - \kappa - i\infty$, where $\kappa \equiv \arg\left\{\tilde{k}(\omega)\right\}, 0 \leq \kappa < \pi/2$, as depicted in Fig. 7.5.

7.4.2   Weyl's Proof

The proof that the angular spectrum integral appearing in Eq. (7.141) is independent of the transformation angles λ and δ, and, subject to certain conditions on the analyticity of the spectral function $U(k_x, k_y)$, reduces to the expression given in Eq. (7.142) is now given based upon the classic proof by Weyl [8] in 1919. This important proof begins with the spatial angular spectrum integral given in Eq. (7.111) under the general coordinate transformation given in Eqs. (7.136) and (7.137), which may be written in the general form

$$I(\mathbf{r}, \lambda, \delta) \equiv \int_0^{2\pi} d\beta \int_C d\alpha \, \frac{U(k_x, k_y)}{\gamma} e^{i\tilde{\mathbf{k}}^{\pm}\cdot\mathbf{r}} \frac{\partial(k_x, k_y)}{\partial(\alpha, \beta)}, \qquad (7.144)$$

where C is the contour extending from $\alpha = 0$ to $\alpha = \pi/2 - \kappa - i\infty$ that is described by Eq. (7.125). This particular form of the angular spectrum integral takes into account the polar coordinate form of the domain of integration for k_x and k_y as expressed in Eq. (7.134) when the polar axis is along k_z, which is then modified by the transformation to an arbitrary polar axis direction specified by the transformation angles λ and δ through the Jacobian

$$J(k_x, k_y; \alpha, \beta) \equiv \frac{\partial(k_x, k_y)}{\partial(\alpha, \beta)}. \qquad (7.145)$$

The proof then entails showing that $I(\mathbf{r}, \lambda, \delta)$ is independent of the parameters λ and δ by determining the partial derivatives $\partial I/\partial\lambda$ and $\partial I/\partial\delta$ and showing that they both vanish.

From Eq. (7.115), the spectral amplitude function appearing in Eq. (7.144) is given by

$$U(k_x, k_y) = f(k_x, k_y, \gamma), \qquad (7.146)$$

where

$$f(k_x, k_y, k_z) = \int_{-\infty}^{\infty} f(x, y, z) e^{-i(k_x x + k_y y + k_z z)} dx \, dy \, dz \qquad (7.147)$$

is an entire function of complex k_x, k_y, k_z. From the discussion associated with Eq. (7.83), it is seen that the parameters λ and δ, through their specification of the polar axis direction, determine the half-space $|z| > Z$ wherein the integral appearing

Fig. 7.8 Two possible
positive half-spaces and their
associated polar axis
directions

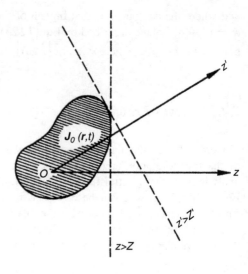

in Eq. (7.147) converges, because the polar axis direction specifies the orientation of the z-axis. Two such possible half-spaces and their associated z-axis directions are depicted in Fig. 7.8 for a current source with compact spatial support.

Define the quantity

$$\psi \equiv \psi(k_x, k_y, k_z) \equiv \frac{1}{k_z} f(k_x, k_y, k_z) e^{i\mathbf{k}\cdot\mathbf{r}}, \qquad (7.148)$$

and let

$$\tilde{\psi} \equiv \psi(k_x, k_y, \gamma). \qquad (7.149)$$

Then the function $\tilde{\psi}$ has an integrable singularity at $\gamma = 0$ and $\tilde{\psi}$ vanishes at $\alpha = \pi/2 - \kappa - i\infty$. With this substitution, the integral $I(\mathbf{r}, \lambda, \delta)$ defined in Eq. (7.144) becomes

$$I(\mathbf{r}, \lambda, \delta) \equiv \int_0^{2\pi} d\beta \int_C d\alpha \, \tilde{\psi} \, J(k_x, k_y; \alpha, \beta). \qquad (7.150)$$

Consider the dependence of $I(\mathbf{r}, \lambda, \delta)$ on the parameter λ. Upon differentiating Eq. (7.150) with respect to λ and assuming that the orders of differentiation and integration can be interchanged, one obtains

$$\frac{\partial I}{\partial \lambda} = \int_0^{2\pi} d\beta \int_C d\alpha \left(\frac{\partial \tilde{\psi}}{\partial \lambda} J + \tilde{\psi} \frac{\partial J}{\partial \lambda} \right) \equiv I_1 + I_2, \qquad (7.151)$$

where

$$I_1 \equiv \int_0^{2\pi} d\beta \int_C d\alpha \, \tilde{\psi} \frac{\partial J}{\partial \lambda}, \tag{7.152}$$

$$I_2 \equiv \int_0^{2\pi} d\beta \int_C d\alpha \, \frac{\partial \tilde{\psi}}{\partial \lambda} J. \tag{7.153}$$

Attention is given first to the integral I_1. The Jacobian appearing here is given by

$$J(k_x, k_y; \alpha, \beta) = \frac{\partial k_x}{\partial \alpha} \frac{\partial k_y}{\partial \beta} - \frac{\partial k_y}{\partial \alpha} \frac{\partial k_x}{\partial \beta},$$

so that

$$\frac{\partial J}{\partial \lambda} = \left(\frac{\partial}{\partial \alpha} \frac{\partial k_x}{\partial \lambda} \right) \frac{\partial k_y}{\partial \beta} + \left(\frac{\partial}{\partial \beta} \frac{\partial k_y}{\partial \lambda} \right) \frac{\partial k_x}{\partial \alpha}$$

$$- \left(\frac{\partial}{\partial \alpha} \frac{\partial k_y}{\partial \lambda} \right) \frac{\partial k_x}{\partial \beta} - \left(\frac{\partial}{\partial \beta} \frac{\partial k_x}{\partial \lambda} \right) \frac{\partial k_y}{\partial \alpha},$$

where it has been assumed that the orders of differentiation can be interchanged. With this substitution, the integral I_1 becomes

$$I_1 = \int_0^{2\pi} d\beta \int_C d\alpha \, \tilde{\psi} \left[\left(\frac{\partial}{\partial \alpha} \frac{\partial k_x}{\partial \lambda} \right) \frac{\partial k_y}{\partial \beta} + \left(\frac{\partial}{\partial \beta} \frac{\partial k_y}{\partial \lambda} \right) \frac{\partial k_x}{\partial \alpha} \right.$$

$$\left. - \left(\frac{\partial}{\partial \alpha} \frac{\partial k_y}{\partial \lambda} \right) \frac{\partial k_x}{\partial \beta} - \left(\frac{\partial}{\partial \beta} \frac{\partial k_x}{\partial \lambda} \right) \frac{\partial k_y}{\partial \alpha} \right]$$

$$= \int_0^{2\pi} d\beta \left\{ \left[\tilde{\psi} \frac{\partial k_x}{\partial \lambda} \frac{\partial k_y}{\partial \beta} - \tilde{\psi} \frac{\partial k_y}{\partial \lambda} \frac{\partial k_x}{\partial \beta} \right]_0^{\pi/2 - \kappa - i\infty} \right.$$

$$\left. - \int_C d\alpha \left[\frac{\partial k_x}{\partial \lambda} \frac{\partial}{\partial \alpha} \left(\tilde{\psi} \frac{\partial k_y}{\partial \beta} \right) - \frac{\partial k_y}{\partial \lambda} \frac{\partial}{\partial \alpha} \left(\tilde{\psi} \frac{\partial k_x}{\partial \beta} \right) \right] \right\}$$

$$+ \int_C d\alpha \left\{ \left[\tilde{\psi} \frac{\partial k_y}{\partial \lambda} \frac{\partial k_x}{\partial \alpha} - \tilde{\psi} \frac{\partial k_x}{\partial \lambda} \frac{\partial k_y}{\partial \alpha} \right]_0^{2\pi} \right.$$

$$\left. - \int_0^{2\pi} d\beta \left[\frac{\partial k_y}{\partial \lambda} \frac{\partial}{\partial \beta} \left(\tilde{\psi} \frac{\partial k_x}{\partial \alpha} \right) - \frac{\partial k_x}{\partial \lambda} \frac{\partial}{\partial \beta} \left(\tilde{\psi} \frac{\partial k_y}{\partial \alpha} \right) \right] \right\} \tag{7.154}$$

upon integration by parts. In order to proceed, special consideration must be allowed for the point $\alpha = \pi/2$ on the contour C in the special case when $\tilde{k}(\omega)$ is real-valued (i.e., when $\kappa = 0$).

As illustrated in Fig. 7.9, in this special limiting case the contour C must be separated into the disjoint contour C_ϵ which excludes an infinitesimally small ϵ-neighborhood about the point $\alpha = \pi/2$ where the slope of the contour is

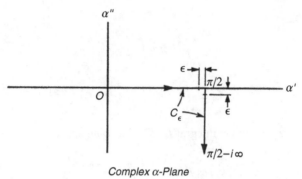

Fig. 7.9 The contour C_ϵ in the complex α-plane in the special case when the wave number $\tilde{k}(\omega)$ is real-valued

discontinuous. Of course, when $\tilde{k}(\omega)$ is complex-valued (i.e., when $0 < \kappa < \pi/2$) the slope of the contour C is continuous along its entirety and this separation is unnecessary. In order to include this special case, Eq. (7.154) must be expanded as

$$
I_1 = \lim_{\epsilon \to 0} \int_0^{2\pi} d\beta \left\{ \left[\tilde{\psi} \frac{\partial k_x}{\partial \lambda} \frac{\partial k_y}{\partial \beta} - \tilde{\psi} \frac{\partial k_y}{\partial \lambda} \frac{\partial k_x}{\partial \beta} \right]_0^{\pi/2-\epsilon} \right.
$$
$$
+ \left[\tilde{\psi} \frac{\partial k_x}{\partial \lambda} \frac{\partial k_y}{\partial \beta} - \tilde{\psi} \frac{\partial k_y}{\partial \lambda} \frac{\partial k_x}{\partial \beta} \right]_{\pi/2-i\epsilon}^{\pi/2-i\infty}
$$
$$
- \int_C d\alpha \left[\frac{\partial k_x}{\partial \lambda} \frac{\partial}{\partial \alpha} \left(\tilde{\psi} \frac{\partial k_y}{\partial \beta} \right) - \frac{\partial k_y}{\partial \lambda} \frac{\partial}{\partial \alpha} \left(\tilde{\psi} \frac{\partial k_x}{\partial \beta} \right) \right] \right\}
$$
$$
+ \lim_{\epsilon \to 0} \int_{C_\epsilon} d\alpha \left\{ \left[\tilde{\psi} \frac{\partial k_y}{\partial \lambda} \frac{\partial k_x}{\partial \alpha} - \tilde{\psi} \frac{\partial k_x}{\partial \lambda} \frac{\partial k_y}{\partial \alpha} \right]_0^{2\pi} \right.
$$
$$
\left. - \int_0^{2\pi} d\beta \left[\frac{\partial k_y}{\partial \lambda} \frac{\partial}{\partial \beta} \left(\tilde{\psi} \frac{\partial k_x}{\partial \alpha} \right) - \frac{\partial k_x}{\partial \lambda} \frac{\partial}{\partial \beta} \left(\tilde{\psi} \frac{\partial k_y}{\partial \alpha} \right) \right] \right\} \quad (7.155)
$$

when $\kappa = 0$.

Consider first the "surface terms" appearing in Eq. (7.155). First of all, because the functions $\tilde{\psi}$, k_x, and k_y are all periodic in β with period 2π, then

$$
\left[\tilde{\psi} \frac{\partial k_y}{\partial \lambda} \frac{\partial k_x}{\partial \alpha} - \tilde{\psi} \frac{\partial k_x}{\partial \lambda} \frac{\partial k_y}{\partial \alpha} \right]_0^{2\pi} = 0, \quad (7.156)
$$

independent of the contour C. Furthermore, from Eq. (7.143) one finds that

$$
\frac{\partial k_x}{\partial \beta} = \tilde{k}(\omega) \sin\alpha \left(\mp \sin\beta \cos\lambda \cos\delta - \cos\beta \sin\delta \right),
$$
$$
\frac{\partial k_y}{\partial \beta} = \tilde{k}(\omega) \sin\alpha \left(\mp \sin\beta \cos\lambda \sin\delta + \cos\beta \cos\delta \right),
$$

so that at $\alpha = 0$,

$$\left.\frac{\partial k_x}{\partial \beta}\right|_{\alpha=0} = \left.\frac{\partial k_y}{\partial \beta}\right|_{\alpha=0} = 0,$$

and because $\tilde{\psi}$ vanishes at $\alpha = \pi/2 - \kappa - i\infty$ for $0 \le \kappa < \pi/2$, then when $\kappa \ne 0$ one finds

$$\left[\tilde{\psi}\frac{\partial k_x}{\partial \lambda}\frac{\partial k_y}{\partial \beta} - \tilde{\psi}\frac{\partial k_y}{\partial \lambda}\frac{\partial k_x}{\partial \beta}\right]_0^{\pi/2-\kappa-i\infty} = 0, \tag{7.157}$$

whereas for $\kappa = 0$ one finds

$$\lim_{\epsilon \to 0}\left\{\left[\tilde{\psi}\frac{\partial k_x}{\partial \lambda}\frac{\partial k_y}{\partial \beta} - \tilde{\psi}\frac{\partial k_y}{\partial \lambda}\frac{\partial k_x}{\partial \beta}\right]_0^{\pi/2-\epsilon} + \left[\tilde{\psi}\frac{\partial k_x}{\partial \lambda}\frac{\partial k_y}{\partial \beta} - \tilde{\psi}\frac{\partial k_y}{\partial \lambda}\frac{\partial k_x}{\partial \beta}\right]_{\pi/2-i\epsilon}^{\pi/2-i\infty}\right\}$$

$$= \lim_{\epsilon \to 0}\left[\tilde{\psi}\frac{\partial k_x}{\partial \lambda}\frac{\partial k_y}{\partial \beta} - \tilde{\psi}\frac{\partial k_y}{\partial \lambda}\frac{\partial k_x}{\partial \beta}\right]_{\alpha=\pi/2-i\epsilon}^{\alpha=\pi/2-\epsilon}. \tag{7.158}$$

This "surface term" will also vanish provided that the quantity to be evaluated is continuous at $\epsilon = 0$, because then the limit as $\epsilon \to 0$ can be taken by setting $\epsilon = 0$. From Eqs. (7.139) and (7.143)

$$\frac{\partial k_x}{\partial \lambda} = \tilde{k}(\omega)\left(\cos\alpha\cos\lambda \mp \sin\alpha\cos\beta\sin\lambda\right)\cos\delta$$

$$= \gamma(\omega)\cos\delta,$$

$$\frac{\partial k_y}{\partial \lambda} = \tilde{k}(\omega)\left(\cos\alpha\cos\lambda \mp \sin\alpha\cos\beta\sin\lambda\right)\sin\delta$$

$$= \gamma(\omega)\sin\delta,$$

and because $\tilde{\psi} = (1/\gamma)f(k_x, k_y, \gamma)e^{i\tilde{\mathbf{k}}^\pm \cdot \mathbf{r}}$, then the final quantity to be evaluated in Eq. (7.158) is indeed continuous at $\epsilon = 0$ and this "surface term" vanishes. Thus, all of the "surface terms" in both Eqs. (7.154) and (7.155) vanish, and these expressions can both be written as

$$I_1 = -\lim_{\epsilon \to 0}\int_0^{2\pi}d\beta\int_C d\alpha\left\{\frac{\partial\tilde{\psi}}{\partial\alpha}\frac{\partial k_x}{\partial\lambda}\frac{\partial k_y}{\partial\beta} + \tilde{\psi}\frac{\partial k_x}{\partial\lambda}\frac{\partial^2 k_y}{\partial\alpha\partial\beta} - \frac{\partial\tilde{\psi}}{\partial\alpha}\frac{\partial k_x}{\partial\beta}\frac{\partial k_y}{\partial\lambda}\right.$$

$$-\tilde{\psi}\frac{\partial^2 k_x}{\partial\alpha\partial\beta}\frac{\partial k_y}{\partial\lambda} + \frac{\partial\tilde{\psi}}{\partial\beta}\frac{\partial k_x}{\partial\alpha}\frac{\partial k_y}{\partial\lambda} + \tilde{\psi}\frac{\partial^2 k_x}{\partial\beta\partial\alpha}\frac{\partial k_y}{\partial\lambda}$$

$$\left.-\frac{\partial\tilde{\psi}}{\partial\beta}\frac{\partial k_x}{\partial\lambda}\frac{\partial k_y}{\partial\alpha} - \tilde{\psi}\frac{\partial k_x}{\partial\lambda}\frac{\partial^2 k_y}{\partial\beta\partial\alpha}\right\}$$

$$= -\lim_{\epsilon \to 0} \int_0^{2\pi} d\beta \int_C d\alpha \left\{ \frac{\partial \tilde{\psi}}{\partial \alpha} \left(\frac{\partial k_x}{\partial \lambda} \frac{\partial k_y}{\partial \beta} - \frac{\partial k_x}{\partial \beta} \frac{\partial k_y}{\partial \lambda} \right) \right.$$
$$\left. + \frac{\partial \tilde{\psi}}{\partial \beta} \left(\frac{\partial k_x}{\partial \alpha} \frac{\partial k_y}{\partial \lambda} - \frac{\partial k_x}{\partial \lambda} \frac{\partial k_y}{\partial \alpha} \right) \right\}, \qquad (7.159)$$

where $C = C_\epsilon$ when $\kappa = 0$, and where the limiting procedure is ignored when $0 < \kappa < \pi/2$.

Because $\psi(k_x, k_y, k_z)$ has an integrable singularity at $k_z = 0$, but is otherwise an analytic function of complex k_z, and is an entire function of complex k_x and k_y, so that $\tilde{\psi} \equiv \psi(k_x, k_y, \gamma)$ is an analytic function on the contour C with an integrable singularity at $\gamma = 0$ (where it also has a branch point),

$$\frac{\partial \tilde{\psi}}{\partial \alpha} = \left[\frac{\partial \psi}{\partial k_x} \frac{\partial k_x}{\partial \alpha} + \frac{\partial \psi}{\partial k_y} \frac{\partial k_y}{\partial \alpha} + \frac{\partial \psi}{\partial k_z} \frac{\partial k_z}{\partial \alpha} \right]_{k_z = \gamma}$$
$$= \left[\frac{\partial \psi}{\partial k_x} \frac{\partial k_x}{\partial \alpha} + \frac{\partial \psi}{\partial k_y} \frac{\partial k_y}{\partial \alpha} + \frac{\partial \psi}{\partial k_z} \frac{\partial \gamma}{\partial \alpha} \right]_{k_z = \gamma}, \qquad (7.160)$$

$$\frac{\partial \tilde{\psi}}{\partial \beta} = \left[\frac{\partial \psi}{\partial k_x} \frac{\partial k_x}{\partial \beta} + \frac{\partial \psi}{\partial k_y} \frac{\partial k_y}{\partial \beta} + \frac{\partial \psi}{\partial k_z} \frac{\partial k_z}{\partial \beta} \right]_{k_z = \gamma}$$
$$= \left[\frac{\partial \psi}{\partial k_x} \frac{\partial k_x}{\partial \beta} + \frac{\partial \psi}{\partial k_y} \frac{\partial k_y}{\partial \beta} + \frac{\partial \psi}{\partial k_z} \frac{\partial \gamma}{\partial \beta} \right]_{k_z = \gamma}, \qquad (7.161)$$

because γ depends upon the parameters α and β in precisely the same manner that k_z does along the contour C. With these substitutions, Eq. (7.159) becomes

$$I_1 = -\lim_{\epsilon \to 0} \int_0^{2\pi} d\beta \int_C d\alpha \left\{ \frac{\partial \psi}{\partial k_x} \frac{\partial k_x}{\partial \alpha} \frac{\partial k_x}{\partial \lambda} \frac{\partial k_y}{\partial \beta} + \frac{\partial \psi}{\partial k_y} \frac{\partial k_x}{\partial \lambda} \frac{\partial k_y}{\partial \alpha} \frac{\partial k_y}{\partial \beta} \right.$$
$$+ \frac{\partial \psi}{\partial k_z} \frac{\partial k_x}{\partial \lambda} \frac{\partial k_y}{\partial \beta} \frac{\partial \gamma}{\partial \alpha} - \frac{\partial \psi}{\partial k_x} \frac{\partial k_x}{\partial \alpha} \frac{\partial k_x}{\partial \beta} \frac{\partial k_y}{\partial \lambda}$$
$$- \frac{\partial \psi}{\partial k_y} \frac{\partial k_x}{\partial \beta} \frac{\partial k_y}{\partial \alpha} \frac{\partial k_y}{\partial \lambda} - \frac{\partial \psi}{\partial k_z} \frac{\partial k_x}{\partial \beta} \frac{\partial k_y}{\partial \lambda} \frac{\partial \gamma}{\partial \alpha}$$
$$+ \frac{\partial \psi}{\partial k_x} \frac{\partial k_x}{\partial \beta} \frac{\partial k_x}{\partial \alpha} \frac{\partial k_y}{\partial \lambda} + \frac{\partial \psi}{\partial k_y} \frac{\partial k_x}{\partial \alpha} \frac{\partial k_y}{\partial \beta} \frac{\partial k_y}{\partial \lambda}$$
$$+ \frac{\partial \psi}{\partial k_z} \frac{\partial k_x}{\partial \alpha} \frac{\partial k_y}{\partial \lambda} \frac{\partial \gamma}{\partial \beta} - \frac{\partial \psi}{\partial k_x} \frac{\partial k_x}{\partial \beta} \frac{\partial k_x}{\partial \lambda} \frac{\partial k_y}{\partial \alpha}$$
$$\left. - \frac{\partial \psi}{\partial k_y} \frac{\partial k_x}{\partial \lambda} \frac{\partial k_y}{\partial \beta} \frac{\partial k_y}{\partial \alpha} - \frac{\partial \psi}{\partial k_z} \frac{\partial k_x}{\partial \lambda} \frac{\partial k_y}{\partial \alpha} \frac{\partial \gamma}{\partial \beta} \right\}_{k_z = \gamma}$$

$$I_1 = -\lim_{\epsilon \to 0} \int_0^{2\pi} d\beta \int_C d\alpha \left\{ \frac{\partial \psi}{\partial k_x} \frac{\partial k_x}{\partial \lambda} \left(\frac{\partial k_x}{\partial \alpha} \frac{\partial k_y}{\partial \beta} - \frac{\partial k_x}{\partial \beta} \frac{\partial k_y}{\partial \alpha} \right) \right.$$

$$+ \frac{\partial \psi}{\partial k_y} \frac{\partial k_y}{\partial \lambda} \left(\frac{\partial k_x}{\partial \alpha} \frac{\partial k_y}{\partial \beta} - \frac{\partial k_x}{\partial \beta} \frac{\partial k_y}{\partial \alpha} \right)$$

$$+ \frac{\partial \psi}{\partial k_z} \left[\frac{\partial k_x}{\partial \lambda} \left(\frac{\partial k_y}{\partial \beta} \frac{\partial \gamma}{\partial \alpha} - \frac{\partial k_y}{\partial \alpha} \frac{\partial \gamma}{\partial \beta} \right) \right.$$

$$\left. \left. + \frac{\partial k_y}{\partial \lambda} \left(\frac{\partial k_x}{\partial \alpha} \frac{\partial \gamma}{\partial \beta} - \frac{\partial k_x}{\partial \beta} \frac{\partial \gamma}{\partial \alpha} \right) \right] \right\}_{k_z = \gamma}$$

$$= -\lim_{\epsilon \to 0} \int_0^{2\pi} d\beta \int_C d\alpha \left\{ \left(\frac{\partial \psi}{\partial k_x} \frac{\partial k_x}{\partial \lambda} + \frac{\partial \psi}{\partial k_y} \frac{\partial k_y}{\partial \lambda} \right) J(k_x, k_y; \alpha, \beta) \right.$$

$$+ \frac{\partial \psi}{\partial k_z} \left[\frac{\partial k_x}{\partial \lambda} \left(\frac{\partial k_y}{\partial \beta} \frac{\partial \gamma}{\partial \alpha} - \frac{\partial k_y}{\partial \alpha} \frac{\partial \gamma}{\partial \beta} \right) \right.$$

$$\left. \left. + \frac{\partial k_y}{\partial \lambda} \left(\frac{\partial k_x}{\partial \alpha} \frac{\partial \gamma}{\partial \beta} - \frac{\partial k_x}{\partial \beta} \frac{\partial \gamma}{\partial \alpha} \right) \right] \right\}_{k_z = \gamma}.$$

$$(7.162)$$

The last two terms in this expression are evaluated in the following manner. Differentiation of the expression $\gamma^2(\omega) = \tilde{k}^2(\omega) - k_x^2 - k_y^2$ with ω held fixed yields $\gamma d\gamma = -k_x dk_x - k_y dk_y$, so that

$$\frac{\partial \gamma}{\partial \alpha} = -\frac{1}{\gamma} \left(k_x \frac{\partial k_x}{\partial \alpha} + k_y \frac{\partial k_y}{\partial \alpha} \right),$$

$$\frac{\partial \gamma}{\partial \beta} = -\frac{1}{\gamma} \left(k_x \frac{\partial k_x}{\partial \beta} + k_y \frac{\partial k_y}{\partial \beta} \right).$$

One then has that

$$\frac{\partial k_y}{\partial \beta} \frac{\partial \gamma}{\partial \alpha} - \frac{\partial k_y}{\partial \alpha} \frac{\partial \gamma}{\partial \beta} = -\frac{1}{\gamma} \frac{\partial k_y}{\partial \beta} \left(k_x \frac{\partial k_x}{\partial \alpha} + k_y \frac{\partial k_y}{\partial \alpha} \right)$$

$$+ \frac{1}{\gamma} \frac{\partial k_y}{\partial \alpha} \left(k_x \frac{\partial k_x}{\partial \beta} + k_y \frac{\partial k_y}{\partial \beta} \right)$$

$$= -\frac{k_x}{\gamma} \left(\frac{\partial k_x}{\partial \alpha} \frac{\partial k_y}{\partial \beta} - \frac{\partial k_x}{\partial \beta} \frac{\partial k_y}{\partial \alpha} \right) = -\frac{k_x}{\gamma} J(k_x, k_y; \alpha, \beta),$$

and

$$
\frac{\partial k_x}{\partial \alpha}\frac{\partial \gamma}{\partial \beta} - \frac{\partial k_x}{\partial \beta}\frac{\partial \gamma}{\partial \alpha} = -\frac{1}{\gamma}\frac{\partial k_x}{\partial \alpha}\left(k_x\frac{\partial k_x}{\partial \beta} + k_y\frac{\partial k_y}{\partial \beta}\right)
$$

$$
+\frac{1}{\gamma}\frac{\partial k_x}{\partial \beta}\left(k_x\frac{\partial k_x}{\partial \alpha} + k_y\frac{\partial k_y}{\partial \alpha}\right)
$$

$$
= -\frac{k_y}{\gamma}\left(\frac{\partial k_x}{\partial \alpha}\frac{\partial k_y}{\partial \beta} - \frac{\partial k_x}{\partial \beta}\frac{\partial k_y}{\partial \alpha}\right) = -\frac{k_y}{\gamma}J(k_x,k_y;\alpha,\beta).
$$

With these final substitutions, Eq. (7.162) becomes

$$
I_1 = -\lim_{\epsilon \to 0}\int_0^{2\pi} d\beta \int_C d\alpha \left\{ \left(\frac{\partial \psi}{\partial k_x}\frac{\partial k_x}{\partial \lambda} + \frac{\partial \psi}{\partial k_y}\frac{\partial k_y}{\partial \lambda} \right) \right.
$$

$$
\left. -\frac{1}{\gamma}\frac{\partial \psi}{\partial k_z}\left(k_x\frac{\partial k_x}{\lambda} + k_y\frac{\partial k_y}{\lambda}\right) \right\}_{k_z=\gamma} J(k_x,k_y;\alpha,\beta),
$$

and because

$$
-\gamma\frac{\partial \gamma}{\partial \lambda} = k_x\frac{\partial k_x}{\partial \lambda} + k_y\frac{\partial k_y}{\partial \lambda},
$$

one finally obtains the result

$$
I_1 = -\lim_{\epsilon \to 0}\int_0^{2\pi} d\beta \int_C d\alpha \left\{ \frac{\partial \psi}{\partial k_x}\frac{\partial k_x}{\partial \lambda} + \frac{\partial \psi}{\partial k_y}\frac{\partial k_y}{\partial \lambda} + \frac{\partial \psi}{\partial k_z}\frac{\partial \gamma}{\partial \lambda} \right\}_{k_z=\gamma} J(k_x,k_y;\alpha,\beta)
$$

$$
= -\lim_{\epsilon \to 0}\int_0^{2\pi} d\beta \int_C d\alpha \frac{\partial \tilde{\psi}}{\partial \lambda} J(k_x,k_y;\alpha,\beta) = -I_2, \tag{7.163}
$$

upon comparison with Eq. (7.153). Hence, from Eq. (7.151) one obtains the desired result that $I(\mathbf{r},\lambda,\delta)$ is independent of the angular parameter λ. Similarly, by changing λ to δ everywhere in this proof one also obtains that $I(\mathbf{r},\lambda,\delta)$ is independent of the angular parameter δ. Hence

$$
\frac{\partial I}{\partial \lambda} = \frac{\partial I}{\partial \delta} = 0, \tag{7.164}
$$

and this completes Weyl's proof for a dispersive and absorptive medium. As a consequence, Eq. (7.142) is valid for any choice of the angles λ and δ such that the point of observation of the field is within the region of convergence of the integral.

Fig. 7.10 Choice of the polar
axis in Weyl's integral

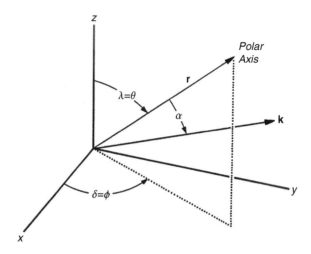

7.4.3   Weyl's Integral Representation

If one chooses the polar axis to be along the observation direction that is specified
by the field point position vector $\mathbf{r} = \mathbf{r}(r, \phi, \theta)$, then the transformation angles δ
and λ are chosen equal to the azimuthal angle $\delta = \phi$ and the angle of declination
$\lambda = \theta$ of the observation direction. With this choice the positive sign is used in front
of the integral appearing in Eq. (7.142) and

$$\tilde{\mathbf{k}}^{\pm}(\omega) \cdot \mathbf{r} = \tilde{k}(\omega) r \cos \alpha, \qquad (7.165)$$

where α (when it is real-valued) is the angle between the polar axis and the
wave vector (when it too is real-valued), as illustrated in Fig. 7.10. With these
substitutions, the angular spectrum integral in Eq. (7.142) becomes

$$u(\mathbf{r}) = \tilde{k}(\omega) \int_0^{2\pi} d\beta \int_C \sin \alpha d\alpha \, U(k_x, k_y) e^{i\tilde{k}(\omega)r \cos \alpha}. \qquad (7.166)$$

Consider now the special case when the spectral amplitude function $U(k_x, k_y)$
appearing in Eq. (7.166) is unity; viz.

$$U(k_x, k_y) = 1.$$

With this substitution, Eq. (7.166) becomes [with the change of variable $\xi = \cos \alpha$
so that $\alpha = 0$ corresponds to $\xi = 1$ and $\alpha = \pi/2 - \kappa - i\infty$ corresponds to
$\xi = (\infty)e^{i(\pi/2-\kappa)}$]

$$u(\mathbf{r}) = 2\pi \tilde{k}(\omega) \int_C \sin \alpha \, e^{i\tilde{k}(\omega)r \cos \alpha} d\alpha$$

$$= -2\pi \tilde{k}(\omega) \int_1^{\infty e^{i(\pi/2 - \kappa)}} e^{i\tilde{k}(\omega)r\xi} d\xi$$

$$= -2\pi \tilde{k}(\omega) \frac{e^{i\tilde{k}(\omega)r\xi}}{i\tilde{k}(\omega)r} \Big|_{\xi=1}^{\xi=\infty e^{i(\pi/2-\kappa)}} = -2\pi i \frac{e^{i\tilde{k}(\omega)r}}{r},$$

where the antiderivative vanishes at the upper limit of integration independently of $\kappa(\omega) \equiv \arg\left\{\tilde{k}(\omega)\right\}$. Substitution of these two expressions in Eq. (7.111) then yields the general result

$$\frac{e^{i\tilde{k}(\omega)r}}{r} = \frac{i}{2\pi} \int_{-\infty}^{\infty} \frac{1}{\gamma(\omega)} e^{i\tilde{\mathbf{k}}^{\pm}(\omega)\cdot\mathbf{r}} dk_x dk_y \qquad (7.167)$$

known as *Weyl's integral* [8], which expresses a spherical wave in terms of a superposition of plane waves in the dispersive medium. Because

$$\tilde{\mathbf{k}}^{\pm}(\omega) \cdot \mathbf{r} = k_x x + k_y y \pm \gamma(\omega)z,$$

then on the plane $z = 0$ there is no distinction between $\tilde{\mathbf{k}}^+(\omega)$ and $\tilde{\mathbf{k}}^-(\omega)$. Notice that the term $\gamma(\omega)z$ provides exponential decay for the convergence of Weyl's integral. On the plane $z = 0$ the only decay factor appearing in the integrand of the integral in Eq. (7.167) is due to the term $\gamma^{-1}(\omega)$, which goes to zero as γ goes off to infinity, but not with sufficient rapidity for absolute convergence requirements to be satisfied. Consequently, Weyl's integral (7.167) converges everywhere except on the plane $z = 0$, and it is conditionally convergent everywhere on the plane $z = 0$ except at the origin where it is divergent, as depicted in Fig. 7.11 for (a) the general angular spectrum representation given in Eqs. (7.105)–(7.108), and (b) Weyl's integral given in Eq. (7.167).

An alternate representation of Weyl's integral may be directly obtained from Eq. (7.142) with $U(k_x, k_y) = 1$, so that in place of Eq. (7.167) one obtains

$$\frac{e^{i\tilde{k}(\omega)r}}{r} = \frac{i\tilde{k}(\omega)}{2\pi} \int_{-\pi}^{\pi} d\beta \int_C \sin\alpha \, d\alpha \, e^{i(k_x x + k_y y \pm \gamma(\omega)z)},$$

where the change in the integration domain from $\beta = 0 \to 2\pi$ to $\beta = -\pi \to \pi$ is justified because it results in the same range of values for k_x and k_y. From Eqs. (7.116), (7.117) and (7.133),

$$k_x = \tilde{k}(\omega) \sin\alpha \cos\beta,$$

$$k_y = \tilde{k}(\omega) \sin\alpha \sin\beta,$$

$$\gamma(\omega) = \tilde{k}(\omega) \cos\alpha,$$

and the above expression becomes

$$\frac{e^{i\tilde{k}(\omega)r}}{r} = \frac{i\tilde{k}(\omega)}{2\pi} \int_{-\pi}^{\pi} d\beta \int_C \sin\alpha\, d\alpha\, e^{i\tilde{k}(\omega)(x\sin\alpha\cos\beta + y\sin\alpha\sin\beta \pm z\cos\beta)},$$

(7.168)

which is the *polar coordinate form of Weyl's integral representation*, where the positive sign appearing in the exponential of the integrand is employed when $z > 0$ and the negative sign when $z < 0$. This form of the representation definitively shows that Weyl's integral expresses a spherical wave as a superposition of inhomogeneous plane waves, where the elements of the ordered triple $(\sin\alpha\cos\beta, \sin\alpha\sin\beta, \pm\cos\beta)$ with $\alpha \in C$ and $\beta = -\pi \to \pi$ are the complex direction cosines of the elementary plane wave normals.

7.4.4   Sommerfeld's Integral Representation

Consider the change of variable

$$\zeta = \tilde{k}(\omega)\sin\alpha$$

(7.169)

in the polar coordinate form (7.168) of Weyl's integral representation, so that as $\alpha = \alpha' + i\alpha''$ varies over $0 \to \pi/2 - \kappa - i\infty$ on the contour C, the variable ζ varies over $0 \to \tilde{k}(\omega)e^{-i\kappa(\omega)}(\infty) = k \cdot \infty$. Define the quantity

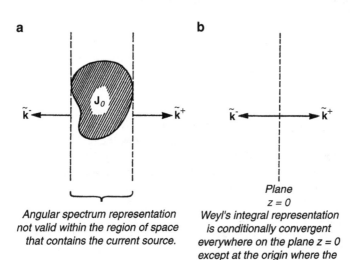

Fig. 7.11 Spatial frequency regions of validity of (**a**) the angular spectrum of plane waves representation and (**b**) Weyl's integral representation of a spherical wave

$$\Gamma(\omega) \equiv \left[\zeta^2 - \tilde{k}^2(\omega)\right]^{1/2}$$

$$= -i\tilde{k}(\omega)\cos\alpha = -i\gamma(\omega), \qquad (7.170)$$

where the branch of the square root is determined such that $\Re\{\Gamma(\omega)\} \geq 0$ as α varies over the contour C as follows:

$$\Re\{\Gamma(\omega)\} = \Re\left\{-i\tilde{k}(\omega)\cos(\alpha' + i\alpha'')\right\}$$

$$= -k(\omega)\left(\cos\kappa\,\sin\alpha'\,\sinh\alpha'' - \sin\kappa\,\cos\alpha'\,\cosh\alpha''\right).$$

Because $0 \leq \kappa < \pi/2$ so that both $\cos\kappa$ and $\sin\kappa$ are nonnegative, and because, as α varies over the contour C, $\sin\alpha'$ varies over $0 \rightarrow \cos\kappa$, $\sinh\alpha''$ varies over $0 \rightarrow -\infty$, $\cos\alpha'$ varies over $1 \rightarrow \sin\kappa$, and $\cosh\alpha''$ varies over $1 \rightarrow +\infty$, then $\Re\{\Gamma(\omega)\} = Ak(\omega)$ where A varies over $0 \rightarrow +\infty$ as α varies over C. In addition, along the contour C

$$\Im\{\zeta\} = \Im\left\{\tilde{k}(\omega)\sin\alpha\right\}$$

$$= k(\omega)\Im\left\{(\cos\kappa + i\sin\kappa)(\sin\alpha'\,\cosh\alpha'' + i\cos\alpha'\,\sinh\alpha'')\right\}$$

$$= k(\omega)\cos\kappa\,\cosh\alpha''(\cos\alpha'\,\tanh\alpha'' + \tan\kappa\,\sin\alpha') = 0$$

from Eq. (7.125), so that ζ is real-valued as α varies over C.

With substitution from Eqs (7.169) and (7.170), the polar coordinate form (7.168) of Weyl's integral representation becomes

$$\frac{e^{i\tilde{k}(\omega)r}}{r} = \frac{1}{2\pi}\int_0^\infty i\,\frac{\zeta d\zeta}{\tilde{k}(\omega)\cos\alpha}\int_{-\pi}^\pi d\beta\,e^{i\zeta(x\cos\beta + y\sin\beta)}e^{\mp\Gamma z}$$

$$= \frac{1}{2\pi}\int_0^\infty \frac{1}{\Gamma(\omega)}e^{\mp\Gamma(\omega)z}\zeta d\zeta\int_{-\pi}^\pi d\beta\,e^{i\zeta(x\cos\beta + y\sin\beta)},$$

where the upper sign choice is employed for $z > 0$ and the lower sign choice for $z < 0$. If one now transforms the spatial coordinates (x, y, z) in the Cartesian representation to cylindrical polar coordinates (r, φ, z), where $x = r\cos\varphi$, $y = r\sin\varphi$ with $r = \sqrt{x^2 + y^2}$ the cylindrical coordinate representation of the spherical radius r, then the above expression becomes

$$\frac{e^{i\tilde{k}(\omega)r}}{r} = \frac{1}{2\pi}\int_0^\infty \frac{1}{\Gamma(\omega)}e^{\mp\Gamma(\omega)z}\zeta d\zeta\int_{-\pi}^\pi e^{i\zeta r\cos(\beta - \varphi)}d\beta,$$

and consequently

$$\frac{e^{i\tilde{k}(\omega)r}}{r} = \int_0^\infty \frac{1}{\Gamma(\omega)}e^{\mp\Gamma(\omega)z}J_0(\zeta r)\zeta d\zeta, \qquad (7.171)$$

where $J_0(\xi)$ denotes the Bessel function of the first kind of order zero. This result may be written in a more explicit form as

$$\frac{e^{i\tilde{k}(\omega)r}}{r} = \int_0^\infty \frac{1}{\left[\zeta^2 - \tilde{k}^2(\omega)\right]^{1/2}} e^{\mp\left[\zeta^2 - \tilde{k}^2(\omega)\right]^{1/2} z} J_0(\zeta r)\zeta\, d\zeta. \tag{7.172}$$

It is then seen that the path of integration extends from the origin along the positive, real ζ-axis with a suitable indentation into the complex ζ-plane about the branch point at $\zeta = k$ when $\tilde{k} = \tilde{k}(\omega)$ assumes a real value k.

If one now uses the relation [18] $J_0(\xi) = \left(H_0^{(1)}(\xi) - H_0^{(1)}(-\xi)\right)/2$ which expresses the Bessel function of the first kind of order zero in terms of the Hankel function of the first kind (or Bessel function of the third kind) of order zero, then the spherical wave representation given in Eq. (7.171) becomes

$$\frac{e^{i\tilde{k}(\omega)r}}{r} = \frac{1}{2}\left\{ \int_0^\infty \frac{1}{\Gamma} e^{\mp\Gamma z} H_0^{(1)}(\zeta r)\zeta\, d\zeta - \int_0^\infty \frac{1}{\Gamma} e^{\mp\Gamma z} H_0^{(1)}(-\zeta r)\zeta\, d\zeta \right\}$$

$$= \frac{1}{2}\left\{ \int_0^\infty \frac{1}{\Gamma} e^{\mp\Gamma z} H_0^{(1)}(\zeta r)\zeta\, d\zeta + \int_{-\infty}^0 \frac{1}{\Gamma} e^{\mp\Gamma z} H_0^{(1)}(\zeta r)\zeta\, d\zeta \right\}$$

under the transformation $\zeta \to -\zeta$ in the second integral. One then finally obtains the classical result [19, 20],

$$\frac{e^{i\tilde{k}(\omega)r}}{r} = \frac{1}{2}\int_0^\infty \frac{1}{\Gamma(\omega)} e^{\mp\Gamma(\omega)z} H_0^{(1)}(\zeta r)\zeta\, d\zeta$$

$$= \frac{1}{2}\int_{-\infty}^\infty \frac{1}{\left[\zeta^2 - \tilde{k}^2(\omega)\right]^{1/2}} e^{\mp\left[\zeta^2 - \tilde{k}^2(\omega)\right]^{1/2} z} H_0^{(1)}(\zeta r)\zeta\, d\zeta,$$

$$\tag{7.173}$$

which is known as *Sommerfeld's integral representation*, where the upper sign choice appearing in the exponential is selected for $z > 0$ and the lower sign choice for $z < 0$. The contour of integration now extends along the real axis from $-\infty$ to $+\infty$ in the complex ζ-plane with the exception of a suitable indentation into the upper half-plane at the origin in order to avoid the logarithmic branch point that the Hankel function exhibits at that point, in addition to any suitable indentation about the branch point at $\zeta = k$ when $\tilde{k}(\omega)$ assumes a real value k. Finally, notice that in this representation the polar axis is fixed along the z-axis.

It is not necessary that the contour of integration appearing in Eqs. (7.171)–(7.173) lie along the real ζ-axis, but rather that the contour for each representation must only lie within a specific domain of analyticity defined by each integrand. For the representation given in Eqs. (7.171) and (7.172) it is readily seen that the path of integration must lie within the strip of analyticity

$$-\Im\left\{\tilde{k}(\omega)\right\} < \Im\{\zeta\} < \Im\left\{\tilde{k}(\omega)\right\} \qquad (7.174)$$

that is defined by the branch points at complex $\zeta = \pm\tilde{k}(\omega)$. Because of the logarithmic branch point that the Hankel function $H_0^{(1)}(\xi)$ exhibits at the origin, the path of integration for Sommerfeld's integral representation (7.173) must lie within the strip of analyticity

$$0 < \Im\{\zeta\} < \Im\left\{\tilde{k}(\omega)\right\}. \qquad (7.175)$$

Of course, there are permissible deformations of the contour of integration outside the strip of analyticity through the use of analytic continuation. Indeed, the path along the real ζ-axis that is described in the discussion following Eq. (7.173) is a trivial example of one such path.

7.4.5   Ott's Integral Representation

The major difficulty in both the application and extension of the polar coordinate form (7.168) of Weyl's integral representation is the simple fact that there are two integrations to perform. This is somewhat offset by the generality afforded by the arbitrary choice of direction for the polar axis that is not provided by Sommerfeld's integral representation (7.173) which entails only a single integration. In order to partially overcome these two complications, Ott [21] applied the transformation given in Eq. (7.169) with the definition in Eq. (7.170) to Sommerfeld's integral representation (7.173). Under this transformation; viz.,

$$\zeta = \tilde{k}(\omega)\sin\alpha,$$
$$d\zeta = \tilde{k}(\omega)\cos\alpha\, d\alpha,$$

the path of integration along the real ζ-axis is transformed to the contour C_0 that is described by Eq. (7.125) and extends from $-\pi/2 + \kappa + i\infty$ through the origin to $\pi/2 - \kappa - i\infty$ in the complex α-plane, as illustrated in Fig. 7.12, where $\kappa(\omega) \equiv \arg\left\{\tilde{k}(\omega)\right\}$. Under this transformation the representation appearing in Eq. (7.173) becomes

$$\frac{e^{i\tilde{k}(\omega)r}}{r} = i\frac{\tilde{k}(\omega)}{2}\int_{C_0} H_0^{(1)}\left(\tilde{k}(\omega)r\sin\alpha\right) e^{\pm i\tilde{k}(\omega)z\cos\alpha}\sin\alpha\, d\alpha, \qquad (7.176)$$

which is known as *Ott's integral representation*. This integral representation of a spherical wave forms the basis of Baños' classical research [20] on dipole radiation in the presence of a conducting half-space.

Fig. 7.12 Contour of
integration for Ott's integral
representation

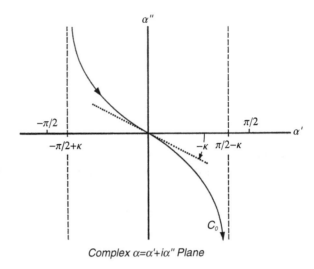

Complex $\alpha = \alpha' + i\alpha''$ Plane

7.5   Angular Spectrum Representation of Multipole Wave Fields

Multipole expansions of both the scalar optical and vector electromagnetic fields are widely employed in both classical and quantum electrodynamics [22] as well as in optics [23]. The genesis of this type of expansion may be found in the early research of scattering by a spherical particle due to Clebsch [24], Mie [25], Debye [26], and Bromwich [27]. In 1903, E. T. Whittaker [28] presented the first multipole expansion of a source-free scalar wave field. This was accomplished by first expressing the solution of the homogeneous scalar wave equation as a linear superposition of homogeneous plane waves (known as a Whittaker-type expansion [12]). Expansion of the plane wave amplitude functions in a series of spherical harmonics then resulted in Whittaker's multipole expansion of the free-field. Whittaker's classic analysis was later extended by A. J. Devaney and E. Wolf [29] to the electromagnetic field produced by a localized charge–current distribution. The analysis presented here closely follows their method of analysis[2] as generalized to a temporally dispersive HILL medium.

The general description begins with the time-domain form of the macroscopic Maxwell's equations [see Eqs. (7.9)–(7.12)] in a temporally dispersive HILL medium (a simple dispersive medium), as described by the constitutive relations given in Eqs. (7.3)–(7.5) without any wave vector dependence. Here $\mathbf{J}(\mathbf{r}, t) = \mathbf{J}_c(\mathbf{r}, t) + \mathbf{J}_{ext}(\mathbf{r}, t)$ is the total current density given by the sum of the *conduction current density* $\mathbf{J}_c(\mathbf{r}, t)$ and the *external current density* $\mathbf{J}_{ext}(\mathbf{r}, t)$. The total current

---

[2]See also A. J. Devaney's Ph.D. thesis [30] in 1971, which became required reading by many later graduate students (including myself) at The Institute of Optics.

density is related to the charge density through the equation of continuity $\nabla \cdot \mathbf{J}(\mathbf{r}, t) = -\partial \varrho(\mathbf{r}, t)/\partial t$. Because $\varrho_c = 0$ [see the discussion following Eq. (5.10)], then $\nabla \cdot \mathbf{J}_c(\mathbf{r}, t) = 0$ and the conduction current density is solenoidal. Furthermore, the externally supplied charge and current sources satisfy the continuity equation

$$\nabla \cdot \mathbf{J}_{ext}(\mathbf{r}, t) = -\frac{\partial \varrho_{ext}(\mathbf{r}, t)}{\partial t}.$$

The temporal Fourier transform of Maxwell's equations then results in, with substitution from the Fourier transform of the constitutive relations for a temporally dispersive HILL medium, the set of frequency domain field equations

$$\nabla \cdot \tilde{\mathbf{E}}(\mathbf{r}, \omega) = \|4\pi\| \frac{\tilde{\varrho}_{ext}(\mathbf{r}, \omega)}{\epsilon(\omega)}, \tag{7.177}$$

$$\nabla \times \tilde{\mathbf{E}}(\mathbf{r}, \omega) = \left\|\frac{1}{c}\right\| i\omega\mu(\omega)\tilde{\mathbf{H}}(\mathbf{r}, \omega), \tag{7.178}$$

$$\nabla \cdot \tilde{\mathbf{H}}(\mathbf{r}, \omega) = 0, \tag{7.179}$$

$$\nabla \times \tilde{\mathbf{H}}(\mathbf{r}, \omega) = \left\|\frac{1}{c}\right\| i\omega\epsilon_c(\omega)\tilde{\mathbf{E}}(\mathbf{r}, \omega) + \left\|\frac{4\pi}{c}\right\| \tilde{\mathbf{J}}_{ext}(\mathbf{r}, \omega), \tag{7.180}$$

where $\epsilon_c(\omega) \equiv \epsilon(\omega) + \|4\pi\| i\sigma(\omega)/\omega$ is the complex permittivity.

Upon taking the curl of Faraday's law (7.178) and using the relations given in Gauss' law (7.177) and Ampère's law (7.180), there results

$$\left(\nabla^2 + \tilde{k}^2(\omega)\right) \tilde{\mathbf{E}}(\mathbf{r}, \omega) = -\frac{\|4\pi\|}{\epsilon(\omega)} \left(i\frac{\omega\epsilon(\omega)\mu(\omega)}{\|c^2\|}\tilde{\mathbf{J}}_{ext}(\mathbf{r}, \omega) - \nabla\tilde{\varrho}_{ext}(\mathbf{r}, \omega)\right),$$

where $\tilde{k}(\omega)$ is the complex wavenumber defined in Eq. (7.114). For the purpose of this analysis, one now defines the *complex-valued wavenumber*

$$\tilde{k}_0(\omega) \equiv \lim_{\sigma(\omega)\to 0} \tilde{k}(\omega) = \frac{\omega}{\|c\|} (\mu(\omega)\epsilon(\omega))^{1/2} \tag{7.181}$$

as the nonconducting limit of the complex wavenumber $\tilde{k}(\omega)$. With this identification, the above expression for the temporal frequency transform of the electric field intensity becomes

$$\left(\nabla^2 + \tilde{k}^2(\omega)\right) \tilde{\mathbf{E}}(\mathbf{r}, \omega) = -\frac{\|4\pi\|}{\epsilon(\omega)} \left(i\frac{\tilde{k}_0^2(\omega)}{\omega}\tilde{\mathbf{J}}_{ext}(\mathbf{r}, \omega) - \nabla\tilde{\varrho}_{ext}(\mathbf{r}, \omega)\right), \tag{7.182}$$

where $\tilde{\varrho}_{ext}(\mathbf{r}, \omega) = \nabla \cdot \tilde{\mathbf{J}}_{ext}(\mathbf{r}, \omega)/(i\omega)$. In a similar fashion, the curl of Ampère's law (7.180), together with Faraday' law (7.178), yields

$$\left(\nabla^2 + \tilde{k}^2(\omega)\right) \tilde{\mathbf{H}}(\mathbf{r}, \omega) = - \left\| \frac{4\pi}{c} \right\| \nabla \times \tilde{\mathbf{J}}_{ext}(\mathbf{r}, \omega). \tag{7.183}$$

The electric and magnetic field vectors described by Maxwell's equations in a dispersive HILL medium then satisfy this pair of *reduced wave equations*, and conversely, the solutions of Eqs. (7.182) and (7.183) which behave as outgoing spherical waves at infinity satisfy Maxwell's equations (7.177)–(7.180) in all of space [29].

7.5.1 Multipole Expansion of the Scalar Optical Wave Field Due to a Localized Source Distribution

Consider the particular solution of the reduced scalar wave equation

$$\left(\nabla^2 + \tilde{k}^2(\omega)\right) \tilde{\psi}(\mathbf{r}, \omega) = -\|4\pi\| \tilde{\rho}(\mathbf{r}, \omega) \tag{7.184}$$

for a given spectral source distribution $\tilde{\rho}(\mathbf{r}, \omega)$ that is assumed to be a continuous function of position and is confined to a finite region about the origin (i.e. has compact support). In particular, let $\tilde{\rho}(\mathbf{r}, \omega) \equiv 0$ for all $r \equiv |\mathbf{r}| > R$, where $R > 0$ is some real constant, independent of the angular frequency ω. The well-known particular solution of Eq. (7.184) that behaves as an outgoing spherical wave at infinity is given by

$$\tilde{\psi}(\mathbf{r}, \omega) = \frac{\|4\pi\|}{4\pi} \int_{r' \leq R} \tilde{\rho}(\mathbf{r}', \omega) \frac{e^{i\tilde{k}(\omega)|\mathbf{r}-\mathbf{r}'|}}{|\mathbf{r} - \mathbf{r}'|} d^3 r'. \tag{7.185}$$

The spherical wave kernel appearing in this integral transform may be expressed as a superposition of plane waves through Weyl's integral [see Eqs. (7.167) and (7.168)]

$$\frac{e^{i\tilde{k}(\omega)r}}{r} = \frac{i}{2\pi} \int_{-\infty}^{\infty} \frac{1}{\gamma(\omega)} e^{i\tilde{\mathbf{k}}^{\pm}(\omega) \cdot \mathbf{r}} dk_x dk_y$$

$$= \frac{i\tilde{k}(\omega)}{2\pi} \int_{-\pi}^{\pi} d\beta \int_C \sin\alpha \, d\alpha \, e^{i\tilde{k}(\omega)(x \sin\alpha \cos\beta + y \sin\alpha \sin\beta \pm z \cos\beta)}, \tag{7.186}$$

where $\tilde{\mathbf{k}}^{\pm}(\omega) \cdot \mathbf{r} = k_x x + k_y y \pm \gamma(\omega) z$ with $\gamma(\omega) \equiv \left(\tilde{k}^2(\omega) - k_x^2 - k_y^2\right)^{1/2}$, and where the contour C, illustrated in Fig. 7.5, extends from $\alpha = 0$ to $\alpha = $

$\pi/2 - \kappa - i\infty$, where $\kappa \equiv \arg\{\tilde{k}(\omega)\}$ with $0 \leq \kappa \leq \pi/2$. The positive sign appearing in the exponential in Eq. (7.186) is employed when $z > 0$ while the negative sign is used when $z < 0$. The polar coordinate form of Weyl's integral expresses a spherical wave as a superposition of inhomogeneous plane waves, the ordered triple $(\sin\alpha\cos\beta, \sin\alpha\sin\beta, \pm\cos\alpha)$ describing the complex direction cosines of the elementary plane wave normals.

Weyl's integral representation of the spherical wave appearing in the integrand of Eq. (7.185) is

$$\frac{e^{i\tilde{k}(\omega)|\mathbf{r}-\mathbf{r}'|}}{|\mathbf{r}-\mathbf{r}'|} = \frac{i\tilde{k}(\omega)}{2\pi} \int_{-\pi}^{\pi} d\beta \int_C \sin\alpha \, d\alpha \, e^{i\tilde{k}(\omega)\mathbf{s}\cdot(\mathbf{r}-\mathbf{r}')}, \tag{7.187}$$

where $\hat{\mathbf{s}} \equiv \hat{\mathbf{1}}_x \sin\alpha\cos\beta + \hat{\mathbf{1}}_y \sin\alpha\sin\beta \pm \hat{\mathbf{1}}_z \cos\alpha$ is a complex-valued vector whose components are the complex direction cosines associated with the inhomogeneous plane wave normals. Substitution of this result in Eq. (7.185) and interchanging the order of integration then results in the expression

$$\tilde{\psi}(\mathbf{r}, \omega) = \|4\pi\| \frac{i\tilde{k}(\omega)}{8\pi^2} \int_{-\pi}^{\pi} d\beta \int_C \sin\alpha \, d\alpha \, \hat{\psi}(\hat{\mathbf{s}}, \omega) e^{i\tilde{k}(\omega)\hat{\mathbf{s}}\cdot\mathbf{r}} \tag{7.188}$$

for the particular solution of the reduced wave equation (7.184), where the spectral amplitude function $\hat{\psi}(\hat{\mathbf{s}}, \omega)$ is defined in terms of the temporal frequency spectrum of the source distribution by

$$\hat{\psi}(\hat{\mathbf{s}}, \omega) = \int_{r' \leq R} \tilde{\rho}(\mathbf{r}', \omega) e^{-i\tilde{k}(\omega)\hat{\mathbf{s}}\cdot\mathbf{r}'} d^3 r'. \tag{7.189}$$

Just as in Whittaker's representation [28] of the source-free wave field, the *Devaney–Wolf representation* [29] expresses the radiated spectral wave field $\tilde{\psi}(\mathbf{r}, \omega)$ as a superposition of plane waves; however, while the Whittaker representation contains only homogeneous plane waves, the Devaney–Wolf representation contains both homogeneous and inhomogeneous plane wave components. In the nondispersive limit, the inhomogeneous components become evanescent as the contour C approaches the limiting case illustrated in Fig. 7.9 comprised of the straight line segment $\alpha = 0 \to \pi/2$ along the real axis followed by the straight line $\alpha = \pi/2 \to \pi/2 - i\infty$ with fixed real part and imaginary part varying from 0 to $-\infty$. Integral representations of this general type describe the wave field as an angular spectrum of plane waves [12, 29].

It is seen from Eq. (7.189) that the spectral amplitude function $\hat{\psi}(\hat{\mathbf{s}}, \omega)$ is related to the spatiotemporal Fourier transform of the source distribution

$$\tilde{\rho}(\mathbf{k}, \omega) = \int_{r' \leq R} \tilde{\rho}(\mathbf{r}', \omega) e^{-i\mathbf{k}\cdot\mathbf{r}'} d^3 r' \tag{7.190}$$

with $\mathbf{k} = \tilde{k}(\omega)\hat{\mathbf{s}}$, so that

$$\hat{\psi}(\hat{\mathbf{s}}, \omega) = \tilde{\rho}\left(\tilde{k}(\omega)\hat{\mathbf{s}}, \omega\right). \tag{7.191}$$

Because the integral appearing on the right side of Eq. (7.190) extends over a finite space domain, and since $\tilde{\rho}(\mathbf{r}, \omega)$ has been assumed to be a continuous function of the spatial coordinates, it then follows[3] that $\tilde{\rho}(\mathbf{k}, \omega)$ is the boundary value along the real k_x', k_y', k_z' axes of an entire function of the complex variables $k_x = k_x' + i k_x'', k_y = k_y' + i k_y'', k_z = k_z' + i k_z''$. The relation given in Eq. (7.191) is then seen to be valid when $\hat{\mathbf{s}}$ is complex. Furthermore, $\hat{\psi}(\hat{\mathbf{s}}, \omega)$ is the boundary value along the contour C of an entire function of the three complex variables $s_x = s_x' + i s_x'', s_y = s_y' + i s_y'', s_z = s_z' + i s_z''$.

After Whittaker [28], the spectral amplitude function $\hat{\psi}(\hat{\mathbf{s}}, \omega)$ is now expanded in a series of *spherical harmonics*

$$Y_\ell^m(\theta, \varphi) \equiv i^\ell C_\ell^m P_\ell^m(\cos\theta)e^{im\phi} \tag{7.192}$$

as

$$\hat{\psi}(\hat{\mathbf{s}}, \omega) = \sum_{\ell=0}^{\infty} \sum_{m=-\ell}^{\ell} (-i)^\ell a_\ell^m(\omega) Y_\ell^m(\alpha, \beta), \tag{7.193}$$

where the expansion coefficients $a_\ell^m(\omega)$, otherwise known as the *multipole moments*, are given by the projections of $\hat{\psi}(\hat{\mathbf{s}}, \omega)$ onto the spherical harmonics as

$$a_\ell^m(\omega) = i^\ell \int_{-\pi}^{\pi} d\beta \int_0^{\pi} d\alpha \, \sin(\alpha)\hat{\psi}(\hat{\mathbf{s}}, \omega) Y_\ell^{m*}(\alpha, \beta). \tag{7.194}$$

Here

$$P_\ell^m(u) \equiv \frac{d^m}{du^m} P_\ell(u) \tag{7.195}$$

defines the associated Legendre polynomial of degree ℓ and order m in terms of the Legendre polynomials through Rodriques' formula

$$P_\ell(u) \equiv \frac{1}{2^\ell \ell!} \frac{d^\ell}{du^\ell} (u^2 - 1)^\ell. \tag{7.196}$$

---

[3] As noted by Devaney and Wolf [29], the result that the Fourier transform of a continuous function possessing compact support is a boundary value along the real axis of an entire function follows from a theorem on analytic functions defined by definite integrals that is given in Sect. 5.5 of Copson [4]. The multidimensional form of this result is provided by the Plancherel–Pólya theorem (see page 352 of Ref. [31]).

Finally, the normalization constants appearing in Eq. (7.192) are given by

$$C_\ell^m \equiv (-1)^m (-i)^\ell \sqrt{\frac{(2\ell+1)(\ell-m)!}{4\pi(\ell+m)!}}. \tag{7.197}$$

As noted by Devaney and Wolf [29], although the multipole moments $a_\ell^m(\omega)$ at a fixed value of the angular frequency ω, defined in Eq. (7.194), "depend explicitly only on those spectral amplitudes $\hat{\psi}(\hat{\mathbf{s}}, \omega)$ which are associated with real $\hat{\mathbf{s}}$ (i.e., those corresponding to homogeneous plane waves in the angular spectrum representation)," the fact that $\hat{\psi}(\hat{\mathbf{s}}, \omega)$ is the boundary value of an entire function implies that the expansion appearing in Eq. (7.193) is valid for all unit vectors associated with the complex contour C depicted in Fig. 7.5.

The multipole moments $a_\ell^m(\omega)$, given in Eq. (7.194), can also be expressed in terms of the spectral source distribution $\tilde{\rho}(\mathbf{r}, \omega)$ through substitution from Eq. (7.189). This substitution, followed by an interchange of the order of integration, gives

$$a_\ell^m(\omega) = \int_{r' \leq R} d^3 r'\, \rho(\mathbf{r}', \omega) i^\ell \int_{-\pi}^{\pi} d\beta \int_0^\pi d\alpha\ \sin(\alpha) Y_\ell^{m*}(\alpha, \beta) e^{-i\tilde{k}(\omega)\hat{\mathbf{s}}\cdot\mathbf{r}'}.$$

Whittaker [28] showed that the homogeneous plane wave expansion of the spectral multipole field

$$\Lambda_\ell^m(\mathbf{r}, \boldsymbol{\omega}) \equiv j_\ell\left(\tilde{k}(\omega)r\right) Y_\ell^m(\theta, \phi), \tag{7.198}$$

where $j_\ell(\zeta)$ denotes the spherical Bessel function of order ℓ, is given by [29]

$$\Lambda_\ell^m(\mathbf{r}, \boldsymbol{\omega}) = (-i)^\ell \frac{1}{4\pi} \int_{-\pi}^{\pi} d\beta \int_0^\pi d\alpha\ \sin(\alpha) Y_\ell^m(\alpha, \beta) e^{i\tilde{k}(\omega)\hat{\mathbf{s}}\cdot\mathbf{r}}, \tag{7.199}$$

where the ordered-triple (r, θ, ϕ) denotes the spherical polar coordinates of the field point $\mathbf{r}$ with reference to the same coordinate axis system as the complex-valued vector $\hat{\mathbf{s}}$. With this identification, the above expression for the multipole moments becomes

$$a_\ell^m(\omega) = \int_{r' \leq R} \rho(\mathbf{r}', \omega) \Lambda_\ell^{m*}(\mathbf{r}, \boldsymbol{\omega}) d^3 r', \tag{7.200}$$

which is the desired result.

Substitution of the expansion given in Eq. (7.193) for $\hat{\psi}(\hat{\mathbf{s}}, \omega)$ into the expression (7.188) for the radiated spectral wave field $\hat{\psi}(\mathbf{r}, \omega)$, followed by an interchange of the order of integration and summation, then results in the series expansion [29]

$$\tilde{\psi}(\mathbf{r}, \omega) = \tilde{k}(\omega) \sum_{\ell=0}^{\infty} \sum_{m=-\ell}^{\ell} a_\ell^m(\omega) \tilde{\Pi}_\ell^m(\mathbf{r}, \boldsymbol{\omega}), \tag{7.201}$$

where

$$\tilde{\Pi}_\ell^m(\mathbf{r}, \omega) = -\frac{\|4\pi\|}{8\pi^2}(-i)^{\ell+1}\int_{-\pi}^{\pi}d\beta\int_C\sin\alpha\,d\alpha\,Y_\ell^m(\alpha, \beta)e^{i\tilde{k}(\omega)\hat{\mathbf{s}}\cdot\mathbf{r}}. \qquad (7.202)$$

As first shown by Erdélyi [32] and later proved by Devaney and Wolf [29], the quantity appearing on the right-hand side of Eq. (7.202) is precisely the angular spectrum of plane waves representation of the spectral scalar multipole field of degree ℓ and order m, given by

$$\tilde{\Pi}_\ell^m(\mathbf{r}, \omega) = h_\ell^{(+)}(\tilde{k}(\omega)r)Y_\ell^m(\theta, \phi), \qquad (7.203)$$

where the ordered-triple (r, θ, ϕ) denotes the spherical polar coordinates of the field point $\mathbf{r}$ and where $h_\ell^{(+)}(\zeta)$ denotes the spherical Hankel function of the first kind of order ℓ. The expansion given in Eq. (7.201) is then seen to be the *multipole expansion of the spectral scalar wave field* $\tilde{\psi}(\mathbf{r}, \omega)$ and the coefficients $a_\ell^m(\omega)$ appearing in that expansion are the corresponding *multipole moments*. Finally, unlike the angular spectrum of plane waves representation which is valid only in the two half-spaces $z > R$ and $z < -R$, the multipole expansion of the same wave field is valid for all $r > R$ (that is, for all points outside the source region).

7.5.1.1   Proof of the Validity of the Angular Spectrum Representation of the Scalar Multipole Field

Proof that the spectral multipole field appearing in Eq. (7.203) is given by the angular spectrum representation given in Eq. (7.202) is now presented based upon the proof given by Devaney and Wolf [29]. The proof begins with Erdélyi's result [32] that any scalar multipole field $\tilde{\Pi}_\ell^m(\mathbf{r}, \omega)$ of order $m \geq 0$ can be generated by a spherical wave $h_0^{(+)}(\tilde{k}(\omega)r) = e^{i\tilde{k}r}/\tilde{k}r$, which is itself proportional to the lowest order scalar multipole field given by $\tilde{\Pi}_\ell^m(\mathbf{r}, \omega) = (1/\sqrt{4\pi})h_0^{(+)}(\tilde{k}(\omega)r)$, through application of the differential operator relation

$$\tilde{\Pi}_\ell^m(\mathbf{r}, \omega) = C_\ell^m\left\{\left[\frac{1}{i\tilde{k}(\omega)}\left(\frac{\partial}{\partial x} + i\frac{\partial}{\partial y}\right)\right]^m P_\ell^m\left(\frac{1}{i\tilde{k}(\omega)}\frac{\partial}{\partial z}\right)\right\}\frac{e^{i\tilde{k}(\omega)r}}{\tilde{k}(\omega)r} \qquad (7.204)$$

for $m \geq 0$, where

$$P_\ell^m\left(\frac{1}{i\tilde{k}}\frac{\partial}{\partial z}\right) \equiv \frac{d^m}{du^m}P_z(u)\Big|_{u=(1/i\tilde{k})\partial/\partial\ell}. \qquad (7.205)$$

Here $P_\ell(u)$ denotes the Legendre polynomial of degree ℓ and the normalization coefficients C_ℓ^m are as defined in Eq. (7.197). The identity

$$\tilde{\Pi}_\ell^{-|m|}(r, \theta, \phi) = (-1)^{|m|}\tilde{\Pi}_\ell^{|m|}(r, \theta, -\phi) \qquad (7.206)$$

is used when $m < 0$, where (r, θ, ϕ) are the spherical polar coordinates of the position vector $\mathbf{r}$.

Substitution of Weyl's integral representation (7.186) of a spherical scalar wave into Eq. (7.204), followed by an interchange of the order of integration and differentiation [justified when $|z| > 0$ so that the double integral in Weyl's integral is uniformly convergent], then results in the expression

$$\tilde{\Pi}_\ell^m(\mathbf{r}, \omega) = \int_{-\pi}^{\pi} d\beta \int_C d\alpha \, \sin(\alpha) \, F(\alpha, \beta) e^{i\tilde{k}(\omega)\hat{\mathbf{s}} \cdot \mathbf{r}}, \qquad (7.207)$$

where

$$F(\alpha, \beta) e^{i\tilde{k}(\omega)\hat{\mathbf{s}} \cdot \mathbf{r}}$$

$$= \frac{i}{2\pi} C_\ell^m \left\{ \left[\frac{1}{i\tilde{k}(\omega)} \left(\frac{\partial}{\partial x} + i \frac{\partial}{\partial y} \right) \right]^m P_\ell^m \left(\frac{1}{i\tilde{k}(\omega)} \frac{\partial}{\partial z} \right) \right\} e^{i\tilde{k}(\omega)\hat{\mathbf{s}} \cdot \mathbf{r}}.$$

Because $\hat{\mathbf{s}} \cdot \mathbf{r} = x \sin\alpha \cos\beta + y \sin\alpha \sin\beta + z \cos\alpha$, the above expression yields

$$F(\alpha, \beta) = \frac{i}{2\pi} C_\ell^m \sin^m \alpha \, e^{im\beta} P_\ell^{(m)}(\cos\alpha). \qquad (7.208)$$

Because $\sin^m \alpha \, P_\ell^{(m)}(\cos\alpha) = P_\ell^m(\cos\alpha)$, then

$$F(\alpha, \beta) = \frac{i}{2\pi} C_\ell^m P_\ell^m(\cos\alpha) e^{im\beta}$$

$$= \frac{i}{2\pi} (-i)^\ell Y_\ell^m(\alpha, \beta), \qquad (7.209)$$

where $Y_\ell^m(\alpha, \beta) = i^\ell C_\ell^m P_\ell^m(\cos\alpha) e^{im\beta}$ is the spherical harmonic of degree ℓ and order m [see Eq. (7.192)]. Substitution of this result in Eq. (7.207) then yields the desired result

$$\tilde{\Pi}_\ell^m(\mathbf{r}, \omega) = \frac{i}{2\pi} (-i)^\ell \int_{-\pi}^{\pi} d\beta \int_C d\alpha \, \sin(\alpha) \, Y_\ell^m(\alpha, \beta) e^{i\tilde{k}(\omega)\hat{\mathbf{s}} \cdot \mathbf{r}}, \qquad (7.210)$$

valid when $|z| > 0$ and $m \geq 0$. The result is also valid when $z = 0$, except at the origin, in the limiting sense [29]

$$\tilde{\Pi}_\ell^m(\mathbf{r}, \omega) \Big|_{z=0} = \frac{i}{2\pi} (-i)^\ell \lim_{|z| \to 0} \left\{ \int_{-\pi}^{\pi} d\beta \int_C d\alpha \, \sin(\alpha) \, Y_\ell^m(\alpha, \beta) e^{i\tilde{k}(\omega)\hat{\mathbf{s}} \cdot \mathbf{r}} \right\}.$$
$$(7.211)$$

For the angular spectrum representation of $\tilde{\Pi}_\ell^m(\mathbf{r}, \omega)$ when $m < 0$, the analysis begins with the expression $\hat{\mathbf{s}} \cdot \mathbf{r} = r(\sin\theta \sin\alpha \cos(\beta + \phi) + \cos\theta \cos\alpha)$ for the dot product between the complex vector $\hat{\mathbf{s}}$ along the normals to the inhomogeneous plane wave phase fronts and the position vector $\mathbf{r}$ at the field point. With this substitution, Eq. (9.124) becomes

$$\tilde{\Pi}_\ell^{|m|}(\mathbf{r}, \omega) = \frac{i}{2\pi}(-i)^\ell \int_{-\pi}^{\pi} d\beta \int_C d\alpha \, \sin(\alpha) \, Y_\ell^{|m|}(\alpha, \beta)$$
$$\times e^{i\tilde{k}(\omega)r[\sin\theta \sin\alpha \cos(\beta+\phi)+\cos\theta \cos\alpha]}.$$

Substitution of this expression into the identity given in Eq. (7.210) then yields

$$\tilde{\Pi}_\ell^{-|m|}(\mathbf{r}, \omega) = (-1)^{|m|}\frac{i}{2\pi}(-i)^\ell \int_{-\pi}^{\pi} d\beta \int_C d\alpha \, \sin(\alpha) \, Y_\ell^{|m|}(\alpha, \beta)$$
$$\times e^{i\tilde{k}(\omega)r[\sin\theta \sin\alpha \cos(\beta-\phi)+\cos\theta \cos\alpha]}.$$

With change of the variable of integration from β to $-\beta$ in this expression and use of the relation $Y_\ell^{-|m|}(\alpha, \beta) = (-1)^{|m|}Y_\ell^{|m|}(\alpha, -\beta)$ there results

$$\tilde{\Pi}_\ell^{-|m|}(\mathbf{r}, \omega) = \frac{i}{2\pi}(-i)^\ell \int_{-\pi}^{\pi} d\beta \int_C d\alpha \, \sin(\alpha) \, Y_\ell^{-|m|}(\alpha, \beta) e^{i\tilde{k}(\omega)r\hat{\mathbf{s}}\cdot\mathbf{r}}.$$

With the final identification that $-|m| = m$ with $m < 0$, the above result becomes formally identical to that given in Eq. (7.208), establishing the result that Eq. (7.210) is valid for $-\ell \leq m \leq \ell$, which then completes the proof.

7.5.1.2 Far-Zone Behavior of the Multipole Expansion

An important feature of the angular spectrum representation of a multipole field is that it directly yields the far-zone behavior of the radiated wave field, and hence, the *radiation pattern*, defined as the angular distribution of the radiated far-zone relative field strength at a fixed distance r from the source. In order to determine this, consider the behavior of $\tilde{\psi}(\mathbf{r}, \omega)$ as $k(\omega)r \to \infty$ along a fixed direction that is specified by the unit vector $\hat{\mathbf{1}}_r \equiv \mathbf{r}/r$, where $k(\omega) \equiv |\tilde{k}(\omega)|$, so that

$$|\mathbf{r} - \mathbf{r}'| \sim r - \mathbf{r}' \cdot \hat{\mathbf{1}}_r. \tag{7.212}$$

Substitution of this asymptotic approximation into Eq. (7.185) then yields

$$\tilde{\psi}(\hat{\mathbf{1}}_r r, \omega) \sim \frac{\|4\pi\|}{4\pi}\tilde{\rho}\left(\hat{\mathbf{1}}_r\tilde{k}(\omega), \omega\right)\frac{e^{i\tilde{k}(\omega)r}}{r} \tag{7.213}$$

as $k(\omega)r \to \infty$, where $\tilde{\rho}(\mathbf{k}, \omega)$ is the spatiotemporal Fourier transform of the source charge distribution $\rho(\mathbf{r}, t)$. However, from Eq. (7.191), $\tilde{\rho}\left(\hat{\mathbf{1}}_r \tilde{k}(\omega), \omega\right)$ is equal to the spectral amplitude function $\hat{\psi}\left(\hat{\mathbf{1}}_r, \omega\right)$, so that the above asymptotic approximation may be written as

$$\tilde{\psi}(\hat{\mathbf{1}}_r r, \omega) \sim \frac{\|4\pi\|}{4\pi} \hat{\psi}\left(\hat{\mathbf{1}}_r, \omega\right) \frac{e^{i\tilde{k}(\omega)r}}{r}, \qquad (7.214)$$

as $k(\omega)r \to \infty$. This important result shows that the radiation pattern of the scalar wave field is exactly given by the spectral amplitude function $\hat{\psi}\left(\hat{\mathbf{1}}_r, \omega\right)$, where $\hat{\mathbf{1}}_r$ is a real unit vector specifying the observation direction. The radiation pattern of the scalar wave field in any given direction $\hat{\mathbf{1}}_r$ from the source is then seen to be given by the complex amplitude of the plane wave component in the angular spectrum representation that is propagating in the direction $\hat{\mathbf{1}}_r$. In particular, with substitution from Eq. (7.193), this asymptotic approximation becomes

$$\tilde{\psi}(\hat{\mathbf{1}}_r r, \omega) \sim \frac{\|4\pi\|}{4\pi} \frac{e^{i\tilde{k}(\omega)r}}{r} \sum_{\ell=0}^{\infty} \sum_{m=-\ell}^{\ell} (-i)^\ell a_\ell^m(\omega) Y_\ell^m(\theta, \phi) \qquad (7.215)$$

as $k(\omega)r \to \infty$, where the ordered pair (θ, ϕ) denotes the spherical polar coordinates of the unit vector $\hat{\mathbf{1}}_r$ along the direction of observation.

Finally, as pointed out by Devaney and Wolf [29]:

The angular spectrum representation (7.188), and the multipole expansion (7.199), are *mode expansions* in the sense that they express the field $\tilde{\psi}(\mathbf{r}, \omega)$ in terms of certain elementary fields (plane wave fields and multipole fields, respectively), each of which satisfies the same equation as does $\tilde{\psi}(\mathbf{r}, \omega)$ outside the source region, namely the Helmholtz equation $\left(\nabla^2 + \tilde{k}^2(\omega)\right) \tilde{\psi}(\mathbf{r}, \omega) = 0$. The range of validity of each of the two expansions is different.

The angular spectrum expansion represents $\tilde{\psi}(\mathbf{r}, \omega)$ outside the strip $|z| \le R$, the multipole expansion represents it outside the sphere $r \le R$. The expansion coefficients in the two representations are related by Eqs. (7.191) and (7.192).

7.5.2 Multipole Expansion of the Electromagnetic Wave Field Generated by a Localized Charge–Current Distribution in a Dispersive Dielectric Medium

The temporal frequency spectra of the electric and magnetic field vectors in a simple (HILL) nonconducting dispersive dielectric medium are found to satisfy the reduced wave equations [see Eqs. (7.182) and (7.183) with $\sigma(\omega) = 0$]

$$\left(\nabla^2 + \tilde{k}^2(\omega)\right)\tilde{\mathbf{E}}(\mathbf{r}, \omega) = -\|4\pi\|\left(i\eta(\omega)\frac{\tilde{k}(\omega)}{\|c\|}\tilde{\mathbf{J}}_{ext}(\mathbf{r}, \omega) - \frac{1}{\epsilon(\omega)}\nabla\tilde{\varrho}_{ext}(\mathbf{r}, \omega)\right),$$

(7.216)

$$\left(\nabla^2 + \tilde{k}^2(\omega)\right)\tilde{\mathbf{H}}(\mathbf{r}, \omega) = -\left\|\frac{4\pi}{c}\right\|\nabla \times \tilde{\mathbf{J}}_{ext}(\mathbf{r}, \omega),$$

(7.217)

with $\tilde{\varrho}_{ext}(\mathbf{r}, \omega) = \nabla \cdot \tilde{\mathbf{J}}_{ext}(\mathbf{r}, \omega)/(i\omega)$, where $\tilde{k}(\omega) \equiv (\omega/\|c\|)(\mu(\omega)\epsilon(\omega))^{1/2}$ is the complex wavenumber for a nonconducting dispersive medium (i.e., for a dispersive dielectric), and where $\eta(\omega) \equiv (\mu(\omega)/\epsilon(\omega))^{1/2}$ is the complex intrinsic impedance of the medium. As in the scalar case, it is assumed that the external charge $\tilde{\varrho}_{ext}(\mathbf{r}, \omega)$ and current $\tilde{\mathbf{J}}_{ext}(\mathbf{r}, \omega)$ sources are both continuous and continuously differentiable functions of position and that both identically vanish when $r > R$ (i.e., have compact support), where $R \geq 0$ is some real constant. The electromagnetic field vectors $\left\{\tilde{\mathbf{E}}(\mathbf{r}, \omega), \tilde{\mathbf{H}}(\mathbf{r}, \omega)\right\}$ generated by the spectral charge–current distribution $\left\{\tilde{\varrho}_{ext}(\mathbf{r}, \omega), \tilde{\mathbf{J}}_{ext}(\mathbf{r}, \omega)\right\}$ are accordingly identified with those particular solutions of the reduced wave equations (7.216) and (7.217) that behave as outgoing spherical waves at infinity.

These reduced wave equations show that each individual Cartesian component of the temporal frequency domain electric and magnetic field vectors of the electromagnetic wave field $\left\{\tilde{\mathbf{E}}(\mathbf{r}, \omega), \tilde{\mathbf{H}}(\mathbf{r}, \omega)\right\}$ satisfy inhomogeneous Helmholtz equations of the form given in Eq. (7.184). The results obtained in the preceding subsection for the multipole expansion of the scalar wave field then apply to each of these Cartesian components. Specifically, Eqs. (7.188) and (7.189) show that the electric and magnetic field vectors have the angular spectrum of plane waves representations

$$\tilde{\mathbf{E}}(\mathbf{r}, \omega) = \|4\pi\|\frac{i\tilde{k}(\omega)}{8\pi^2}\int_{-\pi}^{\pi} d\beta \int_C \sin\alpha \, d\alpha \, \hat{\mathbf{E}}(\hat{\mathbf{s}}, \omega)e^{i\tilde{k}(\omega)\hat{\mathbf{s}}\cdot\mathbf{r}},$$

(7.218)

$$\tilde{\mathbf{H}}(\mathbf{r}, \omega) = \|4\pi\|\frac{i\tilde{k}(\omega)}{8\pi^2}\int_{-\pi}^{\pi} d\beta \int_C \sin\alpha \, d\alpha \, \hat{\mathbf{H}}(\hat{\mathbf{s}}, \omega)e^{i\tilde{k}(\omega)\hat{\mathbf{s}}\cdot\mathbf{r}},$$

(7.219)

where the electric and magnetic spectral amplitude vectors $\left\{\hat{\mathbf{E}}(\hat{\mathbf{s}}, \omega), \hat{\mathbf{H}}(\hat{\mathbf{s}}, \omega)\right\}$ are given by

$$\hat{\mathbf{E}}(\hat{\mathbf{s}}, \omega) = \int_{r'\leq R}\left(i\eta(\omega)\frac{\tilde{k}(\omega)}{\|c\|}\tilde{\mathbf{J}}_{ext}(\mathbf{r}, \omega) - \frac{1}{\epsilon(\omega)}\nabla\tilde{\varrho}_{ext}(\mathbf{r}, \omega)\right)e^{-i\tilde{k}(\omega)\hat{\mathbf{s}}\cdot\mathbf{r}'}d^3r',$$

(7.220)

$$\hat{\mathbf{H}}(\hat{\mathbf{s}}, \omega) = \frac{1}{\|c\|}\int_{r'\leq R}\left(\nabla \times \tilde{\mathbf{J}}_{ext}(\mathbf{r}, \omega)\right)e^{-i\tilde{k}(\omega)\hat{\mathbf{s}}\cdot\mathbf{r}'}d^3r',$$

(7.221)

respectively. These two spectral amplitude functions may be expressed in a simpler form in terms of the spatiotemporal Fourier transforms of the charge and current densities, given by

$$\tilde{\tilde{\varrho}}_{ext}(\mathbf{k}, \omega) = \int_{r' \leq R} \tilde{\varrho}_{ext}(\mathbf{r}', \omega) e^{-i\mathbf{k}\cdot\mathbf{r}'} d^3 r', \qquad (7.222)$$

$$\tilde{\tilde{\mathbf{J}}}_{ext}(\mathbf{k}, \omega) = \int_{r' \leq R} \tilde{\mathbf{J}}_{ext}(\mathbf{r}', \omega) e^{-i\mathbf{k}\cdot\mathbf{r}'} d^3 r', \qquad (7.223)$$

respectively, with the result

$$\hat{\mathbf{E}}(\hat{\mathbf{s}}, \omega) = -\frac{i}{\|c\|} \eta(\omega) \tilde{k}(\omega) \hat{\mathbf{s}} \times \left(\hat{\mathbf{s}} \times \tilde{\tilde{\mathbf{J}}}_{ext}(\tilde{k}(\omega)\hat{\mathbf{s}}, \omega) \right), \qquad (7.224)$$

$$\hat{\mathbf{H}}(\hat{\mathbf{s}}, \omega) = \frac{i}{\|c\|} \tilde{k}(\omega) \hat{\mathbf{s}} \times \tilde{\tilde{\mathbf{J}}}_{ext}(\tilde{k}(\omega)\hat{\mathbf{s}}, \omega). \qquad (7.225)$$

The spectral amplitude field vectors are then seen to satisfy the orthogonality relations

$$\hat{\mathbf{E}}(\hat{\mathbf{s}}, \omega) = -\eta(\omega) \hat{\mathbf{s}} \times \hat{\mathbf{H}}(\hat{\mathbf{s}}, \omega), \qquad (7.226)$$

$$\hat{\mathbf{H}}(\hat{\mathbf{s}}, \omega) = \frac{1}{\eta(\omega)} \hat{\mathbf{s}} \times \hat{\mathbf{E}}(\hat{\mathbf{s}}, \omega), \qquad (7.227)$$

$$\hat{\mathbf{s}} \cdot \hat{\mathbf{E}}(\hat{\mathbf{s}}, \omega) = \hat{\mathbf{s}} \cdot \hat{\mathbf{H}}(\hat{\mathbf{s}}, \omega) = 0. \qquad (7.228)$$

Consequently, at each angular frequency ω and direction $\hat{\mathbf{s}}$, the spectral amplitude vector fields appearing in the integrands of Eqs. (7.218) and (7.219) are monochromatic plane waves satisfying the source-free Maxwell's equations throughout all of space. The angular spectrum of plane waves representation given in Eqs. (7.218) and (7.219) is then seen to be a mode expansion of the electromagnetic field, valid throughout its domain of validity given by $|z| > R$.

In order to obtain the multipole expansions of the monochromatic electromagnetic field vectors $\left\{ \tilde{\mathbf{E}}(\mathbf{r}, \omega), \tilde{\mathbf{H}}(\mathbf{r}, \omega) \right\}$ in the spectral frequency domain, one first expands the spectral amplitude field vectors $\left\{ \hat{\mathbf{E}}(\hat{\mathbf{s}}, \omega), \hat{\mathbf{H}}(\hat{\mathbf{s}}, \omega) \right\}$ in terms of the *vector spherical harmonic functions* $\mathbf{Y}_\ell^m(\alpha, \beta)$, defined in terms of the ordinary spherical harmonic functions $Y_\ell^m(\alpha, \beta)$ by the relation [29]

$$\mathbf{Y}_\ell^m(\alpha, \beta) \equiv \mathcal{L}_s Y_\ell^m(\alpha, \beta), \qquad (7.229)$$

where $\mathcal{L}_s$ is the *orbital angular momentun operator*

$$\mathcal{L}_s \equiv -i \left(\hat{\mathbf{u}}_\beta \frac{\partial}{\partial \alpha} - \frac{1}{\sin \alpha} \hat{\mathbf{u}}_\alpha \frac{\partial}{\partial \beta} \right), \qquad (7.230)$$

with $\hat{\mathbf{u}}_\alpha$ and $\hat{\mathbf{u}}_\beta$ denoting unit vectors in the positive α and β directions, respectively. The vector spherical harmonics can be shown [33] to be everywhere tangent to the unit sphere, viz.

$$\hat{\mathbf{s}} \cdot \mathbf{Y}_\ell^m(\alpha, \beta) = 0, \qquad (7.231)$$

form an orthogonal set in the sense that

$$\int_{-\pi}^{\pi} d\beta \int_0^{\pi} d\alpha \, \sin\alpha \mathbf{Y}_\ell^{m*}(\alpha, \beta) \cdot \mathbf{Y}_{\ell'}^{m'}(\alpha, \beta) = \ell(\ell+1)\delta_{\ell,\ell'}\delta_{m,m'}, \qquad (7.232)$$

and, taken together with the associated vector spherical harmonic functions $\hat{\mathbf{s}} \times \mathbf{Y}_\ell^m(\alpha, \beta)$, form a complete orthogonal basis [29] for all sufficiently well-behaved vector functions $\mathbf{F}(\hat{\mathbf{s}})$ defined on and tangential to the unit sphere $|\hat{\mathbf{s}}| = 1$, so that $\hat{\mathbf{s}} \cdot \mathbf{F}(\hat{\mathbf{s}}) = 0$. In particular, the spectral amplitude field vectors defined by Eqs. (7.224) and (7.225) in terms of the spatiotemporal spectrum of the current source, may be expanded as [see Eq. (7.193)]

$$\hat{\mathbf{E}}(\hat{\mathbf{s}}, \omega) = \sum_{\ell=1}^{\infty} \sum_{m=-\ell}^{\ell} (-i)^\ell \left[a_\ell^m(\omega)\hat{\mathbf{s}} \times \mathbf{Y}_\ell^m(\alpha, \beta) + b_\ell^m(\omega)\mathbf{Y}_\ell^m(\alpha, \beta) \right], \qquad (7.233)$$

$$\hat{\mathbf{H}}(\hat{\mathbf{s}}, \omega) = \frac{1}{\eta(\omega)} \sum_{\ell=1}^{\infty} \sum_{m=-\ell}^{\ell} (-i)^\ell \left[-a_\ell^m(\omega)\mathbf{Y}_\ell^m(\alpha, \beta) + b_\ell^m(\omega)\hat{\mathbf{s}} \times \mathbf{Y}_\ell^m(\alpha, \beta) \right],$$

$$(7.234)$$

where the first expansion has been written in a form that is analogous to the scalar multipole expansion given in Eq. (7.193), the second expansion then following from the orthogonality relations given in Eqs. (7.226)–(7.228). Notice that the summations over ℓ now begin with $\ell = 1$ rather than with $\ell = 0$ as there is no vector spherical harmonic of degree zero.

Application of the orthogonality relation in Eq. (7.232) to the multipole expansions given in Eqs. (7.233) and (7.234) then yields the pair of expressions

$$a_\ell^m(\omega) = -\eta(\omega)\frac{i^\ell}{\ell(\ell+1)} \int_{-\pi}^{\pi} d\beta \int_0^{\pi} d\alpha \, \sin(\alpha) \hat{\mathbf{H}}(\hat{\mathbf{s}}, \omega) \cdot \mathbf{Y}_\ell^{m*}(\alpha, \beta), \quad (7.235)$$

$$b_\ell^m(\omega) = \frac{i^\ell}{\ell(\ell+1)} \int_{-\pi}^{\pi} d\beta \int_0^{\pi} d\alpha \, \sin(\alpha) \hat{\mathbf{E}}(\hat{\mathbf{s}}, \omega) \cdot \mathbf{Y}_\ell^{m*}(\alpha, \beta), \qquad (7.236)$$

for the *multipole moments* as projections of the spectral amplitude field vectors onto the vector spherical harmonics. One may also express $a_\ell^m(\omega)$ and $b_\ell^m(\omega)$ in terms of the spatio-temporal transform of the source current density by substituting the expressions given in Eqs. (7.224) and (7.225) into the above relations with the result [29]

$$a_\ell^m(\omega) = -\frac{\eta(\omega)\tilde{k}(\omega)}{\|c\|}\frac{i^{\ell+1}}{\ell(\ell+1)}\int_{-\pi}^{\pi}d\beta\int_0^{\pi}d\alpha\,\sin(\alpha)$$

$$\times\left[\hat{\mathbf{s}}\times\tilde{\mathbf{J}}_{ext}(\tilde{k}(\omega)\hat{\mathbf{s}},\omega)\right]\cdot\mathbf{Y}_\ell^{m*}(\alpha,\beta),$$

(7.237)

$$b_\ell^m(\omega) = -\frac{\eta(\omega)\tilde{k}(\omega)}{\|c\|}\frac{i^{\ell+1}}{\ell(\ell+1)}\int_{-\pi}^{\pi}d\beta\int_0^{\pi}d\alpha\,\sin(\alpha)$$

$$\times\left\{\hat{\mathbf{s}}\times\left[\hat{\mathbf{s}}\times\tilde{\mathbf{J}}_{ext}(\tilde{k}(\omega)\hat{\mathbf{s}},\omega)\right]\right\}\cdot\mathbf{Y}_\ell^{m*}(\alpha,\beta).$$

(7.238)

Because $\hat{\mathbf{s}}\cdot\mathbf{Y}_\ell^m(\alpha,\beta) = 0$ [see Eq. (7.231)], Eq. (7.237) can be rewritten as [29]

$$a_\ell^m(\omega) = -\frac{\eta(\omega)\tilde{k}(\omega)}{\|c\|}\frac{i^{\ell+1}}{\ell(\ell+1)}\int_{-\pi}^{\pi}d\beta\int_0^{\pi}d\alpha\,\sin(\alpha)$$

$$\times\left\{\hat{\mathbf{s}}\times\left[\hat{\mathbf{s}}\times\tilde{\mathbf{J}}_{ext}(\tilde{k}(\omega)\hat{\mathbf{s}},\omega)\right]\right\}\cdot\left[\hat{\mathbf{s}}\times\mathbf{Y}_\ell^{m*}(\alpha,\beta)\right].$$

(7.239)

Because $-\mathbf{k}\times\left[\mathbf{k}\times\tilde{\mathbf{J}}(\mathbf{k},\omega)\right]/|\mathbf{k}|^2$ is the spatio-temporal Fourier transform $\tilde{\mathbf{J}}_T(\mathbf{k},\omega)$ of the transverse part $\mathbf{J}_T(\mathbf{r},t)$ of the current distribution, then Eqs. (7.238) and (7.239) show that [29] all of the multipole moments $a_\ell^m(\omega)$ and $b_\ell^m(\omega)$, and consequently the electromagnetic field vectors external to the source region, depend only on those Fourier components of the transverse part of the source current distribution for which $\tilde{k}(\omega) = (\omega/\|c\|)n(\omega)$, where $n(\omega)\equiv(\epsilon_c(\omega)\mu(\omega))^{1/2}$ is the complex index of refraction of the medium.

Substitution of the expansions in Eqs. (7.233) and (7.234) for the spectral amplitude vector fields $\hat{\mathbf{E}}(\hat{\mathbf{s}},\omega)$ and $\hat{\mathbf{H}}(\hat{\mathbf{s}},\omega)$ into the angular spectrum of plane waves representation of the temporal frequency spectra of the electric and magnetic field vectors given in Eqs. (7.218) and (7.219), followed by an interchange in the order of integration and summation, results in the series expansions [29]

$$\tilde{\mathbf{E}}(\mathbf{r},\omega) = \sum_{\ell=1}^{\infty}\sum_{m=-\ell}^{\ell}\left[a_\ell^m(\omega)\tilde{\mathbf{E}}_{\ell,m}^e(\mathbf{r},\omega) + b_\ell^m(\omega)\tilde{\mathbf{E}}_{\ell,m}^h(\mathbf{r},\omega)\right],\quad(7.240)$$

$$\tilde{\mathbf{H}}(\mathbf{r},\omega) = \sum_{\ell=1}^{\infty}\sum_{m=-\ell}^{\ell}\left[a_\ell^m(\omega)\tilde{\mathbf{H}}_{\ell,m}^e(\mathbf{r},\omega) + b_\ell^m(\omega)\tilde{\mathbf{H}}_{\ell,m}^h(\mathbf{r},\omega)\right],\quad(7.241)$$

where

$$\tilde{\mathbf{E}}^e_{\ell,m}(\mathbf{r}, \omega) = \eta(\omega)\tilde{\mathbf{H}}^h_{\ell,m}(\mathbf{r}, \omega)$$

$$\equiv \|4\pi\|(-i)^\ell \frac{i\tilde{k}(\omega)}{8\pi^2} \int_{-\pi}^{\pi} d\beta \int_C d\alpha\, \sin\alpha \left[\hat{\mathbf{s}} \times \mathbf{Y}^m_\ell(\alpha, \beta)\right] e^{i\tilde{k}(\omega)\hat{\mathbf{s}}\cdot\mathbf{r}},$$

(7.242)

$$\tilde{\mathbf{E}}^h_{\ell,m}(\mathbf{r}, \omega) = -\eta(\omega)\tilde{\mathbf{H}}^e_{\ell,m}(\mathbf{r}, \omega)$$

$$\equiv \|4\pi\|(-i)^\ell \frac{i\tilde{k}(\omega)}{8\pi^2} \int_{-\pi}^{\pi} d\beta \int_C d\alpha\, \sin\alpha\, \mathbf{Y}^m_\ell(\alpha, \beta) e^{i\tilde{k}(\omega)\hat{\mathbf{s}}\cdot\mathbf{r}}.$$

(7.243)

Devaney and Wolf [29] have shown that the integrals appearing in Eqs. (7.242) and (7.243) are precisely the electromagnetic multipole fields

$$\tilde{\mathbf{E}}^e_{\ell,m}(\mathbf{r}, \omega) = \eta(\omega)\tilde{\mathbf{H}}^h_{\ell,m}(\mathbf{r}, \omega) = \nabla \times \{\nabla \times [\mathbf{r}\Pi^m_\ell(\mathbf{r}, \omega)]\}, \quad (7.244)$$

$$\tilde{\mathbf{E}}^h_{\ell,m}(\mathbf{r}, \omega) = -\eta(\omega)\tilde{\mathbf{H}}^e_{\ell,m}(\mathbf{r}, \omega) = i\tilde{k}(\omega)\nabla \times [\mathbf{r}\Pi^m_\ell(\mathbf{r}, \omega)], \quad (7.245)$$

where $\Pi^m_\ell(\mathbf{r}, \omega)$ is the scalar multipole field defined in Eqs. (7.202) and (7.203). The component vector field quantities $\left\{\tilde{\mathbf{E}}^e_{\ell,m}(\mathbf{r}, \omega), \tilde{\mathbf{H}}^e_{\ell,m}(\mathbf{r}, \omega)\right\}$ are then seen to be the temporal frequency spectra of the electric and magnetic fields generated by an electric multipole while the component vector field quantities $\left\{\tilde{\mathbf{E}}^h_{\ell,m}(\mathbf{r}, \omega), \tilde{\mathbf{H}}^h_{\ell,m}(\mathbf{r}, \omega)\right\}$ are the corresponding fields generated by a magnetic multipole, each of degree ℓ and order m. The pair of expressions in Eqs. (7.240) and (7.241) are then the *multipole expansions* of the temporal frequency domain electric and magnetic field vectors generated by a spatially localized charge–current source distribution (i.e., a charge–current source with compact spatial support), the expansion coefficients $a^m_\ell(\omega)$ and $b^m_\ell(\omega)$ given in Eqs. (7.237) and (7.238) then being the *electric and magnetic multipole moments*, respectively. The temporal frequency spectrum of any electromagnetic multipole wave field then satisfies the homogeneous Maxwell equations for all $r > 0$, valid for all $r > R$ (i.e., valid everywhere external to the source region).

Substitution of the expressions in Eqs. (7.222) and (7.223) for the spatio-temporal Fourier transforms of the charge and current densities into the expressions appearing in Eqs. (7.237) and (7.238) for the multipole moments results in

$$a^m_\ell(\omega) = \frac{\eta(\omega)\tilde{k}(\omega)}{\|c\|} \frac{i^{\ell+1}}{\ell(\ell+1)} \int_{r'\leq R} d^3r'\, \tilde{\mathbf{J}}_{ext}(\mathbf{r}', \omega)$$

$$\cdot \int_{-\pi}^{\pi} d\beta \int_0^{\pi} d\alpha\, \sin(\alpha)\, \hat{\mathbf{s}} \times \mathbf{Y}^{m*}_\ell(\alpha, \beta) e^{-i\tilde{k}(\omega)\hat{\mathbf{s}}\cdot\mathbf{r}'}, \quad (7.246)$$

$$b_\ell^m(\omega) = \frac{\eta(\omega)\tilde{k}(\omega)}{\|c\|} \frac{i^{\ell+1}}{\ell(\ell+1)} \int_{r'\le R} d^3r' \, \tilde{\mathbf{J}}_{ext}(\mathbf{r}', \omega)$$

$$\cdot \int_{-\pi}^{\pi} d\beta \int_0^{\pi} d\alpha \, \sin(\alpha) \, \mathbf{Y}_\ell^{m*}(\alpha, \beta) e^{-i\tilde{k}(\omega)\hat{\mathbf{s}}\cdot\mathbf{r}'}, \qquad (7.247)$$

where the relation $\hat{\mathbf{s}} \times (\hat{\mathbf{s}} \times \mathbf{Y}_\ell^m) = -\mathbf{Y}_\ell^m$, a consequence of the fact that the vector spherical harmonic functions are everywhere tangential to the unit sphere $|\mathbf{s}| = 1$, has been used. Application of the two identities

$$ik\nabla \times \left[\mathbf{r}\Lambda_\ell^m(\mathbf{r})\right] = (-i)^\ell \frac{k}{4\pi} \int_{-\pi}^{\pi} d\beta \int_0^{\pi} d\alpha \, \sin(\alpha) \mathbf{Y}_\ell^m(\alpha, \beta) e^{ik\hat{\mathbf{s}}\cdot\mathbf{r}},$$

$$(7.248)$$

$$\nabla \times \left\{\nabla \times \left[\mathbf{r}\Lambda_\ell^m(\mathbf{r})\right]\right\} = (-i)^\ell \frac{k}{4\pi} \int_{-\pi}^{\pi} d\beta \int_0^{\pi} d\alpha \, \sin(\alpha)\hat{\mathbf{s}} \times \mathbf{Y}_\ell^m(\alpha, \beta) e^{ik\hat{\mathbf{s}}\cdot\mathbf{r}},$$

$$(7.249)$$

where $\Lambda_\ell^m(\mathbf{r}) \equiv j_\ell(kr)Y_\ell^m(\theta, \phi)$ is the *source-free multipole field*, the above pair of expressions for the multipole moments becomes [29]

$$a_\ell^m(\omega) = \frac{4\pi i \eta(\omega)\tilde{k}(\omega)}{\|c\|\ell(\ell+1)\tilde{k}^*(\omega)} \int_{r'\le R} \tilde{\mathbf{J}}_{ext}(\mathbf{r}', \omega) \cdot \left\{\nabla \times \left[\nabla \times \left(\mathbf{r}'\Lambda_\ell^{m*}(\mathbf{r}')\right)\right]\right\} d^3r'$$

$$(7.250)$$

$$b_\ell^m(\omega) = \frac{4\pi i \eta(\omega)\tilde{k}(\omega)}{\|c\|\ell(\ell+1)} \int_{r'\le R} \tilde{\mathbf{J}}_{ext}(\mathbf{r}', \omega) \cdot \left[\nabla \times \left(\mathbf{r}'\Lambda_\ell^{m*}(\mathbf{r}')\right)\right] d^3r'. \qquad (7.251)$$

Notice that $\eta(\omega)\tilde{k}(\omega) = (\omega/\|c\|)\mu(\omega)$, where $\eta(\omega) \equiv \sqrt{\mu(\omega)/\epsilon(\omega)}$ is the complex intrinsic impedance of the nonconducting medium, and that $\tilde{k}(\omega)/\tilde{k}^*(\omega) = e^{i2\psi(\omega)}$, where $\psi(\omega)$ is the phase of the complex wave number. In free space $[\eta(\omega) = 1$ (in cgs units) and $\tilde{k}^*(\omega)/\tilde{k}(\omega) = 1]$, the above pair of expressions for the multipole moments simplify to the well-known expressions given in graduate-level electromagnetics texts (see, for example, Chap. 9 of [34]). The pair of expressions given in Eqs. (7.250)–(7.255) then provide the generalization of the multipole moments to the dispersive, attenuative, nonconducting HILL medium case. The generalization of these expressions to the dispersive, conducting case remains to be given.

7.5.2.1   Proof of the Validity of the Angular Spectrum Representation of the Electromagnetic Multipole Field

Proof that the spectral electromagnetic multipole field given in Eqs. (7.244) and (7.245) is given by the angular spectrum representation in Eqs. (7.242) and (7.243) is now presented based on the proof by Devaney and Wolf [29]. Consider first the vector quantity

$$i\tilde{k}(\omega)\left\{\nabla \times \left[\mathbf{r}\tilde{\Pi}_{\ell}^{m}(\mathbf{r}, \omega)\right]\right\} = i\tilde{k}(\omega)\left\{\tilde{\Pi}_{\ell}^{m}(\mathbf{r}, \omega)\nabla \times \mathbf{r} - \mathbf{r} \times \nabla\tilde{\Pi}_{\ell}^{m}(\mathbf{r}, \omega)\right\}$$

$$= \tilde{k}(\omega)\mathcal{L}_{r}\tilde{\Pi}_{\ell}^{m}(\mathbf{r}, \omega), \tag{7.252}$$

because $\nabla \times \mathbf{r} = \mathbf{0}$, and where

$$\mathcal{L}_{r} \equiv -i\mathbf{r} \times \nabla$$

$$= -i\left(\hat{\mathbf{1}}_{\phi}\frac{\partial}{\partial\theta} - \frac{1}{\sin\theta}\hat{\mathbf{1}}_{\theta}\frac{\partial}{\partial\phi}\right)$$

is the *orbital angular momentum operator* in $\mathbf{r}$-space [cf. Eq. (7.230)]. Substitution of the definition given in Eq. (7.203) for the spectral scalar multipole field $\tilde{\Pi}_{\ell}^{m}(\mathbf{r}, \omega)$ into Eq. (7.252) then gives

$$i\tilde{k}(\omega)\left\{\nabla \times \left[\mathbf{r}\tilde{\Pi}_{\ell}^{m}(\mathbf{r}, \omega)\right]\right\} = \tilde{k}(\omega)h_{\ell}^{(+)}(\tilde{k}(\omega)r)\mathbf{Y}_{\ell}^{m}(\theta, \phi), \tag{7.253}$$

where

$$\mathbf{Y}_{\ell}^{m}(\theta, \phi) \equiv \mathcal{L}_{r}Y_{\ell}^{m}(\theta, \phi) \tag{7.254}$$

is the vector spherical harmonic function of degree ℓ and order m [cf. Eq. (7.229)].

The vector spherical harmonic functions $\mathbf{Y}_{\ell}^{m}(\theta, \phi)$ may be expressed as a linear combination of the ordinary spherical harmonic functions $Y_{\ell}^{m}(\theta, \phi)$ as [33]

$$\mathbf{Y}_{\ell}^{m}(\theta, \phi) = \mathbf{e}_{-}a_{-}Y_{\ell}^{m+1}(\theta, \phi) + \mathbf{e}_{+}a_{+}Y_{\ell}^{m-1}(\theta, \phi) + \hat{\mathbf{1}}_{z}mY_{\ell}^{m}(\theta, \phi),$$

$$\tag{7.255}$$

where

$$a_{-} \equiv \sqrt{(\ell - m)(\ell + m + 1)}, \tag{7.256}$$

$$a_{+} \equiv \sqrt{(\ell + m)(\ell - m + 1)}, \tag{7.257}$$

$$\mathbf{e}_{-} \equiv \frac{1}{2}\left(\hat{\mathbf{1}}_{x} - i\hat{\mathbf{1}}_{y}\right), \tag{7.258}$$

$$\mathbf{e}_{+} \equiv \frac{1}{2}\left(\hat{\mathbf{1}}_{x} + i\hat{\mathbf{1}}_{y}\right). \tag{7.259}$$

Substitution of Eq. (7.255) into Eq. (7.253) then gives

$$i\tilde{k}(\omega)\left\{\nabla \times \left[\mathbf{r}\tilde{\Pi}_{\ell}^{m}(\mathbf{r}, \omega)\right]\right\}$$

$$= \tilde{k}(\omega)\left[\mathbf{e}_{-}a_{-}\tilde{\Pi}_{\ell}^{m+1}(\mathbf{r}, \omega) + \mathbf{e}_{+}a_{+}\tilde{\Pi}_{\ell}^{m-1}(\mathbf{r}, \omega) + m\tilde{\Pi}_{\ell}^{m}(\mathbf{r}, \omega)\right].$$

$$\tag{7.260}$$

Each of the spectral scalar multipole fields appearing on the right-hand side of the above equation are now expressed in terms of the angular spectrum representation given in Eq. (7.202) with the result

$$
i\tilde{k}(\omega)\left\{\nabla \times \left[\mathbf{r}\tilde{\Pi}_\ell^m(\mathbf{r}, \omega)\right]\right\}
$$
$$
= -\frac{\|4\pi\|}{8\pi^2}\tilde{k}(\omega)(-i)^{\ell+1}\int_{-\pi}^{\pi} d\beta \int_C \sin\alpha d\alpha\, \mathbf{G}(\alpha, \beta)e^{i\tilde{k}(\omega)\hat{\mathbf{s}}\cdot\mathbf{r}},
$$

$$(7.261)$$

where

$$
\mathbf{G}(\alpha, \beta) = \mathbf{e}_-a_-Y_\ell^{m+1}(\alpha, \beta) + \mathbf{e}_+a_+Y_\ell^{m-1}(\alpha, \beta) + \hat{\mathbf{1}}_z m Y_\ell^m(\alpha, \beta).
$$

$$(7.262)$$

Comparison of this expression with that given in Eq. (7.255) immediately shows that $\mathbf{G}(\alpha, \beta) = \mathbf{Y}(\alpha, \beta)$, so that

$$
i\tilde{k}(\omega)\left\{\nabla \times \left[\mathbf{r}\tilde{\Pi}_\ell^m(\mathbf{r}, \omega)\right]\right\}
$$
$$
= -\frac{\|4\pi\|}{8\pi^2}\tilde{k}(\omega)(-i)^{\ell+1}\int_{-\pi}^{\pi} d\beta \int_C \sin\alpha d\alpha\, \mathbf{Y}_\ell^m(\alpha, \beta)e^{i\tilde{k}(\omega)\hat{\mathbf{s}}\cdot\mathbf{r}},
$$

$$(7.263)$$

which is precisely the angular spectrum representation (7.243) of the pair of electromagnetic multipole fields given in Eq. (7.245). Finally, application of the curl operator to the expression in Eq. (7.263), noting that the curl operator may be taken inside both integrals on the right-hand side since that double integral converges uniformly for $|z| > 0$, results in

$$
\nabla \times \left\{\nabla \times \left[\mathbf{r}\tilde{\Pi}_\ell^m(\mathbf{r}, \omega)\right]\right\}
$$
$$
= -\frac{\|4\pi\|}{8\pi^2}\tilde{k}(\omega)(-i)^{\ell+1}\int_{-\pi}^{\pi} d\beta \int_C \sin\alpha d\alpha\, \hat{\mathbf{s}} \times \mathbf{Y}_\ell^m(\alpha, \beta)e^{i\tilde{k}(\omega)\hat{\mathbf{s}}\cdot\mathbf{r}},
$$

$$(7.264)$$

which is precisely the angular spectrum representation (7.242) of the pair of electromagnetic multipole fields given in Eq. (7.244). This then completes the proof.

7.5.2.2   Far-Zone Behavior of the Electromagnetic Multipole Field

As in the scalar wave case, the angular spectrum representation of an electromagnetic multipole field directly yields the far-zone behavior of the radiated wave field.

Application of the asymptotic approximation given in Eq. (7.214) to each of the Cartesian components of each spectral field vector $\tilde{\mathbf{E}}(\mathbf{r}, \omega)$ and $\tilde{\mathbf{H}}(\mathbf{r}, \omega)$ then yields

$$\tilde{\mathbf{E}}(\hat{\mathbf{1}}_r r, \omega) \sim \frac{\|4\pi\|}{4\pi} \hat{\mathbf{E}}(\hat{\mathbf{1}}_r, \omega) \frac{e^{i\tilde{k}(\omega)r}}{r}, \qquad (7.265)$$

$$\tilde{\mathbf{H}}(\hat{\mathbf{1}}_r r, \omega) \sim \frac{\|4\pi\|}{4\pi} \hat{\mathbf{H}}(\hat{\mathbf{1}}_r, \omega) \frac{e^{i\tilde{k}(\omega)r}}{r}, \qquad (7.266)$$

as $k(\omega)r \to \infty$ along the fixed direction specified by the unit vector $\hat{\mathbf{1}}_r = \mathbf{r}/r$. The radiation pattern of the radiated electromagnetic wave field is thus seen to be described by the spectral amplitude field vectors $\hat{\mathbf{E}}(\hat{\mathbf{1}}_r, \omega)$ and $\hat{\mathbf{H}}(\hat{\mathbf{1}}_r, \omega)$, where $\hat{\mathbf{1}}_r$ specifies the observation direction relative to the source coordinate system. From Eqs. (7.224) and (7.225), this far-zone asymptotic approximation of the radiated wave field may be expressed in terms of the spatio-temporal Fourier–Laplace transform of the source current distribution as

$$\tilde{\mathbf{E}}(\hat{\mathbf{1}}_r r, \omega) \sim -\frac{i}{4\pi} \left\| \frac{4\pi}{c} \right\| \eta(\omega)\tilde{k}(\omega) \left[\hat{\mathbf{1}}_r \times \left(\hat{\mathbf{1}}_r \times \tilde{\mathbf{J}}_{ext}(\hat{\mathbf{1}}_r, \omega) \right) \right] \frac{e^{i\tilde{k}(\omega)r}}{r}, \quad (7.267)$$

$$\tilde{\mathbf{H}}(\hat{\mathbf{1}}_r r, \omega) \sim \frac{i}{4\pi} \left\| \frac{4\pi}{c} \right\| \tilde{k}(\omega) \left(\hat{\mathbf{1}}_r \times \tilde{\mathbf{J}}_{ext}(\hat{\mathbf{1}}_r, \omega) \right) \frac{e^{i\tilde{k}(\omega)r}}{r}, \qquad (7.268)$$

as $k(\omega)r \to \infty$. Alternatively, the far-zone asymptotic behavior of the radiated electromagnetic wave field may be expressed in terms of the multipole moments $a_\ell^m(\omega)$ and $b_\ell^m(\omega)$ through Eqs. (7.233) and (7.234) as

$$\tilde{\mathbf{E}}(\hat{\mathbf{1}}_r, \omega) \sim \frac{\|4\pi\|}{4\pi} \frac{e^{i\tilde{k}(\omega)r}}{r}$$

$$\times \sum_{\ell=1}^{\infty} \sum_{m=-\ell}^{\ell} (-i)^\ell \left[a_\ell^m(\omega)\hat{\mathbf{1}}_r \times \mathbf{Y}_\ell^m(\theta, \phi) + b_\ell^m(\omega)\mathbf{Y}_\ell^m(\theta, \phi) \right],$$

$$(7.269)$$

$$\hat{\mathbf{H}}(\hat{\mathbf{1}}_r, \omega) = \frac{\|4\pi\|}{4\pi\eta(\omega)} \frac{e^{i\tilde{k}(\omega)r}}{r}$$

$$\times \sum_{\ell=1}^{\infty} \sum_{m=-\ell}^{\ell} (-i)^\ell \left[-a_\ell^m(\omega)\mathbf{Y}_\ell^m(\theta, \phi) + b_\ell^m(\omega)\hat{\mathbf{1}}_r \times \mathbf{Y}_\ell^m(\theta, \phi) \right],$$

$$(7.270)$$

as $k(\omega)r \to \infty$, where (θ, ϕ) are the spherical polar coordinates of the unit vector $\hat{\mathbf{1}}_r$.

As noted by Devaney and Wolf [29], since the radiation pattern of the electromagnetic field is given by either of the spectral amplitude field vectors $\hat{\mathbf{E}}(\hat{\mathbf{1}}_r, \omega)$ or $\hat{\mathbf{H}}(\hat{\mathbf{1}}_r, \omega)$ for all real values of the unit direction vector $\hat{\mathbf{1}}_r$, then Eqs. (7.235) and (7.236), taken together with the orthogonality relations given in Eqs. (7.226)–(7.228), may be interpreted as specifying the multipole moments $a_\ell^m(\omega)$ and $b_\ell^m(\omega)$ in terms of the radiation pattern of the radiated electromagnetic wave field. Hence [29], "all the multipole moments and ... the electromagnetic field at all points outside the sphere $r > R$, are completely specified by the radiation pattern."

7.5.2.3   Radiated Power and the Angular Distribution of Multipole Radiation

Consider first the time-averaged power radiated by the localized charge–current source. From Eq. (5.208), the *time-average Poynting vector* of the radiated time-harmonic field component with angular frequency ω is given by the real part of the *complex Poynting vector* [cf. Eq. (5.209)]

$$\tilde{\mathbf{S}}(\mathbf{r}, \omega) = \frac{1}{2} \left\| \frac{c}{4\pi} \right\| \tilde{\mathbf{E}}(\mathbf{r}, \omega) \times \tilde{\mathbf{H}}^*(\mathbf{r}, \omega). \tag{7.271}$$

Substitution from Eqs. (7.265) and (7.266) into this expression results in the far-zone behavior

$$\tilde{\mathbf{S}}(\mathbf{r}, \omega) \sim \frac{\|4\pi c\|}{32\pi^2} \hat{\mathbf{E}}(\hat{\mathbf{1}}_r, \omega) \times \hat{\mathbf{H}}^*(\hat{\mathbf{1}}_r, \omega) \frac{e^{-2\alpha(\omega)r}}{r^2} \tag{7.272}$$

as $k(\omega)r \to \infty$ along the $\hat{\mathbf{1}}_r$-direction, where $\alpha(\omega) \equiv \Im\left\{\tilde{k}(\omega)\right\}$ is the plane wave attenuation factor in the dispersive medium. The time-average power radiated by the localized charge–current source is then given by the surface integral of the radial component of the time-average Poynting vector over a sphere Σ of radius r as $k(\omega)r \to \infty$, so that

$$\langle P \rangle = \Re\left\{ \int_\Sigma \tilde{\mathbf{S}}(\mathbf{r}, \omega) \cdot \hat{\mathbf{1}}_r r^2 d\Omega \right\}$$
$$\sim \frac{\|4\pi c\|}{32\pi^2} e^{-2\alpha(\omega)r} \Re\left\{ \left[\int_{-\pi}^{\pi} d\phi \int_0^\pi \sin\theta \, d\theta \left[\hat{\mathbf{E}}(\hat{\mathbf{1}}_r, \omega) \times \hat{\mathbf{H}}^*(\hat{\mathbf{1}}_r, \omega) \right] \cdot \hat{\mathbf{1}}_r \right\}.$$

Application of the orthogonality relation given in Eq. (7.227) to the integrand in the above expression then results in

$$\langle P \rangle \sim \frac{\|4\pi c\|}{32\pi^2} e^{-2\alpha(\omega)r} \Re\left\{ \frac{1}{\eta^*(\omega)} \int_{-\pi}^{\pi} d\phi \int_0^\pi \sin\theta \, d\theta \left[\hat{\mathbf{E}}(\hat{\mathbf{1}}_r, \omega) \cdot \hat{\mathbf{E}}^*(\hat{\mathbf{1}}_r, \omega) \right] \right\}.$$

$$\tag{7.273}$$

The time-average radiated power may also be expressed in terms of the magnetic field intensity vector as

$$\langle P \rangle \sim \frac{\|4\pi c\|}{32\pi^2} e^{-2\alpha(\omega)r} \Re \left\{ \eta(\omega) \int_{-\pi}^{\pi} d\phi \int_0^{\pi} \sin\theta \, d\theta \left[\hat{\mathbf{H}}(\hat{\mathbf{1}}_r, \omega) \cdot \hat{\mathbf{H}}^*(\hat{\mathbf{1}}_r, \omega) \right] \right\}. \tag{7.274}$$

Substitution of the multipole expansion (7.233) for $\hat{\mathbf{E}}(\hat{\mathbf{1}}_r, \omega)$ into Eq. (7.273) [or, equivalently, substitution of the multipole expansion (7.234) for $\hat{\mathbf{H}}(\hat{\mathbf{1}}_r, \omega)$ into Eq. (7.274)] with application of the orthogonality relation (7.232) for the vector spherical harmonics then yields the well-known result [29]

$$\langle P \rangle \sim \frac{\|4\pi c\|}{32\pi^2} \Re \left\{ \frac{1}{\eta^*(\omega)} \right\} e^{-2\alpha(\omega)r} \sum_{\ell=1}^{\infty} \sum_{m=-\ell}^{\ell} \ell(\ell+1) \left[|a_\ell^m(\omega)|^2 + |b_\ell^m(\omega)|^2 \right] \tag{7.275}$$

expressing the time-average radiated power $\langle P \rangle$ in terms of the electric and magnetic multipole moments of the spectral amplitude field vectors as $k(\omega)r \to \infty$. This result then shows that, if both the electric and magnetic multipole moments $a_\ell^m(\omega)$ and $b_\ell^m(\omega)$ of a source comprised of a set of incoherently superimposed multipoles of fixed order ℓ are all independent of m, independent of the angular frequency ω, then the far-field angular distribution of the radiated power by that source will be isotropic.

The time-averaged far-field power radiated per unit solid angle is given by [34]

$$\frac{d\langle P \rangle}{d\Omega} = \Re \left\{ \tilde{\mathbf{S}}(\mathbf{r}, \omega) \cdot \hat{\mathbf{1}}_r r^2 \right\}, \tag{7.276}$$

so that, with substitution from Eqs. (7.265) and (7.266) followed by application of the orthogonality relation in Eq. (7.227) and then substitution of the multipole expansion in Eq. (7.233), one obtains

$$\frac{d\langle P \rangle}{d\Omega} \sim \frac{\|4\pi c\|}{32\pi^2} e^{-2\alpha(\omega)r} \Re \left\{ \frac{1}{\eta^*(\omega)} \sum_{m=-\ell}^{\ell} (-i)^\ell \Big| a_\ell^m(\omega) \hat{\mathbf{1}}_r \times \mathbf{Y}_\ell^m(\theta, \phi) \right.$$

$$\left. + b_\ell^m(\omega) \mathbf{Y}_\ell^m(\theta, \phi) \Big|^2 \right\} \tag{7.277}$$

as $k(\omega)r \to \infty$. Notice that the electric $[a_\ell^m(\omega)]$ and magnetic $[b_\ell^m(\omega)]$ multipoles of any fixed order (ℓ, m) possess the same angular dependence at any far-field distance r but have orthogonal polarization states. Notice also that the relative angular distribution of the multipole radiation due to a pulsed source can change

Fig. 7.13 Dipole radiation power patterns $\left|\mathbf{Y}_1^0(\theta,\phi)\right|^2 = \frac{3}{8\pi}\sin^2\theta$ (solid curve) and $\left|\mathbf{Y}_1^{\pm 1}(\theta,\phi)\right|^2 = \frac{3}{16\pi}\left(1+\cos^2\theta\right)$ (dashed curve). Since each radiation pattern depicted here is independent of the azimuthal angle ϕ, the three-dimensional pattern is given by the corresponding surface of revolution about the polar z-axis

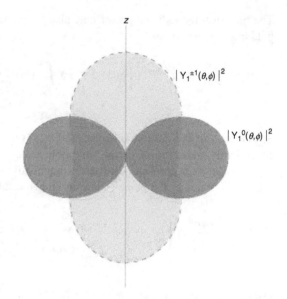

with observation distance r in a dispersive medium due to the frequency dependent attenuation factor $e^{-2\alpha(\omega)r}$ appearing in this expression.

For a pure electric multipole of order (ℓ, m) the above expression [Eq. (7.277)] for the time-averaged far-field power radiated per unit solid angle becomes

$$\frac{d\langle P\rangle}{d\Omega} \sim \frac{\|4\pi c\|}{32\pi^2}\Re\left\{\frac{1}{\eta^*(\omega)}\right\}\left|a_\ell^m(\omega)\right|^2\left|\mathbf{Y}_\ell^m(\theta,\phi)\right|^2 e^{-2\alpha(\omega)r} \qquad (7.278)$$

as $k(\omega)r \to \infty$, with a similar expression for the magnetic case. By comparison, the far-zone electric field strength for this pure multipole is given by [from Eq. (7.269)]

$$\tilde{E}(\hat{\mathbf{1}}_r,\omega) \sim \frac{\|4\pi\|}{4\pi}\left|a_\ell^m(\omega)\right|\left|\mathbf{Y}_\ell^m(\theta,\phi)\right|e^{-\alpha(\omega)r} \qquad (7.279)$$

as $k(\omega)r \to \infty$, with a similar expression for the magnetic field strength. A relative angular plot of either of these two expressions [(7.278) or (7.279)] for given values of ℓ and m provides a radiation or antenna pattern[4] for that particular order multipole, the latter [Eq. (7.279)] being a field strength pattern and the former [Eq. (7.173)] a power pattern. Since both involve relative values, the power pattern

---

[4]From the *IEEE Standard Definitions of Terms for Antennas* (IEEE Std 145-1983), a *radiation pattern* or *antenna pattern* is defined as "a mathematical function or a graphical representation of the radiation properties of the antenna as a function of space coordinates. In most cases, the radiation pattern is determined in the far-field region and is represented as a function of the directional coordinates. Radiation properties include power flux density, radiation intensity, field strength, directivity phase or polarization."

Fig. 7.14 Quadrupole
radiation power patterns
$\left|Y_2^0(\theta,\phi)\right|^2 =$
$\frac{15}{8\pi}\sin^2\theta\cos^2\theta$ (solid curve),
$\left|Y_2^{\pm1}(\theta,\phi)\right|^2 =$
$\frac{5}{16\pi}\left(1 - 3\cos^2\theta + 4\cos^4\theta\right)$
(dashed curve), and
$\left|Y_2^{\pm2}(\theta,\phi)\right|^2 =$
$\frac{5}{16\pi}\left(1 - \cos^4\theta\right)$ (dotted
curve). Since each radiation
pattern depicted here is
independent of the azimuthal
angle ϕ, the
three-dimensional pattern is
given by the corresponding
surface of revolution about
the polar z-axis

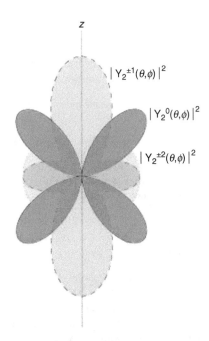

for a pure multipole is just the square of the corresponding field pattern. The power
patterns for dipole radiation ($\ell = 1$, $m = 0, \pm1$) are depicted in Fig. 7.13, the
dipole distribution for $m = 0$ being a linear dipole oscillating along the z-axis at the
origin and the dipole distributions for $m = \pm1$ being comprised of a pair of linear
dipoles in the equatorial plane $\pi/2$ out of phase with each other, one at the origin
along the x-axis and the other at the origin along the y-axis. The power patterns for
quadrupole radiation ($\ell = 2$, $m = 0, \pm1, \pm2$) are depicted in Fig. 7.14. Application
to antenna synthesis (the design of an antenna in order to provide a given far-field
pattern) is an inverse problem of considerable practical engineering interest.

7.6 Applications

The rigorous analysis presented in this chapter provides a basis for describing
pulsed, ultrawideband electromagnetic radiation from both primary sources (antennas) as well as secondary sources (scatterers) embedded in a temporally dispersive
HILL medium. The physical requirement of causality has played a central role
throughout this analysis. The latter topic on secondary sources has direct application
to inverse scattering [35–39] in biomedical imaging (tumor detection), remote
sensing and material identification (ground- and foliage-penetrating radar), and the
topic on primary sources has application to ultrawideband antenna design [40–43]
and undersea communications [44]. These applications are considered in more detail
in Volume 2 of this work.

Problems

7.1 Show that the electric displacement vector $\mathbf{D}(\mathbf{r}, t)$ and magnetic intensity vector $\mathbf{H}(\mathbf{r}, t)$ are invariant under the gauge transformation given in Eqs. (7.57) and (7.58).

7.2 Evaluate the solid angle integral

$$\oint_{4\pi} e^{\pm i \mathbf{F} \cdot \mathbf{G}} \, d\Omega(\mathbf{G}),$$

where $d\Omega(\mathbf{G})$ is the differential element of solid angle of the variable vector $\mathbf{G}$ about the fixed vector $\mathbf{F}$.

7.3 With the result from the previous problem, determine the Fourier spectrum of a function $f(r)$ that depends only upon the magnitude of the position vector $\mathbf{r}$ and not upon the direction.

7.4 Determine the gauge function $\Lambda(\mathbf{r}, t)$ that is required to perform the gauge transformation from the radiation potentials $\{\mathbf{A}, \varphi\}$ given in Eqs. (7.52) and (7.55) for a general temporally dispersive HILL medium with frequency-dependent dielectric permittivity $\epsilon(\omega)$, magnetic permeability $\mu(\omega)$, and electric conductivity $\sigma(\omega)$, to the potentials $\{\mathbf{A}', \varphi'\}$ in the Lorenz gauge, where

$$\nabla \cdot \mathbf{A}'(\mathbf{r}, t) + \frac{\|c\|}{c^2} \frac{\partial \varphi'(\mathbf{r}, t)}{\partial t} = 0.$$

Obtain explicit expressions for the new potentials $\{\mathbf{A}', \varphi'\}$ under this gauge transformation and verify that they do indeed satisfy the Lorenz condition.

7.5 Show that the angular spectrum representations given in Eqs. (7.107) and (7.108) of the vector and scalar potentials yield the angular spectrum representations given in Eqs. (7.105) and (7.106) of the electromagnetic field vectors for the radiation field in a temporally dispersive HILL medium.

7.6 Use the angular spectrum representation in Eq. (7.107) to determine the vector potential field $\mathbf{A}(\mathbf{r}, t)$ in the right half-space region $z > Z > 0$ for $t > 0$ when the external current source is a point source confined to the origin, given by

$$\mathbf{J}_0(\mathbf{r}, t) = \hat{\mathbf{1}} \delta(\mathbf{r}) f(t),$$

where $f(t)$ is a real-valued function of time that vanishes for $t < 0$ with Fourier–Laplace transform $F(\omega)$, $\hat{\mathbf{1}}$ is a unit vector in an arbitrary fixed direction, and where $\tilde{k}(\omega) = \omega/v$, where v is a real, positive-valued constant. Use Weyl's integral given in Eq. (7.167) to perform the required spatial frequency integration.

7.7 Use the angular spectrum representation given in Eq. (7.107) to determine the vector potential field $\mathbf{A}(\mathbf{r}, t)$ in the right half-space region $z > Z > 0$ for $t > 0$ when the external current source is the temporal impulse

$$\mathbf{J}_0(\mathbf{r}, t) = \hat{\mathbf{1}} f(r)\delta(t),$$

where $f(r)$ is a real-valued, spherically symmetric function of time that vanishes for $r > Z > 0$ with Fourier transform $F(k)$, $\hat{\mathbf{1}}$ is a unit vector in an arbitrary fixed direction, and where $\tilde{k}(\omega) = \omega/v$, where v is a real, positive-valued constant.

7.8 Consider a current source that is represented by a pulsed current sheet that begins to radiate at time $t = 0$ and that is uniformly distributed throughout the entire xy-plane, being completely embedded in a temporally dispersive HILL medium with complex permittivity $\epsilon_c(\omega)$, with

$$\mathbf{J}_0(\mathbf{r}, t) = \hat{\mathbf{1}}_x J_0 \delta'(z) F(t),$$

where $F(t) = 0$ for $t < 0$, $\delta'(z)$ is the derivative of the delta function, and J_0 is a measure of the current source strength.

(a) Determine both the vector and scalar potentials of the radiation field produced by this current source.
(b) From this pair of potentials, determine both the electric and magnetic field vectors for this radiation field.

7.9 Derive the angular spectrum of plane waves representation of the vector and scalar potential fields $\mathbf{A}(\mathbf{r}, t)$ and $\varphi(\mathbf{r}, t)$ that is equivalent to that given in Eqs. (6.107) and (6.108) in a temporally dispersive HILL medium that is valid when the current source $\mathbf{J}_0(\mathbf{r}, t)$ begins to radiate at time $t = t_0$ (where t_0 can be negative) instead of at $t = 0$. Determine the form that this representation takes when the ω-integration is taken over C ($\omega = ia - \infty \to ia + \infty$) instead of being taken over C_+ ($\omega = ia \to ia + \infty$).

7.10 Consider the current source distribution

$$\mathbf{J}_0(\mathbf{r}, t) = \begin{cases} \mathbf{f}(\mathbf{r}) \cos{(\omega_0 t)}; & t > t_0 \\ 0; & t < t_0 \end{cases}$$

for real-valued ω_0, where $\mathbf{f}(\mathbf{r})$ is a continuous, bounded, real-valued function of position such that

$$\mathbf{f}(\mathbf{r}) = \mathbf{0}; \quad |\mathbf{r}| > R,$$

so that the current source has compact support in three-dimensional space. Assume that $v(\omega)$ is real-valued.

(a) Determine the space–time domain (i.e., the values of **r** and t) for which the ω-integral appearing in the angular spectrum for the vector potential field derived in Problem 6.9 can be converted into a contour integral that is closed at $|\omega| = \infty$ in the upper-half of the complex ω-plane.

(b) Determine the space–time domain for which the ω-integral appearing in the angular spectrum representation for this vector potential field can be converted into a contour integral that is closed at $|\omega| = \infty$ in the lower-half of the complex ω-plane.

(c) Show that in a certain specified space–time domain where the vector potential field is nonzero, it can be expressed in the form

$$\mathbf{A}(\mathbf{r}, t) = \mathbf{A}_0(\mathbf{r}, t) + \mathbf{A}_{t_0}(\mathbf{r}, t),$$

where $\mathbf{A}_0(\mathbf{r}, t)$ is independent of t_0 and is an expansion of time-harmonic plane waves all with angular frequency ω_0. Derive explicit integral expressions for both $\mathbf{A}_0(\mathbf{r}, t)$ and $\mathbf{A}_{t_0}(\mathbf{r}, t)$. Finally, discuss the behavior of $\mathbf{A}_{t_0}(\mathbf{r}, t)$ as $t_0 \to -\infty$.

7.11 Derive the equations of transformation to an arbitrary polar axis given in Eqs. (7.136) and (7.137) that are obtained from the Cartesian coordinate representation of the wave vector $\tilde{\mathbf{k}}$ through the following succession of rotations about the origin: (i) the k_x and k_y coordinate axes are rotated counterclockwise through the angle δ about the k_z-axis, forming the (k'_x, k'_y, k'_z)-coordinate system; (ii) the k'_x and k'_z coordinate axes are rotated counterclockwise through the angle λ about the k'_y-axis, bringing the k'_z-axis into alignment with the arbitrarily chosen polar axis along the $\hat{k}_z$-axis and forming the final $(\hat{k}_x, \hat{k}_y, \hat{k}_z)$ coordinate system.

References

1. J. J. Stamnes, *Radiation and Propagation of Light in Crystals*. PhD thesis, The Institute of Optics, University of Rochester, Rochester, New York, 1974.
2. J. A. Stratton, *Electromagnetic Theory*. New York: McGraw-Hill, 1941.
3. L. Lorenz, "On the identity of the vibrations of light with electrical currents," *Philos. Mag.*, vol. 34, pp. 287–301, 1867.
4. E. T. Copson, *An Introduction to the Theory of Functions of a Complex Variable*. London: Oxford University Press, 1935. p. 110.
5. E. T. Whittaker and G. N. Watson, *Modern Analysis*. London: Cambridge University Press, fourth ed., 1963. Section 6.222.
6. E. T. Copson, *An Introduction to the Theory of Functions of a Complex Variable*. London: Oxford University Press, 1935. Section 6.1.
7. H. B. Phillips, *Vector Analysis*. New York: John Wiley & Sons, 1933.
8. H. Weyl, "Ausbreitung elektromagnetischer Wellen über einem ebenen Leiter," *Ann. Physik (Leipzig)*, vol. 60, pp. 481–500, 1919.
9. E. T. Whittaker, "x," *Math. Ann.*, vol. 57, pp. 333–355, 1902.
10. G. C. Sherman, A. J. Devaney, and L. Mandel, "Plane-wave expansions of the optical field," *Opt. Commun.*, vol. 6, pp. 115–118, 1972.

11. G. C. Sherman, J. J. Stamnes, A. J. Devaney, and É. Lalor, "Contribution of the inhomogeneous waves in angular-spectrum representations," *Opt. Commun.*, vol. 8, pp. 271–274, 1973.

12. A. J. Devaney and G. C. Sherman, "Plane-wave representations for scalar wave fields," *SIAM Rev.*, vol. 15, pp. 765–786, 1973.

13. P. C. Clemmow, *The Plane Wave Spectrum Representation of Electromagnetic Fields*. Oxford: Pergamon, 1966.

14. J. J. Stamnes, *Waves in Focal Regions: Propagation, Diffraction and Focusing of Light, Sound and Water Waves*. Bristol, UK: Adam Hilger, 1986.

15. M. Nieto-Vesperinas, *Scattering and Diffraction in Physical Optics*. New York: Wiley-Interscience, 1991.

16. T. B. Hansen and A. D. Yaghjian, *Plane-Wave Theory of Time-Domain Fields*. New York: IEEE, 1999.

17. A. J. Devaney, *Mathematical Foundations of Imaging, Tomography and Wavefield Inversion*. Cambridge: Cambridge University Press, 2012.

18. G. N. Watson, *A Treatise on the Theory of Bessel Functions*. Cambridge: Cambridge University Press, second ed., 1958. Sect. 3.62, Eq. (5).

19. A. Sommerfeld, "Über die Ausbreitung der Wellen in der drahtlosen Telegraphie," *Ann. Phys. (Leipzig)*, vol. 28, pp. 665–737, 1909.

20. A. Baños, *Dipole Radiation in the Presence of a Conducting Half-Space*. Oxford: Pergamon, 1966. Sect. 2.12.

21. H. Ott, "Reflexion und Brechung von Kugelwellen; Effekte 2. Ordnung," *Ann. Phys.*, vol. 41, pp. 443–467, 1942.

22. W. Heitler, *The Quantum Theory of Radiation*. Oxford: Clarendon Press, 1954.

23. L. Mandel and E. Wolf, *Optical Coherence and Quantum Optics*. Cambridge: Cambridge University Press, 1995.

24. R. F. A. Clebsch, "Über die Reflexionan einer Kugelfläche," *J. Reine Angew. Math.*, vol. 61, pp. 195–262, 1863.

25. G. Mie, "Beiträge zur Optik truber Medien, speziell kollaidaler Metallosungen," *Ann. Phys. (Leipzig)*, vol. 25, pp. 377–452, 1908.

26. P. Debye, "Der lichtdruck auf Kugeln von beliegigem Material," *Ann. Phys. (Leipzig)*, vol. 30, pp. 57–136, 1909.

27. T. J. I. Bromwich, "The scattering of plane electric waves by spheres," *Phil. Trans. Roy. Soc. Lond.*, vol. 220, p. 175, 1920.

28. E. T. Whittaker, "On the partial differential equations of mathematical physics," *Math. Ann.*, vol. 57, pp. 333–355, 1903.

29. A. J. Devaney and E. Wolf, "Multipole expansions and plane wave representations of the electromagnetic field," *J. Math. Phys.*, vol. 15, no. 11, p. 234, 1974.

30. A. J. Devaney, *A New Theory of the Debye Representation of Classical and Quantized Electromagnetic Fields*. PhD thesis, The Institute of Optics, University of Rochester, 1971.

31. B. A. Fuks, *Introduction to the Theory of Analytic Functions of Several Complex Variables*. Providence: Amer. Math. Soc., 1963.

32. A. Erdélyi, "Zur Theorie der Kugelwellen," *Physica (The Hague)*, vol. 4, pp. 107–120, 1937.

33. E. L. Hill, "The theory of vector spherical harmonics," *Am. J. Phys.*, vol. 22, pp. 211–214, 1954.

34. J. D. Jackson, *Classical Electrodynamics*. New York: John Wiley & Sons, third ed., 1999.

35. R. S. Beezley and R. J. Krueger, "An electromagnetic inverse problem for dispersive media," *J. Math. Phys.*, vol. 26, no. 2, pp. 317–325, 1985.

36. G. Kristensson and R. J. Krueger, "Direct and inverse scattering in the time domain for a dissipative wave equation. I. Scattering operators," *J. Math. Phys.*, vol. 27, no. 6, pp. 1667–1682, 1986.

37. G. Kristensson and R. J. Krueger, "Direct and inverse scattering in the time domain for a dissipative wave equation. II. Simultaneous reconstruction of dissipation and phase velocity profiles," *J. Math. Phys.*, vol. 27, no. 6, pp. 1683–1693, 1986.

38. J. P. Corones, M. E. Davison, and R. J. Krueger, "Direct and inverse scattering in the time domain via invariant embedding equations," *J. Acoustic Soc. Am.*, vol. 74, pp. 1535–1541, 1983.

39. G. Kristensson and R. J. Krueger, "Direct and inverse scattering in the time domain for a dissipative wave equation. Part III. Scattering operators in the presence of a phase velocity mismatch," *J. Math. Phys.*, vol. 28, pp. 360–370, 1987.

40. H. E. Moses and R. T. Prosser, "Initial conditions, sources, and currents for prescribed time-dependent acoustic and electromagnetic fields in three dimensions," *IEEE Trans. Antennas Prop.*, vol. 24, no. 2, pp. 188–196, 1986.

41. H. E. Moses and R. T. Prosser, "Exact solutions of the three-dimensional scalar wave equation and Maxwell's equations from the approximate solutions in the wave zone through the use of the Radon transform," *Proc. Roy. Soc. Lond. A*, vol. 422, pp. 351–365, 1989.

42. H. E. Moses and R. T. Prosser, "Acoustic and electromagnetic bullets: Derivation of new exact solutions of the acoustic and Maxwell's equations," *SIAM J. Appl. Math.*, vol. 50, no. 5, pp. 1325–1340, 1990.

43. H. E. Moses and R. T. Prosser, "The general solution of the time-dependent Maxwell's equations in an infinite medium with constant conductivity," *Proc. Roy. Soc. Lond. A*, vol. 431, pp. 493–507, 1990.

44. R. W. P. King and T. T. Wu, "The propagation of a radar pulse in sea water," *J. Appl. Phys.*, vol. 73, no. 4, pp. 1581–1590, 1993.

Chapter 8
The Angular Spectrum Representation of Pulsed Electromagnetic and Optical Beam Fields in Temporally Dispersive Media

The integrals which we have obtained are not only general expressions which satisfy the differential equation, they represent in the most distinct manner the natural effect which is the object of the phenomenon... when this condition is fulfilled, the integral is, properly speaking, the equation of the phenomenon; it expresses clearly the character and progress of it... Baron Jean–Baptiste–Joseph Fourier (1822).

A completely general representation of the propagation of a freely propagating electromagnetic wave field into the half-space $z \geq z_0 > Z$ of a homogeneous, isotropic, locally linear, temporally dispersive medium is now considered. The term "freely propagating" is used here[1] to indicate that there are no externally supplied charge or current sources for the field present in this half-space, the field source residing somewhere in the region $z \leq Z$.

It is unnecessary to know what this source is provided that the pair $\{\mathbf{E}_0, \mathbf{B}_0\}$ of electromagnetic field vectors are known functions of time and the transverse position vector $\mathbf{r}_T = \hat{\mathbf{1}}_x x + \hat{\mathbf{1}}_y y$ on the plane $z = z_0^+$, as illustrated in Fig. 8.1. The rigorous formal solution of this planar boundary value problem for the electromagnetic field in the half-space $z \geq z_0^+$ forms the basis of investigation for a wide class of pulsed electromagnetic beam field problems in both optics and electrical engineering.

---

[1]A freely propagating field is fundamentally different from a source-free field because the former has an angular spectrum representation that contains both homogeneous and inhomogeneous plane wave components whereas the latter contains only homogeneous plane wave components in lossless media [1–6]. A generalized description of source-free fields appropriate for dispersive attenuative media is given in the final section of this chapter.

© Springer Nature Switzerland AG 2019
K. E. Oughstun, *Electromagnetic and Optical Pulse Propagation*, Springer Series in Optical Sciences 224, https://doi.org/10.1007/978-3-030-20835-6_8

Fig. 8.1 Geometry of the planar electromagnetic boundary value problem. The initial electromagnetic field vectors
$\mathbf{E}(\mathbf{r}_T, z_0, t) = \mathbf{E}_0(\mathbf{r}_T, t)$,
$\mathbf{B}(\mathbf{r}_T, z_0, t) = \mathbf{B}_0(\mathbf{r}_T, t)$ are specified on the plane $z = z_0^+$ for all time t. The problem is to determine their resultant space–time evolution throughout the positive half-space $z > z_0$ for all time t

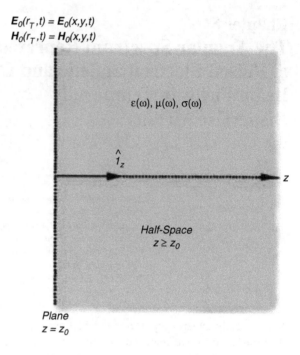

8.1 The Angular Spectrum Representation of the Freely Propagating Electromagnetic Field

Consider an electromagnetic wave field that is propagating into the half-space $z \geq z_0^+ > Z > 0$ and let the electric and magnetic field vectors on the plane $z = z_0^+$, given by the boundary values

$$\mathbf{E}(\mathbf{r}_T, z_0, t) = \mathbf{E}_0(\mathbf{r}_T, t), \qquad (8.1)$$

$$\mathbf{B}(\mathbf{r}_T, z_0, t) = \mathbf{B}_0(\mathbf{r}_T, t), \qquad (8.2)$$

be known functions of time and the transverse position vector $\mathbf{r}_T \equiv \hat{\mathbf{1}}_x x + \hat{\mathbf{1}}_y y$ in the plane $z = z_0$, as indicated by the 0 subscript, as depicted in Fig. 7.1. It is assumed here that the two-dimensional spatial Fourier transform in the transverse coordinates as well as the temporal Fourier–Laplace transform of each field vector on the plane $z = z_0^+$ exists, where

$$\tilde{\mathbf{E}}_0(\mathbf{k}_T, \omega) = \int_{-\infty}^{\infty} dt \int_{-\infty}^{\infty} dx\, dy\, \mathbf{E}_0(\mathbf{r}_T, t) e^{-i(\mathbf{k}_T \cdot \mathbf{r}_T - \omega t)}, \qquad (8.3)$$

$$\mathbf{E}_0(\mathbf{r}_T, t) = \frac{1}{4\pi^3} \Re \left\{ \int_{C_+} d\omega \int_{-\infty}^{\infty} dk_x dk_y\, \tilde{\mathbf{E}}_0(\mathbf{k}_T, \omega) e^{i(\mathbf{k}_T \cdot \mathbf{r}_T - \omega t)} \right\}, \qquad (8.4)$$

and

$$\tilde{\mathbf{B}}_0(\mathbf{k}_T, \omega) = \int_{-\infty}^{\infty} dt \int_{-\infty}^{\infty} dx dy \, \mathbf{B}_0(\mathbf{r}_T, t) e^{-i(\mathbf{k}_T \cdot \mathbf{r}_T - \omega t)}, \tag{8.5}$$

$$\mathbf{B}_0(\mathbf{r}_T, t) = \frac{1}{4\pi^3} \Re \left\{ \int_{C_+} d\omega \int_{-\infty}^{\infty} dk_x dk_y \, \tilde{\mathbf{B}}_0(\mathbf{k}_T, \omega) e^{i(\mathbf{k}_T \cdot \mathbf{r}_T - \omega t)} \right\}, \tag{8.6}$$

where $\mathbf{k}_T \equiv \hat{\mathbf{1}}_x k_x + \hat{\mathbf{1}}_y k_y$ is the transverse wave vector. If the initial time dependence of the electromagnetic field vectors on the plane $z = z_0^+$ is such that both the electric $\mathbf{E}_0(\mathbf{r}, t)$ and magnetic $\mathbf{B}_0(\mathbf{r}, t)$ field vectors vanish for all $t < t_0$ for some finite value of t_0, then the time–frequency transform pairs appearing in Eqs. (8.3)–(8.6) are Laplace transformations and the contour C_+ is the straight line path $\omega = \omega' + ia$ with a being a constant greater than the abscissa of absolute convergence for the initial time evolution of the wave field (see Appendix C); if not, then they are Fourier transformations.

From Eqs. (7.105) and (7.106), the electromagnetic field vectors for the radiation field throughout the positive half-space $z > Z > 0$ are given by

$$\mathbf{E}(\mathbf{r}, t) = \frac{\|4\pi\|}{(2\pi)^3} \Re \left\{ \int_{C_+} \frac{d\omega}{\omega \epsilon_c(\omega)} \right.$$

$$\left. \int_{-\infty}^{\infty} dk_x dk_y \, \frac{\tilde{\mathbf{k}}^+ \times \left(\tilde{\mathbf{k}}^+ \times \tilde{\mathbf{J}}_0(\tilde{\mathbf{k}}^+, \omega) \right)}{\gamma(\omega)} e^{i\left(\tilde{\mathbf{k}}^+ \cdot \mathbf{r} - \omega t \right)} \right\}, \tag{8.7}$$

$$\mathbf{B}(\mathbf{r}, t) = -\frac{\|4\pi/c\|}{(2\pi)^3} \Re \left\{ \int_{C_+} d\omega \, \mu(\omega) \right.$$

$$\left. \int_{-\infty}^{\infty} dk_x dk_y \, \frac{\tilde{\mathbf{k}}^+ \times \tilde{\mathbf{J}}_0(\tilde{\mathbf{k}}^+, \omega)}{\gamma(\omega)} e^{i\left(\tilde{\mathbf{k}}^+ \cdot \mathbf{r} - \omega t \right)} \right\}, \tag{8.8}$$

where

$$\tilde{\mathbf{k}}^+(\omega) = \hat{\mathbf{1}}_x k_x + \hat{\mathbf{1}}_y k_y + \hat{\mathbf{1}}_z \gamma(\omega) \tag{8.9}$$

is the complex wave vector for electromagnetic wave propagation into the positive half-space $z > Z > 0$ with the associated complex wavenumber

$$\tilde{k}(\omega) \equiv \left(\tilde{\mathbf{k}}^+(\omega) \cdot \tilde{\mathbf{k}}^+(\omega) \right)^{1/2}$$

$$= k_0 n(\omega) = \frac{\omega}{\|c\|} [\mu(\omega)\epsilon_c(\omega)]^{1/2}, \tag{8.10}$$

and where $\gamma(\omega)$ is defined as the principal branch of the expression [cf. Eqs. (7.94)–(7.96)]

$$\gamma(\omega) = \left[\tilde{k}^2(\omega) - k_T^2\right]^{1/2}, \tag{8.11}$$

with $k_T^2 \equiv k_x^2 + k_y^2$. On the plane $z = z_0^+ > Z$, $\mathbf{r} = \mathbf{r}_T + \hat{\mathbf{1}}_z z_0$ and the electric field vector given in Eq. (8.7) becomes, with Eq. (8.1),

$$\mathbf{E}(\mathbf{r}_T, t) = \frac{\|4\pi\|}{(2\pi)^3}\Re\left\{\int_{C_+}\frac{d\omega}{\omega\epsilon_c(\omega)}\right.$$

$$\left.\int_{-\infty}^{\infty}dk_x dk_y\,\frac{\tilde{\mathbf{k}}^+ \times \left(\tilde{\mathbf{k}}^+ \times \tilde{\mathbf{J}}_0(\tilde{\mathbf{k}}^+, \omega)\right)}{\gamma(\omega)}e^{i\gamma(\omega)z_0}e^{i(\mathbf{k}_T \cdot \mathbf{r}_T - \omega t)}\right\}. \tag{8.12}$$

Comparison of this expression with that given in Eq. (8.4) shows that the spatiotemporal spectrum of the electric field vector on the plane $z = z_0^+$ is

$$\tilde{\mathbf{E}}_0(\mathbf{k}_T, \omega) = \frac{\|4\pi\|}{2\omega\epsilon_c(\omega)}\frac{\tilde{\mathbf{k}}^+ \times \left(\tilde{\mathbf{k}}^+ \times \tilde{\mathbf{J}}_0(\tilde{\mathbf{k}}^+, \omega)\right)}{\gamma(\omega)}e^{i\gamma(\omega)z_0}. \tag{8.13}$$

Consequently, on any plane $z \geq z_0^+$, $\mathbf{r} = \mathbf{r}_T + \hat{\mathbf{1}}_z z$, the electric field vector is given by

$$\mathbf{E}(\mathbf{r}, t) = \frac{1}{4\pi^3}\Re\left\{\int_{C_+}d\omega\int_{-\infty}^{\infty}dk_x dk_y\,\tilde{\mathbf{E}}_0(\mathbf{k}_T, \omega)e^{i\gamma(\omega)(z-z_0)}e^{i(\mathbf{k}_T \cdot \mathbf{r}_T - \omega t)}\right\}, \tag{8.14}$$

which is the desired angular spectrum representation of the propagated electric field vector in terms of the spectrum of its planar boundary value at $z = z_0^+$. Similarly, the magnetic induction vector on the plane $z = z_0^+ > Z$ is given by, with substitution from Eq. (8.2),

$$\mathbf{B}(\mathbf{r}_T, t) = -\frac{\|4\pi/c\|}{(2\pi)^3}\Re\left\{\int_{C_+}d\omega\,\mu(\omega)\right.$$

$$\left.\int_{-\infty}^{\infty}dk_x dk_y\,\frac{\tilde{\mathbf{k}}^+ \times \tilde{\mathbf{J}}_0(\tilde{\mathbf{k}}^+, \omega)}{\gamma(\omega)}e^{i\gamma(\omega)z_0}e^{i(\mathbf{k}_T \cdot \mathbf{r}_T - \omega t)}\right\}. \tag{8.15}$$

Comparison of this expression with that given in Eq. (8.6) shows that the spatiotemporal spectrum of the magnetic induction field vector on the plane $z = z_0^+$ is given by

$$\tilde{\mathbf{B}}_0(\mathbf{k}_T, \omega) = -\frac{1}{2} \left\| \frac{4\pi}{c} \right\| \mu(\omega) \frac{\tilde{\mathbf{k}}^+ \times \tilde{\mathbf{J}}_0(\tilde{\mathbf{k}}^+, \omega)}{\gamma(\omega)} e^{i\gamma(\omega)z_0}. \tag{8.16}$$

Consequently, the magnetic induction field vector on any plane $z \geq z_0^+$ is

$$\mathbf{B}(\mathbf{r}, t) = \frac{1}{4\pi^3} \Re \left\{ \int_{C_+} d\omega \int_{-\infty}^{\infty} dk_x dk_y \, \tilde{\mathbf{B}}_0(\mathbf{k}_T, \omega) e^{i\gamma(\omega)(z-z_0)} e^{i(\mathbf{k}_T \cdot \mathbf{r}_T - \omega t)} \right\}, \tag{8.17}$$

which is the desired angular spectrum representation of the propagated magnetic induction field vector in terms of the spectrum of its planar boundary value at $z = z_0^+$. Taken together, Eqs. (8.14) and (8.17) constitute the *angular spectrum representation of the freely propagating electromagnetic wave field* for all $z \geq z_0^+ > Z > 0$.

The spatio-temporal spectra of the electromagnetic field vectors at the plane $z = z_0^+$ cannot be chosen independently of each other because both are ultimately determined from the same radiation source situated in the half-space $z < z_0$. Indeed, substitution of Eq. (8.16) into Eq. (8.13) immediately yields the pair of relations

$$\tilde{\mathbf{E}}_0(\mathbf{k}_T, \omega) = -\frac{\|c\|}{\omega\mu(\omega)\epsilon_c(\omega)} \tilde{\mathbf{k}}^+ \times \tilde{\mathbf{B}}_0(\mathbf{k}_T, \omega), \tag{8.18}$$

$$\tilde{\mathbf{B}}_0(\mathbf{k}_T, \omega) = \frac{\|c\|}{\omega} \tilde{\mathbf{k}}^+ \times \tilde{\mathbf{E}}_0(\mathbf{k}_T, \omega). \tag{8.19}$$

In addition, the *transversality condition*

$$\tilde{\mathbf{k}}^+ \cdot \tilde{\mathbf{E}}_0(\mathbf{k}_T, \omega) = \tilde{\mathbf{k}}^+ \cdot \tilde{\mathbf{B}}_0(\mathbf{k}_T, \omega) = 0 \tag{8.20}$$

is satisfied. These three equations are precisely the relations that hold between both field vectors and the associated wave vector for a time-harmonic electromagnetic plane wave field in a temporally dispersive HILL medium [cf. Eqs. (5.61)–(5.63)]. The integrands appearing in the angular spectrum representations given in Eqs. (8.14) and (8.17), viz.

$$\tilde{\mathbf{E}}_0(\mathbf{k}_T, \omega) e^{i[\mathbf{k}_T \cdot \mathbf{r}_T + \gamma(\omega)(z-z_0) - \omega t]} = \tilde{\mathbf{E}}_0(\mathbf{k}_T, \omega) e^{i\left[\tilde{\mathbf{k}}^+(\omega) \cdot \left(\hat{\mathbf{1}}_x x + \hat{\mathbf{1}}_y y + \hat{\mathbf{1}}_z(z-z_0)\right) - \omega t\right]},$$

$$\tilde{\mathbf{B}}_0(\mathbf{k}_T, \omega) e^{i[\mathbf{k}_T \cdot \mathbf{r}_T + \gamma(\omega)(z-z_0) - \omega t]} = \tilde{\mathbf{B}}_0(\mathbf{k}_T, \omega) e^{i\left[\tilde{\mathbf{k}}^+(\omega) \cdot \left(\hat{\mathbf{1}}_x x + \hat{\mathbf{1}}_y y + \hat{\mathbf{1}}_z(z-z_0)\right) - \omega t\right]},$$

are then seen to correspond to a time-harmonic electromagnetic plane wave field that is propagating away from the plane $z = z_0^+$ at each angular frequency ω and transverse wave vector $\mathbf{k}_T = \hat{1}_x k_x + \hat{1}_y k_y$ that is present in the initial spectral amplitude vectors $\left\{ \tilde{\mathbf{E}}_0(\mathbf{k}_T, \omega), \tilde{\mathbf{B}}_0(\mathbf{k}_T, \omega) \right\}$ at that plane, with but one significant difference: the wave vector components k_x and k_y are always real-valued and independent of ω whereas $\gamma(\omega) = \left[\tilde{k}^2(\omega) - k_x^2 - k_y^2 \right]^{1/2}$ is, in general, complex-valued. Hence, each spectral plane wave component is attenuated in the positive z-direction alone; this is just a mathematical consequence of the evaluation of the k_z-integral when the angular spectrum representation was derived in Sect. 7.3. The change in the amount of attenuation with propagation distance in different directions between any two parallel planes $z = $ constant with $z > z_0$ is accounted for solely by the dependence of γ upon k_x and k_y.

8.1.1 Geometric Form of the Angular Spectrum Representation

The plane wave spectral components appearing in the angular spectrum representation given in Eqs. (8.14) and (8.17) of the propagated electromagnetic wave field into the positive half-space $z \geq z_0^+$ may be cast into a more geometric form by setting

$$k_x = \tilde{k}(\omega)p, \tag{8.21}$$

$$k_y = \tilde{k}(\omega)q, \tag{8.22}$$

$$\gamma(\omega) = \tilde{k}(\omega)m, \tag{8.23}$$

where $\tilde{k}(\omega) = k_0 n(\omega)$ is the complex wave number that is given in Eq. (8.10), $k_0 = \omega/c$ is the wave number in vacuum, and $n(\omega) = (c/\|c\|) [\mu(\omega)\epsilon_c(\omega)]^{1/2}$ is the complex index of refraction of the temporally dispersive HILL medium [cf. Eq. (5.32)]. The expression given in Eq. (8.11) then requires that

$$m = \left(1 - p^2 - q^2 \right)^{1/2}, \tag{8.24}$$

where the principal branch of the square root expression is to be taken, as defined in Eqs. (7.94)–(7.96). With these substitutions, the spatial phase term appearing in the exponential factors of both Eqs. (8.14) and (8.17) becomes

$$\mathbf{k}_T \cdot \mathbf{r}_T + \gamma(\omega)(z - z_0) = \tilde{k}(\omega) \left[px + qy + m(z - z_0) \right].$$

The ordered triple of complex numbers (p, q, m) is then seen to be the set of complex direction cosines of the complex wave vector $\tilde{\mathbf{k}}^+ = \hat{\mathbf{1}}_x k_x + \hat{\mathbf{1}}_y k_y + \hat{\mathbf{1}}_z \gamma$, because

$$\tilde{\mathbf{k}}^+ \cdot \mathbf{r} = k_x x + k_y y + \gamma z$$
$$= \tilde{k}(\omega) \left(\frac{k_x}{\tilde{k}(\omega)} x + \frac{k_y}{\tilde{k}(\omega)} y + \frac{\gamma(\omega)}{\tilde{k}(\omega)} z \right)$$
$$= \tilde{k}(\omega) (px + qy + mz).$$

That these generalized direction cosines are, in general, complex-valued follows directly from the fact that, for a causally dispersive medium the wavenumber $\tilde{k}(\omega) = \beta(\omega) + i\alpha(\omega)$ is, in general, complex-valued where $\beta(\omega)$ is the (real-valued) time-harmonic plane wave propagation factor and $\alpha(\omega)$ is the (real-valued) time-harmonic plane wave attenuation factor, given in Eqs. (5.69) and (5.70). Because k_x and k_y must both be real-valued quantities, then the generalized direction cosines p and q must, in general, be complex-valued, so that, with $p = p' + ip''$ and $q = q' + iq''$,

$$k_x = \tilde{k}(\omega)p = (\beta(\omega) + i\alpha(\omega))(p' + ip'')$$
$$= (\beta(\omega)p' - \alpha(\omega)p'') + i(\beta(\omega)p'' + \alpha(\omega)p'), \tag{8.25}$$
$$k_y = \tilde{k}(\omega)p = (\beta(\omega) + i\alpha(\omega))(q' + iq'')$$
$$= (\beta(\omega)q' - \alpha(\omega)q'') + i(\beta(\omega)q'' + \alpha(\omega)q'). \tag{8.26}$$

Hence, in order that both k_x and k_y are real-valued for all values of the angular frequency ω, it is required that

$$p'' = -\frac{\alpha(\omega)}{\beta(\omega)} p', \tag{8.27}$$

$$q'' = -\frac{\alpha(\omega)}{\beta(\omega)} q'. \tag{8.28}$$

With these substitutions, Eqs. (8.25) and (8.26) become, respectively,

$$k_x = \beta(\omega) \left(1 + \frac{\alpha^2(\omega)}{\beta^2(\omega)} \right) p', \tag{8.29}$$

$$k_y = \beta(\omega) \left(1 + \frac{\alpha^2(\omega)}{\beta^2(\omega)} \right) q'. \tag{8.30}$$

Hence, in order to determine the manner in which p' and q' must be required to vary so that k_x and k_y both vary from $-\infty$ to $+\infty$ for all $\omega = \omega' + ia$, $\omega' \geq 0$, it is necessary to know the behavior of $\beta(\omega)$ along the contour C_+ in the complex

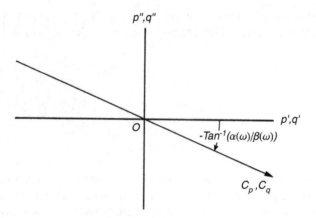

Fig. 8.2 Contours of integration given by the straight lines $C_{p,q}$ in the complex direction cosine plane

ω-plane. Because $\beta(\omega)$ is proportional to the real part of the complex index of refraction which (for the medium models considered in this book, unless otherwise noted; see Problem 8.2) is positive for all $\omega \in C_+$, then p' and q' must both vary from $-\infty$ to $+\infty$. Furthermore, because $\alpha(\omega) \geq 0$ for all $\omega \in C_+$ (for the medium models considered in this book), then p'' varies from $+\infty$ to $-\infty$ as p' varies from $-\infty$ to $+\infty$, and q'' varies from $+\infty$ to $-\infty$ as q' varies from $-\infty$ to $+\infty$. Hence, the contour C_w that $w = w' + iw''$, $w = p, q$ varies over in the complex w-plane is a straight line path through the origin at the angle $-\arctan(\alpha(\omega)/\beta(\omega))$ to the real axis, as illustrated in Fig. 8.2. Notice that C_p and C_q both depend upon the angular frequency ω. When the medium is lossless, α vanishes and the contour lies along the real axis; this may occur, for example, at certain frequency values in a dispersive medium or at all frequency values in a nondispersive medium.

With p'' and q'' as respectively given in Eqs. (8.27) and (8.28), the expression given in Eq. (8.24) for the complex direction cosine $m = (1 - p^2 - q^2)^{1/2}$ becomes

$$m(\omega) = \left[1 - (p' + ip'')^2 - (q' + iq'')^2\right]^{1/2}$$

$$= \left[1 - \left(1 - \frac{\alpha^2(\omega)}{\beta^2(\omega)}\right)\left(p'^2 + q'^2\right) + 2i\frac{\alpha(\omega)}{\beta(\omega)}\left(p'^2 + q'^2\right)\right]^{1/2}. \quad (8.31)$$

In order to evaluate this expression with the appropriate branch choice as set forth in Eqs. (7.94)–(7.96), let

$$m(\omega) \equiv \zeta^{1/2}(\omega) \quad (8.32)$$

with $\zeta = \zeta' + i\zeta''$, where

$$\zeta'(\omega) \equiv \Re\{\zeta(\omega)\} = 1 - \left(1 - \frac{\alpha^2(\omega)}{\beta^2(\omega)}\right)\left(p'^2 + q'^2\right), \tag{8.33}$$

$$\zeta''(\omega) \equiv \Im\{\zeta(\omega)\} = 2\frac{\alpha(\omega)}{\beta(\omega)}\left(p'^2 + q'^2\right). \tag{8.34}$$

Because $\alpha(\omega)/\beta(\omega) \geq 0$ for all $\omega \in C_+$, then

$$\zeta''(\omega) \geq 0, \quad \forall \omega \in C_+. \tag{8.35}$$

Furthermore,

$$\zeta'(\omega) > 0, \quad \text{when } p'^2 + q'^2 < \frac{1}{1 - \alpha^2(\omega)/\beta^2(\omega)}, \tag{8.36}$$

$$\zeta'(\omega) = 0, \quad \text{when } p'^2 + q'^2 = \frac{1}{1 - \alpha^2(\omega)/\beta^2(\omega)}, \tag{8.37}$$

$$\zeta'(\omega) < 0, \quad \text{when } p'^2 + q'^2 > \frac{1}{1 - \alpha^2(\omega)/\beta^2(\omega)}. \tag{8.38}$$

It is then seen that the argument of $\zeta(\omega)$ satisfies the inequality

$$0 \leq \arg\{\zeta(\omega)\} \leq \pi, \tag{8.39}$$

and hence, that the appropriate branch of the argument of $m = \zeta^{1/2}$ satisfies the inequality

$$0 \leq \arg\{m(\omega)\} \leq \frac{\pi}{2}, \tag{8.40}$$

for all $\omega \in C_+$, as illustrated in the sequence of diagrams given in Fig. 8.3. As a consequence, the real and imaginary parts of the complex direction cosine $m(\omega)$ satisfy the inequalities

$$m'(\omega) = \Re\{m(\omega)\} \geq 0, \tag{8.41}$$

$$m''(\omega) = \Im\{m(\omega)\} \geq 0, \tag{8.42}$$

for all $\omega \in C_+$.

Explicit expressions for both $m'(\omega)$ and $m''(\omega)$ are obtained in the following manner. First of all, because $p^2 + q^2 + m^2 = 1$, with $p = p' + ip''$, $q = q' + iq''$, $m = m' + im''$ and substitution from Eqs. (8.27)–(8.28) for p'' and q'', there results

$$\left(1 - \frac{\alpha^2}{\beta^2}\right)\left(p'^2 + q'^2\right) + m'^2 - m''^2 - 2i\left[\frac{\alpha}{\beta}\left(p'^2 + q'^2\right) - m'm''\right] = 1.$$

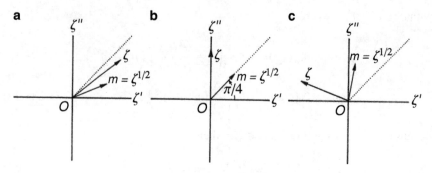

Fig. 8.3 Proper values of the complex direction cosine $m(\omega) = (1-p^2-q^2)^{1/2}$ with $p = p'+ip''$ and $q = q'+iq''$ when (**a**) $p'^2+q'^2 < 1/(1-\alpha^2/\beta^2)$, (**b**) $p'^2+q'^2 = 1/(1-\alpha^2/\beta^2)$, and (**c**) $p'^2+q'^2 > 1/(1-\alpha^2/\beta^2)$

Upon equating real and imaginary parts in the above expression, one obtains the pair of simultaneous equations

$$\left(1-\frac{\alpha^2}{\beta^2}\right)\left(p'^2+q'^2\right)+m'^2-m''^2 = 1, \tag{8.43}$$

$$m'm'' = \frac{\alpha}{\beta}\left(p'^2+q'^2\right). \tag{8.44}$$

Substitution of m'' from the second relation into the first then yields the simple quartic equation

$$m'^4+\left[\left(1-\frac{\alpha^2}{\beta^2}\right)\left(p'^2+q'^2\right)-1\right]m'^2-\frac{\alpha^2}{\beta^2}\left(p'^2+q'^2\right)^2 = 0.$$

The general solution for m'^2 is then given by

$$m'^2 = \frac{1}{2}\left\{\left[1-\left(1-\frac{\alpha^2}{\beta^2}\right)\left(p'^2+q'^2\right)\right]\right.$$
$$\left.\pm\left[1-2\left(1-\frac{\alpha^2}{\beta^2}\right)\left(p'^2+q'^2\right)+\left(1+\frac{\alpha^2}{\beta^2}\right)^2\left(p'^2+q'^2\right)^2\right]^{1/2}\right\}.$$

Because m' is real-valued, then $m'^2 \geq 0$ and only the positive sign choice in the above expression is appropriate. Hence

$$m' = \frac{1}{\sqrt{2}}\left\{\left[1 - \left(1 - \frac{\alpha^2}{\beta^2}\right)\left(p'^2 + q'^2\right)\right]\right.$$

$$\left. \pm \left[1 - 2\left(1 - \frac{\alpha^2}{\beta^2}\right)\left(p'^2 + q'^2\right) + \left(1 + \frac{\alpha^2}{\beta^2}\right)^2 \left(p'^2 + q'^2\right)^2\right]^{1/2}\right\}^{1/2},$$

$$(8.45)$$

and m'' is given by

$$m'' = \frac{\alpha\left(p'^2 + q'^2\right)}{\beta m'}, \qquad (8.46)$$

provided that $\alpha \neq 0$. When $\alpha = 0$, Eq. (8.43) becomes

$$m'^2 - m''^2 = 1 - \left(p'^2 + q'^2\right)$$

and Eq. (8.44) states that either $m' = 0$ or $m'' = 0$. Then $m'' = 0$ and

$$m' = \left[1 - \left(p'^2 + q'^2\right)\right]^{1/2} \qquad (8.47)$$

when $p'^2 + q'^2 \leq 1$, whereas $m' = 0$ and

$$m'' = \left[\left(p'^2 + q'^2\right) - 1\right]^{1/2} \qquad (8.48)$$

when $p'^2 + q'^2 > 1$.

With these results, the complex phase term appearing in the plane wave propagation factor $e^{i\tilde{\mathbf{k}}^+ \cdot \mathbf{r}}$ may be expressed as

$$\tilde{\mathbf{k}}^+(\omega) \cdot \mathbf{r} = k_x x + k_y y + \gamma(\omega)\Delta z$$

$$= \beta(\omega)\left(1 + \frac{\alpha^2(\omega)}{\beta^2(\omega)}\right)\left(p'x + q'y\right) + \left(\beta(\omega)m' - \alpha(\omega)m''\right)\Delta z$$

$$+ i\left(\alpha(\omega)m' + \beta(\omega)m''\right)\Delta z,$$

so that

$$e^{i\tilde{\mathbf{k}}^+(\omega) \cdot \mathbf{r}} = e^{-\left(\alpha(\omega)m' + \beta(\omega)m''\right)\Delta z}$$

$$\times e^{i\beta(\omega)\left[\left(1 + \alpha^2(\omega)/\beta^2(\omega)\right)\left(p'x + q'y\right) + \left(m' - \left(\alpha(\omega)/\beta(\omega)\right)m''\right)\Delta z\right]},$$

$$(8.49)$$

where $\Delta z \equiv z - z_0$. If $p' = q' = 0$, then $e^{i\tilde{\mathbf{k}}^+ \cdot \mathbf{r}}$ represents the spatial part of a time-harmonic homogeneous plane wave with angular frequency ω because the surfaces of constant amplitude coincide with the surfaces of constant phase, given by $\Delta z = $ constant. If $\alpha(\omega) = 0$, then the expression in Eq. (8.49) represents the spatial part of a time-harmonic homogeneous plane wave when $p'^2 + q'^2 \leq 1$ (in which case $m'' = 0$), whereas it represents an evanescent wave when $p'^2 + q'^2 > 1$ (in which case $m' = 0$) because the surfaces of constant amplitude are then orthogonal to the equiphase surfaces. In general, $\alpha(\omega) \neq 0$ and either $p' \neq 0$ or $q' \neq 0$; the expression in Eq. (8.49) then represents the spatial part of a time-harmonic inhomogeneous plane wave with angular frequency ω because the surfaces of constant amplitude $\Delta z = $ constant are different from the surfaces of constant phase, given by

$$\left(1 + \frac{\alpha^2(\omega)}{\beta^2(\omega)}\right)(p'x + q'y) + \left(m' - \frac{\alpha(\omega)}{\beta(\omega)}m''\right)\Delta z = \text{constant}. \qquad (8.50)$$

The inhomogeneous plane wave phase fronts described in Eq. (8.50) propagate in the direction specified by the real-valued vector

$$\mathbf{s} \equiv \left(1 + \frac{\alpha^2(\omega)}{\beta^2(\omega)}\right)(p'\hat{\mathbf{1}}_x + q'\hat{\mathbf{1}}_y) + \left(m' - \frac{\alpha(\omega)}{\beta(\omega)}m''\right)\hat{\mathbf{1}}_z \qquad (8.51)$$

with magnitude

$$s = \left[\left(1 + \frac{\alpha^2(\omega)}{\beta^2(\omega)}\right)^2(p'^2 + q'^2) + \left(m' - \frac{\alpha(\omega)}{\beta(\omega)}m''\right)^2\right]^{1/2}, \qquad (8.52)$$

which is, in general, not equal to unity. The plane phase fronts or cophasal surfaces of the inhomogeneous plane wave described in Eq. (8.49) are then seen to propagate in the direction specified by the set of real-valued directions cosines

$$\left\{\frac{1}{s}\left(1 + \frac{\alpha^2(\omega)}{\beta^2(\omega)}\right)p', \frac{1}{s}\left(1 + \frac{\alpha^2(\omega)}{\beta^2(\omega)}\right)q', \frac{1}{s}\left(m' - \frac{\alpha(\omega)}{\beta(\omega)}m''\right)\right\}, \qquad (8.53)$$

as illustrated in Fig. 8.4. These inhomogeneous plane wave phase fronts advance into the positive half-space $\Delta z > 0$ when the inequality

$$m' - \frac{\alpha(\omega)}{\beta(\omega)}m'' > 0 \qquad (8.54)$$

is satisfied. Substitution of the relation given in Eq. (8.46) into this expression then yields the equivalent inequality

$$m'\left(1 - \frac{\alpha^2(\omega)}{\beta^2(\omega)}\frac{(p'^2 + q'^2)}{m'^2}\right) > 0.$$

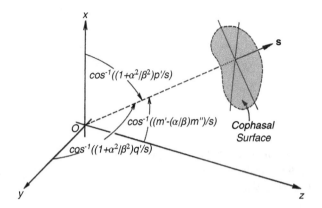

Fig. 8.4 Inhomogeneous plane wave phase front propagating in the direction s

Because $m' > 0$ when $\alpha(\omega) \neq 0$, then the inequality given in Eq. (8.54) will be satisfied when

$$\frac{\left(p'^2 + q'^2\right)}{m'^2} < \frac{\beta^2(\omega)}{\alpha^2(\omega)}, \qquad (8.55)$$

where the right-hand side of this inequality depends solely upon the dispersive properties of the medium. For a nondispersive medium, $\alpha(\omega) = 0$ for all ω and the inequality given in Eq. (8.54) simply becomes $m' > 0$, which is satisfied by the homogeneous plane waves propagating into the positive half-space occupied by a loss-free medium.

Consider now obtaining the conditions (if indeed any do exist in the general case) under which the inequality given in Eq. (8.55) is satisfied. In order to address this, return to the quartic equation in m' following Eq. (8.44); viz.

$$m'^4 + \left[\left(1 - \frac{\alpha^2}{\beta^2}\right)\left(p'^2 + q'^2\right) - 1\right]m'^2 - \frac{\alpha^2}{\beta^2}\left(p'^2 + q'^2\right)^2 = 0.$$

Two inequalities may then be obtained from this equation that are dependent upon the sign of the coefficient of m'^2, as follows.

1. If this coefficient is negative [i.e., if $p'^2 + q'^2 < 1/\left(1 - \alpha^2/\beta^2\right)$] which occurs when $0 \leq \arg\{m(\omega)\} < \pi/4$, then the quartic equation given above implies that

$$\frac{p'^2 + q'^2}{m'^2} < \frac{\beta(\omega)}{\alpha(\omega)}. \qquad (8.56)$$

Hence, if $\beta(\omega) \geq \alpha(\omega)$, then the inequality appearing in Eq. (8.56) implies that the inequality appearing in Eq. (8.55) is satisfied. On the other hand, if $\alpha(\omega) >$

$\beta(\omega)$, then the inequality appearing in Eq. (8.55) may still be satisfied, but there is no guarantee from this method of argument.

2. If this coefficient is positive [i.e., if $p'^2 + q'^2 > 1/(1 - \alpha^2/\beta^2)$] which occurs when $\pi/4 < \arg\{m(\omega)\} \leq \pi/2$, then the quartic equation given above implies that

$$\frac{p'^2 + q'^2}{m'^2} > \frac{\beta(\omega)}{\alpha(\omega)}. \tag{8.57}$$

If $\beta(\omega) > \alpha(\omega)$ then the inequality appearing in Eq. (8.55) may be satisfied, but there is no guarantee from this method of argument. However, it is definitely not satisfied if $\alpha(\omega) \leq \beta(\omega)$.

Notice that when $\alpha(\omega) = 0$, the inequality specified in case 1 reduces to the inequality $p'^2 + q'^2 < 1$ which specifies the homogeneous plane wave components in a loss-free medium. On the other hand, when $\alpha(\omega) = 0$, the inequality specified in case 2 becomes $p'^2 + q'^2 > 1$ which specifies the evanescent plane wave components in a loss-free medium.

The geometric form of the angular spectrum representation of the freely propagating electromagnetic field is then given by

$\mathbf{E}(x, y, z, t)$

$$= \frac{1}{4\pi^3} \Re\left\{ \int_{C_+} d\omega\, e^{-i\omega t} \int_{C_p} \int_{C_q} \tilde{\mathbf{E}}_0(p, q, \omega) e^{i\tilde{k}(\omega)(px + qy + m\Delta z)} \tilde{k}^2(\omega) dp\, dq \right\},$$

$$\tag{8.58}$$

$\mathbf{B}(x, y, z, t)$

$$= \frac{1}{4\pi^3} \Re\left\{ \int_{C_+} d\omega\, e^{-i\omega t} \int_{C_p} \int_{C_q} \tilde{\mathbf{B}}_0(p, q, \omega) e^{i\tilde{k}(\omega)(px + qy + m\Delta z)} \tilde{k}^2(\omega) dp\, dq \right\},$$

$$\tag{8.59}$$

where $p = p' + ip''$ varies over the contour C_p and $q = q' + iq''$ varies over the contour C_q, as illustrated in Fig. 8.2. Because the spatiotemporal frequency spectra of the field vector boundary values are related [cf. Eqs. (8.18) and (8.19)], then the pair of relations given in Eqs. (8.58) and (8.59) for the propagated field vectors in the half-space $z \geq z_0$ may be expressed in terms of either boundary value alone, that is, in terms of either $\tilde{\mathbf{E}}_0(p, q, \omega)$ or $\tilde{\mathbf{B}}_0(p, q, \omega)$. Similar results for the formal solution of such boundary value problems for time-harmonic (or monochromatic) wave propagation may be found in the published work of Bouwkamp [7] and Goodman [8] for the scalar optical field, in Carter [9], Stamnes [10], and Nieto-Vesperinas [11] for the diffraction theory of the electromagnetic field, and in Devaney [12] for the mathematical theory of direct and inverse imaging, tomography, and wave-field inversion.

8.1.2 The Angular Spectrum Representation and Huygen's Principle

It is well known that the solution of the planar boundary value problem that is considered in this chapter can also be obtained through a superposition of spherical waves. The solution in terms of spherical waves has its physical origin in Huygen's principle [13] and has been (and continues to be) a central theme in classical diffraction theory. It is consequently of some importance to establish the connection between the angular spectrum of plane waves representation and the mathematical embodiment of Huygen's principle as found in classical diffraction theory. It is clear that Weyl's integral, as given in Eq. (7.167), which expresses a spherical wave in terms of a superposition of plane waves, provides this connection, which is now derived following the treatment due to Sherman [1, 2].

The derivation begins with the angular spectrum representation of the freely propagating electromagnetic wave field that is given in Eqs. (8.58)–(8.59) where $\Delta z = z - z_0 \geq 0$ is the normal propagation distance between the observation plane at $z \geq z_0^+$ and the input plane at $z = z_0^+$. In this spatio-temporal spectral representation

$$\tilde{\mathbf{E}}_0(p, q, \omega) = \int_{-\infty}^{\infty} \tilde{\mathbf{E}}_0(x, y, \omega) e^{-i\tilde{k}(\omega)(px+qy)} dxdy, \tag{8.60}$$

$$\tilde{\mathbf{B}}_0(p, q, \omega) = \int_{-\infty}^{\infty} \tilde{\mathbf{B}}_0(x, y, \omega) e^{-i\tilde{k}(\omega)(px+qy)} dxdy, \tag{8.61}$$

where

$$\tilde{\mathbf{E}}_0(x, y, \omega) = \int_{-\infty}^{\infty} \mathbf{E}_0(x, y, t) e^{i\omega t} dt, \tag{8.62}$$

$$\tilde{\mathbf{B}}_0(x, y, \omega) = \int_{-\infty}^{\infty} \mathbf{B}_0(x, y, t) e^{i\omega t} dt, \tag{8.63}$$

are the temporal frequency spectra of the initial field vectors on the plane $z = z_0^+$. Substitution of Eq. (8.60) into Eq. (8.58) results in the expression

$$\mathbf{E}(x, y, z, t) = \frac{1}{4\pi^3} \Re \left\{ \int_{C_+} d\omega \, e^{-i\omega t} \int_{-\infty}^{\infty} \int_{-\infty}^{\infty} dx'dy' \, \tilde{\mathbf{E}}_0(x', y', \omega) \right.$$
$$\left. \times \int_{C_p} \int_{C_q} e^{i\tilde{k}(\omega)[p(x-x')+q(y-y')+m\Delta z]\tilde{k}^2(\omega)} dpdq \right\},$$

$$\tag{8.64}$$

and substitution of Eq. (8.61) into Eq. (8.59) yields

$$\mathbf{B}(x,y,z,t) = \frac{1}{4\pi^3}\Re\left\{\int_{C_+} d\omega\, e^{-i\omega t}\int_{-\infty}^{\infty}\int_{-\infty}^{\infty} dx'dy'\,\tilde{\mathbf{B}}_0(x',y',\omega)\right.$$
$$\left.\times \int_{C_p}\int_{C_q} e^{i\tilde{k}(\omega)[p(x-x')+q(y-y')+m\Delta z]}\tilde{k}^2(\omega)dpdq\right\}.$$

(8.65)

The monochromatic spatial impulse response function for the normal propagation distance $\Delta z = z - z_0$ in the temporally dispersive HILL medium at the angular frequency ω is defined here as

$$h(\xi,\eta;\Delta z,\omega) \equiv \frac{1}{(2\pi)^2}\int_{C_p}\int_{C_q} e^{i\tilde{k}(\omega)[p\xi+q\eta+m\Delta z]}\tilde{k}^2(\omega)dpdq,$$

(8.66)

where $\xi = x - x'$ and $\eta = y - y'$ when applied to Eqs. (8.64) and (8.65). Notice that this impulse response function is space-invariant (or isoplanatic) and that it depends upon the angular frequency ω through the complex wave number $\tilde{k}(\omega)$. With the set of relations given in Eqs. (8.21)–(8.23), this expression may be rewritten as

$$h(\xi,\eta;\zeta,\omega) = \frac{1}{(2\pi)^2}\int_{-\infty}^{\infty}\int_{-\infty}^{\infty} e^{i\gamma(\omega)\zeta}e^{i(k_x\xi+k_y\eta)}dk_xdk_y$$
$$= \mathcal{F}^{-1}\left\{e^{i\gamma(\omega)\zeta}\right\},$$

(8.67)

where $\gamma(\omega) = \left[\tilde{k}^2(\omega) - k_x^2 - k_y^2\right]^{1/2}$. The quantity $e^{i\gamma(\omega)\zeta}$ is seen to be the spatial transfer function of the linear dispersive system at the angular frequency ω. With the definition given in Eq. (8.66), the preceding pair of expressions for the propagated field vectors becomes

$$\mathbf{E}(x,y,z,t)$$
$$= \frac{1}{\pi}\Re\left\{\int_{C_+} d\omega\, e^{-i\omega t}\int_{-\infty}^{\infty}\int_{-\infty}^{\infty}\tilde{\mathbf{E}}_0(x',y',\omega)h(x-x',y-y';\Delta z,\omega)dx'dy'\right\}$$

(8.68)

$$\mathbf{B}(x,y,z,t)$$
$$= \frac{1}{\pi}\Re\left\{\int_{C_+} d\omega\, e^{-i\omega t}\int_{-\infty}^{\infty}\int_{-\infty}^{\infty}\tilde{\mathbf{B}}_0(x',y',\omega)h(x-x',y-y';\Delta z,\omega)dx'dy'\right\}$$

(8.69)

and the spatial part of each propagated field vector is given by the two-dimensional convolution of the spatial part of the corresponding initial field vector on the plane $z = z_0^+$ with the spatial impulse response function at each value of the angular frequency ω. Notice that the material dispersion is contained entirely within the spatial impulse response function.

Weyl's integral given in Eq. (7.167) expresses a monochromatic spherical wave in terms of a superposition of monochromatic plane waves as

$$\frac{e^{i\tilde{k}(\omega)r}}{r} = \frac{i}{2\pi} \int_{-\infty}^{\infty} \int_{-\infty}^{\infty} \frac{1}{\gamma(\omega)} e^{i\mathbf{k}^{\pm}(\omega)\cdot\mathbf{r}} dk_x dk_y, \qquad (8.70)$$

where $\mathbf{k}^{\pm}(\omega) = \hat{\mathbf{1}}_x k_x + \hat{\mathbf{1}}_y k_y \pm \hat{\mathbf{1}}_z \gamma(\omega)$, $\mathbf{r} = \hat{\mathbf{1}}_x x + \hat{\mathbf{1}}_y y + \hat{\mathbf{1}}_z z$, and where $r = |\mathbf{r}| = +\sqrt{x^2 + y^2 + z^2}$. The positive sign choice is taken in the positive half-space $z > 0$ whereas the negative sign choice is taken in the negative half-space $z < 0$. With this result in mind, the relation given in Eq. (8.66) may be expressed as

$$\begin{aligned}
h(x, y; \Delta z, \omega) &= \frac{1}{2\pi^2} \int_{-\infty}^{\infty} \int_{-\infty}^{\infty} e^{i\mathbf{k}^+(\omega)\cdot\mathbf{r}^+} dk_x dk_y \\
&= \frac{1}{2\pi^2} \int_{-\infty}^{\infty} \int_{-\infty}^{\infty} e^{i(k_x x + k_y y + \gamma(\omega)\Delta z)} dk_x dk_y \\
&= -\frac{1}{2\pi} \frac{\partial}{\partial z} \left(\frac{i}{2\pi} \int_{-\infty}^{\infty} \int_{-\infty}^{\infty} \frac{1}{\gamma(\omega)} e^{i\mathbf{k}^+(\omega)\cdot\mathbf{r}^+} dk_x dk_y \right), \quad (8.71)
\end{aligned}$$

where $\mathbf{r}^+ \equiv \hat{\mathbf{1}}_x x + \hat{\mathbf{1}}_y y + \hat{\mathbf{1}}_z \Delta z$ with $\Delta z = z - z_0 > 0$.

The interchange of the order of the differentiation and integration operations used in the derivation of the final expression in Eq. (8.71) is justified by the following argument due to Lalor [14]. Define the pair of functions

$$f(z) \equiv \int_{C_p} \int_{C_q} \frac{1}{m} e^{i\tilde{k}(px+qz+mz)} dp dq,$$

$$g(z) \equiv i\tilde{k} \int_{C_p} \int_{C_q} e^{i\tilde{k}(px+qz+mz)} dp dq,$$

where it is desired to show that $g(z) = \partial f(z)/\partial z$ or, equivalently, that

$$\int_{z_0}^{z} g(z') dz' = f(z) - f(z_0).$$

Consider then the integral

$$\int_{z_0}^{z} g(z') dz' = i\tilde{k} \int_{z_0}^{z} dz' \int_{C_p} \int_{C_q} e^{i\tilde{k}(px+qy+mz')} dp dq.$$

Because the Lebesgue integrability of the function $\exp\left[i\tilde{k}(px + qy + mz')\right]$ is ensured by the existence of the integral

$$I = \int_{z_0}^{z} dz' \int_{C_p} \int_{C_q} \left| e^{i\tilde{k}(px+qy+mz')} \right| dpdq$$

for $z' > 0$, then by Fubini's theorem [15], the order of integration in the above expression may be interchanged, so that

$$\int_{z_0}^{z} g(z')dz' = \int_{C_p} \int_{C_q} \frac{1}{m} \left[e^{i\tilde{k}(px+qy+mz')} \right]_{z_0}^{z} dpdq$$

$$= f(z) - f(z_0),$$

as was to be shown.

With substitution of the identity expressed by Weyl's integral [Eq. (8.70)] in Eq. (8.71), one finally obtains the important result that

$$h(x - x', y - y'; \Delta z, \omega) = -\frac{1}{2\pi} \frac{\partial}{\partial z} \left(\frac{e^{i\tilde{k}(\omega)R}}{R} \right), \qquad (8.72)$$

where $R \equiv +\sqrt{(x - x')^2 + (y - y')^2 + (\Delta z)^2}$. This then identifies the spatial impulse response function defined in Eq. (8.66) in terms of the normal derivative of the "free-space" Green's function

$$G(R, \omega) \equiv \frac{e^{i\tilde{k}(\omega)R}}{R} \qquad (8.73)$$

that plays a central role in the mathematical embodiment of Huygen's principle, as given by the integral theorem of Helmholtz and Kirchhoff [13]. Substitution of Eqs. (8.72) and (8.73) into the pair of relations given in Eqs. (8.68) and (8.69) then yields

$$\mathbf{E}(x, y, z, t) = \frac{-1}{2\pi^2} \Re \left\{ \int_{C_+} d\omega \, e^{-i\omega t} \int_{-\infty}^{\infty} \int_{-\infty}^{\infty} \tilde{\mathbf{E}}_0(x', y', \omega) \frac{\partial G(R, \omega)}{\partial z} dx'dy' \right\},$$

$$(8.74)$$

$$\mathbf{B}(x, y, z, t) = \frac{-1}{2\pi^2} \Re \left\{ \int_{C_+} d\omega \, e^{-i\omega t} \int_{-\infty}^{\infty} \int_{-\infty}^{\infty} \tilde{\mathbf{B}}_0(x', y', \omega) \frac{\partial G(R, \omega)}{\partial z} dx'dy' \right\}.$$

$$(8.75)$$

The spatial integrals appearing here are just the *first Rayleigh–Sommerfeld diffraction integrals* [16] of classical optics. The solution in terms of the normal derivatives of the initial field vectors, which yields the *second Rayleigh–Sommerfeld diffraction integrals*, is left as an exercise (see Problem 8.3).

8.2   Polarization Properties of the Freely Propagating Electromagnetic Wave Field

Of considerable interest to the description of the propagation characteristics of a general electromagnetic wave field are the polarization properties of its electric and magnetic field vectors. The standard treatment [13] of the polarization state of an electromagnetic wave is restricted to the idealized case of a time-harmonic plane wave field. This restriction has, in part, been removed by Nisbet and Wolf [17] for the case of a linearly polarized, time-harmonic wave field with arbitrary spatial form. A complete extension [18] of this treatment to a general pulsed electromagnetic wave field is then directly accomplished through the angular spectrum of plane waves representation for the freely propagating field. Not surprisingly, the coherence properties of an electromagnetic field is also involved, a topic of fundamental importance to both optics and electromagnetics that was inspired by the early Nobel-prize winning work of Dennis Gabor (1971 Nobel Prize in Physics) and begun in earnest by Emil Wolf in the early 1950s with Max Born (1954 Nobel Prize in Physics) at Edinburgh University [19].

There are three fundamental, interrelated parts for the complete description of an electromagnetic wave field: the first is quite obviously Maxwell's unifying theory for the interrelated electric and magnetic field vectors (which naturally contains Gauss's, Faraday's, and Ampère's laws); the second is Stokes' theory of polarization and it's extensions; and the third is Wolf's scalar theory of optical coherence [13, 19, 20] and it's subsequent extension to electromagnetic wave theory [21]. The first two parts (Maxwell's equations and polarization theory) are usually developed in an idealized manner with the third part (coherence theory) incorporating physical reality into the overall theory. The *degree of coherence* in either the space–time domain or the space-frequency domain then provides the fundamental measure of this intrinsic wave-field property. Although the genesis of this property occurred within the mathematical framework of scalar optical field theory [13, 19, 20, 22], which in turn is based on Young's double-slit interference experiment with partially coherent light, it is the more recently developed electromagnetic coherence theory that fully accounts for the "intrinsic vectorial nature of the electromagnetic field" [23]. Because of this, electromagnetic coherence and electromagnetic polarization are found to be intimately connected [24, 25]. However, as important as this fundamental property is, its complete development is beyond the scope of this text.

From the pair of expressions given in Eqs. (8.14) and (8.17), the angular spectrum of plane waves representation of a freely propagating electromagnetic wave field may be written as

$$E(\mathbf{r}, t) = \frac{1}{\pi} \Re \left\{ \int_{C_+} \tilde{E}(\mathbf{r}, \omega) e^{-i\omega t} d\omega \right\}, \tag{8.76}$$

$$B(\mathbf{r}, t) = \frac{1}{\pi} \Re \left\{ \int_{C_+} \tilde{B}(\mathbf{r}, \omega) e^{-i\omega t} d\omega \right\}, \tag{8.77}$$

where the (complex-valued) temporal frequency spectrum of each field vector has the angular spectrum representation

$$\tilde{E}(\mathbf{r}, \omega) = \frac{1}{(2\pi)^2} \int_{-\infty}^{\infty} \int_{-\infty}^{\infty} \tilde{E}_0(\mathbf{k}_T, \omega) e^{i\tilde{\mathbf{k}}^+ \cdot \mathbf{r}^+} dk_x dk_y$$

$$\equiv \mathbf{p}_e(\mathbf{r}, \omega) + i\mathbf{q}_e(\mathbf{r}, \omega), \tag{8.78}$$

$$\tilde{B}(\mathbf{r}, \omega) = \frac{1}{(2\pi)^2} \int_{-\infty}^{\infty} \int_{-\infty}^{\infty} \tilde{B}_0(\mathbf{k}_T, \omega) e^{i\tilde{\mathbf{k}}^+ \cdot \mathbf{r}^+} dk_x dk_y$$

$$\equiv \mathbf{p}_m(\mathbf{r}, \omega) + i\mathbf{q}_m(\mathbf{r}, \omega). \tag{8.79}$$

Here $\mathbf{p}_j(\mathbf{r}, \omega)$ and $\mathbf{q}_j(\mathbf{r}, \omega)$ denote the real and imaginary parts, respectively, of the temporal frequency spectrum domain form of the appropriate field vector, as indicated by the subscript $j = e, m$. With this substitution, the pair of relations given in Eqs. (8.76) and (8.77) becomes

$$E(\mathbf{r}, t) = \frac{1}{\pi} \Re \left\{ \int_{C_+} \tilde{V}_e(\mathbf{r}, t; \omega) d\omega \right\}, \tag{8.80}$$

$$B(\mathbf{r}, t) = \frac{1}{\pi} \Re \left\{ \int_{C_+} \tilde{V}_m(\mathbf{r}, t; \omega) d\omega \right\}, \tag{8.81}$$

where

$$\tilde{V}_j(\mathbf{r}, t; \omega) \equiv \left[\mathbf{p}_j(\mathbf{r}, \omega) + i\mathbf{q}_j(\mathbf{r}, \omega) \right] e^{-i\omega t}, \tag{8.82}$$

for $j = e, m$, are complex-valued vector fields that describe the spatial properties of each monochromatic field component appearing in the propagated field representation given in Eqs. (8.80) and (8.81).

8.2.1   Polarization Ellipse for Complex Field Vectors

With the analysis of Born and Wolf [13] as a guide, consider the behavior of the time-harmonic complex vector field

$$\tilde{V}(\mathbf{r}, t) \equiv \left[\mathbf{p}(\mathbf{r}) + i\mathbf{q}(\mathbf{r}) \right] e^{-i\omega t} \tag{8.83}$$

at a fixed point $\mathbf{r} = \mathbf{r}_0$ in space. Here $\tilde{\mathbf{V}}(\mathbf{r}, t)$ represents either $\tilde{\mathbf{V}}_e(\mathbf{r}, t)$ or $\tilde{\mathbf{V}}_m(\mathbf{r}, t)$. As time varies the end point of the vector $\tilde{\mathbf{V}}(\mathbf{r}_0, t)$ describes a curve in the plane that is specified by the pair of (real-valued) vector $\mathbf{p}(\mathbf{r}_0)$ and $\mathbf{q}(\mathbf{r}_0)$. Furthermore, because $\tilde{\mathbf{V}}(\mathbf{r}_0, t)$ is periodic in time at any fixed point $\mathbf{r} = \mathbf{r}_0$ in space, this curve must then be closed. Now let

$$\mathbf{p}(\mathbf{r}_0) + i\mathbf{q}(\mathbf{r}_0) \equiv (\mathbf{a} + i\mathbf{b})e^{i\varphi}, \tag{8.84}$$

where φ is as yet unspecified. The real vectors $\mathbf{a}$ and $\mathbf{b}$ may then be expressed in terms of $\mathbf{p}(\mathbf{r}_0)$, $\mathbf{q}(\mathbf{r}_0)$, and φ as

$$\mathbf{a} = \mathbf{p}(\mathbf{r}_0) \cos \varphi + \mathbf{q}(\mathbf{r}_0) \sin \varphi, \tag{8.85}$$

$$\mathbf{b} = -\mathbf{p}(\mathbf{r}_0) \sin \varphi + \mathbf{q}(\mathbf{r}_0) \cos \varphi. \tag{8.86}$$

Notice that $\mathbf{a}$, $\mathbf{b}$, and φ are all functions of the (fixed) position vector $\mathbf{r}_0$. Consider now choosing the angle φ such that the vectors $\mathbf{a}$ and $\mathbf{b}$ are orthogonal so that $\mathbf{a} \cdot \mathbf{b} = 0$ and

$$\tan (2\varphi) = \frac{2\mathbf{p}(\mathbf{r}_0) \cdot \mathbf{q}(\mathbf{r}_0)}{\mathbf{p}^2(\mathbf{r}_0) - \mathbf{q}^2(\mathbf{r}_0)}. \tag{8.87}$$

The parameters that specify the spatial properties of the complex vector field $\tilde{\mathbf{V}}(\mathbf{r}, t)$ at any fixed point $\mathbf{r} = \mathbf{r}_0$ may now be considered to be the five independent components of the orthogonal vectors $\mathbf{a}$ and $\mathbf{b}$ and the associated phase factor φ, instead of the six independent components of the vectors $\mathbf{p}$ and $\mathbf{q}$. With substitution from Eq. (8.84), the expression given in Eq. (8.83) for the monochromatic complex vector field $\tilde{\mathbf{V}}(\mathbf{r}, t)$ at the fixed point $\mathbf{r} = \mathbf{r}_0$ becomes

$$\begin{aligned} \tilde{\mathbf{V}}(\mathbf{r}_0, t) &\equiv \tilde{\mathbf{V}}^{(r)}(\mathbf{r}_0, t) + i\tilde{\mathbf{V}}^{(i)}(\mathbf{r}_0, t) \\ &= \left[\mathbf{a} \cos (\omega t - \varphi) + \mathbf{b} \sin (\omega t - \varphi) \right] \\ &\quad - i \left[\mathbf{a} \sin (\omega t - \varphi) - \mathbf{b} \cos (\omega t - \varphi) \right]. \end{aligned} \tag{8.88}$$

If a Cartesian coordinate system is now defined with origin at the fixed field point $\mathbf{r} = \mathbf{r}_0$ and with the x- and y-coordinate directions chosen along the direction of the vectors $\mathbf{a} = \mathbf{a}(\mathbf{r}_0)$ and $\mathbf{b} = \mathbf{b}(\mathbf{r}_0)$, respectively, then the components of the real part of $\tilde{\mathbf{V}}(\mathbf{r}_0, t)$ with respect to this coordinate system are given by

$$\begin{aligned} \tilde{V}_x^{(r)}(\mathbf{r}_0, t) &= a(\mathbf{r}_0) \cos (\omega t - \varphi(\mathbf{r}_0)), \\ \tilde{V}_y^{(r)}(\mathbf{r}_0, t) &= b(\mathbf{r}_0) \sin (\omega t - \varphi(\mathbf{r}_0)), \\ \tilde{V}_z^{(r)}(\mathbf{r}_0, t) &= 0, \end{aligned} \tag{8.89}$$

and the components of the imaginary part of $\mathbf{V}(\mathbf{r}_0, t)$ with respect to this coordinate system are given by

$$\tilde{V}_x^{(i)}(\mathbf{r}_0, t) = a(\mathbf{r}_0) \cos \left(\omega t - \varphi(\mathbf{r}_0) + \pi/2\right),$$
$$\tilde{V}_y^{(i)}(\mathbf{r}_0, t) = b(\mathbf{r}_0) \sin \left(\omega t - \varphi(\mathbf{r}_0) + \pi/2\right), \qquad (8.90)$$
$$\tilde{V}_z^{(i)}(\mathbf{r}_0, t) = 0,$$

where $a(\mathbf{r}_0) \equiv |\mathbf{a}(\mathbf{r}_0)|$ and $b(\mathbf{r}_0) \equiv |\mathbf{b}(\mathbf{r}_0)|$. Both of the above two sets of equations describe an ellipse in time, called the *polarization ellipse*, that is given by

$$\frac{\left(\tilde{V}_x^{(j)}(\mathbf{r}_0, t)\right)^2}{a^2} + \frac{\left(\tilde{V}_y^{(j)}(\mathbf{r}_0, t)\right)^2}{b^2} = 1; \quad j = r, i, \qquad (8.91)$$

with semi-axes of lengths a and b along the x- and y-coordinate axes, respectively. Notice that the real and imaginary parts of $\tilde{\mathbf{V}}(\mathbf{r}_0, t)$ are $\pi/2$ out of phase as they trace out the polarization ellipse. The real vectors $\mathbf{p}(\mathbf{r}_0)$ and $\mathbf{q}(\mathbf{r}_0)$ are then seen to form a pair of conjugate semi-diameters of the polarization ellipse, as illustrated in Fig. 8.5.

The semi-axis lengths a and b of the polarization ellipse described by the complex vector field at the fixed point $\mathbf{r} = \mathbf{r}_0$ are readily obtained from Eqs. (8.85)–(8.87) as

$$a^2 = \frac{1}{2}\left[p^2 + q^2 + \left(\left(p^2 - q^2\right)^2 + 4\,(\mathbf{p} \cdot \mathbf{q})^2\right)^{1/2}\right], \qquad (8.92)$$

$$b^2 = \frac{1}{2}\left[p^2 + q^2 - \left(\left(p^2 - q^2\right)^2 + 4\,(\mathbf{p} \cdot \mathbf{q})^2\right)^{1/2}\right]. \qquad (8.93)$$

Fig. 8.5 The polarization ellipse at a fixed point in space

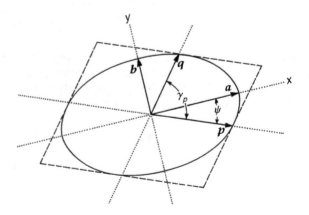

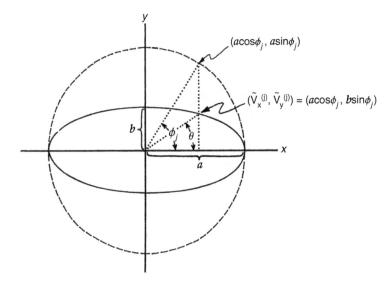

Fig. 8.6 The geometric relation between the eccentric angle ϕ_j and the polar angle θ of the point $\left(\tilde{V}_x^{(j)}, \tilde{V}_y^{(j)}\right)$ on the polarization ellipse

In order to determine the angle ψ between the vectors **a** and **p**, depicted in Fig. 8.5, one begins by expressing the equation of the polarization ellipse in parametric form as

$$\tilde{V}_x^{(j)}(\mathbf{r}_0, t) = a \cos \phi_j, \tag{8.94}$$

$$\tilde{V}_y^{(j)}(\mathbf{r}_0, t) = b \sin \phi_j, \tag{8.95}$$

for $j = r, i$, where $\phi_j = \phi_j(\mathbf{r}_0, t)$ denotes the eccentric angle that is depicted in Fig. 8.6 for the case when $a \geq b$ (this inequality is assumed to hold throughout the remaining analysis). From the geometry of the figure it is seen that the eccentric angle ϕ_j is related to the polar angle of the point $\left(\tilde{V}_x^{(j)}, \tilde{V}_y^{(j)}\right)$ on the vibration ellipse by

$$\tan \theta = \frac{\tilde{V}_y^{(j)}}{\tilde{V}_x^{(j)}} = \frac{b}{a} \tan \phi_j. \tag{8.96}$$

Comparison of the pair of expressions in Eq. (8.94) and (8.95) with the expressions given in Eqs. (8.89) and (8.90) shows that for the vibration ellipse of the real part $\mathbf{V}^{(r)}(\mathbf{r}_0, t)$ of the complex field vector

$$\phi_r(\mathbf{r}_0, t) = \omega t - \varphi(\mathbf{r}_0), \tag{8.97}$$

whereas for the imaginary part

$$\phi_i(\mathbf{r}_0, t) = \omega t - \varphi(\mathbf{r}_0) + \pi/2, \tag{8.98}$$

at any fixed point $\mathbf{r} = \mathbf{r}_0$ in the positive half-space. From Eq. (8.83) it follows that $\tilde{\mathbf{V}}^{(r)}(\mathbf{r}, t) = \mathbf{p}(\mathbf{r})$ when $t = 0$, and $\tilde{\mathbf{V}}^{(i)}(\mathbf{r}, t) = \mathbf{p}(\mathbf{r})$ when $\omega t = -\pi/2$, so that the eccentric angle of $\mathbf{p}$ is $-\varphi$; from Eq. (8.96), the angle ψ between the vectors $\mathbf{p}$ and $\mathbf{a}$ (see Fig. 8.5) is then given by

$$\tan \psi = \frac{b}{a} \tan \varphi. \tag{8.99}$$

Finally, if γ_p denotes the angle between the vectors $\mathbf{p}$ and $\mathbf{q}$, as illustrated in Fig. 8.5, and if β is an auxiliary angle defined as

$$\tan \beta \equiv \frac{q}{p}, \tag{8.100}$$

then Eq. (7.87) becomes

$$\tan (2\varphi) = \frac{2pq}{p^2 - q^2} \cos \gamma_p = \tan (2\beta) \cos \gamma_p. \tag{8.101}$$

Summarizing these standard results, if the vectors $\mathbf{p} = \mathbf{p}(\mathbf{r}_0)$ and $\mathbf{q} = \mathbf{q}(\mathbf{r}_0)$ are given at the fixed point $\mathbf{r} = \mathbf{r}_0$, where γ_p is the angle between these two vectors and β is the auxiliary angle defined in Eq. (8.100) as the inverse tangent of the ratio of the magnitudes of these two vectors, then the principal semi-axes of the vibration ellipse at the point $\mathbf{r}_0$ are given by the relations in Eqs. (8.92) and (8.93) and the angle ψ that the major axis makes with the vector $\mathbf{p}$ is given by Eq. (8.99) where the phase factor φ is found from Eq. (8.101).

From Eqs. (8.82)–(8.84), each time-harmonic vector field component appearing in the propagated field representation given in Eqs. (8.80) and (8.81) is of the form

$$\tilde{\mathbf{V}}(\mathbf{r}, t; \omega) = \big(\mathbf{a}(\mathbf{r}, \omega) + i\mathbf{b}(\mathbf{r}, \omega)\big)e^{i(\varphi(\mathbf{r},\omega) - \omega t)} \tag{8.102}$$

which is the complex representation of a time-harmonic wave that is propagating in the direction specified by $\nabla\varphi(\mathbf{r})$. There are then two possible senses in which the polarization ellipse is traced out, corresponding to left-handed and right-handed polarizations. Curiously enough, there are two opposing definitions of this polarization sense: one in optics (physics) and one in electrical engineering.

Consider first the definition in optics. If the sign of the scalar triple product $[\mathbf{a}, \mathbf{b}, \nabla\varphi] = (\mathbf{a} \times \mathbf{b}) \cdot \nabla\varphi$ is positive, then to an observer looking in a direction that is opposite to that in which the wave field is propagating (looking into the light), the end point of the field vector describes its ellipse in the counterclockwise sense and the polarization is said to be left-handed, as illustrated in part (a) of Fig. 8.7. On the other hand (pun intended), if the sign of the scalar triple product $[\mathbf{a}, \mathbf{b}, \nabla\varphi]$

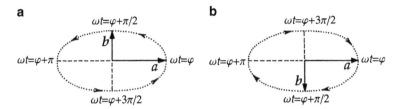

Fig. 8.7 Left-handed (**a**) and right-handed (**b**) polarization ellipses in optics. The polarization senses illustrated here are reversed in electrical engineering

is negative, then the polarization ellipse is described in the clockwise sense and the polarization is said to be right-handed, as illustrated in part (b) of the figure. In electrical engineering the polarization sense is defined for an observer looking in the direction that the wave field is propagating (looking along the direction of propagation). Left-handed then becomes right-handed and right-handed becomes left-handed when one goes from an optics to an electrical engineering point of view, and vice versa.

8.2.2 Propagation Properties of the Polarization Ellipse

Consider now the relation between the initial and propagated polarization properties of the pulsed electromagnetic wave field in the positive half-space $z \geq z_0^+$. From the expressions given in Eqs. (8.78), (8.79) and (8.84), the complex representation of the polarization ellipse for the propagated electric and magnetic field vectors is seen to be given by

$$\left(\mathbf{a}_e(\mathbf{r}^+, \omega) + i\mathbf{b}_e(\mathbf{r}^+, \omega)\right) e^{i\varphi_e(\mathbf{r}^+, \omega)}$$
$$= \frac{1}{(2\pi)^2} \int_{-\infty}^{\infty} \int_{-\infty}^{\infty} \tilde{\mathbf{E}}_0(\mathbf{k}_T, \omega) e^{i\mathbf{k}^+ \cdot \mathbf{r}^+} dk_x dk_y,$$

(8.103)

$$\left(\mathbf{a}_m(\mathbf{r}^+, \omega) + i\mathbf{b}_m(\mathbf{r}^+, \omega)\right) e^{i\varphi_m(\mathbf{r}^+, \omega)}$$
$$= \frac{1}{(2\pi)^2} \int_{-\infty}^{\infty} \int_{-\infty}^{\infty} \tilde{\mathbf{B}}_0(\mathbf{k}_T, \omega) e^{i\mathbf{k}^+ \cdot \mathbf{r}^+} dk_x dk_y,$$

(8.104)

where $\mathbf{r}^+ \equiv \hat{\mathbf{1}}_x x + \hat{\mathbf{1}}_y y + \hat{\mathbf{1}}_z (z - z_0)$ is the position vector of the field observation point in the positive half-space $z \geq z_0$. Notice that with the relations given in Eqs. (8.18) and (8.19), these two equations may be expressed solely in terms of either $\tilde{\mathbf{E}}_0(\mathbf{k}_T, \omega)$ or $\tilde{\mathbf{B}}_0(\mathbf{k}_T, \omega)$, if desired, where [cf. Eqs. (8.3)–(8.6)]

$$\tilde{\mathbf{E}}_0(\mathbf{k}_T, \omega) = \int_{-\infty}^{\infty} \int_{-\infty}^{\infty} \tilde{\mathbf{E}}_0(\mathbf{r}_T, \omega) e^{-i\mathbf{k}_T \cdot \mathbf{r}_T} dx dy, \tag{8.105}$$

$$\tilde{\mathbf{B}}_0(\mathbf{k}_T, \omega) = \int_{-\infty}^{\infty} \int_{-\infty}^{\infty} \tilde{\mathbf{B}}_0(\mathbf{r}_T, \omega) e^{-i\mathbf{k}_T \cdot \mathbf{r}_T} dx dy, \tag{8.106}$$

with

$$\tilde{\mathbf{E}}_0(\mathbf{r}_T, \omega) = \int_{-\infty}^{\infty} \mathbf{E}_0(\mathbf{r}_T, t) e^{i\omega t} dt, \tag{8.107}$$

$$\tilde{\mathbf{B}}_0(\mathbf{r}_T, \omega) = \int_{-\infty}^{\infty} \mathbf{B}_0(\mathbf{r}_T, t) e^{i\omega t} dt, \tag{8.108}$$

with $\mathbf{r}_T = \hat{\mathbf{1}}_x x + \hat{\mathbf{1}}_y y$ and $\mathbf{k}_T = \hat{\mathbf{1}}_x k_x + \hat{\mathbf{1}}_y k_y$. As in Eqs. (8.80) and (8.81), the initial field vectors on the $z = z_0^+$ plane may be expressed as

$$\mathbf{E}_0(\mathbf{r}, t) = \frac{1}{\pi} \Re \left\{ \int_{C_+} \tilde{\mathbf{V}}_e^{(0)}(\mathbf{r}, t; \omega) d\omega \right\}, \tag{8.109}$$

$$\mathbf{B}_0(\mathbf{r}, t) = \frac{1}{\pi} \Re \left\{ \int_{C_+} \tilde{\mathbf{V}}_m^{(0)}(\mathbf{r}, t; \omega) d\omega \right\}, \tag{8.110}$$

where

$$\tilde{\mathbf{V}}_e(\mathbf{r}, t; \omega) = \tilde{\mathbf{E}}_0(\mathbf{r}_T, \omega) e^{-i\omega t}$$
$$= \left[\mathbf{p}_e^{(0)}(\mathbf{r}_T, \omega) + i \mathbf{q}_e^{(0)}(\mathbf{r}_T, \omega) \right] e^{-i\omega t}, \tag{8.111}$$

$$\tilde{\mathbf{V}}_m(\mathbf{r}, t; \omega) = \tilde{\mathbf{B}}_0(\mathbf{r}_T, \omega) e^{-i\omega t}$$
$$= \left[\mathbf{p}_m^{(0)}(\mathbf{r}_T, \omega) + i \mathbf{q}_m^{(0)}(\mathbf{r}_T, \omega) \right] e^{-i\omega t}, \tag{8.112}$$

are complex vectors that describe the spatial properties of each monochromatic field component that is present in the initial wave field. With the complex representation of the polarization ellipse that is given in Eq. (8.84), one then has that

$$\tilde{\mathbf{E}}_0(\mathbf{r}_T, \omega) = \mathbf{p}_e^{(0)}(\mathbf{r}_T, \omega) + i \mathbf{q}_e^{(0)}(\mathbf{r}_T, \omega)$$
$$= \left[\mathbf{a}_e^{(0)}(\mathbf{r}_T, \omega) + i \mathbf{b}_e^{(0)}(\mathbf{r}_T, \omega) \right] e^{i\varphi_e^{(0)}(\mathbf{r}_T, \omega)}, \tag{8.113}$$

$$\tilde{\mathbf{B}}_0(\mathbf{r}_T, \omega) = \mathbf{p}_m^{(0)}(\mathbf{r}_T, \omega) + i \mathbf{q}_m^{(0)}(\mathbf{r}_T, \omega)$$
$$= \left[\mathbf{a}_m^{(0)}(\mathbf{r}_T, \omega) + i \mathbf{b}_m^{(0)}(\mathbf{r}_T, \omega) \right] e^{i\varphi_m^{(0)}(\mathbf{r}_T, \omega)}. \tag{8.114}$$

Substitution of Eq. (8.105) with (8.113) into Eq. (8.103) [or (8.106) with (8.114) into (8.104)] then yields the general expression

$$\left[\mathbf{a}_j(\mathbf{r}^+, \omega) + i\mathbf{b}_j(\mathbf{r}^+, \omega) \right] e^{i\varphi_j^{(0)}(\mathbf{r}^+, \omega)}$$

$$= \frac{1}{(2\pi)^2} \int_{-\infty}^{\infty} \int_{-\infty}^{\infty} \left\{ \left[\mathbf{a}_j^{(0)}(\mathbf{r}_T, \omega) + i\mathbf{b}_j^{(0)}(\mathbf{r}_T, \omega) \right] e^{i\varphi_m^{(0)}(\mathbf{r}_T, \omega)} \right.$$

$$\left. \times \int_{-\infty}^{\infty} \int_{-\infty}^{\infty} e^{i\left(\tilde{\mathbf{k}}^+ \cdot \mathbf{r}^+ - \mathbf{k}_T \cdot \mathbf{r}_T \right)} dk_x dk_y \right\} dx' dy'.$$

$$(8.115)$$

The integral appearing in the brackets of this expression is recognized as the *monochromatic spatial impulse response function*, defined in Eq. (8.66), for the normal propagation distance $\Delta z = z - z_0$ in the positive half-space $z \geq z_0^+$ of a temporally dispersive HILL medium at the angular frequency ω; viz.,

$$h(x - x', y - y'; \Delta z, \omega) = \frac{1}{(2\pi)^2} \int_{-\infty}^{\infty} \int_{-\infty}^{\infty} e^{i\left(\tilde{\mathbf{k}}^+ \cdot \mathbf{r}^+ - \mathbf{k}_T \cdot \mathbf{r}_T \right)} dk_x dk_y$$

$$= \frac{1}{(2\pi)^2} \int_{-\infty}^{\infty} \int_{-\infty}^{\infty} e^{i[k_x(x-x') + k_y(y-y') + \gamma(\omega)\Delta z]} dk_x dk_y,$$

$$(8.116)$$

where $\gamma = \gamma(\omega)$ is defined in Eq. (8.11) and should not be confused with the angle γ_p defined in Eq. (8.101) for the polarization ellipse (see Fig. 8.5). With this identification, Eq. (8.115) becomes

$$\left[\mathbf{a}_j(\mathbf{r}^+, \omega) + i\mathbf{b}_j(\mathbf{r}^+, \omega) \right] e^{i\varphi_j^{(0)}(\mathbf{r}^+, \omega)}$$

$$= \int_{-\infty}^{\infty} \int_{-\infty}^{\infty} \left[\mathbf{a}_j^{(0)}(\mathbf{r}_T, \omega) + i\mathbf{b}_j^{(0)}(\mathbf{r}_T, \omega) \right] e^{i\varphi_j^{(0)}(\mathbf{r}_T, \omega)}$$

$$\times h(x - x', y - y'; \Delta z, \omega) dx' dy'.$$

$$(8.117)$$

Hence, the polarization properties of each monochromatic component present in the propagated wave field are given by the convolution of the initial polarization behavior on the plane $z = z_0^+$ with the spatial impulse response function for the normal propagation distance $\Delta z = z - z_0$ in the temporally dispersive HILL medium at the angular frequency ω. As a consequence, if the polarization properties of the initial wave field vary from point to point over the plane at $z = z_0^+$, then the polarization properties of the propagated wave field will also, in general, vary from point to point throughout the positive half-space $\Delta z \geq 0$. In a strict sense, the state of polarization refers to the electromagnetic field vector behavior at a particular point in space and, in general, varies from point to point for each monochromatic component present in the wave field. Moreover, the frequency dependence allows for the state of polarization to vary in time at any fixed observation point in the positive half-space when the wave field is pulsed in time.

For a uniformly polarized field vector over the plane at $z = z_0^+$ it is required that

$$\left[\mathbf{a}_j^{(0)}(\mathbf{r}_T, \omega) + i\mathbf{b}_j^{(0)}(\mathbf{r}_T, \omega)\right] e^{i\varphi_j^{(0)}(\mathbf{r}_T, \omega)} = W_j^{(0)}(\mathbf{r}_T, \omega)\left(\hat{\mathbf{a}}_j^{(0)} + i\hat{\mathbf{b}}_j^{(0)}\right) e^{i\hat{\varphi}_j^{(0)}},$$
(8.118)

where $\hat{\mathbf{a}}_j^{(0)}$ and $\hat{\mathbf{b}}_j^{(0)}$ are both constant vectors and $\hat{\varphi}_j^{(0)}$ is a scalar constant. The only spatial variation in the particular field vector (either electric $j = e$ or magnetic $j = m$) at the plane $z = z_0^+$ appears in the spectral field amplitude function $W_j^{(0)}(\mathbf{r}_T, \omega) = W_j^{(0)}(x, y, \omega)$. With this substitution, Eq. (8.123) for the propagated polarization ellipse becomes

$$\left[\mathbf{a}_j(\mathbf{r}^+, \omega) + i\mathbf{b}_j(\mathbf{r}^+, \omega)\right] e^{i\varphi_j^{(0)}(\mathbf{r}^+, \omega)} = W_j(\mathbf{r}^+, \omega)\left(\hat{\mathbf{a}}_j^{(0)} + i\hat{\mathbf{b}}_j^{(0)}\right) e^{i\hat{\varphi}_j^{(0)}},$$
(8.119)

where

$$W_j(\mathbf{r}^+, \omega) = \int_{-\infty}^{\infty}\int_{-\infty}^{\infty} W_j^{(0)}(\mathbf{r}_T, \omega)h(x - x', y - y'; \Delta z, \omega)dx'dy',$$
(8.120)

and the polarization state for this field vector remains unchanged throughout the positive half-space $z \geq z_0^+$. Notice that the necessary condition specified in Eq. (8.118) requires that both vectors $\hat{\mathbf{a}}_j^{(0)}$ and $\hat{\mathbf{b}}_j^{(0)}$ and the phase constant $\hat{\varphi}_j^{(0)}$ are independent of the angular frequency ω. If this is not the case, then the polarization state of the wave field vector considered will, in general, evolve with time at a given fixed point in space.

Two special cases of considerable importance in regard to the polarization state at a fixed point in space are the linearly polarized and circularly polarized wave fields. For a linearly polarized field vector at the point $\mathbf{r} = \mathbf{r}_0$, the minor axis of the polarization ellipse vanishes so that $b_j = 0$ and Eq. (8.93) then requires that

$$p_j^2 q_j^2 = (\mathbf{p}_j \cdot \mathbf{q}_j)^2,$$
(8.121)

and the angle γ_p between $\mathbf{p}$ and $\mathbf{q}$ is either 0 or π (see Fig. 8.5). The complex representation of the field vector at this point in space is then given by

$$\tilde{\mathbf{V}}_j(\mathbf{r}_0, t) = \mathbf{a}_j(\mathbf{r}_0)e^{i(\varphi_j(\mathbf{r}_0 - \omega t)}.$$
(8.122)

For a circularly polarized field vector at the point $\mathbf{r} = \mathbf{r}_0$, the vectors $\mathbf{a}_j$ and $\mathbf{b}_j$ are indeterminate and consequently, the angle φ_j is also indeterminate. For this to be the case, Eq. (8.87) then requires that

$$\mathbf{p}_j \cdot \mathbf{q}_j = 0 \quad \wedge \quad p_j^2 - q_j^2 = 0,$$
(8.123)

so that the vectors $\mathbf{p}_j = \mathbf{p}_j(\mathbf{r}_0)$ and $\mathbf{q}_j = \mathbf{q}_j(\mathbf{r}_0)$ are orthogonal and of equal magnitude. The complex representation of the field vector at this point in space is then given by

$$\tilde{\mathbf{V}}_j(\mathbf{r}_0, t) = \sqrt{2}\mathbf{p}_j(\mathbf{r}_0)e^{\pm i\pi/4}e^{-i\omega t}, \tag{8.124}$$

where the sign choice depends upon the polarization sense. From this expression it is seen that any given state of circular polarization may be decomposed into the superposition of two properly phased and orthogonally oriented linearly polarized fields. Furthermore, any given state of elliptic polarization may be decomposed into the superposition of a left-handed and right-handed circularly polarized field with the same angular frequency but with unequal amplitudes.

8.2.3   Relation Between the Electric and Magnetic Polarizations of an Electromagnetic Wave

From Eqs. (8.78), (8.79) and (8.102), the complex representation of the polarization ellipses for the temporal frequency spectra of the electric and magnetic field vectors of an electromagnetic wave field are respectively given by

$$\tilde{\mathbf{E}}(\mathbf{r}, \omega) = \mathbf{p}_e(\mathbf{r}, \omega) + i\mathbf{q}_e(\mathbf{r}, \omega)$$
$$= \left[\mathbf{a}_e(\mathbf{r}, \omega) + i\mathbf{b}_e(\mathbf{r}, \omega)\right]e^{i\varphi_e(\mathbf{r}, \omega)} \equiv \tilde{\mathbf{e}}(\mathbf{r}, \omega)e^{i\varphi_e(\mathbf{r}, \omega)}, \tag{8.125}$$

$$\tilde{\mathbf{B}}(\mathbf{r}, \omega) = \mathbf{p}_m(\mathbf{r}, \omega) + i\mathbf{q}_m(\mathbf{r}, \omega)$$
$$= \left[\mathbf{a}_m(\mathbf{r}, \omega) + i\mathbf{b}_m(\mathbf{r}, \omega)\right]e^{i\varphi_m(\mathbf{r}, \omega)} \equiv \tilde{\mathbf{b}}(\mathbf{r}, \omega)e^{i\varphi_m(\mathbf{r}, \omega)}, \tag{8.126}$$

where the complex field vectors defined by

$$\tilde{\mathbf{e}}(\mathbf{r}, \omega) \equiv \mathbf{a}_e(\mathbf{r}, \omega) + i\mathbf{b}_e(\mathbf{r}, \omega), \tag{8.127}$$

$$\tilde{\mathbf{b}}(\mathbf{r}, \omega) \equiv \mathbf{a}_m(\mathbf{r}, \omega) + i\mathbf{b}_m(\mathbf{r}, \omega), \tag{8.128}$$

have been introduced here for notational convenience and are not to be confused with the microscopic electric and magnetic field vectors. Throughout the positive half-space $z \geq z_0^+$, the temporal frequency spectra of the field vectors satisfy the field equations [cf. Eqs. (5.24)–(5.27)]

$$\nabla \cdot \tilde{\mathbf{E}}(\mathbf{r}, \omega) = 0, \tag{8.129}$$

$$\nabla \times \tilde{\mathbf{E}}(\mathbf{r}, \omega) = \left\|\frac{1}{c}\right\| i\omega \tilde{\mathbf{B}}(\mathbf{r}, \omega), \tag{8.130}$$

$$\nabla \cdot \tilde{\mathbf{B}}(\mathbf{r}, \omega) = 0, \tag{8.131}$$

$$\nabla \times \tilde{\mathbf{B}}(\mathbf{r}, \omega) = - \left\| \frac{1}{c} \right\| i\omega\mu(\omega)\epsilon_c(\omega)\tilde{\mathbf{E}}(\mathbf{r}, \omega). \tag{8.132}$$

Substitution of the complex representations for the electric and magnetic field vectors given in Eqs. (8.125) and (8.126), respectively, into the pair of divergence relations given above then yields the pair of relations

$$i\tilde{\mathbf{e}}(\mathbf{r}, \omega) \cdot \varphi_e(\mathbf{r}, \omega) + \nabla \cdot \tilde{\mathbf{e}}(\mathbf{r}, \omega) = 0, \tag{8.133}$$

$$i\tilde{\mathbf{b}}(\mathbf{r}, \omega) \cdot \varphi_m(\mathbf{r}, \omega) + \nabla \cdot \tilde{\mathbf{b}}(\mathbf{r}, \omega) = 0, \tag{8.134}$$

and substitution into the pair of curl relations gives

$$\tilde{\mathbf{B}}(\mathbf{r}, \omega) = \tilde{\mathbf{b}}(\mathbf{r}, \omega)e^{i\varphi_m(\mathbf{r}, \omega)}$$

$$= \frac{\|c\|}{\omega} \left[\nabla\varphi_e(\mathbf{r}, \omega) \times \tilde{\mathbf{e}}(\mathbf{r}, \omega) - i\nabla \times \tilde{\mathbf{e}}(\mathbf{r}, \omega) \right] e^{i\varphi_e(\mathbf{r}, \omega)}, \tag{8.135}$$

$$\tilde{\mathbf{E}}(\mathbf{r}, \omega) = \tilde{\mathbf{e}}(\mathbf{r}, \omega)e^{i\varphi_e(\mathbf{r}, \omega)}$$

$$= \frac{\|c\|}{\omega\mu(\omega)\epsilon_c(\omega)} \left[-\nabla\varphi_m(\mathbf{r}, \omega) \times \tilde{\mathbf{b}}(\mathbf{r}, \omega) + i\nabla \times \tilde{\mathbf{b}}(\mathbf{r}, \omega) \right] e^{i\varphi_m(\mathbf{r}, \omega)}.$$

$$\tag{8.136}$$

The polarization properties of one field vector may then be determined directly from the polarization state of the other field vector at each point of space from this final pair of relations. For example, if the electric field vector is linearly polarized with $\tilde{\mathbf{e}}(\mathbf{r}, \omega) = \mathbf{a}_e(\mathbf{r}, \omega)$, where $\mathbf{a}_e$ is a real-valued vector field, then

$$\tilde{\mathbf{b}}(\mathbf{r}, \omega)e^{i\varphi_m(\mathbf{r}, \omega)} = \frac{\|c\|}{\omega} \left[\nabla\varphi_e(\mathbf{r}, \omega) \times \mathbf{a}_e(\mathbf{r}, \omega) - i\nabla \times \mathbf{a}_e(\mathbf{r}, \omega) \right] e^{i\varphi_e(\mathbf{r}, \omega)},$$

and the magnetic field vector will, in general, be elliptically polarized provided that $\mathbf{a}_e$ is spatially dependent; however, if $\mathbf{a}_e$ is spatially independent so that the electric field vector is uniformly linearly polarized throughout space, then $\nabla \times \mathbf{a}_e = \mathbf{0}$ and the magnetic field vector is also linearly polarized, but with an orientation that may vary from point to point in space.

The pair of expressions given in Eqs. (8.135) and (8.136) shows that the temporal frequency spectra of the electric and magnetic field vectors are not, in general, instantaneously orthogonal to each other, because

$$\tilde{\mathbf{E}}(\mathbf{r}, \omega) \cdot \tilde{\mathbf{B}}(\mathbf{r}, \omega) = -i\frac{\|c\|}{\omega}\tilde{\mathbf{e}}(\mathbf{r}, \omega) \cdot \left[\nabla \times \tilde{\mathbf{e}}(\mathbf{r}, \omega) \right] e^{i2\varphi(\mathbf{r}, \omega)}, \tag{8.137}$$

which does not, in general, vanish. As previously noted, the state of polarization strictly refers to the electromagnetic wave field behavior at a specific point in space and, in general, varies from point to point in the field. Even in the special case when the electric field vector is everywhere linearly polarized in some fixed direction, the magnetic field vector will, in general, be elliptically polarized, as was first shown by Nisbet and Wolf [17]. As a consequence, in contrast with the orthogonality relation [cf. Eqs. (7.32) and (7.33)]

$$\tilde{\mathbf{E}}(\mathbf{k}, \omega) \cdot \tilde{\mathbf{B}}(\mathbf{k}, \omega) = 0 \qquad (8.138)$$

for the spatiotemporal spectra of the field vectors, the temporal frequency spectra of the electric and magnetic field vectors are not, in general, instantaneously orthogonal to each other. However, the long-time average of the real-valued quantity $\tilde{\mathbf{E}}^{(r)}(\mathbf{r}, t; \omega) \cdot \tilde{\mathbf{B}}^{(r)}(\mathbf{r}, t; \omega)$, with $\tilde{\mathbf{E}}^{(r)}(\mathbf{r}, t; \omega) \equiv \Re\left\{\tilde{\mathbf{E}}(\mathbf{r}, \omega)e^{-i\omega t}\right\}$ and $\tilde{\mathbf{B}}^{(r)}(\mathbf{r}, t; \omega) \equiv \Re\left\{\tilde{\mathbf{B}}(\mathbf{r}, \omega)e^{-i\omega t}\right\}$ does indeed vanish, because [cf. Eq. 5.206)]

$$\left\langle \tilde{\mathbf{E}}^{(r)}(\mathbf{r}, t; \omega) \cdot \tilde{\mathbf{B}}^{(r)}(\mathbf{r}, t; \omega) \right\rangle = \frac{1}{2}\Re\left\{\tilde{\mathbf{E}}(\mathbf{r}, \omega) \cdot \tilde{\mathbf{B}}^*(\mathbf{r}, \omega)\right\}$$

$$= \frac{\|c\|}{2\omega}\Re\left\{\tilde{\mathbf{e}}(\mathbf{r}, \omega) \cdot \left[\nabla\varphi_e(\omega) \times \tilde{\mathbf{e}}^*(\mathbf{r}, \omega)\right]\right.$$

$$\left. + i\tilde{\mathbf{e}}(\mathbf{r}, \omega) \cdot \left[\nabla \times \tilde{\mathbf{e}}^*(\mathbf{r}, \omega)\right]\right\}$$

$$= \frac{\|c\|}{2\omega}\Re\left\{i\tilde{\mathbf{e}}(\mathbf{r}, \omega) \cdot \left[\nabla \times \tilde{\mathbf{e}}^*(\mathbf{r}, \omega)\right]\right\}$$

$$= 0. \qquad (8.139)$$

The fact that the quantity $\tilde{\mathbf{e}} \cdot (\nabla \times \tilde{\mathbf{e}}^*)$ is real-valued, used in the above proof, directly follows from the vector differential identity $\nabla \cdot (\mathbf{v} \times \mathbf{w}) = \mathbf{w} \cdot (\nabla \times \mathbf{v}) - \mathbf{v} \cdot (\nabla \times \mathbf{w})$. With $\mathbf{v} = \tilde{\mathbf{e}}$ and $\mathbf{w} = \tilde{\mathbf{e}}^*$ as well as the fact that $\tilde{\mathbf{e}} \times \tilde{\mathbf{e}}^* = \mathbf{0}$, this identity yields the result $\tilde{\mathbf{e}} \cdot (\nabla \times \tilde{\mathbf{e}}^*) = \tilde{\mathbf{e}}^* \cdot (\nabla \times \tilde{\mathbf{e}})$, and consequently

$$\tilde{\mathbf{e}} \cdot (\nabla \times \tilde{\mathbf{e}}^*) = \frac{1}{2}\left[\tilde{\mathbf{e}} \cdot (\nabla \times \tilde{\mathbf{e}}^*) + \tilde{\mathbf{e}}^* \cdot (\nabla \times \tilde{\mathbf{e}})\right]$$

$$= \frac{1}{2}\left\{\tilde{\mathbf{e}} \cdot (\nabla \times \tilde{\mathbf{e}}^*) + \left[\tilde{\mathbf{e}} \cdot (\nabla \times \tilde{\mathbf{e}}^*)\right]^*\right\}$$

$$= \Re\left\{\tilde{\mathbf{e}} \cdot (\nabla \times \tilde{\mathbf{e}}^*)\right\}.$$

Thus, the temporal frequency spectra of the electric and magnetic field vectors are, on the average, mutually orthogonal.

$$\tilde{\tilde{\mathbf{E}}}(\mathbf{k}, \omega) \cdot \tilde{\tilde{\mathbf{B}}}(\mathbf{k}, \omega) = 0$$

$$\Updownarrow \mathcal{F}_{\mathbf{k}}$$

$$\tilde{\mathbf{E}}(\mathbf{r}, \omega) \cdot \tilde{\mathbf{B}}(\mathbf{r}, \omega) \neq 0$$

$$\left\langle \tilde{\mathbf{E}}^{(r)}(\mathbf{r}, t; \omega) \cdot \tilde{\mathbf{B}}^{(r)}(\mathbf{r}, t; \omega) \right\rangle = 0$$

$$\Updownarrow \mathcal{F}_{\omega}$$

$$\mathbf{E}(\mathbf{r}, t) \cdot \mathbf{B}(\mathbf{r}, t) \neq 0$$

$$\langle \mathbf{E}(\mathbf{r}, t) \cdot \mathbf{B}(\mathbf{r}, t) \rangle \neq 0.$$

As a summary of these results, indicated in the above set of relations, the spatio-temporal frequency spectra $\tilde{\tilde{\mathbf{E}}}(\mathbf{k}, \omega)$ and $\tilde{\tilde{\mathbf{B}}}(\mathbf{k}, \omega)$ of the electric and magnetic field vectors, respectively, are always mutually orthogonal. The temporal frequency spectral wave components $\tilde{\mathbf{E}}^{(r)}(\mathbf{r}, t; \omega) \equiv \Re \left\{ \tilde{\mathbf{E}}(\mathbf{r}, \omega) e^{-i\omega t} \right\}$ and $\tilde{\mathbf{B}}^{(r)}(\mathbf{r}, t; \omega) \equiv \Re \left\{ \tilde{\mathbf{B}}(\mathbf{r}, \omega) e^{-i\omega t} \right\}$ of these field vectors are not, in general, orthogonal, but are mutually orthogonal in the long-time average sense. Finally, the electric $\mathbf{E}(\mathbf{r}, t)$ and magnetic $\mathbf{B}(\mathbf{r}, t)$ field vectors are not, in general, orthogonal and neither is their long-time average.

8.2.4 The Uniformly Polarized Wave Field

For an electromagnetic wave field that is uniformly polarized in the electric field vector, Eqs. (8.118), (8.125) and (8.126) require that the initial field vectors on the plane at $z = z_0^+$ are given by

$$\tilde{\mathbf{E}}_0(\mathbf{r}_T, \omega) = \tilde{E}_0(\mathbf{r}_T, \omega)(\mathbf{a}_e + i\mathbf{b}_e)e^{i\varphi_e}, \tag{8.140}$$

$$\tilde{\mathbf{B}}_0(\mathbf{r}_T, \omega) = \tilde{B}_0(\mathbf{r}_T, \omega)\big(\mathbf{a}_m(\mathbf{r}_T, \omega) + i\mathbf{b}_m(\mathbf{r}_T, \omega)\big)e^{i\varphi_m(\mathbf{r}_T, \omega)}, \tag{8.141}$$

where $\mathbf{a}_e$ and $\mathbf{b}_e$ are fixed vectors and where φ_e is a scalar constant. These two field vectors are not independent and must be oriented such that Eqs. (8.135) and (8.136) are satisfied. In particular, Eq. (8.135) requires that

$$\tilde{B}_0(\mathbf{r}_T, \omega)\big(\mathbf{a}_m(\mathbf{r}_T, \omega) + i\mathbf{b}_m(\mathbf{r}_T, \omega)\big)e^{i\varphi_m(\mathbf{r}_T, \omega)}$$

$$= -i\frac{\|c\|}{\omega}\big(\nabla \tilde{E}_0(\mathbf{r}_T, \omega)\big) \times (\mathbf{a}_e + i\mathbf{b}_e)e^{i\varphi_e}, \tag{8.142}$$

from which it is seen that the temporal frequency spectra of the electric and magnetic field vectors are orthogonal at the plane $z = z_0^+$. Although not necessary, the complex vectors $(\mathbf{a}_e + i\mathbf{b}_e)$ and $(\mathbf{a}_m + i\mathbf{b}_m)$ are chosen to be normalized, so that

$$|\mathbf{a}_j + i\mathbf{b}_j|^2 = a_j^2 + b_j^2 = 1 \tag{8.143}$$

for $j = e, m$. The spatio-temporal frequency spectra of the initial field vectors at the plane $z = z_0^+$ are then given by

$$\tilde{\tilde{\mathbf{E}}}_0(\mathbf{k}_T, \omega) = (\mathbf{a}_e + i\mathbf{b}_e)e^{i\varphi_e} \int_{-\infty}^{\infty} \int_{-\infty}^{\infty} \tilde{E}_0(\mathbf{r}_T, \omega)e^{-i\mathbf{k}_T \cdot \mathbf{r}_T} dx dy$$

$$= (\mathbf{a}_e + i\mathbf{b}_e)e^{i\varphi_e} \tilde{\tilde{E}}_0(\mathbf{k}_T, \omega), \tag{8.144}$$

$$\tilde{\tilde{\mathbf{B}}}_0(\mathbf{k}_T, \omega) = i\frac{\|c\|}{\omega}(\mathbf{a}_e + i\mathbf{b}_e)e^{i\varphi_e} \times \int_{-\infty}^{\infty} \int_{-\infty}^{\infty} \left(\nabla \tilde{E}_0(\mathbf{r}_T, \omega)\right)e^{-i\mathbf{k}_T \cdot \mathbf{r}_T} dx dy$$

$$= \frac{\|c\|}{\omega}\tilde{\mathbf{k}}^+ \times (\mathbf{a}_e + i\mathbf{b}_e)e^{i\varphi_e} \tilde{\tilde{E}}_0(\mathbf{k}_T, \omega), \tag{8.145}$$

where

$$\tilde{\tilde{E}}_0(\mathbf{k}_T, \omega) = \int_{-\infty}^{\infty} \int_{-\infty}^{\infty} \tilde{E}_0(\mathbf{r}_T, \omega)e^{-i\mathbf{k}_T \cdot \mathbf{r}_T} dx dy. \tag{8.146}$$

In addition, the space–time form of the initial field vectors at the plane $z = z_0^+$ is given by

$$\mathbf{E}_0(\mathbf{r}_T, t) = \frac{1}{\pi}\Re\left\{(\mathbf{a}_e + i\mathbf{b}_e)e^{i\varphi_e}\int_{C_+} \tilde{E}_0(\mathbf{r}_T, \omega)e^{-i\omega t} d\omega\right\}, \tag{8.147}$$

$$\mathbf{B}_0(\mathbf{r}_T, t) = \frac{\|c\|}{\pi}\Re\left\{i(\mathbf{a}_e + i\mathbf{b}_e)e^{i\varphi_e}\times\int_{C_+} \frac{\nabla \tilde{E}_0(\mathbf{r}_T, \omega)}{\omega}e^{-i\omega t} d\omega\right\}. \tag{8.148}$$

The propagated field vectors are found from the angular spectrum representation given in Eqs. (8.14) and (8.17) with substitution from Eqs. (8.144) and (8.145) as

$$\mathbf{E}(\mathbf{r}, t) = \frac{1}{4\pi^3}\Re\left\{(\mathbf{a}_e + i\mathbf{b}_e)e^{i\varphi_e}\right.$$

$$\left.\times\int_{C_+} d\omega \int_{-\infty}^{\infty} \int_{-\infty}^{\infty} \tilde{\tilde{E}}_0(\mathbf{k}_T, \omega)e^{i(\tilde{\mathbf{k}}^+ \cdot \mathbf{r} - \omega t)} dk_x dk_y\right\}, \tag{8.149}$$

$$\mathbf{B}(\mathbf{r}, t) = -\frac{\|c\|}{4\pi^3}\Re\left\{(\mathbf{a}_e + i\mathbf{b}_e)e^{i\varphi_e}\right.$$

$$\left.\times\int_{C_+} \frac{d\omega}{\omega} \int_{-\infty}^{\infty} \int_{-\infty}^{\infty} \tilde{\mathbf{k}}^+(\omega)\tilde{\tilde{E}}_0(\mathbf{k}_T, \omega)e^{i(\tilde{\mathbf{k}}^+ \cdot \mathbf{r} - \omega t)} dk_x dk_y\right\}. \tag{8.150}$$

Hence, the propagated field vectors are seen to be everywhere orthogonal in the positive half-space $z \geq z_0^+$ when the electric field vector is uniformly polarized. However, the magnetic field vector is not, in general, uniformly polarized because of its directional dependence on the complex wave vector $\tilde{\mathbf{k}}^+(\omega)$ which causes the orientation of the magnetic field polarization ellipse to vary in both space and time. In order that the magnetic field vector also be uniformly polarized the spectral quantity $\tilde{\mathbf{k}}^{+}(\omega)\tilde{\bar{E}}_0(\mathbf{k}_T, \omega)$ must have a fixed direction independent of ω; this in turn implies that the initial field amplitude $\tilde{E}_0(\mathbf{r}_T, \omega)$ must be independent of one transverse coordinate direction.

For an electromagnetic wave field that is uniformly polarized in the magnetic field vector the set of relations given in Eqs. (8.140) and (8.141) is replaced by

$$\tilde{\bar{E}}_0(\mathbf{r}_T, \omega) = \tilde{E}_0(\mathbf{r}_T, \omega)\big(\mathbf{a}_e(\mathbf{r}_T, \omega) + i\mathbf{b}_e(\mathbf{r}_T, \omega)\big)e^{i\varphi_e(\mathbf{r}_T, \omega)}, \qquad (8.151)$$

$$\tilde{\bar{B}}_0(\mathbf{r}_T, \omega) = \tilde{B}_0(\mathbf{r}_T, \omega)(\mathbf{a}_m + i\mathbf{b}_m)e^{i\varphi_m}, \qquad (8.152)$$

where $\mathbf{a}_m$ and $\mathbf{b}_m$ are fixed vectors and where φ_m is a scalar constant. The propagated wave field vectors throughout the positive half-space in this case are then found to be given by

$$\mathbf{E}(\mathbf{r}, t) = -\frac{\|c\|}{4\pi^3}\Re\bigg\{(\mathbf{a}_m + i\mathbf{b}_m)e^{i\varphi_m}$$

$$\times \int_{C_+} \frac{d\omega}{\omega\mu(\omega)\epsilon_c(\omega)} \int_{-\infty}^{\infty}\int_{-\infty}^{\infty} \tilde{\mathbf{k}}^{+}(\omega)\tilde{\bar{B}}_0(\mathbf{k}_T, \omega)e^{i(\tilde{\mathbf{k}}^{+}\cdot\mathbf{r}-\omega t)}dk_x dk_y\bigg\}.$$

$$(8.153)$$

$$\mathbf{B}(\mathbf{r}, t) = \frac{1}{4\pi^3}\Re\bigg\{(\mathbf{a}_m + i\mathbf{b}_m)e^{i\varphi_m}$$

$$\times \int_{C_+} d\omega \int_{-\infty}^{\infty}\int_{-\infty}^{\infty} \tilde{\bar{B}}_0(\mathbf{k}_T, \omega)e^{i(\tilde{\mathbf{k}}^{+}\cdot\mathbf{r}-\omega t)}dk_x dk_y\bigg\}.$$

$$(8.154)$$

The propagated field vectors are again seen to be everywhere orthogonal in the positive half-space, however, the electric field vector is now not, in general, uniformly polarized.

In order that both the electric and magnetic field vectors be uniformly polarized throughout the positive half-space $z \geq z_0^+$, Eqs. (8.150) and (8.154) must both be satisfied throughout that region, so that

$$\Re\bigg\{\int_{C_+} d\omega \int_{-\infty}^{\infty}\int_{-\infty}^{\infty} dk_x dk_y\, e^{i(\tilde{\mathbf{k}}^{+}\cdot\mathbf{r}-\omega t)}\Big[\tilde{\bar{B}}_0(\mathbf{k}_T, \omega)(\mathbf{a}_m + i\mathbf{b}_m)e^{i\varphi_m}$$

$$+ \frac{\|c\|}{\omega}\tilde{\bar{E}}_0(\mathbf{k}_T, \omega)(\mathbf{a}_e + i\mathbf{b}_e)e^{i\varphi_e} \times \tilde{\mathbf{k}}^{+}(\omega)\Big]\bigg\} = \mathbf{0},$$

and consequently

$$\tilde{\mathbf{B}}_0(\mathbf{k}_T, \omega)(\mathbf{a}_m + i\mathbf{b}_m)e^{i\varphi_m} = -\|c\|\frac{\tilde{k}(\omega)}{\omega}\tilde{\mathbf{E}}_0(\mathbf{k}_T, \omega)(\mathbf{a}_e + i\mathbf{b}_e)e^{i\varphi_e} \times \hat{\mathbf{s}},$$

$$(8.155)$$

where $\hat{\mathbf{s}} = \hat{\mathbf{1}}_x p + \hat{\mathbf{1}}_y q + \hat{\mathbf{1}}_z m$ is a real-valued unit vector that is defined by the direction of the complex-valued wave vector as $\tilde{\mathbf{k}}^+(\omega) = \tilde{k}(\omega)\hat{\mathbf{s}}$. Notice that the relation given in Eq. (8.161) also follows from the relation given in Eq. (8.19); viz., $\tilde{\mathbf{B}}_0(\mathbf{k}_T, \omega) = (\|c\|/\omega)\tilde{\mathbf{k}}^+(\omega) \times \tilde{\mathbf{E}}_0(\mathbf{k}_T, \omega)$, with $\tilde{\mathbf{E}}_0(\mathbf{k}_T, \omega)$ given by Eq. (8.146) and $\tilde{\mathbf{B}}_0(\mathbf{k}_T, \omega)$ given by the two-dimensional spatial transform of Eq. (8.152). The orientation of the magnetic polarization ellipse is then seen to depend on the orientation of the electric polarization ellipse through the factor $(\mathbf{a}_e + i\mathbf{b}_e) \times \hat{\mathbf{s}}$ which depends upon the direction of the complex wave vector $\tilde{\mathbf{k}}^+(\omega) = \tilde{k}(\omega)\hat{\mathbf{s}}$. Because it is required that both polarization ellipses are fixed in both space and time, the direction of the unit vector $\hat{\mathbf{s}}$ must then be appropriately constrained and this, in turn, constrains the coordinate dependency of the field vectors themselves.

As an example, consider an electromagnetic wave that is uniformly polarized in the electric field vector such that it is linearly polarized along the x-axis in a nondispersive (and hence, nonabsorptive) medium. In that case, $\mathbf{a}_e = \hat{\mathbf{1}}_x$, $\mathbf{b}_e = \mathbf{0}$, and $\varphi_e = 0$, so that

$$(\mathbf{a}_m + i\mathbf{b}_m)e^{i\varphi_m} = \left(\hat{\mathbf{1}}_x p + \hat{\mathbf{1}}_y q + \hat{\mathbf{1}}_z\sqrt{1 - p^2 - q^2}\right) \times \hat{\mathbf{1}}_x$$

$$= \hat{\mathbf{1}}_y\sqrt{1 - p^2 - q^2} - \hat{\mathbf{1}}_z q,$$

from which it is seen that there are two possibilities in order to maintain the requirement that the magnetic field is uniformly polarized. Either $q = q_0$ is fixed, in which case p may be allowed to vary such that either $p^2 \leq 1 - q_0^2$ or $p^2 > 1 - q_0^2$, or else q is allowed to vary in which case p must vary in such a fashion that $m = 0$, that is, such that $p^2 = 1 - q^2$. This latter situation in which $m = 0$ for all possible allowed values of p and q precludes propagation into the positive half-space $\Delta z > 0$ and so is of no interest here. Hence, $q = q_0$ and the field itself must then be independent of the y-coordinate.

As another example, consider an electromagnetic wave that is uniformly polarized in the electric field vector such that it is circularly polarized in the xy-plane in a nondispersive medium. In that case, $\mathbf{a}_e = \hat{\mathbf{1}}_x$, $\mathbf{b}_e = \hat{\mathbf{1}}_y$ and $\varphi_e = 0$, so that

$$(\mathbf{a}_m + i\mathbf{b}_m)e^{i\varphi_m} = \left(\hat{\mathbf{1}}_x p + \hat{\mathbf{1}}_y q + \hat{\mathbf{1}}_z\sqrt{1 - p^2 - q^2}\right) \times \left(\hat{\mathbf{1}}_x + i\hat{\mathbf{1}}_y\right)$$

$$= \hat{\mathbf{1}}_y\sqrt{1 - p^2 - q^2} - \hat{\mathbf{1}}_z q - i\left(\hat{\mathbf{1}}_x\sqrt{1 - p^2 - q^2} - \hat{\mathbf{1}}_z p\right)$$

and p and q must both be fixed (i.e., $p = p_0$ and $q = q_0$) in order for the magnetic polarization ellipse to be uniform. Because $a_m^2 = 1 - p_0^2$ and $b_m^2 = 1 - q_0^2$ the magnetic field is elliptically polarized in general and is circularly polarized when $p_0 = q_0$. In either case, the field itself is independent of both the x- and y-coordinates.

This then establishes the following result:

Theorem 8.1 *The requirement that both the electric and magnetic field vectors of an electromagnetic wave field be uniformly polarized throughout the positive half-space $z \geq z_0^+$ requires that the field be independent of some coordinate direction.*

8.3   Real Direction Cosine Form of the Angular Spectrum Representation

The plane wave spectral components appearing in the angular spectrum representation given in Eqs. (8.14) and (8.17) may be cast into a more explicit geometric form [26] by setting[2] [compare with the transformation relations given in Eqs. (8.21)–(8.23)]

$$k_x = \tilde{k}(\omega)e^{-i\psi(\omega)}p, \tag{8.156}$$

$$k_y = \tilde{k}(\omega)e^{-i\psi(\omega)}q, \tag{8.157}$$

$$\gamma(\omega) = \tilde{k}(\omega)e^{-i\psi(\omega)}m, \tag{8.158}$$

where

$$\psi(\omega) \equiv \arg\left\{\tilde{k}(\omega)\right\} = \arctan\left(\frac{\alpha(\omega)}{\beta(\omega)}\right) \tag{8.159}$$

is the phase angle of the complex wavenumber $\tilde{k}(\omega) = \beta(\omega) + i\alpha(\omega)$. With

$$k(\omega) \equiv \left|\tilde{k}(\omega)\right| = \left(\beta^2(\omega) + \alpha^2(\omega)\right)^{1/2} \tag{8.160}$$

denoting the magnitude of the complex wavenumber $\tilde{k}(\omega) = k(\omega)e^{i\psi(\omega)}$, the set of transformation relations given in Eqs. (8.156)–(8.158) may be expressed as

$$k_x = k(\omega)p, \tag{8.161}$$

$$k_y = k(\omega)q, \tag{8.162}$$

[2]Notice that the same notation that was used in Sect. 8.1.1 is being used here. Care must then be exercised not to mix these results.

$$\gamma(\omega) = k(\omega)m. \tag{8.163}$$

Because both k_x and k_y are real-valued, then both direction cosines p and q must also be real-valued. In addition, because $\gamma(\omega) = \left(\tilde{k}^2(\omega) - k_T^2\right)^{1/2}$ with $k_T^2 = k_x^2 + k_y^2$, then

$$m(\omega) = \left(e^{i2\psi(\omega)} - \left(p^2 + q^2\right)\right)^{1/2}, \tag{8.164}$$

which is, in general, complex-valued. As in Eq. (8.32), let $m(\omega) \equiv \zeta^{1/2}(\omega)$ with $\zeta = \zeta' + i\zeta''$, where

$$\zeta'(\omega) \equiv \Re\{\zeta(\omega)\} = \cos\left(2\psi(\omega)\right) - \left(p^2 + q^2\right), \tag{8.165}$$

$$\zeta''(\omega) \equiv \Im\{\zeta(\omega)\} = \sin\left(2\psi(\omega)\right). \tag{8.166}$$

From the real and imaginary parts of the expression $\tilde{k}(\omega) = (\omega/c)n(\omega)$ one obtains the pair of expressions

$$\beta(\omega) = \frac{1}{c}\left(\omega' n_r(\omega) - \omega'' n_i(\omega)\right), \tag{8.167}$$

$$\alpha(\omega) = \frac{1}{c}\left(\omega' n_i(\omega) + \omega'' n_r(\omega)\right), \tag{8.168}$$

with $\omega = \omega' + i\omega''$. Along the real angular frequency axis, $\beta(\omega') = (\omega'/c)n_r(\omega')$ and $\alpha(\omega') = (\omega'/c)n_i(\omega')$, both of which are nonnegative for all $\omega' \geq 0$. If both $\beta(\omega)$ and $\alpha(\omega)$ are to be nonnegative for all ω on the contour C_+ [see the discussion following Eqs. (8.3)–(8.6)], then the pair of inequalities

$$\omega' n_r(\omega) \geq \omega'' n_i(\omega), \tag{8.169}$$

$$\omega' n_i(\omega) + \omega'' n_r(\omega) \geq 0, \tag{8.170}$$

must be satisfied $\forall \omega \in C_+$. Substitution of the second inequality into the first yields the trivial inequality $\omega'^2 + \omega''^2 \geq 0$ so that only one of the inequalities appearing in Eqs. (8.169) and (8.170) needs to be assumed, the other then being satisfied automatically.

Because of its necessity, it is assumed here that the plane wave propagation factor $\beta(\omega)$ is nonnegative $\forall \omega \in C_+$. It then follows that $0 \leq \psi(\omega) \leq \pi/2 \; \forall \omega \in C_+$. Consequently, $\sin(2\psi(\omega)) \geq 0$ so that $\zeta''(\omega) \geq 0$ on C_+. On the other hand, $\cos(2\psi(\omega)) > 0$ when $0 \leq \psi(\omega) \leq \pi/4$, $\cos(2\psi(\omega)) = 0$ when $\psi(\omega) = \pi/4$, and $\cos(2\psi(\omega)) < 0$ when $\pi/4 \leq \psi(\omega) \leq \pi/2$, so that $\zeta'(\omega) \geq 0$ when $(p^2 + q^2) \leq \cos(2\psi(\omega))$ and $\zeta'(\omega) < 0$ when $(p^2 + q^2) > \cos(2\psi(\omega)) \; \forall \omega \in C_+$.

Consequently,

$$p^2 + q^2 < \cos\left(2\psi(\omega)\right) \Rightarrow \zeta'(\omega) > 0 \quad \wedge \quad \zeta''(\omega) \geq 0$$

$$\Rightarrow 0 \leq \arg\left\{\zeta(\omega)\right\} < \frac{\pi}{2}$$

$$\Rightarrow 0 \leq \arg\left\{m(\omega)\right\} < \frac{\pi}{4}, \qquad (8.171)$$

$$p^2 + q^2 = \cos\left(2\psi(\omega)\right) \Rightarrow \zeta'(\omega) = 0 \quad \wedge \quad \zeta''(\omega) \geq 0$$

$$\Rightarrow \arg\left\{\zeta(\omega)\right\} = \frac{\pi}{2}$$

$$\Rightarrow \arg\left\{m(\omega)\right\} = \frac{\pi}{4}, \qquad (8.172)$$

$$p^2 + q^2 > \cos\left(2\psi(\omega)\right) \Rightarrow \zeta'(\omega) < 0 \quad \wedge \quad \zeta''(\omega) \geq 0$$

$$\Rightarrow \frac{\pi}{2} < \arg\left\{\zeta(\omega)\right\} \leq \pi$$

$$\Rightarrow \frac{\pi}{4} < \arg\left\{m(\omega)\right\} < \frac{\pi}{2}, \qquad (8.173)$$

for all $\omega \in C_+$, as illustrated in Fig. 8.8. As a consequence, the real and imaginary parts of the direction cosine $m(\omega) \equiv m'(\omega) + im''(\omega)$ satisfy the respective inequalities

$$m'(\omega) \equiv \Re\left\{m(\omega)\right\} \geq 0, \qquad (8.174)$$

$$m''(\omega) \equiv \Im\left\{m(\omega)\right\} \geq 0, \qquad (8.175)$$

for all $\omega \in C_+$. Explicit expressions for both $m'(\omega)$ and $m''(\omega)$ are obtained from the square of Eq. (8.164) as

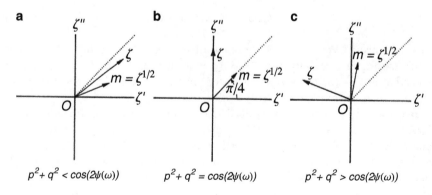

Fig. 8.8 Inequalities satisfied by the direction cosines p and q on the contour C_+

$$m'(\omega) = \left\{ \frac{1}{2}\left(\cos\left(2\psi(\omega)\right) - \left(p^2 + q^2\right) \right) \right.$$

$$\left. + \frac{1}{2}\left[\left(\cos\left(2\psi(\omega)\right) - \left(p^2 + q^2\right) \right)^2 + \frac{1}{\sin^2\left(2\psi(\omega)\right)} \right]^{1/2} \right\}^{1/2},$$

$$\text{(8.176)}$$

provided that $\sin\left(2\psi(\omega)\right) \neq 0$, and

$$m''(\omega) = \frac{\sin\left(2\psi(\omega)\right)}{2m'(\omega)}, \qquad \text{(8.177)}$$

provided that $m'(\omega) \neq 0$. If $m'(\omega) = 0$, then $\sin\left(2\psi(\omega)\right) = 0$ and either $\psi(\omega) = 0$, in which case $(p^2 + q^2) \geq 1$ and

$$m''(\omega) = \left(p^2 + q^2 - 1\right)^{1/2}, \qquad \text{(8.178)}$$

or $\psi(\omega) = \pi/2$, in which case

$$m''(\omega) = \left(p^2 + q^2 + 1\right)^{1/2}, \qquad \text{(8.179)}$$

for all values of p and q. Finally, for the special case of a lossless medium, $\psi(\omega) = 0$ and

$$m(\omega) = m'(\omega) = \sqrt{1 - (p^2 + q^2)} \qquad \text{(8.180)}$$

if $(p^2 + q^2) \leq 1$, whereas

$$m(\omega) = im''(\omega) = i\sqrt{(p^2 + q^2) - 1} \qquad \text{(8.181)}$$

if $(p^2 + q^2) > 1$.

With these results, the plane wave propagation factor appearing in the angular spectrum representation given in Eqs. (8.14) and (8.17) becomes [cf. Eq. (8.49)]

$$e^{i\tilde{\mathbf{k}}^+(\omega)\cdot\mathbf{r}} = e^{-k(\omega)m''(\omega)\Delta z}e^{ik(\omega)\left(px+qy+m'(\omega)\Delta z\right)}, \qquad \text{(8.182)}$$

with $\Delta z \equiv z - z_0 \geq 0$. If $p = q = 0$, then $\exp\left(i\tilde{\mathbf{k}}^+(\omega)\cdot\mathbf{r}\right)$ represents the spatial part of a homogeneous plane wave because the surfaces of constant amplitude coincide with the surfaces of constant phase given by $\Delta z =$ constant. If $\alpha(\omega) = 0$, then Eq. (8.182) represents the spatial part of a homogeneous plane wave when $(p^2 + q^2) \leq 1$ (in which case $m'' = 0$), whereas it represents the spatial part of an evanescent wave when $(p^2 + q^2) > 1$ (in which case $m' = 0$) and the surfaces of constant amplitude are orthogonal to the cophasal surfaces. In general, $\alpha(\omega) \neq 0$ and if either $p \neq 0$ or $q \neq 0$, then the expression appearing in Eq. (8.182) represents the spatial part of an inhomogeneous plane wave of angular frequency ω because the surfaces of constant amplitude $\Delta z =$ constant are different from the surfaces of constant phase $(px + qy + m'(\omega)\Delta z) =$ constant. These inhomogeneous plane wave phase fronts then propagate in the direction that is specified by the unit vector [cf. Eq. (8.51)]

$$\hat{\mathbf{s}} = \hat{\mathbf{1}}_x \frac{p}{s} + \hat{\mathbf{1}}_y \frac{q}{s} + \hat{\mathbf{1}}_z \frac{m'(\omega)}{s}, \qquad (8.183)$$

with

$$s = \sqrt{p^2 + q^2 + m'^2(\omega)}, \qquad (8.184)$$

as depicted in Fig. 8.9 [cf. Fig. 8.4)]. Because $m'(\omega) \geq 0$ for all real values of p and q, as well as for all $\omega \in C_+$, then those inhomogeneous plane wave components with $m'(\omega) > 0$ have phase fronts that advance into the positive half-space $\Delta z > 0$.

From Eq. (8.176) it is seen that the inequality

$$m'(\omega)\Big|_{(p^2+q^2)>\cos(2\psi(\omega))} < m'(\omega)\Big|_{(p^2+q^2)<\cos(2\psi(\omega))} \qquad (8.185)$$

is satisfied when $\sin(2\psi(\omega)) \neq 0$, so that

$$m''(\omega)\Big|_{(p^2+q^2)>\cos(2\psi(\omega))} > m''(\omega)\Big|_{(p^2+q^2)<\cos(2\psi(\omega))}, \qquad (8.186)$$

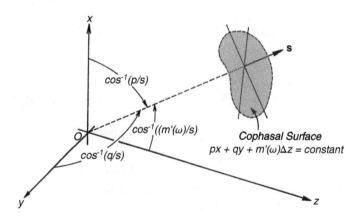

Fig. 8.9 Inhomogeneous plane wave phase front propagating in the direction specified by the unit vector $\hat{\mathbf{s}} = \hat{\mathbf{1}}_x p/s + \hat{\mathbf{1}}_y q/s + \hat{\mathbf{1}}_z m'(\omega)/s$

provided that $m'(\omega) \neq 0$. Consequently, the amplitude attenuation associated with the inhomogeneous plane wave spectral components in the exterior region $\mathcal{R}_> \equiv \{(p,q) | (p^2 + q^2) > \cos(2\psi(\omega))\}$ is typically larger than that associated with the inhomogeneous plane wave spectral components in the interior region $\mathcal{R}_< \equiv \{(p,q) | (p^2 + q^2) < \cos(2\psi(\omega))\}$.

With these results, the angular spectrum of plane waves representation given in Eqs. (8.14) and (8.17) becomes

$$\mathbf{U}(\mathbf{r},t) = \frac{1}{4\pi^3} \Re \left\{ \int_{C_+} d\omega e^{-i\omega t} \int_{\mathcal{R}_< \cup \mathcal{R}_>} dp dq \, k^2(\omega) \tilde{\mathbf{U}}_0(p,q,\omega) \right.$$
$$\left. \times e^{-k(\omega)m''(\omega)\Delta z} e^{ik(\omega)(px+qy+m'(\omega)\Delta z)} \right\}$$

(8.187)

for $\Delta z \geq 0$. Here $\mathbf{U}(\mathbf{r},t)$ represents either the electric field vector $\mathbf{E}(\mathbf{r},t)$, in which case $\tilde{\mathbf{U}}_0(p,q,\omega)$ represents the Fourier–Laplace transform $\tilde{\mathbf{E}}_0(p,q,\omega)$ of the boundary value for that field vector, or the magnetic field vector $\mathbf{B}(\mathbf{r},t)$, in which case $\tilde{\mathbf{U}}_0(p,q,\omega)$ represents the Fourier–Laplace transform $\tilde{\mathbf{B}}_0(p,q,\omega)$ of the boundary value for that field vector.

8.3.1 Electromagnetic Energy Flow in the Angular Spectrum Representation

A quantity of critical interest here is the time-average flow of energy associated with the individual plane wave components appearing in the angular spectrum representation of the freely-propagating electromagnetic wave field. This is given by the real part of the complex Poynting vector

$$\tilde{\mathbf{S}}(\mathbf{r}; \mathbf{k}_T, \omega) \equiv \frac{1}{2} \left\| \frac{c}{4\pi} \right\| \tilde{\mathbf{E}}(\mathbf{r}; \mathbf{k}_T, \omega) \times \tilde{\mathbf{H}}^*(\mathbf{r}; \mathbf{k}_T, \omega),$$

where [see Eqs. (8.18) and (8.19)]

$$\tilde{\mathbf{E}}(\mathbf{r}; \mathbf{k}_T, \omega) = \tilde{\mathbf{E}}_0(\mathbf{k}_T, \omega) e^{i\tilde{\mathbf{k}}^+(\omega) \cdot \Delta \mathbf{r}},$$
$$\tilde{\mathbf{H}}(\mathbf{r}; \mathbf{k}_T, \omega) = \tilde{\mathbf{H}}_0(\mathbf{k}_T, \omega) e^{i\tilde{\mathbf{k}}^+(\omega) \cdot \Delta \mathbf{r}}$$
$$= \frac{\|c\|}{\mu(\omega)\omega} \tilde{\mathbf{k}}^+(\omega) \times \tilde{\mathbf{E}}_0(\mathbf{k}_T, \omega) e^{i\tilde{\mathbf{k}}^+(\omega) \cdot \Delta \mathbf{r}},$$

with $\Delta \mathbf{r} \equiv \hat{\mathbf{1}}_x x + \hat{\mathbf{1}}_y y + \hat{\mathbf{1}}_z \Delta z$. Since $\tilde{\mathbf{k}}^+(\omega) = k(\omega)\left(\hat{\mathbf{1}}_x p + \hat{\mathbf{1}}_y q + \hat{\mathbf{1}}_z m(\omega)\right)$, so that $\left(\tilde{\mathbf{k}}^+(\omega)\right)^* = k(\omega)\left(\hat{\mathbf{1}}_x p + \hat{\mathbf{1}}_y q + \hat{\mathbf{1}}_z m^*(\omega)\right)$, then the complex Poynting vector for each individual plane wave spectral component at the real-valued angular frequency ω becomes

$$
\begin{aligned}
\tilde{\mathbf{S}}(\mathbf{r}; \mathbf{k}_T, \omega) &= \left\|\frac{c^2}{4\pi}\right\| \frac{k(\omega)}{2\mu^*(\omega)\omega} \left|\tilde{\mathbf{E}}_0(\mathbf{k}_T, \omega)\right|^2 e^{-2k(\omega)m''(\omega)\Delta z} \\
&\quad \times \left[\hat{\mathbf{1}}_x p + \hat{\mathbf{1}}_y q + \hat{\mathbf{1}}_z \left(m'(\omega) - im''(\omega)\right)\right] \\
&= \left\|\frac{c^2}{4\pi}\right\| \frac{k(\omega)}{2\,|\mu(\omega)|^2\,\omega} \left|\tilde{\mathbf{E}}_0(\mathbf{k}_T, \omega)\right|^2 e^{-2k(\omega)m''(\omega)\Delta z} \\
&\quad \times \Bigg\{\mu_r(\omega)\left(\hat{\mathbf{1}}_x p + \hat{\mathbf{1}}_y q + \hat{\mathbf{1}}_z m'(\omega)\right) + \mu_i(\omega)\hat{\mathbf{1}}_z m''(\omega) \\
&\qquad +i\left[\mu_i(\omega)\left(\hat{\mathbf{1}}_x p + \hat{\mathbf{1}}_y q + \hat{\mathbf{1}}_z m'(\omega)\right) - \mu_r(\omega)\hat{\mathbf{1}}_z m''(\omega)\right]\Bigg\},
\end{aligned}
$$

$$(8.188)$$

where $\mu_r(\omega) \equiv \Re\{\mu(\omega)\}$ and $\mu_i(\omega) \equiv \Im\{\mu(\omega)\}$, and where ω is taken to be real-valued. The time-average Poynting vector associated with each individual plane wave spectral component at the real-valued angular frequency ω is then obtained by taking the real part of the above expression, with the result

$$
\begin{aligned}
\left\langle \tilde{\mathbf{S}}_\omega(\mathbf{r}, t; p, q)\right\rangle &= \left\|\frac{c^2}{4\pi}\right\| \frac{k(\omega)}{2\,|\mu(\omega)|^2\,\omega} \left|\tilde{\mathbf{E}}_0(\mathbf{k}_T, \omega)\right|^2 e^{-2k(\omega)m''(\omega)\Delta z} \\
&\quad \times \left\{\mu_r(\omega)\left(\hat{\mathbf{1}}_x p + \hat{\mathbf{1}}_y q\right) + \hat{\mathbf{1}}_z \left[\mu_r(\omega)m'(\omega) + \mu_i(\omega)m''(\omega)\right]\right\}.
\end{aligned}
$$

$$(8.189)$$

The direction of the time-average flow of electromagnetic energy for a given plane wave spectral component in the pulsed beam wave field is then seen to depend upon the sign of the quantity $\left[\mu_r(\omega)m'(\omega) + \mu_i(\omega)m''(\omega)\right]$. If this quantity is positive, then the energy flow at that angular frequency is directed into the positive half-space $z > z_0$, while if it is negative the time-average energy flow is directed towards the negative half-space.

8.3.2   Homogeneous and Evanescent Plane Wave Contributions to the Angular Spectrum Representation

Consider the monochromatic wave field

$$
U_\omega(\mathbf{r}, t) = \Re\left\{\tilde{U}(\mathbf{r}, \omega)e^{-i\omega t}\right\}
$$

$$(8.190)$$

whose complex (phasor) field amplitude is given by the angular spectrum representation

$$\tilde{U}(\mathbf{r}, \omega) = \frac{1}{4\pi^2} \int_{-\infty}^{\infty} \tilde{U}_0(\mathbf{k}_T, \omega) e^{i\gamma(\omega)\Delta z} e^{i(k_x x + k_y y)} dk_x dk_y \qquad (8.191)$$

with $\gamma(\omega) \equiv \left(\tilde{k}^2(\omega) - k_x^2 - k_y^2 \right)^{1/2}$. This representation is expressed in terms of a superposition of monochromatic plane waves whose amplitude and phase are given by the spectral distribution function

$$\tilde{U}_0(\mathbf{k}_T, \omega) = \tilde{U}_0(k_x, k_y, \omega)$$

$$= \int_{-\infty}^{\infty} \tilde{U}_0(\mathbf{r}'_T, \omega) e^{-i\mathbf{k}_T \cdot \mathbf{r}'_T} dx' dy'$$

$$= \int_{-\infty}^{\infty} \tilde{U}_0(x', y', \omega) e^{-i(k_x x' + k_y y')} dx' dy'. \qquad (8.192)$$

In free-space, $\tilde{k}(\omega) = k = \omega/c$ and the spectral components satisfying the inequality $k_x^2 + k_y^2 \leq k^2$ represent homogeneous plane waves while the spectral components satisfying the opposite inequality $k_x^2 + k_y^2 > k^2$ represent evanescent plane waves. If the medium is dispersive (and hence absorptive), then $\tilde{k}(\omega) = \beta(\omega) + i\alpha(\omega)$ is complex-valued and this straightforward separation is not possible. Because of this, *the development in the remainder of this subsection is henceforth restricted to free-space.*

In free-space, it is found convenient to separate the domain of integration $\mathcal{D} = \mathcal{R}^2$ in the angular spectrum representation given in Eq. (8.191) into the two parts

$$\mathcal{D}_H \equiv \left\{ (k_x, k_y) \,\middle|\, k_x^2 + k_y^2 \leq k^2 \right\}, \qquad (8.193)$$

$$\mathcal{D}_E \equiv \left\{ (k_x, k_y) \,\middle|\, k_x^2 + k_y^2 > k^2 \right\}, \qquad (8.194)$$

where $\mathcal{D} = \mathcal{D}_H \cup \mathcal{D}_E$. The complex field amplitude then separates into the two parts

$$\tilde{U}_H(\mathbf{r}, \omega) \equiv \frac{1}{4\pi} \int_{\mathcal{D}_H} \tilde{U}_0(k_x, k_y, \omega) e^{i\sqrt{k^2 - k_x^2 - k_y^2}\,\Delta z} e^{i(k_x x + k_y y)} dk_x dk_y, \quad (8.195)$$

$$\tilde{U}_E(\mathbf{r}, \omega) \equiv \frac{1}{4\pi} \int_{\mathcal{D}_E} \tilde{U}_0(k_x, k_y, \omega) e^{-\sqrt{k_x^2 + k_y^2 - k^2}\,\Delta z} e^{i(k_x x + k_y y)} dk_x dk_y, \quad (8.196)$$

where

$$\tilde{U}(\mathbf{r}, \omega) = \tilde{U}_H(\mathbf{r}, \omega) + \tilde{U}_E(\mathbf{r}, \omega). \qquad (8.197)$$

Here $\tilde{U}_H(\mathbf{r}, \omega)$ represents the homogeneous wave contribution and $\tilde{U}_E(\mathbf{r}, \omega)$ represents the evanescent wave contribution to the wave field $\tilde{U}(\mathbf{r}, \omega)$. The stationary phase approximation of the asymptotic behavior of these two-dimensional integrals is given in Appendix G.

Substitution of Eq. (8.192) into Eq. (8.196) gives

$$\tilde{U}_E(\mathbf{r}, \omega) = -\frac{1}{2\pi} \int_{-\infty}^{\infty} \tilde{U}_0(x', y', \omega) I(x - x', y - y', \Delta z) dx' dy', \qquad (8.198)$$

with

$$I(\xi, \eta, z) \equiv -\frac{1}{2\pi} \int_{D_E} e^{-\sqrt{k_x^2 + k_y^2 - k^2} z} e^{i(k_x \xi + k_y \eta)} dk_x dk_y, \qquad (8.199)$$

which is related to the evanescent part of the monochromatic spatial impulse response function $h(\xi, \eta; z, \omega)$ defined in Eq. (8.66). The exact evaluation of this integral representation of the evanescent wave contribution forms the main focus of this subsection.

The integral appearing in Eq. (8.199) may be expressed in a more convenient form through the change of variable

$$k_x = \kappa \cos \alpha, \quad k_y = \kappa \sin \alpha, \qquad (8.200)$$

where $\kappa = k \to \infty$ and $\alpha = 0 \to 2\pi$. With the definition of the angle γ through the pair of relations [not to be confused with the complex wave number component $\gamma(\omega)$ defined in Eq. (8.11)]

$$\cos \gamma = \frac{\xi}{\sqrt{\xi^2 + \eta^2}}, \quad \sin \gamma = \frac{\eta}{\sqrt{\xi^2 + \eta^2}}, \qquad (8.201)$$

the transformed integral expression in Eq. (8.199) becomes

$$I(\xi, \eta, z) \equiv -\frac{1}{2\pi} \int_k^{\infty} \int_0^{2\pi} e^{-\sqrt{\kappa^2 - k^2} z} e^{i\kappa \sqrt{\xi^2 + \eta^2} \cos(\alpha - \gamma)} \kappa \, d\kappa \, d\alpha.$$

The integral over α is immediately recognized as the integral representation of the zero-order Bessel function of the first kind, where

$$J_0(\zeta) = \frac{1}{2\pi} \int_0^{2\pi} e^{-i\zeta \cos(\alpha - \gamma)} d\alpha, \qquad (8.202)$$

so that

$$I(\xi, \eta, z) \equiv -\int_k^{\infty} e^{-\sqrt{\kappa^2 - k^2} z} J_0\left(\kappa \sqrt{\xi^2 + \eta^2}\right) \kappa \, d\kappa. \qquad (8.203)$$

Under the final change of variable $\zeta = \sqrt{\kappa^2 - k^2}$, so that $\kappa d\kappa = \zeta d\zeta$, the above expression becomes

$$I(\xi, \eta, z) \equiv -\int_0^\infty e^{-z\zeta} J_0\left(\sqrt{\xi^2 + \eta^2}\sqrt{k^2 + \zeta^2}\right)\zeta d\zeta. \qquad (8.204)$$

With the addition theorem for Bessel functions (see Sect. 8.531(1) of Ref. [27])

$$J_0\left(\sqrt{a^2 + b^2 - 2ab\cos\theta}\right) = J_0(a)J_0(b) + 2\sum_{n=1}^\infty J_n(a)J_n(b)\cos(n\theta),$$

the Bessel function appearing in Eq. (8.204) may be expressed as

$$J_0\left(\sqrt{\xi^2 + \eta^2}\sqrt{k^2 + \zeta^2}\right)$$

$$= J_0\left(k\sqrt{\xi^2 + \eta^2}\right)J_0\left(\zeta\sqrt{\xi^2 + \eta^2}\right)$$

$$+2\sum_{n=1}^\infty J_n\left(k\sqrt{\xi^2 + \eta^2}\right)J_n\left(\zeta\sqrt{\xi^2 + \eta^2}\right)\cos(n\pi/2)$$

$$= -J_0\left(k\sqrt{\xi^2 + \eta^2}\right)J_0\left(\zeta\sqrt{\xi^2 + \eta^2}\right)$$

$$+2\sum_{n=0}^\infty (-1)^n J_{2n}\left(k\sqrt{\xi^2 + \eta^2}\right)J_{2n}\left(\zeta\sqrt{\xi^2 + \eta^2}\right).$$

Substitution of this result in Eq. (8.204) and interchanging the order of integration and summation then results in the expression

$$I(\xi, \eta, z) = J_0\left(k\sqrt{\xi^2 + \eta^2}\right)\int_0^\infty J_0\left(\zeta\sqrt{\xi^2 + \eta^2}\right)e^{-z\zeta}\zeta d\zeta$$

$$-2\sum_{n=0}^\infty (-1)^n J_{2n}\left(k\sqrt{\xi^2 + \eta^2}\right)\int_0^\infty J_{2n}\left(\zeta\sqrt{\xi^2 + \eta^2}\right)e^{-z\zeta}\zeta d\zeta.$$

$$(8.205)$$

With the integral identity (see Sect. 6.621(4) of Ref. [27])

$$\int_0^\infty J_{2m}\left(\zeta\sqrt{x^2 + y^2}\right)e^{-z\zeta}\zeta d\zeta$$

$$= -\frac{\partial}{\partial z}\left\{\frac{1}{\sqrt{x^2 + y^2 + z^2}}\left[\frac{\sqrt{x^2 + y^2 + z^2} - z}{\sqrt{x^2 + y^2}}\right]^{2m}\right\}$$

the above expression becomes

$$I(\xi, \eta, z) = -J_0\left(k\sqrt{\xi^2 + \eta^2}\right)\frac{\partial}{\partial z}\left\{\frac{1}{\sqrt{\xi^2 + \eta^2 + z^2}}\right\}$$

$$+ 2\sum_{n=0}^{\infty}(-1)^n J_{2n}\left(k\sqrt{\xi^2 + \eta^2}\right)$$

$$\times\frac{\partial}{\partial z}\left\{\frac{1}{\sqrt{\xi^2 + \eta^2 + z^2}}\left[\frac{\sqrt{\xi^2 + \eta^2 + z^2} - z}{\sqrt{\xi^2 + \eta^2}}\right]^{2n}\right\},$$

$$(8.206)$$

which involves an infinite summation of even-order Bessel functions. With the definition of the Lommel functions [28]

$$U_m(u, v) \equiv \sum_{n=0}^{\infty}(-1)^n \left(\frac{u}{v}\right)^{m+2n} J_{m+2n}(v), \qquad (8.207)$$

$$V_m(u, v) \equiv \sum_{n=0}^{\infty}(-1)^n \left(\frac{v}{u}\right)^{m+2n} J_{-m-2n}(v), \qquad (8.208)$$

which arise naturally in the diffraction theory of light through a circular aperture [13], the expression in Eq. (8.206) may be written in the form

$$I(\xi, \eta, z)$$

$$= \frac{\partial}{\partial z}\left\{\frac{-J_0\left(k\sqrt{\xi^2 + \eta^2}\right) + 2U_0\left(k\sqrt{\xi^2 + \eta^2 + z^2} - kz, k\sqrt{\xi^2 + \eta^2}\right)}{\sqrt{\xi^2 + \eta^2 + z^2}}\right\},$$

$$(8.209)$$

which is originally due to Bertilone [29].

The evanescent wave contribution to the time-harmonic (phasor) wave field $\tilde{U}(\mathbf{r}, \omega)$ is then given by

$$\tilde{U}_E(\mathbf{r}, \omega) = -\frac{1}{2\pi}\int_{-\infty}^{\infty}\tilde{U}_0(x', y', \omega)I(x - x', y - y', \Delta z)dx'dy', \qquad (8.210)$$

with

$$I(x - x', y - y', \Delta z) = \frac{\partial}{\partial z}\left\{\frac{2U_0(kR - k\Delta z, k\rho) - J_0(k\rho)}{R}\right\}, \qquad (8.211)$$

where $R = \sqrt{(x - x')^2 + (y - y')^2 + (\Delta z)^2}$ is the distance between the initial field point at $\mathbf{r}' = (x', y', z_0)$ and the observation point at $\mathbf{r} = (x, y, z_0)$, and where $\rho = \sqrt{(x - x')^2 + (y - y')^2}$.

For a pulsed wave field in vacuum, the spatio-temporal pulse evolution separates into homogeneous and evanescent wave contributions as

$$U(\mathbf{r}, t) = U_H(\mathbf{r}, t) + U_E(\mathbf{r}, t), \tag{8.212}$$

with

$$U_H(\mathbf{r}, t) = \frac{1}{\pi} \Re \int_0^\infty \tilde{U}_H(\mathbf{r}, \omega) e^{-i\omega t} d\omega, \tag{8.213}$$

$$U_E(\mathbf{r}, t) = \frac{1}{\pi} \Re \int_0^\infty \tilde{U}_E(\mathbf{r}, \omega) e^{-i\omega t} d\omega. \tag{8.214}$$

Substitution of the expression given in Eq. (8.210) with Eq. (8.211) into the above Fourier integral representation of the evanescent wave contribution to the pulsed wave field then yields the expression

$$U_E(\mathbf{r}, t) = -\frac{1}{2\pi^2} \int_{-\infty}^\infty \frac{\partial}{\partial z} \left\{ \frac{G(x', y'; x, y, z, t)}{R} \right\} dx' dy', \tag{8.215}$$

where

$$G(x', y'; x, y, z, t) \equiv \Re \int_0^\infty \tilde{U}_0(x', y', t')$$

$$\times \left\{ 2U_0((R - \Delta z)\omega/c, \rho\omega/c) - J_0(\rho\omega/c) \right\} e^{-i\omega t'} d\omega. \tag{8.216}$$

Substitution of the Fourier integral relation

$$\tilde{U}_0(x', y', \omega) = \int_{-\infty}^\infty U_0(x', y', t') e^{i\omega t'} dt'$$

into the above expression, followed by an interchange of the integration order, then gives [taking note of the fact that $U_0(x', y', t')$ is real-valued]

$$G(x'y'; x, y, z, t) = c \int_{-\infty}^\infty U_0(x', y', t')$$

$$\times \left\{ \int_0^\infty \left[2U_0(k(R - \Delta z), k\rho) - J_0(k\rho) \right] \right.$$

$$\left. \times \cos(ck|t' - t|) dk \right\} dt', \tag{8.217}$$

where the variable of integration in the inner integral has been changed from ω to $k = \omega/c$.

Consider first the evaluation of the k-integral in Eq. (8.217). Expansion of the zeroth-order Lommel function appearing in that integral in terms of an infinite series of integer-order Bessel functions through the defining relation given in Eq. (8.207) results in

$$\int_0^\infty U_0(k(R - \Delta z), k\rho) \cos(ck|t' - t|)dk$$

$$= \sum_{n=0}^\infty \left(\frac{R - \Delta z}{\rho}\right)^{2n} \int_0^\infty \cos(ck|t' - t|) J_{2n}(k\rho)dk.$$

From Sects. 6.771(10) and 8.940(1) of Ref. [27], the cosine transform of the integer order Bessel function is given by

$$\int_0^\infty J_{2n}(a\zeta)\cos(b\zeta)d\zeta = \begin{cases} \dfrac{(-1)^n}{\sqrt{a^2 - b^2}}\cos(2n\arccos(b/a)), & 0 \le b < a \\ 0, & 0 < a < b, \end{cases}$$

so that, together with the summation (see Sect. 1.447(2) of [27])

$$\sum_{n=0}^\infty \xi^n \cos(n\vartheta) = \frac{1 - \xi\cos\vartheta}{1 - 2\xi\cos\vartheta + \xi^2}, \qquad |\xi| < 1,$$

the above integral becomes

$$\int_0^\infty U_0(k(R - \Delta z), k\rho)\cos(ck|t' - t|)dk$$

$$= \begin{cases} \dfrac{1}{\sqrt{\rho^2 - c^2(t - t')^2}} \sum_{n=0}^\infty \left(\dfrac{R - \Delta z}{\rho}\right)^{2n}\cos(n\vartheta), & 0 \le c|t' - t| < \rho \\ 0, & 0 \le \rho < c|t' - t| \end{cases}$$

$$= \begin{cases} \dfrac{1}{\sqrt{\rho^2 - c^2(t - t')^2}}\left(\dfrac{1 - \xi\cos\vartheta}{1 - 2\xi\cos\vartheta + \xi^2}\right), & 0 \le c|t' - t| < \rho \\ 0, & 0 \le \rho < c|t' - t|, \end{cases} \tag{8.218}$$

where

$$\xi \equiv \left(\frac{R - \Delta z}{\rho}\right)^{2n}, \tag{8.219}$$

$$\vartheta \equiv 2\arccos\left(\frac{c|t' - t|}{\rho}\right). \tag{8.220}$$

Finally, the remaining k-integral appearing in Eq. (8.217) is given by

$$\int_0^\infty J_0(k\rho)\cos(ck|t'-t|)dk = \begin{cases} \dfrac{1}{\sqrt{\rho^2 - c^2(t'-t)^2}}, & 0 \le c|t'-t| < \rho \\ 0, & 0 \le \rho < c|t'-t|. \end{cases} \quad (8.221)$$

Combination of Eqs. (8.218) and (8.221) then gives the desired integration

$$\int_0^\infty \left[2U_0(k(R-\Delta z), k\rho) - J_0(k\rho)\right]\cos(ck|t'-t|)dk$$

$$= \begin{cases} \dfrac{1}{\sqrt{\rho^2 - c^2(t'-t)^2}}\left[2\left(\dfrac{1-\xi\cos\vartheta}{1-2\xi\cos\vartheta+\xi^2}\right) - 1\right], & 0 \le c|t'-t| < \rho \\ 0, & 0 \le \rho < c|t'-t|. \end{cases}$$

$$(8.222)$$

Notice that the integration results given in Eqs. (8.218), (8.221) and (8.222) are improperly defined when $\rho = 0$. As a consequence, the final result given in Eq. (8.222) has physical meaning here only in the context of the integrand of Eq. (8.217) where one may treat the integration in the generalized function sense [29]. Finally, notice that

$$2\left(\frac{1-\xi\cos\vartheta}{1-2\xi\cos\vartheta+\xi^2}\right) - 1 = \frac{R\Delta z}{R^2 - c^2(t'-t)^2}, \quad (8.223)$$

from Eqs. (8.219)–(8.220).

Substitution of the integration result given in Eq. (8.222) with Eq. (8.223) into Eq. (8.217) then yields the following result due to Bertilone [29]:

$$\frac{G(x', y'; x, y, z, t)}{R} = c\int_{t-\rho/c}^{t+\rho/c} \frac{U_0(x', y', t')}{\sqrt{\rho^2 - c^2(t'-t)^2}}\left(\frac{\Delta z}{R^2 - c^2(t'-t)^2}\right)dt'.$$

$$(8.224)$$

The final integral appearing in this expression is nonsingular in the positive half-space $\Delta z > 0$ for any properly behaved initial field behavior $U_0(x, y, t)$ at the plane $z = z_0$. Substitution of this result in Eq. (8.215) then yields the exact expression

$$U_E(\mathbf{r}, t) = -\frac{c}{2\pi^2}\int_{-\infty}^{\infty}\int_{t'=t-\rho/c}^{t+\rho/c} \frac{U_0(x', y', t')}{\sqrt{\rho^2 - c^2(t'-t)^2}}$$

$$\times \frac{\partial}{\partial z}\left\{\frac{\Delta z}{R^2 - c^2(t'-t)^2}\right\}dx'dy'dt'$$

$$(8.225)$$

for the evanescent contribution to the propagated pulsed wave field.

In contrast with the total field, the evanescent field is found to be non-causal since $U_E(\mathbf{r}, t)$ is expressed by Eq. (8.225) in terms of the field behavior $U_0(x', y', t')$ on the $z = z_0$ plane at the future times from $t' = t - \rho/c$ to $t' = t + \rho/c$. Because the total (causal) wave field is given as the sum of the homogeneous and evanescent parts, the homogeneous part must then possess the precise non-causal behavior to exactly cancel the non-causal contribution from the evanescent part. Such nonphysical behavior was first reported by Sherman, Stamnes, Devaney and Lalor [30] who concluded that "for many physical problems, it is inappropriate to give physical significance to $U_H(\mathbf{r}, t)$ and $U_E(\mathbf{r}, t)$ independently." Similar remarks would then appear likely to apply to the separate inhomogeneous plane wave contributions from the interior and exterior regions $\mathcal{R}_<$ and $\mathcal{R}_>$, respectively [see Eqs. (8.156)–(8.186)], when the medium is dispersive; however, a rigorous proof of this extension of Bertilone's proof to dispersive, attenuative media remains to be given.

8.4 Pulsed Electromagnetic Beam Fields and Source-Free Fields

The angular spectrum of plane waves representation of the freely propagating electromagnetic field given in Eqs. (8.14) and (8.17) expresses that wave field throughout the positive half-space $z \geq z_0$ as a superposition of both homogeneous and inhomogeneous plane waves. In the idealized limit of a lossless medium, $\psi(\omega) = 0$ and one obtains homogeneous plane wave components in the interior circular domain $\mathcal{R}_< \equiv \{(p, q)|(p^2 + q^2) < 1\}$ and evanescent plane wave components in the exterior domain $\mathcal{R}_> \equiv \{(p, q)|(p^2 + q^2) > 1\}$. Because there is no time-average energy flow from the evanescent plane wave components into the half-space $z > z_0$, an electromagnetic beam field in a lossless medium is typically (perhaps casually) defined, in part, by the requirement that its angular spectrum does not contain any evanescent wave components, and consequently that its field can be represented by an angular spectrum that contains only homogeneous plane wave components [9]. Because the homogeneous plane wave spectral components do not attenuate when propagating through such a lossless medium, this condition ensures that all of the angular spectrum components of the beam field remain present in their initial proportion throughout the positive half-space $z \geq z_0$. Wave fields in lossless media that only contain homogeneous plane wave components in their angular spectrum representation are known as *source-free wave fields* [4, 6].

Because such a distinction cannot be made when the medium is attenuative, Sherman's definition [4, 6] of source-free wave fields needs to be generalized in order to treat that more general case. Because of the inequality given in Eq. (8.186), wave fields in dispersive attenuative media that contain only inhomogeneous plane wave components with direction cosines that satisfy the inequality $(p^2 + q^2) < \cos(2\psi(\omega))$ for all $\omega \in C_+$ in their angular spectrum representation are called source-free wave fields. In the limit as the material loss goes to zero at all frequencies, these inhomogeneous plane wave components become homogeneous plane wave components with direction cosines satisfying the inequality $(p^2 + q^2) < 1$. Specifically, one has the following definition for a scalar wave field that is described by the scalar version of the angular spectrum representation given in Eq. (8.187):

Definition 8.1 The scalar wave field $\tilde{U}(\mathbf{r}, \omega)$ is said to be source-free if and only if $\tilde{U}_0(p, q, \omega) \in L^2$ vanishes almost everywhere for $(p^2 + q^2) > \cos(2\psi(\omega))$ for all $\omega \in C_+$.

An electromagnetic beam field in a dispersive lossy medium may then be defined, in part, by the requirement that each of its scalar components is a generalized source-free wave field, so that

$$\tilde{\mathbf{E}}_0(p, q, \omega) = \tilde{\mathbf{B}}_0(p, q, \omega) = \mathbf{0}, \quad \forall (p, q) \ni p^2 + q^2 > \cos(2\psi(\omega)). \quad (8.226)$$

Each propagated field vector is then given by the angular spectrum of plane wave representation appearing in Eq. (8.187) taken only over the interior circular region $\mathcal{R}_<$. For reasons of notational simplicity, the properties of source-free wave fields are presented here for the scalar case. The results then apply directly to the electromagnetic wave-field case.

8.4.1 General Properties of Source-Free Wave Fields

With the classic analysis of Sherman [6] as a guide, consider the monochromatic scalar wave function $\tilde{U}(\mathbf{r}, \omega)$ that satisfies the scalar Helmholtz equation [cf. Eqs. (5.54) and (5.55)]

$$\left(\nabla^2 + \tilde{k}^2(\omega)\right) \tilde{U}(\mathbf{r}, \omega) = 0 \quad (8.227)$$

throughout the positive half-space $z \geq z_0$ and which has the boundary value

$$\lim_{\Delta z \to 0} \left\{\tilde{U}(\mathbf{r}, \omega)\right\} = \tilde{U}_0(\mathbf{r}_T, \omega), \quad (8.228)$$

where $\Delta z \equiv z - z_0$. This wave function then has the angular spectrum of plane waves representation

$$\tilde{U}(\mathbf{r}, \omega) = \frac{1}{4\pi^2} \int_{-\infty}^{\infty} \int_{-\infty}^{\infty} \tilde{U}_0(p, q, \omega) H(p, q, \omega; x, y, \Delta z) k^2(\omega) dp dq$$

$$(8.229)$$

throughout the positive half-space $\Delta z \geq 0$, where

$$\tilde{U}_0(p,q,\omega) = \int_{-\infty}^{\infty}\int_{-\infty}^{\infty} \tilde{U}_0(x,y,\omega)e^{-ik(px+qy)}dxdy, \qquad (8.230)$$

$$\tilde{U}_0(x,y,\omega) = \frac{1}{(2\pi)^2}\int_{-\infty}^{\infty}\int_{-\infty}^{\infty} \tilde{U}_0(p,q,\omega)e^{ik(px+qy)}k^2dpdq, \qquad (8.231)$$

form a two-dimensional Fourier transform pair, and where

$$H(p,q,\omega;x,y,\Delta z) \equiv G(p,q,\omega,\Delta z)e^{ik(\omega)(px+qy)} \qquad (8.232)$$

is the kernel appearing in the angular spectrum integral in Eq. (8.191) [cf. Eq. (8.182)] with

$$G(p,q,\omega,\Delta z) \equiv e^{ik(\omega)m(\omega)\Delta z}$$
$$= e^{ik(\omega)\Delta z[e^{i2\psi(\omega)}-(p^2+q^2)]^{1/2}}$$
$$= e^{i\tilde{k}(\omega)\Delta z\left[1-\left(\underline{p}^2+\underline{q}^2\right)\right]^{1/2}}. \qquad (8.233)$$

The complex-valued variables $\underline{p} \equiv pe^{-i\psi}$ and $\underline{q} \equiv qe^{-i\psi}$ have been introduced here in order to simplify the notation.[3] The conditions under which the scalar wave field given by Eq. (8.229) is the valid representation of the propagated wave field throughout the half-space $z > z_0$ that is due to the boundary value given in Eq. (8.228) has been considered in detail by Lalor [14], Montgomery [31], Sherman and Bremermann [5], and Sherman [4, 6] for a lossless nondispersive medium. A sketch of the proof of the validity of this angular spectrum representation of the solution to the Helmholtz equation given in Eq. (8.227) that has the boundary value given in Eq. (8.228) with complex wave number $\tilde{k}(\omega) = (\omega/c)n(\omega)$ is now given.

8.4.1.1 Validity of the Angular Spectrum Representation

The method of proof presented here follows that given by Sherman [6] for a lossless medium. For all $\omega \in C_+$, assume that $\tilde{U}_0(\mathbf{r}_T,\omega) = \tilde{U}_0(x,y,\omega) \in L^2$ and let $\tilde{U}(\mathbf{r},\omega) = \tilde{U}(x,y,z,\omega)$ be given by the angular spectrum representation in Eq. (8.229) for all $\Delta z = z - z_0 > 0$ with $\tilde{U}_0(p,q,\omega) = \mathcal{F}\{\tilde{U}_0(x,y,\omega)\}$ being given by the two-dimensional Fourier transform specified in Eq. (8.230). Because $H(p,q,\omega;x,y,\Delta z) \in L^2$, then the function formed by the product $\tilde{U}_0(p,q,\omega)H(p,q,\omega;x,y,\Delta z)$ is in L^1, and hence, $\tilde{U}(x,y,z,\omega)$ exists and is

[3]Notice that $\underline{p}$ and $\underline{q}$ are identical to the complex direction cosines p and q used in Sect. 8.1 [cf. Eqs. (8.21)–(8.23) and Eqs. (8.156)–(8.158)].

bounded for all $\Delta z > 0$. The spatial partial derivatives of the product function $\tilde{U}_0(p, q, \omega)H(p, q, \omega; x, y, \Delta z)$ satisfy the inequalities

$$\left| \frac{\partial^{j+m+n}}{\partial x^j \partial y^m \partial z^n} \tilde{U}_0(p, q, \omega)H(p, q, \omega; x, y, \Delta z) \right| \leq M(p, q, \delta, \omega) \qquad (8.234)$$

for all $\Delta z \geq \delta > 0$, where δ is an arbitrary positive number, and where

$$M(p, q, \delta, \omega) \equiv \left| \left(k^{j+m+n} p^j q^m e^{i2\psi(\omega)} - \left(p^2 + q^2 \right) \right)^{n/2} \tilde{U}_0(p, q, \omega)G(p, q, \omega, \delta) \right|$$
$$(8.235)$$

is independent of z. Because the right-hand side of Eq. (8.235) is the product of two functions that are in L^2, then it is in L^1. Hence, by Lebesgue's theorem of bounded convergence [32], the integrand in Eq. (8.229) is sufficiently well-behaved that $\tilde{U}(\mathbf{r}, \omega) = \tilde{U}(x, y, z, \omega)$ is continuous with respect to each spatial variable and has continuous partial derivatives with respect to each spatial variable, where

$$\frac{\partial^{j+m+n}}{\partial x^j \partial y^m \partial z^n} \tilde{U}(x, y, z, \omega)$$

$$= \frac{1}{4\pi^2} \int_{-\infty}^{\infty} \int_{-\infty}^{\infty} \frac{\partial^{j+m+n}}{\partial x^j \partial y^m \partial z^n} \left(\tilde{U}_0(p, q, \omega)H(p, q, \omega; x, y, \Delta z) \right) k^2(\omega) dp dq$$
$$(8.236)$$

for all real values of x and y and for all $\Delta z \geq \delta > 0$. The proof that the partial differentiation operation can be interchanged in its order with the integration operation in obtaining the expression on the right-hand side of Eq. (8.236) may be found in a generalization of a theorem by McShane [33] from a single to a multiple integral; although that theorem is stated only for a single differentiation, it may be repeatedly applied $(j + m + n)$ times to obtain the result given in Eq. (8.236).

If one applies this result in order to determine the Laplacian of $\tilde{U}(x, y, z, \omega)$, one obtains the result given in Eq. (8.227). Hence, for all $\omega \in C_+$, the wave function $\tilde{U}(\mathbf{r}, \omega) = \tilde{U}(x, y, z, \omega)$ that is given in Eq. (8.229) satisfies the scalar Helmholtz equation (8.227) with complex wavenumber $\tilde{k}(\omega) = k(\omega)e^{i\psi(\omega)}$ for all $\Delta z \geq \delta > 0$ with δ arbitrarily small.

Consider now the proof of Eq. (8.228) when the limiting procedure is taken as a limit in the mean for all $\tilde{U}(\mathbf{r}, \omega) = \tilde{U}(x, y, z, \omega) \in L^2$. In that case it must be shown that

$$\lim_{\Delta z \to 0} \left\{ I(\Delta z, \omega) \right\} = 0, \qquad (8.237)$$

where

$$I(\Delta z, \omega) \equiv \int_{-\infty}^{\infty} \int_{-\infty}^{\infty} \left| \tilde{U}(x, y, \Delta z, \omega) - \tilde{U}_0(x, y, \omega) \right|^2 dxdy \qquad (8.238)$$

for $\Delta z > 0$. The two-dimensional spatial Fourier transform of the quantity $\left(\tilde{U}(x, y, \Delta z, \omega) - \tilde{U}_0(x, y, \omega) \right)$ is equal to $\tilde{\tilde{U}}_0(p, q, \omega)(G(p, q, \omega, \Delta z) - 1)$ for $\Delta z > 0$ so that by Parseval's theorem

$$I(\Delta z, \omega) = \int_{-\infty}^{\infty} \int_{-\infty}^{\infty} \left| \tilde{\tilde{U}}_0(p, q, \omega)(G(p, q, \omega, \Delta z) - 1) \right|^2 dpdq \qquad (8.239)$$

for $\Delta z > 0$. Because $|G(p, q, \omega, \Delta z) - 1| \leq 2$ for all $\Delta z > 0$, then

$$\left| \tilde{\tilde{U}}_0(p, q, \omega)(G(p, q, \omega, \Delta z) - 1) \right|^2 \leq 4 \left| \tilde{\tilde{U}}_0(p, q, \omega) \right|^2$$

for $\Delta z > 0$. One can then apply Lebesgue's theorem of bounded convergence [32] to obtain

$$\lim_{\Delta z \to 0} \{I(\Delta z, \omega)\} = \int_{-\infty}^{\infty} \int_{-\infty}^{\infty} \left\{ \lim_{\Delta z \to 0} \left[\left| \tilde{\tilde{U}}_0(p, q, \omega) \right|^2 |G(p, q, \omega, \Delta z) - 1|^2 \right] \right\}$$

$$\times k^2(\omega) dpdq$$

$$= 0. \qquad (8.240)$$

This then proves the following generalization of a theorem due to Sherman.[4]

Theorem 8.2 (Sherman's First Theorem) *For all angular frequencies $\omega \in C_+$, let $\tilde{U}_0(\mathbf{r}_T, \omega) = \tilde{U}_0(x, y, \omega) \in L^2$ with two-dimensional Fourier transform $\tilde{\tilde{U}}_0(p, q, \omega) = \mathcal{F}\left\{ \tilde{U}_0(x, y, \omega) \right\}$, and let $\tilde{U}(\mathbf{r}, \omega) = \tilde{U}(x, y, z, \omega)$ be given by the angular spectrum of plane waves representation in Eq. (8.229) for all $\Delta z = z - z_0 > 0$. Then for all $\Delta z > 0$, $\tilde{U}(\mathbf{r}, \omega) = \tilde{U}(x, y, z, \omega)$ is bounded and has continuous partial derivatives of all orders with respect to each spatial coordinate variable. Moreover, $\tilde{U}(\mathbf{r}, \omega) = \tilde{U}(x, y, z, \omega)$ satisfies the scalar Helmholtz equation with complex wave number $\tilde{k}(\omega) = k(\omega)e^{i\psi(\omega)}$ for all $\Delta z > 0$ and has the boundary value*

$$\lim_{\Delta z \to 0} \{\tilde{U}(x, y, z, \omega)\} = \tilde{U}_0(x, y, \omega). \qquad (8.241)$$

From Schwartz's inequality it follows that if a function $F(p, q) \in L^2$ vanishes almost everywhere outside a bounded region in (p, q)-space, then $F(p, q) \in$

---

[4] See Theorem 1 of [6].

L^1. Hence, if $\tilde{U}(\mathbf{r}, \omega)$ is source-free, then $\tilde{U}_0(p, q, \omega) \in L^1$. The following three theorems due to Sherman[5] then remain valid with only minor modifications (included here) in a dispersive attenuative medium.

Theorem 8.3 (Sherman's Second Theorem) *If the scalar wave field $\tilde{U}(\mathbf{r}, \omega)$ is source-free, then the relation given in Eq. (8.229) can be used to extend $\tilde{U}(\mathbf{r}, \omega)$ into the region $z \leq z_0$ to obtain a bounded solution of the Helmholtz equation (8.227) for all space.*

The extension of the source-free wave field $\tilde{U}(\mathbf{r}, \omega)$ through the use of the integral representation given in Eq. (8.229) means only that $\tilde{U}(\mathbf{r}, \omega)$ is defined by Eq. (8.229) for all z. This theorem shows that it is not necessary to restrict attention to the positive half-space $z \geq z_0$ when considering a source-free wave field. The proof of this theorem is essentially inchanged from that given by Sherman [6] in the lossless case.

As a consequence of this theorem, source-free wave fields may always be considered to extend throughout all space. Because source-free fields satisfy the homogeneous Helmholtz equation given in Eq. (8.227) throughout all space, there are then no sources of the field anywhere in any finite reach of space; any source of the field must then be located infinitely far away. Because a source-free field is bounded for all x, y, z, such infinitely removed sources can then have no mathematical significance. Because of this property, Sherman [4, 6] regarded source-free fields to be wave fields that extended throughout all space with no sources anywhere, as reflected in the name "source-free." This type of field is clearly a special case of a freely propagating wave field that has no sources in the positive half-space $z \geq z_0$.

Theorem 8.4 (Sherman's Third Theorem) *For a given boundary value $\tilde{U}_0(\mathbf{r}_T, \omega) = \tilde{U}_0(x, y, \omega) \in L^2$ for all $\omega \in C_+$, the field $\tilde{U}(\mathbf{r}, \omega) = \tilde{U}(x, y, z, \omega)$ is source-free if and only if $\tilde{U}_0(x, y, \omega)$ is equal almost everywhere to a function $f_0(x, y, \omega)$ that can be extended to the entire space of two complex variables $X = x + ix''$, $Y = y + iy''$ as an entire function $f_0(X, Y, \omega)$ such that*

$$|f_0(X, Y, \omega)| \leq Ae^{k'(\omega)(x''^2 + y''^2)^{1/2}} \tag{8.242}$$

for all $\omega \in C_+$, where x'' and y'' are real variables and A is a positive constant.

This theorem, a consequence of the Plancherel–Pólya theorem [34], allows one to determine directly from the boundary value $\tilde{U}_0(x, y, \omega)$ whether the resultant field $\tilde{U}(\mathbf{r}, \omega) = \tilde{U}(x, y, z, \omega)$ is source-free without having to evaluate the Fourier transform $\tilde{U}_0(p, q, \omega) = \mathcal{F}\left\{\tilde{U}_0(x, y, \omega)\right\}$.

---

[5] See Theorems 2 through 4 of [6].

Proof In order to prove the sufficiency part of Theorem 8.4, let the field $\tilde{U}(\mathbf{r}, \omega) = \tilde{U}(x, y, z, \omega)$ be source-free for all $\omega \in C_+$, so that, by definition, $\tilde{U}_0(p, q, \omega) = \mathcal{F}\left\{\tilde{U}_0(x, y, \omega)\right\}$ vanishes almost everywhere for $(p^2 + q^2) > \cos(2\psi(\omega))$ for all $\omega \in C_+$. If one then lets $x'' = r \cos\zeta$, $y'' = r \sin\zeta$, then it follows directly from the necessity portion of the proof of the Plancherel–Pólya theorem [34] that $\tilde{U}_0(\mathbf{r}_T, \omega) = \tilde{U}_0(x, y, \omega)$ is equal almost everywhere to a function $f_0(x, y, \omega)$ that can be extended to the entire space of two complex variables $X = x + ix''$, $Y = y + iy''$ as an entire function $f_0(X, Y, \omega)$ such that

$$|f_0(X, Y, \omega)| \le Ae^{rK_f(\zeta)}, \tag{8.243}$$

where $K_f(\zeta)$ is the least upper bound of the quantity $k(p \cos\zeta + q \sin\zeta$ with fixed ζ for (p, q) in the region $(p^2 + q^2) \le \cos(2\psi(\omega))$ for all $\omega \in C_+$, where k is the transform parameter appearing in Eqs. (8.230) and (8.231). With $p = R \cos\xi$ and $q = R \sin\xi$ with $R \le \cos^{1/2}(2\psi(\omega))$, then $K_f(\zeta)$ is the least upper bound of the quantity $kR \cos(\zeta - \xi)$, and hence

$$K_f(\zeta) = k \cos^{1/2}(2\psi(\omega)) \equiv k'(\omega), \tag{8.244}$$

which, when substituted in Eq. (8.243), yields the result given in Eq. (8.242).

In order to prove the necessity part of the theorem, let $\tilde{U}_0(x, y, \omega) = f_0(x, y, \omega)$ almost everywhere, where $f_0(x, y, \omega)$ can be extended to an entire function $f_0(X, Y, \omega)$ of the two complex variables $X = x + ix''$, $Y = y + iy''$ that satisfies the inequality given in Eq. (8.242). Consider the function $h_\omega(\zeta, x, y)$ that is defined by the limit

$$h_\omega(\zeta, x, y) \equiv \lim_{r \to \infty}\left\{\frac{1}{r}\ln|f_0(x + ir \cos\zeta, y + ir \sin\zeta|\right\}, \tag{8.245}$$

where $x'' = r \cos\zeta$, $y'' = r \sin\zeta$, as before, and let

$$h_\omega(\zeta) \equiv \sup\{h_\omega(\zeta, x, y)\} \tag{8.246}$$

be the least upper bound of $h_\omega(\zeta, x, y)$ for all x and y with fixed ζ for all $\omega \in C_+$. By the Plancherel–Pólya theorem [34], the Fourier transform $F_0(p, q, \omega) = \mathcal{F}\{f_0(x, y, \omega)\}$ vanishes almost everywhere outside a convex region $\mathcal{D}_f$ in (p, q)–space, where

$$k(p \cos\zeta + q \sin\zeta) \le h_\omega(\zeta) \tag{8.247}$$

for all $(p, q) \in \mathcal{D}_f$. Because $f_0(X, Y, \omega)$ satisfies the inequality given in Eq. (8.242), then

$$h_\omega(\zeta) \le \lim_{r \to \infty}\left\{\frac{1}{r}\ln\left|Ae^{k'r}\right|\right\} = k'. \tag{8.248}$$

With $p = R \cos \xi$ and $q = R \sin \xi$, the above two inequalities yield the inequality

$$R \cos (\zeta - \xi) \le \cos^{1/2} (2\psi(\omega)), \tag{8.249}$$

so that $F_0(p, q, \omega)$ vanishes almost everywhere outside the circular region $R^2 = (p^2 + q^2) \le \cos (2\psi(\omega))$. Finally, because the Fourier transforms of two functions that are equal almost everywhere are themselves equal almost everywhere, then $\tilde{U}_0(p, q, \omega)$ vanishes almost everywhere for $(p^2 + q^2) > \cos (2\psi(\omega))$.

Theorem 8.5 (Sherman's Fourth Theorem) *The angular spectrum representation given in Eq. (8.229) can be used to extend the wave field $\tilde{U}(\mathbf{r}, \omega) = \tilde{U}(x, y, z, \omega)$ to the entire space of three complex variables $X = x + ix''$, $Y = y + iy''$, $Z = z + iz''$ as an entire function $\tilde{U}(X, Y, Z, \omega)$ such that for constant $z'' = \Im\{Z\}$,*

$$\left| \tilde{U}(X, Y, Z, \omega) \right| \le B e^{k(x''^2 + y''^2)^{1/2}}, \tag{8.250}$$

where x'', y'', z'' are real variables and B is a positive constant that can depend on z'', if and only if $\tilde{U}(\mathbf{r}, \omega)$ is source-free.

Proof In order to prove the necessity part of this theorem, assume that $\tilde{U}(\mathbf{r}, \omega) = \tilde{U}(x, y, z, \omega)$ is a source-free wave field and consider the representation

$$\tilde{U}(X, Y, Z, \omega) = \frac{1}{4\pi^2} \int_{R_<} \tilde{U}_0(p, q, \omega) H(p, q, \omega; X, Y, Z) k^2(\omega) dp dq, \tag{8.251}$$

where $X = x + ix''$, $Y = y + iy''$, and $Z = z + iz''$. From Eqs. (8.232) and (8.233)

$$\left| \tilde{U}(X, Y, Z, \omega) \right| \le \frac{1}{4\pi^2} \int_{R_<} \left| \tilde{U}_0(p, q, \omega) \right| e^{-k(\omega)m''(\omega)z} e^{-k(\omega)m'(\omega)z''} k^2(\omega) dp dq,$$

where $m'(\omega) \equiv \Re\{m(\omega)\} \ge 0$ and $m''(\omega) \equiv \Im\{m(\omega)\} \ge 0$ for all $\omega \in C_+$. In any bounded region $\mathcal{R}$ of X, Y, Z-space, the product of the two exponentials appearing in the integrand of the above equation is bounded by a positive constant M that depends upon the region $\mathcal{R}$, so that

$$\left| \tilde{U}(X, Y, Z, \omega) \right| \le \frac{M}{4\pi^2} \int_{R_<} \left| \tilde{U}_0(p, q, \omega) \right| k^2(\omega) dp dq = M A. \tag{8.252}$$

Hence, the integral defined in Eq. (8.251) exists for all X, Y, Z in any bounded region of X, Y, Z-space. Because the integrand appearing in Eq. (8.251) is an entire function of complex X, Y, Z, then $\tilde{U}(X, Y, Z, \omega)$ is an entire function of X, Y, Z, provided that $\tilde{U}(\mathbf{r}, \omega) = \tilde{U}(x, y, z, \omega)$ is source-free [35]. It then follows from the sufficiency proof of Theorem 8.5 that $\tilde{U}(X, Y, Z, \omega)$ satisfies the inequality given in Eq. (8.250) for constant $z'' \equiv \Im\{Z\}$.

In order to prove the sufficiency part of this theorem, assume that $\tilde{U}(\mathbf{r}, \omega) = \tilde{U}(x, y, z, \omega)$ may be extended to the entire space of three complex variables X, Y, Z as an entire function $\tilde{U}(X, Y, Z, \omega)$ for all $\omega \in C_+$. If one then replaces $f_0(x, y, \omega)$ with $\tilde{U}(x, y, Z, \omega)$ in the necessity part of the proof of Theorem 8.6 and keeps $z'' = \Im\{Z\}$ constant throughout, it then follows that the quantity $\tilde{U}_0(p, q, \omega)G(p, q, \omega, \Delta z)$ vanishes almost everywhere for $p^2 + q^2 > \cos(2\psi(\omega))$. Because $G(p, q, \omega, \Delta z)$ does not vanish for any finite values of p and q [see Eq. (8.233)], then $\tilde{U}_0(p, q, \omega)$ vanishes almost everywhere for $p^2 + q^2 > \cos(2\psi(\omega))$.

An immediate consequence of this theorem is that a source-free wave field $\tilde{U}(\mathbf{r}, \omega) = \tilde{U}(x, y, z, \omega)$ has all of the properties that are associated with an entire function in complex variable theory. In particular, $\tilde{U}(\mathbf{r}, \omega) = \tilde{U}(x, y, z, \omega)$ can be expanded in each spatial variable in a Taylor series that converges for all values of x, y, and z. In addition, the specification of the field throughout a finite region of space is sufficient to determine the field throughout all space through the process of analytic continuation. This result is of fundamental consequence to diffraction and scattering theory [10, 11] as well as to inverse optics [13, 36].

8.4.1.2   The Sherman Expansion of Source-Free Wave Fields

Because the complex wave number $\tilde{k}(\omega)$ is analytic along the contour C_+, then for a source-free wave field one may represent the function $G(p, q, \omega, \Delta z)$ defined in Eq. (8.233) by its Taylor series expansion

$$G(\underline{p}, \underline{q}, \omega, \Delta z) = \sum_{r=0}^{\infty} \sum_{s=0}^{\infty} \frac{G^{(2r,2s)}(0, 0, \omega, \Delta z)}{(2r)!(2s)!} \underline{p}^{2r} \underline{q}^{2s}, \tag{8.253}$$

where

$$G^{(m,n)}(\underline{p}, \underline{q}, \omega, \Delta z) \equiv \frac{\partial^{m+n} G(\underline{p}, \underline{q}, \omega, \Delta z)}{\partial \underline{p}^m \partial \underline{q}^n}. \tag{8.254}$$

Substitution of this Taylor series expansion into the angular spectrum representation given in Eq. (8.229) then yields

$$\tilde{U}(\mathbf{r}, \omega) = \frac{1}{4\pi^2} \sum_{r=0}^{\infty} \sum_{s=0}^{\infty} \frac{G^{(2r,2s)}(0, 0, \omega, \Delta z)}{(2r)!(2s)!}$$

$$\times \int_{-\infty}^{\infty} \int_{-\infty}^{\infty} \underline{p}^{2r} \underline{q}^{2s} \tilde{U}_0(\underline{p}, \underline{q}, \omega) e^{i\tilde{k}(\omega)\left(\underline{p}x + \underline{q}y\right)} \tilde{k}^2(\omega) dp dq. \tag{8.255}$$

Because

$$\tilde{U}_0^{(m,n)}(x, y, \omega) = \frac{\partial^{m+n}\tilde{U}_0(x, y, \omega)}{\partial x^m \partial y^n}$$

$$= \frac{1}{4\pi^2} \int_{-\infty}^{\infty} \int_{-\infty}^{\infty} \tilde{U}_0(\underline{p}, \underline{q}, \omega) \left(i\tilde{k}(\omega)\right)^{m+n} \underline{p}^m \underline{q}^n$$

$$\times e^{i\tilde{k}(\omega)\left(\underline{p}x+\underline{q}y\right)} \tilde{k}^2(\omega) dp dq,$$

(8.256)

one finally obtains the Sherman expansion of the source-free wave field

$$\tilde{U}(\mathbf{r}, \omega) = \sum_{r=0}^{\infty} \sum_{s=0}^{\infty} \frac{G^{(2r,2s)}(0, 0, \omega, \Delta z)}{(2r)!(2s)!\left(i\tilde{k}(\omega)\right)^{2(r+s)}} \tilde{U}_0^{(2r,2s)}(x, y, \omega).$$ (8.257)

This series expansion explicitly displays the contribution to the transverse spatial behavior of the propagated wave field due to each even-ordered spatial partial derivative of the boundary value given in Eq. (8.228).

Notice that the factor $(\tilde{k}(\omega))^{-2(r+s)}$ appearing in the above series expansion is misleading because the same factor with an opposite-signed exponent is contained in the partial derivative $G^{(2r,2s)}(0, 0, \omega, \Delta z)$. From Eq. (8.233)

$$G^{(m,n)}(\underline{p}, \underline{q}, \omega, \Delta z) \equiv \frac{\partial^{m+n} G(\underline{p}, \underline{q}, \omega, \Delta z)}{\partial \underline{p}^m \partial \underline{q}^n}$$

$$= \left(i\tilde{k}(\omega)\Delta z\right)^{m+n} e^{i\tilde{k}(\omega)\Delta z\left(1-\underline{p}^2-\underline{q}^2\right)^{1/2}} \varphi^{(m,n)}(\underline{p}, \underline{q}),$$

(8.258)

where

$$\varphi^{(m,n)}(\underline{p}, \underline{q}) \equiv \frac{\partial^{m+n}}{\partial \underline{p}^m \partial \underline{q}^n} \left(1 - \underline{p}^2 - \underline{q}^2\right)^{1/2}.$$ (8.259)

With these substitutions, Sherman's expansion (8.257) becomes

$$\tilde{U}(\mathbf{r}, \omega) = \sum_{r=0}^{\infty} \sum_{s=0}^{\infty} \frac{(\Delta z)^{2(r+s)}}{(2r)!(2s)!} \varphi^{(2r,2s)}(0, 0) \tilde{U}_0^{(2r,2s)}(x, y, \omega) e^{i\tilde{k}(\omega)\Delta z}.$$

(8.260)

In addition, the quantity $\varphi^{(2r,2s)}(0,0)$ appearing in this form of the Sherman expansion is given by

$$\varphi^{(2r,2s)}(0,0) = \frac{\partial^{2(r+s)}}{\partial \underline{p}^{2r}\partial \underline{q}^{2s}}\left(1 - \underline{p}^2 - \underline{q}^2\right)^{1/2}\bigg|_{\underline{p}=\underline{q}=0}$$

$$= \theta(2r)\theta(2s), \tag{8.261}$$

where $\theta(0) = 1$ and

$$\theta(2m) \equiv (-1)^m \frac{1}{2}\left(\frac{1}{2}-1\right)\left(\frac{1}{2}-2\right)\cdots\left(\frac{1}{2}-m+1\right)\frac{(2m)!}{m!} \tag{8.262}$$

for integer values of $m > 0$.

With these results, one finally obtains the spatial series representation for either the electric or magnetic field vector of the pulsed, source-free electromagnetic wave field [cf. Eq. (8.187)] as [26]

$$\mathbf{U}(\mathbf{r},t) = \sum_{r=0}^{\infty}\sum_{s=0}^{\infty}\frac{\theta(2r)\theta(2s)}{(2r)!(2s)!}(\Delta z)^{2(r+s)}$$

$$\times \frac{1}{\pi}\Re\left\{\int_{C_+}\tilde{\mathbf{U}}_0^{(2r,2s)}(x,y,\omega)e^{i\left(\tilde{k}(\omega)\Delta z - \omega t\right)}d\omega\right\}, \tag{8.263}$$

where

$$\tilde{\mathbf{U}}_0^{(2r,2s)}(x,y,\omega) \equiv \frac{\partial^{2(r+s)}}{\partial x^{2r}\partial y^{2s}}\int_{-\infty}^{\infty}\mathbf{U}_0(x,y,t)e^{i\omega t}dt$$

$$= \int_{-\infty}^{\infty}\mathbf{U}_0^{(2r,2s)}(x,y,t)e^{i\omega t}dt. \tag{8.264}$$

The representation given in Eq. (8.263) explicitly displays the temporal evolution of a pulsed electromagnetic beam field through a single contour integral taken over the even-order spatial partial derivatives of the temporal frequency spectrum of the boundary value for the appropriate field vector at the plane $z = z_0$. This representation is exact provided that the wave field is source-free. For a plane wave field propagating in the positive z-direction, only the $r = s = 0$ term is non-vanishing, in which case Eq. (8.263) reduces to a single contour integral and the pulse evolution is independent of the transverse position in the field. For a beam field, however, the temporal pulse evolution will, in general, be dependent upon the transverse position in the field through the even-ordered spatial derivatives of the transverse beam field profile at the plane $z = z_0$.

8.4.1.3 Validity of the Sherman Expansion

A sketch of the proof of the validity of the Sherman expansion given in Eq. (8.267) for a source-free wave field in a dispersive, attenuative medium is now given, following the method of proof given by Sherman [6].

Definition 8.2 Let $\mathcal{B}$ denote the space of complex-valued functions $f(x, y)$ that can be extended to the entire space of two complex variables $X = x + ix''$, $Y = y + iy''$ as an entire function $f(X, Y)$ such that

$$|f(X, Y)| \leq Ae^{(k-\delta k)(|X|^2 + |Y|^2)^{1/2}} \tag{8.265}$$

for some $\delta k > 0$, where x, y, x'', and y'' are all real variables and where A is a positive real constant.

From the theory of generalized functions [35, 37], every function $f(x, y) \in \mathcal{B}$ can be considered to be a distribution (a generalized function in the space $\mathcal{D}'$) with Fourier transform $F(p, q)$ that is an ultradistribution (a generalized function in the space $\mathcal{Z}'$ which, in turn, is defined as the set of all generalized functions in the space $\mathcal{Z}$ of slowly increasing entire functions [37]). One can determine $F(p, q)$ by taking the Fourier transform of the two-dimensional Taylor series expansion

$$f(x, y) = \sum_{m=0}^{\infty} \sum_{n=0}^{\infty} a_{mn} x^m y^n \tag{8.266}$$

with $a_{mn} = f^{(m,n)}(0, 0)/(m!n!)$, term by term to obtain

$$F(p, q) = \int_{-\infty}^{\infty} \int_{-\infty}^{\infty} f(x, y) e^{-ik(px+qy)} dx dy$$

$$= \left(\frac{2\pi}{k}\right)^2 \sum_{m=0}^{\infty} \sum_{n=0}^{\infty} \frac{a_{mn}}{(-ik)^{m+n}} \delta^{(m)}(p) \delta^{(n)}(q), \tag{8.267}$$

where $\delta^{(m)}(p)$ denotes the mth-order derivative of the Dirac delta function [see Appendix A]. If the relation in Eq. (8.260) is now multiplied by a function $\Psi(p, q) \in \mathcal{Z}$ and the result is integrated over all space, one can integrate the series expansion on the right-hand side term by term to obtain [5]

$$\int_{-\infty}^{\infty} \int_{-\infty}^{\infty} F(p, q) \Psi(p, q) dp dq = \left(\frac{2\pi}{k}\right)^2 \sum_{m=0}^{\infty} \sum_{n=0}^{\infty} \frac{a_{mn}}{(-ik)^{m+n}} \Psi^{(m,n)}(0, 0). \tag{8.268}$$

The convergence of the series appearing on the right-hand side of this expression is guaranteed by generalized function theory for all functions $f(x, y) \in \mathcal{B}$ and functions $\Psi(p, q) \in \mathcal{Z}$.

Following the method of proof given by Sherman [6], the series expansion appearing in Eq. (8.261) is now used to extend the class of generalized functions that $F(p, q)$ is defined onto one that is much broader than $\mathcal{Z}$. The method begins with the following two definitions.

Definition 8.3 Let $\Psi(p, q, X, Y, Z)$ denote a function of the two real variables p, q and the three complex parameters X, Y, Z that is defined for all p, q, X, Y, Z. The function $\Psi(p, q, X, Y, Z)$ is said to be in $\mathcal{C}$ for X, Y, Z in a region $\mathcal{R}$ if and only if

1. For fixed values of p and q such that $(p^2 + q^2) < \cos\theta$ for real $\theta \in [0, \pi/2)$, the quantity $|\Psi(p, q, X, Y, Z)|$ is bounded for all $X, Y, Z \in \mathcal{R}$;
2. For fixed values of $X, Y, Z \in \mathcal{R}$, the function $|\Psi(p, q, X, Y, Z)|$ can be extended to the space of two complex variables $P = p + ip''$, $Q = q + iq''$ as a function that is analytic for $|P|^2 + |Q|^2 < \cos\theta$.

Notice that no restrictions are placed on the behavior of the function $\Psi(p, q, X, Y, Z) \in \mathcal{C}$ outside the circle $|P|^2 + |Q|^2 = \cos\theta$ or outside the region $\mathcal{R}$.

Definition 8.4 The function $\Psi_\nu(p, q, X, Y, Z)$ is said to converge to zero as $\nu \to \infty$ in the sense of $\mathcal{C}$ if and only if, for all $X, Y, Z \in \mathcal{R}$, $\Psi_\nu(p, q, X, Y, Z) \to 0$ as $\nu \to \infty$ uniformly for all p, q such that $(p^2 + q^2) \le \cos(\theta) + \delta$ for some $\delta > 0$.

The following theorems due to Sherman [6] then remain valid in the temporally dispersive medium case.

Theorem 8.6 *Let $\Psi(p, q, X, Y, Z) \in \mathcal{C}$ for X, Y, Z in a region $\mathcal{R}$, let $f(x, y) \in \mathcal{B}$, and let a_{mn} be given by*

$$a_{mn} = \frac{1}{m!n!} f^{(m,n)}(0, 0) \qquad (8.269)$$

for all integer values of m and n with $a_{00} = f(0, 0)$. Then the series

$$S(X, Y, Z) = \left(\frac{2\pi}{k}\right)^2 \sum_{m=0}^{\infty} \sum_{n=0}^{\infty} \frac{a_{mn}}{(-ik)^{m+n}} \Psi^{(m,n)}(0, 0, X, Y, Z) \qquad (8.270)$$

converges absolutely and uniformly for all $X, Y, Z \in \mathcal{R}$.

Corollary 8.1 *Let $\tilde{U}_0(x, y, \omega) \in \mathcal{B}$ possess the two-dimensional Taylor series expansion that is given by*

$$\tilde{U}_0(x, y, \omega) = \sum_{m=0}^{\infty} \sum_{n=0}^{\infty} a_{mn}(\omega) x^m y^n \qquad (8.271)$$

with

$$a_{mn}(\omega) = \frac{1}{m!n!}\tilde{U}_0^{(m,n)}(0,0,\omega).\tag{8.272}$$

Then the series

$$\sum_{m=0}^{\infty}\sum_{n=0}^{\infty}a_{mn}(\omega)\tilde{U}_{mn}(x,y,z,\omega) \equiv \sum_{m=0}^{\infty}\sum_{n=0}^{\infty}\frac{a_{mn}(\omega)}{(ik)^{m+n}}H^{(m,n)}(0,0,\omega;x,y,\Delta z)$$

$$\tag{8.273}$$

is absolutely and uniformly convergent for all $x, y, \Delta z$.

Theorem 8.7 *Let $\tilde{U}_0(x,y,\omega) \in \mathcal{B}$. Then the series that is obtained by replacing x, y, z with $X = x + ix'', Y = y + iy'', Z = z + iz''$ in the series*

$$\tilde{U}(x,y,z,\omega) = \sum_{m=0}^{\infty}\sum_{n=0}^{\infty}\frac{a_{mn}(\omega)}{(ik)^{m+n}}H^{(m,n)}(0,0,\omega;x,y,\Delta z)\tag{8.274}$$

converges uniformly to an entire function $\tilde{U}(X,Y,Z,\omega)$ of three complex variables X, Y, Z. As a consequence, the function $\tilde{U}(x,y,z,\omega)$ given above can be extended to the space of three complex variables X, Y, Z as an entire function $\tilde{U}(X,Y,Z,\omega)$.

Corollary 8.2 *Let $\tilde{U}_0(x,y,\omega) \in \mathcal{B}$ and let $\tilde{U}(x,y,z,\omega)$ be given by Eq. (8.274). Then for all x, y, z, the function $\tilde{U}(x,y,z,\omega)$ is continuous and has continuous partial derivatives with respect to x, y, z of all orders that can be obtained simply by differentiating the series in Eq. (8.274) term by term.*

Theorem 8.8 *Let $\tilde{U}_0(x,y,\omega) \in \mathcal{B}$ and let $\tilde{U}(x,y,z,\omega)$ be given by Eq. (8.274). Then $\tilde{U}(x,y,z,\omega)$ satisfies the Helmholtz equation given in Eq. (8.227) for all x, y, z and satisfies the boundary value given in Eq. (8.228).*

Proof Application of the preceding corollary to the expression given in Eq. (8.274) yields, for the Laplacian of $\tilde{U}(x,y,z,\omega)$,

$$\nabla^2\tilde{U}(x,y,z,\omega) = \sum_{m=0}^{\infty}\sum_{n=0}^{\infty}\frac{a_{mn}(\omega)}{(ik)^{m+n}}\nabla^2 H^{(m,n)}(0,0,\omega;x,y,\Delta z)$$

$$= \sum_{m=0}^{\infty}\sum_{n=0}^{\infty}\frac{a_{mn}(\omega)}{(ik)^{m+n}}\left(i\tilde{k}(\omega)\right)^2 H^{(m,n)}(0,0,\omega;x,y,\Delta z)$$

$$= -\tilde{k}^2(\omega)\tilde{U}(x,y,z,\omega)$$

and $\tilde{U}(x, y, z, \omega)$ satisfies the Helmholtz equation. Furthermore, because $\tilde{U}(x, y, z, \omega)$ is continuous, one then has that

$$\tilde{U}(x, y, z_0, \omega) = \lim_{\Delta z \to 0} \left\{ \tilde{U}(x, y, z, \omega) \right\}$$

$$= \sum_{m=0}^{\infty} \sum_{n=0}^{\infty} \frac{a_{mn}(\omega)}{(ik)^{m+n}} H^{(m,n)}(0, 0, \omega; x, y, 0)$$

$$= \sum_{m=0}^{\infty} \sum_{n=0}^{\infty} a_{mn}(\omega) x^m y^n = \tilde{U}_0(x, y, \omega)$$

and $\tilde{U}(x, y, z, \omega)$, as given by Eq. (8.274), satisfies the boundary value given in Eq. (8.228).

Theorem 8.9 (Sherman's Fifth Theorem) *Let $\tilde{U}_0(x, y, \omega) \in \mathcal{B}$ as well as $\tilde{U}_0(x, y, \omega) \in L^2$. Then the series given in Eq. (8.274) converges to the same function $\tilde{U}(\mathbf{r}, \omega) = \tilde{U}_0(x, y, z, \omega)$ that is given by the integral representation in Eq. (8.229).*

Proof Following the method of proof given by Sherman [6], let $\tilde{U}_0(x, y, \omega) \in \mathcal{B}$ as well as $\tilde{U}_0(x, y, \omega) \in L^2$. Because $\tilde{U}_0(x, y, \omega) \in \mathcal{B}$, then $\tilde{U}_0(X, Y, \omega)$ is an entire function of the complex variables $X = x + ix''$, $Y = y + iy''$, and satisfies the inequality given in Eq. (8.265). As a consequence, $\tilde{U}_0(x, y, \omega)$ is continuous for all x, y and $\tilde{U}_0(X, Y, \omega)$ satisfies the inequality [cf. Eq. (8.249)]

$$\left| \tilde{U}_0(X, Y, \omega) \right| \leq A e^{(k-\delta k)(|x''|^2 + |y''|^2)^{1/2}}$$

for some $\delta k > 0$. According to the Plancherel–Pólya theorem [34], the two-dimensional spatial Fourier transform $\tilde{U}_0(p, q, \omega)$ of $\tilde{U}_0(x, y, \omega)$ vanishes almost everywhere for $(p^2 + q^2) > \left(\cos^{1/2}(2\psi(\omega)) - \delta k/k \right)^2$, as shown in Eq. (8.249). As a consequence, $\tilde{U}_0(p, q, \omega) \in L^1$ and

$$\tilde{U}_0(x, y, \omega) = \frac{1}{(2\pi)^2} \int_{-\infty}^{\infty} \int_{-\infty}^{\infty} \tilde{U}_0(p, q, \omega) e^{ik(px+qy)} k^2 dp dq \qquad (8.275)$$

almost everywhere. Because both $\tilde{U}_0(x, y, \omega)$ and the right-hand side of Eq. (8.275) are continuous for all x, y, then Eq. (8.275) must be valid everywhere. The partial derivatives of the integrand

$$I(p, q; x, y) \equiv k^2 \tilde{U}_0(p, q, \omega) e^{ik(px+qy)}$$

appearing in Eq. (8.275) satisfy the inequality

$$\left| \frac{\partial^{m+n}}{\partial x^m \partial y^n} I(p, q; x, y) \right| \leq \left| (ikp)^m (ikq)^n k^2 \tilde{\tilde{U}}_0(p, q, \omega) \right|,$$

where the right-hand side of this inequality is independent of the spatial coordinate variables x, y, z and is in L^1. The orders of differentiation and integration may then be interchanged when using the Fourier integral representation given in Eq. (8.275) to determine the spatial partial derivatives of $\tilde{U}_0(x, y, \omega)$, so that

$$\tilde{U}_0^{(m,n)}(x, y, \omega) = \frac{1}{(2\pi)^2} \int_{-\infty}^{\infty} \int_{-\infty}^{\infty} (ik)^{m+n} \tilde{\tilde{U}}_0(p, q, \omega) p^m q^n e^{ik(px+qy)} k^2 dpdq.$$

Substitution of this expression into the equation $a_{mn} = \tilde{U}_0^{(m,n)}(0, 0, \omega)/(m!n!)$ for the coefficients appearing in the two-dimensional Taylor series expansion of $\tilde{U}_0(x, y, \omega)$ then gives

$$a_{mn}(\omega) = \left(\frac{k}{2\pi} \right)^2 \int_{-\infty}^{\infty} \int_{-\infty}^{\infty} \frac{(ik)^{m+n}}{m!n!} \tilde{\tilde{U}}_0(p, q, \omega) p^m q^n dpdq, \qquad (8.276)$$

which, upon substitution into Eq. (8.274), yields

$$\tilde{U}(x, y, z, \omega) = \sum_{m=0}^{\infty} \sum_{n=0}^{\infty} \frac{1}{m!n!} H^{(m,n)}(0, 0, \omega; x, y, \Delta z)$$

$$\times \left(\frac{k(\omega)}{2\pi} \right)^2 \int_{-\infty}^{\infty} \int_{-\infty}^{\infty} \tilde{\tilde{U}}_0(p, q, \omega) p^m q^n dpdq.$$

With the interchange of the order of the double summation and integration in this equation, one finally obtains

$$\tilde{U}(x, y, z, \omega) = \frac{1}{(2\pi)^2} \int_{-\infty}^{\infty} \int_{-\infty}^{\infty} \tilde{\tilde{U}}_0(p, q, \omega)$$

$$\times \left[\sum_{m=0}^{\infty} \sum_{n=0}^{\infty} \frac{p^m q^n}{m!n!} H^{(m,n)}(0, 0, \omega; x, y, \Delta z) \right] k^2(\omega) dpdq.$$

$$(8.277)$$

Because $\tilde{\tilde{U}}_0(p, q, \omega)$ is non-vanishing for $p^2 + q^2 \leq \left(\cos^{1/2}(2\psi(\omega)) - \delta k/k \right)^2$, the interchange of the order of the operations that led to Eq. (8.276) is justified if the double series appearing in the integrand is absolutely and uniformly convergent for all x, y, z and $p^2 + q^2 > \left(\cos^{1/2}(2\psi(\omega)) - \delta k/k \right)^2$. This condition is satisfied by this double-series summation because it is just the Taylor series expansion for

$H(p, q, \omega; x, y, \Delta z)$ in the variables p, q and because $H(P, Q, \omega; X, Y, \Delta Z)$ is an analytic function of the complex variables $P = p + ip''$, $Q = q + iq''$ in the region $|P|^2 + |Q|^2 \leq \left(\cos^{1/2}(2\psi(\omega)) - \delta k/k\right)^2$ for all $\delta k > 0$ and all $X, Y, \Delta Z$. Hence, the expression given in Eq. (8.277) reduces to the angular spectrum representation given in Eq. (8.229) and the theorem is proved.

Corollary 8.3 *Let* $\tilde{U}(\mathbf{r}, \omega) = \tilde{U}(x, y, z, \omega)$ *as given by Eq. (8.229) be source-free with a square-integrable angular spectrum* $\tilde{U}_0(p, q, \omega)$ *that vanishes almost everywhere outside the circle* $p^2 + q^2 = \left(\cos^{1/2}(2\psi(\omega)) - \delta k/k\right)^2$ *for some* $\delta k > 0$. *Then* $\tilde{U}(\mathbf{r}, \omega)$ *is given by the series in Eq. (8.274).*

8.4.2 Separable Pulsed Beam Fields

A problem of special interest is that in which the spatial and temporal properties of a pulsed electromagnetic beam field are separable. Although this separability property is frequently assumed in the open literature, the general conditions under which it is valid are rarely, if ever, addressed. For that special case, assume that the spatial and temporal properties of the initial field vectors at the plane $z = z_0$ are separable in the sense that [26]

$$\mathbf{E}_0(\mathbf{r}_T, t) = \mathbf{E}_0(x, y, t) = \hat{\mathbf{E}}_0(x, y) f(t), \tag{8.278}$$

$$\mathbf{B}_0(\mathbf{r}_T, t) = \mathbf{B}_0(x, y, t) = \hat{\mathbf{B}}_0(x, y) g(t), \tag{8.279}$$

so that

$$\tilde{\mathbf{E}}_0(p, q, \omega) = \hat{\tilde{\mathbf{E}}}_0(p, q, k) \tilde{f}(\omega), \tag{8.280}$$

$$\tilde{\mathbf{B}}_0(p, q, \omega) = \hat{\tilde{\mathbf{B}}}_0(p, q, k) \tilde{g}(\omega). \tag{8.281}$$

One now has the spatial Fourier transform pair relations

$$\hat{\tilde{U}}_0(p, q, k) = \int_{-\infty}^{\infty} \int_{-\infty}^{\infty} \hat{U}_0(x, y) e^{-ik(px+qy)} dx dy, \tag{8.282}$$

$$\hat{U}_0(x, y) = \frac{1}{(2\pi)^2} \int_{-\infty}^{\infty} \int_{-\infty}^{\infty} \hat{\tilde{U}}_0(p, q, k) e^{ik(px+qy)} k^2 dp dq \tag{8.283}$$

for both $\hat{\mathbf{E}}_0(x, y)$ and $\hat{\mathbf{B}}_0(x, y)$, and the separate Fourier–Laplace transform pair relations

$$\tilde{h}(\omega) = \int_{-\infty}^{\infty} h(t) e^{i\omega t} dt, \tag{8.284}$$

$$h(t) = \frac{1}{\pi} \Re \left\{ \int_{C_+} \tilde{h}(\omega) e^{-i\omega t} d\omega \right\} \tag{8.285}$$

for both $f(t)$ and $g(t)$. It is seen from Eqs. (8.18) and (8.19) that these functions cannot be independently chosen because their spectra must satisfy the relation

$$\tilde{\mathbf{B}}_0(p, q, k)\tilde{g}(\omega) = (\mu(\omega)\epsilon_c(\omega))^{1/2} \hat{\mathbf{s}} \times \tilde{\mathbf{E}}_0(p, q, k)\tilde{f}(\omega), \tag{8.286}$$

where $\hat{\mathbf{s}} = \hat{\mathbf{1}}_x p + \hat{\mathbf{1}}_y q + \hat{\mathbf{1}}_z m(\omega)$. The separation of this relation into spatial and temporal frequency components as

$$\tilde{\mathbf{B}}_0(p, q, k) = \hat{\mathbf{s}} \times \tilde{\mathbf{E}}_0(p, q, k), \tag{8.287}$$

$$\tilde{g}(\omega) = (\mu(\omega)\epsilon_c(\omega))^{1/2} \tilde{f}(\omega), \tag{8.288}$$

does not provide a unique solution of Eq. (8.286) except in special cases.

Consider now the propagation of such a separable pulsed wave field in a (fictitious) dispersive medium that is lossless. Substitution of the generic form of the seperability condition given in Eqs. (8.280) and (8.281) into the angular spectrum representation given in Eq. (8.187) then yields

$$\mathbf{U}(\mathbf{r}, t) = \frac{1}{4\pi^3} \Re \left\{ \int_{C_+} d\omega\, \tilde{h}(\omega) e^{-i\omega t} \int_{\mathcal{R}_< \cup \mathcal{R}_>} k^2(\omega) dp dq \right.$$
$$\left. \times \tilde{\mathbf{U}}_0(p, q, k) G(p, q, \omega, \Delta z) e^{ik(\omega)(px+qy)} \right\}, \tag{8.289}$$

where $\mathcal{R}_< = \{(p, q)| p^2 + q^2 < 1\}$ defines the homogeneous plane wave domain, $\mathcal{R}_> = \{(p, q)| p^2 + q^2 > 1\}$ defines the evanescent plane wave domain, and where [see Eq. (8.233)]

$$G(p, q, \omega, \Delta z) \equiv e^{ik(\omega)m\Delta z}$$
$$= e^{ik(\omega)\Delta z(1-p^2-q^2)^{1/2}} \tag{8.290}$$

is the propagation factor for a plane wave progressing in the direction specified by the direction cosines (p, q, m).

Because the evanescent plane wave components in the angular spectrum representation do not provide any time-average energy flow into the positive half-space $z > z_0$ [see Eq. (8.189)], an electromagnetic beam field is defined, in part, by the requirement that its angular spectrum does not contain any evanescent plane wave components [9], and consequently can be represented by an angular spectrum that contains only homogeneous plane wave components. Because the homogeneous plane wave spectral components do not attenuate with propagation distance in a lossless medium, this condition ensures that all of the angular spectrum components of the beam field are maintained in their initial proportion throughout the positive half-space $z \geq z_0$. As described in Sect. 8.4.1, such wave fields are known as *source-free wave fields* [4, 6] and, as such, possess several rather unique properties.

If the initial spatial frequency spectrum $\tilde{\mathbf{U}}_0(p, q, k)$ of a separable pulsed beam wave field contains only homogeneous plane wave components, then the inner integration domain $\mathcal{R}_<$ can be extended to all of p, q-space and the propagation factor $G(p, q, \omega, \Delta z)$ may then be represented by its Maclaurin's series expansion given in Eq. (8.253) with (8.254). Substitution of this expansion in Eq. (9.289) and interchanging the order of integration and summation then yields

$$
\mathbf{U}(\mathbf{r}, t) = \frac{1}{4\pi^3} \Re \left\{ \sum_{r=0}^{\infty} \sum_{s=0}^{\infty} \int_{C_+} d\omega \, \frac{G^{(2r,2s)}(0, 0, \omega, \Delta z)}{(2r)!(2s)!} \tilde{h}(\omega) e^{-i\omega t} \right.
$$
$$
\left. \times \int_{-\infty}^{\infty} \int_{-\infty}^{\infty} \tilde{\mathbf{U}}_0(p, q, k) p^{2r} q^{2s} e^{ik(\omega)(px+qy)} k^2(\omega) dp dq \right\}.
$$

(8.291)

Because [see Eq. (8.256)]

$$
\hat{\mathbf{U}}_0^{(m,n)}(x, y) \equiv \frac{\partial^{m+n} \hat{\mathbf{U}}_0(x, y)}{\partial x^m \partial y^n}
$$
$$
= \frac{1}{2\pi^2} \int_{-\infty}^{\infty} \int_{-\infty}^{\infty} \tilde{\mathbf{U}}_0(p, q, k) (ik(\omega))^{m+n} p^m q^n
$$
$$
\times e^{ik(\omega)(px+qy)} k^2(\omega) dp dq,
$$

(8.292)

one then obtains

$$
\mathbf{U}(\mathbf{r}, t) = \frac{1}{\pi} \Re \left\{ \sum_{r=0}^{\infty} \sum_{s=0}^{\infty} \int_{C_+} d\omega \, \tilde{h}(\omega) e^{-i\omega t} \right.
$$
$$
\left. \times \frac{G^{(2r,2s)}(0, 0, \omega, \Delta z)}{(2r)!(2s)!(ik(\omega))^{2(r+s)}} \hat{\mathbf{U}}_0^{(2r,2s)}(x, y) \right\},
$$

(8.293)

for all $\Delta z \geq 0$. The spatial series appearing in the above expression is just *Sherman's expansion* [4, 6]. As was noted following Eq. (8.257), the factor $(ik(\omega))^{-2(r+s)}$ appearing in the above expansion is misleading because the same factor with an opposite-signed exponent is contained in the partial derivative $G^{(2r,2s)}(0, 0, \omega, \Delta z)$. From Eq. (8.290) it is found that

$$G^{(m,n)}(p, q, \omega, \zeta) = (ik(\omega)\zeta)^{m+n} e^{ik(\omega)\zeta(1-p^2-q^2)^{1/2}} \varphi^{(m,n)}(p, q),$$ (8.294)

where

$$\varphi^{(m,n)}(p, q) \equiv \frac{\partial^{m+n}}{\partial p^m \partial q^n} \left(1 - p^2 - q^2\right)^{1/2}.$$ (8.295)

With these substitutions, the above expression for the propagated, separable wave field becomes

$$\mathbf{U}(\mathbf{r}, t) = \sum_{r=0}^{\infty} \sum_{s=0}^{\infty} \frac{(\Delta z)^{2(r+s)}}{(2r)!(2s)!} \varphi^{(2r,2s)}(0, 0)$$

$$\times \frac{1}{\pi} \Re \left\{ \int_{C_+} \hat{\mathbf{U}}_0^{(2r,2s)}(x, y) \tilde{h}(\omega) e^{i(k(\omega)\Delta z - \omega t)} d\omega \right\},$$

(8.296)

for all $\Delta z \geq 0$. The transverse spatial variation of both the source-free electric and magnetic field vectors at any plane $z > z_0$ is again seen to depend solely upon the transverse spatial variation of the field at the initial plane at $z = z_0$ through all of the even-order spatial derivatives of the corresponding field vector at that plane. At first impression, it would appear that this transverse spatial variation is independent of the wavenumber $k(\omega)$, and hence of the angular frequency ω, so that the function $\hat{\mathbf{U}}_0^{(2r,2s)}(x, y)$ could be taken out from under the integral in Eq. (8.296). This observation deserves a more careful examination.

If $\hat{\mathbf{U}}_0^{(2r,2s)}(x, y)$ is indeed independent of the wavenumber $k(\omega)$, then the propagated source-free pulsed beam field is given by the product of two separate factors, one describing the transverse spatial variation and the other describing the longitudinal spatio-temporal variation, each factor dependent upon the propagation distance $\Delta z = z - z_0$. This apparent wavenumber (or frequency) independence is a direct consequence of the two-dimensional Maclaurin series expansion of the propagation kernel $G(p, q, \Delta z) \equiv e^{ikm\Delta z}$, with $m = \sqrt{1 - p^2 - q^2}$, whose explicit dependence upon $k(\omega)$ is then transferred to the transverse spatial derivatives of $\hat{\mathbf{U}}_0(x, y)$. With this in mind, Eq. (8.296) may be rewritten as

$$\mathbf{U}(\mathbf{r}, t) = \hat{\mathbf{U}}(x, y) \frac{1}{\pi} \Re \left\{ \int_{C_+} \tilde{h}(\omega) e^{i(k(\omega)\Delta z - \omega t)} d\omega \right\},$$ (8.297)

with

$$\hat{\mathbf{U}}(x, y) \equiv \sum_{r=0}^{\infty} \sum_{s=0}^{\infty} \frac{(\Delta z)^{2(r+s)}}{(2r)!(2s)!} \varphi^{(2r,2s)}(0, 0) \hat{\mathbf{U}}_0^{(2r,2s)}(x, y).$$ (8.298)

Unfortunately, except in trivial cases (such as for a plane wave), this series representation converges extremely slowly so that an exceedingly large (perhaps even infinite) number of terms is required. The integral representation corresponding to this series expansion of the transverse field distribution is found to be given by

$$\hat{\mathbf{U}}(x, y) = \frac{1}{4\pi^2} \int_{-\infty}^{\infty} \int_{-\infty}^{\infty} \tilde{\hat{\mathbf{U}}}_0(v_x, v_y) e^{i(v_x x + v_y y + (\gamma_v - v)\Delta z)} dv_x dv_y, \qquad (8.299)$$

where $\gamma_v \equiv \left(v^2 - v_x^2 - v_y^2\right)^{1/2}$ with

$$\tilde{\hat{\mathbf{U}}}_0(v_x, v_y) = \int_{-\infty}^{\infty} \int_{-\infty}^{\infty} \hat{\mathbf{U}}_0(x, y) e^{-i(v_x x + v_y y)} dx dy. \qquad (8.300)$$

Notice that this integral representation of the source-free transverse field vector $\hat{\mathbf{U}}(x, y)$ that is defined by the infinite summation given in Eq. (8.298) is independent of the value of the spatial frequency v appearing in the propagation factor in the integrand of Eq. (8.299) only in the limit as $v \to \infty$. The reason that this is so is found in the infinite double summation of Eq. (8.298). The transverse field variation is expressed there in terms of all of the even-order spatial derivatives of $\hat{\mathbf{U}}_0(x, y)$, and this, in turn, requires that the transverse spatial variation of this initial field structure be known all the way down to an infinitesimal scale, that is, as $1/v \to 0$. This, of course, is just the geometrical optics limit.

This rather curious result deserves further explanation. First of all, notice that the field must be source-free. This means that the wave field does not contain any evanescent field components in a lossless medium. This, in turn, means that the wave number component [see Eq. (9.12)]

$$\gamma(\omega) = \left(k^2(\omega) - k_T^2\right)^{1/2}$$

is real-valued for all points (k_x, k_y) at which $\tilde{\hat{\mathbf{U}}}_0(k_x, k_y)$ is nonzero, where $k_T^2 = k_x^2 + k_y^2$. In general, for each value of ω, the value of the wave number $k(\omega)$ defines the transition circle $k_T^2 = k^2(\omega)$ in $k_x k_y$-space between homogeneous and evanescent plane wave components. This wave number value then appears as the upper limit of integration for the homogeneous plane wave contribution to the angular spectrum representation and this, in turn, sets an upper limit on the spatial frequency scale (or equivalently, a lower limit on the spatial scale) for the transverse spatial structure of the wave field. However, for a source-free wave field, this upper limit is replaced by infinity as the homogeneous wave propagation factor $G(k_x, k_y, \omega, \Delta z)$ is replaced by its Maclaurin's series expansion. The result is an expression for the transverse field variation $\hat{\mathbf{U}}(x, y)$ that is indeed independent of the wave number.

The results presented here then show that, except in special cases (such as for a plane wave), if the initial wave field is separable in the sense defined in

Eqs. (8.278) and (8.279), then the propagated wave field will not, in general, remain separable unless it is strictly source-free. In that idealized case, the wave field remains separable throughout its propagation and its transverse spatial variation is independent of the wavenumber (or frequency). This then demonstrates how a seemingly innocent assumption (in this case, the assumption that the wave field contains only homogeneous wave components) can lead to extreme results when they are taken to their logical limit.

If the initial pulse is strictly quasimonochromatic (in which case [20] $\Delta\omega/\omega_c \ll 1$, where $\Delta\omega$ is the pulse spectrum bandwidth centered at ω_c), then the propagated field is, to some degree of approximation, separable with the spatial frequency ν being taken as the wavenumber $k(\omega_c)$ evaluated at this characteristic angular frequency of the pulse. Except in special cases, this approximation does not hold in the ultrawideband case. In that case, the expression (8.297) for the propagated wave field should be written as

$$\mathbf{U}(\mathbf{r}, t) = \frac{1}{\pi}\Re\left\{\int_{C_+} \hat{\mathbf{U}}(x, y)\tilde{h}(\omega)e^{i(k(\omega)\Delta z - \omega t)}d\omega\right\}, \tag{8.301}$$

with $\hat{\mathbf{U}}(x, y)$ given by Eq. (8.299) with $\nu = k(\omega)$.

8.5 Paraxial Approximation of the Angular Spectrum of Plane Waves Representation

If the real direction cosines p and q are both sufficiently small in magnitude in comparison to unity, then the complex direction cosine $m(\omega)$ may be approximated by the first few terms in its Maclaurin series as

$$m(\omega) = e^{i\psi(\omega)}\left(1 - \left(p^2 + q^2\right)e^{-i2\psi(\omega)}\right)^{1/2}$$

$$\approx e^{i\psi(\omega)} - \frac{1}{2}\left(p^2 + q^2\right)e^{-i\psi(\omega)}. \tag{8.302}$$

With this approximation, the propagation kernel $G(p, q, \omega)$ appearing in the angular spectrum representation in Eq. (8.187) becomes

$$G(p, q, \omega) \equiv e^{ik(\omega)m(\omega)\Delta z} \tag{8.303}$$

$$\approx e^{i\tilde{k}(\omega)\Delta z}e^{-\tilde{k}^*(\omega)\Delta z(p^2+q^2)/2}, \tag{8.304}$$

where the superscript asterisk denotes complex conjugation. The general spatial phase dispersion due to propagation is then replaced by quadratic phase dispersion in this paraxial approximation, so named because of its general validity in the paraxial region about the z-axis described by the pair of inequalities $p \ll 1$ and $q \ll 1$.

Substitution of the paraxial approximation to the propagation kernel given in Eq. (8.302) into the angular spectrum representation given in Eq. (8.187) then results in the expression

$$\mathbf{U}(\mathbf{r}, t) = \frac{1}{\pi} \Re \left\{ \int_{C_+} \tilde{\mathbf{U}}(\mathbf{r}, \omega) e^{i\left(\tilde{k}(\omega)\Delta z - \omega t\right)} d\omega \right\}, \tag{8.305}$$

where

$$\tilde{\mathbf{U}}(\mathbf{r}, \omega) \approx -i \frac{\tilde{k}(\omega)}{2\pi \Delta z} \int_{-\infty}^{\infty} \tilde{\mathbf{U}}_0(x', y', \omega) e^{i\frac{\tilde{k}(\omega)}{2\Delta z}\left((x-x')^2+(y-y')^2\right)} dx' dy' \tag{8.306}$$

is one form of the Fresnel–Kirchhoff diffraction integral [8, 13]. The same result is obtained from the first Rayleigh–Sommerfeld diffraction integral representation [see Eqs. (8.74) and (8.75)] in the Fresnel approximation wherein the distance $R \equiv \sqrt{(x - x')^2 + (y - y')^2 + \Delta z^2}$ between the source and field points is approximated as $R \approx \Delta z + (x-x')^2/(2\Delta z) + (y-y')^2/(2\Delta z)$. In this approximation, the spherical secondary wavelets embodied in the Huygens–Fresnel principle have been replaced by parabolic secondary wavelets. Since the Fresnel approximation is valid in the vanishingly small diffraction angle limit $(x - x')/\Delta z \ll 1$, $(y - y')/\Delta z \ll 1$ and results in the expression given in Eqs. (8.191) and (8.192), it is then equivalent to the paraxial approximation.

Because $\tilde{k}(\omega) = \beta(\omega) + i\alpha(\omega)$, the Fresnel–Kirchhoff diffraction integral appearing in Eq. (8.305) may be rewritten in the more revealing form

$$\tilde{\mathbf{U}}(\mathbf{r}, \omega) \approx -i \frac{\tilde{k}(\omega)}{2\pi \Delta z} \int_{-\infty}^{\infty} \tilde{\mathbf{U}}_0(x', y', \omega) e^{-\frac{\alpha(\omega)}{2\Delta z}\left((x-x')^2+(y-y')^2\right)}$$

$$\times e^{i\frac{\beta(\omega)}{2\Delta z}\left((x-x')^2+(y-y')^2\right)} dx' dy', \tag{8.307}$$

which explicitly displays the manner in which the material attenuation influences the monochromatic beam-field diffraction. Specifically, material attenuation decreases the effects of diffraction relative to the geometrical optics contribution. Nevertheless, care must be exercised when this paraxial representation is applied to pulsed beam problems, particularly when the initial pulse is ultrawideband, since the results may not be strictly causal.

8.5.1 Accuracy of the Paraxial Approximation

Limits on the accuracy of the paraxial approximation have been reported for pulsed beam propagation phenomena in lossless dispersive media by Melamed and Felsen

[38, 39]. Surprisingly, the situation is simplified when material loss is present [40]. The results are best illustrated through a detailed numerical example. Because of its central role in bioelectromagnetics as well as in foliage and ground penetrating radar, a dispersive material of central interest here is water. The Rocard–Powles model for the dielectric permittivity of triply-distilled water at 25 °C [see Eq. (4.196) of Sect. 4.4.3] is given by Eq. (4.197) which accurately describes the complex-valued relative permittivity over the frequency domain $0 \leq \omega \leq 1 \times 10^{13}$ r/s extending from the static into the low infrared region of the electromagnetic spectrum, as illustrated in Fig. 4.6. The resultant frequency dispersion of the real and imaginary parts of the complex wave number $\tilde{k}(\omega) = \beta(\omega) + i\alpha(\omega)$ is depicted in Fig. 8.10. Notice the peak in the absorption at $\omega \approx 6 \times 10^{13}$ r/s, which is then followed by a monotonic decrease in attenuation for larger frequencies. At the lower angular frequency ($\omega_c = 2\pi \times 10^7$ r/s) indicated in Fig. 8.10, the absorption is small [$\alpha(\omega_c) = 4.66 \times 10^{-4}$ m^{-1}] and $\cos(2\psi(\omega_c)) = 1.0000$ so that the complex direction cosine $m(\omega_c)$ is practically the same as that for an ideal lossless medium. This is evident from the pair of graphs presented in Fig. 8.11 which depict the

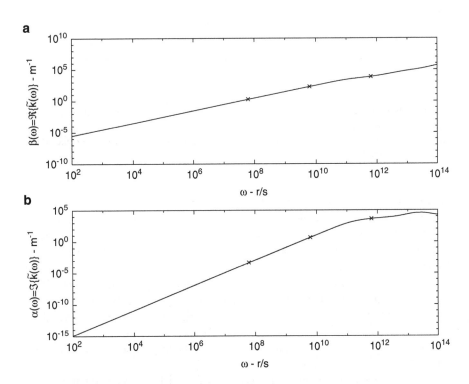

Fig. 8.10 Angular frequency dependence of the (**a**) real and (**b**) imaginary parts of the complex wave number $\tilde{k}(\omega) = \beta(\omega + i\alpha(\omega)$ for the Rocard–Powles–Debye model of triply-distilled water at 25 °C. The values marked with a cross ($\times$) in each diagram indicate the real and imaginary values of the wave number at the angular frequency values $\omega_c = 2\pi \times 10^7$ r/s in the HF, $\omega_c = 2\pi \times 10^9$ r/s in the UHF, and $\omega_c = 2\pi \times 10^{11}$ r/s in the EHF regions of the electromagnetic spectrum

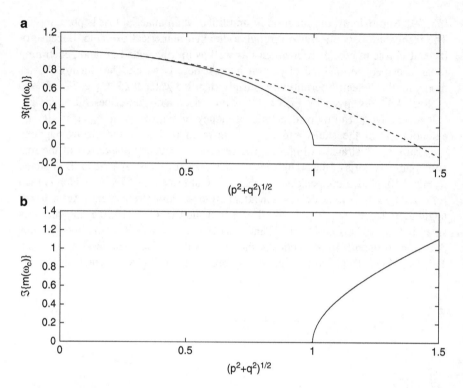

Fig. 8.11 Behavior of the (**a**) real and (**b**) imaginary parts of the complex direction cosine $m(\omega_c) = \left\{ exp\left[i2\psi(\omega_c)\right] - (p^2 + q^2) \right\}^{1/2}$ at $\omega_c = 2\pi \times 10^7$ r/s as functions of $\sqrt{p^2 + q^2}$. The solid curves depict the exact behavior while the dashed curves describe the behavior obtained with the paraxial approximation

behavior of the real and imaginary parts of $m(\omega_c)$ as functions of $\sqrt{p^2 + q^2}$. The dashed curves in this figure depict the behavior of the paraxial approximation given in Eq. (8.302). The real part of this paraxial approximation is seen to be accurate for $\sqrt{p^2 + q^2} \leq 0.6$, while the imaginary part, which is approximately zero for all real values of p and q, is accurate for all $\sqrt{p^2 + q^2} \leq 1$.

The above behavior is reflected in Fig. 8.12 which depicts the behavior of the magnitude and phase of the propagation kernel $G(p, q, \omega_c)$, as well as in Fig. 8.13 which depicts the behavior of the real and imaginary parts of $G(p, q, \omega_c)$ as functions of $\sqrt{p^2 + q^2}$ at the propagation distance $\Delta z = 0.1$ m. These results demonstrate the well-known result that the paraxial approximation is accurate only when $\sqrt{p^2 + q^2} < 1$ in a lossless medium.

At the intermediate angular frequency ($\omega_c = 2\pi \times 10^9$ r/s) indicated in Fig. 8.14, the absorption is moderate [$\alpha(\omega_c) = 4.65$ m^{-1}] and $\cos(2\psi(\omega_c)) = 0.9988$. The complex direction cosine $m(\omega_c)$ is then a slightly smoothed version of that for an ideal lossless medium. The accuracy of the paraxial approximation is seen to be slightly improved over that for the low-loss case illustrated in Fig. 8.11, and

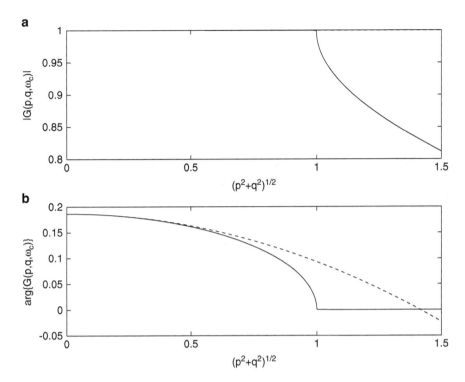

Fig. 8.12 Behavior of the (**a**) magnitude and (**b**) phase of the propagation kernel $G(p, q, \omega_c)$ at $\omega_c = 2\pi \times 10^7$ r/s as functions of $\sqrt{p^2 + q^2}$ at the propagation distance $\Delta z = 0.1$ m. The solid curves depict the exact behavior while the dashed curves describe the behavior obtained with the paraxial approximation

this slight improvement in accuracy is reflected in the paraxial approximation of the propagation kernel $G(p, q, \omega_c)$ whose magnitude and phase are depicted in Fig. 8.15 and whose real and imaginary parts are depicted in Fig. 8.16 for the propagation distance $\Delta z = 0.1$ m. Notice that both the amplitude and phase of the paraxial approximation of $G(p, q, \omega_c)$ lose accuracy as the quantity $\sqrt{p^2 + q^2}$ approaches unity from below in this moderate-absorption case, as reflected in the nearly identical behavior exhibited by the real and imaginary parts of $G(p, q, \omega_c)$.

At the upper angular frequency ($\omega_c = 2\pi \times 10^{11}$ r/s) indicated in Fig. 8.10, the absorption is large [$\alpha(\omega_c) = 4.27 \times 10^3$ m^{-1}] and $\cos(2\psi(\omega_c)) = 0.4615$. The complex direction cosine $m(\omega_c)$ is then a smoothed version of that for an ideal lossless medium, as seen in Fig. 8.17. The accuracy of the paraxial approximation has improved significantly over that for the previous two cases, particularly in the interior region $\mathcal{R}_<$ where $(p^2 + q^2) \leq \cos(2\psi(\omega_c))$, and remains reasonably accurate out to twice this value. Because of the high value of the absorption at this frequency, the paraxial approximation of the propagation kernel $G(p, q, \omega_c)$, whose magnitude and phase are depicted in Fig. 8.18 and whose real and imaginary parts are depicted

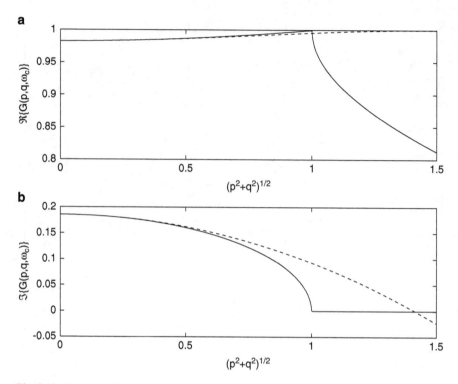

Fig. 8.13 Behavior of the (**a**) real and (**b**) imaginary parts of the propagation kernel $G(p, q, \omega_c)$ at $\omega_c = 2\pi \times 10^7$ r/s as functions of $\sqrt{p^2 + q^2}$ at the propagation distance $\Delta z = 0.1$ m. The solid curves depict the exact behavior while the dashed curves describe the behavior obtained with the paraxial approximation

in Fig. 8.19 for the propagation distance $\Delta z = 0.1$ m, is nearly identical to the exact behavior for all values of p and q for which $|G(p, q, \omega_c)|$ is essentially nonzero.

Finally, notice that the accuracy of the paraxial approximation of $G(p, q, \omega_c)$ increases as the propagation distance increases. This important property is illustrated through comparison of Figs. 8.16 (for $\Delta z = 0.1$ m), 8.20 (for $\Delta z = 1.0$ m) and 8.21 (for $\Delta z = 10.0$ m) for the intermediate frequency case ($\omega_c = 2\pi \times 10^9$ r/s) where the absorption is moderate [$\alpha(\omega_c) = 4.65$ m^{-1} so that $z_d = 0.215$ m]. As the propagation distance increases, both the real and imaginary parts of the paraxial approximation of the propagation kernel are seen to improve in accuracy. At the largest propagation distance considered, the error is almost exclusively due to a small p, q-dependent phase error.

These numerical results show that the accuracy of the paraxial approximation improves both as the material attenuation increases as well as when the propagation distance into the attenuative medium increases. This trend indicates that the inhomogeneous plane wave components in the exterior domain $\mathcal{R}_> = \{(p, q) | (p^2 + q^2) > \cos(2\psi(\omega))\}$ become entirely negligible in comparison with

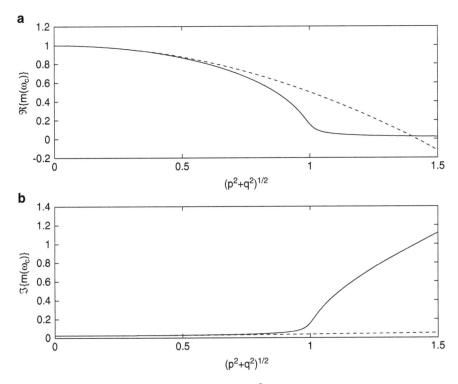

Fig. 8.14 Same as in Fig. 8.11 but with $\omega_c = 2\pi \times 10^9$ r/s

the homogeneous and inhomogeneous plane wave components in the interior domain $\mathcal{R}_< = \{(p,q)|(p^2+q^2) < \cos{(2\psi(\omega))}\}$ as the propagation distance Δz typically exceeds a single absorption depth $z_d \equiv \alpha^{-1}(\omega_c)$ at some characteristic frequency ω_c of the initial pulse.

8.5.2 Gaussian Beam Propagation

An example of central importance in both optics and microwave theory considers the spatial propagation properties of a Gaussian beam wave field. Let the initial transverse field behavior at the plane $z = z_0$ be described as

$$\hat{U}_0(x,y) = A_0 e^{-\left(\sigma_x^2 x^2 + \sigma_y^2 y^2\right)}, \qquad (8.308)$$

with fixed amplitude A_0 and constants σ_x and σ_y that set the initial beam widths in the x- and y-directions, respectively. With use of the integral identity

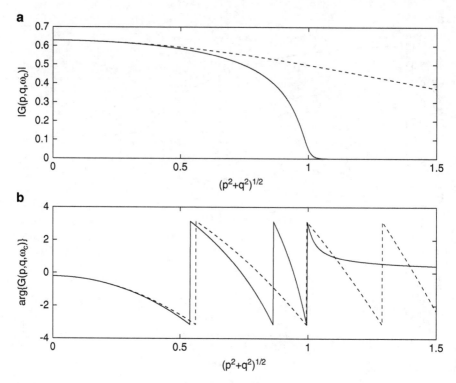

Fig. 8.15 Same as in Fig. 8.12 but with $\omega_c = 2\pi \times 10^9$ r/s

$$\int_{-\infty}^{\infty} e^{-\alpha\zeta^2} e^{\pm i\beta\zeta} d\zeta = \left(\frac{\pi}{\alpha}\right)^{1/2} e^{-\beta^2/4\alpha}, \quad \alpha > 0, \tag{8.309}$$

the initial field spectrum [see Eq. (8.300)] is found to be given by

$$\tilde{\tilde{U}}_0(\nu_x, \nu_y) = A \frac{\pi}{\sigma_x \sigma_y} e^{-\left(\nu_x^2/4\sigma_x^2 + \nu_y^2/4\sigma_y^2\right)}. \tag{8.310}$$

The propagated field distribution is then obtained from Eq. (8.299) as

$$\hat{U}(x, y) = \frac{A}{4\pi\sigma_x\sigma_y} \int_{-\infty}^{\infty}\int_{-\infty}^{\infty} e^{-\left(\nu_x^2/4\sigma_x^2 + \nu_y^2/4\sigma_y^2\right)} e^{i\left(\nu_x x + \nu_y y + (\gamma_\nu - \nu)\Delta z\right)} d\nu_x d\nu_y. \tag{8.311}$$

With the paraxial approximation

$$\gamma_\nu - \nu = \nu\left[\left(1 - \nu_x^2/\nu^2 - \nu_y^2/\nu^2\right)^{1/2} - 1\right] \approx -\frac{\nu_x^2 + \nu_y^2}{2\nu} \tag{8.312}$$

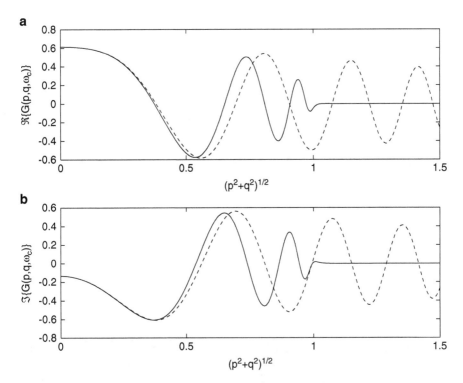

Fig. 8.16 Same as in Fig. 8.13 but with $\omega_c = 2\pi \times 10^9$ r/s

of the propagation factor, valid when $(v_x^2 + v_y^2)/v^2 \ll 1$, the propagated field given in Eq. (8.311) may be separated into the product of a pair of two-dimensional Gaussian beam fields as

$$\hat{U}(x, y) \approx A\hat{U}(x)\hat{U}(y), \qquad (8.313)$$

with

$$\hat{U}(x) \equiv \frac{1}{2\pi^{1/2}\sigma_x} \int_{-\infty}^{\infty} e^{-\left(v_x^2/4\sigma_x^2 + i\,\Delta z/2v\right)} e^{iv_x x} dv_x$$

$$= \frac{1}{\left(1 + 2i\sigma_x^2 \Delta z/v\right)^{1/2}} e^{-x^2/\left(1/\sigma_x^2 + 2i\,\Delta z/v\right)}, \qquad (8.314)$$

and with an exactly analogous expression for $\hat{U}(y)$. Notice that this result depends explicitly on the value of the spatial frequency v; this is due to the paraxial approximation given in Eq. (8.312) that was used in obtaining Eq. (8.314). In the geometric optics limit as $v \to \infty$, $\hat{U}(x) \to e^{-\sigma_x^2 x^2}$.

The argument of the exponential factor appearing in Eq. (8.314) can be separated into real and imaginary parts as

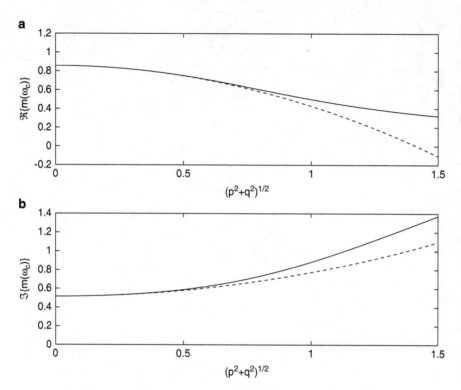

Fig. 8.17 Same as in Fig. 8.11 but with $\omega_c = 2\pi \times 10^{11}$ r/s

$$-\frac{x^2}{1/\sigma_x^2 + 2i\,\Delta z/v} = -\frac{\sigma_x^2}{1 + 4\sigma_x^4(\Delta z)^2/v^2}x^2 + i\frac{2\Delta z/v}{1/\sigma_x^4 + 4(\Delta z)^2/v^2}x^2$$

$$= -\frac{1}{w_x^2(\Delta z)}x^2 + i\frac{v}{2R_x(\Delta z)}x^2, \qquad (8.315)$$

where

$$w_x(\Delta z) \equiv w_{x0}\left(1 + \left(\frac{2\delta z}{w_{x0}^2 v}\right)^2\right)^{1/2} \qquad (8.316)$$

is the beam radius (or "spot size") at the e^{-1} amplitude point in the x-direction with beam waist

$$w_{x0} \equiv w_x(0) = \frac{1}{\sigma_x}, \qquad (8.317)$$

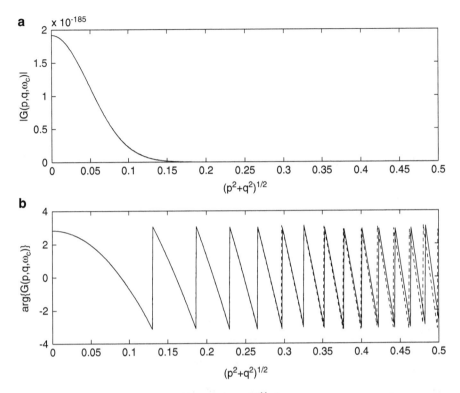

Fig. 8.18 Same as in Fig. 8.12 but with $\omega_c = 2\pi \times 10^{11}$ r/s

and where

$$R_x(\Delta z) \equiv \Delta z \left(1 + \left(\frac{w_{x0}^2 v}{2\Delta z} \right)^2 \right) \tag{8.318}$$

is the radius of curvature of the phase front[6] in the xz-plane in the paraxial approximation. Finally, the square root factor appearing in Eq. (8.314) may be written as

$$\frac{1}{\left(1 + 2i\sigma_x^2 \Delta z/v \right)^{1/2}} = \frac{e^{-i\frac{1}{2} \arctan \left(2\Delta z/(w_{x0}^2 v) \right)}}{\left(1 + 4(\Delta z)^2/(w_{x0}^4 v^2) \right)^{1/4}}$$

---

[6]Notice that $z = \sqrt{R^2 - (x^2 + y^2)} \approx R - (x^2 + y^2)/(2R)$ in the parabolic approximation of a spherical wave front with radius R, as appears in the imaginary part of Eq. (8.315) due to the paraxial approximation made in Eq. (8.312).

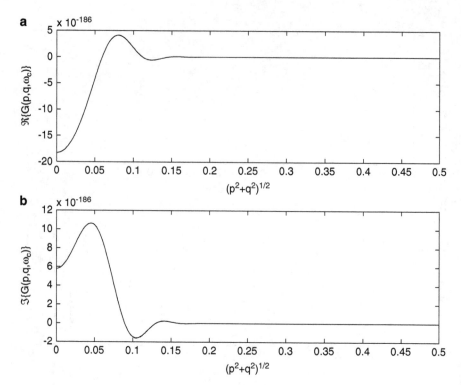

Fig. 8.19 Same as in Fig. 8.13 but with $\omega_c = 2\pi \times 10^{11}$ r/s

$$= \sqrt{\frac{w_{x0}}{w_x(\Delta z)}} e^{-i\psi_x(\Delta z)/2}, \tag{8.319}$$

where

$$\psi_x(\Delta z) \equiv \arctan\left(\frac{2\Delta z}{w_{x0}^2 \nu}\right) \tag{8.320}$$

describes the phase shift with propagation distance Δz away from the beam waist at $z = z_0$.

With these identifications, the two-dimensional Gaussian beam field given in Eq. (8.314) becomes

$$\hat{U}(x) \approx \sqrt{\frac{w_{x0}}{w_x(\Delta z)}} e^{-(x/w_x(\Delta z))^2} e^{(i/2)[(\nu/R_x(\Delta z))x^2 - \psi_x(\Delta z)]}, \tag{8.321}$$

with an analogous expression for $\hat{U}(y)$. The Gaussian beam wave front is a plane wave at the beam waist [$R_x(0) = R_y(0) = \infty$], and for large $|\Delta z|$ it approaches a

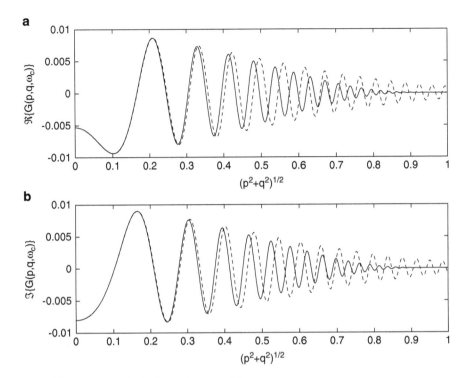

Fig. 8.20 Same as in Fig. 8.15 but with $\Delta z = 1.0\,\text{m}$

spherical wave front with center at the point $\mathbf{r}_0 = (0, 0, z_0)$ and with radius equal
to Δz, and in between these two limits it is, in general, an astigmatic wave front.
The divergence angle θ_x of the Gaussian beam in the xz-plane is obtained from the
limiting behavior of the expression $\tan \theta_x = w_x(\Delta z)/R_x(\Delta z)$ as $|\Delta z| \to \infty$ with
the result

$$\tan \theta_x = \frac{2}{\nu w_{x0}}. \tag{8.322}$$

Finally, the collimated beam length or Rayleigh range $2z_R$ is defined as the distance
over which the beam radius remains less than or equal to $\sqrt{2}$ of its value at the beam
waist, so that $w(z_R) \equiv \sqrt{2}w_0$, with solution

$$z_R = \frac{1}{2}\nu w_0^2. \tag{8.323}$$

The Gaussian beam approximates a collimated beam over the Rayleigh range
$|\Delta z| \leq z_R$ on either side of the beam waist, whereas outside of this range it behaves
more like a converging spherical wave when $\Delta z < z_R$ and a diverging spherical
wave when $\Delta z > z_R$.

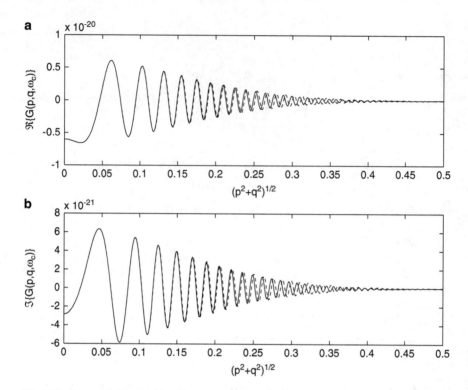

Fig. 8.21 Same as in Fig. 8.16 but with $\Delta z = 10.0\,\text{m}$

All of the paraxial beam parameters for a Gaussian beam are explicitly dependent upon the spatial frequency parameter ν which, in effect, sets an upper limit to the spatial frequency scale. With $\nu = k = 2\pi/\lambda$, these Gaussian beam parameters assume their usual form in the paraxial approximation as [41, 42]

$$w_x(\Delta z) = w_{x0}\left(1 + \left(\frac{\Delta z}{z_R}\right)^2\right)^{1/2}, \tag{8.324}$$

$$R_x(\Delta z) = \Delta z\left(1 + \left(\frac{z_R}{\Delta z}\right)^2\right), \tag{8.325}$$

$$\psi_x(\Delta z) = \arctan\left(\frac{\Delta z}{z_R}\right), \tag{8.326}$$

$$\tan\theta_x = \frac{\lambda}{\pi w_{x0}}, \tag{8.327}$$

$$z_R = \frac{\pi}{\lambda}w_{x0}^2. \tag{8.328}$$

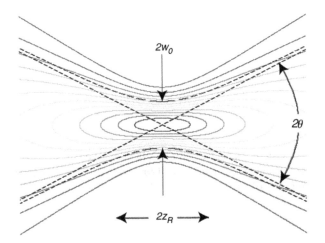

Fig. 8.22 Contour plot of the Gaussian beam amplitude with propagation distance Δz about the beam waist when $w_0 = 20\lambda$, illustrating the relationship of the beam waist $2w_0$, Rayleigh range $2z_R$, and angular divergence 2θ parameters with the main features of the propagation pattern

The physical relation of these parameters to the Gaussian beam propagation pattern they describe is illustrated in the contour plot depicted in Fig. 8.22 for the case when $w_0/\lambda = 20$. The symmetric pair of upper and lower dashed curves describe the beam half-width values $\pm w(\Delta z)$ which asymptotically approach the pair of dashed lines at $\pm\theta$ as $\Delta z \to \pm\infty$. Notice that if the spot size parameter w_0 is decreased from the value used in computing this field pattern, the Rayleigh range will decrease and the angular divergence will increase, resulting in a more divergent field behavior, whereas if it is increased, the Rayleigh range will be lengthened and the divergence angle narrowed, resulting in a more collimated beam behavior.

8.5.3 Asymptotic Behavior

Finally, consider the asymptotic behavior as $vR \to \infty$ of the integral representation of the transverse field $\hat{U}(x, y)$ defined in Eq. (8.209). Under the change of variable $v_x = pv$, $v_y = qv$, this representation may be expressed as

$$\hat{U}(x, y) = \frac{v^2}{4\pi^2} e^{-iv\Delta z} \int_{-\infty}^{\infty} \int_{-\infty}^{\infty} \tilde{\hat{U}}_0(p, q) e^{iv(px+qy+m\Delta z)} dpdq, \qquad (8.329)$$

where $m = \left(1 - p^2 - q^2\right)^{1/2}$. With $x_0 = y_0 = 0$ and fixed direction cosines $\xi_1 = x/R$, $\xi_2 = y/r$, $\xi_3 = \Delta z/R$, and with $\mathbf{V}(p, q, m) = m\tilde{\hat{U}}_0(p, q)$ [see Appendix G, Eq. (1)], the asymptotic approximation of $\hat{U}(x, y)$ is obtained from Eq. (G.86) as

$$\hat{U}(x, y) \sim -i \frac{\nu \Delta z}{2\pi R^2} \tilde{\hat{U}}_0(x/R, y/R) e^{i\nu R} e^{-i\nu \Delta z (z_0/R-1)}, \tag{8.330}$$

as $\nu R \to \infty$. With $\nu = k = 2\pi/\lambda$, the above expression becomes

$$\hat{U}(x, y) \sim -i \frac{\Delta z}{\lambda R^2} \tilde{\hat{U}}_0(x/R, y/R) e^{i 2\pi R/\lambda} e^{-i 2\pi (\Delta z/\lambda)(z_0/R-1)}, \tag{8.331}$$

as $R/\lambda \to \infty$.

As an example, for the Gaussian beam wave field given in Eq. (8.308) with initial beam widths (or beam waists) $w_{x0} = 1/\sigma_x$ and $w_{y0} = 1/\sigma_y$, the initial field spectrum given in Eq. (8.310) becomes

$$\tilde{\hat{U}}_0(x/R, y/R) = A\pi w_{x0} w_{y0} e^{-(\nu^2/4R^2)(w_{x0}^2 x^2 + w_{y0}^2 y^2)}.$$

Substitution of this expression in Eq. (8.331) then yields

$$\hat{U}(x, y) \sim -i\pi A \frac{w_{x0} w_{y0}}{\lambda R^2/\Delta z} e^{-(\pi/\lambda R)^2 (w_{x0}^2 x^2 + w_{y0}^2 y^2)} e^{i 2\pi R/\lambda} e^{-i 2\pi (\Delta z/\lambda)(z_0/R-1)},$$

as $R/\lambda \to \infty$. An analogous expression is obtained from Eqs. (8.313) and (8.321) in the limit as $\Delta z \approx R \to \infty$ (see Problem 8.12).

8.6   The Inverse Initial Value Problem

This chapter concludes with a concise description of the solution to the following time-dependent inverse source problem: Determine the charge sources $\varrho(\mathbf{r}, t)$ and currents $\mathbf{J}(\mathbf{r}, t)$ that are zero everywhere except over the time interval $-T < t < T$ such that for $t > T$ they produce prescribed solutions of the source-free (or homogeneous) Maxwells' equations

$$\nabla \times \mathbf{E}(\mathbf{r}, t) = - \left\| \frac{1}{c} \right\| \mu_0 \frac{\partial \mathbf{H}(\mathbf{r}, t)}{\partial t}, \tag{8.332}$$

$$\nabla \times \mathbf{H}(\mathbf{r}, t) = \left\| \frac{1}{c} \right\| \epsilon_0 \frac{\partial \mathbf{E}(\mathbf{r}, t)}{\partial t}, \tag{8.333}$$

$$\nabla \cdot \mathbf{E}(\mathbf{r}, t) = \nabla \cdot \mathbf{H}(\mathbf{r}, t) = 0, \tag{8.334}$$

in free-space. Although the solution to the inverse problem has been shown to be nonunique [43, 44], Moses and Prosser [45] have shown that by specifying a partial time-dependence in the wave-zone of the radiation field, a unique solution can be found for each such specification.

The early analysis of Moses and Prosser [45] begins with the *Bateman–Cunningham form of Maxwell's equations*, which is obtained in the following manner. First of all, notice that the pair of curl relations in Eqs. (8.332) and (8.333) may be expressed as

$$\nabla \times \mathbf{E}(\mathbf{r}, t) = -\frac{1}{c}\eta_0 \frac{\partial \mathbf{H}(\mathbf{r}, t)}{\partial t}, \tag{8.335}$$

$$\eta_0 \nabla \times \mathbf{H}(\mathbf{r}, t) = \frac{1}{c}\frac{\partial \mathbf{E}(\mathbf{r}, t)}{\partial t}, \tag{8.336}$$

where $\eta_0 \equiv \sqrt{\mu_0/\epsilon_0}$ is the intrinsic impedance of free space . If one then defines the complex vector field $\boldsymbol{\psi}(\mathbf{r}, t)$ as

$$\boldsymbol{\psi}(\mathbf{r}, t) \equiv \mathbf{E}(\mathbf{r}, t) - i\eta_0 \mathbf{H}(\mathbf{r}, t), \tag{8.337}$$

Maxwell's equations may be expressed in complex form as

$$\nabla \times \boldsymbol{\psi}(\mathbf{r}, t) = -i\frac{1}{c}\frac{\partial \boldsymbol{\psi}(\mathbf{r}, t)}{\partial t}, \tag{8.338}$$

$$\nabla \cdot \boldsymbol{\psi}(\mathbf{r}, t) = 0, \tag{8.339}$$

which is the Bateman–Cunningham form of Maxwell's equations when t is replaced by ct. The general solution of the these equations may then be expressed in terms of the eigenfunctions of the curl operator [46].

In a subsequent refinement of their work that is based upon the Radon transform [47], Moses and Prosser [48] have shown that there always exists a vector function of position $\mathbf{G}(\mathbf{r})$ such that for any causal, finite energy solution $\{\mathbf{E}(\mathbf{r}, t), \mathbf{H}(\mathbf{r}, t)\}$ of Maxwell's equations,

$$\lim_{r \to \infty} (r\mathbf{E}(\mathbf{r}, t)) = \mathbf{G}(r - ct, \theta, \phi), \tag{8.340}$$

$$\lim_{r \to \infty} (r\mathbf{H}(\mathbf{r}, t)) = \eta_0 \hat{\mathbf{r}} \times \mathbf{G}(r - ct, \theta, \phi), \tag{8.341}$$

with

$$\hat{\mathbf{r}} \cdot \mathbf{G}(\mathbf{r}) = 0, \tag{8.342}$$

where $\hat{\mathbf{r}} \equiv \mathbf{r}/r$ is the unit vector in the direction of the position vector $\mathbf{r} = (r, \theta, \phi)$ with spherical polar coordinates (r, θ, ϕ) such that $x = r \sin\theta \cos\phi$, $y = r \sin\theta \sin\phi$, $z = r \cos\theta$ with $0 \leq \theta \leq \pi$ and $0 \leq \phi < 2\pi$. Causality then requires that the radiated field vectors $\mathbf{E}(\mathbf{r}, t)$ and $\mathbf{H}(\mathbf{r}, t)$ are obtained from their respective initial temporal field behaviors. The pair of asymptotic limits given in Eqs. (8.340) and (8.341) then show that, with the exception of the $1/r$ factor, all causal, finite energy solutions of Maxwell's equations propagate outward to infinity like one-dimensional electromagnetic waves along rays specified by the polar angles θ and ϕ. Furthermore, the exact electromagnetic field vectors are given in terms of the wave zone vector field $\mathbf{G}(\mathbf{r})$ as [48]

Fig. 8.23 Space–time
domain (as indicated by the
shaded region) of an
electromagnetic bullet that is
defined in the interior region
of the cone $0 \le \theta < \Theta_c$ and
radial space–time region
$a + ct < r < b + ct$

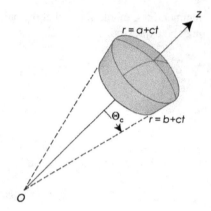

$$\mathbf{E}(\mathbf{r}, t) = \frac{1}{2\pi} \int_0^{\pi} \sin\theta' d\theta' \int_0^{2\pi} d\phi' \frac{\partial}{\partial p} \mathbf{G}(\mathbf{r} \cdot \hat{\mathbf{u}} - ct, \theta', \phi'), \qquad (8.343)$$

$$\mathbf{H}(\mathbf{r}, t) = \frac{1}{2\pi} \int_0^{\pi} \sin\theta' d\theta' \int_0^{2\pi} d\phi' \hat{\mathbf{u}} \times \frac{\partial}{\partial p} \mathbf{G}(\mathbf{r} \cdot \hat{\mathbf{u}} - ct, \theta', \phi'), \quad (8.344)$$

where $p = \mathbf{r} \cdot \hat{\mathbf{u}} - ct$ and $\hat{\mathbf{u}} = (\sin\theta' \cos\phi', \sin\theta' \sin\phi', \cos\theta')$.

So-called 'electromagnetic bullets' may then be constructed by requiring that the
vector function $\mathbf{G}(r - ct, \theta, \phi) = \mathbf{G}(r - ct, \theta)$ describing the far-zone behavior be
independent of the azimuthal angle ϕ and that it identically vanishes both outside
of the cone $0 \le \theta < \Theta_c$ as well as outside of the radial space–time region $a +
ct < r < b + ct$, where $a < b$, as depicted in Fig. 8.23. The solution of the
inverse source problem then provides the source distribution required to produce
this radiated electromagnetic bullet.

Unfortunately, the preceding analysis of the inverse source problem due to Moses
and Prosser [45, 48], while of considerable historical importance, is only applicable
to sources with a specific separable space–time dependence. In addition, these
solutions have been shown, in general, not to be minimum-energy solutions. These
limitations have been addressed by Marengo, Devaney, and Ziolkowski [49] who
reconsider the time-dependent inverse source problem with far-field data based upon
a limited-view Radon transform, a problem analogous to a limited-view computed
tomography reconstruction. Their analysis considers the inverse source problem for
the inhomogeneous scalar wave equation

$$\left(\nabla^2 - \frac{1}{c^2} \frac{\partial^2}{\partial t^2}\right) U(\mathbf{r}, t) = -\|4\pi\| Q(\mathbf{r}, t), \qquad (8.345)$$

where the radiating source $Q(\mathbf{r}, t)$, which is assumed to be localized within a
simply-connected region $\mathcal{D} \in \mathfrak{R}^3$, is to be determined from measurements of the
radiated field [see Eq. (3.81)]

$$U(\mathbf{r}, t) = \frac{\|4\pi\|}{4\pi} \int_{-\infty}^{\infty} dt' \int_{-\infty}^{\infty} d^3 r' \frac{Q(\mathbf{r}', t')}{|\mathbf{r} - \mathbf{r}'|} \delta\left(t' - t + |\mathbf{r} - \mathbf{r}'|/c\right) \tag{8.346}$$

for all $\mathbf{r} \notin \mathcal{D}$ and all $t \in \mathfrak{R}$. Because of the existence of nonradiating sources within the spatial support $\mathcal{D}$ of the source [43, 44], whose fields identically vanish for all $\mathbf{r} \notin \mathcal{D}$, this inverse source problem does not admit a unique solution unless certain *a priori* constraints are applied, as done by Moses and Prosser [45, 48], but not without introducing unnecessary limitations on the solution.

The asymptotic behavior of the radiated field given in Eq. (8.346) is given by

$$U(\mathbf{r}, t) \sim \frac{\|4\pi\|}{4\pi} \frac{F(\hat{\mathbf{s}}, \tau)}{r} \tag{8.347}$$

as $r \to \infty$, where $\hat{\mathbf{s}} \equiv \mathbf{r}/r$ is a unit vector that specifies the direction of observation of the radiated wave field, $\tau \equiv t - r/c$ is the retarded time, and where

$$F(\hat{\mathbf{s}}, \tau) = \int_{-\infty}^{\infty} dt' \int_{-\infty}^{\infty} d^3 r' Q(\mathbf{r}', t') \delta(t' - \tau - \mathbf{r}' \cdot \hat{\mathbf{s}}/c). \tag{8.348}$$

If the time-domain radiation pattern $F(\hat{\mathbf{s}}, \tau)$ is known for all times $\tau \in \mathfrak{R}$ over all observation directions $\hat{\mathbf{s}} \in \mathcal{S}^2$, where $\mathcal{S}^2 \in \mathfrak{R}^3$ is the unit sphere, the radiated wave field $U(\mathbf{r}, t)$ can be then determined [50] everywhere outside the source domain $\mathcal{D}$; however, a unique determination of $U(\mathbf{r}, t)$ for $\mathbf{r} \notin \mathcal{D}$ is not possible if the radiation pattern $F(\hat{\mathbf{s}}, \tau)$ is known only over a discrete set of directions $\hat{\mathbf{s}}$.

The analysis of the inverse source problem with far-field data taken over a discrete set of directions has been given by Marengo, Devaney, and Ziolkowski [49] using a limited-view Radon transform inversion. Their motivation in solving this important problem was (a) the direct problem of synthesizing the minimum energy (minimum L^2 norm) source $Q(\mathbf{r}, t)$ with prescribed spatial support $\mathcal{D}$ that produces a given far-field radiation pattern $F(\hat{\mathbf{s}}, \tau)$, as for an electromagnetic bullet, and (b) the inverse problem of reconstructing an unknown source using far-field data, as for source (target) identification and interrogation. These two interrelated direct and inverse problems are now treated in the following subsections based upon their analysis.

8.6.1   The Direct Problem

Define the source function $\varrho(\chi)$ as

$$\varrho(\chi) \equiv Q(\mathbf{r}, t), \tag{8.349}$$

where $\chi \equiv (\chi_0, \chi_1, \chi_2, \chi_3)$ with $\chi_0 = ct$, $\chi_1 = x$, $\chi_2 = y$, $\chi_3 = z$, and define the unit vector $\hat{v}_s \equiv (v_{s0}, v_{s1}, v_{s2}, v_{s3})$ with $v_{s0} = 1/\sqrt{2}$, $v_{s1} = -s_x/\sqrt{2}$, $v_{s2} =$

$-s_y/\sqrt{2}$, $v_{s3} = -s_z/\sqrt{2}$, where s_x, s_y, s_z are the components of the unit vector $\hat{\mathbf{s}} = (s_x, s_y, s_z)$. In addition, define the *masking function* $M(\chi)$ as

$$M(\chi) \equiv \begin{cases} 1, & \text{if } \chi \in \mathcal{D} \\ 0, & \text{if } \chi \notin \mathcal{D} \end{cases}, \tag{8.350}$$

where $\mathcal{D}$ denotes the space–time region containing the source. With these definitions, the expression (8.348) for the radiation pattern becomes

$$F(\hat{\mathbf{s}}, \tau) = \frac{1}{\sqrt{2}} \int_{-\infty}^{\infty} M(\chi)\varrho(\chi)\delta\left(c\tau/\sqrt{2} - \chi \cdot \hat{\boldsymbol{v}}_s\right) d^4\chi. \tag{8.351}$$

This result may then be expressed in terms of the four-fold Radon transform (see Appendix H)

$$\mathcal{R}[\varrho](\hat{\boldsymbol{v}}, \xi) = \int_{-\infty}^{\infty} \varrho(\chi)\delta(\xi - \chi \cdot \hat{\boldsymbol{v}})d^4\chi \tag{8.352}$$

of the function $\varrho(\chi)$ over the set of hyperplanes $\chi \cdot \hat{\boldsymbol{v}} - \xi = 0$, where $\hat{\boldsymbol{v}}$ is a unit vector defining the orientation of the hyperplane in the four-dimensional Radon domain and where the real-valued parameter ξ defines the distance of the hyperplane to the origin. The expression given in Eq. (8.351) for the time-domain radiation pattern may then be expressed as

$$F(\hat{\mathbf{s}}, \tau) = \frac{1}{\sqrt{2}}\mathcal{R}[M\varrho](\hat{\boldsymbol{v}}_s, c\tau/\sqrt{2}). \tag{8.353}$$

Hence, the radiation pattern $F(\hat{\mathbf{s}}, \tau)$ along a fixed direction $\hat{\mathbf{s}}$ is given by the projection of the source function $\varrho(\chi)$ onto the line taken along the direction of the unit vector $\hat{\boldsymbol{v}}_s$, depicted in Part (a) of Fig. 8.24. It then follows from Eq. (8.353) that, for a prescribed radiation pattern $F(\hat{\mathbf{s}}, \tau)$, the data necessary to reconstruct the source function $\varrho(\chi)$ consist of the Radon projections along radial lines that are tangent to the generalized four-dimensional light cone with apex at the origin, as depicted in Part (b) of Fig. 8.24 in the three-dimensional case. In this manner, Marengo, Devaney, and Ziolkowski have shown that [49]

> Radon projections onto directions outside the light cone cannot be inferred from $F(\hat{\mathbf{s}}, \tau)$, making the inversion nonunique. The inverse source problem with far-field data reduces to finding source functions consistent with radon projections provided *only* for directions that lie tangentially on the surface of the light cone (nonradiating sources are those whose Radon projections onto those directions vanish ...).

The frequency-domain radiation pattern

$$\tilde{F}(\hat{\mathbf{s}}, \omega) = \int_{-\infty}^{\infty} F(\hat{\mathbf{s}}, \tau)e^{i\omega\tau} d\tau \tag{8.354}$$

Fig. 8.24 (a)
Three-dimensional depiction
of the Radon transform
representation of the radiation
pattern along the fixed
direction ŝ due to the source
distribution $\varrho(\chi)$ in a
"viewing" direction specified
by the unit vector ŝ. (b) The
relation of the unit vector ŝ to
the light cone in
three-dimensional
space–time. (After Fig. 1 in
[49].)

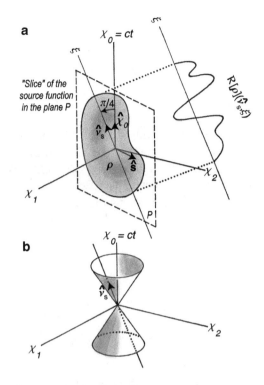

is, with substitution from Eq. (8.351), found to be given by

$$\tilde{F}(\hat{s}, \omega) = \frac{1}{c} \int_{-\infty}^{\infty} M(\chi)\varrho(\chi)e^{i(\sqrt{2}\omega/c)\hat{v}_s \cdot \chi}\, d^4\chi, \tag{8.355}$$

which is just the spatio-temporal Fourier transform of the masked source function
$M(\chi)\varrho(\chi)$ evaluated on the surface of a generalized cone centered at the origin in
the four-dimensional $(\omega/c, \mathbf{k})$ Fourier domain. It is then seen that [49]

> while $F(\hat{s}, \tau)$ provides information about Radon projections of $\varrho(\chi)$ *only* along lines
> tangent to the light cone ..., $\tilde{F}(\hat{s}, \omega)$ provides information about the fourfold Fourier
> transform of $\varrho(\chi)$ *only* over the surface of an analogous generalized cone in the [four-
> dimensional] Fourier domain. Thus $\tilde{F}(\hat{s}, \omega)$ alone is insufficient to uniquely determine $\varrho(\chi)$
> through Fourier inversion, confirming the nonunique nature of the inverse source problem."

8.6.2   The Inverse Problem

The development of the minimum energy solution to the inverse source problem
from far-field data is now given. By 'minimum energy' solution is meant the
minimum L^2 norm $\int |\varrho(\chi)|^2 d^4\chi$ solution $\varrho_{Emin}(\chi) \equiv Q_{Emin}(\mathbf{r}, t)$. Only the general
formulation of the solution is given here. More specific details with applications to

discrete samples may be found in the paper by Marengo, Devaney, and Ziolkowski [49] as well as in related work cited there.

The general solution begins by expressing Eq. (8.351) for the radiation pattern in the form of a linear mapping as [49]

$$F = P\varrho, \tag{8.356}$$

where $\varrho(\chi) \in X$ with $X \equiv L^2\{\mathfrak{R}^4\}$ denoting the Hilbert space of L^2 functions of the variable $\chi \in \mathfrak{R}^4$, $F(\hat{s}, \tau) \in \mathcal{Y}$ with $\mathcal{Y} = \mathcal{T} \times \mathcal{S}$ denoting the Hilbert space that is formed from the direct product of the space $\mathcal{T} \equiv L^2\{\mathfrak{R}\}$ of L^2 functions of the variable $\tau \in \mathfrak{R}$ with the space $\mathcal{S} \equiv L^2\{S^2\}$ of L^2 functions of the unit vector $\hat{s}$, where S^2 denotes the unit sphere, and where $P : X \rightarrow \mathcal{Y}$ is a linear transform mapping the space X into the space $\mathcal{Y}$. Inner products in these Hilbert spaces are then defined in the usual way [51].[7]

With the definition $\langle P^\dagger F, \varrho\rangle_X = \langle F, P\varrho\rangle_\mathcal{Y}$ of the adjoint $P^\dagger$ of a linear operator P, substitution of Eq. (8.351) then shows that

$$\left(P^\dagger F\right)(\chi) = M(\chi)\frac{1}{c}\int_{S^2} F\left(\hat{s}, (\sqrt{2}/c)\hat{v}_s \cdot \chi\right) d^2s$$
$$= M(\chi)\frac{1}{c}\int_{S^2} F\left(\hat{s}, t - \hat{s} \cdot \mathbf{r}/c\right) d^2s, \tag{8.357}$$

which is associated with the operation of *backprojection* (see [49] and the references contained therein). The appearance of the masking function in this expression guarantees that $P^\dagger$ maps $\mathcal{Y} \rightarrow X$, so that $PP^\dagger$ maps $\mathcal{Y} \rightarrow \mathcal{Y}$ and $P^\dagger P$ maps $X \rightarrow X$. The final form of Eq. (8.357) shows that $\left(P^\dagger F\right)(\chi)$ is given by the product of the masking function $M(\chi)$ with a linear superposition of time-dependent plane waves with amplitudes given by the far-field pattern $F(\hat{s}, \tau)$. In particular, the quantity $\left(P^\dagger F\right)(\chi)$ is seen to be a source-free field (see Sect. 8.4.1) that is defined by a plane wave expansion truncated within the space–time domain $\mathcal{D}$ containing the source [see Eq. (8.350)]. It then follows from Sherman's first theorem (Theorem 8.2) that $\left(P^\dagger F\right)(\chi)$ satisfies the homogeneous wave equation

$$\left(\nabla^2 - \frac{1}{c^2}\frac{\partial^2}{\partial t^2}\right)\left(P^\dagger F\right)(\chi) = 0, \tag{8.358}$$

for all $\chi \in \mathcal{D}$ excluding the boundary $\partial\mathcal{D}$ of $\mathcal{D}$.

---

[7]For example, the inner product of two functions $\varrho_1(\chi) \in X$ and $\varrho_2(\chi) \in X$ is defined as

$$\langle\varrho_1, \varrho_2\rangle_X \equiv \int \varrho_1^*(\chi)\varrho_2(\chi)d^4\chi,$$

where the superscript asterisk (*) denotes the complex conjugate.

The unique, minimum L^2 norm solution to the inverse source problem expressed in Eq. (8.356) is found to be [52, 53]

$$\varrho_{Emin}(\chi) = \left(P^\dagger \bar{F}\right)(\chi), \tag{8.359}$$

where $\bar{F}(\hat{s}, \tau)$ denotes the *filtered time-domain radiation pattern* that is defined by the relation

$$\left(P P^\dagger \bar{F}\right)(\hat{s}, \tau) = F(\hat{s}, \tau). \tag{8.360}$$

If the far-field data measurements are noise-free, then Eqs. (8.359) and (8.360) provide the normal solution to the inverse source problem. However, in the inevitable presence of noise in the data, some regularization procedure [52] (such as Tikhonov–Phillips regularization) may be used to generate approximate minimum energy solutions to the inverse source problem [54]. The noise-free solution is only considered here.

Taken together, Eqs. (8.358) and (8.359) give

$$\left(\nabla^2 - \frac{1}{c^2}\frac{\partial^2}{\partial t^2}\right)\varrho_{Emin}(\chi) = 0 \tag{8.361}$$

for all $\chi \in \mathcal{D}$ with $\chi \notin \partial\mathcal{D}$. Notice that the frequency domain analog of this result, which is implicitly contained in the inverse source problem solutions due to Bleistein and Cohen [44] and Devaney and Porter [55, 56] and others, is the usual form employed in this methodology.

For the case of a time-independent masking function $M(\chi) = M(\mathbf{r})$, Eqs. (8.351), (8.357), and (8.360) yield

$$
\begin{aligned}
\left(P P^\dagger \bar{F}\right)(\hat{s}, \tau) &= \frac{1}{c}\int_{-\infty}^{\infty} dt' \int_{-\infty}^{\infty} d^3 r'\, M(\mathbf{r}')\delta(\tau - t' + \mathbf{r}' \cdot \hat{s}/c) \\
&\quad \times \int_{S^2} d^2 s'\, \bar{F}(\hat{s}', t' - \mathbf{r}' \cdot \hat{s}'/c) \\
&= \frac{1}{c}\int_{S^2} d^2 s' \int_{-\infty}^{\infty} dt''\, \bar{F}(\hat{s}', t'') \\
&\quad \times \int_{-\infty}^{\infty} d^3 r'\, M(\mathbf{r}')\delta\left(t'' - \tau - (\hat{s} - \hat{s}') \cdot \mathbf{r}'/c\right) \\
&= \frac{1}{c}\int_{S^2} d^2 s'\, \bar{F}(\hat{s}', \tau) \otimes h(\hat{s} - \hat{s}', \tau), \tag{8.362}
\end{aligned}
$$

where the symbol $\otimes$ here denotes the *temporal convolution operation*

$$f(\tau) \otimes g(\tau) = \int_{-\infty}^{\infty} f(\tau - t)g(t)dt, \tag{8.363}$$

and where

$$h(\hat{\mathbf{s}} - \hat{\mathbf{s}}', \tau) \equiv \int_{-\infty}^{\infty} d^3 r' \, M(\mathbf{r}') \delta \left(\tau + (\hat{\mathbf{s}} - \hat{\mathbf{s}}') \cdot \mathbf{r}'/c \right). \tag{8.364}$$

Taken together, Eqs. (8.360) and (8.362) show that

$$F(\hat{\mathbf{s}}, \tau) = \frac{1}{c} \int_{S^2} d^2 s' \, \bar{F}(\hat{\mathbf{s}}', \tau) \otimes h(\hat{\mathbf{s}} - \hat{\mathbf{s}}', \tau). \tag{8.365}$$

Upon comparison of Eq. (8.364) with Eq. (8.351), the quantity $h(\hat{\mathbf{s}} - \hat{\mathbf{s}}', \tau)$ is identified as the time-domain radiation pattern due to the space–time separable source

$$Q(\mathbf{r}, t) = M(\mathbf{r}) \delta(t - \hat{\mathbf{s}}' \cdot \mathbf{r}/c), \tag{8.366}$$

which is uniformly distributed over the spatial support $M(\mathbf{r})$ and is impulsively excited with a progressive time delay $t' = \hat{\mathbf{s}}' \cdot \mathbf{r}/c$. With this understanding, the integral of the convolution given in Eq. (8.365) states that [49]

> the unfiltered data $F(\hat{\mathbf{s}}, \tau)$ are equal to the sum over all available directions $\hat{\mathbf{s}}'$ of the time-domain radiation pattern of a source that consists of a uniform distribution of point radiators (within the spatial support of the sought-after source) all of which are excited, with a progressive time delay $\mathbf{r} \cdot \hat{\mathbf{s}}'/c$, by the same time signature, the latter being precisely - for a given $\hat{\mathbf{s}}'$ - the time signature of the filtered data $\bar{F}(\hat{\mathbf{s}}', \tau)$.

That is, along each fixed direction $\hat{\mathbf{s}}'$, the filtered data $\bar{F}(\hat{\mathbf{s}}', \tau)$ may be interpreted as the excitation that must be applied at each point of the uniform source distribution $Q(\mathbf{r}, t) = M(\mathbf{r}) \delta(t - \hat{\mathbf{s}}' \cdot \mathbf{r}/c)$ in order to obtain the original, unfiltered data $F(\hat{\mathbf{s}}, \tau)$ through a linear superposition of the time-domain radiation patterns due to each of these synthetic sources.

In order to compute the minimum energy source from Eqs. (8.357) and (8.359), the relation appearing in Eq. (8.365) must be inverted. Upon taking the temporal Fourier transform of Eq. (8.365), one obtains

$$\tilde{F}(\hat{\mathbf{s}}, \omega) = \frac{1}{c} \int_{S^2} d^2 s' \, \tilde{\bar{F}}(\hat{\mathbf{s}}', \omega) \tilde{h}(\hat{\mathbf{s}} - \hat{\mathbf{s}}', \omega), \tag{8.367}$$

where [from the temporal Fourier transform of Eq. (8.364)]

$$\tilde{h}(\hat{\mathbf{s}} - \hat{\mathbf{s}}', \omega) = \int_{-\infty}^{\infty} M(\mathbf{r}') e^{-i(\omega/c)\mathbf{r}' \cdot (\hat{\mathbf{s}} - \hat{\mathbf{s}}')} d^3 r'. \tag{8.368}$$

The minimum energy solution is then obtained in the following manner. First, numerically (or, if possible, analytically) solve the filtering operation specified in Eq. (8.367) to determine $\tilde{\bar{F}}(\hat{\mathbf{s}}, \omega)$ along each specified direction $\hat{\mathbf{s}}$. Then, upon taking the inverse temporal Fourier transform of $\tilde{\bar{F}}(\hat{\mathbf{s}}, \omega)$, recover the filtered data $\bar{F}(\hat{\mathbf{s}}, \tau)$.

Finally, use the first form of Eq. (8.357) to backproject this filtered data $\bar{F}(\hat{s}, \tau)$, as required by the reconstruction formula given in Eq. (8.359). Further details may be found in the paper by Marengo, Devaney, and Ziolkowski [49] and the references contained therein.

8.7 Summary

This rather lengthy chapter has provided fundamental material necessary for the study of dispersive pulsed beam propagation as well as extending this theory into several related topics of interest. Of significant interest here is the question concerning separable pulsed beam fields and the full implications of the commonly used paraxial approximation. In particular, it was shown here that, with the exception of trivial cases (such as a pulsed, plane wave field), space–time separability for a pulsed, beam wave field is strictly valid only in the geometrical optics limit as $\omega \to \infty$.

Problems

8.1 Solve the relation given in Eq. (8.13) for the applied current density $\tilde{\mathbf{J}}_0(\tilde{\mathbf{k}}^+, \omega)$ of the source.

8.2 Redo the analysis presented from Eq. (8.29) through Eq. (8.42) for a negative index medium where $n'(\omega) < 0$ over some frequency domain.

8.3 Obtain the solution equivalent to that given in Eqs. (8.74) and (8.75) expressed in terms of the normal derivatives of the initial field vectors at the plane $z = z_0^+$. The spatial integrals appearing in this representation are the *second Rayleigh–Sommerfeld diffraction integrals* of classical optics.

8.4 For a uniformly polarized wave field in the electric field vector, show that for the magnetic field vector to also be uniformly polarized, the spectral quantity $\tilde{\mathbf{k}}^+(\omega)\tilde{\tilde{E}}_0(\mathbf{k}_T, \omega)$ must have a fixed direction independent of ω. Show that this in turn implies that the initial field amplitude $\tilde{E}_0(\mathbf{r}_T, \omega)$ must be independent of one transverse coordinate direction.

8.5 Show that the temporal frequency transform of the electromagnetic field vectors given in Eqs. (8.14) and (8.17) satisfies the vector Helnmholtz equations $\left(\nabla^2 + \tilde{k}^2(\omega)\right)\tilde{\mathbf{E}}(\mathbf{r}, \omega) = \mathbf{0}$ and $\left(\nabla^2 + \tilde{k}^2(\omega)\right)\tilde{\mathbf{B}}(\mathbf{r}, \omega) = \mathbf{0}$.

8.6 Provide the conditions required for the separation given in Eqs. (8.287) and (8.288) to be a unique solution of Eq. (8.286).

8.7 Derive the expression

$$\tilde{K}(x - x', y - y', z) = \frac{z}{R^3}(ikR - 1)e^{ikR}$$

for the monochromatic impulse response function $\tilde{K}(x, y, z) = -2\pi \tilde{U}(\mathbf{r}, \omega)$ [i.e., the scalar wave-field produced in the positive half-space $z > z_0$ when $\tilde{U}_0(x', y', \omega) = \delta(x', y')$] for the planar boundary value problem in free-space, where $R = \sqrt{(x - x')^2 + (y - y')^2 + (z - z_0)^2}$. This function is also referred to as the scalar dipole field. Using the relation

$$2\frac{\partial \mathsf{U}_0(u, v)}{\partial u} = \left[\left(\frac{v}{u}\right)^2 - 1\right]\mathsf{U}_1(u, v) - \frac{v}{u}J_1(v),$$

show that the evanescent wave contribution to the scalar dipole field is given by

$$\tilde{K}_E(x - x', y - y', z) = \frac{z}{R^3}\Bigg[J_0(k\rho) - 2\mathsf{U}_0(kR - k\Delta z, k\rho)$$

$$+ \frac{kR\rho}{\Delta z}J_1(k\rho) - 2kR\mathsf{U}_1(kR - k\Delta z, k\rho)\Bigg].$$

The homogeneous plane wave contribution to the scalar dipole field is then given by

$$\tilde{K}_H(x - x', y - y', z) = \tilde{K}(x - x', y - y', z) - \tilde{K}_E(x - x', y - y', z).$$

Show that, along the axis of the dipole

$$\tilde{K}(0, 0, z) = \left\{\frac{ik^2}{kz} - \frac{k^2}{(kz)^2}\right\}e^{ikz},$$

$$\tilde{K}_E(0, 0, z) = -\frac{k^2}{(kz)^2},$$

$$\tilde{K}_H(0, 0, z) = \frac{ik^2}{kz}\left\{e^{ikz/2} - \frac{\sin(kz/2)}{kz/2}\right\}e^{ikz/2}.$$

Notice that the homogeneous wave contribution along the dipole axis is negligible in comparison to the evanescent contribution when $kz \ll 1$, while the opposite is true when $kz \gg 1$.

8.8 Determine the spectral amplitude $\tilde{U}_s(p, q, \omega)$ of the time-harmonic spherical wave $\tilde{U}_s(\mathbf{r}, \omega) = e^{ikr}/r$ radiating away from the origin. With this result obtain explicit expressions for the homogeneous and evanescent components of this spherical wave field along the z-axis (i.e. when $p = q = 0$).

8.9 Derive explicit expression for the endpoints α_1 and α_4 of the contour C depicted in Fig. G.2.

8.10 Let $\tilde{U}(\mathbf{r}, \omega) = \tilde{U}(r, \vartheta, \varphi, \omega)$ be a solution of the homogeneous scalar Helmholtz equation

$$\left(\nabla^2 + k^2\right)\tilde{U}(\mathbf{r}, \omega) = 0,$$

and let $\tilde{U}(r, \vartheta, \varphi, \omega)$ have an asymptotic expansion of the form

$$\tilde{U}(r, \vartheta, \varphi, \omega) = \frac{e^{ikr}}{kr} \sum_{n=0}^{N-1} \frac{B_n(\vartheta, \varphi, \omega)}{(kr)^n} + \mathcal{O}\left\{(kr)^{-N}\right\}$$

for arbitrarily large N as $kr \to \infty$ with fixed $k > 0$ and fixed direction cosines $\xi_1 = \sin\vartheta\cos\varphi$, $\xi_2 = \sin\vartheta\sin\varphi$, $\xi_3 = \cos\vartheta$. With the assumption that the asymptotic expansions of the partial derivatives of $\tilde{U}(\mathbf{r}, \omega) = \tilde{U}(r, \vartheta, \varphi, \omega)$ with respect to r, ϑ, φ up to the second order can be obtained by differentiating the above asymptotic expansion term by term, derive the general recursion relation between the coefficients $B_{n+1}(\vartheta, \varphi, \omega)$ and $B_n(\vartheta, \varphi, \omega)$. In particular, show that

$$B_1(\vartheta, \varphi, \omega) = \frac{i}{2}\mathcal{L}^2 B_0(\vartheta, \varphi, \omega),$$

where $\mathcal{L}^2$ is the differential operator defined by

$$\mathcal{L}^2 \equiv -\frac{1}{\sin\vartheta}\frac{\partial}{\partial\vartheta}\left(\sin\vartheta\frac{\partial}{\partial\vartheta}\right) - \frac{1}{\sin^2\vartheta}\frac{\partial^2}{\partial\varphi^2}.$$

8.11 Show that the integral representation of the source-free transverse field vector $\hat{U}(x, y)$ given in Eq. (8.299) has the series expansion given in Eq. (8.298).

8.12 Obtain the limiting behavior as $\Delta z \approx R \to \infty$ of the Gaussian beam field given in Eqs. (8.313) and (8.330).

8.13 Derive the asymptotic approximation given in Eqs. (8.347) and (8.348) of the far-field behavior of the radiation field given in Eq. (8.346).

8.14 Show that the expression given in Eq. (8.357) for the quantity $(P^\dagger F)(\chi)$ follows from the definition of the adjoint $P^\dagger$ of the linear operator P and Eq. (8.351) for the radiation pattern $F(\hat{s}, \tau)$ with $\tau \equiv t - r/c$.

References

1. G. C. Sherman, "Application of the convolution theorem to Rayleigh's integral formulas," *J. Opt. Soc. Am.*, vol. 57, pp. 546–547, 1967.
2. G. C. Sherman, "Integral-transform formulation of diffraction theory," *J. Opt. Soc. Am.*, vol. 57, pp. 1490–1498, 1967.
3. J. R. Shewell and E. Wolf, "Inverse diffraction and a new reciprocity theorem," *J. Opt. Soc. Am.*, vol. 58, no. 12, pp. 1596–1603, 1968.
4. G. C. Sherman, "Diffracted wave fields expressible by plane-wave expansions containing only homogeneous waves," *Phys. Rev. Lett.*, vol. 21, no. 11, pp. 761–764, 1968.
5. G. C. Sherman and H. J. Bremermann, "Generalization of the angular spectrum of plane waves and the diffraction transform," *J. Opt. Soc. Am.*, vol. 59, no. 2, pp. 146–156, 1969.
6. G. C. Sherman, "Diffracted wave fields expressible by plane-wave expansions containing only homogeneous waves," *J. Opt. Soc. Am.*, vol. 59, pp. 697–711, 1969.
7. C. J. Bouwkamp, "Diffraction theory," *Rept. Prog. Phys.*, vol. 17, pp. 35–100, 1954.
8. J. W. Goodman, *Introduction to Fourier Optics*. New York: McGraw-Hill, 1968.
9. W. H. Carter, "Electromagnetic beam fields," *Optica Acta*, vol. 21, pp. 871–892, 1974.
10. J. J. Stamnes, *Waves in Focal Regions: Propagation, Diffraction and Focusing of Light, Sound and Water Waves*. Bristol, UK: Adam Hilger, 1986.
11. M. Nieto-Vesperinas, *Scattering and Diffraction in Physical Optics*. New York: Wiley-Interscience, 1991.
12. A. J. Devaney, *Mathematical Foundations of Imaging, Tomography and Wavefield Inversion*. Cambridge: Cambridge University Press, 2012.
13. M. Born and E. Wolf, *Principles of Optics*. Cambridge: Cambridge University Press, seventh (expanded) ed., 1999.
14. E. Lalor, "Conditions for the validity of the angular spectrum of plane waves," *J. Opt. Soc. Am.*, vol. 58, pp. 1235–1237, 1968.
15. H. L. Royden, *Real Analysis*. New York: Macmillan, second ed., 1968. p. 269.
16. A. Sommerfeld, *Optics*, vol. IV of *Lectures in Theoretical Physics*. New York: Academic, 1964. paperback edition.
17. A. Nisbet and E. Wolf, "On linearly polarized electromagnetic waves of arbitrary form," *Proc. Camb. Phil. Soc.*, vol. 50, pp. 614–622, 1954.
18. K. E. Oughstun, "Polarization properties of the freely-propagating electromagnetic field of arbitrary spatial and temporal form," *J. Opt. Soc. Am. A*, vol. 9, no. 4, pp. 578–584, 1992.
19. E. Wolf, "Recollections of Max Born," in *Tribute to Emil Wolf: Science and Engineering Legacy in Physical Optics* (T. P. Jannson, ed.), ch. 2, pp. 29–49, Bellingham, WA, USA: SPIE Press, 2004.
20. L. Mandel and E. Wolf, *Optical Coherence and Quantum Optics*. Cambridge: Cambridge University Press, 1995. Ch. 3.
21. E. Wolf, *Introduction to the Theory of Coherence and Polarized Light*. Cambridge: Cambridge University Press, 2007.
22. M. Born and E. Wolf, *Principles of Optics: Electromagnetic Theory of Propagation, Interference and Diffraction of Light*. New York: Pergamon Press, 1 ed., 1959.
23. T. Voipio, T. Setälä, and A. T. Friberg, "Statistical similarity and complete coherence of electromagnetic fields in time and frequency domains," *J. Opt. Soc. Am. A*, vol. 32, no. 5, pp. 741–750, 2015.
24. C. Brosseau, "What polarization of light is: The contribution of Emil Wolf," in *Tribute to Emil Wolf: Science and Engineering Legacy in Physical Optics* (T. P. Jannson, ed.), pp. 51–93, Bellingham, WA, USA: SPIE Press, 2004.
25. A. Friberg, "Electromagnetic theory of optical coherence," in *Tribute to Emil Wolf: Science and Engineering Legacy in Physical Optics* (T. P. Jannson, ed.), ch. 4, pp. 95–113, Bellingham, WA, USA: SPIE Press, 2004.

26. K. E. Oughstun, "The angular spectrum representation and the Sherman expansion of pulsed electromagnetic beam fields in dispersive, attenuative media," *Pure Appl. Opt.*, vol. 7, no. 5, pp. 1059–1078, 1998.

27. I. S. Gradshteyn and I. M. Ryzhik, *Table of Integrals, Series, and Products.* New York: Academic Press, 1980.

28. G. N. Watson, *A Treatise on the Theory of Bessel Functions.* Cambridge: Cambridge University Press, 1922.

29. D. C. Bertilone, "The contribution of homogeneous and evanescent plane waves to the scalar optical field: exact diffraction formulae," *J. Modern Optics*, vol. 38, no. 5, pp. 865–875, 1991.

30. G. C. Sherman, J. J. Stamnes, A. J. Devaney, and É. Lalor, "Contribution of the inhomogeneous waves in angular-spectrum representations," *Opt. Commun.*, vol. 8, pp. 271–274, 1973.

31. W. D. Montgomery, "Algebraic formulation of diffraction applied to self imaging," *J. Opt. Soc. Am.*, vol. 58, no. 8, pp. 1112–1124, 1968.

32. E. C. Titchmarsh, *The Theory of Functions.* London: Oxford University Press, 1937. Section 10.5.

33. E. J. McShane, *Integration.* Princeton, NJ: Princeton University Press, 1944. p. 217.

34. W. Kaplan, *Introduction to Analytic Functions.* Reading, MA: Addison-Wesley, 1966. p. 171.

35. H. Bremermann, *Distributions, Complex Variables, and Fourier Transforms.* Reading, MA: Addison-Wesley, 1965. Ch. 8.

36. T. B. Hansen and A. D. Yaghjian, *Plane-Wave Theory of Time-Domain Fields.* New York: IEEE, 1999.

37. I. M. Gel'fand and G. E. Shilov, *Generalized Functions*, vol. I. New York: Academic, 1964. Ch. 2.

38. T. Melamed and L. B. Felsen, "Pulsed-beam propagation in lossless dispersive media. I. A numerical example," *J. Opt. Soc. Am. A*, vol. 15, pp. 1277–1284, 1998.

39. T. Melamed and L. B. Felsen, "Pulsed-beam propagation in lossless dispersive media. II. Theory," *J. Opt. Soc. Am. A*, vol. 15, pp. 1268–1276, 1998.

40. K. E. Oughstun, "Asymptotic description of pulse ultrawideband electromagnetic beam field propagation in dispersive, attenuative media," *J. Opt. Soc. Am. A*, vol. 18, no. 7, pp. 1704–1713, 2001.

41. H. Kogelnik, "Imaging of optical modes - Resonators with internal lenses," *Bell Syst. Tech. J.*, vol. 44, pp. 455–494, 1965.

42. H. Kogelnik and T. Li, "Laser beams and resonators," *Proc. IEEE*, vol. 54, no. 10, pp. 1312–1329, 1966.

43. A. J. Devaney and E. Wolf, "Radiating and nonradiating classical current distributions and the fields they generate," *Phys. Rev. D*, vol. 8, pp. 1044–1047, 1973.

44. N. Bleistein and J. K. Cohen, "Nonuniqueness in the inverse source problem in acoustics and electromagnetics," *J. Math. Phys.*, vol. 18, pp. 194–201, 1977.

45. H. E. Moses and R. T. Prosser, "Initial conditions, sources, and currents for prescribed time-dependent acoustic and electromagnetic fields in three dimensions," *IEEE Trans. Antennas Prop.*, vol. 24, no. 2, pp. 188–196, 1986.

46. H. E. Moses, "Eigenfunctions of the curl operator, rotationally invariant Helmholtz theorem, and applications to electromagnetic theory and fluid mechanics," *SIAM J. Appl. Math.*, vol. 21, pp. 114–144, 1971.

47. H. E. Moses and R. T. Prosser, "A refinement of the Radon transform and its inverse," *Proc. Roy. Soc. Lond. A*, vol. 422, pp. 343–349, 1989.

48. H. E. Moses and R. T. Prosser, "Exact solutions of the three-dimensional scalar wave equation and Maxwell's equations from the approximate solutions in the wave zone through the use of the Radon transform," *Proc. Roy. Soc. Lond. A*, vol. 422, pp. 351–365, 1989.

49. E. A. Marengo, A. J. Devaney, and R. W. Ziolkowski, "New aspects of the inverse source problem with far-field data," *J. Opt. Soc. Am. A*, vol. 16, pp. 1612–1622, 1999.

50. E. Heyman and A. J. Devaney, "Time-dependent multipoles and their application for radiation from volume source distributions," *J. Math. Phys.*, vol. 37, pp. 682–692, 1996.

51. T. F. Jordan, *Linear Operators for Quantum Mechanics.* New York: John Wiley & Sons, 1969.

52. M. Bertero, "Linear inverse and ill-posed problems," in *Advances in Electronics and Electron Physics* (P. W. Hawkes, ed.), pp. 1–120, New York: Academic Press, 1989.

53. D. N. G. Roy, *Methods of Inverse Problems in Physics*.   Boca Raton, Fla: CRC Press, second ed., 1991.

54. R. W. Deming and A. J. Devaney, "A filtered backpropagation algorithm for GPR," *J. Env. Eng. Geo.*, vol. 0, pp. 113–123, 1996.

55. R. P. Porter and A. J. Devaney, "Generalized holography and computational solutions to inverse source problems," *J. Opt. Soc. Am.*, vol. 72, pp. 1707–1713, 1982.

56. A. J. Devaney and R. P. Porter, "Holography and the inverse source problem. Part II: inhomogeneous media," *J. Opt. Soc. Am. A*, vol. 2, pp. 2006–2011, 1985.

Chapter 9
Free Fields in Temporally Dispersive Media

There is nothing in the world except empty curved space. Matter,
charge, electromagnetism, and other fields are only
manifestations of the curvature of space. John Wheeler (1957).

If there are no sources of an electromagnetic field present anywhere in space during a period of time, then that field is said to be a *free-field* during that time. The detailed properties of such free-fields were first studied in detail by Sherman, Devaney and Mandel [1], Sherman, Stamnes, Devaney and Lalor [2], and Devaney and Sherman [3] in the early 1970s. Such fields are of interest because they form the simplest type of wave phenomena encountered in both electromagnetics and optics.[1]

Without any loss of generality, attention is restricted here to the field behavior for all nonnegative time $t \geq 0$ during which it is assumed that there are no sources for the field anywhere in all of space; viz.,

$$\varrho(\mathbf{r}, t \geq 0) = 0, \tag{9.1}$$

$$\mathbf{J}(\mathbf{r}, t \geq 0) = \mathbf{0}. \tag{9.2}$$

Any sources that produced the field were nonvanishing only during negative times $t < 0$. It is unnecessary to know what these sources were provided that the initial values

$$\mathbf{D}(\mathbf{r}, 0) = \mathbf{D}_0(\mathbf{r}), \tag{9.3}$$

$$\mathbf{B}(\mathbf{r}, 0) = \mathbf{B}_0(\mathbf{r}), \tag{9.4}$$

---

[1]It is somewhat ironic that this is where I began my studies into this topic while a graduate student at The Institute of Optics at The University of Rochester under the free-range guidance of my research advisor Prof. George C. Sherman, inspired by the unbridled enthusiasm of Tony Devaney, the Nordic wisdom of Jakob Stamnes, and the guiding-star direction of Prof. Emil Wolf. Much of the development presented in this chapter has been taken, with permission, from George Sherman's Advanced Physical Optics class notes when extended to temporally dispersive HILL media.

© Springer Nature Switzerland AG 2019
K. E. Oughstun, *Electromagnetic and Optical Pulse Propagation*, Springer Series in Optical Sciences 224, https://doi.org/10.1007/978-3-030-20835-6_9

for the electric displacement and magnetic induction field vectors are known, where $\mathbf{D}_0(\mathbf{r})$ and $\mathbf{B}_0(\mathbf{r})$ as well as their first time derivatives $\dot{\mathbf{D}}_0(\mathbf{r})$ and $\dot{\mathbf{B}}_0(\mathbf{r})$ are prescribed vector functions of position $\mathbf{r} \in \mathcal{R}^3$. That the initial values for the magnetic intensity vector $\mathbf{H}(\mathbf{r}, t)$ and the conduction current density vector $\mathbf{J}_c(\mathbf{r}, t)$ are not needed is a consequence of the fact that these initial values are not all independent, as is now shown.

9.1 Laplace–Fourier Representation of the Free Field

The electromagnetic field in a homogeneous, isotropic, locally linear, temporally dispersive medium occupying all of space with no externally supplied charge or current sources is described by the set of *source-free Maxwell's equations*

$$\nabla \cdot \mathbf{E}(\mathbf{r}, t) = 0, \tag{9.5}$$

$$\nabla \times \mathbf{E}(\mathbf{r}, t) = - \left\| \frac{1}{c} \right\| \frac{\partial \mathbf{B}(\mathbf{r}, t)}{\partial t}, \tag{9.6}$$

$$\nabla \cdot \mathbf{B}(\mathbf{r}, t) = 0, \tag{9.7}$$

$$\nabla \times \mathbf{H}(\mathbf{r}, t) = \left\| \frac{1}{c} \right\| \frac{\partial \mathbf{D}(\mathbf{r}, t)}{\partial t} + \left\| \frac{4\pi}{c} \right\| \mathbf{J}_c(\mathbf{r}, t), \tag{9.8}$$

with the constitutive (or material) relations

$$\mathbf{D}(\mathbf{r}, t) = \mathbf{D}_0(\mathbf{r}) + \int_0^t \hat{\epsilon}(t - t') \mathbf{E}(\mathbf{r}, t') dt', \tag{9.9}$$

$$\mathbf{H}(\mathbf{r}, t) = \mathbf{H}_0(\mathbf{r}) + \int_0^t \hat{\mu}^{-1}(t - t') \mathbf{B}(\mathbf{r}, t') dt', \tag{9.10}$$

$$\mathbf{J}_c(\mathbf{r}, t) = \mathbf{J}_{c0}(\mathbf{r}) + \int_0^t \hat{\sigma}(t - t') \mathbf{E}(\mathbf{r}, t') dt'. \tag{9.11}$$

With use of the Laplace transform relation [see Eq. (C.13) in Appendix C]

$$\mathcal{L}\left\{\dot{f}(\mathbf{r}, t)\right\} = -f(\mathbf{r}, 0) - i\omega\mathcal{L}\left\{f(\mathbf{r}, t)\right\}, \tag{9.12}$$

the temporal Laplace transform of the source-free Maxwell's equations given in Eqs. (9.5)–(9.8) yields the set of temporal frequency domain relations

$$\nabla \cdot \tilde{\mathbf{E}}(\mathbf{r}, \omega) = 0, \tag{9.13}$$

$$\nabla \times \tilde{\mathbf{E}}(\mathbf{r}, \omega) = \left\| \frac{1}{c} \right\| \left(i\omega\tilde{\mathbf{B}}(\mathbf{r}, \omega) + \mathbf{B}_0(\mathbf{r}) \right), \tag{9.14}$$

$$\nabla \cdot \tilde{\mathbf{B}}(\mathbf{r}, \omega) = 0, \tag{9.15}$$

$$\nabla \times \tilde{\mathbf{H}}(\mathbf{r}, \omega) = -\left\|\frac{1}{c}\right\| \left(i\omega \tilde{\mathbf{D}}(\mathbf{r}, \omega) + \mathbf{D}_0(\mathbf{r}) \right) + \left\|\frac{4\pi}{c}\right\| \tilde{\mathbf{J}}_c(\mathbf{r}, \omega), \tag{9.16}$$

where

$$\tilde{\mathbf{H}}(\mathbf{r}, \omega) = \int_0^\infty \mathbf{H}(\mathbf{r}, t) e^{i\omega t} dt$$

$$= \frac{i}{\omega} \mathbf{H}_0(\mathbf{r}) + \mu^{-1}(\omega) \tilde{\mathbf{B}}(\mathbf{r}, \omega), \tag{9.17}$$

$$\tilde{\mathbf{D}}(\mathbf{r}, \omega) = \int_0^\infty \mathbf{D}(\mathbf{r}, t) e^{i\omega t} dt$$

$$= \frac{i}{\omega} \mathbf{D}_0(\mathbf{r}) + \epsilon(\omega) \tilde{\mathbf{E}}(\mathbf{r}, \omega), \tag{9.18}$$

$$\tilde{\mathbf{J}}_c(\mathbf{r}, \omega) = \int_0^\infty \mathbf{J}_c(\mathbf{r}, t) e^{i\omega t} dt$$

$$= \frac{i}{\omega} \mathbf{J}_{c0}(\mathbf{r}) + \sigma(\omega) \tilde{\mathbf{J}}_c(\mathbf{r}, \omega), \tag{9.19}$$

with substitution from Eqs. (9.9)–(9.11), respectively. Substitution of these expressions in Eq. (9.16) then gives

$$\nabla \times \tilde{\mathbf{B}}(\mathbf{r}, \omega) = -\frac{\|c\|}{v^2(\omega)} i\omega \tilde{\mathbf{E}}(\mathbf{r}, \omega) - \frac{i}{\omega} \mu(\omega) \left(\nabla \times \mathbf{H}_0(\mathbf{r}) - \left\|\frac{4\pi}{c}\right\| \mathbf{J}_{c0}(\mathbf{r}) \right), \tag{9.20}$$

where

$$v^2(\omega) \equiv \frac{\|c^2\|}{\mu(\omega)\epsilon_c(\omega)} \tag{9.21}$$

is the square of the complex velocity [cf. Eq. (7.30)]. Here

$$\epsilon_c(\omega) \equiv \epsilon(\omega) + i\|4\pi\| \frac{\sigma(\omega)}{\omega} \tag{9.22}$$

is the complex permittivity of the dispersive medium [cf. Eq. (5.28)]. The initial values $\mathbf{H}_0(\mathbf{r})$, $\mathbf{D}_0(\mathbf{r})$, and $\mathbf{J}_{c0}(\mathbf{r})$ are related through Ampère's law (9.8) at time $t = 0$ as

$$\nabla \times \mathbf{H}_0(\mathbf{r}) - \left\|\frac{4\pi}{c}\right\| \mathbf{J}_{c0}(\mathbf{r}) = \left\|\frac{1}{c}\right\| \dot{\mathbf{D}}_0(\mathbf{r}), \tag{9.23}$$

where

$$\dot{\mathbf{D}}_0(\mathbf{r}) \equiv \frac{\partial \mathbf{D}(\mathbf{r}, t)}{\partial t}\bigg|_{t=0}$$

$$= \frac{1}{2\pi} \int_C (-i\omega)\epsilon(\omega)\tilde{\mathbf{E}}(\mathbf{r}, \omega)d\omega. \qquad (9.24)$$

With these results, Eq. (9.20) becomes

$$\nabla \times \tilde{\mathbf{B}}(\mathbf{r}, \omega) = -\frac{\|c\|}{v^2(\omega)} \left(i\omega\tilde{\mathbf{E}}(\mathbf{r}, \omega) + \frac{i}{\omega\epsilon_c(\omega)}\dot{\mathbf{D}}_0(\mathbf{r}) \right), \qquad (9.25)$$

which directly reduces to the form obtained when the medium is nonconducting and nondispersive.[2]

The three-dimensional spatial Fourier transform of the set of temporal frequency domain relations given in Eqs. (9.13)–(9.15) and (9.25) results in the set of spatio-temporal frequency domain equations [see Appendix E]

$$\mathbf{k} \cdot \tilde{\tilde{\mathbf{E}}}(\mathbf{k}, \omega) = 0, \qquad (9.26)$$

$$\mathbf{k} \times \tilde{\tilde{\mathbf{E}}}(\mathbf{k}, \omega) = \left\|\frac{1}{c}\right\| \left(\omega\tilde{\tilde{\mathbf{B}}}(\mathbf{k}, \omega) - i\tilde{\mathbf{B}}_0(\mathbf{k}) \right), \qquad (9.27)$$

$$\mathbf{k} \cdot \tilde{\tilde{\mathbf{B}}}(\mathbf{k}, \omega) = 0, \qquad (9.28)$$

$$\mathbf{k} \times \tilde{\tilde{\mathbf{B}}}(\mathbf{k}, \omega) = -\frac{\|c\|}{v^2(\omega)} \left(\omega\tilde{\tilde{\mathbf{E}}}(\mathbf{k}, \omega) + \frac{1}{\omega\epsilon_c(\omega)}\tilde{\mathbf{D}}_0(\mathbf{k}) \right). \qquad (9.29)$$

The set of partial differential equations given in Eqs. (9.5)–(9.8) with the constitutive relations given in Eqs. (9.9)–(9.11) have thus been replaced by the set of algebraic vector relations given in Eqs. (9.26)–(9.29) which are now directly solved for the Laplace–Fourier spectra $\tilde{\tilde{\mathbf{E}}}(\mathbf{k}, \omega)$ and $\tilde{\tilde{\mathbf{B}}}(\mathbf{k}, \omega)$ of the electromagnetic field vectors in terms of the initial values of these field vectors and the appropriate medium response.

The vector product of $\mathbf{k}$ with the relation in Eq. (9.27) gives, after use of the vector identity $\mathbf{k} \times \left(\mathbf{k} \times \tilde{\tilde{\mathbf{E}}}(\mathbf{k}, \omega) \right) = \left(\mathbf{k} \cdot \tilde{\tilde{\mathbf{E}}}(\mathbf{k}, \omega) \right)\mathbf{k} - k^2\tilde{\tilde{\mathbf{E}}}(\mathbf{k}, \omega)$ for the vector triple product and the orthogonality relation $\mathbf{k} \cdot \tilde{\tilde{\mathbf{E}}}(\mathbf{k}, \omega) = 0$ given in Eq. (9.26),

$$-k^2\tilde{\tilde{\mathbf{E}}}(\mathbf{k}, \omega) = \left\|\frac{1}{c}\right\| \left(\omega\mathbf{k} \times \tilde{\tilde{\mathbf{B}}}(\mathbf{k}, \omega) - i\mathbf{k} \times \tilde{\mathbf{B}}_0(\mathbf{k}) \right),$$

---

[2]In the limiting, idealized case of a nonconducting, nondispersive medium, $\epsilon_c(\omega) = \epsilon$, $v = \|c\|/n_r$ where $n_r = \sqrt{\epsilon\mu}$ is the real-valued index of refraction of the homogeneous, isotropic, locally linear medium, and $\dot{\mathbf{D}}_0(\mathbf{r}) = \epsilon\dot{\mathbf{E}}_0(\mathbf{r}) = -i\omega\epsilon\mathbf{E}_0(\mathbf{r})$.

where $k^2 \equiv \mathbf{k} \cdot \mathbf{k}$. With substitution from Eq. (9.29), the above expression becomes

$$-k^2 \tilde{\tilde{\mathbf{E}}}(\mathbf{k}, \omega) = -\frac{1}{v^2(\omega)} \left(\omega^2 \tilde{\tilde{\mathbf{E}}}(\mathbf{k}, \omega) + \frac{1}{\epsilon_c(\omega)} \tilde{\mathbf{D}}_0(\mathbf{k}) \right) - \frac{i}{\|c\|} \mathbf{k} \times \tilde{\mathbf{B}}_0(\mathbf{k}),$$

with solution

$$\tilde{\tilde{\mathbf{E}}}(\mathbf{k}, \omega) = i \frac{i \tilde{\mathbf{D}}_0(\mathbf{k})/\epsilon_c(\omega) - \left(v^2(\omega)/\|c\|\right) \mathbf{k} \times \tilde{\mathbf{B}}_0(\mathbf{k})}{\omega^2 - k^2 v^2(\omega)}. \tag{9.30}$$

In an analogous manner for the magnetic induction field vector, the vector product of $\mathbf{k}$ with the relation in Eq. (9.29) gives, after use of the vector identity $\mathbf{k} \times \left(\mathbf{k} \times \tilde{\tilde{\mathbf{B}}}(\mathbf{k}, \omega) \right) = \left(\mathbf{k} \cdot \tilde{\tilde{\mathbf{B}}}(\mathbf{k}, \omega) \right) \mathbf{k} - k^2 \tilde{\tilde{\mathbf{B}}}(\mathbf{k}, \omega)$ for the vector triple product and the orthogonality relation $\mathbf{k} \cdot \tilde{\tilde{\mathbf{B}}}(\mathbf{k}, \omega) = 0$ given in Eq. (9.28),

$$-k^2 \tilde{\tilde{\mathbf{B}}}(\mathbf{k}, \omega) = -\frac{\|c\|}{v^2(\omega)} \left(\omega \mathbf{k} \times \tilde{\tilde{\mathbf{E}}}(\mathbf{k}, \omega) + \frac{1}{\omega \epsilon_c(\omega)} \mathbf{k} \times \tilde{\mathbf{D}}_0(\mathbf{k}) \right).$$

With substitution from Eq. (9.27), the above expression becomes

$$-k^2 \tilde{\tilde{\mathbf{B}}}(\mathbf{k}, \omega) = -\frac{\omega}{v^2(\omega)} \left(\omega \tilde{\tilde{\mathbf{B}}}(\mathbf{k}, \omega) - i \tilde{\mathbf{B}}_0(\mathbf{k}) \right) - \frac{\|c\|}{\omega \epsilon_c(\omega) v^2(\omega)} \mathbf{k} \times \tilde{\mathbf{D}}_0(\mathbf{k}),$$

with solution

$$\tilde{\tilde{\mathbf{B}}}(\mathbf{k}, \omega) = i \frac{\omega \tilde{\mathbf{B}}_0(\mathbf{k}) + i \left(\|c\|/\omega \epsilon_c(\omega) \right) \mathbf{k} \times \tilde{\mathbf{D}}_0(\mathbf{k})}{\omega^2 - k^2 v^2(\omega)}. \tag{9.31}$$

The inverse Laplace–Fourier transform of the spatiotemporal spectral field vectors given in Eqs. (9.30) and (9.31) then yields the *Laplace–Fourier integral representation of the free-field*

$$\mathbf{E}(\mathbf{r}, t) = \frac{i}{(2\pi)^4} \int_C d\omega \int_{-\infty}^{\infty} d^3k \frac{\frac{i}{\epsilon_c(\omega)} \tilde{\mathbf{D}}_0(\mathbf{k}) - \frac{v^2(\omega)}{\|c\|} \mathbf{k} \times \tilde{\mathbf{B}}_0(\mathbf{k})}{\omega^2 - k^2 v^2(\omega)} e^{i(\mathbf{k} \cdot \mathbf{r} - \omega t)},$$

$$\tag{9.32}$$

$$\mathbf{B}(\mathbf{r}, t) = \frac{i}{(2\pi)^4} \int_C d\omega \int_{-\infty}^{\infty} d^3k \frac{\omega \tilde{\mathbf{B}}_0(\mathbf{k}) + i \frac{\|c\|}{\omega \epsilon_c(\omega)} \mathbf{k} \times \tilde{\mathbf{D}}_0(\mathbf{k})}{\omega^2 - k^2 v^2(\omega)} e^{i(\mathbf{k} \cdot \mathbf{r} - \omega t)},$$

$$\tag{9.33}$$

valid for all $t \geq 0$.

Although it appears that the initial value problem for the source-free field has been solved and that Eqs. (9.32) and (9.33) represent that solution, that is not yet the case. Specifically, it is not yet known as to whether this Laplace–Fourier integral representation forms a solution. All that is known is that *if* there is a solution with the properties required to ensure that the above analysis is valid (e.g., if the field vectors and initial values are square-integrable), then that solution must then be given by Eqs. (9.32) and (9.33), and hence, there is only one such solution. All that remains to be proven is the existence of the solution.

9.1.1 Plane Wave Expansion of the Free Field in a Nondispersive Nonconducting Medium

In order to determine whether the relations given in Eqs. (9.32) and (9.33) represent the solution to the initial value problem stated in Eqs. (9.1)–(9.4) as well as to gain further insight into the properties of such free fields, the special case of a nondispersive nonconducting medium is considered, for then the ω-integrals in Eqs. (9.32) and (9.33) can be directly evaluated. For a nondispersive nonconducting medium, Eqs. (9.32) and (9.33) become

$$\mathbf{E}(\mathbf{r}, t) = \frac{i}{(2\pi)^4} \int_C d\omega \int_{-\infty}^{\infty} d^3k \frac{\omega \tilde{\mathbf{E}}_0(\mathbf{k}) - \frac{v^2}{\|c\|} \mathbf{k} \times \tilde{\mathbf{B}}_0(\mathbf{k})}{\omega^2 - k^2 v^2} e^{i(\mathbf{k}\cdot\mathbf{r} - \omega t)},$$

(9.34)

$$\mathbf{B}(\mathbf{r}, t) = \frac{i}{(2\pi)^4} \int_C d\omega \int_{-\infty}^{\infty} d^3k \frac{\omega \tilde{\mathbf{B}}_0(\mathbf{k}) + \|c\| \mathbf{k} \times \tilde{\mathbf{E}}_0(\mathbf{k})}{\omega^2 - k^2 v^2} e^{i(\mathbf{k}\cdot\mathbf{r} - \omega t)},$$

(9.35)

for all $t \geq 0$. Attention is then given to integrals of the form

$$I_{C_j} \equiv \int_{C_j} \frac{\alpha \omega + \beta}{\omega^2 - v^2 k^2} e^{-i\omega t} d\omega$$

(9.36)

for $j = 1, 2, 3, 4$ for positive t, where $v = c/n_r$ is the real phase velocity with n_r denoting the real index of refraction of the nondispersive nonconducting medium. Here α and β are functions that are both independent of ω, k is fixed, and C_j are the separate portions of the closed contour C in the complex ω-plane depicted in Fig. 9.1: C_1 is the straight line contour parallel to the ω'-axis and extending from $-R + ia$ to $R + ia$ in the upper-half plane, C_2 is the straight line segment from $R + ia$ to R, C_3 is the semicircular arc $\omega = Re^{i\theta}$ with θ varying from 0 to $-\pi$ in the clockwise direction, and C_4 is the straight line segment from $-R$ to $-R + ia$. The integral about the entire closed contour C' is then given by the sum of the

Fig. 9.1 Contour of
integration
$C' = C_1 \cup C_2 \cup C_3 \cup C_4$ in
the complex ω-plane

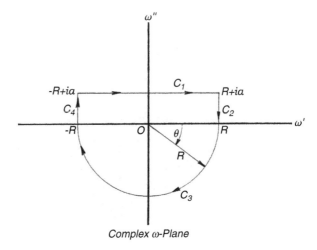

Complex ω-Plane

integrals over each of the contours C_j for $j = 1, 2, 3, 4$, so that

$$I_{C'} = \sum_{j=1}^{4} I_{C_j}, \tag{9.37}$$

where $I_{C'}$ may also be evaluated by application of the residue theorem. If the
integrals over C_2, C_3, and C_4 vanish in the limit as $R \rightarrow \infty$, then $I_{C_1} = I_{C'}$
with $C_1 \rightarrow C$ in that limit, where C is the ω integration contour appearing in the
integral representation given in Eqs. (9.34) and (9.35).

On each contour C_j, the magnitude of the integral I_{C_j} is bounded in the manner

$$|I_{C_j}| \leq \int_{C_j} \left| \frac{\alpha\omega + \beta}{\omega^2 - v^2 k^2} \right| \left| e^{-i(\omega' + i\omega'')t} \right| |d\omega| = \int_{C_j} \left| \frac{\alpha\omega + \beta}{\omega^2 - v^2 k^2} \right| e^{\omega'' t} |d\omega| \,,$$

where

$$\left| \frac{\alpha\omega + \beta}{\omega^2 - v^2 k^2} \right| = \left[\frac{|\alpha|^2 |\omega|^2 + |\beta|^2 + 2\Re\{\alpha\beta\omega\}}{|\omega^2|^2 + v^4 k^4 - 2v^2 k^2 \Re\{\omega^2\}} \right]^{1/2}$$

$$= \left[\frac{|\alpha|^2 \left(\omega'^2 + \omega''^2 \right) + |\beta|^2 + 2\Re\left\{\alpha\beta(\omega' + i\omega'')\right\}}{\omega'^4 + \omega''^4 + 2\omega'^2 \omega''^2 + v^4 k^4 - 2v^2 k^2 \left(\omega'^2 - \omega''^2 \right)} \right]^{1/2} \,. \tag{9.38}$$

Consider first the contour integrals over the vertical line segments C_2 and C_4. On
either of these contours, $0 \leq \omega'' \leq a$ and $\omega' = \pm R$, where the upper sign choice is

taken for C_2 and the lower sign choice for C_1, so that

$$e^{\omega'' t} \le e^{at},$$

and [from Eq. (9.38)]

$$\left| \frac{\alpha\omega + \beta}{\omega^2 - v^2 k^2} \right| = \left[\frac{|\alpha|^2 \left(R^2 + \omega''^2 \right) + |\beta|^2 + 2\Re \left\{ \alpha\beta(R + i\omega'') \right\}}{R^4 + \omega''^4 + 2R^2\omega''^2 + v^4 k^4 - 2v^2 k^2 \left(R^2 - \omega''^2 \right)} \right]^{1/2}$$

$$= \left[\frac{\mathcal{O}(R^2)}{\mathcal{O}(R^4)} \right]^{1/2} = \mathcal{O}\left(\frac{1}{R} \right)$$

as $R \to \infty$. Consequently, as R becomes infinitely large

$$\left| I_{C_{2,4}} \right| \le \int_0^a \left| \frac{\alpha\omega + \beta}{\omega^2 - v^2 k^2} \right| e^{\omega'' t} d\omega''$$

$$\le \frac{M}{R} e^{at} \int_0^a d\omega'' = \frac{Ma}{R} e^{at} \to 0 \tag{9.39}$$

as $R \to \infty$ for all $t \ge 0$.

Consider next the integral over the semicircular contour C_3, along which $\omega = Re^{-i\theta} = R(\cos\theta - i\sin\theta)$ with $\theta = 0 \to \pi$, so that $d\omega = -iRe^{-i\theta}d\theta$ and $|d\omega| = Rd\theta$. For $0 \le \theta \le \pi/2$, the inequality $2\theta/\pi \le \sin\theta$ applies, so that

$$e^{\omega'' t} = e^{-Rt\sin\theta} \le e^{-(2/\pi)Rt\theta}.$$

In addition, from Eq. (9.38)

$$\left| \frac{\alpha\omega + \beta}{\omega^2 - v^2 k^2} \right| = \left[\frac{|\alpha|^2 R^2 + |\beta|^2 + 2\Re \left\{ \alpha\beta R(\cos\theta - i\sin\theta) \right\}}{R^4 + v^4 k^4 - 2v^2 k^2 R^2(\cos^2\theta - \sin^2\theta)} \right]^{1/2}$$

$$= \left[\frac{\mathcal{O}(R^2)}{\mathcal{O}(R^4)} \right]^{1/2} = \mathcal{O}\left(\frac{1}{R} \right)$$

as $R \to \infty$. Consequently, as R becomes infinitely large

$$\left| I_{C_3} \right| \le \int_0^\pi \left| \frac{\alpha\omega + \beta}{\omega^2 - v^2 k^2} \right| e^{-Rt\sin\theta} Rd\theta$$

$$\le 2\frac{M}{R} \int_0^{\pi/2} e^{-Rt\sin\theta} Rd\theta$$

$$\le 2M \int_0^{\pi/2} e^{-(2/\pi)Rt\theta} d\theta = \frac{M\pi}{Rt} \left(1 - e^{-Rt} \right) \to 0$$

as $R \to \infty$ for all $t > 0$. For $t = 0$ it is noted that the integral expressions given in Eqs. (9.34) and (9.35) actually represent the quantities $U(t)\mathbf{E}(\mathbf{r}, t)$ and $U(t)\mathbf{B}(\mathbf{r}, t)$, where $U(t)$ denotes the Heaviside unit step function. Because these quantities are discontinuous at $t = 0$, one must take the limit as $t \to 0^+$. Because $|I_{C_3}| \to 0$ as $R \to \infty$ for all $t > 0$, then $|I_{C_3}| \to 0$ as $R \to \infty$ and $t \to 0^+$. Thus

$$|I_{C_3}| \to 0 \tag{9.40}$$

as $R \to \infty$ for all $t \geq 0$.

With Eq. (9.37), these limiting results then show that

$$\lim_{R \to \infty} I_{C_1} = \lim_{R \to \infty} \oint_{C'} \frac{\alpha \omega + \beta}{\omega^2 - v^2 k^2} e^{-i\omega t} d\omega, \tag{9.41}$$

where $C' = C_1 \cup C_2 \cup C_3 \cup C_4$ is the entire closed contour depicted in Fig. 9.1. This contour integral over C' is now evaluated using the residue theorem. The integrand has two simple poles at $\omega = \pm vk$ with residues

$$\text{Residue} \atop (\omega = +vk) = \frac{\alpha vk + \beta}{2vk} e^{-ivkt},$$

$$\text{Residue} \atop (\omega = -vk) = \frac{-\alpha vk + \beta}{-2vk} e^{ivkt} = \frac{\alpha vk - \beta}{-2vk} e^{ivkt}. \tag{9.42}$$

For fixed k, these are the only poles enclosed by the contour C' for all $R > \sqrt{v^2 k^2 + \delta^2}$ for arbitrarily small $\delta > 0$. Application of the residue theorem to the right-hand side of Eq. (9.41) then yields

$$\lim_{R \to \infty} I_{C_1} = -i\pi \left[\left(\alpha + \frac{\beta}{vk} \right) e^{-ivkt} + \left(\alpha - \frac{\beta}{vk} \right) e^{ivkt} \right] \tag{9.43}$$

for all $t \geq 0$. Application of the result given in Eq. (9.43) to the integral expressions given in Eqs. (9.34) and (9.35) then results in

$$\mathbf{E}(\mathbf{r}, t) = \frac{1}{16\pi^3} \int_{-\infty}^{\infty} \left\{ \left[\tilde{\mathbf{E}}_0(\mathbf{k}) - \frac{v}{\|c\|} \hat{\mathbf{s}} \times \tilde{\mathbf{B}}_0(\mathbf{k}) \right] e^{i(\mathbf{k} \cdot \mathbf{r} - vkt)} \right.$$

$$\left. + \left[\tilde{\mathbf{E}}_0(\mathbf{k}) + \frac{v}{\|c\|} \hat{\mathbf{s}} \times \tilde{\mathbf{B}}_0(\mathbf{k}) \right] e^{i(\mathbf{k} \cdot \mathbf{r} + vkt)} \right\} d^3 k, \tag{9.44}$$

$$\mathbf{B}(\mathbf{r},t) = \frac{1}{16\pi^3} \int_{-\infty}^{\infty} \left\{ \left[\tilde{\mathbf{B}}_0(\mathbf{k}) + \frac{\|c\|}{v}\hat{\mathbf{s}} \times \tilde{\mathbf{E}}_0(\mathbf{k}) \right] e^{i(\mathbf{k}\cdot\mathbf{r}-vkt)} \right.$$

$$\left. + \left[\tilde{\mathbf{B}}_0(\mathbf{k}) - \frac{\|c\|}{v}\hat{\mathbf{s}} \times \tilde{\mathbf{E}}_0(\mathbf{k}) \right] e^{i(\mathbf{k}\cdot\mathbf{r}+vkt)} \right\} d^3k,$$

$$(9.45)$$

where $\hat{\mathbf{s}} \equiv \mathbf{k}/k$. Notice that the exponential factor $e^{i(\mathbf{k}\cdot\mathbf{r}-vkt)}$ represents a plane wave traveling in the direction $\hat{\mathbf{s}} = \mathbf{k}/k$, while $e^{i(\mathbf{k}\cdot\mathbf{r}+vkt)}$ represents a plane wave traveling in the opposite direction given by $-\hat{\mathbf{s}}$. Upon combining similar terms in this pair of expressions for the free field, one obtains

$$\mathbf{E}(\mathbf{r},t) = \frac{1}{(2\pi)^3} \int_{-\infty}^{\infty} \left[\tilde{\mathbf{E}}_0(\mathbf{k}) \cos(vkt) + i\frac{v}{\|c\|}\hat{\mathbf{s}} \times \tilde{\mathbf{B}}_0(\mathbf{k}) \sin(vkt) \right] e^{i\mathbf{k}\cdot\mathbf{r}} d^3k,$$

$$(9.46)$$

$$\mathbf{B}(\mathbf{r},t) = \frac{1}{(2\pi)^3} \int_{-\infty}^{\infty} \left[\tilde{\mathbf{B}}_0(\mathbf{k}) \cos(vkt) - i\frac{\|c\|}{v}\hat{\mathbf{s}} \times \tilde{\mathbf{E}}_0(\mathbf{k}) \sin(vkt) \right] e^{i\mathbf{k}\cdot\mathbf{r}} d^3k,$$

$$(9.47)$$

for $t > 0$.

The free field representation given in Eqs. (9.44) and (9.45) may also be simplified by noting that the integrand factors involving e^{ivkt} are the complex conjugates of the corresponding integrand factors involving e^{-ivkt}. For example, let

$$\mathbf{E}_{\pm}(\mathbf{r},t) \equiv \int_{-\infty}^{\infty} \left[\tilde{\mathbf{E}}_0(\mathbf{k}) \pm \frac{v}{\|c\|}\hat{\mathbf{s}} \times \tilde{\mathbf{B}}_0(\mathbf{k}) \right] e^{i(\mathbf{k}\cdot\mathbf{r}\pm vkt)} d^3k,$$

so that

$$\mathbf{E}(\mathbf{r},t) = \frac{1}{2(2\pi)^3}\left[\mathbf{E}_+(\mathbf{r},t) + \mathbf{E}_-(\mathbf{r},t) \right].$$

Because the initial values $\mathbf{E}_0(\mathbf{r})$ and $\mathbf{B}_0(\mathbf{r})$ for the field vectors are both real, then their spatial frequency spectra satisfy the symmetry relations $\tilde{\mathbf{E}}_0^*(\mathbf{k}) = \tilde{\mathbf{E}}_0(-\mathbf{k})$ and $\tilde{\mathbf{B}}_0^*(\mathbf{k}) = \tilde{\mathbf{B}}_0(-\mathbf{k})$ for real-valued $\mathbf{k}$, so that

$$\mathbf{E}_+^*(\mathbf{r},t) = \int_{-\infty}^{\infty} \left[\tilde{\mathbf{E}}_0^*(\mathbf{k}) + \frac{v}{\|c\|}\hat{\mathbf{s}} \times \tilde{\mathbf{B}}_0^*(\mathbf{k}) \right] e^{-i(\mathbf{k}\cdot\mathbf{r}+vkt)} d^3k$$

$$= \int_{-\infty}^{\infty} \left[\tilde{\mathbf{E}}_0(-\mathbf{k}) + \frac{v}{\|c\|}\hat{\mathbf{s}} \times \tilde{\mathbf{B}}_0(-\mathbf{k}) \right] e^{-i(\mathbf{k}\cdot\mathbf{r}+vkt)} d^3k.$$

Under the change of variable $\mathbf{k}' = -\mathbf{k}$ so that $k' = k$ and $\hat{\mathbf{s}}' = -\hat{\mathbf{s}}$, the above expression becomes

$$
\mathbf{E}_+^*(\mathbf{r}, t) = -\int_\infty^{-\infty} \left[\tilde{\mathbf{E}}_0(\mathbf{k}') - \frac{v}{\|c\|} \hat{\mathbf{s}}' \times \tilde{\mathbf{B}}_0(\mathbf{k}') \right] e^{i(\mathbf{k}'\cdot\mathbf{r} - vk't)} d^3 k'
$$

$$
= \int_{-\infty}^{\infty} \left[\tilde{\mathbf{E}}_0(\mathbf{k}') - \frac{v}{\|c\|} \hat{\mathbf{s}}' \times \tilde{\mathbf{B}}_0(\mathbf{k}') \right] e^{i(\mathbf{k}'\cdot\mathbf{r} - vk't)} d^3 k'
$$

$$
= \mathbf{E}_-(\mathbf{r}, t),
$$

so that

$$
\mathbf{E}(\mathbf{r}, t) = = \frac{1}{2(2\pi)^3} \left[\mathbf{E}_-^*(\mathbf{r}, t) + \mathbf{E}_-(\mathbf{r}, t) \right]
$$

$$
= \frac{1}{(2\pi)^3} \Re \{ \mathbf{E}_-(\mathbf{r}, t) \},
$$

with an analogous expression holding for $\mathbf{B}(\mathbf{r}, t)$. The integral representations appearing in Eqs. (9.44) and (9.45) may then be expressed as

$$
\mathbf{E}(\mathbf{r}, t) = \frac{1}{(2\pi)^3} \Re \int_{-\infty}^{\infty} \left[\tilde{\mathbf{E}}_0(\mathbf{k}) - \frac{v}{\|c\|} \hat{\mathbf{s}} \times \tilde{\mathbf{B}}_0(\mathbf{k}) \right] e^{i(\mathbf{k}\cdot\mathbf{r} - vkt)} d^3 k, \quad (9.48)
$$

$$
\mathbf{B}(\mathbf{r}, t) = \frac{1}{(2\pi)^3} \Re \int_{-\infty}^{\infty} \left[\tilde{\mathbf{B}}_0(\mathbf{k}) + \frac{\|c\|}{v} \hat{\mathbf{s}} \times \tilde{\mathbf{E}}_0(\mathbf{k}) \right] e^{i(\mathbf{k}\cdot\mathbf{r} - vkt)} d^3 k, \quad (9.49)
$$

for $t > 0$.

9.1.2 Uniqueness of the Plane Wave Expansion of the Initial Value Problem

Assume that $\tilde{\mathbf{E}}_0(\mathbf{k})$ and $\tilde{\mathbf{B}}_0(\mathbf{k})$ have the necessary properties to ensure that the spatial and temporal partial derivatives of the plane wave expansions given in Eqs. (9.48) and (9.49) for the free field vectors $\mathbf{E}(\mathbf{r}, t)$ and $\mathbf{B}(\mathbf{r}, t)$ can be obtained by differentiating the integrands in these expressions. The divergence of Eq. (9.48) then gives

$$
\nabla \cdot \mathbf{E}(\mathbf{r}, t) = \frac{1}{(2\pi)^3} \Re \int_{-\infty}^{\infty} i\mathbf{k} \cdot \left[\tilde{\mathbf{E}}_0(\mathbf{k}) - \frac{v}{\|c\|} \hat{\mathbf{s}} \times \tilde{\mathbf{B}}_0(\mathbf{k}) \right] e^{i(\mathbf{k}\cdot\mathbf{r} - vkt)} d^3 k,
$$

$$
= \frac{1}{(2\pi)^3} \Re \left\{ i \int_{-\infty}^{\infty} \mathbf{k} \cdot \tilde{\mathbf{E}}_0(\mathbf{k}) e^{i(\mathbf{k}\cdot\mathbf{r} - vkt)} d^3 k \right\}. \quad (9.50)
$$

Hence, $\nabla \cdot \mathbf{E}(\mathbf{r}, t)$ will vanish in accordance with the Maxwell–Gauss relation (9.5) in a source-free HILL medium if $\mathbf{k} \cdot \tilde{\mathbf{E}}_0(\mathbf{k}) = 0$, which is satisfied if and only if

$$\nabla \cdot \mathbf{E}_0(\mathbf{r}) = 0. \tag{9.51}$$

Therefore, $\mathbf{E}(\mathbf{r}, t)$ will be divergenceless if $\mathbf{E}_0(\mathbf{r})$ is divergenceless. Similarly, the divergence of Eq. (9.49) gives

$$\nabla \cdot \mathbf{B}(\mathbf{r}, t) = \frac{1}{(2\pi)^3} \Re \int_{-\infty}^{\infty} i\mathbf{k} \cdot \left[\tilde{\mathbf{B}}_0(\mathbf{k}) - \frac{v}{\|c\|} \hat{\mathbf{s}} \times \tilde{\mathbf{E}}_0(\mathbf{k}) \right] e^{i(\mathbf{k}\cdot\mathbf{r}-vkt)} d^3k,$$

$$= \frac{1}{(2\pi)^3} \Re \left\{ i \int_{-\infty}^{\infty} \mathbf{k} \cdot \tilde{\mathbf{B}}_0(\mathbf{k}) e^{i(\mathbf{k}\cdot\mathbf{r}-vkt)} d^3k \right\}. \tag{9.52}$$

Hence, $\nabla \cdot \mathbf{B}(\mathbf{r}, t)$ will vanish in accordance with the Maxwell–Gauss relation (9.7) in a source-free nonconducting, nondispersive HILL medium if $\mathbf{k} \cdot \tilde{\mathbf{B}}_0(\mathbf{k}) = 0$, which is satisfied if and only if

$$\nabla \cdot \mathbf{B}_0(\mathbf{r}) = 0. \tag{9.53}$$

Therefore, $\mathbf{B}(\mathbf{r}, t)$ will be divergenceless if $\mathbf{B}_0(\mathbf{r})$ is divergenceless.

The curl of Eq. (9.48) gives

$$\nabla \times \mathbf{E}(\mathbf{r}, t) = \frac{1}{(2\pi)^3} \Re \int_{-\infty}^{\infty} i\mathbf{k} \times \left[\tilde{\mathbf{E}}_0(\mathbf{k}) - \frac{v}{\|c\|} \hat{\mathbf{s}} \times \tilde{\mathbf{B}}_0(\mathbf{k}) \right] e^{i(\mathbf{k}\cdot\mathbf{r}-vkt)} d^3k,$$

$$= \frac{1}{(2\pi)^3} \Re \left\{ i \int_{-\infty}^{\infty} \left[\mathbf{k} \times \tilde{\mathbf{E}}_0(\mathbf{k}) + \frac{v}{\|c\|} k \tilde{\mathbf{B}}_0(\mathbf{k}) \right] e^{i(\mathbf{k}\cdot\mathbf{r}-vkt)} d^3k \right\}, \tag{9.54}$$

because $\mathbf{k} \times \left(\hat{\mathbf{s}} \times \tilde{\mathbf{B}}_0(\mathbf{k}) \right) = \hat{\mathbf{s}} \left(\mathbf{k} \cdot \tilde{\mathbf{B}}_0(\mathbf{k}) \right) - \tilde{\mathbf{B}}_0(\mathbf{k})(\mathbf{k} \cdot \hat{\mathbf{s}}) = -k\tilde{\mathbf{B}}_0(\mathbf{k})$. In addition, the partial derivative of Eq. (9.49) with respect to time gives

$$\frac{\partial \mathbf{B}(\mathbf{r}, t)}{\partial t} = \frac{1}{(2\pi)^3} \Re \int_{-\infty}^{\infty} (-ivk) \left[\tilde{\mathbf{B}}_0(\mathbf{k}) + \frac{\|c\|}{v} \hat{\mathbf{s}} \times \tilde{\mathbf{E}}_0(\mathbf{k}) \right] e^{i(\mathbf{k}\cdot\mathbf{r}-vkt)} d^3k$$

$$= -\|c\| \frac{1}{(2\pi)^3} \Re \left\{ i \int_{-\infty}^{\infty} \left[\mathbf{k} \times \tilde{\mathbf{E}}_0(\mathbf{k}) + \frac{v}{\|c\|} k \tilde{\mathbf{B}}_0(\mathbf{k}) \right] e^{i(\mathbf{k}\cdot\mathbf{r}-vkt)} d^3k \right\}$$

$$= -\|c\| \nabla \times \mathbf{E}(\mathbf{r}, t), \tag{9.55}$$

after comparison with Eq. (9.54). Hence, the Maxwell–Faraday relation (9.6) is satisfied if the initial value $\mathbf{B}_0(\mathbf{r})$ satisfies the relation given in Eq. (9.53). Similarly, the curl of Eq. (9.49) gives

$$\nabla \times \mathbf{B}(\mathbf{r}, t) = \frac{1}{(2\pi)^3} \Re \int_{-\infty}^{\infty} i\mathbf{k} \times \left[\tilde{\mathbf{B}}_0(\mathbf{k}) + \frac{\|c\|}{v} \hat{\mathbf{s}} \times \tilde{\mathbf{E}}_0(\mathbf{k}) \right] e^{i(\mathbf{k}\cdot\mathbf{r} - vkt)} d^3k$$

$$= \frac{1}{(2\pi)^3} \Re \left\{ i \int_{-\infty}^{\infty} \left[\mathbf{k} \times \tilde{\mathbf{B}}_0(\mathbf{k}) - \frac{\|c\|}{v} k \tilde{\mathbf{E}}_0(\mathbf{k}) \right] e^{i(\mathbf{k}\cdot\mathbf{r} - vkt)} d^3k \right\},$$

(9.56)

because $\mathbf{k} \times \left(\hat{\mathbf{s}} \times \tilde{\mathbf{E}}_0(\mathbf{k}) \right) = \hat{\mathbf{s}} \left(\mathbf{k} \cdot \tilde{\mathbf{E}}_0(\mathbf{k}) \right) - \tilde{\mathbf{E}}_0(\mathbf{k}) \left(\mathbf{k} \cdot \hat{\mathbf{s}} \right) = -k \tilde{\mathbf{E}}_0(\mathbf{k})$. In addition, the partial derivative of Eq. (9.48) with respect to time gives

$$\frac{\partial \mathbf{E}(\mathbf{r}, t)}{\partial t} = \frac{1}{(2\pi)^3} \Re \int_{-\infty}^{\infty} (-ivk) \left[\tilde{\mathbf{E}}_0(\mathbf{k}) - \frac{v}{\|c\|} \hat{\mathbf{s}} \times \tilde{\mathbf{B}}_0(\mathbf{k}) \right] e^{i(\mathbf{k}\cdot\mathbf{r} - vkt)} d^3k$$

$$= \frac{v^2}{\|c\|} \frac{1}{(2\pi)^3} \Re \left\{ \int_{-\infty}^{\infty} \left[\mathbf{k} \times \tilde{\mathbf{B}}_0(\mathbf{k}) - \frac{\|c\|}{v} k \tilde{\mathbf{E}}_0(\mathbf{k}) \right] e^{i(\mathbf{k}\cdot\mathbf{r} - vkt)} d^3k \right\}$$

$$= \frac{\|c\|}{\epsilon \mu} \nabla \times \mathbf{B}(\mathbf{r}, t) \tag{9.57}$$

after comparison with Eq. (9.56), where $v = \|c\| / \sqrt{\epsilon \mu}$. Hence, the Maxwell–Ampère relation (9.8) in a source-free nonconducting, nondispersive HILL medium is satisfied if the initial value $\mathbf{E}_0(\mathbf{r})$ satisfies the relation given in Eq. (9.51).

Maxwell's equations in a homogeneous, isotropic, locally linear (HILL) non-conducting, nondispersive medium with no charge or current sources are therefore satisfied by the plane wave expansions given in Eqs. (9.48) and (9.49), as well as in Eqs. (9.44)–(9.45) and Eqs. (9.46)–(9.47), provided that the initial values $\mathbf{E}_0(\mathbf{r})$ and $\mathbf{B}_0(\mathbf{r})$ are solenoidal, satisfying the relations given in Eqs. (9.51) and (9.53), respectively. That is, the expressions given in Eqs. (9.48) and (9.49) for $\mathbf{E}(\mathbf{r}, t)$ and $\mathbf{B}(\mathbf{r}, t)$, respectively, represent a solution to the initial value problem if the initial values $\mathbf{E}_0(\mathbf{r})$ and $\mathbf{B}_0(\mathbf{r})$ are both divergenceless. It now remains to be determined whether these expressions for $\mathbf{E}(\mathbf{r}, t)$ and $\mathbf{B}(\mathbf{r}, t)$ satisfy the initial values. Because the integral expressions given in Eqs. (9.48) and (9.49), as well as in Eqs. (9.44)–(9.45) and (9.46)–(9.47), actually represent the vector field quantitites $U(t)\mathbf{E}(\mathbf{r}, t)$ and $U(t)\mathbf{B}(\mathbf{r}, t)$, where $U(t)$ is the Heaviside unit step function, they are then discontinuous at $t = 0$. For that reason, one must consider their limiting behavior as $t \rightarrow 0^+$. In order to determine the limiting value for the electric field vector, construct the difference vector

$$\mathbf{I}_E(\mathbf{r}, t) \equiv (2\pi)^3 \left(\mathbf{E}(\mathbf{r}, t) - \mathbf{E}_0(\tilde{\mathbf{r}}) \right).$$

With substitution from Eq. (9.46) and the definition $\tilde{\mathbf{E}}_0(\mathbf{k}) = \mathcal{F}\{\mathbf{E}_0(\mathbf{r})\}$, one finds that

$$
|\mathbf{I}_E(\mathbf{r}, t)| = \left| \int_{-\infty}^{\infty} \left[\tilde{\mathbf{E}}_0(\mathbf{k}) \cos{(vkt)} + i \frac{v}{\|c\|} \hat{\mathbf{s}} \times \tilde{\mathbf{B}}_0(\mathbf{k}) \sin{(vkt)} \right] e^{i\mathbf{k}\cdot\mathbf{r}} d^3 k \right.
$$

$$
\left. - \int_{-\infty}^{\infty} \tilde{\mathbf{E}}_0(\mathbf{k}) e^{i\mathbf{k}\cdot\mathbf{r}} d^3 k \right|
$$

$$
= \left| \int_{-\infty}^{\infty} \left[\tilde{\mathbf{E}}_0(\mathbf{k})\big(\cos{(vkt)} - 1\big) + i \frac{v}{\|c\|} \hat{\mathbf{s}} \times \tilde{\mathbf{B}}_0(\mathbf{k}) \sin{(vkt)} \right] e^{i\mathbf{k}\cdot\mathbf{r}} d^3 k \right|
$$

$$
\leq \int_{-\infty}^{\infty} \left\{ \left| \tilde{\mathbf{E}}_0(\mathbf{k}) \right| |\cos{(vkt)} - 1| + \frac{v}{\|c\|} \left| \hat{\mathbf{s}} \times \tilde{\mathbf{B}}_0(\mathbf{k}) \right| |\sin{(vkt)}| \right\} \left| d^3 k \right|.
$$

The right-hand side of the final inequality in the above expression can be made as small as desired by choosing $t > 0$ sufficiently small. Hence, $\mathbf{I}_E(\mathbf{r}, t)$ tends to $\mathbf{0}$ as $t \to 0^+$, and consequently

$$
\lim_{t \to 0^+} \mathbf{E}(\mathbf{r}, t) = \mathbf{E}_0(\mathbf{r}) \tag{9.58}
$$

when $\mathbf{E}(\mathbf{r}, t)$ is given by any of the expressions appearing in Eqs. (9.32), (9.44), (9.46), or (9.48). In a similar fashion, the difference vector

$$
\mathbf{I}_B(\mathbf{r}, t) \equiv (2\pi)^3 \big(\mathbf{B}(\mathbf{r}, t) - \mathbf{B}_0(\mathbf{r})\big)
$$

is found to satisfy the inequality

$$
|\mathbf{I}_B(\mathbf{r}, t)| \leq \int_{-\infty}^{\infty} \left\{ \left| \tilde{\mathbf{B}}_0(\mathbf{k}) \right| |\cos{(vkt)} - 1| - \frac{\|c\|}{v} \left| \hat{\mathbf{s}} \times \tilde{\mathbf{E}}_0(\mathbf{k}) \right| |\sin{(vkt)}| \right\} \left| d^3 k \right|,
$$

where the right-hand side can be made as small as desired by taking $t > 0$ sufficiently small. Hence, $\mathbf{I}_B(\mathbf{r}, t)$ tends to $\mathbf{0}$ as $t \to 0^+$, and consequently

$$
\lim_{t \to 0^+} \mathbf{B}(\mathbf{r}, t) = \mathbf{B}_0(\mathbf{r}) \tag{9.59}
$$

when $\mathbf{B}(\mathbf{r}, t)$ is given by any of the expressions appearing in Eqs. (9.33), (9.45), (9.47), or (9.49).

The following result due to Devaney [4] has thus been established: If the initial field values $\mathbf{E}_0(\mathbf{r})$ and $\mathbf{B}_0(\mathbf{r})$ satisfy the respective conditions given in Eqs. (9.51) and (9.53) of being divergenceless and have the properties required to ensure that the above analysis is valid, then there is one and only one solution for $t > 0$ to the source-free Maxwell's equations in a nonconducting, nondispersive HILL medium filling all of space satisfying the initial conditions $\mathbf{E}(\mathbf{r}, 0) = \mathbf{E}_0(\mathbf{r})$ and $\mathbf{B}(\mathbf{r}, 0) = \mathbf{B}_0(\mathbf{r})$.

9.2 Transformation to Spherical Coordinates in k-Space

The expressions given in Eqs. (9.48) and (9.49) for the free field vectors $\mathbf{E}(\mathbf{r}, t)$ and $\mathbf{B}(\mathbf{r}, t)$ for $t > 0$ are each in the form of a superposition of monochromatic plane waves and are appropriately called plane wave expansions. As the variables of integration change, both the frequency and the direction of propagation of the plane waves change. In order to explicitly display this, a change of variables to spherical coordinates in **k**-space is made, illustrated in Fig. 9.2, where

$$k_x = \frac{\omega}{c} \sin \alpha \cos \beta, \tag{9.60}$$

$$k_y = \frac{\omega}{c} \sin \alpha \sin \beta, \tag{9.61}$$

$$k_z = \frac{\omega}{c} \cos \alpha, \tag{9.62}$$

with spherical polar coordinate variables that vary over the respective domains $0 \leq \omega < \infty, 0 \leq \alpha \leq \pi, 0 \leq \beta \leq 2\pi$. The Jacobian of this transformation then gives

$$d^3 k = \frac{\partial(k_x, k_y, k_z)}{\partial(\omega, \alpha, \beta)} d\alpha d\beta d\omega$$

$$= \left(\frac{\omega}{c}\right)^2 \sin \alpha \, d\alpha d\beta d\left(\frac{\omega}{c}\right), \tag{9.63}$$

where

$$k = \frac{\omega}{c}. \tag{9.64}$$

Fig. 9.2 Spherical coordinates in **k**-space

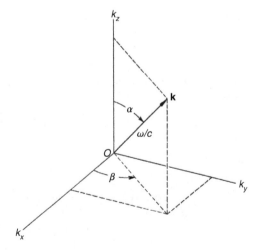

Under this transformation, the plane wave expansions given in Eqs. (9.48) and (9.49) become

$$\mathbf{E}(\mathbf{r}, t) = \frac{1}{(2\pi c)^3} \Re \int_0^\infty \omega^2 d\omega \int_0^\pi \sin\alpha d\alpha \int_0^{2\pi} d\beta$$

$$\times \left[\tilde{\mathbf{E}}_0(\mathbf{k}) - \frac{v}{\|c\|} \hat{\mathbf{s}} \times \tilde{\mathbf{B}}_0(\mathbf{k}) \right] e^{i(\mathbf{k}\cdot\mathbf{r} - (v/c)\omega t)},$$

$$(9.65)$$

$$\mathbf{B}(\mathbf{r}, t) = \frac{1}{(2\pi c)^3} \Re \int_0^\infty \omega^2 d\omega \int_0^\pi \sin\alpha d\alpha \int_0^{2\pi} d\beta$$

$$\times \left[\tilde{\mathbf{B}}_0(\mathbf{k}) + \frac{\|c\|}{v} \hat{\mathbf{s}} \times \tilde{\mathbf{E}}_0(\mathbf{k}) \right] e^{i(\mathbf{k}\cdot\mathbf{r} - (v/c)\omega t)},$$

$$(9.66)$$

for all $t > 0$. This pair of expressions for the free field vectors are plane wave expansions with the integration taken over the angular frequency ω and over the propagation direction specified by the polar angles α and β explicitly displayed. In terms of the element of solid angle $d\Omega(\mathbf{k})$ in the direction of $\mathbf{k}$, where

$$d\Omega(\mathbf{k}) \equiv d\Omega = \sin\alpha \, d\alpha d\beta, \qquad (9.67)$$

these expressions can be further simplified as

$$\mathbf{E}(\mathbf{r}, t) = \frac{1}{(2\pi c)^3} \Re \int_0^\infty \omega^2 d\omega \int_{4\pi} d\Omega \left[\tilde{\mathbf{E}}_0(\mathbf{k}) - \frac{v}{\|c\|} \hat{\mathbf{s}} \times \tilde{\mathbf{B}}_0(\mathbf{k}) \right] e^{i(\mathbf{k}\cdot\mathbf{r} - (v/c)\omega t)},$$

$$(9.68)$$

$$\mathbf{B}(\mathbf{r}, t) = \frac{1}{(2\pi c)^3} \Re \int_0^\infty \omega^2 d\omega \int_{4\pi} d\Omega \left[\tilde{\mathbf{B}}_0(\mathbf{k}) + \frac{\|c\|}{v} \hat{\mathbf{s}} \times \tilde{\mathbf{E}}_0(\mathbf{k}) \right] e^{i(\mathbf{k}\cdot\mathbf{r} - (v/c)\omega t)},$$

$$(9.69)$$

for all $t > 0$. Finally, if one introduces the new field vector

$$\tilde{\mathcal{E}}(\mathbf{k}) \equiv \frac{1}{(2\pi)^3} \left[\tilde{\mathbf{E}}_0(\mathbf{k}) - \frac{v}{\|c\|} \hat{\mathbf{s}} \times \tilde{\mathbf{B}}_0(\mathbf{k}) \right], \qquad (9.70)$$

in which case

$$\frac{\|c\|}{v} \hat{\mathbf{s}} \times \tilde{\mathcal{E}}(\mathbf{k}) = \frac{1}{(2\pi)^3} \left[\frac{\|c\|}{v} \hat{\mathbf{s}} \times \tilde{\mathbf{E}}_0(\mathbf{k}) - \hat{\mathbf{s}} \times \left(\hat{\mathbf{s}} \times \tilde{\mathbf{B}}_0(\mathbf{k}) \right) \right]$$

$$= \frac{1}{(2\pi)^3} \left[\tilde{\mathbf{B}}_0(\mathbf{k}) + \frac{\|c\|}{v} \hat{\mathbf{s}} \times \tilde{\mathbf{E}}_0(\mathbf{k}) \right], \qquad (9.71)$$

then the expressions given in Eqs. (9.48)–(9.49) and (9.68)–(9.69) become

$$\mathbf{E}(\mathbf{r},t) = \Re \int_{-\infty}^{\infty} \tilde{\mathcal{E}}(\mathbf{k}) e^{i(\mathbf{k}\cdot\mathbf{r}-vkt)} d^3k$$

$$= \frac{1}{c^3}\Re \int_0^{\infty} \omega^2 d\omega \int_{4\pi} d\Omega \,\tilde{\mathcal{E}}(\mathbf{k}) e^{i(\mathbf{k}\cdot\mathbf{r}-(v/c)\omega t)}, \tag{9.72}$$

$$\mathbf{B}(\mathbf{r},t) = \frac{\|c\|}{v}\Re \int_{-\infty}^{\infty} \left(\hat{\mathbf{s}}\times\tilde{\mathcal{E}}(\mathbf{k})\right) e^{i(\mathbf{k}\cdot\mathbf{r}-vkt)} d^3k$$

$$= \frac{\|c\|}{c^3 v}\Re \int_0^{\infty} \omega^2 d\omega \int_{4\pi} d\Omega \,\left(\hat{\mathbf{s}}\times\tilde{\mathcal{E}}(\mathbf{k})\right) e^{i(\mathbf{k}\cdot\mathbf{r}-(v/c)\omega t)}, \tag{9.73}$$

for all $t > 0$.

9.2.1   Plane Wave Representations and Mode Expansions

The preceding relations express the solution to the initial value problem in a form that is particularly convenient for a given mode expansion. A mode expansion is a superposition of particular solutions (called modes) to the differential equations governing a physical problem. In the present case the governing equations are Maxwell's equations in a nonconducting, nondispersive HILL medium with no sources, and the modes are the real parts of the complex, monochromatic electromagnetic plane waves

$$\tilde{\mathbf{E}}(\mathbf{r},t,\mathbf{k}) = \tilde{\mathcal{E}}(\mathbf{k}) e^{i(\mathbf{k}\cdot\mathbf{r}-vkt)},$$

$$\tilde{\mathbf{B}}(\mathbf{r},t,\mathbf{k}) = \frac{\|c\|}{v}\hat{\mathbf{s}}\times\tilde{\mathcal{E}}(\mathbf{k}) e^{i(\mathbf{k}\cdot\mathbf{r}-vkt)}.$$

These modes are especially convenient to work with because the dynamics of monochromatic plane waves are particularly simple.

Mode expansions are convenient to work with in the study of the interaction of an electromagnetic wave field with any linear system [5]. The field incident on the system can always be represented as a linear superposition of the modes of the system. The manner in which each mode interacts with the system can then be investigated independently of the other modes. The final result is then obtained by superimposing the results obtained for each mode separately.

One factor that complicates the application of the plane wave mode expansion given in Eqs. (9.72) and (9.73) is that each integrand includes plane waves that are traveling in all directions. As an illustration, let the initial field vectors $\mathbf{E}_0(\mathbf{r})$ and $\mathbf{B}_0(\mathbf{r})$ be confined to the region $z < 0$ at $t = 0$, both field vectors identically vanishing throughout the half-space $z > 0$. The wave field is then incident upon some linear system (e.g., a telescope) in the region $z > 0$ for $t > 0$. In order to

obtain an expression for the output from the linear system for $t > 0$, one can express
the incident wave field by the plane wave expansion given in Eqs. (9.72) and (9.73),
pass each plane wave component through the linear system, and then superimpose
the results to obtain the total output from the system. However, in this approach, one
must include plane wave components that are traveling through the linear system
towards the plane $z = 0$ (backwards through the telescope) as well as components
that are traveling away from that plane (forwards through the telescope).

Another feature of the plane wave expansion given in Eqs. (9.72) and (9.73)
worth noting is that the free field vectors $\mathbf{E}(\mathbf{r}, t)$ and $\mathbf{B}(\mathbf{r}, t)$ are expressed as the
real parts of the complex wave fields $\mathbf{E}_c(\mathbf{r}, t)$ and $\mathbf{B}_c(\mathbf{r}, t)$, respectively, which are
themselves expressed as

$$\mathbf{E}_c(\mathbf{r}, t) = \int_0^\infty \tilde{\mathbf{E}}_c(\mathbf{r}, \omega)e^{-i\omega t}\,d\omega, \tag{9.74}$$

$$\mathbf{B}_c(\mathbf{r}, t) = \int_0^\infty \tilde{\mathbf{B}}_c(\mathbf{r}, \omega)e^{-i\omega t}\,d\omega. \tag{9.75}$$

Functions of this form are called *analytic signals* [6]. If $\tilde{\mathbf{E}}_c(\mathbf{r}, \omega)$ and $\tilde{\mathbf{B}}_c(\mathbf{r}, \omega)$
are sufficiently well behaved, then $\mathbf{E}_c(\mathbf{r}, t' + it'')$ and $\mathbf{B}_c(\mathbf{r}, t' + it'')$ are analytic
functions of complex $t = t' + it''$ for $t'' < 0$.

9.2.2 Polar Coordinate Axis Along the Direction of Observation

The preceding plane wave representations of the free field are expressed in spherical
polar coordinates with the polar axis chosen to be along the z-axis of the original
Cartesian coordinate system. Somewhat simpler expressions for the free field
vectors $\mathbf{E}(\mathbf{r}, t)$ and $\mathbf{B}(\mathbf{r}, t)$ can be obtained by expressing the components k_x, k_y, k_z
in terms of a new spherical coordinate system whose polar axis is along the direction
of the observation point described by the position vector $\mathbf{r}$. The new variables of
integration are then given by

$$k'_x = \frac{\omega}{c}\sin\alpha'\cos\beta', \tag{9.76}$$

$$k'_y = \frac{\omega}{c}\sin\alpha'\sin\beta', \tag{9.77}$$

$$k'_z = \frac{\omega}{c}\cos\alpha', \tag{9.78}$$

which vary over the domain $0 \le \omega < \infty$, $0 \le \alpha' \le \pi$, $0 \le \beta' \le 2\pi$, where k_z' is the component of **k** along the field observation point position vector **r**, so that

$$k_z' = \mathbf{k} \cdot \mathbf{r} = k \cos \alpha'. \tag{9.79}$$

The (k_x', k_y', k_z') coordinate system can be obtained from the (k_x, k_y, k_z) coordinate system through a sequence of three successive rotations through the Euler angles ϕ, θ, and ψ as follows [7]. The initial (k_x, k_y, k_z) coordinate system is first rotated through the angle ϕ counterclockwise about the k_z-axis, resulting in the intermediate (ξ, η, ζ) coordinate system. This intermediate coordinate system is then rotated counterclockwise through the angle θ about the ξ-axis, producing the new intermediate (ξ', η', ζ') coordinate system. Finally, this intermediate coordinate system is rotated counterclockwise through the angle ψ about the ζ'-axis, resulting in the desired (k_x', k_y', k_z') coordinate system. Because only the k_z'-axis is specified by the position vector **r**, the k_x'- and k_y'-axes are completely arbitrary as is the angle ψ. The transformation matrix from the (k_x, k_y, k_z) coordinate system to the (k_x', k_y', k_z') system is then given by

$$\underline{A} = \begin{pmatrix} \cos\psi & \sin\psi & 0 \\ -\sin\psi & \cos\psi & 0 \\ 0 & 0 & 1 \end{pmatrix} \begin{pmatrix} 1 & 0 & 0 \\ 0 & \cos\theta & \sin\theta \\ 0 & -\sin\theta & \cos\theta \end{pmatrix} \begin{pmatrix} \cos\phi & \sin\phi & 0 \\ -\sin\phi & \cos\phi & 0 \\ 0 & 0 & 1 \end{pmatrix}$$

$$= \begin{pmatrix} (\cos\phi\cos\psi - \cos\theta\sin\phi\sin\psi) & (\sin\phi\cos\psi + \cos\theta\cos\phi\sin\psi) & (\sin\theta\sin\psi) \\ (-\cos\phi\sin\psi - \cos\theta\sin\phi\cos\psi) & (-\sin\phi\sin\psi + \cos\theta\cos\phi\cos\psi) & (\sin\theta\cos\psi) \\ \sin\phi\sin\theta & -\cos\phi\sin\theta & \cos\theta \end{pmatrix},$$

$$\tag{9.80}$$

so that

$$\begin{pmatrix} k_x' \\ k_y' \\ k_z' \end{pmatrix} = \underline{A} \begin{pmatrix} k_x \\ k_y \\ k_z \end{pmatrix}, \tag{9.81}$$

which may be written in matrix notation as $\mathbf{k}' = \underline{A}\mathbf{k}$. Because the transformation matrix $\underline{A}$ is orthogonal, the inverse transformation $\mathbf{k} = \underline{A}^{-1}\mathbf{k}'$ from the (k_x', k_y', k_z') coordinate system to the (k_x, k_y, k_z) system is given by the transpose $\underline{A}^T$ of the transformation matrix $\underline{A}$ as

$$\underline{A}^{-1} = \underline{A}^T$$

$$= \begin{pmatrix} (\cos\phi\cos\psi - \cos\theta\sin\phi\sin\psi) & (-\cos\phi\sin\psi - \cos\theta\sin\phi\cos\psi) & (\sin\phi\sin\theta) \\ (\sin\phi\cos\psi + \cos\theta\cos\phi\sin\psi) & (-\sin\phi\sin\psi + \cos\theta\cos\phi\cos\psi) & (-\cos\phi\sin\theta) \\ \sin\theta\sin\psi & \sin\theta\cos\psi & \cos\theta \end{pmatrix},$$

$$\tag{9.82}$$

Upon expressing the components k_x, k_y, k_z in terms of the new spherical polar coordinates ω/c, α', β' which have their polar axis along the direction of the position vector $\mathbf{r}$ of the field observation point, where

$$\begin{pmatrix} k_x \\ k_y \\ k_z \end{pmatrix} = \underline{A}^{-1} \begin{pmatrix} k'_x \\ k'_y \\ k'_z \end{pmatrix} = \frac{\omega}{c} \underline{A}^{-1} \begin{pmatrix} \sin\alpha'\cos\beta' \\ \sin\alpha'\sin\beta' \\ \cos\alpha' \end{pmatrix}, \qquad (9.83)$$

it is found that

$$k_x = \frac{\omega}{c}\big[\sin\alpha'\cos\phi\cos(\beta'+\psi)$$
$$- \sin\alpha'\cos\theta\sin\phi\cos(\beta'+\psi) + \cos\alpha'\sin\theta\sin\phi \big], \quad (9.84)$$
$$k_y = \frac{\omega}{c}\big[\sin\alpha'\sin\phi\cos(\beta'+\psi)$$
$$+ \sin\alpha'\cos\theta\cos\phi\sin(\beta'+\psi) - \cos\alpha'\sin\theta\cos\phi \big], \quad (9.85)$$
$$k_z = \frac{\omega}{c}\big[\sin\alpha'\sin\theta\sin(\beta'+\psi) + \cos\theta\cos\alpha' \big], \qquad (9.86)$$

where the real angle ψ is completely arbitrary.

Under this transformation, the plane wave expansions given in Eqs. (9.65) and (9.66) become

$$\mathbf{E}(\mathbf{r},t) = \frac{1}{(2\pi c)^3}\Re\int_0^\infty \omega^2 d\omega \int_0^\pi \sin\alpha' d\alpha' \int_0^{2\pi} d\beta'$$
$$\times \Big[\tilde{\mathbf{E}}_0(\mathbf{k}) - \frac{v}{\|c\|}\hat{\mathbf{s}}\times\tilde{\mathbf{B}}_0(\mathbf{k}) \Big] e^{i(\omega/c)(r\cos\alpha'-vt)}, \qquad (9.87)$$

$$\mathbf{B}(\mathbf{r},t) = \frac{1}{(2\pi c)^3}\Re\int_0^\infty \omega^2 d\omega \int_0^\pi \sin\alpha' d\alpha' \int_0^{2\pi} d\beta'$$
$$\times \Big[\tilde{\mathbf{B}}_0(\mathbf{k}) + \frac{\|c\|}{v}\hat{\mathbf{s}}\times\tilde{\mathbf{E}}_0(\mathbf{k}) \Big] e^{i(\omega/c)(r\cos\alpha'-vt)}, \qquad (9.88)$$

for all $t > 0$, where the angular dependence of the quantity $\mathbf{k}\cdot\mathbf{r}$ has been explicitly displayed. This representation of the plane wave expansion of the free field has the same properties as that for the representation given in Eqs. (9.65) and (9.66). Notice that the solution to the initial value problem can be expressed in other ways that exhibit different properties of the free field; perhaps the most important form is Poisson's solution [8] which is equivalent to the plane wave expansions considered here.

The extension of this analysis of the free field to a dispersive HILL medium remains to be accomplished. Because the residue analysis presented in Sect. 9.1.1 requires that the analytic properties of the integrand be known throughout the complex ω-plane, a specific model for the material dispersion must then be employed. Different models will then result in different results and it is essential that the model be causal in order that the results have true physical meaning.

9.3 Propagation of the Free Electromagnetic Field

The plane wave representation of the free field given in Eqs. (9.72) and (9.73) may be rewritten as

$$
\mathbf{E}(\mathbf{r}, t) = \frac{1}{2c^3} \left\{ \int_{4\pi} d\Omega \int_0^\infty \omega^2 d\omega \, \tilde{\mathcal{E}}(\mathbf{k}) e^{i(\mathbf{k} \cdot \mathbf{r} - (v/c)\omega t)} \right.
$$
$$
\left. + \int_{4\pi} d\Omega \int_0^\infty \omega^2 d\omega \, \tilde{\mathcal{E}}^*(\mathbf{k}) e^{-i(\mathbf{k} \cdot \mathbf{r} - (v/c)\omega t)} \right\}, \tag{9.89}
$$

$$
\mathbf{B}(\mathbf{r}, t) = \frac{\|c\|}{2c^3 v} \left\{ \int_{4\pi} d\Omega \int_0^\infty \omega^2 d\omega \left(\hat{\mathbf{s}} \times \tilde{\mathcal{E}}(\mathbf{k}) \right) e^{i(\mathbf{k} \cdot \mathbf{r} - (v/c)\omega t)} \right.
$$
$$
\left. + \int_{4\pi} d\Omega \int_0^\infty \omega^2 d\omega \left(\hat{\mathbf{s}} \times \tilde{\mathcal{E}}^*(\mathbf{k}) \right) e^{-i(\mathbf{k} \cdot \mathbf{r} - (v/c)\omega t)} \right\}, \tag{9.90}
$$

for all $t > 0$. Because the initial values $\mathbf{E}_0(\mathbf{r})$ and $\mathbf{B}_0(\mathbf{r})$ are real-valued, then their spatial Fourier transforms satisfy the respective symmetry relations $\tilde{\mathbf{E}}_0(-\mathbf{k}) = \tilde{\mathbf{E}}_0^*(\mathbf{k})$ and $\tilde{\mathbf{B}}_0(-\mathbf{k}) = \tilde{\mathbf{B}}_0^*(\mathbf{k})$, so that $\tilde{\mathcal{E}}(-\mathbf{k}) = \tilde{\mathcal{E}}^*(\mathbf{k})$. With this result, Eqs. (9.89) and (9.90) become

$$
\mathbf{E}(\mathbf{r}, t) = \frac{1}{2c^3} \left\{ \int_{4\pi} d\Omega \int_0^\infty \omega^2 d\omega \, \tilde{\mathcal{E}}(\mathbf{k}) e^{i(\mathbf{k} \cdot \mathbf{r} - (v/c)\omega t)} \right.
$$
$$
\left. + \int_{4\pi} d\Omega \int_0^\infty \omega^2 d\omega \, \tilde{\mathcal{E}}(-\mathbf{k}) e^{-i(\mathbf{k} \cdot \mathbf{r} - (v/c)\omega t)} \right\},
$$

$$
\mathbf{B}(\mathbf{r}, t) = \frac{\|c\|}{2c^3 v} \left\{ \int_{4\pi} d\Omega \int_0^\infty \omega^2 d\omega \left(\hat{\mathbf{s}} \times \tilde{\mathcal{E}}(\mathbf{k}) \right) e^{i(\mathbf{k} \cdot \mathbf{r} - (v/c)\omega t)} \right.
$$
$$
\left. + \int_{4\pi} d\Omega \int_0^\infty \omega^2 d\omega \left(\hat{\mathbf{s}} \times \tilde{\mathcal{E}}(-\mathbf{k}) \right) e^{-i(\mathbf{k} \cdot \mathbf{r} - (v/c)\omega t)} \right\}.
$$

With the change of variable $\omega \rightarrow -\omega$ in the second term of both of these expressions, noting that $\mathbf{k} \rightarrow -\mathbf{k}$ under this transformation, one finally obtains

$$\mathbf{E}(\mathbf{r}, t) = \frac{1}{2c^3} \int_{4\pi} d\Omega \int_{-\infty}^{\infty} \omega^2 d\omega \, \tilde{\mathcal{E}}(\mathbf{k}) e^{i(\mathbf{k}\cdot\mathbf{r}-(v/c)\omega t)}, \tag{9.91}$$

$$\mathbf{B}(\mathbf{r}, t) = \frac{\|c\|}{2c^3 v} \int_{4\pi} d\Omega \int_{-\infty}^{\infty} \omega^2 d\omega \left(\hat{\mathbf{s}} \times \tilde{\mathcal{E}}(\mathbf{k})\right) e^{i(\mathbf{k}\cdot\mathbf{r}-(v/c)\omega t)}, \tag{9.92}$$

for all $t > 0$. The spatio-temporal behavior described by this representation is now considered for several special cases.

9.3.1   Initial Field Values Confined Within a Sphere of Radius R

Let the initial values $\mathbf{E}_0(\mathbf{r})$ and $\mathbf{E}_0(\mathbf{r})$ be confined to the interior of a sphere of radius $R > 0$ centered at the origin, so that

$$\mathbf{E}_0(\mathbf{r}) = \mathbf{B}_0(\mathbf{r}) = \mathbf{0}; \quad \forall \mathbf{r} \ni |\mathbf{r}| > R, \tag{9.93}$$

and

$$\tilde{\mathbf{E}}_0(\mathbf{k}) = \int_{|\mathbf{r}|<R} \mathbf{E}_0(\mathbf{r}) e^{-i\mathbf{k}\cdot\mathbf{r}} d^3r, \tag{9.94}$$

$$\tilde{\mathbf{B}}_0(\mathbf{k}) = \int_{|\mathbf{r}|<R} \mathbf{B}_0(\mathbf{r}) e^{-i\mathbf{k}\cdot\mathbf{r}} d^3r. \tag{9.95}$$

If $\mathbf{E}_0(\mathbf{r})$ and $\mathbf{B}_0(\mathbf{r})$ are bounded for all $\mathbf{r}$ such that $|\mathbf{r}| < R$, then

$$\frac{\partial \tilde{\mathbf{E}}_0(\mathbf{k})}{\partial \omega} = \int_{|\mathbf{r}|<R} \mathbf{E}_0(\mathbf{r}) \frac{\partial}{\partial \omega} \left(e^{-i\mathbf{k}\cdot\mathbf{r}}\right) d^3r, \tag{9.96}$$

$$\frac{\partial \tilde{\mathbf{B}}_0(\mathbf{k})}{\partial \omega} = \int_{|\mathbf{r}|<R} \mathbf{B}_0(\mathbf{r}) \frac{\partial}{\partial \omega} \left(e^{-i\mathbf{k}\cdot\mathbf{r}}\right) d^3r. \tag{9.97}$$

The spectra $\tilde{\mathbf{E}}_0(\mathbf{k})$ and $\tilde{\mathbf{B}}_0(\mathbf{k})$ of the initial field vectors, and consequently the spectral function $\tilde{\mathcal{E}}(\mathbf{k})$, are then entire functions of ω because derivatives with respect to ω of all orders exist. Consider then the behavior of $\tilde{\mathbf{E}}_0(\mathbf{k})$ and $\tilde{\mathbf{B}}_0(\mathbf{k})$ on the circle $|\omega| = $ constant in the complex ω-plane. First, the inequalities

$$\left|\tilde{\mathbf{E}}_0(\mathbf{k})\right| \leq \int_{|\mathbf{r}|<R} |\mathbf{E}_0(\mathbf{r})| \left|e^{-i(\omega/c)\hat{\mathbf{s}}\cdot\mathbf{r}}\right| \left|d^3r\right|,$$

$$\left|\tilde{\mathbf{B}}_0(\mathbf{k})\right| \leq \int_{|\mathbf{r}|<R} |\mathbf{B}_0(\mathbf{r})| \left|e^{-i(\omega/c)\hat{\mathbf{s}}\cdot\mathbf{r}}\right| \left|d^3r\right|,$$

directly follow from Eqs. (9.94) and (9.95). Because

$$\left|e^{-i(\omega/c)\hat{\mathbf{s}}\cdot\mathbf{r}}\right| = \left|e^{-i((\omega'+i\omega'')/c)\hat{\mathbf{s}}\cdot\mathbf{r}}\right|$$

$$= e^{(\omega''/c)\hat{\mathbf{s}}\cdot\mathbf{r}},$$

then these inequalities become

$$\left|\tilde{\mathbf{E}}_0(\mathbf{k})\right| \leq \int_{|\mathbf{r}|<R} |\mathbf{E}_0(\mathbf{r})| \, e^{(\omega''/c)\hat{\mathbf{s}}\cdot\mathbf{r}} \left|d^3r\right|,$$

$$\left|\tilde{\mathbf{B}}_0(\mathbf{k})\right| \leq \int_{|\mathbf{r}|<R} |\mathbf{B}_0(\mathbf{r})| \, e^{(\omega''/c)\hat{\mathbf{s}}\cdot\mathbf{r}} \left|d^3r\right|.$$

In addition, one has that $\max_{|\mathbf{r}|<R} \{\hat{\mathbf{s}}\cdot\mathbf{r}\} = R$ and $\min_{|\mathbf{r}|<R} \{\hat{\mathbf{s}}\cdot\mathbf{r}\} = -R$, so that, for all ω'',

$$\left|\tilde{\mathbf{E}}_0(\mathbf{k})\right| \leq \int_{|\mathbf{r}|<R} |\mathbf{E}_0(\mathbf{r})| \, e^{|\omega''|R/c} \left|d^3r\right|$$

$$\leq M_1 e^{|\omega''|R/c}, \tag{9.98}$$

$$\left|\tilde{\mathbf{B}}_0(\mathbf{k})\right| \leq \int_{|\mathbf{r}|<R} |\mathbf{B}_0(\mathbf{r})| \, e^{|\omega''|R/c} \left|d^3r\right|$$

$$\leq M_2 e^{|\omega''|R/c}, \tag{9.99}$$

because $\mathbf{E}_0(\mathbf{r})$ and $\mathbf{B}_0(\mathbf{r})$ are bounded for all $\mathbf{r}$ such that $|\mathbf{r}| < R$.

Therefore, if the initial values $\mathbf{E}_0(\mathbf{r})$ and $\mathbf{B}_0(\mathbf{r})$ are bounded for all $\mathbf{r}$ such that $|\mathbf{r}| < R$ and identically vanish for all $\mathbf{r}$ such that $|\mathbf{r}| > R$, then both $\tilde{\mathbf{E}}_0(\mathbf{k})$ and $\tilde{\mathbf{B}}_0(\mathbf{k})$ can grow no faster than $M' e^{|\omega''|R/c}$, where $M' = \sup(M_1, M_2)$.[3] Consequently,

$$\left|\tilde{\boldsymbol{\mathcal{E}}}(\mathbf{k})\right| = \frac{1}{(2\pi)^3} \left|\tilde{\mathbf{E}}_0(\mathbf{k}) - \frac{v}{\|c\|}\hat{\mathbf{s}} \times \tilde{\mathbf{B}}_0(\mathbf{k})\right|$$

$$\leq \frac{1}{(2\pi)^3} \left\{\left|\tilde{\mathbf{E}}_0(\mathbf{k})\right| + \left|\frac{v}{\|c\|}\right| \left|\tilde{\mathbf{B}}_0(\mathbf{k})\right|\right\}$$

$$\leq \frac{1}{(2\pi)^3} \left\{M_1 + \left|\frac{v}{\|c\|}\right| M_2\right\} e^{|\omega''|R/c}$$

$$\leq M e^{|\omega''|R/c}, \tag{9.100}$$

---

[3] This result is part of the Paley–Wiener theorem.

so that the integrands in Eqs. (9.91) and (9.92) satisfy the inequality

$$
\left| \omega^2 \tilde{\mathcal{E}}(\mathbf{k}) e^{i(\mathbf{k}\cdot\mathbf{r}-(v/c)\omega t)} \right| = \left| \omega^2 \hat{\mathbf{s}} \times \tilde{\mathcal{E}}(\mathbf{k}) e^{i(\mathbf{k}\cdot\mathbf{r}-(v/c)\omega t)} \right|
$$

$$
= \left| \omega^2 \right| \left| \tilde{\mathcal{E}}(\mathbf{k}) \right| \left| e^{i(\omega/c)(\hat{\mathbf{s}}\cdot\mathbf{r}-vt)} \right|
$$

$$
\leq \left| \omega^2 \right| M e^{|\omega''|(R/c)} e^{-(\omega''/c)(\hat{\mathbf{s}}\cdot\mathbf{r}-vt)} \tag{9.101}
$$

in the complex ω-plane. Consequently, in order for the integrands in the plane wave expansions given in Eqs. (9.91) and (9.92) to possess exponential attenuation, the inequality

$$
\hat{\mathbf{s}} \cdot \mathbf{r} > R + vt \tag{9.102}
$$

must be satisfied when $\omega'' > 0$, whereas the inequality

$$
\hat{\mathbf{s}} \cdot \mathbf{r} < vt - R \tag{9.103}
$$

must be satisfied when $\omega'' < 0$.

The integrands appearing in the plane wave expansions given in Eqs. (9.91) and (9.92) possess exponential decay for those values of $\hat{\mathbf{s}} = \mathbf{k}/k$, $\mathbf{r}$, and t satisfying the inequalities given in Eqs. (9.102) and (9.103). By Jordan's lemma, if $\left| \tilde{\mathbf{E}}_0(\mathbf{k}) \right|$ and $\left| \tilde{\mathbf{B}}_0(\mathbf{k}) \right|$ go to zero faster than $1/|\omega|^2$ as $\omega \to \infty$, then the integral over ω in Eqs. (9.91) and (9.92) vanishes for these values of $\hat{\mathbf{s}}$, $\mathbf{r}$, and t. That is, all modes propagating in the direction $\hat{\mathbf{s}}$ appearing in the plane wave expansion for the free field vectors $\mathbf{E}(\mathbf{r}, t)$ and $\mathbf{B}(\mathbf{r}, t)$ will not contribute to the free field at the point P with position vector $\mathbf{r}$ if the conditions specified in Eqs. (9.102) and (9.103) are satisfied.

A geometrical construction is given in Figs. 9.3 and 9.4 that illustrates the directions that the modes either do or do not contribute to the field at the point P with position vector $\mathbf{r}$ from the origin O as a function of time. Figure 9.3 depicts the construction of three successive cones at the instants $t_0 = 0$, $t_1 > t_0$, and $t_2 > t_1$ with generators $\hat{\mathbf{s}}_0$, $\hat{\mathbf{s}}_1$, and $\hat{\mathbf{s}}_2$, respectively, such that the inequality given in Eq. (9.102) is satisfied in the interior of a given cone at that instant of time. Any plane wave with direction $\hat{\mathbf{s}}$ lying inside the first cone with generator $\hat{\mathbf{s}}_0$ satisfies the inequality $\hat{\mathbf{s}} \cdot \mathbf{r} > R$, any plane wave with direction $\hat{\mathbf{s}}$ lying inside the second cone with generator $\hat{\mathbf{s}}_1$ satisfies the inequality $\hat{\mathbf{s}} \cdot \mathbf{r} > R + vt_1$, and any plane wave with direction $\hat{\mathbf{s}}$ lying inside the third cone with generator $\hat{\mathbf{s}}_1$ satisfies the inequality $\hat{\mathbf{s}} \cdot \mathbf{r} > R + vt_2$. These plane wave modes then do not contribute to the free field at P at these instants of time. Figure 9.4 depicts the same type of construction at the same instants of time for the inequality given in Eq. (8.103). Notice that the time instant t_1 has been chosen to be given by $t_1 = R/v$ when the cone becomes a plane. For earlier times these cones open to the left and for later times they open to the

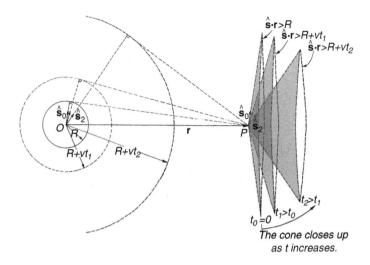

Fig. 9.3 Wave vector regions satisfying the inequality in Eq. (9.102)

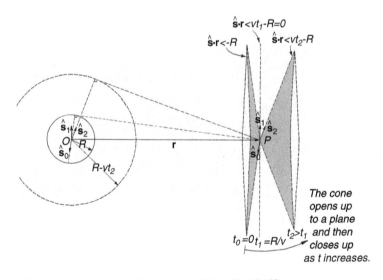

Fig. 9.4 Wave vector regions satisfying the inequality in Eq. (9.103)

right in the figure. Taken together, this pair of diagrams shows that, at each instant of time $t \geq t_0 = 0$, the only plane wave modes that contribute to the free field at the point P are those whose propagation directions $\hat{\mathbf{s}}$ lie in the space between the respective cones described by the pair of inequalities in Eqs. (9.102) and (9.103). It is then seen that when $t \geq R/v$, only those plane wave modes whose direction $\hat{\mathbf{s}}$ has a positive component along the position vector $\mathbf{r}$ to the observation point P (i.e., such that $\hat{\mathbf{s}} \cdot \mathbf{r} > 0$) contribute to the field at P, and that when

$$t \geq \frac{|\mathbf{r}| + R}{v}, \tag{9.104}$$

no plane wave modes contribute and the field at P vanishes.

9.3.2   Initial Field Values Confined Inside a Closed Convex Surface

Let the initial field values $\mathbf{E}_0(\mathbf{r})$ and $\mathbf{B}_0(\mathbf{r})$ identically vanish outside a simply connected region V with convex surface S, so that

$$\tilde{\mathbf{E}}_0(\mathbf{k}) = \int_V \mathbf{E}_0(\mathbf{r}) e^{-i\mathbf{k}\cdot\mathbf{r}} d^3 r, \tag{9.105}$$

$$\tilde{\mathbf{B}}_0(\mathbf{k}) = \int_V \mathbf{B}_0(\mathbf{r}) e^{-i\mathbf{k}\cdot\mathbf{r}} d^3 r, \tag{9.106}$$

and let $\mathbf{E}_0(\mathbf{r})$ and $\mathbf{B}_0(\mathbf{r})$ both be continuous in V and have continuous partial derivatives up through the third order in V. Let the surface S be described by the relation

$$S = \{(x, y, z) |\, F(x, y, z) = 0\}, \tag{9.107}$$

where the equation $F(x, y, z) = 0$ can be solved for x, y, and z as

$$z = \left\{ \begin{matrix} Z_+(x, y) \\ Z_-(x, y) \end{matrix} \right\}, \tag{9.108}$$

$$y = \left\{ \begin{matrix} Y_+(x, z) \\ Y_-(x, z) \end{matrix} \right\}, \tag{9.109}$$

$$x = \left\{ \begin{matrix} X_+(y, z) \\ X_-(y, z) \end{matrix} \right\}, \tag{9.110}$$

with $Y_+(x) \equiv$ maximum value of $Y_+(x, z)$ obtained when z takes on all possible values on S with fixed x on S, $Y_-(x) \equiv$ minimum value of $Y_-(x, z)$ obtained when z takes on all possible values on S with fixed x on S, $X_+ \equiv$ maximum value of $X_+(y, z)$ obtained when y and z take on all possible values on S, and $X_- \equiv$ minimum value of $X_-(y, z)$ obtained when y and z take on all possible values on S.

With these identifications, Eqs. (9.105) and (9.106) can be written as

$$\tilde{\mathbf{E}}_0(\mathbf{k}) = \int_{X_-}^{X_+} dx \int_{Y_-(x)}^{Y_+(x)} dy \int_{Z_-(x,y)}^{Z_+(x,y)} dz\, \mathbf{E}_0(\mathbf{r}) e^{-i\mathbf{k}\cdot\mathbf{r}}, \tag{9.111}$$

$$\tilde{\mathbf{B}}_0(\mathbf{k}) = \int_{X_-}^{X_+} dx \int_{Y_-(x)}^{Y_+(x)} dy \int_{Z_-(x,y)}^{Z_+(x,y)} dz\, \mathbf{B}_0(\mathbf{r}) e^{-i\mathbf{k}\cdot\mathbf{r}}. \tag{9.112}$$

The multiple integration appearing in Eq. (9.111) may be evaluated in a stepwise manner as follows: First, the innermost integral is evaluated as

$$\mathbf{I}_1(x, y) = \int_{Z_-(x,y)}^{Z_+(x,y)} \mathbf{E}_0(\mathbf{r}) e^{-ik_z z} dz$$

$$= \frac{i}{k_z}\left\{\left[\mathbf{E}_0(\mathbf{r}) e^{-ik_z z}\right]_{Z_-(x,y)}^{Z_+(x,y)} - \int_{Z_-(x,y)}^{Z_+(x,y)} \frac{\partial \mathbf{E}_0(\mathbf{r})}{\partial z} e^{-ik_z z} dz\right\},$$

and the second integral may be evaluated in the same manner as

$$\mathbf{I}_2(x) = \int_{Y_-(x)}^{Y_+(x)} \mathbf{I}_1(x, y) e^{-ik_y y} dy$$

$$= \frac{i}{k_y}\left\{\left[\mathbf{I}_1(x, y) e^{-ik_y y}\right]_{Y_-(x)}^{Y_+(x)} - \int_{Y_-(x)}^{Y_+(x)} \frac{\partial \mathbf{I}_1(x, y)}{\partial y} e^{-ik_y y} dy\right\}.$$

In the same manner, the outermost integral in Eq. (9.111) may be evaluated as

$$\tilde{\mathbf{E}}_0(\mathbf{k}) = \int_{X_-}^{X_+} \mathbf{I}_2(x) e^{-ik_x x} dx$$

$$= \frac{i}{k_x}\left\{\left[\mathbf{I}_2(x) e^{-ik_x x}\right]_{X_-}^{X_+} - \int_{X_-}^{X_+} \frac{\partial \mathbf{I}_2(x)}{\partial x} e^{-ik_x x} dx\right\}$$

$$= -\frac{i}{k_x k_y k_z}\left\{\left[\left[\mathbf{E}_0(\mathbf{r}) e^{-i\mathbf{k}\cdot\mathbf{r}}\right]_{z=Z_-(x,y)}^{z=Z_+(x,y)}\right.\right.$$

$$\left. - \int_{Z_-(x,y)}^{Z_+(x,y)} \frac{\partial \mathbf{E}_0(\mathbf{r})}{\partial z} e^{-i\mathbf{k}\cdot\mathbf{r}} dz\right]_{y=Y_-(x)}^{y=Y_+(x)}$$

$$- \int_{Y_-(x)}^{Y_+(x)}\left[\left[\frac{\partial \mathbf{E}_0(\mathbf{r})}{\partial y} e^{-i\mathbf{k}\cdot\mathbf{r}}\right]_{z=Z_-(x,y)}^{z=Z_+(x,y)}\right.$$

$$\left.\left. - \int_{Z_-(x,y)}^{Z_+(x,y)} \frac{\partial^2 \mathbf{E}_0(\mathbf{r})}{\partial y \partial z} e^{-i\mathbf{k}\cdot\mathbf{r}} dz\right] dy\right\}_{x=X_-}^{x=X_+}$$

$$+ \frac{i}{k_x k_y k_z} \int_{X_-}^{X_+} \left\{ \left[\left[\frac{\partial \mathbf{E}_0(\mathbf{r})}{\partial x} e^{-i\mathbf{k}\cdot\mathbf{r}} \right]_{z=Z_-(x,y)}^{z=Z_+(x,y)} \right. \right.$$

$$\left. - \int_{Z_-(x,y)}^{Z_+(x,y)} \frac{\partial^2 \mathbf{E}_0(\mathbf{r})}{\partial x \partial z} e^{-i\mathbf{k}\cdot\mathbf{r}} dz \right]_{y=Y_-(x)}^{y=Y_+(x)}$$

$$- \int_{Y_-(x)}^{Y_+(x)} \left[\left[\frac{\partial^2 \mathbf{E}_0(\mathbf{r})}{\partial x \partial y} e^{-i\mathbf{k}\cdot\mathbf{r}} \right]_{z=Z_-(x,y)}^{z=Z_+(x,y)} \right.$$

$$\left. \left. - \int_{Z_-(x,y)}^{Z_+(x,y)} \frac{\partial^3 \mathbf{E}_0(\mathbf{r})}{\partial x \partial y \partial z} e^{-i\mathbf{k}\cdot\mathbf{r}} dz \right] dy \right\} dx,$$

$$(9.113)$$

after substitution from the preceding two integrations for $\mathbf{I}_2(x)$ and $\mathbf{I}_1(x, y)$. An analogous expression holds for the spatial frequency spectrum $\tilde{\mathbf{B}}_0(\mathbf{k})$ of the magnetic field. Because $k_x = (\omega/c) \sin\alpha \cos\beta$, $k_y = (\omega/c) \sin\alpha \sin\beta$, and $k_z = (\omega/c) \cos\alpha$, and because $\hat{\mathbf{s}} \equiv \mathbf{k}/k = (c/\omega)\mathbf{k}$, then the inequality

$$\left| e^{-i\mathbf{k}\cdot\mathbf{r}} \right| = \left| e^{-i(\omega/c)\hat{\mathbf{s}}\cdot\mathbf{r}} \right| = e^{(\omega''/c)\hat{\mathbf{s}}\cdot\mathbf{r}} \le e^{|\omega''|R/c} \qquad (9.114)$$

is satisfied for all $\mathbf{r}$ such that $|\mathbf{r}| \le R$. For each complex vector component of $\tilde{\mathbf{E}}_0(\mathbf{k})$, the inequality

$$\left| \tilde{E}_{0j}(\mathbf{k}) \right| = \left[\tilde{E}_{0j}(\mathbf{k}) \tilde{E}_{0j}^*(\mathbf{k}) \right]^{1/2} \le \frac{\mathcal{K} e^{|\omega''|R/c}}{(\omega/c)^3 \sin^2\alpha \cos\alpha \sin\beta \cos\beta} \qquad (9.115)$$

is satisfied, where $j = x, y, z$. Then, for α and β both bounded away from both 0 and $\pi/2$, one has the inequality

$$\left| \tilde{E}_{0j}(\mathbf{k}) \right| \le \mathcal{K}_1 \frac{e^{|\omega''|R/c}}{\omega^3}, \qquad (9.116)$$

and similarly

$$\left| \tilde{B}_{0j}(\mathbf{k}) \right| \le \mathcal{K}_2 \frac{e^{|\omega''|R/c}}{\omega^3}, \qquad (9.117)$$

where $\mathcal{K}_1$ and $\mathcal{K}_2$ are positive constants. However, the integrals appearing in Eqs. (9.91) and (9.92) are both taken over 4π steradians. The apparent "divergence" in the inequality when either α or β is equal to either 0 or $\pi/2$ is clearly dependent upon the choice of polar axes. That is, this "divergence" exists for each particular set of coordinate axes so that the overlap produced for two carefully selected coordinate axes effectively eliminates the problem because the solution must be independent of the choice of axes. Consequently, the inequalities appearing in Eqs. (9.116) and (9.117) hold over 4π steradians. It then follows that

$$
\begin{aligned}
\left|\tilde{\mathcal{E}}(\mathbf{k})\right| &= \frac{1}{(2\pi)^3} \left|\tilde{\mathbf{E}}_0(\mathbf{k}) - \frac{v}{\|c\|}\hat{\mathbf{s}} \times \tilde{\mathbf{B}}_0(\mathbf{k})\right| \\
&\leq \frac{1}{(2\pi)^3} \left[\left|\tilde{\mathbf{E}}_0(\mathbf{k})\right| + \frac{|v|}{\|c\|}\left|\tilde{\mathbf{B}}_0(\mathbf{k})\right|\right] \\
&\leq \frac{1}{(2\pi)^3} \left[\mathcal{K}_1 + \frac{|v|}{\|c\|}\mathcal{K}_2\right] \frac{1}{|\omega|^3} e^{|\omega''|R/c} \\
&\leq \frac{\mathcal{M}}{|\omega|^3} e^{|\omega''|R/c},
\end{aligned}
\tag{9.118}
$$

where $\mathcal{M}$ is another positive constant. The integrands appearing in Eqs. (9.91) and (9.92) then satisfy the inequality

$$
\begin{aligned}
\left|\omega^2\hat{\mathbf{s}} \times \tilde{\mathcal{E}}(\mathbf{k})e^{i(\mathbf{k}\cdot\mathbf{r}-(v/c)\omega t)}\right| &= \left|\omega^2\tilde{\mathcal{E}}(\mathbf{k})e^{i(\mathbf{k}\cdot\mathbf{r}-(v/c)\omega t)}\right| \\
&= \left|\omega^2\right|\left|\tilde{\mathcal{E}}(\mathbf{k})\right|\left|e^{i(\omega/c)(\hat{\mathbf{s}}\cdot\mathbf{r}-vt)}\right| \\
&\leq \frac{\mathcal{M}}{|\omega|} e^{|\omega''|R/c}e^{-(\omega''/c)(\hat{\mathbf{s}}\cdot\mathbf{r}-vt)},
\end{aligned}
\tag{9.119}
$$

in the complex ω-plane. One then obtains the same conditions, given in Eqs. (9.102) and (9.103), for exponential decay in the free field. The results of the previous subsection (Sect. 9.3.2) then apply here where R now represents the radius of the smallest sphere that completely encloses the simply connected region V.

9.3.3 Propagation of the Free Electromagnetic Wave Field

Consider the earlier form of the plane wave expansion of the free field given in Eqs. (9.46) and (9.47) as

$$
\mathbf{E}(\mathbf{r}, t) = \frac{1}{(2\pi)^3} \int_{-\infty}^{\infty} \left[\tilde{\mathbf{E}}_0(\mathbf{k}) \cos(vkt) + i\frac{v}{\|c\|}\hat{\mathbf{s}} \times \tilde{\mathbf{B}}_0(\mathbf{k}) \sin(vkt)\right] e^{i\mathbf{k}\cdot\mathbf{r}} d^3k,
\tag{9.120}
$$

$$
\mathbf{B}(\mathbf{r}, t) = \frac{1}{(2\pi)^3} \int_{-\infty}^{\infty} \left[\tilde{\mathbf{B}}_0(\mathbf{k}) \cos(vkt) - i\frac{\|c\|}{v}\hat{\mathbf{s}} \times \tilde{\mathbf{E}}_0(\mathbf{k}) \sin(vkt)\right] e^{i\mathbf{k}\cdot\mathbf{r}} d^3k,
\tag{9.121}
$$

for $t > 0$. From the identity

$$\frac{\sin(AB)}{AB} = \frac{1}{4\pi} \int_{4\pi} e^{i\mathbf{A}\cdot\mathbf{B}} d\Omega(\mathbf{A}),$$

where $d\Omega(\mathbf{A})$ denotes the differential element of solid angle about the direction of the vector $\mathbf{A}$, one finds that

$$\frac{\sin(vkt)}{vkt} = \frac{1}{4\pi} \int_{4\pi} e^{i\mathbf{k}\cdot\mathbf{v}t} d\Omega(\mathbf{v}),$$

and

$$\cos(vkt) = \frac{d}{dt}\left(\frac{\sin(vkt)}{vk}\right)$$

$$= \frac{1}{4\pi}\frac{d}{dt}\left[t\int_{4\pi} e^{i\mathbf{k}\cdot\mathbf{v}t} d\Omega(\mathbf{v})\right].$$

Substitution of these results into the plane wave expansions given in Eqs. (9.120) and (9.121) then gives

$$\mathbf{E}(\mathbf{r},t) = \frac{1}{32\pi^4} \int_{4\pi} d\Omega(\mathbf{v}) \int_{-\infty}^{\infty} d^3k \left\{ \tilde{\mathbf{E}}_0(\mathbf{k}) \frac{d}{dt}\left[te^{i\mathbf{k}\cdot(\mathbf{r}+\mathbf{v}t)}\right] \right.$$

$$\left. + i\frac{v^2}{\|c\|}t\mathbf{k} \times \tilde{\mathbf{B}}_0(\mathbf{k})e^{i\mathbf{k}\cdot(\mathbf{r}+\mathbf{v}t)} \right\}$$

$$= \frac{1}{4\pi} \int_{4\pi} d\Omega(\mathbf{v}) \left\{ \frac{d}{dt}\left[t\mathbf{E}_0(\mathbf{r}+\mathbf{v}t)\right] + \frac{v^2}{\|c\|}t\nabla \times \mathbf{B}_0(\mathbf{r}+\mathbf{v}t) \right\}$$

$$= \frac{1}{4\pi} \left\{ \frac{\partial}{\partial t} \int_{4\pi} t\mathbf{E}_0(\mathbf{r}+\mathbf{v}t)d\Omega(\mathbf{v}) + \frac{v^2}{\|c\|}t \int_{4\pi} \nabla \times \mathbf{B}_0(\mathbf{r}+\mathbf{v}t)d\Omega(\mathbf{v}) \right\},$$

$$(9.122)$$

$$\mathbf{B}(\mathbf{r},t) = \frac{1}{32\pi^4} \int_{4\pi} d\Omega(\mathbf{v}) \int_{-\infty}^{\infty} d^3k \left\{ \tilde{\mathbf{B}}_0(\mathbf{k}) \frac{d}{dt}\left[te^{i\mathbf{k}\cdot(\mathbf{r}+\mathbf{v}t)}\right] \right.$$

$$\left. - i\|c\|t\mathbf{k} \times \tilde{\mathbf{E}}_0(\mathbf{k})e^{i\mathbf{k}\cdot(\mathbf{r}+\mathbf{v}t)} \right\}$$

$$= \frac{1}{4\pi} \int_{4\pi} d\Omega(\mathbf{v}) \left\{ \frac{d}{dt}\left[t\mathbf{B}_0(\mathbf{r}+\mathbf{v}t)\right] - \|c\|t\nabla \times \mathbf{E}_0(\mathbf{r}+\mathbf{v}t) \right\}$$

$$= \frac{1}{4\pi} \left\{ \frac{\partial}{\partial t} \int_{4\pi} t\mathbf{B}_0(\mathbf{r}+\mathbf{v}t)d\Omega(\mathbf{v}) - \|c\|t \int_{4\pi} \nabla \times \mathbf{E}_0(\mathbf{r}+\mathbf{v}t)d\Omega(\mathbf{v}) \right\},$$

$$(9.123)$$

for $t > 0$.

Fig. 9.5 The spherical
surface $|\mathbf{r} - \mathbf{r}'| = vt =$
constant about the field
observation point P

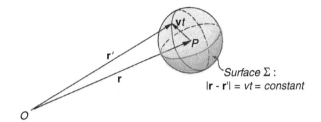

Surface Σ :
$|\mathbf{r} - \mathbf{r}'| = vt = $ constant

Consider now rewriting this pair of integral representations as surface integrals over the spherical surface Σ illustrated in Fig. 9.5, given by $|\mathbf{r} - \mathbf{r}'| = vt = $ constant. A differential element of area on the surface Σ is given by

$$dA' = |\mathbf{r} - \mathbf{r}'|^2 d\Omega(\mathbf{v})$$
$$= (vt)^2 d\Omega(\mathbf{v}),$$

so that

$$d\Omega(\mathbf{v}) = \frac{dA'}{vt|\mathbf{r} - \mathbf{r}'|}. \tag{9.124}$$

With these identifications, the relations given in Eqs. (9.122) and (9.123) become

$$\mathbf{E}(\mathbf{r}, t) = \frac{1}{4\pi v}\left\{ \frac{\partial}{\partial t} \int_\Sigma \frac{\mathbf{E}_0(\mathbf{r}')}{|\mathbf{r} - \mathbf{r}'|} dA' + \frac{v^2}{\|c\|} \int_\Sigma \frac{\nabla \times \mathbf{B}_0(\mathbf{r}')}{|\mathbf{r} - \mathbf{r}'|} dA' \right\}, \tag{9.125}$$

$$\mathbf{B}(\mathbf{r}, t) = \frac{1}{4\pi v}\left\{ \frac{\partial}{\partial t} \int_\Sigma \frac{\mathbf{B}_0(\mathbf{r}')}{|\mathbf{r} - \mathbf{r}'|} dA' - \|c\| \int_\Sigma \frac{\nabla \times \mathbf{E}_0(\mathbf{r}')}{|\mathbf{r} - \mathbf{r}'|} dA' \right\}, \tag{9.126}$$

for $t > 0$, where Σ is the surface $|\mathbf{r} - \mathbf{r}'| = vt$. In the scalar case, this is just Poisson's solution to the initial value problem.

Let the initial field values $\mathbf{E}_0(\mathbf{r})$ and $\mathbf{B}_0(\mathbf{r})$ as well as their spatial derivatives vanish outside of a simply connected region V with closed outer surface S. Because the initial field vectors $\mathbf{E}_0(\mathbf{r}')$ and $\mathbf{B}_0(\mathbf{r}')$ and their spatial derivatives are then identically zero when the point $\mathbf{r}' = \mathbf{r} + \mathbf{v}t$ is outside V, the free field vectors $\mathbf{E}(\mathbf{r}, t)$ and $\mathbf{B}(\mathbf{r}, t)$ also vanish when this occurs. That is, given an observation point P that is exterior to the initial value region V, the free field will be zero at P until the instant of time t_1 when the sphere Σ_1 of radius vt_1 drawn about P as center first intersects the region V, as depicted in Fig. 9.6. Furthermore, the free field at P will return to zero at a later time t_2 when the sphere Σ_2 of radius vt_2 with center at P completely encloses the region V for the first time, as also depicted in Fig. 9.6. The only time when the free field at P is nonzero is when the sphere Σ with radius vt centered at P intersects the surface S that surrounds the region containing the initial field values.

Fig. 9.6 Spherical regions
about the field observation
point P that determine the
time instants t_1 and t_2
between which the initial
values confined to within V
contribute to the free field
at P

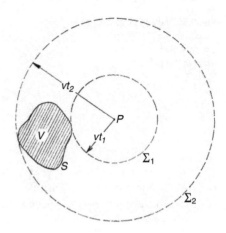

As a consequence, the surface S of the region V can be considered to be a
wavefront that propagates both "inward" and "outward" according to the laws of
geometrical optics from the time $t = 0$ through positive times t, as illustrated
in the sequence of diagrams in Fig. 9.7. At $t = 0$ the field vectors $\mathbf{E}(\mathbf{r}, t)$ and
$\mathbf{B}(\mathbf{r}, t)$ are equal to their respective initial values $\mathbf{E}_0(\mathbf{r})$ and $\mathbf{B}_0(\mathbf{r})$, both of which
identically vanish outside the simply connected region V with closed outer surface
S, as illustrated by the shaded region in part (a) of Fig. 9.7.

At a slightly later time $t = t_1 > 0$, the "wavefront" surface S will have
propagated both inward and outward through a distance vt_1, where $v = c/\sqrt{\epsilon\mu}$
is the constant phase velocity in the nondispersive HILL medium. If the time t_1 is
short enough that there is no overlap between the inward propagated surface and
any other points of the initial surface S, as depicted in part (b) of the figure, then
the free field vectors will identically vanish everywhere outside the region V' that is
enclosed by the outward propagated surface S'. At a later time $t_2 > t_1$ such that the
inward propagating "wavefront" surface overlaps points on the initial surface S, as
illustrated in part (c) of the figure, the free field vectors will then vanish everywhere
exterior to the annular region V'' with closed surface S''. The outer portion of S''
propagates with velocity v in the direction of the outward normal vector $\hat{\mathbf{n}}$ to the
initial surface S and consequently retains the shape of that initial surface. The inner
portion of S'' propagates with velocity v in the direction of the inward normal $-\hat{\mathbf{n}}$
to S so that its shape will be the inversion of S, as illustrated.

Problems

9.1 Derive the relation given in Eq. (9.25) with Eq. (9.16) as the starting point.

9.2 Derive the identity

$$\frac{\sin (AB)}{AB} = \frac{1}{4\pi} \int_{4\pi} e^{i\mathbf{A}\cdot\mathbf{B}} d\Omega (\mathbf{A}),$$

Fig. 9.7 Inward and outward propagation of the geometric "wavefront" from the surface S. The shaded regions indicate the space where the free field is nonvanishing and the unshaded regions indicate the space where the free field identically vanishes

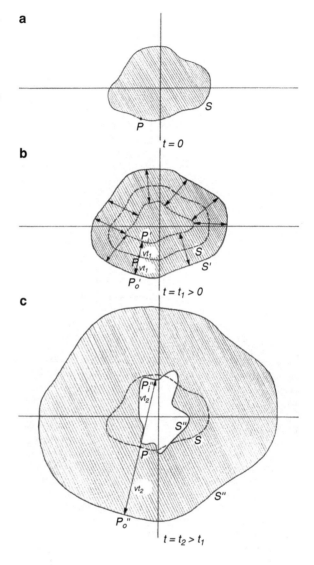

where $d\Omega(\mathbf{A})$ denotes the differential element of solid angle about the direction of the vector $\mathbf{A}$.

9.3 Determine the electric field vector $\mathbf{E}(\mathbf{r}, t)$ for $t > 0$ when $\tilde{\mathcal{E}}(\mathbf{k}) \equiv (1/2\pi)^3[\tilde{\mathbf{E}}_0(\mathbf{k}) - (v/\|c\|)\hat{\mathbf{s}} \times \tilde{\mathbf{B}}_0(\mathbf{k})] = (1/2\pi)^3 \hat{\mathbf{1}}\delta(k - \omega_0/v)$, where $\hat{\mathbf{1}}$ is a fixed unit vector in some specified direction.

9.4 Let the initial values for the electric and magnetic field vectors be given by $\mathbf{E}_0(\mathbf{r}) = \nabla \times \nabla \times \hat{\mathbf{1}}_z f(r)$ and $\mathbf{B}_0(\mathbf{r}) = \nabla \times \hat{\mathbf{1}}_z F(r)$, where $f(r)$ and $F(r)$ are sufficiently well-behaved scalar functions of $r \equiv |\mathbf{r}|$ alone. Use the plane wave

Fig. 9.8 Illustration for
Problem 9.5

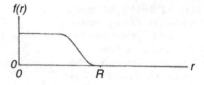

expansion given in Eq. (8.48) to determine the electric field $\mathbf{E}(\mathbf{r}, t)$ for $t > 0$.
Simplify the solution so that the only integral it contains is $u(r) = \int r F(r) dr$.

9.5 Let $\Pi(r, t)$ be a solution of the scalar wave equation

$$\nabla^2 \Pi(r, t) - \frac{1}{v^2} \ddot{\Pi}(r, t) = 0$$

for $t > 0$ with the initial values $\Pi(r, 0) = f(r)$ and $\dot{\Pi}(r, 0) = F(r)$, where $f(r)$
and $F(r)$ are scalar functions of $r \equiv |\mathbf{r}|$ alone.

(a) Let $F(r) = 0$ and let $f(r)$ vary with r as illustrated in Fig. 9.8, where $f(r) = 0$
    for $r > R$. Sketch the behavior of $\Pi(r, t)$ as a function of r for fixed t such that
    $t > R/v$.
(b) Let $f(r) = 0$ for all r and let $F(r) = 1$ for $r < R$ and $F(r) = 0$ for $r > R$.
    Determine $\Pi(r, t)$ for all $t > 0$.
(c) Let $F(r) = -\frac{v}{r} \frac{\partial}{\partial r} [r f(r)]$. Determine $\Pi(r, t)$ for all $t > 0$ in terms of the
    function $f(r)$.

References

1. G. C. Sherman, A. J. Devaney, and L. Mandel, "Plane-wave expansions of the optical field,"
   Opt. Commun., vol. 6, pp. 115–118, 1972.
2. G. C. Sherman, J. J. Stamnes, A. J. Devaney, and É. Lalor, "Contribution of the inhomogeneous
   waves in angular-spectrum representations," *Opt. Commun.*, vol. 8, pp. 271–274, 1973.
3. A. J. Devaney and G. C. Sherman, "Plane-wave representations for scalar wave fields," *SIAM
   Rev.*, vol. 15, pp. 765–786, 1973.
4. A. J. Devaney, *A New Theory of the Debye Representation of Classical and Quantized
   Electromagnetic Fields*. PhD thesis, The Institute of Optics, University of Rochester, 1971.
5. J. W. Goodman, *Introduction to Fourier Optics*. New York: McGraw-Hill, 1968.
6. M. Born and E. Wolf, *Principles of Optics*. Cambridge: Cambridge University Press, seventh
   (expanded) ed., 1999.
7. H. Goldstein, *Classical Mechanics*. Reading, MA: Addison-Wesley, 1950. Chapter 4.
8. N. S. Koshlyakov, M. M. Smirnov, and E. B. Gliner, *Differential Equations of Mathematical
   Physics*. Amsterdam: North-Holland, 1964. Ch. VI, Sect. 3.

Appendix A
The Dirac Delta Function

A.1 The One-Dimensional Dirac Delta Function

The *Dirac delta function* [1] in one-dimensional space may be defined by the pair
of equations

$$\delta(x) = 0; \quad x \neq 0, \tag{A.1}$$

$$\int_{-\infty}^{\infty} \delta(x)\, dx = 1. \tag{A.2}$$

It is clear from this definition that $\delta(x)$ is not a function in the ordinary mathematical
sense, because if a function is zero everywhere except at a single point and
the integral of this function over its entire domain of definition exists, then the
value of this integral is necessarily also equal to zero. Because of this, it is more
appropriate to regard $\delta(x)$ as a functional quantity with a certain well-defined
symbolic meaning. For example, one can consider a sequence of functions $\delta(x, \varepsilon)$
that, with increasing values of the parameter ε, differ appreciably from zero only
over a decreasing x-interval about the origin and which are such that

$$\int_{-\infty}^{\infty} \delta(x, \varepsilon)\, dx = 1 \tag{A.3}$$

for all values of ε. Although it may be tempting to try to interpret the Dirac delta
function as the limit of such a sequence of well-defined functions $\delta(x, \varepsilon)$ as $\varepsilon \to \infty$,
it must be recognized that this limit need not exist for all values of the independent
variable x. However, the limit

$$\lim_{\varepsilon \to \infty} \int_{-\infty}^{\infty} \delta(x, \varepsilon)\, dx = 1 \tag{A.4}$$

© Springer Nature Switzerland AG 2019
K. E. Oughstun, *Electromagnetic and Optical Pulse Propagation*, Springer Series
in Optical Sciences 224, https://doi.org/10.1007/978-3-030-20835-6

must exist. As a consequence, one may interpret any operation that involves the delta function $\delta(x)$ as implying that this operation is to be performed with a function $\delta(x, \varepsilon)$ of a suitable sequence and that the limit as $\varepsilon \to \infty$ is to be taken at the conclusion of the calculation. The particular choice of the sequence of functions $\delta(x, \varepsilon)$ is immaterial, provided that any oscillations near the origin $x = 0$ are not too violent [2]. For example, each of the following functions forms a sequence with respect to the parameter ε that satisfies the required properties:

$$\delta(x, \varepsilon) = \frac{\varepsilon}{\sqrt{\pi}} e^{-\varepsilon^2 x^2} \quad \& \quad \delta(x, \varepsilon) = \mathrm{rect}_{1/\varepsilon}(x) \quad \& \quad \delta(x, \varepsilon) = \frac{\varepsilon}{\pi} \mathrm{sinc}(\varepsilon x)$$

in Cartesian coordinates, and

$$\delta(r, \varepsilon) = \mathrm{circ}_{1/\varepsilon}(r) \quad \& \quad \delta(r, \varepsilon) = \frac{\varepsilon}{r} J_1(2\pi \varepsilon r)$$

in polar coordinates, where $\mathrm{rect}_{1/\varepsilon}(x) \equiv \varepsilon/2$ when $|x| < 1/\varepsilon$ and is zero otherwise, $\mathrm{circ}_{1/\varepsilon}(r) \equiv \varepsilon^2/\pi$ when $r < 1/\varepsilon$ and is zero otherwise, and $\mathrm{sinc}(x) \equiv \sin(x)/x$ when $x \neq 0$ and $\mathrm{sinc}(0) \equiv \lim_{x \to 0} \sin(x)/x = 1$.

Let $f(x)$ be a continuous and sufficiently well-behaved function of $x \in (-\infty, \infty)$ and consider the value of the definite integral

$$\int_{-\infty}^{\infty} f(x)\delta(x - a)dx = \lim_{\varepsilon \to \infty} \int_{-\infty}^{\infty} f(x)\delta(x - a, \varepsilon)dx.$$

When the parameter ε is large, the value of the integral appearing on the right-hand side of this equation depends essentially on the behavior of $f(x)$ in the immediate neighborhood of the point $x = a$ alone, and the error that results from the replacement of $f(x)$ by $f(a)$ may be made as small as desired by taking ε sufficiently large. Hence

$$\lim_{\varepsilon \to \infty} \int_{-\infty}^{\infty} f(x)\delta(x - a, \varepsilon)dx = f(a) \lim_{\varepsilon \to \infty} \int_{-\infty}^{\infty} \delta(x - a, \varepsilon)dx,$$

so that

$$\int_{-\infty}^{\infty} f(x)\delta(x - a)dx = f(a). \tag{A.5}$$

This result is referred to as the *sifting property* of the delta function. Notice that, for this result to hold, the domain of integration need not be extended over all $x \in (-\infty, \infty)$; it is only necessary that the domain of integration contain the point $x = a$

in its interior, so that

$$\int_{a-\Delta_1}^{a+\Delta_2} f(x)\delta(x-a)dx = f(a), \tag{A.6}$$

where $\Delta_{1,2} > 0$. It is then seen that $f(x)$ need only be continuous at $x = a$.
   The above results may be written symbolically as

$$f(x)\delta(x-a) = f(a)\delta(x-a), \tag{A.7}$$

the meaning of such a statement being that the two sides yield the same result when integrated over any domain containing the point $x = a$. For the special case when $f(x) = x^k$ with $k > 0$ and $a = 0$, Eq. (A.7) yields

$$x^k\delta(x) = 0, \quad \forall k > 0. \tag{A.8}$$

Theorem A.1 (Similarity Relationship (Scaling Law)) *For all* $a \neq 0$

$$\delta(ax) = \frac{1}{|a|}\delta(x). \tag{A.9}$$

Proof In order to prove this relationship one need only compare the integrals of $f(x)\delta(ax)$ and $f(x)\delta(x)/|a|$ for any sufficiently well-behaved continuous function $f(x)$. For the first integral one has (for any $a \neq 0$)

$$\int_{-\infty}^{\infty} f(x)\delta(ax)dx = \pm\frac{1}{a}\int_{-\infty}^{\infty} f(y/a)\delta(y)dy = \frac{1}{|a|}f(0),$$

where the upper or lower sign choice is taken accordingly as $a > 0$ or $a < 0$, respectively, and for the second integral one obtains

$$\int_{-\infty}^{\infty} f(x)\frac{1}{|a|}\delta(x)dx = \frac{1}{|a|}f(0).$$

Comparison of these two results then shows that $\delta(ax) = \delta(x)/|a|$, as was to be proved.

For the special case $a = -1$, Eq. (A.9) yields

$$\delta(-x) = \delta(x), \tag{A.10}$$

so that the delta function is an even function of its argument.

Theorem A.2 (Composite Function Theorem) *If $y = f(x)$ is any continuous function of x with simple zeroes at the points x_i [i.e., $y = 0$ at $x = x_i$ and $f'(x_i) \neq 0$] and no other zeroes, then*

$$\delta(f(x)) = \sum_i \frac{1}{|f'(x_i)|} \delta(x - x_i).$$ (A.11)

Proof Let $g(x)$ be any sufficiently well-behaved continuous function and let $\{x_i\}$ denote the set of points at which $y = 0$. Under the change of variable $x = f^{-1}(y)$ one has that

$$\int_{-\infty}^{\infty} g(x)\delta(f(x))dx = \int_R g\left(f^{-1}(y)\right) \delta(y) \frac{1}{|f'(f^{-1}(y))|} dy$$

$$= \sum_{x_i} g\left(f^{-1}(0)\right) \frac{1}{|f'(f^{-1}(0))|} = \sum_i g(x_i) \frac{1}{|f'(x_i)|},$$

where R denotes the range of $f(x)$. In addition,

$$\int_{-\infty}^{\infty} g(x) \left\{ \sum_i \frac{1}{|f'(x_i)|} \delta(x - x_i) \right\} dx = \sum_i \frac{1}{|f'(x_i)|} \int_{-\infty}^{\infty} g(x)\delta(x - x_i)dx$$

$$= \sum_i g(x_i) \frac{1}{|f'(x_i)|}.$$

Comparison of these two expressions then proves the theorem.

As an example, consider the function $f(x) = x^2 - a^2$ which has simple zeroes at $x = \pm a$. Then $|f'(\pm a)| = 2|a|$ so that, for $a \neq 0$,

$$\delta(x^2 - a^2) = \frac{1}{2|a|} (\delta(x - a) + \delta(x + a)).$$

An additional relationship of interest that employs the Dirac delta function is

$$\int_{-\infty}^{\infty} \delta(\xi - x)\delta(x - \eta)dx = \delta(\xi - \eta),$$ (A.12)

which is seen to be an extension of the sifting property to the delta function itself. This equation then implies that if both sides are multiplied by a continuous function of either ξ or η and the result integrated over all values of either ξ or η, respectively, an identity is obtained. That is, because

$$\int_{-\infty}^{\infty} f(\xi)\delta(\xi - \eta)d\xi = f(\eta),$$

and

$$\int_{-\infty}^{\infty} f(\xi) \left[\int_{-\infty}^{\infty} \delta(\xi - x)\delta(x - \eta)dx \right] d\xi$$

$$= \int_{-\infty}^{\infty} \left[\int_{-\infty}^{\infty} f(\xi)\delta(\xi - x)d\xi \right] \delta(x - \eta)dx$$

$$= \int_{-\infty}^{\infty} f(x)\delta(x - \eta)dx = f(\eta)$$

then the expression in Eq. (A.12) follows. In a similar manner, because

$$\int_{-\infty}^{\infty} g(\eta) \left[\int_{-\infty}^{\infty} \delta(\xi - x)\delta(x - \eta)dx \right] d\eta$$

$$= \int_{-\infty}^{\infty} \left[\int_{-\infty}^{\infty} g(\eta)\delta(x - \eta)d\eta \right] \delta(\xi - x)dx$$

$$= \int_{-\infty}^{\infty} g(x)\delta(\xi - x)dx = g(\xi)$$

and

$$\int_{-\infty}^{\infty} g(\eta)\delta(\xi - \eta)d\eta = g(\xi),$$

then the expression in Eq. (A.12) is again obtained.

Consider next what interpretation may be given to the derivatives of the delta function, accomplished through use of the function sequence $\delta(x, \varepsilon)$. Consider then the ordinary integral $\int_{-\infty}^{\infty} f(x)\delta'(x, \varepsilon)dx$ which may be evaluated by application of the method of integration by parts with $u = f(x)$ and $dv = \delta'(x, \varepsilon)dx$, so that

$$\int_{-\infty}^{\infty} f(x)\delta'(x, \varepsilon)dx = f(\infty)\delta(\infty, \varepsilon) - f(-\infty)\delta(-\infty, \varepsilon) - \int_{-\infty}^{\infty} f'(x)\delta(x, \varepsilon)dx.$$

Upon proceeding to the limit as $\varepsilon \to \infty$, the first two terms appearing on the right-hand side of this equation both vanish because

$$\lim_{\varepsilon \to \infty} \delta(\pm\infty, \varepsilon) = 0, \tag{A.13}$$

with the result

$$\int_{-\infty}^{\infty} f(x)\delta'(x)dx = -f'(0). \tag{A.14}$$

Upon repeating this procedure n times for the nth-order derivative of the delta function, one obtains the general result

$$\int_{-\infty}^{\infty} f(x)\delta^{(n)}(x)dx = (-1)^n f^{(n)}(0). \tag{A.15}$$

As a special case of Eq. (A.14), let $f(x) = x$ so that

$$\int_{-\infty}^{\infty} x\delta'(x)dx = -1 = -\int_{-\infty}^{\infty} \delta(x)dx,$$

and one then has the equivalence

$$x\delta'(x) = -\delta(x). \tag{A.16}$$

Because $\delta(x)$ is an even function and x is an odd function, it then follows that $\delta'(x)$ is an odd function of its argument; that is

$$\delta'(-x) = -\delta'(x). \tag{A.17}$$

The generalization of Eq. (A.16) may be directly obtained from Eq. (A.15) by letting $f(x) = x^n$. In that case, $f^{(n)}(x) = n!$ and this relation gives

$$\int_{-\infty}^{\infty} x^n\delta'(x)dx = (-1)^n n! = (-1)^n n! \int_{-\infty}^{\infty} \delta(x)dx,$$

and one then has the general equivalence

$$x^n\delta^{(n)}(x) = (-1)^n n!\delta(x). \tag{A.18}$$

The even-order derivatives of the delta function are even functions and the odd-order derivatives are odd functions of the argument.

It is often convenient to express the Dirac delta function in terms of the *Heaviside unit step function* $U(x)$ defined by the relations $U(x) = 0$ when $x < 0$, $U(x) = 1$ when $x > 0$. Consider the behavior of the derivative of $U(x)$. If, as before, a superscript prime denotes differentiation with respect to the argument, one obtains formally upon integration by parts (with the limits $-x_1 < 0$ and $x_2 > 0$),

$$\int_{-x_1}^{x_2} f(x)U'(x)dx = [f(x)U(x)]_{-x_1}^{x_2} - \int_{-x_1}^{x_2} f'(x)U(x)dx$$

$$= f(x_2) - \int_{0}^{x_2} f'(x)dx$$

$$= f(x_2) - [f(x_2) - f(0)] = f(0),$$

where $f(x)$ is any continuous function. Upon setting $x = y - a$ and $f(x) = f(y - a) = F(y)$ and then proceeding to the limits as $-x_1 \to -\infty$ and $x_2 \to +\infty$, the above result becomes

$$\int_{-\infty}^{\infty} F(y)U'(y-a)dy = F(a),$$

and the derivative $U'(x)$ is seen to satisfy the sifting property given in Eq. (A.5). In particular, with $F(y) = 1$ and $a = 0$, this expression becomes

$$\int_{-\infty}^{\infty} U'(y)dy = 1,$$

and $U'(x)$ also satisfies the property given in Eq. (A.2) which serves to partially define the delta function. Moreover, $U'(x) = 0$ for all $x \neq 0$ and property (A.1) is also satisfied. Hence, one may identify the derivative of the unit step function with the delta function, so that

$$\delta(x) = \frac{dU(x)}{dx}. \tag{A.19}$$

In addition, it is seen that [from Eqs. (A.6) and (A.19)]

$$U(x) = \int_{-\infty}^{x} \delta(\xi)d\xi. \tag{A.20}$$

The Dirac delta function may also be introduced through the use of the Fourier integral theorem [3], which may be written as

$$f(a) = \int_{-\infty}^{\infty} dv \int_{-\infty}^{\infty} dx \, f(x)e^{i2\pi v(x-a)} \tag{A.21}$$

for any sufficiently well-behaved, continuous function $f(x)$. Define the function sequence

$$K(x - a, \varepsilon) \equiv \int_{-\varepsilon}^{\varepsilon} e^{i2\pi v(x-a)}dv = \frac{\sin(2\pi(x-a)\varepsilon)}{\pi(x-a)} \tag{A.22}$$

with limit

$$K(x - a) \equiv \lim_{\varepsilon \to \infty} K(x - a, \varepsilon). \tag{A.23}$$

Strictly speaking, this limit does not exist in the ordinary sense when $x \neq a$; however, the limit does exist and has the value zero when $x \neq a$ if it is interpreted in the sense of a Cesáro limit [4]. Upon inversion of the order of integration, Eq. (A.21) may be formally rewritten as

$$f(a) = \int_{-\infty}^{\infty} f(x)K(x-a)dx,$$ (A.24)

which should be interpreted as meaning that

$$f(a) = \lim_{\varepsilon \to \infty} \int_{-\infty}^{\infty} f(x)K(x-a,\varepsilon)dx.$$ (A.25)

Thus, the function $K(x-a)$ satisfies the sifting property (A.5) of the delta function. If one sets $f(x)=1$ and $a=0$ in Eq. (A.24), there results $\int_{-\infty}^{\infty} K(x)dx = 1$ and $K(x)$ satisfies the property given in Eq. (A.2) which serves to partially define the delta function. Because $K(x) = \lim_{\varepsilon \to \infty} K(x,\varepsilon) = 0$ when $x \neq 0$, so that the property given in Eq. (A.1) is also satisfied, one then obtains from Eq. (A.23) the relation

$$\delta(x) = \int_{-\infty}^{\infty} e^{i2\pi vx}dv.$$ (A.26)

That is, the Dirac delta function may be regarded as the Fourier transform of unity. The reciprocal relation follows from Eq. (A.25) upon setting $f(x) = \exp(i2\pi vx)$ and $a=0$, so that

$$1 = \int_{-\infty}^{\infty} \delta(x)e^{-i2\pi vx}dx,$$ (A.27)

which also follows directly from the sifting property given in Eq. (A.5). Notice that this relation by itself is not sufficient to imply the validity of Eq. (A.26).

A.2   The Dirac Delta Function in Higher Dimensions

The definition of the Dirac delta function may easily be extended to higher-dimensional spaces. In particular, consider three-dimensional vector space in which case the defining relations given in Eqs. (A.1) and (A.2) become

$$\delta(\mathbf{r}) = 0; \quad \mathbf{r} \neq 0,$$ (A.28)

$$\int_{-\infty}^{\infty} \delta(\mathbf{r})d^3r = 1.$$ (A.29)

The function

$$\delta(\mathbf{r}) \equiv \delta(x, y, z)$$

$$\equiv \delta(x)\delta(y)\delta(z), \tag{A.30}$$

where $\mathbf{r} = \hat{\mathbf{1}}_x x + \hat{\mathbf{1}}_y y + \hat{\mathbf{1}}_z z$ is the position vector with components (x, y, z) clearly satisfies Eqs. (A.28) and (A.29) and so defines a three-dimensional Dirac delta function. The sifting property given in Eq. (A.5) then becomes

$$\int_{-\infty}^{\infty} f(\mathbf{r})\delta(\mathbf{r} - \mathbf{a})d^3 r = f(\mathbf{a}), \tag{A.31}$$

and the similarity relationship or scaling law given in Eq. (A.9) now states that

$$\delta(a\mathbf{r}) = \frac{1}{|a|^3}\delta(\mathbf{r}), \tag{A.32}$$

where a is a scalar constant. The Fourier transform pair relationship expressed in Eqs. (A.26) and (A.27) becomes

$$\delta(\mathbf{r}) = \frac{1}{(2\pi)^3} \int_{-\infty}^{\infty} e^{i\mathbf{k}\cdot\mathbf{r}} d^3 k, \tag{A.33}$$

$$1 = \int_{-\infty}^{\infty} \delta(\mathbf{r})e^{-i\mathbf{k}\cdot\mathbf{r}} d^3 r, \tag{A.34}$$

where $\mathbf{k} = \hat{\mathbf{1}}_x k_x + \hat{\mathbf{1}}_y k_y + \hat{\mathbf{1}}_z k_z = 2\pi(\hat{\mathbf{1}}_x v_x + \hat{\mathbf{1}}_y v_y + \hat{\mathbf{1}}_z v_z)$.

The generalization of the three-dimensional Dirac delta function to more general coordinate systems requires more careful attention. Suppose that a function $\Delta(\mathbf{r})$ is given in Cartesian coordinates as

$$\Delta(\mathbf{r}) = \delta(x)\delta(y)\delta(z) \tag{A.35}$$

and it is desired to express $\Delta(\mathbf{r})$ in terms of the orthogonal curvilinear coordinates (u, v, w) defined by

$$u = f_1(x, y, z), \quad v = f_2(x, y, z), \quad w = f_3(x, y, z), \tag{A.36}$$

where f_1, f_2, f_3 are continuous, single-valued functions of x, y, z with a unique inverse $x = f_1^{-1}(u, v, w), y = f_2^{-1}(u, v, w), z = f_3^{-1}(u, v, w)$. That is, an expression for $\Delta(\mathbf{r})$ is desired in terms of the coordinate variables (u, v, w) that satisfies the relation

$$\int_{-\infty}^{\infty} \Delta(\mathbf{r} - \mathbf{r}')\varphi(u, v, w)dV = \varphi(u', v', w'), \tag{A.37}$$

where dV is the differential volume element in u, v, w-space and (u', v', w') is the point corresponding to (x', y', z') under the coordinate transformation given in Eq. (A.36). If the point $\mathbf{r} = (x, y, z)$ is varied from $\mathbf{r}$ to $\mathbf{r} + \delta\mathbf{r}_1$ by changing the coordinate variable u to $u + \delta u$ while keeping v and w fixed, then

$$\delta\mathbf{r}_1 = \frac{\partial\mathbf{r}}{\partial u}\delta u.$$

Similarly, if the point $\mathbf{r} = (x, y, z)$ is varied from $\mathbf{r}$ to $\mathbf{r} + \delta\mathbf{r}_2$ by changing the coordinate variable v to $v + \delta v$ while keeping u and w fixed, then

$$\delta\mathbf{r}_2 = \frac{\partial\mathbf{r}}{\partial v}\delta v.$$

The parallelogram with sides $\delta\mathbf{r}_1$ and $\delta\mathbf{r}_2$ then has area

$$\delta A = |\delta\mathbf{A}| = |\delta\mathbf{r}_1 \times \delta\mathbf{r}_2| = \left|\frac{\partial\mathbf{r}}{\partial u} \times \frac{\partial\mathbf{r}}{\partial v}\right|\delta u\delta v. \tag{A.38}$$

If the point $\mathbf{r} = (x, y, z)$ is now varied from $\mathbf{r}$ to $\mathbf{r} + \delta\mathbf{r}_3$ by changing the coordinate variable w to $w + \delta w$ while keeping u and v fixed, then

$$\delta\mathbf{r}_3 = \frac{\partial\mathbf{r}}{\partial w}\delta w,$$

and the volume of the parallelepiped with edges $\delta\mathbf{r}_1, \delta\mathbf{r}_2$, and $\delta\mathbf{r}_3$ is then given by

$$\delta V = |\delta\mathbf{r}_3 \cdot (\delta\mathbf{r}_1 \times \delta\mathbf{r}_2)| = \left|\frac{\partial\mathbf{r}}{\partial w} \cdot \left(\frac{\partial\mathbf{r}}{\partial u} \times \frac{\partial\mathbf{r}}{\partial v}\right)\right|\delta u\delta v\delta w. \tag{A.39}$$

The quantity

$$J\left(\frac{x, y, z}{u, v, w}\right) \equiv \frac{\partial(x, y, z)}{\partial(u, v, w)} \equiv \frac{\partial\mathbf{r}}{\partial w} \cdot \left(\frac{\partial\mathbf{r}}{\partial u} \times \frac{\partial\mathbf{r}}{\partial v}\right) \tag{A.40}$$

is recognized as the *Jacobian of the coordinate transformation* of x, y, z with respect to u, v, w. With this result for the differential element of volume, Eq. (A.37) becomes

$$\int_{-\infty}^{\infty} \Delta(\mathbf{r} - \mathbf{r}')\varphi(u, v, w)\left|J\left(\frac{x, y, z}{u, v, w}\right)\right| du\,dv\,dw = \varphi(u', v', w'), \tag{A.41}$$

from which it is immediately seen that

$$\delta(u)\delta(v)\delta(w) = \left|J\left(\frac{x, y, z}{u, v, w}\right)\right|\delta(x)\delta(y)\delta(z). \tag{A.42}$$

Because this transformation is assumed to be single-valued, then

$$\delta(x)\delta(y)\delta(z) = \frac{\delta(u)\delta(v)\delta(w)}{\left| J\left(\frac{x,y,z}{u,v,w}\right)\right|} = \left| J\left(\frac{u,v,w}{x,y,z}\right)\right| \delta(u)\delta(v)\delta(w), \qquad (A.43)$$

where $J(u, v, w/x, y, z)$ is the *Jacobian of the inverse transformation*.

Consider finally the description of a function $\varrho(\mathbf{r})$ that vanishes everywhere in three-dimensional space except on a surface S and is such that

$$\int_{-\infty}^{\infty} \varrho(\mathbf{r})\varphi(\mathbf{r})d^3r = \int_S \varsigma(\mathbf{r})\varphi(\mathbf{r})d^2r, \qquad (A.44)$$

where $\varsigma(\mathbf{r})$ is the value of $\varrho(\mathbf{r})$ on the surface S, that is, when $\mathbf{r} \in S$. Choose orthogonal curvilinear coordinates (u, v, w) such that $w = w_0$ describes the surface S for some constant w_0, in which case ∇w is parallel to the normal to the surface S, and is such that ∇u and ∇v are both perpendicular to the normal to the surface S. The differential element of area of the surface S is then given by Eq. (A.38). Furthermore, both $\partial \mathbf{r}/\partial w$ and $(\partial \mathbf{r}/\partial u) \times (\partial \mathbf{r}/\partial v)$ are normal to S so that

$$J\left(\frac{x, y, z}{u, v, w}\right) = \left|\frac{\partial \mathbf{r}}{\partial w}\right| \left|\frac{\partial \mathbf{r}}{\partial u} \times \frac{\partial \mathbf{r}}{\partial v}\right|. \qquad (A.45)$$

With this result, Eq. (A.44) may be written as

$$\int_{-\infty}^{\infty} \varrho(\mathbf{r})\varphi(\mathbf{r}) \left|\frac{\partial \mathbf{r}}{\partial w}\right| \left|\frac{\partial \mathbf{r}}{\partial u} \times \frac{\partial \mathbf{r}}{\partial v}\right| du\,dv\,dw = \int_S \varsigma(\mathbf{r})\varphi(\mathbf{r})d^2r,$$

which, with Eq. (A.38), may be expressed as

$$\int_{-\infty}^{\infty} \varrho(\mathbf{r})\varphi(\mathbf{r}) \left|\frac{\partial \mathbf{r}}{\partial w}\right| d^2r\,dw = \int_S \varsigma(\mathbf{r})\varphi(\mathbf{r})d^2r. \qquad (A.46)$$

From this result it then follows that

$$\varrho(\mathbf{r}) = \frac{\varsigma(\mathbf{r})}{\left|\frac{\partial \mathbf{r}}{\partial w}\right|}\delta(w - w_0), \qquad (A.47)$$

which is the solution of Eq. (A.44). This result can be simplified somewhat by noting that when the variable w is varied while u and v are held fixed, then the changes in $\mathbf{r}$ and w are related by $\delta w = \nabla w \cdot \delta\mathbf{r}$, so that

$$\frac{\partial \mathbf{r}}{\partial w} \cdot \nabla w = 1.$$

Moreover, because both $\partial\mathbf{r}/\partial w$ and ∇w are normal to the surface S described by $w = w_0$, then

$$\left|\frac{\partial\mathbf{r}}{\partial w}\right| |\nabla w| = 1,$$

and, as a result, Eq. (A.47) becomes

$$\varrho(\mathbf{r}) = |\nabla w|\, \varsigma(\mathbf{r})\delta(w - w_0) \tag{A.48}$$

as the solution to Eq. (A.44).

References

1. P. A. M. Dirac, *The Principles of Quantum Mechanics*. Oxford: Oxford University Press, 1930. Sect. 15.
2. M. J. Lighthill, *Introduction to Fourier Analysis and Generalized Functions*. London, England: Cambridge University Press, 1970.
3. E. C. Titchmarsh, *Introduction to the Theory of Fourier Integrals*. London: Oxford University Press, 1939. Ch. I.
4. B. van der Pol and H. Bremmer, *Operational Calculus Based on the Two-Sided Laplace Integral*. London: Cambridge University Press, 1950. pp. 100–104.

Appendix B
Helmholtz' Theorem

Because

$$\nabla^2 \left(\frac{1}{R}\right) = -4\pi \delta(\mathbf{R}) \tag{B.1}$$

where $\mathbf{R} = \mathbf{r} - \mathbf{r}'$ with magnitude $R = |\mathbf{R}|$ and where $\delta(\mathbf{R}) = \delta(\mathbf{r} - \mathbf{r}') = \delta(x - x')\delta(y - y')\delta(z - z')$ is the three-dimensional Dirac delta function (see Appendix A), then any sufficiently well-behaved vector function $\mathbf{F}(\mathbf{r}) = \mathbf{F}(x, y, z)$ can be represented as

$$\mathbf{F}(\mathbf{r}) = \int_V \mathbf{F}(\mathbf{r}')\delta(\mathbf{r} - \mathbf{r}')\, d^3r' = -\frac{1}{4\pi} \int_V \mathbf{F}(\mathbf{r}')\nabla^2 \left(\frac{1}{R}\right) d^3r'$$

$$= -\frac{1}{4\pi}\nabla^2 \int_V \frac{\mathbf{F}(\mathbf{r}')}{R}\, d^3r', \tag{B.2}$$

the integration extending over any region V that contains the point $\mathbf{r}$. With the identity $\nabla \times \nabla \times = \nabla\nabla \cdot - \nabla^2$, Eq. (B.2) may be written as

$$\mathbf{F}(\mathbf{r}) = \frac{1}{4\pi}\nabla \times \nabla \times \int_V \frac{\mathbf{F}(\mathbf{r}')}{R}\, d^3r' - \frac{1}{4\pi}\nabla\nabla \cdot \int_V \frac{\mathbf{F}(\mathbf{r}')}{R}\, d^3r'. \tag{B.3}$$

Consider first the divergence term appearing in this expression. Because the vector differential operator ∇ does not operate on the primed coordinates, then

$$\frac{1}{4\pi}\nabla \cdot \int_V \frac{\mathbf{F}(\mathbf{r}')}{R}\, d^3r' = \frac{1}{4\pi} \int_V \mathbf{F}(\mathbf{r}') \cdot \nabla \left(\frac{1}{R}\right) d^3r'. \tag{B.4}$$

© Springer Nature Switzerland AG 2019
K. E. Oughstun, *Electromagnetic and Optical Pulse Propagation*, Springer Series in Optical Sciences 224, https://doi.org/10.1007/978-3-030-20835-6

Moreover, the integrand appearing in this expression may be expressed as

$$\mathbf{F}(\mathbf{r}') \cdot \nabla \left(\frac{1}{R} \right) = -\mathbf{F}(\mathbf{r}') \cdot \nabla' \left(\frac{1}{R} \right)$$

$$= -\nabla' \cdot \left(\frac{\mathbf{F}(\mathbf{r}')}{R} \right) + \frac{1}{R} \nabla' \cdot \mathbf{F}(\mathbf{r}'), \tag{B.5}$$

where the superscript prime on the vector differential operator ∇' denotes differentiation with respect to the primed coordinates alone. Substitution of Eq. (B.5) into Eq. (B.4) and application of the divergence theorem to the first term then yields

$$\frac{1}{4\pi} \nabla \cdot \int_V \frac{\mathbf{F}(\mathbf{r}')}{R} d^3 r' = -\frac{1}{4\pi} \int_V \nabla' \cdot \left(\frac{\mathbf{F}(\mathbf{r}')}{R} \right) d^3 r' + \frac{1}{4\pi} \int_V \frac{\nabla' \cdot \mathbf{F}(\mathbf{r}')}{R} d^3 r'$$

$$= -\frac{1}{4\pi} \oint_S \frac{1}{R} \mathbf{F}(\mathbf{r}') \cdot \hat{n} d^2 r' + \frac{1}{4\pi} \int_V \frac{\nabla' \cdot \mathbf{F}(\mathbf{r}')}{R} d^3 r'$$

$$= \phi(\mathbf{r}), \tag{B.6}$$

which is the desired form of the scalar potential $\phi(\mathbf{r})$ for the vector field $\mathbf{F}(\mathbf{r})$. Here S is the surface that encloses the regular region V and contains the point $\mathbf{r}$.

For the curl term appearing in Eq. (B.3) one has that

$$\frac{1}{4\pi} \nabla \times \int_V \frac{\mathbf{F}(\mathbf{r}')}{R} d^3 r' = -\frac{1}{4\pi} \int_V \mathbf{F}(\mathbf{r}') \times \nabla \left(\frac{1}{R} \right) d^3 r'$$

$$= \frac{1}{4\pi} \int_V \mathbf{F}(\mathbf{r}') \times \nabla' \left(\frac{1}{R} \right) d^3 r'. \tag{B.7}$$

Moreover, the integrand appearing in the final form of the integral in Eq. (B.7) may be expressed as

$$\mathbf{F}(\mathbf{r}') \times \nabla' \left(\frac{1}{R} \right) = \frac{\nabla' \times \mathbf{F}(\mathbf{r}')}{R} - \nabla' \times \left(\frac{\mathbf{F}(\mathbf{r}')}{R} \right), \tag{B.8}$$

so that

$$\frac{1}{4\pi} \nabla \times \int_V \frac{\mathbf{F}(\mathbf{r}')}{R} d^3 r' = \frac{1}{4\pi} \int_V \frac{\nabla' \times \mathbf{F}(\mathbf{r}')}{R} d^3 r' - \frac{1}{4\pi} \int_V \nabla' \times \left(\frac{\mathbf{F}(\mathbf{r}')}{R} \right) d^3 r'$$

$$= \frac{1}{4\pi} \int_V \frac{\nabla' \times \mathbf{F}(\mathbf{r}')}{R} d^3 r' + \frac{1}{4\pi} \oint_S \frac{1}{R} \mathbf{F}(\mathbf{r}') \times \hat{n} d^2 r'$$

$$= \mathbf{A}(\mathbf{r}), \tag{B.9}$$

which is the desired form of the vector potential $\mathbf{A}(\mathbf{r})$.

The relations given in Eqs. (B.3), (B.6), and (B.9) then show that

$$\mathbf{F}(\mathbf{r}) = -\nabla \phi(\mathbf{r}) + \nabla \times \mathbf{A}(\mathbf{r}), \tag{B.10}$$

which may also be written as

$$\mathbf{F}(\mathbf{r}) = \mathbf{F}_\ell(\mathbf{r}) + \mathbf{F}_t(\mathbf{r}), \tag{B.11}$$

where

$$\mathbf{F}_\ell(\mathbf{r}) = -\nabla \phi(\mathbf{r})$$
$$= -\frac{1}{4\pi} \nabla \int_V \frac{\nabla' \cdot \mathbf{F}(\mathbf{r}')}{|\mathbf{r} - \mathbf{r}'|} d^3 r' + \frac{1}{4\pi} \nabla \oint_S \frac{\mathbf{F}(\mathbf{r}')}{|\mathbf{r} - \mathbf{r}'|} \cdot \hat{\mathbf{n}} d^2 r' \tag{B.12}$$

is the *longitudinal or irrotational part of the vector field* (where $\nabla \times \mathbf{F}_\ell(\mathbf{r}') = \mathbf{0}$), and where

$$\mathbf{F}_t(\mathbf{r}) = \nabla \times \mathbf{A}(\mathbf{r})$$
$$= \frac{1}{4\pi} \nabla \times \nabla \times \int_V \frac{\mathbf{F}(\mathbf{r}')}{|\mathbf{r} - \mathbf{r}'|} d^3 r'$$
$$= \frac{1}{4\pi} \nabla \times \int_V \frac{\nabla' \times \mathbf{F}(\mathbf{r}')}{|\mathbf{r} - \mathbf{r}'|} d^3 r' + \frac{1}{4\pi} \nabla \times \oint_S \frac{\mathbf{F}(\mathbf{r}')}{|\mathbf{r} - \mathbf{r}'|} \times \hat{\mathbf{n}} d^2 r' \tag{B.13}$$

is the *transverse or solenoidal part of the vector field* (where $\nabla \cdot \mathbf{F}_t(\mathbf{r}') = 0$).

If the surface S recedes to infinity and if the vector field $\mathbf{F}(\mathbf{r})$ is regular at infinity, then the surface integrals appearing in the above expressions vanish and Eqs. (B.12) and (B.13) become

$$\mathbf{F}_\ell(\mathbf{r}) = -\nabla \phi(\mathbf{r})$$
$$= -\frac{1}{4\pi} \nabla \int_V \frac{\nabla' \cdot \mathbf{F}(\mathbf{r}')}{|\mathbf{r} - \mathbf{r}'|} d^3 r', \tag{B.14}$$
$$\mathbf{F}_t(\mathbf{r}) = \nabla \times \mathbf{A}(\mathbf{r})$$
$$= \frac{1}{4\pi} \nabla \times \int_V \frac{\nabla' \times \mathbf{F}(\mathbf{r}')}{|\mathbf{r} - \mathbf{r}'|} d^3 r'. \tag{B.15}$$

Taken together, the above results constitute what is known as Helmholtz' theorem [1].

Theorem B.1 (Helmholtz' Theorem) *Let $\mathbf{F}(\mathbf{r})$ be any continuous vector field with continuous first partial derivatives. Then $\mathbf{F}(\mathbf{r})$ can be uniquely expressed in terms* *of the negative gradient of a scalar potential $\phi(\mathbf{r})$ and the curl of a vector potential* $\mathbf{A}(\mathbf{r})$, *as embodied in Eqs. (B.10)–(B.13).*

Reference

1. H. B. Phillips, *Vector Analysis*. New York: John Wiley & Sons, 1933.

Appendix C
The Effective Local Field

The average electric field intensity acting on a given molecule in a dielectric is called the *local* or *effective* field. In a linear isotropic dielectric, the spatially averaged induced molecular charge separation is directly proportional to, and in the same direction as, the local field at that molecular site so that the average induced molecular dipole moment is given by

$$\langle\langle\tilde{\mathbf{p}}(\mathbf{r}, \omega)\rangle\rangle = \epsilon_0 \alpha(\omega)\langle\langle\tilde{\mathbf{E}}_{eff}(\mathbf{r}, \omega)\rangle\rangle \qquad (C.1)$$

at the fixed angular frequency ω of the applied time-harmonic field [see Eq. (4.171)]. The *molecular polarizability* $\alpha(\omega)$ characterizes the frequency-dependent linear response of the molecules comprising the dielectric to the applied electric field.

The effective local field may be determined by removing the molecule under consideration while maintaining all of the remaining molecules in their time-averaged polarized states (time-averaging being used only to remove the effects of thermal fluctuations), the spatially averaged electric field intensity then being calculated in the cavity left vacant by that removed molecule [1–3]. Let V_m denote the volume of that single molecular cavity region. The effective local field at that molecular site is then given by the difference

$$\langle\langle\tilde{\mathbf{E}}_{eff}(\mathbf{r}, \omega)\rangle\rangle = \frac{1}{V_m}\int_{V_m}\tilde{\mathbf{e}}(\mathbf{r}', \omega)d^3 r' - \frac{1}{V_m}\int_{V_m}\tilde{\mathbf{e}}_m(\mathbf{r}', \omega)d^3 r', \qquad (C.2)$$

where $\tilde{\mathbf{e}}(\mathbf{r}', \omega)$ is the total microscopic electric field at the point $\mathbf{r}' \in V_m$, and where $\tilde{\mathbf{e}}_m(\mathbf{r}', \omega)$ is the electric field due to the charge distribution of the molecule under consideration evaluated at the point $\mathbf{r}' \in V_m$.

If the dielectric material is locally homogeneous (i.e., its dielectric properties at any point in the material are essentially constant over a macroscopically small but microscopically large region), then the first integral appearing in Eq. (C.2) is essentially the macroscopic electric field defined in Eq. (4.6). In particular, if the

© Springer Nature Switzerland AG 2019
K. E. Oughstun, *Electromagnetic and Optical Pulse Propagation*, Springer Series in Optical Sciences 224, https://doi.org/10.1007/978-3-030-20835-6

Fig. C.1 Spherical cavity region V of radius R with point charge q situated a distance $r' < R$ from the center O

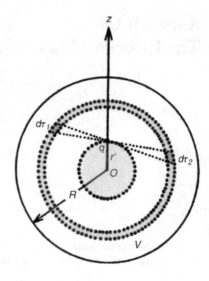

weighting function $w(\mathbf{r}')$ is taken to be given by $1/V_m$ when $\mathbf{r}' \in V_m$ and 0 otherwise, then Eq. (4.1) gives

$$\tilde{\mathbf{E}}(\mathbf{r}, \omega) \equiv \langle\langle \tilde{\mathbf{e}}(\mathbf{r}, \omega)\rangle\rangle = \frac{1}{V_m} \int_{V_m} \tilde{\mathbf{e}}(\mathbf{r} - \mathbf{r}', \omega) d^3 r', \qquad (C.3)$$

where $\mathbf{r} \in V_m$. With the change of variable $\mathbf{r}'' = \mathbf{r} - \mathbf{r}'$, the first integral appearing in Eq. (C.2) is then obtained.

For the second integral appearing in Eq. (C.2), consider determining the average electric field intensity inside a sphere of radius R containing a point charge q that is located a distance r' from the center O of the sphere, as illustrated in Fig. C.1. The z-axis is chosen to be along the line from the center O of the sphere passing through the point charge q. With this choice, symmetry shows that the average field over the spherical volume must be along the z-axis. The average electric field in V is then given by the scalar quantity

$$\bar{e}_z = \frac{1}{V} \int_V \tilde{e}_z d^3 r, \qquad (C.4)$$

where V is the volume of the spherical region. It is convenient to separate this volume integral into two parts, one taken over the spherical shell V_1 between the radii r' and R, and the other taken over the inner sphere V_2 of radius r', so that

$$\bar{e}_z = \frac{1}{V} \int_{V_1} \tilde{e}_z d^3 r + \frac{1}{V} \int_{V_2} \tilde{e}_z d^3 r. \qquad (C.5)$$

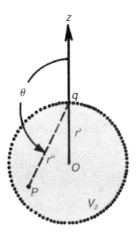

Fig. C.2 Spherical polar coordinate system about the point charge q for the integral over the inner spherical region V_2 with center at O and radius r'

The integral over the spherical shell region V_1 vanishes because of the equal and opposite contributions arising from the pair of volume elements $d\tau_1$ and $d\tau_2$ that are intercepted by the element of solid angle $d\Omega$, as depicted in Fig. C.1. Because the magnitude of $\tilde{e}_z$ decreases with the square of the distance from the point charge q, whereas the volume element $d\tau = r^2 d\Omega$ increases with the square of this distance, their product remains constant. For positive q, $\tilde{e}_z$ is positive at $d\tau_1$ and it is negative at $d\tau_2$, whereas for negative q, $\tilde{e}_z$ is negative at $d\tau_1$ and positive at $d\tau_2$. In either case, the two contributions to the integral over V_1 cancel and that integral then vanishes.

The integral of $\tilde{e}_z$ over the inner volume V_2 is then equal to the same integral over the entire spherical region V. In order to evaluate this final volume integral, choose spherical polar coordinates (r'', θ, φ) with origin at the point charge q, as illustrated in Fig. C.2. At any point $P \in V_2$,

$$\tilde{e}_z = \tilde{\mathbf{e}} \cdot \hat{\mathbf{1}}_z = \frac{\|4\pi\|}{4\pi\epsilon_0} \frac{q}{r''^2} \cos(\theta), \tag{C.6}$$

so that

$$\int_V \tilde{e}_z d^3r = \frac{\|4\pi\|}{4\pi\epsilon_0} q \int_0^{2\pi} d\varphi \int_{\pi/2}^{\pi} \sin(\theta)\cos(\theta) \left[\int_0^{-2r'\cos(\theta)} dr'' \right] d\theta$$

$$= -\frac{\|4\pi\|}{\epsilon_0} qr' \int_{\pi/2}^{\pi} \cos^2(\theta)\sin(\theta)d\theta = -\frac{\|4\pi\|}{3\epsilon_0} qr'. \tag{C.7}$$

The average electric field intensity inside the sphere due to the point charge q is then given by

$$\bar{e}_z = \frac{1}{(4/3)\pi R^3} \int_V \tilde{e}_z d^3r = -\frac{\|4\pi\|}{4\pi\epsilon_0} \frac{qr'}{R^3}. \tag{C.8}$$

The electric dipole moment of the point charge q with reference to the center O of the spherical region is given by $\tilde{\mathbf{p}} \equiv qr'\hat{\mathbf{1}}_z$, so that Eq. (C.8) may be written in the general form

$$\bar{\mathbf{e}} = -\frac{\|4\pi\|}{4\pi\epsilon_0 R^3}\tilde{\mathbf{p}}. \tag{C.9}$$

The second integral appearing in Eq. (C.2) is then given by

$$\frac{1}{V_m}\int_{V_m}\tilde{\mathbf{e}}_m(\mathbf{r}', \omega)d^3r' = -\frac{\|4\pi\|}{4\pi\epsilon_0 r_m^3}\tilde{\mathbf{p}}_m(\mathbf{r}, \omega), \tag{C.10}$$

where $\tilde{\mathbf{p}}_m$ is the dipole moment of the molecule under consideration. Because $N = 1/V_m$ is the local volume density of molecules, then with the assumption that all of the local molecules have parallel and equal polarization vectors, the macroscopic polarization is given by

$$\tilde{\mathbf{P}}(\mathbf{r}, \omega) = N\tilde{\mathbf{p}}_m(\mathbf{r}, \omega)$$

so that the spatially averaged self-field of the molecule is given by

$$\frac{1}{V_m}\int_{V_m}\tilde{\mathbf{e}}_m(\mathbf{r}', \omega)d^3r' = -\frac{\|4\pi\|}{3\epsilon_0}\tilde{\mathbf{P}}(\mathbf{r}, \omega). \tag{C.11}$$

With these substitutions, Eq. (C.2) for the *spatially averaged effective field* becomes

$$\langle\langle\tilde{\mathbf{E}}_{eff}(\mathbf{r}, \omega)\rangle\rangle = \tilde{\mathbf{E}}(\mathbf{r}, \omega) + \frac{\|4\pi\|}{3\epsilon_0}\tilde{\mathbf{P}}(\mathbf{r}, \omega), \tag{C.12}$$

and the local field is larger than the macroscopic electric field. This expression for the effective local field was first derived by Lorentz [1] who used a somewhat different definition of the local field as the field value at the center of the molecule rather than that averaged over the molecular volume.

References

1. H. A. Lorentz, *The Theory of Electrons*. Leipzig: Teubner, 1906. Ch. IV.
2. C. Kittel, *Introduction to Solid State Physics*. New York: John Wiley & Sons, fourth ed., 1971. Ch. 13.
3. J. D. Jackson, *Classical Electrodynamics*. New York: John Wiley & Sons, third ed., 1999.

Appendix D
Magnetic Field Contribution to the Lorentz Model of Resonance Polarization

With the complete Lorentz force relation as the driving force, the equation of motion of a harmonically bound electron is given by

$$\frac{d^2\mathbf{r}_j}{dt^2} + 2\delta_j \frac{d\mathbf{r}_j}{dt} + \omega_j^2 \mathbf{r}_j = -\frac{q_e}{m}\left(\mathbf{E}_{\text{eff}}(\mathbf{r}, t) + \frac{1}{\|c\|}\frac{d\mathbf{r}_j}{dt} \times \mathbf{B}_{\text{eff}}(\mathbf{r}, t)\right), \qquad \text{(D.1)}$$

where $\mathbf{E}_{\text{eff}}(\mathbf{r}, t)$ is the *effective local electric field intensity* and $\mathbf{B}_{\text{eff}}(\mathbf{r}, t)$ is the *effective local magnetic induction field*, and where $\mathbf{r}_j = \mathbf{r}_j(\mathbf{r}, t)$ describes the displacement of the electron from its equilibrium position. Here q_e denotes the magnitude of the charge and m the mass of the harmonically bound electron with undamped resonance frequency ω_j and phenomenological damping constant δ_j. The temporal Fourier integral representation of the electric and magnetic field vectors of the effective local plane electromagnetic wave is given by

$$\tilde{\mathbf{E}}_{\text{eff}}(\mathbf{r}, \omega) = \int_{-\infty}^{\infty} \mathbf{E}_{\text{eff}}(\mathbf{r}, t)e^{i\omega t}\, dt, \qquad \text{(D.2)}$$

$$\tilde{\mathbf{B}}_{\text{eff}}(\mathbf{r}, \omega) = \int_{-\infty}^{\infty} \mathbf{B}_{\text{eff}}(\mathbf{r}, t)e^{i\omega t}\, dt = \frac{\|c\|}{\omega}\mathbf{k}(\omega) \times \tilde{\mathbf{E}}_{\text{eff}}(\mathbf{r}, \omega), \qquad \text{(D.3)}$$

where $\mathbf{k}(\omega)$ is the *wave vector* of the plane wave field with magnitude given by the *wavenumber* $k(\omega) = \omega/c$ because the effective local field is essentially a microscopic field. With the temporal Fourier integral representation

$$\tilde{\mathbf{r}}_j(\omega) = \int_{-\infty}^{\infty} \mathbf{r}_j(t)e^{i\omega t}\, dt, \qquad \text{(D.4)}$$

© Springer Nature Switzerland AG 2019
K. E. Oughstun, *Electromagnetic and Optical Pulse Propagation*, Springer Series in Optical Sciences 224, https://doi.org/10.1007/978-3-030-20835-6

the dynamical equation of motion (D.1) becomes

$$\left(\omega^2 + 2i\delta_j\omega - \omega_j^2\right)\tilde{\mathbf{r}}_j = \frac{q_e}{m}\left(\tilde{\mathbf{E}}_{\text{eff}} - i\tilde{\mathbf{r}}_j \times \left(\mathbf{k} \times \tilde{\mathbf{E}}_{\text{eff}}\right)\right)$$

$$= \frac{q_e}{m}\left[\left(1 + i\tilde{\mathbf{r}}_j \cdot \mathbf{k}\right)\tilde{\mathbf{E}}_{\text{eff}} - i\left(\tilde{\mathbf{r}}_j \cdot \tilde{\mathbf{E}}_{\text{eff}}\right)\mathbf{k}\right], \quad \text{(D.5)}$$

with formal solution

$$\tilde{\mathbf{r}}_j = \frac{q_e}{m}\left[\frac{1 + i\tilde{\mathbf{r}}_j \cdot \mathbf{k}}{\omega^2 - \omega_j^2 + 2i\delta_j\omega}\tilde{\mathbf{E}}_{\text{eff}} - i\frac{\tilde{\mathbf{r}}_j \cdot \tilde{\mathbf{E}}_{\text{eff}}}{\omega^2 - \omega_j^2 + 2i\delta_j\omega}\mathbf{k}\right]. \quad \text{(D.6)}$$

The electron displacement vector may then be expressed as a linear combination of the orthogonal pair of vectors $\mathbf{k}$ and $\tilde{\mathbf{E}}_{\text{eff}}$

$$\tilde{\mathbf{r}}_j = a_j\tilde{\mathbf{E}}_{\text{eff}} + b_j\mathbf{k}, \quad \text{(D.7)}$$

where, because of the transversality relation $\mathbf{k} \cdot \tilde{\mathbf{E}}_{\text{eff}} = 0$,

$$a_j = \frac{\tilde{\mathbf{r}}_j \cdot \tilde{\mathbf{E}}_{\text{eff}}}{\tilde{E}_{\text{eff}}^2}, \quad b_j = \frac{c^2}{\omega^2}\tilde{\mathbf{r}}_j \cdot \mathbf{k}. \quad \text{(D.8)}$$

The pair of scalar products appearing in the above expression may be evaluated from Eq. (D.6) as

$$\tilde{\mathbf{r}}_j \cdot \mathbf{k} = -i\frac{q_e}{m}\frac{\tilde{\mathbf{r}}_j \cdot \tilde{\mathbf{E}}_{\text{eff}}}{\omega^2 - \omega_j^2 + 2i\delta_j\omega}k^2,$$

$$\tilde{\mathbf{r}}_j \cdot \tilde{\mathbf{E}}_{\text{eff}} = \frac{q_e}{m}\frac{1 + i\tilde{\mathbf{r}}_j \cdot \mathbf{k}}{\omega^2 - \omega_j^2 + 2i\delta_j\omega}\tilde{E}_{\text{eff}}^2.$$

Substitution of the second relation into the first then yields

$$\tilde{\mathbf{r}}_j \cdot \mathbf{k} = -i\frac{(q_e/mc)^2\tilde{E}_{\text{eff}}^2\omega^2}{(\omega^2 - \omega_j^2 + 2i\delta_j\omega)^2 - (q_e/mc)^2\tilde{E}_{\text{eff}}^2\omega^2}, \quad \text{(D.9)}$$

and substitution of this result into the second relation gives

$$\tilde{\mathbf{r}}_j \cdot \tilde{\mathbf{E}}_{\text{eff}} = \frac{q_e/m}{\omega^2 - \omega_j^2 + 2i\delta_j\omega}$$

$$\times \left[1 + \frac{(q_e/mc)^2\tilde{E}_{\text{eff}}^2\omega^2}{(\omega^2 - \omega_j^2 + 2i\delta_j\omega)^2 - (q_e/mc)^2\tilde{E}_{\text{eff}}^2\omega^2}\right]\tilde{E}_{\text{eff}}^2. \quad \text{(D.10)}$$

The coefficients a_j and b_j appearing in Eq. (D.7) are then given by

$$a_j = \frac{q_e/m}{\omega^2 - \omega_j^2 + 2i\delta_j\omega}$$

$$\times \left[1 + \frac{(q_e/mc)^2 \tilde{E}_{\text{eff}}^2 \omega^2}{(\omega^2 - \omega_j^2 + 2i\delta_j\omega)^2 - (q_e/mc)^2 \tilde{E}_{\text{eff}}^2 \omega^2} \right], \tag{D.11}$$

$$b_j = -i \frac{(q_e/m)^2 \tilde{E}_{\text{eff}}^2}{(\omega^2 - \omega_j^2 + 2i\delta_j\omega)^2 - (q_e/mc)^2 \tilde{E}_{\text{eff}}^2 \omega^2}, \tag{D.12}$$

respectively.

The local (or microscopic) induced dipole moment $\tilde{\mathbf{p}}_j \equiv -q_e \tilde{\mathbf{r}}_j$ for the jth Lorentz oscillator type is then given by [compare with Eq. (4.209)]

$$\tilde{\mathbf{p}}_j(\mathbf{r}, \omega) = -q_e \left(a_j \tilde{\mathbf{E}}_{\text{eff}}(\mathbf{r}, \omega) + b_j \mathbf{k} \right). \tag{D.13}$$

If there are N_j Lorentz oscillators per unit volume of the jth type, then the macroscopic polarization induced in the medium is given by the summation over all oscillator types of the spatially averaged locally induced dipole moments as

$$\tilde{\mathbf{P}}(\mathbf{r}, \omega, \tilde{E}_{\text{eff}}) = \sum_j N_j \langle\langle \tilde{\mathbf{p}}_j(\mathbf{r}, \omega) \rangle\rangle$$

$$= \left\langle\!\left\langle \tilde{\mathbf{E}}_{\text{eff}}(\mathbf{r}, \omega) \right\rangle\!\right\rangle \sum_j N_j \alpha_{j\perp}(\omega, \tilde{E}_{\text{eff}}) + \mathbf{k} \sum_j N_j \alpha_{j\parallel}(\omega, \tilde{E}_{\text{eff}}).$$
$$\tag{D.14}$$

Here

$$\alpha_{j\perp}(\omega, \tilde{E}_{\text{eff}}) \equiv \alpha_{j\perp}^{(0)}(\omega) + \alpha_{j\perp}^{(2)}(\omega, \tilde{E}_{\text{eff}})$$

$$= \frac{-q_e^2/m}{\omega^2 - \omega_j^2 + 2i\delta_j\omega}$$

$$\times \left[1 + \frac{(q_e/mc)^2 \tilde{E}_{\text{eff}}^2 \omega^2}{(\omega^2 - \omega_j^2 + 2i\delta_j\omega)^2 - (q_e/mc)^2 \tilde{E}_{\text{eff}}^2 \omega^2} \right] \tag{D.15}$$

is defined here as the *perpendicular component of the atomic polarizability*, with $\alpha_{j\perp}^{(0)}(\omega) \equiv \alpha_{j\perp}(\omega, 0) = \alpha_j(\omega)$, where $\alpha_j(\omega)$ is the classical expression

(4.211) for the atomic polarizability when magnetic field effects are neglected, and

$$\alpha_{j\parallel}(\omega, \tilde{E}_{\text{eff}}) \equiv i \frac{(q_e^3/m^2)\tilde{E}_{\text{eff}}^2}{(\omega^2 - \omega_j^2 + 2i\delta_j\omega)^2 - (q_e/mc)^2\tilde{E}_{\text{eff}}^2\omega^2} \tag{D.16}$$

is the *parallel component of the atomic polarizability*, where $\alpha_{j\parallel}(\omega, 0) = 0$.

The atomic polarizability is then seen to be nonlinear in the local electric field strength when magnetic field effects are included. However, numerical calculations [1] show that these nonlinear terms are entirely negligible for effective field strengths that are typically less than $\sim 10^{12}$ V/m and that they begin to have a significant contribution for field strengths that are typically greater than $\sim 10^{15}$ V/m for a highly absorptive material.

The physical origin of the nonlinear term considered here is due to the diamagnetic effect that appears in the analysis of the interaction of an electromagnetic field with a charged particle in the quantum theory of electrodynamics [2, 3]. The Hamiltonian for this coupled system is given by [see Eqs. (XIII.71)–(XIII.72) of Messiah [2]]

$$\mathcal{H} = \mathcal{H}_0 - \frac{q_e}{2mc}\mathbf{H}\cdot\mathbf{L} + \frac{q_e^2}{8mc^2}H^2\sum_{j=1}^{Z}r_{j\perp}^2,$$

in Gaussian units, where $\mathcal{H}_0$ is the Hamiltonian of the center of mass system of the isolated atom with Z spinless electrons, $\mathbf{H}(\mathbf{r})$ is the magnetic field intensity vector with magnitude $H \equiv |\mathbf{H}|$, $r_\perp$ is the projection of the position vector $\mathbf{r}$ on the plane perpendicular to $\mathbf{H}(\mathbf{r})$, and $\mathbf{L} = \sum_{j=1}^{Z}(\mathbf{r}_j \times \mathbf{p}_j)$ is the total angular momentum of the Z atomic electrons. The third term in the above expression for the Hamiltonian is the main factor in atomic diamagnetism. The order of magnitude of this factor is given by $\sim (Zq_e^2/12mc^2)H^2\langle r^2\rangle$, where $\langle r^2\rangle \sim 1 \times 10^{-16}$ cm^2 for a bound electron. The ratio of this quantity to the level distance $\mu_B H$, where $\mu_B \equiv q_e\hbar/2mc$ is the Bohr magneton, is found [2] to be $\sim 10^{-9}ZH$ gauss. For a single electron atom ($Z = 1$), the diamagnetic effect will become significant when $H \geq 10^9$ gauss, which corresponds to an electric field strength $E \geq 10^9$ esu, or equivalently $E \geq 3 \times 10^{13}$ V/m, in agreement with the preceding classical result that the nonlinear effects in the Lorentz model become significant for an applied field strength between 10^{12} V/m and 10^{15} V/m.

In addition, nonlinear optical effects are found to dominate the linear response when the local field strength becomes comparable to the Coulomb field of the atomic nucleus [4]. As an estimate of this field strength, if the distance between the nucleus and the bound electron is taken to be given by the Bohr radius $a_0 \equiv \hbar^2/mq_e^2 \approx 5.29 \times 10^{-9}$ cm, where $\hbar \equiv h/2\pi$ and h is Planck's constant, the electric field strength is $E \approx 5.13 \times 10^{11}$ V/m, in general agreement with the preceding estimates.

References

1. K. E. Oughstun and R. A. Albanese, "Magnetic field contribution to the Lorentz model," *J. Opt. Soc. Am. A*, vol. 23, no. 7, pp. 1751–1756, 2006.
2. A. Messiah, *Quantum Mechanics*, vol. II. Amsterdam: North-Holland, 1962.
3. C. Cohen-Tannoudji, J. Dupont-Roc, and G. Grynberg, *Photons and Atoms: Introduction to Quantum Electrodynamics*. New York: John Wiley & Sons, 1989. Section III.D.
4. R. W. Boyd, *Nonlinear Optics*. San Diego: Academic, 1992. Ch. 1.

Appendix E
The Fourier–Laplace Transform

The complex temporal frequency spectrum of a vector function $\mathbf{f}(\mathbf{r}, t)$ of both position $\mathbf{r}$ and time t that vanishes for $t < 0$ is of central importance to the solution of problems in time-domain electromagnetics and optics and is considered here in some detail following, in part, the treatment by Stratton [1]. The *Laplace transform* of $\mathbf{f}(\mathbf{r}, t)$ with respect to the time variable t is defined here as

$$\mathcal{L}\{\mathbf{f}(\mathbf{r}, t)\} \equiv \int_0^\infty \mathbf{f}(\mathbf{r}, t) e^{i\omega t} \, dt, \tag{E.1}$$

which is simply a Fourier transform with complex angular frequency ω that is taken over only the positive time interval. Let $\mathbf{f}'(\mathbf{r}, t)$ be another vector function of both position and time such that

$$\mathbf{f}'(\mathbf{r}, t) = \mathbf{f}(\mathbf{r}, t); \quad t > 0, \tag{E.2}$$

but which may not vanish for $t \leq 0$. The Laplace transform given in Eq. (E.1) may then be written as

$$\mathcal{L}\{\mathbf{f}(\mathbf{r}, t)\} = \int_{-\infty}^\infty U(t) \mathbf{f}'(\mathbf{r}, t) e^{i\omega t} \, dt, \tag{E.3}$$

where $U(t) = 0$ for $t < 0$ and $U(t) = 1$ for $t > 0$ is the Heaviside unit step function. For real ω, the Laplace transform of $\mathbf{f}(\mathbf{r}, t)$ is then seen to be equal to the Fourier transform of $U(t)\mathbf{f}'(\mathbf{r}, t)$, viz.

$$\mathcal{L}\{\mathbf{f}(\mathbf{r}, t)\} = \mathcal{F}_\omega\{U(t)\mathbf{f}'(\mathbf{r}, t)\}; \quad \text{for real } \omega, \tag{E.4}$$

© Springer Nature Switzerland AG 2019
K. E. Oughstun, *Electromagnetic and Optical Pulse Propagation*, Springer Series in Optical Sciences 224, https://doi.org/10.1007/978-3-030-20835-6

where the subscript ω indicates that it is the Fourier transform variable. The inverse Fourier transform of this equation then gives

$$U(t)\mathbf{f}'(\mathbf{r}, t) = \mathcal{F}^{-1}\{\mathcal{L}\{\mathbf{f}(\mathbf{r}, t)\}\} \tag{E.5}$$

for real ω.

For complex ω, let $\omega = \omega' + i\omega''$ where $\omega' \equiv \Re\{\omega\}$ and $\omega'' \equiv \Im\{\omega\}$. The Laplace transform given in Eq. (E.3) then becomes

$$\mathcal{L}\{\mathbf{f}(\mathbf{r}, t)\} = \int_{-\infty}^{\infty} \left[U(t)\mathbf{f}'(\mathbf{r}, t)e^{-\omega'' t} \right] e^{i\omega' t} dt$$

$$= \mathcal{F}_{\omega'}\left\{ U(t)\mathbf{f}'(\mathbf{r}, t)e^{-\omega'' t} \right\}. \tag{E.6}$$

The inverse Fourier transform of this expression then yields

$$U(t)\mathbf{f}'(\mathbf{r}, t)e^{-\omega'' t} = \mathcal{F}_{\omega'}^{-1}\{\mathcal{L}\{\mathbf{f}(\mathbf{r}, t)\}\}$$

$$= \frac{1}{2\pi} \int_{-\infty}^{\infty} \mathcal{L}\{\mathbf{f}(\mathbf{r}, t)\}e^{-i\omega' t} d\omega', \tag{E.7}$$

which may be rewritten as

$$U(t)\mathbf{f}'(\mathbf{r}, t) = \frac{1}{2\pi} \int_{-\infty}^{\infty} \mathcal{L}\{\mathbf{f}(\mathbf{r}, t)\}e^{-i(\omega' + i\omega'')t} d\omega'$$

$$= \frac{1}{2\pi} \int_{-\infty + i\omega''}^{\infty + i\omega''} \mathcal{L}\{\mathbf{f}(\mathbf{r}, t)\}e^{-i\omega t} d\omega$$

$$= \frac{1}{2\pi} \int_{C} \mathcal{L}\{\mathbf{f}(\mathbf{r}, t)\}e^{-i\omega t} d\omega. \tag{E.8}$$

Here C denotes the straight line contour $\omega = \omega' + i\omega''$ with ω'' fixed and ω' varying over the real domain from $-\infty$ to $+\infty$. Because $\mathbf{f}(\mathbf{r}, t) = U(t)\mathbf{f}'(\mathbf{r}, t)$, Eqs. (E.1) and (E.8) then define the Laplace transform pair relationship

$$\tilde{\mathbf{f}}(\mathbf{r}, \omega) \equiv \mathcal{L}\{\mathbf{f}(\mathbf{r}, t)\} = \int_{0}^{\infty} \mathbf{f}(\mathbf{r}, t)e^{i\omega t} dt, \tag{E.9}$$

$$\mathbf{f}(\mathbf{r}, t) \equiv \mathcal{L}^{-1}\{\tilde{\mathbf{f}}(\mathbf{r}, \omega)\} = \frac{1}{2\pi} \int_{C} \tilde{\mathbf{f}}(\mathbf{r}, \omega)e^{-i\omega t} d\omega, \tag{E.10}$$

where $\tilde{\mathbf{f}}(\mathbf{r}, \omega)$ is the complex temporal frequency spectrum of $\mathbf{f}(\mathbf{r}, t)$ with $\omega = \omega' + i\omega''$. Notice that $\omega'' = \Im\{\omega\}$ plays a passive role in the Laplace transform operation because it remains constant in both the forward and inverse transformations. Nevertheless, its presence can be important because the factor $e^{-\omega'' t}$ appearing in the integrand of the transformation (E.9) may serve as a convergence factor when $\omega'' > 0$. In particular,

$$\tilde{\mathbf{f}}(\mathbf{r}, \omega) = \int_0^\infty \mathbf{f}(\mathbf{r}, t) e^{-\omega'' t} e^{i\omega' t} dt \qquad (\text{E.11})$$

is just the Fourier transform $\mathcal{F}_{\omega'}\{\mathbf{f}(\mathbf{r}, t) e^{-\omega'' t}\}$. The Fourier transform of $\mathbf{f}(\mathbf{r}, t)$ alone is

$$\mathcal{F}_{\omega'}\{\mathbf{f}(\mathbf{r}, t)\} = \int_0^\infty \mathbf{f}(\mathbf{r}, t) e^{i\omega' t} dt,$$

which exists provided that $\mathbf{f}(\mathbf{r}, t)$ is absolutely integrable; viz.,

$$\lim_{T \to \infty} \int_0^T |\mathbf{f}(\mathbf{r}, t)| \, dt < \infty.$$

If $\mathbf{f}(\mathbf{r}, t)$ does not vanish properly at infinity, then the above integral fails to converge and the existence of the Fourier transform $\mathcal{F}_{\omega'}\{\mathbf{f}(\mathbf{r}, t)\}$ is not guaranteed. However, if there exists a real number γ such that

$$\lim_{T \to \infty} \int_0^T \left| \mathbf{f}(\mathbf{r}, t) e^{-\gamma t} \right| dt < \infty, \qquad (\text{E.12})$$

then $\mathbf{f}(\mathbf{r}, t)$ is transformable for all $\omega'' \geq \gamma$ and its temporal frequency spectrum is given by the Laplace transform (E.9). The lower bound γ_a of all of the values of γ for which the inequality appearing in Eq. (E.12) is satisfied is called the *abscissa of absolute convergence* for the function $\mathbf{f}(\mathbf{r}, t)$.

The Laplace transform of the time derivative $\partial \mathbf{f}(\mathbf{r}, t)/\partial t$ can be related to the Laplace transform of $\mathbf{f}(\mathbf{r}, t)$ through integration by parts as

$$\mathcal{L}\left\{ \frac{\partial \mathbf{f}(\mathbf{r}, t)}{\partial t} \right\} = \int_0^\infty \frac{\partial \mathbf{f}(\mathbf{r}, t)}{\partial t} e^{i\omega t} dt$$

$$= \left[\mathbf{f}(\mathbf{r}, t) e^{i\omega t} \right]_0^\infty - i\omega \int_0^\infty \mathbf{f}(\mathbf{r}, t) e^{i\omega t} dt$$

$$= -\mathbf{f}(\mathbf{r}, 0) - i\omega \mathcal{L}\{\mathbf{f}(\mathbf{r}, t)\}, \qquad (\text{E.13})$$

where the fact that $|\mathbf{f}(\mathbf{r}, t) e^{i\omega t}| = |\mathbf{f}(\mathbf{r}, t)| e^{-\omega'' t}$ must vanish as $t \to \infty$ for all $\omega'' \geq \gamma_a$ has been used in obtaining the final form of Eq. (E.13).

For the appropriate form of the convolution theorem for the Laplace transform, consider determining the function whose Laplace transform is equal to the product

$\tilde{f}_1(\omega)\tilde{f}_2(\omega) = \mathcal{L}\{f_1(t)\}\mathcal{L}\{f_2(t)\}$ so that

$$
\begin{aligned}
\mathcal{L}^{-1}\left\{\tilde{f}_1(\omega)\tilde{f}_2(\omega)\right\} &= \frac{1}{2\pi}\int_C \tilde{f}_1(\omega)\tilde{f}_2(\omega)e^{-i\omega t}\,d\omega \\
&= \frac{1}{2\pi}\int_C \tilde{f}_1(\omega)\left[\int_0^\infty f_2(\tau)e^{i\omega\tau}\,d\tau\right]e^{-i\omega t}\,d\omega \\
&= \int_0^\infty d\tau\cdot f_2(\tau)\left[\frac{1}{2\pi}\int_C \tilde{f}_1(\omega)e^{-i\omega(t-\tau)}\,d\omega\right],
\end{aligned}
$$

and consequently

$$
\mathcal{L}^{-1}\left\{\tilde{f}_1(\omega)\tilde{f}_2(\omega)\right\} = \int_0^\infty f_1(t-\tau)f_2(\tau)U(t-\tau)\,d\tau,
$$

where the unit step function $U(t-\tau)$ is explicitly included in this expression to emphasize the fact that $f_1(t)$ vanishes for $t<0$. Because $U(t-\tau)$ vanishes for negative values of its argument, the upper limit of integration in τ must be t and the above equation becomes

$$
\mathcal{L}^{-1}\left\{\tilde{f}_1(\omega)\tilde{f}_2(\omega)\right\} = \int_0^t f_1(t-\tau)f_2(\tau)\,d\tau, \tag{E.14}
$$

which may be rewritten as

$$
\mathcal{L}\left\{\int_0^\infty f_1(t-\tau)f_2(\tau)\,d\tau\right\} = \mathcal{L}\{f_1(t)\}\,\mathcal{L}\{f_2(t)\}, \tag{E.15}
$$

which is the convolution theorem for the Laplace transform.

The spatiotemporal Fourier–Laplace transform of a vector function $\mathbf{F}(\mathbf{r},t)$ of both position $\mathbf{r}$ and time t that vanishes for $t<0$ is defined here by the pair of relations

$$
\tilde{\tilde{\mathbf{F}}}(\mathbf{k},\omega) \equiv \mathcal{F}\mathcal{L}\{\mathbf{F}(\mathbf{r},t)\} = \int_{-\infty}^\infty d^3r \int_0^\infty dt\cdot \mathbf{F}(\mathbf{r},t)e^{-i(\mathbf{k}\cdot\mathbf{r}-\omega t)}, \tag{E.16}
$$

$$
\mathbf{F}(\mathbf{r},t) \equiv \mathcal{F}^{-1}\mathcal{L}^{-1}\{\tilde{\tilde{\mathbf{F}}}(\mathbf{k},\omega)\} = \frac{1}{(2\pi)^4}\int_{-\infty}^\infty d^3k \int_C d\omega\cdot \tilde{\tilde{\mathbf{F}}}(\mathbf{k},\omega)e^{i(\mathbf{k}\cdot\mathbf{r}-\omega t)}, \tag{E.17}
$$

where $\mathbf{k} = \hat{\mathbf{1}}_x k_x + \hat{\mathbf{1}}_y k_y + \hat{\mathbf{1}}_z k_z$ and $\mathbf{r} = \hat{\mathbf{1}}_x x + \hat{\mathbf{1}}_y y + \hat{\mathbf{1}}_z z$. Because,

$$
\frac{\partial \mathbf{F}(\mathbf{r},t)}{\partial x_j} = \frac{1}{(2\pi)^4}\int_{-\infty}^\infty d^3k \int_C d\omega\cdot ik_j\tilde{\tilde{\mathbf{F}}}(\mathbf{k},\omega)e^{i(\mathbf{k}\cdot\mathbf{r}-\omega t)},
$$

from Eq. (E.17), then the transforms of the first spatial derivatives of $\mathbf{F}(\mathbf{r}, t)$ are given by

$$\mathcal{FL}\left\{\frac{\partial \mathbf{F}(\mathbf{r}, t)}{\partial x_j}\right\} = ik_j \tilde{\tilde{\mathbf{F}}}(\mathbf{k}, \omega), \tag{E.18}$$

where $x_1 = x$, $x_2 = y$, $x_3 = z$ and $k_1 = k_x$, $k_2 = k_y$, $k_3 = k_z$. With this result, the spatiotemporal transform of the divergence of the vector field $\mathbf{F}(\mathbf{r}, t)$ is found to be

$$\mathcal{FL}\{\nabla \cdot \mathbf{F}(\mathbf{r}, t)\} = i\mathbf{k} \cdot \tilde{\tilde{\mathbf{F}}}(\mathbf{k}, \omega), \tag{E.19}$$

and the spatiotemporal transform of the curl of $\mathbf{F}(\mathbf{r}, t)$ is given by

$$\mathcal{FL}\{\nabla \times \mathbf{F}(\mathbf{r}, t)\} = i\mathbf{k} \times \tilde{\tilde{\mathbf{F}}}(\mathbf{k}, \omega). \tag{E.20}$$

The spatiotemporal transforms of higher-order spatial derivatives of $\mathbf{F}(\mathbf{r}, t)$ may then be obtained through repeated application of the above relations.

Reference

1. J. A. Stratton, *Electromagnetic Theory*. New York: McGraw-Hill, 1941.

Appendix F
Reversible and Irreversible, Recoverable and Irrecoverable Electrodynamic Processes in Dispersive Dielectrics

A critical re-examination of the Barash and Ginzburg result [1], presented in Sect. 5.2.2, for the electromagnetic energy density and evolved heat in a dispersive dissipative medium is presented here based upon the analysis by Glasgow et al. [2, 3] for a simple dispersive dielectric [$\mu(\omega) = \mu_0$, $\sigma(\omega) = 0$]. In that case the total electromagnetic energy density in the coupled field–medium system is given by [cf. Eqs. (5.186) and (5.187)]

$$\mathcal{U}'(\mathbf{r}, t) = \left\| \frac{1}{4\pi} \right\| \frac{1}{2} \left(\epsilon_0 E^2(\mathbf{r}, t) + \mu_0 H^2(\mathbf{r}, t) \right) + \int_{-\infty}^{t} \mathbf{E}(\mathbf{r}, t') \cdot \frac{\partial \mathbf{P}(\mathbf{r}, t')}{\partial t'} dt'.$$

(F.1)

The first two terms on the right-hand side of this equation describe the *electromagnetic field energy density in the absence of the dispersive medium*

$$\mathcal{U}_{em}(\mathbf{r}, t) \equiv \left\| \frac{1}{4\pi} \right\| \frac{1}{2} \left(\epsilon_0 E^2(\mathbf{r}, t) + \mu_0 H^2(\mathbf{r}, t) \right),$$

(F.2)

and the last term describes the *electromagnetic energy density in the coupled field–medium system*

$$\mathcal{U}_{int}(\mathbf{r}, t) \equiv \int_{-\infty}^{t} \mathbf{E}(\mathbf{r}, t') \cdot \frac{\partial \mathbf{P}(\mathbf{r}, t')}{\partial t'} dt',$$

(F.3)

which is defined by Glasgow et al. [2] as the *interaction energy*. By comparison, Barash and Ginzburg [1] separated the interaction energy into two parts: one part which represents the electromagnetic energy that is reactively stored in the dispersive medium, and another part, $\mathcal{Q}(\mathbf{r}, t)$, which represents the dissipation of

© Springer Nature Switzerland AG 2019
K. E. Oughstun, *Electromagnetic and Optical Pulse Propagation*, Springer Series in Optical Sciences 224, https://doi.org/10.1007/978-3-030-20835-6

electromagnetic energy in the medium. Notice that

$$\frac{\partial \mathcal{U}'(\mathbf{r}, t)}{\partial t} = \frac{\partial \mathcal{U}(\mathbf{r}, t)}{\partial t} + \mathcal{Q}(\mathbf{r}, t)$$

$$= \frac{\partial \mathcal{U}_{em}(\mathbf{r}, t)}{\partial t} + \frac{\partial \mathcal{U}_{int}(\mathbf{r}, t)}{\partial t}, \tag{F.4}$$

where the first form of this separation is due to Barash and Ginzburg [1] [cf. Eq. (5.193)], and the second form is due to Glasgow et al. [2], where $\mathcal{U}(\mathbf{r}, t) = \mathcal{U}_{em}(\mathbf{r}, t) + \mathcal{U}_{rev}(\mathbf{r}, t)$ and $\mathcal{U}_{int}(\mathbf{r}, t) = \mathcal{U}_{rev}(\mathbf{r}, t) + \mathcal{U}_{irrev}(\mathbf{r}, t)$.

The *Helmholtz free energy* (also known as the work function) $\psi \equiv U - TS$ is defined [4] as the difference between the internal energy U of the system and the heat energy TS, where T is the absolute temperature and S the entropy. Von Helmholtz (1882) called this ψ function the *free energy* of a system, because [4] "its change in a reversible isothermal process equals the energy that can be 'freed' in the process and converted to mechanical work." The decrease in ψ is then equal to the maximum work done on the system in a reversible isothermal process. Glasgow et al. [2] generalize this Helmholtz free energy to the irreversible, dissipative, nonequilibrium case by defining the *dynamical free energy* as "the work the system can do on, or return to, an external agent," which naturally reduces to the Helmholtz free energy in the reversible case. With regard to Eq. (5.271), this dynamical free energy $\mathcal{U}(\mathbf{r}, t)$ is given by the sum of the field energy term $\mathcal{U}_{em}(\mathbf{r}, t)$ and the reactive (or reversible) energy term $\mathcal{U}_{rev}(\mathbf{r}, t)$. The evolved heat energy $\mathcal{Q}(\mathbf{r}, t)$ then corresponds to the irreversible (or latent) part of the interaction energy.

The interaction energy defined in Eq. (5.270) can be expressed in terms of the instantaneous (or causal) spectrum of the electric field vector,[1] defined as [5]

$$\tilde{\mathbf{E}}_t(\mathbf{r}, \omega) \equiv \frac{1}{2\pi} \int_{-\infty}^{t} \mathbf{E}(\mathbf{r}, t') e^{i\omega t'} dt', \tag{F.5}$$

in the following manner due to Peatross, Ware, and Glasgow [6]. From Eq. (4.95), the macroscopic polarization density of a temporally dispersive HILL medium can be expressed as

$$\mathbf{P}(\mathbf{r}, t) = \int_{-\infty}^{\infty} \mathbf{E}(\mathbf{r}, t') G(t - t') dt' \tag{F.6}$$

---

[1] Notice that this instantaneous spectrum is equal to the Fourier transform of $\mathbf{E}(\mathbf{r}, t') U(t - t')$, where $U(t)$ is the Heaviside unit step-function.

with Green's function

$$G(t) \equiv \frac{\epsilon_0}{2\pi} \int_{-\infty}^{\infty} \chi_e(\omega) e^{-i\omega t} d\omega. \tag{F.7}$$

This Green's function may be expressed as the sum of two terms, the first associated with the real part $\chi_e'(\omega) \equiv \Re\{\chi(\omega)\}$ and the second with the imaginary part $\chi_e''(\omega) \equiv \Im\{\chi(\omega)\}$ of the electric susceptibility, as

$$G(t) = G'(t) + G''(t), \tag{F.8}$$

where

$$G'(t) \equiv \frac{\epsilon_0}{2\pi} \int_{-\infty}^{\infty} \chi_e'(\omega) e^{-i\omega t} d\omega, \tag{F.9}$$

$$G''(t) \equiv \frac{\epsilon_0}{2\pi} \int_{-\infty}^{\infty} \chi_e''(\omega) e^{-i\omega t} d\omega. \tag{F.10}$$

Causality is introduced through the fact that the real and imaginary parts of the electric susceptibility satisfy the Plemelj formulae given in Eqs. (4.155) and (4.156), viz.

$$\chi_e'(\omega) = \frac{1}{\pi} \mathcal{P} \int_{-\infty}^{\infty} \frac{\chi_e'(\omega')}{\omega' - \omega} d\omega', \tag{F.11}$$

$$\chi_e''(\omega) = -\frac{1}{\pi} \mathcal{P} \int_{-\infty}^{\infty} \frac{\chi_e'(\omega')}{\omega' - \omega} d\omega'. \tag{F.12}$$

Substitution of Eq. (5.278) into (5.276) then yields

$$G'(t) = \frac{\epsilon_0}{2\pi^2} \int_{-\infty}^{\infty} d\omega' \, \chi_e''(\omega) \mathcal{P} \int_{-\infty}^{\infty} \frac{e^{-i\omega t}}{\omega' - \omega} d\omega. \tag{F.13}$$

Because

$$\mathcal{P} \int_{-\infty}^{\infty} \frac{e^{-i\omega t}}{\omega' - \omega} d\omega = \begin{cases} i\pi e^{-i\omega' t}, & t > 0 \\ -i\pi e^{-i\omega' t}, & t < 0 \end{cases}, \tag{F.14}$$

then Eq. (5.280) shows that $G'(t) = G''(t)$ for $t > 0$ and $G'(t) = -G''(t)$ for $t < 0$, so that

$$G(t) = \begin{cases} 2G''(t), & t > 0 \\ 0, & t < 0 \end{cases}. \tag{F.15}$$

Substitution of this result in Eq. (5.273) then gives

$$\mathbf{P}(\mathbf{r}, t) = 2 \int_{-\infty}^{t} \mathbf{E}(\mathbf{r}, t') G''(t - t') dt'$$

$$= i \frac{\epsilon_0}{\pi} \int_{-\infty}^{\infty} d\omega \, \chi_e''(\omega) e^{-i\omega t} \int_{-\infty}^{t} dt' \, \mathbf{E}(\mathbf{r}, t') e^{i\omega t'}, \qquad (F.16)$$

where causality is now explicitly expressed in the upper limit of integration, in agreement with the expression given in Eq. (4.95). The time derivative of this expression is then given by

$$\frac{\partial \mathbf{P}(\mathbf{r}, t)}{\partial t} = \frac{\epsilon_0}{\pi} \left[\int_{-\infty}^{\infty} d\omega \, \omega \chi_e''(\omega) e^{-i\omega t} \int_{-\infty}^{t} dt' \, \mathbf{E}(\mathbf{r}, t') e^{i\omega t'} + i \mathbf{E}(\mathbf{r}, t) \int_{-\infty}^{\infty} \chi_e''(\omega) d\omega \right].$$
$$(F.17)$$

Because $\chi_e''(\omega)$ is an odd function of ω, the final integral in the above expression is equal to zero and consequently

$$\frac{\partial \mathbf{P}(\mathbf{r}, t)}{\partial t} = \frac{\epsilon_0}{\pi} \int_{-\infty}^{\infty} d\omega \, \omega \chi_e''(\omega) e^{-i\omega t} \int_{-\infty}^{t} dt' \, \mathbf{E}(\mathbf{r}, t') e^{i\omega t'}$$

$$= 2\epsilon_0 \int_{-\infty}^{\infty} \omega \chi_e''(\omega) \tilde{\mathbf{E}}_t(\mathbf{r}, \omega) e^{-i\omega t} d\omega. \qquad (F.18)$$

Substitution of this result into Eq. (5.270) then yields

$$\mathcal{U}_{int}(\mathbf{r}, t) = 2\epsilon_0 \int_{-\infty}^{\infty} d\omega \, \omega \chi_e''(\omega) \int_{-\infty}^{t} \mathbf{E}(\mathbf{r}, t') \cdot \tilde{\mathbf{E}}_{t'}(\mathbf{r}, \omega) e^{-i\omega t'} dt'.$$

The time derivative of the complex conjugate of Eq. (5.275) results in the identification $\partial \tilde{\mathbf{E}}_t^*(\mathbf{r}, \omega)/\partial t = (1/2\pi) \mathbf{E}(\mathbf{r}, t) e^{-i\omega t}$, so that the above expression for the interaction energy becomes

$$\mathcal{U}_{int}(\mathbf{r}, t) = 4\pi\epsilon_0 \int_{-\infty}^{\infty} d\omega \, \omega \chi_e''(\omega) \int_{-\infty}^{t} \tilde{\mathbf{E}}_{t'}(\mathbf{r}, \omega) \cdot \frac{\partial \tilde{\mathbf{E}}_{t'}^*(\mathbf{r}, \omega)}{\partial t'} dt'.$$

Because $\mathcal{U}_{int}(\mathbf{r}, t)$ is a real-valued quantity, the above expression may be written as

$$\mathcal{U}_{int}(\mathbf{r}, t) = 2\pi\epsilon_0 \int_{-\infty}^{\infty} d\omega \, \omega \chi_e''(\omega)$$

$$\times \int_{-\infty}^{t} \left[\tilde{\mathbf{E}}_{t'}(\mathbf{r}, \omega) \cdot \frac{\partial \tilde{\mathbf{E}}_{t'}^*(\mathbf{r}, \omega)}{\partial t'} + \tilde{\mathbf{E}}_{t'}^*(\mathbf{r}, \omega) \cdot \frac{\partial \tilde{\mathbf{E}}_{t'}(\mathbf{r}, \omega)}{\partial t'} \right] dt'.$$

$$= 2\pi\epsilon_0 \int_{-\infty}^{\infty} d\omega\, \omega \chi_e''(\omega) \int_{-\infty}^{t} \frac{\partial \left|\tilde{\mathbf{E}}_{t'}(\mathbf{r},\omega)\right|^2}{\partial t'} dt'$$

$$= 2\pi\epsilon_0 \int_{-\infty}^{\infty} \omega \chi_e''(\omega) \left|\tilde{\mathbf{E}}_t(\mathbf{r},\omega)\right|^2 d\omega. \tag{F.19}$$

It is then seen that

$$\mathcal{U}_{int}(\mathbf{r},t) \geq 0 \tag{F.20}$$

for all time t, where $\mathcal{U}_{int}(\mathbf{r},-\infty) = 0$, in agreement with the Landau–Lifshitz result [7] for the asymptotic heat that $\mathcal{U}_{int}(\mathbf{r},+\infty) \geq 0$. This generalization [6] of the Landau–Lifshitz result shows that the interaction energy can never run a deficit; that is, at any instant of time t the work that the electromagnetic field does on the medium always exceeds the work that the medium does against the field.

F.1   Reversible and Irreversible Electrodynamic Processes

The energy dissipated and consequently lost to a dielectric medium from some physically realizable electromagnetic field since time $t = -\infty$ may be expressed as

$$\int_{\mathcal{D}} \mathcal{U}_{int}(\mathbf{r},\infty) d^3r = \int_{\mathcal{D}} \left(\mathcal{U}_{int}(\mathbf{r},\infty) - \mathcal{U}_{int}(\mathbf{r},-\infty)\right) d^3r,$$

because $\mathcal{U}_{int}(\mathbf{r},-\infty) = 0$, where $\mathcal{D}$ is some fixed region contained within the dielectric body. The quantity $\int_{\mathcal{D}}(\mathcal{U}_{int}(\mathbf{r},\infty) - \mathcal{U}_{int}(\mathbf{r},t))d^3r$ then represents the electromagnetic energy lost to the medium in the region $\mathcal{D}$ over the future time interval $t' > t$ given the present state of the medium at time t that has been established by the electromagnetic field over the past time interval $t' < t$. When this quantity is negative, its opposite

$$\int_{\mathcal{D}} \left(\mathcal{U}_{int}(\mathbf{r},t) - \mathcal{U}_{int}(\mathbf{r},\infty)\right) d^3r > 0$$

may then be interpreted as representing a reversible process of energy that is "borrowed" from the electromagnetic field and returned to it at a later time as $t \to \infty$. The value of this energy difference $\int_{\mathcal{D}}(\mathcal{U}_{int}(\mathbf{r},t) - \mathcal{U}_{int}(\mathbf{r},\infty))d^3r$ will be different for different future fields, the physically proper value then being given by an appropriate extremum principle. To that end, the interaction energy density given in Eq. (5.286) is denoted as

$$\mathcal{U}_{int}[E](\mathbf{r},t) \equiv 2\pi\epsilon_0 \int_{-\infty}^{\infty} \omega \chi_e''(\omega) |\tilde{\mathbf{E}}_t(\mathbf{r},\omega)|^2 d\omega, \tag{F.21}$$

where

$$\tilde{E}_t(\mathbf{r}, \omega) \equiv \frac{1}{2\pi} \int_{-\infty}^{t} \mathbf{E}(\mathbf{r}, t') e^{i\omega t'} dt', \tag{F.22}$$

from Eq. (5.272). The reversible and irreversible energy densities of the electromagnetic field in a simple temporally dispersive dielectric $[\mu = \mu_0$ and $\sigma(\omega) = 0]$ are then given by the following definitions due to Broadbent, Hovhannisyan, Clayton, Peatross, and Glasgow [2]

Definition F.1 (Reversible Electromagnetic Energy Density) The reversible energy density $\mathcal{U}_{rev}(\mathbf{r}, t) = \mathcal{U}_{rev}[E](\mathbf{r}, t)$ of the electromagnetic field at time t in a simple temporally dispersive dielectric is given by the supremum (least upper bound) of values that the quantity $\mathcal{U}_{int}(\mathbf{r}, t) - \mathcal{U}_{int}(\mathbf{r}, \infty)$ can attain over all possible alternative future fields; that is,

$$\mathcal{U}_{rev}[E](\mathbf{r}, t) \equiv \sup_{E_f} \left\{ \mathcal{U}_{int}[E](\mathbf{r}, t) - \mathcal{U}_{int}[EU(t - t') + E_f](\mathbf{r}, \infty) \right\}, \tag{F.23}$$

where $U(t)$ denotes the Heaviside unit step function, and where E_f denotes the possible future electric fields with $E_f(\mathbf{r}, t') = 0$ for all $t' < t$.

Notice that the quantity $\mathcal{U}_{int}[EU(t - t') + E_f](\mathbf{r}, t) = \mathcal{U}_{int}[E](\mathbf{r}, t)$ in the above expression, where t' denotes the "past" time variable appearing in the integrand of Eq. (5.289).

Definition F.2 (Irreversible Electromagnetic Energy Density) The irreversible energy density $\mathcal{U}_{irrev}(\mathbf{r}, t) = \mathcal{U}_{irrev}[E](\mathbf{r}, t)$ of the electromagnetic field at time t in a simple temporally dispersive dielectric is given by the complement of the reversible energy density as the infimum (greatest lower bound) of values that the quantity $\mathcal{U}_{int}(\mathbf{r}, \infty)$ can attain over all possible alternate future fields; that is,

$$\mathcal{U}_{irrev}[E](\mathbf{r}, t) \equiv \inf_{E_f} \left\{ \mathcal{U}_{int}[EU(t - t') + E_f](\mathbf{r}, \infty) \right\}. \tag{F.24}$$

Notice that

$$\mathcal{U}_{int}[E](\mathbf{r}, t) = \mathcal{U}_{rev}[E](\mathbf{r}, t) + \mathcal{U}_{irrev}[E](\mathbf{r}, t) \tag{F.25}$$

as required by conservation of energy.

This formulation of the irreversible energy density is in a form that is now appropriate for the calculus of variations. Because any future field solution $\mathbf{E}_f(\mathbf{r}, t')$ vanishes for all $t' < t$, as required in Definition F.1 above, then the temporal Fourier transform of the concatenated electric field quantity $\mathbf{E}(\mathbf{r}, t') U(t - t') + \mathbf{E}_f(\mathbf{r}, t')$ is given by

$$\mathcal{F}\{\mathbf{E}(\mathbf{r}, t')U(t - t') + \mathbf{E}_f(\mathbf{r}, t')\}$$

$$= \frac{1}{2\pi} \int_{-\infty}^{t} \mathbf{E}(\mathbf{r}, t')e^{i\omega t'} dt' + \frac{1}{2\pi} \int_{t}^{\infty} \mathbf{E}_f(\mathbf{r}, t')e^{i\omega t'} dt'$$

$$= \tilde{\mathbf{E}}_t(\mathbf{r}, \omega) + \tilde{\mathbf{E}}_f(\mathbf{r}, \omega). \tag{F.26}$$

The variational derivative of the irreversible electromagnetic energy density defined in Eq. (5.291) with respect to future possible fields then gives, with Eq. (5.286),

$$\delta_{E_f}\{\mathcal{U}_{int}[EU(t - t') + E_f](\mathbf{r}, \infty)\}$$

$$= \delta_{\tilde{E}_f} \int_{-\infty}^{\infty} \omega \chi_e''(\omega) \left| \tilde{\mathbf{E}}_t(\mathbf{r}, \omega) + \tilde{\mathbf{E}}_f(\mathbf{r}, \omega) \right|^2 d\omega = 0. \tag{F.27}$$

Glasgow et al. [2] then define the new electric field vector functions

$$\mathbf{E}_+(\mathbf{r}, t') \equiv \mathbf{E}_f(\mathbf{r}, t' + t), \tag{F.28}$$

$$\mathbf{E}_-(\mathbf{r}, t') \equiv \mathbf{E}(\mathbf{r}, t' + t)U(-t'), \tag{F.29}$$

whose temporal Fourier transforms

$$\tilde{\mathbf{E}}_+(\mathbf{r}, \omega; t) = \frac{1}{2\pi} \int_{-\infty}^{\infty} \mathbf{E}_f(\mathbf{r}, t' + t)e^{i\omega t'} dt'$$

$$= e^{-i\omega t} \frac{1}{2\pi} \int_{-\infty}^{\infty} \mathbf{E}_f(\mathbf{r}, \tau)e^{i\omega \tau} d\tau = e^{-i\omega t} \tilde{\mathbf{E}}_f(\mathbf{r}, \omega), \tag{F.30}$$

$$\tilde{\mathbf{E}}_-(\mathbf{r}, \omega; t) = \frac{1}{2\pi} \int_{-\infty}^{\infty} \mathbf{E}(\mathbf{r}, t' + t)U(-t')e^{i\omega t'} dt'$$

$$= e^{-i\omega t} \int_{-\infty}^{t} \mathbf{E}(\mathbf{r}, \tau)e^{i\omega \tau} d\tau = e^{-i\omega t} \tilde{\mathbf{E}}_t(\mathbf{r}, \omega), \tag{F.31}$$

are analytic in the upper and lower half-planes, respectively. Substitution of these field quantities in Eq. (5.294) with the variation now taken with respect to $\tilde{E}_+$ then results in the expression

$$\int_{-\infty}^{\infty} \omega \chi_e''(\omega) \left(\tilde{\mathbf{E}}_-(\mathbf{r}, \omega; t) + \tilde{\mathbf{E}}_+(\mathbf{r}, \omega; t) \right) \delta\tilde{\mathbf{E}}_+^*(\mathbf{r}, \omega; t)d\omega = 0. \tag{F.32}$$

Because $\tilde{\mathbf{E}}_+(\mathbf{r}, \omega; t)$ is analytic in the upper half of the complex ω-plane, then so also is its variation $\delta\tilde{\mathbf{E}}_+(\mathbf{r}, \omega; t)$. The variation $\delta\tilde{\mathbf{E}}_+^*(\mathbf{r}, \omega; t)$ is then analytic in the

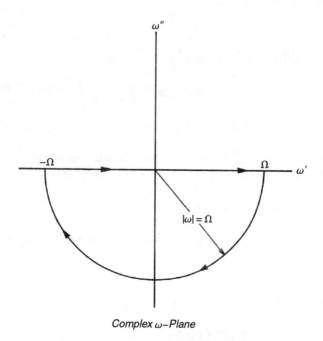

Complex ω–Plane

Fig. F.1 Completion of the contour of integration in the lower half of the complex ω-plane with a semicircular arc of radius $|\omega| = \Omega \to \infty$

lower half plane. The relation given in Eq. (5.299) cannot then be satisfied unless the remainder of the integrand is also analytic in the lower half plane. If the integrand in Eq. (5.299) is analytic in the lower half-plane and has magnitude that goes to zero sufficiently fast as $|\omega| \to \infty$ in the lower half-plane (Fig. F.1), then by completion of the contour with a semicircle in the lower half-plane centered at the origin with radius $\Omega \to \infty$, as illustrated in Fig. 5.7, application of Cauchy's theorem results in Eq. (5.299).

This result then provides a general variational approach for the determination of the irreversible electromagnetic energy density in a simple dispersive dielectric [2]. However, this result does not contradict the Barash and Ginzburg result [1] presented in Sect. 5.2.2 because complete knowledge of the temporal frequency dependence of the imaginary part of the electric susceptibility is required, such as that provided by an analytic model.

As an illustration, consider a single-resonance Lorentz model dielectric with resonance angular frequency ω_0, phenomenological damping constant δ, and plasma frequency $b = \sqrt{(\|4\pi\|/\epsilon_0)Nq_e^2/m}$. In their analysis, Glasgow et al. [2] take the electric susceptibility for this causal model medium to be given by the expression

$$\chi_e(\omega) = -\|4\pi\| \frac{b^2}{(\omega - (\omega_0 - i\delta))(\omega + (\omega_0 + i\delta))}, \tag{F.33}$$

with a denominator that differs from the usual expression $(\omega^2 + 2i\delta\omega - \omega_0^2)$ by the quantity δ^2 [cf. Eq. (4.211)]. The reversible and irreversible energy densities are then found as

$$\mathcal{U}_{rev}[E](\mathbf{r}, t) = \left\| \frac{1}{4\pi} \right\| \frac{\epsilon_0}{2} \frac{b^2}{\omega_0^2} \left[\Re \left\{ (\omega_0 - i\delta)\tilde{\mathbf{E}}_-(\mathbf{r}, \omega_0 - i\delta; t) \right\}^2 \right.$$

$$\left. + \left(\omega_0^2 + \delta^2 \right) \Im \left\{ \tilde{\mathbf{E}}_-(\mathbf{r}, \omega_0 - i\delta; t) \right\}^2 \right], \quad \text{(F.34)}$$

$$\mathcal{U}_{irrev}[E](\mathbf{r}, t) = \left\| \frac{1}{4\pi} \right\| 2\epsilon_0 \frac{\delta b^2}{\omega_0^2} \int_{-\infty}^{t} \Re \left\{ (\omega_0 - i\delta)\tilde{\mathbf{E}}_-(\mathbf{r}, \omega_0 - i\delta; t') \right\}^2 dt'.$$

$$\text{(F.35)}$$

From the constitutive relation $\tilde{\mathbf{P}}(\mathbf{r}, \omega) = \epsilon_0 \chi_e(\omega) \tilde{\mathbf{E}}(\mathbf{r}, \omega)$ relating the macroscopic polarization and electric field spectra in a temporally dispersive HILL medium, it is found that [2]

$$\Re \left\{ (\omega_0 - i\delta)\tilde{\mathbf{E}}_-(\mathbf{r}, \omega_0 - i\delta; t) \right\} = \frac{\omega_0}{b^2} \dot{\mathbf{P}}(\mathbf{r}, t), \quad \text{(F.36)}$$

$$\Im \left\{ \tilde{\mathbf{E}}_-(\mathbf{r}, \omega_0 - i\delta; t) \right\} = -\frac{\omega_0}{b^2} \mathbf{P}(\mathbf{r}, t). \quad \text{(F.37)}$$

With substitution of these two expressions, Eqs. (5.301) and (5.302) yield

$$\mathcal{U}_{rev}[E](\mathbf{r}, t) = \left\| \frac{1}{4\pi} \right\| \frac{\epsilon_0}{2} \left[\frac{1}{2b^2} \dot{P}^2(\mathbf{r}, t) + \frac{\omega_0^2}{2b^2} P^2(\mathbf{r}, t) \right], \quad \text{(F.38)}$$

$$\mathcal{U}_{irrev}[E](\mathbf{r}, t) = \left\| \frac{1}{4\pi} \right\| \epsilon_0 \frac{\delta}{b^2} \int_{-\infty}^{t} \dot{P}^2(\mathbf{r}, t') dt', \quad \text{(F.39)}$$

where the δ^2 term in Eq. (5.305) has been accordingly neglected. Equation (5.305) shows that the reversible energy density for a single Lorentz oscillator type is given by the sum of its kinetic and potential energy densities.

F.2 Recoverable and Irrecoverable Electrodynamic Processes

Based upon the expression given in Eq. (5.286) for the electromagnetic interaction energy density $\mathcal{U}_{int}(\mathbf{r}, t)$, it is seen that as $t \to \infty$, any given point in the (finitely extended) dielectric medium will have interacted with the entire electromagnetic pulse (provided that it has compact temporal support) so that $\mathcal{U}_{int}(\mathbf{r}, \infty)d^3r \equiv \lim_{t \to \infty} \mathcal{U}_{int}(\mathbf{r}, t)d^3r$ specifies the total amount of electromagnetic energy that was dissipated into the medium at that point due the entire pulse evolution at that point.

However, as the electromagnetic pulse propagates through the dielectric medium, a continuous interchange of electromagnetic energy takes place between the field and the atoms and molecules comprising the medium. This real-time energy exchange, for example, forms the physical basis for understanding both "slow" and "fast" light phenomena [8]. In particular, the phenomenon of "slow" light occurs when energy from the leading edge of the pulse is reactively stored in the medium and subsequently returned to the trailing edge of the pulse [9] as occurs in self-induced transparency [10]. The phenomenon of "fast" light, on the other hand, involves the distortion of the pulse envelope shape due to the unrecoverable dissipation of the leading edge energy such that the peak in the pulse envelope is shifted forward in time [11]. This forward temporal shift of the peak amplitude point in a pulse may then result in an apparent superluminal pulse envelope velocity; however, because the peak amplitude points in the dynamical pulse evolution are not causally related, this apparent superluminal velocity does not violate the Special Theory of Relativity.

The analysis of this real-time energy dissipation requires a modified interpretation to that given in the preceding subsection, as reflected in the following definitions due to Glasgow, Meilstrup, Peatross, and Ware [3]:

Definition F.3 (Recoverable Electromagnetic Energy Density) The recoverable energy density $\mathcal{U}_{rev}[E](t)$ of the electromagnetic field interaction at time t is the supremum of the energy density that the dielectric body can subsequently return to the field under the influence of a well-chosen future field; that is,

$$\mathcal{U}_{rec}[E](t) \equiv \sup_{E_+^t} \left\{ \mathcal{U}_{int}[E](t) - \mathcal{U}_{int}[E_-^t + E_+^t](+\infty) \right\}$$

$$= \mathcal{U}_{int}[E](t) - \inf_{E_+^t} \left\{ \mathcal{U}_{int}[E_-^t + E_+^t](+\infty) \right\}, \qquad (F.40)$$

where E_-^t denotes the fixed past field evolution with $E_-^t(t') = E(t)U(t - t')$ so that $E_-^t(t') = E(t)$ for $t' \leq t$ and $E_-^t(t') = 0$ for $t' > t$, and where E_+^t denotes a variable future electric field with $E_+^t(t') = 0$ for all $t' < t$.

Definition F.4 (Irrecoverable Electromagnetic Energy Density) The irrecoverable energy density $\mathcal{U}_{irrec}[E](t)$ of the electromagnetic field interaction at time t in a simple temporally dispersive dielectric is given by the complement of the recoverable energy density at time t as the infimum of the amount of energy that must be eventually dissipated by the dielectric body under the influence of a well-chosen future field, that is,

$$\mathcal{U}_{irrec}[E](t) \equiv \inf_{E_+^t} \left\{ \mathcal{U}_{int}[E_-^t + E_+^t](+\infty) \right\}. \qquad (F.41)$$

Both the recoverable and irrecoverable electromagnetic energy densities are naturally causal because [3] "the field evolution after time t does not affect the energy and loss at that time", where

$$\mathcal{U}_{rec}[E](t) = \mathcal{U}_{rec}[E_-^t + E_+^t](t) = \mathcal{U}_{rec}[E_-^t](t),$$
$$\mathcal{U}_{irrec}[E](t) = \mathcal{U}_{irrec}[E_-^t + E_+^t](t) = \mathcal{U}_{irrec}[E_-^t](t).$$

In addition, these two definitions satisfy the conservation of energy expression

$$\mathcal{U}_{int}[E](t) = \mathcal{U}_{rec}[E](t) + \mathcal{U}_{irrec}[E](t) \tag{F.42}$$

at each instant of time t. As stated in their paper [3], the supremum in Eq. (5.307) and the infimum in Eq. (5.308)

> are accomplished by treating the past field as fixed and allowing the future field to vary as necessary to accomplish the extrema. Thus $\mathcal{U}_{rec}[E](t)$ gives the largest portion of $\mathcal{U}_{int}(t)$ that could possibly be converted back to field energy after t, given a fixed past field. Conversely, $\mathcal{U}_{irrec}[E](t)$ represents the smallest amount of eventual loss $\mathcal{U}_{int}(+\infty)$ that is required for fields that share a common history prior to t.... both extrema are accomplished by the same future field.

The difference between the recoverable and reversible energy densities, as well as that between the irrecoverable and irreversible energy densities, may be explained in the following manner [12]: Let $\mathcal{U}_-$ be the "*creation energy*", defined as the minimum electromagnetic energy required to create a state of a dispersive dissipative dielectric. This is the energy that is required to bring the dielectric from its quiescent state in the distant past at $t = -\infty$ to its present state at time t. The *recoverable energy* defined in Def. 5.3 is the maximum energy that can be extracted from the dispersive dissipative dielectric, starting from the present time t and ending up in its original quiescent state in the distant future at $t = +\infty$. Because the dielectric body is dissipative, it then requires a non-negative influx of electromagnetic energy in order for the dielectric to be taken from its quiescent state at $t = -\infty$ to a new state at time t and then returned to its quiescent state at $t = +\infty$. Expressed mathematically,

$$\mathcal{U}_- - \mathcal{U}_{rec} \geq 0 \quad \Longrightarrow \quad \mathcal{U}_- \geq \mathcal{U}_{rec}, \tag{F.43}$$

stating that "it takes no less energy to create a state than is extractable from it."[2] The *irrecoverable energy* in a dispersive dissipative dielectric is defined in Def. 5.4 as the work that has been done on the dielectric from its quiescent state in the distant past at $t = -\infty$ to the present time t in order to establish the dielectric state at time t minus the energy that can be recovered, so that

$$\mathcal{U}_{irrec} = W - \mathcal{U}_{rec}, \tag{F.44}$$

---

[2] Scott Glasgow, personal correspondence (August 25, 2016).

where W is the work that has been expended (up to the time t) in order to establish that dielectric state. Notice that W depends upon the path taken to establish the present dielectric state, and hence, is not a state variable. In contrast, the *irreversible energy* in a dispersive dissipative dielectric is given by the work W expended to establish the dielectric state at time t minus the optimal creation energy $\mathcal{U}_-$ that could have been done in establishing this state, so that

$$\mathcal{U}_{irrev} = W - \mathcal{U}_-. \tag{F.45}$$

Because of the inequality in Eq. (5.310), one then has that

$$\mathcal{U}_{irrev} \leq \mathcal{U}_{irrec}, \tag{F.46}$$

where both $\mathcal{U}_{irrev}$ and $\mathcal{U}_{irrec}$ are non-negative because the creation energy $\mathcal{U}_-$ is the minimum over all possible values of W; that is, $\mathcal{U}_-$ is the minimum energy that could have been used to establish the dielectric state at time t. In summary, as stated by Glasgow (see footnote 15):

> the irreversible energy is a measure of the remorse you feel in having created the given dielectric state inefficiently... The irrecoverable energy on the other hand is a measure of the work you won't get back from the dielectric in the future despite even best efforts to create the state. Neither of these notions of loss is a state variable, but very much path dependent.

The minimum possible value of the irrecoverable energy $\mathcal{U}_{irrec}$, given by

$$\min\left\{W - \mathcal{U}_{rec}\right\} = \mathcal{U}_- - \mathcal{U}_{rec} \geq 0, \tag{F.47}$$

is a state variable and is a measure of the intrinsic "lossiness" of any given dielectric state.

In a nonconducting dielectric medium, the differential form of Poynting's theorem is given by Eq. (5.176) as $\partial \mathcal{U}'(\mathbf{r}, t)/\partial t = -\nabla \cdot \mathbf{S}(\mathbf{r}, t)$. Integration of this expression over the region of space $\mathcal{D}$ occupied by the dielectric body followed by application of the divergence theorem to the right-hand side then yields

$$\frac{\partial}{\partial t} \int_{\mathcal{D}} \mathcal{U}'(\mathbf{r}, t) d^3 r = -\oint_{\mathcal{S}} \mathbf{S}(\mathbf{r}, t) \cdot \hat{\mathbf{n}} d^2 r,$$

where $\mathcal{S}$ is the surface enclosing $\mathcal{D}$. If the integration domain is extended to all of space, then the surface integral of the normal component of the Poynting vector vanishes because the electromagnetic field vectors go to zero faster than r^{-1} as $r = |\mathbf{r}|$ recedes to infinity, so that

$$\frac{\partial}{\partial t} \int_{-\infty}^{\infty} \mathcal{U}'(\mathbf{r}, t) d^3 r = 0 \implies \int_{-\infty}^{\infty} \mathcal{U}'(\mathbf{r}, t) d^3 r = W, \tag{F.48}$$

where $\mathcal{W}$ is a constant independent of the time t. From Eq. (5.271), one has that $\mathcal{U}'(\mathbf{r}, t) = \mathcal{U}_{em}(\mathbf{r}, t) + \mathcal{U}_{int}(\mathbf{r}, t)$, and hence

$$\frac{\partial}{\partial t} \int_{-\infty}^{\infty} \mathcal{U}_{em}(\mathbf{r}, t)d^3r + \frac{\partial}{\partial t} \int_{-\infty}^{\infty} \mathcal{U}_{int}[E](\mathbf{r}, t)d^3r = 0. \qquad (\text{F.49})$$

Define the *total electromagnetic field energy* as [3]

$$\mathcal{E}[E](t) \equiv \int_{-\infty}^{\infty} \mathcal{U}_{em}(\mathbf{r}, t)d^3r, \qquad (\text{F.50})$$

so that, with substitution from Eq. (5.315)

$$\mathcal{E}[E](t) = \int_{-\infty}^{\infty} \mathcal{U}'(\mathbf{r}, t)d^3r - \int_{-\infty}^{\infty} \mathcal{U}_{int}(\mathbf{r}, t)d^3r$$

$$= \mathcal{W} - \int_{-\infty}^{\infty} \mathcal{U}_{int}(\mathbf{r}, t)d^3r. \qquad (\text{F.51})$$

The *net change in the total electromagnetic field energy* after the so-called "distinguished" time t separating the past (fixed) field evolution from the future (possible) field evolution is then defined as

$$\Delta_t \mathcal{E}[E](t) \equiv \mathcal{E}[E](\infty) - \mathcal{E}[E](t), \qquad (\text{F.52})$$

so that

$$\Delta_t \mathcal{E}[E](t) = \int_{-\infty}^{\infty} \left(\mathcal{U}_{int}[E](t) - \mathcal{U}_{int}[E](\infty) \right)d^3r$$

$$= \int_{-\infty}^{\infty} \mathcal{U}_{rec}[E](t)d^3r - \int_{-\infty}^{\infty} \left(\mathcal{U}_{int}[E](\infty) - \mathcal{U}_{irrec}[E](t) \right)d^3r. $$

$$(\text{F.53})$$

Because $\mathcal{U}_{int}[E](\infty) \geq \mathcal{U}_{irrec}[E](t)$, one finally obtains the inequality set by the *global recoverable energy* [3]

$$\Delta_t \mathcal{E}[E](t) \leq \int_{-\infty}^{\infty} \mathcal{U}_{rec}[E](t)d^3r, \qquad (\text{F.54})$$

stating that the net change in the total field energy in all of space after the distinguished instant of time t is always either equal to or bounded above by the recoverable energy in all of space at that instant of time t. However, the implications of this result are tempered by the simplifying assumption used in obtaining Eq. (5.315) that tacitly requires the dielectric material to fill all of space. One may, nevertheless, think of the result as applying to each point in the infinitely extended medium provided that the electromagnetic field behavior is the same at each point in the medium; that is, provided that it is an ideal time-harmonic uniform plane wave.

For a multiple-resonance Lorentz model dielectric with susceptibility [see Eq. (4.219)]

$$\chi_e(\omega) = \sum_j \chi_j(\omega) = \sum_j \frac{f_j \omega_p^2}{\omega_j^2 - 2i\delta_j\omega - \omega^2}, \tag{F.55}$$

where f_j is the oscillator strength of the dipole line and $\omega_p = \|4\pi\| N q_e^2/m$ is the plasma frequency, the electromagnetic energy density in the coupled field–medium system, or interaction energy $\mathcal{U}_{int}(t)$, defined in Eq. (5.270), may be separated into reversible and irreversible parts as [cf. Eqs. (5.305) and (5.306)]

$$\mathcal{U}_{rev}[E](\mathbf{r}, t) = \left\| \frac{1}{4\pi} \right\| \frac{\epsilon_0}{2} \sum_j \left\{ \frac{1}{f_j \omega_p^2} \dot{P}_j^2(\mathbf{r}, t) + \frac{\omega_j^2}{f_j \omega_p^2} P_j^2(\mathbf{r}, t) \right\}, \tag{F.56}$$

$$\mathcal{U}_{irrev}[E](\mathbf{r}, t) = \left\| \frac{1}{4\pi} \right\| 2\epsilon_0 \sum_j \frac{\delta_j}{f_j \omega_p^2} \int_{-\infty}^{t} \dot{P}_j^2(\mathbf{r}, t')dt'. \tag{F.57}$$

Glasgow et al. [3] refer to these two parts of the interaction energy as the collected energy $\mathcal{U}_e[E](\mathbf{r}, t)$ and loss $\mathcal{U}_\ell[E](\mathbf{r}, t)$ terms, respectively, stating that they are "model-dependent", as described in the following:

> Barash and Ginzburg point out that the parametrization of a given χ is generally not unique. Furthermore, two distinct parametrizations for the same χ generally result in different energy allocations between $\mathcal{U}_\ell$ and $\mathcal{U}_e$ for a given field. Thus, in the Barash and Ginzburg approach one can change the fraction of energy that is allocated as "lost" at a given time by changing the parametrization of χ, even though the choice of parametrization has no effect on physically measurable quantities.

Glasgow et al. [3] go on to state that "this is unsatisfactory at the physical level," the purpose of their approach being to "develop concepts of recoverable energy and loss that depend only on χ and the electric field, and not on the parametrization of χ." It is important to note here that the same can be said for the mathematical description of any physical system, from mechanics to thermodynamics to electromagnetism to circuit theory. As stated by Maxwell [13] in the introductory chapter on the "Nature of Physical Science" in his text on *Matter and Motion*, "Physical science is that department of knowledge which relates to the order of nature, or, in other words, to the regular succession of events." The physical sciences rely upon the interplay between the development of physical models to describe natural phenomena and experimental observation and measurement of that phenomena. If different physical models (based upon fundamental physical laws) used to describe a physical system predict different results, then the model that most closely agrees with experiment is likely the best one, and there may even be a better one. Such is the case with the Lorentz, Drude, Debye, and Rocard–Powles–Debye models described in Chap. 4.

The difference between the irrecoverable energy of Definition 5.4 and the irreversible loss given in Eq. (F.57) for a multiple-resonance Lorentz model dielectric is found to be given by [3]

$$\mathcal{U}_{irrec}[E](\mathbf{r}, t) - \mathcal{U}_{irrev}[E](\mathbf{r}, t) = \mathcal{U}_\ell[E^t_- + E^t_+](\mathbf{r}, t), \tag{F.58}$$

where

$$\mathcal{U}_\ell[E^t_- + E^t_+](\mathbf{r}, t) \equiv \left\| \frac{1}{4\pi} \right\| 2\epsilon_0 \inf_{E^t_+} \sum_j \frac{\delta_j}{f_j \omega_p^2} \int_{-\infty}^t \dot{P}_j^2[E^t_- + E^t_+](\mathbf{r}, t')dt' \tag{F.59}$$

is the *inevitable future loss* that will occur using the Lorentz model to describe the dispersive properties of the dielectric. That this model "determines how loss is allocated between past and future" [3] is of no real consequence because its applicability is restricted to an ideal time-harmonic uniform plane wave.

References

1. Y. S. Barash and V. L. Ginzburg, "Expressions for the energy density and evolved heat in the electrodynamics of a dispersive and absorptive medium," *Usp. Fiz. Nauk.*, vol. 118, pp. 523–530, 1976. [English translation: Sov. Phys.-Usp. vol. 19, 163–270 (1976)].
2. C. Broadbent, G. Hovhannisyan, M. Clayton, J. Peatross, and S. A. Glasgow, "Reversible and irreversible processes in dispersive/dissipative optical media: Electro-magnetic free energy and heat production," in *Ultra-Wideband, Short-Pulse Electromagnetics 6* (E. L. Mokole, M. Kragalott, and K. R. Gerlach, eds.), pp. 131–142, New York: Kluwer Academic, 2003.
3. S. Glasgow, M. Meilstrup, J. Peatross, and M. Ware, "Real-time recoverable and irrecoverable energy in dispersive-dissipative dielectrics," *Phys. Rev. E*, vol. 75, pp. 16616-1–16616-12, 2007.
4. F. W. Sears, *An Introduction to Thermodynamics, the Kinetic Theory of Gases, and Statistical Mechanics*. Reading, MA: Addison-Wesley, second ed., 1953.
5. J. H. Eberly and K. Wódkiewicz, "The time-dependent physical spectrum of light," *J. Opt. Soc. Am.*, vol. 67, no. 9, pp. 1252–1261, 1977.
6. J. Peatross, M. Ware, and S. A. Glasgow, "Role of the instantaneous spectrum on pulse propagation in causal linear dielectrics," *J. Opt. Soc. Am. A*, vol. 18, no. 7, pp. 1719–1725, 2001.
7. L. D. Landau and E. M. Lifshitz, *Electrodynamics of Continuous Media*. Oxford: Pergamon, 1960. Ch. IX.
8. R. W. Boyd and D. J. Gauthier, "Slow and Fast Light," in *Progress in Optics* (E. Wolf, ed.), vol. 43, pp. 497–530, Amsterdam: Elsevier, 2002.
9. C. G. B. Garrett and D. E. McCumber, "Propagation of a Gaussian light pulse through an anomalous dispersion medium," *Phys. Rev. A*, vol. 1, pp. 305–313, 1970.
10. L. Allen and J. H. Eberly, *Optical Resonance & Two-Level Atoms*. Wiley, 1975.
11. R. Y. Chiao and A. M. Steinberg, "Tunneling times and superluminality," in *Progress in Optics* (E. Wolf, ed.), vol. 37, pp. 345–405, Amsterdam: Elsevier, 1997.

12. S. Glasgow, J. Corson, and C. Verhaaren, "Dispersive dielectrics and time reversal: Free energies, orthogonal spectra, and parity in dissipative media," *Phys. Rev. E*, vol. 82, p. 011115, 2010.
13. J. C. Maxwell, *Matter and Motion*. London: Sheldon Press, 1925.

Appendix G
Stationary Phase Approximations
of the Angular Spectrum Representation

The temporal frequency spectrum $\tilde{\mathbf{U}}(\mathbf{r}, \omega)$ of a pulsed wave field $\mathbf{U}(\mathbf{r}, t)$ may be expressed in the positive half-space $z > 0$ of a nonabsorptive (and hence, nondispersive) medium by the real direction cosine form of the angular spectrum of plane waves representation as (see Sect. 8.3.2)

$$\tilde{\mathbf{U}}(\mathbf{r}, \omega) = \tilde{\mathbf{U}}_H(\mathbf{r}, \omega) + \tilde{\mathbf{U}}_E(\mathbf{r}, \omega) \tag{G.1}$$

with [see Eqs. (8.195) and (8.196)]

$$\tilde{\mathbf{U}}_J(\mathbf{r}, \omega) = \int_{\mathcal{D}_J} \tilde{\tilde{\mathbf{U}}}(p, q, \omega) e^{ik(px+qy+mz)} dp\, dq \tag{G.2}$$

for $z > 0$ with $J = H, E$. Here $k = \omega/c$ is real-valued and

$$m = +\left(1 - p^2 - q^2\right)^{1/2}, \quad (p, q) \in \mathcal{D}_H = \left\{(p, q) | 0 \le p^2 + q^2 \le 1\right\}, \tag{G.3}$$

$$m = +i\left(p^2 + q^2 - 1\right)^{1/2}, \quad (p, q) \in \mathcal{D}_E = \left\{(p, q) | p^2 + q^2 > 1\right\}. \tag{G.4}$$

For a sufficiently well behaved spatiotemporal frequency spectrum $\tilde{\tilde{\mathbf{U}}}(p, q, \omega)$, the spectral wave field $\tilde{\mathbf{U}}(\mathbf{r}, \omega)$ is found [1, 2] to satisfy the homogeneous Helmholtz equation

$$\left(\nabla^2 + k^2\right) \tilde{\mathbf{U}}(\mathbf{r}, \omega) = \mathbf{0}, \tag{G.5}$$

© Springer Nature Switzerland AG 2019
K. E. Oughstun, *Electromagnetic and Optical Pulse Propagation*, Springer Series in Optical Sciences 224, https://doi.org/10.1007/978-3-030-20835-6

and the Sommerfeld radiation condition [3, page 250, Eq. (1d)]

$$\lim_{r \to \infty} r \left(\frac{\partial \tilde{U}(\mathbf{r}, \omega)}{\partial r} - ik\tilde{U}(\mathbf{r}, \omega) \right) = \mathbf{0} \qquad (G.6)$$

in the half-space $z > 0$, which guarantees that there are only outgoing waves at infinity. The spectral wave field component $\tilde{U}_H(\mathbf{r}, \omega)$ is expressed in Eq. (8.195) as a superposition of homogeneous plane waves that are propagating in directions specified by the real-valued direction cosines (p, q, m), illustrated in Fig. 8.9, whereas the spectral wave field component $\tilde{U}_E(\mathbf{r}, \omega)$ is expressed in Eq. (8.196) as a superposition of evanescent plane waves that are propagating in directions perpendicular to the z-axis as specified by the real-valued direction cosines $(p, q, 0)$ and exponentially decaying at different rates in the z-direction. Because the behavior of the evanescent plane wave components is so vastly different from that for the homogeneous plane wave components, the two component spectral wave fields $\tilde{U}_H(\mathbf{r}, \omega)$ and $\tilde{U}_E(\mathbf{r}, \omega)$ are usually treated separately, as done here.

In view of the fact that the integral appearing in the angular spectrum representation in Eq. (G.2) cannot be evaluated analytically (except in special cases), meaningful approximations are then required. Typically, the most appropriate approximations in the majority of applications are those that are valid as the distance

$$R = \sqrt{(x - x_0)^2 + (y - y_0)^2 + (z - z_0)^2} \qquad (G.7)$$

from the fixed point (x_0, y_0, z_0) becomes large in comparison with the wavelength $\lambda = 2\pi/k = 2\pi c/\omega$ of the spectral amplitude wave field. This approximation is provided by the asymptotic expansion of $\tilde{U}(\mathbf{r}, \omega)$, $\tilde{U}_H(\mathbf{r}, \omega)$, and $\tilde{U}_E(\mathbf{r}, \omega)$ as $kR \to \infty$ with fixed k and fixed observation direction specified by the (fixed) direction cosines (ξ_1, ξ_2, ξ_3) defined by

$$\xi_1 \equiv \frac{x - x_0}{R}, \quad \xi_2 \equiv \frac{y - y_0}{R}, \quad \xi_3 \equiv \frac{z - z_0}{R}. \qquad (G.8)$$

The analysis presented here follows that given by Sherman, Stamnes, and Lalor [4] in 1975 who consider the asymptotic behavior of $\tilde{U}(\mathbf{r}, \omega)$ as the point of observation (x, y, z) recedes to infinity in a fixed direction (ξ_1, ξ_2, ξ_3) with positive z-component $\xi_3 > 0$ through the point (x_0, y_0, z_0), as well as when the point of observation recedes to infinity in a fixed direction $(\xi_1, \xi_2, 0)$ perpendicular to the z-axis through a fixed point (x_0, y_0, z_0) in the positive half-space $z > 0$, all obtained through an extension of the method of stationary phase.

G.1   The Method of Stationary Phase

The method of stationary phase was originally developed by G. G. Stokes [5] in 1857 and Lord Kelvin [6] in 1887 for the asymptotic approximation of Fourier transform type integrals of the form

$$f(\lambda) = \int_a^b g(t)e^{i\lambda h(t)}dt \tag{G.9}$$

as $\lambda \to \infty$, where a, b, $g(t)$, $h(t)$, λ and t are all real-valued. The functions $h(\zeta)$ and $g(\zeta)$ are assumed to be analytic functions of the complex variable $\zeta = t + i\tau$ in some domain containing the closed interval $[a, b]$ along the real axis [7]. The exponential kernel $e^{i\lambda h(t)}$ appearing in the integrand is then purely oscillatory so that as λ becomes large, its oscillations become very dense and destructive interference occurs almost everywhere. The exceptions occur at any stationary phase point t_j defined by

$$h'(t_j) \equiv 0, \tag{G.10}$$

because the phase term $\lambda h(t)$ is nearly constant in a neighborhood of each such point, as well as from the lower and upper end points $t = a$ and $t = b$ of the integration domain because the effects of destructive interference will be incomplete there.

Assume that there is a single interior stationary phase point at $t = t_0$ where $a < t_0 < b$, and is such that $h''(t_0) \neq 0$. In a neighborhood about this stationary phase point, $h(t)$ may be represented by its Taylor series expansion

$$h(t) = h(t_0) + \frac{1}{2}h''(t_0)(t - t_0)^2 + \cdots \tag{G.11}$$

so that[1] $h(t) - h(t_0) = \mathcal{O}\left\{(t - t_0)^2\right\}$, whereas in a neighborhood of any other point $\tau \in [a, b]$, $h(t) - h(\tau) = \mathcal{O}\left\{(t - \tau)\right\}$. The expansion given in Eq. (G.11) then suggests the change of variable

$$h(t) - h(t_0) \equiv \pm s^2, \tag{G.12}$$

where $+s^2$ is used when $h''(t_0) > 0$ while $-s^2$ is used when $h''(t_0) < 0$. For values of t in a neighborhood of t_0, Eq. (G.12) may be inverted with use of the Taylor series

---

[1]The order symbol $\mathcal{O}$ is defined as follows. Let $f(z)$ and $g(z)$ be two functions of the complex variable z that possess limits as $z \to z_0$ in some domain $\mathcal{D}$. Then $f(z) = \mathcal{O}(g(z))$ as $z \to z_0$ iff there exist positive constants K and δ such that $|f(z)| \leq K|g(z)|$ whenever $0 < |z - z_0| < \delta$.

expansion given in Eq. (G.11) as

$$t - t_0 = \left[\frac{2}{|h''(t_0)|} \right]^{1/2} s + \mathcal{O}\left\{ s^2 \right\}, \qquad (G.13)$$

valid when either $h''(t_0) > 0$ or $h''(t_0) < 0$. With the Taylor series expansion

$$g(t) = g(t_0) + g'(t_0)(t - t_0) + \cdots, \qquad (G.14)$$

the integral in Eq. (G.9) becomes

$$\begin{aligned}
f(\lambda) &= \int_a^b g(t) e^{i\lambda h(t)} dt \\
&\sim \left[\frac{2}{|h''(t_0)|} \right]^{1/2} g(t_0) e^{i\lambda h(t_0)} \int_{-s_1}^{s_2} e^{\pm i\lambda s^2} \left[1 + \mathcal{O}\left\{ s \right\} \right] ds \\
&\sim \left[\frac{2}{|h''(t_0)|} \right]^{1/2} g(t_0) e^{i\lambda h(t_0)} \int_{-\infty}^{\infty} e^{\pm i\lambda s^2} ds + \mathcal{O}\left\{ \int_{-\infty}^{\infty} s^2 e^{\pm i\lambda s^2} ds \right\}
\end{aligned}$$

as $\lambda \to \infty$, so that

$$\int_a^b g(t) e^{i\lambda h(t)} dt \sim \left[\frac{2\pi}{\lambda |h''(t_0)|} \right]^{1/2} g(t_0) e^{i\lambda h(t_0)} e^{\pm i\pi/4} + \mathcal{O}\left\{ \lambda^{-1} \right\} \qquad (G.15)$$

as $\lambda \to \infty$, where $+\pi/4$ corresponds to the case when $h''(t_0) > 0$ while $-\pi/4$ corresponds to $h''(t_0) < 0$.

The end point contributions to the asymptotic behavior of this integral are obtained through a straightforward integration by parts as

$$\begin{aligned}
\int_a^b g(t) e^{i\lambda h(t)} dt &= \left. \frac{g(t)}{i\lambda h'(t)} \right|_a^b - \int_a^b \frac{d}{dt} \left(\frac{g(t)}{i\lambda h'(t)} \right) e^{i\lambda h(t)} dt \\
&\sim \frac{1}{i\lambda} \left[\frac{g(b)}{h'(b)} e^{i\lambda h(b)} - \frac{g(a)}{h'(a)} e^{i\lambda h(a)} \right] + \mathcal{O}\left\{ \lambda^{-2} \right\}, \quad (G.16)
\end{aligned}$$

as $\lambda \to \infty$. Combination of Eqs. (G.15) and (G.16) then results in the asymptotic approximation

$$\begin{aligned}
\int_a^b g(t) e^{i\lambda h(t)} dt &\sim \left[\frac{2\pi}{\lambda |h''(t_0)|} \right]^{1/2} g(t_0) e^{i\lambda h(t_0)} e^{\pm i\pi/4} \\
&\quad + \frac{1}{i\lambda} \left[\frac{g(b)}{h'(b)} e^{i\lambda h(b)} - \frac{g(a)}{h'(a)} e^{i\lambda h(a)} \right] + \mathcal{O}\left\{ \lambda^{-2} \right\}
\end{aligned}$$

$$(G.17)$$

as $\lambda \to \infty$.

As an example, consider the asymptotic behavior of the Fourier integral of a continuous function $f(t)$ that identically vanishes outside of the finite domain $[\alpha, \beta]$; i.e. has compact support. Since $h(t) = t$ does not possess any stationary phase points, the asymptotic behavior is due solely to the endpoints. Repeated integration by parts N times then results in the expression

$$\int_\alpha^\beta f(t)e^{-i\omega t}\,dt = \sum_{n=0}^{N-1}\left(\frac{-i}{\omega}\right)^{n+1}\left\{f^{(n)}(\alpha)e^{-i\omega\alpha} - f^{(n)}(\beta)e^{-i\omega\beta}\right\} + R_N(\omega),$$

where

$$R_N(\omega) \equiv \left(\frac{-i}{\omega}\right)^N\int_\alpha^\beta f^{(N)}(t)e^{-i\omega t}\,dt.$$

By the Riemann–Lebesgue lemma [8], if $f^{(N)}(t)$ is continuous in the interval $[\alpha, \beta]$, then the integral appearing in the remainder term $R_N(\omega)$ tends to zero in the limit as $\omega \to \infty$, so that $R_N(\omega) = \mathcal{O}\{\omega^{-N}\}$. The above result is then the asymptotic series as $\omega \to \infty$ of the Fourier transform of the Nth-order continuous function $f(t)$ with compact support $[\alpha, \beta]$.

G.2 Generalization to Angular Spectrum Integrals

The generalized asymptotic approximation presented here for the angular spectrum integral appearing in Eq. (G.2) is based upon an earlier extension of the method of stationary phase to multiple integrals of the form

$$I(kR) = \int_{\mathcal{D}} g(p, q)e^{ikRf(p,q)}\,dp\,dq, \tag{G.18}$$

as $kR \to \infty$. In this extended method of stationary phase, originally due to Focke [9], Braun [10], Jones and Kline [11], and Chako [12], the functions $f(p,q)$ and $g(p,q)$ are taken to be independent of kR and to be sufficiently smooth in the domain $\mathcal{D}$ and the quantity $kRf(p,q)$ is required to be real-valued for all $(p,q) \in \mathcal{D}$. Heuristically, as kR becomes large, small variations in (p,q) result in rapid oscillations of the exponential factor appearing in the integrand of Eq. (G.18) and this, in turn, results in destructive interference in the integration about that point in the integration domain $\mathcal{D}$ so that most of the domain $\mathcal{D}$ provides only a negligible contribution to the integral $I(kR)$. Significant contributions to $I(kR)$ come from the respective neighborhoods of specific types of critical points that are ordered according to their relative asymptotic order as follows:

Critical Point of the First Kind: An interior point $(p_s, q_s) \in \mathcal{D}$ where

$$\frac{\partial f(p,q)}{\partial p}\bigg|_{(p_s,q_s)} = \frac{\partial f(p,q)}{\partial q}\bigg|_{(p_s,q_s)} = 0, \qquad\qquad \text{(G.19)}$$

at which point the phase function $f(p, q)$ is stationary, called an *interior station-ary phase point* because the exponential factor in the integrand of Eq. (G.18) does not oscillate at that point, is a critical point of the first kind because it results in an asymptotic contribution to the integral appearing in Eq. (G.18) that is of order $\mathcal{O}\{(kR)^{-1}\}$ as $kR \to \infty$.

Critical Point of the Second Kind: A boundary point $(p_b, q_b) \in \partial\mathcal{D}$ on the curve that forms the boundary of the integration domain $\mathcal{D}$ at which $\partial f/\partial s|_{(p_b,q_b)} = 0$, where ds describes a differential element of arc length along the boundary curve, is a critical point of the second kind because it yields a contribution of order $\mathcal{O}\{(kR)^{-3/2}\}$ as $kR \to \infty$.

Critical Point of the Third Kind: A corner point $(p_c, q_c) \in \partial\mathcal{D}$ on the curve that forms the boundary of the integration domain $\mathcal{D}$ is a critical point of the third kind because the slope of the boundary curve is discontinuous, resulting in a contribution of order $\mathcal{O}\{(kR)^{-2}\}$ as $kR \to \infty$.

Additional critical points can arise at any singularities of the function $g(p, q)$ appearing in the integrand of Eq. (G.18). Unfortunately, this heuristic description does not lead to a rigorous asymptotic description of integrals of the form given in Eq. (G.18).

Although these rigorous extensions of the stationary phase method have little resemblance to the heuristic stationary phase argument presented above, they do provide justification of that heuristic argument provided that certain restrictions (as specified above) are placed on both $f(p, q)$ and $g(p, q)$. Unfortunately, these restrictions are not satisfied by the angular spectrum integrals given in Eq. (G.2). In particular, the restriction that the functions $f(p, q)$ and $g(p, q)$ must possess a finite number of continuous partial derivatives in $\mathcal{D}$ except pos-sibly at a finite number of isolated integrable singularities of $g(p, q)$ is not satisfied by the angular spectrum integral given in Eq. (G.2) on the unit circle $p^2 + q^2 = 1$ that forms the boundary between $\mathcal{D}_H$ and $\mathcal{D}_E$. In addition, the requirement that $kRf(p, q)$ is real-valued in $\mathcal{D}$ cannot be applied to either $\tilde{\mathbf{U}}(\mathbf{r}, \omega)$ or $\tilde{\mathbf{U}}_E(\mathbf{r}, \omega)$ when $\xi_3 > 0$ since m is imaginary in $\mathcal{D}_E$. Nevertheless, if the evanescent contribution $\tilde{\mathbf{U}}_E(\mathbf{r}, \omega)$ is indeed negligible in comparison to the homogeneous contribution $\tilde{\mathbf{U}}_H(\mathbf{r}, \omega)$ then, at the very least, the leading term in the asymptotic expansion of $\tilde{\mathbf{U}}(\mathbf{r}, \omega)$ can be obtained from the asymptotic expansion of $\tilde{\mathbf{U}}_H(\mathbf{r}, \omega)$. This is not always the case, however, as shown by the following example due to Sherman et al. [4]. A time-harmonic spherical wave

$$\tilde{U}_s(\mathbf{r}, \omega) = \frac{e^{ikr}}{r} = \frac{e^{ik(x^2+y^2+z^2)^{1/2}}}{(x^2 + y^2 + z^2)^{1/2}}$$

radiating away from the origin has spectral amplitude [13]

$$\tilde{U}_s(p, q, \omega) = \frac{ik}{2\pi m} = \frac{ik}{2\pi (1 - p^2 - q^2)^{1/2}}.$$

When the point of observation lies along the z-axis, $p = q = 0$ and the integrals in Eq. (G.2) may be directly evaluated to obtain

$$\tilde{U}_{SH}(\mathbf{r}, \omega) = \frac{e^{ikz}}{z} - \frac{1}{z},$$

$$\tilde{U}_{SE}(\mathbf{r}, \omega) = \frac{1}{z}.$$

It is then seen that the homogeneous and evanescent wave contributions are of the same order in $1/z$, so that, in this special case, the evanescent contribution to the wave field cannot be neglected in comparison to the homogeneous contribution. This result is shown to be a direct consequence of the singular behavior of $\tilde{U}_s(p, q, \omega) = ik/2\pi m$ on the unit circle $p^2 + q^2 = 1$.

In the generalized asymptotic method due to Sherman, Stamnes and Lalor [4] that is considered here, specific restrictions are placed on the spectral amplitude function $\tilde{U}(p, q, \omega)$. In particular, $\tilde{U}(p, q, \omega) \in \mathcal{T}_N$, where $\mathcal{T}_N$ is defined for positive, even integer N as the set of all spectral functions $\tilde{U}(p, q, \omega)$ that are independent of the spatial coordinates x, y, z and that satisfy the following conditions for all ω:

1. $\tilde{U}(p, q, \omega)$ can be expressed in the form (with the dependence on the angular frequency ω suppressed)

$$\tilde{U}(p, q, \omega) = \frac{V(p, q, m)}{m}, \qquad (G.20)$$

   where m is as defined in Eqs. (G.3) and (G.4), and where $V(p, q, m) \equiv |V(p, q, m)|$ is bounded for all p, q.
2. $V(p, q, s)$ is a complex, continuous vector function of the three independent variables p, q, s that are defined (a) for all real p and q, (b) for real $s \in [0, 1]$, and (c) for pure imaginary $s = i\sigma$ with $\sigma > 0$.
3. $V(p, q, s)$ possesses continuous, bounded partial derivatives up to the order N with respect to p, q, s for all p, q, s within its domain of definition.

Notice that condition 1 ensures the convergence of the angular spectrum integrals appearing in Eq. (G.2) for all $z > 0$, while conditions 2 and 3 are required for the derivation of the asymptotic approximations of these integrals.

G.3   Approximations Valid Over a Hemisphere

The asymptotic approximation of $\tilde{\mathbf{U}}(\mathbf{r}, \omega) = \tilde{\mathbf{U}}(x, y, z, \omega)$ that is valid as the point of observation $\mathbf{r} = (x, y, z)$ recedes to infinity in a fixed direction with $z > 0$ through the point $\mathbf{r}_0 = (x_0, y_0, z_0)$ is considered first. This is done in two steps. First, the method of stationary phase is extended using the technique due to Focke [9] and Chako [12] in order to show that when $\tilde{\tilde{\mathbf{U}}}(p, q, \omega) \in \mathcal{T}_N$ has an interior stationary phase point at $(p_s, q_s) = (\xi_1, \xi_2)$, only an arbitrarily small neighborhood of that stationary phase point contributes to the asymptotic behavior as $R \to \infty$ up to terms of order $\mathcal{O}\{(kR)^{-N}\}$. A standard application of the method of stationary phase is then used to obtain the asymptotic contribution due to this interior stationary phase point.

G.3.1   Extension of the Method of Stationary Phase

The method of approach used by Focke [9] and Chako [12] employs a neutralizer function $v(p, q)$ to isolate the interior stationary point. Notice that since the point of observation $\mathbf{r} = (x, y, z)$ recedes to infinity in a fixed direction with $z > 0$ through the point $\mathbf{r}_0 = (x_0, y_0, z_0)$, then the direction cosine values ξ_1, ξ_2, ξ_3 are constant with $\xi_3 > 0$, and consequently, the stationary phase point $(p_s, q_s) = (\xi_1, \xi_2)$ is fixed within the interior of the region $\mathcal{D}_H$. Define Ω_1 and Ω_2 to be two arbitrarily small neighborhoods of the point (p_s, q_s), both of which are completely contained within the interior of $\mathcal{D}_H$, where $\Omega_1 \subset \Omega_2$ is a proper subset of Ω_2. Let $v(p, q)$ be a real, continuous function of the independent variables p and q with continuous partial derivatives of all orders for all real values of p and q that satisfies the property

$$0 \le v(p, q) \le 1, \tag{G.21}$$

with

$$v(p, q) = \begin{cases} 1, & \text{when } (p, q) \in \Omega_1 \\ 0, & \text{when } (p, q) \notin \Omega_1 \end{cases}. \tag{G.22}$$

The existence of such a continuous function for arbitrary regions Ω_1 and Ω_2 with $\Omega_1 \subset \Omega_2$ has been given by Bremmerman [14]. Additional details about the explicit form of this neutralizer function $v(p, q)$ are unnecessary.

The principal result of this section is expressed in the following theorem due to Sherman, Stamnes and Lalor [4].

Theorem G.1 *Let $\tilde{\tilde{\mathbf{U}}}(p, q, \omega) \in \mathcal{T}_N$ for some positive even integer N. Then for $z > 0$, the spectral wave field $\tilde{\mathbf{U}}(\mathbf{r}, \omega) = \tilde{\mathbf{U}}(x, y, z, \omega)$ given in Eqs. (G.1) and (G.2)* *with k a positive real constant satisfies*

$$\tilde{U}(x, y, z, \omega) = \tilde{U}_0(x, y, z) + \mathcal{R}(x, y, z), \tag{G.23}$$

where[2]

$$\tilde{U}_0(x, y, z) = \int_{\mathcal{D}_H} v(p, q)\tilde{\tilde{U}}(p, q, \omega)e^{ik(px+qy+mz)}dpdq, \tag{G.24}$$

and where $\mathcal{R}(x, y, z) = \mathcal{O}\left\{(kR)^{-N}\right\}$ as $kR \to \infty$ uniformly with respect to ξ_1 and ξ_2 for all real ξ_1, ξ_2 such that $\delta < \sqrt{1 - \xi_1^2 - \xi_2^2} \le 1$ for any positive constant $\delta < 1$.

Notice that the dependence on the angular frequency ω has been suppressed in the above expressions. In addition, the statement that $\mathcal{R}(x, y, z) = \mathcal{O}\left\{(kR)^{-N}\right\}$ as $kR \to \infty$ uniformly with respect to ξ_1 and ξ_2 means that $\mathcal{R}(x, y, z)$ is bounded as $|\mathcal{R}(x, y, z)| \le M(kR)^{-N}$ as $kR \to \infty$ with M a positive constant independent of ξ_1 and ξ_2. This theorem then directly leads to the result that the asymptotic behavior of the spectral wave field $\tilde{U}(\mathbf{r}, \omega)$ of order lower than $(kR)^{-N}$ as $kR \to \infty$ is completely determined by an arbitrarily small neighborhood of the point (p_s, q_s). In particular, if $\tilde{\tilde{U}}(p, q, \omega) \in \mathcal{J}_N$ when N is arbitrarily large, then the asymptotic expansion of $\tilde{U}(\mathbf{r}, \omega)$ is equal to the asymptotic expansion of $\tilde{U}_0(x, y, z)$. The proof of this theorem [4] is based upon a modification of the proof of Theorem 1 by Chako [12] or Theorem 3 by Focke [9].

Proof Construct three neutralizer functions $v_j(p, q)$, $j = 1, 2, 3$, each of which is a real continuous function of p and q with continuous partial derivatives of all orders for all p, q and which satisfy

$$0 \le v_j(p, q) \le 1$$

with

$$v_1(p, q) + v_2(p, q) + v_3(p, q) = 1.$$

Choose positive constants C_1, C_2, C_3, C_4 such that

$$\sqrt{\xi_1^2 + \xi_2^2} < C_1 < C_2 < 1 < C_3 < C_4,$$

with C_1 sufficiently large that the Ω_2-neighborhood of the point (p_s, q_s) is completely contained within the region $\sqrt{p^2 + q^2} < C_1$, as illustrated in Fig. G.1. Since $\xi_1^2 + \xi_2^2 = 1 - \xi_3^2$ with $\xi_3 > \delta > 0$, the constants C_1, C_2, C_3, C_4 clearly exist.

---

[2]Notice the typographical error in Eq. (2.5) of Ref. [4] where the spectral amplitude function $\tilde{\tilde{U}}(p, q, \omega)$ was inadvertently omitted from that equation.

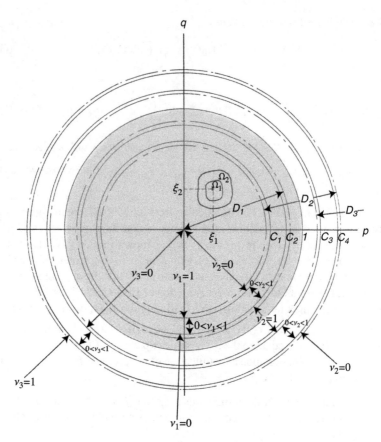

Fig. G.1 Graphical illustration of the notation used in the proof of Theorem 1. The shaded area indicates the region where $m = \sqrt{1 - p^2 - q^2}$ is real-valued, and the unshaded area indicates the region where $m = i\sqrt{p^2 + q^2 - 1}$ is pure imaginary

With these four constants specified, the three neutralizer functions $v_j(p, q)$, $j = 1, 2, 3$ are now required to satisfy the conditions

$$v_1(p, q) = \begin{cases} 1, & \text{for } \sqrt{p^2 + q^2} \leq C_1 \\ 0, & \text{for } \sqrt{p^2 + q^2} \geq C_2 \end{cases},$$

$$v_2(p, q) = \begin{cases} 1, & \text{for } C_2 \leq \sqrt{p^2 + q^2} \leq C_3 \\ 0, & \text{for } \sqrt{p^2 + q^2} \leq C_1 \ \& \ \sqrt{p^2 + q^2} \geq C_4 \end{cases},$$

$$v_3(p, q) = \begin{cases} 1, & \text{for } \sqrt{p^2 + q^2} \geq C_4 \\ 0, & \text{for } \sqrt{p^2 + q^2} \leq C_3 \end{cases}.$$

Finally, three overlapping regions D_1, D_2, D_3 in the p, q-plane are defined as

$$D_1 \equiv \left\{ (p, q) \,\middle|\, \sqrt{p^2 + q^2} \leq C_2 + \varepsilon \right\},$$

$$D_2 \equiv \left\{ (p, q) \,\middle|\, C_1 - \varepsilon \leq \sqrt{p^2 + q^2} \leq C_4 + \varepsilon \right\},$$

$$D_3 \equiv \left\{ (p, q) \,\middle|\, C_3 - \varepsilon \leq \sqrt{p^2 + q^2} \right\},$$

where $\varepsilon > 0$ is a positive constant that satisfies the three inequalities

$$\varepsilon < 1 - C_2,$$

$$\varepsilon < C_3 - 1,$$

$$\varepsilon < C_1 - \sqrt{\xi_1^2 + \xi_2^2}.$$

Each of these three regions is illustrated in Fig. G.1. Features of note that are important for this proof are the following:

- each of the neutralizer functions $v_j(p, q)$, $j = 1, 2, 3$, vanishes both for points $(p, q) \notin D_j$ and for points (p, q) in some neighborhood of the boundary ∂D_j of D_j,
- the stationary phase point (p_s, q_s) lies within the interior of D_1 and is exterior to both regions D_2 and D_3,
- the direction cosine $m = +\sqrt{1 - p^2 - q^2}$ is real-valued for all points $(p, q) \in D_1$,
- and $m = +i\sqrt{p^2 + q^2 - 1}$ is pure imaginary for all points $(p, q) \in D_3$.

Separate the angular spectrum integral for $\tilde{\mathbf{U}}(\mathbf{r}, \omega)$ that is given in Eqs. (G.1) and (G.2) into three parts as

$$\tilde{\mathbf{U}}(\mathbf{r}, \omega) = \tilde{\mathbf{U}}_1(\mathbf{r}, \omega) + \tilde{\mathbf{U}}_2(\mathbf{r}, \omega) + \tilde{\mathbf{U}}_3(\mathbf{r}, \omega),$$

where

$$\tilde{\mathbf{U}}_j(\mathbf{r}, \omega) \equiv \int_{D_j} v_j(p, q) \tilde{\mathbf{U}}(p, q, \omega) e^{ik(px + qy + mz)} \, dp \, dq$$

for $j = 1, 2, 3$. Each of these integrals is now separately treated.

Consider first the asymptotic approximation of the integral representation of $\tilde{\mathbf{U}}_1(\mathbf{r}, \omega)$ which is in a form that is appropriate for the method of stationary phase. Since the stationary phase point (p_s, q_s) is the only critical point of significance in the domain D_1, and since the critical points of the phase function $\mathbf{k} \cdot \mathbf{r} = k(px + qy + mz)$ on the boundary ∂D_1 of D_1 do not contribute to the integral because of the neutralizer function $\nu_1(p, q)$, and since the region Ω_2 which contains the neighborhood Ω_1 of the stationary phase point (p_s, q_s) lies entirely within D_1, then it follows from the proof of Theorem 1 in Chako [12] that

$$\tilde{U}_1(\mathbf{r}, \omega) = \tilde{U}_0(x, y, z) + \mathcal{R}(x, y, z),$$

where $\mathcal{R}(x, y, z) = \mathcal{O}\left\{(kR)^{-N}\right\}$ as $kR \to \infty$ with fixed $k > 0$, uniformly with respect to ξ_1, ξ_2 for all real ξ_1 and ξ_2 such that $\delta < \sqrt{1 - \xi_1^2 - \xi_2^2} \leq 1$ for any positive constant δ with $0 < \delta < 1$.

Consider next the asymptotic approximation of the integral representation of $\tilde{U}_3(\mathbf{r}, \omega)$. In this case, the quantity $m = +i\sqrt{p^2 + q^2 - 1}$ is pure imaginary with $im \leq -\sqrt{(C_3 - \varepsilon)^2 - 1} < 0$ for all $(p, q) \in D_3$. For a sufficiently large value of $R = \sqrt{(x - x_0)^2 + (y - y_0)^2 + (z - z_0)^2}$, a positive constant a then exists such that for $z > a$,

$$\left|\tilde{U}_3(\mathbf{r}, \omega)\right| \leq \int_{D_3} \left|\tilde{\tilde{U}}(p, q, \omega)\right| e^{ikmz} dp dq$$

$$\leq e^{-k(z-a)\sqrt{(C_3-\varepsilon)^2-1}} \int_{D_3} \left|\tilde{\tilde{U}}(p, q, \omega)\right| e^{ikma} dp dq.$$

Since $\tilde{\tilde{U}}(p, q, \omega) \in \mathcal{T}_N$, it then follows from the first condition [see Eq. (1)] in the definition of $\mathcal{T}_N$ that the final integral appearing in the above inequality converges to a finite, nonnegative constant M_3 that is independent of the direction cosines ξ_1, ξ_2, ξ_3. Since $z = z_0 + \xi_3 R$ [see Eq. (G.8)] and since $\xi_3 > 0$, then the above inequality becomes

$$\left|\tilde{U}_3(\mathbf{r}, \omega)\right| \leq M_3 e^{-k[(z_0-a)+R\delta]\sqrt{(C_3-\varepsilon)^2-1}},$$

independent of ξ_1, ξ_2. Hence, $\left|\tilde{U}_3(\mathbf{r}, \omega)\right| = \mathcal{O}\left\{(kR)^{-N}\right\}$ uniformly with respect to ξ_1, ξ_2 as $kR \to \infty$ for all positive even integer values N.

Consider finally the asymptotic approximation of the integral representation of $\tilde{U}_2(\mathbf{r}, \omega)$, which turns out to be much more involved than the previous two cases. The analysis begins with the change of variables

$$p = \sin \alpha \cos \beta,$$
$$q = \sin \alpha \sin \beta,$$

with Jacobian $J(p, q/\alpha, \beta) = \sin \alpha \cos \beta$. In addition,

$$m = \left(1 - p^2 - q^2\right)^{1/2} = \cos \alpha,$$

where α is, in general, complex as the region D_2 extends from inside the inner region where m is real-valued into the outer region where m is pure imaginary, as illustrated in Fig. G.1. In addition, let the angles ϑ, φ describe the intersection point on the unit sphere with center at the origin $\mathbf{r} = \mathbf{0}$ in coordinate space with the line through the origin that is parallel to the direction of observation $\mathbf{r} = (x, y, z)$ as it recedes to infinity through the fixed point $\mathbf{r}_0 = (x_0, y_0, z_0)$ with $z > 0$, where

$$\xi_1 = \sin \vartheta \, \cos \varphi,$$
$$\xi_2 = \sin \vartheta \, \sin \varphi,$$
$$\xi_3 = \cos \vartheta,$$

with $0 \le \vartheta < \pi/2$ and $0 \le \varphi < 2\pi$. Under these two changes of variables, the integral for $\tilde{\mathbf{U}}_2(\mathbf{r}, \omega)$ becomes

$$\tilde{\mathbf{U}}_2(\mathbf{r}, \omega) = \int_0^{2\pi} \int_C A(\alpha, \beta) e^{ikR[\sin \vartheta \, \sin \alpha \, \cos (\beta - \varphi) + \cos \vartheta \, \cos \alpha]} d\alpha d\beta,$$

where

$$A(\alpha, \beta) \equiv v_2(p, q) V(p, q, m) e^{ik(px_0 + qy_0 + mz_0)} \sin \alpha,$$

with $V(p, q, m)$ defined in Eq. (1). The contour of integration C in the complex α-plane, which extends from $\alpha_1 = \arcsin \left(\sqrt{C_1 - \varepsilon} \right)$ along the real α'-axis to $\pi/2$ and then to the endpoint at $\alpha_4 = \pi/2 - i \cosh^{-1} \left(\sqrt{C_4 + \varepsilon} \right)$, depicted in Fig. G.2, then results in a complete, single covering of the original integration domain D_2.

Notice that, in contrast with the original form of the integral, the phase

$$\Phi \equiv \sin \vartheta \, \sin \alpha \, \cos (\beta - \varphi) + \cos \vartheta \, \cos \alpha$$

appearing in the above transformed integral for $\tilde{\mathbf{U}}_2(\mathbf{r}, \omega)$ is analytic and the amplitude function $A(\alpha, \beta)$ is continuous with continuous partial derivatives with respect to α and β up to order N over the entire integration domain D_2. However, the contour of integration C is now complex, as illustrated in Fig. G.2, and this results in the phase Φ being complex-valued when $\alpha = \alpha' + i\alpha''$ varies from $\pi/2$ to α_4 along C.

As α varies over the contour C with the angle β held fixed, the phase function Φ varies over a simple curve $C(\beta)$ of finite length in the complex α-plane. Since the partial derivative

$$\left(\frac{\partial \Phi}{\partial \alpha} \right)_\beta = \sin \vartheta \, \cos \alpha \, \cos (\beta - \varphi) - \cos \vartheta \, \sin \alpha$$

Fig. G.2 Contour of
integration C for the
α-integral in $\check{U}_2(\mathbf{r}, \omega)$

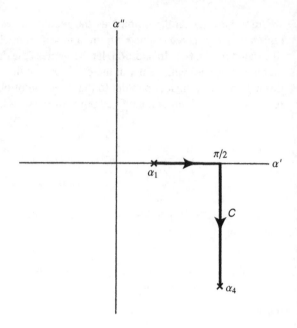

is an entire function of complex α for all β, then

$$\left(\frac{\partial \alpha}{\partial \Phi}\right)_\beta = \left[\left(\frac{\partial \Phi}{\partial \alpha}\right)_\beta\right]^{-1}$$

is an analytic function of complex α for all α, β provided that $\left(\frac{\partial \Phi}{\partial \alpha}\right)_\beta \neq 0$. Since $\cos \vartheta = \xi_3 > \delta$, then $\left(\frac{\partial \Phi}{\partial \alpha}\right)_\beta \neq 0$ when $\cos \alpha = 0$. It then follows that the zeros of $\left(\frac{\partial \Phi}{\partial \alpha}\right)_\beta$ occur for those values of α that satisfy the relation

$$\tan \alpha = \tan \vartheta \cos (\beta - \varphi).$$

On the portion of the contour C over which α is real, $\alpha_1 \leq \alpha \leq \pi/2$, and since $\cos \vartheta = \xi_3 \leq C_1 - \varepsilon$ (see Fig. G.1), then $\vartheta \leq \alpha \leq \pi/2$ and one obtains the inequality $\tan \alpha > \tan \vartheta \cos (\beta - \varphi)$. On the portion of the contour C over which α is complex, $\Re\{\alpha\} = \pi/2$ and $\tan \alpha = -i \tanh (\Im\{\alpha\})$ is pure imaginary whereas $\tan \vartheta \cos (\beta - \varphi)$ is real. As a consequence, the quantity $\left(\frac{\partial \Phi}{\partial \alpha}\right)_\beta$ has no zeros for all α on the contour C with $0 \leq \beta < 2\pi$, and this in turn implies that $\left(\frac{\partial \alpha}{\partial \Phi}\right)_\beta = \left[\left(\frac{\partial \Phi}{\partial \alpha}\right)_\beta\right]^{-1}$ is an analytic function of complex $\alpha \in C$ when $0 \leq \beta < 2\pi$. The change of variable of integration in the α-integral of $\tilde{U}_2(\mathbf{r}, \omega)$ to the variable Φ can then be made with the result

$$\tilde{U}_2(\mathbf{r}, \omega) = \int_0^{2\pi} \int_{C(\beta)} A\big(\alpha(\Phi), \beta\big) \left(\frac{\partial \alpha}{\partial \Phi}\right)_\beta e^{ikR\Phi} d\Phi d\beta.$$

As a consequence of the prescribed properties for both the function $\mathbf{V}(p, q, m) = m\tilde{\mathbf{U}}(p, q, \omega)$ [cf. Eq. (1)] and the neutralizer function $v_2(p, q)$, the function $A(\alpha, \beta)$ possesses N continuous partial derivatives with respect to the variable $\alpha \in C$ for all fixed $\beta \in [0, 2\pi)$.[3] Because the quantity $\left(\frac{\partial \alpha}{\partial \Phi}\right)_\beta$ is an analytic function of $\alpha \in C$, the product $A(\alpha, \beta) \left(\frac{\partial \alpha}{\partial \Phi}\right)_\beta$ also possesses N continuous partial derivatives taken with respect to α along the contour C with fixed $\beta \in [0, 2\pi)$. Notice that differentiation with respect to Φ along the contour $C(\beta)$ with fixed β is equivalent to differentiation with respect to α along C with fixed β followed by multiplication by the quantity $\left(\frac{\partial \alpha}{\partial \Phi}\right)_\beta$. Because $\left(\frac{\partial \alpha}{\partial \Phi}\right)_\beta$ is analytic with respect to α, the quantity $A(\alpha, \beta) \left(\frac{\partial \alpha}{\partial \Phi}\right)_\beta$ has N continuous partial derivatives with respect to $\Phi \in C(\beta)$ for all fixed $\beta \in [0, 2\pi)$. The Φ-integral

$$I(kR, \beta) = \int_{C(\beta)} A\big(\alpha(\Phi), \beta\big) \left(\frac{\partial \alpha}{\partial \Phi}\right)_\beta e^{ikR\Phi} d\Phi$$

appearing in the integral expression for $\tilde{U}_2(\mathbf{r}, \omega)$ is now integrated by parts N times by integrating the exponential factor $e^{ikR\Phi}$ each time and differentiating the remaining factor, with the result [15]

$$I(kR, \beta) = L_N(\Phi_4) - L_N(\Phi_1) + \mathcal{R}_N(kR, \beta),$$

where

$$L_N(\Phi_j) = \sum_{n=0}^{N-1} i^{n-1} \left\{ \frac{\partial^n}{\partial \Phi^n} \left[A\big(\alpha(\Phi), \beta\big) \left(\frac{\partial \alpha}{\partial \Phi}\right)_\beta \right] \right\}_\beta \Bigg|_{\Phi=\Phi_j} \frac{e^{ikR\Phi_j}}{(kR)^{n+1}},$$

with $\Phi_j = \sin\vartheta \sin\alpha_j \cos(\beta - \varphi) + \cos\vartheta \cos\alpha_j$ for $j = 1, 4$, where α_1 and α_4 denote the endpoints of the contour C (see Fig. G.2), and where

$$\mathcal{R}_N(kR, \beta) = (-ikR)^{-N} \int_{C(\beta)} \left\{ \frac{\partial^N}{\partial \Phi^N} \left[A\big(\alpha(\Phi), \beta\big) \left(\frac{\partial \alpha}{\partial \Phi}\right)_\beta \right] \right\}_\beta e^{ikR\Phi} d\Phi.$$

---

[3]Notice that the derivatives with respect to α must be taken along the contour C since both $V(p, q, m)$ and $v_2(p, q)$ are defined only for real-valued p and q. That is, variation of α along the contour C corresponds to p, q varying along the real axis.

As a consequence of the prescribed properties for the neutralizer function $v_2(p, q)$, the function $A(\alpha, \beta)$ and all N of its partial derivatives $\frac{\partial^n A(\alpha, \beta)}{\partial \alpha^n}$, $n = 0, 1, \ldots, N - 1$, taken along the contour C, vanish at the endpoints Φ_1 and Φ_4 of the contour C. Each of the N partial derivatives $\frac{\partial^n A(\alpha, \beta)}{\partial \Phi^n}$ taken along the contour $C(\beta)$ then also vanish at the endpoints Φ_1 and Φ_4, so that

$$L_N(\Phi_1) = L_N(\Phi_4) = 0.$$

Furthermore, because the integrand in the above expression for the remainder term $\mathcal{R}_N(kR, \beta)$ is continuous along the contour $C(\beta)$ and as this contour is of finite length for all ϑ, φ, and β, then that integral is bounded by some positive constant M_2 independent of ϑ, φ, and β, so that

$$|\mathcal{R}_N(kR, \beta)| \leq M_2(kR)^{-N}.$$

Consequently,

$$\left| \tilde{U}_2(\mathbf{r}, \omega) \right| \leq 2\pi M_2(kR)^{-N},$$

so that $\tilde{U}_2(\mathbf{r}, \omega) = \mathcal{O}\left\{ (kR)^{-N} \right\}$ uniformly with respect to ξ_1, ξ_2 as $kR \to \infty$ with fixed $k > 0$.

   In summary, following the method of proof given by Sherman, Stamnes, and Lalor [4], it has been established that the three terms $\tilde{U}_j(\mathbf{r}, \omega)$, $j = 1, 2, 3$, whose sum gives $\tilde{U}(\mathbf{r}, \omega)$, satisfy the order relations

$$\tilde{U}_1(\mathbf{r}, \omega) = \tilde{U}_0(x, y, z) + \mathcal{O}\left\{ (kR)^{-N} \right\},$$

$$\tilde{U}_2(\mathbf{r}, \omega) = \mathcal{O}\left\{ (kR)^{-N} \right\},$$

$$\tilde{U}_3(\mathbf{r}, \omega) = \mathcal{O}\left\{ (kR)^{-N} \right\},$$

uniformly with respect to ξ_1, ξ_2 as $kR \to \infty$ with fixed $k > 0$. This then completes the proof of the theorem.

G.3.2   Asymptotic Approximation of $\tilde{U}(\mathbf{r}, \omega)$

Theorem G.1 establishes that any terms of order lower than $(kR)^{-N}$ in the asymptotic behavior of $\tilde{U}(\mathbf{r}, \omega)$ as $kR \to \infty$ with fixed $k > 0$ must be contributed by the term [see Eq. (G.24)]

$$\tilde{\mathbf{U}}_0(x, y, z) = \int_{\mathcal{D}_H} v(p, q)\tilde{\mathbf{U}}(p, q, \omega)e^{ik(px+qy+mz)}dpdq. \tag{G.25}$$

This integral representation can be expressed in the form of the integral $I(kR)$ in Eq. (G.18) with phase function

$$f(p, q) = \xi_1 p + \xi_2 q + \xi_3 m \tag{G.26}$$

and amplitude function (now a vector)

$$\mathbf{g}(p, q) = v(p, q)\tilde{\mathbf{U}}(p, q, \omega)e^{ik(px_0+qy_0+mz_0)}. \tag{G.27}$$

Because $v(p, q) = 0$ when $(p, q) \notin \Omega_2$, the domain of integration $\mathcal{D}_H$ appearing in Eq. (G.25) can be replaced by a domain $\mathcal{D}$ that is taken as any region within the interior of the region $\mathcal{D}_H$ that contains the subregion Ω_2 within its interior (see Fig. G.3). With that choice, the phase function $f(p, q)$ defined in Eq. (G.26) is real-valued and infinitely differentiable in $\mathcal{D}$, and the amplitude function $g(p, q)$ has continuous, bounded partial derivatives up to order N in $\mathcal{D}$. The integral in Eq. (G.25) then satisfies all of the conditions required for the direct application of the stationary phase method due to Braun [10]. Because there is only the single critical point [the interior stationary phase point $(p, q) = (\xi_1, \xi_2)$] in $\mathcal{D}$, and because the integrand and its derivatives all vanish on the boundary of the domain $\mathcal{D}$, application of the method of stationary phase due to Braun [10] then gives[4]

$$\tilde{\mathbf{U}}(x, y, z) = \frac{e^{ikR}}{kR} \sum_{n=0}^{\frac{N}{2}-1} \frac{\mathbf{B}_n(\vartheta, \varphi)}{(kR)^n} + \mathcal{O}\left\{(kR)^{-\frac{N}{2}}\right\}, \tag{G.28}$$

as $kR \to \infty$ with fixed $k > 0$, positive z, and fixed direction cosines ξ_1, ξ_2, and with $\tilde{\mathbf{U}}(p, q, \omega) \in \mathcal{T}_N$, where the coefficients $\mathbf{B}_n(\vartheta, \varphi)$ are independent of R. Notice that, as required by the definition of the set $\mathcal{T}_N$ of spectral functions introduced just prior to Eq. (1), N is a positive, even integer. Finally, notice that the order of the remainder term in Eq. (G.28) is a refinement of Braun's estimate as reported by Sherman, Stamnes, and Lalor [4].

Application of the result given in Sect. 5.1 of the paper by Jones and Kline [11] yields the zeroth-order term

$$\mathbf{B}_0(\vartheta, \varphi) = -2\pi i \mathbf{V}(\xi_1, \xi_2, \xi_3)e^{ik(\xi_1 x_0+\xi_2 y_0+\xi_3 z_0)}, \tag{G.29}$$

---

[4]Although the stationary phase result given by Braun [10] is for a scalar field, the result may be directly generalized to a vector field by separately applying it to each of the scalar components of that vector field.

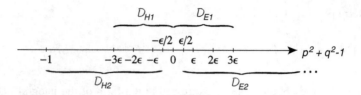

Fig. G.3 Illustration of the integration regions $\mathcal{D}_{J1}$ and $\mathcal{D}_{J2}$, $J = H, E$, appearing in Eqs. (G.2)–(G.4)

where [cf. Eq. (1)] $\mathbf{V}(p, q, m) \equiv m\tilde{\mathbf{U}}(p, q, \omega)$ with $m = +\sqrt{1 - p^2 - q^2}$, and where [see Eqs. (G.7) and (G.8)] $\xi_1 = (x - x_0)/R$, $\xi_2 = (y - y_0)/R$, $\xi_3 = (z - z_0)/R$ with $R = \sqrt{(x - x_0)^2 + (y - y_0)^2 + (z - z_0)^2}$ are related to the angles ϑ and φ through the relations

$$\xi_1 = \sin \vartheta \cos \varphi,$$
$$\xi_2 = \sin \vartheta \sin \varphi, \tag{G.30}$$
$$\xi_3 = \cos \vartheta,$$

with $0 \le \vartheta < \pi/2$ and $0 \le \varphi < 2\pi$.

By making use of the fact that the spectral wave field $\tilde{\mathbf{U}}(\mathbf{r}, \omega)$ satisfies the Helmholtz equation (G.5) for $z > 0$, Sherman [16] has shown that if a solution $\tilde{\mathbf{U}}(\mathbf{r}, \omega)$ of the Helmholtz equation has an asymptotic expansion of the form given in Eq. (G.28) for arbitrarily large (even) values of N, and if the asymptotic expansion of each of the partial derivatives of $\tilde{\mathbf{U}}(\mathbf{r}, \omega)$ with respect to the spherical coordinate variables R, ϑ, φ up to order 2 can be obtained by term by term differentiation of the asymptotic expansion given in Eq. (G.28), then the coefficients $\mathbf{B}_n(\vartheta, \varphi)$ satisfy the recursion formula (see Problem 8.9)

$$(n + 1)\mathbf{B}_{n+1}(\vartheta, \varphi) = \frac{i}{2}\left[\mathcal{L}^2 - n(n + 1)\right]\mathbf{B}_n(\vartheta, \varphi), \tag{G.31}$$

where $\mathcal{L}^2$ is the differential operator defined by

$$\mathcal{L}^2 \equiv -\frac{1}{\sin \vartheta}\frac{\partial}{\partial \vartheta}\left(\sin \vartheta \frac{\partial}{\partial \vartheta}\right) - \frac{1}{\sin^2 \vartheta}\frac{\partial^2}{\partial \varphi^2}. \tag{G.32}$$

Since the partial derivatives of the spectral wave field $\tilde{\mathbf{U}}(\mathbf{r}, \omega)$ given in Eqs. (8.195) and (8.196) satisfy these requirements, then the coefficients $\mathbf{B}_{n+1}(\vartheta, \varphi)$ appearing in the asymptotic expansion (G.28) of this wave field satisfy the recursion formula given in Eq. (G.31) provided that $\tilde{\mathbf{U}}(p, q, \omega) \in \mathcal{T}_\infty$. Since the functional dependence of the coefficients $\mathbf{B}_{n+1}(\vartheta, \varphi)$ on the spatiotemporal spectrum $\tilde{\mathbf{U}}(p, q, \omega)$ of the wave field is independent of N, then Eq. (G.31) also holds for $\tilde{\tilde{U}}(p, q, \omega) \in \mathcal{T}_N$ with N finite.

Sherman's recursion formula (G.31) then provides a straightforward method for obtaining higher-order terms in the asymptotic approximation of the spectral wave field $\tilde{U}(\mathbf{r}, \omega)$ from the first term in the expansion (G.28). For example, $\mathbf{B}_1(\vartheta, \varphi) = (i/2)\mathcal{L}^2 \mathbf{B}_0(\vartheta, \varphi)$, so that if $\tilde{\tilde{U}}(p, q, \omega) \in \mathcal{T}_N$ with $N \geq 6$,

$$\tilde{U}(\mathbf{r}, \omega) = -2\pi i \frac{e^{ikR}}{kR} \left(1 + \frac{i}{2kR}\mathcal{L}^2\right) \mathbf{V}(\xi_1, \xi_2, \xi_3)e^{ik(\xi_1 x_0 + \xi_2 y_0 + \xi_3 z_0)}$$

$$+ \mathcal{O}\left\{(kR)^{-3}\right\} \tag{G.33}$$

as $kR \to \infty$ with fixed $k > 0$ and fixed ξ_1, ξ_2 with $z > 0$.

G.4 Approximations Valid on the Plane $z = z_0$

The asymptotic approximation of $\tilde{U}(\mathbf{r}, \omega) = \tilde{U}(x, y, z, \omega)$ that is valid as the point of observation $\mathbf{r} = (x, y, z)$ recedes to infinity in a fixed direction perpendicular to the z-axis with $z > 0$ through the point $\mathbf{r}_0 = (x_0, y_0, z_0)$ is now considered. This case is excluded in Sect. G.3 because the method of analysis used there cannot be extended to include the special case when $\xi_3 = 0$. As in Theorem G.1, it is again assumed that $\tilde{\tilde{U}}(p, q, \omega) \in \mathcal{T}_N$ for some positive even integer N and that $z = z_0 > 0$.

Upon setting $z = z_0$ in Eq. (G.2) an integral of the form given in Eq. (G.18) is obtained, viz.

$$I(kR) = \int_{\mathcal{D}} g(p, q)e^{ikRf(p,q)}dpdq, \tag{G.34}$$

with

$$f(p, q) = \xi_1 p + \xi_2 q, \tag{G.35}$$

$$g(p, q) = \tilde{\tilde{U}}(p, q, \omega)e^{ik(px_0 + qy_0 + mz_0)}, \tag{G.36}$$

and $\mathcal{D} = \mathcal{D}_J$, $J = H, E$. The phase function $f(p, q)$ is then real and analytic in both regions $\mathcal{D}_H$ and $\mathcal{D}_E$. However, the amplitude function $g(p, q)$ need not have any of its partial derivatives with respect to p and q exist on the unit circle $p^2 + q^2 = 1$. In order to circumvent this problem, Sherman, Stamnes, and Lalor [4] isolate that unit circle using a neutralizer function as follows. Let $v_0(p, q)$ be a real, continuous function with continuous partial derivatives of all orders for all p, q that satisfies the inequality

$$0 \le v_0(p, q) \le 1, \tag{G.37}$$

with

$$v_0(p, q) = \begin{cases} 1 & \text{for} & 1 - \varepsilon \le p^2 + q^2 \le 1 + \varepsilon, \\ 0 & \text{for either } p^2 + q^2 \le 1 - 2\varepsilon \text{ or } p^2 + q^2 \ge 1 + 2\varepsilon, \end{cases} \tag{G.38}$$

where $\varepsilon > 0$ is a positive constant with $\varepsilon < 1/3$. The integral in Eq. (G.2) can then be written in the form

$$\tilde{\mathbf{U}}_J(x, y, z_0, \omega) = \tilde{\mathbf{U}}_{J1}(x, y, z_0, \omega) + \tilde{\mathbf{U}}_{J2}(x, y, z_0, \omega) \tag{G.39}$$

with $J = H, E$, where

$$\tilde{\mathbf{U}}_{J1}(x, y, z_0, \omega) = \int_{\mathcal{D}_{J1}} v_0(p, q)\tilde{\mathbf{U}}(p, q, \omega)e^{ik(px+qy+mz_0)}\,dpdq, \tag{G.40}$$

$$\tilde{\mathbf{U}}_{J2}(x, y, z_0, \omega) = \int_{\mathcal{D}_{J2}} [1 - v_0(p, q)]\tilde{\mathbf{U}}(p, q, \omega)e^{ik(px+qy+mz_0)}\,dpdq. \tag{G.41}$$

The various regions of integration in the p, q-plane appearing in these integral expressions are defined as

$$\mathcal{D}_{H1} \equiv \left\{ (p, q) \,|\, 1 - 3\varepsilon \le p^2 + q^2 \le 1 \right\},$$

$$\mathcal{D}_{H2} \equiv \left\{ (p, q) \,|\, 0 \le p^2 + q^2 \le 1 - \varepsilon/2 \right\},$$

$$\mathcal{D}_{E1} \equiv \left\{ (p, q) \,|\, 1 \le p^2 + q^2 \le 1 + 3\varepsilon \right\},$$

$$\mathcal{D}_{E2} \equiv \left\{ (p, q) \,|\, p^2 + q^2 \le 1 + \varepsilon/2 \right\},$$

and are depicted in Fig. G.3. The asymptotic approximation of each of the integrals appearing in Eqs. (G.40) and (G.41) is now separately considered.

G.4.1   The Region $\mathcal{D}_{H2}$

The integrand appearing in the expression (9.225) for $\tilde{\mathbf{U}}_{H2}(x, y, z_0, \omega)$ is found [4] to satisfy all of the conditions required in Theorem 1 of Chako [12]. Because the region $\mathcal{D}_{H2}$ does not contain any critical points of that integrand, and because the amplitude function $[1 - v_0(p, q)]\tilde{\mathbf{U}}(p, q, \omega)$ and all N of its partial derivatives with respect to p and q vanish on the boundary of $\mathcal{D}_{H2}$, it then follows that

$$\tilde{U}_{H2}(x, y, z_0, \omega) = \mathcal{O}\left\{(kR)^{-N}\right\} \tag{G.42}$$

as $kR \to \infty$ with fixed $k > 0$.

G.4.2   The Region $\mathcal{D}_{E2}$

The integrand appearing in the expression (G.41) for $\tilde{U}_{E2}(x, y, z_0, \omega)$ is also found
[4] to satisfy the conditions in Theorem 1 of Chako [12], but the integration domain
$\mathcal{D}_{E2}$ appearing in Eq. (G.41) extends to infinity whereas Chako's proof is given only
for a finite domain of integration. Its extension to the case of an infinite domain has
been given by Sherman, Stamnes, and Lalor [4] who consider the same integral
but with the region of integration now given by $1 + \varepsilon/2 \le p^2 + q^2 \le K$, where
$K > 1 + \varepsilon/2$ is an arbitrary constant that is allowed to go to infinity. This change
then introduces two complications into Chako's proof. For the first, the amplitude
function $\tilde{U}(p, q, \omega)$ and its partial derivatives with respect to p and q do not, in
general, vanish on the boundary $p^2 + q^2 = K$ for finite K, but they do vanish in
the limit as $K \to \infty$ because each of these boundary terms contains the exponential
factor $e^{-z_0\sqrt{K-1}}$ with $z_0 > 0.[5]$ The second complication occurs in the remainder
integral after integration by parts N times, which is now taken over an infinite
domain. However, all that is required in the proof is that this integral is convergent
and this is guaranteed by the presence of the exponential factor $e^{-z_0\sqrt{p^2+q^2-1}}$ in
the integrand. Hence, the result of the theorem due to Chako applies in this case, so
that

$$\tilde{U}_{E2}(x, y, z_0, \omega) = \mathcal{O}\left\{(kR)^{-N}\right\} \tag{G.43}$$

as $kR \to \infty$ with fixed $k > 0$.

G.4.3   The Region $\mathcal{D}_{H1}$

Attention is now turned to obtaining the asymptotic approximation of $\tilde{U}_{H1}(x, y, z_0, \omega)$ as $kR \to \infty$ with fixed $k > 0$. Under the change of variable $p = \sin\alpha\cos\beta$,

---

[5]Notice that the special case when $z_0 = 0$ can be treated only if additional restrictions are placed on
the behavior of the spectral amplitude function $\tilde{U}(p, q, \omega)$ and its partial derivatives with respect
to p and q in the limit as $p^2 + q^2 \to \infty$.

$q = \sin \alpha \sin \beta$ the integral representation appearing in Eq. (G.40) becomes

$$\tilde{U}_{H1}(x, y, z_0, \omega) = \int_{\beta_0}^{\beta_0 + 2\pi} \int_{\alpha'_1}^{\pi/2} A'(\alpha, \beta) e^{ikR \sin \alpha \cos (\beta - \varphi)} d\alpha d\beta, \qquad (G.44)$$

with fixed direction cosines $\xi_1 = \sin \vartheta \cos \varphi$, $\xi_2 = \sin \vartheta \sin \varphi$ [see Eq. (G.8)] with $0 \leq \vartheta < \pi/2$ and $0 \leq \varphi < 2\pi$. Here β_0 is an arbitrary real constant, $\alpha'_1 \equiv \arcsin \sqrt{1 - 3\varepsilon} = \arccos \sqrt{3\varepsilon}$, and

$$A'(\alpha, \beta) \equiv v_0(p, q) V(p, q, m) e^{ik(px_0 + qy_0 + mz_0)} \sin \alpha, \qquad (G.45)$$

with $m = (1 - p^2 - q^2)^{1/2} = \cos \alpha$ and $V(p, q, m)$ is as defined in Eq. (1). Since β_0 is an arbitrary constant, it is chosen for later convenience to be given by $\beta_0 = \varphi - \pi/4$. The integral appearing in Eq. (G.44) is now in the form of the integral given in Eq. (G.34) with phase and amplitude functions

$$f(\alpha, \beta) = \sin \alpha \cos (\beta - \varphi), \qquad (G.46)$$

$$g(\alpha, \beta) = A'(\alpha, \beta), \qquad (G.47)$$

respectively. Because both of these functions satisfy all of the requirements necessary for application of the method of stationary phase, this method may now be directly applied.

The critical points of the integral in Eq. (G.44) occur at the stationary phase points that are defined by the condition [cf. Eq. (G.19)]

$$\left. \frac{\partial f}{\partial \alpha} \right|_{(\alpha_s, \beta_s)} = \left. \frac{\partial f}{\partial \beta} \right|_{(\alpha_s, \beta_s)} = 0, \qquad (G.48)$$

so that both $\cos \alpha \cos (\beta - \varphi) = 0 \Rightarrow \alpha = \pi/2 \vee \beta = \varphi + \pi/2, \varphi + 3\pi/2$ and $\sin \alpha \sin (\beta - \varphi) = 0 \Rightarrow \beta = \varphi, \varphi + \pi$. The stationary phase points are then

$$\text{Point a:} \quad (\alpha_s, \beta_s) = (\pi/2, \varphi),$$

$$\text{Point b:} \quad (\alpha_s, \beta_s) = (\pi/2, \varphi + \pi),$$

both of which occur on the boundary of the integration region $\mathcal{D}_{H1}$, as illustrated in Fig. G.4. Additional critical points occur at the two corners $(\pi/2, \beta_0)$ and $(\pi/2, \beta_0 + 2\pi)$. Because of the above choice that $\beta_0 = \varphi - \pi/4$, neither of the stationary phase points coincide with a boundary corner of the integration domain, as depicted in Fig. G.4.

Consider first the contribution from the two corner critical points which are just an artifact of the change of variables of integration. Because their location along the line $\alpha = \pi/2$ is completely determined by the choice of the constant β_0, it is expected that, taken together, they do not contribute to the asymptotic behavior of

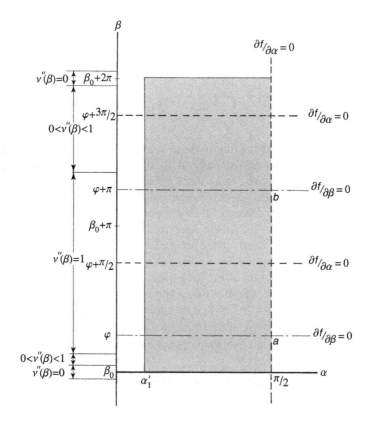

Fig. G.4 Illustration of the integration region $\mathcal{D}_{H1}$ appearing in Eq. (G.44) and the location of the critical points appearing in that integral

$\tilde{\mathbf{U}}_{H1}(x, y, z_0, \omega)$ as $kR \to \infty$. The verification of this expected result is given by the following argument [4]. Construct a periodic neutralizer function $v''(\beta)$ with period 2π that is a real, continuous function of β with continuous derivatives of all orders, and is such that

$$v''(\beta) = 1, \quad \text{when} \quad \varphi - \pi/8 \le \beta \le \varphi + 9\pi/8,$$

and $v''(\beta) = 0$ in some neighborhood of β_0, as depicted on the left side of Fig. G.4. The integral representation (G.44) of the field $\tilde{\mathbf{U}}_{H1}(x, y, z_0, \omega)$ may then be expressed as

$$\tilde{U}_{H1}(x, y, z_0, \omega) = \int_{\beta_0}^{\beta_0+2\pi} \int_{\alpha_1'}^{\pi/2} v''(\beta) A'(\alpha, \beta) e^{ikR \sin\alpha \cos(\beta-\varphi)} d\alpha d\beta$$

$$+ \int_{\beta_0}^{\beta_0+2\pi} \int_{\alpha_1'}^{\pi/2} \left(1 - v''(\beta)\right) A'(\alpha, \beta) e^{ikR \sin\alpha \cos(\beta-\varphi)} d\alpha d\beta.$$

Because the integrand vanishes (due to the neutralizer function) at the corner points, the only critical points in the first integral are the stationary phase points labeled a and b in Fig. G.4. Moreover, because $v''(\beta) = 1$ in a separate neighborhood of each of the stationary phase points, the asymptotic behavior of this first integral is identical to the stationary phase point contributions to the asymptotic behavior of $\tilde{U}_{H1}(x, y, z_0, \omega)$ as $kR \to \infty$. On the other hand, because the quantity $(1 - v''(\beta))$ vanishes at both of the stationary phase points, the only critical points of importance in the second integral are the corner points. Because the integrand in this second integral is a periodic function of β with period 2π, the integrand can be made to vanish at the corner points simply by changing the region of integration so that it extends from $\varphi - 9\pi/8$ to $\varphi + 7\pi/8$. It then follows from Theorem 1 of the paper [12] by Chako that the second integral is $\mathcal{O}\left\{(kR)^{-N}\right\}$ as $kR \to \infty$.

The field $\tilde{U}_{H1}(x, y, z_0, \omega)$ may then be expressed as

$$\tilde{U}_{H1}(x, y, z_0, \omega) = \tilde{U}_{H1}^{(a)}(x, y, z_0, \omega) + \tilde{U}_{H1}^{(b)}(x, y, z_0, \omega) + \mathcal{O}\left\{(kR)^{-N}\right\},$$
$$(G.49)$$

where $\tilde{U}_{H1}^{(a)}(x, y, z_0, \omega)$ and $\tilde{U}_{H1}^{(b)}(x, y, z_0, \omega)$ denote the separate contributions to the asymptotic behavior of $\tilde{U}_{H1}(x, y, z_0, \omega)$ from the stationary phase points a and b, respectively. In order to obtain the separate asymptotic approximations of both $\tilde{U}_{H1}^{(a)}(x, y, z_0, \omega)$ and $\tilde{U}_{H1}^{(b)}(x, y, z_0, \omega)$, the asymptotic expansion due to a boundary stationary phase point of the hyperbolic type must be used [4]. This has been shown to be of the same form as that for a boundary stationary phase point of the elliptic type, which has been given in Sects. 3.3–3.4 of Bremmerman [14]. Thus

$$\tilde{U}_{H1}^{(j)}(x, y, z_0, \omega) = \tilde{U}_{H1}^{(j_1)}(x, y, z_0, \omega) + \tilde{U}_{H1}^{(j_2)}(x, y, z_0, \omega),$$
$$(G.50)$$

with

$$\tilde{U}_{H1}^{(j_1)}(x, y, z_0, \omega) = \frac{e^{\pm ikR}}{kR} \sum_{n=0}^{\frac{N}{2}} \frac{B_{Hn}^{(j_1)}(\varphi)}{(kR)^n} + o\left\{(kr)^{-N/4}\right\},$$
$$(G.51)$$

$$\tilde{U}_{H1}^{(j_2)}(x, y, z_0, \omega) = \frac{e^{\pm ikR}}{(kR)^{3/2}} \sum_{n=0}^{\frac{N}{2}-1} \frac{B_{Hn}^{(j_2)}(\varphi)}{(kR)^n} + o\left\{(kr)^{-N/4}\right\},$$
$$(G.52)$$

as $kR \rightarrow \infty$ with fixed $k > 0$, where $\mathbf{B}_{Hn}^{(j_1)}(\varphi)$ and $\mathbf{B}_{Hn}^{(j_2)}(\varphi)$ are both independent of R. The upper sign in the exponential appearing in Eqs. (G.51) and (G.52) is used when $j = a$ and the lower sign is used when $j = b$. Explicit expressions for the coefficients $\mathbf{B}_{Hn}^{(j_1)}(\varphi)$ and $\mathbf{B}_{Hn}^{(j_2)}(\varphi)$ are given in Appendix II of the paper [4] by Sherman, Stamnes, and Lalor.

G.4.4   The Region $\mathcal{D}_{E1}$

Consider finally obtaining the asymptotic approximation of $\tilde{\mathbf{U}}_{E1}(x, y, z_0, \omega)$ as $kR \rightarrow \infty$ with fixed $k > 0$. With the change of variable defined by

$$p = \cosh \mu \cos \lambda, \tag{G.53}$$

$$q = \cosh \mu \sin \lambda, \tag{G.54}$$

so that

$$m = +i \sinh \mu, \tag{G.55}$$

with Jacobian $J(p, q/\mu, \lambda) = \sinh \mu \cosh \mu$, the integral appearing in Eq. (G.40) becomes

$$\tilde{\mathbf{U}}_{E1}(x, y, z_0, \omega) = \int_{\beta_0}^{\beta_0 + 2\pi} \int_0^{\mu_1} A''(\mu, \lambda) e^{ikR \cosh(\mu) \cos(\lambda - \varphi)} d\mu d\lambda, \tag{G.56}$$

where $\mu_1 \equiv \sinh^{-1}\left(\sqrt{3\epsilon}\right) = \cosh^{-1}\left(\sqrt{1 + 3\epsilon}\right)$ and

$$A''(\mu, \lambda) \equiv -i\nu'(p, q) V(p, q, m) e^{ik(px_0 + qy_0 + mz_0)} \cosh \mu, \tag{G.57}$$

with p and q as given in Eqs. (G.53) and (G.54). The integral appearing in Eq. (G.56) is now in the form of the integral in Eq. (G.34) with amplitude and phase functions

$$g(\mu, \lambda) = A''(\mu, \lambda), \tag{G.58}$$

$$f(\mu, \lambda) = \cosh(\mu) \cos(\lambda - \varphi), \tag{G.59}$$

respectively.

As was found for the integral in Eq. (G.44) for the spectral field component $\tilde{\mathbf{U}}_{H1}(x, y, z_0, \omega)$, all of the critical points lie on the boundary $\partial \mathcal{D}_{E1}$ of the integration domain. Just as in that case, the critical points on the boundary $\mu = \mu_1$ and at the corners $(\mu, \lambda) = (0, 0)$ and $(\mu, \lambda) = (0, 2\pi)$ do not contribute to the asymptotic behavior of the integral in Eq. (G.56). The asymptotic behavior of the spectral field component $\tilde{\mathbf{U}}_{E1}(x, y, z_0, \omega)$ is then completely determined by the contributions from the stationary phase points (a) at $(\mu_s, \lambda_s) = (0, \varphi)$ and (b) at $(\mu_s, \lambda_s) = (0, \varphi + \pi)$. As in Eq. (G.49), the field $\tilde{\mathbf{U}}_{E1}(x, y, z_0, \omega)$ is expressed as

$$\tilde{\mathbf{U}}_{E1}(x, y, z_0, \omega) = \tilde{\mathbf{U}}_{E1}^{(a)}(x, y, z_0, \omega) + \tilde{\mathbf{U}}_{E1}^{(b)}(x, y, z_0, \omega) + \mathcal{O}\left\{(kR)^{-N}\right\},$$
(G.60)

where $\tilde{\mathbf{U}}_{E1}^{(a)}(x, y, z_0, \omega)$ and $\tilde{\mathbf{U}}_{E1}^{(b)}(x, y, z_0, \omega)$ denote the separate contributions to the asymptotic behavior of $\tilde{\mathbf{U}}_{H1}(x, y, z_0, \omega)$ from the stationary phase points (a) and (b), respectively. An analysis similar to that following Eq. (G.49) then yields [4]

$$\tilde{\mathbf{U}}_{E1}^{(j)}(x, y, z_0, \omega) = \tilde{\mathbf{U}}_{E1}^{(j_1)}(x, y, z_0, \omega) + \tilde{\mathbf{U}}_{E1}^{(j_2)}(x, y, z_0, \omega),$$
(G.61)

with

$$\tilde{\mathbf{U}}_{E1}^{(j_1)}(x, y, z_0, \omega) = \frac{e^{\pm ikR}}{kR} \sum_{n=0}^{\frac{N}{2}} \frac{\mathbf{B}_{En}^{(j_1)}(\varphi)}{(kR)^n} + o\left\{(kr)^{-N/4}\right\},$$
(G.62)

$$\tilde{\mathbf{U}}_{E1}^{(j_2)}(x, y, z_0, \omega) = \frac{e^{\pm ikR}}{(kR)^{3/2}} \sum_{n=0}^{\frac{N}{2}-1} \frac{\mathbf{B}_{En}^{(j_2)}(\varphi)}{(kR)^n} + o\left\{(kr)^{-N/4}\right\},$$
(G.63)

as $kR \to \infty$ with fixed $k > 0$, where $\mathbf{B}_{En}^{(j_1)}(\varphi)$ and $\mathbf{B}_{En}^{(j_2)}(\varphi)$ are both independent of R. The upper sign in the exponential appearing in Eqs. (G.62) and (G.63) is used when $j = a$ and the lower sign is used when $j = b$.

G.4.5   Relationship Between the Coefficients $\mathbf{B}_{Hn}^{(j_1)}(\varphi)$, $\mathbf{B}_{Hn}^{(j_2)}(\varphi)$ and $\mathbf{B}_{En}^{(j_1)}(\varphi)$, $\mathbf{B}_{En}^{(j_2)}(\varphi)$

The various relationships between the coefficients $\mathbf{B}_{Hn}^{(j_1)}(\varphi)$ and $\mathbf{B}_{Hn}^{(j_2)}(\varphi)$ in the asymptotic expansion given in Eqs. (G.51) and (G.52) for $\tilde{\mathbf{U}}_{H1}(x, y, z_0, \omega)$ and the coefficients $\mathbf{B}_{En}^{(j_1)}(\varphi)$ and $\mathbf{B}_{En}^{(j_2)}(\varphi)$ in the asymptotic expansion given in Eqs. (G.62) and (G.63) for $\tilde{\mathbf{U}}_{E1}(x, y, z_0, \omega)$ are given by Jones and Kline [11], as modified by Sherman, Stamnes, and Lalor [4], as[6]

$$\mathbf{B}_{En}^{(a_1)}(\varphi) = +\mathbf{B}_{Hn}^{(a_1)}(\varphi),$$
(G.64)

$$\mathbf{B}_{En}^{(a_2)}(\varphi) = -\mathbf{B}_{Hn}^{(a_2)}(\varphi),$$
(G.65)

---

[6]Explicit expressions for all of these coefficients are unnecessary for the final asymptotic expansion and so are not given here. The interested reader should consult Appendices I–III in the 1976 paper [4] by Sherman, Stamnes, and Lalor.

$$\mathbf{B}_{En}^{(b_1)}(\varphi) = -\mathbf{B}_{Hn}^{(b_1)}(\varphi), \tag{G.66}$$

$$\mathbf{B}_{En}^{(b_2)}(\varphi) = -\mathbf{B}_{Hn}^{(b_2)}(\varphi). \tag{G.67}$$

The pair of relations in Eqs. (G.66) and (G.67) then show that the contribution of the point b to $\tilde{\mathbf{U}}_{E1}(x, y, z_0, \omega)$ is equal in magnitude but opposite in sign with the contribution of the point b to $\tilde{\mathbf{U}}_{H1}(x, y, z_0, \omega)$. Consequently, the point b does not contribute to the asymptotic behavior of the field quantity $\tilde{\mathbf{U}}_{H1}(x, y, z_0, \omega) + \tilde{\mathbf{U}}_{E1}(x, y, z_0, \omega)$ with order lower than $\mathcal{O}\left\{(kR)^{-N/4}\right\}$ that arises from the sum of the second terms on the right hand side of Eqs. (G.49) and (G.60). In a similar manner, Eq. (G.65) implies that the contribution of the point a to the series involving inverse half powers of (kR) in the expansion of $\tilde{\mathbf{U}}_{E1}(x, y, z_0, \omega)$ is equal in magnitude but opposite in sign to the same contribution of the point a to $\tilde{\mathbf{U}}_{H1}(x, y, z_0, \omega)$. Taken together with the previous result, this implies that the asymptotic expansion of the field quantity $\tilde{\mathbf{U}}_{H1}(x, y, z_0, \omega) + \tilde{\mathbf{U}}_{E1}(x, y, z_0, \omega)$ does not include terms with half powers of (kR) of order lower than $\mathcal{O}\left\{(kR)^{-N/4}\right\}$. Finally, the relation given in Eq. (G.64) implies that the field components $\tilde{\mathbf{U}}_{H1}(x, y, z_0, \omega)$ and $\tilde{\mathbf{U}}_{E1}(x, y, z_0, \omega)$ contribute equally to the remaining terms in the asymptotic expansion of the total field $\tilde{\mathbf{U}}(x, y, z_0, \omega)$ involving only inverse integral powers of (kR). It then follows from this result together with Eqs. (G.1), (G.39), (G.43), and (G.51) that

$$\tilde{\mathbf{U}}(x, y, z_0, \omega) = 2 \frac{e^{ikR}}{kR} \sum_{n=0}^{\frac{N}{2}} \frac{\mathbf{B}_{Hn}^{(a_1)}(\varphi)}{(kR)^n} + o\left\{(kr)^{-N/4}\right\}, \tag{G.68}$$

as $kR \to \infty$ with fixed $k > 0$ and $z = z_0$.

The coefficients $\mathbf{B}_{Hn}^{(a_1)}(\varphi)$ appearing in the asymptotic expansion (G.68) are found [4] to be given by the limiting expression

$$\mathbf{B}_{Hn}^{(a_1)}(\varphi) = \frac{1}{2} \lim_{\theta \to \frac{\pi}{2}} \mathbf{B}_n(\vartheta, \varphi) \tag{G.69}$$

of the coefficients $\mathbf{B}_n(\vartheta, \varphi)$ appearing in the asymptotic expansion (G.28). It then follows that the results of this section can be combined with those of Sect. G.3.1 to yield the asymptotic expansion

$$\tilde{\mathbf{U}}(x, y, z_0, \omega) = \frac{e^{ikR}}{kR} \sum_{n=0}^{\frac{N}{2}} \frac{\mathbf{B}_n(\vartheta, \varphi)}{(kR)^n} + o\left\{(kr)^{-N/4}\right\}, \tag{G.70}$$

as $kR \to \infty$ with fixed $k > 0$ and $z_0 > 0$ that is uniformly valid with respect to the angles ϑ and φ in their respective domains $0 \le \vartheta \le \pi/2$ and $0 \le \varphi \le 2\pi$. The zeroth-order coefficient $\mathbf{B}_0(\vartheta, \varphi)$ is given in Eq. (G.29) and the first-order coefficient is given by $\mathbf{B}_1(\vartheta, \varphi) = (i/2)\mathcal{L}^2\mathbf{B}_0(\vartheta, \varphi)$, where the differential operator $\mathcal{L}^2$ is defined in Eq. (G.32).

G.5 Asymptotic Approximations of $\tilde{U}_H(\mathbf{r}, \omega)$ and $\tilde{U}_E(\mathbf{r}, \omega)$

The respective dominant terms in the asymptotic expansion of the homogeneous field component $\tilde{U}_H(\mathbf{r}, \omega)$ and in the asymptotic expansion of the evanescent field component $\tilde{U}_E(\mathbf{r}, \omega)$ are now derived for each of the three direction cosine regions $0 < \xi_3 < 1$, $\xi_3 = 0$, and $\xi_3 = 1$.

G.5.1 Approximations Valid Over the Hemisphere $0 < \xi_3 < 1$

In order to obtain the asymptotic approximation of the homogeneous wave contribution $\tilde{U}_H(\mathbf{r}, \omega)$ to the total spectral wave field, the notation and definitions used in Eqs. (G.37)–(G.41) are applied with z_0 replaced by z. The only critical point of the integral for $\tilde{U}_{H2}(\mathbf{r}, \omega)$, where the integrand is nonzero, is the interior stationary phase point $(p_s, q_s) = (\xi_1, \xi_2)$. The asymptotic expansion of $\tilde{U}_{H2}(\mathbf{r}, \omega)$ is then seen to be identical with the asymptotic expansion of $\tilde{U}(\mathbf{r}, \omega)$ given in Eqs. (G.28) and (G.33).

It then follows from the above result together with Eq. (G.1) that the asymptotic expansion of the evanescent wave contribution $\tilde{U}_E(\mathbf{r}, \omega)$ to the total wave field is identical to that of $-\tilde{U}_{H1}(\mathbf{r}, \omega)$. The asymptotic expansion of the integral expression for $\tilde{U}_{H1}(\mathbf{r}, \omega) = \tilde{U}_{H1}(x, y, z, \omega)$ is treated in the same manner as was employed for $\tilde{U}_{H1}(x, y, z_0, \omega)$ in Sect. G.3.2. The change of integration variables given by $p = \sin \alpha \cos \beta$, $q = \sin \alpha \sin \beta$ that was used in the proof of Theorem G.1 is first made, resulting in the integral [cf. Eq. (G.44)]

$$\tilde{U}_{H1}(x, y, z, \omega) = \int_{\beta_0}^{\beta_0+2\pi} \int_{\alpha_1'}^{\pi/2} A'(\alpha, \beta) e^{ikRf(\alpha,\beta)} d\alpha d\beta, \qquad (G.71)$$

with phase function [cf. Eq. (G.46)]

$$f(\alpha, \beta) = \sin \vartheta \sin \alpha \cos (\beta - \varphi) + \cos \vartheta \cos \alpha. \qquad (G.72)$$

The location of the critical points for this phase function are the same as those depicted in Fig. G.4. As in that case, the points a and b are the only critical points that contribute to the asymptotic behavior of the integral under consideration. However, unlike that case, the saddle points here are not ordinary stationary phase points where both $\partial f/\partial \alpha$ and $\partial f/\partial \beta$ vanish, but rather they are boundary stationary phase points where the isotimic contours $f(\alpha, \beta) = $ constant are tangent to the boundary of the integration domain; in the present case they are both points on the boundary line $\alpha = \pi/2$ where $\partial f/\partial \beta = 0$. The resultant asymptotic expansion is then found to be [4, 10, 11]

$$\tilde{\mathbf{U}}_{H1}(x, y, z, \omega) = \frac{\sqrt{2\pi/\sin\vartheta}\,e^{i\pi/4}}{(kR)^{3/2}\cos\vartheta}\left[\mathbf{V}(\xi_1', \xi_2', 0)e^{ik(x_0\xi_1' + y_0\xi_2')}e^{ikR\sin\vartheta}\right.$$

$$\left. + i\mathbf{V}(-\xi_1', -\xi_2', 0)e^{-ik(x_0\xi_1' + y_0\xi_2')}e^{-ikR\sin\vartheta}\right] + \mathcal{O}\left\{(kR)^{-2}\right\}$$

$$\text{(G.73)}$$

as $kR \to \infty$ with fixed $k > 0$ and $N \geq 8$, where

$$\xi_1' = \xi_1/\sin\vartheta = \cos\varphi, \tag{G.74}$$

$$\xi_2' = \xi_2/\sin\vartheta = \sin\varphi. \tag{G.75}$$

The resultant asymptotic expansions for $\tilde{\mathbf{U}}_H(\mathbf{r}, \omega)$ and $\tilde{\mathbf{U}}_E(\mathbf{r}, \omega)$ are then given by (for $N \geq 8$)

$$\tilde{\mathbf{U}}_H(x, y, z, \omega) = -2\pi i\frac{e^{ikR}}{kR}\mathbf{V}(\xi_1, \xi_2, \xi_3)e^{ik(\xi_1 x_0 + \xi_2 y_0 + \xi_3 z_0)}$$

$$+ \tilde{\mathbf{U}}_{H1}(x, y, z, \omega) + \mathcal{O}\left\{(kR)^{-2}\right\}, \tag{G.76}$$

$$\tilde{\mathbf{U}}_E(x, y, z, \omega) = -\tilde{\mathbf{U}}_{H1}(x, y, z, \omega) + \mathcal{O}\left\{(kR)^{-2}\right\}, \tag{G.77}$$

as $kR \to \infty$ with fixed ξ_1, ξ_2, and $k > 0$.

These results then show that the evanescent wave contribution $\tilde{\mathbf{U}}_E(\mathbf{r}, \omega)$ is of higher order in $(kR)^{-1}$ than the homogeneous wave contribution $\tilde{\mathbf{U}}_H(\mathbf{r}, \omega)$. This then provides a rigorous justification for neglecting $\tilde{\mathbf{U}}_E(\mathbf{r}, \omega)$ in comparison to $\tilde{\mathbf{U}}_H(\mathbf{r}, \omega)$ as $kR \to \infty$ with fixed $k > 0$ when $0 < \xi_3 < 1$. In addition, the evanescent wave contribution $\tilde{\mathbf{U}}_E(\mathbf{r}, \omega)$ is of even higher order in $(kR)^{-1}$ in comparison to the homogeneous wave contribution $\tilde{\mathbf{U}}_H(\mathbf{r}, \omega)$ as $kR \to \infty$ in the special case when $\mathbf{V}(p, q, 0) = \mathbf{0}$.

G.5.2 Approximations Valid on the Plane $z = z_0$

The asymptotic behavior of the spectral wave field on the plane $z = z_0$ (or, equivalently, when $\xi_3 = 0$) is directly obtained from the analysis presented in Sect. G.3.2, the present analysis focusing on obtaining explicit expressions for the dominant terms in the separate homogeneous and evanescent wave contributions. The asymptotic expansion of the homogeneous spectral wave contribution $\tilde{\mathbf{U}}_H(x, y, z_0, \omega)$ is obtained from Eqs. (G.39), (G.42), and (G.49)–(G.52), taken together with the series expressions [4] for the coefficients $\mathbf{B}_{Hn}^{j1}(\varphi)$ and $\mathbf{B}_{Hn}^{j2}(\varphi)$, with the result (for $N \geq 8$)

$$\tilde{\mathbf{U}}_H(x, y, z_0, \omega) = -\frac{i\pi}{kR}\Big[\mathbf{V}(\xi_1, \xi_2, 0)e^{ikR}e^{ik(x_0\xi_1+y_0\xi_2)}$$
$$-\mathbf{V}(-\xi_1, -\xi_2, 0)e^{-ikR}e^{-ik(x_0\xi_1+y_0\xi_2)}\Big] + \mathcal{O}\left\{(kR)^{-3/2}\right\}$$

(G.78)

as $kR \to \infty$ with fixed ξ_1, ξ_2, and $k > 0$. Similarly, the asymptotic expansion of the evanescent spectral wave contribution $\tilde{\mathbf{U}}_E(x, y, z_0, \omega)$ is obtained from Eqs. (G.39), (G.43), (G.60), (G.64), and (G.66) with the result (for $N \geq 8$)

$$\tilde{\mathbf{U}}_E(x, y, z_0, \omega) = -\frac{i\pi}{kR}\Big[\mathbf{V}(\xi_1, \xi_2, 0)e^{ikR}e^{ik(x_0\xi_1+y_0\xi_2)}$$
$$+\mathbf{V}(-\xi_1, -\xi_2, 0)e^{-ikR}e^{-ik(x_0\xi_1+y_0\xi_2)}\Big] + \mathcal{O}\left\{(kR)^{-3/2}\right\}$$

(G.79)

as $kR \to \infty$ with fixed ξ_1, ξ_2, and $k > 0$. These two expressions then show that the homogeneous and evanescent wave contributions are of the same order in $(kR)^{-1}$ as $kR \to \infty$. As a consequence, the evanescent wave contribution $\tilde{\mathbf{U}}_E(x, y, z_0, \omega)$ cannot, in general, be neglected in comparison to the homogeneous wave contribution $\tilde{\mathbf{U}}_H(x, y, z_0, \omega)$ on the plane $z = z_0$.

G.6   Approximations Valid on the Line $x = x_0$, $y = y_0$

Along the line $x = x_0$, $y = y_0$, one has that $\xi_3 = 0$ and the homogeneous spectral wave contribution can be written as

$$\tilde{\mathbf{U}}_H(x_0, y_0, z, \omega) = \int_0^{2\pi} d\beta \int_0^{\pi/2} d\alpha \, \mathbf{A}(\alpha, \beta)e^{ikz\cos\alpha}, \qquad (G.80)$$

where

$$\mathbf{A}(\alpha, \beta) = \mathbf{V}(p, q, m)e^{ik(px_0+qy_0+mz_0)} \sin\alpha, \qquad (G.81)$$

with $p = \sin\alpha\cos\beta$, $q = \sin\alpha\sin\beta$, and $m = \sqrt{1 - p^2 - q^2} = \cos\alpha$ [see the proof of Theorem G.1]. Because the phase function in Eq. (G.80) does not involve the integration variable β, the asymptotic behavior of the integral can be obtained from the single integral

$$\tilde{\mathbf{U}}_H(x_0, y_0, z, \omega) = \int_0^{\pi/2} \mathbf{A}(\alpha)e^{ikR\cos\alpha}d\alpha, \qquad (G.82)$$

where $R = z - z_0$, with

$$\mathbf{A}(\alpha) \equiv e^{ikz_0 \cos \alpha} \int_0^{2\pi} \mathbf{A}(\alpha, \beta) d\beta. \qquad (G.83)$$

Application of the method of stationary phase for single integrals in Sect. G.1 shows that the only contributions to the asymptotic behavior of the integral in Eq. (G.82) arise from the endpoints of the integral at $\alpha = 0$ and $\alpha = \pi/2$ with the result [see Eq. (G.17)]

$$\tilde{\mathbf{U}}_H(x_0, y_0, z, \omega) = -2\pi i \frac{e^{ikR}}{kR} \mathbf{V}(0, 0, 1) e^{ikz_0} + \frac{i}{kR} \mathbf{A}(\pi/2) + \mathcal{O}\left\{(kR)^{-3/2}\right\}$$

$$(G.84)$$

as $kR \to \infty$ with $k > 0$ and for $N \geq 4$.

Because the first term in Eq. (G.84) is the dominant term in the asymptotic approximation of $\tilde{\mathbf{U}}(x_0, y_0, z, \omega)$, Eq. (G.1) then shows that the asymptotic behavior of the evanescent wave contribution is given by

$$\tilde{\mathbf{U}}_E(x_0, y_0, z, \omega) = -\frac{i}{kR} \mathbf{A}(\pi/2) + \mathcal{O}\left\{(kR)^{-3/2}\right\} \qquad (G.85)$$

as $kR \to \infty$ with $k > 0$. The homogeneous and evanescent wave contributions are again seen to be of the same order in $(kR)^{-1}$ as $kR \to \infty$ so that, in general, $\tilde{\mathbf{U}}_E(x_0, y_0, z, \omega)$ cannot be neglected in comparison to $\tilde{\mathbf{U}}_H(x_0, y_0, z, \omega)$ for large $kR \to \infty$.

G.7 Summary

The stationary phase asymptotic expansions presented in this appendix are all valid as $kR \to \infty$ with fixed wavenumber $k > 0$ and fixed direction cosines $\xi_1 = (x - x_0)/R$, $\xi_2 = (y - y_0)/R$ with $N \geq 12$ in a nonabsorptive (and hence, strictly speaking, nondispersive) medium. Less restrictive conditions on N result in special cases.

The asymptotic behavior of the spectral wave field $\tilde{\mathbf{U}}(\mathbf{r}, \omega)$ is the same for all $\xi_3 = (z - z_0)/R$ such that $0 \leq \xi_3 \leq 1$ and is given by [cf. Eq. (G.70)]

$$\tilde{\mathbf{U}}(\mathbf{r}, \omega) = -2\pi i \frac{e^{ikR}}{kR} \left(1 + \frac{i}{2kR} \mathcal{L}^2\right) \mathbf{V}(\xi_1, \xi_2, \xi_3) e^{ik(\xi_1 x_0 + \xi_2 y_0 + \xi_3 z_0)}$$

$$+ \mathcal{O}\left\{(kR)^{-3}\right\} \qquad (G.86)$$

as $kR \to \infty$ with fixed $k > 0$, where $\mathcal{L}^2$ is the differential operator defined in Eq. (G.32).

The asymptotic behavior of the separate homogeneous and evanescent component wave fields $\tilde{U}_J(\mathbf{r}, \omega)$, $J = H, E$, however, depends on the value of ξ_3, separating into the three cases (a) $0 < \xi_3 < 1$, (b) $\xi_3 = 0$, and (c) $\xi_3 = 1$, as described in Sect. G.5. The results show that the evanescent wave contribution $\tilde{U}_E(\mathbf{r}, \omega)$ is negligible in comparison to the homogeneous wave contribution $\tilde{U}_H(\mathbf{r}, \omega)$ for large $kR \to \infty$ in case (a), but not necessarily in cases (b) and (c).

In some important applications of the angular spectrum representation in electromagnetic wave theory, integral representations of the form given in Eqs. (G.1) and (G.2) are obtained with $\tilde{U}(p, q, \omega) \notin \mathcal{T}_N$ because of the presence of isolated singularities in the integrand. This occurs, for example, in the analysis of the reflection and refraction of a nonplanar wave field (e.g., an electromagnetic beam field) at a planar interface separating two different media [13, 17]. A neutralizer function can then be used to isolate each singularity. Because $\tilde{U}(p, q, \omega) \in \mathcal{T}_N$, the asymptotic approximation of the resultant integral that does not contain any of the singularities can then be obtained using the two-dimensional stationary phase results of Sherman, Stamnes, and Lalor [4] presented here. Each of the remaining integrals contains one of the isolated singularities, $\tilde{U}(p, q, \omega) \notin \mathcal{T}_N$ and so its asymptotic approximation must be obtained using some other technique. In some case, as in the reflection and refraction problem [17], a change of integration variable can result in a transformed integral that is amenable to the stationary phase method presented here. If that is not the case, uniform asymptotic methods [18–20] may then need to be employed.

References

1. C. J. Bouwkamp, "Diffraction theory," *Rept. Prog. Phys.*, vol. 17, pp. 35–100, 1954.
2. E. Lalor, "Conditions for the validity of the angular spectrum of plane waves," *J. Opt. Soc. Am.*, vol. 58, pp. 1235–1237, 1968.
3. A. Sommerfeld, *Optics*, vol. IV of *Lectures in Theoretical Physics*. New York: Academic, 1964. paperback edition.
4. G. C. Sherman, J. J. Stamnes, and É. Lalor, "Asymptotic approximations to angular-spectrum representations," *J. Math. Phys.*, vol. 17, no. 5, pp. 760–776, 1976.
5. G. G. Stokes, "On the discontinuity of arbitrary constants which appear in divergent developments," *Trans. Camb. Phil. Soc.*, vol. X, pp. 106–128.
6. L. Kelvin, "On the waves produced by a single impulse in water of any depth, or in a dispersive medium," *Proc. Roy. Soc.*, vol. XLII, p. 80, 1887.
7. E. T. Copson, *Asymptotic Expansions*. London: Cambridge University Press, 1965.
8. E. T. Whittaker and G. N. Watson, *A Course of Modern Analysis*. New York: MacMillan, 1943. Sect. 9.41.
9. J. Focke, "Asymptotische Entwicklungen mittels der Methode der stationären phase," *Ber. Verh. Saechs. Akad. Wiss. Leipzig*, vol. 101, no. 3, pp. 1–48, 1954.
10. G. Braun, "Zur Methode der stationären Phase," *Acta Phys. Austriaca*, vol. 10, pp. 8–33, 1956.

11. D. S. Jones and M. Kline, "Asymptotic expansion of multiple integrals and the method of stationary phase," *J. Math. Phys.*, vol. 37, pp. 1–28, 1958.
12. N. Chako, "Asymptotic expansions of double and multiple integrals occurring in diffraction theory," *J. Inst. Math. Appl.*, vol. 1, no. 4, pp. 372–422, 1965.
13. A. Baños, *Dipole Radiation in the Presence of a Conducting Half-Space*. Oxford: Pergamon, 1966. Sect. 2.12.
14. H. Bremermann, *Distributions, Complex Variables, and Fourier Transforms*. Reading, MA: Addison-Wesley, 1965. Ch. 8.
15. A. Erdélyi, "Asymptotic representations of Fourier integrals and the method of stationary phase," *SIAM J. Appl. Math.*, vol. 3, pp. 17–27, 1955.
16. G. C. Sherman, "Recursion relations for coefficients in asymptotic expansions of wavefields," *Radio Science*, vol. 8, pp. 811–812, 1973.
17. J. Gasper, "Reflection and refraction of an arbitrary wave incident on a planar interface," M.S. dissertation, The Institute of Optics, University of Rochester, Rochester, NY, 1972.
18. N. Bleistein, "Uniform asymptotic expansions of integrals with stationary point near algebraic singularity," *Com. Pure and Appl. Math.*, vol. XIX, no. 4, pp. 353–370, 1966.
19. N. Bleistein, "Uniform asymptotic expansions of integrals with many nearby stationary points and algebraic singularities," *J. Math. Mech*, vol. 17, no. 6, pp. 533–559, 1967.
20. N. Bleistein and R. Handelsman, *Asymptotic Expansions of Integrals*. New York: Dover, 1975.

Appendix H
The Radon Transform

The Radon transform, named after the Czech mathematician Johann Radon (1887–1956) has its origin in his 1919 paper [1] (which then led to the Riemann–Lebesgueým theorem). For the purpose of this brief development, it is best introduced in connection with the projection-slice theorem in Fourier analysis in the following manner. Consider a two-dimensional scalar (object) function $O(\mathbf{r}) = O(x, y)$, which may be expressed as

$$O(\mathbf{R}) = \int_{\mathcal{D}} O(\mathbf{r}) \delta(\mathbf{R} - \mathbf{r}) d^2 r, \qquad (\text{H.1})$$

where $\mathcal{D}$ is any region containing the point $\mathbf{R}$, and where $\delta(\mathbf{r})$ is the two-dimensional Dirac delta function which has the Fourier integral representation

$$\delta(\mathbf{R} - \mathbf{r}) = \frac{1}{4\pi^2} \int_{-\infty}^{\infty} e^{i\mathbf{k}\cdot(\mathbf{R}-\mathbf{r})} d^2 k. \qquad (\text{H.2})$$

Let $\hat{\mathbf{n}}$ be the unit vector in the direction of the vector $\mathbf{k}$ so that $\mathbf{k} = k\hat{\mathbf{n}}$. With this substitution, Eq. (H.2) becomes

$$\delta(\mathbf{R} - \mathbf{r}) = \frac{1}{4\pi^2} \int_{-\infty}^{\infty} d^2 k \, e^{i\mathbf{k}\cdot\mathbf{R}} e^{-ik\mathbf{r}\cdot\hat{\mathbf{n}}},$$

and since

$$e^{-ik\mathbf{r}\cdot\hat{\mathbf{n}}} = \int_{-\infty}^{\infty} dx \, \delta(x - \mathbf{r}\cdot\hat{\mathbf{n}}) e^{-ikx},$$

© Springer Nature Switzerland AG 2019
K. E. Oughstun, *Electromagnetic and Optical Pulse Propagation*, Springer Series in Optical Sciences 224, https://doi.org/10.1007/978-3-030-20835-6

one then obtains the expression

$$\delta(\mathbf{R} - \mathbf{r}) = \frac{1}{4\pi^2} \int_{-\infty}^{\infty} d^2k\, e^{i\mathbf{k}\cdot\mathbf{R}} \int_{-\infty}^{\infty} dx\, \delta(x - \mathbf{r}\cdot\hat{\mathbf{n}})e^{-ikx}. \tag{H.3}$$

Substitution of this result into Eq. (H.1) then gives

$$O(\mathbf{R}) = \frac{1}{4\pi^2} \int_{-\infty}^{\infty} d^2k\, e^{i\mathbf{k}\cdot\mathbf{R}} \int_{-\infty}^{\infty} dx\, P(\hat{\mathbf{n}}; x)e^{-ikx}, \tag{H.4}$$

where

$$P(\hat{\mathbf{n}}; x) \equiv \int_{\mathcal{D}} O(\mathbf{r})\delta(x - \mathbf{r}\cdot\hat{\mathbf{n}})d^2r \tag{H.5}$$

denotes the projection of the object function $O(\mathbf{r})$ onto the direction defined by the unit vector $\hat{\mathbf{n}}$. The variable x appearing in this expression is now interpreted as the real variable defining the location in the $\hat{\mathbf{n}}$-direction at which the rectilinear line integral through the object function $O(\mathbf{r})$ is taken. As an illustration, the vertical line integral through the object function depicted in Fig. H.1 is given by

$$\int_{\ell} O(x, y')dy' = \int_{\mathcal{D}} O(x', y')\delta(x - x')dx'dy'$$
$$= \int_{\mathcal{D}} O(\mathbf{r})\delta(x - \mathbf{r}\cdot\hat{\mathbf{n}})d^2r,$$

which is precisely the expression defined in Eq. (H.5).

Fig. H.1 Geometry of the vertical line integral through the object function $O(\mathbf{r})$

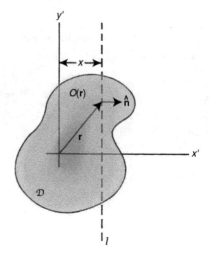

The two-dimensional *Radon transform* of a (sufficiently well-behaved) function $O(\mathbf{r}) = O(x, y)$ is then defined as [2]

$$\mathcal{R}[O](\hat{\mathbf{n}}; x) \equiv \int_{\mathcal{D}} O(\mathbf{r})\delta(x - \mathbf{r} \cdot \hat{\mathbf{n}})d^2r. \tag{H.6}$$

With use of the notation (see Appendix E)

$$\mathcal{F}_k \{f(x)\} = \int_{-\infty}^{\infty} f(x)e^{-ikx}dx, \tag{H.7}$$

$$\mathcal{F}_x^{-1} \left\{ \tilde{f}(k) \right\} = \frac{1}{2\pi} \int_{-\infty}^{\infty} \tilde{f}(k)e^{ikx}dk, \tag{H.8}$$

for the Fourier transform and its inverse, respectively, where x and k are referred to as Fourier conjugate variables, then as a special case of the Fourier integral theorem [3], $\mathcal{F}_x^{-1} \circ \mathcal{F}_k = 1$, that is, the subsequent application of the forward and inverse Fourier transforms (on a sufficiently well-behaved class of functions) results in the identity operator. Eq. (H.4) may then be expressed as

$$O(\mathbf{R}) = \mathcal{F}_{\mathbf{k}}^{-1} \circ \mathcal{F}_k \left\{ P(\hat{\mathbf{n}}; x) \right\}, \tag{H.9}$$

so that

$$\mathcal{F}_{\mathbf{k}} \{O(\mathbf{r})\} = \mathcal{F}_k \left\{ P(\hat{\mathbf{n}}; x) \right\}. \tag{H.10}$$

This important result is known as the *projection slice theorem* (or central slice theorem) [4]. In its N-dimensional generalization, this theorem relates the $(N - 1)$-dimensional Fourier transforms of the projections of an object function onto hyperplanes to the N-dimensional Fourier transform of that original function. Simply stated, this theorem states that the $(N - 1)$-dimensional Fourier transform of a given projection of the function is equal to a "slice" through the N-dimensional Fourier transform of that function.

The Fourier integral representation of the delta function given in Eq. (H.2) involves a vector $\mathbf{k}$ which may be expressed in polar coordinate form as being the vector with magnitude $k \in [0, \infty)$ and with direction specified by a unit vector $\hat{\mathbf{n}}$ which is at an angle $\theta \in [0, 2\pi)$ with respect to some fixed reference direction; one then has that

$$\int_{-\infty}^{\infty} d^2k = \int_0^{2\pi} d\theta \int_0^{\infty} kdk.$$

However, it is equally valid to regard the integration variable in Eq. (H.2) as being a vector $\boldsymbol{\kappa}$ with "magnitude" $\kappa \in (-\infty, \infty)$ and direction specified by the unit vector $\hat{\mathbf{N}}$ which is at an angle $\phi \in [0, \pi)$ with respect to the reference direction; in this case one has that

$$\int_{-\infty}^{\infty} d^2k = \int_0^{\pi} d\phi \int_{-\infty}^{\infty} |\kappa| d\kappa,$$

where the absolute value of κ must be employed since the differential element of area d^2k must always be nonnegative. Just as $\mathbf{k} = k\hat{\mathbf{n}}$, one also has that $\kappa = \kappa\hat{\mathbf{N}}$. There is then a one-to-one correspondence between the points in $\mathbf{k}$-space and the points in κ-space. In addition, $\hat{\mathbf{N}}$ may be treated as a conventional unit vector with its range of directions (with respect to some fixed reference direction) appropriately restricted. The identity appearing in Eq. (H.3) may then be expressed in κ-space as

$$\delta(\mathbf{R} - \mathbf{r}) = \frac{1}{4\pi^2} \int_0^{\pi} d\phi \int_{-\infty}^{\infty} d\kappa \int_{-\infty}^{\infty} dx \, |\kappa| e^{i(\mathbf{R}\cdot\hat{\mathbf{N}}-x)} \delta(x - \mathbf{R}\cdot\hat{\mathbf{N}}). \qquad (\text{H.11})$$

Substitution of this expression into Eq. (H.1) and using the definition of the projection of the object function onto the direction specified by the unit vector $\hat{\mathbf{N}}$ given in Eq. (H.5) then results in the representation of the object function as

$$O(\mathbf{R}) = \frac{1}{4\pi^2} \int_0^{\pi} d\phi \int_{-\infty}^{\infty} d\kappa \, |\kappa| e^{i\kappa\mathbf{R}\cdot\hat{\mathbf{N}}} \mathcal{F}_k \left\{ P(\hat{\mathbf{N}}; x) \right\}. \qquad (\text{H.12})$$

This result then yields the well-known *filtered backprojection algorithm* for object reconstruction [5]. It is said to be "filtered" because of the quantity $|\kappa|$ appearing in the integrand of Eq. (H.12).

The inverse of the two-dimensional Radon transformation given in Eq. (H.6) may be obtained in the following manner [2]. Since $|k| = k \, \text{sgn}(k)$, where $\text{sgn}(k) = +1$ if $k > 0$, $\text{sgn}(k) = 0$ if $k = 0$, and $\text{sgn}(k) = -1$ if $k < 0$ is the signum function , then the representation of the delta function given in Eq. (H.11) may be rewritten as

$$\delta(\mathbf{R}-\mathbf{r}) = \frac{1}{4\pi^2} \int_0^{\pi} d\phi \int_{-\infty}^{\infty} d\kappa \int_{-\infty}^{\infty} dx \, \kappa \, \text{sgn}(\kappa) e^{i(\mathbf{R}\cdot\hat{\mathbf{N}}-x)} \delta(x - \mathbf{R}\cdot\hat{\mathbf{N}}). \qquad (\text{H.13})$$

Since [see Eq. (B.14) of Vol. 1]

$$\int f(x)\delta'(x - a)dx = -\int f'(x)\delta(x - a)dx,$$

where the prime denotes differentiation with respect to the variable x, then with $f'(x) = \kappa \, \text{sgn}(\kappa) e^{i(\mathbf{R}\cdot\hat{\mathbf{N}}-x)}$ so that $f(x) = i \, \text{sgn}(\kappa) e^{i(\mathbf{R}\cdot\hat{\mathbf{N}}-x)}$, the final integral in Eq. (H.13) becomes

$$\int \kappa \, \text{sgn}(\kappa) e^{i(\mathbf{R}\cdot\hat{\mathbf{N}}-x)} \delta(x - \mathbf{R}\cdot\hat{\mathbf{N}})dx$$

$$= -i \int \text{sgn}(\kappa)\delta'(x - \mathbf{R}\cdot\hat{\mathbf{N}}) e^{i(\mathbf{R}\cdot\hat{\mathbf{N}}-x)}dx,$$

and hence

$$\delta(\mathbf{R}-\mathbf{r}) = -\frac{i}{4\pi^2}\int_0^{\pi} d\phi \int_{-\infty}^{\infty} d\kappa \int_{-\infty}^{\infty} dx \ \mathrm{sgn}(\kappa)\delta'(x-\mathbf{R}\cdot\hat{\mathbf{N}})e^{i(\mathbf{R}\cdot\hat{\mathbf{N}}-x)}. \quad \text{(H.14)}$$

With the identity [2]

$$\mathcal{P}\int \frac{f(x)}{x}dx = -\frac{i}{2}\int dx \int_{-\infty}^{\infty} dk \ \mathrm{sgn}(k)e^{ikx} f(x), \quad \text{(H.15)}$$

where $\mathcal{P}$ indicates that the Cauchy principal value of the integral is to be taken, Eq. (H.14) becomes

$$\delta(\mathbf{R} - \mathbf{r}) = -\frac{2}{\pi^2}\mathcal{P}\int_0^{\pi} d\phi \int_{-\infty}^{\infty} dx \ \frac{\delta'(x - \mathbf{R}\cdot\hat{\mathbf{N}})}{\mathbf{R}\cdot\hat{\mathbf{N}} - x}. \quad \text{(H.16)}$$

Substitution of this expression in Eq. (H.1) then yields

$$O(\mathbf{r}) = \frac{1}{2\pi^2}\mathcal{P}\int_0^{\pi} d\phi \int_{-\infty}^{\infty} dx \ \frac{P'(\hat{\mathbf{N}}; x)}{\mathbf{R}\cdot\hat{\mathbf{N}} - x}, \quad \text{(H.17)}$$

which is known as the *inverse Radon transform relationship*, where

$$P'(\hat{\mathbf{N}}; x) = \int_{\mathcal{D}} O(\mathbf{r})\delta'(x - \mathbf{R}\cdot\hat{\mathbf{N}})d^2r \quad \text{(H.18)}$$

is the derivative of the projection with respect to the variable x.

References

1. J. Radon, "Über lineare Funktionaltransformationen und Funktionalgleichungen," *Acad. Wiss. Wien*, vol. 128, pp. 1083–1121, 1919.
2. H. H. Barrett, "The Radon transform and its applications," in *Progress in Optics* (E. Wolf, ed.), vol. XXI, pp. 217–286, Amsterdam: North-Holland, 1984.
3. E. C. Titchmarsh, *Introduction to the Theory of Fourier Integrals*. London: Oxford University Press, 1939. Ch. I.
4. R. M. Mersereau and A. V. Oppenheim, "Digital reconstruction of multidimensional signals from their projections," *Proc. IEEE*, vol. 62, pp. 1319–1338, 1974.
5. A. C. Kak, "Computerized tomography with X-ray, emission, and ultrasound sources," *Proc. IEEE*, vol. 67, pp. 1245–1272, 1979.

Index

© Springer Nature Switzerland AG 2019
K. E. Oughstun, *Electromagnetic and Optical Pulse Propagation*, Springer Series
in Optical Sciences 224, https://doi.org/10.1007/978-3-030-20835-6

CPSIA information can be obtained
at www.ICGtesting.com
Printed in the USA
LVHW051818300820
664592LV00001B/14

9 783030 208837